Environmental Biotechnology: A Biosystems Approach

Environmental Biotechnology: A Biosystems Approach

DANIEL A. VALLERO, PhD

Adjunct Professor of Engineering Ethics, Pratt School of Engineering, Duke University, North Carolina, USA

ELSEVIER

AMSTERDAM • BOSTON • HEIDELBERG • LONDON • NEW YORK • OXFORD
PARIS • SAN DIEGO • SAN FRANCISCO • SINGAPORE • SYDNEY • TOKYO

Academic Press is an imprint of Elsevier

Academic Press is an imprint of Elsevier
32 Jamestown Road, London NW1 7BY, UK
30 Corporate Drive, Suite 400, Burlington, MA 01803, USA
525 B Street, Suite 1800, San Diego, CA 92101-4495, USA

First edition 2010

British Library Cataloguing-in-Publication Data
A catalogue record for this book is available from the British Library

Library of Congress Cataloging-in-Publication Data
A catalog record for this book is available from the Library of Congress

ISBN : 978-0-12-375089-1

For information on all Academic Press publications visit our website at www.elsevierdirect.com

Typeset by TNQ Books and Journals

Printed and bound in United States of America
10 11 12 13   10 9 8 7 6 5 4 3 2 1

To Chloe Jayne Randall

CONTENTS

PREFACE ... ix

CHAPTER 1 Environmental biotechnology: an overview ... 1

CHAPTER 2 A question of balance: using *versus* abusing biological systems ... 45

CHAPTER 3 Environmental biochemodynamic processes ... 99

CHAPTER 4 Systems ... 167

CHAPTER 5 Environmental risks of biotechnologies ... 229

CHAPTER 6 Reducing biotechnological risks ... 275

CHAPTER 7 Applied microbial ecology: bioremediation ... 325

CHAPTER 8 Biotechnological implications: a systems approach ... 401

CHAPTER 9 Environmental risks of biotechnologies: economic sector perspectives ... 443

CHAPTER 10 Addressing biotechnological pollutants ... 491

CHAPTER 11 Analyzing the environmental implications of biotechnologies ... 539

CHAPTER 12 Responsible management of biotechnologies ... 577

APPENDIX 1 Background information on environmental impact statements ... 635
APPENDIX 2 Cancer slope factors ... 641
APPENDIX 3 Verification method for rapid polymerase chain reaction systems to detect biological agents ... 649
APPENDIX 4 Summary of persistent and toxic organic compounds in North America, identified by the United Nations as highest priorities for regional actions ... 651
APPENDIX 5 Sample retrieval from ECOTOX database for Rainbow Trout (*Oncorhynchus mykiss*) exposed to DDT and its metabolites in freshwater ... 663
GLOSSARY ... 679
INDEX ... 737
COLOR PLATE

PREFACE

Environmental biotechnology is a vital component of the scientific and engineering toolkit needed to address environmental problems. Environmental biotechnology embodies more than an explanation of the biological principles underlying environmental engineering. Environmental biotechnology depends on a systematic view of the myriad factors involved when organisms are used to solve society's problems. Thus, both the title and subtitle of this book are important.

A systems approach to biotechnology requires a modicum of understanding of a number of disciplines, especially environmental engineering, systems biology, environmental microbiology, and ecology. This book introduces all of these fields from the perspective of how to apply them to achieve desired environmental outcomes *and* how to recognize and avoid problems in such applications. This approach means that the treatment of these four disciplines is predominantly focused on biotechnology and is not meant to be an exhaustive treatise on any of the four. This book's principal value lies at the intersection of the four disciplines. However, engineering requires specifics, so my intention is that the reader gain a sufficient grasp of each so as to know when more details are needed and when to consult the references at the end of each chapter to seek out these important details.

BIOTECHNOLOGY AT THE INTERSECTION OF DISCIPLINES

Environmental engineering is a broad field, including both *abiotic* and *biotic* solutions to pollution and environmental problems. This book's primary environmental engineering focus is on the biotic solutions, so the reader should consult general environmental engineering texts and specific chemical and physical treatment resources to find abiotic treatment methods to match the biotic approaches discussed here. For example, after reading a discussion of a particular biotechnology, e.g. Chapter 7's exposition of a biofilter used to treat a specific organic pollutant, the reader may be inclined to look up that pollutant to see what other non-biotechnological methods, e.g. pumping and air sparging, have been used in its treatment. This book certainly includes discussions on abiotic techniques in Chapter 10, but limits the discussion to the treating of those pollutants that may result from biotechnologies (e.g. if a hazardous byproduct is produced, it may need to undergo thermal treatment).

Systems biology and molecular biology are addressed insofar as genetic engineering is an important part of environmental biotechnology. An understanding of genetic material and how it can be manipulated either intentionally or unintentionally is crucial to both applications and implications. As in environmental engineering, the discussion is focused less on a theoretical and comprehensive understanding of DNA and RNA for their own sake than would be found in a systems biology text. Again, if the reader needs more information, the references should be consulted and should lead to more specific information. In addition, the book addresses a number of emerging technologies used in environmental assessment, particularly drawing on systems biology, such as the computational methods associated with genomics, proteomics, and the other "omics" systems.

I recall at least one of my professors at the University of Kansas differentiating microbiologists from engineers. Microbiologists are interested in intrinsic aspects of the "bugs," whereas engineers are interested in what the "bugs" can do [1]. I have been careful with the taxonomy of the organisms, but it is not the intent to exhaustively list every microbe of value to environmental biotechnology. When the reader needs more detail on a particular organism and when trying to find other microbes that may work in a biotechnology, the references and notes should help initiate the quest.

More than a few of my ecologist colleagues may cringe when I say that microbes have instrumental value, not intrinsic value, in many environmental biotechnologies. Engineers, including environmental engineers, are focused on outcomes. They design systems to achieve target outcomes within specified ranges of tolerance and acceptability. As such, they say a bacterium is a means, not an end in itself. Ecologists tend to be more interested in the whole system, i.e. the ecosystem. Thus, the microbes, especially those that have been supercharged genetically, must be seen for how they fit within the whole system, not just the part of the system that needs to be remediated. This book, therefore, includes this ecological perspective, especially when addressing potential implications, such as *gene flow* and *biodiversity*. In fact, one of the themes of this book is that engineers must approach even "slam dunk" biotechnologies with whole systems in mind, with considerations of impact in space and time, i.e. a systems approach to biotechnology.

THE SYSTEMS APPROACH

One way to address environmental biotechnology is to ask whether it is "good" or "bad." Of course, the correct answer is that "it depends." According to my colleague at Duke, Jeff Peirce, this is one of the few universally correct statements in engineering. The tough part of such a statement, of course, is deciding to some degree of satisfaction on just what "it depends."

The same biotechnology can be good or bad. It just depends. It depends on risks versus rewards. It depends on what is valued. It depends on reliability and uncertainty of outcome. It depends on short-term versus long-term perspectives. It depends on the degree of precaution needed in a given situation. Mostly, it depends on whether the outcome is ideal, or at a minimum acceptable, based on the consideration of the myriad relationships of all of the factors. Such factors include not only the physical, chemical, biological aspects of a biotechnology, but also those related to sociological and economic considerations. That is, the same technology is good or bad, depending on the results of a systematic perspective.

I would recommend that the question about the dependencies driving the acceptability of a given environmental biotechnology be asked at the beginning of any environmental biotechnology course. I recognize just how tempting it is in teaching an environmental biotechnology course to jump into *how* to use living things to treat pollution, with little thought as to *whether* to use a biotechnology. Perhaps this is because we expect that other perspectives, such as abiotic treatment, will be addressed in courses specifically addressing these technologies, and after having completed courses in every major treatment category, the student will then be able to select the appropriate method for the contaminant at hand. This is much like the need for a really good course in concrete and another excellent course in steel, as a foundation (literally and figuratively) in structural engineering. Such reductionism has served engineering well. In environmental sciences and engineering, the newer views do not lessen the need for similar specific knowledge in the foundational sciences, but in light of the importance of the connections between living things and their surroundings, newer pedagogies are calling for a more systematic view to put these basics into systems that account for variations in complexity and scale.

Biotechnologists are justifiably tempted to keep doing that which has worked in the past. For those in the fields of biological wastewater treatment and hazardous waste biotechnologies, the art of engineering is to move thoughtfully, with some trepidation, from what is known to the realm of the unknown. This microbe was effective in treating contaminant A, so why not acclimate the microbe to a structurally similar compound, e.g. the same molecule with a methyl group or one with an additional ring? Often this works well under laboratory conditions and even in the field, so long as conditions do not change dramatically. Such acclimation was the precursor to more dramatic and invasive forms of *genetic modification*, especially recombinant DNA techniques. This book explores some of the knowns and unknowns of what happens *systematically* when we manipulate the genetic material of an organism. Perhaps, the system is no more influenced by a genetically modified organism than by those that bioengineers have manipulated by letting the organism adapt on its own to the new food source. But, perhaps not.

When I originally proposed the concept for this book, I thought that I would dedicate it almost exclusively to potential *implications* of environmental biotechnologies. I thought that others had done admirable jobs of writing about the *applications.* After delving into the topic in earnest, I came to the conclusion that I was only half right. Indeed, the previous texts in environmental biotechnologies were thorough and expansive. Some did a really good job of laying out the theory and the techniques of environmental biotechnology. However, most were not all that interested in what may go wrong or what happens outside of the specific application. This is not meant to be a criticism, since the authors state upfront that their goal is to enhance the reader's understanding of these applications. The implication, to me at least, is that their work starts after the decision has been made to destroy a certain chemical compound, using the most suitable technique. In this instance "suitable" may be translated to mean "efficient." How rapidly will microbe X degrade contaminant A? How complete is the degradation (e.g. all the way to carbon dioxide and water)? How does microbe X compare in degradation rates to microbes Y and Z? How efficiently will microbe X degrade contaminant A if we tweak its DNA? How broadly can microbe X's degradation be applied to similar compounds?

These are all extremely important questions. Efficiency is an integral but not an exclusive component of effectiveness. Thus, my original contention was half wrong. I could not discuss implications without also discussing applications. I liken this to the sage advice of a former Duke colleague, Senol Utku. He has been a leader in designing adaptive structures that often follow intricate, nonlinear relationships between energy and matter. His students were therefore often eager to jump into nonlinear mathematical solutions, but he had to pull them back to a more complete understanding of linear solutions. He would tell them that it is much like a banana. How can one understand a "non-banana" without first understanding the "banana"? Thus, my systematic treatment of environmental biotechnology requires the explanation of both applications (bananas) and implications (non-bananas).

The term "systems" has become an adjective. For decades, engineers have had systems engineering. We now have systems biology, systems medicine, and even systems chemistry. Early on, systems simply meant a comprehensive approach, such as a life cycle or critical path view. Later, another connotation was that it provided a distinction from compartmental or reductionist perspectives. Now, the systems moniker conveys a computational approach. Lately, subdivisions of the basic sciences have also become systematic in perspective. For example, systems microbiology approaches microorganisms or microbial communities comprehensively by integrating fundamental biological knowledge with genomics and other data to give an integrated representation of how a microbial cell or community operates. This text attempts to address all of these perspectives and more, but all through the lens of the environment.

Along the way, I became aware that there was not a good term that included all of these perspectives. Pioneers in environmental modeling, such as Donald MacKay and Panos Georgopoulos, advanced the field of chemodynamics. In fact, I have drawn heavily from their work. The challenge is how to insert biology into such chemodynamic frameworks.

For many in the environmental sciences and engineering fields, environmental biotechnologies that most readily come to mind are various waste treatment processes, those that often begin with the suffix "bio." Thus, I decided to use the term *biochemodynamics* to refer to the myriad bio-chemo-physical processes and mechanisms at work in environmental biotechnologies. At one point, I even suggested calling this book *Environmental Biochemodynamics*. However, while such a title would distinguish the focus away from abiotic processes, it would leave out some of the important topics covered, such as the societal and feasibility considerations needed in biotechnological decisions.

Environmental biotechnology is all about optimization, so it requires a systematic perspective, at least in its thermodynamic and comprehensive connotations. In particular, biotechnologists are keenly interested in bioremediation of existing contaminants.

To optimize, we must get the most benefit and the least risk by using biology to solve an important problem or fill a vital need. In my research, I discovered a very interesting workshop that took place in 1986 [2]. The workshop was interesting for many reasons. It was held by a regulatory agency, the US Environmental Protection Agency, but predominantly addressed ways to advance environmental biotechnology. In other words, the entity that was chastising polluters was simultaneously looking for ways to support these same polluters financially and scientifically so as to become non-polluters!

Such an approach is not uncommon in its own right, since in the previous decade the same agency had funded research and paid to build wastewater treatment plants to help the same facilities being fined and otherwise reproved for not meeting water quality guidelines and limits. This is a case of the "stick" being followed by the "carrot." The 1986 workshop was actually refreshing, since it was an effort to help scientists come up with ways to push the envelope of technology to complement the growing arsenal of rules and standards for toxic chemicals in the environment.

One of the challenges posed in the mid-1980s was that the National Academy of Sciences had just sketched a schematic to address risks posed by chemicals. It followed a physicochemical structure that consisted of identifying chemical *hazards* and seeing how people may come into contact with these hazards, i.e. *exposure*. The combination of these factors led to what the academy called *risk assessment*. This seemed to work adequately for chemical hazards to one species (*Homo sapiens*), but did not fit quite well with hazards that behave differently than pharmaceuticals, pesticides or other chemical agents, i.e. physical (e.g. UV light) or biological (e.g. microorganisms) hazards. The Academy recently has proposed new schema that may better fit biotechnological risks.

So, indeed, it was good that experts were getting together in 1986 to find new applications of biotechnology to treat and control pollution. However, it appears that even after almost a quarter century some of the challenges have not been addressed, at least not fully. Some of the concerns expressed in 1986 are no longer being expressed widely. The proceedings of the meeting state:

> Federal, State and local regulatory policies pose barriers to field-testing and thereby the development of commercial genetically engineered biotechnology products. Permitting and reporting requirements and the uncertain regulatory climate were identified as additional barriers to the development of the biotechnology control technology [3].

Other concerns persist, as evidenced when the proceedings mention that:

> The public has vague concerns about the risks that may be presented by the use of biotechnology products. The Panelists felt that the public does not usually perceive a distinction between engineered and nonengineered microorganisms and that the public does not understand the scientific basis or applications of biotechnology. These deficiencies pose a barrier to the public's ability to evaluate the issues raised by and the risks associated with biotechnology. ... The

> concerns involve the credibility and capabilities of industry and regulatory agencies to identify and assess potential risks presented by biotechnology and how risks and benefits are balanced in the decision-making process. [4]

A number of biotechnologies still have these credibility problems, most notably those related to food supplies. However, and I am not sure when it happened, at some point in time in the last few decades, environmental biotechnology passed the initial risk test. At least in the United States, there has been some tacit consensus that the environmental advantages of manipulating genetic material in microorganisms to clean up wastes override any environmental and other risks that may result from such modifications. My research did not uncover a specific declaration of this consensus, but it becomes obvious if one compares the uncertainties and questions asked in the 1980s to the research and regulatory agendas today. Interestingly, such a scientific consensus is not universal. For example, some European scientists look at genetically modified organisms (GMOs) of all types with a healthy skepticism.

At least some of the reason for less skepticism toward environmental biotechnology may be the result of the environment in which it emerged. The reader is reminded that in the early 1980s, hazardous waste sites seemed to be cropping up all over the nation. In fact, in the letter to the EPA Administration that transmitted the proceedings mentioned above, the chair, G.E. Omenn, Dean of the School of Public Health and Community Medicine at the University of Washington, stated that:

> The Nation needs alternative technologies to complement present "burn or bury" approaches to chemical pollutants ... Within the microbial treatment arena, improvements are needed, some of which might draw upon genetic engineering methods. [5] The United States does indeed continue to worry a great deal about research that involves genetic manipulation to produce medical and warfare agents, sometimes involving near relatives to the microbes being used in other biotechnologies, including remediation. In fact, the National Institutes of Health have comprehensive guidelines to address physical containment of GMOs and their genetic material. But that is addressed only at research. This begs the question of when is the introduction of genetic material no longer research. History has shown us that when something is introduced into a different, less controlled system, unexpected outcomes are almost always assured. That is, to some extent all environmental biotechnologies can be considered "research."

SEMINAR DISCUSSIONS

These uncertainties and differences in perspective led to my recognition of the need to approach all biotechnologies with a large degree of humility. So, this book includes a "seminar" at the end of each chapter. The seminar addresses a topic about which there is no consensus or where the understanding of the potential outcomes is only now emerging. The topics are those of public concern and of scientific importance. As such, there are many right and wrong answers to the questions posed at the end of each seminar. The seminars are designed for open discussion, so I recommend that a three-step process be used in the classroom or breakout group, depending on the learning environment.

First, the seminar should be read and the references consulted. Second, the students and/or discussion group members write their individual answers to the seminar questions. Third, the class/group openly shares their answers with the whole group with the facilitator ensuring each perspective is shared and with the main points written on a whiteboard or flipchart. Perhaps the major points could be grouped into natural categories (e.g. social concerns, scientific uncertainties, unacceptable risks, etc.), with each member given two votes on which are most

Table P.1 Comparison of benefits and risk from transgenic herbicide-resistant plant

Potential benefits	Potential risks
Simpler weed management based on fewer herbicides	Greater reliance on herbicides for weed control
Decrease in herbicide use	Increase in herbicide use
Less contamination of the ecosystem	More contamination of the water, soil, and air and shift in exposure patterns
Use of environmentally more benign herbicides	Development of resistance in weed species by introgression of the transgenes
Reduction of the need for mechanical soil treatment	Shifts in population of weeds towards more tolerant species
Less crop injury	Increase in volunteer problems in agricultural rotation systems
Improved weed control	Negative effects of herbicides on non-target species

Source: H.A. Kuiper, G.A. Kleter and M.Y. Noordam (2000). Risks of the release of transgenic herbicide-resistant plants with respect to humans, animals, and the environment. *Crop Protection* 19 (8-10): 773–778.

important. The top few problems could then be discussed with regard to possible solutions, including needed research.

For example, the United States has had a fairly strong consensus in support of many biotechnological applications in drug development, industry, and environmental cleanup, but there remains a comparative uneasiness about certain biotechnologies. In the case of food supplies, this may be recognition that the final product may find its way to our kitchen table. It may also be because agriculture systems are very complex, with many steps from seed to table, and are vulnerable to mistakes. Kuiper *et al.*, for example, indicated that humans, animals, and the environment are at some level of risk whenever a GMO, in this case an herbicide-resistant plant, is used. In fact, every decision is a balance between potential benefits and potential risks (see Table P.1) [6]. A key question is why is there a difference between such biotechnologies and the seeming lack of concern about environmental biotechnologies.

REDUCTIONISM VERSUS THE SYSTEMS APPROACH

In times of specialization among and within the sciences, we tend to sharpen our focus, which is usually a good thing. For instance, bioscientists, biotechnologists, and bioengineers often pursue and apply information that meets a particular need. We often isolate our research and interest so tightly that we cannot worry about what is going on in the rest of our own discipline, let alone other disciplines. This baby is usually only well understood by a small cadre of fellow sojourners with a common expertise in highly esoteric subject matter.

I recently discussed with a fellow "seasoned" researcher, who happens to be a world-class microbiologist, the safety and risk of using genetically modified organisms for bioremediation. We both expressed concern that some of the questions that were asked in the late 1970s were still not completely answered. As mentioned, it appears that somewhere along the way, the engineering community dropped these questions, but neither of us could find a clear point in time for such a decision.

Thus, those who apply the physical and biological sciences must decide how they go about using data, making those data into information, and, hopefully add knowledge on how this information, evidence if you will, can best be used to solve the big and mounting problems. Biotechnology provides an excellent illustration of such optimization schemes.

At one end of the spectrum is the total devotion to the application of living things to solve problems; doing whatever gets us to the levels of thermodynamic efficiency we have defined as a performance standard. This means that we can go about unchallenged in modifying genetic material, moving massive amounts of soil and water to bioreactors, and tightly controlling the conditions that give us some predefined metric for efficiency. At the other end is stifling caution that keeps us from designing and using tools based on the state-of-the-science.

Bioremediation, for example, has been greatly improved by understanding the environmental conditions and the microbial processes that lead to more efficiency degradation of some very recalcitrant compounds. As has been standard practice of biological treatment for over a century, we put the microbes to work and use their needs for carbon and energy to do things they would not do with the prodding of an engineer. This logically led to the innate and learned creativity of the bioengineer who began to ask whether we could do something to the "bugs" to make them even more efficient. This gave birth to the bioreactor (first the common tricking filters and their ilk) where we chose the right bugs from their natural habitats, observed how they broke down similar organic material, withheld their natural sources of carbon, exposed them to some new food (our wastes), and patiently and incrementally added enough of the new food so that the endogenous processes found new ways of donation and acceptance of electrons (energy).

In the process, where before a few bugs would take many days to break down such organic matter, our bioreactors could now process millions of gallons of waste per day and release effluent that met what were before thought to be unreachable standards of purity. In 1976, when I started in this business, the gold standard was 20 parts per million (ppm) total suspended solids and 20 ppm biochemical oxygen demand for effluent discharges to the waters of the United States. To my young colleagues, this is like saying that my first PC had 128 kilobytes of random access memory (which it did). These were nevertheless profoundly difficult measures of success.

The next logical step was to treat substances heretofore not considered amenable to biological treatment. The microbes rarely had to rely on these compounds as sources of carbon and energy. There simply were always enough other food sources that were easier to digest; with no need to remove chlorine or to break aromatic rings. So, some time in the late 1970s biological treatment began to emerge as a very viable hazardous waste treatment processes. But, the recalcitrance and variability of chemical composition, as well as the arrival of new DNA techniques made for a logical arranged marriage between the microbes and synthetic organic contaminants.

The need to reconcile reductionist and systematic thinking is ongoing in numerous scientific and design disciplines. For instance, there is an ongoing debate within engineering and design professions concerning the role of evidence in support of the often-stated "form follows function." The postulation is that designers must not only gather physical data, but must add social scientific information and human factors to the mix. This requires asking questions of past users (e.g. clients, patients, subjects, consumers, visitors, policy makers, taxpayers, etc.). From that, a better design will emerge.

The *bioinformatics* challenge is two-fold, however. First, how can reliable information be gathered to address the needs problem at hand? For example, in designing a genetic laboratory, how much of it follows the traditional lab needs for good lab practice (e.g. bench surface area, chemical segregation, storage of hazardous materials, hood design, clean areas, etc.) versus what is specific to the type of genetic research that will be taking place (e.g. tissue preparation, other media needs such as soil, water, and biota handling, genetic material identification apparatus, etc.)? The delta between these two paradigms according to the evidence-based designers, cannot follow the old paradigms, but needs reliable information.

The book attempts to find a balance between rigorous reductionism and the systems approach. The engineering community must avoid being overly myopic in its general acceptance of technologies and designs that work (e.g. bioremediation of oil spills using genetically modified bacteria), while being sufficiently cautious in taking a systematic view (e.g. considering the possible impacts of these modified bacteria in the whole ecosystem, including gene flow and changes in the chemical compounds in the oil that may change their affinity for certain media and compartments in the environment).

STRUCTURE AND PEDAGOGY

This book consists of 12 chapters. They have been designed to provide a primary text for two full semesters of undergraduate study (e.g. Introduction to Environmental Biotechnology; Advanced Environmental Biotechnology). It is also designed to be a resource text for a graduate level seminar in environmental biotechnology (e.g. Environmental Implications of Biotechnology).

Chapter 1 introduces the science that underpins both the applications and implications of environmental biotechnology. It provides the background and historical context of contemporary issues in biotechnology, using the environmental impact assessment process as a teaching and learning vehicle. In particular, the chapter attempts to enhance the chaotic nature of environmental outcomes, i.e. how initial conditions can lead to various outcomes as demonstrated by event and decision trees. The seminar, Antibiotic Resistance and Dual Use, expands the reader's perspectives on the science (e.g. aerosol science and biology) and societal issues associated with current environmental and security issues.

Chapter 2 addresses the various scientific principles involved in environmental biotechnologies. That is, it introduces biochemodynamics. In fact, Table 2.9 is a digest of much of the subject matter addressed in Chapters 3 through 7, so it can be a good resource for exam preparation and review. The seminar discussion, GMOs and Global Climate Change, addresses the pros and cons of whether and how genetic manipulations are a needed tool to address a major environmental problem. The seminar is the book's major discussion of algae, which are becoming increasingly important to biotechnologies.

Chapter 3 provides detailed discussion of each of the processes described in Table 2.9, i.e. the underpinning biochemodynamic processes. This is also the first place where microbial metabolism and growth are discussed in detail. Thus, Chapter 3 may be used as a standalone source to introduce the science of a graduate seminar, or for professors designing their own "coursebook" who need a chapter on the fundamentals of environmental transport and fate. However, I would strongly recommend that such a coursebook include Chapters 4 and 5, since these go into much greater detail on biotransformation and risk, respectively. The seminar topic addresses how well models can predict the transfer of genetic materials. I must admit, I have more questions than answers regarding this topic, so the questions at the end should reveal some real weaknesses in currently available models. As such, I would greatly appreciate the reader's ideas. Please email them to me at dav1@duke.edu.

Chapter 4 is a pivotal chapter. It suggests the need for a systematic perspective. Up to this point, the science being discussed can be seen from numerous perspectives, e.g. how the principles can be applied to clean up a waste site or how these same principles can be used to avoid problems in such a clean up. Chapter 4, however, imposes an onus on the reader to appreciate the chaos. That is why I begin with the lyrics from Sting's song. (My grammar checker hated this quote, incidentally, due to the double negative, but I believe it captures the peril of single-mindedness that our proposed solution is the best solution.) Too often, we exaggerate the expected benefits and ignore the potential risks and downsides of our decisions. As such, Chapter 4 draws from proven tools, e.g. the fugacity models, industrial ecology, and life cycle analysis, and extends them to biotechnologies. Such extensions require a large helping of humility. The seminar topic deals with comparisons of biological agents used for good and ill, asking questions related to when a biological cleanup is successful and whether the introduction of a species to the environment is worth the risks. The comparison of two species of *Bacillus* points to the need to ask whether genetic manipulations are sufficiently understood before introducing new strains to the environment, even for noble causes like bioremediation.

Chapter 5 introduces environmental risk assessment, especially as it relates to biotechnologies. The problem and challenge in writing this chapter is that the lion's share of risk literature addresses chemical risks, rather than biological risks. The scientific community is increasingly aware that microbial risks do not necessarily follow the traditional hazard identification/dose-response, exposure and effects cascade. However, some biotechnological risk indeed is chemical (e.g. the production of toxin). Thus, Chapter 5 introduces the basics of risk assessment (e.g. thresholds, dose-response curves, exposure assessment techniques), but also introduces nuances that may help tie environmental microbiology to environmental engineering risk concepts. The seminar addresses risk tradeoffs, especially when it comes to manipulating genetic material for environmental results.

Chapter 6 addresses ways to reduce and manage risks. In following the risk assessment discussions in Chapter 5, a number of environmental problems are considered with an eye toward ways to address them (e.g. addressing release of antibiotics and microbial resistance, destruction of endocrine disruptors). Managing risks requires an understanding of possible outcomes, so the chapter includes some expansive thinking on what could happen once a microbe enters the environment.

With the help of Drew Gronewold of the US Environmental Protection Agency, Chapter 6 includes a hypothetical scenario using Bayesian techniques. In the interest of full disclosure, one of the great frustrations in writing this book is the lack of reliable quantitative tools to predict outcomes. Unlike risk assessments in the nuclear industry, for example, few decision trees in biotechnology can produce probabilities of outcomes. This is partially because there are so many variables in the ambient environment compared to the controlled conditions of a nuclear power plant. In addition, nuclear power plants are data-rich. Everyone who is potentially exposed to radiation wears a monitoring device that records values that can be aggregated and compared to reliable radiation health effects data (e.g. cancer). In environmental studies, data are scarce and the outcomes are numerous (human health outcomes, ecosystem damage, etc.). The hypothetical scenario at least gives us an opportunity to consider the changes that could occur. Again, I welcome the reader's ideas on how useful this is and how it can be improved.

The Chapter 6 seminar addresses biomimicry. Is it universally acceptable to mimic nature, or does it introduce unexpected risks under certain conditions? The consideration of the botanical pesticides and their derivatives provides an interesting discussion of the often erroneous assumption that *natural* means *safe*. After all, some of the most toxic substances are natural, e.g. the botulinum toxin and aflatoxins. In addition, many of the pyrethroids have been altered chemically so as not to resemble the original botanical.

Chapter 7 most closely resembles traditional environmental biotechnology texts. It is mainly devoted to the application of microbial systems to clean up pollution. Thus, it can be extracted in its entirety for professors and facilitators needing a summary of biological treatment mechanisms and processes. The seminar discussion addresses a currently important topic: how can the disciplines of environmental microbiology be reconciled with bioremediation? In particular, the seminar goes into detail on previous attempts at providing semi-quantitative tools to predict important factors like biodegradation rates. This is a currently important topic, since regulatory agencies around the world are looking for better ways to predict environmental harm *before* a chemical reaches the marketplace. In fact, it appears that the Toxic Substances Control Act may soon be amended to improve such risk prioritization.

Chapter 8 is the mirror image of Chapter 7, as it presents the implications of environmental biotechnologies. The chapter recognizes the value of those applications considered in

Chapter 7, but encourages systematic thinking that must include proactive measures to prevent negative impacts. The seminar discussion addresses the scary problem of long-term transport of microbes and their possible impacts on coral reefs. I chose this seminar for two major reasons. First, coral reefs are complex biological systems that demonstrate how a slight change can substantially alter their condition. Second, the case demonstrates a global scale transport associated with a micro-scale problem. Thus, it is an ideal "teachable moment" to consider scale and complexity involved in a real-world environmental problem.

Chapter 9 is arguably the most expansive part of the book. It addresses the environmental implications of all non-environmental biotechnologies. In fact, many concerns remain about industrial, medical, and especially agricultural biotechnologies. In addition, considering the specific environmental impacts of the technologies, they also provided some lessons for environmental biotechnologists (see for example the discussion box on Hormonally Active Agents, and the case discussion, King Corn or Frankencorn). Also, the discussion of enzymes ties very closely to environmental bioreactors. The seminar topic on vaccines is particularly timely at this writing, since the H1N1 influenza outbreak has dramatically heightened awareness of the risks and benefits associated with vaccines.

Chapter 10 was written with recognition that biotechnologies, just like all technologies, generate pollutants that must be treated. The biodegradable fraction of these pollutants can be treated using those approaches in Chapter 7. However, other abiotic techniques must at times also be deployed. Thus, the chapter includes study designs and assessment approaches that may need to be used to address pollutants generated during biotechnological operations. The seminar topic, in fact, compares and contrasts traditional environmental study designs to those needed for a specific biotechnological project (i.e. gene flow from crops).

Chapters 11 and 12 address the professionalism needed in environmental biotechnological enterprises. This includes ethical and practice considerations. The chapter seminars address the challenges associated with the first canon of all engineering professions, i.e. to hold paramount the safety, health, and welfare of the public. The Chapter 11 seminar explores ways to be inclusive of the public's input and the Chapter 12 seminar delves into ways to approach risk tradeoffs based on a case involving TNT-laden soil.

This book covers a wide range of scientific disciplines, so some terminology may be new or at least used in ways not familiar to most readers. In fact, a number of terms have multiple definitions, depending on the particular subject matter. Thus, readers are encouraged to turn to the Glossary at the end of the book when encountering any term with which they are not fully familiar. Important terms occurring in the Glossary are signaled by the use of italic in the text. The Glossary is quite expansive, since it includes terms used by numerous professions and disciplines involved in environmental biotechnologies. These terms have been gathered from numerous sources, including my own lexicon. A number of sources are mentioned in the endnotes, but some sources have long been forgotten (e.g. past and present colleagues, former mentors, forgotten articles, etc.).

THE CHALLENGE

My first discussions of the idea for this book with the gifted Elsevier editor Christine Minihane included a fear that no single text could capture the entirety of the applications and implications of environmental biotechnologies. Upon its completion, I am even more convinced of this. Early in our discussions, I offered the possibility that we might be able to create an electronic community where the various elements of environmental biotechnology reside on a website where people could update and correct the material in this book, could expand on topics, and add new topics. In addition, new teaching and learning tools, as well as actual case studies could be added and updated as they change [see Discussion Box: Bioreactors to the Rescue]. Finally, community members could share new analytical and quantitative techniques, such as successful uses of decision trees, Bayesian approaches, models, root cause and failure analyses, and other approaches used within and outside of the environmental biotechnology community. If you believe this is a worthwhile endeavor, and especially if you would like to participate, please let me know.

Daniel A. Vallero, PhD

Discussion Box: Bioreactors to the Rescue

There is ample evidence that such biological systems can provide cutting edge solutions needed to protect the environment and public health. A case in point is the U.S. Army's Deployable Aqueous Aerobic Bioreactor (DAAB), which is a portable wastewater treatment system being developed to provide: on-site treatment of wastewater at forward operating bases, rapid response to failures (such as during natural disasters) of treatment works, and a rapidly and readily deployed wastewater treatment system for humanitarian needs during crises [7].

Consider two of the most intractable global challenges: natural disasters and war. As this book goes to final printing, engineers, physicians, and first responders from myriad fields are working feverishly against the devastating and truly tragic tolls taken by the earthquake and aftershocks in Haiti. In addition to the hundreds of thousands who perished during and immediately after the earthquake, millions are and will continue to be at risk of waterborne diseases. As discussed in Chapter 7, environmental biotechnologies must be part of the solution to the aftermath of disasters. In the case of Haiti and in war zones, for example, sustainable and low maintenance systems are being employed. As evidence, the U.S. Army has contracted with Sam Houston State University (SHSU) in Texas to develop a bioreactor that can clean water without the need for external sources of energy or chemical compounds.

The bioreactor uses indigenous soil bacteria which have been collected by scientists at SHSU, who describe the process as consisting of a subset of these bacteria whose genetic material is modified to produce "biofilm that is self-regulating and highly efficient at cleaning wastewater" (See Chapter 7). According to the researchers the process is rapid, "cleaning influent wastewater within 24 hours after set-up to discharge levels that exceed the standards established by the Environmental Protection Agency for municipal wastewater." The sludge production is also manageable, i.e. the original waste volume is decreased by over 90%. This compares to about a month needed for a typical septic tank, which often can only decrease volume by 50% of less [8].

Another important feature of any portable waste system is that it be "scalable." The SHSU developers claim that this system can be used to treat wastes from a single residence to larger scales, such as neighborhoods in Haiti or for an army base in Afghanistan. The keys to sustainable biotechnologies are that they not require intricate operations, that they not depend on scarce materials and energy sources that are difficult to obtain and maintain. Biotechnologies can meet these criteria.

Benefits, as discussed in Chapter 11, can be indirect and difficult to quantify. In this instance, one of indirect but crucial benefits of such of an adaptive biotechnology is an improvement in troop safety. In Afghanistan, for example, clean water has to be trucked precariously due to lack of potable local water supplies. The U.S. Marine Corps' Marine and Energy Assessment Team estimates each soldier requires about 22 gallons of clean water daily, so if the prototypes of sustainable, *in situ* biotechnologies work out, they could translate into 50 fewer military trucks needing to traverse the dangerous terrain [9].

Other applications are also possible, such on tankers and cruise ships, as well as temporary conditions, such as during power outages.

NOTES

1. I have actually softened this view in my paraphrasing. If memory serves, it was closer to "microbiologists like to name the bugs, while we don't care what they are called so much as what they do."
2. US Environmental Protection Agency (1986). The Proceedings of the United States Environmental Protection Agency Workshop on Biotechnology and Pollution Control. Bethesda, Maryland, March 20–21, 1986.

3. Ibid., VIII-2.
4. Ibid., VIII-3.
5. G.E. Omenn (1986). Letter to the Honorable Lee M. Thomas, Administrator, US Environmental Protection Agency. March 25, 1986.
6. H.A. Kuiper, G.A. Kleter and M.Y. Noordam (2000). Risks of the release of transgenic herbicide-resistant plants with respect to humans, animals, and the environment. *Crop Protection* 19 (8-10): 773–778.
7. U.S. Army Corps of Engineers (2010). "Deployable Aqueous Aerobic Bioreactor." Environmental Laboratory. *EL Newsroom;* http://el.erdc.usace.army.mil/news.cfm?List=24; accessed on February 11, 2010.
8. S. Holland (2010). Quoted in "'Revolutionary' Water Treatment Units on their Way to Afghanistan." *Today@Sam.* http://www.shsu.edu/~pin_www/T%40S/2010/RevolutionaryWaterTreatmentUnitsQnTheirWay ToAfghanistan.html; accessed on February 11, 2010.
9. K. Drummond (2010). "Pure Water for Haiti, Afghanistan: Just Add Bacteria." *Wired.Com.* February 10, 2010; http://www.wired.com/dangerroom/2010/02/bacteria-based-water-treatment-headed-to-afghanistan-haiti-next/#ixzzOfFIUsYhH; accessed on February 10, 2010.

CHAPTER 1

Environmental Biotechnology: An Overview

As industrial biotechnology continues to expand in many sectors around the world, it has the potential to be both disruptive and transformative, offering opportunities for industries to reap unprecedented benefits through pollution prevention.

Brent Erickson (2005) [1]

Two of the important topics at the threshold of the 21st century have been the environment and biotechnology. Erickson, representing BIO, the largest biotechnology organization, with more than 1200 members worldwide, succinctly yet optimistically characterized the marriage of environmental issues with the advances in biotechnology. Considered together, they present some of the greatest opportunities and challenges to the scientific community. Biotechnologies offer glimpses to solutions to some very difficult environmental problems, such as improved energy sources (e.g. literally "green" sources like genetically modified algae), elimination and treatment of toxic wastes (e.g. genetically modified bacteria to break down persistent organic compounds in sediments and oil spills), and better ways to detect pollution (e.g. transgenic fish used as indicators by changing different colors in the presence of specific pollutants in a drinking water plant).

Tethered to these arrays of opportunities are some still unresolved and perplexing environmental challenges. Many would say that advances in medical, industrial, agricultural, aquatic, and environmental biotechnologies have been worth the risks. Others may agree, only with the addition of the *caveat*, "so far."

This text is not arguing *whether* biotechnologies are necessary. Indeed, humans have been manipulating genetic material for centuries. The main objective here is that thought be given to possible, often unexpected, environmental outcomes from well-meaning, important, and even necessary biotechnologies. Environmental biotechnology, then, is all about the balance between the applications that provide for a cleaner environment and the implications of manipulating genetic material.

In some ways, this is no different than any environmental assessment. An assessment is only as good as the assumptions and information from which it draws. Good science must underpin environmental decisions. The sciences are widely varied in environmental biotechnology, including most disciplines of physics, chemistry, and biology. Thus, to characterize the risks and opportunities of environmental biotechnology, we must enlist the expertise of engineers, microbiologists, botanists, zoologists, geneticists, medical researchers, geologists, geographers, land use planners, hydrologists, meteorologists, computational experts, systems biologists, and ecologists.

BIOCHEMODYNAMICS

The only way to properly characterize biological systems is by simultaneously addressing chemical reactions, motion, and biological processes. Mass and energy exchanges are taking place constantly within and between cells, and at every scale of an ecosystem or a human population. Thus, biochemodynamics addresses energy and matter as they move (dynamics), change (chemical transformation), and cycle through organisms (biology). A single chemical or organism undergoes biochemodynamics, from its release to its environmental fate (see Figure 1.1).

Since biotechnologies apply the principles of science, the only way to assess them properly is by considering them biochemodynamically. Recently, the environmental community has become increasingly proficient in using biomonitoring to assess ecosystem condition or to determine pathways that have led to xenobiotic body burdens in humans. This has come to be known as exposure reconstruction. In other words, by analyzing concentrations of substances in tissue, the route that led to these concentrations can retrace the pathways, such as those in Figure 1.1.

Reconstruction of *body burden* in an organism that follows the release of a substance to the environment is an example of the biochemodynamic approach. To date, the use of biomonitoring data for environmental assessment has been limited to relatively straightforward

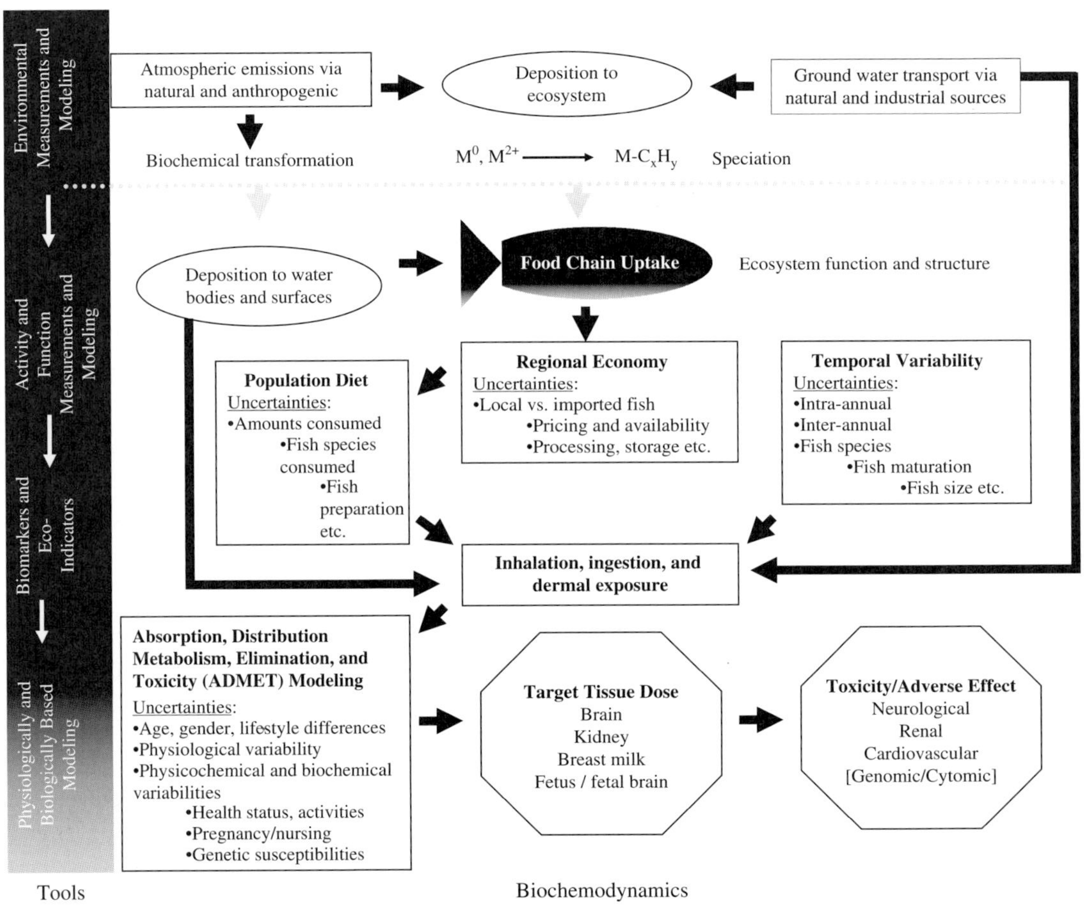

FIGURE 1.1 Biochemodynamic pathways for a substance (in this case a single chemical compound). The fate is mammalian tissue. Various modeling tools are available to characterize the movement, transformation, uptake, and fate of the compound. Similar biochemodynamic paradigms can be constructed for multiple chemicals (e.g. mixtures) and microorganisms. *Source:* Adapted from discussions with D. Mangis, US Environmental Protection Agency in 2007.

exposure scenarios, such as those involving inert and persistent chemicals with relatively long biological half-lives and well-defined sources and pathways of exposure (e.g. the metal lead [p3b] that is inhaled or ingested). More complex scenarios, including multiple chemical, multiple route of entry to the body and multiple pathway exposures, will need to complement biological information with large amounts of chemical and physical data (e.g. multimedia dynamics of the chemical). Table 1.1 provides examples of available population biomarker databases that can complement biomonitoring data.

Assessing biological doses and their effects using exposure measurements constitutes a "forward" analytical approach, whereas estimating or reconstructing exposures from biomarkers invokes an "inverse" methodology. The forward analysis can be accomplished through the direct application of exposure, toxicokinetic, and toxicodynamic models (discussed in Chapter 2), which can be either empirical or mechanistic (i.e. biologically based). Reconstruction requires application of both numerical model inversion techniques and toxicokinetic and/or toxicodynamic models. Physical, chemical, and biological information must be merged into biochemodynamic information to underpin a systematic, environmental assessment.

Physiologically based toxicokinetic (PBTK) and biologically based dose-response (BBDR) models combined with numerical inversion techniques and optimization methods form a biochemodynamic framework to support environmental risk assessment (see Figure 1.2). The inversion approach contrasts with so-called "brute-force sampling," wherein possible factors as evaluated one-by-one. The biochemodynamic approach calls for a systematic evaluation of available methods and computational tools that can be used to "merge" existing forward models and biomarker data [2].

ASSESSING BIOTECHNOLOGICAL IMPACTS

Any consideration of present or future environmental problems requires a systematic perspective. Everything in the environment is interconnected. If we do not ask questions about the possible environmental impacts of biotechnologies and we have no data from which to answer these questions, we may be unpleasantly surprised in time when ecological and human health problems occur. This is doubly bad if such problems could have been prevented with a modicum of foresight. Since this is actually the rationale for environmental impact statements (EISs), they provide a worthwhile framework for the application of biochemodynamics in environmental assessments.

The National Environmental Policy Act (NEPA) was the first of the major pieces of legislation in the United States to ask that the environment be viewed systematically. It was signed into law in 1970 after contentious hearings in the US Congress. NEPA is not really a technical law, but created the environmental impact statement (EIS) and established the Council on Environmental Quality (CEQ) in the Office of the President. Of the two, the EIS represented a sea change in how the federal government was to conduct business. Agencies were required to prepare EISs on any major action that they were considering that could "significantly" affect the quality of the environment. From the outset, the agencies had to reconcile often-competing values, i.e. their mission and the protection of the environment. This ushered in a new environmental ethos that continues today.

Biotechnologies are tailor-made for the assessment process, owing to their complexities and the difficulty of predicting side effects and unexpected outcomes. For example, the US Department of Agriculture's Biotechnology Regulatory Services program and Animal and Plant Health Inspection Service regulate the importation, movement, and potential releases of genetically engineered (GE) organisms, especially plants, insects, and microorganisms that may pose a plant pest risk [3]. The USDA works with the US Environmental Protection Agency (EPA) and the Food and Drug Administration (FDA), since GE organisms are also used for environmental, medical, and industrial applications. Thus, in the United States, the federal

Table 1.1 Examples of biomarker databases available to conduct exposure reconstructions

	OP			PYR			Metals					
Program/Study	Chlorpyrifos	Diazinon	Malathion	Permethrins	Cyfluthrin	Cypermethrin	As	Cd	Cr	Ho\MeHg	Pb	Location: Number of subjects
CHAMACOS (1999–2000) Castornia *et al*., 2003	bd	bd	bd									CA: 600 pregnant women
CTEPP (2000–01) Wilson *et al*., 2004 (*)	ac	ac		ad	ad							NC: OH: 257 children (1.5-5 yr)
MNCPES (1997) Quackenboss *et al*., 2000 (*)	ac	ac	ac									MN: 102 children (3-12 yr)
NHANES-III (1988–94) Hill *et al*., 1995 (*)	c							c			bc	US: 1000 adults (20-59 yr)
NHANES (1999–2000) CDC, 2005 (*)	cd	cd	cd					c		c	bc	US: 9,282 subjects (all ages)
NHANES 2001–02 CDC, 2005 (*)	cd	cd	cd	c	cd	c		c		c	bc	US: 10,477 subjects (all ages)
NHANES 2003–04 (*)	cd	cd	cd	c	cd	c	c	c		c	bc	US: 9,643 subjects (all ages)
NHEXAS-AZ (1995–97) Robertson *et al*., 1999	ac	ac	ac				ac	ac	ac		ac	AZ: 179 subjects (all ages)

NHEXAS-MD (1995–96)	ac	ac	ac	ac	ac		ac	MD: 80 subjects (above 10 yr)
NHEXAS-V (1995–97) Whitmore *et al.*, 1999 (*)			ac	ac	ac	c	ac	EPA Region V: 251 subj. (all ages)

Notes:
* = databases first to be analyzed by Georgopoulos *et al.*
a = Measurements of multimedia concentrations (indoor, outdoor, and personal air; drinking water; duplicate diet; dust; and soil).
b = Partial measurements of environmental concentrations (e.g. outdoor air concentrations; pesticide use; etc.).
c = Specific metabolites.
d = Non-specific metabolites.
Abbreviations: OP: organophosphates; PYR: pyrethroids. CHAMACOS = Center for the Health Assessment of Mothers and Children of Salinas; CTEPP = Children's Total Exposures to Persistent Pesticides and Other Persistent Organic Pollutants; MNCPES = Minnesota Children's Pesticide Exposure Study; NHANES = National Health and Nutrition Examination Survey; and NHEXAS = National Human Exposure Assessment Survey.

Referenced studies:
R. Castorina, A. Bradman, T.E. McKone, D.B. Barr, M.E. Harnly and B. Eskenazi (2003). Cumulative organophosphate pesticide exposure and risk assessment among pregnant women living in an agricultural community: a case study from the CHAMACOS cohort. *Environmental Health Perspectives* 111(13): 1640–1648.
N.K. Wilson, J.C. Chuang, R. Iachan, C. Lyu, S.M. Gordon, M.K. Morgan, *et al.* (2004). Design and sampling methodology for a large study of preschool children's aggregate exposures to persistent organic pollutants in their everyday environments. *Journal of Exposure Analysis and Environmental Epidemiology* 14(3): 260–274.
J.J. Quackenboss, E.D. Pellizzari, P. Shubat, R.W. Whitmore, J.L. Adgate, K.W. Thomas, *et al.* (2003). Design strategy for assessing multi-pathway exposure for children: the Minnesota Children's Pesticide Exposure Study (MNCPES). *Journal of Exposure Analysis and Environmental Epidemiology* 10(2): 145–158.
R.L. Hill, Jr, S.L. Head, S. Baker, M. Gregg, D.B. Shealy, S.L. Bailey, *et al.* (1995). Pesticide residues in urine of adults living in the United States: reference range concentrations. *Environmental Research* 71(2): 99–108.
CDC (2005). Third National Report on Human Exposure to Environmental Chemicals. NCEH Pub. No. 05-0570, Centers for Disease Control and Prevention, Atlanta, Georgia; http://www.cdc.gov/exposurereport/; accessed on August 12, 2009.
G.L. Robertson, M.D. Lebowitz, M.K. O'Rourke, S. Gordon and D. Moschandreas (1999). The National Human Exposure Assessment Survey (NHEXAS) study in Arizona: Introduction and preliminary results. *Journal of Exposure Analysis and Environmental Epidemiology* 9(5): 427–434.
R.W. Whitmore, M.Z. Byron, C.A. Clayton, K.W. Thomas, H.S. Zelon, E.D. Pellizzari, P.J. Lioy and J.J. Quackenboss (1999). Sampling design, response rates, and analysis weights for the National Human Exposure Assessment Survey (NHEXAS) in EPA region 5. *Journal of Exposure Analysis and Environmental Epidemiology* 9(5): 369–380.

Source: P.G. Georgopoulos, A.F. Sasso, S.S. Isukapalli, P.J. Lioy, D.A. Vallero, M. Okino and L. Reiter (2009). Reconstructing population exposures to environmental chemicals from biomarkers: Challenges and opportunities. *Journal of Exposure Science and Environmental Epidemiology* 19: 149–171.

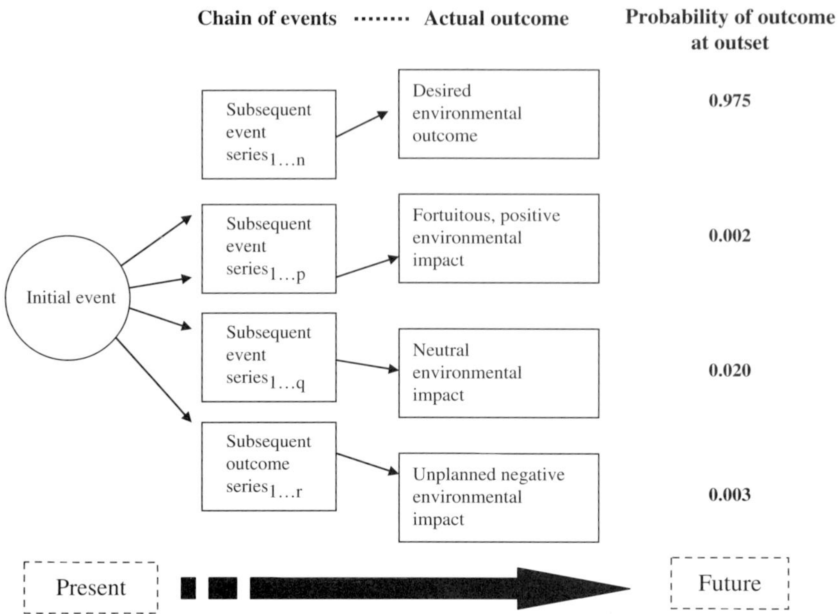

FIGURE 1.2
Hypothetical event tree of possible outcomes from the initial action (e.g. using genetically modified microbes to breakdown a chemical waste in an aquifer).

government seems to be aware of the need to look at possible implications in a systematic way. Under the biotechnology regulations, transgenic plants, insects, mollusks, and microbes are subject to regulation if they potentially pose a plant pest risk. A large number of organisms are included.

Any major action by a federal agency that may significantly affects the human environment falls under NEPA, which means that environmental impacts must be considered prior to undertaking the action. Unless an agency action is categorically excluded from a NEPA-mandated environmental analysis, the agency must analyze the action through the preparation of an environmental assessment (EA), and if needed, an EIS. An action that would result in "less-than-significant" or no environmental impacts can be categorically excluded. For example, a categorical exclusion would apply to the permitting of the "confined release of a GE organism involving a well-known species that does not raise any new issues." However, a categorical exclusion is not an "exemption" from NEPA, merely a determination that an EA or EIS is not necessary.

Many ecologists would argue that *any* action will impact ecosystems, since they are entirely interconnected to other systems and the systems within the ecosystem are influenced by changes among these systems. However, the NEPA catchword is "significant." Scientists, especially statisticians, have difficulty with the use of this term outside of prescribed boundaries, e.g. significant at the 0.05 level (5% likelihood that the outcome occurred due to chance). However, in general use, the term seems to indicate importance or that the action's impacts are substantial.

Environmental assessments consider the need for the proposed action, especially highlighting and evaluating possible alternatives, including a "no-action" alternative. In other words, would the environment be better off if the action is not taken compared to all of the other alternative actions? All of these options are viewed in terms of potential impacts, with a ranking or comparison of the alternatives, and a recommendation to decision makers on how best to implement the proposed program with the least environmental implications. The EA is mainly a step to determine, in a publicly available document, whether to prepare an EIS. If the proposed action lacks a significant impact on the environment, the government agency will issue a Finding of No Significant Impact (FONSI) [4]. If it determines that an aspect of the quality of the human environment may be "significantly affected" by the proposed action, then agency is required to prepare an EIS, which involves a more in-depth inquiry into the proposal and any "reasonable" alternatives to it. For example, the USDA writes an EA before granting permits for introductions of GE organisms that are considered new or novel (the crop species, the trait, or both), with an opportunity for public comment before a permit is granted. The agency also prepares an EA when it decides that a GE plant or microorganism will no longer be regulated. The steps in the EA process are:

Consultation and coordination with other federal, tribal, state, or local agencies;
Public scoping;
Federal Register notices;
Public comments on a draft EA;
Public meetings on a draft EA;
Publication of final EA and FONSI; and
Supplements to a previous EA.

An EIS is more detailed and comprehensive than the EA. Agencies often strive to receive a so-called FONSI, so that they may proceed unencumbered on a mission-oriented project [5]. The process assists in deciding whether to use the NEPA process to improve decision making behind projects of a more narrow scope, such as the deregulation of a specific GE crop. The evaluation includes a discussion of direct, indirect, and cumulative impacts resulting from the adoption of one of several reasonable alternatives, including the no-action alternative. Again, this is evidence of the need for a systematic view of biotechnological environmental impacts.

The EIS may also specify actions that would mitigate any impact of the biotechnology product, that is, possible measures that could reduce any potential impact would be put into place prior to project implementation. Thus, an EIS can only be written by a multidisciplinary team of experts.

The EIS process includes:

Consultation and coordination with other federal, tribal, state, or local agencies, when appropriate;
Scoping;
Federal Register notices;
Public comment on draft EIS;
Public meetings on draft EIS when appropriate;
Publication of a final EIS; and
Supplements to an inadequate EIS, when necessary.

States also have their own environmental assessment processes (see Table 1.2). Like the federal EIS process, the states have their own emphases and concerns about environmental impacts.

The EIS process, when followed properly, is an example of underpinning environmental decisions with reliable biochemodynamic information. For example, in the USDA process, decisions about field-testing of GE crops must ensure that these tests neither pose plant pest risks nor pose significant impacts to the human environment [6]. An incomplete or inadequate assessment will lead to delays and increase the chance of an unsuccessful project, so sound science is needed from the outset of the project design. Even worse, a substandard assessment may allow for hazards and risks down the road.

The final EIS step is the Record of Decision (ROD). This means that someone in the agency will be held accountable for the decisions made and the actions taken. The ROD describes the alternatives and the rationale for final selection of the best alternative. It also summarizes the comments received during the public reviews and how the comments were addressed. Many states have adopted similar requirements for RODs.

Table 1.2 North Carolina's State Environmental Policy Act (SEPA) review process

Step I: Applicant consults/meets with Department of Environment and Natural Resources (DENR) about potential need for SEPA document and to identify/scope issues of concern.
Step II: Applicant submits draft environmental document to DENR.
Environmental document is either an environmental assessment (EA) or an environmental impact statement (EIS).
Step III: DENR-Lead Division reviews environmental document.
Step IV: DENR-Other Divisions review environmental document.
15–25 calendar days. DENR issues must be resolved prior to sending to the Department of Administration – State Clearinghouse (SCH) review.
Step V: DENR-Lead Division sends environmental document and FONSI (a) to SCH.
Step VI: SCH publishes Notice of Availability for environmental document in NC Environmental Bulletin. Copies of environmental document and FONSI are sent to appropriate state agencies and regional clearinghouses for comments.
Interested parties have either 30 (EA) or 45 (EIS) calendar days from the Bulletin publication date to provide comments.
Step VII: SCH forwards copies of environmental document comments to DENR-Lead Division who ensures that applicant addresses comments.
SCH reviews applicant's responses to comments and recommends whether environmental document is adequate to meet SEPA.
Substantial comments may cause applicant to submit revised environmental document to DENR-Lead Division. This will result in repeating of Steps III–VI.
Step VIII: Applicant submits final environmental document to DENR-Lead Division.
Step IX: DENR-Lead Division sends final environmental document and FONSI (in case of EA and if not previously prepared) to SCH.
Environmental Assessment (EA)
Step X: SCH provides letter stating one of the following:

- Document needs supplemental information, or
- Document does not satisfy a FONSI, and an EIS should be prepared, or
- Document is adequate; SEPA is complete.

Environmental Impact Statement (EIS)
Step XI: After lead agency determines the FEIS is adequate, SCH publishes a Record of Decision (ROD) in the NC Environmental Bulletin.

Notes:
PUBLIC HEARING(S) ARE RECOMMENDED (BUT NOT REQUIRED) DURING THE DRAFT STAGE OF DOCUMENT PREPARATION FOR BOTH EA AND EIS.
For an EA, if no significant environmental impacts are predicted, the lead agency (or sometimes the applicant) will submit both the EA and the Finding of No Significant Impact (FONSI) to SCH for review (either early or later in the process).
Finding of No Significant Impact (FONSI): Statement prepared by Lead Division that states proposed project will have only minimal impact on the environment.

The EIS documents were supposed to be a type of "full disclosure" of actual or possible problems if a federal project is carried out. Fully disclosing possible impacts can be likened to Lorenz's view of chaos (*see* Discussion Box: Little Things Matter in a Chaotic World). For example, even a low probability outcome must be considered. In fact, most environmental risk calculations deal with low probabilities (e.g. a one-in-a-million risk of cancer following a lifetime exposure to a certain carcinogen). As shown in Figure 1.2, the fact that the vast majority of outcomes will be the desired effect does not obviate the need to consider all potential outcomes. This hypothetical example has four possible outcomes (in most situations there are myriad outcomes) from one initial event (e.g. the use of a genetically modified microbe to break down a persistent chemical). The good news is that 97.5% of the time, the beneficial outcome is achieved. And, sometimes an unplanned benefit is realized (0.2%). Most of the rest of the outcomes are neither good nor bad (2%). However, on rare occasions, given the complexities and variable environmental conditions, the beneficial outcomes do not occur and negative impacts ensue (0.3%). NEPA and other systematic decision support

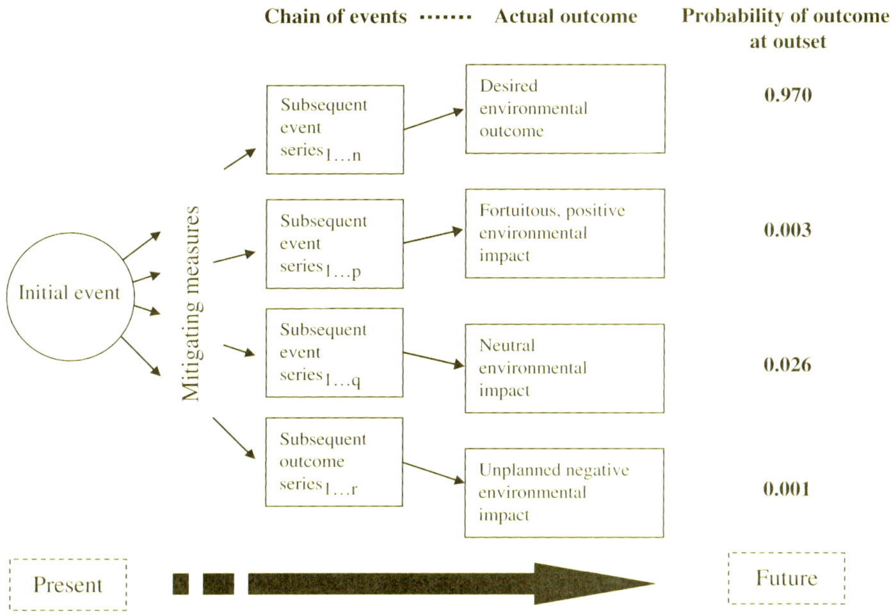

FIGURE 1.3
Hypothetical event tree of possible outcomes from the initial action with the addition of mitigating measures to decrease the likelihood of negative impacts.

systems must help to determine whether 0.3% risk of a negative outcome is acceptable. This depends on the severity, persistence, and extent of the negative impact. For example, if the microbial population of an ecosystem is not in danger of irreversible or long-term damage and the scope of the problem is contained, then this probability of harm may be acceptable. However, if the microbial population changes and there is long-term loss of biodiversity, the risk may not be worth it. For these scenarios, other alternatives must be sought.

Figure 1.3 shows the same hypothetical scenario as that in Figure 1.2, but in this case actions are taken to prevent some of the adverse outcomes. Such mitigating measures [7] can include better matches of microbes to the specific environmental conditions, more frequent and reliable monitoring of the project, and using safer (e.g. non-genetically modified microbes) methods. The likelihood of the adverse outcome has fallen to 0.1%, but the likelihood of the desired outcome has decreased to 97.3% (including the fortuitous benefits).

These may seem like small differences, but environmental decisions often hinge on a few parts per billion or a risk difference of 0.00001 on whether a project is acceptable. Thus, in this hypothetical case, the measure of success has decreased from 97.5% to 97.3%, or a success rate decrease of 0.2%. Sometimes, such a drop affects cleanup rates (e.g. the 0.2% rate translates into another three months before a target cleanup level is achieved). It may also translate into the inability for some microbes to break down certain recalcitrant pollutants. Conversely, the better adverse outcomes may well be worth it, if the 0.1% improvement means less ecosystem effects and fewer releases and exposures to toxic substances.

The reasons given for not taking mitigating measures often have to do with costs and efficiencies. For example, in the scenario described in Figure 1.3, the naturally available microbes may be slower to degrade the compound, so at the same point in time in the future, less of the waste has been detoxified. Even though the possibility of negative outcomes has been cut, so has the removal of the toxic waste. This is an example of a contravening risk and risk tradeoff; that is, we must decide whether a less efficient contaminant cleanup (human health risk) is more important than ecosystem integrity (ecological risk).

In a complete event tree, all of the events following the initial event would need to be considered. This is the only way that the probability of the final outcome (positive, neutral or negative) can be calculated as the result of contingent probabilities down the line. In fact, each of the mitigating measures shown in Figure 1.3 has a specific effect on the ultimate probability of the outcome. Thus, mitigating measures can be seen as interim events with their own contingent probability (e.g. choosing natural attenuation versus enhanced biodegradation lowers the probability of the desired rate of biodegradation, but also lowers the probability of adverse genetic effects on the ecosystem).

The systematic approach considers all of the potential impacts to the environment from any of the proposed alternatives, and compares those outcomes to a "no action" alternative. In the first years following the passage of NEPA many agencies tried to demonstrate that their "business as usual" was in fact very environmentally sound. In other words, the environment would be better off with the project than without it (action is better than no action). Too often, however, an EIS was written to justify the agency's mission-oriented project. One of the key advocates for the need for a national environmental policy, Lynton Caldwell, is said to have referred to this as the federal agencies using an EIS to "make an environmental silk purse from a mission-oriented sow's ear!" [8].

The courts adjudicated some very important laws along the way, requiring federal agencies to take NEPA seriously. Some of the aspects of the "give and take" and evolution of federal agencies' growing commitment to environmental protection was the acceptance of the need for sound science in assessing environmental conditions and possible impacts, and the very large role of the public in deciding on the environmental worth of a highway, airport, dam, waterworks, treatment plant, or any other major project sponsored by or regulated by the federal government. This was a major impetus in the growth of the environmental disciplines since the 1970s. Experts were needed who could not only conduct sound science but who could communicate what their science means to the public.

All federal agencies must adhere to a common set of regulations [9] to "adopt procedures to ensure that decisions are made in accordance with the policies and purposes of the Act." Agencies are required to identify the major decisions called for by their principal programs and make certain that the NEPA process addresses them. This process must be set up in advance, early in the agency's planning stages. For example, if waste remediation or reclamation is a possible action, the NEPA process must be woven into the remedial action planning processes from the beginning with the identification of the need for and possible kinds of actions being considered.

Noncompliance or inadequate compliance with NEPA rules and regulations can lead to severe consequences, including lawsuits, increased project costs, delays, and the loss of the public's trust and confidence, even if the project is designed to improve the environment, and even if the compliance problems seem to be only "procedural." The US EPA is responsible for reviewing the environmental effects of all federal agencies' actions. This authority was written as Section 309 of the Clean Air Act (CAA). The review must be followed with the EPA's public comments on the environmental impacts of any matter related to the duties, responsibilities, and authorities of EPA's administrator, including EISs. The EPA's rating system (see Appendix 1) is designed to determine whether a proposed action by a federal agency is unsatisfactory from the standpoint of public health, environmental quality, or public welfare. This determination is published in the Federal Register and referred to the CEQ.

BIOTECHNOLOGY AND BIOENGINEERING

Biotechnology as an endeavor is not new. Even the term itself is almost a century old. Karl Ereky, a Hungarian engineer, is credited with coining the word "biotechnology" in 1919 when he referred to approaches that recruited the help of living organisms to produce materials. More recently, 1992, the Convention on Biological Diversity settled on defining biotechnology as "any technological application that uses biological systems, living organisms or derivatives thereof, to make or modify products and processes for specific use" [10]. This goes well beyond microbiology, thus, an understanding of biotechnology entails a need to consider the chemical processes at work in living systems.

One of the challenges of this book is to get a sense of the extent to which existing engineering analytical tools can support biotechnological decision making, especially as this applies to potential environmental impacts. One indispensable tool used to assess the complete environmental footprint of a process is the life cycle analysis (LCA).[a] We will address this in much greater detail in subsequent chapters, but it is worth noting now that LCA is more than a particular software package or set of engineering diagrams and charts. It is a way of considering the history and future of a biotechnological enterprise as a complete system with respect to inputs and outputs. As such, it provides a means of demonstrating and evaluating the physical, chemical, and biological systems within a system. That is, the LCA considers the environmental worthiness of any endeavor, including biotechnologies, within the context of first principles of thermodynamics, motion and the other laws and theories that underpin the system. All of the energy and matter inputs must balance with outputs. As such, the outcomes can be studied rationally and objectively.

Numerous cases demonstrate the lack of a life cycle perspective (see Discussion Box: Little Things Matter in a Chaotic World). The systematic nature of LCA extends from these first physical principles to biological principles. Arguably, the two bioengineering disciplines are biomedical and environmental engineering. Both deal directly and indirectly with living things and, as such, with biotechnologies. Both approach biology as a means of understanding and managing risks to living organisms, especially humans. Often, however, the information that goes into LCAs is qualitative, such as the hazards posed directly or indirectly by organisms, especially genetically modified microorganisms (*see* Tables 1.3 and 1.4). Note that such hazards extend to both human populations and ecosystems [11].

Table 1.3 **European Federation of Biotechnology's classes of risks posed by genetically modified microorganisms**

Hazard level	Description of microbial hazard
Least	Never identified as causative agents of disease in humans nor offer any threat to the environment
Hazardous when contained, low human population risk	May cause disease in human and might, therefore, offer a hazard to laboratory workers. They are unlikely to spread in the environment. Prophylactics are available and treatment is effective
Severe when contained, moderate human population risk	Severe threat to the health of laboratory workers, but a comparatively small risk to the population at large. Prophylactics are available and treatment is effective
High human population risk	Severe illness in humans and serious hazard to laboratory workers and to people at large. In general, effective prophylactics are not available and no effective treatment is known
Greatest ecological and human population risk	Most severe threat to the environment, beyond humans. May be responsible for heavy economic losses. Includes several classes, Ep1, Ep2, Ep3 (see Table 1.4 for description) to accommodate plant pathogens

Source: Adapted from: B. Jank, A. Berthold, S. Alber and O. Doblhoff-Dier (1999). Assessing the impacts of genetically modified microorganisms. *International Journal of Life Cycle Analysis* 4 (5): 251–252.

[a] LCA can also be shorthand for life cycle assessment, which for the sake of this discussion, is synonymous with life cycle analysis.

Table 1.4 European Federation of Biotechnology classes of microorganisms causing diseases in plants

European Federation of Biotechnology Class	Description of Microbes in Class
Ep 1	May cause diseases in plants but have only local significance. They may be mentioned in a list of pathogens for the individual countries concerned. Very often they are endemic plant pathogens and do not require any special physical containment. However, it may be advisable to employ good microbiological techniques
Ep 2	Known to cause outbreaks of disease in crops as well as in ornamental plants. These pathogens are subject to regulations for species listed by authorities in the country concerned
Ep 3	Mentioned in quarantine lists. Importation and handling are generally forbidden. The regulatory authorities must be consulted by prospective users

Source: Compiled from: H.L.M. Lelieveld, B. Boon, A. Bennett, G. Brunius, M. Cantley, A. Chmiel, *et al.* (1996). Safe biotechnology. 7. Classification of microorganisms on the basis of hazard. *Applied Microbiology and Biotechnology* 45: 723-729; W. Frommer and the Working Party on Safety in Biotechnology of the European Federation of Biotechnology (1992). Safe biotechnology. 4. Recommendations for safety levels for biotechnological operations with microorganisms that cause diseases in plants. *Applied Microbiology and Biotechnology* 38:139–140; and M. Küenzi and the Working Party on Safety in Biotechnology of the European Federation of Biotechnology (1985). Safe biotechnology. General considerations. *Applied Microbiology and Biotechnology* 21: 1–6.

DISCUSSION BOX

Little Things Matter in a Chaotic World

Most would agree that the Monarch butterfly is a beautiful creature. What if we lost it because of biotechnology, as suggested by a recent report? The report stated that pollen from corn that had been genetically engineered with genetic material from soil bacterium *Bacillus thuringiensis* (Bt) posed a threat of killing Monarch butterfly larvae [12].

The Bt produces a protein that targets insect pests. Scientists "borrow" the genetic material that expresses this protein and insert it into plant species, including corn.

The original report showed only that Bt-containing pollen fed directly to Monarch larvae is toxic but did not include realistic field exposures. Since then, more intensive studies suggest the risk is low enough to be acceptable, given the benefits of insect resistance.

Some studies indicated that corn pollen normally travels in limited distances and that the pollen has a tendency not to accumulate on the favored Monarch food, i.e. milkweed leaves. Also, pollen production usually does not occur at the same time as the active feeding by Monarch larvae. These factors supported the US EPA decision to continue to approve the planting of Bt corn. The question in such decisions is whether the decision was based on sufficient field studies and the possibility of the combination of rare events.

Biotechnology puts living things to work for certain purposes. An excellent example is *phytoremediation*, which utilizes biochemodynamic processes to remove, degrade, transform, or stabilize contaminants that reside in soil and groundwater (see Figure 1.4). Subtle changes in any of these processes can make the difference between a successful remediation effort and a failure. Phytoremediation uses plants to capture the water from plumes of contaminated aquifers. The plants take up the water by the capillary action of their roots, transport it upward through the plant until the water is transpired to the atmosphere. The good news is that many of the contaminants have been biochemically transformed or at least sequestered in the plant tissue.

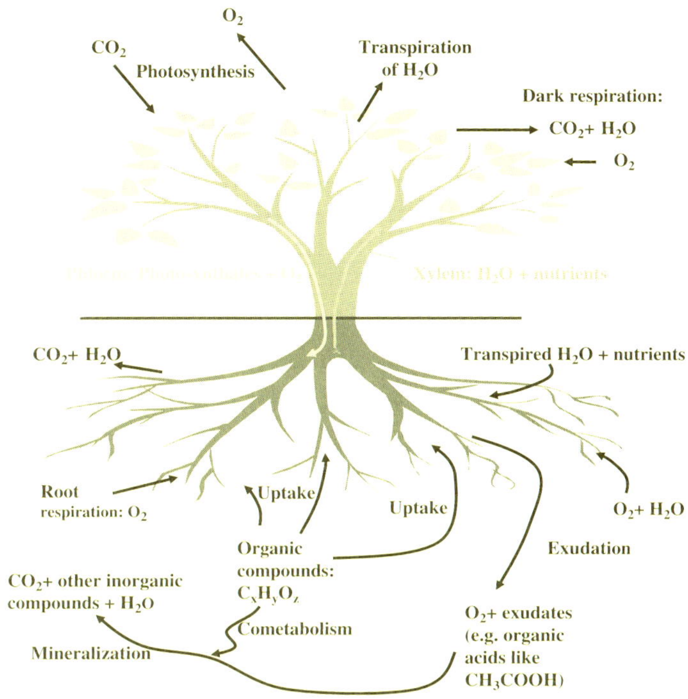

FIGURE 1.4
Biochemodynamic processes at work in phytoremediation (see Chapter 7). [*See color plate section*]
Source: Adapted from R. Kamath, J.A. Rentz, J.L. Schnoor and P.J.J. Alvarez (2004). Phytoremediation of hydrocarbon-contaminated soils: principles and applications. *Studies in Surface Science and Catalysis* 151: 447–478.

Plants do not metabolize organic contaminants to carbon dioxide and water as microbes do. Rather they transform parent compounds into non-phytotoxic metabolites. After uptake by the plant, the contaminant undergoes a series of reactions to convert, conjugate, and compartmentalize the metabolites. Conversion includes oxidation, reduction, and hydrolysis. Conjugation reactions chemically link these converted products (i.e. phase 1 metabolites) to glutathione, sugars, or amino acids, so that the metabolites (i.e. phase 2 metabolites) have increased aqueous solubility and, hopefully, less toxicity than the parent compound. After this conjugation, the compounds are easier for the plant to eliminate or compartmentalize to other tissues. Compartmentalization (phase 3) causes the chemicals to be segregated into vacuoles or bound to the cell wall material, such as the polymers lignin and hemicellulose. Phase 3 conjugates are considered to be bound residues in that laboratory extraction methods have difficulty finding the original parent compounds [13].

Enter the butterfly. It turns out that some of the phytoremediation products of conversion reactions can become more toxic than the parent contaminants when consumed by animals or potentially leached to the environment from fallen leaves. For example, the release of contaminants from conjugated complexes or compartmentalization could occur in the gut of a worm, a snail, or a *butterfly* [14]. This means that there is a distinct possibility of re-introducing the pollutant, by means of the butterfly, into the food chain.

Ironically, the butterfly is also the metaphor for chaos. Edward Lorenz's Butterfly Effect postulates "sensitive dependence upon initial conditions" [15] as a postulate of chaos theory: a small change for good or bad can reap exponential rewards and costs. Lorenz, at a 1963 New York Academy of Sciences meeting, related the comments of a "meteorologist who had remarked that if the theory were correct, one flap of a seagull's wings would be enough to alter the course of the weather forever." Lorenz later revised the seagull example to be that of a butterfly in his 1972 paper "Predictability: Does the Flap of a Butterfly's Wings in Brazil Set Off a Tornado in Texas?" at a meeting of the American Association for the Advancement

(Continued)

of Science in Washington, DC. In both instances Lorenz argued that future outcomes are determined by seemingly small events cascading through time. Engineers and mathematicians struggle with means to explain, let alone predict, such outcomes of so-called "ill-posed" problems. Engineers generally prefer orderly systems and a well-posed problem, that is, one that is uniquely solvable (i.e. a unique solution exists) and one that is dependent upon a continuous application of data. By contrast, an ill-posed problem does not have a unique solution and can only be solved by discontinuous applications of data, meaning that even very small errors or perturbations can lead to large deviations in possible solutions [16].

The importance of seemingly small things within a systematic approach is demonstrated by what happened near the Iron Gate Dam in Europe. This case demonstrates the enormous ecological price that must be paid when biodiversity is destroyed. The case is very interesting in that something that we do not ordinarily consider to be a pollutant or even a limiting ecological factor, silicates, led to major problems.

The Black Sea is the largest enclosed catchment basin, receiving freshwater and sediment inputs from rivers draining half of Europe and parts of Asia. The Danube River, which flows into the Black Sea, receives effluents from eight European countries, and is the largest source of stream-borne nutrients. The sea is highly sensitive to eutrophication, i.e. nutrient enrichment leading to adverse trophic changes, and has experienced change numerous times in recent decades. In less than a decade, the system changed from an extremely biologically diverse one to a system dominated by jellyfish (*Aurelia* and the combjelly *Mnemiopsi*) [17].

These invaders were unintentionally introduced in the mid-1980s, culminating in the fisheries almost completely vanishing by the early 1990s. This collapse was first attributed to unpalatable carnivores that fed on plankton, roe, and larvae. Subsequently, however, the jellyfish takeover was found to result from human perturbations in the coastal ecosystems and in the drainage basins of the rivers, including changing the hydrologic character of out-flowing rivers. The biggest of these was the damming of the Danube in 1972 by the Iron Gates, approximately 1000 km upstream from the Black Sea. In addition, urban and industrial development, heavy use of commercial fertilizers, over-fishing, and the introduction of exotic, invasive organisms (e.g., *Mnemiopsi*) contributed to the problem. After 1970 this change in nutrient concentrations induced phytoplankton blooms during the warm months and changed the dominance to nonsiliceous species that were not a first choice as food for meso-zooplankton. The decreased fish stocks further increased the dominance of the jellyfish, since they competed better than the game fish for the same food. Ironically, since the mid-1990s, the ecosystems have begun to improve, mainly due to increased nutrient (phosphorus and nitrogen) loading. In most situations, we are looking to decrease this loading, to prevent eutrophication. But in this system, the added nutrients have allowed certain plankton and benthic (bottom dwelling) organisms to re-colonize. The abundance of jellyfish has also stabilized, with a concomitant increase in anchovy eggs and larvae.

Nutrient limitation occurs when the presence of a chemical, such as phosphorus or nitrogen, is insufficient to sustain the growth of community or species. Usually, marine systems are nitrogen-limited whereas freshwater plankton systems are phosphorus-limited. Numerous freshwater organisms can fix atmospheric nitrogen but, with minor exceptions, the nitrogen is impeded in marine water. The nutrient requirements differ by species. A disturbance in the ratio of nitrogen, phosphorus, silica, and even iron changes the biotic composition of a particular plankton community.

Often, all four nutrients can be considered as limiting. For instance, the lack of silica limits diatoms. This was observed first in natural blooms off Cape Mendocino in the United States and subsequently observed in the northwest part of the Black Sea, after closing the Iron Gates dam. The case also demonstrates that economics is crucial, since the marine ecosystem improvement directly corresponds to the decline of the economies of Central and Eastern European nations in the 1990s.

So what is the lesson from the butterfly and jellyfish? Small changes for good or bad can produce unexpectedly large effects. Ignoring the biochemodynamic details can lead to big problems down the road.

The smallest living systems, the viruses, bacteria, and other microbes, are amazing biochemical factories. For much of human history, we have treated them as marvelous black boxes, wherein mysterious and elegant processes take place. These processes not only keep the microbes alive, but they provide remarkable proficiencies to adapt to various hostile environments. Some produce spores; many have durability and protracted latency periods; all have the ability to reproduce in large numbers until environmental conditions become more favorable. The various systems that allow for this efficient survival have become increasingly better understood in recent decades, to the point that cellular and subcellular processes of uptake and absorption, nutrient distribution, metabolism and product elimination have been characterized, at least empirically. More recently, the genetic materials of deoxyribonucleic acid (DNA) and the various forms of ribonucleic acids (RNA) have been mapped. As genes have become better understood, so has the likelihood of their being manipulated. Such manipulation is the stuff of biotechnology.

Biotechnology began as a passive and adaptive approach. For example, sanitary engineers noted that natural systems, such as surface waters and soil, were able to break down organic materials. Such processes are now known as biodegradation. When engineers put microbial populations (usually bacteria and fungi) to work to clean up the soil or groundwater, this is called bioremediation. In other words, biological processes are providing a remedy for some very important societal problems, i.e. polluted resources.

As scientists studied these processes, they realized that various genera of microbes had the ability to use detritus on forest floors, suspended organic material in water and organic material adsorbed onto soil particles as sources of energy needed for growth, metabolism and reproduction. The engineers [18] correctly hypothesized that a more concentrated system could be fabricated to do the same thing with society's organic wastes. Thus, trickling filters, oxidation ponds, and other wastewater treatment systems are merely supercharged versions of natural systems.

In a passive biotechnological system, the microbes used are those found in nature, but that have been allowed to acclimate to the organic material that needs to be broken down. The microbial population's preference for more easily and directly derived electron transfer (i.e. energy sources) is overcome only permitting them to come into contact with the chemicals in the waste. In the presence of oxygen the stoichiometry stated simply is:

$$\text{organic matter} + O_2 \xrightarrow{\text{microbes}} CO_2 + H_2O + \text{cell matter biomass} + \text{other end products} \quad (1.1)$$

Thus, the microbes adapt their biological processes to use these formerly unfamiliar compounds as their energy sources and, in the process, break them down into less toxic substances. Ultimately the microbes degrade complex organic wastes to carbon dioxide and water in the presence of molecular oxygen, or methane and water when molecular oxygen is absent, known as aerobic and anaerobic digestion, respectively.

Numerous examples of passive systems have been put to use with the evolution of complex societies. For example, passive biotechnologies were needed to allow for large-scale agriculture, including hybrid crops and nutrient cycling in agriculture and vaccines in medicine.

Very recently, more active systems have been used increasingly to achieve such societal gains, but at an exponentially faster pace. In addition, scientists have developed biotechnologies that bestow products that simply would not exist using passive systems.

ENVIRONMENTAL BIOTECHNOLOGY AS A DISCIPLINE

If having a professional society is an indication that something is a scientific discipline, then environmental biotechnology meets that requirement. The International Society of Environmental Biotechnology [19] promotes interest in environmental biotechnology and offers the exchange of information regarding the development, use, and regulation of biological systems for remediation of contaminated environments (land, air, water), and for environment-friendly processes (green manufacturing technologies and sustainable development). This definition mainly focuses on the application of biotechnology to the natural environment, but is not nearly broad enough to encompass the comprehensive and complex relationship between biotechnology and the environment.

BIOTECHNOLOGY AND SOCIETY

> Even minor tampering with nature is apt to bring serious consequences, as did the introduction of a single chemical (DDT). Genetic engineering is tampering on a monumental scale, and nature will surely exact a heavy toll for this trespass.
>
> **Eva Novotny [20]**

Society will judge the success of all engineering research, including biotechnology, based on its results. If devices and systems are designed that improve and protect life, then engineers are successful. Conversely, if the risks outweigh the benefits, we have failed. Engineering research is largely evaluated based on its risks and reliability. We will discuss these topics in greater detail in Chapter 4.

As scientists, we are asked if we have appropriately considered all of the possible human and ecological impacts, not merely from the way we conduct research, but also in how that research is or will be applied. Movies and books have challenged researchers to consider the possible future impacts of even seemingly benign research, when such research is emergent. It does not particularly matter that dire consequences are only remote and unlikely outcomes. Biotechnology and other emerging technologies present a particular challenge, owing to the numerous areas of uncertainty. Could the design lead to environmental risk and will this risk be distributed proportionately throughout society? Will the consequences be irreversible and widespread? In short, biotechnological research requires a healthy, honest, and objective perspective and a level of commitment to the future, including sustaining and improving environmental quality. Sometimes scientists become so committed to the possible and actual beneficial aspects of their research that they overlook or at least give less weight to possible, negative outcomes (see Figure 1.5).

Biotechnological research embodies practicality. Of course, bioengineering researchers are interested in advancing knowledge, but always with an eye toward practice. Society demands that the state-of-the-science be advanced as rapidly as possible *and* that no dangerous side effects ensue. The engineering practitioner and researcher are also adept at optimizing among numerous variables for the best design outcomes. Emergent areas are associated with some degree of peril. A recent international query of top scientists [21] asked about what are the most pressing technologies needed to help developing countries (see Table 1.5). Each expert was asked the following questions about the specific technologies:

- Impact. How much difference will the technology make in improving health?
- Appropriateness. Will it be affordable, robust, and adjustable to health care settings in developing countries, and will it be socially, culturally, and politically acceptable?
- Burden. Will it address the most pressing health needs?
- Feasibility. Can it realistically be developed and deployed in a time frame of 5–10 years?
- Knowledge gap. Does the technology advance health by creating new knowledge?
- Indirect benefits. Does it address issues such as environmental improvement and income generation that have indirect, positive effects on health?

All of the areas of need identified in Table 1.5 involve biotechnologies, either directly (as in numbers 5 and 8, the need for improved sequencing and genetically modified organisms) or indirectly (e.g. improved environmental tools. Thus, bioengineers are at the forefront of technological progress and will continue to play an increasingly important role in the future.

FIGURE 1.5
Biotechnological research often has a heavy dose of uncertainty. Usually, organisms that have had their genetic material manipulated are less competitive in natural systems. However, a healthy respect for possible, negative implications is needed, since the introduction of any species to an ecosystem or individual organisms can lead to unexpected consequences.

The concomitant societal challenges require that every engineer fully understand the implications and possible drawbacks of these technological breakthroughs. Key among them will be biotechnical advances at smaller scales, well below the cell and approaching the molecular level.

Technological processes at these scales require that engineers improve their grasp of the potential environmental implications. As biotechnologies advance, so will the concomitant societal challenges. Many engineering and science disciplines will be involved, requiring a better appreciation for and ability to predict the implications and possible drawbacks of technological developments. Key among them will be biotechnical advances at smaller scales that approach the molecular level. Biotechnological processes at these scales, for example, require that engineers improve their grasp of the potential environmental implications.

RISKS AND RELIABILITY OF NEW BIOTECHNOLOGIES

There are two major schools of thought on determining whether to pursue biotechnologies: the precautionary approach and the evidence-based approach. The precautionary principle states:

> When an activity raises threats of harm to human health or the environment, precautionary measures should be taken even if some cause-and-effect relationships are not fully established scientifically. [22]

Table 1.5 Ranking by global health experts of top ten biotechnologies needed to improve health in developing countries

Final ranking	Biotechnology
1	Modified molecular technologies for affordable, simple diagnosis of infectious diseases
2	Recombinant technologies to develop vaccines against infectious diseases
3	Technologies for more efficient drug and vaccine delivery systems
4	Technologies for environmental improvement (sanitation, clean water, bioremediation)
5	Sequencing pathogen genomes to understand their biology and to identify new antimicrobials
6	Female-controlled protection against sexually transmitted diseases, both with and without contraceptive effect
7	Bioinformatics to identify drug targets and to examine pathogen–host interactions
8	Genetically modified crops with increased nutrients to counter specific deficiencies
9	Recombinant technology to make therapeutic products (for example, insulin, interferons) more affordable
10	Combinatorial chemistry for drug discovery

Source: Data from survey conducted in A.S. Daar, H. Thorsteinsdóttir, D.K. Martin, A.C. Smith, S. Nast and P.A. Singer (2002). Top ten biotechnologies for improving health in developing countries. *Nature Genetics* 32: 229–232.

The precautionary perspective requires that a threshold of certainty be met before allowing a new technology. The evidence-based perspective allows the technology so long as the evidence supports it. This may seem similar, but in fact, the two viewpoints are quite different with respect to onus. The precautionary approach clearly places the onus on the technologist who wants to use the biotechnology to prove that it is safe under every possible scenario, whereas the evidence-based approach allows the technology if it undergoes a risk assessment that shows that the risks, even those that are unknown, are acceptable.

Risk management is an example of optimization. However, optimizing among variables is not usually straightforward for biotechnology and bio-engineering applications. Optimization models often apply algorithms to arrive at a net benefit/cost ratio, with the selected option being the one with the largest value, i.e. greatest quantity of benefits compared to costs. Steven Kelman of Harvard University was one of the first to articulate the weaknesses and dangers of taking a purely utilitarian approach in managing environmental, safety and health risks [23]. Kelman asserts that in such risk management decisions, a larger benefit/cost ratio does not always point to the correct decision. He also opposes the use of dollars, i.e. monetization of non-marketed benefits or costs, to place a value on environmental resources, health and quality of life. He uses a logical technique of *reductio ad absurdum* (from Greek, η? εις το αδυνατον απαγωγη, "reduction to the impossible") where an assumption is made for the sake of argument, a result found, but it is so absurd that the original assumption must have been wrong [24]. For example, the consequences of an act, whether positive or negative, can extend far beyond the act itself. Kelman gives the example of telling a lie. Using the pure benefit/cost ratio, if the person telling the lie has much greater satisfaction (however that is quantified) than the dissatisfaction of the lie's victim, the benefits would outweigh the cost and the decision would be morally acceptable. At a minimum, the effect of the lie on future lie-telling would have to be factored into the ratio, as would other cultural norms.

Another of Kelman's examples of flaws of utilitarianism is the story of two friends on an Arctic expedition, wherein one becomes fatally ill. Before dying, he asks that the friend return to that very spot on the Arctic ice in 10 years to light a candle in remembrance. The friend promises to do so. If no one else knows of the promise and the trip would be a great inconvenience, the benefit/cost approach instructs him not to go (i.e. the costs of inconvenience outweigh the benefit of the promise because no one else knows of the promise). These examples illustrate that benefit/cost information is valuable, but care must be taken in choosing the factors that go into the ratio, properly weighing subjective and non-quantifiable data, ensuring that the views of those affected by the decision are properly considered, and being mindful of possible conflicts of interest and undue influence of special interests.

From either the precautionary or evidence-based scientific perspective, environmental systems must be viewed comprehensively. There is no way to evaluate a biotechnology without considering the concept of risk [25]. Managing risks to human and ecosystem health is one of the principal bioengineering mandates. Reliability lets us know how well the natural and engineered biosystems are working, not only in their delivery of manufactured products, but also in ensuring that wastes are not released, that exposures are prevented and minimized and that ecosystems are protected now and in the future.

Risk, as it is generally understood, is the chance that some unwelcome event will occur. The operation of an automobile, for example, introduces the driver and passengers to the risk of a crash that can cause damage, injuries, and even death. Understanding the factors that lead to a risk is known as *risk analysis*. The reduction of this risk (for example, by installing airbags and wearing seat belts in the driving example) is *risk management*. Risk management is often differentiated from *risk assessment*, which is comprised of the scientific considerations of a risk. Risk management includes the policies, laws, and other societal aspects of risk.

Enlisting the biological sciences to address society's medical, industrial, agricultural, and environmental challenges must be accompanied by considerations of the interrelationships among factors that put people and the environment at risk, suggesting that biotechnologists are indeed risk analysts. Technologies must be based on the sound application of the physical and social sciences. The public expects safe products and processes, and the public holds biotechnologists accountable for its health, safety, and welfare. Engineers employ systems to reduce and to manage risks. So, bioengineers employ biosystems, to reduce and to manage environmental risks. Risk is indirectly proportional to reliability. Thus, bioengineers seek ways to enhance the reliability of these systems. Like all engineering disciplines, biotechnological design deals directly or indirectly with risk by improving system reliability.

Both risk and reliability are expressed as probabilities. Most everyone, at least intuitively, assesses the risks of new technologies before they use them and, when presented solutions by engineers, makes decisions about the reliability of the designs. People, for good reason, want to be assured that every aspect of a new biotechnology is and will be "safe." But, safety is a relative term. Calling something "safe" integrates a value judgment that is invariably accompanied by uncertainties. The safety of a product or process can be described, at least to some extent, in objective and quantitative terms. Factors of safety are a part of every design. Most of the time, environmental safety is expressed by its opposite term, risk. As such, risk is a good metric against which biotechnological safety is judged [26].

Bioengineering is indeed an "applied social science," so bioengineering must be seen less as a profession that builds things than one that provides useful outcomes. It is tempting at this point to count only those factors that are quantifiable. This includes measures and interventions that protect and enhance human and ecological resources and assets; those goods and services that are valued. Even deciding on value is laden with perspective, (anthropocentric, biocentric, and ecocentric viewpoints are discussed in chapter 12). Here is where the social sciences can help the bioengineer. Social science is concerned with human society and its individual members. Although, most engineers are steeply trained in the physical sciences (those that explain natural phenomena), their professional codes are written as social science mandates. For example, the stated mission of the Biomedical Engineering Society (BES) consists of five key areas, three of which directly address the application of biological sciences to solve and to prevent societal problems:

- Fostering translation of biomedical engineering and technology to industrial and clinical applications through enhancing interactions with industry and clinical medicine ...
- Holding top-quality scientific meetings and publishing the best possible journal for the communication and exchange of state-of-the-art knowledge at the frontier of biomedical engineering and bioengineering.
- Enhancing the impact of biomedical engineering on economy and human health and maximizing the performance of the discipline ... [27]

Biosystem engineering success or failure is in large measure determined by what the engineer does with respect to what the profession "expects." As mentioned, safety is always a fundamental facet of our professional duties. Thus, we need a set of criteria that tells us when designs and projects are sufficiently safe. Four main safety criteria should be applied to test a biotechnology's worthiness [28]:

The design must comply with applicable laws.
The design must adhere to "acceptable engineering practice."
Alternative designs must be sought to see if there are safer practices.
Possible misuse of the product or process must be foreseen.

The first two criteria are easier to follow than the third and fourth. The well-trained bioengineer can look up the physical, chemical, and biological factors to calculate tolerances and factors of safety for specific designs. Laws have authorized the thousands of pages of regulations and guidance that codify when acceptable risk and safety thresholds are crossed; meaning that the design has failed to provide adequate protection. Bioengineering standards of practice go a step further. Biosystematic success or failure here can be difficult to recognize, since the biosciences being applied are rather esoteric and specific to the technology. Only other engineers with explicit expertise in this biology specialty can adequately ascertain the ample margin of safety as dictated by sound engineering principles and practice. Identifying alternatives and predicting misuse requires quite a bit of creativity and imagination. This type of failure falls within the domain of social science.

BEYOND BIOTECHNOLOGICAL APPLICATIONS

Indeed, numerous applications of biotechnology are effectively addressing environmental problems, but implications of using living systems to solve society's ills also constitute some of the greatest challenges. Living organisms can be used to sense and to indicate changes in the environment (so-called bioindicators). Also, living organisms are being put to work to clean up chemically contaminated waste sites (e.g. microbes that break down hazardous chemicals into less toxic byproducts). Yet, merely presenting the positive sides of environmental biotechnology leaves the impression that it is entirely a beneficial resource, without hazards. Conversely, only worrying about the environmental pollution associated with emerging biotechnological advances also paints an unrealistic, overly negative image of biotechnology's impact on society. Therefore, it is important and crucial to treat every aspect of biotechnology objectively and in a scientifically credible manner. There are many sources that either support or condemn biotechnology. The attempt here is to provide a balanced and comprehensive consideration of all aspects of environmental biotechnology.

This begs the question then of what we mean when we say environmental biotechnology. Both words are important. In fact, in spite of the growth of environmental science and engineering over recent decades, there is not complete unanimity as to what is meant by important environmental terms. Pollution, contamination, hazard, risk, and other key environmental terms have various meanings within the environmental scientific community. Thus, after a brief introduction of biotechnologies, we will discuss a number of environmental concepts so that at least in this work, we have agreement on vernacular and basic science.

Terminology

Distinguishing "environmental" science and engineering from other scientific endeavors often hinges on what is important. For example, the general definition of biology is that it encompasses everything to do with the study of living things. Thus, to see what a biologist considers important must come from his or her specialization. Biology must be subdivided into more specific fields. Moving toward our specific focus, we are predominantly interested in environmental biology, which is the study of living things with respect to their surroundings. To most engineers and technologists, any field of biology, no matter the adjective in front of it, is less about application than about understanding the principles and concepts of that field. Therefore, even a more specific focus is still mainly interested in description than in application. That is, biologists tend to be more interested in living things as subject matter, whereas bioengineers are more concerned in living things with respect to solving problems. The bioengineering danger is the view that all things, including living things, are merely objects that should be manipulated to achieve some end. So, a biological process may only be seen as unacceptable if it does not meet the stated objective, e.g. converting a given mass of polychlorinated biphenyls to a given mass of carbon dioxide, water, and chlorine. If an ecosystem's other characteristics are not also part of this measure of success, even a successful bioremediation can still be a failure (intermediate chemical products released to the environment, changes in microbial populations that harm ecosystem diversity, etc.).

Environmental protection is another way to provide focus. For example, the United States Environment Protection Agency (US EPA) has as its mission the protection of public health and the environment, and goes about that mission by addressing problems that can either be categorized as human health or ecological. Some see the need to protect ecosystems due to their *inherent value,* and others see the need to protect ecosystems due to their *instrumental value,* i.e. the goods and services that they provide to humans. Either way, introducing possible stressors to these systems must be considered from an ecological, systematic perspective. After all, ecology is the subdiscipline of biology that concerns itself with how organisms interact with their environments.

In addition to the various perspectives on the environment, there are also varying views about biotechnology. Biotechnology means many things to many people. Every part of the word, "biotechnology" is important. The suffix, "bio" indicates that we are exclusively concerned about living things. However, the role that living things play in this technology varies considerably among even scientists working in various fields of biotechnology. For example, some consider biotechnology to include every aspect of the application of living systems to solving any societal problem. This is the broadest of definitions and arguably tells you the least of all.

Therefore, at least with regard to environmental biotechnology, the definitions need to be more specific. In fact, the areas of bioscience, bioengineering, and biotechnology need to be distinguished from each other: *science* is the explanation of natural phenomena; *engineering* is the application of the sciences to addressing societal problems; *technology* results from scientific and engineering advances.

Eureka!

Hēurēka!

Archimedes, as recorded by Marcus Vitruvius Pollio in *De architectura*, c.25 BC

Discovery is the lifeblood of science. It brings excitement, not to mention investment, to those involved and to the broader society. Biotechnology has brought with it actual and promising improvements in almost every economic sector, from the food supply to medicine to environmental remedies. Although it is tempting to think of biotechnology as a uniquely modern and systematic enterprise, many of the discoveries have been incidental and accidental over centuries. The story of Archimedes' discovering the phenomenon of volumetric displacement demonstrates that discovery takes place within and outside of the laboratory. Or, it may demonstrate that the laboratory extends well beyond the bench (as it did to Archimedes' bathtub). Thus, biotechnologies have found their way through numerous medical, industrial, agricultural and defense systems, not just from earnest scientists trying to find new applications of the biosciences.

Indeed, although biotechnology is a modern term, many of the human advances over the past few millennia can be categorized as biotechnological. For example, humans greatly improved crops by selecting various seed types with characteristics preferred for their taste, quality, and disease resistance. The same processes are used in modern times, only now genetic engineering and other scientifically advanced approaches have speeded up the time needed for selection and have enabled a more precise focus on specific selected attributes of the plants.

Biological systems range in complexity and scale, from subcellular to planetary. The two fields of engineering which are most intimately engaged in the life systems are biomedical and environmental engineering. Nevertheless, all fields of engineering to some extent must address living systems. Indeed, any engineering is the practice whereby scientific information is applied in order to improve the world. And this improvement is measured according to where human society places its values. One big area that is not completely addressed in either environmental textbooks or engineering textbooks is the effect that biotechnological sources have on the environment. This can range in scale from changes in cellular chemical messaging systems to possible long-range transport of biological agents with possible risks at the biome, or even global, scale. This entire domain is the subject of environmental biotechnology.

The biomedical, agricultural science, industrial hygiene, and engineering disciplines must apply science and technology to understand and to enhance a myriad of biological systems. In fact, the systematic approach is the best way to address biological challenges, from an understanding of nano-scale changes to a living cell that leads to cancer to the cycling of greenhouse gases that affect global climate.

"System" is a widely used, yet misapplied term in both scientific and lay communities. In fact, it has currency in various venues, from thermodynamics to fluid dynamics to motion to pharmacokinetics. Thus, scientists and engineers who engage in environmental processes must have a common understanding and application of living systems, i.e. biosystems. Biotechnologies present both challenges and opportunities for environmental science and engineering.

This book will present a comprehensive treatment of actual and potential biotechnologies at the full range of environmental scales. It is also a valuable companion and link to biomedical applications, such as computational toxicology and physiologically based, compartmental models. Such emergent and promising tools are being used in both medical and engineering disciplines. The two principal engineering disciplines engaging in biosystems are biomedical and environmental. Unfortunately, most textbooks do not bridge these perspectives. The author's research and teaching at Duke University has provided him with an all-too-rare opportunity to work in both areas. This book will take advantage of this unique perspective to link and to contrast the biosystem approach in these diverse and essential fields.

Oh no!

> The most exciting phrase to hear in science, the one that heralds new discoveries, is not "Eureka!" (I found it!) but "That's funny …"
>
> **Isaac Asimov, 1920–92**

Scientific discoveries that involve only one experimental or independent variable are the easiest to explain. One can manipulate that single factor, keeping others constant, and watch for changes in the dependent variable. Biological systems seldom allow researchers to hold these other variables constant. In fact, many variables are not even known; only outward manifestations or indicators can be observed. Thus, in a static system, i.e. one that is not changing in space and time, the interaction of variables is complex, but in a dynamic system, the uncertainties can increase dramatically. Thus, in characterizing and applying the biological sciences, biotechnologists are challenged to document what is going on. The outcome of the experiment is explained within the context of its constraints. For dynamic systems, things get astronomically more complicated in time and space, as more variables are introduced and environmental conditions change.

When scientific, engineering, and technological advances occur relatively rapidly, unintended consequences can result. More complex approaches are often associated with a lack of understanding of how they work, at least in the early stages of development. Such "black boxes" can contain steps and processes with unforeseen hazards. In addition to complexity, the rapid dissemination of new approaches in modern times can lead to trouble on a grand scale.

Thus, the complexity and scale of possible environmental impacts must be considered as early as possible in decisions involving emerging biotechnologies. Otherwise, numerous countervailing and downstream risks can present themselves. Such risks may not be readily apparent and may even be unprecedented. For example, the new technology may supplant or completely change design and manufacturing processes. Biotechnology-derived products may impact the use of other products.

Agricultural materials, such as hybrid seeds of cash crops may alter farming practices which, in turn, may adversely affect water and soil quality. For instance, if the biologically altered plant's root has a very different capillary function than the non-genetically modified plant nutrients may be held and released at a rate that is not sustainable. Or, the new capillarity may improve sustainability by enhanced root–soil interactions. Further, the shallower or deeper root system may upset the delicate balance in the water cycle. A shallower root system may translate into a diminished ability of the plants to hold soil in place, thus leading to a loss of top soil. A deeper root system may extract water from lower strata of an aquifer, leading to drawdown.

An interesting example of the complications and connectedness between slight changes to an ecosystem and human health is the effect of change in flora and the Ross River virus. Ross River virus causes a severe illness in human populations, manifested in muscle and joint pain, polyarthritis and lethargy. In Australia, for example, the salt marsh mosquito *Aedes vigilax* and the freshwater *Culex annulirostris* are the main vectors of Ross River, but the virus is also carried by *Aedes notoscriptus* and the brackish water species *Aedes funereus*, with marsupials serving as the main reservoir hosts. The connection between ecosystems and human disease is often indirect and complicated [29] (see Figure 1.6).

Most ecologists rightfully rue the fact that wetlands have been dwindling globally. In fact, the civil engineering profession has been quite successful in advancing wetland construction efforts. Historically, however, wetlands were seen as problematic. Major public and private swamp drainage efforts were undertaken in the 20th century to prevent disease.

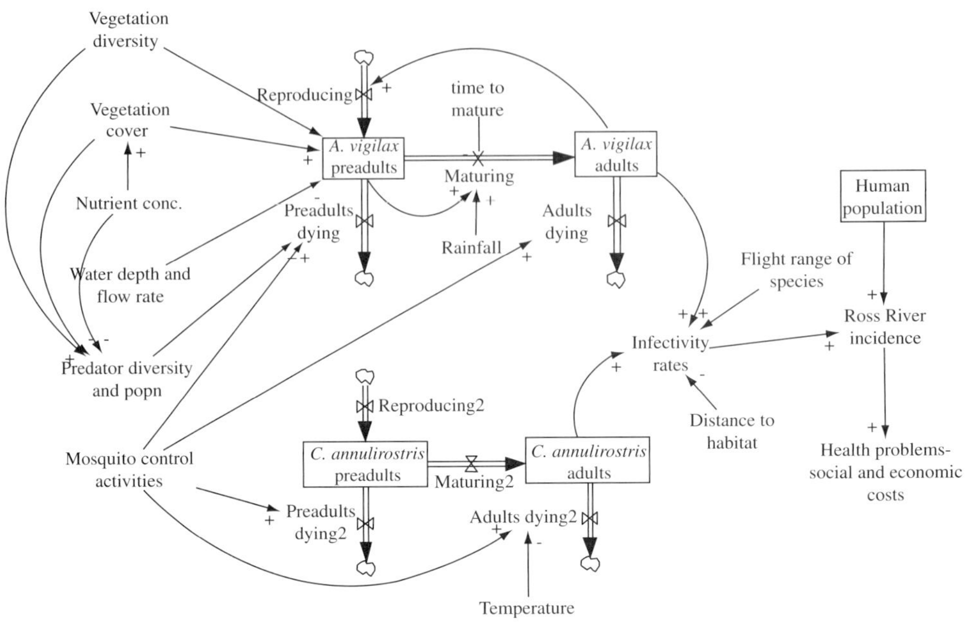

FIGURE 1.6
Conceptual model of factors connecting environmental conditions to mosquito (*Aedes vigilax*) populations to human disease.
Source: M. Cox (2006). Impacts of changes in coastal waterway condition on human well-being. Doctoral dissertation. Centre for Marine Studies, University of Queensland, Queensland, Australia.

Whether a wetland is a contributor to a human disease is a function of location. If the wetland is within mosquito flight range from the nearest human population the risk of an outbreak is increased. This can vary from a few hundred meters to greater than 10 kilometers [30]. Shallow wetlands with shrubs and trees usually support higher mosquito populations than deep pools with steep edges. This may be due to the latter's support of mosquito predators. Thus, if water levels do not fluctuate sizably, mosquito populations can flourish, along with exposures to the virus. Thus, when farmers remove trees with deep roots to grow shallow-rooted crops, these crops do not siphon the water and transpire it through their leaves like trees and shrubs do. Thus, the farmers actually create wetlands, but unlike many wetland construction efforts, these cause problems. The problem is not so much an ecological problem, but an ecosystem event that leads to a human health problem. This highlights that, depending on the circumstances, a change may be adverse or beneficial. Either way, a decision on whether to apply a biotechnology is complex.

THE SCIENCE OF ENVIRONMENTAL BIOTECHNOLOGY

In a span of just a few decades advances and new environmental applications of science, engineering, and their associated technologies have coalesced into a whole new way to see the world. Science is the explanation of the physical world, while engineering encompasses applications of science to achieve results. Thus, what we have learned about the environment by trial and error has incrementally grown into what is now standard practice of environmental science and engineering. This heuristically attained knowledge has come at a great cost in terms of the loss of lives and diseases associated with mistakes, poor decisions (at least in retrospect), and the lack of appreciation of environmental effects.

The "environmental movement" is a relatively young one. The emblematic works of Rachel Carson, Barry Commoner, and others in the 1960s were seen by many as mere straws in the wind. The growing environmental awareness was certainly not limited to the academic and scientific communities. Popular culture was also coming to appreciate the concept of "spaceship earth," i.e. that our planet consisted of a finite life support system and that our air, water, food, soil, and ecosystems were not infinitely elastic in their ability to absorb humanity's willful disregard. The poetry and music of the time expressed these fears and called for a new respect for the environment. The environmental movement was not a unique enterprise, but was interwoven into growing protests about the war in Vietnam, civil rights, and a general discomfort with the "establishment." The petrochemical industry, the military, and capitalism were coming under increased scrutiny and skepticism.

The momentum of the petrochemical revolution following World War II was seemingly unstoppable. However, much of the progress we now take as given was the result of those who agitated against the *status quo* and refused to accept the paradigms of their time. In fact, this book provides evidence of the validity of some of these early environmentalists' causes. A handful of cases were defining moments in the progress in protecting public health and the environment. It seems that every major piece of environmental legislation was preceded by an environmental disaster precipitated from mistakes, mishaps, and misdeeds. Amendments to the Clean Air Act resulted from deadly episodes such as were experienced in Donora, Pennsylvania and London, UK. Hazardous waste legislation came about after public outcries concerning Love Canal in New York state. "Right-to-Know" legislation worldwide grew from the disaster at Bhopal, India. Oil spill and waste contingency plans were strengthened following the Exxon Valdez spill in Alaska. International energy policies changed, with growing anti-nuclear power sentiments, following the near disaster at Three Mile Island in the United States and the actual catastrophe at Chernobyl in Ukraine. Most recently, engineering and public health emergency response planning has been completely revamped in response to the events of September 11, 2001.

Certainly these can all be classified as "environmental" problems, but they represent new, societal paradigms as well. That is the tricky part of dealing with emerging technologies, including biotechnologies. Contemporary society has a way of thrusting problems upon us. Ironically, society demands the promotion of new and better things and processes, simultaneously demanding that scientists, engineers, physicians, and others in the scientific community sufficiently control the consequences of the very same technologies that members of society insist we use. For example, society may demand, reasonably or unreasonably, certain food characteristics (higher nutrition, less fat, attractive color). They may be quite happy with a product, until it is found to have actual or perceived negative characteristics. For example, the public may be pleased that the price of strawberries remains low and the texture of a high quality until they find out that the plants have been genetically altered to resist frost damage. However, this engineered characteristic could have been the principal driver for the lower price and better texture. Likewise, cleanup of polluted waters and sediments can benefit from genetically altered bacteria and fungi to break down some very persistent contaminants, but the public may fear potential problems if these microbes somehow escape their intended use and find their way into unplanned components of the food chain. Prominent and infamous environmental problems have emerged as byproducts of some useful, high-demand enterprise.

BOXES AND ENVELOPES: PUSHING THE BOUNDARIES, CONTAINING THE RISKS

The engineer, more than ever, must balance creativity and innovation with caution and deliberation. Preventing and reducing the risks posed by hazardous substances involves discovery, brilliance, and due diligence. Engineers are simultaneously asked to provide proven products in high and low technologies, while being at the same time encouraged to think outside the box and to push the envelope to find better ways to address the health and ecological perils imposed by these wastes.

The "box" and "envelope" are illustrative metaphors. Bioreactors are boxes in a thermodynamic sense. Physicists would envision a bioreactor as a box (known as a control volume) in which mass and energy enter, within which substances change in terms of energy and mass, and from which the parents and byproducts of these substances exit. The bioreactor is different from other reactors in the fact that biological processes are at the entrance to, within, and at the exit from the control volume. Chemical engineers see this box as any other reactor, nonetheless; that is, the facility is the enclosure in which all the physical and chemical processes are (or should be) controlled. The envelope is also an appropriate bioengineering term. It is the maximum capacity or capability of a system. As new substances are generated or existing mass and energy are released to the environment, the engineer is expected to push beyond existing solutions to reduce the potential for adverse effects that can result from exposures to these wastes. But, prudence dictates that the box and envelope not be contorted in ways that may lead to unacceptable risks. As such, engineering the risks of biotechnology is always in the balance between flux and stability.

Biotechnologies must be addressed across many scales, ranging from the atom to the planet. At the so-called nano-scale, we simultaneously fear the new environmental challenges of nano-materials (e.g. nanoparticles and fullerenes that display properties not as well understood as those of larger scales), while enthusiastically embracing these same nanotechnologies as improvements in ways to measure contaminants and to improve treatment of conventional and toxic wastes. At the planetary scale, we must consider the cumulative effect of contaminants on the atmosphere, the oceans, sensitive biosystems, and human populations. We must also consider economic and political solutions to the problems presented, including multinational perspectives to reduce the need to produce the wastes in the first place. Not only are engineers charged with a mandate to solve and to prevent environmental problems, but we must do such things in a way that does not lead to unacceptable side effects, especially those that affect our vital life systems on this planet. This is embodied in the first canon of engineering practice. We must hold paramount the public's safety, health, and welfare. Engineers have become increasingly adept at distinguishing between what we *can do* and what we *should do.*

RESPONSIBLE BIOENGINEERING

> By three methods we may learn wisdom: First, by reflection, which is noblest; second, by imitation, which is easiest; and third by experience, which is the bitterest.
>
> **Confucius, c.551–479 BC**

Confucius' quote is genuinely applicable to biotechnologies, since it encapsulates the incrementalism of engineering knowledge and wisdom. Engineers spend most of the preparation for the profession learning the scientific principles and applying them to problems; what Confucius might have called "reflection." Next, the newly minted engineer observes and applies lessons from seasoned mentors in increasingly less "safe" venues. And, the engineer hopes not to experience the bitterness of direct failure by adopting practices that have worked for others in the past.

The bioengineering growth relies on a framework of texts, manuals, and handbooks, not only from engineering, but also from allied biological and chemical sciences. Only when experience is added to the mix can wise decisions be made [31]. Engineers who intend to practice must first submit to a rigorous curriculum (approved and accredited by the Accreditation Board for Engineering and Technology), then must sit for the Future Engineers (FE) examination. After some years in the profession (assuming tutelage by and ongoing intellectual osmosis with more experienced professionals), the engineer has sufficiently demonstrated professional strength [32] and sits for the Professional Engineers (PE) exam. Only after passing the PE exam does the National Society for Professional Engineering certify that the engineer is a "professional engineer" and eligible to use the initials PE after his or her name. The engineer is, supposedly, now schooled beyond textbook knowledge. The professional status demarks a transition from knowing the "what" and the "how" to knowing the "why" and the "when." The engineer knows more about why technical and ethical problems require a complete understanding of the facts and possible outcomes (i.e. conditional probabilities). Details and timing are critical attributes of a good engineer. The wise engineer grows to appreciate that the correct answer to many engineering problems is "It depends."

Emergent research, such as that of biotechnology and nanotechnology, continues to become smaller in scale. Many research institutions have numerous *nano-scale* projects (within a range of a few angstroms). Nascent areas of research include ways to link protein engineering with cellular and tissue biomedical engineering applications (e.g. drug delivery and new devices); ultra-dense computer memory; nonlinear dynamics and the mechanisms governing emergent phenomena in complex systems; and state-of-the-art nano-scale sensors (including photonic ones). Complicating the potential societal risks, much of this research frequently employs biological materials and self-assembly devices to design and build some strikingly different kinds of devices. Some of the worst-case scenarios have to do with the replication of the "nano-machines." We need to advance the state-of-the-science to improve the quality of life (e.g. treating cancer, Parkinson's disease, Alzheimer's disease, and improving life expectancies, or cleaning up contaminated hazardous wastes), but in so doing are we introducing new societal risks? In his recent book *Catastrophe: Risk and Response,* Richard Posner, a judge of the US Court of Appeals for the Second Circuit, describes this paradox succinctly:

> Modern science and technology have enormous potential for harm. But they are also bounteous sources of social benefits. The one most pertinent ... is the contribution technology has made to averting both natural and man-made catastrophes, including the man-made catastrophes that technology itself enables or exacerbates. [33]

Posner gives the example of the looming threat of global climate change, caused in part by technological and industrial progress (mainly the internal combustion engine and energy production tied to fossil fuels). Emergent technologies can help to assuage these problems by using alternative sources of energy, such as wind and solar, to reduce global demand for fossil fuels. We will discuss other pending problems, such as the low-probability but highly important outcomes of genetic engineering, e.g. genetically modified organisms (GMOs) used to produce food. There is both a fear that the new organisms will carry with them unforeseen ruin, such as in some way affecting living cells' natural regulatory systems. An extreme viewpoint, as articulated by the renowned physicist Martin Rees, is the growing apprehension about nanotechnology, particularly its current trend toward producing "nanomachines." Biological systems, at the subcellular and molecular levels, could very efficiently produce proteins, as they already do for the own purposes. By tweaking some genetic material at a scale of a few angstroms, parts of the cell (e.g. the ribosome) that manufacture molecules could start producing myriad molecules designed by scientists, such as pharmaceuticals and nanoprocessors for computing. However, Rees is concerned that such assemblers could start self-replicating (like they always have), but without any "shut-off." Some have called this the "gray goo" scenario, i.e. accidentally creating an "extinction technology" from the cell's unchecked ability to exponentially replicate itself if part of their design is to be completely "omnivorous," using all matter as food! No other "life" on earth would exist if this "doomsday" scenario were to occur [34].

Certainly, this is the stuff of science fiction, but it does call attention to the need for vigilance, especially since our track record for becoming aware of the dangers of technologies is so frequently tardy. It also points out that "rare" does not equal "impossible." The events that lead to a rare outcome are all possible. In environmental situations, messing with genetic materials may harm biodiversity, i.e. the delicate balance among species, including trophic states (producer–consumer–decomposer) and predator–prey relationships. Engineers and scientists are expected to push the envelopes of knowledge. We are rewarded for our eagerness and boldness. The Nobel Prize, for example, is not given to the chemist or physicist who has aptly calculated important scientific phenomena, with no new paradigms. It would be rare indeed for engineering societies only to bestow awards on the engineer who for an entire career has used none but proven technologies to design and build structures. This stems from our general approach to contemporary scientific research. We are rugged individualists in a quest to add new knowledge. For example, aspirants seeking PhDs must endeavor to add knowledge to their specific scientific discipline. Scientific journals are unlikely to publish articles that do not at least contain some modicum of originality and newly found information [35]. We award and reward innovation. Unfortunately, there is not a lot of natural incentive for the innovators to stop what they are doing to "think about" possible ethical dilemmas propagated by their discoveries [36].

Thus, biotechnologies call both for pushing the envelopes of possible applications and simultaneously for a rigorous approach to investigate likely scenarios, from the very beneficial to the worst-case ("doomsday") outcomes. This link between fundamental work and outcomes becomes increasingly crucial as such research reaches the marketplace relatively quickly and cannot be confined to the "safety" and rigor of the laboratory and highly controlled scale-ups.

Acceptable risk

Environmental biotechnology thrusts the engineer into uncomfortable places. The frustration for engineers lies in the fact that there is seldom a simple answer to the questions "How healthy is healthy enough?" and "How protected is protected enough?" Managing environmental risks consists of balancing among alternatives. Usually, no single solution to an environmental problem is available. Whether a risk is acceptable is determined by a process of making decisions and implementing actions that flow from these decisions to reduce the adverse outcomes or, at least to lower the chance that negative consequences will occur [37].

Risk managers can expect that whatever risk remains after their project is implemented, those potentially affected will not necessarily be satisfied with that risk. It is difficult to think of any situation where anyone would prefer a project with more risk than with one with less, all other things being equal. It has been said that "acceptable risk is the risk associated with the best of the available alternatives, not with the best of the alternatives which we would hope to have available" [38].

Since risk involves chance, risk calculations are inherently constrained by three conditions:

> The actual values of all important variables cannot be known completely and, thus cannot be projected into the future with complete certainty.
> The physical and biological sciences of the processes leading to the risk can never be fully understood, so the physical, chemical, and biological algorithms written into predictive models will propagate errors in the model.
> Risk prediction using models depends on probabilistic and highly complex processes that make it infeasible to predict many outcomes. [39]

The "go or no go" decision for most engineering designs or projects is based upon some sort of "risk–reward" paradigm, and should be a balance between benefits and costs [40]. This creates the need to have costs and risks significantly outweighed by some societal good. The adverb "significantly" reflects two problems: the uncertainty resulting from the three constraints described above; and the "margin" between good versus bad. Significance is the province of statistics, i.e. it tells us just how certain we are that the relationship between variables cannot be attributed to chance. But, when comparing benefits to costs, we are not all that sure that any value we calculate is accurate. For example, a benefit/cost ratio of 1.3 with confidence levels that give at a range between 1.1 and 1.5 is very different from a benefit/cost ratio of 1.3 with a confidence range between 0.9 and 1.7. The former does not include any values below 1, while the latter does (i.e. 0.9). This value means that even with all of the uncertainties, our calculation shows that the project could be unacceptable. This situation is compounded by the second problem of not knowing the proper margin of safety. That is, we do not know the overall factor of safety to ensure that the decision is prudent. Even a benefit/cost ratio that appears to be mathematically high, i.e. well above 1, may not provide an ample margin of safety given the risks involved.

The likelihood of unacceptable consequences can result from exposure processes, from effects processes or both processes acting together. So, four possible permutations can exist:

Probabilistic exposure with a subsequent probabilistic effect;
Deterministic exposure with a subsequent probabilistic effect;
Probabilistic exposure with a subsequent deterministic effect; or
Deterministic exposure with a subsequent deterministic effect. [41]

A risk outcome is deterministic if the output is uniquely determined by the input. A risk outcome is probabilistic if it is generated by a statistical method, e.g. randomly. Thus, the accuracy of a deterministic model depends on choosing the correct conditions, i.e. those that will actually exist during a project's life, and correctly applying the principles of physics, chemistry, and biology. The accuracy of the probabilistic model depends on choosing the right statistical tools and correctly characterizing the outcomes in terms of how closely the subpopulation being studied (e.g. a community or an ecosystem) resembles those of the population (e.g. do they have the same factors or will there be sufficient confounders to make any statistical inference incorrect?). A way of looking at the difference is that deterministic conditions depend on how well one understands the science underpinning the system, while probabilistic conditions depend on how well one understands the chance of various outcomes (see Table 1.6).

Actually, the deterministic exposure/deterministic effect scenario is not really a risk scenario because there is no "chance" involved. It would be like saying that releasing a 50 kg steel anvil from 1 meter above the earth's surface runs the risk of falling toward the ground! The risk only comes into play when we must determine external consequences of the anvil falling. For example, if an anvil is suspended at the height of one meter by steel wire and used by workers to perform some task (i.e. a deterministic exposure) there is some probability that it may fall (e.g. studies have shown that the wires fail to hold one in 10,000 events, i.e. failure probability of 0.0001), so this would be an example of a deterministic exposure followed by a probabilistic effect (wire failure), i.e. permutation number 2. A biotechnological example would be the potential release of microorganisms. If microbial spores are included in a release from a facility's stack due to a chain of events (e.g. the tanks hold in all but one in 10,000 events), then the failure probability is 0.0001 or 10^{-4}, i.e. the deterministic exposure to the spores could then be calculated from the likelihood that populations would come into contact with them. If that likelihood of contact to a spore suspended in air, re-entrained after setting, from dermal contact, or from food or drinking water, is 0.01 or 10^{-2}, then the overall exposure probability of this scenario is $10^{-4}\,10^{-2} = 10^{-6}$. Note that this is not the risk, but the exposure. The risk is a function of the exposure and the hazard, e.g. the effect from receiving a dose of spores.

Estimating risk using a deterministic approach requires the application of various scenarios, e.g. a very likely scenario, an average scenario or a worst-case scenario. Very likely scenarios are valuable in some situations when the outcome is not life-threatening, or one of severe effects, like cancer. The debate in the public health arena is often between a mean exposure and a worst-case exposure (i.e. maximally exposed and highly sensitive individuals). The latter is more protective, but almost always more expensive and difficult to attain. For example, lowering the emissions of particulate matter (PM) from a power plant stack to protect the mean population of the state from the effects of PM exposure is much easier to achieve than lowering the PM emissions to protect an asthmatic, elderly person living just outside of the power plant property line (see Figure 1.7).

So, then, why not require zero risk from biotechnological materials, or from any hazard for that matter. It becomes an exercise in optimization. Of course, the most protective standards are best, but the feasibility of achieving them can be a challenge. The regulated standard can be very close to zero, especially if one assumes a worst-case scenario for exposure and provides adequate or even more conservative factors of safety. For example, preventing the

Table 1.6 Exposure and effect risk management approaches

	Probabilistic exposure	Deterministic exposure
Probabilistic effect	Contracting the West Nile virus Although many people are bitten by mosquitoes, most mosquitoes do not carry the West Nile virus. There is a probability that a person will be bitten and another, much lower probability, that the bite will transmit the virus. A third probability of this bitten group may be rather high that once bitten by the West Nile virus-bearing mosquito the bite will lead to the actual disease. Another conditional probability exists that a person will die from the disease. So, death from a mosquito bite (probabilistic exposure) leads to a very unlikely death (probabilistic effect)	Occupational exposure to asbestos Exposure to asbestos from vermiculite workers is deterministic because the worker chooses to work at a plant that processes asbestos-containing substances. This is not the same as the person choosing to be exposed to asbestos, only that the exposure results from an identifiable activity. The potential health effects from the exposures are probabilistic, ranging from no effect to death from lung cancer and mesothelioma. These probabilistic effects increase with increased exposures that can be characterized (e.g. number of years in certain jobs, availability of protective equipment and amount of friable asbestos fibers in the air)
Deterministic effect	Death from methyl isocyanate exposure Exposure to a toxic cloud of high concentrations of the gas methyl isocyanate (MIC) is a probabilistic exposure, which is very low for most people. But, for people in the highest MIC concentration plume, such as those in the Bhopal, India, tragedy, death was 100% certain. Lower doses led to other effects, some acute (e.g. blindness) and others chronic (e.g. debilitation that led to death after months or years). The chronic deaths may well be characterized probabilistically, but the immediate poisonings were deterministic (i.e. they were completely predictable based on the physics, chemistry and biology of MIC)	Generating carbon dioxide from combusting methane The laws of thermodynamics dictate that a decision to oxidize methane, e.g. escaping from a landfill where anaerobic digestion is taking place, will lead to the production of carbon dioxide and water (i.e. the final products of complete combustion). Therefore, the engineer should never be surprised when a deterministic exposure (heat source, methane, and oxygen) leads to a deterministic effect (carbon dioxide release to the atmosphere). In other words, the production of carbon dioxide is 100% predictable from the conditions. The debate on what happens after the carbon dioxide is released (e.g. global warming) is the province of probabilistic and deterministic models of these effects

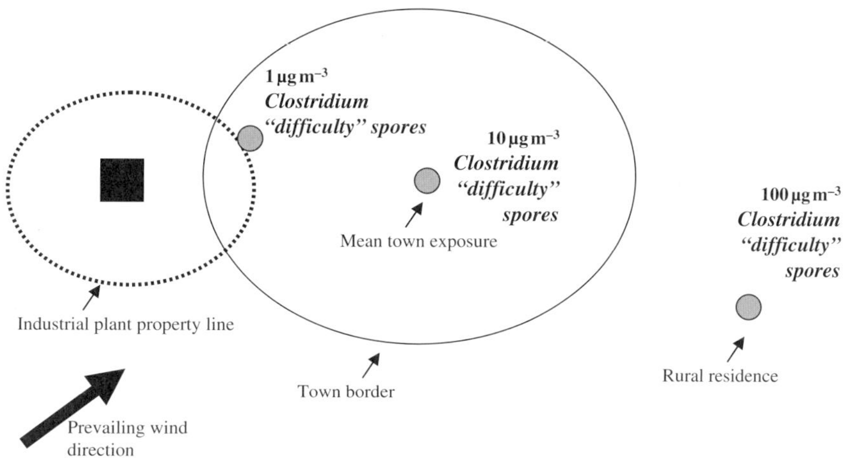

FIGURE 1.7
Difference in control strategies based on concentration of the allowable emissions of a substance to reduce risks to maximally exposed *versus* mean population and a very low exposure scenario. Hypothetical data for fictitious microbe, *Clostridium "difficulty"*, which produces a spore that is an allergen. The concentration would be even lower if the risks are based on a highly sensitive subpopulation (e.g. elderly, infants, asthmatic, or immunocompromised), depending upon the effects elicited by the emitted substance. For example, if the spores are expected to cause cardiopulmonary effects in babies, an additional factor of safety may push the risk-based controls downward by a factor of 10 to achieve 0.1 $\mu g\ m^{-3}$ to protect the maximally exposed, sensitive population.

escape of 99% of microbes from a reactor is much easier and cheaper than preventing the escape of 99.999% microbes. Stated differently, putting controls and fail-safe measures in place to prevent an accidental release 99% of the time is much more feasible than the controls and fail-safe measures needed to prevent an accidental release 99.999% of the time (see Figure 1.8). Depending on the hazard, however, even the 99.999% may not meet engineering and risk standards. Some substances are so toxic, and some biological agents so poorly understood, that prevention of their releases may require numerous redundancies. The risk reduction measures are not only dependent on initial design, but on retrofits, operation and maintenance, and decommissioning of a bioreactor or other biotechnological equipment.

Actual or realistic values are input into the deterministic model. For example, to estimate the risk of tank explosion from rail cars moving through and parked in a community, the number of cars, the flammability and vapor pressure of contents, the ambient temperature, the vulnerability of the tank materials to rupture, and the likelihood of derailment would be assigned numerical values, from which the risk of explosion is calculated. Similarly, a biotechnological deterministic example may estimate the risk of a breach of physical contaminant of a microbe based on adaptability and spore-forming ability of the microbe and environmental conditions, e.g. available nutrients, routes and modes of physical movement by air and water and the physical characteristics of barriers leading to rupture or otherwise fail. A probabilistic approach would require the identification of the initiating events and the plant operational states to be considered; analysis of the adverse outcome using statistical analysis tools, including event trees; application of fault trees for the systems analyzed using the event trees (i.e. reliability analyses – see Chapter 3), collection of probabilistic data (e.g. probabilities of failure and the frequencies of initiating events); and the interpretation of results.

Human beings engage in risk management decisions every day. They must decide throughout whether the risk from particular behaviors is acceptable or whether the potential benefits of a behavior do not sufficiently outweigh the hazards associated with that behavior. In engineering terms, they are optimizing their behaviors based upon a complex set of variables that lead to numerous possible outcomes. A person wakes up and must decide whether to drink

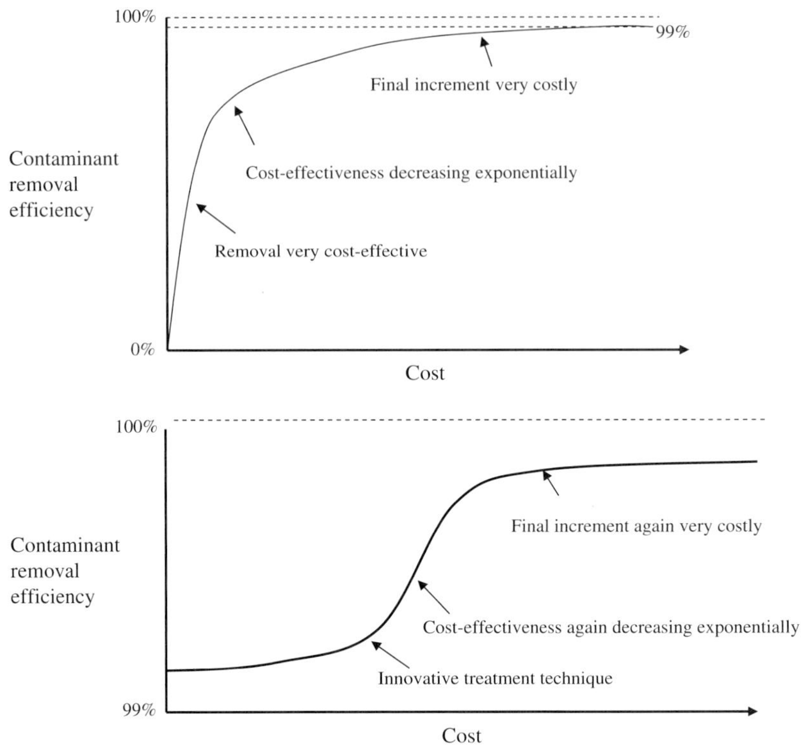

FIGURE 1.8
Prototypical contaminant removal cost-effectiveness curve. In the top diagram, during the first phase a relatively large amount of the contaminant is removed at comparatively low costs. As the concentration in the environmental media decreases, the removal and control costs increase substantially. At an inflexion point, the costs begin to increase exponentially for each unit of contaminant removed, until the curve nearly reaches a steady state where the increment needed to reach complete removal is very costly. The top curve does not recognize innovations that, when implemented, as shown in the bottom diagram, can make a new curve that will again allow for a steep removal of the contaminant until its cost-effectiveness decreases. This concept is known to economists as the law of diminishing returns.
Source: D.A. Vallero (2004). *Environmental Contaminants: Assessment and Control.* Elsevier Academic Press, Burlington, MA.

coffee that contains the alkaloid caffeine. The benefits include the morning "jump-start," but the potential hazards include induced cardiovascular changes in the short term, and possible longer-term hazards from chronic caffeine intake. The decision is also optimized according to other factors, such as sensory input (e.g. a spouse waking earlier and starting the coffee could be a strong positive determinate pushing the decision toward "yes"), habit (more likely to drink a cup if it is part of the morning routine, less likely if it is not), and external influences (e.g. seeing or hearing a commercial suggesting how nice a cup of coffee would be, or conversely reading an article in the morning paper suggesting a coffee-related health risk). This decision includes a "no-action" alternative, along with a number of other available actions. One may choose not to drink coffee or tea. Other examples may include a number of actions, with concomitant risk.

The no-action alternative is not always innocuous. For example, if a person knows that exercise is beneficial but does not act upon this knowledge, the potential for adverse cardiovascular problems is increased. If a person does not ingest an optimal amount of vitamins and minerals, disease resistance may be jeopardized. If a person always stays home to avoid the crowds, no social interaction is possible and the psyche suffers. The management decision in this case may be that the person decided that human-to-human contact, correctly, is a means of transmitting pathogens. But, implementing that decision carries with it another hazard, i.e. social isolation. Likewise, the engineer must take an action only if it provides the optimal solution to the environmental problem, while avoiding unwarranted financial costs, without causing unnecessary disruption to normal activities, and in a manner that is socially acceptable to the community.

In addition, the engineer must weigh and balance any responsibility to represent the client with environmental due diligence. This diligence must be applied to designs and plans for manufacturing processes that limit, reduce or prevent pollution, to ways to reduce risks in operating systems, to the assessment of sites and systems for possible human exposures to hazardous and toxic substances, and to the evaluation of designs systems to reduce or eliminate these exposures. Ultimately, the engineer participates in means to remedy the problem, i.e. to ameliorate health, environmental, and welfare damages. The remedy process varies according to the particular environmental compartment of concern (e.g. water, air, or soil), the characteristics of contaminant of concern (e.g. toxicity, persistence in the environment and likelihood to be accumulate in living tissues), and the specific legislation and regulations covering the project. However, it generally follows a sequence of preliminary studies, screening of possible remedies, selecting the optimal remedy from the reasonable options, and implementing the selected remedy (see Figure 1.9). The evaluation and selection of the best alternative is the stuff of risk management.

Given this seemingly familiar role of the engineer, why then do disasters and injustices occur on our watch? What factors cause the engineer to improperly optimize for the best outcome? In part, failures in risk decision making and management are ethical in nature. Sometimes organizational problems and demands put engineers in situations where the best and most moral decision must be made against the mission as perceived by management. Working within an organization has a way of inculcating the "corporate culture" into professionals. The process is incremental and can "desensitize" employees to acts and policies that an outsider would readily see to be wrong. Much like the proverbial frog placed in water that gradually increases to the boiling point, an engineer can work in gradual isolation, specialization, and compartmentalization that ultimately changes to immoral or improper behavior, such as ignoring key warning signs that a decision to locate the facilitate will have an unfair and disparate impact on certain neighborhoods, that health and safety are being compromised, and that political influence or the "bottom line" of profitability is disproportionately weighted in an engineer's recommendation [42].

Another reason that optimization is difficult is that an engineer must deal with factors and information that may have not been adequately addressed during formal engineering training or even during career development. Although environmental and public health decisions must always give sufficient attention to the toxicity and exposure calculations, these quantitative

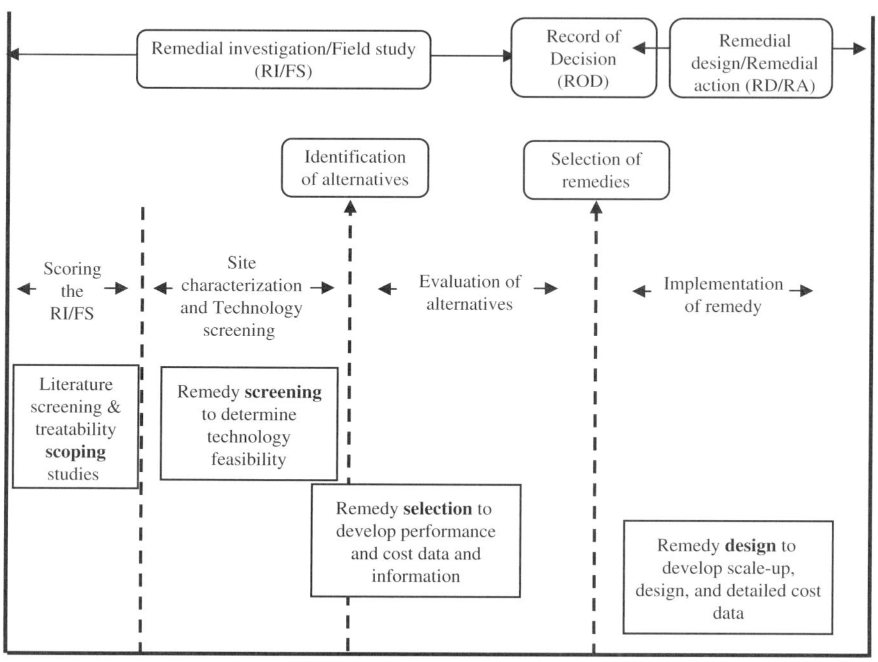

FIGURE 1.9
Steps in cleaning up a contaminated site.
Source: US Environmental Protection Agency (1992). Guide for Conducting Treatability Studies under the Comprehensive Environmental Response, Compensation and Liability Act: Thermal Desorption, EPA/540/R-92/074 B.

results are tempered with feasibility considerations. Thus, engineers' strengths lie to the far left and right of Figure 1.9, but the middle steps, i.e. feasibility and selecting the best alternative, require information that is not "well behaved" in the minds of many engineers. For example, in 1980 the US Congress passed the Comprehensive Environmental Response, Compensation, and Liability Act (CERCLA), commonly referred to as Superfund [43]. The Superfund law authorizes the federal government to respond directly to releases or to the threat of releases of hazardous substances and enables the US Environmental Protection Agency (EPA) to take legal action to force parties responsible for causing the contamination to clean up those sites or reimburse the Superfund for the costs of cleanup. If the responsible parties for site contamination cannot be found or are unwilling or unable to clean up a site, the EPA can use funds from the Superfund to clean up a site.

Bioengineering is "utilitarian"; that is, we are called upon "to produce the most good for the most people" [44]. Bioengineers are also bound to codes of ethics, design criteria, regulations, and standards of practice. Engineers must, to the best of their abilities, consider all possible design outcomes, planned or otherwise. But this is not easy when the most appropriate "benchmarks" for success in biotechnology are moving targets. There are seldom, if ever, clear measures of success and failure. At the beginning of any engineering endeavor, there is no absolute certainty that the project will be a complete success, even if it meets the specifications laid out at the beginning. Sometimes, failure can result in implementing the plan, but other times the failure results from the confluence of events during construction and during the useful life of the project. For example, environmental decisions regarding the level of cleanup at a hazardous waste site can be based on the target health risk expected after cleanup. This remediation target can be based on what has been called the "residential standard." In the United States this was a common measure of success for hazardous waste laws passed in the late 1970s and early 1980s. That is, immediately following the passage of key hazardous waste laws, regulators held the general view that a polluted site needed to be cleaned up to the point that no undue risk remained and the site could be returned to productive use, as if it had never been polluted.

On its face, the residential standard closely follows the steps of hazard identification, dose-response relationships, exposure analysis and effects assessment to characterize risks. However, this could actually lead to a lower level of cleanup in a previously polluted area, which would be unfair to the people living there compared to those living in more pristine areas. Thus, had the engineers designed a cleanup based on the residential standard, their plans may well have been considered failures by today's standards. This also provides an example of why cleanups often must be adaptive to changing conditions [45].

In the United States, the cleanup of a hazardous waste begins with a feasibility study to address nine criteria that will determine the best approach for addressing the contamination, as well as the ultimate level of cleanup:

Overall protection of human health and environment
Compliance with applicable or relevant and appropriate requirements
Long-term effectiveness and permanence
Reduction of toxicity, mobility or volume through treatment
Short-term effectiveness
Ease of implementation
Cost
State acceptance
Community acceptance

The first and fourth criteria are clearly the product of a sound, quantitative risk assessment. The other criteria must also include semi-qualitative and qualitative information. This illustrates the variety of data and information used in evaluating the environmental implications of a biotechnology.

When considering biotechnological advances, or any emerging technology, concerns about downstream impacts are crucial. Possible problems may not present themselves until sufficient time and space has been allowed to elapse. Products that contain dangerous materials like asbestos, lead, mercury, polybrominated compounds, and polychlorinated biphenyls (PCBs) were once considered acceptable and were even required by law or policy to protect the public safety and health, such as asbestos-containing and polybrominated materials to prevent fires, DDT and other persistent pesticides to kill mosquitoes in an effort to prevent disease and methyl tert-butyl ether (MTBE) as a fuel additive to prevent air pollution. Subsequently, these products were all found to cause adverse environmental and health problems, notwithstanding ongoing disagreements within the scientific community about the extent and severity of these and other contaminants. Countless environmental problems are yet to be resolved and others are plagued with incomplete or nonexistent unanimity of thought as to their importance or even whether indeed they are problems, such as the cumulative health and environmental impacts of confined animal feeding operations on microbial populations.

The tradeoff in such cases may be between two competing societal needs, such as a reliable food source and clean water. Sometimes the two needs are mutually exclusive, but others may be met by modifications of one or both solutions (unconfined, local, and more spatially distributed animal operations, along with the public's willingness to accept higher prices for meat). The key to good environmental decision making is that it be informed with reliable information.

SEMINAR TOPIC

Antibiotic Resistance and Dual Use

Molecular biology has both advanced the state of the science of antibiotics and caused them to present vexing problems. Antibiotics are chemicals that interfere with metabolic processes that inhibit the growth of or kill microbes, especially bacteria. The mechanisms of antibiotics vary. Penicillin and vancomycin, for example, cause lysis in gram-positive bacteria (i.e. they are narrow spectrum) by obstructing their ability to synthesize cell walls. Conversely, tetracyclines affect both gram-positive and gram-negative bacteria (i.e. broad spectrum) by impeding their binding to ribosomes, thereby impeding the production of proteins and limiting their activity [46].

The dilemma of *dual use* illustrates the various scientific perspectives involved in a real-world problem. Bacterial resistance to antibiotics is a real and growing problem in the medical community. Tuberculosis, malaria, and ear infections and numerous other diseases are becoming increasingly difficult to treat. For example, tuberculosis cases in the United States were almost completely eliminated shortly after discovery of the pharmaceutical isoniazid in 1940, but the emergence of resistant strains that can only be treated with less effective drugs is very dangerous to public health. The sources are varied. For example, about 2 million patients contract bacterial infections in hospitals each year, known as *nosocomial* infections. Immunocompromised patients are particularly vulnerable to infections of *Staphylococcus aureus*, which is commonly found in hospitals, leading to pneumonia and other ailments. Isolated strains of *S. aureus* are now found to be resistant to previously efficacious antibiotics, including methicillin, oxacillin, penicillin, and amoxicillin. Some strains are even resistant to vancomycin, the antibiotic prescribed by physicians after exhausting other options [47].

A gene in *Yersinia pestis*, the causative agent of plague, was isolated in 2006 and was found to be similar to an *Escherichia coli* gene known to cause multiple types of antibiotic resistance [48]. This non-virulent strain of *Y. pestis* that over-expresses the gene is resistant to numerous common antibiotics, including those commonly prescribed to treat plague infection. Spontaneous antibiotic resistant bacteria have been found to mutate to forms that affect the expression of this gene, indicating a mechanism by which bacteria can acquire resistance [49].

Antibiotic resistance is genetically encoded. Numerous microbes produce antibiotic compounds to protect themselves from other microbes, and as a result, some of these microbes have evolved to be resistant to them. Spontaneous mutations can also produce resistance genes which can be passed on to future generations via genetic exchange processes. For instance, bacteria may transfer a circular strand of DNA (i.e. plasmid) external to its chromosome to another bacterium through conjugation. Bacteria may also get genes that have been released from dead bacteria, incorporating them into their chromosome or plasmid through transformation. A third means is *transduction*, whereby a bacterial virus (i.e. *bacteriophage*) invades the bacterial cell and removes genetic material. When the bacteriophage infects another cell, that gene may be incorporated into the other cell's chromosome or plasmid.

(Continued)

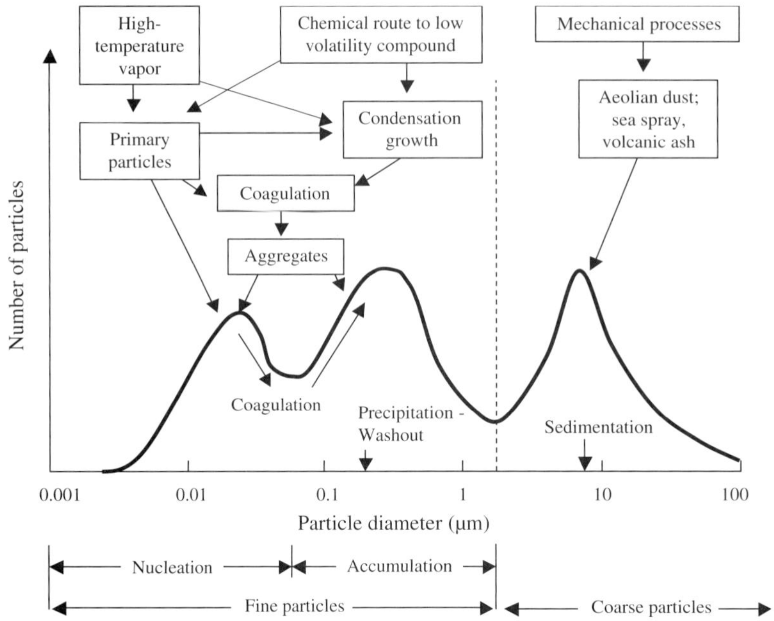

FIGURE 1.10

Prototypical size distribution of tropospheric particles with selected sources and pathways of how the particles are formed. Dashed line is approximately 2.5 μm diameter.

Source: Adapted from United Kingdom Department of Environment, Food, and Rural Affairs, Expert Panel on Air Quality Standards (2004). *Airborne Particles: What Is the Appropriate Measurement on which to Base a Standard? A Discussion Document.* DEFRA, London, UK.

It is quite common in biological laboratories to make antibiotic-resistant bacterial strains by plasmid-insertion of genes to express resistance to a known antibiotic. A major problem with this practice is that this resistance can be transferred from one type of bacteria to another by way of a single gene. For example, two antibiotics, tetracycline and chloramphenicol, are currently used to treat plague infections. This brings us to the dual use problem. Biological weapons could be developed using these antibiotic-resistant pathogens. This is not science fiction. For example, during the 1980s the Soviet Union developed antibiotic-resistant strains of plague, anthrax, tularemia, and glanders bacteria. The Iraqi regime of Saddam Hussein focused on anthrax, botulinum toxin, and aflatoxin [50].

Biological researchers must consider that there are possible uses, possibly not even considered in the quest to provide needed medical advances that may have profound and disastrous dual uses. In addition, the combination of molecular biology and physical sciences is needed. For example, not only the biological practice of gene insertion and ancillary medical research presents problems in security and disease prevention, but so do the means by which the bacteria and other biological materials are transported.

Aerosol science is a major part of most environmental research programs. It addresses the ways that particulate matter moves and changes in the atmosphere and in other environmental compartments. Nebulizers are devices that are used to deliver drugs. Aerosolization has become an important part of biomedical engineering with increasingly sophisticated methods adding to the effectiveness of medical treatment of asthma and other pulmonary diseases. Nebulizers are needed for immediate delivery of deep doses of medicines to the lungs, with research matching the deposition sites in the lung with the type of drug being delivered. Stokes' law states:

$$F_d = 6\pi\mu RV \tag{1.2}$$

where, F_d is the frictional force (Newtons), μ is the fluid's dynamic viscosity (Pascal seconds), R is the radius of the spherical object (meters), and V is the particle's velocity (in m s^{-1}).

Thus, Stokes' law explains the radius of a sphere and the viscosity of a fluid together predict the force needed to move it without settling. So, then, drug delivery to the lungs can be optimized by accounting for the density, morphology, and dynamics of the particle.

Particulate matter (PM) is a common physical classification of particles [51]. The size of a particle is determined by how the particle is formed. For example, combustion can generate very small particles, while coarse particles are often formed by mechanical processes (such as the particles to the left of the dashed line in Figure 1.10 and micrographs in Figures 1.11 and 1.12) and from vehicle exhausts. If particles are sufficiently small and of low mass, they can be suspended in the air

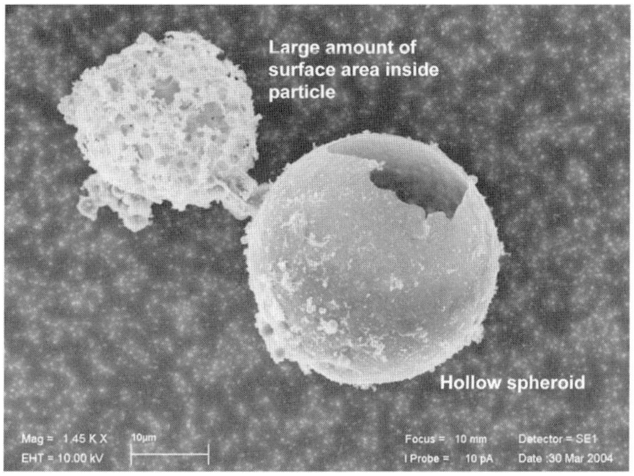

FIGURE 1.11
Scanning electron micrograph of coarse particles emitted from an oil-fired power plant. Diameters of the particles are greater than 20 μm optical diameter. Both particles are hollow, so their aerodynamic diameter is significantly smaller than if they were solid.
Source: Source characterization study by R. Stevens, M. Lynam and D. Proffitt (2004). Photo courtesy of R. Willis, Man Tech Environmental Technology, Inc., 2004; used with permission.

for long periods of time. Larger particles (e.g. >10 μm aerodynamic diameter) are found in smoke or soot (see Figure 1.11), while very small particles (<2.5 μm) may be apparent only indirectly, such as when they diffuse, diffract, absorb, and reflect light (see Figure 1.12).

The term ''aerosol'' is often used synonymously with PM. An aerosol can be a suspension of solid or liquid particles in air, and an aerosol includes both the particles and all vapor or gas phase components of air.

Since very small particles may remain suspended for some time, they can be particularly problematic from a pollutant transport perspective because their buoyancy allows them to travel longer distances. Smaller particles are also challenging because they are associated with numerous health effects (mainly because they can penetrate more deeply into the respiratory system than larger particles).

Generally, the mass of PM falling in two size categories is measured, i.e. ≤2.5 μm diameter, and ≥2.5 μm ≤ 10 μm diameter. These measurements are taken by instruments (see Figure 1.13) with inlets using size exclusion mechanisms to segregate the mass of each size fraction (i.e. ''dichotomous'' samplers). Particles with diameters ≥10 μm are generally of less concern, however they are occasionally measured if a large particulate-emitting source (e.g. a coal mine) is nearby, since these particles rarely travel long distances.

Mass can be determined for a predominantly spherical particle by microscopy, either optical or electron, by light scattering and Mie theory, by the particle's electrical mobility, or by its aerodynamic behavior. However, since most particles are not spherical, PM diameters are often described using an equivalent diameter, i.e., the diameter of a sphere that would have the same fluid properties. Another term, optical diameter, is the diameter of a spherical particle that has an identical refractive index as the particle. Optical diameters are used to calibrate the optical particle sizing instruments, which scatter the same amount of light into the solid angle measured. Diffusion and gravitational settling are also fundamental fluid phenomena used to estimate the efficiencies of PM transport, collection, and removal processes, such as in designing PM monitoring equipment and ascertaining the rates and mechanisms of how particles infiltrate and deposit in the respiratory tract.

Only for very small diameter particles is diffusion sufficiently important that the Stokes diameter is often used. The Stokes diameter for a particle is the diameter of a sphere with the same density and settling velocity as the particle. The Stokes diameter is derived from the aerodynamic drag force caused by the difference in velocity of the particle and the surrounding fluid. Thus, for smooth, spherical particles, the Stokes diameter is identical to the physical or actual diameter. The aerodynamic diameter (D_{pa}) for all particles greater than 0.5 μm can be approximated [52, 53] as the product of the Stokes particle diameter (D_{ps}) and the square root of the particle density (ρ_p):

$$D_{pa} = D_{ps}\sqrt{\rho_p} \qquad (1.3)$$

If the units of the diameters are in μm, the units of density are g cm^{-3}.

Fine particles (<2.5 μm) generally come from industrial combustion processes (such as the particles in Figure 1.12) and from vehicle exhaust. As mentioned, this smaller sized fraction has been closely associated with increased respiratory disease, decreased lung functioning, and even premature death, probably due to their ability to bypass the body's trapping mechanisms, such as cilia in the lungs, and nasal hair filtering. Some of the diseases linked to PM exposure include aggravation of asthma, chronic bronchitis, and decreased lung function.

In addition to health impacts, PM is also a major contributor to reduced visibility, including near national parks and monuments. Also, particles can be transported long distances and serve as vehicles on which contaminants are able to reach water bodies and soils. Acid deposition, for example, can be as dry or wet. Either way, particles play a part in acid rain. In the first, the dry particles enter ecosystems and potentially reduce the pH of receiving waters. In the latter,

(Continued)

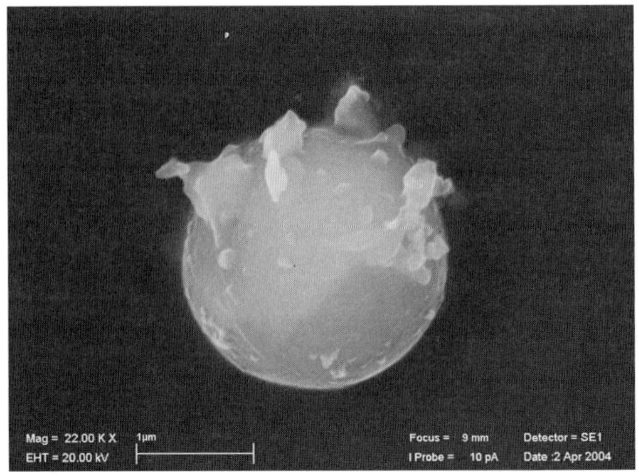

FIGURE 1.12
Scanning electron micrograph of spherical aluminosilicate fly ash particle emitted from an oil-fired power plant. Diameter of the particle is approximately 2.5 μm. Photo courtesy of R. Willis, Man Tech Environmental Technology, Inc., 2004; used with permission.

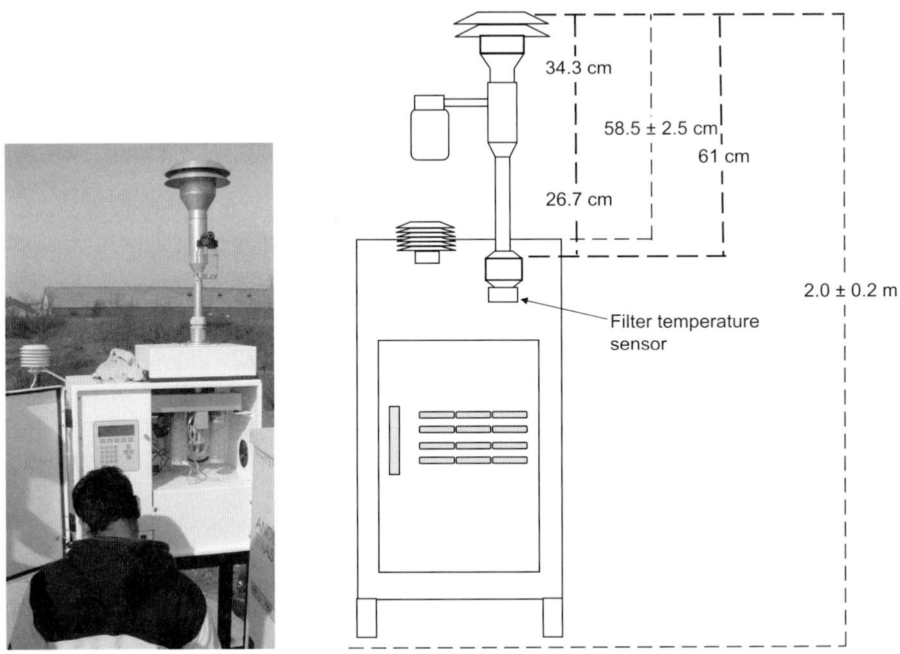

FIGURE 1.13
Photo and schematic of sampling device used to measure particles with aerodynamic diameters ≤2.5 μm. Each sampler has an inlet (top) that takes in particles ≤10 μm. An impacter downstream in the instrument cuts the size fraction to ≤2.5 μm, which is collected on Teflon filter. The filter is weighed before and after collection. The Teflon construction allows for other analyses, e.g. X-ray fluorescence to determine inorganic composition of the particles. Quartz filters would be used if any subsequent carbon analyses are needed. Photo and schematic courtesy of US EPA.

particles are washed out of the atmosphere and, in the process, lower the pH of the rain. The same transport and deposition mechanisms can also lead to exposures to persistent organic contaminants like dioxins and organochlorine pesticides, and heavy metals like mercury that have sorbed in or on particles.

In addition to their inherent toxicity, particles can function as vehicles for transporting and transforming chemical contaminants. For example, compounds that are highly sorptive and that have an affinity for organic matter can use particles as a means for long-range transport. Also, charge differences between the particle and ions (particularly metal cations) will also make particles a means by which contaminants are transported.

The human body and other biological systems have a tremendous capacity for the uptake of myriad types of chemicals and either utilize them to support some bodily function or eliminate them. As analytical capabilities have improved, increasingly lower concentrations of chemicals have been observed in various parts of the body. Some of these chemicals enter the body by inhalation.

The primary function of the human respiratory system is to deliver O_2 to the bloodstream and to remove CO_2 from the body. These two processes occur concurrently as the breathing cycle is repeated. Air containing O_2 flows into the nose and/or mouth and down through the upper airway to the alveolar region, where O_2 diffuses across the lung wall to the bloodstream. The counterflow involves transfer of CO_2 from the blood to the alveolar region and then up the airways and out the nose. Because of the extensive interaction of the respiratory system with the surrounding atmosphere, air pollutants or trace gases can be delivered to the respiratory system.

The anatomy of the respiratory system is shown in Figure 1.14. This system may be divided into three regions: the nasal, tracheobronchial, and pulmonary. The nasal region is composed of the nose and mouth cavities and the throat. The tracheobronchial region begins with the

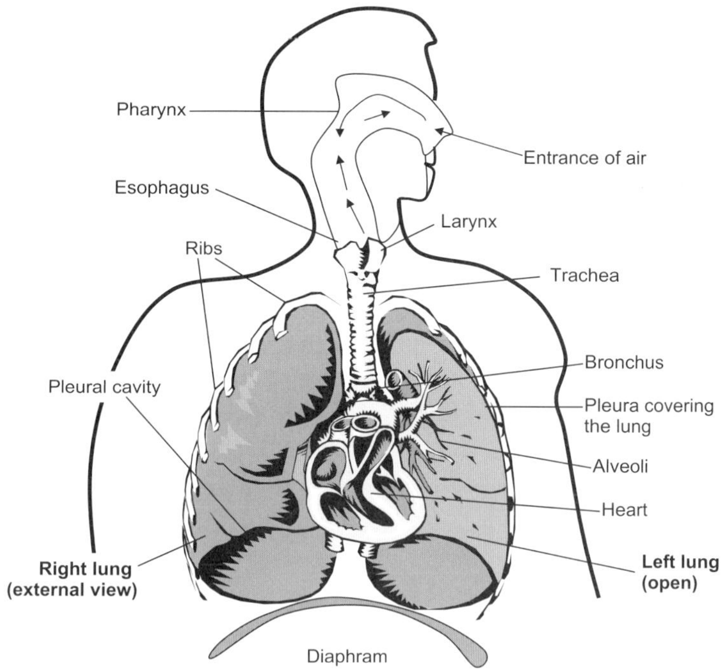

FIGURE 1.14
Anatomy of the human respiratory system.
Source: D.A. Vallero (2008). *Fundamentals of Air Pollution*, 4th Edition. Elsevier Academic Press, Burlington, MA.

trachea and extends through the bronchial tubes to the alveolar sacs. The pulmonary region is composed of the terminal bronchi and alveolar sacs, where gas exchange with the circulatory system occurs. Figure 1.14 illustrates the continued bifurcation of the trachea to form many branching pathways of increasingly smaller diameter by which air moves to the pulmonary region. The trachea branches into the right and left bronchi. Each bronchus divides and subdivides at least 20 times; the smallest units, bronchioles, are located deep in the lungs. The bronchioles end in about 3 million air sacs, the alveoli.

The behavior of particles and gases in the respiratory system is greatly influenced by the region of the lung in which they are located [54]. Air passes through the upper region and is humidified and brought to body temperature by gaining or losing heat. After the air is channeled through the trachea to the first bronchi, the flow is divided at each subsequent bronchial bifurcation until very little apparent flow is occurring within the alveolar sacs. Mass transfer is controlled by molecular diffusion in this final region. Because of the very different flows in the various sections of the respiratory region, particles suspended in air and gaseous air pollutants are treated differently in the lung.

Particle behavior in the lung is dependent on the aerodynamic characteristics of particles in flow streams. In contrast, the major factor for gases is the solubility of the gaseous molecules in the linings of the different regions of the respiratory system. The aerodynamic properties of particles are related to their size, shape, and density. The behavior of a chain type or fiber may also be dependent on its orientation to the direction of flow. The deposition of particles in different regions of the respiratory system depends on their size. The nasal openings permit very large dust particles to enter the nasal region, along with much finer airborne particulate matter. Particles in the atmosphere can range from less than 0.01 μm to more than 50 μm in diameter. The relationship between the aerodynamic size of particles and the regions where they are deposited is shown in Figure 1.15. Larger particles are deposited in the nasal region by impaction on the hairs of the nose or at the bends of the nasal passages. Smaller particles pass through the nasal region and are deposited in the tracheobronchial and pulmonary regions. Particles are removed by impacts with the walls of the bronchi when they are unable to follow the gaseous streamline flow through subsequent bifurcations of the bronchial tree. As the airflow decreases near the terminal bronchi, the smallest particles are removed by Brownian motion, which pushes them to the alveolar membrane.

The respiratory system has several mechanisms for removing deposited aerosols. The walls of the nasal and tracheobronchial regions are coated with a mucous fluid. Nose blowing, sneezing, coughing, and swallowing help remove particles from the upper airways. The tracheobronchial walls have fiber cilia which sweep the mucous fluid upward, transporting particles to the top of the trachea, where they are swallowed. This mechanism is often referred to as the *mucociliary escalator*. In the pulmonary region of the respiratory

(Continued)

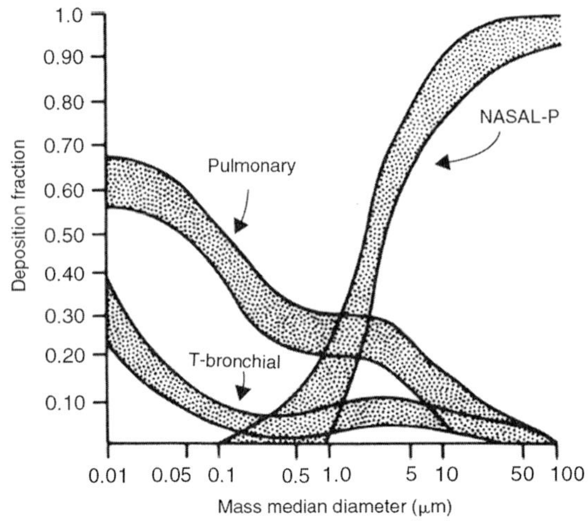

FIGURE 1.15
Particle deposition as a function of particle diameter in various regions of the lung. The nasopharyngeal region consists of the nose and throat; the tracheobronchial (T-bronchial) region consists of the windpipe and large airways; and the pulmonary region consists of the small bronchi and the alveolar sacs.
Source: Task Group on Lung Dynamics (1966). *Health Physics* 12: 173.

system, foreign particles can move across the epithelial lining of the alveolar sac to the lymph or blood systems, or they may be engulfed by scavenger cells called alveolar macrophages. The macrophages can move to the mucociliary escalator for removal. For gases, solubility controls removal from the airstream. Highly soluble gases such as SO_2 are absorbed in the upper airways, whereas less soluble gases such as NO_2 and ozone (O_3) may penetrate to the pulmonary region. Irritant gases are thought to stimulate neuroreceptors in the respiratory walls and cause a variety of responses, including sneezing, coughing, bronchoconstriction, and rapid, shallow breathing. The dissolved gas may be eliminated by biochemical processes or may diffuse to the circulatory system.

Since the location of particle deposition in the lungs is a function of aerodynamic diameter and density, then changing the characteristics of aerosols can greatly affect their likelihood to elicit an effect. Larger particles (>5 μm) tend to deposit before reaching the lungs, especially being captured by ciliated cells that line the upper airway. Moderately sized particles (1–5 μm) are more likely to deposit in the central and peripheral airways and in the alveoli but are often scavenged by macrophages. Particles with an aerodynamic diameter less than 1 μm remain suspended in air and are generally exhaled. Recent studies have shown that large drug particles may be able to evade macrophages past the ciliated cells of the upper respiratory tract and deep into the lungs.

Recently, scientists have taken advantage of these characteristics of the lung mechanisms and of Stokes' law to "improve" the physics of the aerosol so that more medicine is delivered to the target locations, overriding or passing by these mechanisms. Optimal particle slip, shape, and density, as well as porosity of the particles have been changed to deliver "drug particles that were large enough to evade macrophages past the ciliated cells of the upper respiratory tract and deep into the lungs" [55]. Medicine particle design has usually sought a standard size range of 1–5 μm, but a recent study used particles of non-standardized density and aerodynamic diameter. Specifically, the researchers believed that large porous particles would have the mass and dynamics of smaller particles but since they are bigger they would more effectively evade scavenging macrophages in the alveoli. Thus, doses would be less frequent, since more of the medicine would penetrate to the desired, deeper locations in the lungs [56].

Indeed, as the researchers postulated:

> *large porous particles of insulin stayed active in rat lungs for 96 hours, 15 times longer than the longest-acting aerosol on the market. They also found that porous particles embedded with testosterone effectively raised blood hormone levels for extended periods. In the case of estradiol (a potent female sex hormone) delivered as an aerosol into the lungs of rats, bioavailability approached 87%, a much higher percentage than previously achieved. Such enhanced efficiencies permit prescribed medicines to be taken at less frequent intervals and at lower doses, thereby improving convenience. Future applications may include not only obvious pulmonary disorders such as asthma, but also inhalant delivery of insulin, testosterone, estradiols, and monoclonal antibodies for treating viral diseases. [57]*

It did not seem to occur to the researchers that while the porous particle technology holds tremendous promise for improving drug inhalants, they may have introduced a mechanism by which the natural defenses of respiratory system could be overridden. Specifically, any deep lung penetration of spores (e.g. anthrax) would increase their lethality since a greater number of spores would penetrate more deeply into the lungs. Pathogenic microbes would also become more virulent (e.g. pneumonic plague, tularemia, Q fever, smallpox, viral encephalitis, viral hemorrhagic fevers, and botulism). Subsequently, the researchers have recognized the new opportunity for "reverse engineering" of an inhaled drug delivery system, which would increase vulnerabilities in public health and national security.

Seminar Questions

Considering the two biochemodynamic factors of gene transfer and aerosol transport, what are the important steps that should be taken to prevent dual use problems?

What other biochemodynamic factors should be considered for public health protection? ... for preventing terrorist acts?

How can the lessons of antibiotic resistance and aerosol delivery be applied to environmental biotechnology? In particular, what lessons can be learned from these chaotic biomedical systems to prevent a benign entity from evoking an environmental problem?

What are the similarities and differences in considerations of dual use as it applies to environmental engineers, biomedical engineers, agricultural scientists, microbiologists, and medical researchers?

How might a reductionist view differ from a systems biology view in addressing this dual use problem and preventing possible negative outcomes from environmental biotechnologies?

REVIEW QUESTIONS

Consider Figure 1.1 as an illustration of how human populations and ecosystems become connected in terms of exposures to harmful substances. How might the bioengineer use this flow to inform the application of biotechnologies?

When NEPA was passed in 1970, most of the biotechnological revolution was not yet under way. What improvements do you recommend that would address new challenges posed by biotechnologies?

State why you agree or disagree with Kelman's criticism of utilitarian viewpoints in environmental protection. Give two biotechnological examples to support your position.

Give an example of a tradeoff in biotechnology. Does it properly weight environmental considerations?

Why is bioreactor risk never zero? What steps can be taken to improve biotechnological risk?

NOTES AND COMMENTARY

1. B. Erickson (2005). Meeting the Future: A Research Agenda for Sustainability. Highlights of International Workshop. Washington, DC, May 18–20, 2005.
2. P.G. Georgopoulos, A.F. Sasso, S.S. Isukapalli, P.J. Lioy, D.A. Vallero, M. Okino and L. Reiter (2009). Reconstructing population exposures to environmental chemicals from biomarkers: Challenges and opportunities. *Journal of Exposure Science and Environmental Epidemiology* 19: 149–171.
3. This discussion is based on information in: US Department of Agriculture (2006). Animal and Plant Health Inspection Service. BRS Fact Sheet: National Environmental Policy Act and Its Role in USDA's Regulation of Biotechnology. http://www.aphis.usda.gov/publications/biotechnology/content/printable_version/BRS_FS_NEPA_02-06.pdf; accessed August 1, 2009.
4. Pronounced "Fonzy," like the nickname for character Arthur Fonzerelli portrayed by Henry Winkler in the television show *Happy Days*.
5. This is understandable if the agency is in the business of something not directly related to environmental work, but even the natural resources and environmental agencies have asserted that there is no significant impact to their projects. It causes the cynic to ask, then, why are they engaged in any project that has no significant impact? The answer is that the term "significant impact" is really understood to mean "significant adverse impact" to the human environment.
6. Ibid.
7. I learned the meaning of mitigation in my first professional job, i.e. writing an environmental impact statement for a large, coal-fired power plant that needed an EPA permit to release turbine cooling water to the Missouri River. I asked why other cooling approaches, especially cooling towers and lakes, were not being designed into this large, 600+ megawatt facility. After all of the risk assessment and management decisions were made, the power company had to add some features, such as sloughs and co-generation and heated water sharing with a nearby chemical company. Actually, in retrospect, the advice that I received from my senior colleagues was soon vindicated by the federal decision to eliminate all once-through cooling systems shortly after the acceptance of my EIS.

My second EIS was also somewhat controversial, but for different reasons. The EIS was again called for because of wastewater discharge issues, but this time it was a city's wastewater treatment funding under the so-called "Construction Grants" program pursuant to Section 201 of the Federal Water Pollution Control Act Amendments. The city also needed a discharge permit. I sometimes wonder to this day why this facility of the hundreds being constructed in the mid-1970s with federal dollars (often 75% of the total project costs) rose to meet the two EIS metrics; i.e., being a "major federal action" and one with a "significant environmental impact." It must have been the innovation of land application of the wastewater. It was not a typical way of releasing wastes to the environment. The paradigm was, and still is in most instances, that if the waste came to the plant in the form of a wastewater, it was to be released to a water body.

This was a classic case of pleasing the technologists and professionals, and alienating the rest of the public. After all, the nutrients in the wastewater are really fertilizers, so the plants growing in the sprayed fields would benefit. The engineers were happy, the agricultural scientists were happy, and the city planners were happy. We were turning a waste into a resource, after all!

That was my take up to my first public hearing. What a surprise! The questions and indictments went something like this: "What about drift?" "This is sewage, and sewage is loaded with pathogens, can you guarantee that they will not find their way to the air that I breathe or the water that I drink?" "Don't things like viruses pass through systems untreated? So, aren't you just putting this waste out there to infect me?"

I cannot speak for my colleagues, but I was stunned. Why couldn't they see? As the EIS progressed and I learned more about land application, many if not all of the concerns were also being voiced around the world. And the scientific community was ill-equipped to allay the fears. I learned many valuable lessons about environmental assessments in that relatively small town on the High Plains. One was that risk communications should be ever on the mind of the environmental professional. And beyond communications, one should learn to listen to the community members. They live there and they will be affected by the "expert" decisions long after we leave.

Elmo Roper had it right back in 1942. The famous pollster said: " many of us make two mistakes in our judgment of the common man. We overestimate the amount of information he has; and underestimate his intelligence." Roper seemed to be surprised that the general public frequently has too little information to decide on important matters. Roper was even more surprised that in spite of this lack of sufficient information, the common person's "native intelligence generally brings him to a sound conclusion." This is important to keep in mind in dealing with people who will potentially be affected by the recommendations of environmental experts.

8. Quote attributed to Timothy Kubiak, one of Professor Caldwell's former graduate students in Indiana University's Environmental Policy Program. Kubiak has since gone on to become a successful environmental policy maker in his own right, first at EPA and then at the US Fish and Wildlife Service.
9. 40 CFR 1507.3.
10. Albert Sasson (2005). *Medical Biotechnology: Achievements, Prospects and Perceptions.* United Nations University Press, Toyko, Japan.
11. B. Jank, A. Berthold, S. Alber and O. Doblhoff-Dier (1999). Assessing the impacts of genetically modified microorganisms. *International Journal of Life Cycle Analysis* 4 (5): 251–252.
12. O.J. Losey, L. Rainier and M. Carter (1999). Transgenic pollen harms Monarch larvae. *Nature* 399: 214.
13. R. Kamath, J.A. Rentz, J.L. Schnoor and P.J.J. Alvarez (2004). Phytoremediation of hydrocarbon-contaminated soils: principles and applications. *Studies in Surface Science and Catalysis* 151: 447–478.
14. J.M. Yoon, B.T. Oh, C.L. Just and J.L. Schnoor (2002). Uptake and leaching of octahydro-1,3,5,7-tetranitro-1,3,5,7-tetrazocine by hybrid poplar trees. *Environmental Science & Technology* 36 (21): 4649.
15. R.C. Hilborn (1994). *Chaos and Nonlinear Dynamics.* Oxford University Press, Oxford, UK.
16. J. Hadamard (1923). *Lectures on the Cauchy Problem in Linear Partial Differential Equations.* Yale University Press, New Haven, CT.
17. The sources for the Iron Gates discussion are Global Environmental Facility (2005). Project Brief/Danube Regional Project—Phase 1: ANNEX 11 Causes and Effects of Eutrophication in the Black Sea; http://www.gefweb.org/Documents/Council_Documents/GEF_C17/Regional_Danube_Annex_II_Part_2.pdf; accessed April 27, 2005. C. Lancelot, J. Staneva, D. Van Eeckhout, J.-M. Beckers and E. Stanev (2002). Modelling the Danube-influenced Northwestern Continental Shelf of the Black Sea. II: Ecosystem Response to Changes in Nutrient Delivery by the Danube River after its Damming in 1972, Estuarine, Coastal and Shelf Science 54: 473–499.
18. Such engineers were either formally trained in microbiology, like Ross McKinney at MIT, or learned about the microbial world on-the-job.
19. International Society of Environmental Biotechnology: http://www.iseb-web.org/; accessed January 22, 2009.
20. Spokesperson for Scientists for Global Responsibility.
21. A.S. Daar, H. Thorsteinsdóttir, D.K. Martin, A.C. Smith, S. Nast and P.A. Singer (2002). Top ten biotechnologies for improving health in developing countries. *Nature Genetics* 32: 229–232.
22. Wingspread Statement on the Precautionary Principle (1998). Reported in the Science and Environmental Health Network (2000); http://www.sehn.org/precaution.html; accessed August 5, 2009.
23. S. Kelman (1981). Cost–benefit analysis: An ethical critique. *Regulation* 5 (1): 33–40.
24. This is also known as proof by contradiction.
25. For example, Love Canal, Times Beach, Missouri and the Valley of the Drums in Kentucky are major cases that led to regulatory changes.
26. This is not to suggest that there is unanimity, or even consensus, within the scientific community on the role of risk in evaluating new technologies. For example, many scientists and policy makers prefer precaution to evidence-based decision making. Precaution would more likely ban or tightly control a new technology when key

evidence is missing, whereas risk-based decisions tend to allow a technology to go forward so long as the existing and relevant evidence supports its safety and healthy use.
27. Biomedical Engineering Society (2009). BMES Mission and Vision. http://www.bmes.org/mc/page.do?sitePageId=71345&orgId=bes; accessed June 11, 2009.
28. C.B. Fleddermann (1999). Safety and risk. *Engineering Ethics*, Chapter 5. Prentice-Hall, Upper Saddle River, NJ.
29. K. Glass (2005). Ecological mechanisms that promote arbovirus survival: a mathematical model of Ross River virus transmission. *Transactions of the Royal Society of Tropical Medicine and Hygiene* 99(4): 252–260.
30. A. Muhar, P.E.R. Dale, L. Thalib and E. Arito (2000). The spatial distribution of Ross River Virus infections in Brisbane: Significance of residential location and relationships with vegetation types. *Environmental Health and Preventative Medicine* 4(4): 184–189; and R.C. Russell (1999). Constructed wetlands and mosquitoes: Health hazards and management options – an Australian perspective. *Ecological Engineering* 12(1-2): 107–124.
31. For example, in Acts 24:25 and II Peter 1:6 St Peter associates maturation with greater "self-control" or "temperance" (Greek *kratos* for "strength"). Interestingly, he considered knowledge as a prerequisite for temperance. Thus, from a professional point of view, we could take his argument to mean that one can really only understand and appropriately apply scientific theory and principles after one practices them. This is, in fact, the path taken toward the preparation of most professionals. For example, graduates of undergraduate engineering programs may have mastered the mathematical and scientific requirements of the curriculum and, as such, can sit for a Future Engineer (FE) or Engineer-in-Training (EIT) examination. After passing the exam, however, they must also work under the mentorship of a licensed engineer for a specified number of years before being allowed to sit for the Professional Engineering (PE) exam.
32. Ibid., from the Greek *kratos* (strength). Biosystems engineering is complex and complicated, so the nurtured *kratos* has an added measure of importance.
33. R. Posner (2004). *Catastrophe: Risk and Response.* Oxford University Press, New York, NY.
34. M. Rees (2003). *Our Final Hour: A Scientist's Warning: How Terror, Error, and Environmental Disaster Threaten Humankind's Future In This Century - On Earth and Beyond.* Basic Books, New York, NY.
35. Depending on the journal, this can contradict another tenet of scientific research, i.e. the research should be able to be conducted by other researchers, following the methodology described in the article, and derive the same results. However, there is little incentive to replicate research if the likelihood of publication is low. That is, the research is no longer "new" because it was conducted by the original researcher, so the journal may well reject the second, replicate research.
36. However, the engineering profession is beginning to come to grips with this issue. For example, in emergent "macroethical" areas like nanotechnology, neurotechnology, and even sustainable design approaches. For example, see: National Academy of Engineering, 2004, *Emerging Technologies and Ethical Issues in Engineering.* The National Academies Press, Washington, DC.
37. The Royal Society (1992). *Risk: Analysis, Perception and Management.* The Royal Society, London, UK.
38. S.L. Derby and R.L. Keeney (1981). Risk analysis: Understanding "How Safe Is Safe Enough?", *Risk Analysis*, 1 (3): 217–224.
39. Adapted (i.e. added biological considerations) from: M.G. Morgan (1981). Probing the question of technology-induced risk. *IEEE Spectrum* (18) 11: 58–64.
40. Department of the Environment, United Kingdom Government (1994). *Sustainable Development, the UK Strategy.* Cmnd 2426, HMSO, London, UK.
41. Morgan, Probing the question of technology-induced risk.
42. For case analyses where engineers have made such unethical decisions, see: W.M. Evan and M. Manion (2002). *Minding the Machines: Preventing Technological Disasters.* Prentice–Hall PTR, Upper Saddle River, NJ.
43. Comprehensive Environmental Response, Compensation and Liability Act of 1980 (42 USC '9601-9675), December 11, 1980. In 1986, CERCLA was updated and improved under the Superfund Amendments and Reauthorization Act (42 USC 9601 et seq), October 17, 1986.
44. J.S. Mill, 1863, *Utilitarianism.*
45. See M. Martin and R. Schinzinger's (1996). *Ethics in Engineering*, McGraw-Hill, New York, NY, for an excellent discussion of the roles of moral reasoning and ethical theories in engineering decision making.
46. This process follows that called for in: National Research Council (1983). *Risk Assessment in the Federal Government: Managing the Process.* National Academies Press, Washington, DC; and National Research Council (1993). *Issues in Risk Assessment.* National Academies Press, Washington, DC.
47. Federation of American Scientists (2009). Cases in Dual Use: Biological Research. http://www.fas.org/biosecurity/education/dualuse/; accessed on September 20, 2009.
48. Ibid.
49. R.A. Udani and S.B. Levy (2006). MarA-like regulator of multidrug resistance in *Yersinia pestis. Antimicrobial Agents and Chemotherapy* 50 (9): 2971–2975.
50. Federation of American Scientists (2009). Aerosol Delivery Case Study. http://www.fas.org/biosecurity/education/dualuse/; accessed September 20, 2009.
51. Ibid.
52. United Kingdom Department of Environment, Food, and Rural Affairs, Expert Panel on Air Quality Standards (2004). *Airborne Particles: What Is the Appropriate Measurement on which to Base a Standard? A Discussion Document.* DEFRA, London, UK.
53. Aerosol textbooks provide methods to determine the aerodynamic diameter of particles less than 0.5 micrometers. For larger particles gravitational settling is more important and the aerodynamic diameter is often used.

54. American Lung Association (1978). *Health Effects of Air Pollution*. American Lung Association. New York, NY.
55. Federation of American Scientists, Aerosol Delivery Case Study.
56. D.A. Edwards, J. Hanes, G. Caponnetti, J. Hrkach, A. Ben-Jebria, M.-L. Eskew, J. Mintzes, D. Deaver, N. Lotan and R. Langer (1997). Large porous particles for pulmonary drug delivery. *Science* 276: 1868–1871.
57. Federation of American Scientists, Aerosol Delivery Case Study.

CHAPTER 2

A Question of Balance: Using *versus* Abusing Biological Systems

LESSONS FROM ENVIRONMENTAL SYSTEMS

An effective scientist is arguably foremost an ardent observer. The scientist observes the cascade of events that seem to lead to an outcome. The scientist measures and models these events in a laboratory to try to see which factors appear to be most important. In a word, the scientist is modeling and reconstructing nature. Even before the title "scientist" was applied to them, humans have carefully observed traits and characteristics of organisms that could serve a useful purpose. Biotechnology is a striking example of *anthropocentricism*.

For millennia, humans have patiently and laboriously selected those organisms that seemed to have met these purposes and adapted systems to enhance their value. This not only meant selecting individuals from a species of plant or animal that was more hardy, prolific, and nutritious than other individuals, but controlling the conditions around them to achieve better food and fiber. That is, intuitively, humans have understood the complexities of biotic and abiotic systems, especially how certain factors are more sensitive than others in generating some value to their community. Only in the last few decades have the methods for selection and adaptation been abbreviated by applying the tools of systems biology.

Engineers arguably are best known and usually more comfortable when designing with "hard and dry" materials, such as metals, ceramics, and plastics. Inspired by the success of the "soft and wet" natural world, bioengineers and scientists are now designing new practical applications by collapsing the lessons of eons of natural biological adaptation into almost immediate results. In fact, environmental engineers have welcomed the soft and wet in applying biological principles to treat pollution, as have biomedical engineers in their application of physical and chemical principles to address anatomical and physiological challenges in the human body.

Mimicking and adapting biological systems to meet societal needs is systematic, reverse engineering at its highest biological complexity. Over eons of time, molecules assembled into cells, cells organized into higher organisms, which subsequently developed complex life-support biomechanisms. Bioengineers are now analyzing, characterizing, and doing forensics on these natural mechanisms. Recently, this new knowledge has allowed bioengineers to borrow and build from these natural systematic successes to develop new products and provide services, such as improving medicine, agriculture and the environment.

Environmental Biotechnology

Science is chock full of lessons learned from nature, such as following biological design of birds wings to unravel the secrets of flight. God's blueprint for life is simultaneously humbling and inspiring. For millenia, humans have observed natural systems, searching for clues of how they work and adapting these observations to needs. Modern agriculture, for example, has resulted from applying the lessons learned from nature. However, the pace has quickened and the challenges become more daring as even more complicated processes are being reconstructed. For example, Duke University's Center for Biologically Inspired Materials and Material Systems (CBIMMS) "seeks to use some of biology's materials and Lilliputian self-assembly methods to design and build some strikingly different kinds of devices..." including learning "more about how nature did it first, and how to 'engineer around' bioprocesses that can cause disease" [1]

Such bioengineering depends on sophisticated tools, many of which are already being used in the biosciences. First, reliable analytical methods are needed to understand the existing, biological systems (see Discussion Box: Limits of Detection).

DISCUSSION BOX

Limits of Detection

Detection limits are important in environmental biotechnologies. The limit of detection is the lowest concentration or mass that can be differentiated from a blank with statistical confidence, which is a function of sample handling and preparation, sample extraction efficiencies, chemical separation efficiencies, and capacity and specifications of all analytical equipment being used. For example, sampling aerosols around a bioreactor is limited to about 0.002–100 μm aerodynamic diameter, since this is the lower limit of detection using a condensation nucleus counter, and that assumes proper sampling prior to the analysis.

The method detection limit (MDL) is the minimum concentration of a substance that can be measured and reported with 99% confidence that the analyte concentration is greater than zero and is determined from analysis of a sample in a given matrix containing the analyte [2]. The MDL is expected to apply to a wide range of environmental samples including water and wastewater effluent. As such, the MDL for an analytical procedure can vary according to the kind of sample taken and requires a complete, specific, and well-defined analytical method. The minimum limit (ML) is the lowest concentration at which an entire analytical system must give a recognizable signal and acceptable calibration point for the analyte. It is equivalent to the concentration of the lowest calibration standard, so long as the system employs all method-specified sample weights, volumes, and cleanup procedures. The ML is calculated as the product of the MDL by 3.18 and rounding the result to the number nearest to (1, 2, or 5) × 10n, where n is an integer.

The importance of analytical advances related to environmental systems can be easy to overlook, but the recently profound progress in science and engineering owes much to increases in analytical capabilities. As an example, in the 1970s it was not unheard of to have an environmental standard that was below the detection limits of analytical systems. For instance, the author recalls a measurement of mercury (Hg) concentrations in a lake near St Louis, Missouri in which the standard at that time was 20 parts per million (ppm) but the analytical equipment had limits of detection above that level (e.g. 30 ppm). Therefore, the report of the lake's Hg concentration could only reliably say that the levels were above the water quality standard if concentrations were above 30 ppm. So, for example, if the actual Hg concentration were 25 ppm, it would only show up as "non-detected," even though it was really a standard violation. Incidentally, most states have adopted water quality criteria for the protection of aquatic life and human health that fall in the range of 1 to 50 ppt, although current federal methods in the United States do not detect or quantify mercury in this range. A non-detect result using the US Environmental Protection Agency Method 1631, for example, would show only that Hg concentrations are below 200 ppt but would not establish that they are at or below the applicable water quality criterion [3]. No matter, these limits of detection are five orders of magnitude better than just a few decades ago. The current limits of detection for Hg are shown in Table 2.1.Note that these are all below the lowest ambient water quality criterion for Hg of 12 ng L^{-2} (nationwide) [4] and 1.3 ng L^{-2} (Great Lakes) [5].

Table 2.1 Limits of detection for mercury and monomethylmercury

Measurement	Method	Method detection limits
Total mercury in water	EPA 1631	0.121 ng/L (ppt)
Monomethylmercury in water	EPA 1630 with distillation	0.0192 ng/L (ppt)
Total mercury in sediment/soil	EPA 1631	0.302 ng/g (ppb)
Monomethylmercury in sediment/soil	EPA 1630 with extraction	0.0124 ng/g (ppb)
Total mercury in tissue	EPA 1631	0.378 ng/g (ppb)
Monomethylmercury in tissue	EPA 1630 with digestion	1.29 ng/g (ppb)

ppt = parts per trillion; ppb = parts per billion.
Source: Pacific Northwest National Laboratory (2009). Mercury Analytical Laboratory. Determination of total mercury; http://marine.pnl.gov/resources/determination.stm; accessed September 23, 2009.

Detection limits steps [6]:

1. Select an estimated detection limit using one of the following:
   - **(a)** The concentration value corresponding to an instrument signal/noise ranging from 2.5 to 5.
   - **(b)** The concentration equivalent of three times the standard deviation of replicate instrumental measurements of the analyte in reagent water.
   - **(c)** That region of the standard curve where there is a significant change in sensitivity, i.e., a break in the slope of the standard curve.
   - **(d)** Instrumental limitations.
2. Prepare reagent (blank) water that is as free of analyte as possible. Reagent or interference free water is defined as a water sample in which *analyte* and *interferent* concentrations are not detected at the method detection limit of each analyte of interest. Analytes are the compounds that are being investigated, e.g. if a pesticide is being analyzed. The analytes would include the active ingredient and the so-called inert ingredients (so-called, because the term "active" applies only to the particular biocidal mode of action that makes the pesticide efficacious against a particular pest, not to whether a chemical is chemically or biologically active). The interferents are the substances that, when present, may interfere with the particular analytical method and add to incorrect findings. Interferences are defined as systematic errors in the measured analytical signal of an established procedure caused by the presence of interfering chemical species. Thus, the interferent concentration is presupposed to be normally distributed in representative samples of a given matrix.
3. **(a)** If the MDL is to be determined in reagent (blank) water, prepare a laboratory standard (analyte in reagent water) at a concentration which is at least equal to or in the same concentration range as the estimated method detection limit. (Recommend between 1 and 5 times the estimated method detection limit.) Proceed to Step 4.
   (b) If the MDL is to be determined in another sample matrix, analyze the sample. If the measured level of the analyte is in the recommended range of one to five times the estimated detection limit, proceed to Step 4.
   If the measured level of analyte is less than the estimated detection limit, add a known amount of analyte to bring the level of analyte between one and five times the estimated detection limit.

(Continued)

If the measured level of analyte is greater than five times the estimated detection limit, there are two options:

(1) Obtain another sample with a lower level of analyte in the same matrix if possible.

(2) The sample may be used as if for determining the method detection limit if the analyte level does not exceed 10 times the MDL of the analyte in reagent water. The variance of the analytical method changes as the analyte concentration increases from the MDL, hence the MDL determined under these circumstances may not truly reflect method variance at lower analyte concentrations.

4. (a) Take a minimum of seven aliquots of the sample to be used to calculate the MDL and process each through the entire analytical method. Make all computations according to the defined method with final results in the method reporting units. If a blank measurement is required to calculate the measured level of analyte, obtain a separate blank measurement for each sample aliquot analyzed. The average blank measurement is subtracted from the respective sample measurements.

(b) It may be economically and technically desirable to evaluate the estimated method detection limit before proceeding with 4(a). This will: (1) Prevent repeating this entire procedure when the costs of analyses are high and (2) insure that the procedure is being conducted at the correct concentration. It is quite possible that an inflated MDL will be calculated from data obtained at many times the real MDL even though the level of analyte is less than five times the calculated method detection limit. To insure that the estimate of the method detection limit is a good estimate, it is necessary to determine that a lower concentration of analyte will not result in a significantly lower method detection limit. Take two aliquots of the sample to be used to calculate the method detection limit and process each through the entire method, including blank measurements as described above in 4(a). Evaluate these data:

(1) If these measurements indicate the sample is in desirable range for determination of the MDL, take five additional aliquots and proceed. Use all seven measurements for calculation of the MDL.

(2) If these measurements indicate the sample is not in correct range, re-estimate the MDL, obtain new sample as in 3 and repeat either 4(a) or 4(b).

5. Calculate the variance (S^2) and standard deviation (S) of the replicate measurements, as follows:

$$S^2 = \frac{1}{n-1}\left[\sum_{i=1}^{n} x_i^2 - \frac{\left[\sum_{i=1}^{n} X_i\right]^2}{n}\right] \quad S = (S^2)^{\frac{1}{2}} \tag{2.1}$$

Where, X_i; $i = 1$ to n, are the analytical results in the final method reporting units obtained from the n sample aliquots; and Σ refers to the sum of the X values from $i = 1$ to n.

6. (a) Compute the MDL:

$$\text{MDL} = (n-1, 1-\alpha = 0.99)(S) \tag{2.2}$$

Where, MDL = the method detection limit; $t(n - 1, 1 - \alpha = 0.99)$ = the Student's t value appropriate for a 99% confidence level and a standard deviation estimate with n−1 degrees of freedom (see Table 2.2).

(b) The 95% confidence interval estimates for the MDL derived in 6(a) are computed according to the following equations derived from percentiles of the chi-square over degrees of freedom distribution (χ^2/df):

LCL = 0.64 MDL

UCL = 2.20 MDL

Where, LCL and UCL are the lower and upper 95% confidence limits respectively based on seven aliquots.

Table 2.2 **Student's *t* values at the 99% confidence level**

Number of replicates	Degrees of freedom (n – 1)	$t_{(cn-1,0.99)}$
7	6	3.143
8	7	2.998
9	8	2.896
10	9	2.821
11	10	2.764
16	15	2.602
21	20	2.528
26	25	2.485
31	30	2.457
61	60	2.390
00	00	2.326

Source: United States Code of Federal Regulations: 40 CFR 131.6, Appendix B – Definition and procedure for determination of the method detection limit – Revision 1.11. [49 FR 43430, Oct. 26, 1984; 50 FR 694, 696, Jan. 4, 1985, as amended at 51 FR 23703, June 30, 1986].

7. Optional iterative procedure to verify the reasonableness of the estimate of the MDL and subsequent MDL determinations.

(a) If this is the initial attempt to compute MDL based on the estimate of MDL formulated in Step 1, take the MDL as calculated in Step 6, spike the matrix at this calculated MDL and proceed through the procedure starting with Step 4.

(b) If this is the second or later iteration of the MDL calculation, use S^2 from the current MDL calculation and S^2 from the previous MDL calculation to compute the F-ratio. The F-ratio is calculated by substituting the larger S^2 into the numerator S^2A and the other into the denominator S^2B. The computed F-ratio is then compared with the F-ratio found in the table which is 3.05 as follows: if $S^2A/S^2B<3.05$, then compute the pooled standard deviation by the following equation:

$$S_{pooled} = \left[\frac{6S_A^2 + 6S_B^2}{12}\right]^{\frac{1}{2}} \tag{2.3}$$

If $S^2A/S^2B>3.05$, re-spike at the most recent calculated MDL and process the samples through the procedure starting with Step 4. If the most recent calculated MDL does not permit qualitative identification when samples are spiked at that level, report the MDL as a concentration between the current and previous MDL which permits qualitative identification.

(c) Use the Spooled as calculated in 7(b) to compute the final MDL according to the following equation:

$$\text{MDL} = 2.681\ (\text{Spooled}) \tag{2.4}$$

Where 2.681 is equal to $t(12,1-\alpha = 0.99)$.

(d) The 95% confidence limits for MDL derived in 7c are computed according to the following equations derived from percentiles of the chi squared over degrees of freedom distribution.
LCL = 0.72 MDL
UCL = 1.65 MD
Where LCL and UCL are the lower and upper 95% confidence limits respectively based on 14 aliquots.

(Continued)

An example calculation of the MDL is:

A laboratory conducts seven runs of a sample containing 0.1 ng mL^{-1} Hg, with a SD = 0.007.

The standard deviation of 7 runs of 0.1 ng mL^{-1} = 0.007 × Student's *t* value at the 99% confidence level. From Table 2.2, this is found to be 3.143.

Therefore, the MDL = 0.007 ng mL^{-1} × 3.143 = 0.022 ng mL^{-1}.

8. Report the results. The analytical method used must be specifically identified by number or title and the MDL for each analyte expressed in the appropriate method reporting units. If the analytical method permits options that affect the method detection limit, these conditions must be specified with the MDL value. The sample matrix used to determine the MDL must also be identified with MDL value. Report the mean analyte level with the MDL and indicate if the MDL procedure was iterated. If a laboratory standard or a sample that contained a known amount of analyte was used for this determination, also report the mean recovery. If the level of analyte in the sample was below the determined MDL or exceeds 10 times the MDL of the analyte in reagent water, do not report a value for the MDL.

The limit of detection is both an analytical and a sampling threshold. If an instrument can only detect down to 1 ppb, this is an analytical limitation. However, in reality, if the sample has been held for some time, or the sample must be extracted from the soil or trapping device in the field, this is a limit, even if the laboratory can detect down to 1 ppb. Statistical methods for dealing with non-detects are used, but a non-detect should never be reported as 0, since one can only say with confidence that it was not seen. It may not be present, but the scientist or engineer can only report what is known, and that is dictated by the limits of detection.

The performance is expressed in terms of precision, accuracy, specificity, false positives, and false negatives. *Precision* describes how refined and repeatable an operation can be performed, such as the exactness in the instruments and methods used to obtain a result. It is an indication of the uniformity or reproducibility of a result. This can be likened to shooting arrows [7], with each arrow representing a data point. The spread of arrows is equally precise in targets A and B in Figure 2.1. Assuming that the center of the target, i.e. the bull's eye, is the "true value," data set B is more accurate than A. If the archer is consistently missing the bull's eye in the same direction at the same distance, this is an example of *bias* (systematic error). This consistent deviation from the true value can be corrected by calibration and adjustments to equipment (e.g. by running known standards in our analytical equipment). To stay with the archery analogy, the archer would move her sight up and to the right.

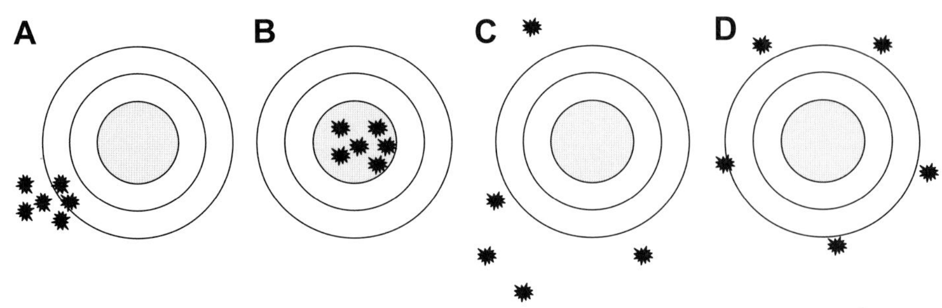

FIGURE 2.1
Precision and accuracy. The bull's eye represents the true value. Targets A and B demonstrate data sets that are precise; targets B and D data sets that are accurate, and targets C and D data sets that are imprecise. Target B is the ideal data set, which is precise and accurate.

Aerosol limits of detection

In the environmental sciences, the use of the term "phase" is nuanced from that of physics. In the earlier Hg example, total mercury in water consists of the various chemical species of Hg (e.g. elemental – Hg^0, alkylated – mono- or dimethylmercury) and whether they are in solution or part of the solids (suspended or sediment). However, in atmospheric sciences the most important phase distribution is that of *vapor* versus *particulate* phase. The distinction between gases and vapors has to do with the physical phase that

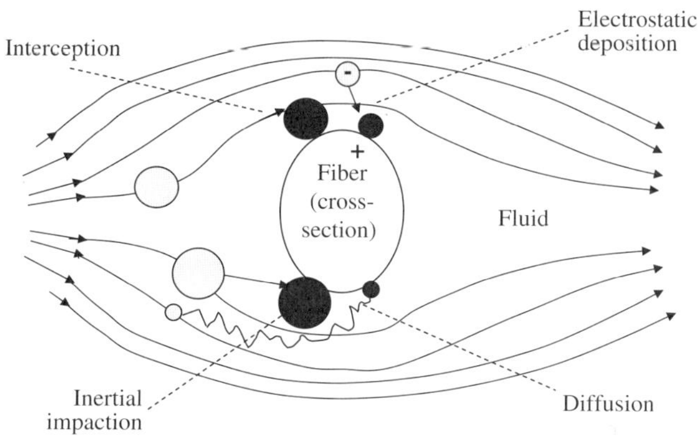

FIGURE 2.2
Mechanical processes important to filtration.
Source: D.A. Vallero (2008). *Fundamentals of Air Pollution*, 4th Edition. Elsevier Academic Press, Burlington, MA; adapted from: K.L. Rubow (2004). Filtration: fundamentals and applications, in *Aerosol and Particle Measurement Short Course.* University of Minnesota, Minneapolis, MN, August 16–18.

a substance would be under environmental conditions, e.g. at standard temperature and pressure. Particulate matter (PM) is an expression of all particles, whether liquid or solid. An aerosol is a liquid or solid particle that is suspended in a gas; in environmental sciences this gas is usually air, but in reactors, stacks, and other non-ambient conditions, this can be various flue gases. Standard atmospheric conditions can be defined as 1 atm pressure (760 mmHg) and 298°K (25 °C + 273) [8].

Since particulate matter is an important vector for genetic material transport (e.g. spore or cysts moved by advection, or genetic material dissolved in or sorbed to aerosols), their measure is a limiting factor in estimating gene transfer, including that of transgenic species.

In the United States, the Clean Air Act established the national ambient air quality standards (NAAQS) for particulate matter (PM) in 1971, requiring measurements of total suspended particulates (TSP) as measured by a high volume sampler, i.e. a device that collected a large range of sizes of particles (aerodynamic diameters up to 50 μm). Smaller particles are more likely to be inhaled than larger particles, so in 1987 the US EPA changed the standard for PM from TSP to PM_{10}, i.e. particle matter $\leq$10 μm diameters [9]. The NAAQS for PM_{10} became a 24-hour average of 150 $\mu g\ m^{-3}$ (not to exceed this level more than once per year), and an annual average of 50 $\mu g\ m^{-3}$ arithmetic mean. However, subsequent research showed the need to protect people breathing even smaller PM in air, since most of the particles that penetrate deeply into the air–blood exchange regions of the lung are quite small. Thus, in 1997, the US EPA added a new fine particle (diameters $\leq$2.5), known as $PM_{2.5}$ [10].

Aerosols are collected using equipment that separates out the size fraction of concern. Filtration is an important technology in every aspect of environmental engineering, i.e. air pollution, wastewater treatment, drinking water, and even hazardous waste and sediment cleanup. Basically, filtration consists of four mechanical processes: 1. diffusion; 2. interception; 3. inertial impaction; and 4. electrostatics (see Figure 2.2).

Diffusion is important only for very small particles ($\leq$0.1 μm diameter) because the Brownian motion allows them to move away in a "random walk" away from the air stream. Interception works mainly for particles with diameters between 0.1 and 1 μm. The particle does not leave the air stream but comes into contact with the filter medium (e.g. a strand of fiberglass). Inertial impaction collects particles that are sufficiently large to leave the air stream by inertia (diameters $\geq$1 μm). Electrostatics consist of electrical interactions between the atoms in the filter and those in the particle at the point of contact (Van der Waal's force), as well as electrostatic attraction (charge differences between particle and filter medium). Other important factors affecting filtration efficiencies include the thickness and pore diameter or the filter, the uniformity of particle

(Continued)

diameters and pore sizes, the solid volume fraction, the rate of particle loading onto the filter (e.g. affecting particle "bounce"), the particle phase (liquid or solid), capillarity and surface tension (if either the particle or the filter media are coated with a liquid), and characteristics of air or other carrier gases, such as velocity, temperature, pressure, and viscosity.

Aerosol measurement is an expression of the mass of particles within bands of particle sizes, i.e. aerometric diameters (e.g. $<2.5\ \mu m$, $>2.5\ \mu m$ but $<10\ \mu m$, and $>10\ \mu m$). Figure 2.3 shows an inlet of the $PM_{2.5}$ sampler that is designed to extract ambient aerosols from the surrounding air stream, remove particles with aerodynamic diameters $>10\ \mu m$, and move the remaining smaller particles to the next stage. Figure 2.4 illustrates the impactor and filter assembly for removing those particles $<10\ \mu m$ but greater than $2.5\ \mu m$ in diameter but allows particles of $2.5\ \mu m$ in diameter to pass and be collected on a filter surface. Particles $<10\ \mu m$ but $<2.5\ \mu m$ are removed downstream from the inlet by a single-stage, single-flow, single-jet impactor assembly. Aerosols are collected on filters that are weighed before and after sampling. This system uses 37 mm diameter glass filters immersed in low volatility, low viscosity diffusion oil. The oil is added to reduce the impact of "bounce", i.e. particle hit the filter and are not reliably collected [11].

The before and after weight difference is the particulate mass in a given air volume, expressed in units of mass (e.g. ng) per units of volume (most often m^3), i.e. $ng\ m^{-3}$. This represents the mass concentration of aerosols in ambient air requires a sensitivity of ± 100 ng for PM_{10} on standard $0.20 \times 0.25\ m^2$ filter and a sensitivity of ± 1 ng for 37 mm or 47 mm diameter Teflon filters. The LD for the balance must be good enough to meet the atmospheric limits of detection of $60\ ng\ m^{-3}$ for PM_{10} and $40\ ng\ m^{-3}$ for $PM_{2.5}$. This assumes a 24-hour sampling period on a standard filter for PM_{10} at a flow rate of $0.0189\ m^3\ sec^{-1}$, and a 24-hour sampling period on a 47 mm diameter filter for $PM_{2.5}$ at a flow rate of $2.78 \times 10^{-4}\ m^{-3}\ sec^{-1}$ [12].

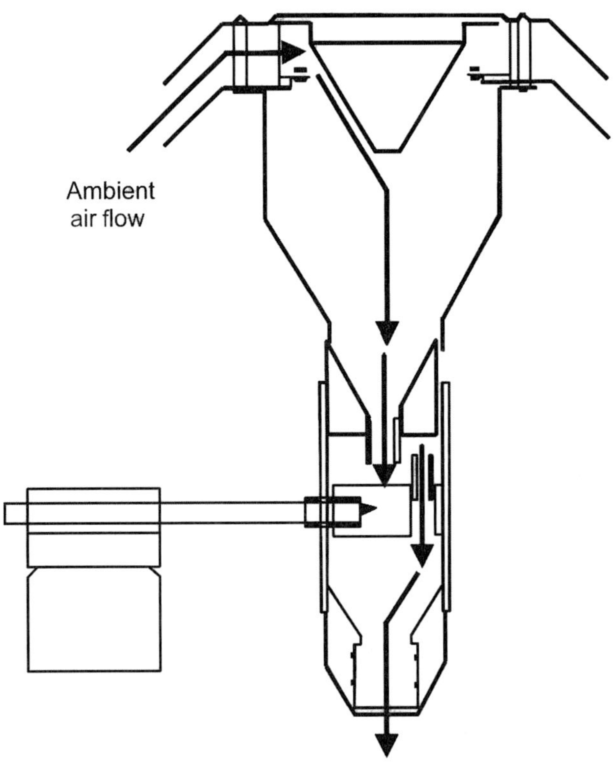

FIGURE 2.3
Flow of air through a sampler inlet head used to collect particulate matter with aerodynamic diameters $<2.5\ \mu m$ ($PM_{2.5}$). WINS = well impactor ninety-six, i.e. the design of the particle impactor specified by the US EPA for reference method samplers for $PM_{2.5}$.
Source: US Environmental Protection Agency (1998). *Quality Assurance Guidance Document 2.12. Monitoring $PM_{2.5}$ in Ambient Air Using Designated Reference or Class I Equivalent Methods.* Research Triangle Park, North Carolina, November 1998.

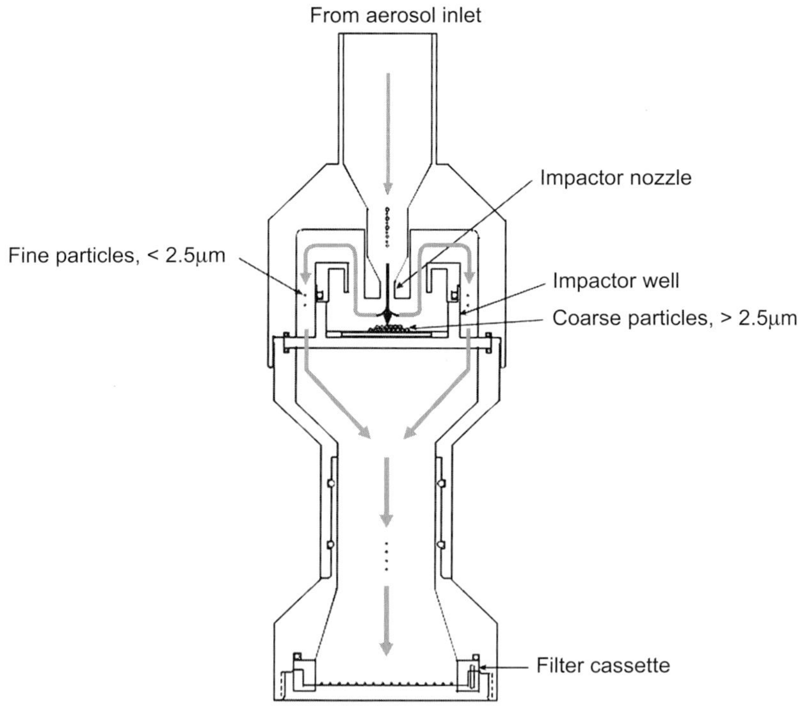

FIGURE 2.4
Flow of air through an impactor well and filter holder used to collect particulate matter with aerodynamic diameters <2.5 µm ($PM_{2.5}$).
Source: US Environmental Protection Agency (1998). *Quality Assurance Guidance Document 2.12. Monitoring* $PM_{2.5}$ *in Ambient Air Using Designated Reference or Class I Equivalent Methods.* Research Triangle Park, North Carolina, November 1998.

Pre-weighing must account the relative humidity (i.e. water in the filter adds mass), not only during weighing but for a substantial amount of time (e.g. 24 hour preceding period), as part of numerous calibration steps.

Precision and accuracy in aerosol measurements are greatly affected by collection efficiency. Accuracy is the same as that defined in the previous section, but precision for aerosols is defined as the random variation among individual measurements of the same property, usually under prescribed identical conditions. For ambient particulate concentration measurements, precision is usually expressed in terms of a standard deviation estimated by collocated sampling or by reweighing filters and comparing reproducibility. Typical precision is within 5% [13].

Microbial limits of detection

The TML applies to chemical substances, but microbial limits also exist. In fact, regulatory agencies are looking for ways to lower levels of detection for microbes. An illustrative example of how to determine the usefulness of an environmental analytical system is the 2004 US Environmental Protection Agency review of a bacterial test kit. The Invitrogen Corporation developed the PathAlert™ detection kits for the bacteria *Francisella tularensis* (*F. tularensis*), *Yersinia pestis* (*Y. pestis*), and *Bacillus anthracis* (*B. anthracis*) as part of the Environmental Technology Verification Program [14]. The results are summarized here, and the method used to validate this test is provided in Appendix 3.

In the verification of the bacterial test, precision was determined from the overall percentage of consistent responses for all the sample sets. Responses were deemed to be consistent if all responses of the four replicates were the same. For *F. tularensis* replicates, 95% of the sample sets (20 out of 21) showed consistent results. Likewise, 95% of the sample sets (20 out of 21) were consistent for *Y. pestis*. For both, the inconsistency resulted from an inconclusive result for a drinking water (DW) replicate for each bacterium.

(Continued)

B. anthracis also showed 95% consistency (20 out of 21) but the one sample set with inconsistent results was the infective dose in a performance test (PT) sample. Here, three of the four samples were inconclusive, but the fourth sample was positive for *B. anthracis*. The infective dose of *B. anthracis* was below the method level of detection (LD) for this bacterium [15].

Accuracy was assessed by evaluating how often the results were positive in the presence of a concentration of contaminant above the method LD. Contaminant-only performance testing (PT) samples were used for this analysis. An overall percent agreement was determined by dividing the number of positive responses by the overall number of analyses of contaminant-only PT samples above the method LD. The results are presented in Table 2.3.

For *F. tularensis*, *Y. pestis*, and *B. anthracis*, all samples at concentration levels above the vendor-stated method LD generated positive responses for each set of replicates, resulting in 100% agreement for the overall accuracy of the detection kit for each bacterium. The infective/lethal dose for *Y. pestis* is 0.28 colony-forming units per milliliter (cfu mL^{-1}) and for *B. anthracis* is 200 cfu mL^{-1}. Both doses were below the method LD and not included in the accuracy calculations for those bacteria.

Specificity is the ability of a test to show a negative response when the contaminant is in fact absent. The specificity rate of this kit was determined by dividing the number of negative responses by the total number of unspiked samples. Unspiked interferent PT samples and unspiked DW samples were used to assess specificity. For *F. tularensis* and *Y. pestis*, one unspiked DW replicate for each bacterium produced an inconclusive response (see Table 2.4) [16].

A *false positive response* is defined as a detectable or positive test response when the agent is not present, in this case it was the interferent PT samples or DW samples which were not spiked. The false positive rate was the frequency of false positive results out of the total number of unspiked samples. The false negative response was defined as a negative response when the sample was spiked with a contaminant at a concentration greater than the method LOD. Spiked PT (contaminant and interferent) samples and spiked DW samples were included in the analysis.

Table 2.3 **Accuracy of bacteria test kit based on percentage of positive results compared to total replicates for each bacterium species**

Bacterium	Concentration range of samples used in accuracy calculations (cfu/mL)	Overall accuracy (Positive results out of total replicates)
F. tularensis	2×10^4 to 5×10^5	100% (20/20)
Y. pestis	2×10^2 to 5×10^3	100% (16/16)
B. anthracis	2×10^4 to 5×10^5	100% (16/16)

Source: US Environmental Protection Agency and Battelle National Laboratory (2004). ETV Joint Verification Statement. Rapid Polymerase Chain Reaction: Detecting Biological Agents and Pathogens in Water; http://www.epa.gov/ordnhsrc/pubs/vsInvitrogen121404.pdf; accessed September 24, 2009.

Table 2.4 **Specificity of bacteria test kit based on percentage of negative results compared to total replicates for each bacterium species**

Bacterium	Overall specificity (Negative results out of total replicates)
F. tularensis	96% (23/24)
Y. pestis	96% (23/34)
B. anthracis	100% (22/22)

Source: US Environmental Protection Agency and Battelle National Laboratory (2004). ETV Joint Verification Statement. Rapid Polymerase Chain Reaction: Detecting Biological Agents and Pathogens in Water; http://www.epa.gov/ordnhsrc/pubs/vsInvitrogen121404.pdf; accessed September 24, 2009.

Conversely, the *false negative rate* was reported as the frequency of false negative results out of the total number of spiked samples for a particular contaminant. The results are presented in Table 2.5. No false positives or false negatives were found for any of the sample matrices for any bacteria for this kit. One replicate for unspiked DW in two different DW samples showed an inconclusive result for *F. tularensis* and one for *Y. pestis*. Two inconclusive results were reported; one for *F. tularensis* in one DW sample and one for *Y. pestis* in a different DW sample (see Table 2.5).

Table 2.5 **False positive and false negative response rates of bacteria test kit**

Bacterium	False positive rate	False negative rate
F. tularensis	0/24	0/60
Y. pestis	0/24	0/56
B. anthracis	0/22	0/56

Source: US Environmental Protection Agency and Battelle National Laboratory (2004). ETV Joint Verification Statement. Rapid Polymerase Chain Reaction: Detecting Biological Agents and Pathogens in Water; http://www.epa.gov/ordnhsrc/pubs/vslnvitrogen121404.pdf; accessed September 24, 2009.

Next, mathematical and engineering skills, including deductive and inductive reasoning and intuition, help to disentangle the biological mechanisms and to arrive and "imagineer" ways that these mechanisms can be applied to existing challenges:

> What we have learned from biologists is that biology operates on the "just good enough" principle. The general perspective is that things aren't optimized. In biology, things work just well enough to survive. As engineers, we can use our optimization techniques to create better structures for specific purposes.
>
> Robert Clark, Former Chair of Duke University's Center for Biologically Inspired Materials and Material Systems [17]

Can we really expect to "improve" nature? After all, as we will discuss later in this chapter, the naturally adapted systems were built from a relatively small set of molecules, from configurations of just a few elements. What if we were able to add some elements and configurations to these elegant systems? What if we could make subtle changes to cellular features to give outcomes that could take generations of adaptive selection to produce?

The bioengineering challenge begins with a careful examination of the natural processes to see how "synthetic chemistries and bio-based synthesis" can create new molecules from the same biological principles but reapply these principles in novel and enhanced ways [18]. In 2001, for example, Duke developed the so-called "liposomes," which are capsules with a thickness of a mere two molecules and about one-hundredth the size of a red blood cell. Such nanoscale capsules can deliver a chemotherapy drug to a tumor and then apply the dose that can be triggered by focused heating for cancer therapy.

In fact the liposomes were inspired by the two-layer fatty membranes that enclose the typical cell. However, the fatty material in the liposomes is chemically modified to be waxier than natural cells. Thus, in a heated environment the liposome material undergoes a change from solid to liquid phase, making the membrane more porous and conducive to mass transfer until the heat is removed. The possibilities are exciting, for example, polymers might be designed and their structures improved in biological systems by emulating the electrical signals between muscles and the nervous system and using atomic force microscopy.

ENVIRONMENTAL BIOMIMICRY

Imitation is said to be the highest form of flattery. Are we paying nature a compliment or are we exhibiting rashness in mimicking natural systems? Organisms are complex, as biomedical, microbiological and environmental sciences and engineering increasingly find as they attempt to unravel animal and human physiological processes. Bumping this up a level of scale, reconstructing natural mechanisms and the systems' biology can also advance environmental biotechnologies. Biotechnology and, more recently, nanotechnology hold much promise. In fact nanotechnology has gained some of the media and scientific attention once held by biotechnology. Numerous articles have been written on how we should apply the lessons learned from biotechnology, so that they are not repeated. One prominent lesson is that observing nature can teach bioengineers how to enhance design by applying its principles. Processes and mechanisms do indeed work quite well in nature, so the bioengineer should study them, find out how they work, and apply them to society's needs. This is biomimicry.

All emergent technologies arise out of a hope for their *applications*, but this hope always comes with a fear of *implications*. On the implications side, there is much uncertainty about possible risks to consumers and the environment. Frankly, not much is known about how physical, chemical, and biological characteristics of materials change at dimensions less than 100 nanometers (that is the threshold between ordinary or "bulk" and "nano" materials that enjoy general agreement of the scientific community).

Some changes are obvious, such as the difference in electromagnetic properties, e.g. at the nanoscale the metal gold is red. Carbon is one of the principal elements that comprise nanomaterials (see Figure 2.5), but so are numerous metallic substances, e.g. titanium dioxide, zinc, cerium, iron, and silver. These and other nano-enabled materials provide numerous environmental applications. Engineers currently fabricate pollution detection and control systems from these materials, including membranes, adsorbents, oxidants, and catalysts. They are also used in analytical equipment, such as real-time chromatographs and pollutant sensors.

The good news is that scientific breakthroughs are happening, with great promise in medical and engineering applications, such as the improved and more targeted drug delivery and better

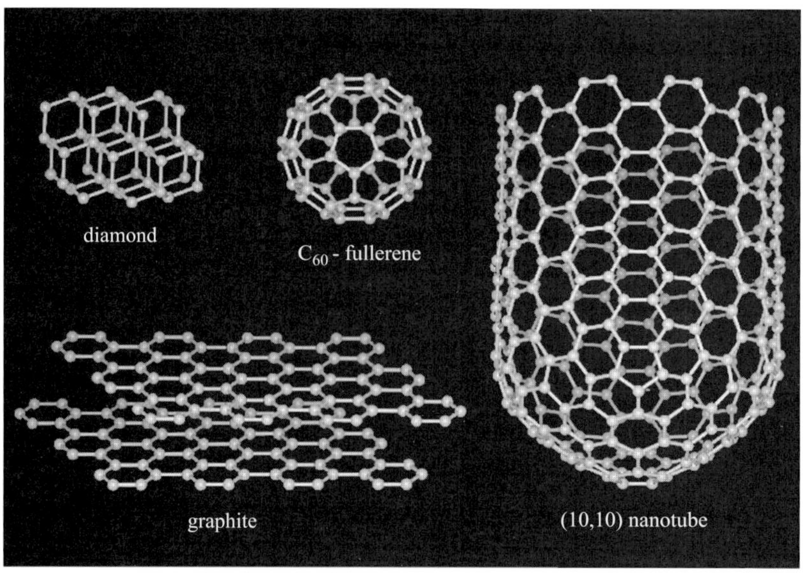

FIGURE 2.5
Structures of four carbon-based nanomaterials. The buckminsterfullerene (or simply, fullerene) consists of 60 bond carbons. The fullerene and the nanotube are nanoparticles. The tubes are elongated fullerenes.
Drawing used with permission from Mark Wiesner.

diagnostics mentioned above. How wonderful it would be if biomedical engineers were able to employ the principles of nanotechnology to target tumor cells without damaging normal, healthy surrounding cells. But these same technologies give us pause in what happens if they are released to the environment. Do they aggregate and become no different than bulk particles that we breathe all the time? Do they change chemically (e.g. hydrolyze) after release, rendering them harmless, or do they take on even more toxic properties?

Thus, nanotechnology, as all systems, must be considered from a life cycle perspective. The pros and cons are complicated and often contravening. For example, nanotechnologies are presently being used to clean up hazardous waste sites. As evidence, nano iron (Fe) particles have been shown to enhance the breakdown of toxic substances in groundwater. In conventional treatment, iron filings have been used to treat groundwater (see Figure 2.6). However, when nano Fe particles are injected, removal rates have improved considerably (Figure 2.7), probably because the nano particles have significantly larger overall surface area, larger numbers of reaction sites on the particles, and the nano particles themselves are more reactive than their bulk-scale counterparts [19]. They may also provide a greater number of nanoscale surfaces on which microbes can reside, allowing contaminants to come into contact with biofilm, and enhancing biodegradation (see Discussion Box: Biochemodynamic Films in Chapter 7).

Applications of innovative, nanoscale technologies will continue to be arrows in the bioengineering quiver. We must be careful to balance the benefits and opportunities with risks.

Scientists continue to improve their ability to characterize systems at increasingly smaller scales. In recent years, nanotechnology has gained some of the media and scientific attention once held by biotechnology. These innovations in genetic engineering apply our growing

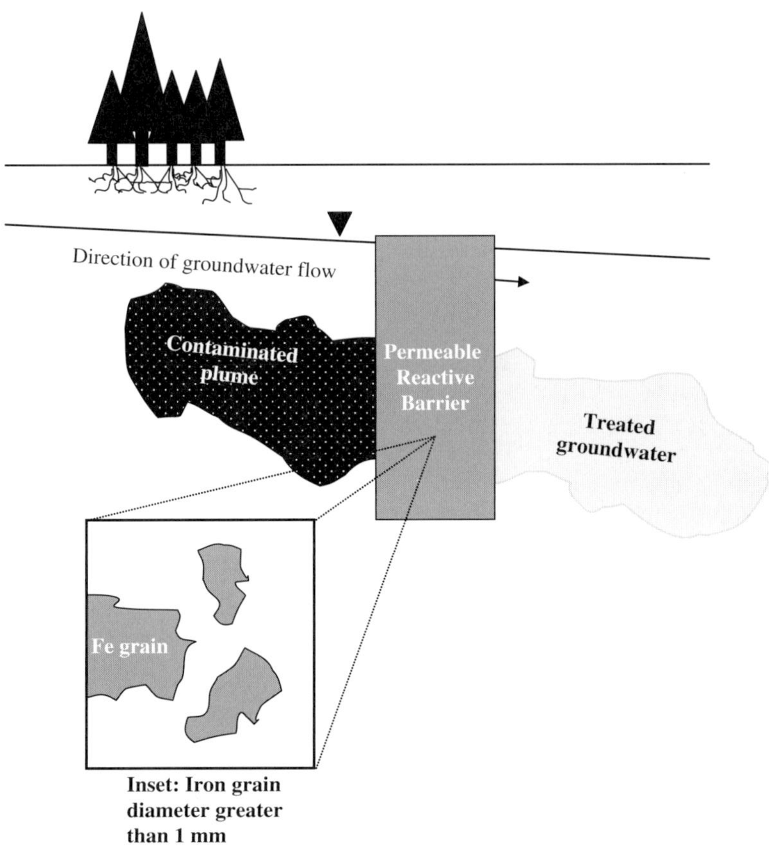

FIGURE 2.6
Conventional onsite (*in situ*) treatment of contaminated groundwater with injected bulk-scale (about 1 mm diameter) iron particles.
Adapted from: P.G. Tratnyek and R.L. Johnson (2006). Nanotechnologies for environmental cleanup. *Nano Today* 1 (2), May.

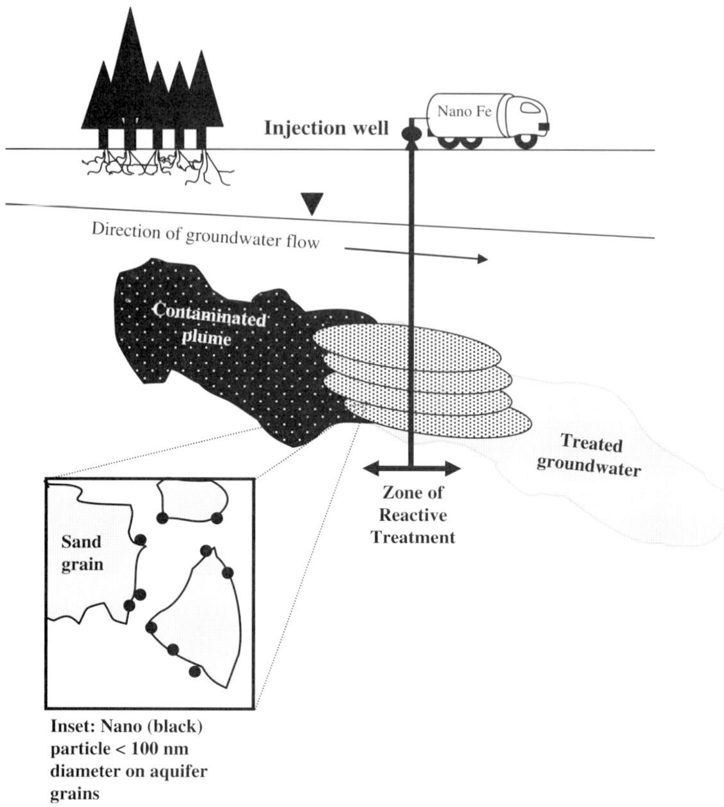

FIGURE 2.7
Nanotechnology-based (*in situ*) treatment of contaminated groundwater by creating a reactive treatment zone resulting from sequential injection of nano-sized Fe to form overlapping zones of nano iron particles sorbed to the grains of native aquifer material (e.g. sand).
Adapted from: P.G. Tratnyek and R.L. Johnson (2006). Nanotechnologies for environmental cleanup. *Nano Today* 1 (2), May.

understanding of DNA to society's need, or if you are more jaundiced in your view of genetic modification, the manipulation of the building blocks of life. In fact, there is much to learn from biotechnology. For example, nature can inform and improve design if we can apply its principles (i.e. biomimicry). However, when scientists manipulate the molecules of life, concerns about ethics and societal risk arise.

ENGINEERED SYSTEMS INSPIRED BY BIOLOGY

A mere handful of elements, arranged in myriad ways, provide the structure for all living systems. These biophile (*bio* – life; *phil* – affinity) elements, shown in Table 2.6, were first identified by Victor Moritz Goldschmidt, considered the father of modern geochemistry [20]. Geochemistry is often associated with the abiotic fields of mineralogy and petrology, yet we can trace the classification of an atom economy to Goldschmidt since taxonomy of the various elements in the periodic table was indeed functional. The group with an affinity for and which form the living systems indeed constitutes the lion's share of matter involved in living tissue.

Biophile elements are enriched in the biosphere as they cycle through living organisms, remains, and natural processing over time (e.g. formation of fossil fuels). Thus, the group not only includes the major biophiles that are found in all living tissue, but also those elements that form bonds with organic carbon. In fact, a mere 11 elements make up 99.8% of the atoms in the human body [21].

Biologists often think of geochemistry as an abiotic field, but perhaps it is not so ironic that geochemistry was the first to differentiate biophiles from other abiotic elements. After all, living systems are always comprised of both biotic and abiotic material, and geochemistry is

Table 2.6 **List of biophile elements**

Type	Element	Mean percentage of atoms in the human body	Mean percentage of atoms in microorganisms
Major biophile	Oxygen	65.0	65.0
	Carbon	18.0	18.0
	Hydrogen	10.0	10.0
	Nitrogen	3.0	3.0
	Phosphorus	1.0	0.4
	Sulfur	0.26	0.3
	Iodine	NR	NR
	Chlorine	0.14	NR
Minor biophile	Boron	NR	NR
	Calcium	1.4	NR
	Magnesium	0.5	NR
	Potassium	0.34	NR
	Sodium	0.14	NR
	Vanadium	NR	NR
	Manganese	NR	NR
	Iron	NR	NR
	Copper	NR	NR

Note: NR = not reported, so if present, only in trace amounts.
Source: C.L. Hollabaugh (2007). Modification of Goldschmidt's geochemical classification of the elements to include arsenic, lead and mercury as biophile elements. In: R. Datta, D. Sarkar and R. Hannigan, eds, *Concepts and Applications in Environmental Geochemistry*, Elsevier, Amsterdam, The Netherlands, pp. 9–32; human body percentages from: U. Lindh (2005). Biological functions of the elements. In: *Essentials of Medical Geology*, Elsevier, Amsterdam, The Netherlands, pp. 115–160; microorganism percentages from: D.R. Schneider and R.J. Billingsley (1990). *Bioremediation: A Desk Manual for the Environmental Professional*, Cahners Publishing Co., Des Plaines, IL.

the chemistry of the earth. The relationships between living organisms and their environment reveal themselves in cycles of matter and energy into and out of the organism. From a biosystematic perspective, this means that the organism itself is a thermodynamic control volume. In turn, the population of organisms is part of the larger control volumes (e.g. a microbe in the intestine; the intestine in the animal; the herd as prey in a habitat; the habitat as part of an ecosystem's structure).

In addition to those elements that comprise most of the tissue mass, i.e. carbon, oxygen, hydrogen, and nitrogen, the other biophiles listed in Table 2.7 add nuances that are the consequence and the cause of the survival of various species. Thus, if these systems and processes can be understood, characterized, emulated, and ultimately enhanced in artificial systems, certain qualities of biological systems can be put to use in agriculture, medicine, engineering, and environmental applications. However, the downsides must also be considered. Countervailing and subsequent risks must be avoided and prevented. What may be good for a few organisms may be bad overall for the system.

One of the means to manage present and future risks is to model the factors, constraints, relationships, synergies, antagonisms, and sensitivities of various combinations of the elements of a system. This can be a thermodynamic model wherein mass and energy balances

Table 2.7 Exposure factors for selected biological warfare agents

Agent	Disease	Transmission	Lethality	Infectivity	Required detection capability[a]
Bacteria					
Bacillus anthracis	Anthrax	Spores in aerosol	High ~ 100%	10,000 organisms	5,000 org/m^3 air
Vibrio cholera	Cholera	Food and water			
		Aerosol	Low with treatment	1 million organisms	500,000 org/L water
Yersinia pestis	Pneumonic plague	Aerosol inhalation	High unless treated	< 100 organisms	< 25 org/m^3 air
Franciscella tularensis	Tularemia (rabbit fever)	Aerosol inhalation	Moderate	1 to 50 organisms	< 25 org/m^3 air
Shigelladysenteriae	Dysentery	Inhalation and ingestion	Moderate	10 to 100 organisms	25 org/m^3 air
					25 org/L water
Rickettsia					
Coxiella burnetti	Q fever	Aerosol inhalation	Very low	10 organisms	5 org/m^3 air
					< 5 org/kg food
Rickettsia rickettsii	Rockey Mountain spotted fever	Vectors	Low	N/A	N/A
Viruses					
Ebola virus	Ebola	Direct contact	High for Zaire strain	N/A	
		Aerosol			
Venezuelan Equine Encephalitis (VEE) virus	Encephalitis	Vectors	Low	N/A	
Yellow fever virus	Yellow fever	Vector/tick	Low	N/A	
Rift Valley fever virus	Rift Valley fever	Vector/ mosquito	Low	N/A	
Variola virus	Smallpox	Aerosol	High to moderate	N/A	
Hanta virus	Hanta	Aerosol	43% in US	N/A	N/A
Dengue fever	Dengue fever	Aedes mosquito	Low to moderate	N/A	

[a]*These numbers were calculated by dividing the infectivity level by 2 m^3 (the amount of air assumed to be breathed in two hours by an active adult) or by 2 L, the amount of water consumed during a day.*

Source: T.E. McKone, B.M. Huey, E. Downing and L.M. Duffy (Eds) (2000). *Strategies to Protect the Health of Deployed U.S. Forces: Detecting, Characterizing, and Documenting Exposures.* National Academies Press, Washington, DC.

are permutated and perturbed to see what happens as a result. What goes into a control volume? What happens after these elements and energies enter the volume? And, what comes out of the control volume as a result? Next, what happens as these changed elemental arrangements (i.e. molecules) re-enter the control volume and as the energy levels change? This is very complicated indeed. Invariably, when working with living systems, characterizing mass and energy is incomplete. Often, only highly constrained and controlled systems can be studied. Natural, living systems are geometrically more complex than anything that can be simulated in a laboratory. Therefore, bioscience has much to learn by iteratively characterizing and understanding these biosystems. So, let us start with a consideration of the cycling of some key biophiles.

ENVIRONMENTAL MICROBIOLOGY

Arguably, much of the interest in microbiology in the environmental sciences began with a focus on pathogenic microbes. Avoidance of disease was the profound interest of medical sciences, which called for improved taxonomical information about microorganisms, as well as ways to identify and to characterize them (e.g. microscopic analysis, staining techniques, and identification of spore formation processes). However, microbiology was also needed to support engineering applications, especially the beneficial use of bacteria to treat wastes, mainly those in surface waters.

Both the pathogenic and beneficial interests led to the incorporation of microbiological sciences into environmental sciences and engineering in the 20th century. It was only in the last part of the century that microbiological techniques were used to identify genetically modified strains of microorganisms (See Discussion Box: Pollutant-Degrading Bacteria: Microbiology Meets Bioengineering).

In recent years there has been a growing concern about microbes, especially the use of bacteria and other organisms in terrorist and warfare applications (see Tables 2.7 and 2.8). Science must consider and advance methods for preparedness, health, and security, such enhancing ways "to anticipate life-threatening exposures" [22]. This calls for improvement in the application of microbiology. These microbiological applications will also improve the characterization of potential exposures to detrimental microorganisms, including the various exposure pathways based on the dimensions of adverse effects, e.g., severity, populations affected and persistence of the effects [23].

Discussion Box

Pollutant-Degrading Bacteria: Microbiology Meets Bioengineering

All bacteria are prokaryotes, which are basically biomolecules surrounded by a membrane and a cell wall. Their DNA is in the form of a single bacterial chromosome and is not bound within a nucleus, as in the more advance eukaryotes (higher plants and animals). Prokaryotic DNA is not associated with histone proteins; thus it is referred to as "naked" DNA.

Taxonomy is important. Most soil bacteria, for example, are nonpathogenic. However, certain species of gram-negative bacilli, e.g. from the genus *Acinetobacter*, have been found to be pathogenic, especially to immunocomprised individuals. Only a few species are likely pathogenic, so distinguishing species is crucial. There are numerous ways to classify bacteria (see Figure 2.8). The Gram's stain is a microbiological technique to classify bacteria on the basis of their forms, sizes, cell shape, and gram reaction. The approach is based upon the Hucker modification of the original Gram's stain method. Gram-positive and gram-negative bacteria are both stained by the primary stain, i.e. crystal violet dye. When iodine is added, a crystal violet–iodine complex is formed within the cell wall. A decolorizing agent extracts

Table 2.8 Characteristics of Selected Biological Toxins

Source	Toxin	LD_{50} (μG/kg)	Required Detection Capability[a]	Notes
Bacteria *Clostridium botulinium*	Botulinium A, B, C, D, E	~ 0.02 (inhalation) 1 (oral)	0.1 mg/m^3 0.02 mg/L (water or food)	Among the most potent toxins known Delayed lethality Persists in food and water Breaks down within 12 hours in air
Clostridium perfringens	Gangrene-causing enzyme	0.1 to 5	0.3 mg/m^3	Delayed action Low mortality, but very debilitating
Clostridium tetani	Tetanus toxin	~ 3	N/A	Delayed action Relatively unstable and heat sensitive
Cornyebacterium diptheria	Diptheria toxin	0.03	N/A	Lethal Rapid acting
Staphylococcus aureus	Staphylococcus enterotoxin A, B, C, D, E (Toxicity is for type B)	0.4 (aerosol ED_{50}) 20 (aerosol LD_{50}) 0.3 (oral ED_{50})	0.058 mg/m^{30} 3 mg/m^3 0.007 mg/L	Rapid acting Symptoms persist for 24 to 48 hours Severely incapacitating Can be lethal Large-scale production feasible Very stable
Dinoflagellates *Gonyaulax tamerensis, Gonyaulax catanella*, and related species	Saxitoxin (shellfish poison)	1 (aerosol inhalation) 7 (oral)	0.01 mg/m^3 (air) 0.2 mg/L	Lethal Rapid acting Soluble in water Relatively persistent
Takifugu poecilonotuss	Tetrodotoxin	1.5 to 3 (inhalation) 30 (oral)	0.3 mg/m^3 (air)	Lethal Rapid acting
Algae *Anacystis* sp.	Anatoxin A (VFDF)	170 to 250 (Ip)[b]		Very fast death factor
Anabanea flosaquae		5,000 (oral) 2,100 (dermal)	100 mg/L(kg) (water or food)	Very rapid acting
Microcystis aeruginosa (*Anacystis cyanea*)	Microcystin (FDF)	25 to 100 (Ip)[b]	~10 mg/m^3 (air) ~2 mg/L (water)	Lethal, rapid acting Fast death factor

Fungi				
Fusarium sp.	Trichothecene mycotoxins ("yellow rain")	25 to 500 (inhalation) 1,600 (oral)	40 mg/m^3 (air) 40 mg/L	Nonlethal, delayed effects Inhalation, ingestion, dermal Very stable Small repeated doses are cumulative
Plants				
Ricinus communis	Ricin	1,000	150 mg/m^3 (air) 20 mg/L (water)	Lethal, delayed action Easily produced Persistent
Animals				
Palythoa (soft corals)	Palytoxin	0.08 to 0.4	0.035 mg/m^3 (air) 0.006 mg/L (water)	Lethal and rapid acting Stable
Conus geographus	Conotoxins	3 to 6	~0.6 mg/m^3 (air)	Water soluble
Conus magnus (fish-hunting cone snails)			~0.1 mg/L (water)	Highly stable Can be used as aerosols
Easily synthesized *Phyllobates aurotaenia* and *Phyllobates terribilis*	Batrachotoxin	0.1 to 0.2	0.015 mg/m^3 (air)	Rapid acting and lethal Very stable Can be synthesized

[a] *Assumes 70-kg adult breathing at a rate of 0.016 m^3/min for 30 minutes for air or the ingestion of 3 L water or 3 kg food by a 70-kg adult.*
[b] *IP refers to intraperitoneal injection dose to mice.*
Source: T.E. McKone, B.M. Huey, E. Downing and L.M. Duffy (Eds) (2000). *Strategies to Protect the Health of Deployed U.S. Forces: Detecting, Characterizing, and Documenting Exposures.* National Academies Press, Washington, DC.

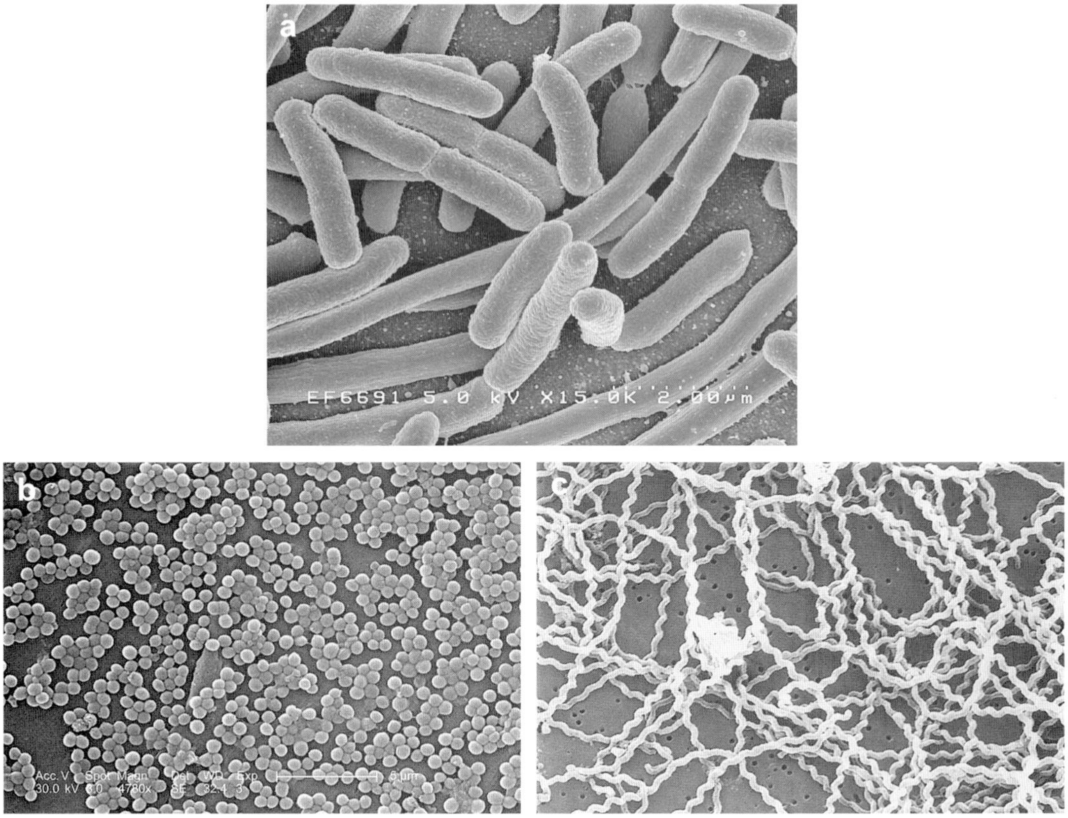

FIGURE 2.8
Common shapes of bacteria. (A) Bacilli or rod-shaped (*Escherichia coli*); (B) cocci or spherical (*Staphylococcus aureus*); and (C) borrelia or spiral-shaped (*Leptospira interrogans*).
Source: University of Florida/Institute of Food and Agricultural Sciences, Department of Fisheries and Aquatic Sciences (2003). *A Beginner's Guide to Water Management – Bacteria.* Information Circular 106.

lipid from the cell wall of gram-negative bacteria, thus increasing the crystal violet–iodine complex, so that the complex diffuses from the cell. Simultaneously, the gram-positive bacteria lose water which leads to a decrease in the porosity of the cell wall, thereby trapping the crystal violet–iodine complex within the cell. The increased porosity means that the counterstain (e.g. safranin) can permeate the cell wall of gram-negative bacteria.

Staining is quite useful in differentiating the types and characteristics of a microbe's cell wall and, as such, often distinguishes pathogenic microbes from beneficial microbes. In particular, the cell wall behavior elicits immune responses in mammals, including humans.

The knowledge base at the confluence of microbiology and engineering has been consistently growing since the 1990s regarding those microorganisms that participate in the degradation of recalcitrant organic compounds. The consensus viewpoint of the bioremediation industry near the end of the 20th century was that all that was required to provide the desired degradation reaction was to manipulate the geochemical conditions, which would lead to the stimulation of microbes having the appropriate metabolic pathways needed to break down the target compound. After this pathway was established, the microbe could be isolated and its metabolic processes understood [24].

Reductive dechlorination of tetrachloroethene (PCE) and trichloroethene (TCE) was recognized [25] in 1983, but the research demonstrated that the degradation rate of each subsequent reductive dechlorination step became slower than the preceding one, meaning that vinyl chloride (VC) would accumulate (see Figure 2.9). This caused engineering researchers to ignore the anaerobic biodegradation approach for PCE and TCE since the more toxic and carcinogenic VC would accumulate (see Appendix 2). Simultaneously, chlorinated ethenes were found to be cometabolized by aerobic bacteria [26]. *Cometabolism* (in this instance, co-oxidation in Phase 1 metabolism) is the fortuitous transformation of a compound by an enzyme synthesized by the cell for metabolism of another compound. The chlorinated ethenes, for example, are able to be degraded by this mechanism. In the late 1990s, PCE was

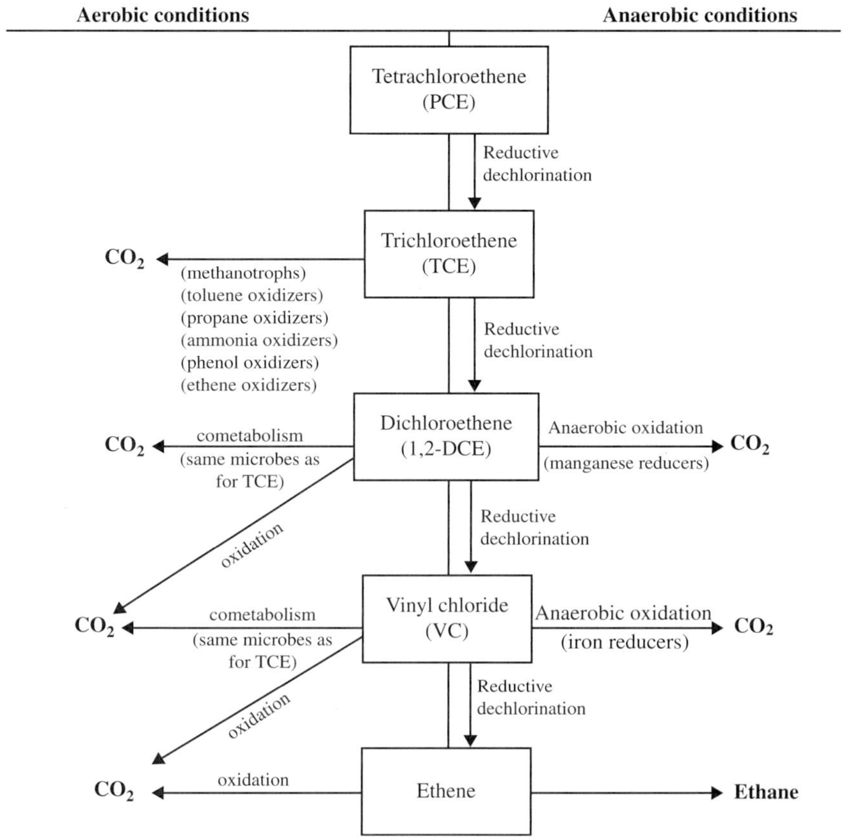

FIGURE 2.9
Degradation pathways of chlorinated ethers. Note that vinyl chloride accumulates as a result of anaerobic digestion, i.e. reductive dechlorination of a degradation product, i.e. 1,2-DCE.
Source: Environmental Security Technology Certification Program (2004). White Paper: Bioaugmentation for Remediation of Chlorinated Solvents: Technology Development, Status, and Research Needs. Prepared by GeoSyntec Consultants.

shown to be cometabolized by *Pseudomonas putida* OX1 [27]. *Pseudomonas putida* and *Methylosinus trichosporium* OB3b and other organisms were found to degrade chloroethenes, chloroethanes, chloromethanes, chloropropanes, and other halogenated organic compounds by cometabolism [28].

The bacterium *Acinetobacter spp.* is one of many that are used in bioremediation. It is a strict aerobe. This means it needs oxygen as its exclusive electron acceptor and cannot survive in the absence of O_2. Conversely, those bacteria that use electron acceptor other than oxygen and survive without O_2, e.g. *Desulfovibrio spp.*, are strict anaerobes. A facultative anaerobe, e.g. *Escherichia coli,* survives in the presence and absence of oxygen. That is, they are able to switch their electron acceptor depending upon the availability of these compounds in the environment. For example, note the numerous oxidizers and reducers in Figure 2.9.

ENVIRONMENTAL BIOCHEMODYNAMICS

Discussions of the environmental aspects of biotechnology must always consider both physical processes associated with chemical reactions and other chemical processes. Thus, environmental physics must be considered mutually with environmental chemistry. Such is the nature of environmental science; all concepts are interrelated.

The interconnectedness between physical, chemical, and biological processes is evident in Table 2.9, which lists some of the most important physical and chemical processes involved in the fate of biotechnological materials. It is important to bear in mind that all environmental processes are a function of both the chemical characteristics of the compartment

Table 2.9 Physical, chemical, and biological processes important to the fate and transport of biotechnological substances in the environment

Process	Description	Physical phases involved	Major mechanisms at work	Outcome of process	Factors included in process
Advection	Transport by turbulent flow; mass transfer	Aqueous, gas	Mechanical	Transport due to mass transfer	Concentration gradients, porosity, permeability, hydraulic conductivity, circuitousness or tortuosity of flow paths
Dispersion	Transport from source	Aqueous, gas	Mechanical	Concentration gradient-driven	Concentration gradients, porosity, permeability, hydraulic conductivity, circuitousness or tortuosity of flow paths
Molecular diffusion	Fick's law (concentration gradient)	Aqueous, gas, solid	Mechanical	Concentration gradient-driven transport	Concentration gradients
Liquid separation	Various fluids of different densities and viscosities are separated within a system	Aqueous	Mechanical	Recalcitrance to due formation of separate gas and liquid phases (e.g., gasoline in water separates among benzene, toluene and xylene)	Polarity, solubility, K_d, K_{ow}, K_{oc}, coefficient of viscosity, density
Density stratification	Distinct layers of differing densities and viscosities	Aqueous	Physical/chemical	Recalcitrance or increased mobility in transport of lighter fluids (e.g. LNAPLs) that float at water table in groundwater, or at atmospheric pressure in surface water	Density (specific gravity)
Migration along flow paths	Faster through large holes and conduits, e.g. path between sand particles in an aquifer	Aqueous, gas	Mechanical	Increased mobility through fractures	Porosity, flow path diameters
Sedimentation	Heavier compounds settle first	Solid	Chemical, physical, mechanical, varying amount of biological	Recalcitrance due to deposition of denser compounds	Mass, density, viscosity, fluid velocity, turbulence (R_N)

Filtration	Retention in mesh	Solid	Chemical, physical, mechanical, varying amount of biological mechanisms	Recalcitrance due to sequestration, destruction and mechanical trapping of compounds in soil micropores	Surface charge, soil, particle size, sorption, polarity
Volatilization	Phase partitioning to vapor	Aqueous, gas	Physical	Increased mobility as vapor phase of contaminant migrates to soil gas phase and atmosphere	P^0, concentration of contaminant, solubility, temperature
Dissolution	Co-solvation, attraction of water molecule shell	Aqueous	Chemical	Various outcomes due to formation of hydrated compounds (with varying solubilities, depending on the species)	Solubility, pH, temperature, ionic strength, activity
Fugacity	Escape from one type of environmental compartment to another	All phases	Physical, but influenced by chemical and biological	Fleeing potential	All partitioning conditions affect fugacity
Absorption	Retention on solid surface	Solid	Chemical, physical, mechanical, varying amount of biological mechanisms	Partitioning of lipophilic compounds into soil organic matter	Polarity, surface charge, Van der Waals attraction, electrostatics, ion exchange, solubility, K_d, K_{ow}, K_{oc}, coefficient of viscosity, density
Adsorption	Retention on solid surface	Solid	Chemical, physical, varying amount of biological mechanism	Recalcitrance due to ion exchanges and charge separations	Polarity, surface charge, Van der Waals attraction, electrostatics, ion exchange, solubility, K_d, K_{ow}, K_{oc}, coefficient of viscosity, density

(*Continued*)

Table 2.9 Physical, chemical, and biological processes important to the fate and transport of biotechnological substances in the environment—cont'd

Process	Description	Physical phases involved	Major mechanisms at work	Outcome of process	Factors included in process
Chemisorption	Retention on surface wherein strength of interaction is stronger than solely physical adsorption, resembling chemical bonding	Solid	Chemical and biochemical in addition to the physical mechanisms	Recalcitrance due to ion exchanges and charge separations	Polarity, surface charge, Van der Waals attraction, electrostatics, ion exchange, solubility, K_d, K_{ow}, K_{oc}, coefficient of viscosity, density, presence of biofilm
Ion exchange	Cations attracted to negatively charged particle surfaces or anions are attracted to positively charged particle surfaces, causing ions on the particle surfaces to be displaced	Solid	Chemical and biochemical in addition to the physical mechanisms	Recalcitrance due to ion exchanges and charge separations	Polarity, surface charge, Van der Waals attraction, electrostatics, ion exchange, solubility, K_d, K_{ow}, K_{oc}, coefficient of viscosity, density, presence of biofilm
Complexation	Reactions with matrix (e.g. soil compounds like humic acid) that form covalent bonds	Solid	Chemical, varying amount of biological	Recalcitrance and transformation due to reactions with soil organic compounds to form residues (bound complexes)	Available oxidants/reductants, soil organic matter content, pH, chemical interfaces, available O_2, electrical interfaces, temperature
Oxidation/reduction	Electron loss and gain	All	Chemical, physical, varying amount of biological	Destruction or transformation due to mineralization of simple carbohydrates to CO_2 and water from respiration of organisms	Available oxidants/reductants, soil organic matter content, pH, chemical interfaces, available O_2, electrical interfaces, temperature
Ionization	Complete co-solvation leading to separation of compound into cations and anions	Aqueous	Chemical	Dissolution of salts into ions	Solubility, pH, temperature, ionic strength, activity

Hydrolysis	Reaction of water molecules with contaminants	Aqueous	Chemical	Various outcomes due to formation of hydroxides (e.g. aluminum hydroxide) (with varying solubilities, depending on the species)	Solubility, pH, temperature, ionic strength, activity
Photolysis	Reaction catalyzed by electromagnetic energy (sunlight)	Gas (major phase)	Chemical, physical	Photo-oxidation of compounds with hydroxyl radical upon release to the atmosphere	Free radical concentration, wavelength and intensity of EM radiation
Bioavailability	Fraction of the total mass of a compound present in a compartment that has the potential of being absorbed by the organism	All phases	Biological and chemical	Uptake, absorption, distribution, metabolism and elimination	*Bioaccumulation* is the process of uptake into an organism from the abiotic compartments. *Bioconcentration* is the concentration of the pollutant within an organism above levels found in the compartment in which the organism lives
Biodegradation	Microbially mediated, enzymatically catalyzed reactions	Aqueous, solid	Chemical, biological	Various outcomes, including destruction and formation of daughter compounds (degradation products) intracellularly and extracellularly	Microbial population (count and diversity), pH, temperature, soil moisture, acclimation potential of available microbes, nutrients, appropriate enzymes in microbes, available and correct electron acceptors (i.e. oxygen for aerobes, others for anaerobes)

(*Continued*)

Table 2.9 Physical, chemical, and biological processes important to the fate and transport of biotechnological substances in the environment—cont'd

Process	Description	Physical phases involved	Major mechanisms at work	Outcome of process	Factors included in process
Cometabolism	Other organic compounds metabolized concurrently by microbes that are degrading principal energy source	Aqueous, but wherever biofilm comes into contact with organic compounds	Biochemical	Coincidental degradation of organic compounds	Enhanced microbial activity, presence of a good energy source (i.e. successful acclimation) and production of enzymes in metabolic pathways
Activation	Metabolic, detoxification process that renders a compound more toxic	Aqueous, gas, solid, tissue	Biochemical	Phase 1 or Phase 2 metabolic oxidation of aromatic compounds (e.g. polycyclic aromatic hydrocarbons) to form more toxic epoxides	
Cellular respiration	Conversion of nutrients' biochemical energy into adenosine triphosphate (ATP), with release of waste products	Catabolic reactions, oxidation of one molecule and reduction of another	Biochemical	Microbial respiration, along with metabolism, degrades organic compounds	Aerobic respiration involves oxygen as final electron acceptor, whereas anaerobic respiration has another final electron acceptor
Fermentation	Cellular energy derived from oxidation of organic compounds	Endogenous electron acceptor, i.e. usually an organic compound. Differs from cellular respiration where electrons are donated to an exogenous electron acceptor, (e.g. O_2) via an electron transport chain	Biochemical	Organic compounds degraded to alcohols and organic acids, ultimately to methane and water	Often an anaerobic process
Enzymatic catalysis	Cell produces biomolecules (i.e. complex proteins) that speed up biochemical reactions	Enzymes reactive site binds substrates by non-covalent interactions	Biological (intracellular in single-celled and multi-celled organisms)	Catalyzed reaction follows three steps: substrate fixation, reaction, desorption of the product	Non-covalent bonding includes hydrogen bonds, dipole–dipole interactions, van der Waals or dispersion forces, π stacking interactions, hydrophobic effect

Metal catalysis	Reactions sped up in the presence of certain metallic compounds (e.g. noble metal oxides in the degradation of nitric acid)	Aqueous, gas, solid, and biotic	Chemical (especially reduction and oxidation)	Same chemical reaction, but faster	Chemical form of metal, pH, temperature
Pharmacokinetics/ toxicokinetics	Rates at which uptaken substances are absorbed, distributed, metabolized and eliminated	Absorption, distribution, metabolism and elimination of a substance by the body, as affected by uptake, distribution, binding, elimination, and biotransformation	Biochemical	Mass balance of substance after uptake	Available detoxification and enzymatic processes in cells
Pharmaco-dynamics/ toxicodynamics	Effects and modes of action of chemicals in an organism	Uptake, movement, binding, and interactions of molecules at their site of action	Biochemical	Fate of compound or its degradates	Affinities of compounds to various tissues

(e.g., substrate, air or water) and those of the substance residing in that compartment. The inherent properties of the substance are influenced and changed by the extrinsic properties of the media in which the pollutant resides in the environment. Thus, Table 2.9 briefly explains both sets of properties.

Environmental chemodynamics is concerned with how chemicals move and change in the environment. Since we are concerned with the bioengineering aspects of transport, transformation and fate, let us add the life sciences to the physicochemical consideration, i.e. *biochemodynamics*. The principles underpinning environmental biochemodynamics are discussed in detail in Chapter 3.

BIOPHILE CYCLING

All living systems on earth consist of molecular arrangements of the elements carbon, oxygen, hydrogen, and most contain nitrogen. These four biophile elements have an affinity for each other so as to form complex organic compounds. In fact, up until less than two centuries ago, such organic compounds were thought only to be able to be produced within natural biological systems. Friedrich Wöhler [29] is credited with synthesizing the first organic compound outside of an organism [30] when he reacted silver isocyanate with ammonium chloride to form urea and silver chloride:

$$AgNCO + NH_4Cl \rightarrow (NH_2)_2CO + AgCl \tag{2.5}$$

All of the earth's creatures are carbon-based, so it is fitting to consider the processes and systems that share this element. Carbon, with its atomic number 12, has four electrons in its outermost shell, so it can either give four away or gain four. This means that it readily forms covalent bonds, and is the main reason so many organic compounds are possible.

Living systems both reduce and oxidize carbon. Reduction is the act of gaining electrons, whereas oxidation is the act of losing electrons from the outermost shell. Reduction often takes place in the absence of molecular oxygen (O_2), such as in the rumen of cattle, in sludge at the bottom of a lagoon, or in buried detritus on the forest floor. Anaerobic bacteria get their energy by reduction, breaking down organic compounds into methane (CH_4) and water.

Conversely, aerobic microbes get their energy from oxidation, forming carbon dioxide (CO_2) and water. Plants absorb CO_2 for photosynthesis, the process whereby plants convert solar energy into biomass and release O_2 as a byproduct. Thus, the essential oxygen is actually the waste product of photosynthesis and is derived from carbon-based compounds. Respiration generates carbon dioxide as a waste product of oxidation that takes place in organisms, so there is a balance between green plants' uptake of CO_2 and release of O_2 in photosynthesis and the uptake of O_2 and release of CO_2 in respiration by animals, microbes and other organisms.

Combined with hydrogen, carbon forms hydrocarbons – which can be good or bad. For example, those released when burning fossil fuels can be toxic and lead to smog, but hydrocarbons are essential as food. They make nature colorful, such as the carotenoids (organic pigment in photosynthetic organisms including algae), and evoke our sense of smell, such as the terpenes produced by variety of pines and other coniferous trees, which are the primary constituents of essential oils in plants and flowers used as natural flavor additives in food. Hydrocarbons also make up medicines and myriad other products that are part of our daily lives.

Combined with oxygen and hydrogen, carbon forms the biochemicals, including sugars, cellulose, lignin, chitins, alcohols, fats, and esters. Combined with nitrogen, carbon forms alkaloids – naturally occurring amines produced by plants and animals, which combine to form proteins. Combined with sulfur, carbon is the source of antibiotics, proteins, and amino acids. Combined with phosphorus and a few other elements, carbon forms ribonucleic acid (RNA) and deoxyribonucleic acid (DNA), the chemical codes of life.

Of course, biotechnologies are rooted in carbon, but so are other new technologies. For example, classes of nanomaterials are often carbon-based, such as carbon-60 (C_{60}). Interestingly, these spherical structures consisting of 60 carbon atoms are called fullerenes or Buckyballs, after the famous designer Buckminster Fuller, in honor of his innovative geodesic domes and spheres. When these fullerenes combine, they link into nanotubes. Also, the functional groups attached to the outside of these nanospheres and tubes determine their usefulness (e.g. ways to keep them from aggregating into larger particles that do not possess the electromagnetic properties needed for medicinal and engineering purposes) and the potential hazards (e.g. increased toxicity and mobility in biosystems, such as the human body or ecosystems).

The most prominent greenhouse gases, carbon dioxide and methane, are carbon-based compounds. These are just two of the carbon compounds that are cycled continuously through the environment (see Figure 2.10).

Figure 2.10 demonstrates the importance of sinks and sources of carbon. For example, if carbon can remain sequestered in the soil, roots, sediment, and other compartments, it is not released to the atmosphere. Thus, it cannot have an impact on the greenhouse effect. Even relatively small amounts of methane and carbon dioxide can profoundly increase the atmosphere's greenhouse potential.

Carbon bonds to itself and to other elements in a myriad of ways, forming single, double, and triple bonds with itself. This makes for millions of possible organic compounds. An organic compound is a compound that includes at least one carbon-to-carbon or carbon-to-hydrogen bond.

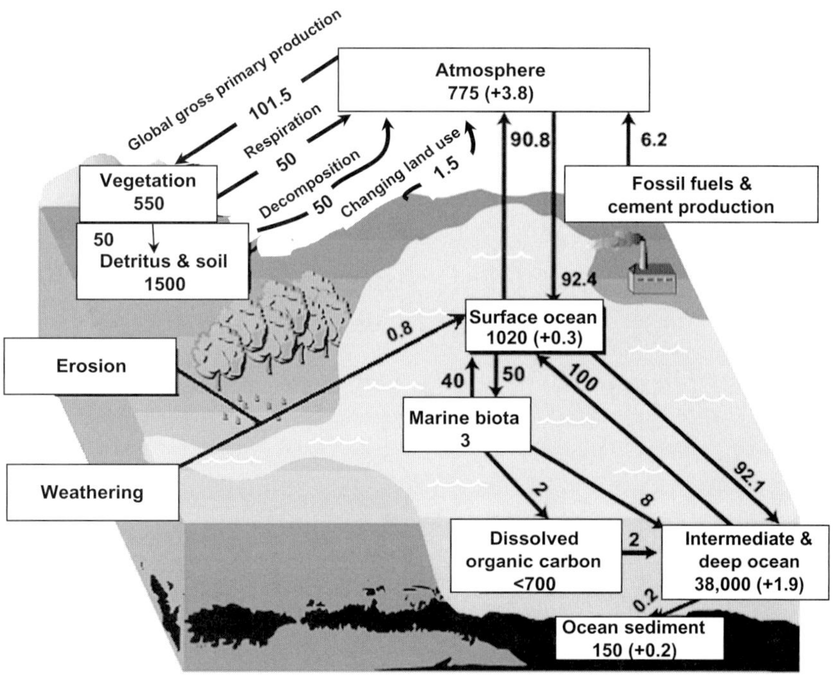

FIGURE 2.10
Global carbon cycle from 1992 to 1997. Carbon pools are boxes, expressed in gigatons (Gt) of carbon (Note: Gt C $= 10^{15}$ g C). Annual increments are expressed in Gt C per year (shown in parentheses). All fluxes indicated by the arrows are expressed in Gt C per year. The inferred net terrestrial uptake of 0.7 Gt C per year considers gross primary production (~101.5), plant respiration (~50), decomposition (~50), and additional removal from the atmosphere directly or indirectly, through vegetation and soil and eventual flow to the ocean through the terrestrial processes of weathering, erosion, and runoff (~0.8). Net ocean uptake (~1.6) considers air/sea exchange (~92.4 gross uptake, −90.8 gross release). As the rate of fossil fuel burning increases and CO_2 is released to the atmosphere it is expected that the fraction of this C remaining in the atmosphere will increase resulting in a doubling or tripling of the atmospheric amount in the coming century.
Source: M. Post. Oak Ridge National Laboratory; http://cdiac.ornl.gov/pns/graphics/c_cycle.htm; accessed January 29, 2009.

Slight changes to an organic molecule can profoundly affect its behavior. For example, there are large ranges of solubility for organic compounds, depending upon the presence of polar groups in their structure. The addition of an alcohol group to the gas ethane to produce ethanol, e.g. by fermentation, changes the phase and increases aqueous solubility. This means that considerations for siting a bioreactor must take into account the physical, chemical, and biological affinities of all the substances, from the raw materials to intermediate compounds to the intended product. It also determines any pollution control devices, for example, if intermediate products have lower vapor pressures than raw materials, they will have to be addressed as air pollutants, whereas all of the liquid phase compounds may be multi-phase pollutants. For example, they may be released through leaks and conduits as liquids, but a portion will also partition to the gas phase (e.g. in the reactor headspace).

Organic compounds can be further classified into two basic groups: aliphatics and aromatics. Hydrocarbons are the most fundamental type of organic compound. They contain only the elements carbon and hydrogen. Hydrocarbons are an important group of pollutants, either as pollutants themselves, e.g. components of toxic compounds (e.g. polycyclic aromatic hydrocarbons), or as precursors of other pollutants, e.g. ozone (O_3), the key ingredient of tropospheric smog. In fact, the presence of hydrocarbons is an important part of the formation of smog. For example, places like Los Angeles that have photochemical oxidant smog problems are looking for ways to reduce the amount of hydrocarbons released to the air.

Aliphatic compounds are classified into a few chemical families. Each carbon normally forms four covalent bonds. Alkanes are hydrocarbons that form chains with each link of the carbon. A single link is CH_4, methane. The carbon chain length increases with the addition of carbon atoms. For example, ethane's structure is:

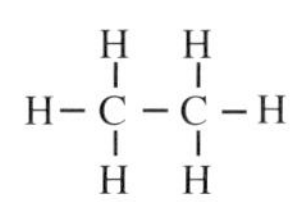

And the prototypical alkane structure is:

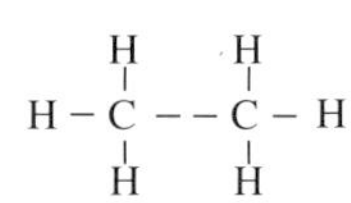

The alkanes contain a single bond between each carbon atom, and include the simplest organic compound, methane (CH_4), and its derivative "chains" such as ethane (C_2H_6) and butane (C_4H_{10}). Alkenes contain at least one double bond between carbon atoms. For example, 1,3-butadiene's structure is $CH_2{=}CH{-}CH{=}CH_2$. The numbers "1" and "3" indicate the position of the double bonds. The alkynes contain triple bonds between carbon atoms, the simplest being ethyne, $CH{\equiv}CH$ which is commonly known as acetylene (the gas used by welders).

The aromatics are all based upon the six-carbon configuration of benzene (C_6H_6). The carbon–carbon bond in this configuration shares more than one electron, so that benzene's structure (Figure 2.11) allows for resonance among the double and single bonds, i.e. the actual benzene bonds flip locations. Benzene is the average of two equally contributing resonance structures.

The term "aromatic" comes from the observation that many compounds derived from benzene were highly fragrant, such as vanilla, wintergreen oil, and sassafras. Aromatic compounds, thus, contain one or more benzene rings. The rings are planar, meaning that they remain in the same geometric plane as a unit. However, in compounds with more than one ring, such as the highly toxic polychlorinated biphenyls (PCBs), each ring is planar, but the rings bound together may or may not be planar. This is actually a very important property for toxic compounds. It has been shown that some planar aromatic compounds are more toxic

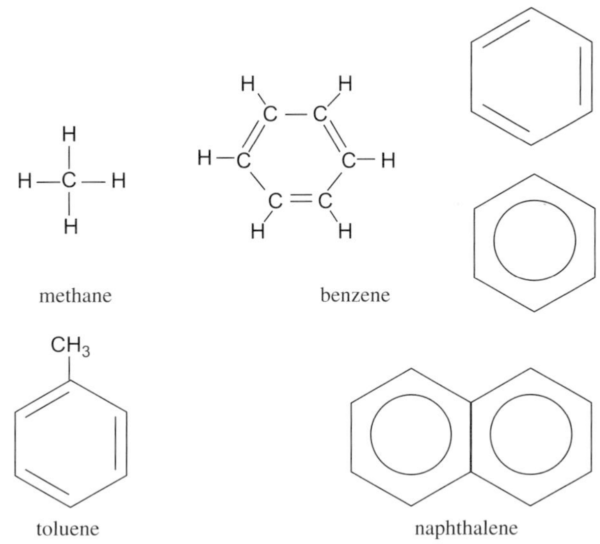

FIGURE 2.11
Fundamental organic compound structures. Methane is the simplest aliphatic structure and benzene is the simplest aromatic structure. Note that the benzene molecule has alternating double and single bonds between the carbon atoms. The double and single bonds flip, i.e. resonate. This is why the benzene ring is also shown as the two structures on the right, which are the commonly used condensed forms for aromatic compounds, such as the solvent toluene and the polycyclic aromatic hydrocarbon naphthalene.

than their non-planar counterparts, possibly because living cells may be more likely to allow planar compounds to bind to them and to produce nucleopeptides that lead to biochemical reactions associated with cellular dysfunctions, such as cancer or endocrine disruption.

Both the aliphatic and aromatic compounds can undergo substitutions of the hydrogen atoms. These substitutions render new properties to the compounds, including changes in solubility, vapor pressure, and toxicity. For example, halogenation (substitution of a hydrogen atom with a halogen) often makes an organic compound much more toxic. For example, trichloroethane is a highly carcinogenic liquid that has been found in drinking water supplies, whereas non-substituted ethane is a gas with relatively low toxicity. This is also why one of the means for treating the large number of waste sites contaminated with chlorinated hydrocarbons and aromatic compounds involves dehalogenation techniques.

The important functional groups that are part of many organic compounds are shown in Table 2.10.

Structures of organic compounds can induce very different physical and chemical characteristics, as well as change the bioaccumulation and toxicity of these compounds. For example, the differences between the estradiol and a testosterone molecule may seem small but they cause significant differences in the growth and reproduction of animals. The very subtle differences between an estrogen and an androgen, female and male hormones respectively, can be seen in these structures. Incremental changes to a simple compound, such as ethane can make for large differences (see Table 2.11). Replacing two or three hydrogen atoms with chlorine atoms makes for differences in toxicities between the nonhalogenated form and the chlorinated form. The same is true for the simplest aromatic, benzene. Substituting a methyl group for one of the hydrogen atoms forms toluene. Replacing a hydrogen atom with a hydroxyl group on a benzene ring yields phenol, with substantially different properties than benzene.

CARBON BIOGEOCHEMISTRY

By far, most carbon-based compounds are organic, but a number of inorganic compounds are also important. In fact, the one that is getting the most attention for its role in climate, carbon dioxide, is an inorganic compound because its carbon atom does not contain a covalent bond

Table 2.10 Structures of organic compounds

Chemical class	Functional group
Alkanes	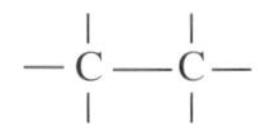
Alkenes	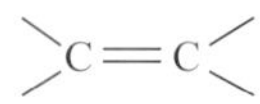
Alkynes	
Aromatics	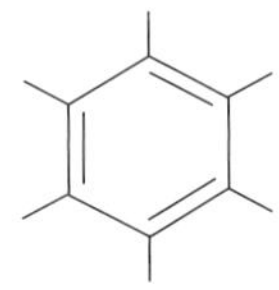
Alcohols	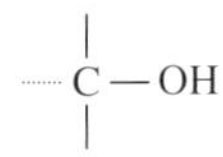
Amines	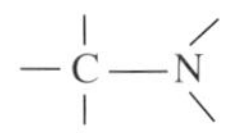
Aldehydes	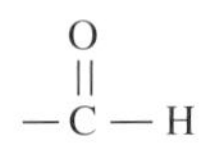
Ether	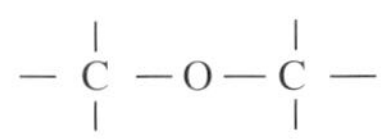
Ketones	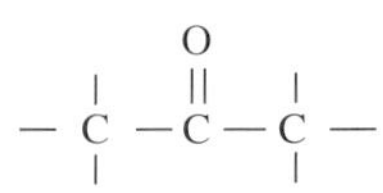
Carboxylic acids	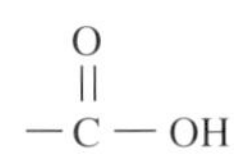
Alkyl halides [31]	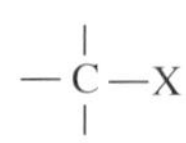
Phenols (aromatic alcohols)	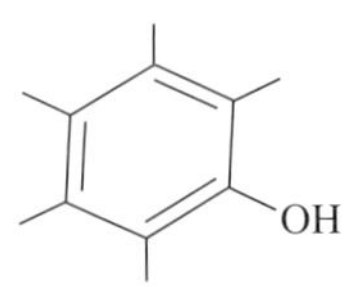
Substituted aromatics (substituted benzene derivatives)	
Nitrobenzene	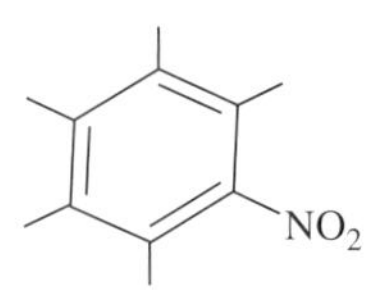

Table 2.10 Structures of organic compounds—cont'd

Chemical class	Functional group
Monosubstituted alkylbenzenes	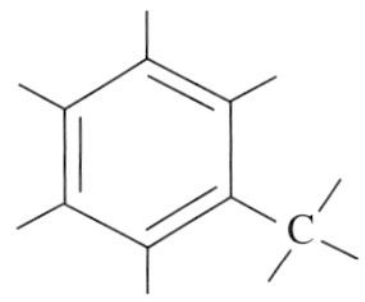
Toluene (Simplest monosubstituted alkylbenzene)	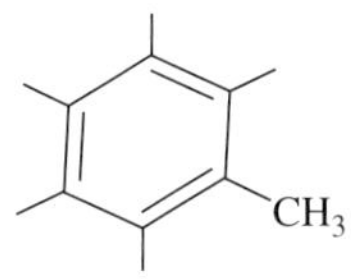
Polysubstituted alkylbenzenes	
1,2-alkylbenzene (also known as *ortho* or *o*-...)	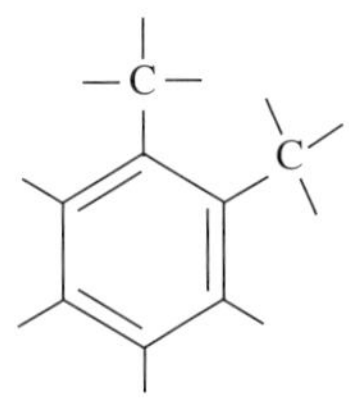
1,2-xylene or *ortho*-xylene (*o*-xylene)	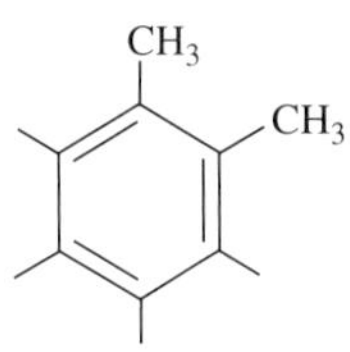
1,3-xylene or *meta*-xylene (*m*-xylene)	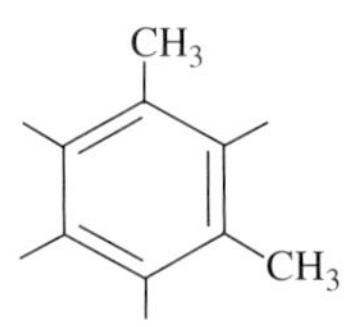
1,4-xylene or *para*-xylene (*p*-xylene)	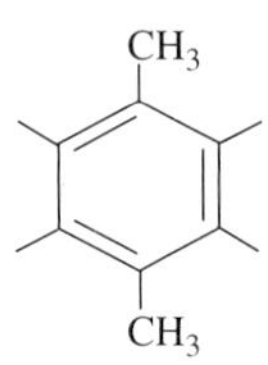
Hydroxyphenols do not follow general nomenclature rules for substituted benzenes	
Catechol (1,2-hydroxiphenol)	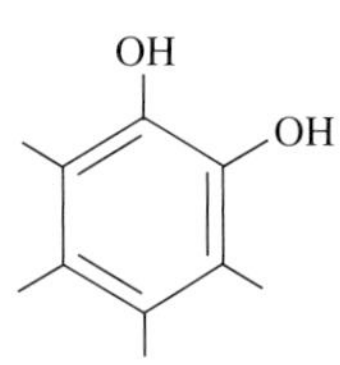

(*Continued*)

Table 2.10 Structures of organic compounds—cont'd

Chemical class	Functional group
Resorcinol (1,3-hydroxiphenol)	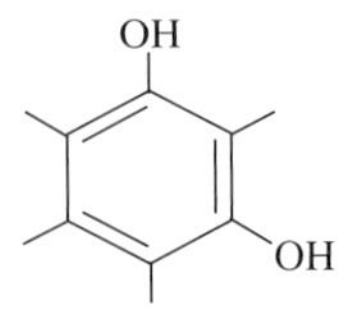
Hydroquinone (1,4-hydroxiphenol)	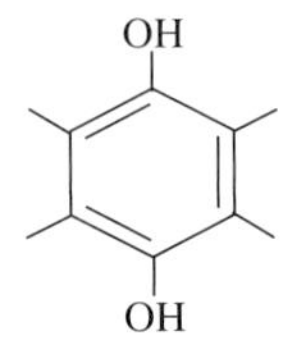

Table 2.11 Incremental differences in molecular structure leading to changes in physicochemical properties and hazards, and the levels of protection by environmental and public health regulators

Compound	Physical state at 25 °C	− log P° solubility in H_2O at 25 °C (mol L^{-1})	−log Vapor pressure at 25 °C (atm)	Worker exposure limits (parts per million)	Regulating agency
Methane, CH_4	Gas	2.8	−2.4	25	Canadian Safety Association
Tetrachloromethane (carbon tetrachloride), CCl_4	Liquid	2.2	0.8	2 short-term exposure limit (STEL) = 60 min	National Institute of Occupation Health Sciences (NIOSH)
Ethane, C_2H_6	Gas	2.7	−1.6	None (simple asphyxiant)	Occupational Safety and Health Administration (OSHA)
Trichloroethane, $C_2H\,Cl_3$	Liquid	2.0	1.0	450 STEL (15 min)	OSHA
Benzene, C_6H_6	Liquid	1.6	0.9	5 STEL	OSHA
Phenol, C_6H_6O	Liquid	0.2	3.6	10 ppm	OSHA
Toluene C_7H_8	Liquid	2.3	1.4	150 STEL	UK Occupational and Environmental Safety Services

with other carbon or hydrogen atoms. Other important inorganic carbon compounds include the pesticides sodium cyanide (NaCN) and potassium cyanide (KCN),and the toxic gas carbon monoxide (CO). Inorganic compounds also include inorganic acids, such as carbonic acid (H_2CO_3) and cyanic acid (HCNO) and compounds derived from reactions with the anions carbonate (CO_3^{2-}) and bicarbonate (HCO_3^-).

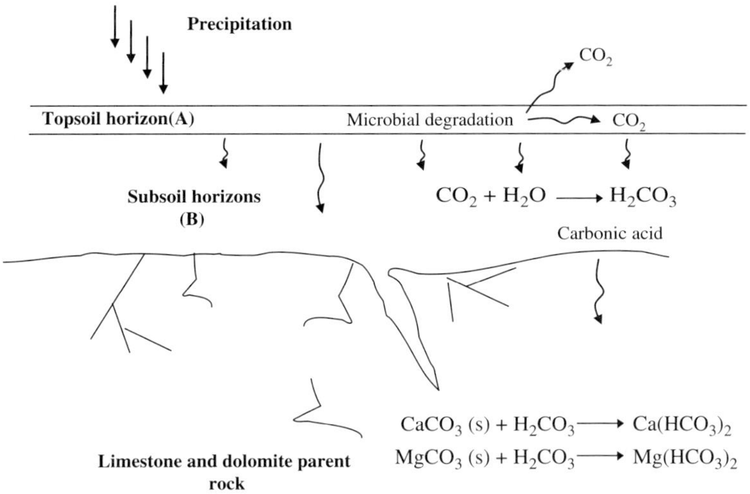

FIGURE 2.12
Biogeochemistry of carbon equilibrium. The processes that release carbonates are responsible for much of the buffering capacity of natural soils against the effects of acid rain.

Many of these forms exist in equilibrium with one another. For example, Figure 2.12 demonstrates the equilibrium among carbonates, bicarbonates, organic compounds, carbonic acid, and carbon dioxide. On a global scale, the mean pH of uncontaminated rain is about 5.6, owing to its dissolution of carbon dioxide, CO_2. As the water droplets fall through the air, the CO_2 in the atmosphere becomes dissolved in the water, setting up an equilibrium condition:

$$CO_2 \text{ (gas in air)} \leftrightarrow CO_2 \text{ (dissolved in the water)} \tag{2.6}$$

The CO_2 in the water reacts to produce hydrogen ions, as

$$CO_2 + H_2O \leftrightarrow H_2CO_3 \rightarrow H^+ + HCO_3^- \tag{2.7}$$

$$HCO_3^- \leftrightarrow 2H^+ + CO_3^{2-} \tag{2.8}$$

Assuming the mean partial pressure CO_2 in the air to be 3.0×10^{-4} atm, it is possible to calculate the pH of water in equilibrium. Such chemistry is always temperature-dependent, so let us assume that the air is 25 °C. We can also assume that the mean concentration of CO_2 in the troposphere is 350 ppm, but this concentration is rising by some estimates at a rate of 1 ppm per year.

Henry's law states that the concentration of a dissolved gas is directly proportional to the partial pressure of that gas above the solution:

$$p_a = K_H[c] \tag{2.9}$$

where K_H = Henry's law constant

p_a = partial pressure of the gas

$[c]$ = molar concentration of the gas

or,

$$p_a = K_H C_W \tag{2.10}$$

where C_W is the concentration of gas in water.

Henry's law, therefore, is a function of a substance's solubility in water and its vapor pressure and expresses the proportionality between the concentration of a dissolved contaminant and its partial pressure in the open atmosphere at equilibrium. That is, the Henry's law constant is an example of an *equilibrium constant*, which is the ratio of concentrations when chemical equilibrium is reached in a reversible reaction, the time when the rate of the forward reaction is the same as the rate of the reverse reaction.

The CO_2 concentration of the water droplet at equilibrium with air is obtained from the partial pressure of Henry's law constant:

$$p_{CO_2} = K_H[CO_2]_{aq} \tag{2.11}$$

The change from carbon dioxide in the atmosphere to carbonate ions in water droplets follows a sequence of equilibrium reactions:

$$CO_{2(g)} \overset{K_H}{\leftrightarrow} CO_{2(aq)} \overset{K_r}{\leftrightarrow} H_2CO_{3(aq)} \overset{Ka1}{\leftrightarrow} HCO^-_{3(aq)} \overset{Ka2}{\leftrightarrow} CO^{2-}_{3(aq)} \tag{2.12}$$

The processes that release carbonates increase the buffering capacity of natural soils against the effects of acidic water (pH <5). Thus, carbonate-rich soils like those of central North America are able to withstand even elevated acid deposition compared to the thin soils in areas such as the Canadian Shield, the New York Finger Lakes region, and much of Scandinavia.

The concentration of carbon dioxide CO_2 is constant, since the CO_2 in solution is in equilibrium with the air that has a constant partial pressure of CO_2. And the two reactions and ionization constants for carbonic acid are:

$$H_2CO_3 + H_2O \leftrightarrow HCO_3^- + H_3O^+ \qquad K_{a1} = 4.3 \times 10^{-7} \tag{2.13}$$

$$HCO_3^- + H_2O \leftrightarrow CO_3^{-2} + H_3O^+ \qquad K_{a2} = 4.7 \times 10^{-11} \tag{2.14}$$

K_{a1} is four orders of magnitude greater than K_{a2}, so the second reaction can be ignored for environmental acid rain considerations. The solubility of gases in liquids can be described quantitatively by Henry's law, so for CO_2 in the atmosphere at 25 °C we can apply the Henry's law constant and the partial pressure to find the equilibrium. The K_H for $CO_2 = 3.4 \times 10^{-2}$ mol L^{-1} atm^{-1}. We can find the partial pressure of CO_2 by calculating the fraction of CO_2 in the atmosphere. Since the mean concentration of CO_2 in the earth's troposphere is 350 ppm by volume in the atmosphere, the fraction of CO_2 must be 350 divided by 1,000,000 or 0.000350 atm.

Thus, the carbon dioxide and carbonic acid molar concentration can now be found:

$$[CO_2] = [H_2CO_3] = 3.4 \times 10^{-2} \text{ mol } L^{-1} atm^{-1} \times 0.000350 \text{ atm} = 1.2 \times 10^{-5} M$$

The equilibrium is $[H_3O^+] = [HCO^-]$. Taking this and our carbon dioxide molar concentration gives us:

$$K_{a1} = 4.3 \times 10^{-7} = \frac{[HCO_3^-][H_3O^+]}{CO_2} = \frac{[H_3O^+]^2}{1.2 \times 10^{-5}}$$

$$[H_3O^+]^2 = 5.2 \times 10^{-12}$$

$$[H_3O^+] = 2.6 \times 10^{-6} M$$

Or, the droplet pH is about 5.6.

Carbon dioxide, with water, is the ultimate product of aerobic microbial respiration, but it is also an important greenhouse gas. From the preceding discussion, a global increase in CO_2 concentrations must also change the mean acidity of precipitation. For example, many models expect a rather constant increase in tropospheric CO_2 concentrations. For example, the increase from the present 350 ppm to 400 ppm tropospheric CO_2 concentrations would be accompanied by a proportional decrease in precipitation pH. The molar concentration can be adjusted using the previous equations:

$$3.4 \times 10^{-2} \text{ mol L}^{-1} \text{ atm}^{-1} \times 0.000400 \text{ atm} = 1.4 \times 10^{-5}\text{M},$$

$$\text{so } 4.3 \times 10^{-7} = \frac{[H_3O^+]^2}{1.4 \times 10^{-5}} \text{ and } [H_3O+]^2$$

$$= 6.0 \times 10^{-12} \text{ and } [H_3O^+] = 3.0 \times 10^{-6}M.$$

Thus, average water droplet pH would be decrease to about 5.5. This means that the incremental increase in atmospheric carbon dioxide can be expected to contribute to greater acidity in natural rainfall. The precipitation rates themselves would also be affected if greenhouse gas concentrations continue to increase, so any changes in atmospheric precipitation rates would also, on average, be expected to be more acidic. The forcing factors for these interrelationships are shown in Figure 2.13. This is an interesting example of how the earth is actually a very large bioreactor. Changing one variable can profoundly change the entire system; in this instance the release of one gas changes numerous physical (e.g. temperature) and chemical (e.g. precipitation pH) factors, which in turn evoke a biological response (biome and ecosystem diversity).

A second-order change may occur between two prominent greenhouse gases: carbon dioxide and methane. The increased amounts of CO_2 will likely affect global temperature, which affects biomes and the kinetics within individual ecosystems. This will in turn change ecological structure, such as tree associations, which may result in changes to canopies and forest floors. Other ecosystem structures will also undergo changes, such as those in wetlands. If the wetlands soils undergo increased reduction then there will be an attendant increase in anaerobic microbial decomposition. The anaerobes will therefore lead to increasing build-up

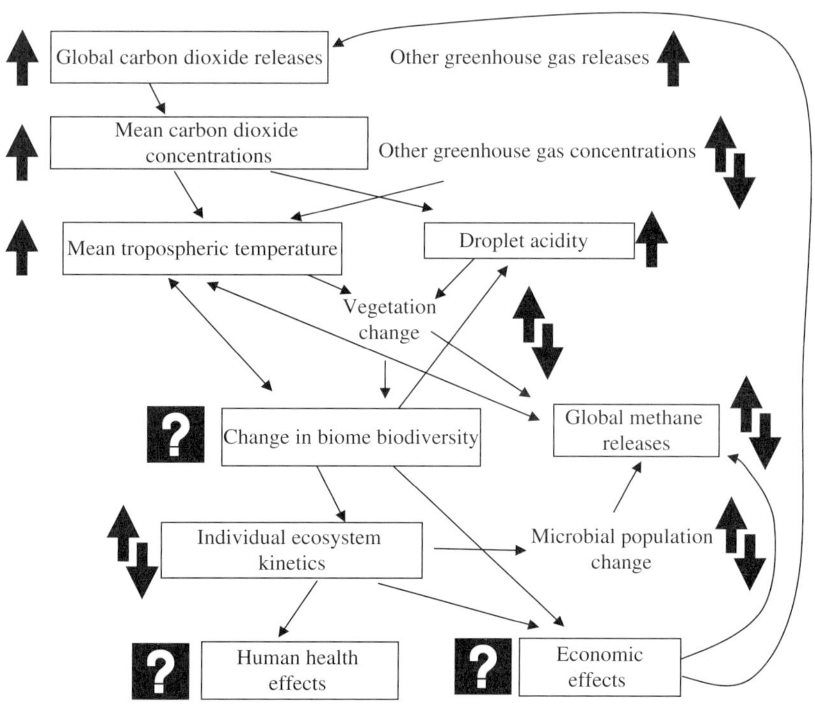

FIGURE 2.13
Systematic view of changes in tropospheric carbon dioxide. Thick arrows indicate whether this factor will increase (up arrow), decrease (down arrow) or will vary depending on the specifics (e.g. some greenhouse gas releases have decreased, e.g. the chlorofluorocarbons, and some gases can cool the atmosphere, e.g. sulfate aerosols). Question mark indicates that the type and/or direction of change are unknown or mixed. Thin arrows connect the factors as drivers toward downstream effects.

of CH_4 in the atmosphere. Owing to methane's strong radiant gas potential, this could likely lead to increasing global temperatures, all other factors being held constant.

However, if greater biological activity and increased photosynthesis is triggered by the increase in CO_2, and wetland depth is decreased, CH_4 global concentrations would fall, leading to less global temperature rise. Conversely, if this increased biological activity and photosynthesis leads to a decrease in forest floor detritus mass, then less anaerobic activity may lead to lower releases of CH_4. In actuality, there will be increases and decreases at various scales, so the net effects on a complex, planetary system is highly uncertain.

It is important to note that CO_2 is not the most important gas associated with pollution leading to acid rain. In fact, the oxides of two other biophile elements, sulfur and nitrogen, have rightly drawn the most attention from the scientific community. These compounds can dramatically decrease the pH of rain. However, the increase in CO_2 means that the pH of rainfall, which is not neutral to begin with, can adversely affect the fish and wildlife in and around surface waters with even lower concentrations of sulfur and nitrogen compounds.

As mentioned, methane (CH_4) is the product of anaerobic decomposition and human food production. Methane also is emitted during the combustion of fossil fuels and cutting and clearing of forests. The concentration of CH_4 in the atmosphere has been steady at about 0.75 for over a thousand years, and then increased to 0.85 ppm in 1900. Since then, in the space of only a hundred years, it has skyrocketed to 1.7 ppm. Methane is removed from the atmosphere by reaction with the hydroxyl radical (OH) as

$$CH_4 + OH + 9O_2 \rightarrow CO_2 + 0.5H_2 + 2H_2O + 5O_3 \tag{2.15}$$

This indicates that the reaction creates carbon dioxide, water vapor, and ozone, all of which are greenhouse gases, so the effect of one molecule of methane is devastating to the production of the greenhouse effect.

The lessons for biotechnology are numerous. For example, our very measures of success, carbon dioxide and methane, are greenhouse gases. So, the bioengineer should expect to be asked why the solution to one problem, e.g. cleaning up a contaminated site, is in the process contributing to another, global climate change. Another lesson is that we can learn from the feedback systems, for example, at which points in the event cascade in Figure 2.13 bioengineering principles can be put to use to ameliorate the problems.

One of the biggest engineering challenges is how to put the biogeochemical cycles to work to reduce the impact of global climate change debate, in light of the seeming paucity of ways to deal with the problem. The National Academy of Engineering has identified the most important challenges to the future of engineering. Both the nitrogen and carbon biogeochemical cycles are explicitly identified among the most pressing engineering needs.

The biogeochemical cycle that extracts nitrogen from the air for its incorporation into plants – and hence food – has become altered by human activity. With widespread use of fertilizers and high-temperature industrial combustion, humans have doubled the rate at which nitrogen is removed from the air relative to pre-industrial times, contributing to smog and acid rain, polluting drinking water, and even worsening global warming. Engineers must design countermeasures for nitrogen cycle problems, while maintaining the ability of agriculture to produce adequate food supplies [32]. Like carbon, nitrogen and the other nutrient elements are essential and toxic, depending on its dose and form. The Academy articulates this challenge:

> The biogeochemical cycle that extracts nitrogen from the air for its incorporation into plants – and hence food – has become altered by human activity. With widespread use of fertilizers and high-temperature industrial combustion, humans have doubled the rate at which nitrogen is removed from the air relative to pre-industrial times, contributing to smog and acid rain, polluting drinking water, and even worsening global warming.

Engineers must design countermeasures for nitrogen cycle problems, while maintaining the ability of agriculture to produce adequate food supplies. [33]

Bioengineers can expect to be increasingly called upon to recommend improvements to food life cycles (e.g. animal feeding operations, farmlands, rangelands and groceries). How can engineering innovation improve the efficiency of various human activities related to nitrogen, from making fertilizer to recycling food wastes? Currently, less than half of the fixed nitrogen generated by farming practices actually ends up in harvested crops. And less than half of the nitrogen in those crops actually ends up in the foods that humans consume. In other words, fixed nitrogen leaks out of the system at various stages in the process – from the farm field to the feedlot to the sewage treatment plant. Engineers not only need to identify the leakage points and devise systems to plug them, i.e. the structural and mechanical solutions, but must engage biological solutions, such as understanding the processes in Table 2.9 that lead to increased nitrogen emissions, and applying this understanding to modify the processes accordingly [34].

Greenhouse gases

The earth acts as a reflector to the sun's rays, receiving the radiation from the sun, reflecting some of it into space (called *albedo*), and adsorbing the rest, only to reradiate this into space as heat. In effect the earth acts as a wave converter, receiving the high energy high frequency radiation from the sun and converting most of it into low energy low frequency heat to be radiated back into space. In this manner, the earth maintains a balance of temperature.

In order to better understand this balance, the light energy and the heat energy have to be defined in terms of their radiation patterns, as shown in Figure 2.14. The incoming radiation (light) wavelength has a maximum at around 0.5 nm and almost all of it is less than 3 nm. The heat energy spectrum, or that energy reflected back into space, has the maximum at about 10 nm and almost all or it at a wavelength higher than 3 nm.

As both the light and heat energy pass through the earth's atmosphere they encounter the aerosols and gases surrounding the earth. These can either allow the energy to pass through, or they can interrupt it by scattering or absorption. If the atoms in the gas molecules vibrate at the same frequency as the light energy, they will absorb the energy and not allow it to pass

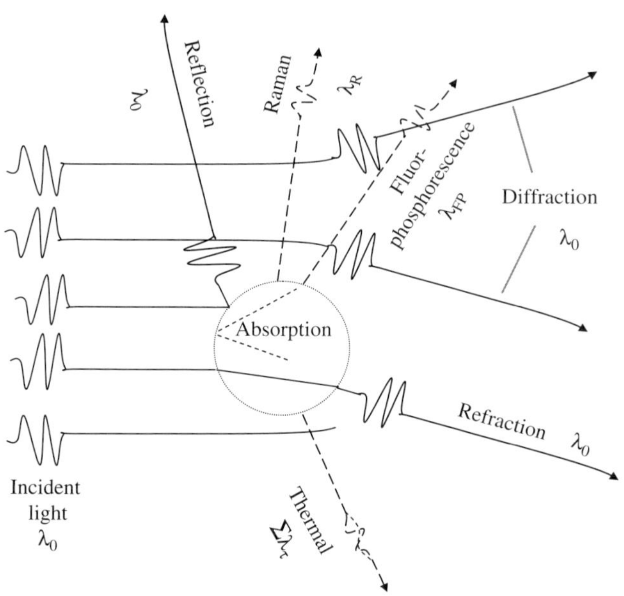

FIGURE 2.14
Patterns for heat and light energy.

through. Aerosols will scatter the light and provide a "shade" for the earth. The incoming radiation is impeded by water vapor and oxygen and ozone, as discussed in the preceding section. Most of the light energy comes through unimpeded.

The heat energy, however, encounters several potential impediments. As it is trying to reach outer space, it finds that water vapor, CO_2, CH_4, O_3, and N_2O all have absorptive wavelengths right in the middle of the heat spectrum. Quite obviously, an increase in the concentration of any of these will greatly limit the amount of heat transmitted into space. These gases are appropriately called *greenhouse gases* because their presence will limit the heat escaping into space, much like the glass of a greenhouse or even the glass in your car limits the amount of heat that can escape, thus building up the temperature under the glass cover.

The effectiveness of a particular gas to promote global warming (or cooling, as is the case with aerosols) is known as *forcing*. The gases of most importance in forcing are listed in Table 2.12.

Carbon dioxide is the product of decomposition of organic material, whether biologically or through combustion. The effectiveness of CO_2 as a global warming gas has been known for over 100 years, but the first useful measurements of atmospheric CO_2 were not taken until 1957. The data from Mauna Loa in Hawaii are exceptionally useful since they show that even in the 1950s the CO_2 concentration had increased from the baseline 280 ppm to 315 ppm, and this has continued to climb over the last 50 years at a constant rate of about 1.6 ppm per year. The most serious problem with CO_2 is that the effects on global temperature due to its greenhouse effect are delayed. Even abrupt decreases in CO_2 emissions will ameliorate but will not reverse the current increasing trend. In fact, even if the United States and Europe decrease their emissions, increases in emissions in developing economics, especially those of China and India, are very likely to increase tropospheric CO_2 concentrations dramatically for decades.

Methane is the product of anaerobic decomposition and human food production. One of the highest producers of methane in the world is New Zealand, which boasts 80 million sheep. Methane also is emitted during the combustion of fossil fuels and cutting and clearing of forests. The concentration of CH_4 in the atmosphere had been steady at about 0.75 for over a thousand years, and then increased to 0.85 ppm in 1900. Since then, in the space of only a hundred years, it has skyrocketed to 1.7 ppm. Methane is removed from the atmosphere by reaction with the hydroxyl radical (OH) as

$$CH_4 + OH + 9O_2 \rightarrow CO_2 + 0.5H_2 + 2H_2O + 5O_3 \qquad (2.16)$$

But in so doing, it creates carbon dioxide, water vapor, and ozone, all of which are greenhouse gases, so the effect of one molecule of methane is devastating to the production of the greenhouse effect.

Halocarbons, or the same gang of suspects in the destruction of atmospheric ozone, are also at work in promoting global warming. The most effective global warming gases are CFC-11 and CFC-12, both of which are no longer manufactured, and the banning of these substances has shown a leveling off in the stratosphere.

Table 2.12 **Relative forcing of increased global temperature (excluding water vapor)**

Greenhouse gas	Percent of relative radiative forcing
Carbon dioxide, CO_2	64
Methane, CH_4	19
Halocarbons (mostly CFCs)	11
Nitrous Oxide, N_2O	6

Increased atmospheric concentrations of nitrous oxide mainly result from human activities, especially the cutting and clearing of tropical forests. The greatest problem with nitrous oxide is that there appear to be no natural removal processes for this gas and so its residence time in the stratosphere is quite long.

The net effect of these global pollutants is still being debated. Various atmospheric models used to predict temperature change over the next hundred years vary widely. They nevertheless agree that some positive change will occur. By the year 2100, even if we do not increase our production of greenhouse gases and international agreements are reached and subsequently followed, the global temperature is likely to be between 0.5 and 1.5 °C warmer than at present. This effect of this on natural systems and dynamics in the oceans and atmosphere could be devastating.

Sequestration

Anticipating the continued use of fossil fuels, engineers have explored technological methods of capturing the carbon dioxide produced from fuel burning, including innovative ways to sequester it in reservoirs in soil, under the Earth's surface, in the oceans, and in biomass [35]. Sequestration is a biosystematic solution since it is an ongoing process on Planet Earth, with myriad interactions between biotic and abiotic factors. The arrows in Figure 2.13 show that carbon compounds, especially CO_2 and CH_4, find their way to the ocean, forests, and other carbon sinks. Human activities can influence biogeochemical processes adversely or beneficially; sequestration is an example of systematic engineering as an intervention against the global buildup of greenhouse gases, especially CO_2.

CARBON SEQUESTRATION IN SOIL

The soil is a great friend of the bioengineer. It is home to *Pseudomonas* and numerous other species that have been used to treat wastes for decades. The very essence of a soil's "value" has been its capacity to support plant life, especially crops. At a minimum, environmental biotechnology must include an understanding of soil properties such as their texture or grain size (see Table 2.13), ion exchange capacities, ionic strength, pH, microbial populations, and soil organic matter content.

Soil is a matrix made up of various components, including organic matter and unconsolidated material. The matrix contains liquids (i.e. "substrate" to the chemist and bioengineer) within its interstices. Much of the substrate in this matrix is water with varying amounts of

Table 2.13 Commonly used soil texture classifications

Name	Size range (mm)
Gravel	>2.0
Very coarse sand	1.0–1.999
Coarse sand	0.500–0.999
Medium sand	0.250–0.499
Fine sand	0.100–0.249
Very fine sand	0.050–0.099
Silt	0.002–0.049
Clay	<0.002

Source: T. Loxnachar, K. Brown, T. Cooper and M. Milford (1999). *Sustaining Our Soils and Society.* American Geological Institute, Soil Science Society of America, USDA Natural Resource Conservation Service, Washington, DC.

solutes. At least for most environmental conditions, air and water are solutions of very dilute amounts of compounds. For example, air's solutes represent small percentages of the solution at the highest (e.g. water vapor) and most other solutes represent parts per million (recall that there is about 350 ppm carbon dioxide). Soil is a conglomeration of all states of matter. Soil is predominantly solid, but frequently has large fractions of liquid (soil water) and gas (soil air, methane, carbon dioxide) that make up the matrix. The composition of each fraction is highly variable. For example, soil gas concentrations are different from those in the atmosphere and change profoundly with depth from the surface. Table 2.14 illustrates the inverse relationship between carbon dioxide and molecular oxygen. Sediment is really an underwater soil. It is a collection of particles that have settled on the bottom of water bodies.

Ecosystems are combinations of these media. For example, a wetland system consists of plants that grow in soil, sediment, and water. The water flows through living and non-living materials. Microbial populations live in the surface water, with aerobic species congregating near the water surface and anaerobic microbes increasing with depth due to the decrease in oxygen levels, due to the reduced conditions. Air is not only important at the water and soil interfaces, but it is a vehicle for nutrients and contaminants delivered to the wetland. The groundwater is fed by the surface water during high water conditions, and feeds the wetland during low water.

So, another way to think about these environmental media is that they are compartments, each with boundary conditions, kinetics and partitioning relationships within a compartment or among other compartments. Chemicals, whether nutrients or contaminants, change as a result of the time spent in each compartment. The bioengineering challenge is to describe, characterize, and predict the behaviors of various chemical species as they move through the media. When something is amiss, the cause and cure lie within the physics, chemistry, and biology of the system.

Soil conservation is an important part of sustainable agriculture and food production, since it entails keeping soil from becoming a pollutant in the surface waters, and its ability to sieve and filter pollutants that would otherwise end up in drinking water. Another perhaps less obvious benefit is that soil is a vast sink for carbon. Soil is lost when land is degraded by deforestation and as a result of inadequate land use and management in sensitive soil systems, especially those in the tropics and sub-tropics, such as slash and burn and other aggressive practices. As is often the case in ecosystems, some of the most valuable ecosystems in terms of the amount of carbon sequestered and oxygen generated are also the most sensitive. Tropical systems, for example, often have some of the thinnest soils due to the rapid oxidation processes that take place in humid, oxidized environments.

Sensitive systems are often given value by society, or at least a certain segment of society (e.g. industry), for a single purpose. Bauxite, for example, is present in tropical soils due to the

Table 2.14 **Composition of two important gases in soil air found experimentally in a soil column**

	Silty clay		Silty clay loam		Sandy loam	
Depth from surface (cm)	O_2 (% volume of air)	CO_2 (% volume of air)	O_2 (% volume of air)	CO_2 (% volume of air)	O_2 (% volume of air)	CO_2 (% volume of air)
30	18.2	1.7	19.8	1.0	19.9	0.8
61	16.7	2.8	17.9	3.2	19.4	1.3
91	15.6	3.7	16.8	4.6	19.1	1.5
122	12.3	7.9	16.0	6.2	18.3	2.1
152	8.8	10.6	15.3	7.1	17.9	2.7
183	4.6	10.3	14.8	7.0	17.5	3.0

Source: V.P. Evangelou (1998). *Environmental Soil and Water Chemistry: Principles and Applications.* John Wiley and Sons, Inc., New York.

physical and chemical conditions of the tropics (aluminum in parent rock material, oxidation, humidity, and ion exchange processes). However, from a life cycle and resource planning perspective, such single-mindedness is folly. The decision to extract bauxite, iron or other materials from sensitive tropical rainforests must be seen in terms of local, regional, and global impacts. With this in mind, international organizations are promoting improved land use systems and land management practices that give both economic and environmental benefits that can be sustained over time.

Keeping soil intact protects biological diversity, improves ecosystem conditions, and increases carbon sequestration. This last-mentioned benefit includes numerous forms of carbon in all physical phases. Soil gases include CO_2 and CH_4. Plant root systems, fungi, and other organisms that are comprised of amino acids, proteins, carbohydrates, and other organic compounds live in the soil. Even inorganic forms of carbon are held in soil, such as the carbonate, bicarbonate, and carbonic acid chemical species that are in soils as a result of chemical reactions with parent rock material, especially limestone and dolomite. When the soils are lost, all of these carbon compounds become available to be released to the atmosphere. Thus, soil conservation is a passive process that can be enhanced to sequester carbon.

ACTIVE SEQUESTRATION

Active sequestration is the application of technologies to send carbon to the sinks, including deep rock formations and the oceans. Such technology can be applied directly to sources. For example, fires from China's coal mines presently release about 1 billion metric tons of CO_2 to the atmosphere every year. Estimates put India's coal mine fire releases to be about 50 million metric tons. This accounts for as much as one percent of all carbon greenhouse releases. This is about the same as the CO_2 released by all of the gasoline-fuel automobiles in the United States. Engineering solutions that reduce these emissions would actively improve the net greenhouse gas global flux.

NITROGEN AND SULFUR BIOCHEMODYNAMICS

The biophile element nitrogen is a foundational element of life. For example, the bases in RNA and DNA, i.e. adenine, cytosine, guanine, thymine (DNA only) and uracil (RNA only), are all nitrogen compounds. From an environmental perspective, it is common to consider compounds of sulfur (S), another biophile, and nitrogen (N) compounds together. Both elements play huge roles in photosynthesis and other processes important to flora and microbes. Thus, S and N compounds are synthesized and manufactured in enormous volumes for agriculture. Along with phosphorus (P) and potassium (K), S and N compounds provide the macro and micronutrients to ensure productive crop yields. Conversely, certain S and N compounds can harm living systems, including the health of humans. A number of S and N compounds can adversely affect the environment and can lead to welfare impacts, such as the corrosion of buildings and other structures and diminished visibility due to the formation of haze. Like carbon, sulfur and nitrogen must be understood from a biogeochemical, systematic perspective. As nutrients, they also demonstrate the concept that pollution is often a resource that is simply in the wrong place.

Another reason to consider the biogeochemistry of sulfur and nitrogen compounds together is because their oxidized species [e.g. sulfur dioxide (SO_2) and nitrogen dioxide (NO_2)] form acids when they react with water. The lowered pH is responsible for many challenges, such as the working of bioreactors to treat wastes or even at the planetary scale, where rainfall becomes more acidic, leading to ecological problems. In addition, many sulfur and nitrogen pollutants result from combustion. While they share many things in common, however, sulfur and nitrogen pollutants actually are very different in their sources and in the processes that can lead to biotechnological challenges.

Sulfur is present in most fossil fuels, usually higher in coal than in crude oil. Prehistoric plant life is the source for most fossil fuels. Most plants contain sulfur as a nutrient and as the plants become fossilized a fraction of the sulfur volatilizes (i.e. becomes a vapor) and is released. However, some sulfur remains in the fossil fuel and can be concentrated because much of the carbonaceous matter is driven off. Thus, the S-content of the coal is available to react with oxygen when the fossil fuel is combusted. In fact, the S-content of coal is an important characteristic in its economic worth; the higher the S-content the less it is worth. So, the lower the sulfur content and volatile constituents and the higher the carbon content, the more valuable the coal. Since combustion is the combination of a substance (fuel) with molecular oxygen (O_2) in the presence of heat (denoted by the Δ above the arrow in the one-way, i.e. irreversible, reaction), the reaction for complete or efficient combustion of a hydrocarbon results in the formation of carbon dioxide and water:

$$(CH)_x + O_2 \xrightarrow{\Delta} CO_2 + H_2O \tag{2.17}$$

Fossil fuels contain other elements which are also oxidized. When sulfur is present, the side reaction forms oxides of sulfur. Thus, sulfur dioxide is formed as:

$$S + O_2 \xrightarrow{\Delta} SO_2 \tag{2.18}$$

Actually, many other oxidized forms of sulfur can form during combustion, so air pollution experts refer to them collectively as SO_x, commonly seen in air pollution literature.

Similarly, nitrogen compounds also form during combustion, but their sources are very different from those of sulfur compounds. Recall that the troposphere, the part of the atmosphere where we live and breathe, is made up mainly of molecular nitrogen (N_2). More than three-fourths of the troposphere is N_2, so the atmosphere itself is the source of much of the N that forms oxides of nitrogen (NO_x). Because N_2 is relatively non-reactive under most atmospheric conditions, it seldom enters into chemical reactions, but under high pressure and at very high temperatures, it will react with O_2:

$$N_2 + O_2 \xrightarrow{\Delta} 2NO \tag{2.19}$$

Where will we find such conditions under which N_2 will react this way? Actually, the answer is sitting in any driveway or garage. The automobile's internal combustion engine is a major source of oxides of nitrogen. As are electricity generating stations that need to heat boilers to make steam to turn turbines to convert mechanical energy into electrical energy. Approximately 90–95% of the nitrogen oxides generated in combustion processes are in the form of nitric oxide (NO), but like the oxides of sulfur, other nitrogen oxides can form, especially nitrogen dioxide (NO_2), so air pollution experts refer to NO and NO_2 collectively as NO_x. In fact, in the atmosphere the emitted NO is quickly converted photochemically to nitrogen dioxide (NO_2). Such high temperature/high pressure conditions exist in internal combustion engines, like those in automobiles and other so-called "mobile sources." Thus, NO_x is one of the major mobile source air pollutants (others include particulate matter, hydrocarbons, carbon monoxide, and in some countries, the heavy metal lead, Pb).

In addition to the atmospheric nitrogen, other sources exist, particularly the nitrogen in fossil fuels. The nitrogen oxides generated from atmospheric nitrogen are known as "thermal NO_x" since they form at high temperatures, such as near burner flames in combustion chambers. Nitrogen oxides that form from the fuel or feedstock are called "fuel NO_x." Unlike the sulfur compounds, a significant fraction of the fuel nitrogen remains in the bottom ash or in unburned aerosols in the gases leaving the combustion chamber, i.e. the fly ash. Nitrogen oxides can also be released from nitric acid plants and other types of industrial processes involving the generation and/or use of nitric acid (HNO_3).

Nitric oxide is a colorless, odorless gas and is essentially insoluble in water. Nitrogen dioxide has a pungent acid odor and is somewhat soluble in water. At low temperatures such as those often present in the ambient atmosphere, NO_2 can form the molecule NO_2-O_2N or simply

N_2O_4 that consists of two identical simpler NO_2 molecules. This is known as a *dimer*. The dimer N_2O_4 is distinctly reddish-brown and contributes to the brown haze that is often associated with photochemical smog incidents.

Both NO and NO_2 are harmful and toxic to humans, although atmospheric concentrations of nitrogen oxides are usually well below the concentrations expected to lead to adverse health effects. The low concentrations are due to the moderately rapid reactions that occur when NO and NO_2 are emitted into the atmosphere. Much of the concern for regulating NO_x emissions is to suppress the reactions in the atmosphere that generate the highly reactive molecule ozone (O_3). Nitrogen oxides play key roles in O_3 formation. Ozone forms photochemically (i.e. the reaction is caused or accelerated by light energy) in the lowest level of the atmosphere, known as the troposphere, where people live. Nitrogen dioxide is the principal gas responsible for absorbing sunlight needed for these photochemical reactions. So, in the presence of sunlight, the NO_2 that forms from the NO incrementally stimulates the photochemical smog-forming reactions because nitrogen dioxide is very efficient at absorbing sunlight in the ultraviolet portion of its spectrum. This is why ozone episodes are more common in summer and in areas with ample sunlight. Other chemical ingredients, i.e. ozone precursors, in O_3 formation include volatile organic compounds (VOCs) and carbon monoxide (CO). Governments around the world regulate the emissions of precursor compounds to diminish the rate at which O_3 forms.

Many compounds contain both nitrogen and sulfur along with the typical organic elements (carbon, hydrogen, and oxygen). The reaction for the combustion of such compounds, in general form, is:

$$C_aH_bO_cN_dS_e + (4a + b - 2c) \rightarrow aCO_2 + \left(\frac{b}{2}\right)H_2O + \left(\frac{d}{2}\right)N_2 + eS \tag{2.20}$$

This reaction demonstrates the incremental complexity as additional elements enter the reaction. In the real world, pure reactions are rare. The environment is filled with mixtures. Reactions can occur in sequence, parallel or both. For example, a feedstock to a municipal incinerator contains myriad types of wastes, from garbage to household chemicals to commercial wastes, and even small (and sometimes) large industrial wastes that may be illegally dumped. For example, the nitrogen-content of typical cow manure is about 5 kg per metric ton (about 0.5%). If the fuel used to burn the waste also contains sulfur along with the organic matter, then the five elements will react according to the stoichiometry of the reaction in Equation 2.18.

Certainly, combustion specifically and oxidation generally are very important processes to bioengineering, when heat sources are involved. Numerous other processes involve nitrogen and sulfur. Among the most important in the environment, oxidation *and* reduction involve various electron acceptances and donations by N and S chemical species. An oxidation–reduction (known as "redox") reaction is the simultaneous loss of an electron (oxidation) by one substance joined by an electron gain (reduction) by another in the same reaction. In oxidation, an element or compound loses, i.e. donates, electrons. Oxidation also occurs when oxygen atoms are gained or when hydrogen atoms are lost. Conversely, in reduction, an element or compound gains, i.e. captures, electrons. Reduction also occurs when oxygen atoms are lost or when hydrogen atoms are gained. The nature of redox reactions means that each oxidation–reduction reaction is a pair of two simultaneously occurring "half-reactions." The formation of sulfur dioxide and nitric oxide by acidifying molecular sulfur is a redox reaction:

$$S(s) + NO_3^-(aq) \rightarrow SO_2(g) + NO(g) \tag{2.21}$$

The designations in parentheses give the physical phase of each reactant and product: "s" for solid; "aq" for aqueous; and "g" for gas.

The oxidation half-reactions for this reaction are:

$$S \rightarrow SO_2 \tag{2.22}$$

$$S + 2H_2O \rightarrow SO_2 + 4H^+ + 4e^- \quad (2.23)$$

The reduction half-reactions for this reaction are:

$$NO_3^- \rightarrow NO \quad (2.24)$$

$$NO_3^- + 4H^+ + 3e^- \rightarrow NO + 2H_2O \quad (2.25)$$

Therefore, the balanced oxidation–reduction reactions are:

$$4NO_3^- + 3S + 16H^+ + 6H_2O \rightarrow 3SO_2 + 16H^+ + 4NO + 8H_2O \quad (2.26)$$

$$4NO_3^- + 3S + 4H^+ \rightarrow 3SO_2 + 4NO + 2H_2 \quad (2.27)$$

Oxidation–reduction reactions are not only responsible for pollution; they are also very beneficial. Redox reactions are part of essential metabolic and respiratory processes. Redox is commonly used to treat wastes, e.g. to ameliorate toxic substances and to treat wastes, by taking advantage of electrons donating and accepting microbes, or by abiotic chemical redox reactions. For example, in drinking water treatment a chemical oxidizing or reducing agent is added to the water under controlled pH. This reaction raises the valence of one reactant and lowers the valence of the other. Thus redox removes compounds that are "oxidizable," such as ammonia, cyanides, and certain metals like selenium, manganese, and iron. It also removes other "reducible" metals like mercury (Hg), chromium (Cr), lead, silver (Ag), cadmium (Cd), zinc (Zn), copper (Cu), and nickel (Ni). Oxidizing cyanide (CN^-) and reducing of Cr^{6+} to Cr^{3+} are two examples where the toxicity of inorganic contaminants can be greatly reduced by redox [36].

A reduced form of sulfur that is highly toxic and an important pollutant is hydrogen sulfide (H_2S). Certain microbes, especially bacteria, reduce N and S, using the N or S as energy sources through the acceptance of electrons. For example, sulfur-reducing bacteria can produce hydrogen sulfide (H_2S), by chemically changing oxidized forms of sulfur, especially sulfates (SO_4). To do so, the bacteria must have access to the sulfur, i.e. it must be in the water, which can be in surface or groundwater, or the water in soil and sediment, especially in the biofilm around particles. These sulfur-reducers are often anaerobes, i.e. bacteria that live in water where concentrations of molecular oxygen (O_2) are deficient. The bacteria remove the O_2 molecule from the sulfate leaving only the sulfur, which in turn combines with hydrogen to form gaseous H_2S. In groundwater, sediment and soil water, H_2S is formed from the anaerobic or nearly anaerobic decomposition of deposits of organic matter, e.g. plant residues. Thus, redox principles can be used to treat H_2S contamination, i.e. the compound can be oxidized using a number of different oxidants (see Table 2.15). Strong oxidizers, like molecular oxygen and hydrogen peroxide, most effectively oxidize the reduced forms of sulfur, nitrogen or any reduced compound.

Table 2.15 Theoretical amounts of various agents required to oxidize 1 mg L^{-1} of sulfide ion

Oxidizing agent	Amount (mg L^{-1}) needed to oxidize 1 mg L^{-1} of S_2^- based on practical observations	Theoretical stoichiometry (mg L^{-1})
Chlorine (Cl_2)	2.0 to 3.0	2.2
Chlorine dioxide (ClO_2)	7.2 to 10.8	4.2
Hydrogen peroxide (H_2O_2)	1.0 to 1.5	1.1
Potassium permanganate ($KMnO_4$)	4.0 to 6.0	3.3
Oxygen (O_2)	2.8 to 3.6	0.5
Ozone (O_3)	2.2 to 3.6	1.5

Source: Water Quality Association (1999). Ozone Task Force Report, *Ozone for POU, POE and Small Water System Applications.* Lisle, IL.

Ionization is also important in environmental reactions. The arrangement of the electrons in the atom's outermost shell, i.e. valence, determines the ultimate chemical behavior of the atom. The outer electrons become involved in transfer to and sharing with shells in other atoms, i.e. forming new compounds and ions. An atom will gain or lose valence electrons to form a stable ion that will have the same number of electrons as the noble gas nearest the atom's atomic number. For example, the nitrogen cycle (Figure 2.15) includes three principal forms that are soluble in water under environmental conditions: the cation (positively charged ion) ammonium (NO_3^-) and the anions (negatively charged ions) nitrate (NH_4^+) and nitrite (NO_2^-). Nitrates and nitrites combine with various organic and inorganic compounds. Once taken into the body, NO_3^- is converted to NO_2^-. Since NO_3^- is soluble and readily available as a nitrogen source for plants (e.g. to form plant tissue such as amino acids and proteins), farmers are the biggest users of NO_3^- compounds in commercial fertilizers (although even manure can contain high levels of NO_3^-).

A serious illness in infants is due to the conversion of nitrate to nitrite by the body, which can interfere with the oxygen-carrying capacity of the blood, known as methemoglobinemia. Especially in small children, when nitrates compete successfully against molecular oxygen, the blood carries methemoglobin (as opposed to healthy hemoglobin), giving rise to clinical symptoms. At 15–20% methemoglobin, children can experience shortness of breath and blueness of the skin (i.e., clinical cyanosis). At 20–40% methemoglobin, hypoxia will result. This acute condition can deteriorate a child's health rapidly over a period of days, especially if the water source continues to be used. Long-term, elevated exposures to nitrates and nitrites can cause an increase in the kidneys' production of urine (diuresis), increased starchy deposits, and hemorrhaging of the spleen [37].

Nutrients, like the compounds of nitrogen and sulfur, are important in every environmental medium. They present a challenge to environmental biotechnology, since certain forms in some scenarios are essential, whereas the same forms cause environmental insults in other scenarios. They can be air pollutants, water pollutants, as well as indicators of eutrophication (i.e. nutrient enrichment), ecological condition, and acid rain. They are some of the best examples of the need for a systematic viewpoint. Nutrients are valuable but, in the wrong place under the wrong conditions, they become pollutants.

The systematic approach requires an understanding of the processes that affect biotechnologies in the environment, either adversely or beneficially.

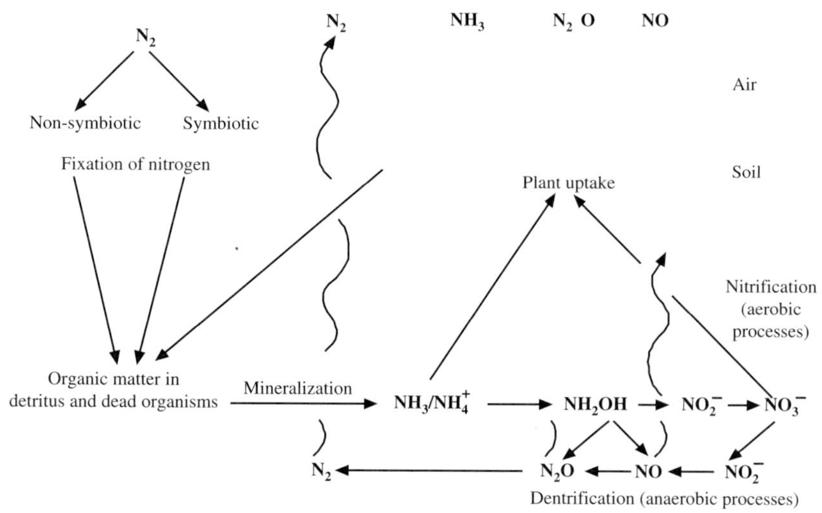

FIGURE 2.15
Biochemical nitrogen cycle.

SEMINAR TOPIC

GMOs and Global Climate Change

Cornell researcher R.J. Herring [38] argues that genetically modified organisms (GMOs) can play a large role in ensuring adequate food supplies. Recombinant DNA technologies can provide crops that are more adaptive and resilient, particularly for "vulnerable farmers and nations ... aggravated by the twin global challenges of climate change and ensuring the sustainability of agriculture" [39].

Oregon State University researcher S. Strauss is convinced that GMOs can:

> *... help rescue major tree species that have been devastated by exotic diseases, such as have occurred for chestnut and elm in the United States, to improve the efficiency of environmental cleanup and to reduce the risks of ecological harm due to the spread of exotic tree varieties. Products such as disease-resistant chestnut and elm should have direct benefits for promoting forest biodiversity by resurrecting key species that support many kinds of organisms in the ecosystems in which they occur. [40]*

Conversely, some of the risks associated with GMOs include public health and ecological concerns, such as irreversible changes to native plants and biodiversity. In fact, some have described genetically modified crops as a "leaky technology," in which pollen and other genetic material is spread by advection, insects, and higher animals, including humans. Pollen is a particle that can be transported long distances (see Figure 2.16). For example, pollen from exotic trees has been observed to be advectively transported from western North America to Greenland (see Figure 2.17).

One question is whether it is ethical, feasible or worth the risk to modify genes to respond to what seems to be an anthropogenic problem. For example, if it is likely that certain adverse outcomes will be on the increase as the climate changes, e.g. malaria and other tropical diseases or increases in fungal infestations of crops, should we be trying to "improve" a population's genetic structure to deal with these changes? We are doing this with plants now, but clearly there is a huge moral difference between plants and humans.

The relationship between energy sources and global climate change is direct. The more fossil fuels that are burned, the greater the amount of greenhouse gas that is emitted to the atmosphere. Thus, new biofuels are being sought and at the top of the list are algae. In fact, the US Department of Energy estimates that 7.5 billion gallons of biodiesel can be generated by algae on 200,000 hectares of desert land [41].

The biological taxonomy of algae includes a wide variety of simple organisms in the Protista kingdom. Algae are actually not plants since they have no roots, leaves or other plant structures but, like plants, they undergo photosynthesis.

Oil-rich microalgae can potentially be the source of high-lipid mass needed to generate biofuels, due to efficient photosynthetic conversion. In addition, microalgal energy avoids the geopolitical problems of competition between fuel and food, as is the case for corn-generated biofuels, for example. An estimated 2 megahectares (Mhs), or about 1.1% of the total cropping area of the United States, is enough to produce algal biomass as the source of 50% of the transportation fuel needs in the US [42]. The large-scale microalgal culturing needed to provide lipids, e.g. triglycerides, which are the basis for biofuels, can come from a number of candidate species of diatoms and microalgae, including *Chlorella spp.*, *Dunaliella salina*, *Spirulina spp.* and *Hematococcus pluvialis* [43]. A large culture system is currently operational using an 800 ha open-pond system to grow the halophile *Dunaliella salina* in a highly saline lake in Australia [44].

Algae farms (see Figure 2.18) depend on surface area (as opposed to volume in many bioreactors), because surface is a critical determinant for capturing sunlight. Their productivity is measured in terms of

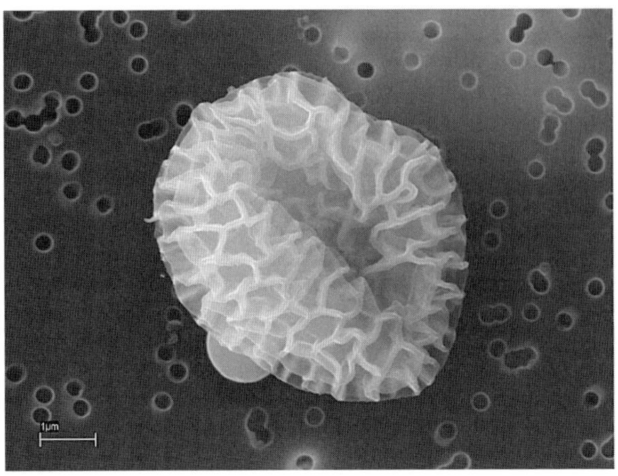

FIGURE 2.16
Scanning electron micrograph of a pollen particle.
Photo courtesy of R. Willis, US Environmental Protection Agency.

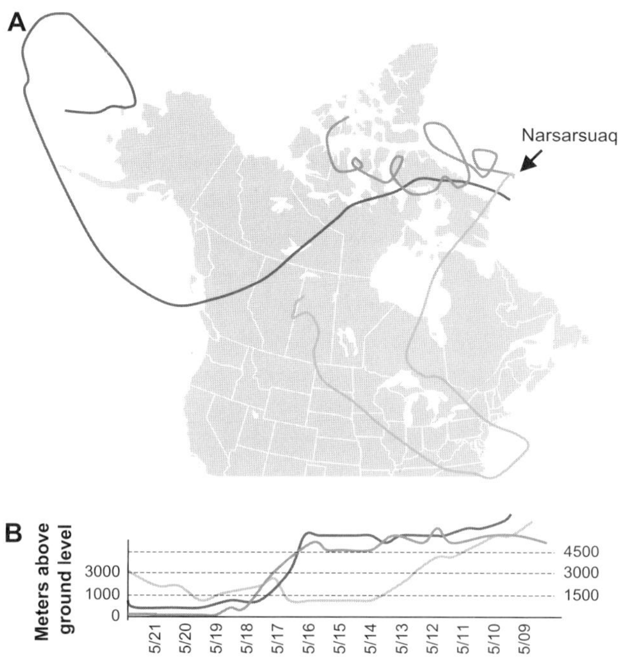

FIGURE 2.17
Long distance pollen transport to Narsarsuaq, Greenland at ground level (red), 1000 m above the ground (blue) and 3000 m (green) on 21 May 2003. (A) Backward trajectories for 21 May 2003. (B) Altitudinal variation (meters) of the three air volumes used in the backward trajectories analysis. [*See color plate section*]

biomass produced per day per unit of available surface area. In addition to replacing fossil fuels as a greener (literally and figuratively) fuel source, algae also can extend carbon capture and sequestration efforts when co-located or located near a large carbon source (e.g. a power plant), as shown in Figure 2.18. Typical coal-fired electricity generating facilities emit flue gas from their stacks containing up to 13% CO_2. At these levels, CO_2 enhances transfer and uptake of CO_2 in the ponds.

Research has been ongoing to find the algae's mechanisms of lipid accumulation, including whether there is a specific "lipid trigger" induced by processes like nitrogen (N) and phosphorus (P) nutrient starvation. This has involved finding ways to produce improved algae strains, starting by finding the genetic variability among algal isolates, as well as using flow cytometry to select "naturally-occurring high lipid individuals, and exploring algal viruses as potential genetic vectors" [45]. In addition to N and P starvation, silica (Si) depletion in diatoms has also been shown to induce lipid accumulation. Nitrogen is a major component of cellular biomolecules, but Si is not, so the Si effect on lipid production is likely to be less complex. The problem with any nutrient starvation strategy is that since lipids are intracellular products, the total lipid productivity is the product of cell lipid content and biomass productivity. Thus, based on the first principles of thermodynamics, the lower biomass compromises the overall lipid/energy productivity.

These challenges showed the importance of enhancing both the biotechnology and the state of the science of biochemistry and molecular biology of diatoms' mechanisms for lipid accumulation, especially isolating and characterizing numerous enzymes in the pathways for synthesizing carbohydrates and lipid. That is where biofuels will gain their carbon and energy, so cloning of the genes that code for these enzymes became a major emphasis for the Department of Energy. This has led to improved bioreactor design, including stages, i.e. a stage that optimizes cell growth and division in a nutrient-sufficient medium followed by a stage that optimizes nutrient-starvation or other physiological stress to induce lipid accumulation.

The advances have also called for genetically engineering of the algae "to manipulate microalgal lipid levels by overexpressing or down-regulating key genes in the lipid or carbohydrate synthetic pathways" [46]. Completing the genome sequences of a number of algae has increased this effort. These include the red alga *Cyanidioschyzon merolae*, the diatoms *Thalassiosira pseudonana* and *Phaeodactylum tricornutum*, and the unicellular green alga *Ostreococcus tauri.* In addition, nuclear transformation of several microalgal species is ongoing, as is chloroplast transformation for green, red, and euglenoid algae. Organelle transformation is advancing (e.g. sequenced plastid, mitochondrial, and nucleomorph genomes), as is a genetic transformation system (e.g. in green algae *Chlamydomonas reinhardtti* and *Volvox carteri*) [47].

Other engineering research has focused on optimizing pond design and using CO_2 from power plant flue gases as carbon sources [48].

Since algae are not grown as a crop and are often seen as opportunistic biota, modifying algal DNA has not received the same skepticism as cash crops, like corn and potatoes. However, algae are important parts of most aquatic environments, so any change to algal

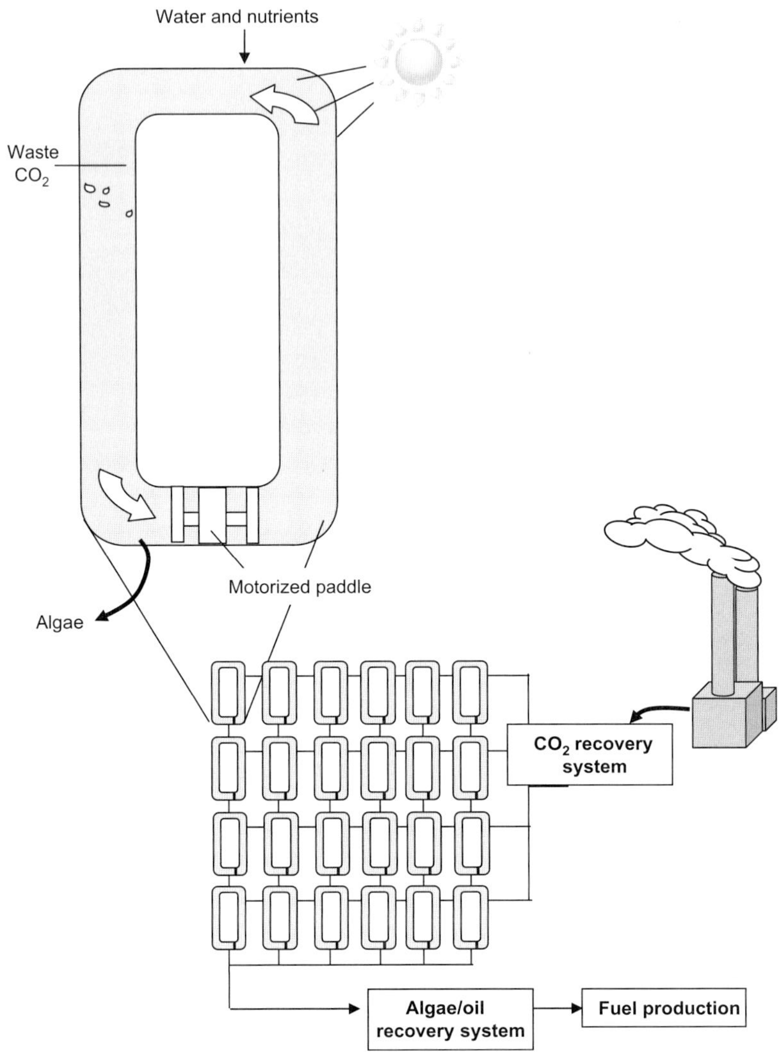

FIGURE 2.18
Algae farm.
Adapted from: National Renewable Energy Laboratory (1998). A Look Back at the US Department of Energy's Aquatic Species Program – Biodiesel from Algae. Report No. NRL/TP-580-24190.

populations will have an impact on food webs and food chains. As such, fish and other seafood will be affected, at least indirectly. For example, if the genetically modified algae are released and allowed to transfer genetic material, this could change delicate relationships between fish and aquatic fauna with the previously natural algae in unknown ways. If one of these changed expressions is an advantage of surviving in warmer waters, these modified algae could completely replace the progenitor species.

In addition to the indirect benefits to potential decreases in atmospheric concentrations of carbon compounds if algae were used as biofuels, scientists from 24 research organizations led by the US Department of Energy (DOE) Joint Genome Institute (and the Monterey Bay Aquarium Research Institute (MBARI) have been decoding genomes of two algal strains, highlighting the genes enabling them to capture carbon and maintain its delicate balance in the oceans. These findings, from a team led by Alexandra Z. Worden of MBARI and published in the April 10 edition of the journal *Science*, will illuminate cellular processes related to algae-derived biofuels being pursued by DOE scientists.

The study sampled two isolates of the photosynthetic algal genus *Micromonas* (see Figure 2.19): one from the South Pacific and the other from the English Channel. The analysis identified approximately 10,000 genes in each isolate that are compressed into genomes that total about 22 million nucleotides. The two isolates of the same species only share about 90% of the same genes [49]. Such information can be used to understand and possibly control to some extent the amount of carbon sequestered in the oceans' algal populations, since these two isolates indicate survival during myriad environmental conditions gene complement may cause them to access and respond to the environment differently.

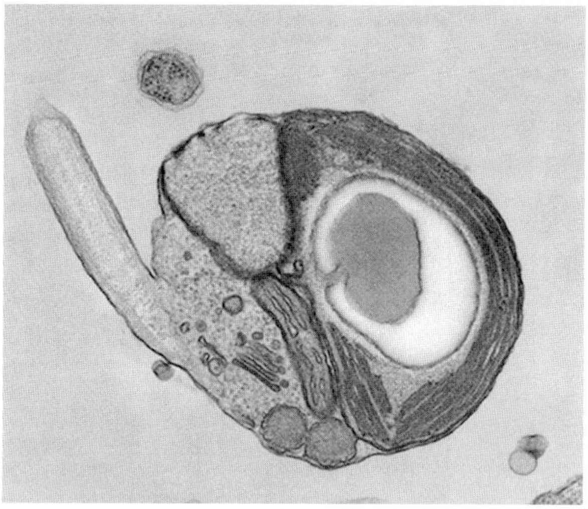

FIGURE 2.19
Transmission electron micrograph of eukaryotic algae, *Micromonas*.
Photo credit: A.Z. Worden, T. Deerinck, M. Terada, J. Obiyashi and M. Ellisman (Monterey Bay Aquarium Research Institute and National Center for Microscopy and Imaging Research).

Seminar Questions

What aspects of global climate change are good candidates for improvement using GMOs?

What are the risks and uncertainties associated with using GMOs to try to ameliorate the effects of global climate change?

How do algae differ from bacteria regarding ecosystem risks associated with genetic modifications? How are these risks similar?

REVIEW QUESTIONS

Give two examples of biomimicry that are providing benefits to environmental quality. Give two examples that may pose a risk to public health and the environment.

What is the difference between a nutrient and a contaminant?

If a laboratory conducts 25 runs of a sample containing 0.01 ng mL^{-1} benzo(a)pyrene with a standard deviation of 0.005, what is the method detection limit for that instrument? If the SD stays the same for the same analyte on the same instrument, but with 50 runs, what is the MDL? Is there a way to estimate the change in the MDL if the laboratory had conducted 100 runs on that instrument, with the same SD? How do the additional runs affect MDL? Can you use this instrument if a project calls for a detection limit of 0.05 ng mL^{-1} b(a)p?

If a filter is removed from an $PM_{2.5}$ monitor and weighed at 10:00 am and again at 11:00 am and the latter weighing is 10% higher than the earlier weighing, what might the reasons be for the increase? Is this acceptable?

If collocated $PM_{2.5}$ samplers have collected samples with measurements 3% of each other, is this sufficiently precise to meet regulatory requirements? Support your answer.

If the flow rate is 2.78×10^{-4} m^{-3} sec^{-1} for 24 hours for all of the measurements in Figure 2.20, what are the $PM_{2.5}$ mass concentration values for the replicates? What is the precision if every other replicate is from collocated Samplers A and B? Do they meet the National Ambient Air Quality Standards?

What are the key processes in Table 2.9 that are at work when a filter is placed on bioreactor's vent?

Filter Lot Number C20102		Analyst J. Armstrong	
Balance Number A44603		QC Supervisor R. Vanderpool	
Presampling Filter Weighing Date 6/30/97		RH 33	Temp 22
Postsampling Filter Weighing Date 8/13/97		RH 38	Temp 21
Filter Number[a]	**Presampling Mass (mg)**	**Postsampling Mass (mg)**	**Net Mass Filter Loading (mg)**
100 mg (WS)	100.000	100.001	0.001
200 mg (WS)	199.999	200.001	0.002
D-110 (LB)	136.546	136.550	0.004
D-111 (LB)	129.999	130.006	0.007
D-112 (LB)	130.633	130.645	0.012
R-700 (FB)	130.896	130.904	0.008
R-701 (FB)	128.339	128.345	0.006
R-702 (FB)	130.929	130.936	0.007
R-691	139.293	139.727	0.434
R-692	136.020	136.455	0.435
R-693	135.818	136.260	0.442
R-694	131.456	131.905	0.449
R-695	137.508	137.973	0.465
R-696	136.098	135.554	0.456
R-697	131.029	131.483	0.454
R-698	125.175	125.641	0.466
R-699	131.165	131.633	0.468
R-691 (R)	139.293	139.730	0.437
100 mg (WS)	100.002	100.001	0.001
200 mg (WS)	199.998	200.000	0.002

[a]Indicate working standard (WS), lab blank (LB), field blank (FB), or replicate (R) measurement here.

FIGURE 2.20

Example of a laboratory data form for aerosol measurements. Information is fictitious. Adapted from: US Environmental Protection Agency (1998). *Quality Assurance Guidance Document 2.12. Monitoring $PM_{2.5}$ in Ambient Air Using Designated Reference or Class I Equivalent Methods.* Research Triangle Park, North Carolina, November 1998.

How might genetic modification of bacteria affect the processes shown in Figures 2.6 and 2.9? Considering possible scenarios for carbon sequestration, what role may microbes play in the long-term storage? In particular, what role may genetically modified organisms play in methanogenesis?

If a coal-fired electricity generating power plant releases 20 tons of carbon dioxide per year, how much surface area and algal mass would your algae farm (*Thalassiosira pseudonana*) need to photosynthesize this CO_2 to O_2? How much oil would be produced from lipid accumulation? State your assumptions. What seems to be the most sensitive variable in this problem?

Compare the energy that can be produced from biofuels from growing 100 ha of corn; 100 ha of switchgrass and 100 ha of pond surface of a *Chlamydomonas reinhardtti* algae farm in Fargo, ND. What would happen if the amount of N and P were increased by 50% in each of these systems? How would the energy balances change if you located the same farms near San Diego, CA? State your assumptions. What seems to be the most sensitive variable in this problem?

How might the drift of genetic material differ for algae compared to bacteria? How might it differ for algae compared to larger plants? Estimate the extent of horizontal gene transfer of a genetically modified strain of *Cyanidioschyzon merolae* compared with that of *Zea mays* based on mechanisms of physical and biological transport. State your assumptions. What seems to be the most sensitive variable in this problem?

If a certain algal species has the chemical composition of $C_{106}H_{263}O_{110}N_{16}P$, what is the mass of each element if you have a mean mass of 1000 kg of algae in your algae farm? If 0.1 mg L^{-1}

N and 0.04 mg L^{-1} is available for algal production, which nutrient is limiting your algal production, N or P (assuming no other macronutrients or micronutrients are limiting)? What would happen if P were decreased by 50%? What if N were decreased by 50%? What if they both were reduced by 25%? How much biofuel can be generated from this algae farm? State your assumptions. What seems to be the most sensitive variable in this problem?

NOTES AND RESOURCES

1. M. Basgall (2002). One upping nature in a quest for new materials. Office of News and Communications, Duke University. http://www.pratt.duke.edu/news/?id=175; accessed August 7, 2009.
2. United States Code of Federal Regulations: 40 CFR 131.6, Appendix B – Definition and procedure for determination of the method detection limit – Revision 1.11.
3. J.A. Hanlon (2007). Office of Wastewater Management, US Environmental Protection Agency. Memorandum to Water Division Directors, Regions 1–10. August 27, 2007.
4. United States Code of Federal Regulations: 40 CFR 132.6, Table 4.
5. United States Code of Federal Regulations: 40 CFR 131.6. The method detection limit, according to 40 CFR 136, Appendix B, is 0.2 ng L^{-1} and the minimum level of quantitation is 0.5 ng L^{-1}.
6. This procedure is taken directly from United States Code of Federal Regulations: 40 CFR 131.6, Appendix B – Definition and procedure for determination of the method detection limit – Revision 1.11.
7. My apologies to the originator of this analogy, who deserves much credit for this teaching device. The target is a widely used way to describe precision and accuracy.
8. US Environmental Protection Agency (1998). *Quality Assurance Guidance Document 2.12. Monitoring $PM_{2.5}$ in Ambient Air Using Designated Reference or Class I Equivalent Methods.* Research Triangle Park, North Carolina, November 1998.
9. The diameter most often used for airborne particle measurements is the "aerodynamic diameter." The aerodynamic diameter (D_{pa}) for all particles greater than 0.5 μm can be approximated as the product of the Stokes particle diameter (D_{ps}) and the square root of the particle density (ρ_p):

$$D_{pa} = D_{ps}\sqrt{\rho_p} \tag{2.28}$$

If the units of the diameters are in μm, the units of density are g cm^{-3}.
The Stokes diameter D_{ps} is the diameter of a sphere with the same density and settling velocity as the particle. The Stokes diameter is derived from the aerodynamic drag force caused by the difference in velocity of the particle and the surrounding fluid. Thus, for smooth, spherical particles, the Stokes diameter is identical to the physical or actual diameter.
Aerosol textbooks provide methods to determine the aerodynamic diameter of particles less than 0.5 μm. For larger particles gravitational settling is more important and the aerodynamic diameter is often used.
10. For information regarding particle matter (PM) health effects and inhalable, thoracic and respirable PM mass fractions see: US Environmental Protection Agency (1996). Air Quality Criteria for Particulate Matter. Technical Report No. EPA/600/P-95/001aF, Washington, DC.
11. Ibid.
12. P. Solomon, G. Norris, M. Landis and M. Tolocka (2001). Chemical analysis methods for atmospheric aerosol components. In: P.A. Baron and K. Willeke (Eds), *Aerosol Measurement: Principles, Techniques, and Applications*, 2nd Edition. Wiley-Intersciences, Inc., Hoboken, NJ.
13. United States Code of Federal Regulations, Part 58, Appendix A, 1997; and US Environmental Protection Agency (1998).
14. US Environmental Protection Agency and Battelle National Laboratory (2004). ETV Joint Verification Statement. Rapid Polymerase Chain Reaction: Detecting Biological Agents and Pathogens in Water; http://www.epa.gov/ordnhsrc/pubs/vsInvitrogen121404.pdf; accessed September 24, 2009.
15. US EPA (2004).
16. Ibid.
17. Quote included in M. Basgall (2002).
18. D. Needham, in M. Basgall (2002).
19. P.G. Tratnyek and R.L. Johnson (2006). Nanotechnologies for environmental cleanup. *Nano Today* 1 (2): May.
20. V.M. Goldschmidt (1923). Geochemische Verteilungsgesetze der Elemente. Skrifter utg. av det Norske Visenskaps-Akademii i Oslo I. *Mat.-Naturv. Klasse* 1–17.
21. C.L. Hollabaugh (2007). Modification of Goldschmidt's geochemical classification of the elements to include arsenic, lead and mercury as biophile elements. In: R. Datta, D. Sarkar and R. Hannigan (Eds), *Concepts and Applications in Environmental Geochemistry.* Elsevier, Amsterdam, The Netherlands, pages 9–32.
22. T.E. McKone, B.M. Huey, E. Downing and L.M. Duffy (Eds) (2000). *Strategies to Protect the Health of Deployed US Forces: Detecting, Characterizing, and Documenting Exposures.* National Academies Press, Washington, DC.
23. Ibid.
24. Environmental Security Technology Certification Program (2004). White Paper: Bioaugmentation for Remediation of Chlorinated Solvents: Technology Development, Status, and Research Needs. Prepared by GeoSyntec Consultants.

25. E.J. Bouwer and P.L. McCarty (1983). Transformation of 1- and 2-carbon halogenated aliphatic organic compounds under methanogenic conditions. *Applied and Environmental Microbiology* 45: 1286–1294.
26. B.Z. Fathepure, J.P. Nengu and S.A. Boyd (1987). Anaerobic bacteria that dechlorinate perchloroethene. *Applied and Environmental Microbiology* 53: 2671–2674.
27. D. Ryoo, H. Shim, K. Canada, P. Barbieri and T.K. Wood (2000). Aerobic degradation of tetrachloroethylene by toluene-o-xylene monooxygenase of *Pseudomonas stutzeri* OX1. *Nature Biotechnology* 77: 5–8.
28. S. Heald and R.O. Jenkins (1994). Trichloroethylene removal and oxidation toxicity mediated by toluene dioxygenase of *Pseudomonas putida*. *Applied and Environmental Microbiology* 60: 4634–4637; R. Oldenhuis, R.L.J.M. Vink, D.B. Jansen and B. Witholt (1989). Degradation of chlorinated aliphatic hydrocarbons by *Methylosinus trichosporium* OB3b expressing soluble methane monooxygenase. *Applied and Environmental Microbiology* 55: 2819–2826; and R.J. Oldenhuis., J.Y. Oedzes, J.J. Vanderwaarde and D.B. Janssen (1991). Kinetics of chlorinated hydrocarbon degradation by *Methylosinus trichosporium* OB3b and toxicity of trichloroethylene. *Applied and Environmental Microbiology* 57: 7–14.
29. Friedrich Wöhler (1828). Ueber künstliche Bildung des Harnstoffs. *Annalen der Physik und Chemie* 37 (1): 330.
30. Kyriacos Costa Nicolaou and Tamsyn Montagnon (2008). *Molecules That Changed the World*. Wiley–VCH, Hoboken, NJ.
31. The letter "X" commonly denotes a halogen, e.g. fluorine, chlorine, or bromine, in organic chemistry. However, in this text, since it is an amalgam of many scientific and engineering disciplines, where "x" often means an unknown variable and horizontal distance on coordinate grids, this rule is sometimes violated. Note that when consulting manuals on the physicochemical properties of organic compounds, such as those for pesticides and synthetic chemistry, the "X" usually denotes a halogen.
32. National Academy of Engineering (2009). Grand Challenges for Engineering: Manage the Nitrogen Cycle. http://www.engineeringchallenges.org/cms/8996/9132.aspx; accessed August 8, 2009.
33. Ibid.
34. See R.H. Socolow (1999). Nitrogen management and the future of food: lessons from the management of energy and carbon. *Proceedings of the National Academy of Sciences of the United States of America* 96: 6001–6008; also, No 4: Human alteration of the nitrogen cycle: Threats, benefits and opportunities, UNESCO-SCOPE Policy Briefs (2007).
35. National Academy of Engineering (2009). Grand Challenges for Engineering: Develop Carbon Sequestration Methods. http://www.engineeringchallenges.org/cms/8996/9077.aspx; accessed August 8, 2009.
36. Redox reactions are controlled in closed reactors with rapid mix agitators. Oxidation-reduction probes are used to monitor reaction rates and product formation. The reactions are exothermic and can be very violent, when the heat of reaction is released, so care must be taken to use only dilute concentrations, along with careful monitoring of batch processes.
37. US Environmental Protection Agency (EPA) (1999). National Primary Drinking Water Regulations: Technical Fact Sheets. Washington, DC: http://www.epa.gov/OGWDW/hfacts.html.
38. R.J. Herring (2008). Opposition to transgenic technologies: ideology, interests, and collaborative framing. *Nature Reviews Genetics* 9: 458–463.
39. Y. Borofsky (2009). Strategic adaptation: How GMOs could help us deal with climate change. *Breakthrough Generation*; http://breakthroughgen.org/2009/06/18/strategic-adaptation-how-gmo%E2%80%99s-could-help-us-deal-with-climate-change; accessed September 21, 2009.
40. Ibid.
41. National Renewable Energy Laboratory (1998). A look back at the US Department of Energy's Aquatic Species Program – Biodiesel from algae. Report No. NRL/TP-580-24190.
42. Y. Chisti (2007). Biodiesel from microalgae. *Biotechnological Advances* 25: 294–306.
43. M.A. Borowitzka (1999). Commercial production of microalgae: ponds, tanks, tubes and fermenters. *Journal of Biotechnology* 70: 313–321.
44. T. Matsunaga, M. Matsumoto, Y. Maeda, H. Sugiyama, R. Sato and T. Tanaka (2009). Characterization of marine microalga, *Scenedesmus* sp. strain JPCC GA0024 toward biofuel production. *Biotechnology Letters* 31: 1367–1372.
45. Ibid.
46. National Renewable Energy Laboratory (1998).
47. T.L. Walker, C. Collet and S. Purton (2005). Algal transgenics in the genomic era. *Journal of Phycology* 41 (6): 1077–1093.
48. Matsunaga *et al.*, Characterization of marine microalga.
49. US Department of Energy (2009). Joint Genome Institute. Genes from tiny algae shed light on big role managing carbon in world's oceans; http://www.jgi.doe.gov/News/news_09_04_09.html; accessed September 24, 2009.

CHAPTER 3

Environmental Biochemodynamic Processes

The first law of thermodynamics states that energy and matter are neither created nor destroyed in any system; but this law applies to any system in its entirety. It applies to closed systems. It applies to an ecosystem. It applies to a human body. The laws of thermodynamics also apply to any closed system, which is one that is not exchanging energy or matter with its surroundings. The ultimate closed system is the universe; but the universe contains cascades of closed systems down to subcellular systems.

Environmental thermodynamics usually involves open systems. For a defined time period, reactors are closed systems to some extent, although even well-insulated reactors exchange heat with their surroundings. So, chemical engineers are more likely to encounter quasi-closed systems than most environmental engineers who help to design incinerators or other reaction chambers, including bioreactors used in biotechnologies. Bioengineers who practice out in the "field," such as remediation projects, ecosystem protection, and restoration and designing pollution control equipment, more often than not must assume they are working with an open system.

CELLULAR THERMODYNAMICS

Each microbe is a thermodynamic system. Two fundamental cell types exist: *prokaryotic* and *eukaryotic*. The more primitive cell type, prokaryotic cells, have no membrane around their nuclear region, thus their deoxyribonucleic acid (DNA) is naked. Prokaryotes include bacteria, mycoplasma, and simple blue-green algae, i.e. *cyanobacteria*. By contrast, eukaryotic cells have double membranes separating the nucleus from the cytoplasm, and numerous internal membranes to set apart their organelles. All animal and plant cells are eukaryotic. Prokaryotic and eukaryotic cells play important roles in the fate of environmental contaminants.

Prokaryotic organisms commonly produce only exact duplicates of themselves, but higher eukaryotic organisms' cells can be differentiated into diverse cell types. Prokaryotic cells, then, have the advantage of simple needs for nutrients. This allows for their being able to break down contaminants via *biotransformation*. This is also a factor in why engineers are able to acclimate bacteria and other prokaryotes to use recalcitrant xenobiotics as carbon and energy sources in processes to treat hazardous wastes and wastewater. In addition, prokaryotes can resist adverse environmental conditions, grow rapidly, and divide geometrically. Thus, they are ideal for environmental treatment scenarios.

Environmental Biotechnology

Production of proteins is the principal chemical compound output of all cells. Eukaryotes start producing proteins in the nucleus, the large, dense structure within the cell. By the mid 20th century, heredity traits were being linked to the rod-like bundles of DNA, known as chromosomes. The nucleus provides the cellular information through chemical messaging systems, including the polypeptides. Genes, comprised of DNA, direct the formation of cells, i.e. what kind they are and what types will be made and differentiated in the organism. So, the nucleus is the location of all messages regarding reproduction and cell division. Molecular DNA consists of bases linked to form a double helix structure. Two bases are joined together by chemical bonds, and attached to chains of chemically bonded sugar and phosphate molecules. A nucleotide is a unit of DNA that is made up of one sugar molecule, one phosphate molecule, and one base Only four bases exist: adenine (A), thymine (T), guanine (G), and cytosine (C). The base A is always joined to T. The base G is always linked to C. Thus, the sequence of bases on one side of the helix (e.g., AGCGT) complements and establishes the sequence (TCGCA) on the other side of the helix. This sequencing allows for billions of possible messages. Unfortunately, it is also the errors in such sequencing that lead to many of the adverse outcomes resulting from exposures to environmental contaminants, such as cancer and birth defects.

Equilibrium is a physical, chemical and biological concept. It is the state of a system at which the energy and mass of that system are distributed in a statistically most probable manner, obeying the laws of conservation of mass, conservation of energy (first law of thermodynamics), and efficiency (second law of thermodynamics). So, if the reactants and products in a given reaction are in a constant ratio, that is the forward reaction and the reverse reactions occur at the same rate, then that system is in equilibrium. Up to the point where the reactions are yet to reach equilibrium, the process is *kinetic*, i.e. the rates of particular reactions are considered.

In environmental situations, an important determination is whether a system's influences and reactions are in balance. The conservation laws require that everything be balanced eventually, but since we only observe systems within finite timeframes and in confined spatial frameworks, we may only be able to see some of the steps *en route* to reaching equilibrium. Thus, for example, it is not uncommon in the environmental literature to see non-equilibrium constants (i.e. kinetic coefficients [1]). Indeed, not all energy is used in a system.

The second law of thermodynamics states that when energy is converted from one form to another, a certain amount of that energy is not available to do work. No physical, chemical or biological process is ever 100% efficient and a certain amount of energy will not be converted to work. Simply stated, total energy is the sum of usable and unusable energy. The usable energy, i.e. the energy that *is* available for work, is the *free energy* in a system. This applies to all energy, but in environmental situations, including metabolism, respiration, photosynthesis, and biodegradation, free energy is crucial in chemical reactions in living systems.

The equilibrium constant for a chemical reaction depends upon the environmental conditions, notably pressure, temperature, and ionic strength of the solution. An example of a thermodynamic equilibrium reaction is chemical precipitation, such as what occurs in water treatment processes [2]. Environmental engineers treating water apply a chemical reaction to remove microbes and chemical compounds that make water unsafe to drink. This is often a *heterogeneous reaction*, i.e. the reagents and products involved include more than one physical state of matter. For an equilibrium reaction to occur between solid and liquid phases, the solution must be saturated and undissolved solids must be present. So, at a high hydroxyl ion concentration (e.g. pH = 10), the solid phase calcium carbonate ($CaCO_3$) in the water reaches equilibrium with divalent calcium (Ca^{2+}) cations and divalent carbonate (CO_3^{2-}) anions in solution. Such reactions are also commonly used to remove other positively charge ions, i.e. *cations* from water.

For gases, the thermodynamic "equation of state" expresses the relationships of pressure (p), volume (V), and thermodynamic temperature (T) in a defined quantity (n) of a substance. For gases, this relationship is defined most simply in the ideal gas law:

$$pV = nRT \tag{3.1}$$

where R = the universal gas constant or molar gas constant = $8.31434\ \mathrm{J\ mol^{-1}\ {}^\circ K^{-1}}$.

Note that the ideal gas law only applies to ideal gases, those that are made up of molecules taking up negligible space, with negligible spaces between the gas molecules. So, for real gases, the equilibrium relationship is:

$$(p + k)(V - nb) = nRT \tag{3.2}$$

where k = factor for the decreased pressure on the walls of the container due to gas particle attractions; nb = volume occupied by gas particles at infinitely high pressure.

Further, the van der Waals equation of state is:

$$k = \frac{n^2 a}{V^2} \tag{3.3}$$

where a is a constant.

Gas reactions, therefore, depend upon partial pressures. The gas equilibrium K_p is quotient of the partial pressures of the products and reactants, expressed as:

$$K_p = \frac{p_C^z p_D^w}{p_A^x p_B^y} \tag{3.4}$$

Thus, K_p can also be expressed as:

$$K_p = K_{eq}(RT)^{\Delta v} \tag{3.5}$$

where Δv is defined as the difference in stoichiometric coefficients.

As mentioned, free energy is the measure of a system's ability to do work, in this case to drive the chemical reactions. In the cell, free energy is released from the chemical bonds. If reactants have greater free energy than the products have, energy is released from the reaction; which means the reaction is *exergonic*. Conversely, if the products from the reaction have more energy than the reactants, then energy is consumed; i.e. it is an *endergonic* reaction.

Equilibrium constants can be ascertained thermodynamically by employing the Gibbs free energy (G) change for the complete reaction. This is expressed as:

$$G = H - TS \tag{3.6}$$

where G is the energy liberated or absorbed in the equilibrium by the reaction at constant T. H is the system's enthalpy and S is its entropy. Enthalpy is the thermodynamic property expressed as:

$$H = U + pV \tag{3.7}$$

where U is the system's internal energy.

The relationship between a change in free energy and equilibria can be expressed by:

$$\Delta G^* = \Delta G_f^{*0} + RT\ln K_{eq} \tag{3.8}$$

where ΔG_f^{*0} = free energy of formation at steady state ($\mathrm{kJ\ gmol^{-1}}$).

In summary, the total energy in living systems is known as enthalpy (H) and the usable energy is known as free energy (G). Living cells need G for all chemical reactions, especially cell growth, cell division and cell metabolism and health. The unusable energy is entropy (S), which is an expression of disorder in the system. Disorder tends to increase as a result of the many conversion steps outside and inside of the cell. In response, the cells have adapted ways of improving efficiencies. Bioengineers have looked for ways to improve these efficiencies even further. Thus, to understand environmental biotechnologies, the processes that underlie microbial metabolism must be characterized.

Importance of free energy in microbial metabolism

Whether in single-cell or multi-cell organisms, the cells must carry out two very basic tasks in order to survive and grow. They must undergo biosynthesis, i.e. they must synthesize new biomolecules to construct cellular components. They must also harvest energy. Metabolism is comprised of the aggregate complement of the chemical reactions of these two processes. Thus, metabolism is the cellular process that derives energy from a cell's surroundings and uses this energy to operate and to construct even more cellular material.

Energy that does chemical work is exemplified by cellular processes. As mentioned, metabolism has two components: catabolism and anabolism (see Figure 3.1). Catabolism consists of reactions that degrade incoming food, i.e. the energy source, such as carbohydrates. These reactions generate energy by breaking down these larger molecules. Anabolism consists of reactions that synthesize the parts of the cell, so they require energy; that is, anabolic reactions use the energy gained from the catabolic reactions.

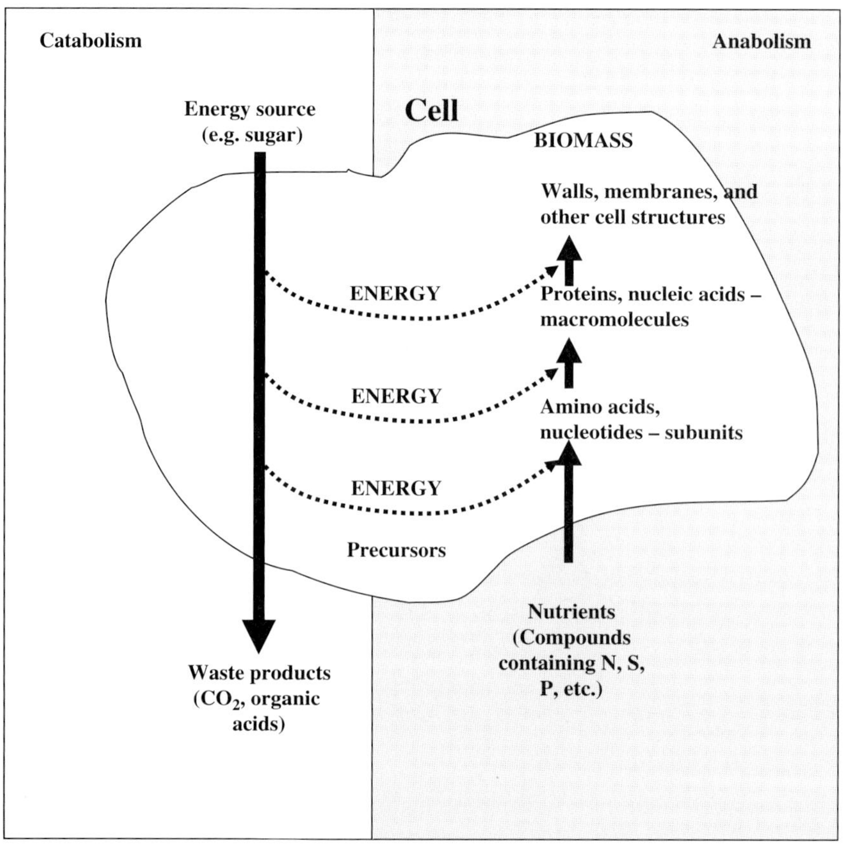

FIGURE 3.1
Cellular metabolism results from catabolic reactions that break down compounds to gain energy that is used to build biomolecules (anabolic metabolism) from nutrients that are taken up by the cell, beginning with simple precursors, then subunits, macromolecules. From these biomolecules, the cellular structures are built.

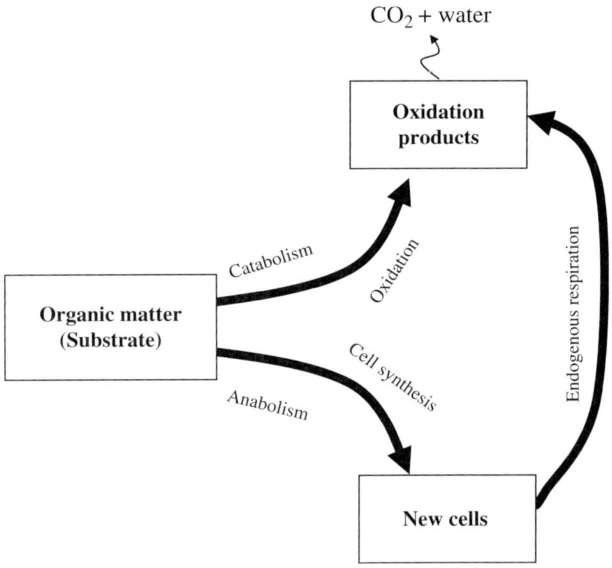

FIGURE 3.2
Microbial oxidation that occurs during the degradation of organic compounds. Both the catabolic and anabolic processes generate oxidation products.

Anabolism and catabolism are two sides of the same metabolic coin, so to speak. Anabolism is synthesizing, whereas catabolism is destroying. But, the only way that anabolism can work to build the cellular components is by the energy released by catabolism's destruction of organic compounds. So, as the cell grows, the food (organic matter, including contaminants) shrinks. Biological treatment takes advantage of these two metabolic functions, depending on the microbial biodegradability of various organic substrates. The microbes use the organic compounds as their exclusive source of energy (catabolism) and their sole source of carbon (anabolism). These life processes degrade the pollutants (see Figure 3.2).

Microbes, e.g. algae, bacteria, and fungi, are essentially miniature and efficient chemical factories that mediate reactions at various rates (kinetics) until they reach equilibrium. These "simple" organisms (and complex organisms alike) need to transfer energy from one site to another to power their machinery needed to stay alive and reproduce. Microbes play a large role in degrading pollutants, whether in natural attenuation, where the available microbial populations adapt to the hazardous wastes as an energy source, or in engineered systems that do the same in a more highly concentrated substrate (see Table 3.1). Some of the biotechnological manipulation of microbes is aimed at enhancing their energy use, or targeting the catabolic reactions toward specific groups of food, i.e. organic compounds.

Again, free energy is an important factor in microbial metabolism. The cell needs it for metabolic processes and bioengineers take advantage of this to use the metabolic pathways to degrade compounds. This occurs in a step-wise progression after the cell comes into contact with the compound. The initial compound, i.e. the parent, is converted into intermediate molecules by the chemical reactions and energy exchanges shown in Figure 3.1. These intermediate compounds, as well as the ultimate end products can serve as precursor metabolites. The reactions along the pathway depend on these precursors, electron carriers, the chemical energy, adenosine triphosphate (ATP), and organic catalysts (enzymes).

The reactant and product concentrations and environmental conditions, especially pH of the substrate, affect the observed ΔG^* values. If a reaction's ΔG^* is a negative value, the free energy is released and the reaction will occur spontaneously, and the reaction is exergonic. If a reaction's ΔG^* is positive, the reaction will not occur spontaneously. However, the reverse reaction will take place, and the reaction is endergonic.

Table 3.1 Genera of microbes shown to be able to degrade a persistent organic contaminant, i.e., crude oil

Bacteria	Fungi
Achromobacter	*Allescheria*
Acinetobacter	*Aspergillus*
Actinomyces	*Aureobasidium*
Aeromonas	*Botrytis*
Alcaligenes	*Candida*
Arthrobacter	*Cephaiosporium*
Bacillus	*Cladosporium*
Beneckea	*Cunninghamella*
Brevebacterium	*Debaromyces*
Coryneforms	*Fusarium*
Erwinia	*Gonytrichum*
Flavobacterium	*Hansenula*
Klebsiella	*Helminthosporium*
Lactobacillus	*Mucor*
Leucothrix	*Oidiodendrum*
Moraxella	*Paecylomyces*
Nocardia	*Penicillium*
Peptococcus	*Phialophora*
Pseudomonas	*Rhodosporidium*
Sarcina	*Rhodotorula*
Spherotilus	*Saccharomyces*
Spirillum	*Saccharomycopisis*
Streptomyces	*Scopulariopsis*
Vibrio	*Sporobolomyces*
Xanthomyces	*Torulopsis*
	Trichoderma
	Trichosporon

Source: US Congress, Office of Technology Assessment (1991). Bioremediation for Marine Oil Spills – Background Paper, OTA-RP-O-70. US Government Printing Office, Washington DC.

Time and energy are limiting factors that determine whether a microbe can efficiently mediate a chemical reaction, so catalytic processes are usually needed. Since an *enzyme* is a biological *catalyst*, these compounds (proteins) speed up the chemical reactions of degradation without themselves being used up. They do so by helping to break chemical bonds in the reactant molecules (see Figure 3.3). By lowering the activation energy needed, a biochemical reaction can be initiated sooner and easier than if the enzymes were not present (Figure 3.4). Indeed,

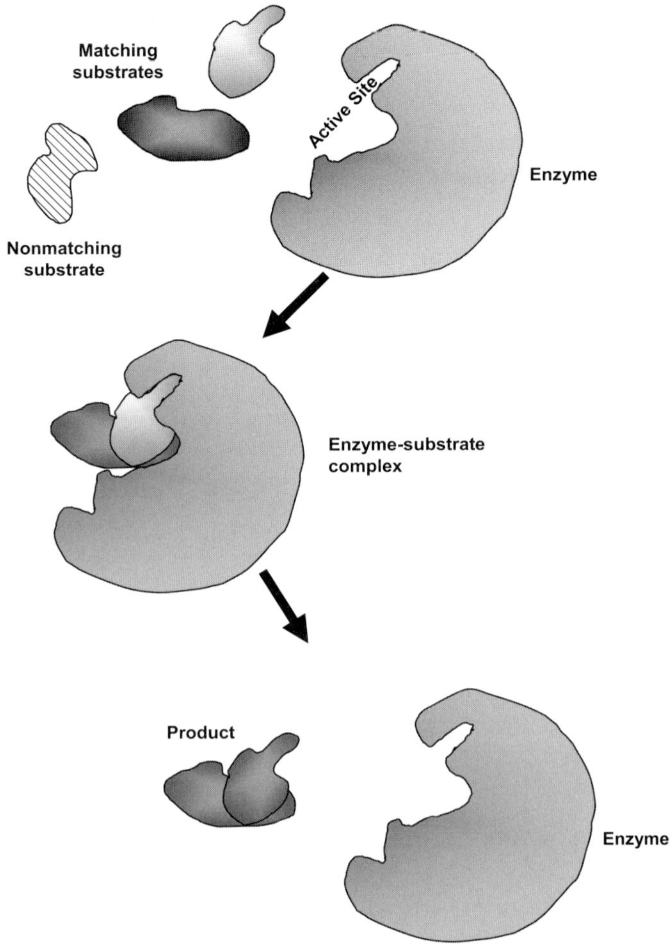

FIGURE 3.3
Substrates specific to the enzyme. The breakdown of the enzyme–substrate complex yields a product and the enzyme becomes available for another catalytic reaction. The nonmatching substrate cannot enter into a complex, so it is not affected by the presence of this particular enzyme. However, another enzyme may have the ''lock'' to match this substrate's ''key.''

enzymes play a very large part in microbial metabolism. They facilitate each step along the metabolic pathway. As catalysts, enzymes reduce the reaction's *activation energy*, which is the minimum free energy required for a molecule to undergo a specific reaction. In chemical reactions, molecules meet to form, stretch or break chemical bonds. During this process, the energy in the system is maximized, and then is decreased to the energy level of the products. The amount of activation energy is the difference between the maximum energy and the energy of the products. This difference represents the energy barrier that must be overcome for a chemical reaction to take place. Catalysts (in this case, microbial enzymes) speed up and increase the likelihood of a reaction by reducing the amount of energy, i.e. the activation energy, needed for the reaction.

Enzymes are usually quite specific. An enzyme is limited in the kinds of substrate that it will catalyze. Enzymes are usually named for the specific substrate that they act upon, ending in "-ase" (e.g. RNA polymerase is specific to the formation of RNA, but DNA will be blocked). Thus, the enzyme is a protein catalyst that has an active site at which the catalysis occurs. The enzyme can bind a limited number of substrate molecules (see Figure 3.3). The binding site is specific, i.e. other compounds do not fit the specific three-dimensional shape and structure of the active site (analogous to a specific key fitting a specific lock). The complex that results, i.e. the enzyme–substrate complex, yields a product and a free enzyme.

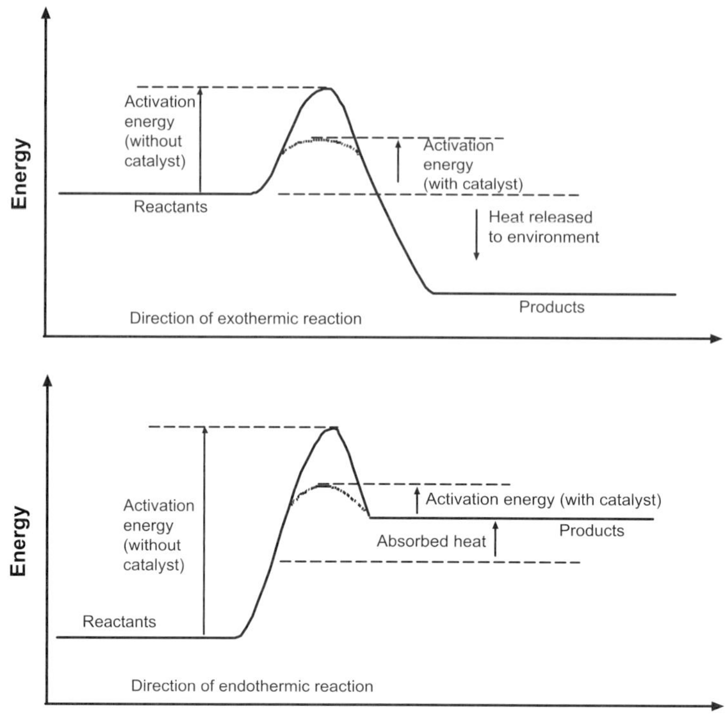

FIGURE 3.4
Effect of a catalyst on an exothermic reaction (*top*) and on an endothermic reaction (*bottom*).

The most common microbial coupling of exergonic and endergonic reactions by means of high-energy molecules to yield a net negative free energy is that of the nucleotide, adenosine triphosphate (ATP) with $\Delta G^* = -12$ to -15 kcal mol^{-1}. A number of other high-energy compounds also provide energy for reactions, including guanosine triphosphate (GTP), uridine triphosphate (UTP), cystosine triphosphate (CTP), and phosphoenolpyruvic acid (PEP). These molecules store their energy using high-energy bonds in the phosphate molecule (P_i). An example of free energy in microbial degradation is the possible first step in acetate metabolism by bacteria:

$$\text{Acetate} + \text{ATP} \rightarrow \text{acetyl-coenzyme A} + \text{ADP} + P_i \tag{3.9}$$

In this case, the P_i represents a release of energy available to the cell. Conversely, to add the phosphate to the two- P_i structure ADP to form the three- P_i ATP requires energy (i.e. it is an *endothermic process*). Thus, the microbe stores energy for later use when it adds the P_i to the ATP.

Dissolution

The measure of the amount of chemical that can dissolve in a liquid is called solubility. It is usually expressed in units of mass of solute (that which is dissolved) in the volume of solvent (that which dissolves). Usually, when scientists use the term "solubility" without any other attributes, they mean the measure of the amount of the solute in water, i.e. aqueous solubility. Otherwise, the solubility will be listed along with the solvent, such as solubility in benzene, solubility in methanol, or solubility in hexane. Solubility may also be expressed in mass per mass or volume per volume, represented as parts per million (ppm), parts per billion (ppb), or parts per trillion (ppt). Occasionally, solubility is expressed as a percent or in parts per thousand, however, this is uncommon for contaminants, and is usually reserved for nutrients and essential gases (e.g. percent carbon dioxide in water or ppt water vapor in the air).

The solubility of a compound is very important to environmental transport, including the transport of genetic material. The diversity of solubilities in various solvents is a fairly reliable indication of where one is likely to find the compound in the environment. For example, the

Table 3.2 **Solubility of tetrachlorodibenzo-*para*-dioxin in water and organic solvents**

Solvent	Solubility (mg L^{-1})	Reference
Water	1.93×10^{-5}	Podoll *et al.* (1986). *Environmental Science & Technology* 20: 490–492
Water	6.90×10^{-4}(25° C)	Fiedler *et al.* (1990). *Chemosphere* 20: 1597–1602
Methanol	10	International Agency for Research on Cancer [3] (IARC)
Lard oil	40	IARC
n-Octanol	50	IARC
Acetone	110	IARC
Chloroform	370	IARC
Benzene	570	IARC
Chlorobenzene	720	IARC
Orthochlorobenzene	1400	IARC

various solubilities of the most toxic form of dioxin, tetrachlorodibenzo-*para*-dioxin (TCDD), are provided in Table 3.2. Based on these solubility differences, if a bioreactor has been operating and releasing dioxins, one would expect TCDD to have a much greater affinity for sediment, organic particles, and the organic fraction of soils. The low water solubilities indicate that dissolved TCDD in the water column should be at only extremely low concentrations. But, as will be seen in the discussion regarding co-solvation, for example, other processes may override any single process, e.g. dissolution, in an environmental system.

POLARITY

A number of physicochemical characteristics of a substance come into play in determining its solubility. One is a substance's *polarity.* The polarity of a molecule is its unevenness in charge. Since the water molecule's oxygen and two hydrogen atoms are aligned so that there is a slightly negative charge at the oxygen end and a slightly positive charge at the hydrogen ends, and since "like dissolves like," polar substances have an affinity to become dissolved in water, and nonpolar substances resist being dissolved in water.

Increasing temperature, i.e. increased kinetic energy, in a system increases the velocity of the molecules, so that intermolecular forces are weakened. With increasing temperature, the molecular velocity becomes sufficiently large so as to overcome all intermolecular forces, so that the liquid boils (vaporizes). Intermolecular forces may be relatively weak or strong. The weak forces in liquids and gases are often called van der Waals forces.

Phase Partitioning

The previous discussion regarding dioxins in various environments is an example of how chemicals may divide their presence and concentrations in various compartments in the environment, i.e. *phase partitioning*. Phase partitioning is also sometimes called "phase distribution."

It is a principal subject matter of equilibrium physics and chemistry (and contrasted with *kinetic* physics and chemistry). If a compound has high aqueous solubility, i.e. it is easily dissolved in water under normal environmental conditions of temperature and pressure, it is *hydrophilic*. If, conversely, a substance is not easily dissolved in water under these conditions, it is said to be *hydrophobic*. Since many contaminants are organic (i.e. consist of molecules containing carbon-to-carbon bonds and/or carbon-to-hydrogen bonds), the solubility can be further differentiated as to whether under normal environmental conditions of temperature and pressure the substance is easily dissolved in organic solvents. If so the substance is said to be *lipophilic* (i.e., readily dissolved in lipids). If, conversely, a substance is not easily dissolved in organic solvents under these conditions, it is said to be *lipophobic*.

This affinity for either water or lipids underpins an important indicator of environmental partitioning; i.e., the octanol–water partition coefficient (K_{ow}). The K_{ow} is the ratio of a substance's concentration in octanol ($C_7H_{13}CH_2OH$) to the substance's concentration in water at equilibrium (i.e., the reactions have all reached their final expected chemical composition in a control volume of the fluid). Octanol is a surrogate for lipophilic solvents in general because it has degrees of affinity for both water and organic compounds, that is, octanol is *amphibilic*. Since the ratio forming the K_{ow} is $[C_7H_{13}CH_2OH]$: $[H_2O]$, then the larger the K_{ow} value, the more lipophilic the substance. Since the water concentration is in the denominator of the K_{ow} calculation, hydrophobic compounds will have relatively large K_{ow} values (small denominator = large number), whereas the K_{ow} of a compound that readily dissolves in water will be relatively small (large denominator = small number). Values for solubility in water and K_{ow} values of some important environmental compounds, along with their densities, are shown in Table 3.3.

Table 3.3 elucidates some additional aspects of solubility and organic/aqueous phase distribution. Water solubility is somewhat inversely related to K_{ow}, but the relationship is uneven. This results from the fact that various organic compounds are likely to have affinities for neither, either, or both the organic and the aqueous phases. Most compounds are not completely associated with either phase; i.e. they have some amount of amphibilicity.

Also, what seem to be minor structural changes to a molecule can make quite a difference in phase partitioning and in density. Even the isomers (i.e., same chemical composition with a different arrangement) vary in their K_{ow} values and densities (note the "1,1" versus "1,2" arrangements of chlorine atoms on 1,1-dichloroethane, and 1,2-dichloroethane, causes the former to have a slightly decreased density but twice the K_{ow} value than the latter. The location

Table 3.3 Solubility, octanol-water partitioning coefficient, and density values for some environmental pollutants

Chemical	Water solubility (mg L^{-1})	K_{ow}	Density (kg m^{-3})
Atrazine	33	724	
Benzene	1780	135	879
Chlorobenzene	472	832	1110
Cyclohexane	60	2754	780
1,1-Dichloroethane	4960	62	1180
1,2-Dichloroethane	8426	30	1240
Ethanol	Completely miscible	0.49	790
Toluene	515	490	870
Vinyl chloride	2790	4	910
Tetrachlorodibenzo-*para*-dioxin (TCDD)	1.9×10^{-4}	6.3×10^{6}	

Source: H.F. Hemond and E.J. Fechner-Levy (2000). *Chemical Fate and Transport in the Environment*. Academic Press, San Diego, CA; TCDD data from the NTP Chemical Repository, National Environmental Health Sciences Institute (2003); and US Environmental Protection Agency (2003). Technical Fact Sheet on Dioxin (2,3,7,8-TCDD).

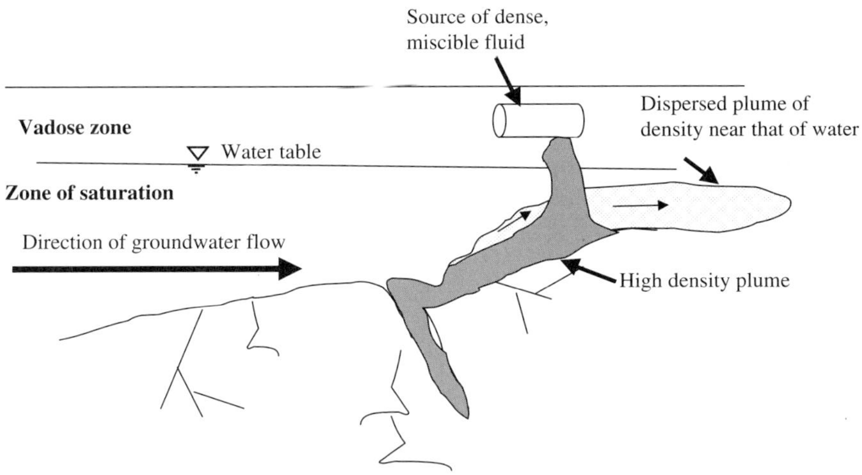

FIGURE 3.5
Hypothetical plume of dense, highly hydrophilic fluid. Based on information provided by M.N. Sara (1991). Groundwater monitoring system design. In: D.M. Nielsen (Ed.), *Practical Handbook of Ground-Water Monitoring.* Lewis Publishers, Chelsea, MI.

of the chlorine atoms alone accounts for a significant difference in water solubility in the two compounds.

The relationship between density and organic/aqueous phase partitioning is very important to pollutant transport, as shown in Figure 3.5. The transport of the nonaqueous phase liquids (NAPLs) through the vadose zone assumes that the NAPLs have extremely high K_{ow} values and extremely low water solubility. That is, they have a greater affinity for lipids than for water. As the aqueous solubility of a substance increases, its flow will increasingly follow the water flow lines.

When a dense, miscible fluid seeps into the zone of saturation, the dense contaminants move downward. The continuation and direction of movement of contaminants upon reaching the bottom of the aquifer are dictated by the shape of the pollutant plume and the slope of the underlying bedrock or other relatively impervious layer, which may be in a direction other than the flow of the groundwater in the aquifer. Solution and dispersion near the boundaries of the plume will have a secondary plume that will generally follow the general direction of groundwater flow. The physics of this system points how that deciding where the plume is heading will entail more than the fluid densities, including solubility and phase partitioning. So, monitoring wells will need to be installed upstream and downstream from the source.

If a source consists entirely of a light, hydrophilic fluid, the plume may be characterized as shown in Figure 3.6. Low-density organic fluids, however, often are highly volatile; i.e., their

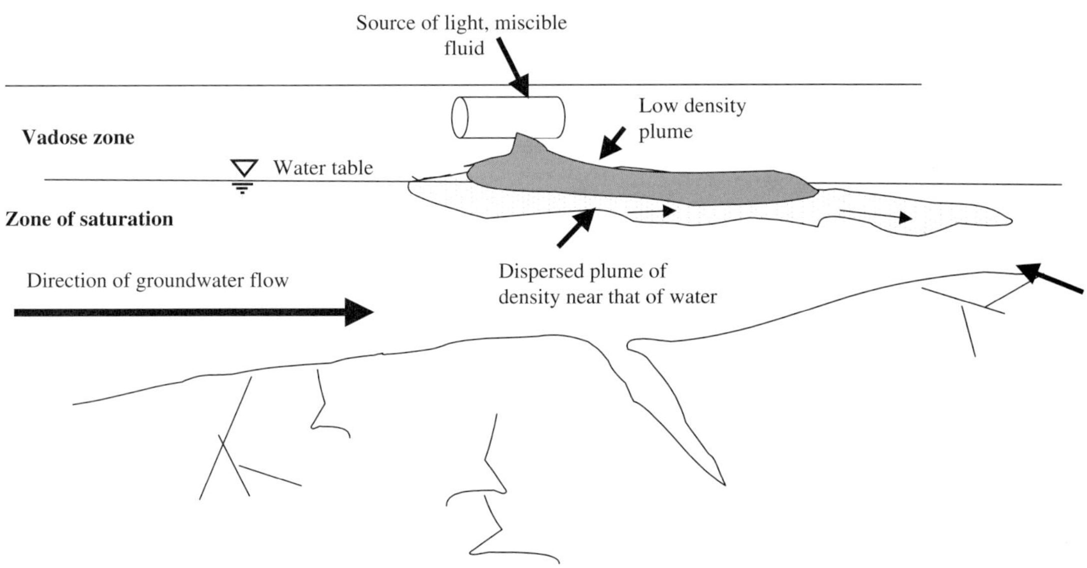

FIGURE 3.6
Hypothetical plume of light, highly hydrophilic fluid.
Based on information provided by: M.N. Sara (1991). Groundwater monitoring system design. In: D.M. Nielsen (Ed.), *Practical Handbook of Ground-Water Monitoring.* Lewis Publishers, Chelsea, MI.

vapor pressures are sufficiently high to change phases from liquid to gas. Thus, vapor pressure is another extremely important physicochemical property of environmental fluids that determines direction and quantity of movement, which must be considered along with density and solubility.

An important process in plume migration is that of *co-solvation*, the process where a substance is first dissolved in one solvent and then the new solution is mixed with another solvent. As mentioned, with increasing aqueous solubility, a pollutant will travel along the flow lines of the ground or surface water. However, even a substance with low aqueous solubility can follow the flow under certain conditions. Even a hydrophobic compound like a chlorinated benzene (called a dense non-aqueous phase liquid, DNAPL), which has very low solubility in pure water, can migrate into and within water bodies if it is first dissolved in an alcohol or an organic solvent (e.g. toluene). So, a DNAPL will migrate downward because its *density* is less than that of water and is transported in the solvent which has undergone co-solvation with the water. Likewise, the ordinarily lipophilic compound can be transported in the vadose zone or upper part of the zone of saturation where it undergoes co-solvation with water and a light non-aqueous phase liquid (LNAPL), e.g. toluene.

THERMODYNAMICS IN ABIOTIC AND BIOTIC SYSTEMS

Biomes, habitats, and individual organisms and their components are thermodynamic systems. In the context of thermodynamics, a system is simply a sector or region in space or some parcel of a sector that has at least one substance that is ordered into phases. Unfortunately, the English language has numerous connotations of "systems". Even scientists have various definitions. For example, a more general understanding of scientists and technicians is that a "system" is a method of organization, e.g. from smaller to larger aggregations. The "ecosystem" and the "organism" are examples of both types of systems. They consist of physical phases and order (e.g. producer–consumer–decomposer; predator–prey; individual–association–community; or cell–tissue–organ–system). They are also a means for understanding how matter and energy move and change within a parcel of matter.

In the previous discussion of cellular metabolism, a distinction was drawn between closed and open systems. Both exist and are important in the environment. Recall that a closed system does not allow material to enter or leave the system (engineers refer to a closed system as a "control mass"). The open system allows material to enter and leave the systems (such a system is known as a *control volume*).

Another thermodynamic concept is that of the *property*. A property is some trait or attribute that can be used to describe a system and to differentiate that system from others. A property must be able to be stated at a specific time independently of its value at any other time and unconstrained by the process that induced the condition (state). An intensive property is independent of the system's mass (such as pressure and temperature). An extensive property is proportionality to the mass of the system (such as density or volume). Dividing the value of an extensive property by the system's mass gives a "specific property," such as *specific heat, specific volume*, or *specific gravity*.

The thermodynamics term for the description of the change of a system from one state (e.g. equilibrium) to another is known as a *process*. Processes may be reversible or irreversible, they may be adiabatic (no gain or loss of heat, so all energy transfers occur through work interactions). Other processes included isometric (constant volume), isothermal (constant temperature), isobaric (constant pressure), isentropic (constant entropy), and isenthalpic (constant enthalpy).

Volatility/solubility/density relationships

Many substances important in biotechnology and that are important in environmental systems exist in various physical states under environmental conditions. In particular,

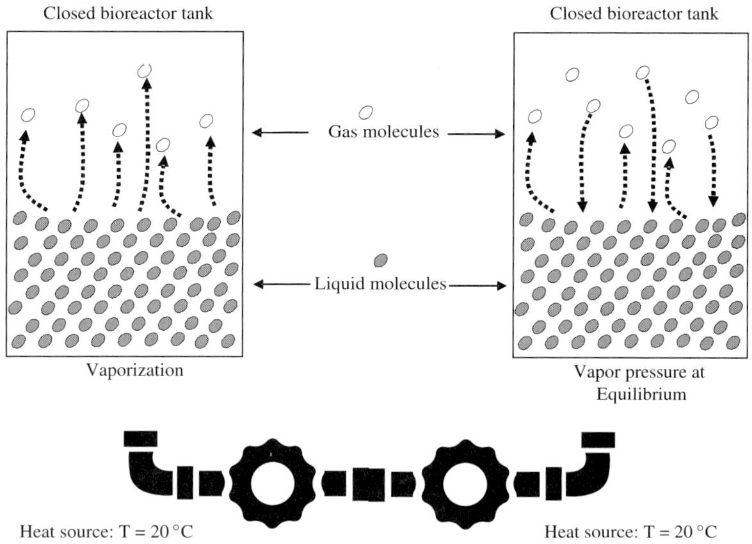

FIGURE 3.7
Bioreactor vapor pressure of a fluid during vaporization and at equilibrium. A portion of a substance in an evacuated, closed container with limited headspace will vaporize. The pressure in the space above the liquid increases from zero and eventually stabilizes at a constant value. This value is what is known as the vapor pressure of that substance. Substances not in a closed container (i.e., infinitely available headspace) will also vaporize, but will continue to vaporize until all of the substance has partitioned to the gas phase.

substances of low molecular weight and certain molecular structures have high enough vapor pressures that they can exist in either the liquid or gas phases under environmental conditions.

The vapor pressure (P^0) of a contaminant in the liquid or solid phase is the pressure that is exerted by its vapor when the liquid and vapor are in dynamic equilibrium (see Figure 3.7). This is actually an expression of the *partial pressure* of a chemical substance in a gas phase that is in equilibrium with the non-gaseous phases. The *ideal gas law* can be used to convert P^0 into moles of vapor per unit volume:

$$\frac{n}{V} = \frac{P^0}{RT} \tag{3.10}$$

where: V = volume of the container

n = number of moles of chemical

R = molar gas constant

$\frac{n}{V}$ is the gas phase concentration (moles L^{-1}) of the chemical.

The P^0 that is published in texts and handbooks is an expression of a chemical in its pure form; that is, P^0 is the force per unit area exerted by a vapor in an equilibrium state with its pure solid, liquid, or solution at a given temperature (see Table 3.4). This situation is seldom encountered in the real world of environmental engineering, so adjustments have to be made to estimates based on published P^0 values.

Therefore, P^0 is a measure of a substance's propensity to evaporate (Figure 3.7), increasing exponentially with an increase in temperature (see Figure 3.8), which means that a statement of P^0 must always be accompanied by a temperature for that P^0. For example, the P^0 of trichloroethene at 21.0 °C is about 7.5 kP, but at 25.5 °C rises to about 9.5 kP [4]. Thus, p^0 is a relative measure of a substance's likely chemical volatility in the environment. As such, P^0 is a component of partitioning coefficients and volatilization rate constants.

Volatile organic compounds (VOCs) have P^0 values greater than 10^{-2} kP; semi-volatile organic compounds (SVOCs) have P^0 values between 10^{-5} and 10^{-2} kP, and the so-called "nonvolatile organic compounds" have P^0 values less than 10^{-5}.

Table 3.4 Vapor pressures at 20 °C for some environmental pollutants

Chemical	Vapor pressure (kP) at 0 °C	Vapor pressure (kP) at 20 °C	Vapor pressure (kP) at 25 °C	Vapor pressure (kP) at 50 °C
Atrazine		4.0×10^{-8}		
Benzene	3.3		12.7	36.2
Chlorobenzene			1.6	
Cyclohexane			13.0	36.3
1,1-Dichloroethane	9.6		30.5	79.2
1,2-Dichloroethane	2.8	9.2	10.6	31.4
Ethanol	1.5		7.9	29.5
Toluene			3.8	
Vinyl chloride	170	344	355	
Tetrachlorodibenzo-*para*-dioxin (TCDD)		4.8×10^{-9}	5.6×10^{-3}	

Sources: Column 2: H. Hemond and E. Fechner-Levy (2000). *Chemical Fate and Transport in the Environment*. Academic Press, San Diego, CA; Columns 3 and 4: D. Lide (Ed.) (1995). *CRC Handbook of Chemistry and Physics*, 76th Edition. CRC, Boca Raton, FL; TCDD data from the NTP Chemical Repository, National Environmental Health Sciences Institute (2003); and US Environmental Protection Agency (2003). Technical Fact Sheet on Dioxin (2,3,7,8-TCDD).

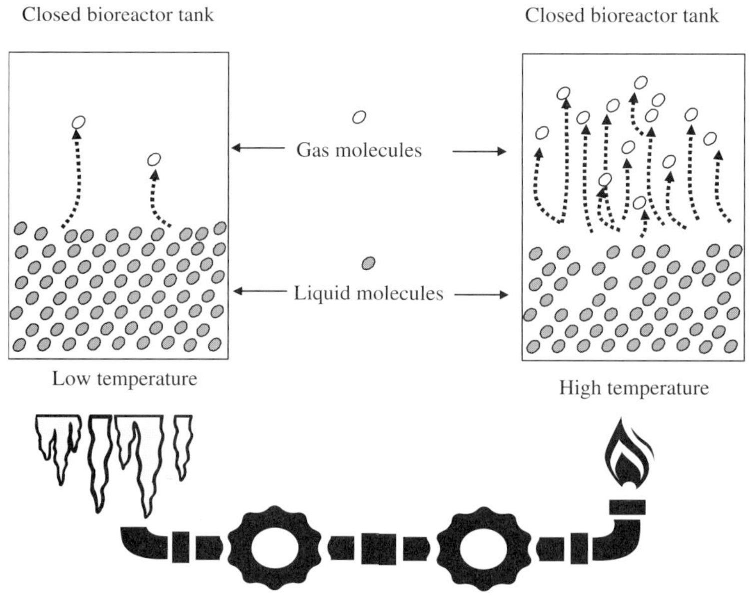

FIGURE 3.8
Bioreactor vapor pressure increases in direct proportion with increasing temperature. Readings of vapor pressure values must always be accompanied by the temperature at which each measured vapor pressure is occurring.

Any substance, depending upon the temperature, can exist in any phase. However, in many environmental contexts, a vapor refers to a substance that is in its gas phase but under typical environmental conditions exists as a liquid or solid under a given set of conditions. Although the pressure in the closed container in Figure 3.8 is constant, molecules of the vapor will continue to condense into the liquid phase and molecules of the liquid will continue to evaporate into the vapor phase. However, the rate of these two processes within each vessel is

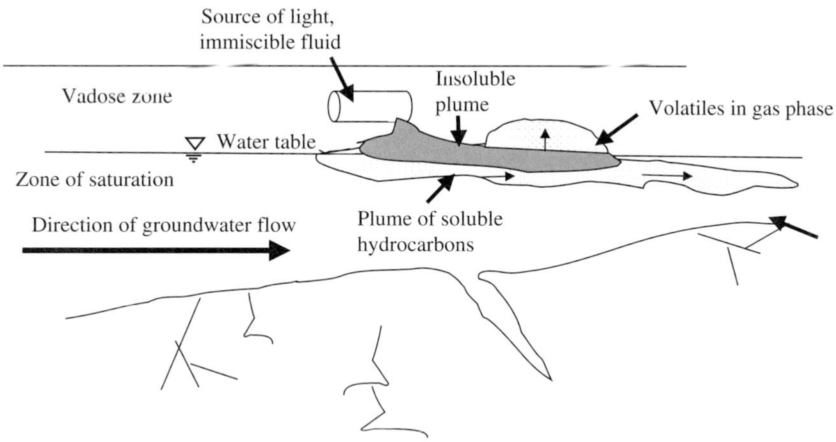

FIGURE 3.9
Hypothetical plume of hydrophobic fluid.
Source: M.N. Sara (1991) Groundwater monitoring system design. In: D.M. Nielsen (Ed.), *Practical Handbook of Ground-Water Monitoring.* Lewis Publishers, Chelsea, MI.

equal, meaning no net change in the amount of vapor or liquid. This is an example of dynamic equilibrium, or *equilibrium vapor pressure.*

At the boiling point temperature, a liquid's vapor pressure is equal to the external pressure. Generally, the higher the substance's vapor pressure at a given temperature, the lower the boiling point. So, compounds with high vapor pressures are classified as "volatile," meaning they form higher concentrations of vapor above the liquid [5]. This means that they are potential air pollutants from storage tanks, etc., and it also means that they can present problems to first responders. For example, if a volatile compound is also flammable, there is a fire and explosion hazard higher than if the substance were less volatile.

In groundwater, if a source of LNAPLs includes a relatively insoluble substance that distributes between liquid and gas phases (see Figure 3.9), the fluid will infiltrate and move along the water table along the top of the zone of saturation, just above the capillary fringe. (Capillarity is discussed in detail in chapter 7.) However, some of the contaminant fluid lags behind the plume and slowly solubilizes in the pore spaces of the soil and unconsolidated material. These more soluble forms of the fluid find their way to the zone of saturation and move with the general groundwater flow. The higher vapor pressures of portions of the plume will lead to upward movement of volatile compounds in the gas phase. Thus, this system has at least three plumes as a result of solubility, density, and vapor pressure of the fluid components and environmental conditions.

Environmental balances

Although many possible outcomes can occur after a substance is released into the environment, the possibilities can fall into three basic categories:

- the chemical may remain where it is released and retain its physicochemical characteristics (at least within a specified time);
- the substance may be transported to another location; or
- the substance may be changed chemically, known as the *transformation* of the chemical.

This is a restatement of the conservation laws mentioned earlier. If we focus on mass within a control volume, it is a statement of *mass balance.* Every molecule of mass moving into and out of the control volume must be accounted for, as well as any chemical changes to the contaminant that take place within the control volume. A control volume may be a nice, neat geometric shape, such as a cube (Figure 3.10A) through which contaminant fluxes are calculated. However, a control volume can also be an organism (e.g. what it eats, metabolizes, and eliminates) or an ecosystem, such as the pond in Figure 3.10B).

Much of the work of environmental assessment is an accounting for the mass on both sides of the mass balance equation. The change in storage of a substance's mass is equal to the difference between the mass of the chemical transported into the system less the mass of the chemical transported out of the system. However, the actual chemical species transported in may be different from what is transported out due to the chemical and biological reactions taking place within the control volume. So, the mass balance equation may be written as:

$$\text{Accumulation or loss of contaminant A} = \text{Mass of A transported in} - \text{Mass of A transported out} \pm \text{Reactions} \quad (3.11)$$

The reactions may be either those that generate chemical A (i.e. *sources*), or those that destroy chemical A (i.e. *sinks*).

The entering mass transported equals the inflow to the system that includes pollutant discharges, transfer from other control volumes and other media (for example, if the control volume is soil, the water and air may contribute mass of chemical A), and formation of chemical A by abiotic chemistry and biological transformation. Conversely, the outflow is the mass transported out of the control volume, which includes uptake, by biota, transfer to other compartments (e.g. volatilization to the atmosphere) and abiotic and biological degradation of chemical A.

The rate of change of mass in a control volume is equal to the rate of chemical A transported in less the rate of chemical A transported out, plus the rate of production from sources, and

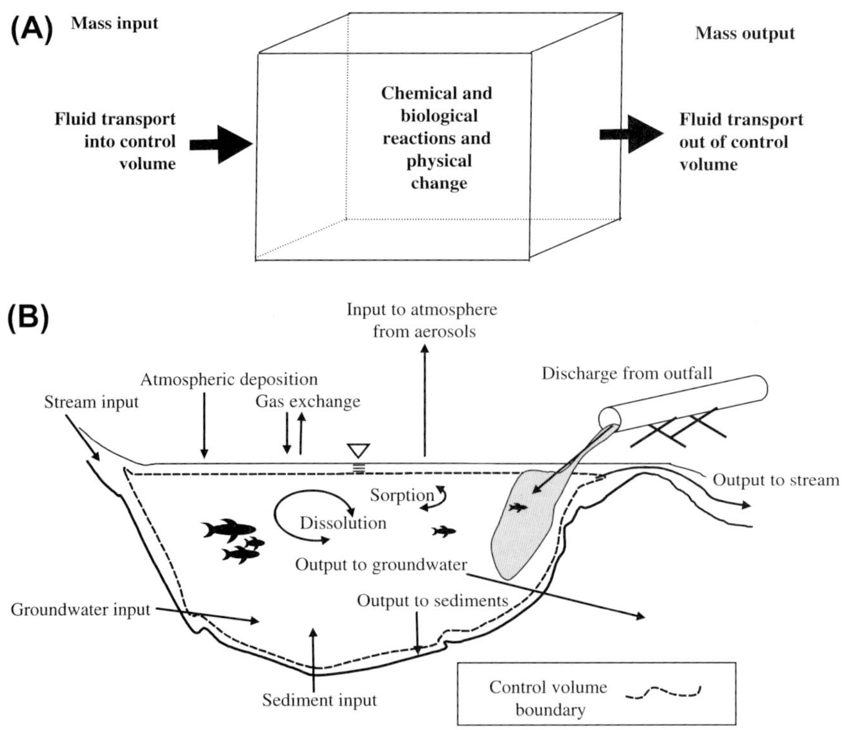

FIGURE 3.10
(A) Control volume of an environmental matrix (e.g. soil, sediment, or other unconsolidated material) or fluid (e.g. water, air, or blood). (B) A pond. Both volumes have equal masses entering and exiting, with transformations and physical changes taking place within the control volume.

minus the rate of elimination by sinks. Stated as a differential equation, the rate of change contaminant A is:

$$\frac{d[A]}{dt} = -v \cdot \frac{d[A]}{dx} + \frac{d}{dx}(D \cdot \frac{d[A]}{dx}) + r \tag{3.12}$$

where: v = fluid velocity

$\frac{d[A]}{dx}$ = concentration gradient of chemical A

r = internal sinks and sources within the control volume

These rates operate at various scales. For example, Eq. 3.12 can be applied from the cell to the planet. It is even the basis for pharmacokinetic and pharmacodynamic modeling (see Discussion Box). Much goes into variable *r*. This leads to a discussion of the transport mechanisms responsible for the movement of a contaminant within a system.

DISCUSSION BOX

Bioengineering Within the Organism: Pharmacodynamics

Two "control volumes" that are commonly considered in environmental exposure and risk assessments are the organism and a defined environmental volume around the organism. Thus, scientists commonly calculate mass balances for the classic control cube and adapt it to the environment (see Figure 3.10A). Not surprisingly, the most studied control volume organism is the human. Humans meet the same criteria as our cube and pond, in that we must fully account for the pollutant mass in and out, as well as the processes that occur within these control volumes.

Exposure assessments in human populations must account for the amount of a contaminant that a person contacts. However, just because a contaminant finds its way to the individual's mouth, nose, or skin, does not necessarily mean the contaminant will harm the individual, or that the contaminant will find its way to a susceptible cell. Organisms have numerous protective mechanisms in their metabolic processes. For example, many compounds are converted to harmless metabolites. An example of such biological activation or *bioactivation*, is likely occurring after an organism is exposed to the toxic compound benzo(a)-pyrene (See Figure 3.20); whereupon within the cell an epoxide is formed as benzo(a)pyrene is metablized – meaning that the likely carcinogen is the epoxide, rather than the parent compound benzo(a)pyrene. Thus, conducting a mass balance in an organism is complicated, within numerous "black boxes". For example, many researchers believe that the epoxide formed trying to metabolize benzo(a)pyrene is a (the) likely carcinogen. Conducting a mass balance in an organism is complicated.

It is sometimes advantageous to look at the human being as a control volume for the mass balance of a contaminant. *Body burden* is the total amount of the contaminant in the body at a given time of measurements. This is an indication of the behavior of the contaminant in the control volume (i.e. the person). Some contaminants accumulate in the body and are stored in fat or bone or they simply are metabolized more slowly and tend to be retained for longer time periods. This concept is at the core of what are known as *physiologically based pharmacokinetic* (PBPK) models. These models attempt to describe what happens to a chemical after it enters the body, showing its points of entry, its distribution (i.e. where it goes after entry), how it is altered by the body, and how it ultimately is eliminated by the body. This is almost identical to the processes that take place in a stream, or a wetland, or other system.

Fugacity

A basic concept of bioengineering is that substances tend to have affinity for certain compartments in abiotic and biotic systems. Thus, bioengineers must understand the specific partitioning relationships that control the "leaving" and "gaining" of compounds among water bodies, within soil and sediment matrices, in and on particles, in the atmosphere, and within organic tissues. In designing biotechnology, the engineer needs to apply these partitioning relationships to estimate and model where a substance will go within a designed system (e.g. a bioreactor) and after it is released. Basic relationships between sorption, solubility, volatilization, and organic carbon–water partitioning are respectively expressed by coefficients of sorption (distribution coefficient, K_D, or solid-water partition coefficient, K_p), dissolution or solubility coefficients, air–water partitioning (and the Henry's law (K_H) constant), and organic carbon–water (K_{oc}).

In biochemodynamics, the environment can be subdivided into finite compartments. Recall that the mass of the contaminant entering and the mass leaving a control volume must be balanced by what remains within the control volume. But, environmental systems are a cascade of control volumes. Within each control volume, an individual compartment may be a gainer or loser of the contaminant, nutrient or other compound mass, but the overall mass must balance.

The generally inclusive term for these compartmental changes is known as *fugacity* or the "fleeing potential" of a substance. It is the propensity of a chemical to escape from one type of environmental compartment to another. Combining the relationships between and among all of the partitioning terms is one means of modeling chemical transport in the environment [6]. This is accomplished by using thermodynamic principles and, hence, fugacity is a thermodynamic term.

The simplest biochemodynamic approach addresses each compartment where a contaminant is found in discrete phases of air, water, soil, sediment, and biota (see Figure 3.11). However, a complicating factor in environmental chemodynamics is that even within a single compartment, a contaminant may exist in various phases (e.g. dissolved in water and sorbed to a particle in the solid phase). Interphase reactions, or the physical interactions of the contaminant at the interface between each compartment, determine the amount of any substance in the environment. Within a compartment, a contaminant may remain unchanged for a designated time period, or it may move physically, or it may be transformed chemically into another substance.

Indeed, in many cases all three mechanisms will take place. A mass fraction will remain unmoved and unchanged. Another fraction remains unchanged but is transported to a different compartment. Another fraction becomes chemically transformed with all products remaining products staying in the compartment where they were generated. And, a fraction of the original contaminant is transformed and then moved to another compartment. So, upon release from a source, the contaminant moves as a result of motion and changes as a result of thermodynamics. We were introduced to fugacity principles in our discussion of the fluid properties K_{ow} and vapor pressure and the partial pressure of gases.

Fugacity requires that at least two phases be in contact with the contaminant. For example, recall that the K_{ow} value is an indication of a compound's likelihood to exist in the organic versus aqueous phase. If a substance is dissolved in water and the water comes into contact with another substance, e.g. octanol, the substance will have a tendency to move from the water to the octanol. Its octanol-water partitioning coefficient reflects just how much of the substance will move until the aqueous and organic solvents (phases) will reach equilibrium. So, for example, in a spill of equal amounts of the polychlorinated biphenyl, decachlorobiphenyl (log K_{ow} of 8.23) and the pesticide chlordane (log K_{ow} of 2.78), the PCB has much greater affinity for the organic phases than does the chlordane (more than five orders of magnitude). This does not mean that a great amount of either of the compounds is likely to stay in the water column, since they are both hydrophobic, but it does mean that they will vary in the time and mass of each contaminant moving between phases. The rate (kinetics) is different, so the time it takes for the PCB and chlordane to reach equilibrium will be different. This can be visualized by plotting the concentration of each compound with time (see Figure 3.12). When the concentrations plateau, the compounds are at equilibrium with their phase.

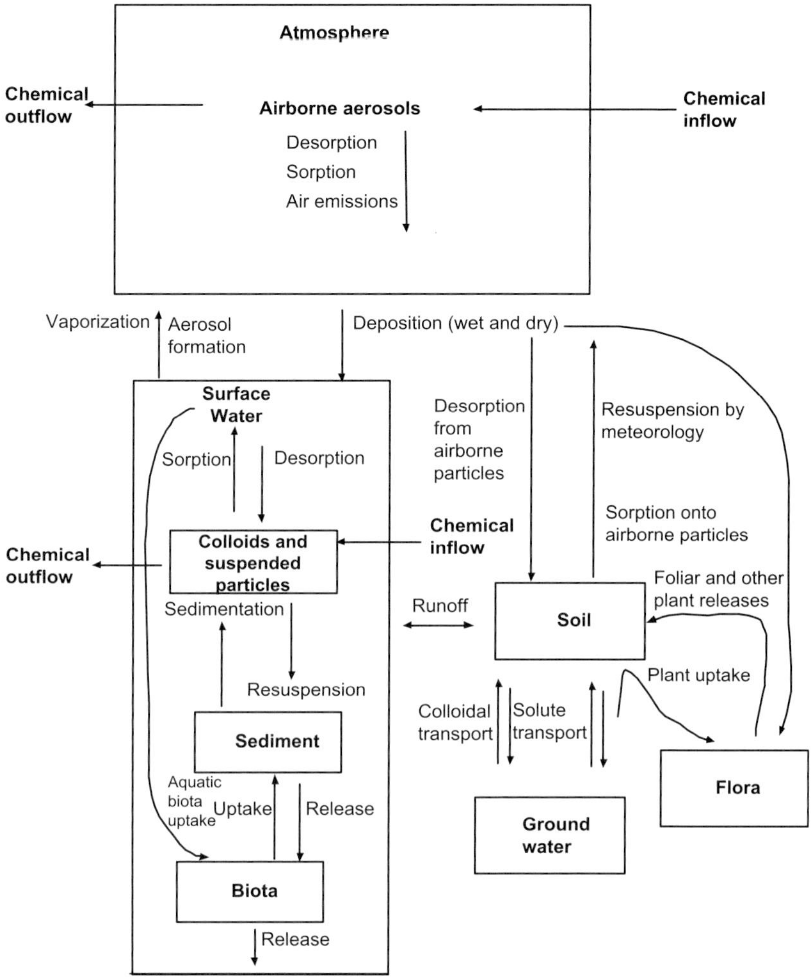

FIGURE 3.11
Simple biochemodynamics of a substance through environmental media, including compartmental uptake, transport, transformation and various fates. Equilibrium constants (e.g. partitioning coefficients) must be developed for each arrow and steady-state conditions may not be assumed, reaction rates and other chemical kinetics must be developed for each arrow and box.

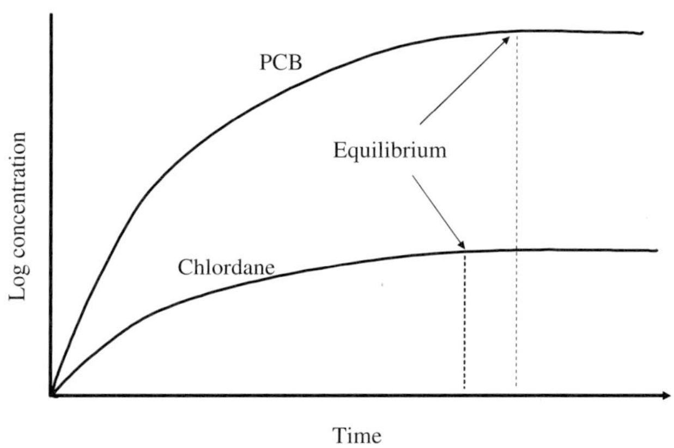

FIGURE 3.12
Bioreactor relative concentrations of a polychlorinated biphenyl (PCB) and chlordane in octanol with time.

Sorption

Sorption is arguably the most important transfer process that determines how bioavailable or toxic a compound will be in surface waters. The physicochemical transfer [7] of a chemical, A, from liquid to solid phase is expressed as:

$$A_{(solution)} + \text{solid} = A - \text{solid} \tag{3.13}$$

The interaction of the solute (i.e. the chemical being sorbed) with the surface of a solid surface can be complex and dependent upon the properties of the chemical and the water. Other fluids are often of such small concentrations that they do not determine the ultimate solid-liquid partitioning. While, it is often acceptable to consider "net" sorption, let us consider briefly the four basic types or mechanisms of sorption:

Adsorption is the process wherein the chemical in solution attaches to a solid surface, which is a common sorption process in clay and organic constituents in soils. This simple adsorption mechanism can occur on clay particles where little carbon is available, such as in groundwater.

Absorption is the process that often occurs in porous materials so that the solute can diffuse into the particle and be sorbed onto the inside surfaces of the particle. This commonly results from short-range electrostatic interactions between the surface and the contaminant.

Chemisorption is the process of integrating a chemical into a porous materials surface via a chemical reaction. In soil, this is usually the result of a covalent reaction between a mineral surface and the contaminant.

Ion exchange is the process by which positively charged ions (cations) are attracted to negatively charged particle surfaces or negatively charged ions (anions) are attracted to positively charged particle surfaces, causing ions on the particle surfaces to be displaced. Particles undergoing ion exchange can include soils, sediment, airborne particulate matter, or even biota, such as pollen particles. Cation exchange has been characterized as being the second most important chemical process on earth, after photosynthesis. This is because the cation exchange capacity (CEC), and to a lesser degree anion exchange capacity (AEC) in tropical soils, is the means by which nutrients are made available to plant roots. Without this process, the atmospheric nutrients and the minerals in the soil would not come together to provide for the abundant plant life on planet earth [8].

These four types of sorption are a mix of physics and chemistry, and are important to biotechnology, since they function at surfaces and are crucial to biofilm and molecular exchanges (see Chapter 7, Discussion Box: Biochemodynamic Films). The first two types of sorption are predominantly controlled by physical factors, and the second two are combinations of chemical reactions and physical processes. Generally, sorption reactions affect three processes [9] in biochemodynamic systems:

The chemical contaminant's transport in water due to distributions between the aqueous phase and particles.

The aggregation and transport of the contaminant as a result of electrostatic properties of suspended solids.

Surface reactions such as dissociation, surface-catalysis, and precipitation of the chemical contaminant.

When a contaminant enters soil, some of the chemical remains in soil solution and some is adsorbed onto the surfaces of the soil particles. Sometimes this sorption is strong due to cations adsorbing to the negatively charged soil particles. In other cases the attraction is weak. Sorption of chemicals on solid surfaces needs to be understood because they hold onto contaminants, not allowing them to move freely with the pore water or the soil solution.

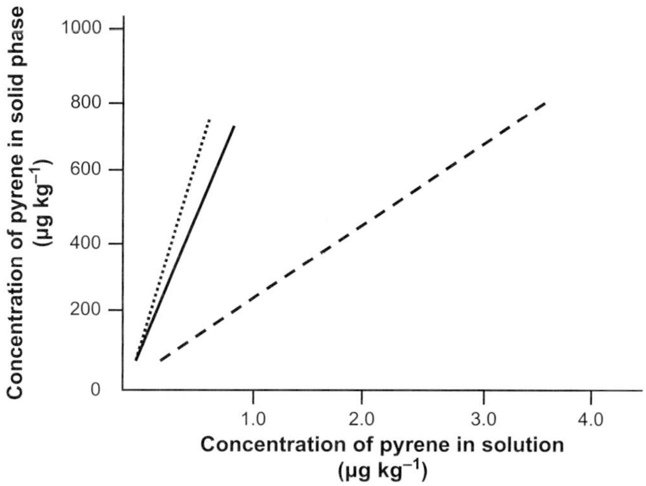

FIGURE 3.13
Three experimentally determined sorption isotherms for the polycyclic aromatic hydrocarbon pyrene.
Source: J. Hassett and W. Banwart (1989). The sorption of nonpolar organics by soils and sediments. In: B. Sawhney and K. Brown (Eds), *Reactions and Movement of Organic Chemicals in Soils.* Soil Science Society of America Special Publication 22, p. 35.

Therefore sorption slows that rate at which substances move downwardly through the soil profile.

Biomolecules and xenobiotic compounds eventually establish a balance between the mass on the solid surfaces and the mass that is in solution. Molecules will migrate from one phase to another to maintain this balance. The properties of both the chemical and the soil (or other matrix) will determine how and at what rates the molecules partition into the solid and liquid phases. These physicochemical relationships, known as *sorption isotherms*, are found experimentally. Figure 3.13 illustrates three isotherms for pyrene from experiments using different soils and sediments.

The *x*-axis shows the concentration of pyrene dissolved in water, and the *y*-axis shows the concentration in the solid phase. Each line represents the relationship between these concentrations for a single soil or sediment. A straight-line segment through the origin represents the data well for the range of concentrations shown. Not all portions of an isotherm are linear, particularly at high concentrations of the contaminant. Linear chemical partitioning can be expressed as:

$$S = K_D C_W \tag{3.14}$$

where: S = concentration of contaminant in the solid phase (mass of solute per mass of soil or sediment)

C_W = concentration of contaminant in the liquid phase (mass of solute per volume of pore water)

K_D = partition coefficient (volume of pore water per mass of soil or sediment) for this contaminant in this soil or sediment.

For many soils and chemicals, the partition coefficient can be estimated using:

$$K_D = K_{OC} OC \tag{3.15}$$

where K_{OC} = organic carbon partition coefficient (volume of pore water per mass of organic carbon) and OC = soil organic matter (mass of organic carbon per mass of soil).

This relationship is a very useful tool for estimating K_D from the known K_{OC} of the contaminant and the organic carbon content of the soil horizon of interest. The actual derivation of K_D is:

$$K_D = C_S (C_W)^{-1} \tag{3.16}$$

where C_S is the equilibrium concentration of the solute in the solid phase and C_W is the equilibrium concentration of the solute in the water.

Therefore, K_D is a direct expression of the partitioning between the aqueous and solid (soil or sediment) phases. A strongly sorbed chemical like a dioxin or the banned pesticide DDT can have a K_D value exceeding 10^6. Conversely, a highly hydrophilic, miscible substance like ethanol, acetone, or vinyl chloride, will have K_D values less than 1. This relationship between the two phases demonstrated by Eq. 3.16 and Figure 3.14 is the *Freundlich sorption isotherm*:

$$C_{sorb} = K_F C^n \tag{3.17}$$

where C_{sorb} is the concentration of the sorbed contaminant; i.e., the mass sorbed at equilibrium per mass of sorbent, and K_F is the Freundlich isotherm constant. The exponent determines the linearity or order of the reaction. Thus, if $n = 1$, then the isotherm is linear; meaning the more of the contaminant in solution, the more would be expected to be sorbed to surfaces. For values of $n < 1$, the amount of sorption is in smaller proportion to the amount of solution and, conversely, for values of $n > 1$, a greater proportion of sorption occurs with less contaminant in solution. These three isotherms are shown in Figure 3.14. Also note that if $n = 1$, then Eq. 3.17 and the Freundlich sorption isotherm are identical.

Research has shown that when organic matter content is elevated in soil and sediment, the amount of a contaminant that is sorbed is directly proportional to the soil/sediment organic matter content. This allows us to convert the K_D values from those that depend on specific soil or sediment conditions to those that are soil/sediment independent sorption constants, K_{OC}:

$$K_{OC} = K_D(f_{OC})^{-1} \tag{3.18}$$

where f_{OC} is the dimensionless weight fraction of organic carbon in the soil or sediment. The K_{OC} and K_D have units of mass per volume. Table 3.5 provides the log K_{OC} values that are calculated from chemical structure and those measured empirically for several organic compounds, and compares them to the respective K_{ow} values.

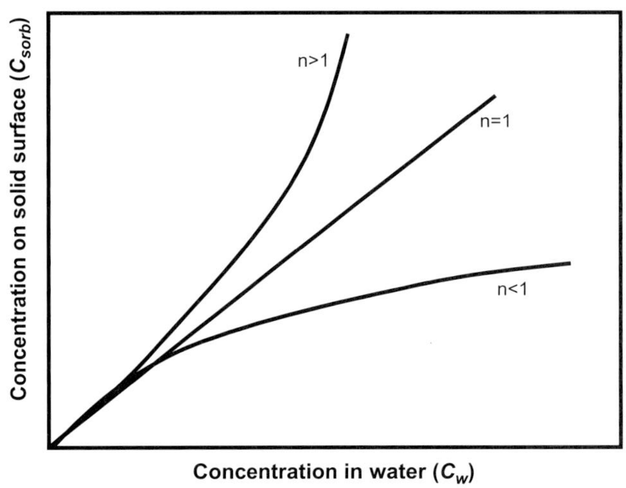

FIGURE 3.14

Hypothetical Freundlich isotherms with exponents (n) less than, equal to, and greater than 1, as applied to the equation $C_{sorb} = K_F C^n$.

Sources: R. Schwarzenbach, P. Gschwend and D. Imboden (1993). *Environmental Organic Chemistry*. John Wiley & Sons, Inc, New York, NY; and H.F. Hemond and E.J. Fechner-Levy (2000). *Chemical Fate and Transport in the Environment*. Academic Press, San Diego, CA.

Table 3.5 Calculated and experimental organic carbon coefficients (K_{ow}) for selected contaminants found at hazardous waste sites

Chemical	log K_{ow}	Calculated		Measured	
		log K_{oc}	K_{oc}	log K_{oc}	K_{oc} (geomean)
Benzene	2.13	1.77	59	1.79	61.7
Bromoform	2.35	1.94	87	2.10	126
Carbon tetrachloride	2.73	2.24	174	2.18	152
Chlorobenzene	2.86	2.34	219	2.35	224
Chloroform	1.92	1.60	40	1.72	52.5
Dichlorobenzene,1,2-*(o)*	3.43	2.79	617	2.58	379
Dichlorobenzene,1,4-*(p)*	3.42	2.79	617	2.79	616
Dichloroethane,1,1-	1.79	1.50	32	1.73	53.4
Dichloroethane,1,2-	1.47	1.24	17	1.58	38.0
Dichloroethylene,1,1-	2.13	1.77	59	1.81	65
Dichloroethylene, trans -1,2-	2.07	1.72	52	1.58	38
Dichloropropane,1,2-	1.97	1.64	44	1.67	47.0
Dieldrin	5.37	4.33	21,380	4.41	25,546
Endosulfan	4.10	3.33	2,138	3.31	2,040
Endrin	5.06	4.09	12,303	4.03	10,811
Ethylbenzene	3.14	2.56	363	2.31	204
Hexachlorobenzene	5.89	4.74	54,954	4.90	80,000
Methyl bromide	1.19	1.02	10	0.95	9.0
Methyl chloride	0.91	0.80	6	0.78	6.0
Methylene chloride	1.25	1.07	12	1.00	10
Pentachlorobenzene	5.26	4.24	17,378	4.51	32,148
Tetrachloroethane,1,1,2,2-	2.39	1.97	93	1.90	79.0
Tetrachloroethylene	2.67	2.19	155	2.42	265
Toluene	2.75	2.26	182	2.15	140
Trichlorobenzene,1,2,4-	4.01	3.25	1,778	3.22	1,659
Trichloroethane,1,1,1-	2.48	2.04	110	2.13	135
Trichloroethane,1,1,2-	2.05	1.70	50	1.88	75.0
Trichloroethylene	2.71	2.22	166	1.97	94.3
Xylene,*o*-	3.13	2.56	363	2.38	241
Xylene,*m*-	3.20	2.61	407	2.29	196
Xylene,*p*-	3.17	2.59	389	2.49	311

Source: US Environmental Protection Agency, 1996, Soil Screening Program.

Volatilization

At its simplest, volatilization is a function of the concentration of a contaminant in solution and the contaminant's partial pressure (see the previous discussion on vapor pressure).

Henry's law states that the concentration of a dissolved gas is directly proportional to the partial pressure of that gas above the solution:

$$p_a = K_H[c] \tag{3.19}$$

where: K_H = Henry's law constant

p_a = Partial pressure of the gas

$[c]$ = Molar concentration of the gas

or,

$$p_A = K_H C_W \tag{3.20}$$

where C_W is the concentration of gas in water.

A proportionality between solubility and vapor pressure can be established for any chemical, since Henry's law is an expression of this proportionality between the concentration of a dissolved contaminant and its partial pressure in the headspace (including the open atmosphere) at equilibrium. A dimensionless version of the partitioning is similar to that of sorption, except that instead of the partitioning between solid and water phases, it is between the air and water phases (K_{AW}):

$$K_{AW} = \frac{C_A}{C_W} \tag{3.21}$$

where C_A is the concentration of gas A in the air.

The relationship between the air/water partition coefficient and Henry's law constant for a substance is:

$$K_{AW} = \frac{K_H}{RT} \tag{3.22}$$

where R is the gas constant (8.21×10^{-2} L atm mol^{-1} K^{-1}) and T is the temperature (°K).

Henry's law relationships work well for most environmental conditions, representing a limiting factor for systems where a substance's partial pressure is approaching zero. At very high partial pressures (e.g. 30 pascals) or at very high contaminant concentrations (e.g. >1000 ppm), Henry's law assumptions cannot be met. Such vapor pressures and concentrations are seldom seen in ambient environmental situations, but may be seen in industrial and other source situations. Thus, in modeling and estimating the tendency for a substance's release in vapor form, Henry's law is a good metric and is often used in compartmental transport models to indicate the fugacity from the water to the atmosphere.

Henry's law constants are highly dependent upon temperature, since both vapor pressure and solubility are also temperature-dependent. So, when using published K_H values, one must compare them isothermically. Also, when combining different partitioning coefficients in a model or study, it is important either to use only values derived at the same temperature (e.g. sorption, solubility, and volatilization all at 20 °C), or to adjust them accordingly. A general adjustment is an increase of a factor of 2 in K_H for each 8 °C temperature increase.

Any sorbed or otherwise bound fraction of the contaminant will not exert a partial pressure, so this fraction should not be included in calculations of partitioning from water to air. For example, it is important to differentiate between the mass of the contaminant in solution (available for the K_{AW} calculation) and that in the suspended solids (unavailable for K_{AW} calculation). This is crucial for many hydrophobic organic contaminants, where they are most likely not to be dissolved in the water column (except as co-solutes), with the largest mass fraction in the water column being sorbed to particles.

The relationship between K_H and K_{ow} is also important. It is often used to estimate the *environmental persistence*, as reflected the chemical *half-life* ($T_{1/2}$) of a contaminant. However, many other variables determine the actual persistence of a compound after its release. Note in the table, for example, that benzene and chloroform have nearly identical values of K_H and K_{ow} yet benzene is far less persistent in the environment. We will consider these other factors in the next chapters, when we discuss abiotic chemical destruction and biodegradation.

With these caveats in mind, however, relative affinity for a substance to reside in air and water can be used to estimate the potential for the substance to partition not only between water and air, but more generally between the atmosphere and biosphere, especially when considering the long-range transport of contaminants (e.g., across continents and oceans) [10]. Such long-range transport estimates make use of both atmospheric $T_{1/2}$ and $K_{H.}$ Also, the relationship between octanol–water and air–water coefficients can also be an important part of predicting a contaminants transport. For example, Figure 3.15 provides some general classifications according various substances' K_{AW} and K_{ow} relationships. Thus, according to Eq. 3.22, this relationship applies to K_H values, since K_{AW} and K_H values are proportional. In general, chemicals in the upper left hand group have a great affinity for the atmosphere, so unless there are contravening factors, this is where to look for them. Conversely, substances with relatively low K_H and K_{ow} values are less likely to be transported long distance in the air.

The air-watering partitioning can also be put to use in the closed system conditions of a bioreactor. For example, two-phase partitioning bioreactors, also known as *biphasic reactors*, take advantage of a contaminant's vapor pressure and affinity for the vapor phase (see Table 3.6).

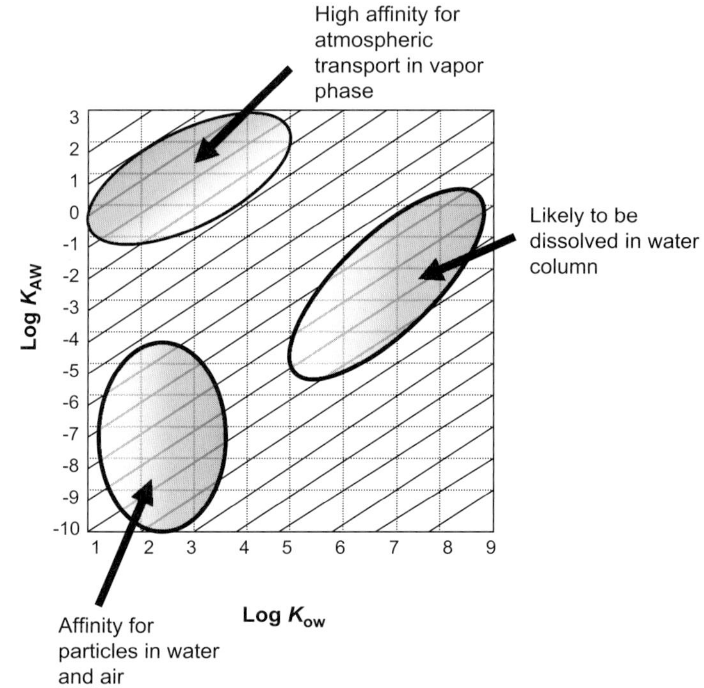

FIGURE 3.15
Relationship between air–water partitioning and octanol–water partitioning and affinity of classes of contaminants for certain environmental compartments under standard environmental conditions.
Source: D. van de Meent, T. McKone, T. Parkerton, M. Matthies, M. Scheringer, F. Wania, *et al.* (1999). Persistence and transport potential of chemicals in a multimedia environment. In *Proceedings of the SETAC Pellston Workshop on Criteria for Persistence and Long-Range Transport of Chemicals in the Environment, 14–19 July 1998, Fairmont Hot Springs, British Columbia, Canada*, Society of Environmental Toxicology and Chemistry, Pensacola, FL.

Table 3.6 Properties and volatile compounds that can be treated in two-phase partitioning bioreactors

VOC/TPPB system	Microorganism	Load[a]	[VOC][b]	RE[c]	EC[d]
Benzene STR[e] with 33%[f] hexadecane	*Alcaligenes xylosoxidans*	140	6	95	133
STR with 33% hexadecane	*Achromobacter xylosoxidans*	1240	60	99	1200
Toluene STR with 33% hexadecane	*Alcaligenes xylosoxidans*	235	9	99	233
STR with 33% hexadecane	*Alcaligenes xylosoxidans*	748	15	97	727
Hexane Biotrickling filter with 5% silicone oil	Activated sludge	100	10	90	90
TPPBs with 10% silicone oil	*Pseudomonas aeruginosa*	180	3	77	140
Fungal TPPBs with 10% silicone oil	*Fusarium solani*	180	3	67	120
Fungal biofilter with 10% silicone oil	*Fusarium solani*	180	3	90	160
Styrene Biotrickling filter with 20% silicone oil	Mixed bacterial culture	555	1	96.8	537

VOC = volatile organic compound.[a] *1 VOC volumetric loading rate (g m^{-3} reactor h^{-1}).*[b] *VOC inlet concentration (g m^{-3}).*[c] *Removal efficiency (%).*[d] *Elimination capacity (g m^{-3} reactor h^{-1}).*[e] *STR=stirred–tank reactor.*[f] *Amount of non-aqueous phase volume added per volume of total working reactor volume (%).*
Source: R. Muñoz, S. Villaverdea, B. Guieysse and S. Revah (2007). Two-phase partitioning bioreactors for treatment of volatile organic compounds. *Biotechnology Advances* 25 (4): 410–422.

In addition to the inherent properties of the compounds being degraded, the bioreactor processes take advantage of aerobic heterotrophic microbes' ability to use these substances as carbon and energy sources. Before this can happen, however, the contaminants and O_2 must first move from the vapor phase to the aqueous phase where they can be metabolized by the microorganisms (see Chapter 7, Discussion Box: Biochemodynamic Films). Thus, even though these are volatile compounds they are actually exclusively treated in the aqueous phase. The volumetric mass transfer rate (mol m^{-3} sec^{-1}) of gaseous substrates (e.g. compounds to be treated, oxygen, and nutrients) to the aqueous phase is:

$$F_{G/A} = K_1 a_{G/A}\left(\frac{S_G}{K_{G/A}} - S_A\right) \tag{3.23}$$

where $K_1 a_{G/A}$ is the global volumetric mass transfer coefficient (hr^{-1}), S_G and S_A are the substrate (e.g. benzene) concentrations (mol m^{-3}) in the bulk gas and aqueous phases, respectively, and $K_{G/A}$ is the substrate partition coefficient (dimensionless) between the gaseous and aqueous phases. $K_{G/A}$ is calculated as follows:

$$K_{G/A} = \frac{S_G}{S_A^*} \tag{3.24}$$

where

S_A^* is the substrate concentration at the gas/aqueous interface (mol m^{-3}).

In addition to the vapor phase and aqueous phase, there is also a non-aqueous (e.g. lipids) phase in the substrate of a bioreactor (see Figure 3.16). The key is getting the microbes and oxygen together, but that can involve moving through at least two phases (Figure 3.16B). This translates into a bioreactor profile similar to the ambient environmental profile in Figure 3.15, except in this case, the compounds fall into sectors based on the octanol–water coefficient (K_{ow}) and $K_{G/A}$. For example, this can predict the relationship between the substrate

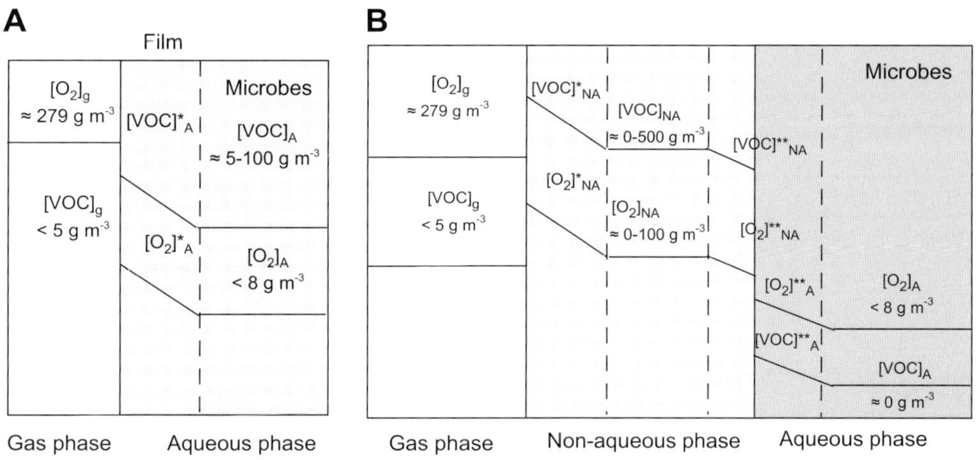

FIGURE 3.16
Concentration profiles of lipophilic substrates (volatile organic compounds – VOCs – and O_2) in a single-phase bioreactor system (A) and in a two-phase partitioning bioreactor (B). [VOC] and $[O_2]$ = concentrations of volatile organic compounds and O_2, respectively, in the treated gas phase ($[\ ]_g$), aqueous phase ($[\ ]_A$), non-aqueous phase ($[\ ]_{NA}$), $[\ ]^*$ and $[\ ]^{**}$ represent the equilibrium concentrations at the gaseous/non-aqueous and non-aqueous/aqueous interfaces, respectively. Note: All concentrations based on air contaminated with 5 g VOC m^{-3}.
Source: R. Muñoz, S. Villaverdea, B. Guieysse and S. Revah (2007). Two-phase partitioning bioreactors for treatment of volatile organic compounds. *Biotechnology Advances* 25 (4): 410–422.

partitioning ($K_{G/A}$) and toxicity to the microbe, e.g. the fungus *Fusarim solani* in a bioreactor (see Figure 3.17).

Persistence is related to partitioning coefficients (see Table 3.7).

The greater the $T_{1/2}$, the more persistent the compound. Persistence is both an intrinsic and extrinsic property of a substance. It is dependent upon the molecular structure of the compound, such as the presence of aromatic rings, certain functional groups, isomeric structures, and especially the number and types of substitutions of hydrogen atoms with

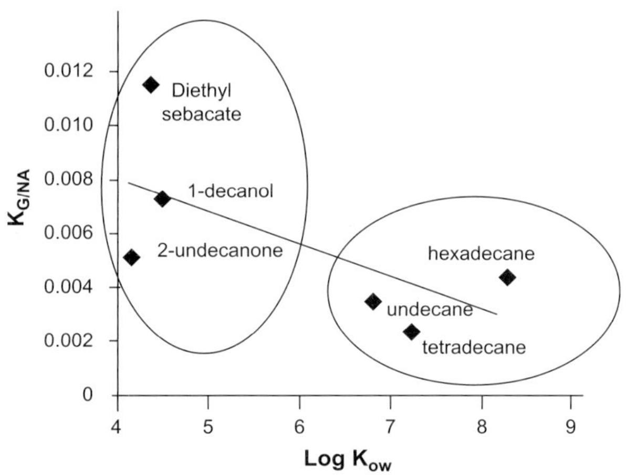

FIGURE 3.17
Gaseous/non-aqueous hexane partition coefficient ($K_{G/NA}$) in organic solvents. The left cluster represents solvents toxic to the fungus, *Fusarim solani*. The right cluster shows biocompatible solvents that were biodegraded by *F. solani*. Note: Silicone oil ($K_{G/NA}$ = 0.0034; unknown log K_{ow}) was the only non-aqueous phase substance tested showing both biocompatible and non-biodegradable characteristics.
Sources: R. Muñoz, S. Villaverdea, B. Guieysse and S. Revah (2007). Two-phase partitioning bioreactors for treatment of volatile organic compounds. *Biotechnology Advances* 25 (4): 410–422; and S. Arriaga, R. Muñoz, S. Hernandez, B. Guieysse and S. Revah (2006). Gaseous hexane biodegradation by *Fusarium solani* in two liquid phase packed-bed and stirred tank bioreactors. *Environmental Science & Technology* 40: 2390–2395.

Table 3.7 Atmospheric persistence compared to octanol–water and Henry's law coefficient

Compound	Half-life (days)	Log K_{ow}	Log K_H
Benzene	7.7	2.1	−0.6
Chloroform	360	1.97	−0.7
DDT	50	6.5	−2.8
Ethyl benzene	1.4	3.14	0.37
Formaldehyde	1.6	0.35	−5.0
Hexachlorobenzene	708	5.5	−3.5
Methyl chloride	470	0.94	−0.44
Methylene chloride	150	1.26	−0.9
PCBs	40	6.4	−1.8
1,1,1 Trichloroethane	718	2.47	0.77

Source: D. Toro and F. Hellweger (1999). Long-range transport and deposition: The role of Henry's law constant. Final report, International Council of Chemical Associations.

halogens (specifically chlorines and bromines). Persistence potential also depends upon the contaminant's relationship to its media. Compound $T_{1/2}$ values are commonly reported for each compartment, so it is possible for a compound to be highly persistent in one medium, yet relatively reactive in another.

Half-lives and rate constants represent identically ordered decay processes, and are inversely related to one another. For example, first-order decay can be expressed in terms of concentration versus time, concentration versus distance, and as biodegradation rates. The first-order rates are: Rate constant = {0.693/T\{1/2\}} and half-life = {0.693/Rate constant}. Thus, a half-life of 2 years is the same as a first-order rate constant of 0.35 year^{-1}, and a half-life of 10 years = a first-order rate constant of 0.0693 (i.e. a slower rate constant is inversely related to a longer half-life. This is an important consideration in estimating the rate at which a contaminant plume will be attenuated, and is commonly used in groundwater studies.

Concentration versus time constants are known as point decay rates (k_{point})m, which are derived from a single concentration value versus time plot, can be used to estimate the length of time that a plume will last. Bulk attenuation rates (k), derived from concentration versus distance plots, are used to see if the contaminant plume is expanding. Biodegradation rates (λ), which are specific to the contaminant and exclude dispersion and other transport mechanisms, can show trends in plume growth or shrinkage. This is now as pollutant attenuation, which follows the rate law. The uses of these rate constants are summarized in Figure 3.18.

The synergy of physical, chemical, and biological processes can be demonstrated by an equation [11] that considers transport (i.e. advection and dispersion) and decay (i.e. biodegradation:

$$C(x,t) = \frac{C_0}{2}\exp\left[\frac{x}{2\alpha_x}\left(1-\sqrt{1+\frac{4\lambda\alpha_x}{v}}\right)\right]\operatorname{erfc}\left(\frac{x-vt\sqrt{1+\frac{4\lambda\alpha_x}{v}}}{2\sqrt{\alpha_x vt}}\right)\operatorname{erf}\left(\frac{Y}{2\sqrt{\alpha_y x}}\right) \quad (3.25)$$

where: C = contaminant concentration

C_0 = initial contaminant concentration

α_x = longitudinal dispersivity

α_y = transverse or horizontal dispersivity

λ = biodegradation rate

t = time

v = retarded velocity of groundwater (v = {seepage velocity/retardation factor}), where the retardation is due to sorption

Y = source width

	Point Decay Rate Constant (k_{point})	Bulk Attenuation Rate Constant (k)	Biodegradation Rate Constant (λ)
USED FOR:	*Plume Duration Estimate.* Used to estimate time required to meet a remediation goal at a particular point within the plume. If wells in the source zone are used to derive k_{point}, then this rate can be used to estimate the time required to meet remediation goals for the entire site. k_{point} should not be used for representing biodegradation of dissolved constituents in groundwater models (use λ as described in the right hand column).	*Plume Trend Evaluation.* Can be used to project how far along a flow path a plume will expand. This information can be used to select the sites for monitoring wells and plan long-term monitoring strategies. Note that k should not be used to estimate how long the plume will persist except in the unusual case where the source has been completely removed, as the source will keep replenishing dissolved contaminants in the plume.	*Plume Trend Evaluation.* Can be used to indicate if a plume is still expanding, or if the plume has reached a dynamic steady state. First calculate λ, then enter λ into a fate and transport model and run the model to match existing data. Then increase the simulation time in the model and see if the plume grows larger than the plume simulated in the previous step. Note that λ should not be used to estimate how long the plume will persist except in the unusual case where the source has been completely removed.
REPRESENTS:	Mostly the change in source strength over time with contributions from other attenuation processes such as dispersion and biodegradation. is not a biodegradation rate as it represents how quickly the source is depleting. In the rare case where the source has been completely removed (for a discussion of source zones, see Wiedemeier et al., 1999), k_{point} will approximate k.	Attenuation of dissolved constituents due to all attenuation processes (primarily sorption, dispersion, and biodegradation).	The biodegradation rate of dissolved constituents once they have left the source. It does not account for attenuation due to dispersion or sorption.
HOW TO CALCULATE:	Plot natural log of concentration vs. time for a single monitoring point and calculate k_{point} = slope of the best-fit line (ASTM, 1998). This calculation can be repeated for multiple sampling points and for average plume concentration to indicate spatial trends in k_{point} as well. 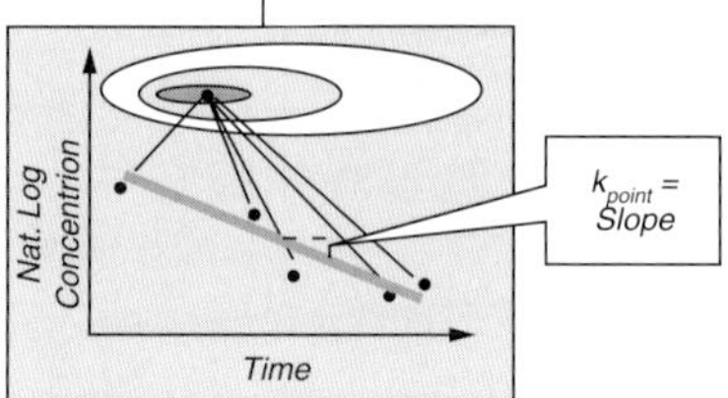 Note this calculation does not account for any changes in attenuation processes, particularly Dual-Equilibrium Desorption (availability) which can reduce the apparent attenuation rate at lower concentrations (e.g., see Kan et al., 1998).	Plot natural log of conc. vs. distance. If the data appear to be first-order, determine the slope of the natural log-transformed data by: 1. Transforming the data by taking natural logs and performing a linear regression on the transformed data, or 2. Plotting the data on a semi-log plot, taking the natural log of the y intercept minus the natural log on the x intercept and dividing by the distance between the two points. Multiply this slope by the contaminant velocity (seepage velocity divided by the retardation factor R) to get k. 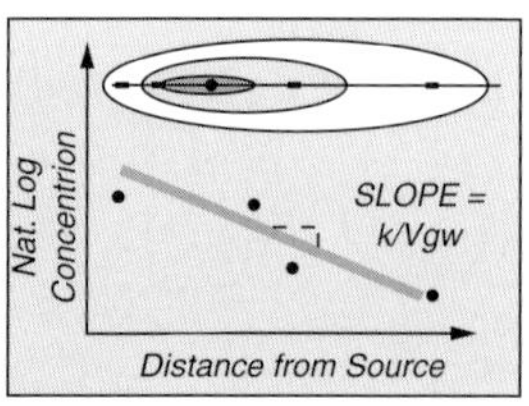	Adjust contaminant concentration by comparison to existing tracer (e.g., chloride, tri-methyl benzenes) and then use method for bulk attenuation rate (see Wiedemeier et al.,1999); or Calibrate a ground-water solute transport computer model that includes dispersion and retardation (e.g., BIOSCREEN, BIOCHLOR, BIOPLUME III, MT3D) by adjusting λ; or Use the method of Buscheck and Alcantar (1995) (plume must be at steady-state to apply this method). Note this method is a hybrid between k and λ as the Buscheck and Alcantar method removes the effects of longitudinal dispersion, but does not remove the effects of transverse dispersion from their λ. 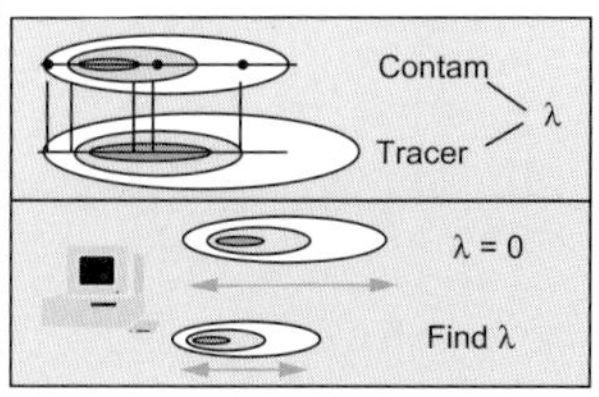

FIGURE 3.18
Steps in applying pollution attenuation rate constants.
Source: C. Newell, H. Rifai, J. Wilson, J. Connor, J. Aziz and M. Suarez (2003). *Ground Water Issue: Calculation and Use of First-Order Rate Constants for Monitored Natural Attenuation Studies.* US Environmental Protection Agency, Ada, OK.

	Point Decay Rate Constant (k_{point})	Bulk Attenuation Rate Constant (k)	Biodegradation Rate Constant (λ)
HOW TO USE:	*To estimate plume lifetime:* The time (t) to reach the remediation goal at the point where k_{point} was calculated is: $$t = \frac{-\text{Ln}\left[\frac{C_{goal}}{C_{stat}}\right]}{k_{point}}$$	*To estimate if a plume is showing relatively little change:* Pick a point in the plume but downgradient of any source zones. Estimate the time needed to decay these dissolved contaminants to meet a remediation goal as these contaminants move downgradient: $$t = \frac{-\text{Ln}\left[\frac{C_{goal}}{C_{stat}}\right]}{k}$$ Calculate the distance L that the dissolved constituents will travel as they are decaying using V_s as the seepage velocity and R is the retardation factor for the contaminant: $$L = \frac{V_s}{R} \cdot t$$ If the plume currently has not traveled this distance L then this rate analysis suggests the plume may expand to that point. If the plume has extended beyond point L, then this rate analysis suggests the plume may shrink in the future. Note that an alternative (and probably easier method) is to merely extrapolate the regression line to determine the distance where the regression line reaches the remediation goal.	*To estimate if a plume is showing relatively little change:* Enter λ in a solute transport model that is calibrated to existing plume conditions. Increase the simulation time (e.g. by 100 years, or perhaps to the year 2525), and determine if the model shows that the plume is expanding, showing relatively little change, or shrinking.
TYPICAL VALUES:	Reid and Reisinger (1999) indicated that the mean point decay rate constant for benzene from 49 gas station sites was 0.46 per year (half-life of 1.5 years). For MTBE they reported point decay rate constants of 0.44 per year (half-life of 1.6 years). In contrast, Peargin (2002) calculated rates from wells that were screened in areas with residual NAPL: the mean decay rate for MTBE was 0.04 per year (half-life of 17 years) the rate for benzene was 0.14 per year (half-life of 5 years). Newell (personal communication) calculated the following median point decay rate constants: 0.33 per year (2.1 year half-life) for 159 benzene plumes at service station sites in Texas; and 0.15 per year (4.7 year half-life) for 37 TCE plumes around the US.	For many BTEX plumes, k will be similar to biodegradation rates λ (on the order of 0.001 to 0.01 per day;) as the effects of dispersion and sorption will be small compared to biodegradation.	For BTEX compounds 0.1 - 1 %/day (half-lives of 700 to 70 days) (Suarez and Rifai, 1999). Chlorinated solvent biodegradation rates may be lower than BTEX biodegradation rates at some sites. For more information about biodegradation rates for a variety of compounds, see Wiedemeier et al., 1999 and Suarez and Rifai, 1999.

Figure 3.18 (Continued)

Figure 3.19 shows the results when the following values are inserted into the equation: $v_s = 100$ ft year^{-1}; $R = 5$; $Y = 40$ ft; $t = 10$ years; and $\alpha_y = 0.1\alpha_x$. Models also need a value for source thickness (b), which is assumed to be 10 ft. Mechanical processes alone or combined underestimate the actual attenuation of contaminant concentrations as compared to when they are combined with biological factors (i.e. source decay and biodegradation). The rate doubles when the decay factors are considered (and the half-life is halved).

Bioavailability

Chemicals move from abiotic compartments into biota. Relatively hydrophobic substances frequently have a strong affinity for fatty tissues (i.e. those containing high K_{ow} compounds).

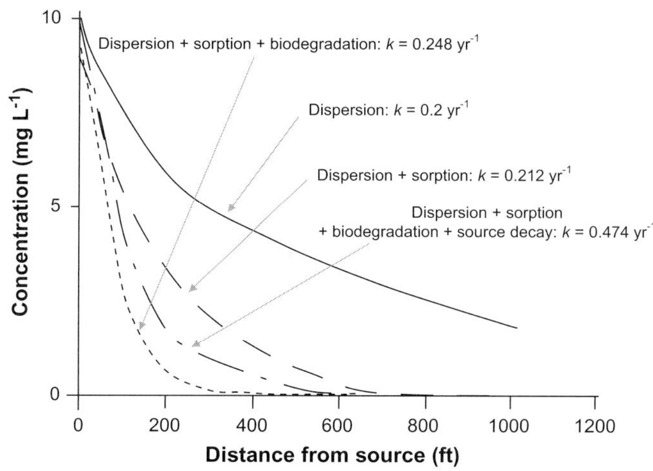

FIGURE 3.19
Effect of incremental contaminant attenuation factors on bulk rate changes to a groundwater plume.
Source: C. Newell, H. Rifai, J. Wilson, J. Connor, J. Aziz and M. Suarez (2003). *Ground Water Issue: Calculation and Use of First-Order Rate Constants for Monitored Natural Attenuation Studies.* US Environmental Protection Agency, Ada, OK.

Therefore, such contaminants can be sequestered and can accumulate in organisms. In other words, certain chemicals are very *bioavailable* to organisms that may readily take them up from the other compartments. Bioavailability is an expression of the fraction of the total mass of a compound present in a compartment that has the potential of being absorbed by the organism. *Bioaccumulation* is the process of uptake into an organism from the abiotic compartments. *Bioconcentration* is the concentration of the pollutant within an organism above levels found in the compartment in which the organism lives. So, for a fish to bioaccumulate DDT, the levels found in the total fish or in certain organs (e.g. the liver) will be elevated above the levels measured in the ambient environment. In fact, DDT is known to bioconcentrate many orders of magnitude in fish. A surface water DDT concentration of 100 parts per trillion in water has been associated with 10 ppm in certain fish species (a concentration of 10,000 times). Thus the straightforward equation for the *bioconcentration factor* (BCF) is the quotient of the concentration of the contaminant in the organism and the concentration of the contaminant in the host compartment. So, for a fish living in water, the BCF is:

$$\mathrm{BCF} = \frac{C_{organism}}{C_w} \tag{3.26}$$

The BCF is applied to an individual organism that represents a genus or some other taxonomical group. However, considering the whole food chain and trophic transfer processes, in which a compound builds up as a result of predator–prey relationships, the term *biomagnification* is used. Some compounds that may not appreciably bioconcentrate within lower trophic state organisms may still become highly concentrated. For example, even if plankton have a small BCF (e.g. 10), if subsequently higher order organisms sequester the contaminant at a higher rate, by the time top predators (e.g. alligators, sharks, panthers, and humans) may suffer from the continuum of biomagnification, with levels many orders of magnitude higher than what is found in the abiotic compartments.

For a substance to bioaccumulate, bioconcentrate, and biomagnify, it must be at least somewhat persistent. If an organism's metabolic and detoxification processes are able to degrade the compound readily, it will not be present (at least in high concentrations) in the organism's tissues. However, if an organism's endogenous processes degrade a compound into a chemical species that is itself persistent, the metabolite or degradation product will bioaccumulate, and may bioconcentrate, and biomagnify. Finally, cleansing or *depuration* will occur if the organism that has accumulated a contaminant enters an abiotic environment that no longer contains the contaminant. However, some tissues have such strong affinities for certain contaminants that the persistence within the organism will remain long after the source of the contaminant is removed. For example, the piscivorous birds, such as the Common Loon (*Gavia immer*), decrease the concentrations of the metal mercury in their bodies by translocating the metal to feathers and eggs. So, every time the birds molt or lay eggs they undergo mercury depuration. Unfortunately, when the birds continue to ingest mercury that has bioaccumulated in their prey (fish), they often have a net increase in tissue Hg concentrations because the bioaccumulation rate exceeds the depuration rate [12].

Bioconcentration can vary considerably in the environment. The extent to which a contaminant builds up in an ecosystem, especially in biota and sediments, is related to the compound's persistence. For example, a highly persistent compound, if nothing else, lasts longer in the environment so there is a greater opportunity for uptake, all other factors being equal. In addition, persistent compounds often possess chemical structures that are also conducive to sequestration by fauna. Such compounds are generally quite often lipophilic, have high K_{ow} values, and usually low vapor pressures. This means that they may bind to the organic molecules in living tissues and may resist elimination and metabolic process, so that they build up over time. However, the bioaccumulation and bioconcentration can vary considerably, both among biota and within the same species of biota. For example, the pesticide mirex has been shown to exhibit bioconcentration factors of 2600 and 51,400 have been observed in pink shrimp and fathead minnows, respectively. The pesticide endrin has shown an even larger interspecies variability in BCF values, with factors ranging more than four orders of magnitude, from 1.4×10^1 to 1.8×10^4, recorded in fish after continuous exposure. Interspecies BCF ranges may also be high. For example, oysters exposed to very low concentrations of the organometallic compound tributyl tin exhibit BCF values ranging from 1000 to 6000 [13].

Even the same compound in a single medium, e.g. a lake's water column or sediment, will show large BCF variability among species of fauna in that compartment. A number of *persistent organic pollutants* (*POPs*) that have been largely banned, some for decades, are still found in environmental samples throughout the world. As might be expected from their partitioning coefficients, they have concentrated in sediment and biota.

PERSISTENT BIOACCUMULATING TOXIC SUBSTANCES

The worst combination of factors is when a compound is persistent in the environment, builds up in organic tissues, and is toxic. Such compounds are referred to as *persistent bioaccumulating toxic* substances (*PBTs*). Recently, the United Nations Environmental Programme (UNEP) reported on the concentrations of the persistent and toxic compounds. Each region of the world was evaluated for the presence of these compounds.

The sources of PBTs are widely varied. Many are intentionally manufactured to serve some public need, such as the control of pests that destroy food and spread disease. Other PBTs are generated as unintended byproducts, such as the products of incomplete combustion. In either case, there are often measures and engineering controls available that can prevent PBT releases, rather than having to deal with them after they have found their way into the various environmental compartments. One of the principal reasons for the concern about the plethora of organic chemicals and heavy metals in the environment has been the connection between exposures to these substances and cancer and other chronic diseases. Intrinsic properties of compounds render them more or less toxic. In addition, physical and chemical properties determine whether the compounds will resist degradation and persist for long time periods and build up in organisms. PBTs are comprised of myriad compounds (see Discussion Box: The Inuit and Persistent Organic Pollutants). One prominent class is the polycyclic aromatic hydrocarbons (PAHs), a family of large, flat compounds with repeating benzene structures, which represent a class of PBTs. The chemical structure, i.e. stereochemistry, renders most PAHs highly hydrophobic, i.e. fat soluble, and difficult for an organism to eliminate (since most blood and cellular fluids are mainly water). This property also enhances the PAHs' ability to

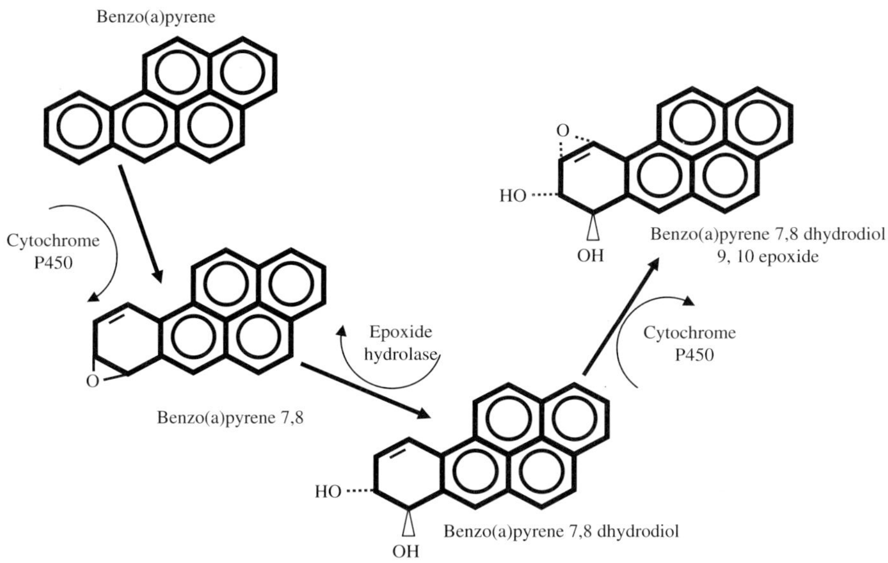

FIGURE 3.20
Biological activation of benzo(a)pyrene to form the carcinogenic active metabolite benzo(a)pyrene 7,8 dihydrodiol 9, 10 epoxide. During metabolism, the biological catalysts (enzymes) cytochrome P450 and epoxide hydrolase are employed to make the molecule more polar, and in the process form diols and epoxides. These metabolites are more toxic than the parent compound.

insert themselves into the deoxyribonucleic acid (DNA) molecule, interfering with transcription and replication. This is why some large organic molecules can be mutagenic and carcinogenic. One of the most toxic PAHs is benzo(a)pyrene, which is found in cigarette smoke, combustion of coal, coke oven emissions, and numerous other processes that use combustion. The compound can become even more toxic when it is metabolized, a process known as biological activation (see Figure 3.20).

DISCUSSION BOX

The Inuit and Persistent Organic Pollutants

Persistent organic pollutants (POPs) include a wide range of substances: industrial chemicals (e.g. PCBs) and byproducts of industrial processes (e.g. hexachlorobenzene – HCB, and chlorinated dioxins), which are unintentionally toxic. Other POP have characteristics that are intentionally toxic, such as insecticides (e.g. DDT) and herbicides (e.g. 2,4-dichlorophenoxyacetic acid; ''2,4-D''), or fungicides (e.g. vinclozolin). Those POPs with substituted chlorines are referred to as organochlorines.

Indigenous peoples and other Arctic populations subsist on traditional food for all or part of their diet. Studies have shown that even very remote Arctic regions have been chronically exposed to POPs, so these subpopulations are vulnerable to be adversely affected. POPs are of particular concern because:

they persist in the environment for long periods of time which allows them to be transported large distances from their sources, are often toxic, and have a tendency to bioaccumulate; many POPs biomagnify in food chains;

many indigenous people in the Arctic depend on traditional diets that are both an important part of their cultural identity and a vital source of nourishment; alternative sources of food often do not exist; however, traditional diets are often high in fat and POPs tend to accumulate in fatty tissue of the animals that are eaten;

most northern residents have not used or directly benefited from the activities associated with the production and use of these chemicals, however indigenous peoples in the Arctic have some of the highest known exposures to these chemicals.

Due to their physicochemical properties, POPs can move many hundreds of kilometers away from their sources, either in the gas phase or attached to particles. They are generally moved by advection, i.e. along with the movement of air masses. Some of the routes of long-range transport of POPs are shown in Figure 3.21).

(Continued)

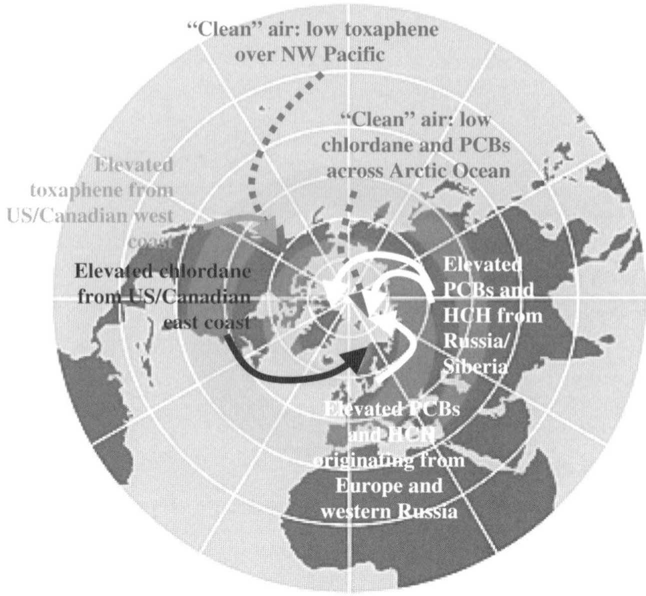

FIGURE 3.21
Long-range transport of persistent organic pollutants in the Arctic regions. Note: HCH – hexachlorocyclohexane; PCBs = polychlorinated buphenyls. [*See color plate section*]
Source: Russian Chairmanship of the Arctic Council (2005). Draft Fact Sheet.

A particularly vulnerable group is the Inuit. Lactating Inuit mother's breast milk, for example, contains elevated levels of PCBs, DDT and its metabolites, chlorinated dioxins and furans, and brominated organics, such as residues from fire retardants, i.e. polybrominated diphenyl ethers (PBDEs), and heavy metals [14]. Risks can vary considerably by age and other factors. For example, infants are particularly vulnerable to some PBTs, because, in general these compounds are lipophilic and find their way to fat reserves in warm-blooded animals. Thus, nursing infants are especially likely to experience unacceptable exposures to these compounds (see Figure 3.22).

These compounds are encountered to varying extents among women in industrially developed as well as in developing nations. Some of the highest levels of contaminants have been detected in the Canadian Inuit, whose diet consists of seal, whale, and other species high on the marine food chain. In the process the Inuit body burden of POPs is quite high [15]. These elevated exposures have led to adverse health effects which have been reported in persons exposed to PCBs who also had evidence of other contaminants in body fluids.

A study of Inuit women from Hudson Bay [16] indicated very high levels of PCBs and dichlorodiphenylethene (DDE) in breast milk; these results prompted an examination of the health status of Inuit newborns [17]. Correlation analysis revealed a statistically significant negative association between male birth length and levels of hexachlorobenzene, mirex, PCBs, and chlorinated dibenzodioxins (CDDs)/CDFs in the fat of mothers' milk. No significant differences were observed between male and female newborns for birth weight, head circumference, or thyroid-stimulating hormone. Immune system effects have also been detected in Inuit infants suspected of receiving elevated levels of PCBs and dioxins during lactation. These babies had a drop in the ratio of the CD4+ (helper) to CD8+ (cytotoxic) T-cells at ages 6 and 12 months (but not at 3 months) [18].

The Inuit situation demonstrates the critical ties between humans and their environment and the importance of physical properties of contaminants (e.g. persistence, bioaccumulation, and toxicity potentials), the conditions of the environment (e.g. the lower Arctic temperatures increase the persistence of many POPs), and the complexities of human activities (e.g. diet and lifestyle) in order to assess risks and, ultimately, to take actions to reduce exposures. The combination of these factors leaves the Inuit in a tragic dilemma. Since they are subsistence anglers and hunters, they depend almost entirely on a tightly defined portion of the earth for food. Their lifestyle and diet dictate dependence on food sources high in POPs.

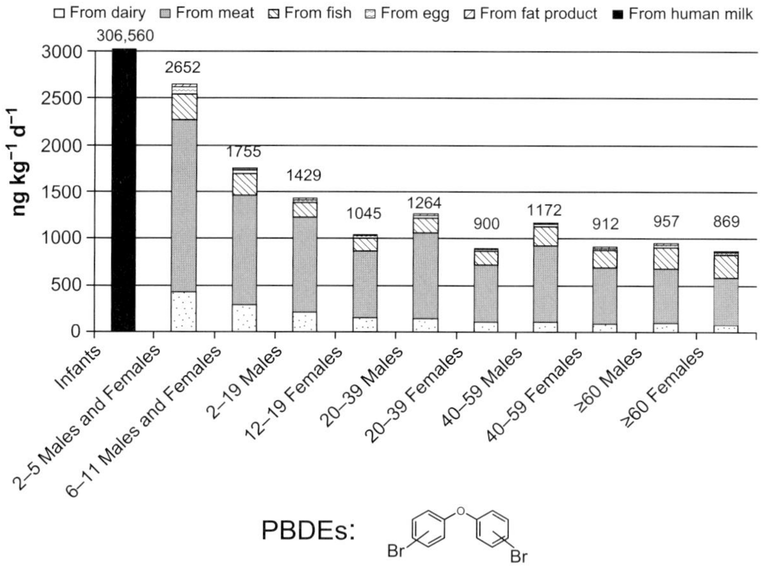

FIGURE 3.22
US population's estimated daily dietary intake of polybrominated diphenyl ethers (PBDE) by age group and food source. Units are picograms per kg (pg kg^{-1}) of body weight per day. In all groups older than 1 year of age, total PBDE intake from meat is significantly higher than from any other food sources. The highest dietary intake values of PBDEs were found in nursing infants (307 ng kg^{-1}) body weight per day, which compares to 1.0 ng kg^{-1} d^{-1} for men or 0.9 ng kg^{-1} d^{-1} for women at ≥60 years of age.
Data from: A. Schecter, O. Päpke, T.R. Harris, K.C. Tung, A. Musumba, J. Olson, and L. Birnbaum (2006). Polybrominated diphenyl ether (PBDE) levels in an expanded market basket survey of U.S. food and estimated PBDE dietary Intake by age and sex. *Environmental Health Perspectives* 114 (10): 1515–1520.

The lesson extends even further, since exposures also include mother's milk. Pediatricians rightly encourage breast feeding for its many attributes, including enhancing the infant's immune system in the critical first weeks after birth. So, in terms of risk tradeoffs, it is dangerous to discourage breast feeding. This lesson not only applies to the Inuit, or even just subsistence farmers, hunters, and anglers, but to all of us. We need to be finding ways to ensure that breast milk everywhere does not contain hazardous levels of PBTs and other contaminants. The only way to do this is to consider the entire life cycle of the pollutants and find ways to prevent their entry into the environment in the first place.

Extrinsic Factors

The greater persistence of POPs in the Arctic regions compared to temperate and tropical regions is a direct result of temperature. Toxicity properties of environmental contaminants are also affected by extrinsic conditions, such as whether the substances are found in the air, water, sediment or soil, along with the conditions of these media (e.g. oxidation–reduction, pH, and grain size). For example, the metal mercury is usually more toxic in reduced and anaerobic conditions because it is more likely to form alkylated organo metallic compounds, like monomethyl mercury and the extremely toxic dimethyl mercury. These reduced chemical species are likely to form when buried under layers of sediment where dissolved oxygen levels approach zero. Ironically, engineers have unwittingly participated in increasing potential exposures to these toxic compounds. With the good intention of attempting to clean up contaminated lakes in the 1970s, engineers recommended and implemented dredging programs. In the process of removing the sediment, however, the metals and other toxic chemicals that had been relatively inert and encapsulated in buried sediment were released to the lake waters. In turn, the compounds were also more likely to find their way to the atmosphere (see Figure 3.23). This is a lesson to engineers to take care to consider the many physical, chemical, and biological characteristics of the compound and the environment where it exists.

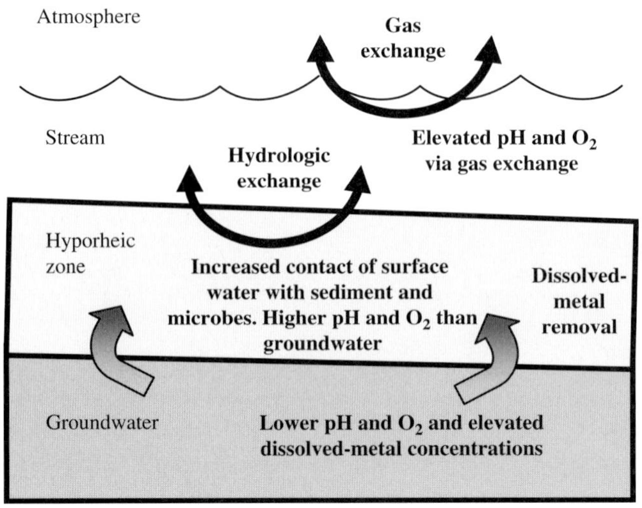

FIGURE 3.23
Exchanges and reactions that can occur in groundwater, sediment, and surface water. Some of the stream water moves into and out of the sediment and in shallow groundwater (i.e. the hyporheic zone). The process can increase the mobility of dissolved metallic compounds.
Source: Adapted from US Geological Survey and D.A. Vallero (2004). *Environmental Contaminants: Assessment and Control.* Elsevier Academic Press, Burlington, MA.

Biochemodynamic persistence and half-life

Substances that remain in the environment long after their release are more likely to continue to cause problems or to be a threat to environmental quality. Persistence is commonly expressed as the chemical half-life ($T_{1/2}$) of a substance, i.e. the time it takes to degrade one-half of the mass. The US Environmental Protection Agency considers a compound to be persistent if it has a $T_{1/2}$ in water, soil or sediment of greater than 60 days, and very persistent if the $T_{1/2}$ is greater than 180 days. In air, the compound is considered persistent if its $T_{1/2}$ is greater than two days. Some of the most notoriously toxic chemicals are also very persistent.

The concept of persistence elucidates the notion of tradeoffs that are frequently needed as part of many responses to environmental insults. It also underlines the important point that good science is necessary but never sufficient to provide an acceptable response to environmental justice issues. Let us consider the pesticide DDT (1,1,1-trichloro-2,2-bis-(4-chlorophenyl)-ethane ($C_{14}H_9C_{15}$)). DDT is relatively insoluble in water (1.2–5.5 mg L^{-1} at 25 °C) and is not very volatile (vapor pressure: 0.02×10^{-5} mmHg at 25 °C) [19]. Looking at the water solubility and vapor pressures alone may lead one to believe that people and wildlife are not likely to be exposed in the air or water. However, the compound is highly persistent in soils, with a $T_{1/2}$ of about 1.1 to 3.4 years, so it may still end up in drinking water in the form of suspended particles or in the air sorbed to fine particles. DDT also exhibits high bioconcentration factors (in the order of 50,000 for fish and 500,000 for bivalves), so once organisms are exposed, they tend to increase body burdens of DDT over their lifetimes. In the environment, the parent DDT is metabolized mainly to DDD (dichlorodiphenyldichlorethane) and DDE (dichloro-diphenyldichloroethylene) [20].

The physicochemical properties of a substance determine how readily it will move among the environmental compartments, i.e. to and from sediment, surface water, soil, groundwater, air, and in the food web, including humans. So, if a substance is likely to leave the water, it is not persistent in water. However, if the compound moves from the water to the sediment, where it persists for long periods of time, it must be considered environmentally persistent. This is an example of how terminology can differ between chemists and engineers. Chemists often define persistence as an intrinsic chemical property of a compound, while engineers see it as both intrinsic and extrinsic (i.e. a function of the media, energy and mass balances, and equilibria). So, engineers usually want to know not only about the molecular weight, functional groups, and ionic form of the compound, but also whether it is found in the air or water, and the condition of the media (e.g. pH, soil moisture, sorption potential, and microbial populations). The movement among phases and environmental compartments is known as partitioning. Many toxic compounds are semi-volatile (i.e. at 20 °C and 101 kPa atmospheric pressure, vapor pressures = 10^{-5} to 10^{-2} kPa), under typical environmental conditions. The low vapor pressures and low aqueous solubilities means they will have low fugacities, i.e. they lack a strong propensity to flee a compartment, e.g. to move from the water to the air.

Even low K_H and K_{AW} compounds, however, can be transported long distances in the atmosphere when sorbed to particles. Fine particles can behave as colloids and stay suspended for extended periods of time, explaining in part why low K_H compounds can be found in the most remote locations relative to their sources, such as in the Arctic regions. This is important, for example, when explaining to indigenous populations why they may be exposed to contaminants that are not produced near them.

If the substrate has sufficient sorption sites (see Figure 3.24), such as many clays and organic matter, the substance may become tightly bound and persistent. The properties of the compound and those of the water, soil, and sediment determine the rate of sorption.

Biochemodynamics includes both chemical persistence and environmental persistence. Henry's law, solubility, vapor pressure, and sorption coefficients for a compound may *prima facie* indicate that the compound is not persistent. However, in real-life scenarios, this may not be the case. For example, there may be a repository of a source of a non-persistent compound that leads to a continuous, persistent exposure of a neighborhood population. Conversely, a compound that is ordinarily not very persistent may become persistent under the right circumstances, e.g. a reactive pesticide that is tracked into a home and becomes entrapped in carpet fibers. The lower rate of photolysis (degradation by light energy) indoors versus outdoors and the sorptive characteristics of the carpet twill, as well as if the pesticide molecule is sorbed to a soil particle embedded in the twill, can lead to dramatically increased environmental half-lives of certain substances. For example, ionic charge can affect the sorption, e.g. a negatively charged site on a particle may increase the persistence of a substance that has an abundance of negatively charged sites (see Figure 3.24).

A potentially important threat looming at present with long-term implications is presented by a suite of chemicals that appear to alter hormonal functions in animals, including mammals. Such chemicals, known as hormonally active agents or endocrine disrupting compounds (or simply endocrine disruptors), have been associated with abnormal spermatogenesis,

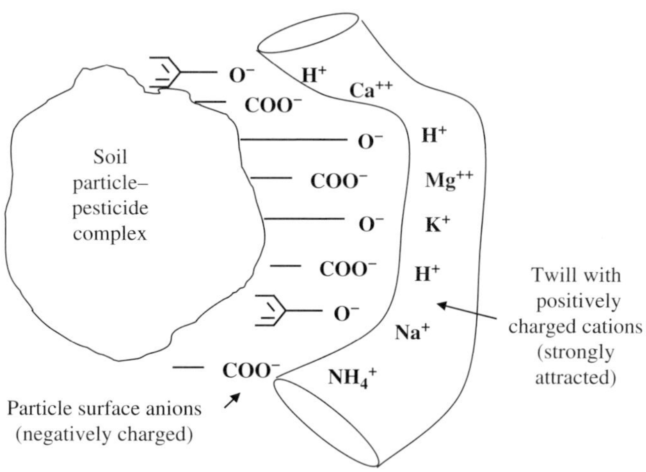

FIGURE 3.24
Negatively charged, pesticide-laden soil particle sorbed to carpet twill that contains numerous sorption sites (positively charged).

feminization of males, masculinization of females, dysfunction of adrenal, pineal and thyroid glands, auto-regulatory problems, and other hormonally related problems. They are diverse in molecular structure (see Table 3.8), come from a myriad of sources and have been detected throughout the environment, i.e. food, water, air, soil and in plant and animal tissues.

Kinetics versus equilibrium

Chemical kinetics is the description of the rate of a chemical reaction [21]. This is the rate at which the reactants are transformed into products. This may take place by abiotic or by biological systems, such as microbial metabolism. Since a rate is a change in quantity that occurs with time, the change we are most concerned with is the change in the concentration of our contaminants into new chemical compounds:

$$\text{Reaction rate} = \frac{\text{change in product concentration}}{\text{corresponding change in time}} \tag{3.27}$$

And,

$$\text{Reaction rate} = \frac{\text{change in reactant concentration}}{\text{corresponding change in time}} \tag{3.28}$$

In environmental degradation, the change in product concentration will be decreasing proportionately with the reactant concentration, so, for substance A the kinetics looks like:

$$\text{Rate} = -\frac{\Delta(A)}{\Delta t} \tag{3.29}$$

The negative sign denotes that the reactant concentration (the parent contaminant), is decreasing. It stands to reason then that the degradation product C resulting from the concentration will be increasing in proportion to the decreasing concentration of the contaminant A, and the reaction rate for Y is:

$$\text{Rate} = \frac{\Delta(C)}{\Delta t} \tag{3.30}$$

By convention, the concentration of the chemical is shown in parentheses to indicate that the system is not at equilibrium. $\Delta(X)$ is calculated as the difference between an initial concentration and a final concentration:

$$\Delta(X) = \Delta(X)_{\text{final}} - \Delta(X)_{\text{initial}} \tag{3.31}$$

Thus, the chemical transformation [22] of one isomer of the compound to another takes place at a certain rate under specific environmental conditions. The rate of reaction at any time is the negative of the slope of the tangent to the concentration curve at that specific time (see Figure 3.25).

For a reaction to occur, the molecules of the reactants must collide. High concentrations of a substance are more likely to collide than low concentrations. Thus, the reaction rate must be a function of the concentrations of the reacting substances. The mathematical expression of this function is known as the "rate law." The rate law can be determined experimentally for any contaminant. Varying the concentration of each reactant independently and then measuring the result will give a concentration curve. Each reactant has a unique rate law (this is one of a contaminant's physicochemical properties).

In a reaction of reactants A and B to yield product C (i.e. $A + B \rightarrow C$), the reaction rate increases in accordance with the increasing concentration of either A or B. If the amount of A is tripled, then the rate of this whole reaction triples. Thus, the rate law for such a reaction is:

$$\text{Rate} = k[\text{A}][\text{B}] \tag{3.32}$$

Table 3.8 Selected compounds found in the environment suspected of adversely affecting hormonal function, based on *in vitro*, *in vivo*, cell proliferation, or receptor-binding studies

Compound[a]	Endocrine effect[b]	Potential source
2,2',3,4',5,5'-Hexachloro-4-biphenylol and other chlorinated biphenylols	Anti-estrogenic	Degradation of PCBs released into the environment
4',7-Dihydroxy daidzein and other isoflavones, flavones, and flavonals	Estrogenic	Natural flora
Aldrin*	Estrogenic	Insecticide
Alkylphenols	Estrogenic	Industrial uses, surfactants
Amitrol*	Thyroid	Thyroid peroxidase inhibitors; inhibits thyroid hormone synthesis
Atrazine*	Neuroendocrine-pituitary (depression of LH surge), testosterone metabolism	Inhibits ligand binding to androgen and estrogen receptors
Bisphenol A and phenolics	Estrogenic	Plastics manufacturing
Chlofentezine*	Thyroid	Enhances secretion of thyroid hormone
DDE (1,1-dichoro-2,2-bis(p-chlorophenyl)ethylene)	Anti-androgenic	DDT metabolite
DDT and metabolites	Estrogenic	Insecticide
Dicofol	Estrogenic or anti-androgenic in top predator wildlife	Insecticide
Dieldrin	Estrogenic	Insecticide
Diethylstilbestrol (DES)	Estrogenic	Pharmaceutical
Endosulfan	Estrogenic	Insecticide
Ethylene thiourea*	Thyroid	Thyroid peroxidase inhibitor
Hydroxy-PCB congeners	Anti-estrogenic (competitive binding at estrogen receptor)	Dielectric fluids
Kepone (Chlorodecone)	Estrogenic	Insecticide
Lindane (γ-hexachlorocyclohexane) and other HCH isomers	Estrogenic and thyroid agonistic	Miticide, insecticide
Linuron*	Androgen	Androgen receptor antagonist
Lutolin, quercetin, and naringen	Anti-estrogenic (e.g. uterine hyperplasia)	Natural dietary compounds
Malathion*	Thyroid antagonist	Insecticide
Methoxychlor	Estrogenic	Insecticide
Nonachlor, trans- *	Estrogenic	Potential estrogen receptor agonist?
Octachlorostyrene*	Thyroid agonist	Electrolyte production
Pentachloronitrobenzene*	Thyroid antagonist	Fungicide, herbicide

(Continued)

Table 3.8 Selected compounds found in the environment suspected of adversely affecting hormonal function, based on *in vitro*, *in vivo*, cell proliferation, or receptor-binding studies—cont'd

Compound[a]	Endocrine effect[b]	Potential source
Pentachlorophenol	Anti-estrogenic (competitive binding at estrogen receptor)	Preservative
Perfluorooctane sulfonate* (PFOS)	Thyroid, reproductive	Suppression of T3,T4; mechanism unknown
Phthalates and their ester compounds	Estrogenic	Plasticizers, emulsifiers
Polychlorinated biphenyls (PCBs)	Estrogenic	Dielectric fluid
Polybrominated Diphenyl Ethers (PBDEs)*	Estrogenic	Fire retardants, including *in utero* exposures
Polycyclic aromatic hydrocarbons (PAHs)	Anti-androgenic (Aryl hydrocarbon-receptor agonist)	Combustion byproducts
Tetrachlorodibenzo-*para*-dioxin and other halogenated dioxins and furans*	Anti-androgenic (Aryl hydrocarbon-receptor agonist)	Combustion and manufacturing (e.g. halogenation) byproduct
Toxaphene	Estrogenic	Animal pesticide dip
Tributyl tin and tin organometallic compounds*	Sexual development of gastropods and other aquatic species	Paints and coatings
Vinclozolin and metabolites	Anti-androgenic	Fungicide
Zineb*	Thyroid antagonist	Fungicide, insecticide
Ziram*	Thyroid antagonist	Fungicide, insecticide

[a]*Not every isomer or congener included in a listed chemical group (e.g. PAHs, PCBs, phenolics, phthalates, and flavinoids) has been shown to have endocrine effects. However, since more than one compound has been associated with hormonal activity, the whole chemical group is listed here.*
[b]*Note that the antagonists' mechanisms result in an opposite net effect. In other words an antiandrogen feminizes and an antiestrogen masculinizes an organism.*
Sources: For full list, study references, study types and cellular mechanisms of action, see Chapter 2 of National Research Council, *Hormonally Active Agents in the Environment*. National Academies Press, Washington, DC, 2000. Source for asterisked (*) compounds is T. Colborn, D. Dumanoski and J.P. Myers, *Our Stolen Future: Are We Threatening Our Fertility, Intelligence and Survival?*, http://www.ourstolenfuture.org/Basics/chemlist.htm; accessed November 10, 2009.

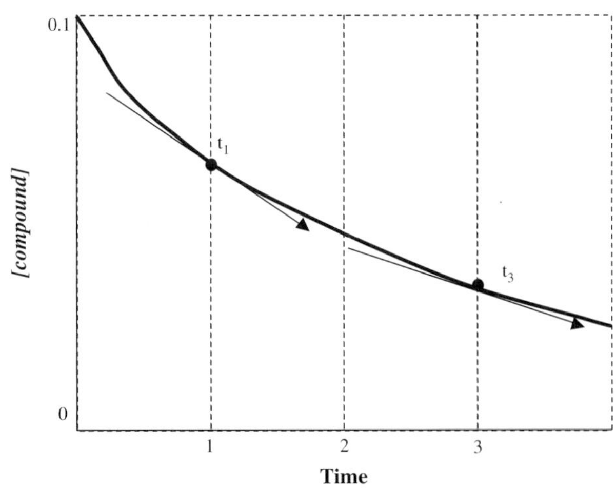

FIGURE 3.25
The kinetics of the transformation of a compound. The rate of reaction at any time is the negative of the slope of the tangent to the concentration curve at that time. The rate is higher at t_1 than at t_3. This rate is concentration-dependent (first-order).

The rate law for the different reaction X + Y → Z, in which the rate is only increased if the concentration of X is increased (changing the Y concentration has no effect on the rate law), must be:

$$\text{Rate} = k[\text{X}] \tag{3.33}$$

Equations 3.29 and 3.30 indicate that the concentrations in the rate law are the concentrations of reacting chemical species at any specific point in time during the reaction. The rate is the velocity of the reaction at that time. The constant k in the equations is the *rate constant*, which is unique for every chemical reaction and is a fundamental physical constant for a reaction, as defined by environmental conditions (e.g. pH, temperature, pressure, type of solvent).

The rate constant is defined as the rate of the reaction when all reactants are present in a 1 molar (M) concentration, so the rate constant k is the rate of reaction under conditions standardized by a unit concentration. By drawing a concentration curve for a contaminant that consists of an infinite number of points at each instant of time, an instantaneous rate can be calculated along the concentration curve. At each point on the curve the rate of reaction is directly proportional to the concentration of the compound at that moment in time. This is a physical demonstration of *kinetic order*. The overall kinetic order is the sum of the exponents (powers) of all the concentrations in the rate law. So for the rate k[A][B], the overall kinetic order is 2. Such a rate describes a second-order reaction because the rate depends on the concentration of the reactant raised to the second power. Other decomposition rates are like k[X], and are first-order reactions because the rate depends on the concentration of the reactant raised to the first power.

The kinetic order of each reactant is the power that its concentration is raised in the rate law. So, k[A][B] is first-order for each reactant and k[X] is first-order X and zero-order for Y. In a zero-order reaction, compounds degrade at a constant rate and are independent of reactant concentration.

A "parent compound" is transformed into "chemical daughters" or "progeny." For example, pesticide kinetics often concerns itself with the change of the active ingredient in the pesticide to its "degradation products."

Fugacity, Z values, and Henry's law

Knowledge about the affinities of a compound for each phase enables predictions of the amount and rate of transformation, transport, and fate. This biochemodynamic behavior as expressed by the partition coefficients can be viewed as a potential; that is, at the time equilibrium is achieved among all phases and compartments, the chemical potential in each compartment has been reached [23].

Chemical concentration and fugacity are directly related via the *fugacity capacity constant* (known as the *Z value*):

$$C_i = Z_i \cdot f \tag{3.34}$$

where: C_i = Concentration of substance in compartment i (mass per volume)

Z_i = Fugacity capacity (time^2 per length^2)

f = Fugacity (mass per length per time^2)

And, at equilibrium, the fugacity of the system of all environmental compartments is:

$$f = \frac{M_{total}}{\sum_i (Z_i \cdot V_i)} \tag{3.35}$$

where: M_{total} = Total number of moles of a substance in all of the environmental system's compartments

V_i = Volume of compartment i where the substance resides

Assuming that a chemical substance will obey the ideal gas law (which is usually acceptable for ambient environmental pressures), then fugacity capacity is the reciprocal of the gas constant (R) and absolute temperature (T). Recall that the ideal gas law states:

$$\frac{n}{V} = \frac{P}{RT} \tag{3.36}$$

where: n = Number of moles of a substance

P = Substance's vapor pressure

Then,

$$P = \frac{n}{V} \cdot RT = f \tag{3.37}$$

And,

$$C_i = \frac{n}{V} \tag{3.38}$$

Therefore,

$$Z_{air} = \frac{1}{RT} \tag{3.39}$$

This relationship allows for predicting the behavior of the substance in the gas phase. The substance's affinity for other environmental media can be predicted by relating the respective partition coefficients to the Henry's law constants. For water, the fugacity capacity (Z_{water}) can be found as the reciprocal of K_H:

$$Z_{water} = \frac{1}{K_H} \tag{3.40}$$

This is the dimensioned version of the Henry's law constant (length^2 per time^2). In sediment, the fugacity capacity is directly proportional to the contaminant's sorption potential, expressed as the solid–water partition coefficient (K_d), and the average sediment density ($\rho_{sediment}$). Sediment fugacity capacity is indirectly proportional to the chemical substance's Henry's law constant:

$$Z_{sediment} = \frac{\rho_{sediment} \cdot K_d}{K_H} \tag{3.41}$$

For biota, particularly fauna and especially fish and other aquatic vertebrates, the fugacity capacity is directly proportional to the density of the fauna tissue (ρ_{fauna}), and the chemical substance's *bioconcentration factor* (BCF), and inversely proportional to the contaminant's Henry's law constant:

$$Z_{fauna} = \frac{\rho_{fauna} \cdot BCF}{K_H} \tag{3.42}$$

As in the case of the sediment fugacity capacity, a higher bioconcentration factor means that the fauna's fugacity capacity increases and the actual fugacity decreases. Again, this is logical, since the organism is sequestering the contaminant and keeping it from leaving if the organism has a large BCF. This is a function of both the species of organism and the characteristics of the contaminant and the environment where the organism resides. So, factors like temperature, pH, and ionic strength of the water and metabolic conditions of the organism will affect BCF and Z_{fauna}. This also helps to explain why published BCF values may have large ranges.

The total biochemodynamic partitioning of the environmental system is merely the aggregation of all of the individual compartmental partitioning. So, the number of moles of the contaminant in each environmental compartment (M_i) is found to be a function of the fugacity, volume, and fugacity capacity for each compartment:

$$M_i = Z_i \cdot V_i \cdot f \quad (3.43)$$

Comparing the respective fugacity capacities for each phase or compartment in an environmental system is useful for a number of reasons. First, if one compartment has a very high fugacity (and low fugacity capacity) for a contaminant, and the source of the contaminant no longer exists, then one would expect the concentrations in that medium to fall rather precipitously with time under certain environmental conditions. Conversely, if a compartment has a very low fugacity, measures (e.g. *in situ* remediation, or removal and abiotic chemical treatment) may be needed to see significant decreases in the chemical concentration of the contaminant in that compartment. Second, if a continuous source of the contaminant exists, and a compartment has a high fugacity capacity (and low fugacity), this compartment may serve as a conduit for delivering the contaminant to other compartments with relatively low fugacity capacities. Third, by definition, the higher relative fugacities of one set of compartments compared to another set in the same ecosystem allow for comparative analyses and estimates of sources and sinks (or "hot spots") of the contaminant, which is an important part of fate, transport, exposure, and risk assessments.

Applying this information allows us to explore fugacity-based, multi-compartmental environmental models. The movement of a contaminant through the environment can be expressed with regard to how equilibrium is achieved in each compartment. The processes driving this movement can be summarized into transfer coefficients or compartmental rate constants, known as *D values* [24]. So, by first calculating the Z values and then equating inputs and outputs of the contaminant to each compartment, we can derive *D* value rate constants. The actual transport process rate (N) is the product of fugacity and the *D* value:

$$N = Df \quad (3.44)$$

And, since the contaminant concentration is Zf, we can substitute and add a first-order rate constant k to give us a first-order rate *D* value (D_R):

$$N = V[c]k = (VZk)f = D_R f \quad (3.45)$$

Although the concentrations are shown as molar concentrations (i.e., in brackets), they may also be represented as mass per volume concentrations [25].

Diffusive and non-diffusive transport processes follow Fick's laws, i.e. diffusive processes, can also be expressed with their own *D* values (D_D), which is expressed by the mass transfer coefficient (K) applied to area A:

$$N = KA[c] = (KAZ)f = D_D f \quad (3.46)$$

Non-diffusive transport (bulk flow or advection) within a compartment with a flow rate (G) has a D value (D_A) and is expressed as:

$$N = G[c] = (GZ)f = D_A f \quad (3.47)$$

This means that a substance is moving through the environment, during its residence time in each phase, is affected by numerous physical transport and chemical degradation and transformation processes. The processes are addressed by models with the respective *D* values, so that the total rate of transport and transformation is expressed as:

$$f(D_1 + D_2 + \ldots D_n) \quad (3.48)$$

Very fast processes have large *D* values, and these are usually the most important when considering the contaminant's behavior and change in the environment.

Models, though imperfect, are important tools for estimating the movement of contaminants in the environment. They do not obviate the need for sound measurements. In fact measurements and models are highly complementary. Compartmental model assumptions must be verified in the field. Likewise, measurements at a limited number of points depend on models to extend their meaningfulness. Having an understanding of the basic concepts of a contaminant transport model, we are better able to explore the principle mechanism for the movement of contaminants throughout the environment.

BIOCHEMODYNAMIC TRANSPORT

Mechanics is the field of physics concerned with the motion and the equilibrium of matter, describing phenomena using Newton's laws. Motion and equilibrium in the environment fall generally within the province of fluid mechanics. Things move at all scales, from molecular to global. Molecular diffusion within sediments, for example, can be an important contaminant transport mechanism. At the other end of the scale, large air masses may be able to transport gases and aerosols in bulk for thousands of kilometers from their sources.

To ensure mass balance, the flux of a substance in a control volume is equal to the mass flux plus the dispersion flux, diffusion flux, as well as source and sink terms. Sources can be the result of a one-time or continuous release of a chemical from a reservoir or result from desorption of the chemical along the way. Sinks can be the result of sorption and surface processes. This means that even if the source is well characterized, sorption takes place in soil, sediment, and biota that will either remove the chemical from the fluid, or under other environmental conditions, the chemical will be desorbed from the soil, sediment or biota. Thus, these interim sources and sinks must be considered in addition to the initial source and final sinks (i.e. the media of the chemical's ultimate fate).

The knowledge of mass balances and partitioning prepares us for three important physical processes responsible for the transport of a contaminant; i.e. advection, dynamic dispersion, and diffusion [26]. First, it is important to consider the means by which a substance finds it way into an environmental system.

Loading

The amount of any substance, e.g. nutrient, pollutant or microbes, that is discharged into a system is known as the *load*. Nutrient loading, pollutant loading, and microorganism loading can be quantified, not only by the gross mass of the substance entering a system, e.g. a water body, but also by the response of that system to the load. Hence the relationship between a substance's entry and its impact on the receiving river, lake, wetland, estuary, ocean or aquifer. However, these receptors can also be compartmentalized into their various media, e.g. soil, sediment, water, and biota.

The first major division of loads is between *point sources* and *nonpoint sources*. This is actually a distinction of convenience. What some may call a point source, others may consider to be a nonpoint source. For example, if a 1 hectare field is near a small creek, this may be a relatively large and dispersed source of a pesticide being released to the creek. However, a 1 ha trickling filter system near a large river may be classified as a point source since the majority of the discharge to the river comes through a conduit (i.e. the *outfall structure*) from the system. However, it is highly likely that pollutants are being released from the treatment facility from sources other than the outfall structure.

Thus, nonpoint sources are generally associated with *runoff*, i.e. multiple sources in a given two-dimensional space that leave these sources and are transported toward a receptor system. These receiving bodies can be above ground (e.g. streams and lakes) or below ground (aquifers). They are usually cascades of systems, such as a contaminant moving atop the soil, with some infiltration. The infiltrated fraction may find its way to the aquifer and moves much more slowly than the runoff above the ground. However, the contaminated groundwater may recharge a stream flow. The same stream may have already been contaminated by the overland flow, so the stream is receiving a continuous flow (groundwater source to stream) and a plug or episodic flow (overland runoff), as shown in Figure 3.26.

For biota, especially bacteria, the nonpoint and point contribution is even more complex than that of chemical compounds. The bacteria themselves may be directly loaded into a stream or other receiving body (Figure 3.26A). In addition, they may accumulate as nonpoint source loads from a combination of sources. As they are transported overland and under the soil surface, the microbes meet hostile and accommodating conditions, which will decrease and increase their numbers, respectively. However, the hostile conditions may lead to the formation of cysts and spores, which are more likely than the bacteria themselves to be transported by advection in the air (see Figure 3.27). When the conditions change to become more accommodating, the spores will generate increased microbial populations. Thus, the actual loading to the body consists of a complicated set of transport and fate processes.

In addition, any loading calculations must take into account background sources (sometimes referred to as "natural" sources). Atmospheric loading is similar, but usually the distinctions are made between *mobile sources* and *stationary sources*. Figures 3.11, 3.21 and 3.26 illustrate that the two systems, hydrologic and atmospheric, join at the interfaces between earth and water bodies and the atmospheric, i.e. terrestrial and tropospheric fluxes. For example, atmospheric deposition is an important source of nutrient loading (e.g. N, P, S, and K) to ecosystems. Ecosystems, e.g. wetlands, load nutrients back into the atmosphere, but often as different chemical species. That is, the ecosystems are acting as control volumes in which reactions are taking place. Many of these reactions are biotic and are mediated by microbes and plants.

The linkages between biota and their environments are influenced by the availability and forms of nutrients. When rain containing nitrates falls to the wetland water and land surfaces, for instance, plant roots take it up and metabolize it into organic forms of nitrogen. Meanwhile, bacteria in sediments reduce it to ammonia. In the opposite direction, the reduced forms are oxidized by other bacteria (e.g. *Nitrosomonas* converts ammonium ions [NH_4^+] to nitrite [NO_2^-], *Nitrobacter* converts the NO_2^- to nitrate [NO_3^-]). Thus, two opposite reactions

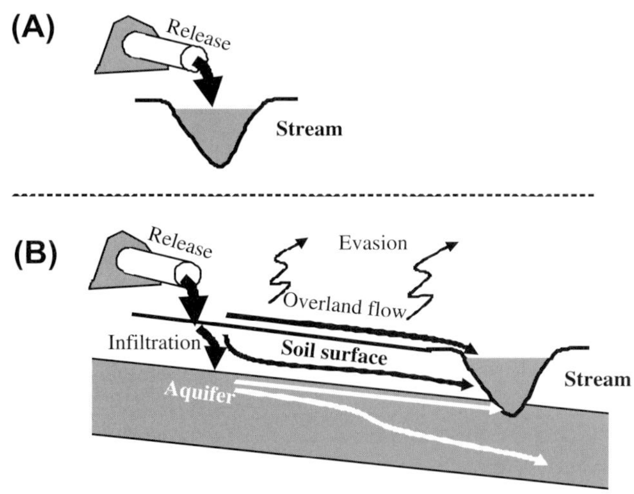

FIGURE 3.26
(A) Point source: material is released directly the receiving body. (B) Nonpoint source: material runs off, infiltrates through soil column, follows groundwater flow lines, and eventually reaches the receiving body. However, along the way, a fraction of the material may be chemically transformed and physically held (e.g. sorbed to soil particles). For volatile compounds, a fraction will also be evaded to the atmosphere. Thus, the net amount of the compound that reaches the water body is a function of numerous factors, including vegetative cover, soil porosity and permeability, groundwater flow rates, soil texture, and Henry's law and other equilibrium coefficients for the released material.

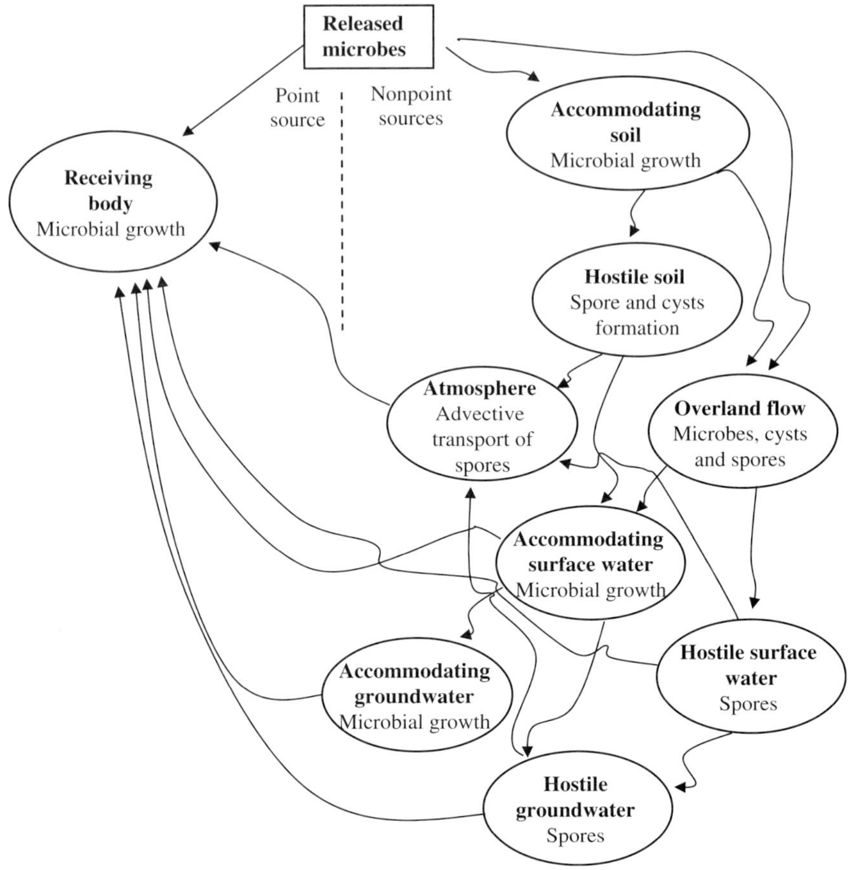

FIGURE 3.27
Difference between point source and nonpoint source releases of microbes. In a point source, the growth, metabolism and formation of cysts and spores depends on the conditions of the receiving water body, which varies in time but not in space. In a nonpoint source scenario, the growth, metabolism, and formation of spores and cysts is a cascade in time and space. In accommodating environments, the microbes will undergo metabolism and growth, which may allow for larger microbial populations delivered to other compartments and eventually to the receiving body. However, various compartments in the flow may be hostile (insufficient nutrients, water, pH, temperature, etc.). These may induce the formation of spores which can be advectively transported in the atmosphere or by the flow of ground and surface waters.

are at play constantly in loading scenarios. In this instance, these are *ammonification* (or *deamination*) and *nitrification:*

$$\text{Ammonification}: \ NH_4^+ + OH^- \Leftrightarrow NH_3 + H_2O \tag{3.49}$$

$$\text{Nitrification}: \ NH_4^+ + 2O_2 \rightarrow NO_3^- + 2H^+ + H_2O \tag{3.50}$$

Note that ammonification is written as an equilibrium reaction, and nitrification is an oxidation reaction. However, the loading depends on numerous environmental conditions, including pH. Thus, under acidic conditions, most of the ammonia nitrogen will ionize to ammonium and under basic conditions the non-ionized ammonia concentrations will increase in proportion to the ammonium. This is important since non-ionized ammonia is very toxic to aquatic biota, whereas ionized species are nutrients needed for plants and algae. Add to this the decomposition of organic matter, and the loading becomes a complicated mix of reactions:

$$\text{Organic nitrogen} + O_2 \rightarrow NH_3 \text{ nitrogen} + O_2 \rightarrow NO_2 \text{ nitrogen} + O_2 \rightarrow NO_3 \text{ nitrogen} + O_2 \tag{3.51}$$

Even this is a gross oversimplification, since the oxidation and reduction depends on the types of bacteria present. For example, the O_2 required for nitrification is theoretically 4.56 mg O_2 mg^{-2} NH_4^+, but this is an *autotrophic* reaction. Thus, O_2 is being produced by the nitrifying bacteria, decreasing the amount needed. However, the growth rate of nitrifying microbes is far less than that of *heterotrophic* microbes decomposing organic wastes. Therefore, when large amounts of organic matter are being degraded, the nitrifiers' growth rate will be sharply limited by the heterotrophs, which means the rate of nitrification will commensurately be decreased [27]. Furthermore, these same responses to conditions exist for all other nutrients in the system, e.g. organic P will be oxidized and oxidized P species reduced, etc.). Thus, the kinetics of nutrient loading are quite complex.

Total maximum daily loading

In the United States, the Clean Water Act (CWA) restricts the total maximum daily load (TMDL) of pollutants released to water bodies that have been deemed to be *impaired* by the state in which technology-based and other engineering controls are not adequate to achieve water quality standards. Thus, the TMDL reflects the amount of a pollutant that can be discharged from point, nonpoint, and natural background sources, including a margin of safety (MOS) for any water quality-limited water body. The TMDL process consists of five steps:

Selection of the pollutant in need of consideration.
Estimation of the water body assimilative capacity (i.e., loading capacity).
Estimation of the pollutant loading from all sources to the water body.
Analysis of current pollutant load and determination of needed reductions to meet assimilative capacity.
Allocation, including a margin of safety, of the allowable pollutant load among the different pollutant sources in a manner that water quality standards are achieved.

A beneficial use that has been impaired is a function of a change in the chemical, physical or biological integrity of surface waters, such as:

- restrictions on fish and wildlife consumption
- tainting of fish and wildlife flavor
- degradation of fish wildlife populations
- fish tumors or other deformities
- bird or animal deformities or reproduction problems
- degradation of benthos
- restrictions on dredging activities
- eutrophication or undesirable algae
- restrictions on drinking water consumption, or taste and odor problems
- beach closings
- degradation of esthetics
- added costs to agriculture or industry
- degradation of phytoplankton and zooplankton populations
- loss of fish and wildlife habitat.

Let us consider two examples of impairments: the Cuyahoga River basin in northeast Ohio and the Rock Creek basin in Maryland. The Cuyahoga area of concern embodies the lower 45 miles of the river from the Ohio Edison Dam to the river's mouth, along with 10 miles of Lake Erie shoreline, and includes 22 miles of urbanized stream between Akron and Cleveland (see Figure 3.28) [28]. The environmental degradation resulted from nutrient loading, toxic substances (including polychlorinated biphenyls (PCBs) and heavy metals), bacterial contamination, habitat change and loss, and sedimentation. Sources for these contaminants include municipal and industrial discharges, bank erosion, commercial/residential development, atmospheric deposition, hazardous waste disposal sites, urban stormwater runoff, combined sewer overflows (CSOs), and wastewater treatment plant bypasses.

In 1994, an advisory about eating fish was issued for Lake Erie and the Cuyahoga River due to elevated polychlorinated biphenyl (PCB) levels in fish tissue. The advisory restricted the

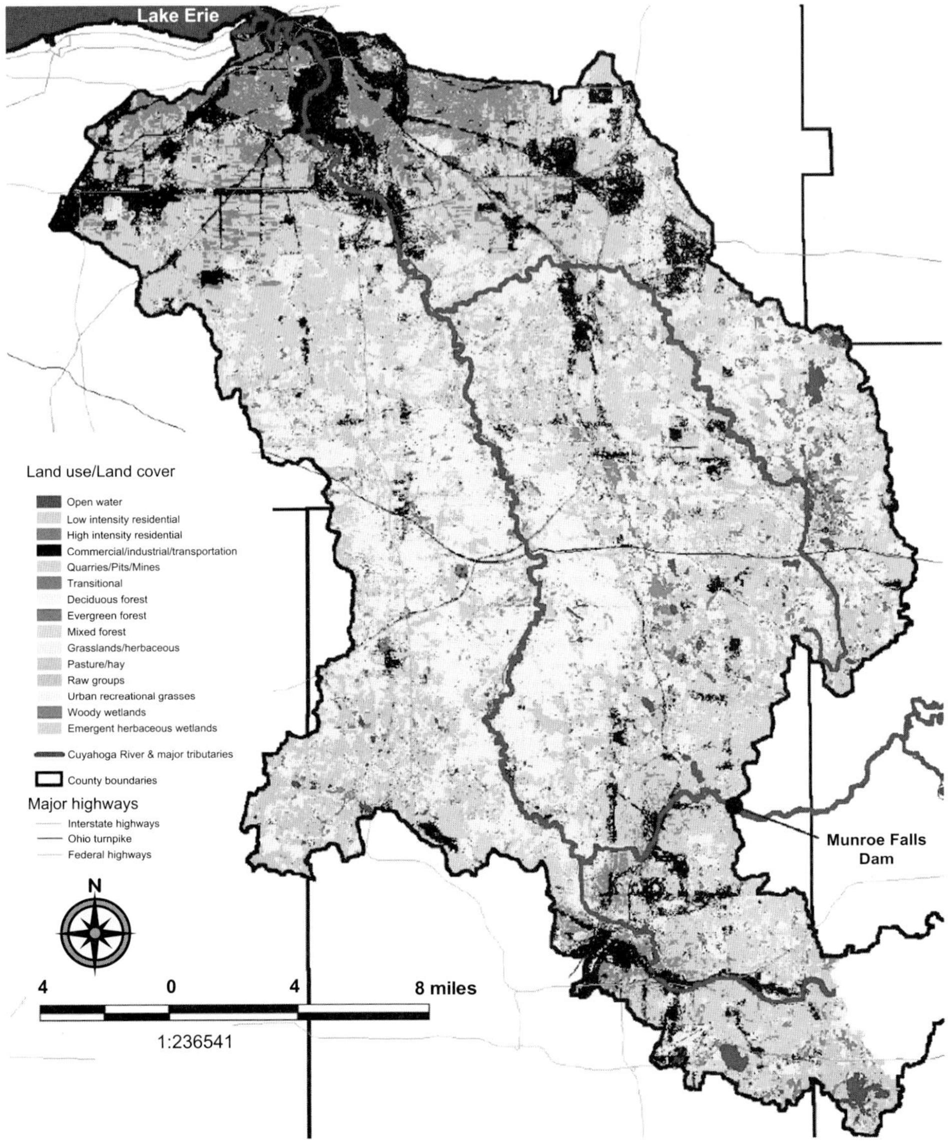

FIGURE 3.28
Land use/ land cover map of the Cuyahoga River basin in northeast Ohio. Drainage is northward to Lake Erie. Approximately 38% of the area is tree covered, 28% is residential, 15% agricultural, 12% industrial, and 3% wetlands. [*See color plate section*]
Source: Ohio Environmental Protection Agency (2003). Lower Cuyahoga River Watershed TMDLs.

consumption of white sucker, carp, brown bullhead, and yellow bullhead in the Cuyahoga River area of concern, and walleye, freshwater drum, carp, steelhead trout, white perch, Coho salmon, Chinook salmon, small mouth bass, white bass, channel catfish, and lake trout in Lake Erie. Beginning at the Ohio Edison Gorge and extending downstream to Lake Erie, measures of fish population conditions ranged from fair to very poor and were below applicable Ohio warm water habitat aquatic life use criteria. Although fish communities have recovered significantly compared to the historically depleted segments of the Cuyahoga River, pollution-tolerant species continue to compose the dominant fish population. Wildlife and fish populations are impaired. Anecdotal information indicates some recovery of Great Blue Heron nesting in the Cuyahoga River watershed. Resident populations of Black-crowned Night Herons have been noted in the navigation channel. The RAP is seeking partners to undertake research in this area in order that an evaluation may be made. Although deformities such as eroded fins, lesions, and external tumors (DELT anomalies) have declined throughout the watershed, significant impairments continue to be found in the headwaters to the nearshore areas of Lake Erie.

Macroinvertebrate populations living at or near the bottom (i.e. benthic organisms) of the Cuyahoga River remain impaired at certain locations, however, there are indications of substantial recovery, ranging from good to marginally good, throughout most free-flowing sections of the river. Some fair and even poor designations are still seen, however.

The US EPA restricts disposal of dredged sediment in most of the Cuyahoga basin due to high concentrations of heavy metals. Only a small amount of the dredged material that contains contaminated sediments is transported and disposed of in a confined disposal facility in the Cleveland area.

The Cuyahoga navigation channel seems to be impaired due to extreme oxygen depletion during summer months. The oxygen demand of sediment oxygen is a factor. Polluted aquifers and surface waters may still be sources for individual supplies and wells. High bacterial counts following rain events periodically adversely affect the two beaches in the area of concern. Swimming advisories are issued after a storm, or if microbial counts exceed certain thresholds. According to some studies, phytoplankton populations in the river are impaired.

Channelization, nonexistent riparian cover, silt, bank reinforcement with concrete and sheet piling, alterations of littoral areas and shorelines, and dredging are contributors to the impairment of fish and wildlife in the area of concern.

Meeting water quality standards in Ohio requires meeting aims at protecting various beneficial uses including recreational activities, aquatic life, and water supply. Ohio's recreational beneficial use attainment is based on fecal coliform or *E. coli* bacteriological criteria. Much of the lower Cuyahoga River is designated for primary contact recreation water where there is an intermediate potential exposure to bacteria and a baseline level of disinfection required.

The general components needed to develop a TMDL are shown in Figure 3.29. The microbial TMDL is expressed as most probable number (MPN) per day and is based on meeting the instream long-term geometric mean of enterococci bacteria. The US EPA's regulations for MPN [29] as well as 40 CFR §130.2(i) also define the TMDL as the "sum of individual wasteload allocations for point sources and load allocations for nonpoint sources and natural background." To determine the MPN, a set of tubes containing enriched broth are inoculated with different amounts of the water sample and incubated at a specific temperature for a predetermined time period. If gas is produced in the tube, a sample of the bacteria in the broth is transferred to additional media to confirm the presence of fecal coliform bacteria. The number of tubes producing gas are converted to express the results of the test as the MPN per 100 mL water, a statistical estimation of the number of bacteria that would give the results shown by the laboratory test. This is a statistical probability number and not an actual empirical result. The test may give higher than actual results because of the built-in 23% positive bias.

The membrane filter technique is an alternate means of expressing MPN, using a measured amount of sample and filtering it through a membrane (pore size = 0.45 μm). Microbes are retained on the membrane and the filter is placed on the surface of a selective agar medium and incubated at a specific temperature for a specified length of time. After incubation, colonies that have formed by the growth of the microbes are counted microscopically at low magnification. The membrane filter technique thus provides an estimate of the number of coliform bacteria that form colonies when cultured (colony-forming units or CFU per 100 mL). Because colonies could be formed from more than one bacterium, the count is merely an estimate. The membrane filtration procedure is generally used in the US for pathogens released to waterways because it is faster and more precise than the MPN technique; however, many local agencies may avoid it because of its complexity and the need for expertise to interpret the findings.

The TMDL for the Cuyahoga basin was determined using a load duration curve, which is calculated from continuous flow data from each gage site. It is a cumulative frequency distribution of the daily mean flow over a gage record period. Next, the load duration curve is generated from the product of flow duration curve and the applicable water quality criterion (see Figure 3.30).

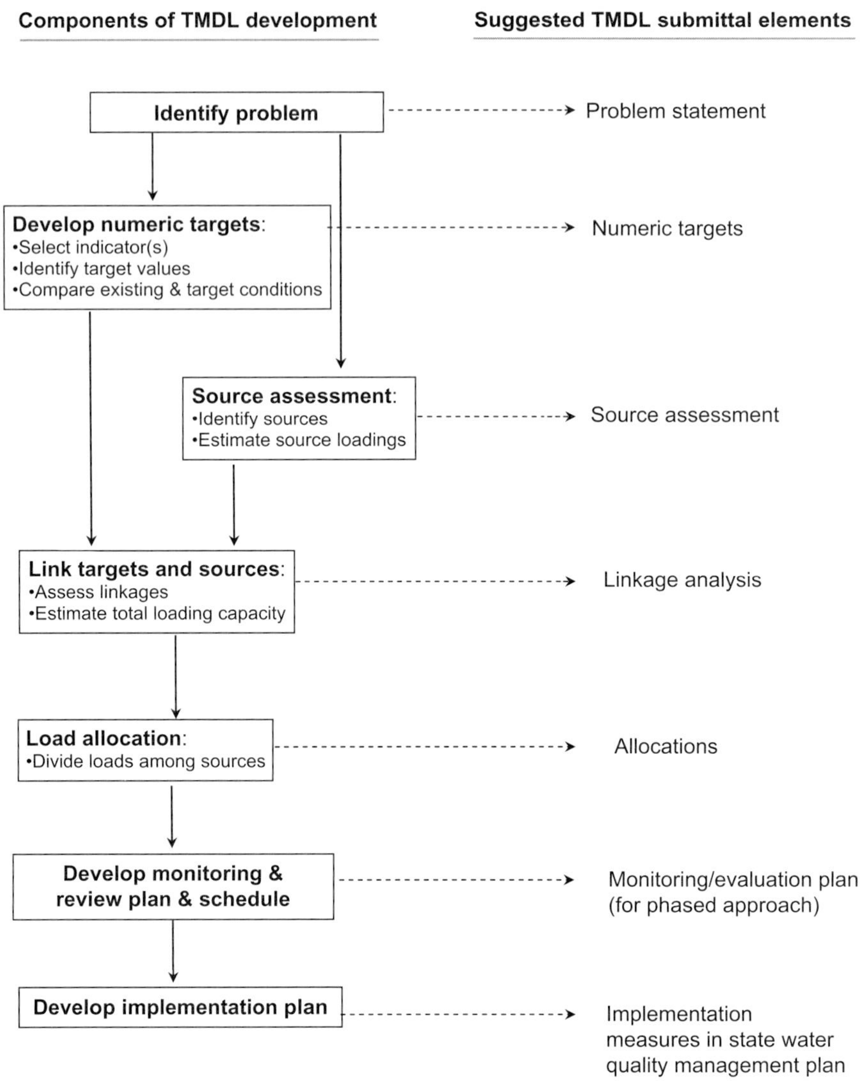

FIGURE 3.29
Steps in developing a total maximum daily load.
Source: US Environmental Protection Agency (2001). Protocol for developing pathogen TMDLs. Report No. EPA 841-R-00-002.

The quantification of the existing load per source is calculated as the in-stream load:

$$\text{In-stream load} = \sum \text{WWTP} + \sum \text{Reservoirs} \sum \text{CSO} + \sum \text{ST} + \text{GW} + \text{RO} - \text{Loss} \quad (3.52)$$

where WWTP = wastewater treatment plants; CSO = combined sewer overflows; ST = septic tanks; GW = groundwater; and RO = runoff.

Thus, the general description of the TMDL is:

$$\text{TMDL} = \sum \text{WLA} + \sum \text{LA} + \text{Background} + \text{Future Growth} + \text{MOS} - \text{Loss} \quad (3.53)$$

where WLA = wasteload allocation of point sources and LA = load allocation for nonpoint sources.

The pollutant reductions required under the TMDL for the Cuyahoga River basin are shown in Table 3.9.

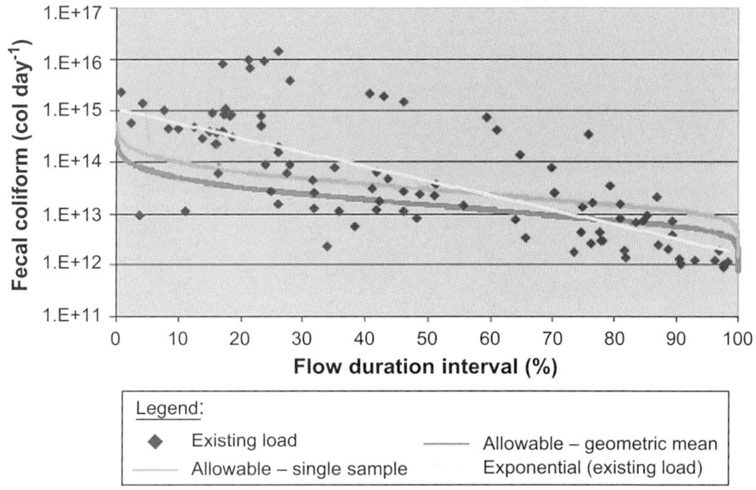

FIGURE 3.30
Fecal coliform TMDLs for a single sample maximum and geometric mean criteria compared to existing load data for the Cuyahoga River at Independence, Ohio, gage station. [*See color plate section*]
Source: Ohio Environmental Protection Agency (2003). Lower Cuyahoga River Watershed TMDLs.

The US Environmental Protection Agency's (EPA) rationale for approving the TMDLs for fecal bacteria in the Rock Creek watershed in Montgomery County, Maryland [30] illustrates how environmental microbiology and bioengineering can be combined to address impairments. The watershed comprises about 19,684 hectares; about 80% of the drainage area lies within Montgomery County, Maryland, and 20% within Washington, District of Columbia. The headwaters of Rock Creek are at Laytonsville, Maryland, flowing through Montgomery County to Washington, DC, eventually reaching the Potomac River. The North Branch of Rock Creek begins at Mount Zion, Maryland, and discharges to Rock Creek in Rockville, Maryland. Two surface impoundments are located in the Rock Creek Watershed: Needwood Lake and Lake Bernard Frank (see Figure 3.31).

The Rock Creek watershed consists of the mainstem of Rock Creek, the North Branch, and the tidal drainage area. The mainstem and North Branch are non-tidal streams, i.e they are

Table 3.9 Source category reductions at Independence, Ohio, needed to meet the TMDL for fecal coliform and phosphorus

	Fecal Coliform			Total Phosphorus		
Source	Existing Ave Load (cfu/year)	Allocated Ave Load (cfu/year)	% Reduction	Existing Ave Load (lb/year)	Allocated Ave Load (lb/year)	% Reduction
Runoff	3.47E + 18	1.42E + 18	59%	219716	113780	48%
Point Sources	9.70E + 13	9.70E + 13	0%	170580	120101	30%
Akron CSOs & Bypass	5.26E + 16	1.05E + 15	98%	34629	4635	87%
Septic	2.27E + 15	1.45E + 15	36%	28831	20181	30%
Lake Rockwell Release	3.22E + 13	3.22E + 13	0%	22043	15847	28%
Groundwater	7.27E + 13	7.27E + 13	0%	12924	12924	0%

The % reduction for TP associated with the Akron CSOs is expected only. The TP removal is incidental to the treatment methods proposed by Akron to treat for fecal coliform.
Source: Ohio Environmental Protection Agency (2003). Lower Cuyahoga River Watershed TMDLs.

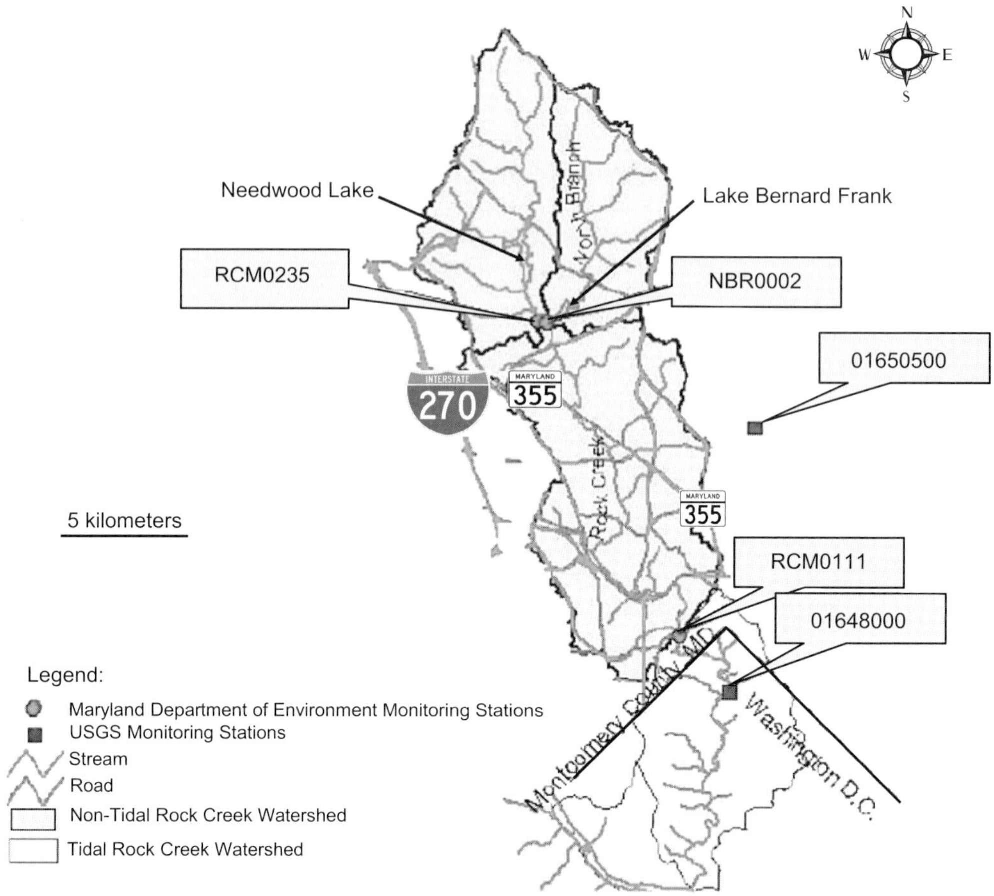

FIGURE 3.31
Rock Creek Watershed.
Source: US Environmental Protection Agency (2007). Decision Rationale: Total Maximum Daily Loads of Fecal Bacteria for the Non-Tidal Rock Creek Basin in Montgomery County, Maryland. Region 3. Philadelphia, Pennsylvania.

free-flowing. The Rock Creek is 33 miles long, with the downstream 9.3 miles flowing through the District of Columbia. Only the last quarter mile of the creek is tidally influenced, with the head of tide located approximately where Pennsylvania Avenue crosses the stream. The creek meets the Potomac River about 108 miles upstream of Chesapeake Bay. Rock Creek drainage area within Montgomery County drainage area is 15,258 hectares.

In 2007, the Maryland Department of the Environment (MDE) prepared the *Total Maximum Daily Loads of Fecal Bacteria for the Non-Tidal Rock Creek Basin in Montgomery County, Maryland*. The document identified nutrients, sediments, and fecal bacteria as the cause of impaired uses in the drainage basin.

The US EPA requires that total maximum daily loads:

Be designed to implement applicable water quality standards.
Include a total allowable load as well as individual wasteload allocations (WLAs) and load allocations (LAs).
Consider the impacts of background pollutant contributions.
Consider critical environmental conditions.
Consider seasonal environmental variations.
Include a margin of safety to account for uncertainty about the relationship between the pollutant and the quality of the receiving waterbody.
Provide reasonable assurance that the TMDLs can be met.
Include public participation.

The total loads allowed is equal to the sum of the individual wasteload allocations (WLAs) for point sources and municipal separate stormwater systems (MS4), plus the land-based load allocations for nonpoint sources set forth below. Also, the TMDLs for fecal bacteria for Rock

Creek are incorporated into the State's water quality management plan. Thus, the general description of the TMDL is:

$$\mathrm{TMDL} = \sum \mathrm{WLA} + \sum \mathrm{LA} + \mathrm{MOS} \quad (3.54)$$

where WLA is the wasteload allocations for point sources in the basin; LA is the load allocation for nonpoint sources; and MOS is the added margin of safety.

The benchmark of success is the state's water quality criteria. For example, in Maryland *E. coli* should not exceed 125 MPN/100 mL steady state geometric mean indicator density; enterococci should not exceed 33 MPN/100 mL in fresh waters and 35 MPN/100 mL in marine water. The steady state geometric mean was calculated from samples collected during steady state flows with at least five representative sampling events. These are shown in Table 3.10. In addition to these sources, the county has a treatment facility that will be allowed to release 60 billion MPN/day according to the provisions of a Clean Water Act (National Pollution Discharge Elimination System) permit. Much of the LA for nonpoint sources will be achieved through best management practices (BMPs).

Neither of these TMDLs specifically identified confined animal feeding operations (CAFOs) as point sources in the WLA fraction of Eqs. 3.52, 3.53, and 3.54. This is interesting and possibly revealing since both basins have substantial land use dedicated to agricultural operations. Perhaps, this fraction is included in the LA (nonpoint sources in Tables 3.9 and 3.10), but CAFOs should be more explicitly identified and calculated, especially when fecal coliform and other microbial populations are part of the contaminants of concern.

Advection

Genetic material can be transported within the hydrosphere and troposphere. In fact, movement within fluids (i.e. air and water) not only is a mechanism for the microbes themselves, but can be the dominant transport mechanism for pollen, cysts, and spores. From a purely physical motion perspective, the most straightforward pollutant transport process arguably is advection ($J_{Advection}$). *Advection* is the transport of matter within the streamlines of a fluid, i.e. with the water or air flow. In terms of total volume and mass of pollutants moved, advection accounts for the lion's share. In fact, another name for advection is bulk transport. During advection, a contaminant is moved along with the fluid or, in the language of environmental science, the *environmental medium*. The contaminant is merely "hitching a ride" on the fluid as it moves through the environment. Environmental fluids move within numerous matrices, such as the flow of air and water between soil particles, within sediment, in unconsolidated materials underground, and in the open atmosphere. Surface water is also an environmental medium in which advection occurs.

Table 3.10 Fecal bacteria summary for Rock Creek, Maryland, TMDL allocations

	Baseline	TMDL	WLA-WWTP	WLA-MS4[2]	LA[3]	MOS[4]
Sub-Watershed	Billion MPN/100 mL enterococci/day					
NBR0002us	1786	37	0	13	24	5% explicit
RCM0235us	497	32	0	12	20	5% explicit
RC0111sub	1672	56	0	35	21	5% explicit
Total	3955	125	0	60	65	

MPN = Most Probable number.
WLA-WWTP = Wasteload allocation for non MS4 systems (municipal or industrial).
WLA-MS4 = Wasteload allocation for MS4 systems.

Advection is considered a passive form of transport because the contaminant moves along with the transporting fluid. That is, the contaminant moves only because it happens to reside in the medium. Advection occurs within a single medium and among media. The rate and direction of transport is completely determined by the rate and direction of the flow of the media.

The simplest bulk transport within one environmental medium or compartment is known as homogeneous advection, where only one fluid is carrying the contaminant. The three-dimensional rate of homogeneous, advective transport is simply the product of the fluid medium's flow rate and the concentration of the contaminant in the medium:

$$N = QC \tag{3.55}$$

where Q is the flow rate of the fluid medium (e.g., $m^3\ sec^{-1}$) and C is the concentration of the chemical contaminant being transported in the medium (e.g., $\mu g\ m^{-3}$).

Therefore, the units for three-dimensional advection are mass per time (e.g. $\mu g\ sec^{-1}$). There is much variability in these rates, so different units will be used for different media. For example, atmospheric transport and large surface waters, like rivers, move large volume plumes, while groundwater systems move very slowly.

Heterogeneous advection refers to those cases where there is a secondary phase present inside the main advective medium. For example, the presence of particulate matter (i.e. suspended solids) in advecting river water, or particles carried by wind. Heterogeneous advection involves more than one transport system within the compartment. For example, the contaminant may be dissolved in the water *and* sorbed to solids that are suspended in the water. Thus, not only the concentration of the dissolved fraction of the contaminant must be known, but also the concentration of chemical in and on the solid particles.

For example, if a system's homogeneous advection of tree pollen is 500 $\mu g\ sec^{-1}$ and suspended particles are moving in the river at a rate of 0.001 $m^3\ sec^{-1}$, and analyses have shown that the suspended particles have an average pollen concentration of 500 $mg\ L^{-1}$, the heterogeneous (total) advective flow of pollen in the river can be estimated if the concentration of particles is assumed to be homogeneous in the river, i.e. the suspended particles can be treated as a homogeneous, advective transport, which can be added to the sorbed pollen fraction for the total stream load:

$$N = \frac{(0.001\ m^3\ sec^{-1})(\ 500\ mg\ L^{-1})(1000L)}{m^3} = 500\ mg\ sec^{-1}$$

Thus, the total advective transport of pollen $= 500\ mg\ sec^{-1} + 50\ \mu g\ sec^{-1} = 500.5\ mg\ sec^{-1}$.

This example illustrates that heterogeneous advection is a common transport mechanism for highly lipophilic compounds or otherwise insoluble matter (e.g. pollen, spores, and cysts with lipid membranes) that are often sorbed to particles as compared to dissolved in water. This may be similar to chemical transport, such as when metals form both lipophilic and hydrophilic species (e.g. ligands), depending upon their speciation. Many biomolecules and other complex organic compounds, such as the PAHs and PCBs, are relatively insoluble in water. Therefore, most of their advective transport is by attaching to particles. In fact, lipophilic organics are likely to have orders of magnitude greater concentrations in suspended matter than is dissolved in the water (recalling the discussion of K_{ow} earlier in this chapter).

Solutes in the groundwater also move in the general direction of groundwater flow, i.e., via advection, with minor control by diffusion. The zone of saturation's pore pressures are different from atmospheric pressure due to *head.* Flow is produced through the pore spaces. There is sufficient difference in head at one location versus another, so the advection follows this *hydraulic gradient* (calculation provided in Figure 3.32).

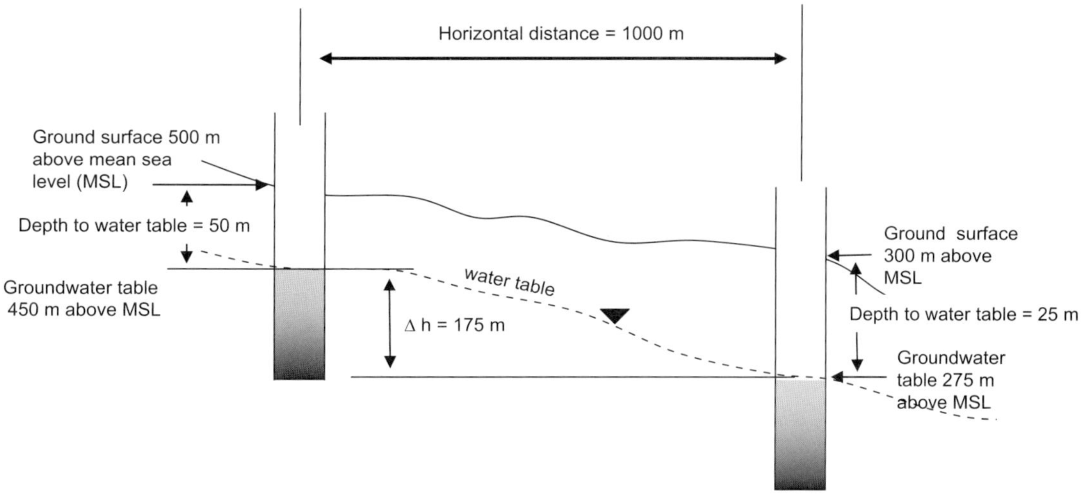

FIGURE 3.32
The hydraulic gradient (K), is the change in hydraulic head (h) over a unit distance. In this case, the horizontal distance is 1000 m. The h is the difference between the upper h (450 m) and the lower h (275 m), thus h = 175 m. So, $K = 175$ m/1000 m, or 0.175 (dimensionless).

Transmissivity is the rate at which water passes through a unit width of the aquifer under a unit hydraulic gradient. It is equal to the hydraulic conductivity multiplied by the thickness of the zone of saturation. It is expressed as volume per time per length such as gallons per day per foot (gal d^{-1} ft^{-1}) or liters per day per meter (L d^{-1} m^{-1}).

Accordingly, solutes in groundwater are predominantly transported by advection.

Another example of advective transport is atmospheric *deposition* of contaminants. The sorption of contaminants to the surface of atmospheric water droplets is known as *wet deposition* and sorption to solid particles is known as *dry deposition*. The process where these contaminants are delivered by precipitation to the earth is advection.

Rather than three-dimensional transport, many advective models are represented by the one-dimensional mass *flux* equation for advection, which can be stated as:

$$J_{Advection} = \bar{v}\eta_e[c] \qquad (3.56)$$

where: $\bar{v}$ = average linear velocity (m s^{-1})

η_e = effective porosity (per cent, unitless)

C = chemical concentration of the solute (kg m^{-3})

Probably the most common application of the flux term is in two dimensions:

$$J_{Advection} = \bar{v}[c] \qquad (3.57)$$

Two-dimensional fluxes are an expression of the transport of a contaminant across a unit area. This rate of this transport is the flux density (see Figure 3.33), which is the contaminant mass moving across a unit area per time. In most environmental applications, fluid velocities vary considerably in time and space (think about calm versus gusty wind conditions, for example). Thus, estimating flux density for advection in a turbulent fluid usually requires a time integration to determine average concentrations of the contaminant. For example, a piece of air monitoring equipment may collect samples every minute, but the model or calculation calls for an hourly value, so the 60 values are averaged to give one integrated concentration of the air pollutant.

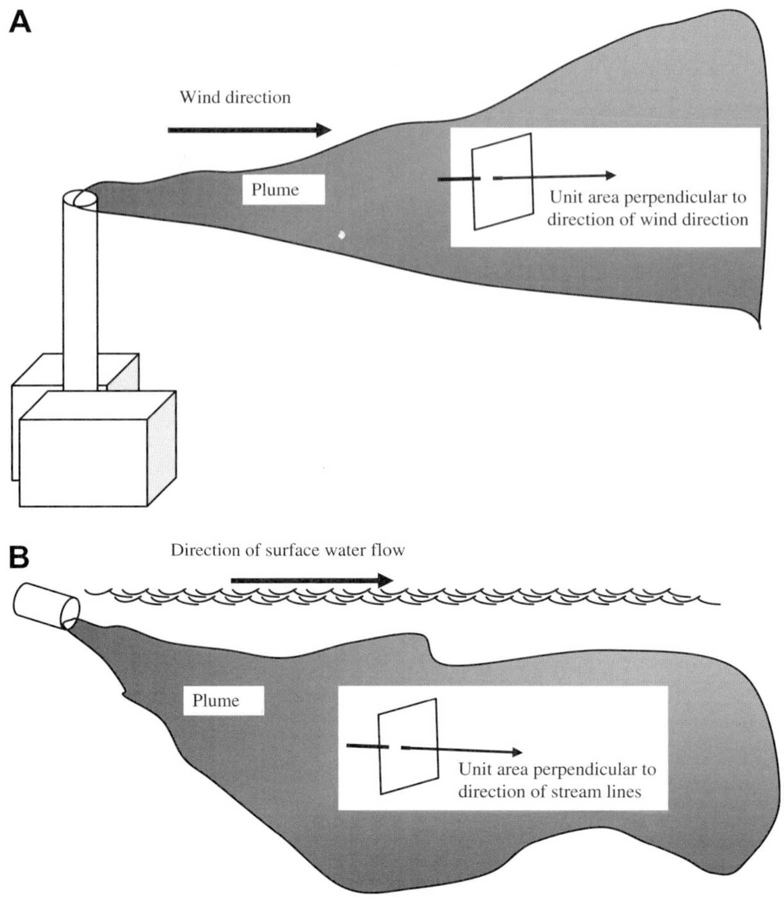

FIGURE 3.33
Determining flux density using an imaginary cross-sectional area across which contaminant flux is calculated in the atmosphere (A) and in surface waters (B).

For example, if the concentration of the pesticide dieldrin is 15 ng L^{-1} in a stream with a velocity of 0.1 m sec^{-1}, the average flux density of the dieldrin as it moves downstream can be calculated as:

$$[\text{dieldrin}] = 15 \text{ ng L}^{-1} = 0.015 \text{ ng m}^{-3}$$

$$J_{Advection} = \bar{v}[c] = (0.1 \text{ m sec}^{-1})(0.015 \text{ ng m}^{-3}) = 0.0015 \text{ ng m}^{-2} = 1.5 \text{ pg m}^{-2}$$

Dispersion

Numerous dispersion processes are at work in environmental biochemodynamic systems. As is the case for diffusion, the type of dispersion can vary according to scale. Contaminant transport literature identifies two principal types, i.e., hydrodynamic dispersion and mechanical dispersion. However, these are actually not mutually exclusive terms. In fact, mechanical dispersion is a factor in dynamic dispersion. See Figures 3.34 and 3.35 for a computationally combined advective and dispersive air transport system.

Such computational approximations are based on first principles of motion and thermodynamics and can be applied to any physical agent. For example, they may be applied to an escaped or intentionally released genetically engineered microbe or its spores. They are also useful in predicting the dispersion of agents in emergency situations.

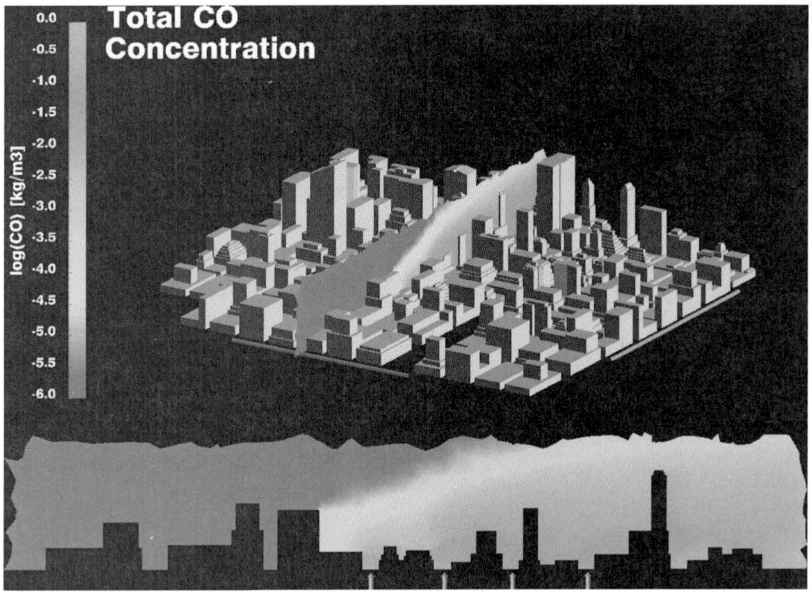

FIGURE 3.34
Profile computational fluid dynamic model depicting an air pollution plume along 59th Street in New York City. Much of the plume is caused by advection by wind through the urban canyons. Dispersion accounts for much of the transport within the street canyons. The vertical profile at the bottom of the figure indicates the dispersion taking place above the buildings as the plume is advected horizontally. The source of the carbon monoxide is a line along the street. [*See color plate section*]
Source: A. Huber (2003). US Environmental Protection Agency.

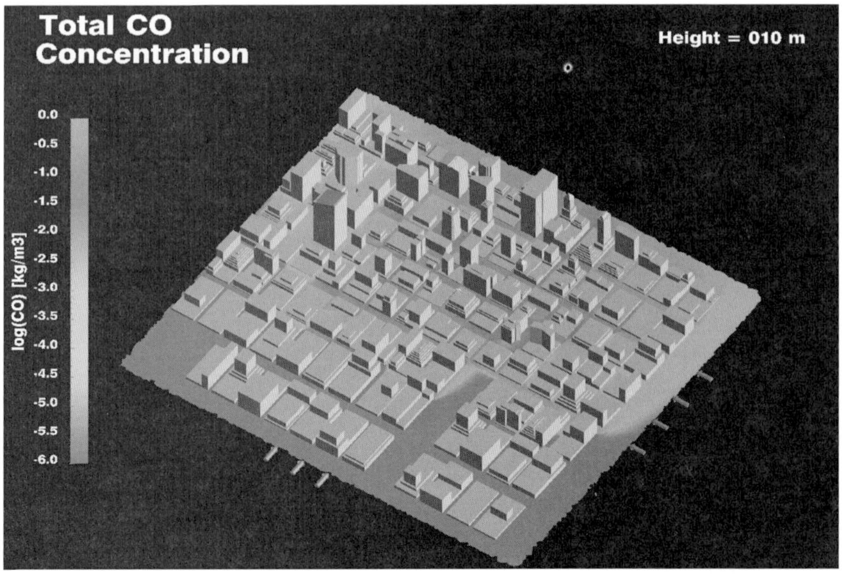

FIGURE 3.35
Plan view of computational fluid dynamic model depicting an air pollution plume along 59th Street in New York City. [*See color plate section*]
Source: A. Huber (2003). US Environmental Protection Agency.

AERODYNAMIC AND HYDRODYNAMIC DISPERSION

The process of a contaminant plume's spread into multiple directions longitudinally is known as dynamic dispersion. If in air, the spreading is known as *aerodynamic dispersion,* and if in water it is *hydrodynamic dispersion.* This spreading results from physical processes that affect the velocity of different molecules in an environmental medium. For example, in aquifers, the process is at work when the contaminant traverses the flow path of the moving groundwater. This is the result of two physical mechanisms: molecular diffusion and mechanical dispersion. Molecular diffusion can occur under both freely flowing and stagnant fluid systems, while mechanical dispersion is of most importance in flowing systems. The units of dynamic dispersion d_d are area per time (e.g. $cm^2 \sec^{-1}$ for groundwater). Dynamic dispersion is expressed as:

$$d_d = \alpha v_x + D_e \tag{3.58}$$

where: a = dispersivity of the porous medium (cm)

v_x = average linear groundwater velocity ($cm \sec^{-1}$)

D_e = diffusion coefficient of the contaminant ($cm^2 \sec^{-1}$)

Mechanical dispersion is the result of the tortuosity of flow paths within an environmental medium. It is especially important in soil and other unconsolidated materials that render circuitous the paths through which the fluid must travel. Thus, when fluids move through spaces in porous media, they cannot move in straight lines, so they tend to spread out longitudinally and vertically. This is what makes mechanical dispersion the dominant mechanism causing hydrodynamic dispersion at the fluid velocities that are often encountered in aquifers and soil.

Since dispersion is the mixing of the pollutant within the fluid body (e.g. aquifer, surface water or atmosphere), the question arises as to whether it is better to calculate the dispersion from physical principles, using a deterministic approach, or to estimate the dispersion using statistics, actually probabilities. The Eularian model bases the mass balance around a differential volume. A Lagrangian model applies statistical theory of turbulence, assuming that turbulent dispersion is a random process described by a distribution function. The Lagrangian model follows the individual random movements of molecules released into the plume, using statistical properties of random motions that are characterized mathematically. Thus, this mathematical approach estimates the movement of a volume of chemical (particle [31]) from one point in the plume to another distinct point during a unit time. Thus, the path each particle takes during this time is an ensemble mean field that relates to the probabilities for particle displacement:

$$\overline{[c]}(x, y, z, t) = M_{Total} P(Dx, t) \tag{3.59}$$

where: $Dx = x_2 - x_1$ = particle displacement

$P(Dx_2, t)$ = probability that the point x_2 will be immersed in the dispersing media at time t

M_{Total} = total mass of particles released at x_1

$\overline{[c]}$ = mean concentration of all released particles = mass of particles the plume $dx \cdot dy \cdot dz$ around x_2

Gaussian dispersion models assume a normal distribution of the plume (see Figure 3.36). This is a common, but at best a first approximation of the actual dispersion within a biochemodynamic system.

In a deterministic approach, the dispersion includes mixing at all scales. At the microscopic scale, the model accounts for frictional effects as the fluid moves through pore spaces, the path length around unconsolidated material (the tortuosity), and the size of the pores. At the larger scales, characteristics of strata and variability in the permeability of the layers must be described. So, a deterministic dispersion flux would be:'

$$J_{Dispersion} = \underline{D} \cdot grad[c] \tag{3.60}$$

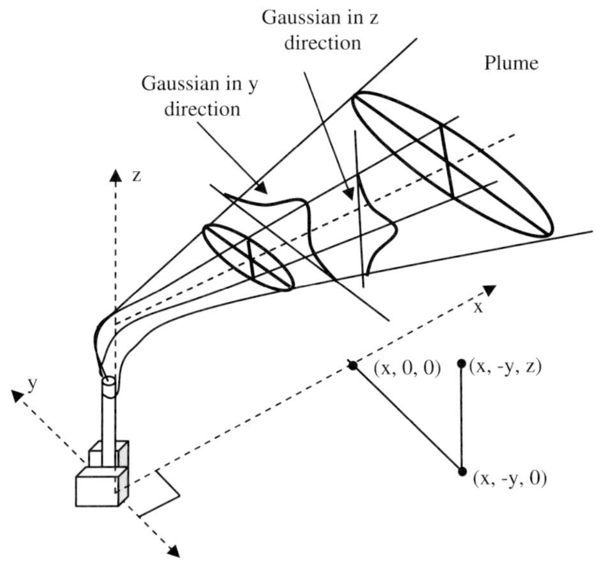

FIGURE 3.36
Gaussian plume. The pollutants are assumed to be distributed vertically and horizontally in a statistically normal manner about the plume center line away from the point of release.

where: $J_{Dispersion}$ = mass flux of solute due to dispersion ($kg\,m^{-2}\,s^{-1}$)

$\underline{D}$ = dispersion tensor ($m\,s^{-1}$)

C = concentration of chemical contaminant ($kg\,m^{-3}$)

The $\underline{D}$ includes coefficients for each direction of dispersion, i.e. longitudinally, horizontally and vertically (D_{xx}, D_{xy}, D_{xz}, D_{yy}, D_{yz}, D_{zz}).

Diffusion

In diffusion, contaminants and other solutes move from higher to lower concentrations in a solution. For example, if a sediment contains methyl mercury (CH_3Hg) in concentrations of 100 ng L^{-1} at a depth of 3 mm and at 10 ng L^{-1}depth of 2 mm, diffusion would account for the upward transport of the CH_3Hg. Diffusion is described by *Fick's laws.* The first law says that the flux of a solute under steady state conditions is a gradient of concentration with distance:

$$J_{Diffusion} = -D\frac{d[c]}{dx} \tag{3.61}$$

where D is a diffusion coefficient (units of area/time), [c] is the molar concentration of the contaminant, and x is distance between the points of contaminant concentration measurements (units of length). Note that the concentration can also be expressed as mass per fluid volume (e.g. mg L^{-1}), in which case, flux is expressed as:

$$J_{Diffusion} = -D\frac{dC}{dx} \tag{3.62}$$

The concentration gradient can also appear in the form:

$$J_{Diffusion} = -d_o i_c \tag{3.63}$$

where d_0 is again the proportionality constant, and:

$$i_c = \frac{\partial[c]}{\partial x} \tag{3.64}$$

The negative sign denotes that the transport is from greater to lesser contaminant concentrations. Fick's second law comes into play when the concentrations are changing with time. The change of concentrations with respect to time is proportional to the second derivative of the concentration gradient:

$$\frac{\partial[c]}{\partial t} = \frac{\partial^2[c]}{\partial x^2} \tag{3.65}$$

All of the diffusion expressed in these equations is one-dimensional, but three-dimensional forms are available and used in models.

Two types of diffusion are important to transport of contaminants: molecular diffusion and turbulent or eddy diffusion. Each Fickian process operates at its own scale. At the molecular level, in surface waters and atmospheric systems, diffusion dominates as a transport mechanism only in a very thin boundary layer between the fluid media. However, in sediments, sludge, and groundwater, this can be an important transport mechanism. Since the concentration gradient (i_c) is the change in concentration (for example, in units of kg m^{-3}) with length (in meters), the units of i_c are kg m^{-4}. Diffusion is therefore analogous to the physical potential field theories (that is, flow is from the direction of high potential to low potential, such as from high pressure to low pressure). This gradient is observed in all phases of matter, solid, liquid or gas. So, molecular diffusion is really only a major factor of transport in porous media, such as soil or sediment, and can be ignored if other processes, such as advection, lead to a flow greater than 2×10^{-5} m s^{-1} [32]. However, it can be an important process for source characterization, since it may be the principal means by which a contaminant becomes mixed in a quiescent container (such as a drum, a buried sediment, or a covered pile) or at the boundaries near clay or artificial liners in landfill systems.

Turbulent motion in fluids is characterized by the formation of eddies of various sizes. These eddies can be modeled according to Fick's first law (concentration gradients), so that the same equations in this chapter applied to molecular diffusion may also be used to estimate the transport of contaminants by eddy diffusion. Like molecular diffusion, eddy diffusion can be modeled in one, two, or three dimensions. One-dimensional models assume that the diffusion coefficient (D) does not change with respect to direction. However, D must be adjusted to the model. This must be done when D is expected to vary with spatial location and time (which it always does, but if the change is not significant, it may be ignored). The coefficient may also be *anisotropic*, that is, it may vary in different directions or vertically in the air or water.

Pollutants can diffuse from biotechnological operations. Assume, for example, that a composting system is operating in a building with a "footprint" of 200 m^2 of untreated soil. The soil air 3 meters beneath the compost has a concentration of 2 μg cm^{-3} total hydrocarbons (THC). If the diffusion coefficient is 0.01 cm^2 sec^{-1} in this particular soil and assuming that the air is well mixed, the flux density of the vapor and the rate of vapor release by molecular diffusion can be calculated as a one-dimensional flux, the vertical concentration (upward on z axis):

$$\frac{dC}{dz} = (2 \times 10^{-6}\ \text{g cm}^{-3})/300\ \text{cm} = 6.7 \times 10^{-9}\ \text{g cm}^{-4}$$

The flux density is:

$$J_{Diffution} = -D\frac{dC}{dx} = (10^{-2}\ \text{cm}^2\ \text{sec}^{-1}) \times (6.7 \times 10^{-9}\ \text{g cm}^{-4}) = 6.7 \times 10^{-11}\ \text{g cm}^{-2}\ \text{sec}^{-1}$$

Applying the flux density to the 200 m^2 (2×10^6 cm^2), the penetration of the vapor into the air:

$$(6.7 \times 10^{-11}\ \text{g cm}^{-2}\ \text{sec}^{-1}) \times (2 \times 10^6\ \text{cm}^2) \times (3600\ \text{sec hr}^{-1}) \times (24\ \text{hr day}^{-1})$$
$$= 11.5\ \text{g day}^{-1}$$

Overall effect of the fluxes, sinks, and sources

Numerous, interrelated biochemodynamic processes determine transport and environmental fate of contaminants. Recall that one of the laws dictating fluid dynamics mentioned at the beginning of this discussion included conservation of mass. This is true, of course, but the molecular structure of the chemical may very well change. The transport depends upon chemical characteristics of the compound (e.g. solubility, vapor pressure, reactivity, and oxidation state) and those of the environment (e.g., presences of microbes, redox potential, ionic strength, and pH). The chemical degradation can be as simple as a first-order decay process (i.e. the degradation of the contaminant concentration C):

$$\frac{\partial C}{\partial t} = -\lambda c \tag{3.66}$$

The degradation (λ) terms are applied to each chemical. The new degradation products call for an iterative approach to the transport and fate of each degradation product can be described. As a new compound is formed, it must go through the same scrutiny for each transport step. This is even more critical if the degradates are toxic. Some are even more toxic than the parent compound.

A model of the expected total flux representing the fate (J_{Fate}) of the contaminant can therefore be:

$$J_{Fate} = J_{Desorption} + J_{Diffusion} + J_{Dilution} + J_{Dispersion} + J_{Advection} - J_{Sorption} - \lambda[c] \tag{3.67}$$

Biochemodynamic transport models

The phase and compartmental distributions can be combined into fugacity-based, chemodynamic transport models. Such models are classified into three types:

Level 1 Model: This model is based on an equilibrium distribution of fixed quantities of contaminants in a closed environment (i.e. conservation of contaminant mass). No chemical or biological degradation, no advection, and no transport among compartments (such as sediment loading or atmospheric deposition to surface waters).

A Level 1 calculation describes how a given quantity of a contaminant will partition among the water, air, soil, sediment, suspended particles, and fauna, but does not taken into account chemical reactions. Early Level 1 models considered an area of 1 km^2 with 70% of the area covered in surface water. Larger areas are now being modeled (e.g., about the size of the state of Ohio).

Level 2 Model: This model relaxes the conservation restrictions of Level 1 by introducing direct inputs (e.g. emissions) and advective sources from air and water. It assumes that a contaminant is being continuously loaded at a constant rate into the control volume, allowing the contaminant loading to reach steady state and equilibrium between contaminant input and output rates. Degradation and bulk movement of contaminants (advection) is treated as a loss term. Exchanges between and among media are not quantitified.

Since the Level 2 approach is a simulation of a contaminant being continuously discharged into numerous compartments and which achieves a steady-state equilibrium, the challenge is to deduce the losses of the contaminant due to chemical reactions and advective (non-diffusive) mechanisms.

Reaction rates are unique to each compound and are published according to reactivity class (e.g. fast, moderate, or slow reactions), which allows modelers to select a class of reactivity for the respective contaminant to insert into transport models. The reactions are often assumed to be first-order, so the model will employ a first-order rate constant for each compartment in the environmental system (e.g. x mol hr^{-1}in water, y mol hr^{-1} in air, z mol hr^{-1}in soil). Much uncertainty is associated with the reactivity class and rate constants, so it is best to use rates published in the literature based upon experimental and empirical studies, wherever possible.

Advection flow rates in Level 2 models are usually reflected by residence times in the compartments. These residence times are commonly set to at one hour in each medium, so the advection rate (G_i) is volume of the compartment divided by the residence time (t):

$$G_i = Vt^{-1} \tag{3.68}$$

Level 3 Model: Same as Level 2, but does not assume equilibrium between compartments, so each compartment has its own fugacity. Mass balance applies to the whole system and each compartment within the system. Includes mass transfer coefficients, rates of deposition and re-suspension of contaminant, rates of diffusion, soil runoff, and area covered. All of these factors are aggregated into an intermedia transport term (D) for each compartment.

The assumption of equilibrium in Level 1 and 2 models is a simplification, and often a gross over-simplification of what actually occurs in environmental systems. When the simplification is not acceptable, kinetics must be included in the model. Numerous diffusive and non-diffusive transport mechanisms are included in Level 3 modeling. For example, values for the various compartments' unique intermedia transport velocity parameters (in length per time dimensions) are applied to all contaminants being modeled (these are used to calculate the D values).

It is important to note that models are only as good as the information and assumptions that go into them. For example, neighborhood-scale effects can modify estimates from transport models or from measurement interpolations (barriers, channeling, local flows, trapping). This applies to all transport models, whether highly computational or simplified (see Figure 3.37). Site-specific differences can greatly affect predicted outcomes, such as the extent of gene flow

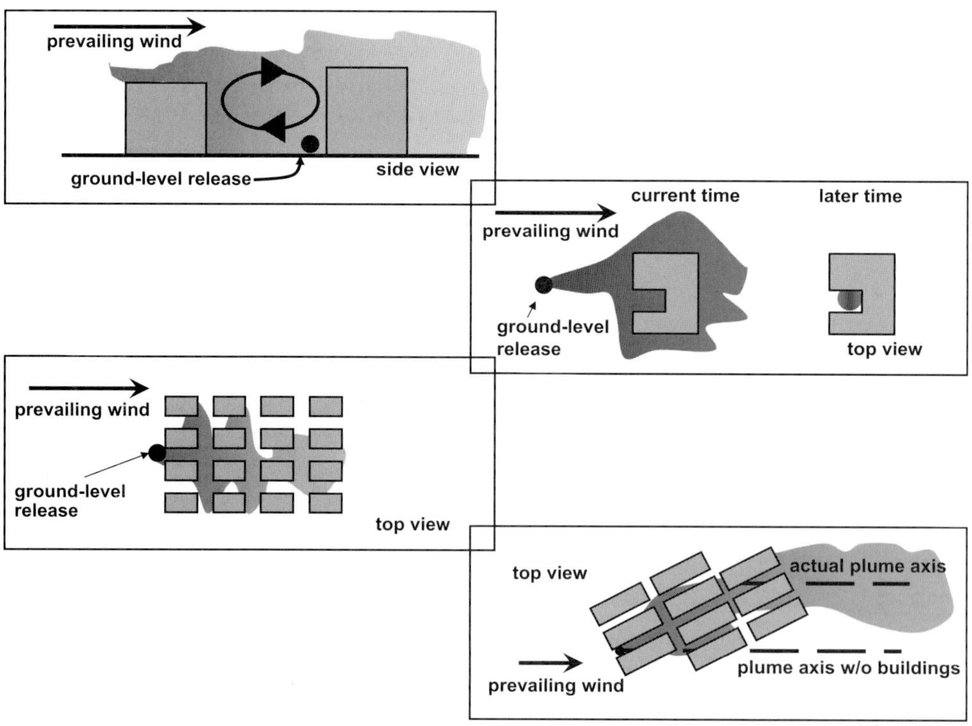

FIGURE 3.37
Scale effects can modify significantly estimates from atmospheric transport models or from monitor interpolations (barriers, channeling, local flows, trapping): there is need for both computational and simplified models.
Source: D.A. Vallero, S.S. Isukapalli, P.G. Georgopoulos and P.J. Lioy (2009). Improved Assessment of Risks from Emergency Events: Application of Human Exposure Measurements. 4th Annual Interagency Workshop: Using Environmental Information to Prepare and Respond to Emergencies. New York, NY, July 17.

or direction and extent of movement after the accidental or intentional release of genetic material after the release of biotechnologically generated agents or byproducts (see Chapter 8, for example).

Thus, numerous physical, chemical, and biological processes are at work within biotechnological systems and in the various receptor systems after products and organisms are released. This calls for a systematic approach.

SEMINAR TOPIC

How Well Can Biochemodynamic Models Predict Transfer of Genetic Materials in the Environment?

The processes that lead to the ultimate fate of biotechnological materials in the environment are complicated. Their relationships with one another are complex. These complexities must be captured in a model for at least two reasons. First, they must document what is going on during a given period of time. Almost always, actual measurement data are not available to characterize the movement and change of materials in the environment. Second, models provide a means of predicting outcomes based on currently available information. Figure 3.38 illustrates the steps that should be taken to develop an environmental model.

The model generators are to the left and the users (stakeholders) are to the right. This illustrates the connection between the scientific bases for biochemodynamic factors discussed in this chapter with their applications to decision making. Each arrow indicates the connection between processes; each factor and process introduces information to the model, but simultaneously adds uncertainty.

Uncertainty lies in estimating the values to be assigned to each compartment of a model, including theoretical uncertainty (e.g. whether Henry's law published values are relevant to a particular scenario) and measurement uncertainty (e.g. limitations in the ability to measure in the real world, such as the need for destructive methods or analytical problems, such as storing samples before they can be analyzed). However, models add another dimension to uncertainty (see Figure 3.39). Uncertainty increases with each addition of information. Combining models adds even more uncertainty.

Uncertainty can be classified as either aleatory or epistemic. *Aleatory uncertainty* is random uncertainty or stochastic uncertainty that is impossible to predict. All environmental systems have inherent, random uncertainty which cannot be reduced by additional observations. On the other had, *epistemic uncertainty* arises from insufficient knowledge about the system. Thus as more reliable information becomes available regarding the processes described in this chapter, one would expect epistemic uncertainty to fall. Thus, from a scalar perspective, any model outcome (Y) is affected by a function of the probability of these two types of uncertainty [33]:

$$Y = h(U, V) \tag{3.69}$$

Where U = all epistemic uncertainties (uncertain parameters), V = aleatory uncertainties (stochastic variables), and h is the computational model considered as a deterministic function of both uncertainties.

An example of aleatory uncertainty would be a forecast of failure to contain a genetically engineered microbe within the physical confinement of a laboratory or a cleanup site, e.g., where the occurrence of failure occurs randomly over time and the actual time of failure cannot be predicted, no matter the size of the data set. Conversely, epistemic uncertainty includes uncertainties inherent to a variable or parameter, as well as uncertainties in the model's algorithms. For example, our model may miss a possible route by which a GMO can be released because we may have wrong information about the microbe's affinity to a certain type of aerosol or soil particle. With increasing information about these relationships, the epistemic uncertainty should decrease and the predictive capability of the model would commensurately increase.

Dispersion of transgenes into natural populations may occur by various mechanisms. In animals, this can occur as a result of (1) vertical gene transfer via matings with feral animals; (2) introduction of invasive species or shifts in metapopulations; and (3) horizontal gene transfer mediated by microbial agents. It is likely that all three processes can occur simultaneously [34].

A means of predicting this transgenic dispersion is by hierarchical holographic modeling (HHM), which addresses large system complexities by:

> *Identifying the components and processes of all sub-systems and suggests ways in which they might interact with each other based on established/supportive information. The technique decomposes the system by looking at it from many different perspectives including, for example, the functions, activities, geo-political boundaries, or structures of the system. HHM can be used in one of two ways—as a hazard identification tool or as a comprehensive analytical modeling tool. The analyst constructs an HHM by first identifying the most appropriate perspectives for the problem in hand. These are used to define the sub-systems which in turn are further decomposed into components, processes, functions or activities, which may or may not overlap with other sub-systems. The analyst can investigate the quantitative properties of the system if the functions, activities, components or processes of the system can be described by a series of overlapping models, subject to overall system constraints. The analyst(s) can also identify hazards by comparing potential interactions between the*

(Continued)

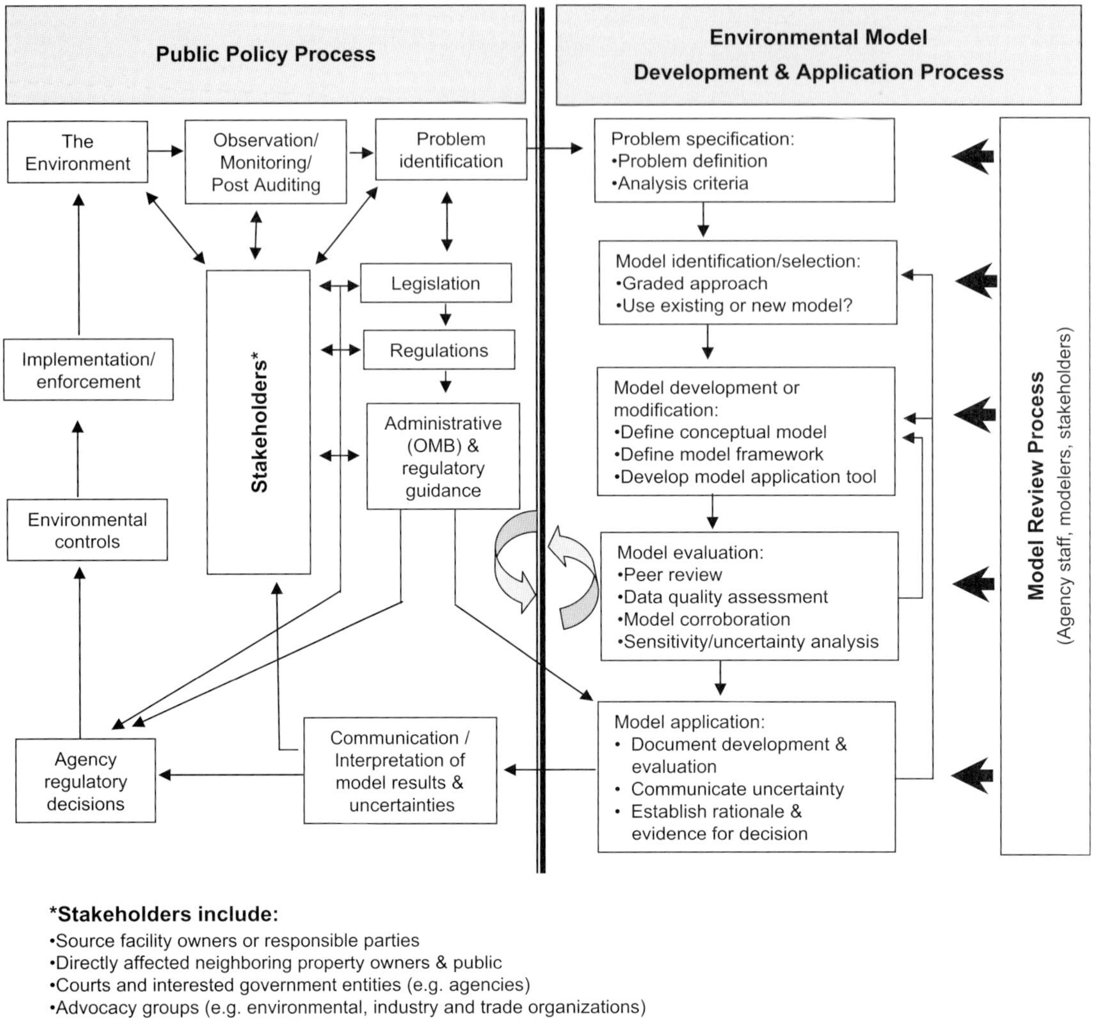

FIGURE 3.38
Steps needed to develop and implement an environmental decision model from inception to completion. These include problem specification; model identification and selection (a site-specific model may be generated *de novo* or based on an existing model framework); model development (including problem- and site-specific model conceptualization, model formulation and design, and model calibration); model evaluation (e.g., based on peer review, data quality assessment, model code verification, model confirmation/corroboration, sensitivity analysis, and uncertainty analysis); model use (diagnostic analysis, solution, and application support for decision making); and review after use.
Source: US Environmental Protection Agency (2006). Science Advisory Board, Regulatory Environmental Modeling, Guidance Review Panel. Review of Agency *Draft Guidance on the Development, Evaluation, and Application of Regulatory Environmental Models* and *Models Knowledge Base.* EPA-SAB-06-009. Letter to Administrator Stephen L. Johnson, August 22, 2006.

> *sub-systems in a qualitative fashion. This is best achieved by a team, whose members are expert in one or more of the chosen perspectives [35].*

This seems reasonable, but what tools are available to these teams to estimate and predict transgene flow after introduction of new genetic material? For example, what if the modified proteins of a transgenic plant are found to cause allergic reactions in a susceptible subpopulation: for example, a gene from a Brazil nut inserted into soybean or corn that triggers a potentially fatal reaction in someone with severe nut allergies. How would the precautionary principle be met, i.e. that a transgenic crop be substantially known not to cause allergies in sensitive subpopulations *before* regulatory approval? How can the potential flow of the allergenic protein be predicted in terms of crop-to-crop genetic drift and human factors (e.g. mislabeling, mishandling)?

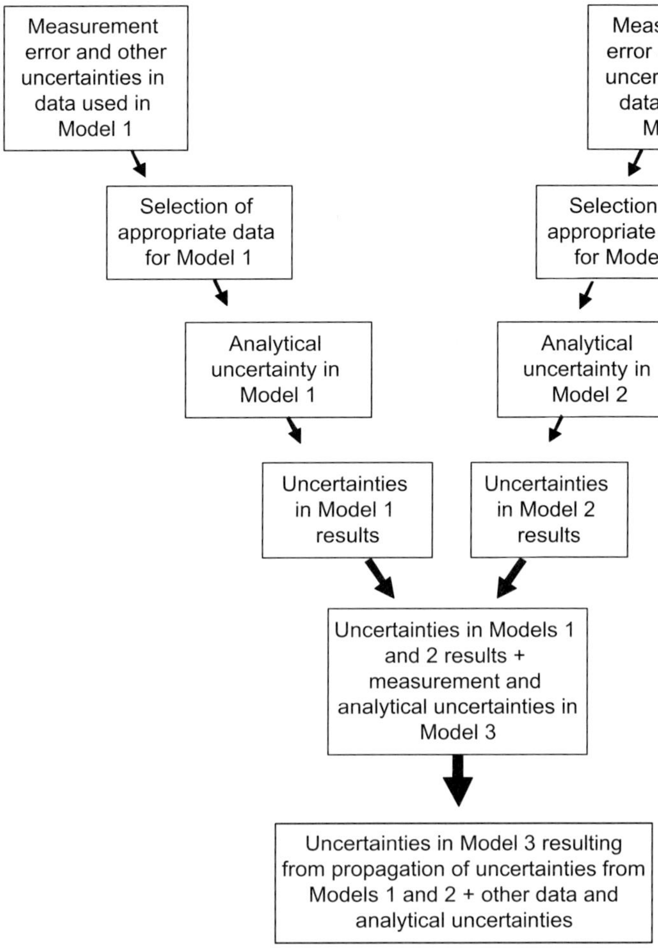

FIGURE 3.39
Propagation of uncertainty in environmental models.

Seminar Questions

How well do presently available models support environmental decision makers who must deal with the uncertainty inherent in environmental models and their application? What additional advice should be given to address uncertainty, and why?

Is the use of methods such as Bayesian networks an effective and practicable way for decision makers to incorporate uncertainty within their decisions and to communicate this uncertainty to stakeholders? If so, how? Are there alternative methods available?

Are there lessons learned from the TMDL process described in this chapter? For example, can the WLA and LA fractions of the TMDL indicate specific genetically modified strains of bacteria in a manner similar to that used to predict and reduce the levels of pathogenic bacteria, nutrients or other pollutants?

REVIEW QUESTIONS

What are the molar concentrations of $[H^+]$ and $[OH^-]$ of rainwater at pH 3.7 at 25 °C? How might these conditions affect the transport and transformation of genetic material in the environment, compared to "normal" rainwater?

How might transport mechanisms (e.g. advection, dispersion, and diffusion) affect spore and other biological material in the environment?

Identify weakness in a Gaussian model. Why are such models used?

Find the properties of two compounds that are part of a biotechnology (e.g. the purification or fermentation processes) with Henry's Law constants that differ by more than 3 logs (See for example, the Risk Assessment Information System: http://rais.ornl.gov/cgi-bin/tox/TOX_select). How do the physical and chemical properties of these two chemicals affect the potential and extent of environmental damage after release?

How does scale affect the choice of model used to predict exposure and risk from a biotechnological agent?

How can confined animal feeding operations be considered in the TMDL process?

An agricultural system is growing genetically modified soybeans. The mean wind speed from the field is 4.5 m sec^{-1}. The average aerodynamic diameter of the aerosols is 2.5 μm. Analyses have shown that these aerosols have an average pollen concentration of 10 mg L^{-1}. What is the heterogeneous (total) advective flow of pollen in the air? Determine whether this is a realistic scenario and, if not, what other information would you need to know to predict advective transport more accurately?

What does the information in the previous question tell you about the likelihood of genetic drift from this field?

Give an example of when is a level 2 model sufficient to estimate gene flow. When would a level 3 model be needed?

NOTES AND COMMENTARY

1. This is one of many counterintuitive and seemingly oxymoronic terms common in environmental sciences, such as "dynamic equilibrium." However, this is the nature of environmental science and engineering. The field is divided between theoretical and empirical concepts. We may know that the sorption or dissolution or volatilization is incomplete, but we have seen similar situations in the laboratory and field so often that we can prepare a "non-equilibrium constant" for a "pseudo-steady state condition."
2. For the calculations and discussions of solubility equilibrium, including this example, see C.C. Lee and S.D. Lin (Eds) (2000). *Handbook of Environmental Engineering Calculations*. McGraw-Hill, New York, NY.
3. Reference for all of the organic solvents: International Agency for Research on Cancer (1977). *Monographs on the Evaluation of the Carcinogenic Risk of Chemicals to Man: 1972 – Present*. World Health Organization, Geneva, Switzerland.
4. T. Boublik, V. Fried and L. Brown (1984). *The Vapour Pressures of Pure Substances*, 2nd Edition. Elsevier, Amsterdam.
5. This is an important aspect of chromatography, as well. The lower boiling point compounds, i.e. the VOCs, usually come off the column first. That is, as the gas chromatograph's oven increases the temperature of the column, the more volatile compounds leave the column first, and they hit the detector first, so their peaks show up before the less volatile compounds. Other chemical factors such as halogenation and sorption affect this relationship, so this is not always the case, but as a general rule, a compound's boiling point is a good first indicator of residence time on a column.
6. Fugacity models are valuable in predicting the movement and fate of environmental contaminants within and among compartments. This discussion is based on work by one of the pioneers in this area, Don MacKay, and his colleagues at the University of Toronto. See, for example, D. MacKay and S. Paterson (1991). Evaluating the fate of organic chemicals: A level III fugacity model. *Environmental Science & Technology* 25: 427–436.
7. W. Lyman (1995). Transport and transformation processes. In: G. Rand (Ed.), *Fundamentals of Aquatic Toxicology: Effects, Environmental Fate, and Risk Assessment*, 2nd Edition. Taylor & Francis, Washington, DC, Chapter 15.
8. Professor Daniel Richter of Duke University's Nicholas School of the Environment has waxed eloquent on this subject.
9. See J. Westfall (1987). Adsorption mechanisms in aquatic surface chemistry. In: *Aquatic Surface Chemistry*. Wiley-Interscience, New York, NY.
10. See D. Mackay and F. Wania (1995). Transport of contaminants to the Arctic: partitioning, processes and models. *The Science of the Total Environment* 160/161: 25–38.
11. This equation, known as the Domenico solution, is found in: C. Newell, H. Rifai, J. Wilson, J. Connor, J. Aziz and M. Suarez (2003). *Ground Water Issue: Calculation and Use of First-Order Rate Constants for Monitored Natural Attenuation Studies*. US Environmental Protection Agency, Ada, OK. The example and associated graphic is also taken from this source.
12. N. Schoch and D. Evers (2002). Monitoring mercury in common loons. New York Field Report, 1998–2000. Report BRI 2001-01 submitted to US Fish Wildlife Service and New York State Department of Environmental Conservation, BioDiversity Research Institute, Falmouth, ME.
13. United Nations Environmental Programme (2002). Chemicals: North American Regional Report, Regionally Based Assessment of Persistent Toxic Substances, Global Environment Facility.
14. B.R. Sonawane (1995). Chemical contaminants in human milk: an overview. *Environmental Health Perspectives* 103 (Supplement 6): 197–205; and K. Hooper and T.A. McDonald (2000). The PBDEs: an emerging environmental challenge and another reason for breast-milk monitoring programs. *Environmental Health Perspectives* 108: 387–392.

15. E. Dewailly, P. Ayotte, S. Bruneau, C. Laliberté, D.C.G. Muir and R.J. Norstrom (1993). Inuit exposure to organochlorines through the aquatic food chain in Arctic Quebec. *Environmental Health Perspectives* 101: 618–620.
16. E. Dewailly, A.J. Nantel, J.P. Weber and F. Meyer (1989). High levels of PCBs in breast milk of Inuit women from Arctic Quebec. *Bulletin of Environmental Contamination Toxicology* 43: 641–646.
17. Dewailly *et al.*, Inuit exposure to organochlorines.
18. E. Dewailly, S. Bruneau, C. Laliberté *et al.* (1993). Breast milk contamination by PCBs and PCDDs/PCDFs in Arctic Quebec: Preliminary results on the immune status of Inuit infants. In: Dioxin '93: 13th International Symposium on Chlorinated Dioxins and Related Compounds; Vienna, Austria, pp. 403–406.
19. The source for the physicochemical properties of DDT and its metabolites is: United Nations Environmental Programme (2002). Chemicals: North American Regional Report, Regionally Based Assessment of Persistent Toxic Substances, Global Environment Facility.
20. The two principal isomers of DDD are: *p,p'*-2,2-bis(4-chlorophenyl)-1,1-dichloroethane; and *o,p'*-1-(2-chlorophenyl)-1-(4-chlorophenyl)-2,2-dichloroethane. The principal isomer of DDE is *p,p'*-1,1'-(2,2-dichloroethenylidene)-bis[4-chlorobenzene].
21. Although "kinetics" in the physical sense and the chemical sense arguably can be shown to share many common attributes, for the purposes of this discussion, it is probably best to treat them as two separate entities. Physical kinetics, as discussed in previous sections, is concerned with the dynamics of material bodies and the energy in a body owing to its motions. Chemical kinetics addresses rates of chemical reactions. The former is more concerned with mechanical dynamics, the latter with thermodynamics.
22. This example was taken from J. Spencer, G. Bodner and L. Rickard (2003). *Chemistry: Structure and Dynamics*, 2nd Edition. John Wiley & Sons, New York, NY.
23. A major source of information in this section is from H.F. Hemond and E.J. Fechner-Levy (2000). *Chemical Fate and Transport in the Environment.* Academic Press, San Diego, CA.
24. The source of the D value discussion is D. MacKay, L. Burns and G. Rand (1995). Fate modeling. In: G. Rand (Ed.), *Fundamentals of Aquatic Toxicology: Effects, Environmental Fate, and Risk Assessment*, 2nd Edition. Taylor & Francis, Washington, DC, Chapter 18.
25. This is the case throughout this text. Bracketed values indicate molar concentrations, but these may always be converted to mass per volume concentration values.
26. The presentation "Groundwater Modelling: Theory of Solute Transport" by Professor W. Schneider, Technische Universität Hamburg–Harburg, was a source for some of the equations used in this section.
27. V. Novotny (2003). *Water Quality*, 2nd Edition. John Wiley & Sons, Inc. Hoboken, NJ.
28. Source of this section is the US Environmental Protection Agency's website: http://www.epa.gov/glnpo/aoc/index.html.
29. United Sates Code of Federal Regulations. 40 CFR §130.2(i).
30. The source for this TMDL discussion is: US Environmental Protection Agency (2007). Decision Rationale: Total Maximum Daily Loads of Fecal Bacteria for the Non-Tidal Rock Creek Basin in Montgomery County, Maryland. Region 3. Philadelphia, Pennsylvania.
31. Science is not always consistent with its terminology. The term "particle" is used in many ways. In dispersion modeling, the term particle usually means a theoretical point that is followed in a fluid. The point represents the path that the pollutant is expected to take. Particle is also used to mean aerosol in atmospheric sciences. Particle is also commonly used to describe unconsolidated materials, such as soils and sediment. The present discussion, for example, accounts for the effects of these particles (e.g. frictional) as the fluid moves through unconsolidated material. The pollutant PM, particle matter, is commonly referred to as "particles." Even the physicist's particle-wave dichotomy comes into play in environmental analysis, as the behavior of light is important in environmental chromatography.
32. This value is taken from W.A. Tucker and L.H. Nelkson (1982). Diffusion coefficients in air and water. *Handbook of Chemical Property Estimation Techniques.* McGraw-Hill, New York, NY. Such low flows are not uncommon in some groundwater systems, or at the boundary between a landfill and the liner, or within a landfill's clay liner.
33. E. Hofer, M. Kloos, B. Krzykacz-Hausmann, J. Peschke and M. Woltereck (2002). An approximate epistemic uncertainty analysis approach in the presence of epistemic and aleatory uncertainties. *Reliability Engineering and System Safety* 77(3): 229–238.
34. G. Linder, E. Little, L. Johnson and C. Vishy (Eds) (2005). Risk and Consequence Analysis Focused on Biota Transfers Potentially Associated with Surface Water Diversions Between the Missouri River and Red River Basins. US Geological Survey.
35. K.R. Hayes (2004). Robust methodologies for ecological risk assessment. Final Report: Inductive Hazard Analysis for GMOs. Department of the Environment, Water, Heritage and the Arts. Canberra, Australia.

CHAPTER 4

Systems

I never saw no miracle of science that didn't go from a blessing to a curse ...

Sting [1]

It is safe to say that the systems approach has been gaining a stronghold in the life sciences. Songwriter Sting's observation is not always the case, but unforeseen consequences have happened frequently enough in bioengineering that we should be mindful of possible adverse outcomes, even when there appears to be consensus on the benefits. The difference between a risk and a benefit, or a "blessing" and a "curse," can only fully be understood when looking at complex systems comprehensively in space and time. Specifically, bioengineering is applying principles learned from advances on two fronts, molecular biology and genetics.

The molecular biology that underpins biotechnology has been evolving toward systems biology for decades. Ecologists, immunologists, and developmental biologists, for example, have been employing non-equilibrium thermodynamics since 1931, performing system analysis since shortly after the discovery of deoxyribonucleic acid (DNA) in 1944 and the articulation of its molecular structure in 1953.

A few years later, molecular biological research began identifying cellular processes, especially feedback regulation in metabolism. Around 1960 recombinant DNA (rDNA) technologies appeared on the scene; about the same time analog simulation and bioenergetics principles were being put to use. Automated DNA sequencing started in the late 1970s, just before early attempts at *in silico* biology began around 1980. The first genome (*Haemophilus influenzae*) was identified in 1995, during the time that high-throughput at the genome scale was beginning to be developed. In the last decade of the 20th century, the human genome was sequenced and genome scale models and analytical techniques were used to develop organism-scale kinetic models [2]. Thus, systems biology and molecular biology are both providing important advances in the systematic analysis of biochemodynamic processes.

There has also been merging of disciplines at the micro-scale. Both microbial ecology and environmental biotechnology have benefited from the exponential growth in knowledge and tools in materials science, bioengineering, computational methods, and microbiology. Microbial ecology strives to characterize and explain microbial communities. These communities are systems that are self-organizing and self-assembling [3].

BIOTECHNOLOGICAL SYSTEMS

Up to this point, the term *system* has been used with its thermodynamic connotation. Another perspective important to biotechnology is the meaning of systems within the context of systems biology. This is the scientific discipline wherein interactions among the components of biological systems take place, including an appreciation of the functions and mechanisms (e.g. enzymatic, metabolic and other pathways). So, then, what is a biological system? This is the antithesis of the reductionist viewpoint that anything that is important to a complex process can be broken down into simpler, more basic entities. Systems indeed consist of these simpler, foundational parts, but to paraphrase Aristotle, the whole system is much more than the aggregation of these individual components. Systems consist of synergies, antagonisms, and other interrelationships of these parts.

These interrelationships and interactions that characterize a biological system require information beyond the descriptive data. Thus, the various "omics" disciplines have been developed to explain the cellular, subcellular, and molecular interactions that affect a biological system:

- Genomics: the systematic study of genomes
- Proteomics: the systematic study of proteins, especially structure and function
- Metabolonomics (also called *metabonomics*): the systematic study of the metabolic status of the whole organism connecting genomics and proteomics with histopathology. Characterizes metabolic pathways after uptake of a compound, as well as the endogenous metabolic products that exist in an organism, but that respond to exogenous agents in various ways
- Transcriptinomics: the systematic study of the messenger RNA (mRNA) produced by cells

These systematic disciplines are now being used in environmental and human health monitoring efforts. Assessing environmental damage from complex systems with various influences and potential adverse outcomes requires a sound characterization of scale and complexity. Assessing biological systems must take place at ascending levels of increasing complexity. First all matter and energy exchanges and conversions within and among organisms and between these organisms and the abiotic environment must be categorized. Next a more refined screening level assessment must be conducted. Ultimately a complete risk assessment can be carried out [4].

Each level requires reliable data. For example, the Organisation for Economic Co-operation and Development has established the Screening Information Data Sets to survey high-production-volume (HPV) chemicals for potential effects, which includes information to support the preliminary screening level assessment of biological systems [5]. For example, the databases that are accessed through this screening level system include information about chemical and biological agents. In addition to inherent physicochemical data, indirect toxicity information and modes of action are available. For example, information on *Bacillus thuringiensis* (Bt) gives information about various proteins and toxins produced from genetically altered organisms, and Bt toxicity to organisms (see Table 4.1), as well as modes of toxicity (see Figure 4.1). The newly configured databases are crucial to data mining and informatics needed to conduct screening and characterization of environmental insults. The data are often collected for purposes other than the level of systematic assessments, e.g. regulatory approval of a new chemical or a registration of a pesticide, so it is imperative that the user is familiar with the limitations in extending such data, i.e. these must be discerned from attributes delineated in the meta-data.

Computational toxicology uses these various tiers of data and the "omics" disciplines, using mathematical and computer models to predict adverse effects and to better understand the mechanism(s) through which a given agent causes harm. As shown in Figure 4.2, these tools computationally link the levels of biological organization as a substance enters a system by uptake and moves through trophic states and food webs, from molecular to population (human or ecosystem). Thus, genomics, proteomics, metabonomics, and the other computational tools will provide systematic insights that are sorely needed in environmental biotechnologies.

Table 4.1 Example of species-specific screening level information available form high-volume chemical database: effects of *Bacillus thuringiensis* (Bt) on fish

Material tested[a]	Species	Concentration	Duration	Results	References
Bta	*Oncorhynchus mykiss*	100 mg/L water	96 h	No-observed-effect level	R.L. Boeri (1991). Acute toxicity of ABG-6305 to the rainbow trout *(Oncorhynchus mykiss)* (Project No. 9107-A). Hampton, NH, Resource Analysts Inc, Enviro Systems Division, pp 1–26 (Unpublished Abbott document)
Btk	*Lepomis macrochirus*	2.9×10^9 cfu/L water[b] 1.2×10^{10} cfu/g diet[c]	32 days	No significant toxicity or pathology	K.P. Christensen (1990). Dipel technical material *(Bacillus thuringiensis* var. *kurstaki)*: Infectivity and pathogenicity to bluegill sunfish *(Lepomis macrochirus)* during a 32-day static renewal test. Wareham, MA, Springborn Laboratories Inc., pp 1–53 (Unpublished Abbott document No. 90-1-3211)
	Oncorhynchus mykiss	2.9×10^9 cfu/L water[b] 1.1×10^{10} cfu/g diet[c]	32 days	20% mortality but not infectivity	K.P. Christensen (1990). Dipel technical material *(Bacillus thuringiensis* var. *kurstaki)*: Infectivity and pathogenicity to rainbow trout *(Oncorhynchus mykiss)* during a 32-day static renewal test. Wareham, MA, Springborn Laboratories Inc., pp 1–57 (Unpublished Abbott document No. 90-2-3219)
	Cyprinodon variegatus	2.6×10^{10} cfu/L water[c] 3.3×10^9 cfu/g diet[c]	30 days	No significant toxicity or pathology	K.P. Christensen (1990). Dipel technical material *(Bacillus thuringiensis* var. *kurstaki)*: Infectivity and pathogenicity to sheepshead minnow *(Cyprinodon variegatus)* during a 30-day static renewal test. Wareham, MA, Springborn Laboratories Inc., vol 2, pp 253–308 (Unpublished Abbott document No. 90-5-3317)
Bti	*Lepomis macrochirus*	1.2×10^{10} cfu/liter water[c] 1.3×10^{10} cfu/g diet[c]	30 days	No significant toxicity or pathology	K.P. Christensen (1990). Vectobac technical material *(Bacillus thuringiensis* var. *israelensis)*: Infectivity and pathogenicity to bluegill sunfish *(Lepomis macrochirus)* during a 30-day static renewal test. Wareham, MA, Springborn Laboratories Inc., pp 1–55 (Unpublished Abbott document No. 90-2-3228)

(*Continued*)

Table 4.1 **Example of species-specific screening level information available form high-volume chemical database: effects of *Bacillus thuringiensis* (Bt) on fish—cont'd**

Material tested[a]	Species	Concentration	Duration	Results	References
	Oncorhynchus mykiss	1.1×10^{10} cfu/L water[c] 1.7×10^{10} cfu/g diet[c]	32 days	No significant toxicity or pathology	K.P. Christensen (1990). Vectobac technical material *(Bacillus thuringiensis* var. *israelensis)*: Infectivity and pathogenicity to rainbow trout *(Oncorhynchus mykiss)* during a 32-day static renewal test. Wareham, MA, Springborn Laboratories Inc., pp 1–55 (Unpublished Abbott document No. 90-2-3242)
	Cyprinodon variegates	1.3×10^{10} cfu/L water[c] 2.1×10^{10} cfu/g diet[c]	30 days	No significant toxicity or pathology	K.P. Christensen (1990). Vectobac technical material *(Bacillus thuringiensis* var. *israelensis)*: Infectivity and pathogenicity to sheepshead minnow *(Cyprinodon variegatus)* during a 30-day static renewal test. Wareham, MA, Springborn Laboratories Inc., pp 1–57 (Unpublished Abbott document No. 90-4-3288)
Btte	*Salmo gairdneri*	100 mg/L water	96 h	No-observed-effect level	D.C. Surprenant (1989). Acute toxicity of *Bacillus thuringiensis* var. *tenebrionis* technical material to rainbow trout *(Salmo gairdneri)* under static renewal conditions. Wareham, MA, Springborn Life Sciences Inc., pp 1–19 (Unpublished Abbott document)
	Oncorhynchus mykiss	1.6×10^{10} cfu/L water[c] 1.34×10^{10} cfu/g diet[c]	30 days	No significant toxicity or pathology	K.P. Christensen (1990). *Bacillus thuringiensis* var. *tenebrionis*: Infectivity and pathogenicity to rainbow trout *(Oncorhynchus mykiss)* during a 30-day static renewal test. Wareham, MA, Springborn Laboratories Inc., pp 1–54 (Unpublished Abbott document No. 90-3-3263)
	Cyprinodon variegatus	9.94×10^{9} cfu/g diet	30 days	No significant toxicity or pathology	K.P. Christensen (1990). *Bacillus thuringiensis* var. *tenebrionis*: Infectivity and pathogenicity to sheepshead minnow *(Cyprinodon variegatus)* during a 30-day static renewal test. Wareham, MA, Springborn Laboratories Inc., pp 1–50 (Unpublished Abbott document No. 90-6-3348)

Notes: a = commercial formulations; b = nominal concentration; and c = measured average concentration. Bta = *Bacillus thuringiensis* subspecies *aizawai;* Bti = *Bacillus thuringiensis* subspecies *israelensis;* Btk = *Bacillus thuringiensis* subspecies *kurstaki;* Btte = *Bacillus thuringiensis* subspecies *tenebrionis.*
Source: United Nations Environment Programme (1999). Environmental Health Criteria 217: *Bacillus thuringiensis.*

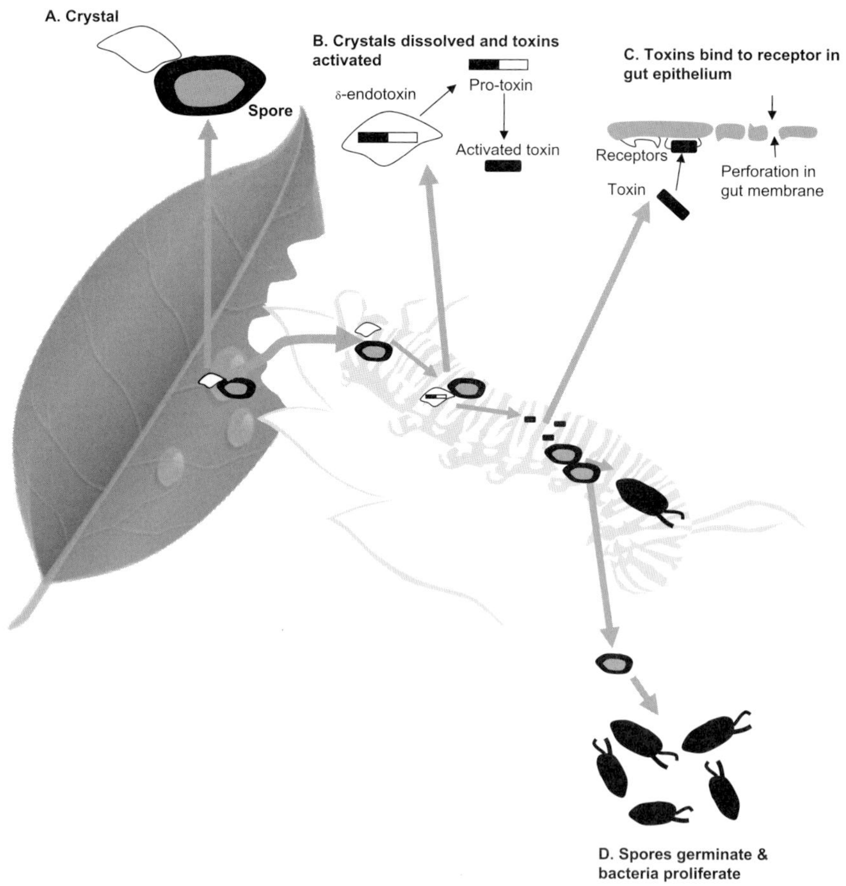

FIGURE 4.1
Example of biological mechanism-related information for screening level risk assessments; in this instance, the mechanism for toxicity of *Bacillus thuringiensis.*
Source: United Nations Environment Programme (1999). Environmental Health Criteria 217: *Bacillus thuringiensis.*

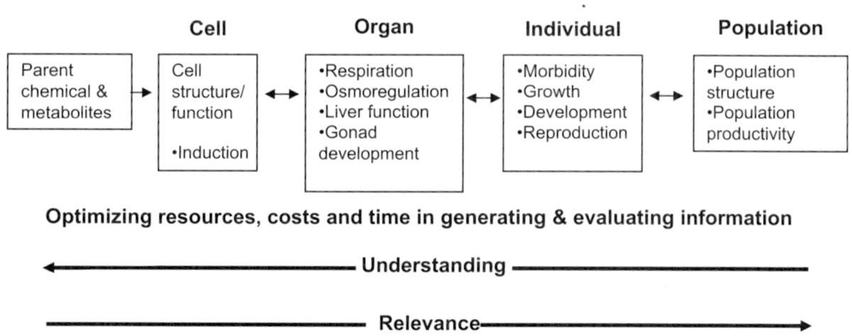

FIGURE 4.2
Critical path from toxicological responses across levels of biological organization would help prioritize risk-based assessment questions and associate data and information needs.
Source: S.P. Bradbury, T.C.J. Feijtel and C.J. Van Leeuwen (2004). Peer reviewed: Meeting the scientific needs of ecological risk assessment in a regulatory context. *Environmental Science & Technology* 38 (23): 463A–470A.

PUTTING BIOLOGY TO WORK

The knowledge about environmental systems has been dramatically increasing for the past five decades. Pollution control and prevention approaches were first aimed at the most basic pollutants, e.g. oxygen demand, solids and pathogenic bacteria in water, particulate matter, carbon monoxide, oxides of sulfur and nitrogen in air. While these continue to be addressed using improved techniques, myriad other pollutants must now be addressed, especially the so-called hazardous and toxic substances.

Even within the environmental professional and scientific communities, the debate continues regarding the adequate level of treatment. Bioengineers and their clients grapple *ad nauseum*. with the question of "how clean is clean?" For example, we can present the same data regarding a contaminated site to two distinguished environmental engineers. One will recommend *in situ* active clean up, such as a pump-and-treat approach, and the other will recommend a passive approach, such as *in situ* natural attenuation, wherein the microbes and abiotic environment is allowed to break down the contaminants over an acceptable amount of time. Still others see the need to "supercharge" the cleanup by enhancing the conditions of the site, such as adding oxygen, moisture and nutrients, to improve the microbial kinetics and the concomitant rates of biodegradation. This is known as bioaugmentation. And, there are those who see the need to remove the wastes and treat them under even greater controlled conditions, i.e. *ex situ* treatment. Both *in situ* and *ex situ* approaches can be enhanced by biotechnology.

Quite likely, all of the options require ongoing monitoring to ensure that the contaminants are in fact breaking down and to determine that they are not migrating away from the site. Different cleanup recommendations result from judgments about the system at hand, notably the initial and boundary conditions, the control volume, the constraints and drivers. The designed solution must be systematic and tailored to the specific waste and environment. For example, a site on Duke University's property was used to bury low-level radioactive waste and spent chemicals. The migration of one of these chemicals, the highly toxic paradioxane, was modeled. The comparison of the effectiveness of active *versus* passive design is shown in Figure 4.3. Is this difference sufficiently significant to justify an active removal and remediation instead of allowing nature to take its course?

Both approaches have risks. Active cleanup potentially exposes workers and the public during removal. There may even be avenues of contamination made possible by the action that would not exist if no action were taken. Conversely, in many cases, without removal of the contaminant, it could migrate to aquifers and surface water that is the source of drinking water,

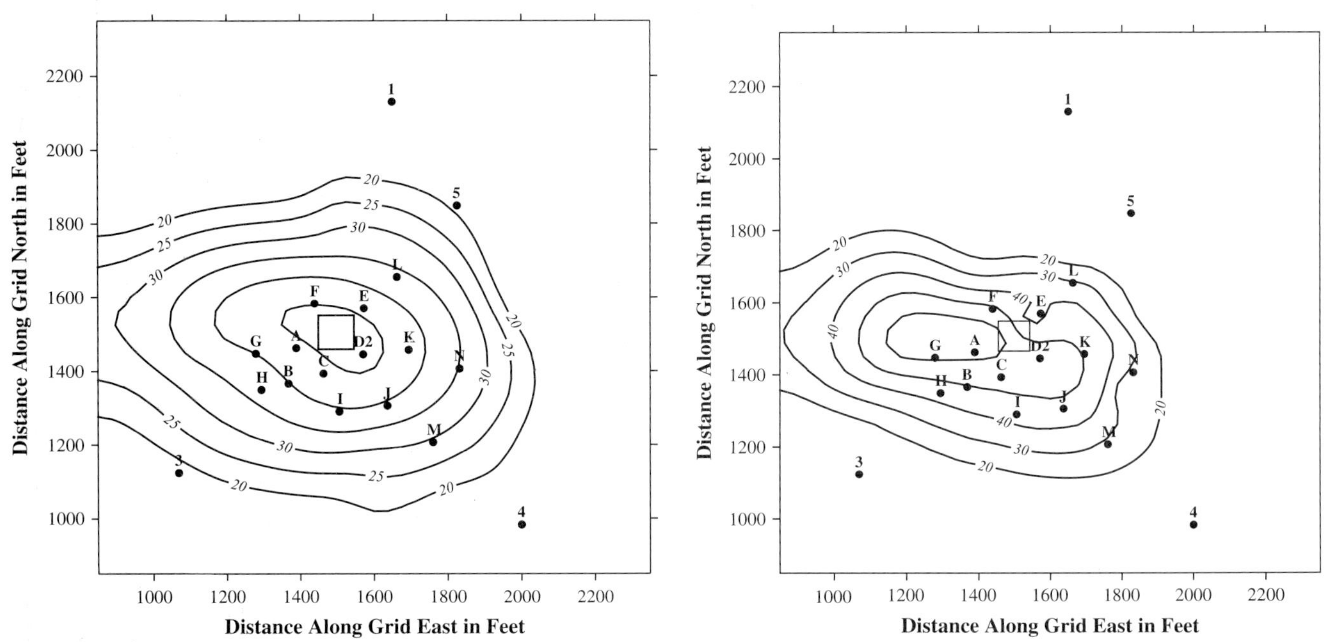

FIGURE 4.3
Duke Forest Gate 11 Waste Site in North Carolina. *Left map:* Modeled paradioxane plume after 50 years of natural attenuation. *Right map:* Paradioxane plume modeled after 10 years of pump and recharge remediation. Numbered points are monitoring wells. The difference in plume size from intervention *versus* natural attenuation is an example of the complexity of risk management decisions, i.e. does the smaller predicted plume justify added costs, possible risk tradeoffs from pumping (e.g. air pollution) and disturbances to soil and vegetation?
Source: M.A. Medina, Jr., W. Thomann, J.P. Holland and Y-C. Lin (2001). Integrating parameter estimation, optimization and subsurface solute transport. *Hydrological Science and Technology* 17, 259–282. Used with permission from first author.

or could remain a hazard for decades if the contaminant is persistent and not amenable to microbial degradation. Thus, engineering is all about risk management. Managing risks requires thoughtful consideration of all options.

From a biotechnological perspective, what would happen if the cleanup went beyond bioaugmentation, e.g. injecting water, oxygen or nutrients to speed up microbial growth? What if a genetically engineered microbe were injected? The risk predictions become more complicated. For example, the plume may not change at all, if the genetically engineered organisms find the environment hostile and do not grow. Thus, only the naturally adapted microbes would survive and continue to degrade the dioxane. Or, the genetically engineered strains may decrease the plume at a faster rate at first, but due to stresses they cause on the microbial population's diversity they then slow down so that the rate of degradation may not be as good as natural attenuation or bioaugmentation alone. The best scenario would be that the genetically engineered microbes continue to degrade the dioxane *and* be compatible with natural biota (or simply do their job and not reproduce and mix their genetic material with native species).

Genetically modified microorganisms (and higher plants and animals, for that matter) can gain an advantage that would allow them to increase in numbers and spread in the environment. The environmental risks will vary according to the characteristics of, and the interactions between, the organism, the trait introduced through the gene, and the environment. Thus, risk assessments need to be conducted on site-specific and case-by-case bases [6].

Obviously, there are numerous other scenarios, and these are oversimplifications of the options available to the bioengineer. The key point is that like almost all environmental interventions, biotechnological applications are not without risk. Risks from genetically modified or engineered microorganisms must be evaluated according to a number of factors. Some important assessment questions [7] include:

- Is the introduced gene unrelated to the species being modified, or is it an extra copy or some modification of the organism's own genetic material?
- Does the new or modified trait allow the organism into which it has been introduced (the "host species") to become toxic or cause disease?
- Will the new or modified trait increase the environmental "fitness" of the host species?
- Is the host species exotic or native to a particular ecosystem, and does it have pest, weed, or native near relatives that may result in gene flow?
- Could the new gene transfer to any other species, either to non-genetically modified individuals of the same species, to closely related species through natural reproductive processes, or to distantly related species through possible (but rare or unlikely) processes or accidents?
- How much of and where will the genetically modified organism (GMO) be released and how will it be managed and monitored?
- Will the GMO persist beyond intended areas and what will be the environmental fate of any new substances produced by the GMO?

These questions need to be sufficiently answered for each site before deeming a genetic engineering enterprise to be a worthy pursuit.

> Resilience is the ability of a system to absorb disturbance and still retain its basic function and structure.
>
> **The Sustainable Development, United Kingdom [8]**

The challenge of bioengineering is to find ways to manage environmental risks that are underpinned by sound science, approaching each project from a "site-wide" perspective that combines health and ecological risks with other factors, e.g. spatial-temporal considerations. This means that whatever residual risk is allowed to remain is based on both traditional risk outcomes (disease, endangered species) *and* future needs (see Figure 4.4). This is the crux of sustainability: good things are sustained, bad things persist.

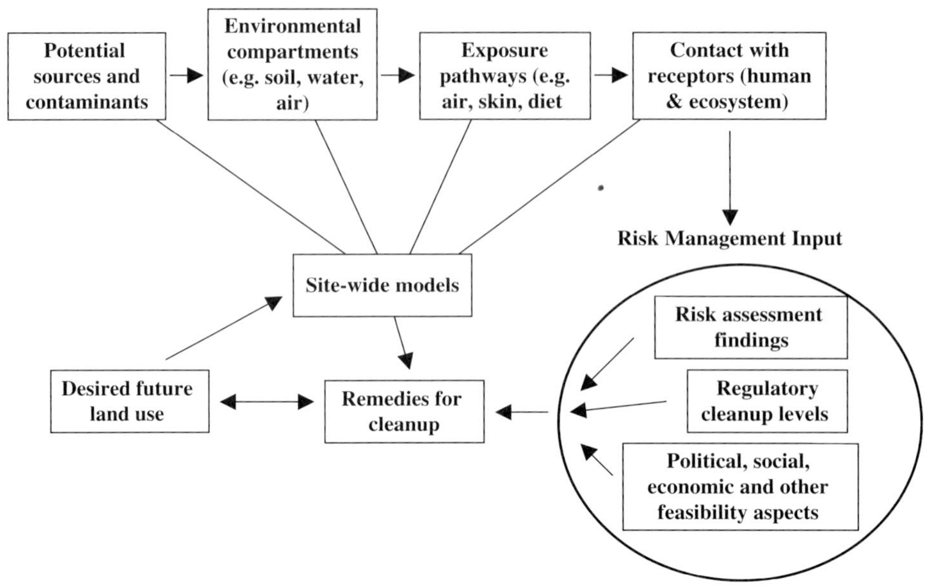

FIGURE 4.4
Site-wide cleanup model based upon targeted risk and future land use.
Source: Adapted from J. Burger, C. Powers, M. Greenberg and M. Gochfeld (2004). The role of risk and future land use in cleanup decisions at the Department of Energy. *Risk Analysis* 24 (6): 1539–1549.

Even a very attractive near-term project may not be so good when viewed from a longer-term perspective. Conversely, a project with seemingly large initial costs may in the long run be the best approach. This opens the door for selecting projects with larger initial risks. Examples of site based risk management have included asbestos and lead remedies, where the workers are subjected to the threat of elevated concentrations of toxicants, but the overall benefits of the action were deemed necessary to protect children. In an integrated engineering project, a risk that is widely distributed in space and time (i.e. numerous building with the looming threat to children's health for decades to come) is avoided in favor of a more concentrated risk that can be controlled (e.g. safety protocols, skilled workers, protective equipment, removal and remediation procedures, manifests and controls for contaminated materials, and ongoing monitoring of fugitive toxicant releases).

This combined risk and land use approach also helps to moderate the challenge of "one size fits all" in environmental cleanup. That is, limited resources may be devoted to other community objectives if the site does not have to be cleaned to the level prescribed by a residential standard. This does not mean that the site can be left to be "hazardous," only that the cleanup level can be based on a land use other than residential, where people are to be protected in their daily lives. For example, if the target land use is similar to the sanitary landfill common to most communities in the United States, the protection of the general public is achieved through measures beyond concentrations of a contaminant. These measures include allowing only authorized and adequately protected personnel in the landfill area, barriers, and leachate collection systems to ensure that contamination is confined within certain areas within the landfill and security devices and protocols (fences, guards, and sentry systems) to limit the opportunities for exposures and risks by keeping people away from more hazardous areas. This can also be accomplished in the private sector. For example, turnkey arrangements can be made so that after the cleanup (private or governmental) meets the risk/land use targets, a company can use the remediated site for commercial or industrial uses. Again, the agreement must include provisions to ensure that the company has adequate measures in place to keep risks to workers and others below prescribed targets, including periodic inspections, permitting, and other types of oversights by governmental entities to ensure compliance with agreements to keep the site clean (i.e. so-called "closure" and "post-closure" agreements).

TRANSFORMING DATA INTO INFORMATION: INDICES

Often, there may be data or models that can be used in risk assessments, but their relationships with each other and thier overall meaning is not evident. An *index* can help to transform environmental data into useful information. The simplest indices are those that have just a few physicochemical variables, such as dissolved oxygen (DO), biochemical oxygen demand (BOD), and specific nutrients (e.g. total nitrogen and phosphorus). As discussed in Chapter 11, these variables are not only important as singular variables, but influence the behavior of the other variables. Even in this simple example, the DO will respond both positively and negatively to increased nutrient levels. All biota have an optimal range of growth and metabolism that varies among species (e.g. algae will add some O_2 with photosynthesis, but use some O_2 for metabolism, whereas the bacteria will generally be net consumers of molecular oxygen). Thus, systematic indices will have a much larger number of variables, and will include biology.

The most widely applied environmental indices that incorporate organisms are those that follow the framework of an *index of biological integrity.* In biological systems, integrity is the capacity of a system to sustain a balanced and healthy community. This means the community of organisms in that system meets certain criteria for species composition, diversity, and adaptability, often compared to a reference site that is a benchmark for integrity.

As such, biological integrity indices are designed to integrate the relationships of chemical and physical parameters with each other and across various levels of biological organization. They are now used to evaluate the integrity of environmental systems using a range of metrics to describe system conditions. They are similar to human "indices" used by physicians where no single biomarker or physical measurement is used, but a variety of markers, with varying weights as to importance, give a reading of a patient's conditions. A low grade fever may not indicate much, but when combined with respiration rates, recent weight loss, and levels of specific liver enzymes, the physician is able to deduce reasons for a patient's symptoms.

Table 4.2 Biological metrics used in the original index of biological integrity (IBI)

Integrity aspect	Biological metric
Species richness and composition	Total number of fish species (total taxa)
	Number of *Catostomidae* species (suckers)
	Number of darter species
	Number of sunfish species
Indicator species metrics	Number of intolerant or sensitive species
	Percent of individuals that are *Lepomis cyanellus* (Centrarchidae)
Trophic function metrics	Percent of individuals that are omnivores
	Percent of individuals that are insectivorous Cyprinidae
	Percent of individuals that are top carnivores or piscivores
Reproductive function metrics	Percent of individuals that are hybrids
Abundance and condition metrics	Abundance or catch per effort of fish
	Percent of individuals that are diseased, deformed, or that have eroded fins, lesions, or tumors (DELTs)

Source: J.R. Karr (1981). Assessment of biotic integrity using fish communities. *Fisheries* 6: 21–27.

Thus, environmental indices combine attributes to determine a system's condition (e.g. diversity and productivity) and to hypothesize stresses. The original index of biotic integrity developed by Karr [9] was based on fish fauna attributes and has provided predictions of how well a system will respond to a combination of stresses. In fact, the index is completely biological, with no direct chemical measurements. However, the metrics (see Table 4.2) are indirect indicators of physicochemical factors (e.g. the abundance of game fish is directly related to dissolved oxygen concentrations, as discussed in Chapter 11). The metrics provide descriptions of a system's structure and function.

An example of the data that are gathered to characterize a system is provided in Table 4.3. The information that is gleaned from these data is tailored to the physical, chemical, and biological conditions of an area (in Table 4.3). In this instance, the information applies exclusively to large spatial regions, so there are quite a few categories of data. However, environmental

Table 4.3 Biological metrics that apply to various regions of North America[a]

Alternative IBI Metrics	Midwestern United States	Central Appalachians	Sacramento-San Joaquin	Colorado Front Range	Western Oregon Ohio	Ohio Headwater Sites	Northeastern United States		Ontario	Central Corn Belt Plain	Wisconsin – Warmwater	Wisconsin – Coldwater	Maryland Coastal Plain	Maryland Non-Tidal
1. **Total Number of Species**	X	X	X	X				X		X			X	X
No. native fish species					X	X	X		X		X			
No. salmomid age classes[b]				X	X									
2. **Number of Darter Species**	X	X		X		X				X	X			
No. sculpin species					X									
No. benthic insectivore species								X						
No. darter and sculpin species							X							
No. darter, sculpin, and madtom species													X	
No. salmonid juveniles (individuals)[b]			X		X			X						
% round-bodied suckers						X[c]								
No. sculpins (individuals)			X											
No. benthic species													X	X
3. **Number of Sunfish Species**	X			X		X				X	X			
No. cyprinid species					X									
No. water column species								X						
No. sunfish and trout species									X					
No. salmonid species			X							X				

Table 4.3 **Biological metrics that apply to various regions of North America[a]—cont'd**

Alternative IBI Metrics	Midwestern United States	Central Appalachians	Sacramento-San Joaquin	Colorado Front Range	Western Oregon Ohio	Ohio Headwater Sites	Northeastern United States		Ontario	Central Corn Belt Plain	Wisconsin – Warmwater	Wisconsin – Coldwater	Maryland Coastal Plain	Maryland Non-Tidal
No. headwater species							X							
% headwater species							X			X				
4. **Number of Sucker Species**	X				X	X		X		X	X			
No. adult trout species[b]			X		X									
No. minnow species				X			X			X				
No. sucker and catifish species									X					
5. **Number of Intolerant Species**	X			X	X	X		X			X	X	X	X
No. sensitive species							X			X				
No. amphibian species			X											
Presence of brook trout									X					
% stenothermal cool and cold water species												X		
% of salmonid ind. as brook trout												X		
6. **% Green Sunfish**	X													
% common carp					X									
% white sucker				X				X						
% tolerant species						X	X			X	X	X	X	X
% creek chub		X												
% dace species									X					
% eastern mundminnow													X	
7. **% Omnivores**	X			X		X	X	X	X	X	X			
% generalist feeders		X												
% generalists, and invertivores														X
8. **% Insectivorous Cyprinids**	X												X	
% insectivore					X			X		X	X		X	X[e]
% specialized insectivores		X												
No. juvenile trout			X											
% insectivorous species						X	X							

(Continued)

Table 4.3 Biological metrics that apply to various regions of North America[a]—cont'd

Alternative IBI Metrics	Midwestern United States	Central Appalachians	Sacramento-San Joaquin	Colorado Front Range	Western Oregon Ohio	Ohio Headwater Sites	Northeastern United States		Ontario	Central Corn Belt Plain	Wisconsin – Warmwater	Wisconsin – Coldwater	Maryland Coastal Plain	Maryland Non-Tidal
9. % **Top Carnivores**	X					X		X	X	X	X	X		
% catchable salmonids					X									
% catchable trout			X				X			X			X	
% pioneering species							X			X			X	
Density catchable wild trout			X											
10. **Number of Individuals (or catch per effort)**	X	X	X	X	X	X[d]	X[d]		X	X	X[d]		X	
Density of individuals								X						X
% abundance of dominant species													X	X
Biomass (per m2)														X[f]
11. % **Hybrids**	X							X						
% introduced species				X	X									
% simple lithophills						X				X	X			X
No. simple lithophills species							X							
% native species			X											
% native wild individuals			X											
% silt-intolerant spawners													X	
12. % **Diseased Individuals (deformities, eroded fins, lesions, and tumors)**	X	X		X	X	X	X	X	X	X	X		X	X

Note: X = metric used in region. Many of these variations are applicable elsewhere.
[a]*Taken from Karr et al. (1986), Leounard and Orth (1986), Moyle et al. (1986), Fausch and Schrader (1987), Hughes and Gammon (1987), Ohio EPA (1987), Miller et al. (1988), Steedman (1988), Simon (1991), Lyons (1992a), Barbour et al. (1995), Simon and Lyons (1995), Hall et al. (1996), Lyons et al. (1996), Roth et al. (1997). For reference, see source publication.*
[b]*Metric suggested by Moyle et al. (1986) or Hughes and Gammon (1987) as a provisional replacement metric in small western salmonid streams.*
[c]*Boat sampling methods only (i.e., larger streams/rivers).*
[d]*Excluding individuals of tolerant species.*
[e]*Non-coastal Plain streams only.*
[f]*Coastal Plain streams only.*
Source: M.T. Barbour, J. Gerritsen, B.D. Snyder and J.B. Stribling (1999). *Rapid Bioassessment Protocols for Use in Streams and Wadeable Rivers: Periphyton, Benthic Macroinvertebrates and Fish*, 2nd Edition. Report No. EPA 841-B-99-002. US Environmental Protection Agency, Office of Water, Washington, DC.

indices are also useful at almost any scale. The information from a biologically based index can be used to evaluate a system, as shown in Figure 4.5.

Systems involve scale and complexities in both biology and chemistry. For example, a fish's direct aqueous exposure (*AE* in μg day^{-1}) is the product of the organism's ventilation volume,

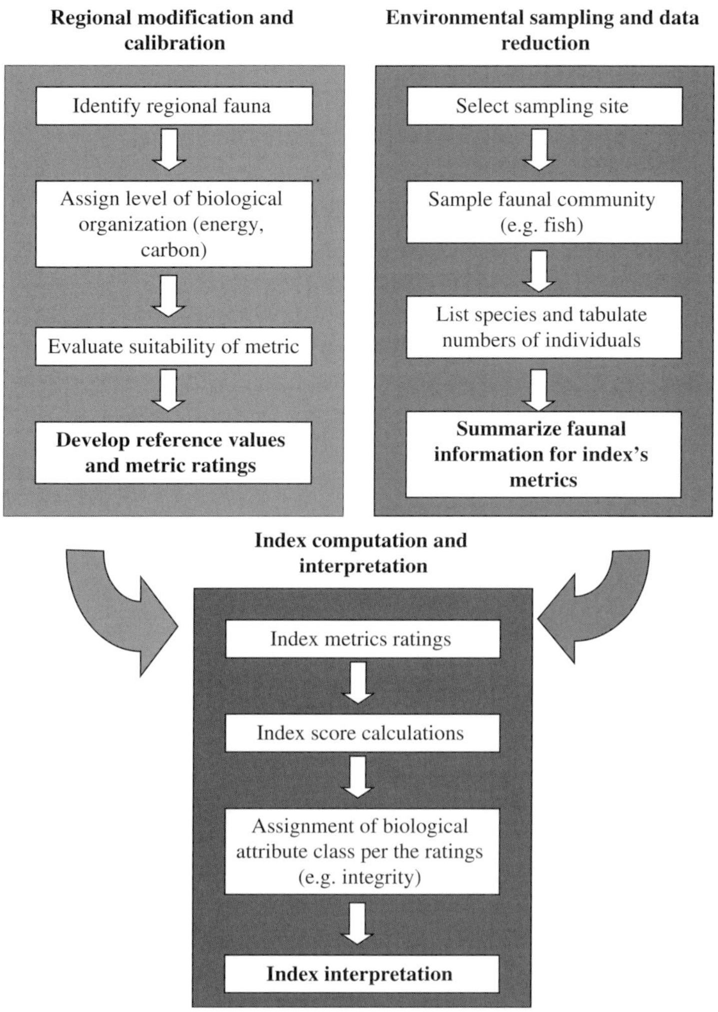

FIGURE 4.5

Sequence of activities involved in calculating and interpreting an Index of Biotic Integrity (IBI).
Source: Adapted from: M.T. Barbour, J. Gerritsen, B.D. Snyder and J.B. Stribling (1999). *Rapid Bioassessment Protocols for Use in Streams and Wadeable Rivers: Periphyton, Benthic Macroinvertebrates and Fish*, 2nd Edition. Report No. EPA 841-B-99-002. US Environmental Protection Agency, Office of Water, Washington, DC; adapted from J.R. Karr (1987). Biological monitoring and environmental assessment: a conceptual framework. *Environmental Management* 11: 249–256.

i.e. the flow Q (in mL day^{-1}), and the compound's aqueous concentration, C_w (μg mL^{-1}). The fish's exposure by its diet (DE, in μg day^{-1}) is the product of its feeding rate, F_w (g wet weight day^{-1}), and the compound's concentration in the fish's prey, C_p (μg g^{-1} wet wt). If the fish's food consists of a single type of prey that is at equilibrium with the water, the fish's aqueous and dietary exposures and the bioconcentration factor (BCF) can be calculated when they are equal:

$$\mathrm{AE} = \mathrm{DE};\ \mathrm{QC_w} = \mathrm{F_wC_p};\ BCF = \frac{Q}{F_w} \tag{4.1}$$

The ventilation-to-feeding ratio for a 1 kg trout has been found [10] to be on the order of $10^{4.3}$ mL g^{-1}. Based on the *quantitative structure activity relationship* (*QSAR*), the BCF of a trout's prey can be assumed to be 0.048 times the octanol-water coefficient (K_{ow}) of a chemical compound. If so, food represents a trout's predominant route of exposure for lipophilic substances in the fish's diet, e.g. compounds with a $K_{ow} > 10^{5.6}$. Exposure must also account for the organism's assimilation of compounds in food, which for very lipophilic compounds will probably account for the majority of exposure compared to that from the water. Even though chemical exchange occurs from both food and water via passive diffusion (Fick's law relationships; see Chapter 3), the uptake from food, unlike direct uptake from water, does not necessarily relax the diffusion gradient into the fish. The difference between digestion and assimilation of food can result in higher contaminant concentrations in the fish's gut.

Predicting expected uptake where the principal route of exchange is dietary can be further complicated by the fact that most fish species exhibit well-defined size-dependent, taxonomic, and temporal trends regarding their prey. Thus, a single bioaccumulation factor (BAF) may not universally be useful for risk assessments for all fish species. Indeed, the BAF may not even apply to different sizes of the same species.

The systematic biological exchange of materials between the organism, in this case various species of fishes, is known as uptake, which can be expressed by the following three differential equations for each age class or cohort of fish [11]:

$$\frac{dB_f}{d_t} = J_g + J_i + J_{bt} \tag{4.2}$$

where, J_g represents the net chemical exchange ($\mu g\ d^{-1}$) across the fish's gills from the water; J_i represents the net chemical exchange ($\mu g\ d^{-1}$) across the fish's intestine from food; and J_{bt} represents the compound's biotransformation rate ($\mu g\ d^{-1}$).

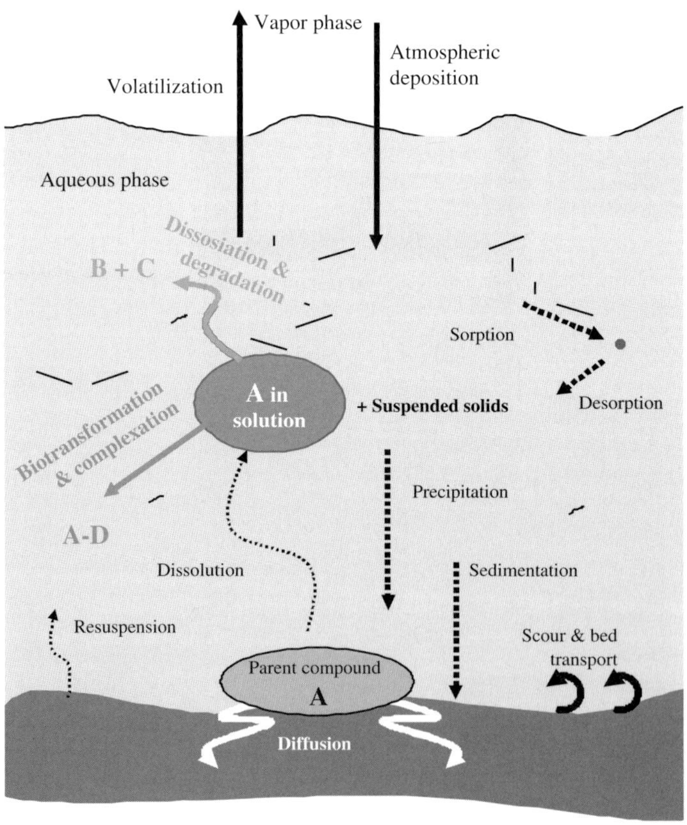

FIGURE 4.6
Transport and transformation phenomena in a water system. The transformation processes, including dissociation and degradation to form metabolites and degradation products (B, C, and D), simultaneously consist of both abiotic (e.g. hydrolysis and photolysis) and biotic (i.e. biodegradation). The parent compound A and its reaction products include molecular diffusion (usually only important in quiescent systems, e.g. sediment layers) and advective processes (see Table 2.1). [*See color plate section*]
Source: Adapted from W.J. Lyman (1995). Transport and Transformation Processes – Chapter 15. In: G. Rand (Ed.), *Fundamentals of Aquatic Toxicology: Effects, Environmental Fate, and Risk Assessment*, 2nd Edition. Taylor & Francis, Washington, DC.

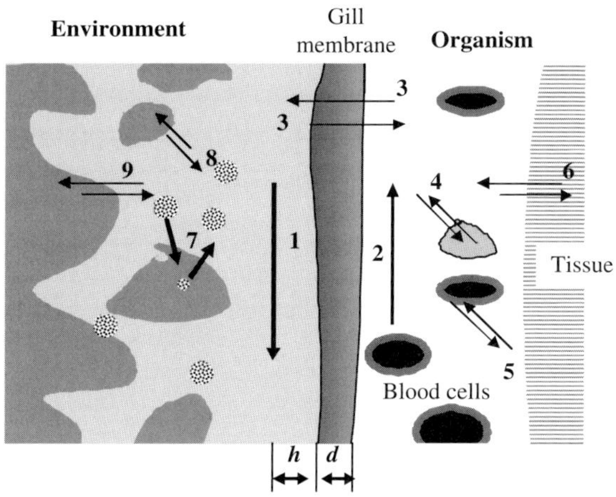

FIGURE 4.7
Transfer of matter as part of the bioaccumulation process in a multiphase system (water, sediment, particles, and biota), as represented by a gill (a lung would be analogous to inhalation in air-breathing organisms, but similar processes occur in dermal and ingestion routes): 1. water flow across membrane; 2. blood flow within organism; 3. chemical flux across membrane; 4. binding by and release from serum proteins; 5. sorption/desorption to blood cells; 6. chemical mass transfer from blood to tissues by perfusion; 7. complexation to and decomplexation from organic carbon in a particulate phase (POC); 8. sorption to and desorption from coarse particulate solids, in addition to internal diffusion within the particles. Note: h = stagnant water (velocity = 0 at interface) layer thickness; and d = diffusion distance across membrane. [*See color plate section*]
Source: Drawn from information provided by A. Spacie, L.S. McCarty and G.M. Rand (1995). Bioaccumulation and bioavailability in multiphase systems – Chapter 16. In: G. Rand (Ed.), *Fundamentals of Aquatic Toxicology: Effects, Environmental Fate, and Risk Assessment*, 2nd Edition. Taylor & Francis, Washington, DC.

$$\frac{dW_d}{d_t} = F_d - E_d - R - EX - SDA \tag{4.3}$$

where, B_f = compound's body burden (μg fish^{-1}) and W_d = dry body weight (g dry wt fish^{-1}) of the average individual within the cohort; and N is the cohort's population density (fish ha^{-1}).

$$\frac{dN}{d_t} = EM - NM - PM \tag{4.4}$$

where, Fd = the fish's feeding; Ed = *egestion* (i.e. expulsion of undigested material); R = routine respiration; EX = excretion; and SDA = specific dynamic action (i.e., the respiratory expenditure in excess of R required to assimilate food). All of these parameters have units of g dry wt day^{-1}.

Numerous processes are involved in environmental systems. These include processes in the environment (see Figure 4.6) and those at the interface between the organism and the environment (see Figure 4.7).

Physiologically based models for fish growth are often formulated in terms of energy content and flow (e.g., kcal fish^{-1} and kcal day^{-1}), Eq. 4.3 is basically the same as such bioenergetic models because energy densities of fish depend on their dry weight [12].

Obviously, feeding depends on the availability of suitable prey, so the mortality of the fish is a function of the individual feeding levels and population densities of its predators. Thus, the fish's dietary exposure is directly related to the organism's feeding rate and the concentrations chemicals in its prey.

CONCENTRATION-BASED MASS BALANCE MODELING [13]

To enhance the understanding of system effects, let us consider a hypothetical compound's transport within a single compartment (surface water). A factory has released the chemical to an estuary with an average depth of 5 m that covers an area of 2 million m^2. The flow rate of water into and out of the estuary is 24,000 m^3 per day. Sediment enters the estuary at a rate of 1 L min^{-1}. Of this, 60% settles to the sediment at the bottom of the estuary and 40% remains suspended and is part of the estuary's outflow. The half-life of the chemical is 300 days. Its evaporation rate gives the chemical a mass transfer coefficient of 0.24 m day^{-1}. The chemical's molecular mass is 100 g mol^{-1}. Its air–water partitioning coefficient, K_{AW}, is 0.01. Its particle to water coefficient (K_{PW}) is 6000 and its bioconcentration factor (i.e. partitioning from the water to the biota) is 9000. The particle (i.e. suspended solids) concentration in the water column is 25 ppm by volume. The volume of aquatic fauna in the estuary is 10 ppm. The factory is releasing the contaminant into the estuary at a rate of 1 kg per day. The background inflow concentration of the contaminant is 10 μg L^{-1}.

From this loading and partitioning information we can calculate the steady state (constant) concentration of contaminant in the estuary's water, particles, and fauna, including loss rates.

First, we must set the total concentration of the contaminant in the water as an unknown value. We can later calculate this value by difference from the total and other known values. We will also convert all units to g h^{-1} for the mass balance.

Contaminant input

$$\text{Discharge rate}(1\ \text{kg day}^{-1}) = \text{nearly } 42\ \text{g h}^{-1}$$

Inflow rate is the flow rate of the estuary times the concentration of the contaminant in the water column =

$$[(24,000\ \text{m}^3\ \text{day}^{-1})/(24\ \text{h day}^{-1})][(10\mu\text{g L}^{-1})(10^{-6}\ \text{g}\ \mu\text{g}^{-1})(1000\ \text{L m}^{-3})] = 10\ \text{g h}^{-1}$$

So, the total input of the contaminant is $42 + 10 = 52$ g h^{-1}

Partitioning between compartments

The total volume of water in the estuary is the average depth time area (5 m × 2,000,000 m^2) = 10^7 m^3.

However, the total volume contains 25 ppm particles and 10 ppm fauna, or:

$$\text{Particle volume} = 25 \times 10^{-6} \times 10^7\ \text{m}^3 = 250\ \text{m}^3$$

and

$$\text{Fauna volume} = 10 \times 10^{-6} \times 10^7\ \text{m}^3 = 100\ \text{m}^3$$

Since the dissolved fraction of the contaminant concentration is $C_{dissolved}$, then the concentration of the contaminant dissolved in the water must be:

$$10^7 \cdot C_{dissolved}$$

And the particle concentration is:

$$250 \cdot K_{PW} \cdot C_{dissolved} = (250 \times 6000)C_{dissolved} = 1.5 \times 10^6 C_{dissolved}$$

And the fauna concentration is:

$$100 \cdot K_{BW} \cdot C_{dissolved} = (100 \times 9000)C_{dissolved} = 9 \times 10^5 C_{dissolved}$$

Or for water, particles, and fauna total:

$$C_{dissolved}(10 + 1.5 + 0.9) \times 10^6 = 12.4 \times 10^6 C_{dissolved}$$

Recall that the total volume must be $10^7 C_W$, so we can use the ratio of the quantities in parentheses for the mass balance:

$$C_{dissolved} = 10/12.4 = 0.81 C_W$$

Sorbed particle concentration $= 1.5/12.4 = 0.12\ C_W$

Bioconcentration $= 0.9/12.4 = 0.07\ C_W$

Thus, 81% of the contaminant is dissolved in the estuary's surface water, 12% is sorbed to particles, and 7% is in the fauna tissue.

The concentration of the contaminant on the particles is therefore $K_{PW}C_{dissolved}$ or 0.81 $K_{PW}C_W = 0.81 \times 6000 = 4860 C_W$. And, the concentration of the contaminant in fauna tissue is 0.81 $K_{BW}C_W = 0.81 \times 9000 = 7290 C_W$.

Outflow

The outflow rate is 24,000 $m^3\ day^{-1} = 1000\ m^3\ h^{-1}$, so the rate of transport of the dissolved contaminant is 1000 $C_{dissolved}$ $g\ h^{-1}$or 810 $C_W\ g\ h^{-1}$.

Sorption is constantly occurring, so there will also be outflow of the contaminant attached to particles (let us assume that the fauna remain in the estuary, or at least that there is no net change in contaminant mass concentrated in the biotic tissue). 40% of the sediment's 1 $L\ min^{-1}$ leaves the estuary; therefore the 0.4 $L\ min^{-1} = 24\ L\ h^{-1}$ of particles containing $4860 C_W\ g\ m^{-3}$. Since, 24 $L\ h^{-1} = 0.024\ m^3\ h^{-1}$, there will be $4860 \times 0.024 = 117\ C_W\ g\ h^{-1}$ contaminant leaving the estuary on the sediment.

Reaction

The product of the estuary water volume, concentration, and rate constant gives the reaction rate. Since the half-life is 300 days (7200 hours), the rate constant is:

$$\frac{Ln(2)}{7200} = 9.6 \times 10^{-5}\ h^{-1}$$

Thus, the reaction rate is $10^7 \times C_W \times 9.6 \times 10^{-5} = 960\ C_W\ g\ h^{-1}$.

Sedimentation

Since the concentration of the contaminant sorbed to particles is $4860 C_W$ and the particle deposition (sedimentation) rate is 60% of the 1 $L\ min^{-1}$ of sediment entering the estuary (i.e. 0.6 $L\ min^{-1} = 36\ L\ h^{-1} = 0.036\ m^3\ h^{-1}$), the contaminant deposition rate is $4860 \times 0.036 C_W = 175 C_W\ g\ h^{-1}$.

Vaporization

The vaporization (evaporation) rate equals the product of the gas's mass transfer coefficient, the estuary's surface area, and the contaminant concentration in water. Thus, for our contaminant, the evaporation rate $= (0.24\ m\ day^{-1})(day\ 24\ h^{-1})(2 \times 10^6\ m^2)(0.81 C_W) = 16200 C_W\ g\ h^{-1}$.

We will assume that no diffusion is taking place between the air and water (i.e., the air contains none of our hypothetical contaminant). If the atmosphere were a source of the contaminant, we would need to add another input term.

Combined process rates

If we assume steady state conditions, we can now combine the calculated rates and set up an equality with our discharge rate (input rate):

Discharge rate = Sum of all process rates

Discharge rate = Dissolved outflow + Sorbed outflow + Reaction + Sedimentation + Vaporization

$$52 = 810C_W + 117C_W + 960C_W + 175C_W + 16200C_W$$

$$52 = 18262C_W$$

$$C_W = 52/18262 = 0.0028\ \text{g m}^{-3} = 0.0028\ \text{mg L}^{-1} = 2.8\ \mu\text{g L}^{-1}$$

So, returning to our calculated rates and substituting C_W, our model shows the following process rates for the hypothetical contaminant in the estuary:

Process	Rate (g h^{-1})	Percent of total
Outflow dissolved in water (810 × 0.0028)	2.3	4%
Outflow sorbed to suspended particles (117 × 0.0028)	0.33	1%
Reaction (960 × 0.0028)	2.7	5%
Sedimentation (175 × 0.0028)	0.49	1%
Vaporization (16,200 × 0.0028)	45.4	89%

So, our model tells us that the largest loss of the contaminant is to the atmosphere. Our contaminant behaves as a volatile compound, since most of its mass (89%) is readily partitioned to the vapor phase. Dissolution and chemical breakdown are also important processes in the mass balance. Sorption and sedimentation are also occurring, but account for far less of the contaminant mass than does volatilization. This means that our contaminant is sufficiently water soluble, sorptive, reactive, and volatile that any monitoring or clean up must account for all compartments in the environment.

To complete our model, let us consider the contaminant concentration in each environmental compartment:

$$\text{Contaminant dissolved in water} = (0.81)(0.0028\ \text{g m}^{-3}) = 0.0023\ \text{g m}^{-3} = 2.3\ \mu\text{g L}^{-1}$$

The concentration on particles is 4860 times the dissolved concentration:

$$\text{Contaminant sorbed to particles} = (4860)(0.0023\ \text{g m}^{-3}) = 11\ \text{g m}^{-3} = 11\ \text{mg L}^{-1}$$

Solid phase media, like soil, sediment, and suspended matter, are usually expressed in weight to weight concentrations. So, if we assume a particle density of 1.5 g cm^{-3} the concentration on particles is about 7.3 mg kg^{-1}.

Also, the suspended solids fraction of contaminants in surface waters is expressed with respect to water volume. Since particles make up 0.000025 of the total volume of the estuary, our contaminant's concentration is $(2.5 \times 10^{-5})(11\ \text{mg L}^{-1}) = 0.000275\ \text{mg L}^{-1}$ or about 275 ng L^{-1} of the water column.

The concentration in the fauna is 7290 times the dissolved concentration:

$$\text{Contaminant concentrated in fauna tissue} = (7290)(0.0023\ \text{g m}^{-3}) = 17\ \text{g m}^{-3}$$
$$= 17\ \text{mg L}^{-1}$$

which is about equal to 17 mg kg^{-1} tissue. Since the fauna volume makes up 10^{-5} of the total volume of the estuary, our contaminant's concentration is $(10^{-5})(17\ \text{mg L}^{-1}) = 1.7 \times 10^{-4}$ mg L^{-1} or 17 μg L^{-1} of the water column.

Since the total mass is 52 g h^{-1}, we have maintained our mass balance. The concentration in each media is an indicator of the relative affinity that our contaminant has for each environmental compartment.

What if the contaminant were less soluble in water and had a higher bioconcentration rate? The calculations indicate that if the contaminant were less soluble, then less mass would be available to be sorbed or bioconcentrated. Keep in mind, however, that this is a mathematical phenomenon and not necessarily a physical one. Yes, the dissolved fraction is used to calculate the mass that moves to the particles and biota, but remember that the coefficients are based upon empirical information. So, the bioconcentration factor that we were given would increase to compensate for the lower dissolved concentration. That is what makes modeling interesting and complex. When one parameter changes, the other parameters must be adjusted. Thus, it is important to keep in mind that no environmental system is completely independent.

FUGACITY, Z VALUES, AND HENRY'S LAW

Before modeling the partitioning of contaminants among the environmental media, let us revisit the relationships of Henry's law constants to equilibrium introduced in Chapter 3. The relative chemical concentrations of a substance in the various compartments and physical phases are predictable from partition coefficients. The more one knows about the affinities of a compound for each phase, the better is one's ability to predict how much and how rapidly a chemical will move. This chemodynamic behavior as expressed by the partition coefficients can be viewed as a potential, that is, at the time equilibrium is achieved among all phases and compartments, the chemical potential in each compartment has been reached [14].

Chemical concentration and fugacity are directly related via the *fugacity capacity constant* (known as the *Z value*):

$$C_i = Z_i \cdot f \tag{4.5}$$

where: C_i = Concentration of substance in compartment i (mass per volume)

Z_i = Fugacity capacity (time^2 per length^2)

f = Fugacity (mass per length per time^2)

And, at equilibrium, the fugacity of the system of all environmental compartments is:

$$f = \frac{M_{total}}{\sum_i (Z_i \cdot V_i)} \tag{4.6}$$

where: M_{total} = Total number of moles of a substance in all of the environmental system's compartments

V_i = Volume of compartment i where the substance resides

If we assume that a chemical substance will obey the ideal gas law (which is usually acceptable for ambient environmental pressures), then fugacity capacity is the reciprocal of the gas constant (R) and absolute temperature (T). Recall that the ideal gas law states:

$$\frac{n}{V} = \frac{P}{RT} \tag{4.7}$$

where: n = Number of moles of a substance

P = Substance's vapor pressure

Then,

$$P = \frac{n}{V} \cdot RT = f \tag{4.8}$$

And,

$$C_i = \frac{n}{V} \tag{4.9}$$

Therefore,

$$Z_{air} = \frac{1}{RT} \tag{4.10}$$

This relationship allows for predicting the behavior of the substance in the gas phase. The substance's affinity for other environmental media can be predicted by relating the respective partition coefficients to the Henry's law constants. For water, the fugacity capacity (Z_{water}) can be found as the reciprocal of K_H:

$$Z_{water} = \frac{1}{K_H} \tag{4.11}$$

This is the dimensioned version of the Henry's law constant (length^2 per time^2)

Fugacity Example 1

What is the fugacity capacity of toluene in water at 20 °C?

Solution:

Since Z_{water} is the reciprocal of the Henry's law constant, which is 6.6×10^{-3} atm m^3 mol^{-1} for toluene, then Z_{water} must be 151.5 mo atm^{-1} m^{-3}.

The fugacity capacity for sediment is directly proportional to the contaminant's sorption potential, expressed as the solid-water partition coefficient (K_d), and the average sediment density ($\rho_{sediment}$). Sediment fugacity capacity is indirectly proportional to the chemical substance's Henry's law constant:

$$Z_{sediment} = \frac{\rho_{sediment} \cdot K_d}{K_H} \tag{4.12}$$

Fugacity Example 2

What is the fugacity capacity of toluene in sediment with an average density of 2400 kg m^{-3} at 20°C in sediment where the K_d for toluene is 1 L kg^{-1}?

Solution:

Since

$$Z_{sediment} = \frac{\rho_{sediment} \cdot K_d}{K_H}$$

then

$$Z_{sediment} = \frac{(2400 \text{ kg m}^{-3}) \cdot (1 \text{ L kg}^{-1}) \cdot (1 \text{ m}^3)}{(6.6 \times 10^{-3} \text{ atm m}^3 \text{ mol}^{-1}) \cdot (1000 \text{ L})}$$

which is for toluene, then $Z_{sediment}$ must be 3.6×10^{-4} mol atm^{-1} m^{-3}.

Note that if the sediment had a higher sorption capacity, for example 1.5 L kg^{-1}, the fugacity capacity constant would be higher (50% times greater in this case).

Conversely, fugacity would decrease by a commensurate amount with increase sorption capacity. This makes physical sense if one keeps in mind that fugacity is the tendency to escape from the medium (in this case, the sediment) and move to another (surface water). So, if the sediment particles are holding the contaminant more tightly due to higher solid-water partitioning, the contaminant is less prone to leave the sediment. If the solid-water partitioning is reduced, i.e. sorption is reduced, the contaminant is freer to escape the sediment and be transported to the water.

The nature of the substrate and matrix material (e.g. texture, clay content, organic matter content, and pore fluid pH) can have a profound effect on the solid-water partition coefficient, and consequently, the $Z_{sediment}$ value.

For biota, particularly fauna and especially fish and other aquatic vertebrate, the fugacity capacity is directly proportional to the density of the fauna tissue (ρ_{fauna}), and the chemical substance's bioconcentration factor (BCF), and inversely proportional to the contaminant's Henry's law constant:

$$Z_{fauna} = \frac{\rho_{fauna} \cdot BCF}{K_H} \tag{4.13}$$

Fugacity Example 3

What is the fugacity capacity of toluene in aquatic fauna which a BCF of 83 L kg^{-1}and tissue density of 1 g cm^{-3} at 20 °C?

Solution:

Since

$$Z_{fauna} = \frac{\rho_{fauna} \cdot BCF}{K_H}$$

then

$$Z_{fauna} = \frac{(1\ \text{g cm}^{-3}) \cdot (83\ \text{L kg}^{-1}) \cdot (1000\ \text{cm}^3) \cdot (\text{kg})}{(6.6 \times 10^{-3}\ \text{atm m}^3\ \text{mol}^{-1}) \cdot (1\ \text{L}) \cdot (1000\ \text{g})}$$

then Z_{fauna} is 0.013 mol atm^{-1} m^{-3}.

As in the case of the sediment fugacity capacity, a higher bioconcentration factor means that the fauna's fugacity capacity increases and the actual fugacity decreases. Again, this is logical, since the organism is sequestering the contaminant and keeping if from leaving if the organism has a large BCF. This is a function of both the species of organism and the characteristics of the contaminant and the environment where the organism resides. So, factors like temperature, pH, and ionic strength of the water and metabolic conditions of the organism will affect BCF and Z_{fauna}. This also helps to explain why published BCF values may have large ranges.

The total partitioning of the environmental system is merely the aggregation of all of the individual compartmental partitioning. So, the moles of the contaminant in each environmental compartment (M_i) is found to be a function of the fugacity, volume, and fugacity capacity for each compartment:

$$M_i = Z_i \cdot V_i \cdot f \tag{4.14}$$

Comparing the respective fugacity capacities for each phase or compartment in an environmental system is useful for a number of reasons. First, if one compartment has a very high fugacity (and low fugacity capacity) for a contaminant, and the source of the contaminant no longer exists, then one would expect the concentrations in that medium to fall rather precipitously with time under certain environmental conditions. Conversely, if a compartment has a very low fugacity, measures (e.g. *in situ* remediation, or removal and abiotic chemical treatment) may be needed to see significant decreases in the chemical concentration of the contaminant in that compartment. Second, if a continuous source of the contaminant exists, and a compartment has a high fugacity capacity (and low fugacity), this compartment may serve as a conduit for delivering the contaminant to other compartments with relatively low fugacity capacities. Third, by definition, the higher relative fugacities of one set of compartments compared to another set in the same ecosystem allow for comparative analyses and estimates of sources and sinks (or "hot spots") of the contaminant, which is an important part of fate, transport, exposure, and risk assessments.

Fugacity Example 4

What is the equilibrium partitioning of 1000 kg of toluene discharged into an ecosystem of 5 × 10^9 m^3 air, 9 × 10^5 m^3 water, and 4.5 m^3 aquatic fauna, with the same K_H, BCF, K_d, and densities for fauna and sediment used in the three previous examples. Assume the temperature is 20 °C and the vapor pressure for toluene is 3.7 × 10^{-2} atm.

Solution:

The first step is to determine the number of moles of toluene released into the ecosystem. Toluene's molecular weight is 92.14, so converting the mass of toluene to moles gives us:

$$\frac{(1000\ \text{kg}) \cdot (1000\ \text{g}) \cdot (1\text{mol})}{(1\ \text{kg}) \cdot (92.14\ \text{g})} = 10853\ \text{mol}$$

The fugacity capacities for each phase are:

$$Z_{air} = \frac{1}{RT} = \frac{1}{0.0821\ \text{L} \cdot \text{atm} \cdot \text{mol}^{-1} \cdot \text{K} \cdot 293^\circ \text{K}} \cdot \frac{1000\ \text{L}}{\text{m}^3} = 41.6\ \text{mol atm}^{-1} m^{-3}$$

$$Z_{water} = \frac{1}{K_H} = \frac{1}{6.6 \times 10^{-3}\ \text{atm m}^3\ \text{mol}^{-1}} = 151.5\ \text{mol atm}^{-1}\ \text{m}^{-3}$$

$$Z_{fauna} = \frac{\rho_{fauna} \cdot BCF}{K_H} = \frac{(1\ \text{g cm}^{-3}) \cdot (83\ \text{L kg}^{-1}) \cdot (1000\ \text{cm}^3) \cdot (\text{kg})}{(6.6 \times 10^{-3}\ \text{atm m}^3\ \text{mol}^{-1}) \cdot (1\ \text{L}) \cdot (1000\ \text{g})} = 0.013\ \text{mol atm}^{-1}\ \text{m}^{-3}$$

The ecosystem fugacity can now be calculated:

$$f = \frac{M_{total}}{\sum_i (Z_i \cdot V_i)} = \frac{10,843 \text{ mol}}{41.6 \cdot 5 \times 10^9 + 151.5 \cdot 9 \times 10^5 + 0.013 \cdot 4.5} = 5.2 \times 10^{-8} \text{ atm}$$

The moles of toluene in each compartment are:

$M_{air} = 5.2 \times 10^{-8} \cdot 5 \times 10^9 \cdot 41.6 = 10{,}816$ mol

$M_{water} = 5.2 \times 10^{-8} \cdot 9 \times 10^5 \cdot 151.5 = 7.1$ mol

$M_{fauna} = 5.2 \times 10^{-8} \cdot 4.5 \cdot 0.013 = 3.0 \times 10^{-9}$ mol

So, the mass of toluene at equilibrium will be predominantly in the air.

The toluene concentration of the air is 10,816 mol divided by the total air volume of 5×10^9 m^3. Since toluene molecular weight is 92.14 grams per mol, then this means the air contains 996,586 grams of toluene, and the air concentration is 199 μg m^{-3}.

The toluene concentration of the water is 7.1 mol divided by the total water volume of $\times$ 10^5 m^3. So, the water contains about 654 grams of toluene, and the water concentration is 727 μg m^{-3}. However, water concentration is usually expressed on a per liter basis, or 727 ng L^{-1}.

The toluene concentration of the aquatic fauna is 3.0×10^{-9} mol divided by the total tissue volume of 4.5 m^3. So, the fish and other vertebrates contain about 276 ng of toluene, and the tissue concentration is 0.06 ng m^{-3}. Thus, even though the largest amount of toluene is found in the air, the highest concentrations are found in the water.

Applying this information allows us to explore fugacity-based, multi-compartmental environmental models. The movement of a contaminant through the environment can be expressed with regard to how equilibrium is achieved in each compartment. The processes driving this movement can be summarized into transfer coefficients or compartmental rate constants, known as *D values* [15]. So, by first calculating the Z values, as we did for toluene in the previous examples, and then equating inputs and outputs of the contaminant to each compartment, we can derive D value rate constants. The actual transport process rate (N) is the product of fugacity and the D value:

$$N = Df \tag{4.15}$$

And, since the contaminant concentration is Zf, we can substitute and add a first-order rate constant k to give us a first-order rate D value (D_R):

$$N = V[c]k = (VZk)f = D_R f \tag{4.16}$$

Although the concentrations are shown as molar concentrations (i.e., in brackets), they may also be represented as mass per volume concentrations, which will be used in our example [16]. Diffusive processes that follow Fick's laws, and can also be expressed with their own D values (D_D), which is expressed by the mass transfer coefficient (K) applied to area A:

$$N = KA[c] = (KAZ)f = D_D f \tag{4.17}$$

Non-diffusive transport (bulk flow or advection) within a compartment with a flow rate (G) has a D value (D_A) is expressed as:

$$N = G[c] = (GZ)f = D_A f \tag{4.18}$$

This means that when a contaminant is moving through the environment, while it is in each phase it is affected by numerous physical transport and chemical degradation and transformation processes. The processes are addressed by models with the respective D values, so that the total rate of transport and transformation is expressed as:

$$f(D_1 + D_2 + \ldots D_n) \tag{4.19}$$

Very fast processes have large D values, and these are usually the most important when considering the contaminants behavior and change in the environment.

FUGACITY-BASED MASS BALANCE MODELING [17]

We can apply a fugacity approach to determine the partitioning of the hypothetical example used earlier in the concentration-based model example, assuming an average temperature of 25 °C. Let us visualize the mass transport of our hypothetical contaminant among the compartments based upon the results of our concentration-based model (see Figure 3.11).

We will use units of mol m^{-3} Pa^{-1} for our Z values.

$$Z_{air} = \frac{1}{RT} = 4.1 \times 10^{-4} \text{ mol m}^{-3} \text{ Pa}^{-1}$$

We can derive the Z_{water} from Z_{air} and the given K_{AW} (0.01):

$$Z_{water} = \frac{Z_{air}}{K_{AW}} = \frac{4.1 \times 10^{-4}}{0.01} = 4.1 \times 10^{-2} \text{ mol m}^{-3} \text{ Pa}^{-1}$$

The $Z_{particles}$ value can be derived from Z_{water} and the given K_{PW} (6000):

$$Z_{particles} = Z_{water} \cdot K_{PW} = (4.1 \times 10^{-2})(6000) = 246 \text{ mol m}^{-3} \text{ Pa}^{-1}$$

The Z_{fauna} value can be derived from Z_{water} and the given K_{BW} (9000):

$$Z_{fauna} = Z_{water} \cdot K_{BW} = (4.1 \times 10^{-2})(9000) = 369 \text{ mol m}^{-3} \text{ Pa}^{-1}.$$

This means that the weighted total Z value (Z_{WT}) for the ecosystem is the sum of these Z values, which we can weight in proportion to their respective volume fractions in the ecosystem:

$$\begin{aligned} Z_{WT} &= Z_{water} + (2.5 \times 10^{-4} Z_{particles}) + (10^{-5} Z_{fauna}) \\ &= (4.1 \times 10^{-2}) + (2.5 \times 10^{-4})(246) + (10^{-5})(369) \\ &= 1.06 \times 10^{-1} \text{ mol m}^{-3} \text{ Pa}^{-1} \end{aligned}$$

The D values (units of mol Pa^{-1} h^{-1}) can be found from the respective flow rates (G) given or calculated in the concentration model example, and the respective Z values:

Outflow in water:

$$D_1 = G_{water} \cdot Z_{water} = 1000 \times 4.1 \times 10^{-2} = 41 \text{ mol Pa}^{-1} \text{ h}^{-1}$$

Outflow sorbed to particles:

$$D_2 = G_{particle} \cdot Z_{particle} = (0.024) \cdot (246) = 5.9 \text{ mol Pa}^{-1} \text{ h}^{-1}$$

Reaction (using rate constant calculated from half-life of contaminant given in the concentration-based model example):

$$D_3 = VZ_{WT}k = (10^7 \times 1.06 \times 10^{-1})(9.6 \times 10^{-5}) = 101.8 \text{ mol Pa}^{-1} \text{ h}^{-1}$$

Sedimentation

$$D_4 = G_{sed}Z_{particle} = (0.036) \cdot (246) = 8.9 \text{ mol Pa}^{-1} \text{ h}^{-1}$$

Vaporization

The hypothetical contaminant's given mass transfer coefficient (k_M) is 0.24 m day^{-1} or 0.01 m h^{-1} (a fairly volatile substance). This mass transfer takes place across the entire surface area of the estuary (A):

$$D_5 = k_M A Z_{water} = (0.01)(2 \times 10^6)(4.1 \times 10^{-2}) = 820 \text{ mol Pa}^{-1}\text{ h}^{-1}$$

Overall mass balance

Now, we can apply these D values to express the overall mass balance of the system according to the contaminant's fugacity in water (f_{water}). Recall that the contaminant's molecular mass is 100 g mol^{-1}, and that we calculated the total input of the contaminant to be 52 g h^{-1}. Thus, the input rate is 0.052 mol h^{-1}:

Contaminant input = $f_{water}\ \Sigma D_i$. So,

$$0.052 = f_{water}D + f_{water}D_2 + f_{water}D_3 + f_{water}D_4 + f_{water}D_5$$

$$0.052 = f_{water}977.6$$

This means that $f_{water} = 5.3 \times 10^{-5}$. Further, we can now calculate the concentrations in all of the media from the derived Z values and the contaminant's f_{water}:

Contaminant dissolved in water =

$$Z_{water} \cdot f_{water} = (4.1 \times 10^{-2})(5.3 \times 10^{-5}) = 2.2 \times 10^{-6} \text{ mol m}^{-3} = 2.2 \times 10^{-4} \text{ g m}^{-3}$$

Contaminant sorbed to suspended particles =

$$Z_{particle} \cdot f_{water} = (246)(5.3 \times 10^{-5}) = 1.3 \times 10^{-1} \text{ mol m}^{-3} = 13 g \text{ m}^{-3}\text{particle}$$

Contaminant in fauna tissue =

$$Z_{fauna} \cdot f_{water} = (369)(5.3 \times 10^{-5}) = 2.0 \times 10^{-1}\text{mol m}^{-3} = 20 g \text{ m}^{-3}\text{tissue}$$

The concentrations derived from the fugacity model are very close to those we derived from the concentration-based model, taking into account rounding. This bears out the relationship between contaminant concentration and the Z and D values.

This model demonstrates the interrelations between and among compartments. In fact, the concentration and fugacity of the contaminant are controlled by the molecular characteristics of the contaminant and the physicochemical characteristics of the environmental compartment. For example, our hypothetical example contaminant's major "forcing function" was the K_{AW} or the mass transfer coefficient for the contaminant leaving the water surface and moving to the atmosphere. In other words, this is one of a number of rate limiting steps that determines where the contaminant ends up.

To demonstrate how one physicochemical characteristic can significantly change the whole system's mass balance, let us reduce the contaminant's mass transfer from a K_{AW} value of 0.24 to 0.024 m day^{-1} (0.001 m h^{-1}). Thus, for our new contaminant, the evaporation rate = 2.4 m day^{-1})(day 24 h^{-1})(2×10^6 m^2)($0.81C_W$) = $1620C_W$ g h^{-1}. So the combined process rates will again be the sum of all process rates:

$$\text{Discharge rate} = \text{Dissolved Outflow} + \text{Sorbed Outflow} + \text{Reaction} + \text{Sedimentation} + \text{Vaporization}$$

$$52 = 810C_W + 117C_W + 960C_W + 175C_W + 1620C_W$$

$$52 = 3682\ C_W$$

$$C_W = 52/3682 = 0.014\ g \text{ m}^{-3} = 0.014 \text{ mg L}^{-1} = 14\ \mu g \text{ L}^{-1}$$

The modeled results for the estuary's process rates for the hypothetical contaminant will change to:

Process	Rate (g h^{-1})	Percent of total
Outflow dissolved in water (810 × 0.014)	11.3	22%
Outflow sorbed to suspended particles (117 × 0.014)	1.6	3%
Reaction (960 × 0.014)	13.4	26%
Sedimentation (175 × 0.014)	2.5	5%
Vaporization (1620 × 0.014)	22.7	44%

Therefore, comparing these values to those derived from the concentration-based example demonstrates a system effect. The change in one parameter, i.e. decreasing the mass transfer of our pollutant to 10% of the original contaminant's vapor pressure, has led to a much more even distribution of the contaminant in the environment. While the air is still the largest repository for the contaminant at equilibrium, its share has fallen sharply (by 45%). And, the fractions dissolved in water and degraded by chemical reactions account for a much larger share of the mass balance (increasing by 18% and 21%, respectively). Sorption and sedimentation's importance has also increased.

Each environmental system will determine the relative importance of the physical and chemical characteristics. The partitioning coefficients will represent the forcing functions accordingly. For example, if a contaminant has a very high BCF, even small amounts will represent high concentrations in the tissues of certain fish. Often, the molecular characteristics of a contaminant that cause it to have a high sorption potential will also render it more lipophilic, so the partitioning between the organic and aqueous phases will also be high. Conversely, the high molecular weight and chemical structures of these same molecules may render them less volatile, so that the water–to–air partitioning may be low. This is not always true, as some very volatile substances are also highly lipophilic (and have high octanol-water partition coefficients) and are quite readily bioconcentrated (having high BCF values). The halogenated solvents are such an example.

Also, it is important that all of these partitioning events are taking place simultaneously. So, a contaminant may have an affinity for a suspended particle, but the particle may consist of organic compounds, including those of living organisms, so sorption, organic-aqueous phase, and bioconcentration partitioning are all taking place together at the same time on the particle. The net result may be that the contaminant stays put on the particle. Researchers are interested in which of these (and other) mechanisms is most accountable for the fugacity. In the real-life environment, however, it often suffices to understand the net effect. That is why there are so many "black boxes" in environmental models. We may have a good experiential and empirical understanding that under certain conditions a contaminant will move or not move, will change or not change, or will elicit or not elicit an effect. We will not usually have a complete explanation of why these things are occurring, but we can be confident that the first principles of science as expressed by the partitioning coefficients will occur unless there is some yet to be explained other factor affecting them. In other words, we will have to live with an amount of uncertainty, but scientists are always looking for ways to increase certainty.

Models are important tools for estimating the movement of contaminants in the environment. They do not obviate the need for sound measurements. In fact measurements and models are highly complementary. Compartmental model assumptions must be verified in the field. Likewise, measurements at a limited number of points depend on models to extend their meaningfulness. Having an understanding of the basic concepts of a contaminant transport model, we are better able to explore the principle mechanism for the movement of contaminants throughout the environment.

BIOLOGY MEETS CHEMISTRY

The relationships between physicochemical attributes and biological metrics are crucial in predictive microbiology. The amounts and forms of chemical species drive the conditions for microbial growth and metabolism. So, if an index is aimed at a total food chain, the sustained health of the top carnivore or other indicator species may be the index's target output. In fact, both matter and energy indices are used for biological systems. For example, the food web structure will influence its *resilience*, i.e. the ease and speed that a perturbed system can return to equilibrium. This is analogous to the engineering concept of *hysteresis*. For example, a biological community can be considered as a simple relationship between active plant tissue, heterotrophic organisms and organic matter from inactive and dead organisms (see Figure 4.8).

Microbial populations comprise key compartments of the food web, including bacterial, fungal and algal communities. For example, in a six-compartment food web model for sea bass (*Dicentrarchus labrax*) [18], three of these compartments are dominated by microbial populations, i.e. the two plankton compartments and the detritus (see Figure 4.9). The exchange between the environment and the organism is usually observed empirically, e.g. water samples. For example, water in the estuary of the River Seine in France had a range of concentrations of various congeners of polychlorinated biphenyls (PCB), as shown in Table 4.4. These concentrations were compared to the concentrations in the tissue of the organisms in the sea bass food web (see Table 4.5).

Most of the PCB congeners appear to biomagnify moving up levels of biological organization. That is, the *D. labrax* concentrations are clearly the highest and the zooplankton concentrations are the lowest for *most* of the congeners. In particular, the bioconcentration rates are high for congeners 101, 118, 149, 153, and 180. However, for less chlorinated congeners, e.g. 28 and 31, this was not the case. This can be explained to some extent by the lipophilicity of the compounds, which is related to a compound's K_{ow} (see Table 4.6). In fact, the less bioconcentrated congeners' K_{ow} are two orders of magnitude lower than those with higher bioconcentrations. This is an example of a model being highly sensitive to a variable, in this instance the compound's affinity for lipids.

Similar systematic relationships exist in other media. For example, Table 4.7 gives the partitioning coefficients for few important pollutants that have been shown to be transported long distances in the atmosphere.

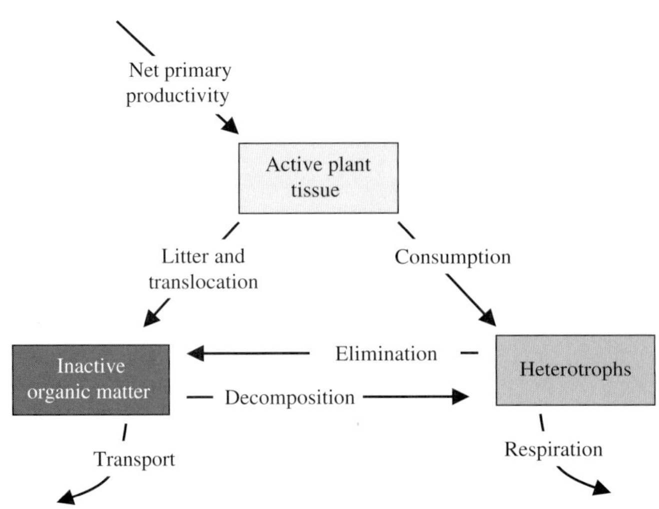

FIGURE 4.8
Transfer of matter and energy within a biological community. [*See color plate section*]
Source: Adapted from M. Begon, J.L. Harper and C.R. Townsend (1996). *Ecology*, 3rd Edition. Blackwell Science, Oxford, UK.

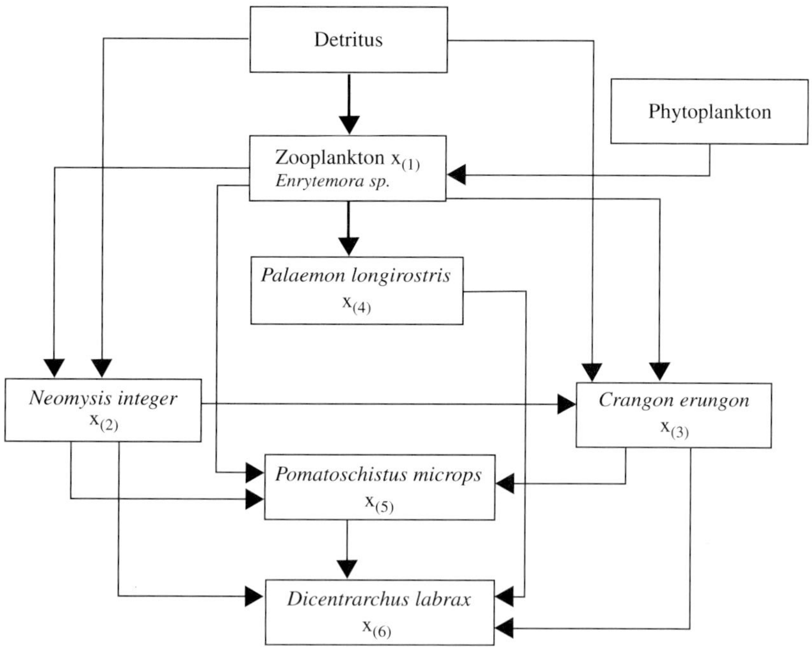

FIGURE 4.9
Six compartment food web model for sea bass (*Dicentrarchus labrax*). Note that the three top compartments are dominated by microorganisms.
Source: V.R. Loizeau, A. Abarnou and A.M. Nesguen (2001). A steady-state model of PCB bioaccumulation in the sea bass (*Dicentrarchus labrax*) food web from the Seine Estuary, France. *Estuaries* 24 (6B): 1074–1087.

Table 4.4 **Mean concentrations of polychlorinated biphenyl congeners (ng g^{-1}) measured in water from the Seine Estuary**

PCB	Concentration (ng L^{-1})
28	0.159
31	0.167
52	0.315
101	0.111
149*	0.120
118	0.075
153	0.075
105	0.015
138	0.072
180	0.040
170*	0.025
194*	0.010

Source: V.R. Loizeau, A. Abarnou and A.M. Nesguen (2001). A steady-state model of PCB bioaccumulation in the sea bass (*Dicentrarchus labrax*) food web from the Seine Estuary, France. *Estuaries* 24 (6B): 1074–1087.

Table 4.5 Mean concentrations of polychlorinated biphenyl congeners ($ng\ g^{-1}$) in six species of aquatic biota from the Seine Estuary (standard deviations)

	31	28	52	101	149	118	153	132	105	138	187	128	180	170	194
Zooplankton	4.2 (0.5)	7.1 (0.6)	14.5 (1.3)	18.5 (2.1)	21.9 (2.3)	12.8 (0.9)	42.3 (5.0)	13.2 (1.4)	8.6 (1.3)	33.6 (3.5)	10.5 (1.1)	3.2 (0.3)	12.3 (1.3)	6.8 (0.7)	1.1 (0.2)
N. integers	6.0 (0.5)	12.5 (1.4)	40.2 (4.2)	65.1 (6.6)	65.2 (6.2)	53.3 (5.4)	119.6 (12.0)	22.2 (2.1)	16.5 (1.5)	94.9 (10.0)	21.7 (2.3)	9.1 (1.0)	59.0 (6.0)	21.4 (2.2)	4.3 (0.5)
P. microps	5.8 (0.6)	9.3 (1.0)	36.5 (3.3)	75.9 (7.7)	74.6 (7.2)	71.5 (6.9)	146.5 (15.0)	42.0 (4.1)	17.2 (1.8)	121.5 (12.8)	32.6 (3.0)	8.7 (0.9)	44.0 (4.6)	15.2 (1.7)	7.9 (0.8)
P. longirostris	2.8 (0.3)	5.6 (0.6)	29.2 (3.1)	22.6 (2.4)	23.2 (2.0)	52.6 (5.4)	96.4 (10.0)	8.1 (0.7)	11.0 (0.9)	75.2 (8.1)	33.2 (2.9)	6.2 (0.6)	51.2 (5.3)	19.2 (2.0)	6.8 (0.7)
C. crangon	2.3 (0.3)	8.4 (0.7)	31.2 (3.3)	22.8 (2.4)	26.5 (2.1)	59.7 (6.1)	156.4 (16.0)	9.4 (1.0)	12.4 (1.4)	131.5 (13.6)	45.5 (5.1)	5.4 (0.6)	81.9 (9.0)	32.7 (2.8)	8.8 (0.9)
Sea bass male (III)	3.5 (0.5)	10.3 (1.5)	44.1 (4.8)	126.5 (13.0)	136.1 (14.0)	144.8 (15.0)	338.8 (35.0)	45.6 (4.1)	40.2 (3.8)	298.7 (27.1)	64.7 (6.6)	12.3 (1.1)	131.3 (12.7)	48.7 (5.0)	5.1 (0.4)

PCB congeners are listed according to their elution time on a gas chromatographic column targeted for semivolatile organic compounds (e.g. DB5).
Source: V.R. Loizeau, A. Abarnou and A.M. Nesguen (2001). A steady-state model of PCB bioaccumulation in the sea bass (*Dicentrarchus labrax*) food web from the Seine Estuary, France. *Estuaries* 24 (6B): 1074–1087.

Table 4.6 Measured biological parameters and log K_{ow} values for polychlorinated biphenyl (PCB) congeners

PCB congener	Log K_{ow}	k_{derm}($cm^3\ g^{-1}\ s^{-1}$)	k_{elim}($\times 10^{-6}\ s^{-1}$)	α (–)
18	5.24	0.181	9.25	0.26
19	5.02	0.250	9.77	0.37
22	5.58	0.190	7.98	0.22
25	5.67	0.230	8.45	0.11
26	5.66	0.191	8.10	0.11
28	5.67	0.172	8.43	0.22
31	5.67	0.174	8.54	0.11
40	5.66	0.203	6.76	0.20
42	5.76	0.193	6.79	0.17
44	5.75	0.177	7.24	0.20
45	5.53	0.192	7.38	0.23
47	5.85	0.206	5.99	0.20
49	5.85	0.213	6.65	0.19
51	5.63	0.275	4.45	0.16
52*	5.84			
64	5.95	0.180	6.65	0.14
74	6.20	0.184	6.41	0.16

Table 4.6 **Measured biological parameters and log K_{ow} values for polychlorinated biphenyl (PCB) congeners—cont'd**

PCB congener	Log K_{ow}	k_{derm}(cm^3 g^{-1} s^{-1})	k_{elim}($\times 10^{-6}$ s^{-1})	α (–)
83	6.26	0.178	5.35	0.15
85	6.30	0.187	5.37	0.16
91	6.13	0.199	5.57	0.15
97	6.29	0.215	5.70	0.11
99	6.39	0.260	4.98	0.15
100	6.23	0.247	6.32	0.11
101	6.38	0.210	5.25	0.15
105	6.65	0.173	5.46	0.16
118	6.33	0.198	5.17	0.14
128	6.74	0.290	4.60	0.16
132	6.58	0.134	4.41	0.12
136	6.22	0.168	4.60	0.13
138*	6.83			
146	6.89	0.177	4.14	0.12
149*	6.67			
153	6.92	0.133	4.48	0.17
170*	7.27			
180*	7.36			
194*	7.80			

Note: k_{derm} is dry tissue based (cm^3 per g dry weight per second); k_{elim} is elimination rate (s^{-1}), α_j is the fractional uptake efficiency (–) of PCB congener j in the digestive tract.
Source of K_{ow} values (except those asterisked*): D.W. Hawker and D.W. Connell (1988). Octanol-water partition coefficients of polychlorinated biphenyl congeners. *Environmental Science & Technology* 22: 382–387. Source of K_{ow} values for asterisked (*) congeners: V.R. Loizeau, A. Abarnou and A.M. Nesguen (2001). A steady-state model of PCB bioaccumulation in the sea bass (*Dicentrarchus labrax*) food web from the Seine Estuary, France. *Estuaries* 24 (6B): 1074–1087. Source of other values: X. Sun, D. Werner and U. Ghosh (2009). Modeling PCB mass transfer and bioaccumulation in a freshwater oligochaete before and after amendment of sediment with activated carbon. *Environmental Science & Technology* 43: 1115–1121.

The kinetics of this community is interdependent. The rate of change of active plant growth and metabolism depends on the input of energy, represented by *net primary productivity* in Figure 4.8. The active plant compartment leads to two outputs consumption of matter (e.g. nutrients) and loss of matter (litter). The rates of change of energy and matter further down the food web depend on subsequent inputs and outputs. The heterotrophs consume living plant biomass and dead organic matter, and then release their own elimination products [19].

These energy and matter kinetics can be input into a system resilience index. This can be useful in estimations of widespread implications and irreversible impacts. For example, resilience of various types of ecosystems has been compared according to the energy needed per unit of active plant tissue (e.g. standing crop). The index would indicate that a system with low total amount of active tissue and a high amount of biomass turnover would be best able to adapt to perturbations. Thus, in Figure 4.10, the pond is predicted to be nearly four orders of magnitude more resilient than a tundra system and three orders more resilient than

Table 4.7 Properties of chemicals used in atmospheric compartmental modeling

Compound	Half-life (days)	Log K_{ow}	Log K_H
Benzene	7.7	2.1	−0.6
Chloroform	360	1.97	−0.7
DDT	50	6.5	−2.8
Ethyl benzene	1.4	3.14	0.37
Formaldehyde	1.6	0.35	−5.0
Hexachlorobenzene	708	5.5	−3.5
Methyl chloride	470	0.94	−0.44
Methylene chloride	150	1.26	−0.9
PCBs	40*	6.4	−1.8
1,1,1 Trichloroethane	718	2.47	0.77

*Note: Polychlorinated biphenyls as a chemical class have long half-lives in the environment, but some congeners are susceptible to photodegradation in the atmosphere, e.g. atmospheric half-lives in days were found by Sinkkonen and Paasivirta (2000) to be: 3 days for PCB 28, 62.5 days for PCB 101, 125 days for PCB 138 and 250 days for PCB 180. All tested PCBs have much shorted half-lives than in water, soil and sediment (≥1.5 orders of magnitude). Source: S. Sinkkonen and J. Paasivirta (2000). Degradation half-life times of PCDDs, PCDFs and PCBs for environmental fate modeling. Chemosphere. 40 (9–11). 943–949.
Source: D. Toro and F. Hellweger (1999). Long-range transport and deposition: The role of Henry's law constant. Final report, International Council of Chemical Associations.

a tropical forest [20]. This would seem to indicate that the implications of a biotechnology, e.g. the introduction of a genetically modified bacterial strain, may be more prolonged with a higher likelihood of irreversibility in systems with lower energy fluxes.

This is where predictive microbiology adds value to an index. Most bacterial growth models have been concerned with the microbial population's response to various physical conditions, especially varying water temperatures, pH or concentrations of chemical substances [21]. In fact, fish food models have attempted to predict quality and shelf life of the organisms after harvest [22]. Conversely, environmental indices may apply the same parameters, but are interested in the fish as indicators of environmental and ecosystem condition, rather than their value as food commodities. This is an example, wherein methodologies used by different scientific communities can be mutually supportive.

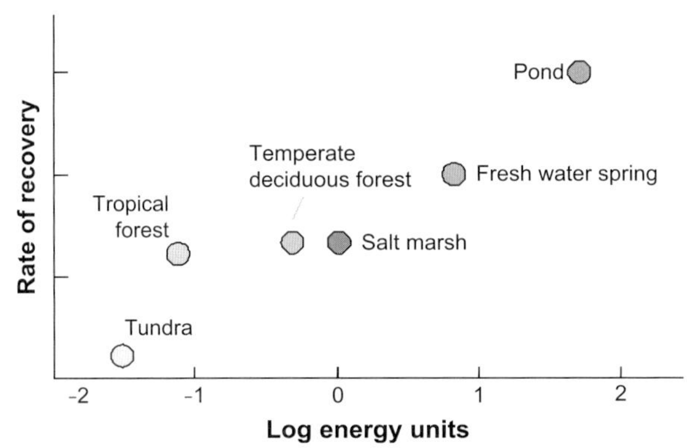

FIGURE 4.10
System resilience index calculated from bioenergetics for six community types. Rate of recovery units are arbitrary; energy units = energy input per unit standing vegetation.
Sources of data: R.V. O'Neill (1976). Ecosystem persistence and heterotrophic regulation. *Ecology* 57: 1244–1253; and M. Begon, J.L. Harper and C.R. Townsend (1996). *Ecology*, 3rd Edition. Blackwell Science, Oxford, UK.

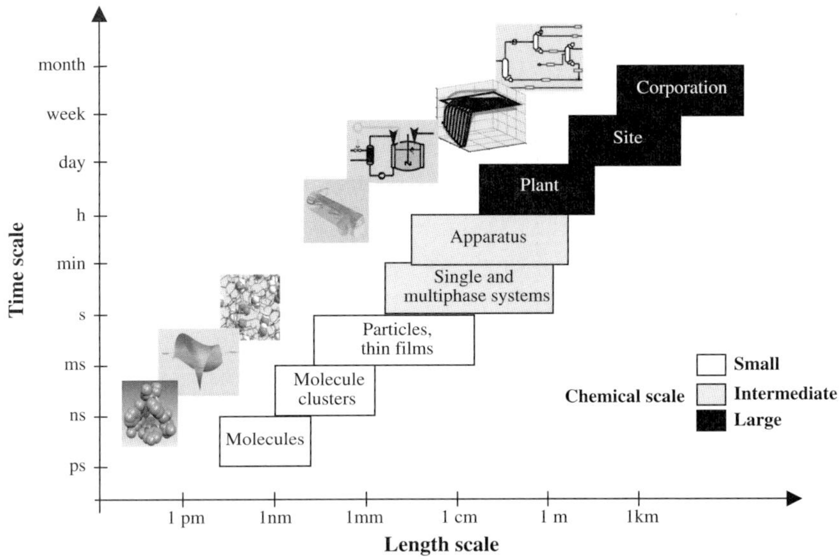

FIGURE 4.11
Scales and complexities of reactors. [*See color plate section*]
Note: ms = millisecond; ns = nanosecond; ps = picosecond.
Source: W. Marquardt, L. von Wedel and B. Bayer (2000). Perspectives on lifecycle process modeling. In: M.F. Malone, J.A. Trainham and B. Carnahan (Eds.), *Foundations of Computer-Aided Process Design*, AIChE Symposium Serial 323, Vol. 96, 192–214.

IMPORTANCE OF SCALE IN BIOSYSTEMS

In Chapters 2 and 3, we embarked on a discussion of thermodynamics. We can visualize biotechnologies as reactors working at various scales in the environment. Engineers are quite familiar with reactors, such as tanks and vats. Reactors not only involve the input of materials and energy, but also reactions. The combination of inputs and reactions within these vessels results in new and often very different forms and amounts of materials and energy that exit. In fact, these new products are the reason for building and operating reactors in the first place. In environmental systems, these thermodynamic behaviors are also occurring, but over a broad domain; having scales ranging from just a few angstroms to global (see Figure 4.11). For example, the processes that lead to a contaminant moving and changing in a bacterium may be very different from those processes at the lake or river scale, which in turn are different from those processes that cause the contaminant's fate as it crosses the ocean. This is simply a manifestation of the first law of thermodynamics, i.e. energy or mass is neither created nor destroyed, only altered in form. This also means that energy and mass within a system must be in balance: what comes in must equal what goes out. These fluxes are measured and yield energy balances within a region in space through which a fluid travels. Recall from Chapter 2 that such a region is known as a control volume, and that the control volumes where these balances occur can take many forms. The first law of thermodynamics frames any biological system, from subcellular to planetary as a reactor where mass and energy enter, change within the control volume, and exit as transformed products. This is the way all environmental biotechnological processes work:

$$\begin{bmatrix} \text{Quantity of} \\ \text{mass per unit volume} \\ \text{in a medium} \end{bmatrix} = [\text{Total flux of mass}] + \begin{bmatrix} \text{Rate of production or loss} \\ \text{of mass per unit volume} \\ \text{in a medium} \end{bmatrix} \quad (4.20)$$

Or, stated mathematically:

$$\frac{dM}{dt} = M_{in} - M_{out} \quad (4.21)$$

where, M = mass, and t = specified time interval.

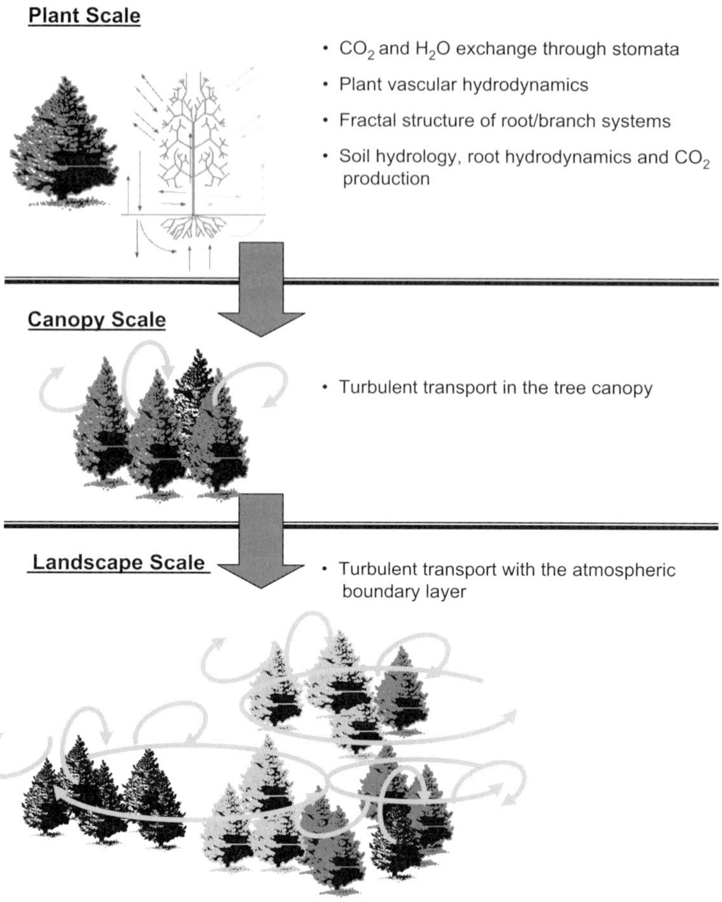

FIGURE 4.12
Three hierarchical scales applied to trees. Although the flow and transport equations do not change, the application of variables, assumptions, boundary conditions, and other factors are scale- and time-dependent. [*See color plate section*] *Source:* G. Katul (2001). Modeling heat, water vapor, and CO_2 transfer across the biosphere–atmosphere interface. Seminar presentation at Pratt School of Engineering, December 1, 2001.

If we are concerned about a specific chemical (e.g. environmental engineers worry about losing good ones, like oxygen, or forming bad ones, like the toxic dioxins), the equation needs a reaction term (R):

$$\frac{dM}{dt} = M_{in} - M_{out} \pm R \tag{4.22}$$

In reality, smaller control volumes assimilate into larger ones. Within reactors are smaller-scale reactors (e.g. within the fish liver, on a soil particle, or in the pollutant plume or a forest, as shown in Figure 4.12). Thus, scale and complexity can vary by orders of magnitude in environmental systems. For example, the human body is a system, but so is the liver, and so are the collections of tissues through which mass and energy flows as the liver performs its function. Each hepatic cell in the liver is a system. At the other extreme, large biomes that make up large parts of the earth's continents and oceans are systems, from the standpoint of biology and thermodynamics.

The interconnectedness of these systems is crucial to understanding biotechnological implications, since mass and energy relationships between and among systems determine the efficiencies of all living systems. For example, if a toxin adversely affects a cell's energy and mass transfer rates, it could have cumulative affect on the tissue and organs of the organism. And, if the organisms that make up a population are less efficient in survival, then the balances needed in the larger systems, e.g. ecosystems and biomes, may be changed, causing problems at the global scale. Viewing this from the other direction, a larger system can be stressed, such as changes in ambient temperature levels or the increased concentrations of contaminants in water bodies and the atmosphere. This results in changes all the way down to the subcellular levels (e.g. higher temperatures or the presence of foreign chemicals at a cell's membrane will change the efficiencies of uptake, metabolism, replication, and survival). Thus, the changes at these submicroscopic scales determine the value of any biotechnology.

The biosystematic viewpoint also includes the interrelationships of the abiotic (non-living) and biotic (living) environments. Biotechnology has been the application of the concept of "trophic state" for much of human history. Organisms, including humans, live within an interconnected network or web of life (see Figure 4.13). In a way this is not any different from the energy and mass budgets of the chemical reactors familiar to chemical engineers. Ecologists attempt to understand the complex interrelationships shown in Figure 4.14, and consider humans to be among the consumers. This systematic view is also valuable for remediation and restoration of disturbed ecosystems, as it not only identifies options, but allows for an assessment of the difficulty in implementing them (see Figure 4.15).

The "feedbacks" in Figure 4.13 are crucial to environmental biotechnology, wherein bioengineers "optimize" the intended products and preserve (limit the effects) on the energy and mass balances. Sometimes the bioengineer must decide that there is no way to optimize both. In this instance, the ethical bioengineer must recommend the "no go" option. That is, the potential downstream costs are either unacceptable or the uncertainties of possible unintended, unacceptable outcomes are too high. Usually, though, the engineer will be able to at least model a number of permutations and optimize solutions from more than two variables (e.g. species diversity, productivity, and sustainability, costs and feasibility, and bioengineered product efficiencies). The challenge is knowing to what extent the model represents the realities as they vary in time and space. Models are inherently uncertain, since they represent something larger than themselves, i.e. they have scalar uncertainties.

As mentioned, the scientific advances needed to provide reliable estimations of the inputs, changes, and outputs of biological systems requires the emerging scientific assessment tools being realized by advances in computational fluid dynamics, computational chemistry; systems biology; molecular, cellular, and biochemical toxicology; and exposure modeling [23].

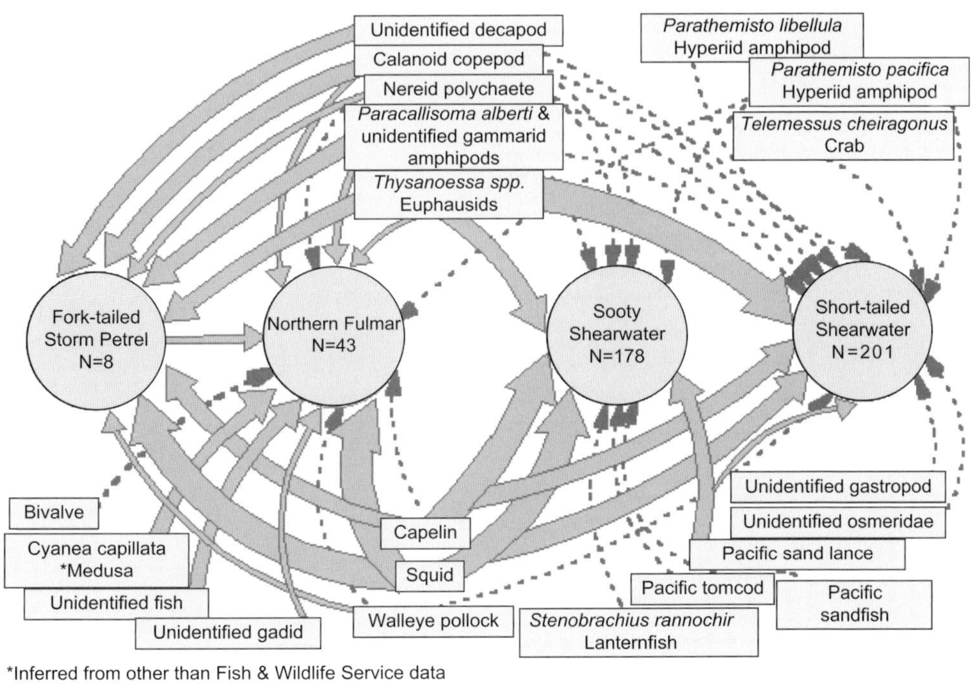

FIGURE 4.13
Flow of energy and mass among invertebrates, fish and seabirds (*Procellariform*) in the Gulf of Alaska. The width of the arrow increases in proportion to the relative flow. Note how some species prefer crustaceans (e.g. copepods and euphausiids), but other species consume larger forage species like squid. [*See color plate section*] *Source:* G.A. Sanger (1983). Diets and food web relationships of seabirds in the Gulf of Alaska and adjacent marine areas. US Department of Commerce, National Oceanic and Atmospheric Administration, OCSEAP Final Report # 45, 631–771.

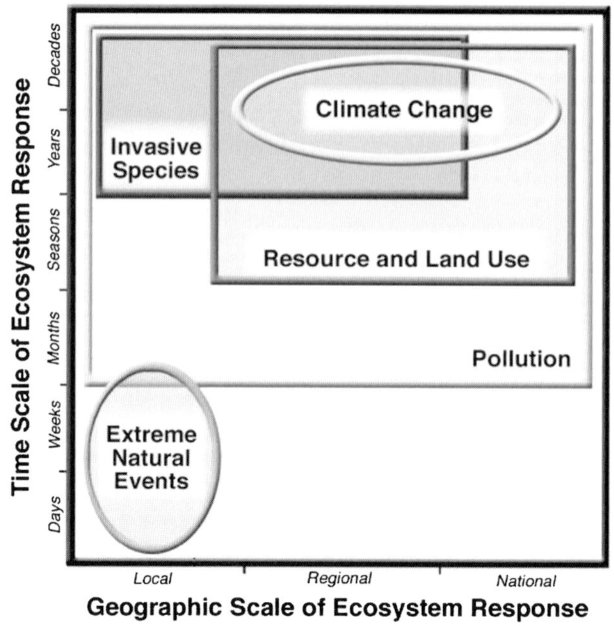

FIGURE 4.14
The response to stressors has temporal and spatial dependencies. Near-field stressors can result from a spill or emergency situation. At the other extreme, global climate change can result from chronic releases of greenhouse gases with expansive (planetary) impacts in direct proportion to significant changes in global climate (temperature increases in the troposphere and oceans, shifting biomes, sea level rise, and migratory patterns). [*See color plate section*]
Source: R. Araujo (2007). US Environmental Protection Agency. Conversation with author.

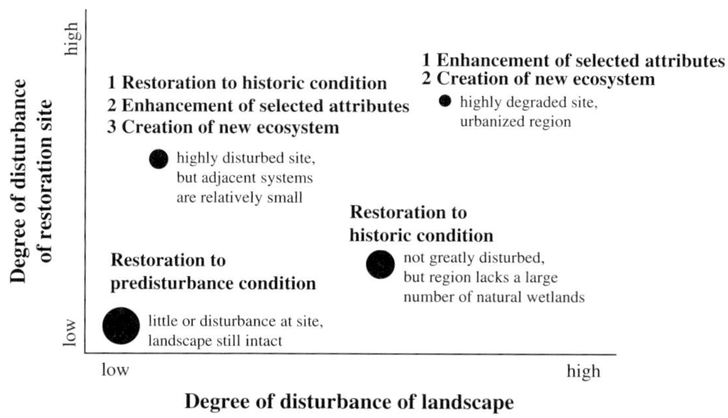

FIGURE 4.15
Restoration strategies applied to Columbia River Estuary ecosystems based on the amount of damage and likelihood of success (size of dot is proportional to relative chance of success).
Source: US Department of Energy (2003). An Ecosystem-Based Approach to Habitat Restoration Projects with Emphasis on Salmonids in the Columbia River Estuary. Final Report (PNNL-14412). Washington, DC.

These tools need to be integrated to characterize the complexities and scalar influences on biological systems (see Figure 4.16). The tools make use of the tiers in biological systems (e.g. trophic states in ecosystems, absorption–distribution–metabolism–elimination in organisms). This integration is important to biochemodynamics. For example, as illustrated in Figure 4.17, results from Step 1 feed into Step 2, which is quantification of dose–response relationships and habitat–response relationships. The cascade of information flows to the next level, e.g. from organism to population. Thus, the response variables in Step 2 are spatially explicit demographic rates of individuals within a population. These demographic rates allow for estimates of population growth rates, population extinction rates, or other population-level outcomes by using population models in Step 3. Step 4 estimates

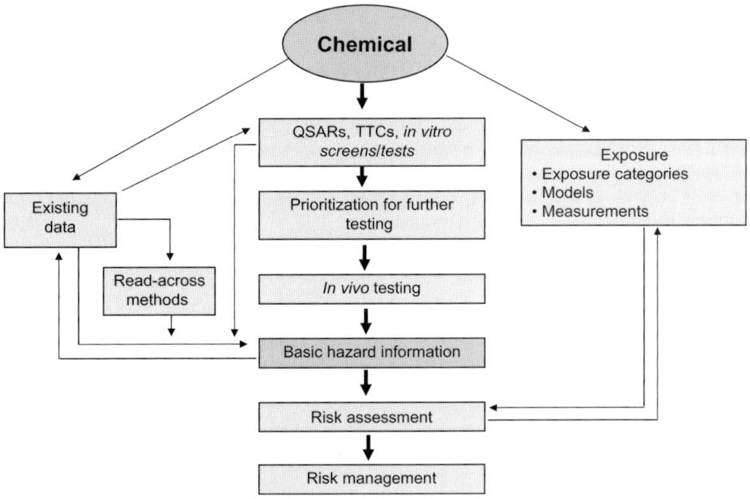

FIGURE 4.16
Framework for integrating environmental exposure information and effects information gained from quantitative structure–activity relationships (QSARs), read-across methods, thresholds of toxicological concern (TTCs), and *in vitro* tests prior to *in vivo* testing to perform risk assessment of chemicals. [*See color plate section*]
Source: S.P. Bradbury, T.C.J. Feijtel, C.J. Van Leeuwen (2004). Peer reviewed: Meeting the scientific needs of ecological risk assessment in a regulatory context. *Environmental Science & Technology* 38 (23): 463A–470A.

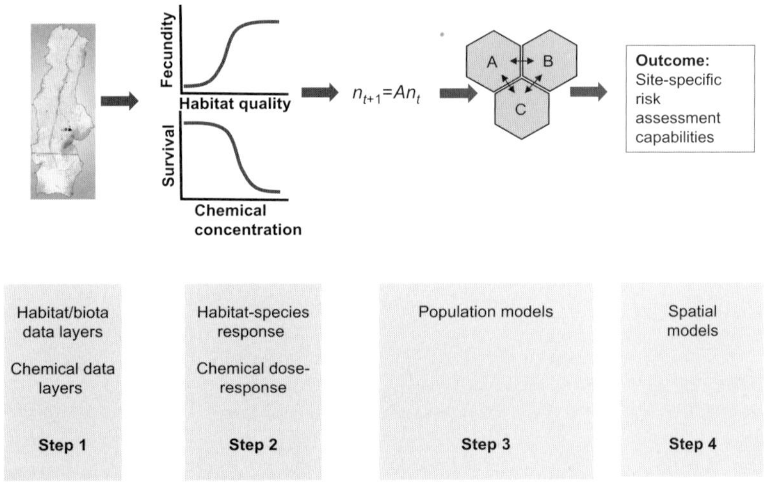

FIGURE 4.17
Conceptual model of spatially explicit population-based risk assessments. Linking databases for species-specific toxicity, demographics, life history, and habitat quality requirements to models that can estimate missing values from existing information will provide the means for projecting population responses for specified species in defined locations. The first of four steps a geographic information system (GIS)-based risk-assessment modeling process is a landscape characterization that requires spatial and temporal characterization of the chemical stressors exposure along with the spatial and temporal characterization of habitat quantity and quality. [*See color plate section*]
Source: S.P. Bradbury, T.C.J. Feijtel, C.J. Van Leeuwen (2004). Peer reviewed: Meeting the scientific needs of ecological risk assessment in a regulatory context. *Environmental Science & Technology* 38 (23): 463A–470A.

habitat-specific population sources and sinks by applying the population dynamics that were derived from Step 3 into the landscape. This systematic assessment provides estimates of risks to biota at the population level posed by chemical, physical, and biological agents, as well as by habitat changes and landscape perturbations [24]. That is, it includes both biotic and abiotic information and systematically analyzes them, moving up the biological tiers.

The computational and "omics" tools can also assist the development of biological effects indicators. The progress of these molecular and biochemical approaches will extend hazard identification and evaluation approaches needed to investigate complex chemical mixtures and to provide increasingly reliable and refined techniques to identify when specific agents are causing adverse effects [25]. Biotechnologies are helping with these causal links. For example, organism response is being used to indicate stresses on biological systems.

SYSTEMS SYNERGIES: BIOTECHNOLOGICAL ANALYSIS

All biotechnological fields require reliable analytical methods. Advancing the state-of-the-science of microscopy and molecular biology in DNA assays will be a particularly fruitful endeavor. Currently, most of this work takes place within each field. Thus, specific methods are described in medical, industrial, agricultural, and environmental journals; often to address a particular need (e.g. identify a specific organism, develop an environmental model, or test a drug). Some specific methods are published for their own sake in analytical journals with interests in chemical, physical, and biological methods, such as environmental and medical chromatography, genomics, pesticide science, and food safety. In fact, most of these methods would likely apply to any of the biotechnological fields.

Organisms are integrators. As they take up, metabolize, and use chemicals they provide fingerprints of conditions that surround them. Scientists often rely on simplistic models to extrapolate from high-dose toxicology data to estimate low-dose response, which frequently renders a finding of low or no adverse health effects at environmentally relevant levels [26]. However, when biota are exposed to a variety of agents, their response can be quite complicated. There is great uncertainty with regard to co-exposures and chemical mixtures. In some instances, when biota come into contact with two or more different compounds, the effects can be more than additive, i.e. the chemicals are eliciting a synergistic effect in the biota. However, there are some instances when co-exposures can be protective, i.e. the effects are antagonistic. Sometimes the effects are mechanistic and physical, such as when a lipophilic compound is found in oily substrate, which allows transfer through skin more readily than the compound where in other substrates. In other situations, the effects are chemical, such as when an otherwise hydrophobic compound is dissolved in an alcohol, which allows it to be transported into more polar substrates (e.g. water-rich targets). Other co-exposures combine these effects with biological factors, such as when active sites on cells are affected by mixtures. Thus, the manner in which an organism integrates a chemical compound is affected by a mix of physical, chemical, and biological factors.

USING BIOINDICATORS

The traditional means of dealing with pollution is to measure the chemical concentrations of contaminants and, if they are outside of the healthy range (e.g. elevated contamination), actions must be taken. However, other ways are available to assess environmental quality. Biological monitoring and assessment techniques have advanced considerably in recent decades. Water quality standards, for example, to protect wildlife and aquatic life began as general guidelines due to limited data and specific research. Improved precision may result in more efficient and effective evaluation of attainment of condition and utilization of restoration resources. Finally, improved precision in uses can enhance demonstrating progress towards management goals. More precise, scientifically defensible biological tools are allowing for improved protection for specific ecosystems.

Distinguishing between natural variability and effects of stressors on ecosystems, along with determining the appropriate level of protection for individual components of ecosystems, are being increasingly addressed. For example, in the United States, a number of states and Native American tribes are presently using biological information to directly assess the condition of their ecological resources [27]. Some states are designating tiered aquatic life uses to clearly articulate and differentiate intended levels of protection with enough specificity to help decision makers to implement their standards to protect a specific site, reach, or watershed, and so that the public adequately understands the goals set to protect ecosystems.

In 2001, the National Research Council (NRC) recommended tiering designated uses as an essential step in setting water quality standards and improving decision making [28]. The NRC considered the Clean Water Act's goals (i.e., "fishable," "swimmable") to be too broad to be used as operational statements of designated use, and recommended greater specificity in defining such uses. For example, rather than stating that a water body needs to be "fishable," the designated use would ideally specify biological characteristics (e.g., cold water fishery) along with other biological conditions need for that particular fish population [29]. Thus, this biological information could be coupled with physical and chemical criteria to show the extent to which the ecological resource, in this case the water body, is meeting its designated use. The NRC described a "position of the criterion" framework, which reflects how representative a criterion is of a designated use according to its position along a conceptual causal pathway (see Figure 4.18). This alignment is comparable to that of performance criteria in other engineering disciplines.

It is also compatible with regulatory review and approval of new and existing products and biocontrols of organisms. For example, if the metabolism of a genetically engineered strain of an organism differs from the non-modified species, this could indicate the potential for ecological and toxicological hazards brought on by the genetic manipulation. Regulatory disapproval of a GE crop may result if the GE crop is determined to be compositionally dissimilar or lack substantial equivalence to progenitor (non-modified) cultivar. This is a first screen. A plant must not vary substantially from the norm [30]. If it does, it could foretell health and environmental problems as the GE crop is used more extensively.

Organisms have distinctive metabolonomes (e.g. hormones, endogenous intermediates and other small-molecule metabolites). For example, Figure 4.19 provides an example of how metabonomic fingerprints can be used to distinguish progenitor from transgenic potatoes.

The presence, condition, and diversity of plants, animals, and other living things can be used to assess the health of a specific ecosystem, such as a stream, lake, estuary, wetland, or forest. Such organisms are referred to as *biological indicators.*

An indicator is in a sense an "integrated" tool that incorporates highly complex information in an understandable manner. A well-known bioindicator is the famous canary in the coal mine. Miners were aware that if they hit a vein that contained "coal gas" (actually high concentrations of methane) they had little time to evacuate before inhalation of the gas would lead to death. However, they realized that due to its small mass, a smaller animal would succumb to the toxic

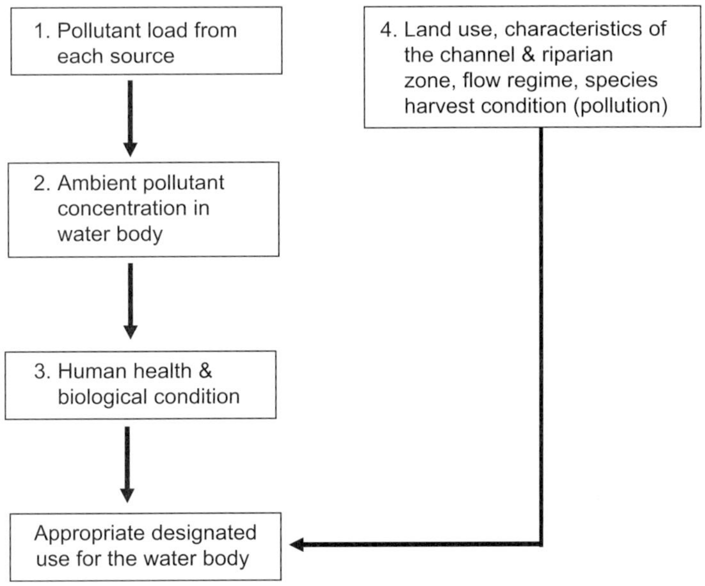

FIGURE 4.18
Types of water quality criteria and their position relative to designated uses.
Sources: US Environmental Protection Agency (2005). Draft Report: Use of Biological Information to Better Define Designated Aquatic Life Uses in State and Tribal Water Quality Standards: Tiered Aquatic Life Uses – August 10, 2005, Washington, DC; and National Research Council.

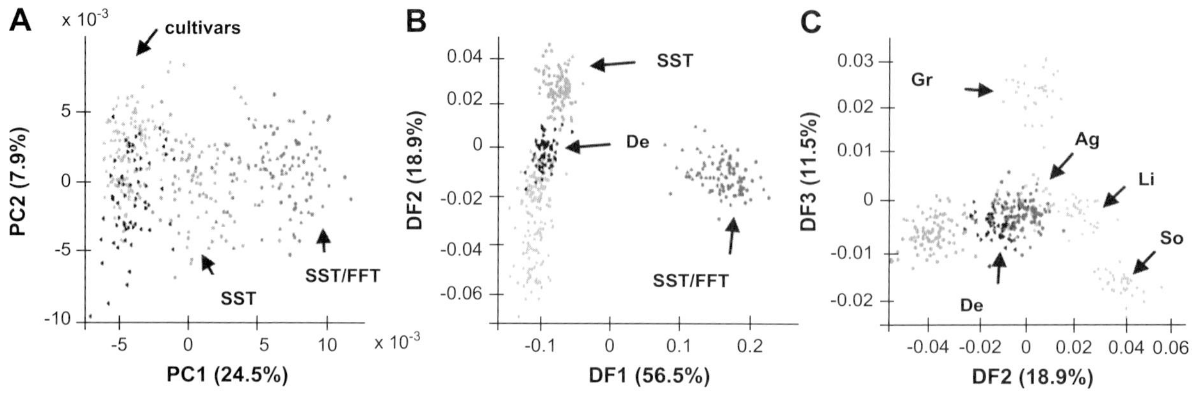

FIGURE 4.19
Flow injection electrospray–mass spectrometric metabolite fingerprints of five conventional potato cultivars and two types of transgenic lines (SST and SST/FFT) analyzed by different multivariate data analysis methods. Computational methods are applied to determine if the endogenous metabolic profiles for the plants are substantially equivalent, in this case, principal component analysis (PC1 and PC2) for regularities and discriminant functions (DF1 and DF2) for differences. If the metabonomic profiles differ substantially, this could indicate unacceptable hazard and risk uncertainties. [*See color plate section*] *Source:* G.S. Catchpole, M. Beckmann, D.P. Enot, M. Mondhe, B. Zywicki, J. Taylor, *et al.* (2005). Hierarchical metabonomics demonstrates substantial compositional similarity between genetically modified and conventional potato crops. *Proceedings of the National Academy of Sciences of the United States of America* 102: 14458–14462.

effects before a human would be affected. The miners did not really care so much how it worked (i.e. the dose–response relationships and routes that will be discussed in the next chapter), they only cared that it worked. Actually the canary is an example of a *bioassay,* which is a test of toxicity or other adverse effect on one or a few organisms to determine the overall expected effect on a system.

Using an indicator or sentry organism can indicate a cumulative effect to relatively low-dose, chronic exposures, which are more common and realistic than laboratory studies that rely on short-duration, high-dose administrations to a relatively small number of test species, from which the results must be mathematically modeled. Living organisms, as natural biological integrators, represent numerous locations (wherever the organism has been) and realistic behaviors (what the organism ordinarily does, so long as the sensor is not disruptive).

Actually, when epidemiological studies are designed to extract fluids to ascertain exposures in a representative group of people, this is an example of *biomonitoring*. That is, whatever these people have been exposed to, if it is analyzed for, will be indicated by the sample, so long as the compound has not been completely metabolized and is present at sufficient concentrations to be detected. Metabolites are also often measured in these studies. For example, a person exposed to nicotine, usually metabolizes the compound rapidly. However, cotinine is a metabolite of nicotine that, when measured, allows the dose of nicotine to be reconstructed. Computational methods are being used for such dose reconstruction. Metabolic pathways yield breakdown products after an organism is exposed to the parent compound. In addition, an organism's endogenous chemicals respond to the exposure to the parent and breakdown products. Thus, there is a combination of the xenobiotic compound and its metabolites, as well as a change in the concentrations of the chemicals that are always produced by the organism. *Metabolonomics* measures the metabolic status of the whole organism, connecting genomics and proteomics (genetic and cellular responses to the xenobiotic exposure, respectively) with histopathology (i.e. the tissue damage). This reveals a "fingerprint" of the organism's response to the uptake and metabolism of a substance. In other words, the omics tools characterize the expected chemical progeny of the parent compound and the profile of the organism's own endogenous compounds as a response to the exposure to the parent xenobiotic compound and its degradation products. In addition, these "omics" can be used at the population and higher trophic levels, so these tools can be useful in biological indication studies.

Indicators of effects are also available. Biological effects at the cellular level range from acute cellular toxicity to changes in the cellular ribonucleic and deoxyribonucleic acid structures, leading to cellular (and tissue) mutations, including cancer. The cells are also homes to chemical signaling processes such as those in the stimulus-response systems in microbes and plants, as well as the endocrine, immune, and neural systems in animals. The presence of enzymes and other chemicals can indicate stress at various biological levels. Metabolonomics are also valuable computational tools for effects studies.

An *ecological indicator* can be a single measure, an index that embodies a number of measures, or a model that characterizes an entire ecosystem or components of that ecosystem. An indicator integrates the physical, chemical, and biological aspects of ecological condition. They are used to determine status and to monitor or predict trends in environmental conditions and possible sources of contamination and stress on systems.

Biocriteria are metrics of a system's biological integrity. A system must be able to support communities of organisms in a balanced manner [31]. One means of determining biological integrity is to compare the current condition of an ecosystem to that of pristine or undisturbed conditions (see Figure 4.20). At a minimum, the threshold for biological integrity is the condition below which a system suffers from dysfunction or impairment. That is, a robust set of bioindicators should be able to show whether the system is above or below a threshold of biological well-being. Such a threshold is known as a reference condition, which ecologists frequently associate with biological integrity. However, most systems have in some way and to some extent been adversely affected by humans. Indeed, the "pristine" system is so rare that environmental scientists more often consider a reference system to be one that is "minimally impaired," i.e. one with high biological integrity, but that has not been untouched by human activities.

Ecosystems and environmental compartments can be degraded by chemical contamination, as well as by physical changes that alter habitats, such as the withdrawal of irrigation water from

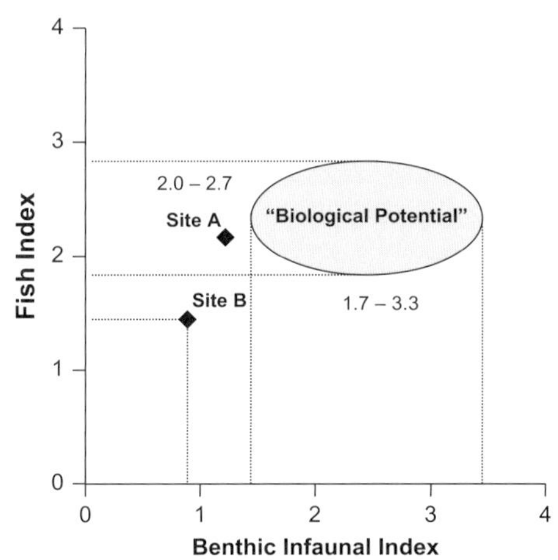

FIGURE 4.20
Graphical representation of a bioassessment comparing actual ecosystem conditions to ideal reference sites. The assessment sites A and B are compared to an ideal biological potential. Site *A* is near its potential, whereas Site *B* deviates from it. The bioassessment can be based on measured and/or modeled information. This includes biocriteria, such as biodiversity, species abundance and productivity. Ecosystem condition information can also be enhanced by complementary physicochemical information, e.g. nutrient cycling and chemical contamination levels. Note: The benthic infaunal index is developed by: (1) defining major habitat types based on classification analysis of benthic species composition and evaluation of the physical characteristics of the resulting site groups; (2) selecting a development data set representative of degraded and undegraded sites in each habitat (3) comparing various benthic attributes between reference sites and degraded sites for each of the major habitat types; (4) selecting the benthic attributes that best discriminated between reference and degraded sites for inclusion in the index; (5) establishing scoring criteria (thresholds) for the selected attributes based on the distribution of values at reference sites; (6) constructing a combined index value for any given sample by assigning an individual score for each attribute, based on the scoring criteria, and then averaging the individual scores; and (7) validating the index with an independent data set.
Source: U.S. Environmental Protection Agency (2000). *Estuarine and Coastal Marine Waters: Bioassessment and Biocriteria Technical Guidance.* Report No. EPA-822-B-00-024. Washington, DC.

Table 4.8 Components of biological integrity

Biospheric elements	Ecosystem processes
Genetics	Mutation, recombination
Individual	Metabolism, growth, reproduction
Population/species	Age-specific birth and death rates Evolution/speciation
Assemblage (community and ecosystem)	Interspecies interactions Energy flow
Landscape	Water cycle Nutrient cycles Population sources and sinks Migration and dispersal

Source: US Environmental Protection Agency.

aquifers and surface waters, over-fishing and overgrazing, and by introducing opportunistic exotic species. Biota are selectively sensitive to all forms of pollution (such as the difference between game and rough fish discussed in the oxygen depletion sections).

Estimating biological integrity requires the application of direct or indirect evaluations of a system's attributes. Indirect evaluations can have the advantage of being cheaper than the direct approaches, but will not often be as robust. An attribute of natural systems to be protected, e.g. a fish population, is an example of an assessment endpoint; whereas an attribute that is quantified with actual measurements, e.g. age classes of the fish population, is known as a measurement endpoint. Reliable and representative assessment and measurement endpoints are needed to reflect a system's biological integrity.

Arguably the most widely used metric for biological integrity is the "Index of Biotic Integrity" (IBI) which consists of 12 attributes in three major groups, i.e., species richness and composition, trophic structure, and fish abundance and condition. The elements of the biosphere are essential to the protection of biological integrity (see Table 4.8).

The ecosystem processes follow the hierarchy of a system's organization, including its various structures and functions. So the metabolism of individual organisms are at one extreme. Population processes, e.g. reproduction, recruitment, dispersal, and speciation are next, while at the highest level of organization, i.e. the communities or ecosystems, processes include nutrient cycling, interspecies interactions, and energy flows. Only a representative amount of biota needs to be sampled. Such selections must aggregate an optimal number of attributes with sufficient precision and sampling efficiency to provide robust indicators of ecosystem health. For example, *benthic* aquatic invertebrates living at the bottom of surface water systems can be very powerful bioindicators since they live in the water for all or most of their lives and remain only in areas suited to their survival (i.e. higher quality conditions). Benthic invertebrates are also relatively easy to collect and identify in the laboratory. They have limited mobility and differ in their ability to tolerate different kinds of pollution, so they are good "sentries" of biological integrity. Since benthic invertebrates can live for more than one year and are limited in their mobility, they can be ideal "integrators" of surface water conditions. These and other "sentry" organisms, analogous to the "canary in the coal mine," integrate or "index" environmental quality (see Discussion Box: Chlorophyll as an Environmental Indicator). When the correct diversity, productivity, and abundance of representative organisms are present, the bioindicators are telling us that the system is healthy (see Figure 4.21).

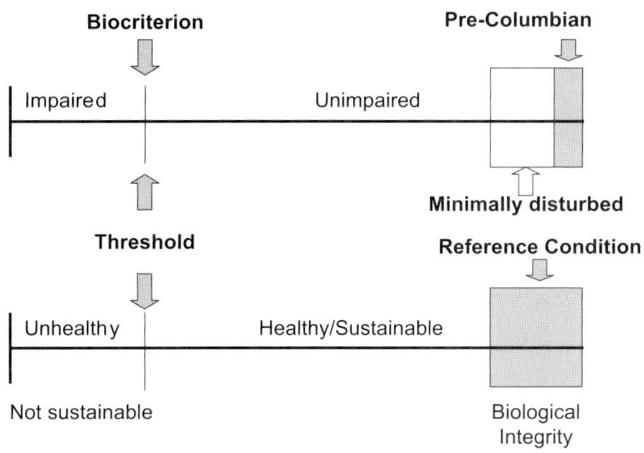

FIGURE 4.21
Need to have biocriteria that match actual ecosystem integrity.
Source: US Environmental Protection Agency (2003). Biological Indicators of Watershed Health. http://www.epa.gov/bioindicators/html/about.html.

DISCUSSION BOX

Chlorophyll as an Environmental Indicator [32]

Chlorophyll is the pigment that gives plants their green color, and is an essential component of photosynthesis whereby plants derive their energy for metabolism, growth, and reproductive processes. Scientists measure the amount of chlorophyll in water as an indirect, yet reliable, indicator of the amount of photosynthesizing taking place in a water body. For example, in a sample collected in a lake or pond, the photosynthetic activity of algae or phytoplankton is indicated by a metric known as chlorophyll. Such a measurement reflects the amount of chlorophyll pigments, both active (alive) or inactive (dead). Thus, chlorophyll can allow the distinction between different life cycles of algal growth (see Figure 4.22). "Chlorophyll *a*" is a measure of the active fraction of the pigments; that is, the portion that was still actively respiring and photosynthesizing at the time of sampling.

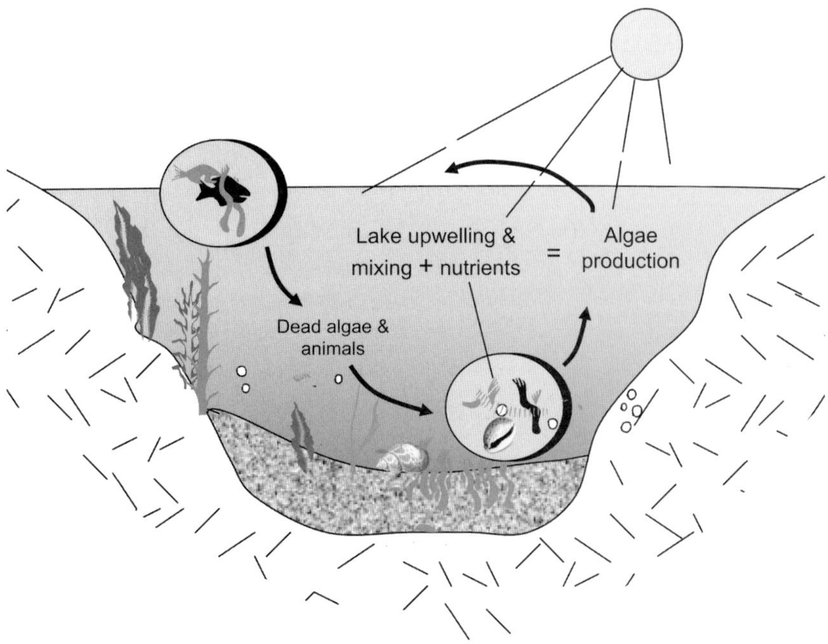

FIGURE 4.22
Algal growth cycle.
Source: State of Washington (2003). Department of Ecology. A Citizen's Guide to Understanding and Monitoring Lakes and Streams.

The amount of algae found in a surface water body will have a large effect on the physical, chemical, and biological mechanisms in the water because the algae produce oxygen when light is present (i.e. in the daytime), and consume oxygen in the dark (nighttime). Algae also expend oxygen when they die and decay. In addition, the decomposition of algae results in the release of nutrients to the lake, which may allow more algae to grow. Thus, the algal and plankton photosynthesis and respiration will affect the water body's pH and suspended solids content. In fact, in lakes the presence of algae in the water column is the principal factor affecting turbidity measurements (e.g. Secchi disk readings).

Algal proliferation can also lead to negative esthetics, such as the "algal blooms" that show up as a greenish scum floating atop ponds and lakes in the summer, as well as the odors associated with the growth. Increasing amounts of sunlight, temperature, and available nutrients with spring warming and summer heat increase algal growth and, therefore, the chlorophyll *a* concentrations. Until limited by the availability of one or more nutrients (especially nitrogen or phosphorus), algae will continue to grow. Strong winds provide mixing of waters, leading to an immediate decrease in algae concentrations as the organisms are distributed throughout the water column. But, winds may also help to release nutrients into the surface water system by agitating nutrients sequestered in bottom sediments, so that a nitrogen or phosphorus limited lake or pond may experience a spike in algal growth following the windy conditions.

The decreasingly available light and reduced temperatures with the onset of fall results in decreasing algal growth. However, in deep lakes that undergo stratification (i.e. different temperatures at different lake levels) a fall algal bloom may occur because the lake mixes with the change of density of the layers due to the temperature differentials at various levels. This allows for more nutrients to be made available to the algae in the water body.

Algal populations, and therefore chlorophyll *a* concentrations, vary greatly with lake depth. Algae must stay within the top portion of the lake where there is sunlight to be able to photosynthesize and stay alive. As they sink below the sunlit portion of the lake, they die. Therefore, few live algae (as measured by chlorophyll *a*) are found at greater depths. Some algae, notably blue–greens (Figure 4.23), have internal "flotation devices" that allow them to regulate their depth and so remain within the top portion of the lake to photosynthesize and reproduce.

Certain algal species, especially the "blue–green" prokaryotes, produce toxins. Usually, the concentrations of toxin are too small to elicit health problems, but should the algal populations become dense, the toxins may exceed safe thresholds. For example, animals have been known to die from consuming water contaminated by algae. The blooms of these algal species usually have a characteristic bluish-green sheen.

FIGURE 4.23
Blue–green algae.
Source: United States National Oceanic and Atmospheric Administration. Coral Reef Information System. Photo: J. Waterbury, Woods Hole/National Aeronautics and Space Administration. Astrobiology Institute: http://www8.nos.noaa.gov/coris; accessed on December 29, 2009.

Since limiting nutrients will limit the number of algae that can grow in surface waters, the best way to address algal problems is to limit the amount of nitrogen and phosphorus entering the water. At one time, *point sources* were a principal source of such contamination, but with greater controls on wastewater treatment plants and other large sources, much of the loading of nutrients to lakes and ponds comes from *nonpoint sources*, such as runoff from farms and septic tanks. Lake management plans, for example, often include a large amount of measures to reduce the amount of nutrients reaching surface waters, including top soil erosion control programs, contour farming, minimum till agriculture, and reduced amounts of fertilizers applied to fields, as well as the banning or strong controls of septic tanks and other nutrient leaching sources in a lake watershed.

Chlorophyll *a* is reported in mass per volume units (usually $\mu g\ L^{-1}$. Many states have no water quality standard for chlorophyll *a*. Concentrations of chlorophyll *a* can vary considerably from one lake to another, even though they may be in the same region. For example, the concentrations for three lakes in the western region of Washington State are shown in Table 4.9. Black Lake would appear to have greater algal growth than do Summit and Blackmans Lakes. Also, Black Lake would appear to be temperature-stratified and experiences fall mixing, allowing for an increase in algal populations in September.

Chlorophyll a: an environmental indicator

Chlorophyll *a* concentrations can be a tool to characterize a lake's trophic status. Though trophic status is not related to any water quality standard, it is a mechanism that can be used to rate a surface water body's productive state. Phytoplankton biomass in aquatic ecosystems can be simply measured as an indicator of water quality and ecosystem condition. Chlorophyll *a* has been established as an indicator of both the potential amount of photosynthesis and of the quantification of phytoplankton biomass [33] and has become a principal measure of the amount of phytoplankton present in a water body. Chlorophyll *a* is also an indirect measure of light penetration [34].

Relatively rapid methods are available for measuring the concentration of chlorophyll *a* in water samples and *in vivo* [35]. Methods are also available to measure chlorophyll *a* with remote sensing and passive multispectral signals associated with phytoplankton [36]. Chlorophyll *a* is a robust indicator of nitrogen and phosphorus enrichment [37]. Reduced water clarity and low dissolved oxygen conditions improve when excess phytoplankton or blooms, measured as chlorophyll *a*, are lowered. Thus, chlorophyll *a* can be a robust indicator of tropic state of a water body. An example is that modeled for lakes in Oregon (Tables 4.10 and 4.11). Along with the chlorophyll *a* readings, visual inspections of surface waters can indicate trophic conditions. Those identified for Chesapeake Bay are shown in Figure 4.24.

Table 4.9 Chlorophyll *a* concentrations ($\mu g\ L^{-1}$) measured in the top stratum (epilimnion) of three lakes in June and September 1989

	Summit Lake	Blackmans Lake	Black Lake
June	1.5	3.3	7.6
September	1.5	3.9	56.2

Source: Coastnet, Oregon State University Extension Sea Grant Program (1996). *Sampling Procedures: A Manual for Estuary Monitoring.*

Table 4.10 Modeled values of seasonal mean and salinity regime-specific chlorophyll *a* concentrations ($\mu g\ L^{-1}$) characterizing trophic conditions to support acceptable dissolved oxygen levels

Season	Tidal-Fresh	Oligohaline	Mesohaline	Polyhaline
Spring	4	5	6	5
Summer	12	7	5	4

Source: Coastnet, Oregon State University Extension Sea Grant Program (1996). *Sampling Procedures: A Manual for Estuary Monitoring.*

Table 4.11 Visual inspection criteria for trophic conditions of a water body

Algal Index value	Category	Description
0	Clear	Conditions vary from no algae to small populations visible to the naked eye
1	Present	Some algae visible to the naked eye but present at low to medium levels
2	Visible	Algae sufficiently concentrated that filaments or balls of algae are visible to the naked eye May be scattered streaks of algae on water surface
3	Scattered surface blooms	Surface mats of algae scattered. May be more abundant in localized areas if winds are calm. Some odor problems
4	Extensive surface blooms	Large portions of the water surface covered by mats of algae. Windy conditions may temporarily eliminate mats, but they will quickly redevelop as winds become calm. Odor problems in localized areas

Source: Coastnet, Oregon State University Extension Sea Grant Program (1996). *Sampling Procedures: A Manual for Estuary Monitoring*.

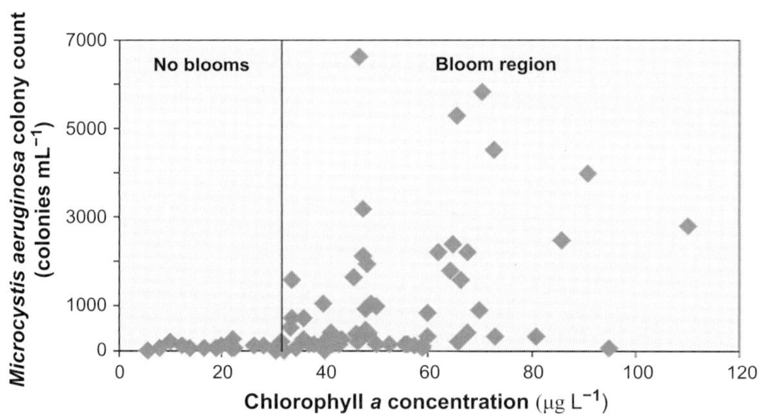

FIGURE 4.24
Colony counts of the algal species *Microcytis aeruginosa* compared to the gradient of chlorophyll *a*, measured in Chesapeake Bay. The vertical line depicts the threshold between bloom and non-bloom conditions (approximately 500 colonies mL^{-1} and 30 μg L^{-1} chlorophyll *a*.
Source: Maryland Department of National Resources (2003). Unpublished data.

BIOSENSORS

Biotechnology takes bioindicators to the next level in detecting environmental insults. So-called "biosensors" make use of biological principles to give information about physicochemical agents that may be present. Such devices can be designed to detect the presence and, with calibration, concentrations of contaminants, or they may be used to sense certain physicochemical properties (solubility, polarity, partitioning, and bioavailability) of a whole sample. Compared to conventional methods, biosensors can improve sensitivity (i.e. the biosensor reliably indicates when the agent or class of compounds is present). Biosensors can also be specific (only respond to stimulus of a single or well-defined set of contaminants) and portable (e.g. results known in the field, with no need to collect samples and return to the lab for "wet chemistry" at the bench). In contrast with chemical or physical analyses, elaborate and expensive instrumentation is not usually necessary when using biosensors.

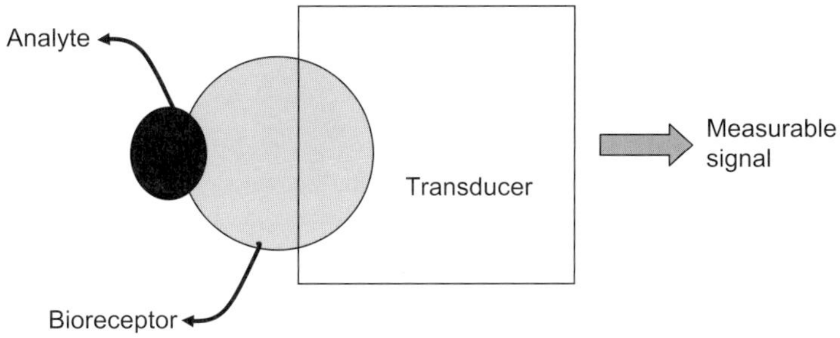

FIGURE 4.25
Schematic of a biosensor. This is a whole cell biosensor if the bioreceptor is a microbe. Otherwise, the bioreceptor can consist of a biomolecule, e.g. an antibody.
Source: Y.H. Lee and R. Mutharasan (2004). Biosensors. In: J.S. Wilson (Ed.), *Sensor Technology Handbook*. Newnes, Burlington, MA.

To date, enzymes, antibodies, subcellular components, and microbes have dominated the types of biological components in biosensors. Enzymes tend to be unstable and expensive to use, so enzyme-based biosensors are more common in medical applications that in environmental biotechnology. Whole microbes are showing promise, as the biological component of biosensors, owing to their diversity and rapid reproduction, in addition to their well-understood culture collections. Whole cell biosensors also take advantage of the biological integration that a microorganism undergoes. As such, the whole cell represents numerous enzymatic reactions, including those involved in cellular respiration and fermentation [38].

The physiological response of immobilized bacteria is a biochemodynamic process. The chemical being detected is transported in a sample to a sensor (e.g. O_2 electrode), from which the biological response is ascertained. This molecular response is what makes a biosensor. The transducer simply provides a specific response to the biochemical activity (see Figure 4.25) [39].

The simplicity of this design provides good specificity, sensitivity, and portability, eliminating the need for expensive instrumentation, except for calibrations at the lab bench. Microbial diversity in the natural environment and the wide availability of microbes in culture collections means that a suitable strain can be matched to the needs of a systematic field study.

Several types of whole-cell bacterial biosensors using recombinant DNA technology are now available. The bacteria are genetically engineered to respond to the presence of chemicals or physiological stresses by synthesizing a reporter protein, such as β-galactosidase, or green fluorescent protein [40]. A biosensor is evaluated on the basis of:

Sensitivity – response of the sensor to per unit change in analyte concentration.
Selectivity – ability of the sensor to respond only to the target analyte. That is, lack of response to other interfering chemicals is the desired feature.
Range – concentration range over which the sensitivity of the sensor is good (also referred to as dynamic range or linearity).
Response time – time needed for the sensor to indicate a certain percentage of its final response due to a step change in analyte concentration.
Reproducibility – accuracy with which the sensor's output is obtained.
Detection limit – lowest concentration of the analyte to which there is a measurable response.
Useful life – time period over which the sensor can be used without significant deterioration in performance characteristics.
Stability – change in its baseline or sensitivity over a fixed time period [41].

Biosensors have been around for some time. Immunoassays, in particular, have been used in environmental applications instruments. The improvements to these biotechnologies will allow for improved environmental and public health assessments.

RELATIONSHIP BETWEEN GREEN ENGINEERING AND BIOTECHNOLOGY

Environmental biotechnology must account for the various spheres of influence in the life cycle, including the technical intricacies involved in manufacturing, using, and decommissioning of a product or system, the infrastructure technologies need to support the product and the social structure in which the product is made and used (see Figure 4.26). This means that no matter how well designed a bioreactor or other biotechnology may be, how carefully microbial and other processes are chosen, how rigorously omics processes are applied, and how high the quality control and assurances, problems will arise if the infrastructure and societal context is not properly characterized and predicted. Each of the spheres in Figure 4.26 affect and are influenced by every concentric sphere.

Decision force fields can be adapted specifically to sustainable designs. For example, if we are primarily concerned about toxic management, we can develop decision force fields based on the various physical and chemical properties of a substance using a multiple objective plot (Figure 4.27). In this plot, two different products can be visually compared in terms of the sustainability, based on toxicity (carcinogenicity), mobility, and partitioning (e.g. sorption, vapor pressure, and Henry's law constants), persistence, and treatability by different methods (e.g. wastewater treatment facilities, pump and treat, etc.). The shape of the curve and the size of the peaks are relative indicators of toxicity and persistence of a potential problem (the inverse of sustainability of healthy conditions).

The plot criteria are selected to provide an estimate of the comparative sustainability of candidate products. It is important to tailor the criteria to the design needs. In the instance of Figure 4.27, this is mainly addressing the toxic hazard and risk of the substances [42]:

Vapor pressure – This sector is a chemical property that tells us the potential of the chemical to become airborne. The low end of the scale is 10^{-8} mmHg; high end is 100 mmHg and above.

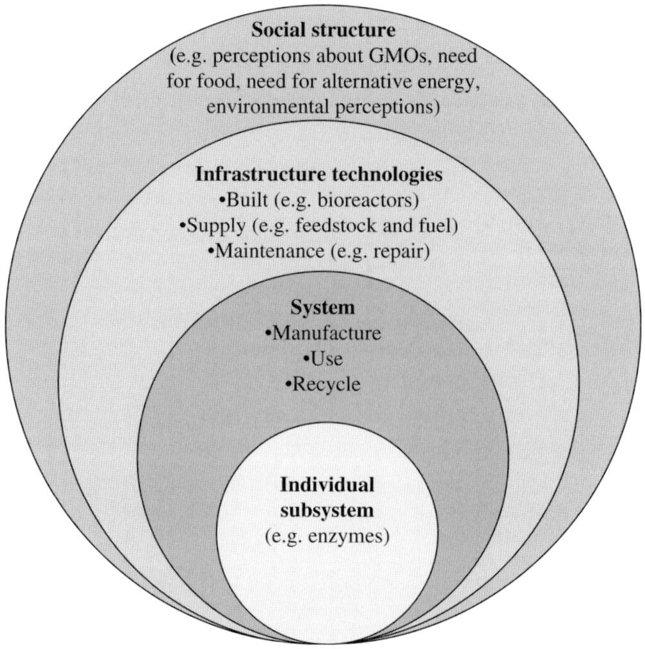

FIGURE 4.26

Spheres or layers of influence in a system. The system consists of interdependencies among each layer. [*See color plate section*]

Source: Adapted from B.R. Allenby and T.E. Graedel (1995). *Industrial Ecology.* Prentice–Hall, New York, NY.

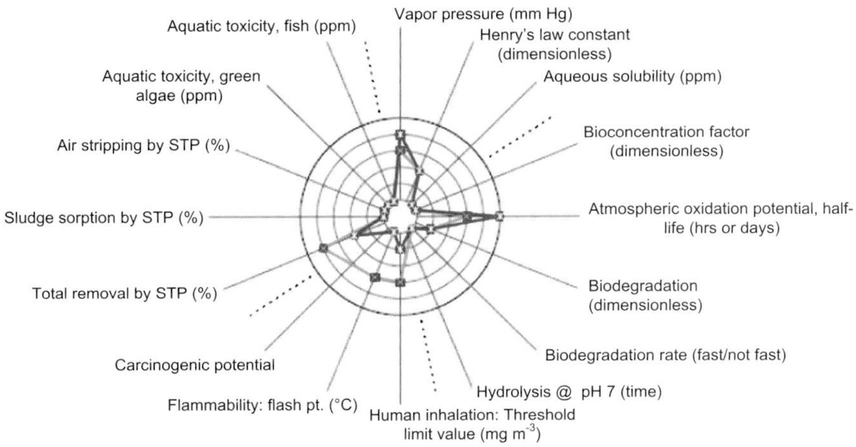

FIGURE 4.27
Hypothetical multiple objective plot of two candidate chemical mixtures to be used in an environmental bioremediation project. Both products appear to have an affinity for the air. Product #1 (open squares) has a larger half-life (i.e. is more persistent), whereas Product #2 (closed squares) is more carcinogenic, flammable, and likely to be taken up by the lungs. Based on these factors, it appears, at least at the screening level, that Product #1 is comparatively better from a sustainability standpoint. [*See color plate section*]
Source: J. Crittenden (used with permission).
Note: STP = sewage treatment plant; ppm = parts per million.

Henry's law – This property tells us how the chemical partitions in air and water. Non-volatile substances have a value of 4×10^{-7} (unitless), moderate volatility is between 4×10^{-4} and 4×10^{-2}, and volatile chemicals are at or above 4. The values are limitless because they are a ratio of concentration in air and water.

Solubility – This property alludes to the potential of the chemical to enter water. Very soluble chemicals are on the order of 10000 ppm and non-soluble entities have a solubility less than 0.1 ppm.

Bioconcentration – This sector specifies the tendency/potential of the chemical to be taken up by biological entities (algae, fish, animals, humans, etc.). A low potential is defined as 250 (unitless) or less, while a high potential is found at 1000 or above.

Atmospheric oxidation, half-life [days] – This property helps to define the fate of the chemical once it enters the atmosphere. A short half-life is desirable as the chemical will have little time to cause adverse effects. A rapid half-life would be on the order of 2 hours or less. A slow half-life is between 1 and 10 days; longer than 10 days is a persistent chemical.

Biodegradation – This sector defines the ability of the environment to break down the chemical. A short biodegradation time is ideal so that the chemical doesn't persist. There are two sectors of biodegradation; one is dimensionless and one has units of time. A biodegradation factor on the order of hours is very quick, whereas a factor on the order of years is long.

Hydrolysis – This describes the potential of the chemical to be broken down into a byproduct and water. It has units of time for a pH of 7. A long hydrolysis time is on the order of many years.

Flammability – This describes the chemical's flash point [degrees C].

Human inhalation – This defines the threshold limit for inhalation of the chemical below which there will be no observed effect in humans. 500 mg m^{-3} and above is a high concentration for which there is little effect. The chemical becomes more of a problem when the limit is 50 mg m^{-3} or less.

Carcinogenicity – This is the potential for the chemical to cause cancer. These data are usually somewhat uncertain due to inaccurate dose-response curves.

Sewage treatment plant (STP) total removal – This is the percent of the chemical that is removed in a wastewater treatment process. 90–100% removal is desirable whereas 0–10% removal describes a chemical that is tough to treat.

STP sludge sorption – This is a percentage of how much of the chemical will adsorb to the sludge in a WWTP. This can be important when the sludge is disposed in a landfill or agriculturally land-applied. 0–10% sorption is ideal so that the chemical doesn't get recycled back to the environment. 90–100% sorption to sludge solids makes disposal difficult.

STP air removal – A percentage of the chemical that is removed to the air from WWT. 0–10% is ideal so that little extra air treatment is needed. 90–100% air removal requires significant air treatment.

Aquatic toxicity (green algae) [ppm] – This sector defines the chemical's toxicity to green algae. A toxic effect on algae can disrupt the entire food chain of an ecosystem. Toxicity is measured on a concentration scale. A low toxicity would be at high concentrations (>100ppm). A high toxicity would be at concentrations on the ppb or ppt scale.

Aquatic toxicity (fish) [ppm] – This defines the toxicity of the chemical to a specific fish species. For example, in the Pacific Northwest, a chemical that is toxic to salmon can cause millions of dollars in economic damage. A low toxicity would be at high concentrations (>100ppm). A high toxicity would be at concentrations on the ppb or ppt scale.

Certainly, green design considers more than toxicity. So, other alternatives for recycling and reuse, avoiding consumer misuse, and disassembly can also be compared with multiple objective plots. The best of these can be considered the benchmark, which is a type of index that conveniently displays numerous factors with appropriate weightings.

Another way to visualize such complex data is the decision matrix. The matrix helps the designer to ensure that all of the right factors are considered in the design phase and that these factors are properly implemented and monitored throughout the project. Integrated engineering approaches require that the engineer's responsibilities extend well beyond the construction, operation, and maintenance stages. Such an approach has been articulated by the American Society of Mechanical Engineers (ASME). The integrated matrix helps DFE to be visualized. This has been recommended by the ASME [43] (see Table 4.12). This allows for the engineer to see the technical and ethical considerations associated with each component of the design, as well as the relationships among these components. For example, health risks, social expectations, and environmental impacts and other societal risks and benefits associated with a device, structure, product or activity can be visualized at various stages of the manufacturing, marketing, and application stages. This yields a number of two-dimensional matrices (see Figure 4.28) for each relevant design component. Further, each respective cell indicates both the importance of that component but the confidence (expressed as scientific certainty) that the engineer can have about the underlying information used to assess the importance (see legend to Figure 4.28).

The matrix approach is qualitative or at best semi-quantitative, but like the multiple objective plots, provides a benchmark for comparing alternatives that would otherwise be incomparable. To some extent, even numerical values can be assigned to each cell to compare them quantitatively, but the results are at the discretion of the analyst, who determines how different areas are weighted. The matrix approach can also focus on design for a more specific measure, such as energy efficiency or product safety, and can be extended to corporate activities as a system.

The key point about benchmarking is the importance of a systematic and prospective viewpoint in design. Whatever tools we can use to help us to model and to predict consequences of available alternatives are an important aspect of green design.

Systematic approaches to bioengineering make use of Design for the Environment (DFE), Design for Disassembly (DFD), and Design for Recycling (DFR) [44]. For example, the concept of a "cap and trade" has been tested and works well for some pollutants. This is a system where companies are allowed to place a "bubble" over a whole manufacturing complex or trade pollution credits with other companies in their industry instead of a "stack-by-stack" and "pipe-by-pipe" approach, i.e. the so-called "command and control" approach. Such policy and regulatory innovations call for some improved technology-based approaches as well as better quality-based approaches, such as leveling out the pollutant loadings and using less expensive technologies to remove the first large bulk of pollutants, followed by higher operation and maintenance (O&M) technologies for the more difficult to treat stacks and pipes. But, the net effect can be a greater reduction of pollutant emissions and effluents than treating each stack or pipe as an independent entity. This is a foundation for most sustainable design approaches, i.e.

Table 4.12 Functions that must be integrated into an engineering design

Baseline studies of natural and built environments
Analyses of project alternatives
Feasibility studies
Environmental impact studies
Assistance in project planning, approval and financing
Design and development of systems, processes and products
Design and development of construction plans
Project management
Construction supervision and testing
Process design
Startup operations and training
Assistance in operations
Management consulting
Environmental monitoring
Decommissioning of facilities
Restoration of sites for other uses
Resource management
Measuring progress for sustainable development

Source: American Society of Mechanical Engineers; http://www.professionalpractice.asme.org/communications/sustainability/2.htm; accessed May 23, 2006.

conducting a life-cycle analysis, prioritizing the most important problems, and matching the technologies and operations to address them. The problems will vary by size (e.g. pollutant loading), difficulty in treating, and feasibility. The easiest ones are the big ones that are easy to treat (so-called "low hanging fruit"). You can do these first with immediate gratification! However, the most intractable problems are often those that are small but very expensive and difficult to treat, i.e. less feasible. Thus, the environmental science requires that expectations be managed from both a technical and an operational perspective, including the expectations of the client, the government, and oneself.

The type of pollution control technology applied depends on the intrinsic characteristics of the contaminants and on the substrate in which they reside. The choice must factor in all of the physical, chemical, and biological characteristics of the contaminant with respect to the matrices and substrates (if soil and sediment) or fluids (air, water, or other solvents) where the contaminants are found. The selected approach must meet criteria for treatability (i.e. the efficiency and effectiveness of a technique in reducing the mobility and toxicity of a waste). The comprehensive remedy must consider the effects each action taken will have on past and proceeding steps.

Eliminating or reducing pollutant concentrations begins with assessing the physical and chemical characteristics of each contaminant, and matching these characteristics with the appropriate treatment technology. All of the kinetics and equilibria, such as solubility, fugacity, sorption, and bioaccumulation factors, will determine the effectiveness of destruction, transformation, removal, and immobilization of these contaminants. For example, Table 4.13 ranks

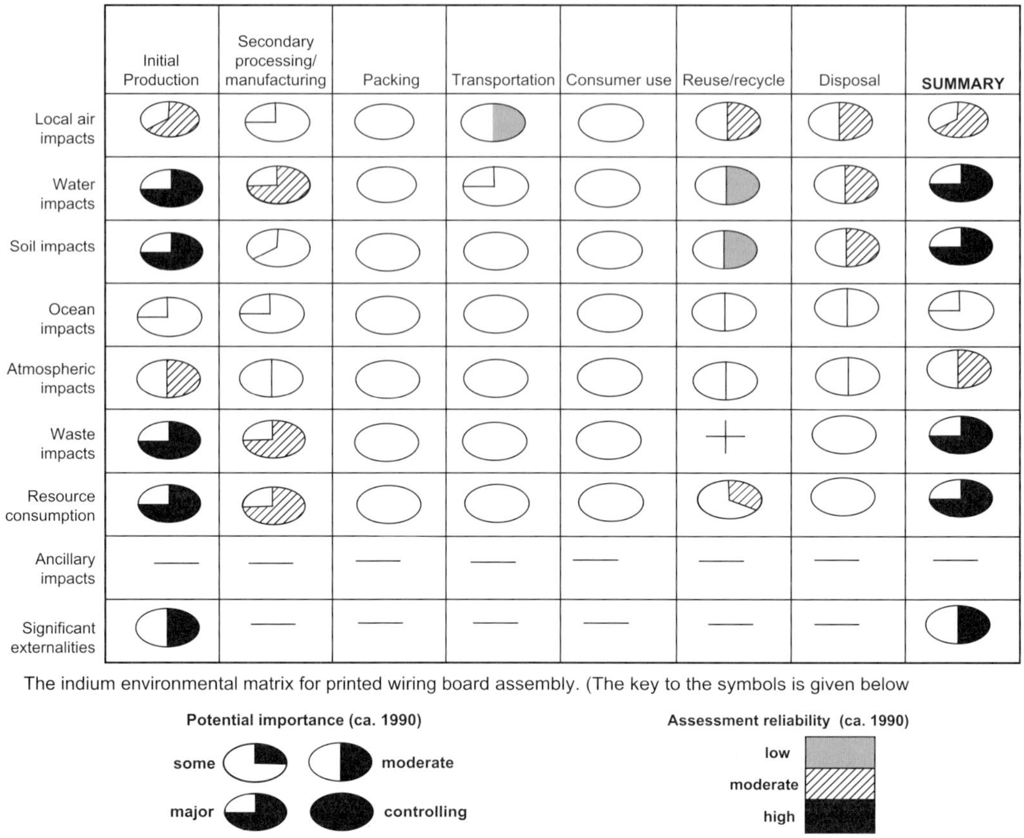

FIGURE 4.28
An example of an integrated engineering matrix; in this instance applied to sustainable designs.
Source: American Society of Mechanical Engineers; http://www.professionalpractice.asme.org/communications/sustainability/2.htm; accessed May 25, 2006.

the effectiveness of selected treatment technologies on organic and inorganic contaminants typically found in contaminated slurries, soils, sludges, and sediments. As shown, there can be synergies (e.g. innovative incineration approaches are available that not only effectively destroy organic contaminants, but in the process also destroys the inorganic cyanic compounds).

Unfortunately, there are also antagonisms among certain approaches, such as the very effective incineration processes for organic contaminants that transform heavy metal species into more toxic and more mobile forms. The increased pressures and temperatures are good for breaking apart organic molecules and removing functional groups that lend them toxicity, but these same factors oxidize or in other ways transform the metals into worse forms. So, when mixtures of organic and inorganic contaminants are targeted, more than one technology may be required to accomplish project objectives, and care must be taken not to trade one problem (e.g. polychlorinated biphenyls, PCBs) for another (e.g. a more mobile species of cadmium).

The characteristics of the soil, sediment, or water will vary the performance of any contaminant treatment or control. For example, sediment, sludge, slurry, and soil characteristics will influence the efficacy of treatment technologies; these include particle size, solids content, and high contaminant concentration (see Table 4.14).

A factor as specific and seemingly mundane as particle size may be the most important limiting characteristic for application of treatment technologies to certain wastes (e.g. contaminated sediments). It reminds us that engineers must continue to be cognizant of minute details. Looking at the tables, we see the peril of "one size fits all thinking." Most treatment technologies work well on sandy soils and sediments. The presence of fine-grained material adversely affects treatment system emission controls because it increases particulate generation during thermal drying, it is more difficult to dewater, and it has greater attraction to the contaminants

Table 4.13 **Effect of the characteristics of the contaminant on decontamination efficiencies**

	Organic contaminants					Inorganic contaminants		
Treatment technology	PCBs	PAHS	Pesticides	Petroleum hydrocarbons	Phenolic compounds	Cyanide	Mercury	Other heavy metals
Conventional incineration	D	D	D	D	D	D	xR	pR
Innovative incineration[a]	D	D	D	D	D	D	xR	I
Pyrolysis[a]	D	D	D	D	D	D	xR	I
Vitrification[a]	D	D	D	D	D	D	xR	I
Supercritical water oxidation	D	D	D	D	D	D	U	U
Wet air oxidation	pD	D	U	D	D	D	U	U
Thermal desorption	R	R	R	R	U	U	xR	N
Immobilization	pI	pI	pI	pI	pI	pI	U	I
Solvent extraction	R	R	R	R	R	pR	N	N
Soil washing[b]	pR	pR	pR	pR	pR	pR	pR	pR
Dechlorination	D	N	pD	N	N	N	N	N
Oxidation[c]	N/D	N/D	N/D	N/D	N/D	N/D	U	xN
Bioremediation[d]	N/pD	N/D	N/D	D	D	N/D	N	N

Note: PCBs – polychlorinated biphenyls
PAHs – polynuclear aromatic hydrocarbons

Primary designation
D = effectively destroys contaminant
R = effectively removes contaminant
I = effectively immobilizes contaminant
N = no significant effect
N/D = effectiveness varies from no effect to highly efficient depending on the type of contaminant within each class
U = effect not known

Prefixes
p = partial
x = may cause release of nontarget contaminant

[a]*This process is assumed to produce a vitrified slag.*
[b]*The effectiveness of soil washing is highly dependent on the particle size of the sediment matrix, contaminant characteristics, and the type of extractive agents used.*
[c]*The effectiveness of oxidation depends strongly on the types of oxidant(s) involved and the target contaminants.*
[d]*The effectiveness of bioremediation is controlled by a large number of variables as discussed in the text.*
Source: US Environmental Protection Agency (2003). Remediation Guidance Document, EPA-905-B94-003 Chapter 7.

Table 4.14 **Effect of particle size, solids content, and extent of contamination on decontamination efficiencies**

	Predominant particle size			Solids content		High contaminant concentration	
Treatment technology	Sand	Silt	Clay	High (slurry)	Low (*in situ*)	Organic compounds	Metals
Conventional incineration	N	X	X	F	X	F	X
Innovative incineration	N	X	X	F	X	F	F
Pyrolysis	N	N	N	F	X	F	F
Vitrification	F	X	X	F	X	F	F
Supercritical water oxidation	X	F	F	X	F	F	X
Wet air oxidation	X	F	F	X	F	F	X
Thermal desorption	F	X	X	F	X	F	N
Immobilization	F	X	X	F	X	X	N
Solvent extraction	F	F	X	F	X	X	N
Soil washing	F	F	X	N	F	N	N
Dechlorination	U	U	U	F	X	X	N
Oxidation	F	X	X	N	F	X	X
Bioslurry process	N	F	N	N	F	X	X
Composting	F	N	X	F	X	F	X
Contained treatment facility	F	N	X	F	X	X	X

Note: F – sediment characteristic favorable to the effectiveness of the process
N – sediment characteristic has no significant effect on process performance
U – effect of sediment characteristic on process is unknown
X – sediment characteristic may impede process performance or increase cost
Source: US Environmental Protection Agency (2003). Remediation Guidance Document, EPA-905-B94-003 Chapter 7.

(particularly clays). Clayey sediments that are cohesive also present materials handling problems in most processing systems. Solids content generally ranges from high, i.e. usually the *in situ* solids content (30–60% solids by weight), to low, e.g. hydraulically dredged sediments (10–30% solids by weight). Treatment of slurries is better at lower solids contents, but this can be achieved even for high solids contents by water addition at the time of processing. It is more difficult to change lower to a higher solids content, but evaporative and dewater approaches, such as those used for municipal sludges, may be employed. Also, thermal and dehalogenation processes are decreasingly efficient as solids content is reduced. More water means increased chemical costs and increased need for wastewater treatment.

We must be familiar with every potential contaminant. Again, a quick review of the tables shows that elevated levels of organic compounds or heavy metals in high concentrations can be drivers in deciding on the appropriate technological solution. Higher total organic carbon (TOC) content favors incineration and oxidation processes. The TOC can be the contaminant of concern or any organic, since they are combustibles with caloric value. Conversely, higher metal concentrations may make a technology less favorable by increasing contaminant mobility of certain metal species following application of the technology.

A number of other factors may affect selection of a treatment technology other than its effectiveness for treatment (some are listed in Table 4.15). Biological processes are used in several of the technologies listed. Some are direct biodegradation processes (e.g. bioslurries

Table 4.15 Selected factors on selecting decontamination and treatment approaches

Treatment technology	Implementability at full scale	Regulatory compliance	Community acceptance	Land requirements	Residuals disposal	Wastewater treatment	Air emissions control
Conventional incineration		✓	✓				✓
Innovative incineration		✓	✓				✓
Pyrolysis		✓					✓
Vitrification	✓	✓					✓
Supercritical water oxidation	✓						
Wet air oxidation							
Thermal desorption					✓	✓	✓
Immobilization				✓			
Solvent extraction					✓	✓	
Soil washing					✓	✓	
Dechlorination							✓
Oxidation	✓						
Bioslurry process	✓						✓
Composting				✓			✓
Contained treatment facility				✓		✓	✓

Note: ✓ – the factor is critical in the evaluation of the technology
Source: US Environmental Protection Agency (2003). Remediation Guidance Document, EPA-905-B94-003 Chapter 7.

and composting), some are subtypes (e.g. bioreactors as a contained treatment facility), and some are components of larger systems (e.g. oxidation and soil washing).

Regulatory compliance and community perception are always a part of decisions regarding an incineration system. Land use considerations, including the acreage needs, are commonly confronted in solidification and solid-phase bioremediation projects (as they are in sludge farming and land application). Disposing of residues following treatment must be part of any process. Treating water effluent and air emissions must be part of the decontamination decision making process.

Note that a good design must account for the entire life cycle of a potential hazard. For example, we must concern ourselves not only about the processes over which we have complete control, such as the manufacturing design process for a product or the treatment of a waste within the company property lines. But, we must think about what happens when a chemical or other stressor enters the environment [45]. We must be able to show how a potential contaminant moves after entering the environment, which is complicated and difficult because there is much variability of chemical and physical characteristics of contaminated media (especially soils and sediments), owing to the strong affinity of most contaminants for fine-grained sediment particles, and due to the limited track record or "scale-up" studies for many treatment technologies. Off-the-shelf models can be used for simple process operations, such as extraction or thermal vaporization applied to single contaminants in relatively pure systems. However, such models have not been appropriately evaluated for a number of other technologies because of the limited database on treatment technologies, such as for contaminated sediments or soils.

Standard engineering practice [46] for evaluating the effectiveness of treatment technologies for any type of contaminated media (solids, liquids, or gases) requires first performing a so-called "treatability" study for a sample that is representative of the contaminated material. The performance data from treatability studies can aid in reliably estimating contaminant concentrations for the residues that remain after treatment, as well as possible waste streams that could be generated by applying a given technology. Treatability studies may be performed at the bench-scale (in the lab) or at pilot-scale level (e.g. a real-world study, but limited in number of contaminants, in spatial extent, or to a specific, highly controlled form of a contaminant, e.g. one pure congener of PCBs, rather than the common mixtures). Most

Table 4.16 Selected waste streams commonly requiring treatability studies

Contaminant loss stream	Treatment technology type						
	Biological	Chemical	Extraction	Thermal desorption	Thermal destruction	Immobilization	Particle separation
Residual solids	X	X	X	X	X	X	X
Wastewater	X	X	X	X			X
Oil/organic compounds			X	X			X
Leachate						X[a]	
Stack gas				X	X		
Adsorption media			X	X			
Scrubber water					X		
Particulates (filter/cyclone)				X	X		

[a]*Long-term contaminant losses must be estimated using leaching tests and contaminant transport modeling similar to that used for sediment placed in a confined disposal facility. Leaching could be important for residual solids for other processes as well.*

Source: US Environmental Protection Agency (2003). Remediation Guidance Document, EPA-905-B94-003 Chapter 7.

treatment technologies include post-treatment or controls for waste streams produced by the processing. The contaminant losses can be defined as the residual contaminant concentrations in the liquid or gaseous streams released to the environment. For technologies that extract or separate the contaminants from the bulk of the sediment, a concentrated waste stream may be produced that requires treatment offsite at a hazardous waste treatment facility, where permit requirements may require destruction and removal efficiencies greater than 99.9999% (i.e. the so-called rule of "six nines"). The other source of loss for treatment technologies is the residual contamination in the sediment after treatment. After disposal, treated wastes are subject to leaching, volatilization, and losses by other pathways. The significance of these pathways depends on the type and level of contamination that is not removed or treated by the treatment process. Various waste streams for each type of technology that should be considered in treatability evaluations are listed in Table 4.16.

Systems are integral to all environmental endeavors. As biochemodynamic tools continue to improve, so will the abilities to assess the risks and rewards of biotechnologies, and hopefully, reap more biotechnological blessings and fewer environmental curses.

SEMINAR TOPIC

Biological Agents: How Clean Is Clean?

When microorganisms are released, they may infect either in humans or animals. Biological threat agents are classified in three categories [47].

Category A are the highest priority agents that:

- **(i)** pose a risk to the national security since they may easily be disseminated;
- **(ii)** are transmitted from person to person;
- **(iii)** may result in high mortality rates; and,
- **(iv)** may cause public panic and require special health preparedness.

This highest threat category includes *Bacillus anthracis* (anthrax), *Francisella tularensis* (tularemia), *Yersinia pestis* (plague), *Variola major* (smallpox), viruses causing viral hemorrhagic fevers and botulinum toxin (botulism) (see Table 4.17).

Category B agents are moderately disseminated and are expected to result in low mortality rates. Category B includes *Coxiella burnetti* (Q-fever), *Brucella* spp. (brucellosis), *Burkholderia* spp. (glanders, melioidosis), viruses causing viral encephalitis, *Rickettsia prowazekii* (typhus fever), and waterborne and food safety threats such as *Vibrio cholera* (cholera), *Shigella* and *Salmonella* spp., respectively, in addition to the toxins ricin, *Staphylococcus* enterotoxin B (SEB) and epsilon toxin of *Clostridium perfringens.*

Agents that cause emerging infectious diseases are included in Category C, such as a range of viruses, e.g. Nipah virus and hantavirus, as well as genetically engineered microbes designed for mass dissemination. Category C agents can be available, easily produced, and may lead to high mortality rates. Thus, these microbes are not presently considered major bioterrorism threats, but could become threats in the future.

Frequently, clinical symptoms may be the first indication of a biological incident, so immediate, reliable, and efficient identification methods, including taxonomic identification, must be available to assist in both clinical and environmental samples. These reliable data will not only support triage, evacuation, and response activities, but ultimately decontamination of the biological threat. The National Research Council has defined decontamination as the process of neutralizing or removing chemical or biological agents from people, structures, articles and/or equipment, and the environment [48]. Effective decontamination requires three elements:

the contaminants involved are correctly identified;

the procedures and equipment are available and are appropriately employed to remove or neutralize the contaminant; and,

the reduction of risk from the contaminant is defensible by scientific and regulatory standards [49].

Detection of biological agents must follow a structured process, such as the one shown in Figure 4.29. In this case, the laboratory is brought on-site, rather than requiring the samples to be transported. This can save time and provide more immediate results to a concerned public.

Anthrax

Bacillus anthracis is a spore-forming bacterium that causes anthrax, which is a zoonotic disease, i.e. it can be transmitted from non-human animals to humans. *B. anthracis* spores remain viable in the environment for years, representing a potential source of infection. Human anthrax exists in three clinical forms: inhalational, gastrointestinal, and cutaneous. Inhalational anthrax results from exposure to *B. anthracis* spores that have aerosolized. Aerosols with $\geq$100 μm aerodynamic diameters usually settle at typical indoor air velocities, but very small particles (i.e., $\leq$5 μm in diameter) can remain suspended for extensive time periods and can move greater distances, increasing their likelihood to be inhaled before impaction into surfaces or settling onto a surface. Since single spores or small clusters of spores of *B. anthracis* have diameters that can range from 5 to 10 μm, they can move with the air stream. In addition, particles can become resuspended. Resuspension rates depend on the spore's size and the sorption properties of the spore's surface [50].

(Continued)

Table 4.17 Potential biological threat agents requiring public health preparedness

Biological agent(s)	Disease
Category A	
Variola major	Smallpox
Bacillus anthracis	Anthrax
Yersinia pestis	Plague
Clostridium botulinum (botulinum toxins)	Botulism
Francisella tularensis	Tularemia
Filoviruses and Arenaviruses (e.g., Ebola virus, Lassa virus)	Viral hemorrhagic fevers
Category B	
Coxiella burnetii	Q fever
Brucella spp.	Brucellosis
Burkholderia mallei	Glanders
Burkholderia pseudomallei	Melioidosis
Alphaviruses (VEE, EEE, WEE[a])	Encephalitis
Rickettsia prowazekii	Typhus fever
Toxins (e.g., ricin, staphylococcal enterotoxin B)	Toxic syndromes
Chlamydia psittaci	Psittacosis
Food safety threats (e.g., *Salmonella spp.*, *Escherichia coli* O157:H7)	
Water safety threats (e.g., *Vibrio cholerae*, *Cryptosporidium parvum*)	
Category C	
Emerging threat agents (e.g., Nipah virus, hantavirus)	

[a]*Venezuelan equine (VEE), eastern equine (EEE), and western equine encephalomyelitis (WEE) viruses.*
Source: US Centers for Disease Control (2002). Report Summary: Public Health Assessment of Potential Biological Terrorism Agents. *Emerging Infectious Diseases* 8 (2): 225–230.

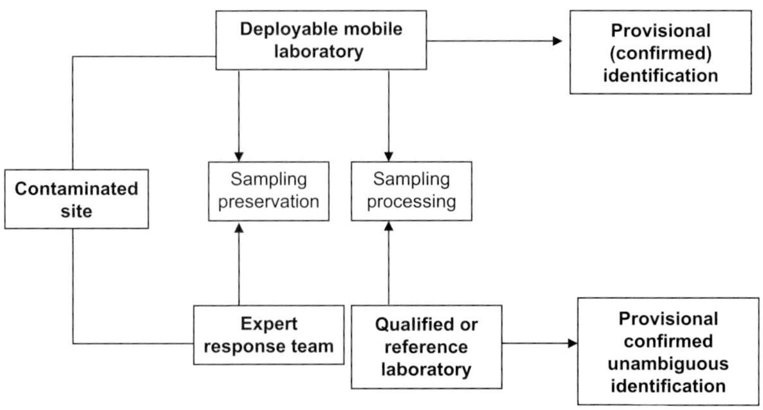

FIGURE 4.29
A general scheme involving sampling and identification of biological threat agents from a biological contaminated site. *Source:* J.M. Blatny, E.M. Fykse, J.S. Olsen, G. Skogan and T. Aarskaug (2008). Identification of biological threat agents in the environment and its challenge. Forsvarets forskningsinstitutt/Norwegian Defense Research Establishment. Report No. FFI-rapport 2008/01371.

On October 5, 2001, a hospital in Boca Raton, Florida notified the Federal Bureau of Investigation (FBI) that a patient had died from inhalational anthrax. The patient had worked at the American Media Incorporated (AMI) facility, which was the first to be targeted through anthrax contaminated mail prior to an incident at NBC News in New York and before the letter to Senator Daschle was received at the Hart Senate Building [51]. Soon after, the AMI building was evacuated and identified as a crime scene as FBI specialists investigated the source of the anthrax. Anthrax spores in powder form were found on the computer keyboard of the deceased as well as in the facility's mail room, indicating that the contamination likely occurred through the mail.

Law enforcement personnel examined the scene to determine whether the situation was suspicious and, if so, they were to contact the nearest of the four Palm Beach Fire Department HAZMAT teams. If the item appeared to be contaminated, people who might have been in contact with it were decontaminated with soap/water and/or a 0.5% bleach solution. The area surrounding the package (e.g. office space, floor, furniture, car, etc.) was also decontaminated with 0.5% hypochlorite solution. Samples of the suspicious material were collected and sent for analysis to a Miami laboratory. However, the Miami lab was overloaded with samples, so samples were queued according to priority, i.e. those that were most suspicious. Potentially contaminated people were told that if any symptoms appeared they were to see their personal physicians for monitoring, but that antibiotics were only needed in the event of a positive test for anthrax exposure. Of the 1000 nasal swabs performed on the likely exposed population, only two people tested positive.

There was a great deal of confusion as to the number of spores needed to infect a person, which slowed the response effort. Labeling the building as a crime scene stalled decontamination efforts and may have contributed to the delay of health officials' treatment of potential victims.

In the Capitol Hill anthrax letter case, both surfaces and air in the buildings were sampled for the presence of anthrax, using wet swabs and wipes for nonporous surfaces and high efficiency particulate arresting (HEPA) vacuuming for porous materials, along with air sampling. Decontamination consisted of removing any anthrax detected in the congressional buildings: fumigating with the antimicrobial pesticide, chlorine dioxide (ClO_2) gas, disinfecting with liquid chlorine dioxide, disinfecting with a neutralizing agent (Sandia foam), and using high efficiency particulate air (HEPA) vacuuming (see Figure 4.30). The ClO_2 fumigant was used to decontaminate parts of the Hart Senate Office Building, along with mail and packages [52].

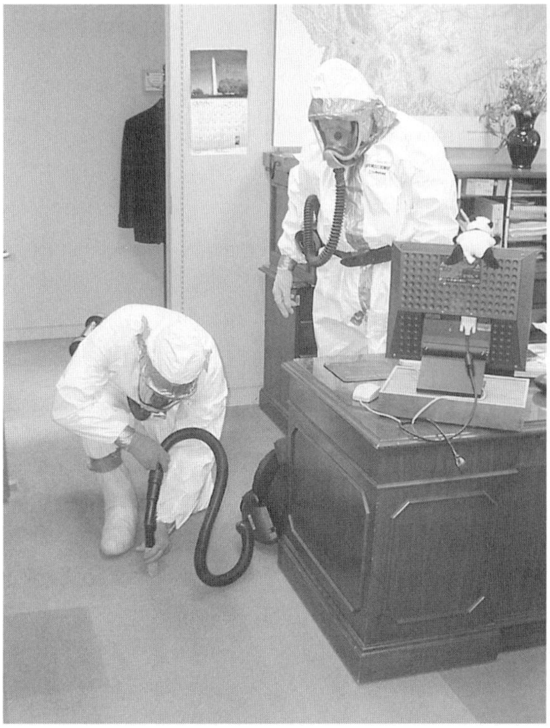

FIGURE 4.30
Decontamination personnel using a high efficiency particulate air (HEPA) vacuum in a congressional office in Washington, DC.
Source: US General Accounting Office (2003). Capitol Hill Anthrax Incident: EPA's Cleanup Was Successful: Opportunities Exist to Enhance Contract Oversight. Report to the Chairman, Committee on Finance, US Senate. Photo by the US Environmental Protection Agency.

These cases illustrate some common problems with using the HAZMAT model for decontamination, including lack of reliable equipment and technologies to determine when contamination exists. As a result, emergency response personnel are at risk and the decontamination of the site can be delayed. In addition, victims that might require immediate attention to alleviate the effects of the contaminant may not receive sufficiently immediate care.

Decontaminating an area or item contaminated by anthrax depends on numerous and variable factors specific to individual locations (see Figure 4.31). No single technology, process, or strategy can be expected to work in every case, so a decontamination plan must consider the following:

- The nature of the contamination, e.g. the strain of anthrax, its entry to the facility, and the physical characteristics that affect the spread of contamination.
- The extent of contamination, e.g. the amount of contamination and possible pathways by which it could have or will spread.
- The objectives of decontamination, e.g. the intended re-use of the facility and building systems and whether items will be decontaminated for re-use or treated for disposal [53].

The likelihood of exposure to anthrax spores is a function of concentration of the spores with time:

$$E = \int_{t=t_1}^{t=t_2} C(t)\,dt \tag{4.23}$$

where E = personal exposure during time period from t_1 to t_2; and $C(t)$ = concentration at interface, at t.

The degree of exposure and the means of protection against exposure vary by the stage of response. During rescue operations, relatively high levels of detection may suffice for chemicals, accompanied by more immediate reporting than in a non-emergency operation, e.g. firefighters will likely work in conditions of high levels of contaminants like polycyclic aromatic hydrocarbons (PAHs) and carbon monoxide, since they are using personal protection equipment (PPE) and since their expertise allows them to allocate appropriate time to rescue (unfortunately, there are examples where their estimates have been wrong). There is less concern about chronic effects (e.g. cancer from PAH exposures) than acute effects (e.g. asphyxiation from CO). During the rescue phase, crime scene, forensics and rescue efforts have primacy over environmental concerns (e.g. levels of dioxins and benzene to protect firefighters with PPE are much higher than a person without protection exposed for 30 years).

Recovery, the next stage, allows for somewhat more time, but still in the first-responder mode of operation. This means that exposure data are being logged so that analyses can be done. The results will all make for better responses in the future and possibly linkages to exposures that may be associated with latent effects. Crime scene forensics are still ongoing (with deference to law enforcement).

In the next stage of response, re-entry, even more time is available for exposure investigations. This stage looks more like a prototypical research protocol, but with the provision that any study should not hinder law enforcement and responder activities and decisions.

Finally, re-habitation must only occur after sufficient decontamination. This stage obviously involves the longest potential exposures, so its

FIGURE 4.31
Decontamination worker inserting a sample in to a vial in the Hart Senate Office Building.
Source: US General Accounting Office (2003). Capitol Hill Anthrax Incident: EPA's Cleanup Was Successful: Opportunities Exist to Enhance Contract Oversight. Report to the Chairman, Committee on Finance, US Senate. Photo by the US Environmental Protection Agency.

exposure metrics are those typically used in risk assessment (e.g. lifetime average daily dose). Conservative approaches are challenged as people want to get back to "normal." However, they should not be allowed to re-enter and re-habitate a contaminated area until it is sufficiently habitable from an exposure perspective.

The extent of contamination and how the contamination spreads are critical considerations in isolating affected areas and selecting appropriate decontamination technologies. For example, if spores are widely dispersed and have traveled through the air, decontamination may involve extensive isolation and fumigation. In contrast, if the contamination is limited to a small area and spores are not likely to become airborne, then minimal isolation and surface decontamination methods alone may suffice.

How clean is clean enough when it comes to biological agent contamination? Are false positives better or worse than false negatives? That is, when it comes to responding to microbial contamination emergencies, how is precaution balanced with efficiency? Commercial field equipment to detect biological agents may produce as many false positives as false negatives. Knowing when a site can be reoccupied is often less than scientifically based. In general, regulating chemical substances that may affect human health and the environment follows certain principles:

- Chemicals should be reviewed against risk-based safety standards based on sound science and protective of human health and the environment.
- Manufacturers should provide regulatory agencies with the necessary information to conclude that new and existing chemicals are safe and do not endanger public health or the environment.
- Regulators must have clear authority to take risk management actions when chemicals do not meet the safety standard, with flexibility to take into account sensitive subpopulations, costs, social benefits, equity and other relevant considerations. Manufacturers and regulators should assess and act on priority chemicals, both existing and new, in a timely manner.
- Green chemistry should be encouraged and provisions assuring transparency and public access to information should be strengthened [54].

However, chemical risk assessment does not directly translate to the risks posed by biological agents. For example, the microbe may induce disease, but other effects can result from the toxins produced by the microbe, or from cysts and spores. These also change the typical pathways, e.g. oral, ingestion, and inhalation, compared to a chemical compound.

Further complications can result when decontamination involves chemical risks. For example, methyl bromine has been shown to be relatively effective for topical disinfection of *B. anthracis.* In addition to being associated with acute and chronic health effects, atmospheric scientists have concluded that use of methyl bromide contributes to the destruction of the ozone layer. Accordingly, under the Montreal Protocol on Substances that Deplete the Ozone Layer and under the Clean Air Act, production of most uses of methyl bromide has been banned in the United States and other countries covered by the Protocol. Revisions to the Clean Air Act in 1998 induced the United States to limit the production and import of methyl bromide to 75% of the 1991 baseline. In 2001, production and import were further reduced to 50% of the 1991 baseline. In 2003, allowable production and import were again reduced to 30% of the baseline, leading to a complete phase-out of production and import in 1995. Beyond 2005, continued production and import of methyl bromide are restricted to critical, emergency, and quarantine and pre-shipment uses:

- *Soil fumigation:* Methyl bromide gas is injected into the soil before a crop is planted. This treatment, which effectively sterilizes the soil, kills the vast majority of soil organisms.
- *Commodity treatment:* Methyl bromide gas is used for post-harvest pest control and can be injected into a chamber or under a tarp containing commodities such as grapes, raisins, cherries, nuts, and imported non-food materials.
- *Structural pest control treatment:* Methyl bromide gas is used to fumigate buildings for termites, warehouses, and food processing facilities for insects and rodents, aircraft for rodents, and ships (and other transportation vehicles) for various pests.
- *Quarantines:* USDA's Animal Plant and Health Inspection Service (APHIS) uses methyl bromide to treat imported commodities as required by quarantine regulations [55].

Anthrax cleanup is not always an emergency situation and may resemble cleanups of chemically contaminated sites. While the state-of-the-science is advancing, improved approaches for detection, early warnings, and decontamination of biological threat agents are needed.

Seminar Questions

How does a biological agent cleanup vary between a reductionist versus a systematic view and when is one view better than the other?

Which physical and chemical properties of *B. anthracis* appear to have the greatest weight in terms of likelihood of exposure? Can these characteristics be extrapolated to other bacteria?

B. anthracis is closely related to *Bacillus thuringiensis.* Does this imply special precautions when using Bt in genetic modifications? Why or why not?

How may systems biology and engineering be used to advance bioindicators and biosensors to assist in emergency response efforts, such as that shown in Figure 4.32?

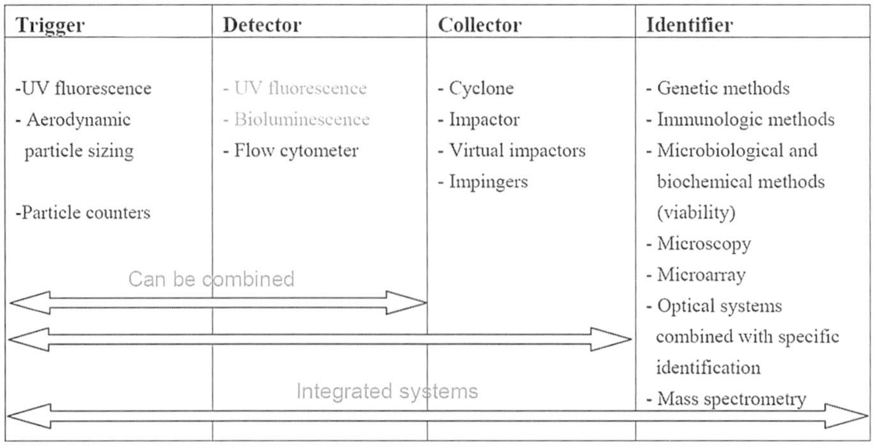

FIGURE 4.32
Technologies used to detect and identify biological threat agents in the air. An integrated detection system must provide sensitive, specific, fast, reliable detection and identification.
Source: J.M. Blatny, E.M. Fykse, J.S. Olsen, G. Skogan and T. Aarskaug (2008). Identification of biological threat agents in the environment and its challenge. Forsvarets forskningsinstitutt/Norwegian Defense Research Establishment. Report No. FFI-rapport 2008/01371.

REVIEW QUESTIONS

Identify at least three systems important to the environment. Explain how closely these adhere to the formal, thermodynamic definitions of systems.

Explain how a past environmental disaster could have been avoided by a greater appreciation of the interconnectedness of environmental systems.

How might "omics" tools be used to enhance environmental decision making?

Draw a decision force field for two products you can buy at a drug store. Apply the criteria from Figure 4.27 to decide which is a better choice from a systematic environmental perspective.

Why would a regulatory agency disapprove a transgenic crop that has a metabonomic profile substantially different from the progenitor?

Estimate the success or failure of a biotechnology (e.g. enhanced size of poultry, resistance to disease, pest resistance) from the standpoint of four control volumes:

The cell
The organism
The population (human or ecosystem)
The earth.

How does scale affect the acceptability of that biotechnology?

How will the "omics" tools help to predict biotechnological artifacts and outcomes?

When is chlorophyll *a* useful as a bioindicator? When is its usefulness limited?

Compare the microbial ecology of algae to that of bacteria. How may the abiotic and biotic conditions required for their growth and metabolism affect their usefulness of bioindicators? How do these conditions affect their usefulness in bioremediation?

Apply the factors in Table 4.2 to an ecosystem near your home. Explain the weighting of each factor and the interrelationships among the factors.

What are the greatest needs in environmental biotechnology that can be met by moving from reductionist to systematic perspectives? What aspects of reductionism must be preserved to ensure sound science?

NOTES AND COMMENTARY

1. Sting. Lyrics from "If I Ever Lose Your Love."
2. H.V. Westerhoff and B.O. Palsson (2004). The evolution of molecular biology into systems biology. *Nature Biotechnology* 22 (10): 1249.
3. B.E. Rittmann, M. Hausner, F. Loffler, N.G. Love, G. Muyzer, S. Okabe, *et al.* (2006). A vista for microbial ecology and environmental biotechnology. *Environmental Science & Technology* 40 (4): 1096–1103.
4. Organisation for Economic Co-operation and Development (1992). *Report of the OECD Workshop on the Extrapolation of Laboratory Aquatic Toxicity Data to the Real Environment*. OECD Environment Monographs No. 59; Paris, France: 59; C.J. Van Leeuwen, J-L.M. Hermens (Eds) (1995). *Risk Assessment of Chemicals: An Introduction*. Kluwer Academic Publishers, Dordrecht, The Netherlands; and C.J. Van Leeuwen *et al.* (1996). *Environmental Toxicology and Pharmacology* (2): 243–299.

5. Organisation for Economic Co-operation and Development (1992). Existing Chemicals Programme, www.oecd.org.
6. Australian Government Department of the Environment, Water, Heritage and the Arts (2009). Assessing risks from GMOs; http://www.environment.gov.au/settlements/biotechnology/assessingrisks.html; accessed August 14, 2009.
7. Ibid.
8. Sustainable Development Commission. London, UK: http://www.sd-commission.org.uk/pages/resilience.html; accessed August 14, 2009.
9. J.R. Karr (1981). Assessment of biotic integrity using fish communities. *Fisheries* 6: 21–27.
10. R.L. Erickson and J.M. McKim (1990). A model for exchange of organic chemicals at fish gills: flow and diffusion limitations. *Aquatic Toxicology* 18: 175–198; and D.J. Stewart, D. Weininger, D.V. Rottiers and T.A. Edsall (1983). An energetics model for lake trout *Salvelinus namaycush*: Application to the Lake Michigan population. *Canadian Journal of Fisheries and Aquatic Sciences* 40: 681–698.
11. M.C. Barber (2008). Bioaccumulation and Aquatic System Simulator (BASS). User's Manual, Version 2.2. Report No. EPA 600/R-01/035, update 2.2, March 2008. US Environmental Protection Agency, Athens, GA.
12. J.A. Kushlan, S.A. Voorhees, W.F. Loftus and P.C. Frohring (1986). Length, mass, and calorific relationships of Everglades animals. *Florida Scientist* 49: 65–79; K.J. Hartman and S.B. Brandt (1995). Estimating energy density of fish. *Transactions of the American Fisheries Society* 124: 347–355; and K. Schreckenbach, R. Knösche and K. Ebert (2001). Nutrient and energy content of freshwater fishes. *Journal of Applied Ichthyology* 17: 142–144.
13. This example is based upon guidance from D. MacKay and S. Paterson (1993). Mathematical models of transport and fate. In: G. Suter (Ed.) (1995). *Ecological Risk Assessment*. Lewis Publishers, Inc., Chelsea, MI; and D. MacKay, L. Burns and G. Rand (1995). Fate modeling – Chapter 18. In: G. Rand (Ed.), *Fundamentals of Aquatic Toxicology: Effects, Environmental Fate, and Risk Assessment*, 2nd Edition. Taylor & Francis, Washington, DC.
14. A major source of information in this section is from H.F. Hemond and E.J. Fechner-Levy (2000). *Chemical Fate and Transport in the Environment*. Academic Press, San Diego, CA.
15. The source of the D value discussion is D. MacKay, L. Burns and G. Rand (1995). Fate modeling – Chapter 18. In: G. Rand (Ed.), *Fundamentals of Aquatic Toxicology: Effects, Environmental Fate, and Risk Assessment*, 2nd Edition. Taylor & Francis, Washington, DC.
16. Bracketed values indicate molar concentrations, but these may always be converted to mass per volume concentration values.
17. This example is also based upon guidance from MacKay and Paterson, Mathematical models of transport and fate.
18. V.R. Loizeau, A. Abarnou and A.M. Nesguen (2001). A steady-state model of PCB bioaccumulation in the sea bass (*Dicentrarchus labrax*) food web from the Seine Estuary, France. *Estuaries* 24 (6B): 1074–1087.
19. M. Begon, J.L. Harper and C.R. Townsend (1996). *Ecology*, 3rd Edition. Blackwell Science, Oxford, UK.
20. R.V. O'Neill (1976). Ecosystem persistence and heterotrophic regulation. *Ecology* 57: 1244–1253.
21. J. Iliopoulou-Georgudaki, C. Theodoropoulos, D. Venieri and M. Lagkadinou (2009). A model predicting the microbiological quality of aquacultured sea bream (*Sparus aurata*) according to physicochemical data: an application in western Greece fish aquaculture. *World Academy of Science, Engineering and Technology* 49: 1–8.
22. K. Koutsoumanis, A. Stamatiou, P. Skandamis and G.J.E. Nychas (2006). Development of a microbial model for the combined effect of temperature and pH on spoilage of ground meat, and validation of the model under dynamic temperature conditions. *Applied and Environmental Microbiology* 72: 124–134; T. Ross and T.A. McMeeking (2003). Modeling microbial growth within food safety risk assessments. *Risk Analysis* 23: 182–197; K. Koutsoumanis and G.J.E. Nychas (2000). Application of a systematic experimental procedure to develop a microbial model for rapid fish shelf-life prediction. *International Journal of Food Microbiology* 60: 171–184; P.S. Taoukis, K. Koutsoumanis and G.J.E. Nychas (1999). Use of time temperature integrators and predictive modelling for shelf life control of chilled fish under dynamic storage conditions. *International Journal of Food Microbiology* 53: 21–31; J.C. Augustin and V. Carlier (2000). Mathematical modelling of the growth rate and lag time for *Listeria monocytogenes*. *International Journal of Food Microbiology* 56: 29–51; and B. Gonzalez-Acosta, Y. Bashan, N. Hernadez-Saavedra, F. Ascencio and G. De la Cruz-Aguero (2006). Seasonal seawater temperature as the major determinant for populations of culturable bacteria in the sediments of an intact mangrove in an arid region. *FEMS Microbiology Ecology* 55: 311–321.
23. Cefic, Europa Bio (2004). European Commission's DG Research. *A European Technology Platform for Sustainable Chemistry*; www.cefic.be.
24. S.P. Bradbury, T.C.J. Feijtel and C.J. Van Leeuwen (2004). Peer reviewed: Meeting the scientific needs of ecological risk assessment in a regulatory context. *Environmental Science & Technology* 38 (23): 463A–470A.
25. US EPA (1993). Methods for Aquatic Toxicity Identification Evaluations: Phase III Toxicity Confirmation Procedures for Samples Exhibiting Acute and Chronic Toxicity; Report No. EPA/600/R-92-081; US Government Printing Office, Washington, DC, 1993; K.T. Ho *et al.* (2002). *Mar. Pollut. Bull.* 44 (4), 286–293; and National Centre for Ecotoxicology and Hazardous Substances (2001). *Direct Toxicity Assessment: Ecotoxicity Test Methods for Effluent and Receiving Water Assessment: Comprehensive Guidance*. Environment Agency, Wallingford, UK.
26. T. Colborn and K. Thayer (2000). Aquatic ecosystems: harbingers of endocrine disruption. *Ecological Applications* 10 (4): 949–957.
27. US Environmental Protection Agency (2005). Draft Report: Use of Biological Information to Better Define Designated Aquatic Life Uses in State and Tribal Water Quality Standards: Tiered Aquatic Life Uses – August 10, 2005, Washington, DC; and US Environmental Protection Agency (2002). Summary of Biological Assessment Programs and Biocriteria Development for States, Tribes, Territories, and Interstate Commissions: Streams and Wadeable Rivers. EPA-822-R-02-048. US Environmental Protection Agency, Washington, DC.

28. National Research Council.
29. Ibid.
30. G.S. Catchpole, M. Beckmann, D.P. Enot, M. Mondhe, B. Zywicki, J. Taylor, *et al.* (2005). Hierarchical metabolomics demonstrates substantial compositional similarity between genetically modified and conventional potato crops. *Proceedings of the National Academy of Sciences of the United States of America* 102: 14458–14462.
31. J. Karr and D. Dudley (1981). Ecological perspectives on water quality goals. *Environmental Management* 5: 55–68.
32. State of Washington, Department of Ecology (2003). A Citizen's Guide to Understanding and Monitoring Lakes and Streams; http://www.ecy.wa.gov/programs/wq/plants/management/joysmanual/chlorophyll.html.
33. See D. Flemer (1969). Continuous measurement of *in vivo* chlorophyll of a dinoflagellate bloom in Chesapeake Bay. *Chesapeake Science* 10: 99–103; and D. Flemer (1969). Chlorophyll analysis as a method of evaluating the standing crop of phytoplankton and primary production. *Chesapeake Science* 10: 301–306.
34. For example, see C. Lorenzen (1972). Extinction of light in the ocean by phytoplankton. *Journal of Conservation* 34: 262–267.
35. See D. Flemer (1969). Continuous measurement of *in vivo* chlorophyll of a dinoflagellate bloom in Chesapeake Bay. *Chesapeake Science* 10: 99–103; and US Environmental Protection Agency (EPA) (1997). *Methods for the Determination of Chemical Substances in Marine and Estuarine Environmental Matrices,* 2nd Edition. Method 446.0. EPA/600/R-97/072. US EPA, Office of Research and Development, Washington, DC.
36. L. Harding, Jr., E. Itsweire and W. Esais (1992). Determination of phytoplankton chlorophyll concentrations in the Chesapeake Bay with aircraft remote sensing. *Remote Sensing of Environment* 40: 79–100.
37. L. Harding, Jr., and E. Perry (1997). Long-term increase of phytoplankton biomass in Chesapeake Bay, 1950–1994. *Marine Ecology Progress Series* 157: 39–52.
38. K. Yagi (2007). Applications of whole-cell bacterial sensors in biotechnology and environmental science. *Applied Microbiology and Biotechnology* 73: 1251–1258.
39. Y.H. Lee and R. Mutharasan (2004). Biosensors. In: J.S. Wilson (Ed.), *Sensor Technology Handbook.* Newnes, Burlington, MA.
40. Yagi, Applications of whole-cell bacterial sensors.
41. Lee and Mutharasan, Biosensors.
42. These criteria were provided by John Crittenden, Arizona State University.
43. American Society of Mechanical Engineers (2005). Sustainability: Engineering Tools; http://www.professionalpractice.asme.org/business_functions/suseng/1.htm; accessed January 10, 2006.
44. See S.B. Billatos (1997). *Green Technology and Design for the Environment.* Taylor & Francis, Washington, DC; and V. Allada (2000). Preparing engineering students to meet the ecological challenges through sustainable product design. *Proceedings of the 2000 International Conference on Engineering Education*, Taipei, Taiwan.
45. US Environmental Protection Agency (2003). Remediation Guidance Document, EPA-905-B94-003 Chapter 7.
46. Ibid.
47. US Centers for Disease Control.
48. National Research Council (1999). *Strategies to Protect the Health of Deployed US Forces*. National Academies Press, Washington, DC.
49. Oak Ridge National Laboratory: B.M. Vogt and J.H. Sorensen (2002). How Clean is Safe? Improving the Effectiveness of Decontamination of Structures and People Following Chemical and Biological Incidents. Report No. ORNL/TM-2002/178. Final Report prepared for the US Department of Energy. Chemical and Biological National Security Program.
50. P.J. Meehan, N.E. Rosenstein, M. Gillen, R.F. Meyer, M.J. Kiefer, S. Deitchman, *et al.* (2004). Responding to detection of aerosolized *Bacillus anthracis* by autonomous detection systems in the workplace. *Morbidity and Mortality Weekly Report* 53: 1–11.
51. Ibid.
52. US General Accounting Office (2003). Capitol Hill Anthrax Incident: EPA's Cleanup Was Successful: Opportunities Exist to Enhance Contract Oversight. Report to the Chairman, Committee on Finance, US Senate.
53. US Department of Labor: Occupational Safety & Health Administration (2009). Etools: Anthrax; http://www.osha.gov/SLTC/etools/anthrax/decon.html; accessed September 30, 2009.
54. Based on: US Environmental Protection Agency (2009). *Essential Principles for Reform of Chemicals Management Legislation*; http://www.epa.gov/oppt/existingchemicals/pubs/principles.html; accessed September 30, 2009.
55. US Environmental Protection Agency (2009). Anthrax spore decontamination using methyl bromide; http://www.epa.gov/pesticides/factsheets/chemicals/methylbromide_factsheet.htm; accessed September 30, 2009.

CHAPTER 5

Environmental Risks of Biotechnologies

Risk is an expression of technological success or failure. Too much risk means the new technology has failed society. Societal expectations of acceptable risk are mandated by the standards and specifications of certifying authorities, such as health codes and regulations, zoning and building codes and regulations, principles of professional engineering and medical practice, design guidebooks, and standards promulgated by international agencies (e.g. ISO, the International Standards Organization) and national standard-setting bodies (e.g. ASTM, the American Society for Testing and Materials).

The risks stemming from operations and products of biotechnologies are not limited to human health, but also involve ecological resources and social welfare. As such, biotechnologies are additionally sanctioned by organizations involved in the life sciences, such as the American Medical Association and regulatory agencies, and are regulated by a variety of public health, food safety, and environmental agencies, such as the US Food and Drug Administration, the US Department of Agriculture, and the US Environmental Protection Agency, and their respective state counterpart agencies.

Most recently, since biotechnologies have the potential for intentional misuse, a number of their research and operational practices are regulated and overseen by homeland security and threat reduction agencies, especially related to microbes that have been or could be used as biological agents in warfare and terrorism.

ESTIMATING BIOTECHNOLOGICAL RISKS

Risk, as it is generally understood, is the chance that some unwelcome event will occur. The understanding of the factors that lead to a risk is known as *risk analysis.* The reduction of this risk (for example, by wearing seat belts in the driving example) is *risk management.* Risk management is often differentiated from *risk assessment,* which is comprised of the scientific considerations of a risk. Risk management includes the policies, laws, and other societal aspects of risk.

There are actually at least two types of risk analysis venues that relate to biotechnologies; those that follow the traditional chemical risk assessment paradigm and those that do not. Those risks that differ from chemical risks may follow certain steps in the risk process, but may need special attention in others (e.g. rather than inducing a disease, the release of a biotechnological product may upset the delicate ecological balances as a result of gene flow). Therefore, the foregoing discussion will follow the general chemical risk paradigm, but differences will be pointed out along the way and especially at the end of the chapter in the seminar discussion.

Environmental Biotechnology

To ascertain possible risks from biotechnologies, the first step is to identify a general hazard (a potential threat) and then to develop a scenario of events that could take place to unleash the potential threat and ultimately lead to an adverse effect. To assess the importance of a given scenario, the severity of the effect and the likelihood that it will occur in that scenario is calculated. This combination of the agent's severity and likelihood of exposure to that agent in a particular scenario constitutes the risk.

The relationship between the severity and probability of a risk follows a general equation [1]:

$$R = f(S, P) \tag{5.1}$$

where risk (R) is a function (f) of the severity (S) and the probability (P) of harm. The risk equation can be simplified to be a product of severity and probability:

$$R = S \times P \tag{5.2}$$

The traditional risk assessment paradigm (see Figure 5.1) is generally a step-wise process. It begins with the identification of a hazard, which is comprised of a summary of an agent's physicochemical properties and routes and patterns of exposure and a review of toxic effects. The tools for hazard identification take into account the chemical structures that are associated with toxicity, metabolic and pharmacokinetic properties, short-term animal and cell tests, long-term animal (*in vivo*) testing, and human studies (e.g. epidemiology, such as longitudinal and case-control studies). These comprise the core components of hazard identification, however additional hazard identification methods have been emerging that increasingly provide improved reliability of characterization and prediction. Risk assessors now can apply biomarkers of genetic damage (i.e. *toxicogenomics*) for more immediate assessments, as well as improved *structure-activity relationships* (SAR), which have incrementally been quantified in terms of stereochemistry and other chemical descriptions, i.e. using *quantitative structure-activity relationships* (QSAR) and computational chemistry. For the most part, however, health-effects research has focused on early indicators of outcome, making it possible to shorten the time between exposure and observation of an effect [2].

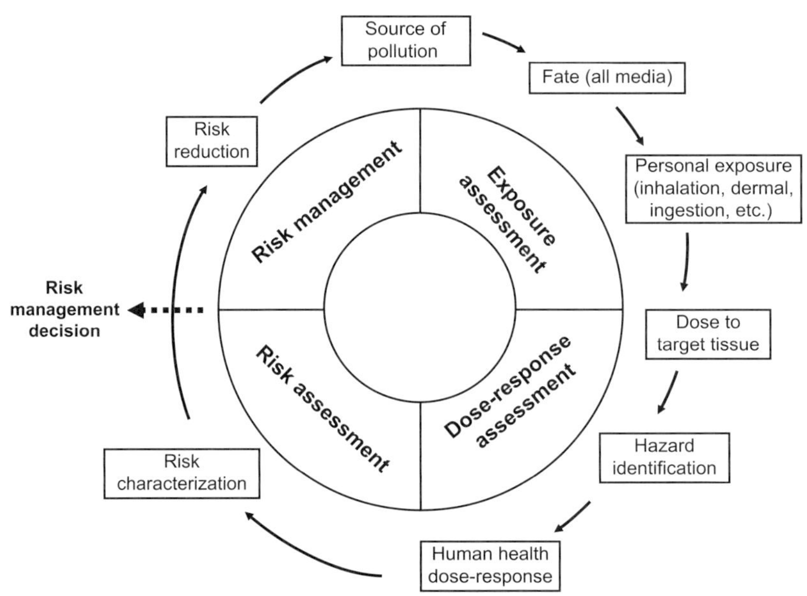

FIGURE 5.1

Risk assessment and management paradigm as employed by environmental agencies in the United States. The inner circle includes the steps recommended by the National Research Council. The outer circle indicates the activities (research and assessment) that are currently used by regulatory agencies to meet these required steps.

Source: NRC (1983). *Risk Assessment in the Federal Government.* National Academy of Sciences, Washington, DC.

Hazards from biological agents are often different than those posed by chemicals. As evidence, the Safety in Biotechnology Working Party of the European Federation of Biotechnology [3] has identified four risk classes for genetically modified organisms:

Risk class 1. No adverse effect, or very unlikely to produce an adverse effect. Organisms in this class are considered to be safe.

Risk class 2. Adverse effects are possible but are unlikely to represent a serious hazard with respect to the value to be protected. Local adverse effects are possible, which can either revert spontaneously (e.g. owing to environmental elasticity and resilience) or be controlled by available treatment or preventive measures. Spread beyond the application area is highly unlikely.

Risk class 3. Serious adverse local effects are likely with respect to the value to be protected, but spread beyond the area of application is unlikely. Treatment and/or preventive measures are available.

Risk class 4. Serious adverse effects are to be expected with respect to the value to be protected, both locally and outside the area of application. No treatment or preventive measures are available.

These classes indicate that even the safest genetic modifications carry some risk and that increasing uncertainty about an organism increases the need for precaution, i.e. decreasing the ability to assume a biological agent in a particular scenario is safe.

The biological agent risk classes are analogous to the hazard classes of various human health risk paradigms, e.g. carcinogen listings or extremely hazardous designations. Thus, a genetically modified microbe can be assigned to the appropriate risk class based on its physical, chemical, and biological properties, "independent of the technique used to select or generate the particular variant, and then scoring it against a set of values to be protected" [4]. Like the hazard identification process for chemicals, the microbe is classified according to inherent properties. It is in the next stage that environmental conditions are taken into account; that is, the characterization of administered dose to various responses in different populations. Both the hazard identification and dose-response information are based on the research that underpins the risk analysis. For microbes, the highest score for any one agent determines the overall risk class for environmental application.

Another similarity of the risk classification of microbes to the chemical hazard identification processes is that it is not uncommon to extrapolate from available knowledge to other microbes with similar characteristics or to yet untested, but similar environmental conditions (e.g. a field study's results in one type of field extrapolated to a different agricultural or environmental remediation setting). In chemical hazard identification, this is accomplished by *structural activity relationships*.

In the United States, ecological risk assessment paradigms have differed from human health risk assessment paradigms. The ecological risk assessment framework (see Figure 5.2) is based mainly on characterizing exposure and ecological effects. Both exposure and effects are considered during problem formulation [5].

Interestingly, the ecological risk framework is driving current thinking in risk assessment. The process shown in the inner circle of Figure 5.1 actually does not target the technical analysis of risk so much as it provides coherence and connections between risk assessment and risk management. In the early 1980s there was confusion and mixing of the two. For example, a share of the criticism of federal response to environmental disasters, such as those in Love Canal, New York, and Times Beach, Missouri, related to the mixing of scientifically sound studies (risk assessment) and decisions on whether to pursue certain actions (risk management). Obviously, this opened the response to charges of political and financial motivation, which was perceived to be overriding science. In fact, the final step of the risk assessment process was referred to as "characterization" to mean that "both quantitative and qualitative

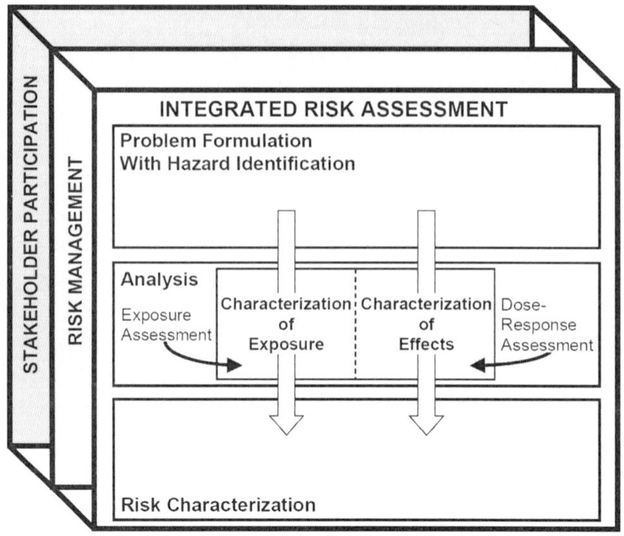

FIGURE 5.2
Framework for integrated human health and ecological risk assessment.
Sources: World Health Organization; and U.S. Environmental Protection Agency (1998).

elements of risk analysis, and of the scientific uncertainties in it, should be fully captured by the risk manager" [6]. In particular, the process allowed for an integration of research with risk assessment, which could underpin risk management decisions. The problem formulation step in the ecological framework has the advantage of providing an analytic-deliberative process early on. That is, it combines sound science with input from various stakeholders inside and outside of the scientific community.

The ecological risk framework calls for the characterization of ecological effects instead of hazard identification used in human health risk assessments. This is because the term "hazard" has been used in chemical risk assessments to connote either intrinsic effects of a stressor or a margin of safety by comparing a health effect with an estimate of exposure concentration. Thus, the term becomes ambiguous when applied to nonchemical hazards, such as those encountered in biological systems. Discussion Box: Risks of Commercializing *Sinorhizobium meliloti*, RMBPC-2 presents an example of a risk assessment of a genetically modified organism.

DISCUSSION BOX

Risks of Commercializing *Sinorhizobium meliloti*, RMBPC-2

An intergeneric microbe is one that is formed by combining genetic material from organisms in different genera. One such microbe, *Sinorhizobium meliloti* (*S. meliloti*) strain RMBPC-2, can be used as a microbial seed inoculant to coat alfalfa (*Medicago spp.*) seeds prior to planting. Research Seeds, Inc. of St Joseph, Missouri, was permitted by the US EPA to manufacture up to a maximum production volume of 500,000 pounds of the microbial seed inoculant during any consecutive 12-month period [7].

The Rhizobia (genera *Rhizobium*, *Sinorhizobium*, and *Bradyrhizobium*) are gram-negative soil bacteria. They are motile, rod-shaped, aerobic, and commonly found on roots (see Figure 5.3). In particular, Rhizobia have a symbiotic relationship with legumes, e.g. *Medicago, Melilotus*, and *Trigonella.* The symbiotic relationship results from the bacteria fixing atmospheric nitrogen, providing ammonium for protein production in the plant. In exchange, the bacteria obtain energy from the plant in the form of photosynthate, specifically dicarboxylates [8].

A member of the alpha subdivision of purple bacteria (i.e. proteobacteria), *S. meliloti* possesses a multipartite genome. These bacteria form growths called nodules on the roots of the legumes, and provide nitrogen in chemical forms that are biologically available to the plants. The benefiting plants return carbon and energy to the Rhizobia.

FIGURE 5.3
The root nodules of a 4-week-old *Medicago italica* inoculated with *Sinorhizobium meliloti.* [*See color plate section*]
Source: Wikipedia photo; *e* http://upload.wikimedia.org/wikipedia/commons/b/b3/Medicago_italica_root_nodules_2.JPG; accessed on October 3, 2009.

Problem formulation with hazard identification

Nitrogen (N) is an essential plant nutrient that, although abundant in the air and in organic matter in the soil, cannot be used directly by plants in the chemical forms provided in these compartments, so it must be fixed by plants via *nodulation*. Conventional methods of making N available to plants have been by adding N-rich fertilizers to the soil or by inoculating seed (i.e. coating the seed) with bacteria able to fix nitrogen. Bacterial N fixation changes the molecular nitrogen (N_2) in the troposphere into an inorganic form that plants can use. In addition, the bacteria leave excess N in the soil, potentially resulting in a net N gain the next growing season. Rhizobia have been used commercially as seed inoculants in the form of seed coatings for over one hundred years. Currently, about 80% of alfalfa grown in the United States is inoculated with rhizobia prior to planting. Nitrogen fixation by *S. meliloti* is specific to the legumes alfalfa, sweet clover, and fenugreek.

Under its authority from the Toxic Substances Control Act (TSCA), Section 5 of TSCA, the US EPA requires that information about the health and environmental effects of new chemical substances (including genetically modified microorganisms) be reviewed before the substances may be used commercially in the United States. This information on new chemical substances or engineered organisms is submitted as a *premanufacture notice* (PMN). Under TSCA, "new" microorganisms are intergeneric.

Regulators began evaluating various intergeneric strains of *Sinorhizobium (Rhizobium) meliloti* beginning in 1987. Research Seeds, Inc. began research on this microbe in 1992, submitting several PMNs for approval to conduct several small- and large-scale research field trials with various strains of these microorganisms, including strain RMBPC-2, which has added genes to regulate the nitrogenase enzyme and genes that increase the delivery of organic acid from the plant to the nodule bacterium. These field trials included strains developed by another company and previously evaluated by the US EPA. These field trials were also subject to a consent order issued by EPA under section 5(e) of TSCA. The order limited the use of the intergeneric strains of *Sinorhizobium meliloti*, including strain RMBPC-2, to specific sites only for research purposes. In 1994, the company submitted a request to commercialize *S. meliloti* strain RMBPC-2. On January 4, 1995, the agency's subcommittee of the Biotechnology Science Advisory Committee (BSAC) met to review the Agency's draft risk assessment. The BSAC conducts scientific peer reviews for risk assessments of certain biotechnology products reviewed by EPA under TSCA. The BSAC submitted its report on March 6, 1995. Based on this review and other information the US EPA amended the order in 1997 to allow limited commercial use of *S. meliloti* strain RMBPC-2. This was the first commercial use of an intergeneric microorganism in the environment approved under TSCA.

The microbial stressor has the potential to split into biological subcomponents, e.g. pathogenicity, altered legume growth resulting from the microbe, as well as chemical subcomponents, e.g. production of toxins,

(continued)

generation of detrimental metabolites, and overproduction of nitrate. As with chemical stressors, characterizing the recombinant microbes to predict their potential adverse effects is crucial to risk assessment. For the recombinant microbe, both the donor and recipient microorganisms must be characterized, including the process by which they are modified. The phenotypic traits of most genetically engineered microbes reviewed under TSCA are encoded and analyzed with a PC-microcomputer version of the "Micro-IS" data system, with the final step for genetically engineered microbe identification being the verification that the microbe's DNA contains the DNA of interest, along with additional vector DNA [9].

Exposure assessment

Sinorhizobium meliloti has been used as a seed inoculant for over one hundred years, with no reported pathogenic effects on humans, animals, or plants associated with the use of these microorganisms. The genetic modifications made to strain RMBPC-2 are not expected to alter these characteristics in the microorganism.

Limited occupational exposure of workers to strain RMBPC-2 during manufacture and processing of the seed inoculant, as well as during inoculant application, was not expected to present substantial risk at the expected exposures due to the low hazard to human health posed by the microorganism.

Effects assessment

Antibiotic resistance

Antibiotic resistance genes like those introduced into strain RMBPC-2 occur commonly in a wide array of naturally occurring microorganisms and are generally much more mobile than those introduced into strain RMBPC-2, due to the stability of the genes' location in this strain. Because of the stability of these genes in strain RMBPC-2, EPA does not believe that strain RMBPC-2 will contribute significantly to the naturally occurring antibiotic resistance gene pool.

According to the risk assessment, as it is used in commerce, strain RMBPC-2 is not found in the same locations in the environment as other microorganisms that are human or animal pathogens. This further reduces the likelihood that such microorganisms might acquire antibiotic resistance characteristics from strain RMBPC-2.

The antibiotics to which strain RMBPC-2 is resistant have few uses in the treatment of human or animal disease, and for the majority of these uses are not the drugs of first choice.

Alfalfa yield

Field tests, lasting up to four years at some sites, have demonstrated that RMBPC-2 is able to significantly increase alfalfa yield under certain conditions (low nitrogen content in the soil and low indigenous rhizobial populations). Overall, RMBPC-2 was shown to perform within the normal range expected of naturally occurring commercial inoculants.

Nodulation

Rhizobial nodulation of legumes is species- and strain-specific. Each rhizobial strain is likely to nodulate only specific legumes (cross-inoculation group). *S. meliloti* is among the most restrictive in nodulation preference, normally nodulating only the legumes alfalfa, sweet clover, and fenugreek. There are isolated reports in the literature, however, that *S. meliloti* may also be able to nodulate a few other leguminous plants such as mesquite.

Due to the potential for *S. meliloti* to inoculate legumes other than alfalfa, the BSAC Subcommittee considered whether to recommend additional testing of strain RMBPC-2 to determine its potential to inoculate legumes outside its cross-inoculation group. EPA believes that there is no hazard associated with the potential for inoculation of leguminous plants outside of the cross-inoculation group for *S. meliloti*, and consequently that such testing is not necessary prior to limited commercial use of strain RMBPC-2.

Environmental persistence, fate, and transport

The scientific advisory committee also considered the need for additional testing of the persistence, dissemination, competitiveness, and genetic stability of strain RMBPC-2. EPA believes that such additional testing is not necessary prior to limited commercial use of the product. The Agency's determination is based on data collected from field trials of *S. meliloti* strain RMBPC-2, as well as trials of similarly modified *S. meliloti* strains and non-modified rhizobia strains. These data support the conclusion that the behavior of strain RMBPC-2 is expected to be consistent with current commercially available rhizobia strains.

Following [10] are the fate study results:

Atmospheric transport: Selective agar plates were mounted on posts located in all four compass directions at various distances – 4, 9, 50, 100, 200, and up to 500 feet – from the perimeter of the test plots on days 0, 1, 2, 3, 4, and 6 after initiation of the strain comparison trial. Additional plates were placed between the four compass points. No colonies appeared on the vast majority of plates regardless of compass direction or distance. A total of 13 colonies appeared on Selective Medium A over a cumulative exposure of 6 hours on day 0 for all compass directions and distances even though a moderate wind blew on the day of application. Later samplings were for 2-hour exposures only. On day 6, the number of colonies on Medium A from the west compass direction (the direction with the highest counts) had dropped from 13 at the 4-foot distance to one colony at both the 100- and 200-foot distances. Overall, little aerial dispersion of the genetically modified microbes occurred. Likewise, aerial dispersion measurements taken at termination, when the fields were being plowed, resulted in no detectable dispersal of inoculant from the test site.

Vertical migration: Movement of the recombinant rhizobia downward through the soil profile past the rhizosphere was measured by plating out soil obtained with a soil-coring device. Twelve-inch cores were taken from control and treated plots in an outside row, immediately adjacent to a plant stalk. The top 2 and bottom 2 inches of the soil core were homogenized and sub-sampled for the presence of added rhizobia. Vertical monitoring used the plant most probable number (MPN) technique for enumeration at various time points up to 312 days. Throughout the season, cell numbers ranged from 7 to $>$138 cells per gram dry soil in the top 2 inches and from 3 to $>$524 cells per gram dry soil in the 10- to 12-inch depth. Rhizobial inoculants also occurred at a depth of 22 to 24 inches. Overall, only minimal movement occurred beyond the root zone. No observed differences occurred in the vertical movement of the recombinant strains versus the wild-type strain.

Horizontal dispersion: The study monitored horizontal movement through the soil by sampling the top 2 inches of the soil surface at a distance of 6 inches away from the edge of the plots in all four compass directions on days 0, 11, and 34. Samples were examined for the presence of three strains: RmSF38 and two recombinants, RMB7101 and RMB7103. Using selective media supplemented with the fluorescent antibody method, samples contained no detectable inoculants. With the more sensitive MPN enumeration technique, counts ranged from 0 to 57 cells per gram dry soil. Consequently, all subsequent analyses used the MPN technique. Up through day 123, cell counts never exceeded 250 cells per gram dry soil, and nearly all counts dropped to 0 by day 159. These results indicate minimal horizontal movement of the rhizobial inoculants throughout the study and no differences in the behavior of the recombinant strains versus the wild type.

Risk characterization

Based on its assessment of available information, the risk assessment determined that the initial commercialization of strain RMBPC-2 presents a low level of risk to health and the environment.
In addition, strain RMBPC-2 has demonstrated a significant advantage over other commercial alfalfa seed inoculants in improving alfalfa yields under certain soil conditions. EPA acknowledges, however, that there are some uncertainties associated with the behavior of this microorganism in the environment, as noted above. Based on its intended use as a seed inoculant, the microorganism is expected to be produced in substantial quantities and to be used in the environment in substantial quantities. The US EPA therefore concluded that it is prudent at this time to limit commercial production of strain RMBPC-2, and to establish a subsequent opportunity for EPA to reexamine the product at a future date and to consider, in light of the information and understanding available at that time, whether additional action is needed to address questions about the behavior of strain RMBPC-2 in the environment.

(Continued)

In 1998, the US EPA issued a significant new use rule (SNUR) under section 5(a)(2) of the TSCA for *S. meliloti* strain RMBPC-2 , that required persons who intend to manufacture, import, or process the bacterium for a significant new use beyond the earlier PMN to notify the agency at least 90 days before commencing these activities [11]. The SNUR was added to provide sufficient time to evaluate the intended use and, if necessary, to prohibit or limit that activity before it can occur.

Ongoing Uncertainties and Concerns

Both effects and fate information had elements of uncertainty. For the greenhouse yield data, uncertainty resulted from the protocol, the alfalfa cultivar relative to the field trials, and the extrapolation to field results. The alfalfa yield in the field may not have reflected the ability of the rhizobia to increase alfalfa growth because the test did not measure total nitrogen in the field soil, and high levels of nitrogen can inhibit nodulation by rhizobia. Heavy weed and leaf hopper infestations also may have confounded the alfalfa yield data [12].

The effects on weedy legumes and other crop legumes in the cross-inoculation group for *R. meliloti* also introduced uncertainty. For the genetically engineered microbes undergoing field testing, no data were available for the risk assessment that would have indicated their competitive ability to nodulate alfalfa relative to native rhizobia. Fate and transport information also introduced uncertainty. Extrapolation from pure laboratory culture and greenhouse studies to the real-world conditions of the field is questionable. The ability to distinguish the released rhizobia from each other and from the indigenous rhizobia was also uncertain.

The RMBPC-2 strain of *S. meliloti* genome is now fully sequenced [13]. In addition to the added genes that regulate nitrogenase enzyme (for nitrogen fixation) and genes that increase the organic acid delivered from the plant to the nodule bacterium, the strain also has the antibiotic resistance marker genes for streptomycin and spectinomycin [14]. The commercial release was permitted in spite of concerns about the impact of the GM microbe on the environment.

Genetically modified *S. meliloti* strains have recently been shown to persist in soil for years, even when separated from the host plant. For example, the recombinant strains that the field trials investigated for persistence – strains RmSF38, RMB7101, and RMB7103– survived at rates of 10^5–10^6 cells per gram dry root into the second year of the field study [15]. Horizontal gene flow to other soil bacteria and microevolution of plasmids has also been observed [16]. Investigations have indicated that modified strains of *S. meliloti* in the arthropod gut can facilitate gene transfer to a number of bacteria [17]. Thus, the initial conclusions in the risk assessment regarding antibiotic resistance and gene flow continue to be an area of uncertainty and concern for genetically engineered strains of *S. meliloti.*

In cases where the outcomes could be substantial and where small changes can lead to very different functions and behaviors from unknown and insufficiently known chemicals or microbes, specific investigations are needed in the laboratory and field, e.g. to estimate competitive advantages and disadvantages compared to natural microbial populations, horizontal transfer of a genetic trait, ability and potential to measure environmental outcomes and the amount of reversibility of the outcomes. Often, the proponents of a biotechnology will have done substantial research on the benefits and operational aspects of the agent, but the regulatory agencies and the public may call for more and better information about unintended and yet-to-be-understood consequences and side effects [18].

The Safety in Biotechnology Working Party of the European Federation of Biotechnology gives an illustrative example to demonstrate this hazard/risk classification, using *Bacillus thuringiensis* (Bt) that has been sprayed onto a cornfield to eliminate the European corn borer. But, if in the process *B. thuringiensis* also kills honey bees, this would be a side effect that would not be tolerable. Furthermore, physical, chemical, and biological factors can influence these effects, e.g. type of application of Bt can influence the amount of drift toward non-target species. Downstream effects can be even more difficult to predict than side effects, since they not only occur within variable space, but also in variable time regimes. For example, risk can arise from both the application method and from the build-up of toxic materials and gene flow following the pesticide drift.

Environmental risks associated with biotechnologies are a function of the interrelationships among factors that put people or ecosystems at risk. Environmental practitioners provide decision makers with thoughtful studies based upon the sound application of the physical sciences and, therefore, are risk assessors by nature. Bioengineers, like all engineers, control factors in their designs. As such, bioengineers are risk managers. Engineers are held responsible for designing safe products and processes, and the public holds us accountable for its health, safety, and welfare. The public expects environmental practitioners to "give results, not excuses" [19], and risk and reliability are accountability measures of their success. Engineers design systems to reduce risk and look for ways to enhance the reliability of these systems. Thus, every biotechnologist deals directly or indirectly with risk and reliability. This can be challenging since risk means different things to different people.

In fact, risk has some very precise definitions within the scientific community. However, the various scientific disciplines have divergent concepts of risk (see Table 5.1). Interestingly, some of the technical definitions differ substantively from the social definitions but others may be more semantic differences or the technical definitions are merely restating the social concepts using mathematical or scientific nomenclature. Biotechnological risk falls into both technical and social categories, since the public has a wide range of perceptions about the value and risks of various biotechnologies.

One means of expressing is by using models. For example, models can be used to estimate exposures. Such models range from "screening-level" to "high-tiered." Screening models included generally over-predict exposures because they are based on conservative default values and assumptions. They provide a first approximation that screens out exposures not likely to be of concern [20]. Conversely, higher-tiered models typically include algorithms that provide specific site characteristics, time–activity patterns, and are based on relatively realistic values and assumptions. Such models require data of higher resolution and quality than the screening models and, in return, provide more refined exposure estimates [21].

Risk involves a stressor, a receptor, and an outcome. The stressor can be physical, chemical or biological. If a water body's temperature increases beyond some threshold value, the fish may die or fail to reproduce. This is an example of physical stressor (heat) on a population of receptors (various fish genera), leading to a deleterious outcome (fish kill). Sometimes, the outcome is not specific to any particular stressor. For example, the fish kill could have resulted from a chemical stressor, such as the release of organic matter into the water body, which was used as food by the aquatic microorganisms, using up the available oxygen. In this case, the stressor is first- and second-order. The first-order stress was the release of organic material; the second-order stress was the decreased dissolved oxygen (DO). The fish kill could also be the result of a biological stressor, such as a dinoflagellate, e.g. *Karenia brevis* (red tide) or *Pfiesteria* piscicida. The outcome in all three scenarios is the same, i.e. a fish kill, but the path to this outcome is different.

The pathway to an unwanted outcome often includes many steps. Biotechnologies can be used to ameliorate hazards, but can also add to hazards. Genetically modified organisms have been used to reduce environmental hazards (e.g. bacteria selected for metabolic and kinetic traits that enhance biodegradation). However, they may also present complications in the risk paradigm. For example, the Ecological Society of America has articulated five specific areas of environmental risks presented by genetically engineered organisms:

creating new or more vigorous pests and pathogens;
exacerbating the effects of existing pests through hybridization with related transgenic organisms;

Table 5.1 Comparison of definitions of risk in technical publications versus social vernacular

Technical definitions of risk (Compiled by the Society of Risk Analysts and recorded in S.M. Macgill, Y.L. Siu (2005). A new paradigm for risk analysis. *Futures* 37: 1105–1131)	1. Possibility of loss, injury, disadvantage or destruction; to expose to hazard or danger; to incur risk of danger
	2. An expression of possible loss over a specific period of time or number of operational cycles
	3. Consequence per unit time = frequency (events per unit time) × magnitude (consequences per event)
	4. Measure of the probability and severity of adverse effects
	5. Conditional probability of an adverse effect (given that the necessary causative events have occurred)
	6. Potential for unwanted negative consequences of an event or activity
	7. Probability that a substance will produce harm under specified conditions
	8. Probability of loss or injury to people and property
	9. Potential for realization of unwanted, negative consequences to human life, health or the environment
	10. Product of the probability of an adverse event times the consequences of that event were it to occur
	11. Function of two major factors: (a) probability that an event, or series of events of various magnitudes, will occur, and (b) the consequences of the event(s)
	12. Probability distribution over all possible consequences of a specific cause which can have an adverse effect on human health, property or the environment
	13. Measure of the occurrence and severity of an adverse effect to health, property or the environment
Social definitions of risk (Compiled in: S.M. Macgill, Y.L. Siu (2005). A new paradigm for risk analysis. *Futures* 37: 1105–1131)	1. Probability of an adverse event amplified or attenuated by degrees of trust, acceptance of liability and/or share of benefit
	2. Opportunity tinged with danger
	3. A code word that alerts society that a change in the expected order of things is being precipitated
	4. Something to worry about/have hope about
	5. An arena for contending discourses over institutional relationships, socio-cultural issues, political and economic power distributions
	6. A threat to sustainability/current lifestyles
	7. Uncertainty
	8. Part of a structure of meaning based in the security of those institutional settings in which people find themselves
	9. The general means through which society envisages its future
	10. Someone's judgment on expected consequences and their likelihood
	11. What people define it to be – something different to different people
	12. Financial loss associated with a product, system or plant
	13. The converse of safety

harm to nontarget species, such as soil organisms, non-pest insects, birds, and other animals;
disruption of biotic communities, including agroecosystems;
irreparable loss or changes in species diversity or genetic diversity within species. [22]

In a way, genetic manipulation is similar to computing. The digital versions of software are so far removed from the analogues that the software functions are different not only in degree but in kind. Thus, the numerous ongoing and potential approaches to genetic engineering are quite different from their "analogues" (i.e. traditional breeding, encompassing viruses, bacteria, algae, fungi, grasses, trees, insects, fish, and shellfish). According to the ESA, these genetically engineered organisms "that present novel traits will need special scrutiny with regard to their environmental effects" [23].

Environmental stressors, including biological agents, can be modeled in a unidirectional and one-dimension fashion, such as the flow depicted in Figure 7.29. Systems biology can provide a conceptual framework that links exposure to environmental outcomes across levels of biological organization (Figure 5.4). Thus, environmental exposure and risk assessment considers coupled networks that span multiple levels of biological organization that can describe the interrelationships within the biological system. Mechanisms can be derived by characterizing and perturbing these networks (e.g. behavioral and environmental factors) [24]. This can apply to a food chain or food web model (see Figure 1.1) or a kinetic model (see Figure 5.5) or numerous other modeling platforms. Computational models are discussed in greater detail in Chapter 7.

To assess the risks associated with genetic manipulations, three questions [25] must be asked:

What are the specific environmental concerns or harm that will or can occur?
What is the probability that the concerns will be realized or that harm will occur?
What are the adverse outcomes (e.g. to health and the environment) when the harm occurs, including how widespread in time and space?

These are deceptively simple questions. They are only easy to answer when an action is clearly wrong with no benefits whatsoever. For example, if a scientist simply wants to genetically

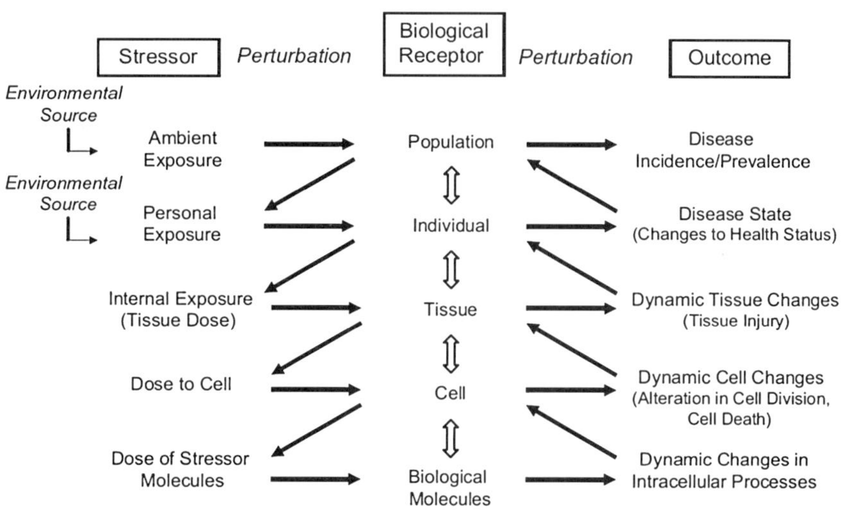

FIGURE 5.4
Systems cascade of exposure-response processes. In this instance, scale and levels of biological organization are used to integrate exposure information with biological outcomes. The stressor (chemical or biological agent) moves both within and among levels of biological organization, reaching various receptors, thereby influencing and inducing outcomes. The outcome can be explained by physical, chemical, and biological processes (e.g. toxicogenomic mode-of-action information).
Source: E.A. Cohen Hubal, A.M. Richard, S. Imran, J. Gallagher, R. Kavlock, J. Blancato and S. Edwards (2008). Exposure science and the US EPA National Center for Computational Toxicology. *Journal of Exposure Science and Environmental Epidemiology.* doi:10.1038/jes.2008.70 [online: November 5, 2008].

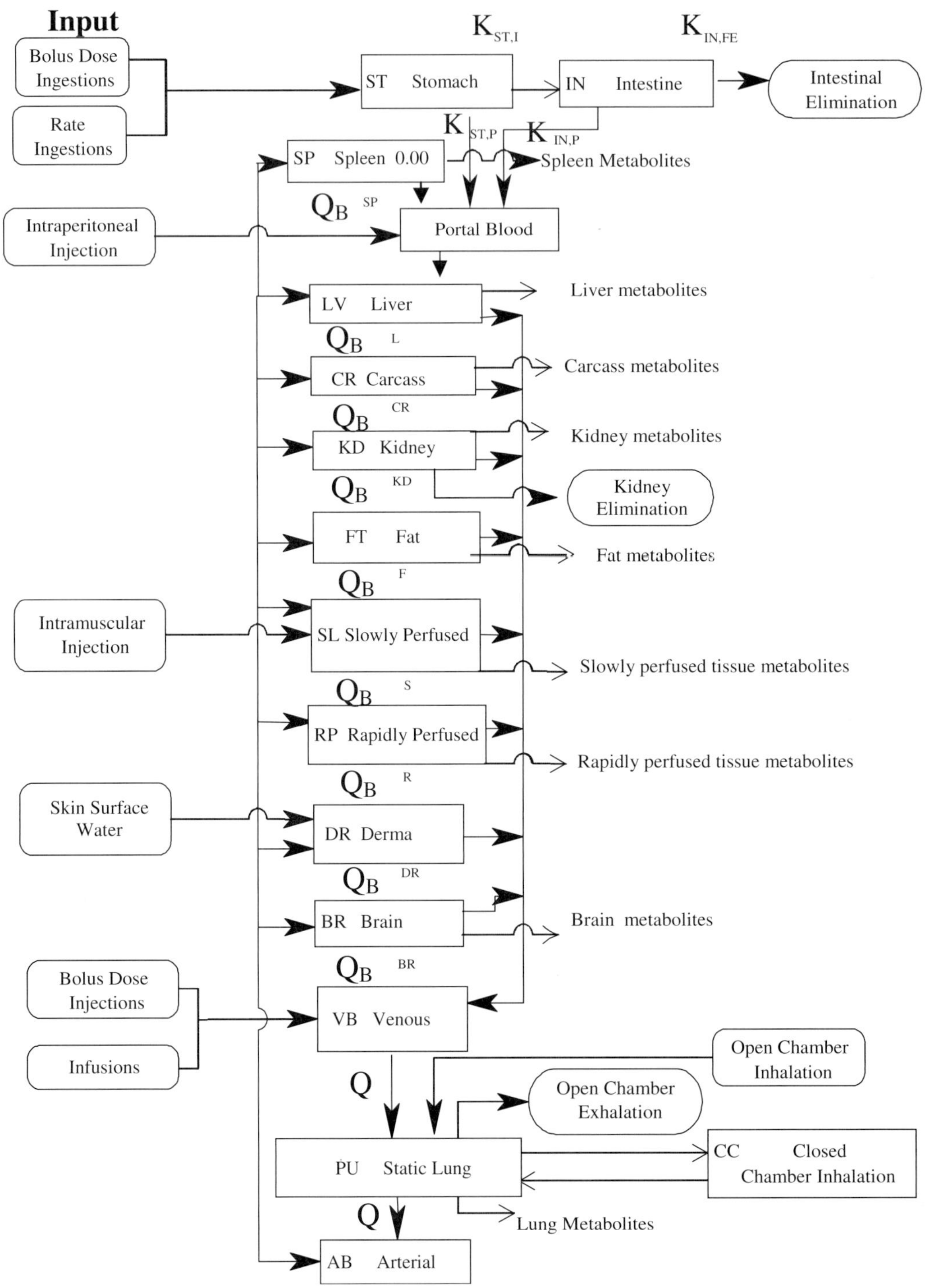

FIGURE 5.5

Toxicokinetic model used to estimate dose as part of an environmental exposure. This diagram represents the static lung, with each of the compartments (brain, carcass, fat, kidney, liver, lung tissue, rapidly and slowly perfused tissues, spleen, and the static lung) having two forms of elimination, an equilibrium binding process, and numerous metabolites. Notes: K = kinetic rate; Q = mass flow; and QB = blood flow. This model can be used for a single chemical and for multiple chemicals. A breathing lung model would consist of alveoli, lower dead space, lung tissue, pulmonary capillaries, and upper dead space compartments. Gastro-intestinal (GI) models allow for multiple circulating compounds with multiple metabolites entering and leaving each compartment, i.e. the GI model consists of the wall and lumen for the stomach, duodenum, lower small intestine, and colon, with lymph pool and portal blood compartments included. Bile flow is treated as an output from the liver to the duodenum lumen. All uptaken substances are treated as circulating. Nonspecific ligand binding, e.g. plasma protein binding, is represented in arterial blood, pulmonary capillaries, portal blood, and venous blood. *Source:* C.C. Dary, P.J. Georgopoulos, D.A. Vallero, R. Tornero-Velez, M. Morgan, M. Okino, *et al.* (2007). Characterizing chemical exposure from biomonitoring data using the exposure related dose estimating model (ERDEM). 17th Annual Conference of the International Society of Exposure Analysis, Durham, NC, October 17, 2007. Adapted from: J.N. Blancato, F.W. Power, R.N. Brown and C.C. Dary (2006). Exposure Related Dose Estimating Model (ERDEM): A Physiologically-Based Pharmacokinetic and Pharmacodynamic (PBPK/PD) Model for Assessing Human Exposure and Risk. Report No. EPA/600/R-06/061. US Environmental Protection Agency, Las Vegas, NV.

engineer a skunk to make it more likely to get rabies, many would oppose such research on the basis of the animal's possible release into the wild and the concomitant increase in rabies in animal and human populations. However, the scientist may argue that he really needs such a skunk because that is the only way to test for rabies vaccines. Whether we agree, this is actually a question of short-term risks versus long-term advancement in medical and other societal knowledge. This is exactly what happened in natural, non-genetic engineering introductions of species, such as kudzu and multiflora roses, which were only later found to have dramatically negative ecological impacts. If we could go back in time and have answers to the three questions above from our full knowledge of the implications observed in the field, these answers would have in all likelihood told the scientists and agricultural extension agents that the risks outweighed the benefits and the decision should be a "no go." Some have argued analogously for genetically modifying microbes to find vaccines or other societal benefits that, in their opinion, override concerns about possible releases of potentially dangerous forms (and the concomitant need to provide containment).

Analogies can be stretched from past events, so long as assumptions and driving factors remains similar and realistic. Consider *biopharming*. This is a play on phonetics; i.e. the "pharm" is short for pharmaceuticals, which sounds like "farm." Biopharming makes use of genetically modified plants and animals, crops and livestock respectively, to produce pharmaceuticals. We are about at the same point in time, just beyond the idea stage, before deciding that kudzu would be a good thing for erosion control. Biopharm crops are being grown on a limited scale in the United States and Europe, and biopharm animals are starting to be raised in New Zealand [26].

Having all the data needed for a *completely* informed risk decision is impossible. Scientific objectivity and humility dictate that risk assessors are upfront about uncertainties. The three questions can only be answered by looking for patterns and analogies from events that are similar to the potential threat being considered. From there, scenarios can be developed to follow various paths to good, bad, and indifferent outcomes. This is known as a decision tree, which is discussed in Chapter 12. Risk assessment is a way to estimate the importance of each scenario, and selecting the one with the most acceptable risk. This is not the same, necessarily, as the one with the most benefits compared to risks, i.e. a benefit to risk ratio or relationship, or benefit to cost ratio or relationship. However, this is indeed one of the more widely used approaches. The challenge is how to quantify many of the benefits and risks, since risk is a function of likelihood and severity of a particular adverse outcome (i.e. harm):

$$R = S \times P \tag{5.3}$$

where R = risk, S = severity of the outcome, and P = probability of the outcome.

Environmental risk is often considered to be the product of the hazard (H) and the exposure (E) to that hazard:

$$R = H \times E \tag{5.4}$$

Biotechnological risk characterization follows four basic steps: hazard identification; dose-response estimation; exposure assessment; and effects assessment.

Biotechnological hazard identification

Anything with the potential to cause harm is a hazard. Some things are inherently hazardous, whereas others are hazardous in one scenario, but essential or desirable in another. Liquid water is a drowning hazard. Ice is a slipping hazard. Sharps, like syringe needles, are infection hazards. Pesticides are health hazards. At least a portion of the hazard is an intrinsic property of a substance, product or process, i.e. a concept of potential harm. For example, a biochemical hazard is an absolute expression of a substance's properties, since all substances have unique physical and chemical properties. These properties can render the substance to be hazardous.

Conversely, Eq. 5.2 shows risk can only occur with exposure. So, if one walks on a street in the summer, the likelihood of slipping on ice is near zero. One's total slipping risk is not necessarily zero (e.g. a person could step on an oily surface or someone could throw ice in one's path). If not in a medical facility, one's infection risk from sharps may be near zero, but a person's total infection risk is not zero (e.g. people may be exposed to the same infection from a person sneezing in their office). If a person does not use pesticides, the pesticide health risk is also lower. However, since certain pesticides are persistent and can remain in the food chain, the person's exposure is not zero. Also, even if a person's pesticide exposure is near zero, that person's cancer risk is not zero, since he or she may be exposed to other cancer hazards and/or may be genetically predisposed to carcinogenesis.

Genetically modified organisms are a special case. In nature, bacteria provide essential functions, such as decomposition of organic matter in soil, sediment, and detritus. However, when these organisms' genetic material is altered, the organisms' endogenous processes may be essentially the same as those in the natural ones, but are in some way enhanced (e.g. faster or able to break down a wider array of chemicals). Other manipulations may cause an organism to do something completely different than its natural functions (e.g. a transgenic species that glows or one that has insecticidal properties). Either way, new hazards may be introduced and must be properly evaluated in a hazard identification process.

Engineers and scientists working in environmental areas consider a number of hazards; the most common is toxicity. Other important environmental hazards are shown in Table 5.2. Hazards can be expressed according to the physical and chemical characteristics, as in Table 5.2, as well as in the ways they may affect living things. For example, Table 5.3 summarizes some of the expressions of biologically based criteria of hazards. Other hazards, such as flammability, are also important to environmental engineering. However, the chief hazard in most environmental situations has been toxicity.

Dose-response

The first means of determining exposure is to identify *dose*, the amount (e.g. mass) of a contaminant that comes into contact with an organism. Dose can be the amount administered to an organism (so-called "applied dose"), the amount of the contaminant that enters the organism ("internal dose"), the amount of the contaminant that is absorbed by an organism over a certain time interval ("absorbed dose"), or the amount of the contaminants or its metabolites that reach a particular "target" organ ("biologically effective dose" or "bio-effective dose"), such as the amount of a hepatotoxin (a chemical that harms the liver) that finds its way to liver cells or a neurotoxin (a chemical that harms the nervous system) that reaches the nerve or other nervous system cells. Theoretically, the higher the concentration of a hazardous substance or microbe that comes into contact with an organism, the greater the expected adverse outcome. The pharmacological and toxicological gradient is the so-called "dose-response" curve (Figure 5.6). Generally, increasing the amount of the dose means a greater incidence of the adverse outcome.

Dose-response assessment generally follows a sequence of five steps [28]:

- fitting the experimental dose-response data from animal and human studies with a mathematical model that fits the data reasonably well;
- expressing the upper confidence limit (e.g. 95%) line equation for the selected mathematical model;
- extrapolating the confidence limit line to a response point just below the lowest measured response in the experimental point (known as the "point of departure"), i.e. the beginning of the extrapolation to lower doses from actual measurements;
- assuming the response is a linear function of dose from the point of departure to zero response at zero dose; and,
- calculating the dose on the line that is estimated to produce the response.

Table 5.2 Hazards defined by the Resource Conservation and Recovery Act

Hazard type	Criteria	Physical/Chemical classes in definition
Corrosivity	A substance with an ability to destroy tissue by chemical reactions	Acids, bases, and salts of strong acids and strong bases. The waste dissolves metals, other materials, or burns the skin. Examples include rust removers, waste acid, alkaline cleaning fluids, and waste battery fluids. Corrosive wastes have a pH of <2.0 or >12.5. The US EPA waste code for corrosive wastes is "D002"
Ignitability	A substance that readily oxidizes by burning	Any substance that spontaneously combusts at 54.3°C in air or at any temperature in water, or any strong oxidizer. Examples are paint and coating wastes, some degreasers, and other solvents. The US EPA waste code for ignitable wastes is "D001"
Reactivity	A substance that can react, detonate or decompose explosively at environmental temperatures and pressures	A reaction usually requires a strong initiator (e.g. an explosive like TNT, trinitrotoluene), confined heat (e.g. saltpeter in gunpowder), or explosive reactions with water (e.g. Na). A reactive waste is unstable and can rapidly or violently react with water or other substances. Examples include wastes from cyanide-based plating operations, bleaches, waste oxidizers, and waste explosives. The US EPA waste code for reactive wastes is "D003"
Toxicity	A substance that causes harm to organisms. Acutely toxic substances elicit harm soon after exposure (e.g. highly toxic pesticides causing neurological damage within hours after exposure). Chronically toxic substances elicit harm after a long period of time of exposure (e.g. carcinogens, immunosuppressants, endocrine disruptors, and chronic neurotoxins)	Toxic chemicals include pesticides, heavy metals, and mobile or volatile compounds that migrate readily, as determined by the Toxicity Characteristic Leaching Procedure (TCLP), or a "TC waste." TC wastes are designated with waste codes "D004" through "D043"

Table 5.3 Biologically-based classification criteria for chemical substances [27]

Criterion	Description
Bioconcentration	The process by which living organisms concentrate a chemical contaminant to levels exceeding the surrounding environmental media (e.g. water, air, soil, or sediment)
Lethal dose (LD)	A dose of a contaminant calculated to expect a certain percentage of a population of an organism (e.g. minnow) exposed through a route other than respiration (dose units are mg [contaminant] kg^{-1} body weight). The most common metric from a bioassay is the lethal dose 50 (LD_{50}), wherein 50% of a population exposed to a contaminant is killed
Lethal concentration (LC)	A calculated concentration of a contaminant in the air that, when respired for 4 hours (i.e. exposure duration = 4 h) by a population of an organism (e.g. rat) will kill a certain percentage of that population. The most common metric from a bioassay is the lethal concentration 50 (LC_{50}), wherein 50% of a population exposed to a contaminant is killed. (Air concentration units are mg [contaminant] L^{-1} air)

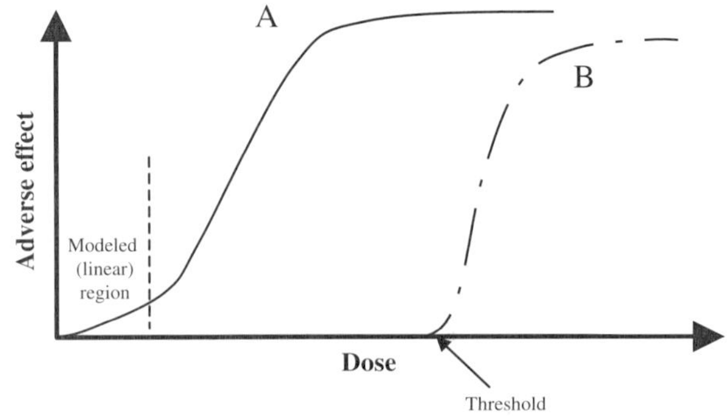

FIGURE 5.6
Prototypical dose-response curves. Curve A represents the "no-threshold" curve, which predicts a response (e.g. cancer) even if exposed to a single molecule ("one-hit model"). As shown, the low end of the curve, i.e. below which experimental data are available, is linear. Thus, Curve A represents a linearized multistage model. Curve B represents toxicity above a certain threshold (no observable adverse effect level (NOAEL) is the level below which no response is expected). Another threshold is the no observable effect concentration (NOEC), which is the highest concentration where no effect on survival is observed ($NOEC_{survival}$) or where no effect on growth or reproduction is observed ($NOEC_{growth}$). Note that both curves are sigmoidal in shape because of the saturation effect at high dose (i.e. less response with increasing dose). *Source:* Adapted from D.A. Vallero (2004). *Environmental Contaminants: Assessment and Control.* Elsevier Academic Press, Burlington, MA.

How does this process track with biotechnologies? For chemicals released into the environment, it is the same. Thus, the risk assessor can use published physical and chemical hazard characteristics for all of the chemicals used in the life cycle of a biotechnology.

The process is useful, but may not be completely applicable to the genetically modified organisms themselves. For example, if a microbe is harmful to a particular type of cell (e.g. a nerve) before and after genetic modification, it may follow the steps just as a neurotoxic chemical. However, if the microbial modifications change microbial populations in an organism or in an ecosystem, the dose-response may become much more complex than a single, abiotic chemical hazard.

The curves in Figure 5.6 represent those generally found for toxic chemicals [29]. Once a substance is suspected of being toxic, the extent and quantification of that hazard is assessed [30]. This step is frequently referred to as a dose-response evaluation because this is when researchers study the relationship between the mass or concentration (i.e. dose) and the damage caused (i.e. response). Many dose-response studies are ascertained from animal studies (*in vivo* toxicological studies), but they may also be inferred from studies of human populations (epidemiology). To some degree, "Petri dish" (i.e. *in vitro)* studies, such as mutagenicity studies like the Ames test [31] of bacteria complement dose-response assessments, but are mainly used for screening and qualitative or, at best, semi-quantitative analysis of responses to substances. The actual name of the test is the "Ames *Salmonella*/ microsome mutagenicity assay" which shows the short-term reverse mutation in histidine dependent *Salmonella* strains of bacteria. Its main use is to screen for a broad range of chemicals that induce genetic aberrations leading to genetic mutations. The process works by using a culture that allows colony formation only by those bacteria whose genes revert to histidine interdependence. As a mutagenic chemical is added to the culture, a biological gradient can usually be determined. That is, the greater the amount of the chemical that is added, the greater the number of microbes, and the larger the size of colonies on the plate. The test is widely used to screen for mutagenicity of new or modified chemicals and mixtures. It is also a "red flag" for carcinogenicity, since cancer is a genetic disease and a manifestation of mutations.

The toxicity criteria include both acute and chronic effects, and include both human and ecosystem effects. These criteria can be quantitative. For example, a manufacturer of a new chemical may have to show that there are no toxic effects in fish exposed to concentrations below 10 mg L^{-1}. If fish show effects at 9 mg L^{-1}, the new chemical would be considered to be toxic.

A contaminant is acutely toxic if it can cause damage with only a few doses. Chronic toxicity occurs when a person or ecosystem is exposed to a contaminant over a protracted period of time, with repeated exposures. The essential indication of toxicity is the dose-response curve. The curves in Figure 5.6 are sigmoidal because toxicity is often concentration-dependent. As the doses increase the response cannot mathematically stay linear (e.g. the toxic effect cannot double with each doubling of the dose). So, the toxic effect continues to increase, but at a decreasing rate (i.e. decreasing slope). Curve A is the classic cancer dose-response, i.e. any amount of exposure to a cancer-causing agent may result in an expression of cancer at the cellular level (i.e. no safe level of exposure). Thus, the curve intercepts the x-axis at 0.

Curve B is the classic noncancer dose-response curve. The steepness of the three curves represents the potency or severity of the toxicity. For example, Curve B is steeper than Curve A, so the adverse outcome (disease) caused by chemical in Curve B is more potent than that of the chemical in Curve A. Obviously, potency is only one factor in the risk. For example, a chemical may be very potent in its ability to elicit a rather innocuous effect, like a headache, and another chemical may have a rather gentle slope (lower potency) for a dreaded disease like cancer.

With increasing potency, the range of response decreases. In other words, as shown in Figure 5.7, a severe response represented by a steep curve will be manifested in greater mortality or morbidity over a smaller range of dose. For example, an acutely toxic contaminant's dose that kills 50% of test animals (i.e. the LD_{50}) is closer to the dose that kills only 5% (LD_5) and the dose that kills 95% (LD_{95}) of the animals. The dose difference of a less acutely toxic contaminant will cover a broader range, with the differences between the LD_{50} and LD_5 and LD_{95} being more extended than that of the more acutely toxic substance.

The major differentiation of toxicity is between carcinogenic and noncancer outcomes. The term "noncancer" is commonly used to distinguish cancer outcomes (e.g. bladder cancer, leukemia, or adenocarcinoma of the lung) from other maladies, such as neurotoxicity, immune system disorders, and endocrine disruption. The policies of many regulatory agencies and international organizations treat cancer differently than noncancer effects, particularly in how the dose-response curves are drawn. As we saw in the dose-response curves, there is no safe dose for carcinogens. Cancer dose-response is almost always a non-threshold curve, i.e. no safe dose is expected while, theoretically at least, noncancer outcomes can have a dose below which the adverse outcomes do not present themselves. So, for all other diseases safe doses of compounds can be established. These are known as reference doses (RfD), usually based on the oral exposure route. If the substance is an air pollutant, the safe dose is known as the reference concentration (RfC), which is calculated in the same manner as the RfD, using units that apply to air (e.g. $\mu g\, m^{-3}$). These references are calculated from thresholds below which no adverse effect is observed in animal and human studies. If the models and data were perfect, the safe level would be the threshold, known as the no observed adverse effect level (NOAEL).

The term "noncancer" has a completely different meaning than the term "anticancer" or "anticarcinogens." Anticancer procedures include radiation and drugs are those that are used to attack tumor cells. Anticarcinogens are chemical substances that work against the processes that lead to cancer, such as antioxidants, and essential substances that help the body's immune, hormonal, and other systems to prevent carcinogenesis.

In the real world, any hazard identification or dose-response research is never perfect and so the data derived from these investigations are often beset with various forms of uncertainty. Chief reasons for this uncertainty include variability among the animals and people being tested, as well as differences in response to the compound by different species (e.g. one species may have decreased adrenal gland activity, while another may show thyroid effects). Whereas this is usually associated with chemical risk, these uncertainties can also be part of microbial data sets. For example, certain immunocompromised subpopulations may respond adversely to microbial exposures that are below thresholds for the general population.

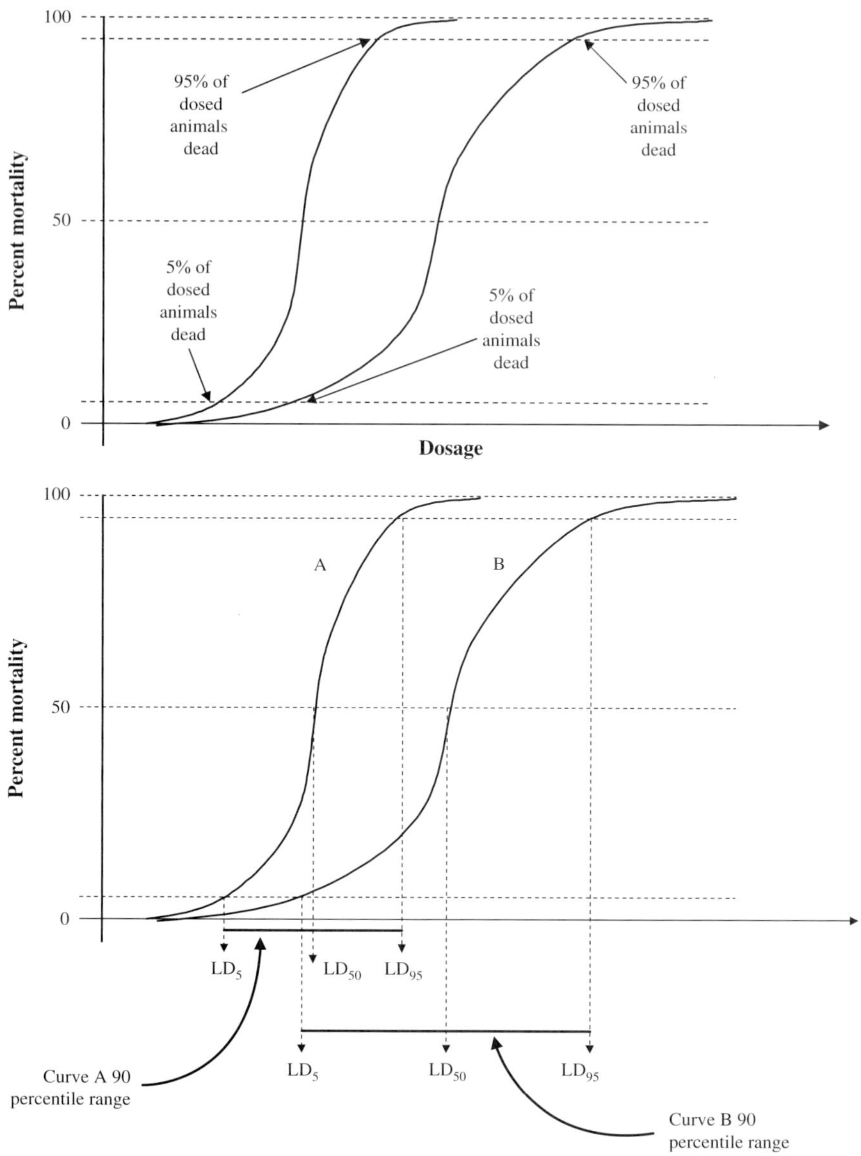

FIGURE 5.7
The greater the potency or severity of response (i.e. steepness of the slope) of dose-response curve the smaller the range of toxic response (90 percentile range shown in bottom graph). Also, note that both curves have thresholds and that curve B is less acutely toxic based upon all three reported lethal doses (LD_5, LD_{50}, and LD_{95}). In fact, the LD_5 for curve A is nearly the same as the LD_{50} for curve B, meaning that at about the same dose, contaminant A kills nearly half the test animals, but contaminant B has only killed 5%. Thus, contaminant A is much more acutely toxic.
Source: D.A. Vallero (2004). *Environmental Contaminants: Assessment and Control.* Elsevier Academic Press, Burlington, MA.

Sometimes, studies only indicate the lowest concentration of a contaminant that causes the effect, i.e. the lowest observed adverse effect level (LOAEL), but the NOAEL is unknown. If the LOAEL is used, one is less certain how close this is to a safe level where no effect is expected. Often, there is temporal incongruence, such as most of the studies taking place in a shorter timeframe than in the real world. Thus, in lieu of long-term human studies, hazards and risks may have to be extrapolated from acute or subchronic studies of the same or similar agents. Likewise, routes and pathways of exposure used to administer the agent to subjects may differ from the likely real world exposures. For example, if the dose of substance in a research study is administered orally, but the pollutant is more likely to be inhaled by humans, this route-to-route extrapolation adds uncertainty. This is particularly problematic for microbial exposures, e.g. inhalational anthrax (*Bacillus anthracis*) is more virulent in human populations than is ingestional anthrax.

Finally, the hazard and exposure data themselves may be weak because the studies from which they have been gathered lack sufficient quality or the precision, accuracy, completeness, and representativeness, or they may not be directly relevant to the risk assessment at hand.

The factors underlying the uncertainties are quantified as specific uncertainty factors (UFs). The uncertainties in the RfD are largely due to the differences between results found in animal testing and expected outcomes in human population. As in other bioengineering operations, a factor of safety must be added to calculations to account for UFs. So, for environmental risk analyses and assessments, the safe level is expressed in the RfD, or in air, the RfC. This is the dose or concentration below which regulatory agencies do not expect a specific unacceptable outcome. Thus, all the uncertainty factors adjust the actual measured levels of no effect (i.e. the threshold values, e.g. NOAELs and LOAELs) in the direction of a zero concentration. This is calculated as:

$$\mathrm{RfD} = \frac{\mathrm{NOAEL}}{\mathrm{UF_{inter}} \times \mathrm{UF_{intra}} \times \mathrm{UF_{other}}} \tag{5.5}$$

The first of the three types of uncertainty is that resulting from the difference between the species tested and *Homo sapiens* (UF_{inter}). Humans may be more or less sensitive than the tested species to a particular compound. The second uncertainty factor is associated with the fact that certain human subpopulations are more sensitive to the effects of a compound than the general human population. These are known as intraspecies uncertainty factors (UF_{intra}). The third type of uncertainty (UF_{other}) results when the available data and science are lacking, such as when a LOAEL is used rather than a NOAEL. That is, data show a dose at which an effect is observed, but the "no effect" threshold has to be extrapolated. Since the UFs are in the denominator, the greater the uncertainties, the closer the safe level (i.e. the RfD) is to zero, i.e. the threshold is divided by these factors. The UFs are usually multiples of 10, although the UF_{other} can range from 2 to 10.

A particularly sensitive subpopulation is children, since they are growing and tissue development is much more prolific than in older years. To address these sensitivities, the Food Quality Protection Act (FQPA) now includes what is known as the "10×" rule. This rule requires that the RfD for products regulated under FQPS, e.g. pesticides, must include an additional factor of 10 for protection of infants, children, and females between the ages of 13 and 50 years (see Figure 5.8). This factor is included in the RfD denominator along with the other three UF values. The RfD that includes the UFs and the 10× protection is known as the population adjusted dose (PAD). A risk estimate that is less than 100% of the acute or chronic PAD does not exceed the Agency's risk concern.

An example of the use of an RfD as a factor of safety can be demonstrated by the US Environmental Protection Agency's decision making regarding the re-registration of the

FIGURE 5.8
The Food Quality Protection Act requires an added protection factor of 10 when calculating factors of safety (reference doses and concentrations) for infants and children, and females of childbearing age.
Photo by the author.

organophosphate pesticide chlorpyrifos. The acute dietary scenario had a NOAEL of 0.5 mg kg^{-1} day^{-1} and the three UF values equaled 100. Thus, the acute RfD = 5×10^{-3} mg kg^{-1} day^{-1} but the more protective acute PAD 5×10^{-4}mg kg^{-1} day^{-1}. The chronic dietary scenario is even more protective, since the exposure is long term. The chronic NOAEL was found to be 0.03 mg kg^{-1} day^{-1}. Thus, the chronic RfD for chlorpyrifos = 3×10^{-4}mg kg^{-1} $day^{-1}$and the more protective acute PAD $\times$ 10^{-5}mg kg^{-1} day^{-1}. Therefore, had the NOAEL threshold been used alone without the safety adjustment of the RfD, the allowable exposure would have been three orders of magnitude higher [32].

Uncertainty can also come from error. Two errors can occur when information is interpreted in the absence of sound science. The first is the *false negative*, or reporting that there is no problem when one in fact exists. The need to address this problem is often at the core of the positions taken by environmental and public health agencies and advocacy groups. They ask questions like:

- What if a biosensor's level of detection is above that needed to show that a contained microbial population is actually being released?
- What if the leak detector registers zero, but in fact toxic substances are being released from the tank?
- What if this substance really does cause cancer but the tests are unreliable?
- What if people are being exposed to a contaminant, but via a pathway other than the ones being studied?
- What if there is a relationship that is different from the laboratory when this substance is released into the "real world," such as the difference between how a chemical behaves in the human body by itself as opposed to when other chemicals are present (i.e. the problem of "complex mixtures")?

The other concern is, conversely, the *false positive*. This can be a major challenge for public health agencies with the mandate to protect people from exposures to environmental contaminants. For example, what if previous evidence shows that an agency had listed a compound as a potential endocrine disruptor, only to find that a wealth of new information is now showing that it has no such effect? This can happen if the conclusions were based upon faulty models, or models that only work well for lower organisms, but subsequently developed models have taken into consideration the physical, chemical, and biological complexities of higher-level organisms, including humans. False positives may force public health officials to devote inordinate amounts of time and resources to deal with so-called "non-problems." False positives also erroneously scare people about an actually or potentially useful product, which can lead to avoidance behaviors (e.g. not using an efficacious drug due to false positives about side effects). False positives, especially when they occur frequently, create credibility gaps between engineers and scientists and the decision makers. In turn the public, those whom we have been charged to protect, lose confidence in us as professionals.

Both false negatives and false positives are rooted in science. Therefore, environmental risk assessment is in need of high quality, scientifically based information. Put in engineering language, the risk assessment process is a "critical path" in which any unacceptable error or uncertainty along the way will decrease the quality of the risk assessment and, quite likely, will lead to a bad environmental decision.

EXPOSURE ESTIMATION

Hazard identification and dose-response comprise the hazard (H) in Eq. 5.4, so next we must consider the second part of the risk equation, exposure (E). An exposure is any contact with an agent. For chemical and biological agents this contact can come about from a number of exposure pathways, i.e. routes taken by a substance, beginning with its source to its endpoint (i.e. a target organ, like the liver, or a location short of that, such as in fat tissues).

Exposure results from sequential and parallel processes in the environment, from release to environmental partitioning to movement through pathways to uptake and fate in the organism (see Figure 5.9). The substances often change to other chemical species as a result of the body's metabolic and detoxification processes. Certainly, genetic modifications can affect such processes. New substances, known as degradation products or metabolites, are produced as cells using the parent compounds as food and energy sources. These metabolic processes, such as hydrolysis and oxidation, are the mechanisms by which chemicals are broken down.

Physical agents, such as electromagnetic radiation, ultraviolet (UV) light, and noise, do not follow this pathway exactly. The contact with these sources of energy can elicit a physiological response that may generate endogenous chemical changes that behave somewhat like the metabolites. For example, UV light may infiltrate and damage skin cells. The UV light helps to promote skin-tumor promotion by activating the transcription factor complex activator protein-1 (AP-1) and enhancing the expression of the gene that produces the enzyme cyclooxygenase-2 (*COX2*). Noise, i.e. acoustical energy, can also elicit physiological responses that affect an organism's chemical messaging systems, i.e. endocrine, immune, and neural. It is possible that genetically modified organisms will respond differently to these physical agents than their unmodified counterparts.

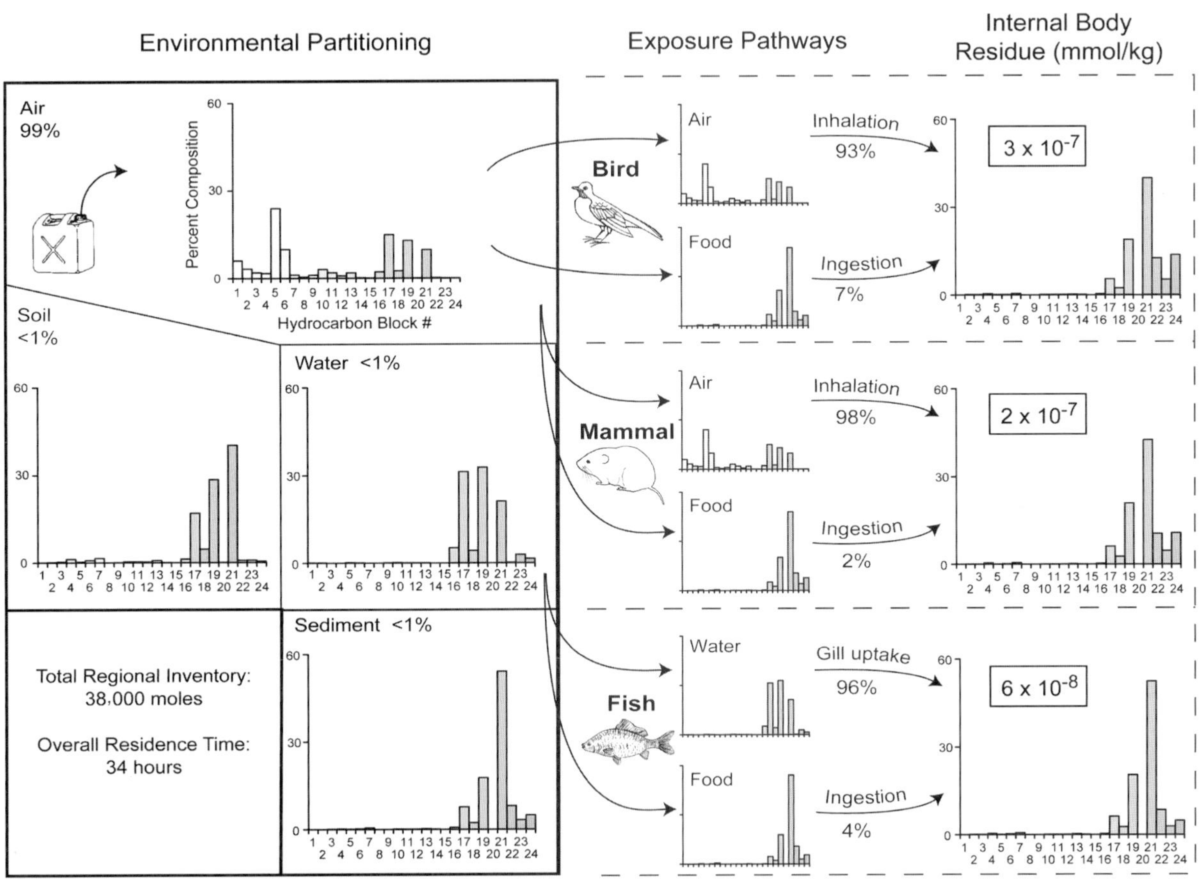

FIGURE 5.9
Processes leading to organismal uptake and fate of chemical and biological agents after release into the environment. In this instance, the predominant sources are air emissions, and predominant pathway of exposure is inhalation. However, due to deposition to surface waters and the agent's affinity for sediment, the ingestion pathways are also important. Dermal pathways, in this case, do not constitute a large fraction of potential exposure.
Source: T. McKone, R. Maddalena, W. Riley, R. Rosenbaum and D. Vallero (2006). Significance of partitioning, kinetics, and uptake at biological exchange surfaces. International Conference on Environmental Epidemiology & Exposure Analysis, Paris, France.

The exposure pathway also includes the manner in which humans and other organisms can come into contact with (i.e. be exposed to) the agent. The pathway has five parts:

The source of contamination (e.g. release from a bioreactor).
An environmental medium and transport mechanism (e.g. pharmaceutical substrate or soil with water moving through it).
A point of exposure (such as a well used for drinking water).
A route of exposure (e.g. inhalation, dietary ingestion, nondietary ingestion, dermal contact, and nasal).
A receptor population (those who are actually exposed or who are where there is a potential for exposure).

If all five parts are present, the exposure pathway is known as a completed exposure pathway. In addition, the exposure may be short-term, intermediate, or long-term. Short-term contact is known as an acute exposure, i.e. occurring as a single event or for only a short period of time (up to 14 days). An intermediate exposure is one that lasts from 14 days to less than one year. Long-term or chronic exposures are greater than one year in duration.

Uncertainty in exposure assessment can increase with increasing scale. For example, determining the exposure for a neighborhood can be more complicated than assessing exposure for each individual in that neighborhood. Even if we do a good job identifying all of the contaminants of concern and its possible source (no small task), we may have little idea of the extent to which the receptor population has come into contact with these contaminants (steps 2 through 4). Thus, assessing exposure involves not only the physical sciences, but the social sciences, e.g. psychology and behavioral sciences. People's activities greatly affect the amount and type of exposures. That is why exposure scientists use a number of techniques to establish activity patterns, such as asking potentially exposed individuals to keep diaries, videotaping, and using telemetry to monitor vital information, e.g. heart and ventilation rates.

General ambient measurements, such as air pollution monitoring equipment located throughout cities, are often not good indicators of actual population exposures. For example, metals and their compounds comprise the greatest mass of toxic substances released into the US environment. This is largely due to the large volume and surface areas involved in metal extraction and refining operations. However, this does not necessarily mean that more people will be exposed at higher concentrations or more frequently to these compounds than to others. A substance that is released, or even if it resides in the ambient environment, is not tantamount to its coming in contact with a *receptor*. Conversely, even a small amount of a substance under the right circumstances can lead to very high levels of exposure (e.g. in biotechnology settings, handling raw materials, and residues for bioreactors).

A recent study by the Lawrence Berkley Laboratory demonstrates the importance of not simply assuming that the released or even background concentrations are a good indicator of actual exposure [33]. The researchers were interested in how sorption may affect microenvironments, so they set up a chamber constructed of typical building materials and furnished with actual furniture like that found in most residential settings. A number of air pollutants were released into the room and monitored (see Figure 5.10). With the chamber initially sealed, the observed decay of xylene, a volatile organic compound, in vapor phase concentrations results from adsorption onto surfaces (walls, furniture, etc.). The adsorption continues for hours, with xylene concentrations reaching a quasi-steady state. At this point the chamber was flushed with clean air to free the vapor phase xylene. The xylene concentrations shortly after the flush began to rise again until reaching a new steady state. This rise must be the result of desorption of the previously sorbed xylene, since the initial source is gone. Sorption is one of the biochemodynamic processes that must be considered to account for differences in the temporal pattern of microenvironmental (e.g. occupational) and ambient concentrations.

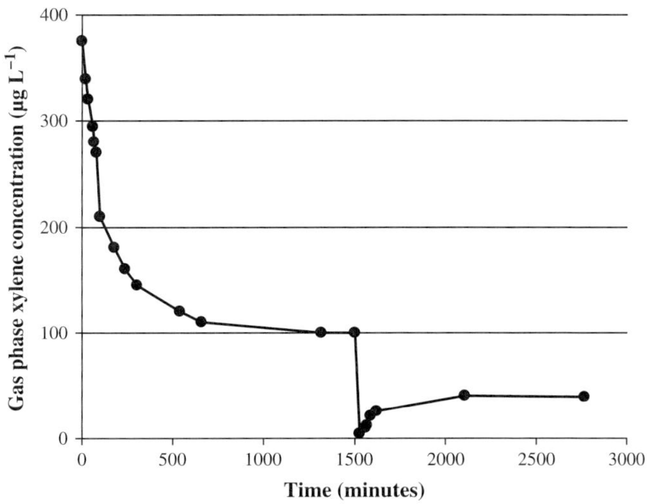

FIGURE 5.10
Concentrations of xylene measured in its vapor phase in a chamber sealed during adsorption and desorption periods. *Source:* Adapted from B. Singer (2003). A tool to predict exposure to hazardous air pollutants. *Environmental Energy Technologies Division News* 4(4): 5.

The simplest quantitative expression of exposure is:

$$E = D/t \tag{5.6}$$

where: E = human exposure during the time period, t (units of concentration [mass per volume] per time) (mg kg^{-1} day)

D = mass of pollutant per body mass (mg kg^{-1})

t = time (day)

Usually, to obtain D, the chemical concentration of a pollutant is measured near the interface of the person and the environment, during a specified time period. This measurement is sometimes referred to as the potential dose (i.e., the chemical has not yet crossed the boundary into the body, but is present where it may enter the person, such as on the skin, at the mouth, or at the nose).

Expressed quantitatively, exposure is a function of the concentration of the agent and time. It is an expression of the magnitude and duration of the contact. That is, exposure to a contaminant is the concentration of that contact in a medium integrated over the time of contact:

$$E = \int_{t=t_1}^{t=t_2} C(t)dt \tag{5.7}$$

where: E = exposure during the time period from t_1 to t_2 , and

$C(t)$ = concentration at the interface between the organism and the environment, at time t.

The concentration at the interface is the potential dose (i.e., the agent has not yet crossed the boundary into the body, but is present where it may enter the receptor). Since the amount of a chemical agent that penetrates from the ambient atmosphere into a control volume affects the concentration term of the exposure equation, a complete mass balance of the contaminant must be understood and accounted for; otherwise exposure estimates will be incorrect. Recall that the mass balance consists of all inputs and outputs, as well as chemical changes to the contaminant:

$$\begin{aligned}\text{Accumulation or loss of contaminant A} = {} & \text{Mass of A transported in} \\ & - \text{Mass of A transported out} \pm \text{Reactions}\end{aligned} \tag{5.8}$$

The reactions may be either those that generate substance A (i.e. *sources*), or those that destroy substance A (i.e. *sinks*). Thus, the amount of mass transported in is the inflow to the system that includes pollutant discharges, transfer from other control volumes and other media (for example, if the control volume is soil, the water and air may contribute mass of chemical A), and formation of chemical A by abiotic chemistry and biological transformation. Conversely, the outflow is the mass transported out of the control volume, which includes uptake, by biota, transfer to other compartments (e.g. volatilization to the atmosphere) and abiotic and biological degradation of chemical A. This means the rate of change of mass in a control volume is equal to the rate of chemical A transported in less the rate of chemical A transported out, plus the rate of production from sources, and minus the rate of elimination by sinks. Stated as a differential equation, the rate of change contaminant A is:

$$\frac{d[A]}{dt} = -v\cdot\frac{d[A]}{dx} + \frac{d}{dx}\left(\Gamma\cdot\frac{d[A]}{dx}\right) + r \qquad (5.9)$$

where: v = fluid velocity

Γ = a rate constant specific to the environmental medium

$d[A]$ = concentration gradient of chemical A

r = internal sinks and sources within the control volume

Reactive compounds can be particularly difficult to measure. For example, many volatile organic compounds in the air can be measured by, first, collection in stainless steel canisters and then analysis by chromatography in the lab. However, some of these compounds, like the carbonyls (notably aldehydes like formaldehyde and acetaldehyde) are prone to react inside the canister, meaning that by the time the sample is analyzed a portion of the carbonyls are degraded (under-reported). Therefore, when sampling highly reactive compounds, other methods should be employed. A common alternative is to trap the compounds with dinitrophenyl hydrazine (DNPH)-treated silica gel tubes that are frozen until being extracted for chromatographic analysis. The combination of suspension in the gel and lower temperatures greatly reduces chemical reactivity. After all, the purpose of the measurement is to see what is in the air, water, soil, sediment, or biota at the time of sampling, so any reactions before the analysis increase measurement error. No matter the sensitivity of sophisticated separation and detection equipment (e.g. gas or liquid chromatography and mass-spectroscopy, respectively), the sampling error can completely detract from the quality of the analysis.

Remember that the chemical that is released may or may not be what the engineer measures. If the released chemical is reactive, some or all of it may have changed into another form (i.e. *speciated*) by the time it is measured. Even relatively non-reactive compounds may speciate between when the sample is collected (e.g. in a water sample, an air canister, a soil core, or a bag) and when the sample is analyzed. In fact, each contaminant has unique characteristics that vary according to the type of media in which it exists, and extrinsic conditions like temperature and pressure. Sample preservation and holding times for the anions according to EPA Method 300.1, Determination of Inorganic Anions in Drinking Water by Ion Chromatography, are shown in Table 5.4. These methods vary according to the contaminant of concern and the environmental medium from which it is collected, so the engineer needs to find and follow the correct methods.

The general exposure Eq. 5.7 is rewritten to address each route of exposure, accounting for chemical concentration and the activities that affect the time of contact. The exposure calculated from these equations is actually the chemical intake (I) in units of concentration (mass per volume or mass per mass) per time, such as mg kg^{-1} day^{-1}:

$$I = \frac{C\cdot CR\cdot EF\cdot ED\cdot AF}{BW\cdot AT} \qquad (5.10)$$

Table 5.4 Preservation and holding times for anion sampling and analysis

PART A : Common anions		
Analyte	Preservation	Holding Time
Bromide	None required	28 days
Chloride	None required	28 days
Fluoride	None required	28 days
Nitrate-N	Cool to 4 °C	48 hours
Nitrite-N	Cool to 4 °C	48 hours
ortho-Phosphate-P	Cool to 4 °C	48 hours
Sulfate	Cool to 4 °C	28 days
PART B : Inorganic disinfection by-products		
Analyte	Preservation	Holding Time
Bromate	50 mg L^{-1} EDA	28 days
Bromide	None required	28 days
Chlorate	50 mg L^{-1} EDA	28 days
Chlorite	50 mg L^{-1} EDA, Cool to 4 °C	14 days

Source: US Environmental Protection Agency, 1997, EPA Method 300.1: Determination of Inorganic Anions in Drinking Water by Ion Chromatography, Revision 1.0.

where: C = chemical concentration of contaminant (mass per volume)

CR = contact rate (mass per time)

EF = exposure frequency (number of events, dimensionless)

ED = exposure duration (time)

These factors are further specified for each route of exposure, such as the lifetime average daily dose (LADD), as shown in Table 5.5. The LADD is obviously based on a chronic, long-term exposure.

Acute and subchronic exposures require different equations, since the exposure duration (ED) is much shorter. For example, instead of LADD, acute exposures to noncarcinogens may use maximum daily dose (MDD) to calculate exposure (see Discussion Box). However, even these exposures follow the general model given in Eq. 5.11.

DISCUSSION BOX

Exposure Calculation

In the process of synthesizing pesticides over an 18-year period, a polymer manufacturer has contaminated the soil on its property with vinyl chloride. The plant closed two years ago but vinyl chloride vapors continue to reach the neighborhood surrounding the plant at an average concentration of 1 mg m^{-3}. Assume that people are breathing at a ventilation rate of 0.5 m^3 h^{-1} (about the average of adult males and females over 18 years of age [34]). The legal settlement allows neighboring residents to evacuate and sell their homes to the company. However, they may also stay. The neighbors have asked for advice on whether to stay or leave, since they have already been exposed for 20 years.

(Continued)

Table 5.5 Equations for calculating lifetime average daily dose (LADD) for various routes of exposure

Route of exposure	Equation LADD (in mg kg^{-1} d^{-1}) =	Definitions
Inhaling aerosols (particulate matter)	$\frac{(C)\cdot(PC)\cdot(IR)\cdot(RF)\cdot(EL)\cdot(AF)\cdot(ED)\cdot(10^{-6})}{(BW)\cdot(TL)}$	*C* = concentration of the contaminant on the aerosol/particle (mg kg^{-1}) *PC* = particle concentration in air (g m^{-3}) *IR* = inhalation rate (m^{-3} h^{-1}) *RF* = respirable fraction of total particulates (dimensionless, usually determined by aerodynamic diameters, e.g. 2.5 μm) *EL* = exposure length (h d^{-1}) *ED* = duration of exposure (d) *AF* = absorption factor (dimensionless) *BW* = body weight (kg) *TL* = typical lifetime (d). 10^{-6} is a conversion factor (kg to mg)
Inhaling vapor phase contaminants	$\frac{(C)\cdot(IR)\cdot(EL)\cdot(AF)\cdot(ED)}{(BW)\cdot(TL)}$	*C* = concentration of the contaminant in the gas phase (mg m^{-3}) Other variables the same as above
Drinking water	$\frac{(C)\cdot(CR)\cdot(ED)\cdot(AF)}{(BW)\cdot(TL)}$	*C* = concentration of the contaminant in the drinking water (mg L^{-1}) *CR* = rate of water consumption (L d^{-1}) *ED* = duration of exposure (d) *AF* = portion (fraction) of the ingested contaminant that is physiologically absorbed (dimensionless) Other variables are the same as above
Contact with soil-borne contaminants	$\frac{(C)\cdot(SA)\cdot(BF)\cdot(FC)\cdot(SDF)\cdot(ED)\cdot(10^{-6})}{(BW)\cdot(TL)}$	*C* = concentration of the contaminant in the soil (mg kg^{-1}) *SA* = skin surface area exposed (cm^{-2}) *BF* = bioavailability (percent of contaminant absorbed per day) *FC* = fraction of total soil from contaminated source (dimensionless) *SDF* = soil deposition, the mass of soil deposited per unit area of skin surface (mg cm^{-1} d^{-1}) Other variables are the same as above

Source: M. Derelanko (1999). Risk assessment. In: M.J. Derelanko and M.A. Hollinger (Eds), *CRC Handbook of Toxicology*. CRC Press, Boca Raton, FL.

Vinyl chloride is highly volatile, so its phase distribution will be mainly in the gas phase rather than the aerosol phase. Although some of the vinyl chloride may be sorbed to particles, we will use only the vapor phase LADD equation, since the particle phase is likely to be relatively small. Also, we will assume that outdoor concentrations are the exposure concentrations. This is unlikely, however, since people spend very little time outdoors compared to indoors, so this may provide an additional factor of safety. To determine how much vinyl chloride penetrates living quarters, indoor air studies would have to be conducted. For a scientist to compare exposures, indoor air measurements should be taken.

Find the appropriate equation in Table 5.5 and insert values for each variable. Absorption rates are published by the EPA and the Oak Ridge National Laboratory's Risk Assessment Information System (http://risk.lsd.ornl.gov/cgi-bin/tox/TOX_select?select=nrad). Vinyl chloride is well absorbed, so we can assume that AF = 1. We will also assume that the person stays in the neighborhood, is exposed at the average concentration 24 hours a day (EL = 24), and that a person lives the remainder of their entire typical lifetime exposed at the measured concentration.

Although the ambient concentrations of vinyl chloride may have been higher when the plant was operating, the only measurements we have are those taken recently. Thus, this is an area of uncertainty that must be discussed with the clients. The common default value for a lifetime is 70 years, so we can assume the longest exposure would be is 70 years (25,550 days). Table 5.6 gives some of the commonly used default values in exposure assessments. If the person is now 20 years of age and has already been exposed for that time, and lives the remaining 50 years exposed at 1 mg m^{-3}:

$$\text{LADD} = \frac{(C)\cdot(IR)\cdot(EL)\cdot(AF)\cdot(ED)}{(BW)\cdot(TL)} = \frac{(1)\cdot(0.5)\cdot(24)\cdot(1)\cdot(25550)}{(70)\cdot(25550)} = 0.2 \text{ mg kg}^{-1}\text{ day}^{-1}$$

If the 20 year old leaves today, the exposure duration would be for the 20 years that the person lived in the neighborhood. Thus, only the ED term would change, i.e. from 25,550 days to 7300 days (i.e. 20 years).

Thus, the LADD falls to 2/7 of its value:

$$\text{LADD} = 0.05 \text{ mg kg}^{-1}\text{ day}^{-1}$$

Table 5.6 Commonly used human exposure factors [35]

Exposure factor	Adult male	Adult female	Child (3–12 years of age) [36]
Body weight (kg)	70	60	15–40
Total fluids ingested (L d^{-1})	2	1.4	1.0
Surface area of skin, without clothing (m^2)	1.8	1.6	0.9
Surface area of skin, wearing clothes (m^2)	0.1–0.3	0.1–0.3	0.05–0.15
Respiration/ventilation rate (L min^{-1}) – Resting	7.5	6.0	5.0
Respiration/ventilation rate (L min^{-1}) – light activity	20	19	13
Volume of air breathed (m^3 d^{-1})	23	21	15
Typical lifetime (years)	70	70	NA
National upper-bound time (90th percentile) at one residence (years)	30	30	NA
National median time (50th percentile) at one residence (years)	9	9	NA

Sources: US Environmental Protection Agency (2003). *Exposure Factor Handbook*; and Agency for Toxic Substances and Disease Registry (2003). ATSDR Public Health Assessment Guidance Manual.

Once the hazard and exposure calculations are done, risks can be characterized risk quantitatively. There are two general ways that such risk characterizations are used in environmental problem solving, i.e. direct risk assessments and risk-based cleanup standards.

DIRECT BIOENGINEERING RISK CALCULATIONS

In its simplest form, risk is the product of the hazard and exposure, but assumptions can greatly affect risk estimates. For example, cancer risk can be defined as the theoretical probability of contracting cancer when continually exposed for a lifetime (e.g. 70 years) to a given concentration of a substance (carcinogen). The probability is usually calculated as an upper confidence limit. The maximum estimated risk may be presented as the number of chances in a million of contracting cancer.

Two measures of risk are commonly reported. One is the individual risk, i.e. the probability of a person developing an adverse effect (e.g. cancer) due to the exposure. This is often reported as a "residual" or increased probability above background. For example, if we want to characterize the contribution of all the power plants in the US to increased cancer incidence, the risk above background would be reported. The second way that risk is reported is population risk, i.e. the annual excess number of cancers in an exposed population. The maximum individual risk might be calculated from exposure estimates based upon a "maximum exposed individual" or MEI. The hypothetical MEI lives an entire lifetime outdoors at the point where pollutant concentrations are highest. Assumptions about exposure will greatly affect the risk estimates. For example, the cancer risk from power plants in the US has been estimated to be 100- to 1000-fold lower for an average exposed individual than that calculated for the MEI.

For cancer risk assessments, the hazard is generally assumed to be the slope factor and the long-term exposure is the lifetime average daily dose:

$$\text{Cancer risk} = SF \times LADD \tag{5.11}$$

Therefore, cancer risk can be calculated if the exposure (LADD) and potency (slope factor) are known (see Discussion Box).

DISCUSSION BOX

Cancer Risk Calculation

Using the lifetime average daily dose value from the vinyl chloride exposure calculation in the previous section, estimate the direct risk to the people living near the abandoned polymer plant. What information needs to be communicated?

Insert the calculated LADD values and the vinyl chloride inhalation slope factor of 3.00×10^{-1} from Appendix 2. The two LADD values under consideration, the cancer risk to the neighborhood exposed for an entire lifetime (exposure duration = 70 years) gives us $0.2\ \text{mg kg}^{-1}\ \text{day}^{-1} \times 0.3\ (\text{mg kg}^{-1}\ \text{day}^{-1})^{-1} = 0.06$. This is an incredibly high risk! The threshold for concern is often 1 in a million (0.000001), while this is a probability of 6%.

Even at the shorter duration period (20 years of exposure instead of 70 years), the risk is calculated as $0.05 \times 0.3 = 0.017$ or nearly a 2% risk. The combination of a very steep slope factor and very high lifetime exposures leads to a very high risk. Vinyl chloride is a liver carcinogen, so unless corrective actions significantly lower the ambient concentrations of vinyl chloride the prudent course of action is that the neighbors accept the buyout and leave the area.

Incidentally, vinyl chloride has relatively high water solubility and can be absorbed to soil particles, so ingestion of drinking water (e.g. people on private wells drawing water from groundwater that has been contaminated) and dermal exposures (e.g. children playing in the soil) are also conceivable. The total risk from a single contaminant like vinyl chloride is equal to the sum of risks from all pathways (e.g. vinyl chloride in the air, water and soil):

$$\text{Total risk} = \sum \text{risks from all exposure pathways} \tag{5.12}$$

Requirements and measures of success are seldom if ever as straightforward as the vinyl chloride example. In fact, the bioengineer would be ethically remiss if the only advice given is to the local community, i.e. whether or not to accept the buyout. Of course, one of the engineering canons is to serve as a "faithful agent" to the clientele. However, the first engineering canon is to hold paramount the health and safety of the public. Thus, the engineer must balance any proprietary information that the client wants to be protected with the need to protect public health. In this case, the engineer must inform the client and prime contractors, for example, that the regulatory agencies need to know that even if the current neighbors are moving, others, including future populations, are threatened.

In other words, a systematic approach is needed, since the current population may be moved from harm's way but remediation is still likely needed to reduce the vinyl chloride concentrations to acceptable levels. This is because biochemodynamic processes are complex. The contaminant may remain in untreated or poorly treated areas, which may later be released (e.g. future excavation, long-term transport from groundwater to surface or from groundwater to drinking water sources). Thus, post-closure monitoring should be designed and operated based on worst-case scenarios.

The risk of adverse outcome other than cancer (so-called "noncancer risk") is generally called the "hazard quotient" (HQ). It is calculated by dividing the maximum daily dose (MDD) by the acceptable daily intake (ADI):

$$\text{Noncancer risk} = \text{HQ} = \frac{\text{MDD}}{\text{ADI}} = \frac{\text{Exposure}}{\text{RfD}} \tag{5.13}$$

Note that this is an index, not a probability, so it is really an indication of relative risk. If the noncancer risk is greater than 1, the potential risk may be significant, and if the noncancer risk is less than 1, the noncancer risk may be considered to be insignificant (see Discussion Box). Thus, the reference dose, RfD, is one type of ADI.

DISCUSSION BOX

Noncancer Risk Calculation

The chromic acid (Cr^{6+}) mists dermal chronic RfD of 6.00×10^{-3} mg kg^{-1} day^{-1}. If the actual dermal exposure of people living near a metal processing plant is calculated (e.g. by intake or LADD) to be 4.00×10^{-3} mg kg^{-1} day^{-1}, calculate the hazard quotient for the noncancer risk of the chromic acid mist to the neighborhood near the plant and interpret the meanings.

From Eq. 5.17:

$$\frac{\text{Exposure}}{\text{RfD}} = \frac{4.00 \times 10^{-3}}{6.00 \times 10^{-3}} = 0.67.$$

Since this is less than 1, one would not expect people chronically exposed at this level to show adverse effects from skin contact. However, at this same chronic exposure, i.e. 4.00×10^{-3} mg kg^{-1} day^{-1}, to hexavalent chromic acid mists via oral route, the RfD is 3.00×10^{-3} mg kg^{-1} day^{-1}, meaning the HQ = 4/3 or 1.3. The value is greater than 1, so we cannot rule out adverse noncancer effects.

If a population is exposed to more than one contaminant, the hazard index (HI) can be used to express the level of cumulative noncancer risk from pollutants 1 through n:

$$\text{HI} = \sum\nolimits_1^n \text{HQ} \tag{5.14}$$

The HI is useful in comparing risks at various locations, e.g. benzene risks in St Louis, Cleveland, and Los Angeles. It can also give the cumulative (additive) risk in a single population exposed to more than one contaminant. For example, if the HQ for benzene is 0.2 (not significant), toluene is 0.5 (not significant), and tetrachloromethane is 0.4 (not significant), the cumulative risk of the three contaminants is 1.1 (potentially significant).

It is desirable to have realistic estimates of the hazard and the exposures in such calculations. However, precaution is the watchword for risk. Estimations of both hazard (toxicity) and exposure are often worst-case scenarios, because the risk calculations can have large uncertainties. Models usually assume effects to occur even at very low doses. Human data are usually gathered from epidemiological studies that, no matter how well they are designed, are fraught with error and variability (science must be balanced with the rights and respect of subjects, populations change, activities may be missed, and confounding variables are ever present). Uncertainties exist in every phase of risk assessment, from the quality of data, to limitations and assumptions in models, to natural variability in environments and populations.

RISK-BASED CLEANUP STANDARDS

For most of the second half of the 20th century, environmental protection was based on two types of controls, i.e. technology-based and quality-based. Technology-based controls are set according to what is "achievable" from the current state of the science and engineering. These are feasibility-based standards. The Clean Air Act has called for "best achievable control technologies (BACT), and more recently for maximally achievable control technologies (MACT). Both standards reflect the reality that even though from an air quality standpoint it would be best to have extremely low levels of pollutants, technologies are not available or are not sufficiently reliable to reach these levels. Requiring unproven or unreliable technologies can even exacerbate the pollution, such as in the early days of wet scrubbers on coal-fired power plants. Theoretically, the removal of sulfur dioxide could be accomplished by venting the power plant flue through a slurry of carbonate, but the technology at the time was unproven and unreliable, allowing all-too-frequent releases of untreated emissions while the slurry systems were being repaired. Selecting a new technology over older proven techniques is unwise if the tradeoff of the benefit of improved treatment over older methods is outweighed by the numerous failures (i.e. no treatment).

Technology-based standards are a part of most environmental programs. Wastewater treatment, groundwater remediation, soil cleaning, sediment reclamation, drinking water supply, air emission controls, and hazardous waste site cleanup all are in part determined by availability and feasibility of control technologies (see Discussion Box: Treatment by Genetic Modification).

DISCUSSION BOX

Treatment by Genetic Modification

Bioengineers consider disinfection to be the primary means of inactivating and destroying pathogens in water. The application of ultraviolet (UV) light is increasingly used for disinfection of drinking water and for treating wastewater. This is accomplished by transferring energy from the UV lamp to the pathogenic microbe's genetic material (DNA and RNA). The appropriate wavelength ranges of UV penetrate the microbe's cell wall, but do not kill it. Rather, the damage to the strands of nucleic acid does not allow the microbe to reproduce properly, so the organism is deactivated [37].

The efficiency of UV disinfection is a function of the water's physicochemical characteristics (see Table 5.7), the intensity of the UV radiation, exposure time for the microbe and the configuration of the reactor.

In effect, the energy from the UV lamp is sufficiently strong and targeted toward the genetic material, thereby unzipping the DNA molecule and preventing the DNA from being rezipped (repaired). Such a repair could take place in the presence of visible light, so a requirement of disinfection is to keep the UV-treated water in the dark for a sufficient period of time to prevent reactivation of the microbe or its oocysts. In other words, at least until the threshold for deactivation is reached, the microbe has been genetically modified. As the bases rejoin to form a new DNA molecule, if the microbe were to survive, it is now a GMO, by definition.

Questions

Is it possible that some of the UV-treated microbes may undergo DNA repair with a resulting transcription of unknown and possibly harmful traits?

Are there differences between the likelihood of survival of microbes for wastewater versus drinking water treatment?

Table 5.7 Physicochemical characteristics impacting ultraviolet disinfection performance

Wastewater Characteristic	Effects on UV Disinfection
Ammonia	Minor effect, if any
Nitrite	Minor effect, if any
Nitrate	Minor effect, if any
Biochemical oxygen demand (BOD)	Minor effect, if any. Although, if a large portion of the BOD is humic and/or unsaturated (or conjugated) compounds, then UV transmittance may be diminished.
Hardness	Affects solubility of metals that can absorb UV light. Can lead to the precipitation of carbonates on quartz tubes.
Humic materials, Iron	High absorbency of UV radiation.
pH	Affects solubility of metals and carbonates.
TSS	Absorbs UV radiation and shields embedded bacteria.

Source: US Environmental Protection Agency (1999). Wastewater Technology Fact Sheet: Ultraviolet Disinfection. Report No. EPA 832-F-99-064.

Quality-based controls are those that are required to ensure that an environmental resource is in good enough condition to support a particular use. For example, a stream may need to be improved so that people can swim in it and so that it can be a source of water supply. Certain streams may need higher levels of protection than others, such as the so-called "wild and scenic rivers." The parameters will vary, but usually include minimum levels of dissolved oxygen and maximum levels of contaminants. The same goes for air quality, where ambient air quality must be achieved, so that concentrations of contaminants listed as National Ambient Air Quality Standards, as well as certain toxic pollutants, are below levels established to protect health and welfare.

Environmental protection in the United States was spearheaded by the US Environmental Protection Agency, created in 1970 and led during its formative years by William Ruckelshaus. Ruckelshaus saw the need for "risk-based" environmental standards and recognized that such standards would receive public support. Risk-based approaches to environmental protection, especially contaminant target concentrations, are designed to require engineering controls and preventive measures to ensure that risks are not exceeded. The risk-based approach actually embodies elements of both technology-based and quality-based standards. The technology assessment helps determine how realistic it will be to meet certain contaminant concentrations, while the quality of the environment sets the goals and means to achieve cleanup. Engineers are often asked, "How clean is clean?" When do we know that we have done a sufficient job of cleaning up a spill or hazardous waste site? It is often not possible to have nondetectable concentrations of a pollutant. Commonly, the threshold for cancer risk to a population is 1 in a million excess cancers. However, one may find that the contaminant is so difficult to remove, that we almost give up on dealing with the contamination and put in measures to prevent exposures, i.e. fencing the area in and prohibiting access. This is often done as a first step in remediation, but is unsatisfying and controversial (and usually politically and legally unacceptable). Thus, even if costs are high and technology unreliable, the engineer must find suitable and creative ways to clean up the mess and meet risk-based standards.

Risk-based target concentrations can be calculated by solving for the target contaminant concentration in the exposure and risk equations. Since risk is the hazard (e.g. slope factor) times the exposure (e.g. LADD), a cancer-risk-based cleanup standard can be found by enumerating the exposure equations (Eqs. 5.6 and 5.7) within the risk equation (in this instance, the drinking water equation from Table 5.5) gives:

$$\text{Risk} = \frac{\text{C}\cdot\text{CR}\cdot\text{EF}\cdot\text{ED}\cdot\text{AF}\cdot\text{SF}}{\text{BW}\cdot\text{AT}} \tag{5.15}$$

and solving for C:

$$\text{C} = \frac{\text{Risk}\cdot\text{BW}\cdot\text{AT}}{\text{CR}\cdot\text{EF}\cdot\text{ED}\cdot\text{AF}\cdot\text{SF}} \tag{5.16}$$

This is the target concentration for each contaminant needed to protect the population from the specified risk, e.g. 10^{-6}. In other words, this is the concentration that must not be exceeded in order to protect a population having an average body weight and over a specified averaging time from an exposure of certain duration and frequency that leads to a risk of 1 in a million. While one-in-a-million added risk is a commonly used benchmark, cleanup may not always be required to achieve this level. For example, if a site is considered to be a "removal" action, i.e. the principal objective is to get rid of a sufficient amount of contaminated soil to reduce possible exposures, that risk reduction target may be as high as one additional cancer per 10,000 (i.e. 10^{-4}).

The decision regarding the actual cleanup level, including whether to approach a contaminated site or facility as a removal or a remedial action, is not risk assessment, but falls within the province of risk management. The bioengineer will have input to the decision, but will not be the only party in that decision. The risk assessment data and information will comprise much of the scientific underpinning of the decision, but legal, economic and other societal drivers will also be considered to arrive at cleanup levels. It is not unusual, for example, for a legal document, such as a consent decree, to prescribe cleanup levels more protective than typical removal or remedial levels.

DISCUSSION BOX

Risk-Based Contaminant Cleanup

A well is the principal water supply for the town of Apple Chill. A study has found that the well contains 80 mg L^{-1} tetrachloromethane (CCl_4). Assuming that the average adult in the town drinks 2 L day^{-1} of water from the well and lives in the town for an entire lifetime, what is the lifetime cancer risk to the population if no treatment is added? What concentration is needed to ensure that the population cancer risk is below 10^{-6}?

The lifetime cancer risk added to Apple Chill's population can be estimated using the LADD and slope factor for CCl_4. In addition to the assumptions given, we will use default values from Table 5.6. We will also assume that people live in the town for their entire lifetimes, and that their exposure duration is equal to their typical lifetime.

Thus, ED and TL terms cancel, leaving the abbreviated LADD =

$$\frac{(\text{C})\cdot(\text{CR})\cdot(\text{AF})}{(\text{BW})}$$

Since we have not specified male or female adults, we will use the average body weight, assuming that there are about the same number of males as females. We look up the absorption factor for CCl_4 and find that it is 0.85, so the adult lifetime exposure is:

$$\text{LADD} = \frac{(80)\cdot(2)\cdot(0.85)}{(65)} = 4.2\ \text{mg kg}^{-1}\ \text{day}^{-1}$$

Using the midpoint value between the default values $\left(\frac{15+40}{2} = 27.5\text{kg}\right)$ for body weight and default CR values (1 L day^{-1}), the children lifetime exposure is:

$\text{LADD} = \frac{(80)\cdot(1)\cdot(0.85)}{(27.5)} = 2.5$ mg kg^{-1} day^{-1} for the first 13 years, and the adult exposure of 4.2 mg kg^{-1} day^{-1} thereafter.

The oral SF for CCl_4 is 1.30×10^{-1} kg day^{-1}, so the added adult lifetime risk from drinking the water is:

$$4.2 \times (1.30 \times 10^{-1}) = 5.5 \times 10^{-1}$$

and, the added risk to children is:

$$2.5 \times (1.30 \times 10^{-1}) = 3.3 \times 10^{-1}.$$

Some subpopulations are more vulnerable to exposures than others. For example, for children, environmental and public health agencies recommend an additional factor of safety beyond what would be used to calculate risks for adults. This is known as the "10×" rule, i.e. children need to be protected 10 times more than adults because they have longer life expectancies (so latency periods for cancer need to be accounted for), and their tissue is developing prolifically and changing. So, in this case, with the added risk, our reported "risk" would be 3.3. While this is statistically impossible (i.e. one cannot have a probability greater than one because it would mean that the outcome is more than 100% likely, which of course is impossible!), it is actually an adjustment to the cleanup concentration. Since the combination of a very high slope of the dose-response curve and a very high LADD increases the risk, children need a measure of protection beyond the general population. This is accomplished either by removing the contaminants from the water or the provision of a new water supply. In any event, the city public works and/or health department should mandate another source of drinking water (e.g. bottled water) immediately.

The cleanup of the water supply to achieve risks below 1 in a million can be calculated from the same information and reordering the risk equation to solve for C:

$$\text{Risk} = \text{LADD} \times \text{SF}$$

$$\text{Risk} = \frac{(\text{C})\cdot(\text{CR})\cdot(\text{AF})\cdot(\text{SF})}{(\text{BW})}$$

$$\text{C} = \frac{(\text{BW})}{(\text{CR})\cdot(\text{AF})\cdot(\text{SF})\cdot\text{Risk}}$$

Based on adult LADD, the well water must be treated so that the tetrachloromethane concentrations are below:

$$\text{C} = \frac{(65)\cdot 10^{-6}}{(2)\cdot(0.85)\cdot(0.13)} = 2.9 \times 10^{-4}\ \text{mg L}^{-1} = 290\ \text{ng L}^{-1}$$

Based on children's LADD, and the additional "10×," the well water must be treated so that the tetrachloromethane concentrations are below:

$$\text{C} = \frac{(27.5)\cdot 10^{-7}}{(1)\cdot(0.85)\cdot(0.13)} = 2.5 \times 10^{-5}\ \text{mg L}^{-1} = 25\ \text{ng L}^{-1}$$

The town must remove the contaminant so that the concentration of CCl_4 in the finished water will be at a level six orders of magnitude less than the untreated well water, i.e. lowered from 80 mg L^{-1} to 25 ng L^{-1}.

Cleanup standards are part of the arsenal needed to manage risks. However, other considerations need to be given to a contaminated site, such as how to monitor the progress in lowering levels and how to ensure that the community stays engaged and is participating in the cleanup actions, where appropriate. Even when the engineering solutions are working well, the engineer must allot sufficient time and effort to these other activities, otherwise skepticism and distrust can arise.

Some general principles that have been almost universally adopted by regulatory agencies, especially those concerned with cancer risks from environmental exposures, are shown in Table 5.8.

Biotechnological risks are difficult to ascertain due to the numerous areas of uncertainty. Many of the risks are not directly tied to a single agent. For example, the release of an organism whose genetic material has been modified may carry downstream risks that are impossible to quantify. The risks to public health, such as cancer and noncancer endpoints, are only one class of outcomes of concern. Others include ecosystem changes from gene flow and other ecological endpoints and opportunity risks. The latter risks include those posed by not allowing biological pesticides that may be safer than abiotically-derived pesticides.

Another consideration is that the overall risk of a biotechnological operation may be safer when compared to the risks posed by its abiotic counterpart. For example, an *in situ* biotreatment process may have less potential of the release of toxic raw materials since the contaminated substrate does not have to be moved. This would result in a decreased likelihood of exposures to chemical toxins. Also, microbial processes may not require the addition of toxic chemicals as part of the treatment process. Additions usually are limited to nutrients, oxygen, and water. In the case of *bioaugmentation*, microbes will also be added.

One of the most difficult risks to quantify is the opportunity risk. That is, how can the loss of the opportunity that a biotechnology affords be compared to the risk it poses? For example, if regulatory agencies are overly cautious about approving a cancer drug that may entail the introduction of carcinogens from possible environmental releases, the risk assessment may indicate that the overall cancer risk is actually improved if the cancers that are prevented and treated with the new drug are part of the risk calculation. Usually, however, it is not that straightforward. For example, what if the cancer risk that is being treated by the drug is relatively benign (e.g. basal cell skin cancer) compared to the more virulent form from the pollutants introduced from the bioreactor (e.g. pancreatic cancer)? Or, what if the outcomes are ecological versus public health? The risk decision and management process must consider these tradeoffs.

Just from the arithmetic, it should be noted that zero risk can be calculated when either the hazard (e.g. toxicity) does not exist or when the exposure to that hazard is zero. A substance found to be associated with cancers based upon animal testing or observations of human populations can be further characterized. Association of two factors, such as the level of exposure to a compound and the occurrence of a disease, does not necessarily mean that one necessarily "causes" the other. Often, after study, a third variable explains the relationship. However, it is important for science to do what it can to link causes with effects. Otherwise, corrective and preventive actions cannot be identified. So, strength of association is a beginning step toward cause and effect. A major consideration in strength of association is the application of sound technical judgment of the weight of evidence. For example, characterizing the weight of evidence for carcinogenicity in humans consists of three major steps [38]:

characterization of the evidence from human studies and from animal studies individually;
combination of the characterizations of these two types of data to show the overall weight of evidence for human carcinogenicity; and,
evaluation of all supporting information to determine if the overall weight of evidence should be changed.

Note that none of these steps is absolutely certain.

Risk information must be presented in a meaningful way, avoiding without overextending the interpretation of the data (see Discussion Box: Biotechnological Communications). A common overextension is to assign a cause when merely an association exists between two factors. However, if all we can say is that the variables are associated, the potentially affected public is likely to want to know more. In particular, the community is likely going to want to know more about what a measurement or modeling result means in terms of a known or

Table 5.8 General principles applied to health and environmental risk assessments conducted by regulatory agencies in the United States

Principle	Explanation
Human data are preferable to animal data	For purposes of hazard identification and dose-response evaluation, epidemiological and other human data better predict health effects than animal models
Animal data can be used in lieu of sufficient, meaningful human data	While epidemiological data are preferred, agencies are allowed to extrapolate hazards and to generate dose-response curves from animal models
Animal studies can be used as basis for risk assessment	Risk assessments can be based upon data from the most highly sensitive animal studies
Route of exposure in animal study should be analogous to human routes	Animal studies are best if from the same route of exposure as those in humans, e.g. inhalation, dermal, or ingestion routes. For example, if an air pollutant is being studied in rats, inhalation is a better indicator of effect than if the rats are dosed on the skin or if the exposure is dietary
Threshold is assumed for non-carcinogens	For non-cancer effects, e.g. neurotoxicity, endocrine dysfunction, and immunosuppression, there is assumed to be a safe level under which no effect would occur (e.g. "no observed adverse effect level", NOAEL, which is preferred, but also "lowest observed adverse effect level," LOAEL)
Threshold is calculated as a reference dose or reference concentration (air)	Reference dose (RfD) or concentration (RfC) is the quotient of the threshold (NOAEL) divided by factors of safety (uncertainty factors and modifying factors; each usually multiples of 10): $\mathrm{RfD} = \frac{\mathrm{NOAEL}}{\mathrm{UF} \times \mathrm{MF}}$
Sources of uncertainty must be identified	Uncertainty factors (UFs) address: Inter-individual variability in testing Interspecies extrapolation LOAEL-to-NOAEL extrapolation Subchronic-to-chronic extrapolation Route-to-route extrapolation Data quality (precision, accuracy, completeness, and representativeness) Modifying factors (MFs) address uncertainties that are less explicit than the UFs
Factors of safety can be generalized	The uncertainty and modifying factors should follow certain protocols, e.g. 10 = for extrapolation from a sensitive individual to a population; 10 = rat-to-human extrapolation, 10 = subchronic-to-chronic data extrapolation) and 10 = LOAEL used instead of NOAEL
No threshold is assumed for carcinogens	There is no safe level of exposure is assumed for cancer causing agents
Precautionary principle applied to cancer model	A linear, no-threshold dose-response model is used to estimate cancer effects at low doses, i.e. to draw the unknown part of the dose-response curve from the region of observation (where data are available) to the region of extrapolation
Precautionary principle applied to cancer exposure assessment	The most highly exposed individual is generally used in the risk assessment (upper-bound exposure assumptions). Agencies are reconsidering this worst-case policy, and considering more realistic exposure scenarios

Source: US Environmental Protection Agency (2001). General Principles for Performing Aggregate Exposure and Risk Assessment. Office of Pesticides Programs, Washington, DC.

suspected adverse affect that has been observed in a community. For example, how do the measured lead (Pb) blood levels relate to reported incidences of learning disabilities of children living in the neighborhood?

Linking associated factors to causes was particularly problematic in early cancer research. Possible causes of cancer were being explored and major research efforts were being directed at myriad physical, chemical, and biological agents. So, there needed to be some manner of

Table 5.9 **Factors to be considered in determining whether exposure to a biomaterial elicits an effect based on Hill's criteria for causality**

Criterion	Description
Strength of association	For exposure to cause and effect, the exposure must be associated with that affect. Strong associations provide more certain evidence of causality than is provided by weak associations. Common epidemiological metrics used in association include risk ratio, odds ratio, and standardized mortality ratio
Consistency	If the chemical exposure is associated with an effect consistently under different studies using diverse methods of study of assorted populations under varying circumstances by different investigators, the link to causality is stronger. For example, if the carcinogenic effect of Chemical X is found in mutagenicity studies, mouse and Rhesus monkey experiments, and human epidemiological studies, there is greater consistency between Chemical X and cancer than if only one of these studies showed the effect
Specificity	The specificity criterion holds that the cause should lead to only one disease and that the disease should result from only this single cause. This criterion appears to be based in the germ theory of microbiology, where a specific strain of bacteria and viruses elicits a specific disease. This is rarely the case in studying most chronic diseases, since a chemical can be associated with cancers in numerous organs, and the same chemical may elicit cancer, hormonal, immunological, and neural dysfunctions
Temporality	Timing of exposure is critical to causality. This criterion requires that exposure to the chemical must precede the effect. For example, in a retrospective study, the researcher must be certain that the manifestation of a disease was not already present before the exposure to the chemical. If the disease were present prior to the exposure, it may not mean that the chemical in question is not a cause, but it does mean that it is not the sole cause of the disease (see Specificity above)
Biologic gradient	This is another essential criterion for chemical risks. In fact, this is known as the ''dose-response'' step in risk assessment. If the level, intensity, duration, or total level of chemical exposure is increased a concomitant, progressive increase should occur in the toxic effect
Biological plausibility	Generally, an association needs to follow a well-defined explanation based on known biological system. However, ''paradigm shifts'' in the understanding of key scientific concepts do change. A noteworthy example is the change in the latter part of the 20th century in the understanding of how the endocrine, immune, neural systems function, from the view that these are exclusive systems to today's perspective that in many ways they constitute an integrated chemical and electrical set of signals in an organism. For example, Candace Pert, a pioneer in endorphin research, has espoused the concept of mind/body, with all the systems interconnected, rather than separate and independent systems
Coherence	The criterion of coherence suggests that all available evidence concerning the natural history and biology of the disease should ''stick together'' (cohere) to form a cohesive whole. By that, the proposed causal relationship should not conflict or contradict information from experimental, laboratory, epidemiologic, theory, or other knowledge sources
Experimentation	Experimental evidence in support of a causal hypothesis may come in the form of community and clinical trials, *in vitro* laboratory experiments, animal models, and natural experiments
Analogy	The term analogy implies a similarity in some respects among things that are otherwise different. It is thus considered one of the weaker forms of evidence

Source: A.B. Hill (1965). The environment and disease: association or causation? *Proceedings of the Royal Society of Medicine, Occupational Medicine* 58: 295.

sorting through findings to see what might be causal and what is more likely to be spurious results. Sir Austin Bradford Hill is credited with articulating key criteria (see Table 5.9) that need to be satisfied to attribute cause and effect in medical research [39].

DISCUSSION BOX

Biotechnical Communications

Bioengineers must produce scientifically sound products and systems in a way that are fair and just. Fairness and justice require inclusion of diverse perspectives, especially of those most directly or indirectly affected by our decisions. Thus, engineers must be able to communicate effectively in order to arrive at adequate designs, to ensure that these technically sound designs are accepted by clients and stakeholder, and to convey sufficient information to users so that the designs are operated and maintained satisfactorily.

Technical communication can be seen as a critical path, where the engineer sends a message and the audience receives it (see Figure 5.11). The means of communication can be either perceptual or interpretive [40]. Perceptual communications are directed toward the senses. Human perceptual communications are similar to those of other animals; that is, we react to sensory information (e.g. reading body language or assigning meaning to gestures, such as a hand held up with palms out, meaning "stop" or smile conveying approval).

Interpretive communications encode messages that require intellectual effort by the receiver to understand the sender's meanings. This type of communication can either be verbal or symbolic. Scientists and engineers draw heavily on symbolic information when communicating amongst themselves. If you have ever mistakenly walked into the seminar where experts are discussing an area of science not familiar to you, using unrecognizable symbols and vernacular, this is an example of symbolic miscommunication. In fact, the experts may be using words and symbols that are used in your area of expertise, but with very different meanings. For example, psychologists speak of "conditioning" with a very different meaning than that of an engineer. The bioengineer and scientist must therefore be aware of the venue of communication to ensure that all stakeholders have a sufficient grasp of the plans, designs, and other aspects of biotechnological operations that will be applied.

In assessing risks, some of Hill's criteria are more important than others. Risk assessments rely heavily on strength of association, e.g. to establish dose-response relationships. Coherence is also very important. Animal and human data should not be extensions of one another and should not disagree. Biological gradient is crucial, since this is the basis for dose-response (the more dose, the greater the biological response).

Temporality is crucial to all scientific research, i.e. the cause must precede the effect. However, this is sometimes difficult to see in some instances, such as when the exposures to suspected agents have been continuous for decades and the health data are only recently available.

The key is that sound bioengineering and scientific judgment, based on the best available and most reliable data, be used to estimate risks. Linking cause and effect is often difficult in environmental matters. It is best for information to be transparent and coherent about the data, including the uncertainties.

Environmental risk by its nature addresses unwanted outcomes. Risk characterization is the stage where the bioengineer pulls together the necessary assumptions, describes the scientific uncertainties, and determines the strengths and limitations of the analyses. The risks are articulated by integrating the analytical results, interpreting adverse outcomes, and describing the uncertainties and weights of evidence.

As mentioned, risk assessment is a process distinct from risk management, where actions are taken to address and reduce the risks. But the two are deeply interrelated and require continuous feedback with each other. Engineers are key players in both efforts.

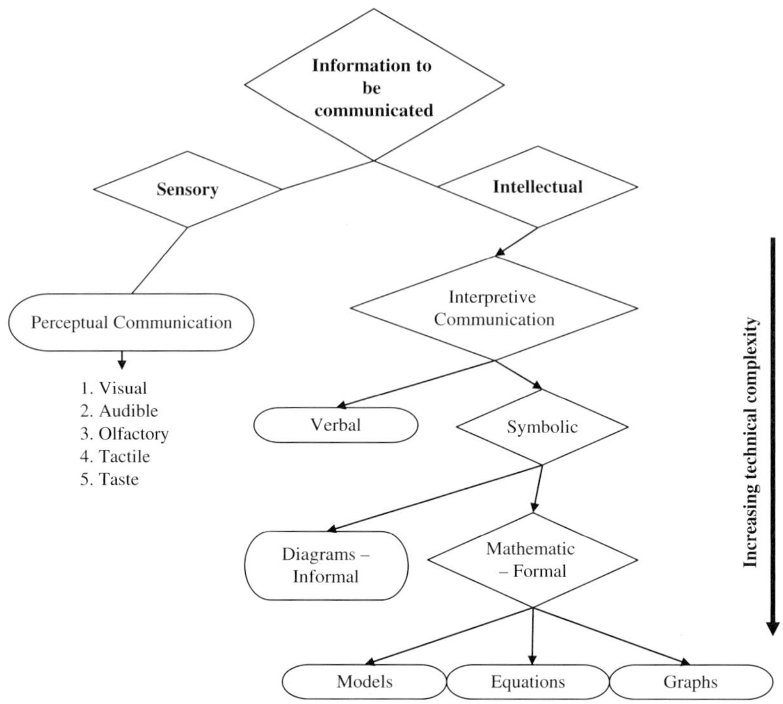

FIGURE 5.11
Risk communication techniques. All humans use perceptual communication, such as observing body language of an engineer or smelling an animal feeding operation. The right side of the figure is the domain of technical communication. Thus, the public may be overwhelmed by perceptive cues or may not understand the symbolic, interpretive language being used by a bioengineer and others in the risk communication process. Thus, the type of communication in a scientific briefing is quite different from that of a public meeting or a briefing for a neighborhood group potentially affected by a risk management decision.
Source: adapted from M. Myers and A. Kaposi (2004). *The First Systems Book: Technology and Management*, 2nd Edition. Imperial College Press, London, UK; and T.R.G. Green (1989). Cognitive dimensions of notations. In: A. Sutcliffe and L. Macaulay (Eds), *People and Computers V*. Cambridge University Press, Cambridge, UK.

Bioengineers need tools to support the risk analysis, but none or all of these tools provide a surefire answer to biotechnological risks. They are too complicated and complex for simple risk calculations. Certainly, components of biotechnologies can indeed be evaluated for their human and ecosystem risks. For example, many of the solvents and other chemicals used in the purification and other steps in bioreactors have been studied sufficiently so that they have well-defined dose-response curves and other hazard metrics.

There are numerous ways to evaluate biotechnological performance. Does it "work" (effectiveness)? Is it the best way to reach the end for which we strive (efficiency)? If it works and if it is the best means of providing the outcome, what is the probability of benefit from the biotechnology applied for a societal (medical, industrial, agricultural or environmental) problem under prescribed conditions. For example, the efficacy of a medical technology is often assessed using case-control studies, wherein controlled trials of an experimental therapy are compared to a control (e.g. case = biotechnology; control = placebo) [41].

Engineers add a few steps. We must consider whether or not the technology will likely continue to "work" (reliability) and, further, we must consider the hazards that can come about as the new technology is used. Risk is a function of likelihood that the hazard will in fact be encountered, so we must also try to predict the adverse implications that society might face (risk characterization). Thus, the "risk" associated with a biotechnology refers to the possibility and likelihood of undesirable and possibly harmful effects. Errors in risk prediction can range from not foreseeing outcomes that are merely annoying (e.g. genetically modified grain that has a color different from the natural form), to those that are devastating (e.g. the release of carcinogens into the environment) [42].

A classic example of such a failure, albeit not a biotechnological one, was the decision making related to the Ford Pinto, a subcompact car produced by Ford Motor Company between 1971 and 1980. The car's fuel tank was poorly placed in such a way that increased the probably of a fire from fuel spillage by a rear collision. A confluence of events, including the poor design, the likelihood of rear-end crashes, and lack of public knowledge of the risk, resulted in injuries and fatalities. The ensuing adverse implications manifested in the series of injuries that resulted from the defect could have been predicted at some level of accuracy. Perhaps they were, and dismissed or overrun by opposing viewpoints in the boardroom. If so, this is an example of detrimental single-mindedness. Financial considerations appear to have trumped safety and health.

Often, the problem is not a question of right versus wrong, but of one right versus another right. In other words, society needs the benefit of the technology, so the *status quo* is not acceptable. However, the benefits must outweigh the risks. That is the challenge of biotechnological risk assessment.

Frankly, one of the frustrations in writing this book is the difficulty in quantifying biotechnological risks. Rather, we often have to settle on semi-quantitative or qualitative risk assessments when we look to the future. Retrospective failure analyses can be more quantitative, since there are forensic techniques available to tease out and assign weights to the factors that led to the outcome. There are even methods to calculate outcomes had other steps been taken.

Much of what we know about the engineering errors associated with today's technology is a manifestation of lessons learned for centuries. If one were to consider this concept of "risk" approximately four thousand years ago in the kingdom of Babylon, the concept takes on a different meaning. Here, the professional "risk" associated with engineering transcended the failures of the product itself, extending to the physical well-being of the engineer himself and his family. In Babylon, should the engineer face failure, his and his family's lives were at stake. Obviously, such enforcement was quite different from today's protocols. This demonstrates that the concept of risk has been evolving and continues to evolve, although for the past century, professional expectations have stabilized compared to earlier times.

SEMINAR TOPIC

Assessing the Risks of Green Transgenesis

Biological science, technology, and engineering do not always proceed along predictable and step-wise paths. In fact, scientists' penchant for linearity is not usually respected by natural systems. Adaptation in these systems can be quite surprising. Consider, for example, our colorful world. Few would tolerate a monochromatic lifestyle. Our products, our gardens, our artwork, our labels, our brochures, and our clothing create a tapestry of colors.

Henry Ford famously quipped that customers could have any color car they wanted, so long as it was black. Nearly a century later, there have been suggestions that new car customers may have any color they want, so long as it is *not* black. Recently, the State of California was reported to be considering banning black paint on new cars because darker paints absorb more incoming solar radiation than lighter-colored paints, meaning that more infrared radiation (heat) is released to the atmosphere. The idea seems to have been dropped for now, probably because it violates the American sanctity of the automobile. In fact, the automobile for some generations has come to represent freedom in many forms, e.g. movement, style, consumer choice.

Color is simply a function of electromagnetic radiation, in this case the visible light spectrum. Arguably, Sir Isaac Newton's short essay entitled "New Theory about Light and Colors" that was published in 1672, and his full-length treatise of 1704, *Opticks*, were seminal works that presaged the advent of modern science in the Renaissance. Thus, color was the subject of Newton's first written work, predating even his masterpiece *Philosophiae Naturalis Principia Mathematica* by 15 years. Newton's numerous breakthroughs in optics include light dispersion using glass prisms [43].

(Continued)

Modern societies have been afforded a kaleidoscope of products chiefly by way of large organic molecules and metallic dyes. Ironically and tragically, a present or abandoned site where commercial dyes have been produced or used can be contaminated with organic xenobiotics, like naphthalene, toluene, xylene, and other petroleum distillate solvents, as well metallic compounds, such as potassium dichromate and the colorful metals, including cadmium and cobalt (for the colors red and blue, respectively).

The challenges for the engineer to date have been how to remove these contaminants from the waste stream during coloration, pigmentation, and dyeing operations, or cleaning up waste sites left behind by these operations. In addition, the coloring-related contaminants have found their way to sanitary landfills and waste combustors, allowing them to re-enter the environment as parent compounds or degradation products.

The risk assessments have been straightforward. The risk is calculated as a function of the amount of these organic and metallic compounds that comes into contact with the receptor population (i.e. the population exposure) and the hazard of each compound (e.g. carcinogenicity, neurotoxicity, etc.).

Waste minimization and pollution prevention can help reduce or even eliminate the use of many of the chemicals used for coloration [44]. We need new manufacturing processes. Some of the answer can come by emulating nature.

But, what if a more sustainable approach to colors could be achieved and what if biotechnology was a means of providing it? Consider the brightly colored birds, and how their plumage can provide intense colorations that are important for mating and other biological functions such as camouflage.

One of the canons of green engineering and sustainable design is avoid being trapped by old paradigms, especially if they are entrenched in fossil fuel dependence, wastefulness and thoughtlessness of what occurs after a product's useful life. Old and new paradigms clash, for example, in how we humans, with our existing technologies, would color the Blue Jay (*Cyanocitta cristata*). In the old paradigm, we well may choose a proven technology, e.g. the application of an aniline dye, a heavy metal or even some cyanide-laden pigment source to obtain the blue plumage. Is that how the Creator decided to color the jay? Pull a feather from the Blue Jay and look at it under a microscope. You will find that the feather lacks any pigment, and some feathers are even transparent! In fact, you could pluck the poor bird bald and look at each "blue" feather and find no color blue. What gives?

In a natural system of optimization, the use of heavy elements and large organic molecules can be a disadvantage. It certainly could dramatically increase the bird's weight, so this would be disadvantageous for flight and survival, especially when another means of obtaining colors is available. So, the bird's plumage does not rely on differential absorption and reflection (i.e. the cobalt compound or blue dye absorbs all other light wavelengths, but reflects in the blue part of the visible light spectrum). These colors come from the feathers' individual "microstructures" made of keratin and air bubbles that create diffraction or scattering (Tyndall effect or Raleigh scattering) of light or by interferences [45] of light by microscopic contours and shapes in the feather. The blue feathers result from differences in the distances traveled by light waves that are reflected. The diffracted light returns the blue color to the mate's (and our) eyes.

And the Blue Jay is not the only bird that uses microstructures to be blue, purple, or of course green. Unlike humans, most birds can see well into the ultraviolet range of the light spectrum, so they possess efficient means of producing colors at the violet to UV range, without pigmentation.

Can we apply Newton's seminal discussion of optics to discover breakthroughs in coloring technology? It could even pave the way for lucrative and more environmentally friendly businesses. Who knows? We may all be wearing transparent, microstructured clothing, using colorful building materials with no pigments and driving cars with microstructure coatings in the near future. We may not have pigments, but we will be as colorful as the Blue Jay. And, when we finally wear out that old shirt, or throw out the box for our new High Definition TV, or junk the old car, we won't have to worry about toxic dyes in the environment. Better yet, what if we designed for disassembly so that we don't have to throw things out at all? What if the function that is driving our design (form) eliminated problems completely long before manufacturing or construction and use?

As we look to natural systems for ways to use colors, we may find some unintended consequences. For example, in our quest to achieve colors, we may try genetic engineering. For example, the author participated in a seminar some years back where the presenter shared the idea of using transgenic fish as indicators of pollution. He was working on selecting genes that change an organism's color in the presence of toxic substances, i.e. a type of biological spectrograph. So, a water supply would pass water by caged fish before being distributed throughout the community. If the water turns red, it may have high levels of unsubstituted aromatics, purple if certain levels of arsenic are present, and so on. This seems attractive as a biosensor system.

What if these colors were available in other species. By inserting the genetic material into our favorite fish, could we not design bioindicator, glowing fish. This is now what is known as a *chimera*, a creature of Greek mythology. It had a serpent's tail, a lion's head, and a goat's body. Chimeric creatures also included the minotaur, which was a combination of a man and bull, and the faun that combined human and goat features. The term has regained usage in modern times, representing organisms that have been genetically engineered by insertion of genetic material from one species into another. The genes of the two species that produce the chimera do not combine within the organism. Instead, the cells of a chimera are a "mosaic of cells of different species" [46].

Transgenic plants provide a special case for risk assessment. Herbicide-resistant crops (HRCs), also known as herbicide-tolerant crops, become resistant to herbicides either directly from genetic modification via transgene technology or by selection in cell or tissue culture for mutations that confer resistance [47]. Risks of HRCs are often associated with glyphosate-resistant crops (GRCs) [48].

The HRCs have been considered sustainable alternatives to commercial herbicide use (which presents classic chemical risks discussed in this chapter). However, the extent to which GRCs have changed herbicides is being debated. Glyphosate is normally often used at rates of a kilogram or more per hectare. Conversely, chemical herbicides, e.g. carfentrazone-ethyl, can be effectively applied at a rate of 10–50 g per ha [49]. Comparisons of herbicide application rates are not necessarily good indicators of environmental and human health risks, since few investigations compare actual exposures, but use the surrogate of the pesticide's active ingredient used per unit area. This is a common problem for all pesticide exposure and risk studies, not just those associated with biotechnologies.

A recent review concluded that the risks associated with gene flow from glyphosate resistance transgenes, particularly the introgression into wild populations, is likely to be minimal, since these expressions offer no advantage in the absence of glyphosate. Conversely, when glyphosate resistance transgenes are linked with genes that provide a fitness advantage in a natural habitat, gene flow could be exacerbated by the elimination of competing plants of the hybrid by the herbicide [50]. Over the long term, this could be the greatest risk of GRCs. Biotechnological research will need to keep advancing to find ways to prevent or mitigate these risks.

Similarly, pathogen-derived resistance (PDR) is a genetic modification technique to control plant viruses. It has been widely used for numerous benefits, against which risks, including environmental risks, must be weighed. Some of the risks are heteroencapsidation, recombination, synergism, gene flow, effect on nontarget organisms, and allergenicity [51]. Heteroencapsidation refers to the encapsidation of the genome of one virus by the coat protein (CP) of a different virus, e.g. in plants that have been infected by more than one virus. Heteroencapsidation also could result from the CP subunits expressed by the transgenic plant rather than from the second viral genome (see Figure 5.12). Because the CP can carry genetic information for pathogenicity and vector specificity, etc., the properties of viruses in transgenic plants may be altered, such as when a virus that does not usually serve as a vector becomes transmissible through heteroencapsidation in a transgenic plant. In addition, a virus could infect an otherwise nonhost plant via heteroencapsidation, followed by vector-mediated transmission. Theoretically, new virus epidemics could result from heteroencapsidation [52].

The question for risk assessors is whether this is sufficient information from which to make sound science-based risk decisions, and if not what are the gaps that need to be filled? In PDR, transgenic plants containing genes or sequences of a parasite can become protected from pathogens that are detrimental to a plant during its life. PDR to plant viruses was first demonstrated in tobacco [53] and tomato [54] plants expressing the coat protein (CP) gene of TMV (*Tobacco mosaic virus*). The agricultural effectiveness of PDR for controlling plant viruses has since been replicated numerous times and is now allowing for resistance against most families of plant viruses in numerous agricultural crops [55].

The risks of such transgenesis are subtle and can defy objective quantification. Chimeras raise numerous concerns, especially when they involve human genetic material. For example, such genetic manipulation engenders the fear of the "slippery slope" [56]. The slippery slope occurs when allowing an act leads to other negative options, foreclosing options that would have helped to prevent problems. Thus, the momentum cannot be overcome and matters keep worsening [57]. For example, if human genetic material is inserted in other species, we have crossed an ethical line that many of us oppose, i.e. compromising and destroying the uniqueness and sanctity of humanity.

Other major ethical issues involved in genetic engineering often center on animal welfare, and risks to human health and the environment. For example, what if a new creature is so different in kind that it has such a competitive advantage (i.e. no effective predators) or an ability to self-replicate that it would pose risks to public health and welfare, in violation of the engineer's first ethical canon. And, how many animals' lives are worth an important discovery? Are we decreasing the genetic diversity of our wildlife or destroying the habitats of other animals?

Motivation for transgenesis is often anthropocentric, such as aiding humans by developing treatments to deadly diseases or methods to assist in the creation of tissues and organs. Opposition includes religious and moral concerns, e.g. the researchers and the biotechnological companies are immorally attempting to "play God" by creating entirely new beings and unnaturally altering the genetic makeup of progeny. Opposition may also be biocentric (concern for other organisms besides us humans) or ecocentric (concern for the systematic impacts to the environment).

This leads to bioethical questions. What impact might research and marketing on these species have on society, such as threats to health? What are the ethical considerations needed on behalf of the creatures themselves? For example, is it ethical to modify a monkey so that its fur glows in the dark? In the case of chimeras, are we not ignoring alarms given by natural systems (e.g. failure of many hybrids to be able to reproduce)?

Mimicking nature can be an effective part of a sustainable design, but biomimicry can be taken too far from an ethical and socially responsible perspective. It becomes a question of optimization, once again. Natural adaptive systems are often an optimization among factors. For example, if heat were the only consideration for designing a polar bear, would its fur be white?

Seminar Questions

Estimate the risks of using transgenic species for a particular environmental purpose (e.g. genetically modified fish to indicate the presence of a contaminant in drinking water).
How might the risk tradeoffs be addressed?

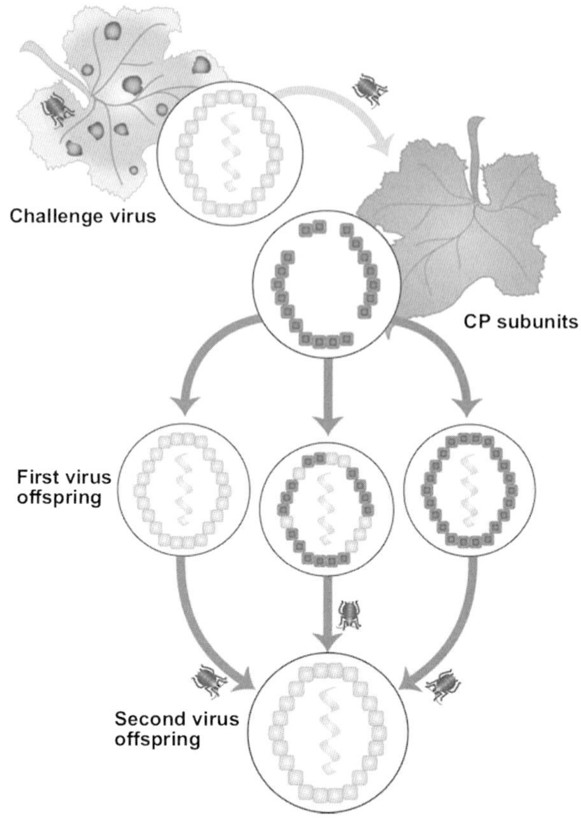

FIGURE 5.12
Heteroencapsidation. In nature, an insect vector acquires a virus (i.e. the *challenge virus*) from an infected plant and transmits it to a transgenic plant expressing a viral CP gene. The genome of the challenge virus can be encapsidated by its own coat protein subunits or those encoded by the transgene (*CP subunits*), either partially or fully (*first virus offspring*). An insect vector can then acquire the newly formed virions and transmit them further into the system. The second virus progeny will be identical to the challenge virus. [*See color plate section*]
Source: M. Fuchs and D. Gonsalves (2007). Decades after their introduction: lessons from realistic field risk assessment studies. *Annual Review of Phytopathology* 45: 173-202.

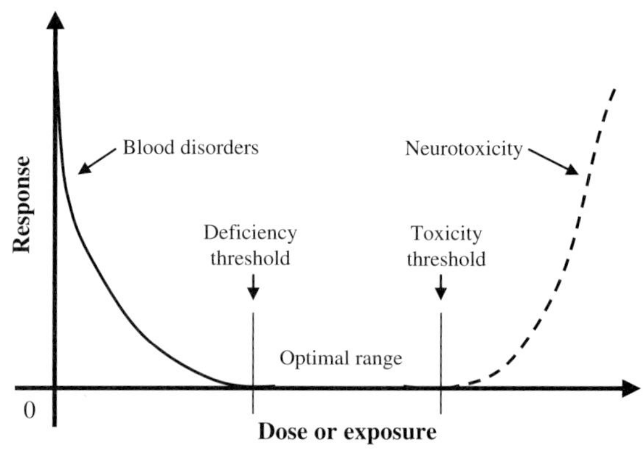

FIGURE 5.13
Dose/exposure-response curve for an essential substance (e.g. a nutrient or vitamin).

REVIEW QUESTIONS

How might assessing the risks of a genetically modified organism differ from conventional chemical risk assessments?

What can biotechnology decision makers learn from the disagreements about the certainties in predicting global climate change?

Describe a possible situation in which a genetically modified organism is physically contained in one system, but the control system being used transfers the genetic material to another medium, such as water or soil, that may or may not be regulated.

What factors influence the accumulation of a chemical in the human body?

Give an example of how communications involving two different scientific disciplines may be unclear regarding biotechnological risks.

What are the major differences in calculating cancer and noncancer risks?

What are the likely exposure pathways for manipulated genetic materials involved in microbial processes? ... in plants? ... in animals? Consider the variability (e.g. by species, by media) and suggest ways that exposure assessments can best be used to predict biotechnological risks.

Give an example of a biological agent that is intrinsically hazardous, another that is not itself hazardous, but that produces toxins that may be hazardous, and another that is both. Construct a risk assessment according to Figures 5.1 and 5.2 for each and discuss uncertainties and gaps in knowledge that must be filled to improve these risk assessments.

NOTES AND COMMENTARY

1. The Safety in Biotechnology Working Party of the European Federation of Biotechnology and O. Doblhoff-Dier (1999). Safe biotechnology 9: Values in risk assessment for the environmental application of microorganisms. *Trends in Biotechnology* 17(8): 307–311.
2. National Academy of Sciences (2002). *Biosolids Applied to Land: Advancing Standards and Practices*. National Academies Press, Washington, DC.
3. Ibid.
4. Ibid.
5. US Environmental Protection Agency (1992). Framework for Ecological Risk Assessment. Report No. EPA/630/R-92/001. Risk Assessment Forum, Washington, DC.
6. National Research Council (2009). *Science and Decisions: Advancing Risk Assessment*. National Academies Press, Washington, DC.
7. The source of this discussion box is: US Environmental Protection Agency (2009). Fact Sheet: Commercialization of *Sinorhizobium (Rhizobium) meliloti*, RMBPC-2; http://www.epa.gov/biotech_rule/pubs/factdft6.htm; accessed October 2, 2009.
8. US Environmental Protection Agency (2003). Risk Assessment Forum. A review of ecological assessment case studies from a risk assessment perspective: Volume II. Report No. EPA/630/R-94/003.
9. Ibid.
10. Ibid.
11. US Code of Federal Regulations (1998). 40 CFR Part 721. *Sinorhizobium meliloti* strain RMBPC—Significant New Use Rule. Final Rule. June 1, 1998.
12. US EPA (2003).
13. The source of these concerns is: Institute of Science in Society (2009). GM microbes invade North America; http://www.i-sis.org.uk/full/GMMINAFull.php; accessed October 3, 2009.
14. Information Systems for Biotechnology (1998). National Biological Impact Assessment Program. *ISB News Report*. May 1998; http://www.nbiap.vt.edu/news/1998/news98.may.html; accessed October 3, 2009.
15. US EPA (2003).
16. J. Morrisey, U. Walsh, A. O'Donnel, Y. Moenne-Laccoz and F. O'Gara (2002). Exploitation of genetically modified inoculants for industrial ecology applications. *Antonie von Leewenhoek* 81: 599–606.
17. A. Hoffman, T. Thimm and C. Tebbe (1999). Fate of plasmid bearing luciferase marker gene tagged bacteria after feeding the soil microarthropod *Onychiurus firmatus* (collembolan). *FEMS Microbiology Ecology* 30: 125–135.
18. Safety in Biotechnology Working Party (1999).
19. C. Mitcham and R.S. Duval (2000). Responsibility in engineering. In: *Engineering Ethics*. Prentice-Hall, Upper Saddle River, NJ, ch. 8.
20. US Environmental Protection Agency (1992). Guidelines for Exposure Assessment. US Environmental Protection Agency, Risk Assessment Forum. Report No. EPA/600/Z-92/001. Washington, DC.
21. Ibid.
22. A. A. Snow, D. A. Andow, P. Gepts, E. M. Hallerman, A. Power, J. M. Tiedje and L. L. Wolfenbarger (2005). Genetically engineered organisms and the environment: Current status and recommendations. *Ecological Applications* 15(2): 377–404.
23. Ibid.
24. E.A. Cohen Hubal, A.M. Richard, S. Imran, J. Gallagher, R. Kavlock, J. Blancato and S. Edwards (2008). Exposure science and the US EPA National Center for Computational Toxicology. *Journal of Exposure Science and Environmental Epidemiology*. doi:10.1038/jes.2008.70 [online: November 5, 2008].

25. S. H. Morris (2006). EU biotech crop regulations and environmental risk: a case of the emperor's new clothes. *Trends in Biotechnology* 24(1): 2–6; and A.J. Conner, T.R. Glare and J.P. Nap (2003). The release of genetically modified crops into the environment. *Plant Journal* 33: 19–46.
26. W. Kaye-Blake, C. Saunders and M. de Aragão Pereira (2008). Potential impacts of biopharming on New Zealand: Results from the Lincoln Trade and Environment Model. Research Report No. 307. Agribusiness and Economics, Lincoln University, Christ Church, NZ.
27. Source: P. Aarne Vesilind, J. Jeffrey Peirce, and Ruth F. Weiner (1993). *Environmental Engineering*, 3rd Edition. Butterworth-Heinemann, Boston, MA.
28. US Environmental Protection Agency (1986). Guidelines for Carcinogen Risk Assessment, Report No. EPA/630/R-00/004. *Federal Register* 51(185):33992–34003, Washington, DC; and R.I. Larsen (2003). An air quality data analysis system for interrelating effects, standards, and needed source reductions: Part 13 – Applying the EPA *Proposed Guidelines for Carcinogen Risk Assessment* to a set of asbestos lung cancer mortality data. *Journal of the Air & Waste Management Association* 53: 1326–1339.
29. J. Duffus and H. Worth (2001). Training program: The Science of Chemical Safety: Essential Toxicology – 4; Hazard and Risk. *IUPAC Educators' Resource Material*, International Union of Pure and Applied Chemistry.
30. Actually, another curve could be shown for essential compounds like vitamins and certain metallic compounds. In such a curve (Figure 5.13), the left-hand side (low dose or low exposure) of the curve would represent deficiency and the right-hand side (high dose or exposure) would represent toxicity, with an optimal, healthy range between these two adverse responses:
Note that the two responses will differ when moving toward lower and higher doses, away from the optimal range. For example, anemia and its related effects may occur at the low end, while neurotoxicity can be an outcome at the high end of exposures. Ideally, exposures within the optimal range has neither effect. Like the other curves, the safe levels of both effects would be calculated and appropriate factors of safety applied. Recall that these curves represent population exposures, rather than individual exposures, thus the ranges will vary according to the sensitivity of a population. The shapes will also vary. For example, for an extremely allergic subpopulation (e.g. nut allergies), the curve may resemble that of the noncancer curve (Curve B in Figure 5.6), with little or no deficiency, since the protein needed from the food source can be obtained in other foods. In this instance, the RfD for the food substance (e.g. the protein eliciting the allergic response) may be quite low due to the numerous uncertainties (the UF and MF values in the denominator will be large). Thus, for practical purposes, an allergic subpopulation's optimal range for this protein source may be virtually zero.
31. For an excellent summary of the theory and practical applications of the Ames test, see K. Mortelmans and E. Zeiger (2000). The Ames *Salmonella*/Microsome Mutagenicity Assay, *Mutation Research* 455: 29–60.
32. US Environmental Protection Agency (2002). Interim Reregistration Eligibility Decision for Chlorpyrifos. Report No. EPA 738-R-01-007. Washington, DC.
33. B. Singer (2003). A tool to predict exposure to hazardous air pollutants. *Environmental Energy Technologies Division News* 4(4): 5.
34. US Environmental Protection Agency (1997). Exposure Factors Handbook, Report No. EPA/600/P-95/002Fa, Washington, DC.
35. These factors are updated periodically by the US EPA in the *Exposure Factor Handbook* at www.epa.gov/ncea/exposfac.htm.
36. The definition of "child" is highly variable in risk assessment. The *Exposure Factors Handbook* uses these values for children between the ages of 3 and 12 years.
37. The source for this discussion box is US Environmental Protection Agency (1999). Wastewater Technology Fact Sheet: Ultraviolet Disinfection. Report No. EPA 832-F-99-064.
38. US Environmental Protection Agency (1986). Guidelines for Carcinogen Risk Assessment, Report No. EPA/630/R-00/004, *Federal Register* 51(185): 33992–34003, Washington, DC.
39. A. Bradford Hill (1965). The environment and disease: association or causation? *Proceedings of the Royal Society of Medicine, Occupational Medicine*, 58: 295; and A. Bradford-Hill (1965). The environment and disease: association or causation? President's Address: *Proceedings of the Royal Society of Medicine* 9: 295–300.
40. The principal sources of this discussion are: M. Myers and A. Kaposi (2004). *The First Systems Book: Technology and Management*, 2nd Edition. Imperial College Press, London, UK; and T.R.G. Green (1989). Cognitive dimensions of notations. In: A. Sutcliffe and L. Macaulay (Eds), *People and Computers V.* Cambridge University Press, Cambridge, UK.
41. See, for example, France Biotech (2009). FROM GMP TO GBP: Fostering bioethics practices (GBP) among the European biotechnology industry. http://cordis.europa.eu/fetch?CALLER=FP6_PROJ&ACTION=D&DOC=16&CAT=PROJ&QUERY=1172750150639&RCN=80077; accessed July 19, 2009.
42. H. Petroski (1985). *To Engineer is Human: The Role of Failure in Successful Design.* St Martin's Press, New York, NY.
43. Source: R. Westfall (1993). *The Life of Isaac Newton.* Cambridge University Press, Boston, MA.
44. I first was made aware of this whole new paradigm driving to Greensboro from Durham, NC, heading to the National Environmental Science and Technology Conference. A radio program was dealing with new ways to think about materials and how we have relied upon old, inefficient means of using technology. Unfortunately, I do not know the name of the expert who was being interviewed, nor even that of the radio program (I tuned in mid-interview), but the discussion was intriguing. I am indebted to this anonymous expert, including his insights about birds and nature.
However, one of the postulations of the expert may not hold. He contended that a reason that nature would not select coloring by heavy metals like cadmium is that flight depends upon lift exceeding drag, so it seems counterintuitive for a natural system to use heavy molecules that would make flight more difficult, i.e. increasing

the drag. I contacted Professor Geoffrey Hill, a respected ornithologist at Auburn University, who informed me that it is not unusual for large molecules to be used as pigments, because they are readily available in the birds' habitats. He cited the example of carotenoid pigments manufactured by carrots and other orange and red plants. Birds ingest the plants and translocate the carotenoids to their feathers. So, it may cost the birds something in flight, but the availability of the pigments can override this need.

By coincidence, the keynote speaker at the conference was Joseph DeSimone, Professor of Chemistry and Chemical Engineering at the University of North Carolina. Professor DeSimone has been recognized as a leader in sustainable industry. He has used supercritical carbon dioxide, for example, as a substitute for hazardous solvents in the dry cleaning industry. Such cleaners are now found throughout the United States and increasingly around the globe. Over 100,000 plants use pressure and polymers to make the CO_2 supercritical. In this form, CO_2 is very efficient at dissolving most organic compounds. Thus a new application of a well understood physical concept is preventing pollution.

45. See R.O. Prum, R.H. Torres, S. Williamson and J. Dyck (1998). Cohered light scattering by blue feather barbs. *Nature* 396: 28–9.
46. T. Seyfer (2007). An overview of chimeras and hybrids. Life Issues.net: Clear Thinking about Crucial Issues; http://www.lifeissues.net/writers/sey/sey_03overview1.html; accessed September 17, 2009.
47. J. Zhang, Y. Zhang and Y. Song (2003). Chloroplast genetic engineering in higher plants. *Acta Botanica Sinica* 45: 509–516.
48. S.O. Duke (Ed.) (1996). *Herbicide-Resistant Crops*. CRC Press, Boca Raton, FL.
49. W.K. Vencill (Ed.) (2002). *Herbicide Handbook*, 8th Edition. Weed Science Society of America, Lawrence, KS.
50. A. L. Cerdeira and S. O. Duke (2006). The current status and environmental impacts of glyphosate-resistant crops: a review. *Journal of Environmental Quality* 35: 1633–1658.
51. M. Fuchs and D. Gonsalves (2007). Decades after their introduction: Lessons from realistic field risk assessment studies. *Annual Review of Phytopathology* 45: 173–202.
52. Ibid.
53. P. Powell-Abel, R. S. Nelson, N. De B Hoffmann, S. G. Rogers, *et al.* (1986). Delay of disease development in transgenic plants that express the tobacco mosaic virus coat protein gene. *Science* 232: 738–743.
54. R. Nelson, S. M. McCormick, X. Delannay, P. Dube, J. Layton, *et al.* (1988). Virus tolerance, plant growth and field performance of transgenic tomato plants expressing the coat protein from tobacco mosaic virus. *Bio/Technology* 6: 403–409.
55. Fuchs and Gonsalves (2007).
56. This discussion draws upon the ideas of Zach Abrams, who conducted undergraduate research on GMOs in my Ethics in Professions course at Duke.
57. An example is the fear that allowing euthanasia will cheapen the sanctity of life. That is, the value of human life changes from a dichotomy (giving and protecting life is good; taking and not protecting life is bad) to a continuum (certain life is good, other life is not so good, and some life is bad). A continuum view can be adjusted depending on conditions and preferences of a society (really sick, old people today, people with bald spots tomorrow?). So, one of the chimera's slippery slopes is that we are taking human DNA and inserting it into nonhuman species, e.g. transplanting human neural cells into the brain of mice indeed does induce the production of neurons in the mouse that are from human origin (see, for example: I. Weissman, Online News Hour with Jim Lehrer transcript, PBS television, July 2005, http://www.pbs.org/newshour/bb/science/july-dec05/ chimeras_weissman-ext.html). In a sense, we now have a chimera not unlike the Greeks' minotaur! We are already going down the slippery slope and gaining speed, in the name of science. Unfortunately, scientists can be some of the most closed-minded people if their research and funding are at risk. Right now, biomedical cutting edge areas that violate many of our moral standards are vigorously defended. Too often, those of us who recommend caution and consideration of the present and likely ethical breaches are labeled Luddites and extremists by the larger scientific community.

CHAPTER 6

Reducing Biotechnological Risks

> Science is built up of facts, as a house is built of stones; but an accumulation of facts is no more a science than a heap of stones is a house.
>
> **Henri Poincaré (1905) Science and Hypothesis**

Risk assessment is the science upon which environmental decisions are made. Risk management decisions must not so much *eliminate* risk, since that is nearly always impossible, but ensure that any risk is *acceptable*. This sets the stage for one of the most important questions in risk-based decisions: What constitutes an acceptable risk?

A convenient standard of biotechnological acceptability is that a risk from an operation, a product or a system should be "as low as reasonably practical" (ALARP), a concept coined by the United Kingdom Health and Safety Commission [1]. The range of possibilities fostered by this standard can be envisioned as a diagram (see Figure 6.1). The upper area (highest actual risk) is clearly where the risk is unacceptable. Below this intolerable level is the ALARP. Risks in this region require measures to reduce risk to the point where costs disproportionately outweigh benefits.

This approach to determining a scientifically and ethically acceptable outcome based upon risks and benefits is one form of *utilitarianism*. The utility of a particular application of a microbial population, for example, is based upon the greatest good that population's growth and metabolism will engender, but this must be compared to the potential harm it may cause. For example, if the microbial population breaks down an organic contaminant that has seeped into the groundwater more efficiently than other available techniques (e.g. pumping out the groundwater and treating it aboveground using air stripping), this would seem acceptable. However, such single-variable assessments are uncommon and can be dangerous. For example, the bioengineer must take into account whether the introduced microbial population's growth and metabolism introduces side effects, such as the production of harmful metabolites, or whether they could change the diversity and condition of neighboring microbial population.

Another aspect of ALARP which is especially important to biotechnological systems is that a margin of safety should be sought. This margin is less a concept of risk assessment than in risk reduction and management, since the margin must be both protective and reasonable [2]. Hence, reaching ALARP necessitates qualitative and/or quantitative measures of the amount of risk reduced and costs incurred with the design decisions:

The ALARP principle is based on the assumption that it is possible to compare *marginal improvements in safety* (marginal risk decreases) with the *marginal costs of the increases in reliability* [3].

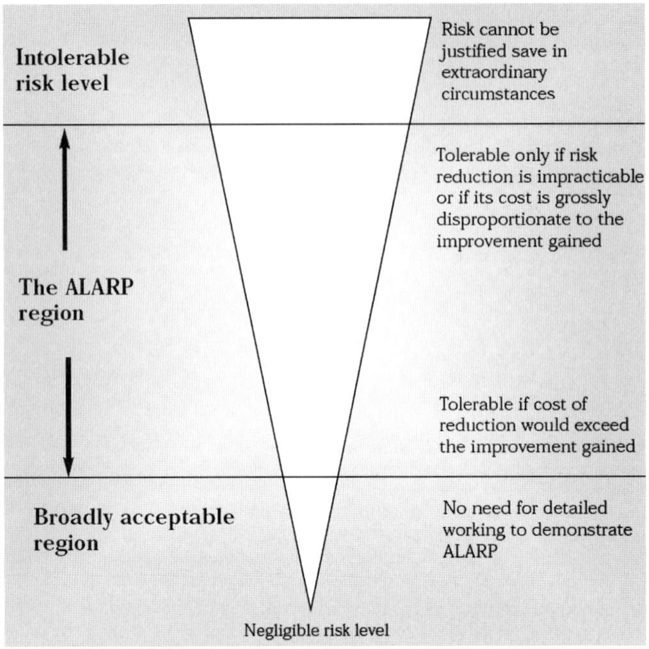

FIGURE 6.1
Three regions of risk tolerance.
Source: United Kingdom Health and Safety Commission (1998); http://www.hse.gov.uk/nuclear/computers.pdf; accessed May 26, 2006.

Acceptable risk must be reconciled with the sound practice of bioengineering. One of the challenges of applying a utilitarian model to engineered systems involving living things is that financial costs and benefits are often easier, or at least more straightforward, to calculate than other (i.e. non-monetized) costs and benefits, and such benefits and costs in biological systems are often not seen right away (see Figure 6.2). In fact, many of the concerns with biological agents is that their impacts may not be measurable until after numerous generations of microbial populations and after the genetic material has moved into other geographic domains (i.e. horizontal gene transfer).

Another problem in assigning value to a biotechnological utility is the issue of costs of *not* designing the biological solution to a problem. For example, the current controversies associated with conducting research at the nanoscale (near the diameter of the hydrogen atom)

FIGURE 6.2
Safety and environmental risks associated with primary and secondary costs. Increased safety and reliability can be gained by considering secondary costs in product and system design. This demonstrates the short-term benefits may lead to long-term costs, especially when costs and impacts in addition to financial measures are considered.
Source: Adapted from M. Martin and R. Schinzinger (1996). *Ethics in Engineering.* McGraw-Hill, New York, NY.

are sometimes rooted in fears of the potential Pandora's box of an unknown technology. If decisions are only made to avoid these problems, society exposes itself to *opportunity risks.* In other words, if we inordinately avoid designs for fear of potential harm, we may forfeit life-saving and enriching technologies. The prominent opportunities may be better known in the medical biotechnological venues, such as drug delivery systems that differentiate tumor tissue from healthy tissue based on electromagnetic differences at the nanoscale. However, this is also true for environmental biotechnologies.

The previously mentioned bioremediation of the organic contaminant in the aquifer would pose an opportunity risk if the pumping and treatment of the water does not do as good a job as a genetically modified microbial population in lowering the concentrations of the contaminant. Thus, if the aquifer is a source of drinking water, more people would be exposed to higher concentrations of the contaminant, so their risks to its hazards (e.g. cancer, nervous system damage, endocrine disruption) would be higher.

Even this academic opportunity cost tradeoff can be complicated by adding just one real-world variable. For instance, what if the introduced microbial population were found to be more allergenic to 10% of the population (e.g. gastric response)? Would this make the pump and treat approach preferable to the bioremediation? Usually, populations are stratified with respect to their outcomes in response to such variables. For example, the allergenicity would not be evenly distributed across the entire populations, but some susceptible subpopulations will respond more sensitively. Thus, the overall population's response is not as important as the response of sensitive subgroups. If 25% of children under the age of 10 years were found to be allergic to the introduced microbes, does this tip the balance? In addition, the severity of the response is crucial. For example, even if the proposed bioremediation option is very efficient in it's treatment of a contaminated aquifer, if the gastric reactions lead to health problems that require hospitalization and other complications, the risk management decision is likely to forgo using that option, or to do so within tightly defined constraints and precautions to prevent exposures to the microbes.

Holding paramount the health, safety, and welfare of the public is any engineer's, including the bioengineer's, primary mandate. However, reaching consensus on the degree of health and environmental risks posed by a biotechnological application is difficult. For starters, the public often exaggerates risks, at least if the metric is the degree to which the risk is based on known scientific facts. Other times, the public ignores seemingly large risks. Thus, abating risks that are in fact quite low could mean unnecessarily complicated and costly measures, whereas not doing enough to address a large, yet relatively uncertain risk is also unacceptable. The tension comes from the possibility that the bioengineer will support an untenable or a less acceptable alternative, i.e. one that in the long run may be more costly and deleterious to the environment or public health.

Part of the divergence in risk-based decision making between scientists and the general public may result from the fact that much of the scientific community has adopted some form of science-based risk assessment, as articulated in Chapter 5. Conversely, the larger population is less inclined to give science such primacy in decision making. Thus, since the risk assessment and risk perception processes differ quite markedly, the two groups can vary substantially in their decisions, as shown in Table 6.1. Not only the final decisions, but the entire critical path taken to reach the decisions will differ. As discussed in Chapter 5, engineers who make biotechnological and environmental decisions follow a path that begins with problem identification, followed by data analyses that ultimately lead to a comprehensive risk characterization. This path includes balances, sometime quantifiable, such as risk–benefit and cost–benefit ratios.

This is not to say the scientists involved in risk assessment cannot and do not engage in perception. In fact, in environmental systems, there is always some "black box" in which purely deductive reasoning cannot be used. It is more a matter of degree and, as mentioned, the primacy given to objective information versus intuition, personal experience and "intangibles" [4]. Perception relies on thought processes, including intuition, personal experiences,

Table 6.1 Differences between environmental risk assessment and risk perception processes

Analytical phase	Risk assessment processes	Risk perception processes
Identifying risk	Physical, chemical, and biological monitoring and measuring the event	Personal awareness
	Deductive reasoning	Intuition
	Statistical inference	
Estimating risk	Magnitude, frequency, and duration calculations	Personal experience
	Cost estimation and damage assessment	Intangible losses and non-monetized valuation
	Economic costs	
Evaluating risk	Cost–benefit analysis Community policy analysis	Personality factors Individual action

Source: Adapted from K. Smith (1992). *Environmental Hazards: Assessing Risk and Reducing Disaster.* Routledge, London, UK.

and personal preferences. Engineers tend to be more comfortable operating in the middle column (using risk assessment processes), while the general public often uses the processes in the far right column. One can liken this to the "left brained" engineer trying to communicate with a "right brained" audience. It can be done, so long as preconceived and conventional approaches do not get in the way.

Recall from the previous chapter that characterizing a biotechnological risk is a function of the hazards and the likelihood of these hazards, distinguishing between actual and perceived hazards and risks. Thus, the biotechnological risk management process must account for a "system of systems," i.e. the environment consists of numerous systems within and among media, pathways, and routes at various levels of biological organization (see Figure 5.4).

RISK QUOTIENT METHOD AND LEVELS OF CONCERN [5]

One means of addressing risks within multiple and embedded systems with an ecosystem is use screening approaches upfront. These generally provide a means of testing for the most important (i.e. sensitive) factors that are increasing risks. The *risk quotient method* is used to address ecosystem risks posed by pesticides by integrating the results of exposure and *ecotoxicity* data. For both acute and chronic endpoints, risk quotients (RQs) are calculated by dividing exposure estimates by ecotoxicity values:

$$RQ = \frac{\text{Exposure}}{\text{Toxicity}} \tag{6.1}$$

This is a deterministic approach, i.e. the RQ is calculated by dividing a point estimate of exposure by a point estimate of effects. This ratio is a simple, screening-level estimate that identifies high- or low-risk situations [6]. The calculation of risk quotients is based upon ecological effects data, pesticide use data, information about the pesticide's environmental transport, transformation and fate, along with estimates of exposure to the pesticide. Applying this method relates an estimated environmental concentration (EEC) to an effect (toxicity) level, such as an LC_{50} [i.e. the concentration of a pesticide where 50% of the organisms die (see Figure 5.6)].

Acute RQs use the LD_{50} or lethal concentration (50%), i.e. the LC_{50}, as the toxicity value, and chronic RQs use the no observed adverse effect concentration (NOAEC) or the no observed adverse effect level (NOAEL) as the toxicity value. The RQs are then compared to regulatory levels of concern (LOCs). These LOCs are criteria used regulators to indicate potential risk to nontarget organisms (e.g. beneficial insects or threaten species) and the need to consider regulatory action. That is, the LOCs are the policy tool for analyzing potential risk to nontarget organisms; so if the RQ exceeds the LOC for a given taxa and exposure duration, there is a potential for adverse effects to nontarget organisms. For example, the US EPA has defined LOCs for acute risk, potential restricted use classification, and for endangered species [7].

Exceeding the criteria indicates that a pesticide, even when used as directed, has the potential to cause adverse effects to nontarget organisms. The LOCs currently address the following *risk presumption* (see Table 6.2) categories:

acute – there is a potential for acute risk; regulatory action may be warranted in addition to restricted use classification;
acute restricted use – the potential for acute risk is high, but this may be mitigated through restricted use classification;

Table 6.2 Risk presumptions for risk quotients and corresponding regulatory levels of concern

Organism	Risk presumption	Risk quotient	Level of concern
Birds[a]	Acute risk	EEC/LC_{50} or LD_{50} ft^{-2} or LD_{50} day^{-1}	0.5
	Acute restricted use	EEC/LC_{50} or LD_{50} ft^{-2} or LD_{50} day^{-1} (or LD_{50} <50 mg kg^{-1})	0.2
	Acute endangered species	EEC/LC_{50} or LD_{50} ft^{-2} or LD_{50} day^{-1}	0.1
	Chronic risk	EEC/NOAEC	1
Wild mammals[a]	Acute risk	EEC/LC_{50} or LD_{50} ft^{-2} or LD_{50} day^{-1}	0.5
	Acute restricted use	EEC/LC_{50} or LD_{50} ft^{-2} or LD_{50} day^{-1} (or LD_{50} <50 mg kg^{-1})	0.2
	Acute endangered species	EEC/LC_{50} or LD_{50} ft^{-2} or LD_{50} day^{-1}	0.1
	Chronic risk	EEC/NOAEC	1
Aquatic animals[b]	Acute risk	EEC/LC_{50} or EC_{50}	0.5
	Acute restricted use	EEC/LC_{50} or EC_{50}	0.1
	Acute endangered species	EEC/LC_{50} or EC_{50}	0.05
	Chronic risk	EEC/NOAEC	1
Terrestrial and semi-aquatic plants	Acute risk	EEC/EC_{25}	1
	Acute endangered species	EEC/EC_{05} or NOAEC	1
Aquatic plants[b]	Acute risk	EEC/EC_{50}	1
	Acute endangered species	EEC/EC_{05} or NOAEC	1

Notes: NOAEC = no observed adverse effect concentration; NOAEL = no observed adverse effect level (NOAEL); LD = dose of a substance that is lethal to a specified percentage of tested animals (in this case, LD_{50} is median lethal dose); EEC = estimated environmental concentration of a substance; EC = effective concentration of a substance required to produce a particular effect in a specified percentage of an animal population (e.g. EC_{05} is the concentration expected to elicit an effect in 5% of the population).
[a]*LD_{50} is the median lethal dose:*

$$LD_{50}\ ft^2 = \frac{M\ ft^{-2}}{LD_{50} \times W};\ LD_{50}day^{-1} = \frac{M_{consumed}\ day^{-1}}{LD_{50} \times W}$$

where, M = mass of substance (mg); $M_{consumed}$ = mass of substance consumed (mg); and W = weight of test animal.

[b]*EEC = (ppm or ppb) in water.*
Source: US Environmental Protection Agency (2007). Appendix E: Risk Quotient Method and LOCs – Risks of Metolachlor Use to Federally Listed Endangered Barton Springs Salamander. Washington, DC.

acute endangered species – the potential for acute risk to endangered species is high, regulatory action may be warranted;
chronic risk – the potential for chronic risk is high, regulatory action may be warranted.

Currently, most regulations do not require risk assessments for chronic risk to plants, acute or chronic risks to nontarget insects, or chronic risk from granular/bait formulations to mammalian or avian species.

The ecotoxicity test values (i.e. *measurement endpoints* – see Table 6.3) applied in the acute and chronic risk quotients are derived from required studies submitted by the registrant. Measurement endpoints are measures of effects that are derived from the results of tests or observational studies and used to estimate the effects on an assessment endpoint of exposure to a stressor. For example, a conventional measure of effect from an acute lethality test is the median lethal concentration (LC_{50}), which might be used to estimate the risk of a fish kill (an assessment endpoint) from exposure to a spill of the tested chemical [8].

Measures of effect and assessment endpoints may be expressed at the same level of biological organization. The same measure of effect may be used, with considerably greater uncertainty, to estimate risks to other levels, e.g. extrapolation from an organismal effect to a population-level assessment endpoint (e.g. the abundance of a certain fish species) or a community-level endpoint (total number of species in the system). Thus, measurement endpoints are the results of tests or observational studies that are used to estimate the effects on an assessment endpoint of exposure to a stressor, e.g. the median lethal concentration (LC_{50}), which might be used to estimate the risk of a fish kill (an assessment endpoint) from exposure to a release of the tested chemical to an ecosystem [9].

Measures of effect and assessment endpoints may be expressed at the same level of organization (organism level in this case). However, the same measure of effect may be used, with considerably greater uncertainty, to estimate risks to a population-level assessment endpoint (abundance of a fish species) or a community-level endpoint (number of species).

Test values may also be derived from data published in the open literature (e.g. the ECOTOX database [10]; see Appendix 5). Examples of ecotoxicity values derived from short-term laboratory studies that assess acute effects include the LC_{50} for fish and birds and the LD_{50} for birds and mammals. In addition, ecotoxicity can be reflected by an effect concentration (EC) that is a statistically or graphically estimated concentration expected to cause one or more

Table 6.3 **Example assessment and measurement endpoints applicable to ecological restoration of surface waters: the endpoints are to provide information on the structure and function of natural elements of ecosystems**

Ecosystem component	Assessment endpoint	Measurement endpoint
Physical habitat	Increase habitat suitability for rainbow trout by 50%	Habitat Suitability Index for rainbow trout
Hydrology	Increase minimum stream flow to 10 $ft^3\ sec^{-1}$	Stream flow
Population	Establish trout population of 50 kg ha^{-1}	Kg trout per ha
Ammonia	Eliminate exceedences of the water quality standard for ammonia	Concentration of ammonia
Ambient toxicity	Eliminate acute and chronic ambient toxicity	96-h LC_{50} values and 7-day IC_{25} values for ambient stream water to fathead minnows and the benthic invertebrate *Hyallela azteca*

Note: IC_{25} = *in vitro* 25% maximal inhibitory concentration, i.e. the point estimate of the chemical concentration that would cause a 25% reduction in a non-lethal biological measurement, such as growth or reproduction.
Source: US Environmental Protection Agency (1995). *Ecological Restoration.* Report No. EPA 841-F-95-007. Washington, DC.

specified effects in a specified percentage of a group of organisms under site-specific conditions. For instance, the EC_{50} can be used for aquatic plants and aquatic invertebrates) and the EC_{25} can be applied to terrestrial plants. Examples of toxicity test effect levels derived from the results of long-term laboratory studies that assess chronic effects are the LOAEL (lowest observed adverse effect level) for birds, fish, and aquatic invertebrates and the NOAEL for birds, fish, and aquatic invertebrates. However, the NOAEL is generally used as the ecotoxicity test value in assessing chronic effects.

The RQ is an ecosystem risk metric, but could be used more widely in risk assessments since it is a direct function of both exposure and hazard. For instance, changes in assessment and measurement endpoints could indicate systematic problems in human populations. If a bird population, for example, experiences a decline in the Habitat Suitability Index, this could be an indicator of pesticide drift that could effect human populations. Changes in RQ could also be an early warning system for gene flow of invasive strains or species of organisms (e.g. microbial changes may lead to changes in higher levels of biological organization).

BIOSYSTEMATIC INTERVENTION

A convenient place to start in finding ways to reduce the environmental risks and to find bioengineering intervention options is to approach the risks from the perspective of one environmental medium, i.e. water. That is, we will consider decision making regarding a single medium-multisystem risk. This reductionist approach has value, since much environmental data is "media-specific" and most regulations are written for a single environmental compartment. This should be the beginning of the intervention, not the end, because the environment is a multimedia, multicompartment system.

The water compartment is complex, ranging from oceans to the water surrounding a clay particle. The public, and the bioengineer for that matter, consider water to be more than simply H_2O. First, water in the environment always contains solutes and suspended matter. The only place that it does not, is in the laboratory, e.g. de-ionized water. In fact, depending on the laboratory practices and quality assurance, we are often only able to say that the impurities do not exist at level above detection limits.

Even "polished," potable or highly treated water contains matter other than H_2O. No drinking water is completely free of substances, and microbial populations live in natural waters. The chemical composition and microbial ecology can reflect the source of the water. If a water supply is from an igneous rock formation, many of the rock's elements will have leached into the water. The same goes for metamorphic and sedimentary rock formations. The amount is directly related to each constituent's aqueous solubility, but is also affected by the quality of the water that originally infiltrated the aquifer, as well as by the characteristics of the soil through which the water percolated.

Water's value is derived from more than its chemical composition. As evidence, the quality of a community's source of drinking water, recreational water bodies, and other landscape considerations helps to define a community. Take the example of a community that believes its water supply is tainted because when clothing is washed it has a reddish-brown residue. Something to keep in mind is that geology and soil type are highly variable, even within the same formation. So, one person's water quality may differ considerably from that of her neighbor only a few hundred meters away. Just because the neighborhood is defined by the county and state as a unit, it does not mean it is homogeneous. The neighborhood has been defined for other reasons (e.g. property purchases, zoning, subdivision regulations, and accessibility to roads and other facilities), so expecting uniform environmental quality is unrealistic.

Again, this illustrates the stark difference between risk perception and risk assessment. For example, the discoloration of the water can be due to the presence of iron or manganese. Indeed, it will ruin a white shirt, but may not be associated with health problems. However, depending on the formation, if iron and manganese are present, so may a series of other metals, such as copper, zinc, lead, mercury, and arsenic (actually a metalloid). In addition, naturally occurring deposits that contain brownish-red iron oxide deposits may indicate the presence of other, more toxic heavy metals or may also contain other harmful substances, notably asbestos. So, pardon the pun, but discoloration could be a proverbial "red flag" (or at least a brownish-red one!).

Often measurements of environmental conditions depend on indicators. An indicator must be both sensitive (i.e. we can see it and it is telling us that something is wrong) and specific (i.e. when we see it we know that it means one or only a few things). But iron in water is neither a sensitive nor a specific indicator of cancer-causing substances in water. It is not even a good indicator of water hardness, since the two ions that cause hardness, calcium and magnesium, may or may not substantially co-occur with iron in natural waters.

The challenge in managing risks of biotechnologies is dealing with uncertainties. For example, a growing concern of water pollution is the presence of medicines that pass through waste treatment systems and find their way into drinking water supplies, allowing microbes to gain resistance against antibiotic drugs. Basquero *et al.* [11] put this problem within a biotechnological context:

An important part of the dispersal and evolution of antibiotic-resistant bacterial organisms depends on water environments. In water, bacteria from different origins (human, animal, environmental) are able to mix, and resistance evolves as a consequence of promiscuous exchange and shuffling of genes, genetic platforms, and genetic vectors.

This quote embodies the far-reaching nature of biotechnological risk management. The public already perceives microbes to be a hazard, even though the vast majority of species are beneficial. Add to this anxiety concerning the complexity and possible large-scale distribution of "Andromeda strains" and "gray goo", and the already established dangers of cross-resistance and "super bugs" – bacteria that are resistant and tolerant of synthetic antibiotics. This yields a heightened skepticism of what the bioengineer is proposing (see Case Study: Genetic Biocontrols of Invaders).

CASE STUDY

Genetic Biocontrols of Invaders

Opportunistic species wreak havoc on ecosystems. They are usually introduced into ecosystems where no natural predators can check the newly arrived species' numbers and geographic range. Introduced species cause us to rethink our concept of pollution. Like so many other issues in this book, the type of pollution caused when an opportunistic, invasive organism colonizes and outlives its welcome is a systematic one. The threat is not usually to a single species, although this could be very important if a species were already threatened or endangered prior to the invasion. It is usually a problem of the whole ecosystem. Something is out of balance.

The concept of invasion is not one that is consistently applied. For example, the very diligent Invasive Species Specialist Group (ISSG) has reluctantly listed the 100 "worst" invasive species in the world (see Table 6.4). In ISSG's own words, the task is difficult:

> *Species and their interactions with ecosystems are very complex. Some species may have invaded only a restricted region, but have a huge probability of expanding, and causing further great damage (for example, see* Boiga irregularis, *the Brown Tree Snake). Other species may already be globally widespread, and causing cumulative but less visible damage. Many biological families or genera contain large numbers of invasive species, often with similar impacts; in these cases one representative species was chosen. The one hundred species aim to collectively illustrate the range of impacts caused by biological invasion. [12]*

Invasive species are organisms that are not native to an ecosystem. They are problematic when they cause harm, such as loss of diversity and other environmental damage, economic problems, or even human health concerns. These organisms can be any biota, that is, microbes, plants, and animals, but usually at least their presence and impact are in part due to human activity.

With few or no predators, these non-native invasive species can consume the food sources much faster than their competitors. In North America, the Great Lakes basin is particularly vulnerable to invasive species. For example, about 170 non-native invasive species have been identified in the Laurentian Great Lakes drainage basin. Lake Erie watershed alone has 132 species, including: algae

Table 6.4 Worst invasive fish species as rated by the Global Invasive Species Database

Genus and species	Common names
Asterias amurensis	Flatbottom seastar, Japanese Seastar, Japanese starfish, Nordpazifischer Seestern, North Pacific seastar, northern Pacific seastar, purple-orange seastar
Clarias batrachus	alimudan, cá trê tráng, cá trèn trang, clarias catfish, climbing perch, freshwater catfish, Froschwels, hito, htong batukan, ikan keling, ikan lele, Ito, kawatsi, keli, klarievyi som, koi, konnamonni, kug-ga, leleh, magur, mah-gur, mangri, marpoo, masarai, mungri, nga-khoo, pa douk, paltat, pantat, pla duk, pla duk dam, pla duk dan, pla duk nam jued, pla duk nam juend, Thai hito, Thailand catfish, trey andaing roueng, trey andeng, walking catfish, wanderwels, Yerivahlay
Cyprinus carpio	carp, carpa, carpat, carpe, carpe, carpe commune, carpeau, carpo, cerpyn, ciortan, ciortanica, ciortocrap, ciuciulean, common carp, crap, crapcean, cyprinos, escarpo, Europäischer Karpfen, European carp, German carp, grass carp, grivadi, ikan mas, kapoor-e-maamoli, kapor, kapr obecný, karp, karp, karp, karp, karp, karp dziki a. sazan, karpa, karpar, karpe, Karpe, karpen, karper, karpfen, karpion, karppi, kerpaille, koi, koi carp, korop, krap, krapi, kyprinos, læderkarpe, lauk mas, leather carp, leekoh, lei ue, mas massan, mirror carp, olocari, pa nai, pba ni, pla nai, ponty, punjabe gad, rata pethiya, saran, Saran, sarmão, sazan, sazan baligi, scale carp, sharan, skælkarpe, soneri masha, spejlkarpe, sulari, suloi, tikure, trey carp samahn, trey kap, ulucari, weißfische, wild carp, wildkarpfen
Gambusia affinis	Barkaleci, Dai to ue, Gambusia, Gambusie, Gambusino, Gambuzia, Gambuzia pospolita, Gambuzija, guayacon mosquito, Isdang canal, Kadayashi, Koboldkärpfling, Kounoupopsaro, Live-bearing tooth-carp, Mosquito fish, Obyknovennaya gambuziya, pez mosquito, San hang ue, Silberkärpfling, tes, Texaskärpfling, Topminnow, western mosquitofish, Western mosquitofish
Lates niloticus	chengu, mbuta, nijlbaars, nilabborre, Nilbarsch, nile perch, perca di nilo, perche du nil, persico del nilo, sangara, Victoria perch, victoriabaars, victoriabarsch
Micropterus salmoides	achigã, achigan, achigan à grande bouche, American black bass, bas dehanbozorg, bas wielkogeby, bass, bass wielkgebowy, biban cu gura mare, black bass, bol'sherotyi chernyi okun', bolsherotnyi amerikanskii tscherny okun, buraku basu, fekete sügér, forelbaars, forellenbarsch, green bass, green trout, großmäuliger Schwarzbarsch, huro, isobassi, khorshid Mahi Baleh Kuchak, lakseabbor, largemouth bass, largemouth black bass, lobina negra, lobina-truche, northern largemouth bass, okounek pstruhový, okuchibasu, Öringsaborre, Ørredaborre, ostracka, ostracka lososovitá, perca americana, perche d'Amérique, perche noire, perche truite, persico trota, stormundet black bass, stormundet ørredaborre, tam suy lo ue, zwarte baars
Oncorhynchus mykiss	pstrag teczowy, rainbow trout, redband trout, Regenbogenforelle, steelhead trout, trucha arco iris, truite arc-en-ciel
Oreochromis mossambicus	blou kurper, common tilapia, fai chau chak ue, Java tilapia, kawasuzume, kurper bream, malea, mojarra, mosambik-maulbrüter, Mozambikskaya tilapiya, Mozambique cichlid, Mozambique mouth-breeder, Mozambique mouthbrooder, Mozambique tilapia, mphende, mujair, nkobue, tilapia, tilapia del Mozambique, tilapia du Mozambique, tilapia mossambica, tilapia mozámbica, trey tilapia khmao, weißkehlbarsch, wu-kuo yu

Source: The IUCN/SSC Invasive Species Specialist Group (ISSG) (http://www.issg.org).

(20 species), submerged plants (8 species), marsh plants (39 species), trees/shrubs (5 species), bacteria (3 species), mollusks (12 species), oligochaetes (9 species), crustaceans (9 species), other invertebrates (4 species), and fishes (23 species). The increase has been attributed for the most part to switching from solid to water ballast in cargo ships, along with the opening of the St Lawrence Seaway in 1959. Among the non-native invasive fish species recently invading Lake Erie is the Chinese bighead carp, *Hypophthalmichthys nobilis* [13].

One of the means of addressing invasive species is by using *biocontrols*, i.e. using one living organism to limit and eliminate

(Continued)

another organism's effects. Introducing sterility has been used to control fish populations. The traditional sterilization approach change's the invasive fish's ploidy (i.e. affecting the complete set of chromosomes), by modifying the chromosome development using pressure, thermal or chemical shock at the point of fertilization so as to disrupt the egg's normal extrusion of a polar body containing a haploid set of maternal chromosomes (see Figure 6.3).

In particular, inducing triploidy has been used effectively to produce sterile offspring, by disrupting sexual reproduction. The retained polar body produces an embryo with two haploid chromosome sets from the female (rather than the normal one haploid set) as well as a third set from the male [14]. The odd sets of chromosomes appear to interfere with the mechanics of pairing of homologous chromosomes during each cell division, which disrupts normal gamete development. The resulting triploid varies from the normal diploid number of chromosomes. However, the major weakness of this method of sterilization is it can be incomplete, that is some of the offspring are not triploid. This can be ameliorated somewhat by producing tetraploids in the first generation.

An important environmental risk of this type of sterilization is the escape of these genetically modified organisms into the wild populations, since triploids still have sufficient levels of sex hormones to elicit normal courtship and spawning. Thus, the entire wild relative species' reproductive success would be in jeopardy, which is exacerbated if the wild relatives are endangered or threatened. In addition, given the large numbers of fish that escape fish farms, if many triploid transgenic fish enter the environment recurrently, some could survive to adversely affect fish diversity [15].

New genetic engineering methods have been developed to induce fish sterility, particularly the application of interference RNA (RNAi), to produce what are known as "sterile ferals" [16]. The RNAi method involves the insertion of a transgene that blocks the expression of an endogenous gene needed to develop gametes and embryos. The blocker expression is controlled by inducible promoter. The pros and

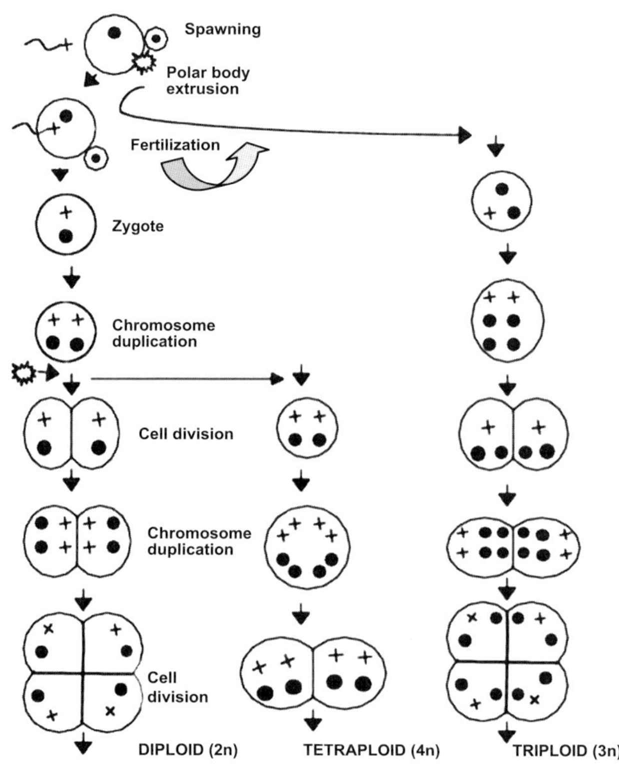

FIGURE 6.3
Steps in gamete fertilization and cell division that lead to the normal diploid fish embryo. Induction of triploidy or tetraploidy occurs by shocks (thermal, chemical or pressure) at a specific time after fertilization. Notes: ✲ denotes the point at which the shock is applied; ● denotes one haploid chromosome set derived from the female parent; + ● denotes one haploid chromosome set derived from the male.
Source: National Research Council (2004). *Biological Confinement of Genetically Engineered Organisms.* National Academies Press, Washington, DC.

cons of tripoidal versus transgenic sterilization of fish are provided in Table 6.5. This indicates that the transgenic approach has all of the ecological concerns of traditional techniques, but with additional uncertainties, such as the presence of repressor chemicals in the water.

A biotechnological concern is the possible spread of deleterious genetic material. Controlling non-native fish species invasions with transgenic fish of the same species as the targeted species is designed to be disruptive to the fish life cycle (see Figure 6.4). These disruptions can occur from before the embryonic stage to juvenile development.

Table 6.5 Comparison of traditional triploid sterilization to transgenic sterilization of fish

Method of sterilization	Strengths	Weakness	Other ecological, social, and regulatory considerations
Triploid sterilization via temperature, chemical or pressure shock to newly fertilized egg (Also called chromosome set manipulation or ploidy manipulation)	1. Methods well developed 2. Limitations understood 3. Ready to use for some species 4. Relatively low cost 5. Shorter research and development time than transgenic methods	1. Triploid induction not 100% effective for treated eggs 2. Requires individual screening to cull failures 3. Potential for mosaic individuals (mix of diploid and triploid)	1. Do sterile individuals retain active reproductive hormone levels and normal courtship behavior? 2. How many modified fish would need to be stocked and at what frequency? 3. Would predation of and competition with native species outweigh benefits gained by stocking sterile non-native fish? 4. Need to adapt methods to biology of each species 5. May be more socially acceptable and more feasible from a regulatory standpoint than transgenic methods
Transgenic sterilization (Involves recombinant DNA techniques; also called genetic modification or genetic engineering)	1. Capability to control sterility expression via repressor or inducer molecules 2. Can build in reduntant sterilization methods by stacking transgenes that affect different stages of development	1. Costly to develop 2. Long research and development period for each transgenic line 3. Limitations not fully understood 4. Probable limits to gene stacking 5. May require individual screening to ensure success	1. See considerations 1–4 for triploid sterilization 2. How stable is the transgene expression? 3. How complete is the induced sterility? 4. Unexpected presence of repressor molecule (e.g. tetracycline) in natural waters might repress expression of sterility genes 5. May not be as socially acceptable as non-transgenic sterilization methods

Source: A.R. Kapuscinski and T.J. Patronski (2005). Genetic methods for biological control of non-native fish in the Gila River basin: Development and testing of methods, potential environmental risks, regulatory requirements, multi-stakeholder deliberation, and cost estimates. Contract report to the US Fish and Wildlife Service (USFWS agreement number 201813N762). University of Minnesota, Institute for Social, Economic and Ecological Sustainability, St Paul, Minnesota. Minnesota Sea Grant Publication F 20.

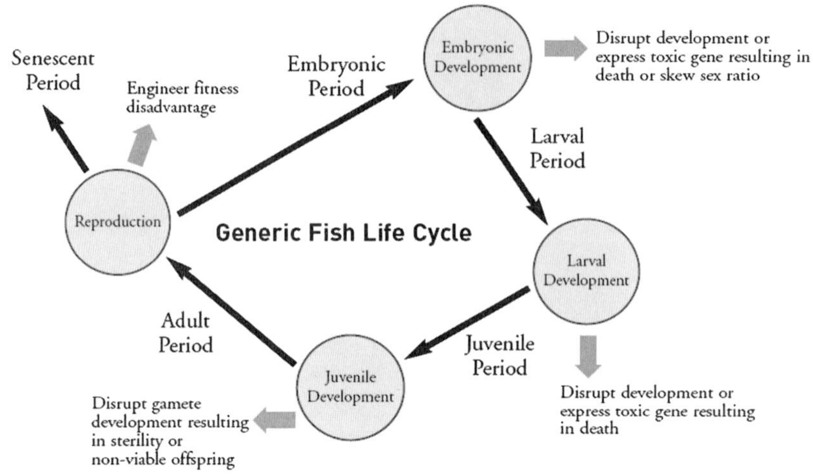

FIGURE 6.4

Stages of a fish life cycle that present opportunity for disruption that leads to sterilization via transgenic processes. [*See color plate section*]

Source: A.R. Kapuscinski and T.J. Patronski (2005). Genetic methods for biological control of non-native fish in the Gila River basin: Development and testing of methods, potential environmental risks, regulatory requirements, multi-stakeholder deliberation, and cost estimates. Contract report to the US Fish and Wildlife Service (USFWS agreement number 201813N762). University of Minnesota, Institute for Social, Economic and Ecological Sustainability, St Paul, Minnesota. Minnesota Sea Grant Publication F 20.

A number of endogenous genes' expression can be disrupted in fish, including the arotamase, estrogen receptor, gonadotropin-releasing hormone, protamine, vitellogenein, and growth-regulating genes. Among the questions that need to be answered are:

How many transgenic fish would have to be released and at what frequency and spatial distribution?

Would transgenic fish show heightened predation on or competition with wild, non-target native fish species?

If so, would this raise ecological risks to a level that outweighs the benefits of the intended biological control?

What risks are associated with a transgenic fish escape to areas of its native distribution and how could these risks be managed?

Would there be significant risks to human health if a transgenic fish was caught and eaten?

Type I and type II errors are likely. A type I error would be to decide not to release a transgenic fish for biological control on the basis of predicted risks that in fact do not exist, allowing the exacerbation of the non-native species problem. A type II error would be to decide to release the transgenic fish based on the assessment of existing data, only to find in time the occurrence of unexpected damage.

Species can hybridize with closely related species, so there is a potential of transgene spread to near relatives of non-target species. The amount of hazard and risk is a function of the proximity and availability of these near relatives, their proficiency to hybridize, and their fertility and viability of their offspring.

Transgenic species also have the potential for unexpected expressions such as increased toxicity or allergenicity for transgenic compared to native strains of the same species. This could lead to human health effects. Table 6.6 summarizes the potential hazards of fish sterilization biotechnologies. The bottom line is that the decision to use these technologies for even a worthwhile environmental effort (dealing with invaders) must be based on a thorough identification of these hazards and a reliable assessment of the risks that these technologies may elicit.

Certainly, new tools are being developed to help with such risk communications. For example, computational methods and natural products chemistry can help bioscientists and engineers test impacts before they reach living systems. That is, *in silico* techniques are preceding *in vivo* and *in vitro* testing. This can help to prevent the production unwanted byproducts all along the medical critical path, including toxic byproducts, as well as microbial processes, e.g. prevention of cross-resistance, antibiotic pass-through treatment facilities, and production of super bugs. (See Discussion Box: Biochemodynamics of Pharmaceuticals).

Table 6.6 **Some of the hazards presented by using genetic biocontrols. Shaded rows apply only to transgenic fish**

Decreasing likelihood (in general)	Hazard	Potential harm
↓	Density-dependent compensation for X years	Wipe out endangered fish before biocontrol effect prevails
	Failure in intended trait change	Increased number of fit non-natives increases disruption of native fish
	Transgene side effect on trait that enhances predation, competition or alters another non-target behavior	Increased disruption of native fish biocontrol effect prevails
	Pest replacement, once the target species is removed	Another pest species may be released from competition/predation and become a greater pest to native fish
	Transgene spread to native range of species	Depress or extirpate native populations
	Transgene spread to closely related species via hybridization	Harm to non-target species and communities
	Transgene spread to fish caught for eating	Harm to human health
	Horizontal gene transfer to non-target species	Depress populations of non-target species

Source: A.R. Kapuscinski and T.J. Patronski (2005). Genetic methods for biological control of non-native fish in the Gila River basin: Development and testing of methods, potential environmental risks, regulatory requirements, multi-stakeholder deliberation, and cost estimates. Contract report to the US Fish and Wildlife Service (USFWS agreement number 201813N762). University of Minnesota, Institute for Social, Economic and Ecological Sustainability, St Paul, Minnesota. Minnesota Sea Grant Publication F 20.

DISCUSSION BOX

Biochemodynamics of Pharmaceuticals

The endocrine, immune, and neurological systems are intertwined. The electrochemical signals they send control an organism's normal functioning, growth, development, and reproduction; so even small disturbances at the wrong time may lead to long-lasting, irreversible effects. An organism is particularly vulnerable during highly sensitive times of development, such as prenatal and pubescent periods, when small changes in endocrine status may have delayed consequences that may not appear until much later in adult life or even in future generations. Numerous studies indicate that pharmaceuticals, personal care products, and their metabolites are present in our nation's water bodies.

The most infamous case of multigenerational endocrine disruption is arguably that of diethylstilbestrol (DES), a synthetic hormone that was prescribed to pregnant women from 1940 to 1971 as a treatment of the mothers to prevent miscarriages. Unfortunately, DES has been subsequently classified as a *known carcinogen*. The major concern was not with the treatment of the mothers, but with the *in utero* exposure that led to a high incidence of cervical cancers in the daughters of the treated mothers.

While DES is generally recognized as a pharmaceutical problem it is, at a minimum, an environmental indicator of the potential problems of newly introduced chemicals. Also, an emerging concern, particularly about human and animal pharmaceuticals, is their "pass-through" into the environment. For example, drugs used in combined animal feeding operations (CAFOs) have been found in waters downstream, even after treatment. This is problematic in at least two ways. First, the drugs and their metabolites (after passing through the animals) may themselves be hormonally active or may suppress immune systems. Second, antibiotics are being introduced to animals in large quantities, giving the targeted pathogens an opportunity to develop resistance and rendering less effective.

(Continued)

Even more troubling is the phenomenon of *cross-resistance*. For example, the US Food and Drug Administration recently proposed withdrawing the approval of enrofloxacin, in the treatment of poultry in CAFOs. Enrofloxacin is one of the antibacterials known as fluoroquinolones, which have been used to treat humans since 1986 [17]. Fluoroquinolone drugs keep chickens and turkeys from dying from *Escherichia coli* (*E. coli*) infection, usually contracted from the animals' own droppings. The pharmaceutical may be an effective prophylactic treatment for *E. coli*, but another genus, *Campylobacter*, appears to be able to build resistance (see Figure 6.5). People who consume poultry products contaminated with fluoroquinolone-resistant *Campylobacter* increase substantially their risk infection by a strain of *Campylobacter* that, for biochemodynamic reasons, is increasingly difficult to treat. Worse yet, the whole class of reliable fluoroquinolone drugs is at risk of losing their efficaciousness, since the cross-resistance can carry over to drugs with similar structures.

Antibiotic resistance results from recombinatorial events, i.e. genetic exchanges among organisms inside populations and communities. The so-called "genetic reactors" in which antibiotic resistance evolves are shown in Figure 6.6. The level 1 reactor consists of the human and animal microbial populations (>500 species) in which the microbes exert their actions (intentional pharmaceutical actions). The level 2 reactor includes the dense nodes of microbial exposures and exchange, i.e. the aggregation of susceptible

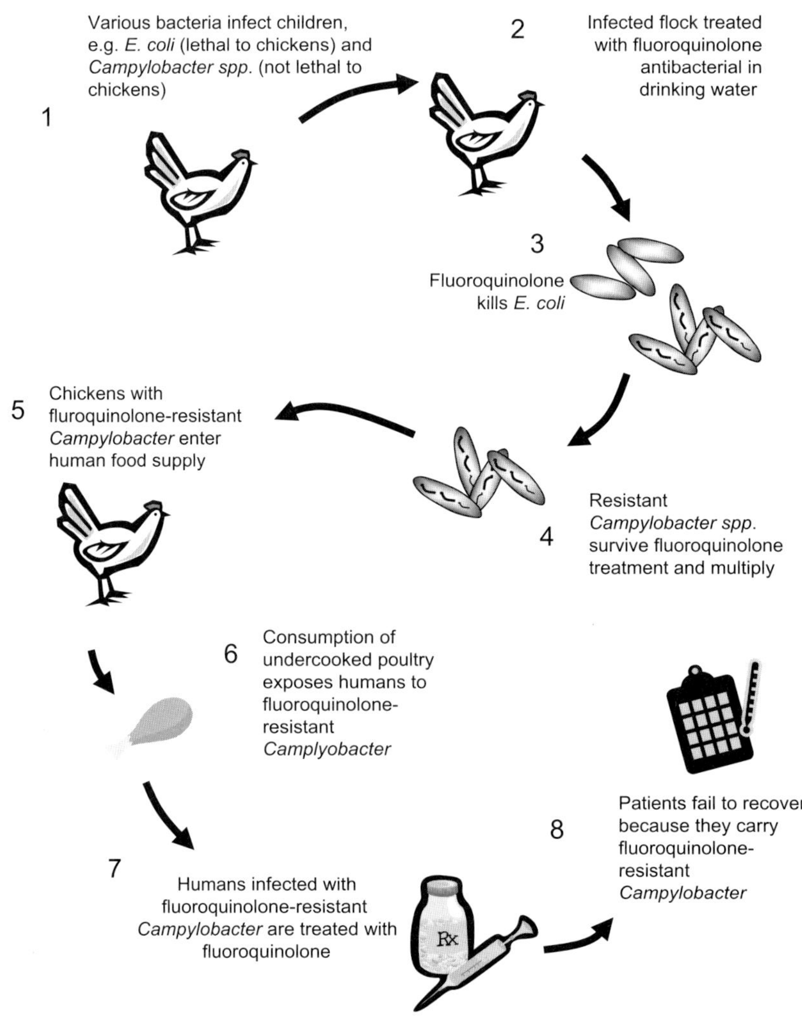

FIGURE 6.5
Steps in the cross-resistance of *Camplyobacter* to fluoroquinolone drugs.
Source: US Food and Drug Administration, L. Bren (2001). Antibiotic resistance from down on the farm. *FDA Veterinarian* 16 (1): 2–4. Graphic by R. Gordon.

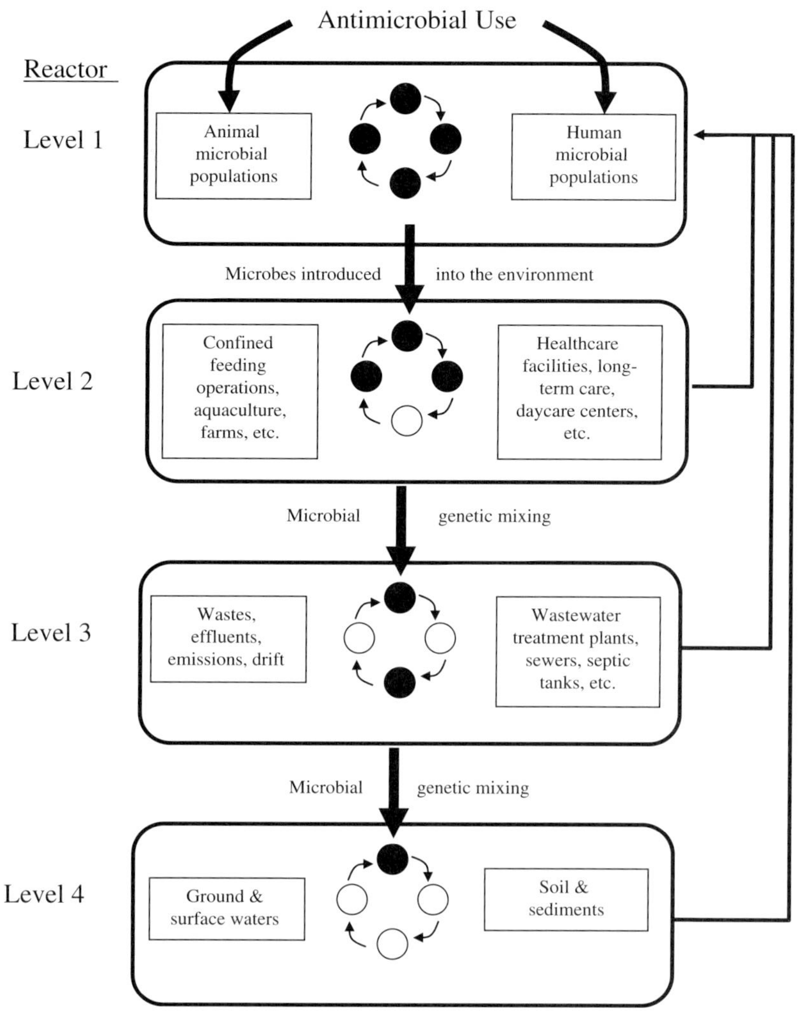

FIGURE 6.6
Genetic reactors that breed antibiotic resistance via genetic exchange and recombination. In the lower level reactors (1 and 2), human and animal microbial populations (filled circles) mix with environmental microbial populations (clear circles), which increases genetic variation, allowing new resistance mechanism in the microbial populations, whereupon these new strains, with the potential for greater resistance, are re-introduced to the human and animal environments (feedback arrows). Therefore, even if the human populations have not yet used an antibiotic, if a similar form is used in animals, the genetic adaptations may allow for resistant strains of bacteria to find their way into human populations, rendering a new antibiotic less efficacious.
Source: Adapted from F. Baquero, J.L. Martínez and R. Cantón (2008). Antibiotics and antibiotic resistance in water environments. *Biotechnology* 19: 260–265.

subpopulations in hospitals, long-term care facilities, feeding, and farming operations, etc. The level 3 reactor consists of the wastes and biological residues, e.g. lagoons, wastewater treatment plants, compost piles, septic tanks, etc. wherein microbes from many different individuals can assimilate and exchange genetic material. The level 4 reactor includes the various environmental media (soil, surface or groundwater environments) in which the microbes from the levels 1 through 3 reactors mix and counteract with organisms in the environment [18].

Carry-over and cross-resistance have been observed in numerous classes of drugs, including synthetic penicillin. Exacerbating the problem, the use of drugs is not limited to treating diseases. In fact, large quantities of antibiotics have been used as growth promoters in CAFOs, so the probability of cross-resistance is further increased, as shown in Figures 6.5 and 6.6.

(Continued)

Computational and green chemistry approaches can "tweak" molecules to see what impacts might result down the road. In some cases the molecule is never actually synthesized, but is a virtual molecule that can be run through various biochemodynamic scenarios as a first screen on possible hazards. These tools can also help with predicting hazards, exposures, and risks to humans and other organisms by improving the synthetic chemistry, e.g. preventing the formulation of chiral and enantiomer compounds that are resistant to natural biodegradation (e.g. left-hand chirals may be much more easily broken down than right-hand chirals of the same compound; also, one chiral may be toxic and the other efficacious). This may also apply to genetically modified organisms. For example, subtle changes to a microbe's DNA can foster unexpected changes in ecosystem competition, which might be predicted using proteomics and other computational tools.

Potential contaminant mixtures and co-exposures are not only important in water systems, but the flow and change of genetic material, platforms, and vectors may take place in numerous other environmental systems. In fact, any environmental risk assessment must consider such genetic changes during problem formulation, when establishing assessment and measurement endpoints, and when calculating RQs. Indeed, each such subsystem has unique considerations. For example, the atmosphere may be more important as a transport medium than as habitat for ecological risks, but is an important medium in terms of inhalation exposure route in humans and animals for airborne pollutants. For plant and microbial populations, the atmosphere connects the sources to the receptors. Soil and sediment differ from air and water in that they exist as matrices of all physical phases, making sorption and dissolution important chemodynamic considerations.

Bioengineers and other technical professionals must reduce uncertainty in the science that they apply in risk assessments. Without a readily understandable "benchmark" of environmental measurements, people can be left with possible misunderstandings, ranging from a failure to grasp a real environmental problem that exists to perceiving a problem even when things are better than or do not differ significantly from those of the general population. Two errors can occur when information is interpreted in the absence of sound science. The first mentioned above is the *false negative,* or reporting that there is no problem when one in fact exists. The need to address this problem is often at the core of the positions taken by environmental and public health agencies and advocacy groups. They ask questions like:

- What if the epidemiological study designed by the county health department shows no harm, but in fact toxic substances are in our drinking water?
- What if the substances that have been identified really do cause cancer but the tests are unreliable?
- What if people are being exposed to a biomaterial, but via a pathway other than the ones being studied (e.g. exposed to genetically modified organism in water when showering or a chemical released by the GMO is transformed into toxic compounds when the water is used for cooking)?
- What if there is a relationship that is different from the laboratory when this substance is released into the "real world," such as the difference between how a biomaterial behaves in the human body by itself as opposed to when other chemicals and microbes are present (i.e. the problem of "complex mixtures" and biological interactions)?

The other concern is, conversely, the *false positive.* This can be a major challenge for public health agencies with the mandate to protect people from exposures to environmental contaminants. We do not want to unnecessarily halt the march of progress in biotechnologies. For example, what if previous evidence shows that an agency had listed a compound transformed after release as a carcinogen, only to find that a wealth of new information is now showing that it has no such effect? This can happen if the conclusions were based upon faulty models, or models that only work well for lower organisms, but subsequently developed models have taken into consideration the physical, chemical, and biological complexities of higher-level organisms, including humans. False positives may force public health officials to devote inordinate amounts of time and resources to deal with so-called "non-problems." False positives also erroneously scare people about potentially useful products [19]. False positives, especially when they occur frequently, create credibility gaps between engineers and scientists and the decision makers. In turn the public loses confidence in science and scientific professionals. Risk assessments need to be based on high quality, scientifically based information. Put in engineering language, the risk assessment process is a "critical path" in which any unacceptable error or uncertainty along the way will decrease the quality of the risk assessment and will increase the likelihood of bad decisions.

While we almost never reach complete scientific consensus when it comes to ascribing cause, technical professionals rely on "weight of evidence," much as juries do in legal matters. The difference is that a single data point in science can undo mountains of contravening evidence. Statisticians tell us that we must reduce both type I and type II errors:

- A *type I error* is rejecting a hypothesis when it is true.
- A *type II error* is failing to reject the hypothesis when it is false.

So, this is simply another way to say that we wish to avoid false negatives and false positives. The challenge of environmental risks posed by many biotechnologies is that we are unlikely to have a sufficient amount of data to allay criticisms that we have either large type I or type II errors. There is an appearance or actuality of false negatives should an engineer or epidemiologist hint at a "clean bill of health." Those who may have to pay for corrective actions from which they receive no direct benefit may complain that the tests are flawed or misinterpreted, i.e. giving false positives (e.g. just because you find 100 ppb of a biomaterial in water does not mean any of the cancers were "caused" by exposures to this biomaterial).

Various groups frequently disagree strongly about the "facts" underlying risk premises, such as whether the data really show that these dosages "cause" cancer or whether they are just coincidental associations. Or, they may agree that they cause cancer, but not at the rate associated with realistic risk scenarios. Or, they may disagree with the scientific approach, e.g. the models are not appropriate to estimate and to predict risks based on the concentrations of chemical X to which people would be exposed (e.g. a conservative model may show high exposures and another model, with less protective algorithms, such as faster deposition rates, may show very low exposures). Or, they may argue that even if the algorithms of the models are valid, the measurements taken are not representative of real-world exposures.

Biotechnological decisions also involve regulatory and policy uncertainties and arguments about the level of protection. For example, should public health be protected so that only one additional cancer would be expected in a population of a million or one in ten thousand? If the former (10^{-6} cancer risk) were required, the plant would have to lower emissions of chemical X far below the levels that would be required for the latter (10^{-4} cancer risk). Or, do the agricultural, medical or industrial benefits outweigh any environmental costs? This is actually an argument about the value. Granted, tools for comparing benefits against costs are crude, but those comparing one benefit against another are even more uncertain. For example, even life is monetized, i.e. a dollar value is placed quite frequently on a prototypical human life, or even expected remaining lifetimes. These are commonly addressed in actuarial and legal circles. For example, risk assessor Paul Schlosser states:

> The processes of risk assessment, risk management, and the setting of environmental policy have tended to carefully avoid any direct consideration of the value of human life. A criticism is that if we allow some level of risk to persist in return for economic benefits, this is putting a value on human life (or at least health) and that this is

> inappropriate because a human life is invaluable – its value is infinite. The criticism is indeed valid; these processes sometimes do implicitly put a finite, if unstated, value on human life.
>
> A bit of reflection, however, reveals that in fact we put a finite value on human life in many aspects of our society. One example is the automobile. Each year, hundreds of thousands of US citizens are killed in car accidents. This is a significant risk. Yet we allow the risk to continue, although it could be substantially reduced or eliminated by banning cars or through strict, nation-wide speed limits of 15 or 20 mph. But we do not ban cars and allow speeds of 65 mph on major highways because we derive benefits, largely economic, from doing so. Hence, our car "policy" sets a finite value on human life.
>
> You can take issue with my car analogy because, when it comes to cars, it is the driver who is taking the risk for his or her own benefit, while in the case of chemical exposure, risk is imposed on some people for the benefit of others. This position, however, is different from saying that a human life has infinite value. This position says that a finite value is acceptable if the individual in question derives a direct benefit from that valuation. In other words, the question is then one of equity in the risk–benefit tradeoff, and the fact that we do place a finite value on life is not of issue. [20]

Another way to address this question is to ask, "How much are we willing to spend to save a human life?" Table 6.7 provides one group's estimates of the costs to save one human life. From what I can gather from the group that maintains the website sharing this information, they are opposed to much of the "environmentalist agenda," and their bias colors these data. However, their method of calculating the amount of money is fairly straightforward. If nothing else, the amounts engender discussions about possible risk tradeoffs since the money may otherwise be put to more productive use.

Schlosser asks, "How much is realistic?" He argues that a line must be drawn between realistic and absurd expenditures. He states:

> In some cases, risk assessment is not used for a risk–benefit analysis, but for comparative risk analysis. For example, in the case of water treatment one can ask: is the risk of cancer from chlorination by-products greater than the risk of death by cholera if we do not chlorinate? Similar, if a government agency has only enough funds to clean up one of two toxic waste sites in the near future, it would be prudent to clean up the site which poses the greatest risk. In both of these cases, one is seeking the

Table 6.7 Regulation cost of saving one life

Activity	Cost ($US)
Auto passive restraint/seat belt standards	100,000.00
Aircraft seat cushion flammability standard	400,000.00
Alcohol and drug control standards	400,000.00
Auto side door support standards	800,000.00
Trenching and excavation standards	1,500,000.00
Asbestos occupational exposure limit	8,300,000.00
Hazardous waste listing for petroleum refining sludge	27,600,000.00
Cover/remove uranium mill tailings (inactive sites)	31,7000,000.00
Asbestos ban	110,700,000.00
Diethylstilbestrol (DES) cattle feed ban	124,800,000.00
Municipal solid waste landfill standards (proposed)	19,107,000,000.00
Atrazine/Alachlor drinking water standard	92,069,700,000.00
Hazardous waste listing for wood preserving chemicals	5,700,000,000,000.00

Source: P.M. Schlosser (1997). Risk assessment: the two-edged sword. http://pw1.netcom.com/~drpauls/just.html; accessed on August 25, 2009.

course of action which will save the greatest number of lives, so this does not implicitly place a finite value on human life. (In the second example, the allocation of finite funds to the government agency does represent a finite valuation, but the use of risk assessment on how to use those funds does not.) [21]

Human beings are fallible, thus they are not always the best assessors or predictors of value. So, how do arguments about where to place value and the arguments made by Schlosser and Feldman fit with environmental decision making? The first step in using risk information is evaluating its usefulness in cause–effect relationships.

CHEMICAL INDICATORS OF BIOLOGICAL AGENTS

One of the challenges to reducing biotechnological risks is that the risks are often a combination of biological and chemical species. For example, a bioreactor uses abiotic chemicals that may be hazardous. The same bioreactor makes use of microbes that can be pathogenic or otherwise be hazardous. The microbes may also produce chemical toxins. The microbes may also generate spores and other biological substances. Thus, a cacophony of hazards can result from a single biotechnological operation.

Risk assessment must be based on a sufficient amount of reliable information. The ability to gather the data from which this information is derived varies. Often, chemical analysis is more straightforward than biological analysis. For example, identifying genetic material is highly complex and prone to uncertainty. Spores of the same fungus may appear quite similar. Pollen can be very diverse in morphology, even when analyzed by electron microscopy. Thus, chemical indicators may be useful in exposure and risk assessments. For example, rather than qualitative descriptions of lichens, soil biota, fungi, etc., a specific chemical compound may be measured. Ratios of various compounds can also be used to indicate specific genera (see Table 6.8).

Table 6.8 Chemical compounds (i.e. saccharides) commonly found in atmospheric aerosols

Compound	Source
Primary sugars (mono- and disaccharides)	
Arabinose	Lichens
Fructose	Lichens Soil biota
Galactose	Soil biota
Glucose	Fungi Lichens Soil biota Wood burning
Mannose	Soil biota
Xylose	Soil biota
Maltose (monohydrated)	Soil biota
Sucrose	Plants Soil biota
Mycose (Trehalose)	Yeast Bacteria, fungi Soil biota

(Continued)

Table 6.8 Chemical compounds (i.e. saccharides) commonly found in atmospheric aerosols—cont'd

Compound	Source
Sugar alcohols	
Arabitol	Fungi, lichens
Erythritol	Lichens Soil biota
Glycerol	Soil biota
Inositol	Soil biota
Mannitol	Fungal spores Fungi Lichens Soil biota
Sorbitol	Bacteria Lichens Soil biota
Xylitol	Fruits, berries, hardwood Soil biota
Anhydrosugars	
Galactosan (1,6-anhydro-β-d-galactopyranose)	Wood burning
Levoglucosan (1,6-anhydro-β-d-glucose, 1,6-anhydro-β-d-glucopyranose)	Wood burning
Mannosan (1,6-anhydro-β-d-mannopyranose)	Wood burning
1,6-Anhydrogluco-furanose	Wood burning

Source: A. Caseiro, I.L. Marr, M. Claeys, A. Kasper-Giebl, H. Puxbaum and C.A. Pio (2007). Determination of saccharides in atmospheric aerosol using anion-exchange high-performance liquid chromatography and pulsed-amperometric detection. *Journal of Chromatography A* 1171 (1-2): 37–45.

Particulate matter is measured throughout the world since it is a well-documented pollutant. Often, the matter consists of carbonaceous material, which is categorized as either elemental carbon or organic carbon. Elemental carbon (EC) consists of common residues of combustion, e.g. soot; whereas organic carbon (OC) is the carbon that has combined with other elements to form complex compounds, including those emitted by plants and most human activities.

The ratio (OC:EC) is useful in determining possible sources of air pollution. The OC fraction can be directly emitted as a particle (known as a primary aerosol) or as a gas that subsequently changes phase to a liquid or solid (i.e. a secondary aerosol). The fact that both primary OC and EC are predominantly emitted from combustion sources means that EC can be used as a tracer for primary combustion-generated OC. The formation of secondary organic aerosol (SOA) increases the ambient concentration of OC and the ambient OC:EC ratio. Thus when the OC:EC ratios exceed a certain expected primary emission ratio, SOA is probably being formed.

The logic of the OC:EC ratio might be applied to other biogenic sources, such as an indication of spores. Most of the biogenic material, e.g. pollen and spores, is in the so-called coarse fraction (particles with aerodynamic diameter >10 μm), along with metals, soil dust, sea salt, and nitrate. Thus, these particles could be analyzed chemically for the compounds in the left column of Table 6.6, compared to organic carbon content and mapped in two dimensions. This could directly indicate the movement of biomass and indirectly indicate the gene flow (movement of pollen and spores). This is been done to some extent with spores from *Cladosporium spp., Aspergillus spp., Penicillium spp.,* and *Alternaria spp.* [22].

RISK CAUSES

Zero risk can only occur when either the hazard does not exist or the exposure to that hazard is zero. Association of two factors, such as the level of exposure to a compound and the occurrence of a disease, does not necessarily mean that one necessarily "causes" the other. Often, after study, a third variable explains the relationship. However, it is important for science to do what it can to link causes with effects. Otherwise, corrective and preventive actions cannot be identified. So, strength of association is a beginning step toward cause and effect (See Discussion Box: Sir Austin Bradford Hill, A Pioneer in Causality). A major consideration in strength of association is the application of sound technical judgment of the weight of evidence. For example, characterizing the weight of evidence for carcinogenicity in humans consists of three major steps [23]:

characterization of the evidence from human studies and from animal studies individually;
combination of the characterizations of these two types of data to show the overall weight of evidence for human carcinogenicity; and,
evaluation of all supporting information to determine if the overall weight of evidence should be changed.

Note that none of these steps is absolutely certain.

Students are rightfully warned in their introductory statistics courses not to confuse association with causality. One can have some very strong statistical associations that are not causal. For example, if one were to observe ice-cream eating in Kansas City and counted the number of people wearing shorts, one would find a strong association between shorts-wearing and ice-cream-eating. Does wearing shorts cause more people to eat more ice-cream? In fact, both findings are caused by a third variable, ambient temperature. Hotter temperatures drive more people to wear shorts *and* to eat more ice-cream.

People have a keen sense of observation, especially when it has to do with the health and safety of their families and neighborhoods. They can "put 2 and 2 together." Sometimes, it seems that as engineers we are asked to tell them that 2 + 2 does *not* equal 4. That cluster of cancers in town may have nothing to do with the green gunk that is flowing out of the abandoned building's outfall. But in their minds, the linkage is obvious.

Sound science requires reliable data and appropriate means of interpreting these data. If causality is assigned when, in fact, none really exists, this not only misrepresents the findings of a given study, but can lead to erroneous assumptions in future investigations. The medical community in the mid-twentieth century was wrestling with how to move beyond simple associations to causality, as cancer was increasingly understood. Although factually correct, simply saying that study after study indicated an association between exposures to a particular physical, chemical, and biological agent would be intellectually unsatisfying and a likely barrier to scientific advancement. What was needed was an agreed-upon set of guidelines to help researchers to determine just how likely an agent was in *causing cancer.* Possible causes of cancer were being explored and major research efforts were being directed at myriad physical, chemical, and biological agents. Thus, there needed to be some manner of sorting through findings to see what might be causal and what would more likely be spurious. Sir Austin Bradford Hill (see Discussion Box) is credited with articulating key criteria that need to be satisfied to attribute cause and effect in medical research (see Chapter 3) [24]. The factors to be considered in determining whether exposure to a chemical or microbial agent might elicit an environmental effect include:

- Criterion 1: Strength of Association
- Criterion 2: Consistency
- Criterion 3: Specificity
- Criterion 4: Temporality
- Criterion 5: Biologic Gradient
- Criterion 6: Plausibility
- Criterion 7: Coherence
- Criterion 8: Experimentation
- Criterion 9: Analogy

DISCUSSION BOX

Sir Austin Bradford Hill, a Pioneer in Causality

In 1965 English epidemiologist and statistician Sir Austin Bradford Hill made a major contribution to the science of epidemiology and risk assessment by publishing his famous paper in which he recommended nine guidelines for establishing the relationship between environmental exposure and effect. Hill meant for the guidelines to be just that – guidelines, and not an absolute test for causality. A situation does not have to meet Hill's nine criteria to be shown to be causally related.

In the introduction to his paper, Hill acknowledges this by suggesting that there will be circumstances where not all of the nine criteria need to be met before action is taken. He recommended that action may need to be taken when the circumstances warrant. In his opinion, in some cases "the whole chain may have to be unraveled" or in other situations "a few links may suffice."

The case of the 1853 cholera epidemic in London, concluded by John Snow to be water-borne, and controlled by the removal of the pump handle, is a classic example where only a few links were understood.

The risk manager must be able to evaluate objectively the data on which the risk assessments are based. In assessing risks, some of Hill's criteria are more important than others. Risk assessments rely heavily on strength of association, e.g. to establish dose-response relationships. Coherence is also very important. Animal and human data findings, for example, should agree. Biological gradient is crucial, since this is the basis for dose-response (the larger the dose, the greater the biological response).

Temporality is essential since the cause must precede the effect. However, this is sometimes difficult to see in some instances, such as when the exposures to suspected agents have been continuous for decades and the health data are only recently available. Or, in the case of biomaterials, exposures to unmodified organisms may be slightly different from genetically modified organisms, so the actual beginning of exposure to the genetically engineered material is not yet completely established.

Linking cause and effect is often difficult. With emergent biotechnologies, the best we can do is to be upfront and clear about the uncertainties and the approaches we use. Environmental risk by nature addresses outcomes that may never materialize anywhere, what Aristotle called "probable impossibilities." This includes both adverse and beneficial outcomes. From a statistical perspective, it is extremely likely that cancer will not be eliminated during our lifetimes (a benefit that is probably impossible). But, the efforts to date have shown great progress toward reducing risks from several forms of cancer. This risk reduction can be attributed to a number of factors, including changes in behavior (smoking cessation, dietary changes, and improved lifestyles), source controls (less environmental releases of cancer-causing agents), and the reformulation of products (substitution of chemicals in manufacturing processes).

Risk characterization is the stage where the scientist summarizes the necessary assumptions, describes the uncertainties, and determines the strengths and limitations of the analyses. The risks are articulated by integrating the analytical results, interpreting adverse outcomes, and describing the uncertainties and weights of evidence.

Risk assessment is a process distinct from risk management, where actions are taken to address and reduce the risks. But the two are deeply interrelated and require continuous feedback with each other. In addition, risk communication between the scientific community and the lay public further complicates the implementation of the risk assessment and management processes. What really sets risk assessment apart from the actual management and policy decisions is that the risk assessment must follow the rigors of the method.

Risk assessments must be objective. Biotechnologists must avoid preconceptions. Frequently, the researcher is quite convinced of the value of the research and searches for facts to support it. The general public expects that its scientists' arguments are based in first principles. We must be careful that this "advocacy science" or, as some might call it, "junk science," does not find its way into biotechnology. There is a canon that is common in most engineering codes that tells us we need to be "faithful agents." This, coupled with an expectation of competency, requires us to be faithful to the first principles of science. This is not to say that biotechnologists have the luxury to ignore the wishes of sponsors and other influential actors, but since the scientists are the ones with their careers and reputations riding on these decisions, they must clearly state when an approach is scientifically unjustifiable.

Unfortunately, many scientific bases for decisions are far removed from first principles. For example, we know how fluids move through conduits (with thanks to Bernoulli *et al.*), but other factors come into play when we estimate how a biomaterial moves through very small vessels (e.g. intercellular transport) and how these change in the chaotic conditions of the environment. The combination of synergies and antagonisms at the molecular and cellular scales makes for uncertainty. Combine this with uncertainties about the effects of enzymes and other catalysts in the cell and we propagate even greater uncertainties. So, at the meso-scale (e.g. a wastewater treatment plant) an engineer may be fairly confident about the application of first principles of contaminant transport, but the biomechanical engineer looking at the same contaminant at the nanoscale is not so confident. In the void of certainty, e.g. at the molecular scale, some irrational arguments are made about what does or does not happen. Biotechnologists had better be prepared for some off-the-wall ideas of how the world works. New hypotheses for biological reactions will be put forward. Some will be completely unjustifiable by physical and biological principles, but they will sound sufficiently plausible to those not immersed in one's area of biotechnological specialization. The challenge of the scientist is to sort through this morass without becoming closed-minded. After all, many scientific breakthroughs have been considered crazy when first proposed (recalling Copernicus, Einstein, Bohr, and Hawking, to name a few). But, even more were in fact wrong and unsupportable upon scientific scrutiny.

In addition to the differences in scientific training and understanding, the disconnections between the public and the scientist can also result from perspective. If neighbors of a biotechnological laboratory live with fear everyday, scientists must respect that fear, even if the scientist and engineer correctly see little risk. A daunting challenge is that systems engineering and biotechnologies are changing so rapidly. We live in a time when misreading the slightest nuance of a design or operation can be the difference between success and failure. Often we are not immediately even sure whether we are succeeding or failing. In fact, we may not know whether the fruition of our ideas in many emerging technologies will be for good or ill until well after the research. The explosion in emergent technologies is like having countless chests before us, some are treasure troves, but others are a Pandora's box [25].

Anything that changes rapidly is difficult to measure. Those who are engaged in the research and the practice of new technologies may not recognize the hazards and pitfalls. In fact, those directly involved in the advancement may be the worst at appraising its worth and estimating its risks. The innovators have a built-in conflict of interest, which works against the ability to serve as society's honest brokers of emerging technologies.

H.A.L. Fisher observed that "there can be no generalizations, only one safe rule for the historian: that he should recognize in the development of human destinies the play of the contingent and the unforeseen ..." and that "the ground gained in one generation may be lost by the next" [26]. Fisher's advice and admonition to the historian holds for the biotechnology. It envisages the famous counsel of George Santayana [27]:

> Progress, far from consisting in change, depends on retentiveness. ... Those who cannot remember the past are condemned to repeat it.

What we remember can save us from much despair and needless waste and failure in the long run. We forget important events at our own peril. But, in the instance of an emerging technology, what is it exactly that we are supposed to remember? It is quite illustrative to consider what the experts were saying over a decade ago about biotechnology. Were they correct? Did the scientific community follow through on their recommendations? Let us consider a few observations.

One overarching theme in the 1980s seemed to be the overall positive view of biotechnology by those engaged in the supporting research. In fact, scientists often took on the role of apologists and advocates. They even opposed the public perceptions that an unknown and untested technology should not be rushed upon the scene without a thorough risk assessment. For example, the US Environmental Protection Agency sponsored a workshop in 1986 on biotechnology and pollution control [28]. The workshop's goals were "to examine the barriers to and incentives for commercialization; to develop recommendations for promoting, evaluating, and regulating field testing; and to identify strategies to foster the development and commercialization of biotechnology control products." The workshop materials complained that environmental biotechnologies were lagging behind other sectors (e.g. agricultural and medical).

This was the perspective of the agency responsible for regulating the products derived from biotechnological enterprises! Certainly, the need for better pollution remediation was and is a major concern. The workshop rightly pointed out this need. However, the almost exclusive focus on the benefits of biotechnologies completely eclipsed concerns about possible downstream impacts.

It is one thing to fail, but quite another not to refuse to learn from our failures. As engineers, we must consider the reasons and events that led to the failure in hopes that corrective actions and preventive measures are put in place to avoid their reoccurrence. This is not easy and is almost always complicated, especially for systems as elegant and complex as the human species.

The difference between success and failure often hinges on very subtle hints in the bioengineering process. Every failure results from a unique series of events. Human factors must always be considered in any design implementation. Often, seemingly identical situations lead to very different conclusions. In fact, the mathematics and statistics of failure analysis are some of the most complicated, relying on nonlinear approaches and chaos, making use of nontraditional statistical methods, such as Bayesian theory [29]. This is not an excuse to acquiesce. To the contrary, the complexities require that engineering creativity and imagination can and must be used to envisage possible failures and to take steps to ensure success (see Case Study: Managing Risks by Distinguishing between Progenitor and Genetically Modified Microbes). The good news is that computational and other tools are improving,

CASE STUDY

Managing Risks by Distinguishing between Progenitor and Genetically Modified Microbes

> *Factors to be considered in determining the level of containment include agent factors such as: virulence, pathogenicity, infectious dose, environmental stability, route of spread, communicability, operations, quantity, availability of vaccine or treatment, and gene product effects such as toxicity, physiological activity, and allergenicity. Any strain that is known to be more hazardous than the parent (wild-type) strain should be considered for handling at a higher containment level. Certain attenuated strains or strains that have been demonstrated to have irreversibly lost known virulence factors may qualify for a reduction of the containment level compared to the Risk Group assigned to the parent strain …*
>
> *National Institutes of Health, Guidelines for Research Involving Recombinant DNA Molecules, April 2002 Revisions [30]*

Granted, the NIH guidelines on genetic modification are mainly focused on research and development experimentation involving recombinant DNA, but the guidelines do address "physical containment guidelines for large-scale (greater than 10 liters of culture) research or production involving viable organisms containing recombinant DNA molecules." Bioremediation projects are generally experimental in nature, since the bioengineer must match never before seen conditions to generally available, scientifically credible approaches. Interestingly, a number of the genera of microorganisms listed by NIH as biohazardous agents (see Table 6.9) have been used in bioremediation projects. In particular, they are listed as agents that are associated with human disease that is rarely serious and for which preventive or therapeutic interventions are *often* available (see Table 6.10). However, the larger the project, the greater the number of variables and potential for breaches of physical containment. For instance, do the unique environmental conditions in a bioremediation project increase the potential for the agents becoming more persistent, more toxic or more virulent (e.g. more allergenic)?

It is beneficial to use the *Actinobacillus* species to breakdown toxic compounds like phenol [31] or to improve strains of *Rhodococcus* using rDNA technology [32], for example, but are there contravening outcomes and risks that are not fully understood?

Human and ecosystem risk assessments must clearly explain all uncertainty and variability of every component of the assessment. In particular, risk assessment documents must include explanations of how uncertainty and variability propagate through the causal chain from hazard to exposure to risk and, ultimately to the risk decision. For example, if a decision must be made about limits to microbial growth that leads to fish kills (e.g. algal or dinoflagellate, such as blooms from *Pfisteria spp.*), risk assessments must document the quality and relevance of data and model quality throughout the entire system.

When data are limited, one option is to use a graphical structure to present the critical path from various factors, their interactions, and possible outcomes. From these, the systems can be perturbed (virtually) through space and time through a sequence of conditional probabilities. It may seem intuitive that the components of this system are interrelated and, furthermore, that observations about one relationship (such as whether or not a shellfish microbial concentration value leads to a lethal response in humans) should strengthen or weaken other relationships in the system. However, most modeling frameworks do not allow information to be passed both "forwards" and "backwards" through a system. Bayesian inference procedures, which involve identifying a prior probability distribution for system parameters, and updating those distributions through evidence (expressed through a likelihood function) to yield posterior probability distributions, provide an elegant solution to this problem.

Common statistical approaches to environmental predictions fall into three categories, each involving a different approach to quantifying the likelihood of an event relative to a set of all possible events. The first approach can be thought of as using a priori beliefs, which, in the case of a single roll of a six-sided die, might reflect an expectation that the die is fair, and therefore that the probability of each of the six possible outcomes (that is, 1, 2, . . . , 6) is exactly 1/6. A second approach is based on empirical evidence, in which the meaning of the underlying probability of events is based entirely on data. In the case of the six-sided die, this approach might involve rolling the die repeatedly and estimating the probability of each outcome to be its observed relative frequency. This is seldom possible in environmental situations since the variables and outcomes are so complex and the data about them so limited.

(Continued)

Table 6.9 Basis for the classification of biohazardous agents by risk group by the National Institutes of Health

Risk Group 1 (RG1)	Agents that are not associated with disease in healthy adult humans
Risk Group 2 (RG2)	Agents that are associated with human disease that is rarely serious and for which preventive or therapeutic interventions are *often* available
Risk Group 3 (RG3)	Agents that are associated with serious or lethal human disease for which preventive or therapeutic interventions *may be* available (high individual risk but low community risk)
Risk Group 4 (RG4)	Agents that are likely to cause serious or lethal human disease for which preventive or therapeutic interventions are *not usually* available (high individual risk and high community risk)

Source: National Institutes of Health (2002). Guidelines for Research Involving Recombinant DNA Molecules; http://oba.od.nih.gov/oba/rac/guidelines_02/APPENDIX_B.htm#AppxB_Tbl1; accessed August 29, 2009.

Table 6.10 Bacterial agents deemed by the National Institutes of Health (Risk Group 2) to be associated with human disease that is rarely serious and for which preventive or therapeutic interventions are *often* available[a]

Bacteria genera
Acinetobacter baumannii (formerly *Acinetobacter calcoaceticus*)
Actinobacillus
Actinomyces pyogenes (formerly *Corynebacterium pyogenes*)
Aeromonas hydrophila
Amycolata autotrophica
Archanobacterium haemolyticum (formerly *Corynebacterium haemolyticum*)
Arizona hinshawii – all serotypes
Bacillus anthracis
Bartonella henselae, B. quintana, B. vinsonii
Bordetella, including *B. pertussis*
Borrelia recurrentis, B. burgdorferi
Burkholderia (formerly *Pseudomonas* species) except those listed in Appendix B-III-A (RG3)
Campylobacter coli, C. fetus, C. jejuni
Chlamydia psittaci, *C. trachomatis*, *C. pneumoniae*
Clostridium botulinum, Cl. chauvoei, Cl. haemolyticum, Cl. histolyticum, Cl. novyi, Cl. septicum, Cl. tetani
Corynebacterium diphtheriae, C. pseudotuberculosis, C. renale
Dermatophilus congolensis
Edwardsiella tarda
Erysipelothrix rhusiopathiae
Escherichia coli – all enteropathogenic, enterotoxigenic, enteroinvasive and strains bearing K1 antigen, including *E. coli* O157:H7
Haemophilus ducreyi, H. influenzae
Helicobacter pylori
Klebsiella – all species except *K. oxytoca* (RG1)
Legionella, including *L. pneumophila*
Leptospira interrogans – all serotypes
Listeria
Moraxella

Table 6.10 Bacterial agents deemed by the National Institutes of Health (Risk Group 2) to be associated with human disease that is rarely serious and for which preventive or therapeutic interventions are *often* available[a]—cont'd

Mycobacterium (except those listed in NIH Guidelines, Appendix B-III-A (RG3)), including *M. avium* complex, *M. asiaticum, M. bovis* BCG vaccine strain, *M. chelonei, M. fortuitum, M. kansasii, M. leprae, M. malmoense, M. marinum, M. paratuberculosis, M. scrofulaceum, M. simiae, M. szulgai, M. ulcerans, M. xenopi*
Mycoplasma, except *M. mycoides* and *M. agalactiae* which are restricted animal pathogens
Neisseria gonorrhoeae, *N. meningitidis*
Nocardia asteroides, N. brasiliensis, N. otitidiscaviarum, N. transvalensis
Rhodococcus equi
Salmonella, including *S. arizonae, S. cholerasuis, S. enteritidis, S. gallinarum-pullorum, S. meleagridis, S. paratyphi*, A, B, C, *S. typhi*, *S. typhimurium*
Shigella, including *S. boydii, S. dysenteriae,* type 1, *S. flexneri, S. sonnei*
Sphaerophorus necrophorus
Staphylococcus aureus
Streptobacillus moniliformis
Streptococcus, including *S. pneumoniae, S. pyogenes*
Treponema pallidum, T. carateum
Vibrio cholerae, V. parahemolyticus, V. vulnificus
Yersinia enterocolitica

[a]*The group also includes fungi, parasitic agents, and viruses that are not listed here but are available in: National Institutes of Health (2002). Guidelines for Research Involving Recombinant DNA Molecules; http://oba.od.nih.gov/oba/rac/guidelines_02/APPENDIX_B.htm#AppxB_Tbl1); accessed August 29, 2009.*

Bayesian statistics is the third approach, which provides a mechanism for combining a priori beliefs with potentially sparse empirical evidence to derive a posterior probability distribution. Bayes' theorem is:

$$P(A|B) = \frac{P(B|A)P(A)}{P(B)} \tag{6.2}$$

where $P(A)$ and $P(B)$ represent the marginal probabilities of events A and B, respectively, while $P(A \mid B)$ and $P(B \mid A)$ represent the conditional probabilities of event A given that event B has occurred, and of event B given that event A has occurred, respectively. The probability $P(A|B)$, in a Bayesian framework, is referred to as the posterior probability of event A, given that event B has occurred. In this context, Bayes' theorem states that the posterior probability of event A (that is, the probability of event A given that event B has occurred) equals the likelihood [written $P(B \mid A)$] times the prior probability distribution of event A [that is, $P(A)$], divided by the marginal distribution of event B. In this way, the prior probability distribution, the likelihood, and the posterior probability distribution both provide the framework for and serve as the necessary elements of a Bayesian statistical problem. In more practical terms, Bayes' theorem allows a priori beliefs about the probability of an event (or an environmental condition, or some other metric in the cause-to-event-to-outcome chains) to be combined with measurements. This combination provides a new and more robust posterior probability distribution (see Figure 6.7).

With advances in microbiology, omics, rDNA identification, and computational methods, it is reasonable to assume that methods will allow distinctions between the abundance of progenitor and genetically modified variations in the same species in water bodies, soil, and other ecosystems. As mentioned in previous chapters, microbes are genetically modified to express desired traits, such as better survival rates in hostile environments (e.g. toxic wastes), ability to degrade recalcitrant compounds, i.e. low biodegradability, and to alter microbe–environment interactions (e.g. biofilms between the soil particles and microbes).

These indeed are desirable traits for the task at hand, e.g. to breakdown wastewater and contaminated soil. However, at least two scenarios in the causal chain present possible uncertainties and

(Continued)

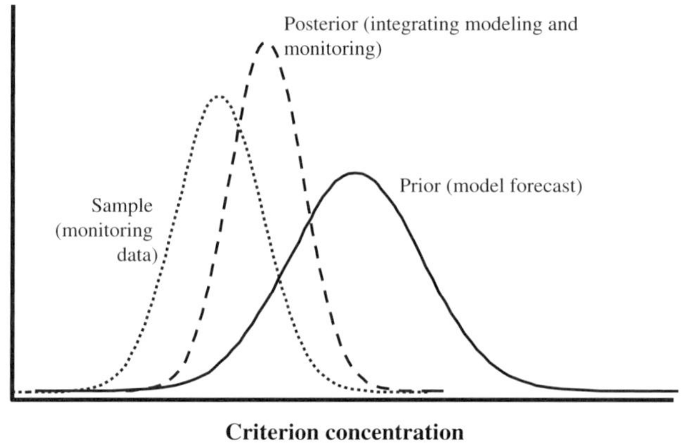

FIGURE 6.7
Bayesian approach combining measurements in the environment with a priori modeled information, yielding a more robust posterior distribution than would be possible with only contingent probabilities.
Source: Based on conversations with K. Reckhow, Duke University, Nicholas School of the Environment.

variabilities that must be considered. First, was it possible that in the process of selection and expression of these desirable traits that another undesirable trait was unintentionally transcripted and expressed? For example, are the more aggressive microbes now able to survive in environments other than the targeted one (horizontal gene flow)? Genetically modified organisms usually do not survive well beyond the first generation (i.e. "suicidal" properties result from the modified transcription). However, if they are better able to survive beyond the first generation (which is a typical situation) and reproduce with this selected trait, this can adversely affect the biochemodynamic balances in an ecosystem (e.g. biodiversity and species abundance ratios are adversely affected) [33].

The second scenario, which modifies the first scenario, considers whether and the extent to which the surviving microbes have a trait that is adverse to ecosystems and human health. For example, if the microbes are food for shellfish and have a higher toxicity to the shellfish, this would increase shellfish mortality, which adversely affects the food chain as the toxic agent moves throughout the food web and/or as the food (shellfish) at a lower trophic level is decreased, so that the predator–prey and other relationships change, altering the entire system.

A human health problem would occur if the genetically modified microbes (e.g. bacteria) are more virulent than the progenitors and/or they are more likely to manifest human endpoints (e.g. higher carcinogenicity, allergenicity, or antibiotic resistance). The aquatic fate, transport, and ingestion of disease-causing pathogenic organisms of fecal origin is affected by numerous potential sources of uncertainty including the fraction of viable organisms entering the aquatic environment, the rate of uptake (and subsequent survival) of those organisms in shellfish tissue, and the subsequent ingestion rate in humans. Intrinsic variability may also affect each of these potential indicators of human and environmental health risk, making it difficult or impossible to infer relationships in the system that cannot be observed (i.e. "black boxes") or for which limited measurement data and empirical evidence do not allow for a direct analysis of outcomes.

Staying with the shellfish scenario, measurements of shellfish toxicity, based on a particular metric of shellfish response to a particular pathogen, may provide information regarding not only the distinction of different pathogen strains in the aquatic environment, but also transmission rates of those pathogens into the shellfish tissue itself. Similarly, the outbreak of human disease (or similar epidemiological endpoint, e.g. allergenicity) in response to a particular exposure level to contaminated shellfish may provide additional evidence of pathogen strain partitioning, and of the relationship between toxicity levels in shellfish and impacts on safe levels of consumption by humans. These relationships are presented graphically in Figure 6.8.

In the hypothetical example in Figure 6.9, the a priori assumption is that a particular pathogen load entering an aquatic environment includes both genetically modified and progenitor strains, albeit less than 0.002 GMO strain abundance, i.e. 99.8% is non-genetically modified. The genetically modified strains are intentionally high to demonstrate the partitioning. However, as mentioned, this can be determined to some extent with microbial survival laboratory and field studies. Similar initial prior distributions are also assumed for the concentration of pathogens in shellfish (for a given *in situ* concentration and pathogen strain partitioning scenario) as well as the toxicity response of shellfish to a particular concentration, and the associated concentration level in humans ingesting the shellfish.

The Bayesian approach can enhance the predictions and improve risk management: for example, if after observing relationships between

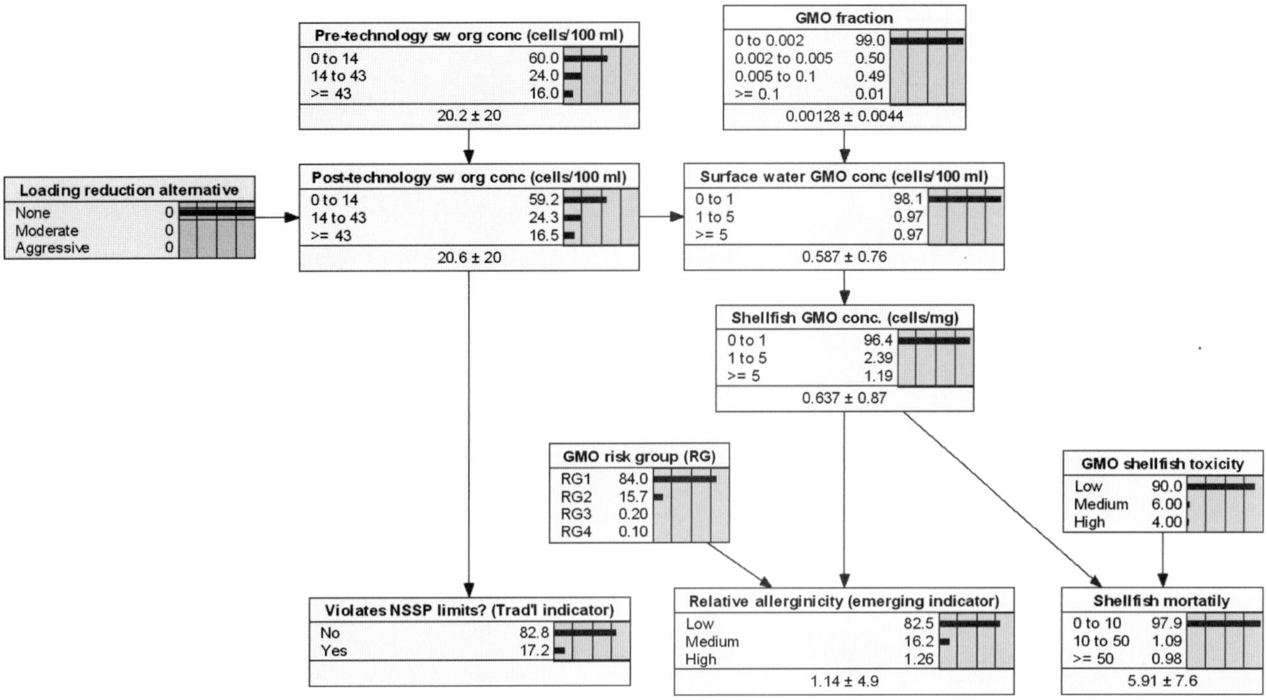

FIGURE 6.8

Cause–event–effect chain reflecting a priori assumptions for horizontal gene flow of bacteria injected into an aquifer to treat recalcitrant pollutants, with possible impacts on microbial survival in a shellfish population. This a priori model assumes, in particular, that there is a 99% chance that genetically-modified organisms (GMOs) make up less than 0.2% of the overall organism population (as indicated by the "GMO fraction" node). This overall organism population is, in turn, impacted by the treatment technology applied. In this model, no additional loading reduction measures are employed, and therefore pre- and post-technology concentration probability distributions are nearly identical (differences are attributed to the probabilistic nature of the underlying model). The uptake of GMOs by shellfish is reflected in the resulting distribution of shellfish tissue concentration values intervals: 0–1; 1–5; and >5 cells per mg shellfish tissue (hypothetical scenario and hypothetical data). The shellfish tissue concentration can propagate into adverse ecosystem (measured through shellfish mortality) and adverse human health (measured through relative allergenicity) impacts. In this a priori scenario, GMO impacts on shellfish health are expected to be relatively minor (with a slight increase in relative mortality), while GMO impacts on humans are also expected to be minor, depending on the GMO risk group. Relative allergenicity and shellfish mortality are emerging indicators of human and ecosystem health risks, although current wastewater management infrastructure decisions are based on meeting more traditional surface water quality standards such as those recommended by the National Shellfish Sanitation Program (NSSP). However, risk assessments will improve as methods for distinguishing GMOs from progenitor strains, especially when microbial virulence and persistence diverges between genetically modified and non-modified strains.
Source: Collaboration between D. Gronewold, US Environmental Protection Agency, and D. Vallero.

some of these variables, the bioengineer obtains an updated understanding of the system response and, in particular, a potentially robust approach to forecasting impacts on human health based on *in situ* concentrations or, if available, better evidence of shellfish toxicity. The decision framework allows the bioengineer, clients, and regulators to see what would happen if risk reduction actions were taken, i.e. decision option selection and optimization. For example, hydraulic and hydrological conditions could be changed so that less of the microbes would move off-site. Or, as shown in Figure 6.9, the bioengineer injects a smaller concentration of genetically modified (including host killing or suicidal expressions) microbes based on the downstream events (improved shellfish microbial concentrations and virulence). This intervention affects the critical path from in situ microbial populations to the off-site movement of microbes to shellfish uptake to the human food supply to impacts on human health. The critical path is usually branched. For example the shellfish uptake can also lead to impacts on ecosystems, including changes in conditions as evidenced by altered diversity, productivity, and sustainability. These critical paths and options are otherwise unobservable and, in most cases, difficult to quantify through empirical evidence alone.

Actually, the techniques available to the bioengineer to manage and reduce the risk of genetically modified microbes are improving and increasingly available. Numerous recombinant DNA techniques have been developed for microbes used in biodegradation of environmental contaminants and for synthesizing small molecules [34]. These include gene expression control mechanisms, mechanisms to contain and

(Continued)

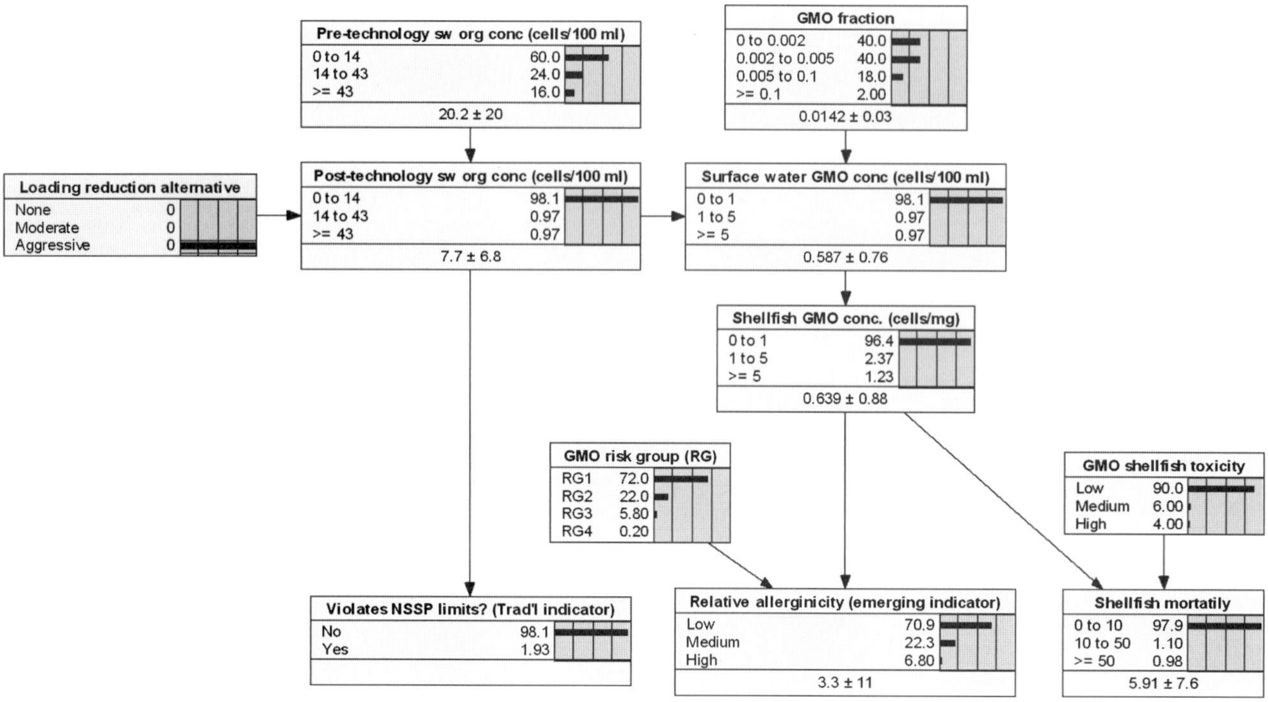

FIGURE 6.9

Cause–event–effect chain reflecting updated probability distributions for horizontal gene flow of bacteria injected into an aquifer to treat recalcitrant pollutants, with possible impacts on microbial survival in a shellfish population. The scenario differs from that in Figure 6.8 in that the municipality, responding to expected water quality standard violation frequencies, has adopted a more aggressive treatment technology, resulting in an expected water quality standard violation frequency of about 2% (compared to the pre-reduction scenario, which was expected to violate standards about 17% of the time). The scenario also reflects updated (i.e. posterior) probability distributions of genetically-modified organism (GMO) partitioning within the overall microbial population (reflected by the ''GMO fraction'' node) and the relative impacts of the GMO on human health (reflected by the ''relative allergenicity'' node). Thus, whereas technology improvements appear to provide adequate protection of human health under traditional metrics, the introduction of a GMO may set the stage for a new human and ecosystem health risk paradigm which must consider new relationships and sources of uncertainty. *Source:* Collaboration between D. Gronewold, US Environmental Protection Agency and D. Vallero.

control persistence of genetically modified microbes, site-directed and random mutagenesis applications to amplify the substrate range or activity of biodegradative enzymes, and methods for monitoring and tracking genetically engineered microorganisms after introduction, during bioremediation, and as they leave the containment.

The question that bioengineers must answer is whether the risks of the bioremediation project or other environmental biotechnological application have been properly addressed. The answer can only be answered with uncertainty. Have the microbes been sufficiently physically and biologically constrained? Is there enough certainty about the risks associated with the microbes to use them on a large scale (NIH's 10 liter culture, for example)? And are the effects to human health and the ecosystem condition worth these risks?

This method provides a risk-based approach for environmental and public health protection. For example, if the outcomes (bottom boxes of Figures 6.8 and 6.9) need to protect a populations so that only 1 in a million (10^{-6}) would be expected to be exposed to lethal levels of these microbes, the bioengineer can work backwards from this concentration and optimize the variables accordingly.

This is a single variable example. Obviously, microbial ecology is much more complex. Thus, numerous sources of data and information are often linked by various models. Let us take the example of controlling nutrient loading to prevent fish kills. There are numerous pathways for nutrients to reach a water body (see Figure 6.10). The events that lead to a fish kill from nutrient loading are shown in Figure 6.11.

The events and conditions are interconnected by a series of probabilities. That is, the dependencies are described by conditional probability distributions, e.g. the low dissolved oxygen is related to the oxygen demand in the sediment and the degree that the water is mixed, i.e. the temperature is stratified. Thus, this particular

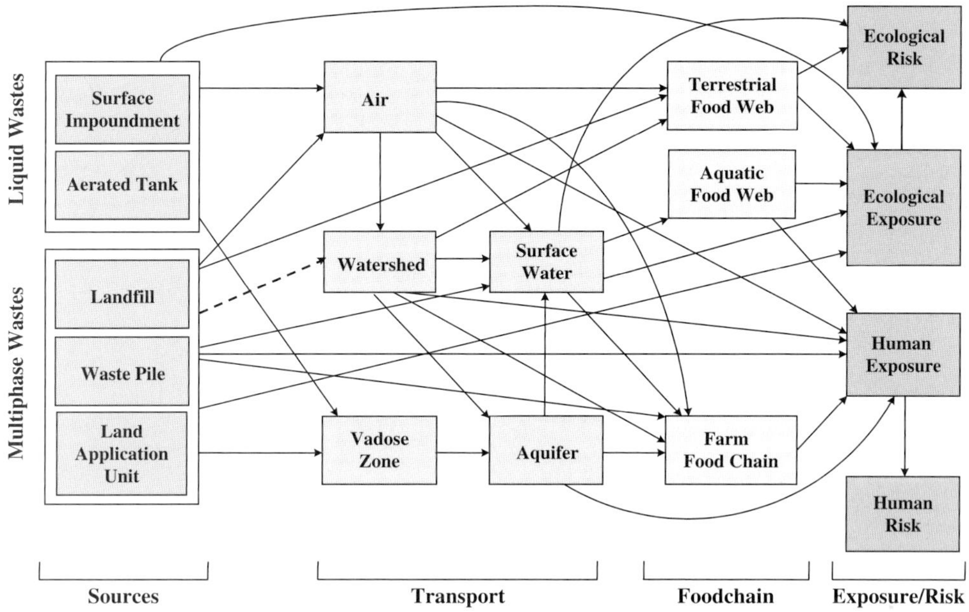

FIGURE 6.10
Environmental transport pathways. Compounds (nutrients, contaminants) and microbes move through the environment wherein they may reside in any of the boxes, before their fate, which is where risks occur.
Source: D.A. Vallero, K.H. Reckhow and A.D. Gronewold (2007). Application of multimedia models for human and ecological exposure analysis. International Conference on Environmental Epidemiology and Exposure. October 17, 2007. Durham, NC. Graphic provided by US Environmental Protection Agency, Athens, GA.

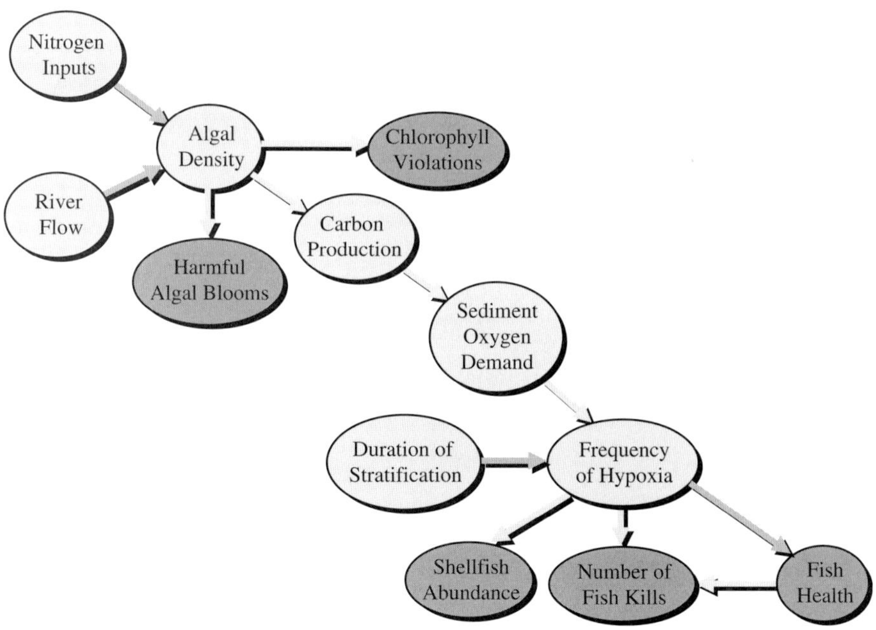

FIGURE 6.11
Flow of events and conditions leading to fish kills.
Source: D.A. Vallero, K.H. Reckhow and A.D. Gronewold (2007). Application of multimedia models for human and ecological exposure analysis. International Conference on Environmental Epidemiology and Exposure. October 17, 2007, Durham, NC.

conditional probability distribution is found as P(hypoxia | sediment oxygen demand | stratification). All of the interrelationships of the variables must have similar conditional probability distributions, which in turn have their own probabilities with other distributions which must be linked with models (see Figure 6.12). This can be important for genetically modified organisms and gene flow. For example, the conditional probability for survival may differ from progenitor strains and must be run through the causal chain, leading to a different outcome (more or less likelihood of a fish kill based on the differential response of various fish genera to the genetically modified microbes or the differential response of genetically modified versus non-modified fish species).

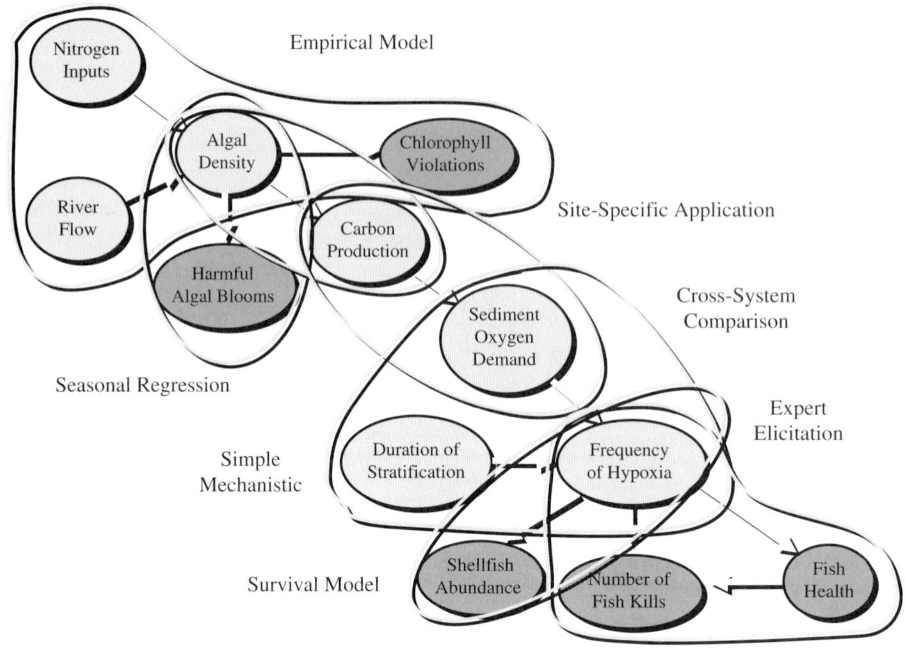

FIGURE 6.12
Linkages of data and models to arrive at estimate of fish kills due to nitrogen loading to a water basin. Each of the groups requires a conditional probability distribution, which is joined with the others to calculate the ultimate outcome (i.e. species abundance and overall system health).
Source: D.A. Vallero, K.H. Reckhow and A.D. Gronewold (2007). Application of multimedia models for human and ecological exposure analysis. International Conference on Environmental Epidemiology and Exposure. October 17, 2007. Durham, NC.

FAILURE: HUMAN FACTORS ENGINEERING

Human factors engineering (HFE) addresses the interface between humans and things, especially how we use these things. Human factor studies address all of the elements at this interface:

- systems performance;
- problems encountered in information presentation, detection, and recognition;
- related action controls;
- workspace arrangement; and
- skills required [35].

Any of these factors can contribute to failure. Sometimes the use itself is problematic, such as the repetitive actions of people using manufacturing or office equipment. From a bioethical perspective, HFE is needed to determine how devices fail and to suggest ways that such failures can be prevented. A design that does not properly account for problems in usage by medical practitioners or by the patients themselves is morally unacceptable. For example, a device is unacceptable if it does not account for variability in dexterity and finger length of practitioners in the operating room. Likewise, a pump delivering medication to a patient is also unacceptable if it requires an inordinate degree of user sophistication, especially if operator error can lead to major medical complications, especially if there is an available alternate design that is more user-friendly.

Concern with emerging technological developments is not the exclusive domain of neo-Luddites. Thoughtful people are simply asking for protections and commitments to precaution before going into an unbridled mode of research and applications. Some would also argue that research in many areas, including genetic engineering and human enhancement, has already passed the point of precaution, so that any attempts to control or even moderate the societal risk with more precaution is akin to changing a tire on a bus as it moves down the highway. It is physically possible, but certainly not very probable. You probably will have to stop the bus for a while. Knowing when to stop the bus requires some objective measure of success and failure.

Utility as a measure of success

Quantitative types, like most engineers, have a strong affinity for objective measures of successes. Thus, we gravitate to *usefulness* as a measure of success. Such utility *is indeed part* of any successful engineering enterprise. After all, engineers are expected to provide reasonable and useful products. In a word, we look for utility. It turns out that one of the major ethical constructs is *utilitarianism*. The first two dictionary [36] definitions of utilitarianism (Latin *utilis*, useful) are relevant to bioethics:

> The belief that the value of a thing or an action is determined by its utility.
> The ethical theory … that all action should be directed toward achieving the greatest happiness for the greatest number of people.

Since utilitarianism is based on outcome, it is a form of consequentialism. Engineers are expected to provide products, so such a theory is attractive. Furthermore, some of the tools used to determine the morality of a decision are objective and quantifiable. Chief among these is the benefit-to-cost (B/C) ratio. It is an attractive metric due to its simplicity and seeming transparency. To determine whether a project is worthwhile, one need only add up all of the benefits and put them in the numerator and all of the costs (or risks) and put them in the denominator. If the ratio is greater than 1, its benefits exceed its costs.

One obvious problem is that some costs and benefits are much easier to quantify than others. Some, like those associated with quality of life, are nearly impossible to quantify accurately. Further, the comparison of doing anything with doing nothing cannot always be captured with a benefit/cost ratio. Opportunity costs and risks are associated with taking no action (e.g. loss of an opportunity to apply an emerging technology may mean delay or nonexistent treatment of diseases). Simply comparing the *status quo* to costs and risks associated with a new technology may be biased toward no action. But the costs (time and money) are not the only reasons for avoiding action. The availability of a new monitoring device may invite more tests or drugs that, if not managed properly, could interfere with quality of life of patients, could carry its own risks, and could add costs to the public and the community (e.g. "preventive medicine" to avoid liability), with little actual benefit to anyone. So, it is not simply a matter of benefits *versus* cost, it is often one risk being traded for another.

Often, addressing contravening risk is a matter of a judgment with very little supporting data. It requires a good bit of intuition, which is a proven analytical tool in engineering. However, the greater the number of contravening risks that are possible, the more complicated such optimization routines become. Risk tradeoff is a very common phenomenon in everyday life. For example, local governments enforce building codes to protect health and safety. Often times, these added protections are associated with indirect, countervailing risks. For example, the costs of construction may increase safety risks via "income" and "stock" effects. The income effect results from pulling money away from family income to pay the higher mortgages, making it more difficult for the family to buy other items or services that would have protected them. The stock effect results when the cost of the home is increased, families have to wait to purchase a new residence, so they are left in substandard housing longer [37]. Such countervailing risks are common in biotechnological decisions. Consumers and scientists alike demand immediate and state-of-the-art technologies be pursued. Simultaneously, they may resist increased risks from income and stock effects by imposing what they believe to be costly and questionable testing. If a technology exists, it will probably be used (supply creates demand). Thus, the bioengineer is frequently asked to optimize two or more conflicting variables in many situations.

Indeed, abating risks that are in fact quite low *could* mean unnecessarily complicated and costly measures and precautions. It may also mean choosing the less acceptable alternative, i.e. one that in the long run may be more costly and deleterious to the public health. The risk assessment and risk perception processes differ markedly. Assessment relies on problem identification, data analysis, and risk characterization, including cost–benefit ratios. Perception relies on thought processes, including intuition, personal experiences, and personal preferences. Everyone has a way of identifying, estimating, and evaluating risks. However, engineers tend to be more comfortable operating in the middle column (using risk assessment processes), while the general public often use the processes in the far right column. One can liken this to the "left brained" engineer trying to communicate with a "right brained" audience. It can be done, so long as preconceived and conventional approaches do not get in the way.

The reasons for failure vary widely, but can be broadly categorized as mistakes, mishaps, and misdeeds. The terms all include the prefix "mis-" that is derived from Old English, "to miss." This type of failure applies to numerous ethical failures. However, the prefix "mis-" can connote something that is done "poorly," i.e. a mistake. It may also mean that an act leads to an accident because the original expectations were overtaken by events, i.e. a mishap. This is an all-too-common shortcoming of professionals, i.e. not upholding the levels of technical competence called for by their field. Medical and engineering codes of ethics, for example, include tenets and principles related to competence, such as only working in one's area of competence or specialty. Finally, "mis-" can suggest that an act is immoral or ethically impermissible, i.e. a misdeed. Interestingly, the theological derivation for the word "sin" (Greek: *hamartano*) means that when one a person has missed the mark, i.e. the goal of moral goodness and ethical uprightness, that person has sinned or has behaved immorally by failing to abide by an ethical principle, such as honesty and justice. Bioethical failures have come about by all three means. The lesson from Santayana is that we must learn from all of these past failures. Learning must be followed by new thinking and action, including the need to forsake what has not worked and shift toward what needs to be done.

Engineering failure can be categorized into five types. Whether the failure is deemed unethical is determined by the type of failure and the circumstances contributing to the failure.

Failure Type 1: Mistakes and miscalculations

Sometimes engineers make mistakes and their works fail due to their own miscalculations, such as when parentheses are not closed in computer code, leading to errors in predicting pharmacokinetic behavior of a drug. Some failures occur when engineers do not correctly estimate the corrosivity that occurs during sterilization of devices (e.g. not properly accounting for fatigue of materials resulting from high temperature and pressure of an autoclave). Such mistakes are completely avoidable if the physical sciences and mathematics are properly applied.

Failure Type 2: Extraordinary natural circumstances

Failure can also occur when factors of safety are exceeded due to extraordinary natural occurrences. Engineers can, with fair accuracy, predict the probability of failure due to natural forces like wind loads and they design the structures for some maximum loading, but these natural forces can be exceeded. Engineers design for an acceptably low probability of failure – not for 100% safety and zero risk. However, tolerances and design specifications must be defined as explicitly as possible.

The tolerances and factors of safety have to match the consequences. A failure rate of 1% may be acceptable for a household compost pile, but it is grossly inadequate for bioreactor performance. And, the failure rate of devices may spike up dramatically during an extreme natural event (e.g. power surges during storms). Equipment failure is but one of the factors that lead to uncontrolled environmental releases. Conditional probabilities of failure should be known. That way, back-up systems can be established in the event of extreme natural events, like hurricanes, earthquakes, and tornados. If appropriate, contingency planning and design considerations are factored into operations, the engineer's device may still fail, but the failure would be considered reasonable under the extreme circumstances.

Failure Type 3: Critical path

No engineer can predict all of the possible failure modes of every structure or other engineered device, and unforeseen situations can occur. A classical, microbial case is the Holy Cross College football team hepatitis outbreak in 1969 [38]. A confluence of events occurred that resulted in water becoming contaminated when hepatitis virus entered a drinking water system. Modeling such a series of events would probably only happen in scenarios with relatively high risks associated agents and conditions that had previously led to an adverse outcome.

In this case, a water pipe connecting the college football field with the town passed through a golf course. Children had opened a water spigot on the golf course, splashed around in the pool they created, and apparently discharged hepatitis virus into the water. A low pressure was created in the pipe when a house caught on fire and water was pumped out of the water pipes. This low pressure sucked the hepatitis-contaminated water into the water pipe. The next morning the Holy Cross football team drank water from the contaminated water line and many came down with hepatitis. The case is memorable because it was so highly unlikely – a combination of circumstances that were impossible to predict. Nevertheless, the job of engineers is to do just that, to try to predict the unpredictable and thereby to protect the health, safety, and welfare of the public.

This is an example of how engineers can fail, but may not be blamed for the failure, since such a set of factors had not previously led to an adverse action. If the public or their peers agree that the synergies, antagonisms, and conditional probabilities of the outcome could not reasonably be predicted, the engineer is likely to be forgiven. However, if a reasonable person deems that a competent engineer *should* have predicted the outcome, the engineer is to that extent accountable.

Indeed, there is always a need to consider risks by analogy, especially when related to complex, biological systems. Many complex situations are so dynamic and multifaceted that there is never an exact precedent for the events and outcomes for any real-world scenario. For example, every bioremediation project will differ from every other such project, but there are analogous situations related to previous projects that can be applied to a particular project. Are the same strains of microbes being used? Are the physical conditions similar, such as soil texture, and biological conditions, such as microbial ecology and plant root systems, ambient temperatures and daily season variabilities, to those in previous studies? Are structurally similar compounds being degraded? Are the volumes of wastes and concentrations similar?

There are numerous examples of ignoring analogies to previous situations that led to adverse outcomes. Although it was not a biotechnological operation, the tragic industrial accident at Bhopal, India, illustrates this type of engineering failure. Perhaps the biggest air pollution disaster of all time occurred in Bhopal in 1984 when a toxic cloud drifted over the city from the Union Carbide pesticide plant. If 12,000 death claims have been filed, based on the fact that many poor and unrepresented people died as a result of the exposures to the plume, this gas leak could realistically have been responsible for the premature death of 20,000 people and permanent injuries to about 120,000. Failure is often described as an outcome when not applying the science correctly (e.g. a mathematical error and an incorrect extrapolation of a physical principle). Another type of failure results from misjudgments of human systems. Bhopal had both.

Although the pesticide manufacturing plant in Bhopal was not a biotechnological operation it can demonstrate the chain of events that can lead to failure [39]. The plant, up until its closing, had produced the insecticides Sevin and Cararyl since 1969, using the intermediate product methyl isocyanate (MIC) in its gas phase. The MIC was produced by the reaction:

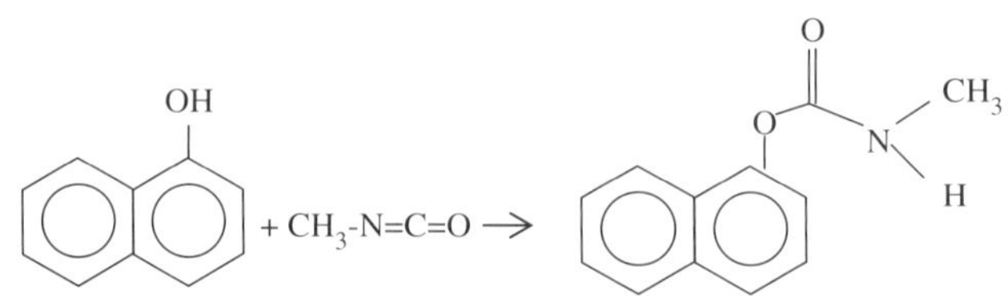

(6.3)

This process was highly cost-effective, involving only a single reaction step. The schematic of MIC processing at the Bhopal plant is shown in Figure 6.13.

MIC is highly water reactive (see Table 6.11); i.e. it reacts violently with water, generating a very strong exothermic reaction that produces carbon dioxide. When MIC vaporizes it becomes a highly toxic gas that, when concentrated, is highly caustic and burns tissues. This can lead to scalding of nasal and throat passages, blinding, and loss of limbs, as well as death.

On December 3, 1984, the Bhopal plant operators became concerned that a storage tank containing MIC was showing signs of overheating and had begun to leak. Introduction of water to the tank resulted in a highly exothermic reaction between the water and MIC,

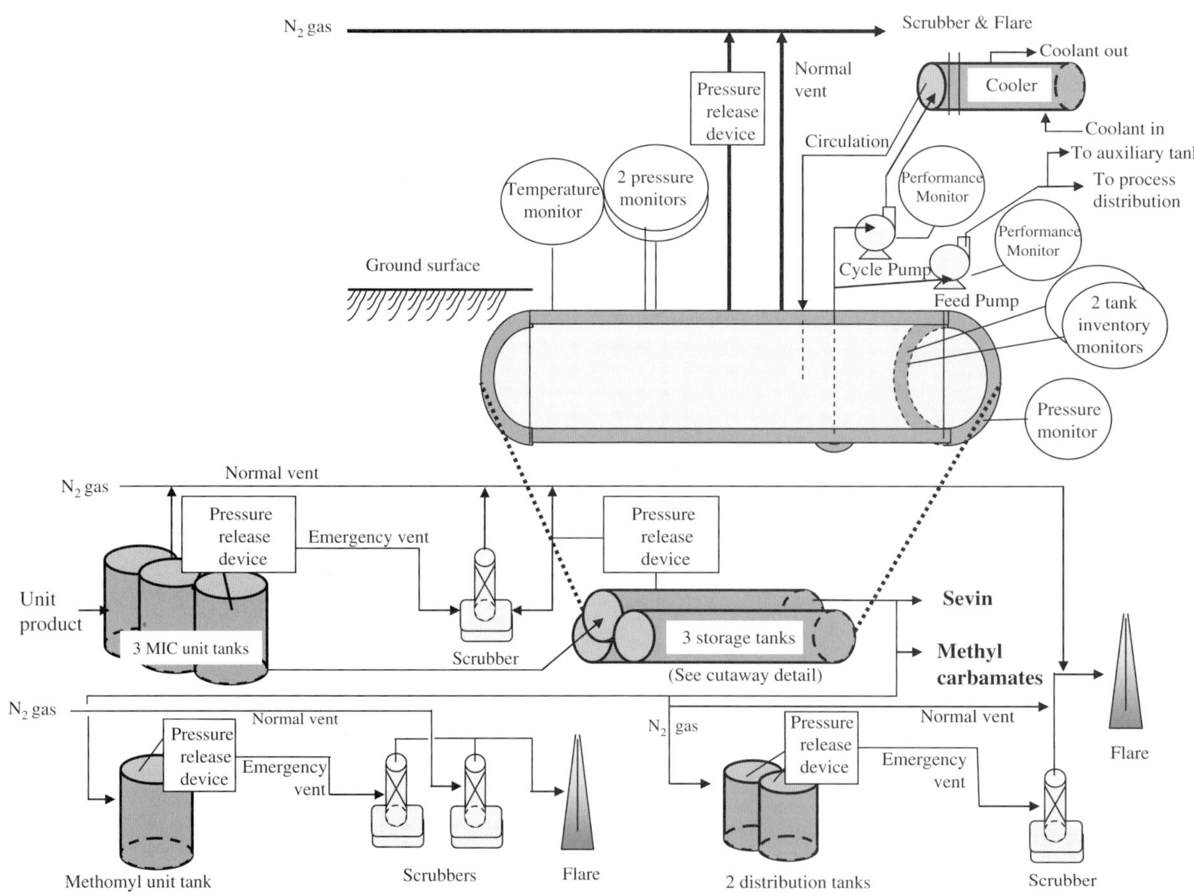

FIGURE 6.13

Schematic of methyl isocyanate processes at the Bhopal, India plant (*c.* 1984).

Source: W. Worthy (1985). Methyl isocyanate: the chemistry of a hazard. *Chemical Engineering News* 63 (66): 29.

Table 6.11 **Properties of methyl isocyanate (MIC)**

Common name	Isocyanic acid, methylester and methyl carbylamine
Molecular mass	57.1
Properties	Melting point : −45 °C; boiling point: 43–45 °C Volatile liquid Pungent odor Reacts violently with water and is highly flammable MIC vapor is denser than air and will collect and stay at low areas The vapor mixes well with air and explosive mixtures are formed May polymerize due to heating or under the influence of water and catalysts Decomposes on heating and produces toxic gases like hydrogen cyanide, nitrogen oxides, and carbon monoxide
Uses	Used in the production of synthetic rubber, adhesives, pesticides, and herbicide intermediates. It is also used for the conversion of aldoximes to nitriles
Side effects	MIC is extremely toxic by inhalation, ingestion, and skin absorption. Inhalation of MIC causes cough, dizziness, shortness of breath, sore throat, and unconsciousness. It is corrosive to the skin and eyes. Short-term exposures also lead to death or adverse effects like pulmonary edema (respiratory inflammation), bronchitis, bronchial pneumonia, and reproductive effects. The Occupational Safety and Health Administration's permissible exposure limit to MIC over a normal 8-hour workday or a 40-hour workweek is 0.05 mg m^{-3}

Sources: US Chemical Safety and Hazards Board, http://www.chemsafety.gov/lib/bhopal.0.1.htr; Chapman and Hall, *Dictionary of Organic Chemistry*, Volume 4, 5th Edition, Mack Printing Company, USA, 1982; and T.W Graham, *Organic Chemistry*, 6th Edition, John Wiley & Son, Inc., Canada, 1996.

generating CO_2. As the gas production increased, tank pressure increased rapidly. The leak rapidly increased in size, and within one hour of the first leakage the tank exploded and released approximately 80,000 lbs (4×10^4 kg) of MIC into the atmosphere.

The Indian government had required that the plant be operated exclusively by Indian workers, so Union Carbide agreed to train them, including flying them to a sister plant in West Virginia for hands-on sessions. In addition, the company required that US engineering teams make periodic on-site inspections for safety and quality control, but these ended in 1982, when the plant decided that these costs were too high. So, instead, the US contingency was responsible only for budgetary and technical controls, but not safety. The last US inspection in 1982 warned of many hazards, including a number that have since been implicated as contributing to the leak and release.

From 1982 to 1984, safety measures declined, attributed to high employee turnover, improper and inadequate training of new employees, and low technical savvy in the local workforce. On-the-job experiences were often substituted for reading and understanding safety manuals. (Remember, this was a pesticide plant.) In fact, workers would complain of typical acute symptoms of pesticide exposure, such as shortness of breath, chest pains, headaches, and vomiting, yet they would typically refuse to wear protective clothing and equipment. The refusal in part stemmed from the lack of air conditioning in this subtropical climate, where masks and gloves can be uncomfortable.

Indian, rather than the more stringent US, safety standards were generally applied at the plant after 1982. This likely contributed to overloaded MIC storage tanks (company manuals cite a maximum of 60% fill).

The release lasted about two hours, after which the entire quantity of MIC had been released. The highly reactive MIC arguably could have reacted and become diluted beyond a certain safe distance. However, over the years tens of thousands of squatters had taken up residence just outside of the plant property, hoping to find work or at least take advantage of the plant's water and electricity. The squatters were not notified of hazards and risks associated with the pesticide manufacturing operations, accept by a local journalist who posted signs saying: "Poison Gas. Thousands of Workers and Millions of Citizens are in Danger."

This is a class instance of a "confluence of events" that led to a disaster. More than a few mistakes were made. The failure analysis found the following:

- The tank that initiated the disaster was 75% full of MIC at the outset.
- A standby overflow tank for the storage tank contained a large amount of MIC at the time of the incident.
- A required refrigeration unit for the tank was shut down 5 months prior to the incident, leading to a three- to four-fold increase in tank temperatures over expected temperatures.
- One report stated that a disgruntled employee unscrewed a pressure gauge and inserted a hose into the opening (knowing that it would do damage, but probably not nearly the scale of what occurred).
- A new employee was told by a supervisor to clean out connectors to the storage tanks, so the worker closed the valves properly, but did not insert safety discs to prevent the valves from leaking. In fact, the worker knew the valves *were* leaking, but they were the responsibility of the maintenance staff. Also the second-shift supervisor position had been eliminated.
- When the gauges started to show unsafe pressures, and even when the leaking gases started to sting mucous membranes of the workers, they found that evacuation exits were not available. There had been no emergency drills or evacuation plans.
- The primary fail-safe mechanism against leaks was a vent-gas scrubber, i.e. normally, this release of MIC would have been sorbed and neutralized by sodium hydroxide (NaOH) in the exhaust lines, but on the day of the disaster, the scrubbers were not working. (The scrubbers were deemed unnecessary, since they had never been needed before.)
- A flare tower to burn off any escaping gas that would bypass the scrubber was not operating because a section of conduit connecting the tower to the MIC storage tank was under repair.
- Workers attempted to mediate the release by spraying water 100 ft high, but the release occurred at 120 feet.

Thus, according to the audit, many checks and balances were in place, but the cultural considerations were ignored or given low priority, such as, when the plant was sited, the need to recognize the differences in land use planning and buffer zones in India compared to Western nations, or the difference in training and oversight of personnel in safety programs [40]. Every engineer and environmental professional needs to recognize that much of what we do is affected by geopolitical realities and that we work in a global economy. This means that we must understand how cultures differ in their expectations of environmental quality. One cannot assume that a model that works in one setting will necessarily work in another without adjusting for differing expectations. Bhopal demonstrated the consequences of ignoring these realities.

In addition, many of the dual use and bioterrorism fears are analogous to the lack of due diligence at Bhopal. For example, the risk categories discussed in Chapter 5 may indicate the need for extra care in using similar strains and species in biotechnologies (e.g. substantial differences and similarities in various strains of *Bacillus spp.*). Just because no direct analogy exists does not mean that even a slight change in conditions may elicit unexpected and unwanted outcomes.

Characterizing as many contingencies and possible outcomes in the critical path is an essential part of many biohazards. The Bhopal incident provides this lesson. For example, engineers working with bioreactors and genetically modified materials must consider all possible avenues of release. They must ensure that fail-safe mechanisms are in place *and* are operational. Quality assurance officers note that testing for an unlikely but potentially devastating event is difficult. Everyone in the decision chain must be "on board." The fact that no incidents have yet to occur (thankfully) means that no one really knows what will happen in such an event. That is why health and safety training is a critical part of the engineering process.

Failure Type 4: Negligence

Engineers also have to protect the public from its members' own carelessness. The case of the woman trying to open a 2-liter soda bottle by turning the aluminum cap the wrong way with a pipe wrench, and having the cap fly off and into her eye, is a famous example of unpredictable ignorance. She sued for damages and won, with the jury agreeing that the design engineers should have foreseen such an occurrence. (The new plastic caps have interrupted threads that cannot be stripped by turning in the wrong direction.)

In the design of water treatment plants, engineers are taught to design the plants so that it is easy to do the right thing, and very difficult to do the wrong thing. Pipes are color-coded, valves that should not be opened or closed are locked, and walking distances to areas of high operator maintenance are minimized and protected. This is called making the treatment plant "operator proof." This is not a rap exclusively on operators of bioreactors and other biotechnological projects and operations. In fact, such standard operating procedures (SOPs) are crucial in any operation that involves repeated actions and a flow of activities. Hospitals, laboratories, factories, schools, and other institutions rely on SOPs [41]. When they are not followed, people's risks are increased. Biosystem engineers recognize that if something can be done incorrectly, sooner or later it will, and that it is their job to minimize such possibilities. That is, both risk and reliability are functions of time.

Risk is a function of time because it is a part of the exposure equation, i.e. the more time one spends in contact with a substance, the greater is the exposure. In contrast, reliability is the extent to which something can be trusted. A system, process or item is reliable so long as it performs the designed function under the specified conditions during a certain time period. In most engineering applications, reliability means that what we design will not fail prematurely. Or, stated more positively, reliability is the mathematical expression of success; that is, reliability is the probability that a system that is in operation at time *0* (t_0) will still be operating until the designed life (time $t = (t_t)$). As such, it is also a measure of the engineering accountability. People in neighborhoods near the biotechnological facility want to know if it will work and will not fail. This is especially true for those facilities that may affect the environment, such as landfills and power plants. Likewise, when environmental cleanup is being proposed, people want to know how certain the engineers are that the cleanup will be successful.

Time shows up again in the so-called *hazard rate*, i.e. the probability of a failure per unit time. Hazard rate may be a familiar term in environmental risk assessments, but many engineers may recognize it as a *failure density*, or *f*(t). This is a function of the likelihood that an adverse outcome will occur, but note that it is not a function of the severity of the outcome. The *f*(t) is not affected by whether the outcome is very severe (such as pancreatic cancer and loss of an entire species) or relatively benign (muscle soreness or minor leaf damage). The likelihood that something will fail at a given time interval can be found by integrating the hazard rate over a defined time interval:

$$P\{t_1 \leq T_f \leq t_2\} = \int_{t_1}^{t_2} f(t)dt \tag{6.4}$$

where T_f = time of failure.

Thus, the reliability function *R*(t), of a system at time t is the cumulative probability that the system has not failed in the time interval from t_0 to t_t:

$$R(t) = P\{T_f \geq t\} = 1 - \int_0^t f(x)dx \tag{6.5}$$

Engineers must be humble, since everything we design *will* fail. We can improve reliability by extending the time (increasing t_t), thereby making the system more resistant to failure. For example, proper engineering design of a landfill barrier can decrease the flow of contaminated water between the contents of the landfill and the surrounding aquifer, e.g. a velocity of a few microns per decade. However, the barrier does not completely eliminate failure, i.e. $R(t) = 0$; it simply protracts the time before the failure occurs (increases Tf) [42].

While disclosure and labeling are absolutely necessary parts of reliability in bioengineering, they are wholly insufficient to prevent accidents. Another such example occurred in the early 1970s, when jet-powered airliners were replacing propeller aircraft. The fueling system at airports was not altered, and the same trucks fueled both types of craft. The nozzle fittings for both types of fuels were therefore the same. A tragic accident occurred near Atlanta, where jet fuel was mistakenly loaded into a Martin 404 propeller craft. The engines failed on takeoff, resulting in fatalities. A similar accident occurred in 1974 in Botswana with a DC-4 and again near Fairbanks, Alaska, with a DC-6 [43]. The fuel delivery systems had to be modified so that it was impossible to put jet fuel into a propeller-driven airplane and vice versa. An example of how this can be done is the modification of the nozzles used in gasoline stations. The orifice in the gas tank of vehicles running on unleaded fuel is now too small to take the nozzles used for either leaded fuel or for diesel fuel [44]. By analogy, bioengineers must recognize that no amount of signs or training could prevent such tragedies.

Failure Type 5: Lack of imagination

Every time something fails, whether a manufactured product (such as a tire blowout) or a constructed facility (such as a bridge collapse), it is viewed as an engineering failure. The job of engineers historically has been to predict the problems that can occur, and to design so as to minimize these events; protecting people from design errors, natural forces, unforeseen events, and ignorance/carelessness.

This objective of engineers changed on 11 September 2001 with the attack on the United States by terrorists. Today the engineer is also required to protect the health, safety, and welfare of the public from acts of terrorism. It has never occurred to most engineers that they have a responsibility to protect people from those who would want to intentionally harm other people or to destroy public facilities intentionally. This is a totally new failure mode in engineering. Such failures can be considered to be "intentional accidents," or failures resulting from intentional actions [45].

Engineers now find themselves in the position of having to address these "intentional accidents." Military engineers of course have had to design against such destructive actions since the days of moats and castles, but those were all structures built explicitly to withstand attack. Civilian engineers have never had to think in these terms, but are now asked to design structures for this contingency. Engineering and engineers have a new challenge – to prevent such "accidents" on civilian targets by terrorists bent on harm to the public. Engineering response to terrorism or even to unexpected misuse of a system (e.g. a device, chemical or microbe), therefore, requires a two-pronged approach – technical and social. That is, the engineer's responsibility goes beyond assumptions about physics, and requires imagination as to possible abuses, misuses and failures of even properly designed systems. This is the venue of human factors engineering to the extreme, where the engineer must now consider possibilities that were previously *almost unthinkable* (for example, see Chapter 12's discussion of risk homeostasis and the theory of offsetting behaviour).

BIOTERRORISM: BAD BIOTECHNOLOGY

There are numerous examples of biotechnological success. Sometimes, however, a success in one venue can lead to problems elsewhere. A genetically modified microbe can lead to intended results in one situation, but in another venue could be pathogenic. There is a distinct likelihood that a number of the designs intended for beneficence will be converted and adapted for malevolence. Anti-terrorist technology is emergent and it is not something for which most engineers have yet developed an aptitude. As evidence, few resources that would protect us from low technology assault have been developed to date. Traditionally, our research funds have been spent either to enhance our own health or to develop sophisticated weapons for countering threats from similarly technically sophisticated enemies. For example, the use-inspired and applied research budgets of the National Institutes of Health and the Department of Defense greatly surpass the budgets for basic science research efforts of the National Science Foundation.

Research that is clearly basic is often supported for its possible, yet tenuous, public benefits. Interestingly, the term *dual use* has two different connotations. The first, which grew in popularity during the Cold War and space missions, is any science, engineering, and technology designed to provide both military and civilian benefits. Better pots and pans, microwave ovens, and DVD players can be touted as having been procreated out of huge, publicly funded military programs. The second definition, which in many instances seems to challenge the first, is any research or technology that simultaneously benefits and places society at risk.

Recently, concerns about terrorism and national security have piqued the public's interest about the research and technology that possibly fits the second definition. For example, in July 2006, the Congressional Research Service reported:

> An issue garnering increased attention is the potential for life sciences research intended to enhance scientific understanding and public health to generate results that could be misused to advance biological weapon effectiveness.
>
> Such research has been called "dual-use" research because of its applicability to both biological countermeasures and biological weapons. The federal government is a major source of life sciences research funding. Tension over the need to maintain homeland security and support scientific endeavor has led to renewed consideration of federal policies of scientific oversight. Balancing effective support of the research enterprise with security risks generated by such research has proven a complex challenge. Policies considered to address science and security generate tensions between federal funding agencies and federal funding recipients. To minimize these tensions while maximizing effective oversight of research, insight and advice from disparate stakeholders is generally considered essential [46].

A real-life case of the conflict in biosystematic dual use recently came to light when a study pointed out specific deficiencies in the protection of the milk supply and its vulnerability to widespread contamination by the *botulinum* toxin. The manuscript from the study was submitted to, and ultimately published in, *Proceedings of the National Academy of Sciences* [47]. The US Department of Health and Human Services had strongly opposed the publication on the grounds that it could encourage and instruct would-be bioterrorists (a dramatic example of dual use's second definition) [48]. One mediating aspect of this debate is that the authors of the article are not funded by the federal government. Had they received federal assistance, the US government likely would have had a greater onus and stronger position to veto the text.

In fact, the authors and their advocates made two arguments that favored publication. First, the information could be helpful to agricultural and food security decision makers charged with protecting the milk supply (analogous to publishing vulnerabilities with a hope this information will drive homeowners to make the necessary changes to improve home safety). The second argument was that the information had already been made readily available via the Internet and other sources.

The case illustrates the quandary of risk avoidance. Like many other engineering writers, the author has been confronted with a decision about just how much one should say about certain vulnerabilities, even though the information being shared was indeed readily available and not confidential or secret in any way. However, engineers are trained "to connect the dots" in ways that many are not, so even if the source information is readily available to anyone interested, we know where to look and how to assimilate the information into new knowledge.

This is the *bottom-up* design process. Engineers do this all the time. So, there is an additional onus on engineers undertaking endeavors in sensitive subject matter to take care to avoid giving new knowledge to those who intend to use it nefariously. And, engineers must be diligent not to fall victim to our own rationalizations, i.e. that we are doing it for the public good or the advancement of the state-of-the-science, if in fact our real intentions are to improve our own lot (e.g. a gold-standard publication, a happy client, or public recognition). There is nothing wrong and much right about improving one's own lot, but engineers are in a position of trust and must hold paramount the public's safety, health, and welfare. And, as the famous physicist Richard Feynman reminded us:

> Science is a long history of learning how not to fool ourselves. [49]

And:

> Science is a way of trying not to fool yourself. The first principle is that you must not fool yourself, and you are the easiest person to fool. [50]

Researchers and practitioners engaged in biotechnology need a healthy dose of realism and a critical eye toward our own justifications. Therefore, better technology cannot be the only, or maybe not even a primary, engineering response to the threat of terrorism. Terrorists with money and skill can get around our technology, and they can use our own technology against us. Even if we were good at anti-terror research, this would eventually fail to protect us.

The argument is not against the use of anti-terrorism technology but to suggest that such technology is only part of the answer. It is not possible for engineers to prevent all such deeds, just as it is not possible for engineers to make anything 100% safe from other kinds of failure.

One of the most interesting, albeit confusing, typologies of the uncertainties involved in risk management decisions was expounded by the former US Secretary of Defense, Donald Rumsfeld [51]:

> As we know, there are known knowns. There are things we know we know. We also know there are known unknowns. That is to say we know there are some things we do not know. But there are also unknown unknowns, the ones we don't know we don't know.

Acceptable risk for a given biotechnology, as most risks, includes uncertainties and ambiguity. The bioengineer must do what is possible to identify and weigh the factors that can lead to adverse environmental hazards and risks, but the risk decision will need to be monitored and re-evaluated in light of the chaotic nature of both the products of genetic manipulations and the environmental systems into which they are released.

SEMINAR TOPIC

The Risks and Opportunities in Mimicking Nature

The field of pest control is immense, and many problems impatiently await a solution. A new territory has opened up for the synthetics chemist, a territory which is still unexplored and difficult, but which holds out the hope that in time further progress will be made.

Paul Müller (Nobel Laureate for the synthesis of DDT), Nobel Lecture, December 11, 1948

The more our world functions like the natural world, the more likely we are to endure on this home that is ours, but not ours alone.

Janine Benyus, President, Biomimicry Institute

Few technological advances have been more polarized than those associated with the manufacture and use of pesticides. In fact, up to just a few years ago, a large volume of pesticides in use to control insects were direct derivatives from organophosphate (OP) chemicals that have been used in warfare agents known as *nerve gases.*

Up to 1998, an estimated 10,000 tons of the active ingredient in chlorpyrifos, a prominent OP pesticide, was being applied to 8 million acres annually in the United States. In 2000, the US EPA revised the human health risk assessment and entered into an agreement with the registrants to eliminate and phase out certain uses of chlorpyrifos, mainly to address food, drinking water, residential, and non-residential uses, posing the greatest risks to children. The mitigation contained in the agreement also reduced certain worker/applicator and ecological exposures by eliminating use sites and reducing application rates [52].

The mode of action of OP pesticides is cholinesterase inhibition, not surprising given their history as nerve agents. Thus, the toxic effects in humans include overstimulation of the nervous system, leading to nausea, dizziness, confusion, and at very high exposures, respiratory paralysis and death. Previous chlorpyrifos exposure assessments [53] have found:

- Dietary exposures from eating food crops treated with chlorpyrifos are below the level of concern for the entire US population, including infants and children. Drinking water risk estimates based on screening models and monitoring data from both ground and surface water for acute and chronic exposures are generally not of concern.
- Residential post-application exposures may occur after termiticide use in residential structures.
- Occupational exposures of concern include mixing/loading liquids for aerial/chemigation and groundboom application, mixing wettable powder for groundboom application, aerial application, and application by backpack sprayer, high-pressure handwand, and hand-held sprayer or duster. Generally, these risks can be mitigated by a combination of additional personal protective equipment and engineering controls, and by reductions in application rates.

Based upon available hazard and exposure data, chlorpyrifos risk quotients (RQs) indicate that a single application of chlorpyrifos poses ecosystem risks, i.e. to small mammals, birds, fish, and aquatic invertebrate species for nearly all registered outdoor uses. Multiple pesticide applications increase the steady state concentrations of OP pesticides with a concomitant increase in risks to wildlife and prolong exposures to toxic concentrations.

To reduce these risks, regulators have recommended a number of measures, including increased retreatment intervals, reduced seasonal maximum amounts applied per acre, and no-spray setback zones around water bodies.

Pesticides also make up the lion's share of the highly persistent organochlorine compounds, many of which have remained at measurable concentrations in the environment decades after being banned or heavily restricted in the world marketplace. The so-called Dirty Dozen is a list of the 12 most notorious persistent organic pollutants (POPs), includes nine organochlorine pesticides. The United States, member states of the EU, and numerous other countries signed a United Nations treaty in Stockholm, Sweden in May 2001 (i.e. the Stockholm Convention) agreeing to reduce or eliminate the production, use, and/or release of these POPs (see Table 6.12). A summary of the biochemodynamic properties of these and other high priority POPs is provided in Appendix 4.

Perhaps nature can point us to alternative types of pesticides with less risks to nontarget species, including humans. In recent decades, for example, scientists have been developing synthetic pyrethroids of the natural-based pesticide pyrethrum, made from extracts from plants in the chrysanthemum family. The flower has the advantage of constant production of low doses of pyrethrum. To replicate this mode of action using a typical pesticide application technique, e.g. spraying, new formulations have been developed to be more toxic and longer-lasting than pyrethrum. The upside is that they can be more efficacious as insecticides, but the downside is that "longer-lasting" is just a synonym for persistence and "more efficacious" is often another name for more toxic. Thus, the increased persistence increases the likelihood of risk and the increased potency increases the hazard. Since these are the two components of risk, the human and ecological risk of the synthetic pyrethroids is increased compared to nature's formulation, pyrethrum.

The natural pyrethrins have low aqueous solubility and high lipophilicity. They degrade rapidly, especially when exposed to sunlight. Pyrethroids are manufactured chemicals similar in structure to the pyrethrins, but usually more toxic to insects, as well as to mammals (see Tables 6.13 and 6.14) and are more persistent in the environment than pyrethrins.

Those pesticides classified as pyrethroids are actually comprised of as many as eight different molecules with the same chemical formula that have their atoms joined together in the same sequence, but with a different arrangement of the atoms, i.e. *stereoisomers*. Each isomer has different biochemodynamic properties insecticidal potencies, as well as different toxicities [54].

Permethrin, *resmethrin*, and *sumithrin* are synthetic pyrethroids commonly applied to kill adult mosquitoes. Permethrin is currently registered and sold in a number of products, e.g. household insect foggers and sprays, yard tick and flea sprays, pet flea dips and sprays, termite treatments, agricultural and livestock products, and mosquito abatement products. Resmethrin is used to control flying and crawling insects and to control insects on ornamental plants on animals, and to kill mosquitoes. It is toxic to fish, so resmethrin is a registered use pesticide (RUP), available for use only under the direction of certified pesticide applicators. Sumithrin is used to control adult mosquitoes and as an insecticide in transport vehicles, as well as an insecticide and miticide for nonfood areas [55].

So, then, how do the risks from these synthesized versions of a natural pesticide compare to the OP and organochlorine pesticides? There are two important types of exposure that need to be considered in such risk assessment: aggregate exposure and cumulative exposure. *Aggregate exposure* is comprised of contact to one substance through every exposure route (e.g. inhalation, dermal, dietary ingestion and non-dietary ingestion). *Cumulative exposure* is comprised of contact with multiple hazardous substances through every exposure route. Thus, looking at these three pesticide classes, it is best to measure and model exposures to every individual compound in the class via the numerous pathways and routes, i.e. cumulative exposure.

Seminar Questions

Can the risks from pyrethroids be predicted based upon their chemical structure, i.e. using structure activity relationships?

What are the uncertainties associated with predicting biochemodynamic factors from a natural product and its synthesized counterparts (e.g. halogen substitutions; see Figure 6.14).

Table 6.12 The Dirty Dozen persistent organic pollutant

Pollutant	Global historical use/Source	Overview of US status
Aldrin and dieldrin	Insecticides used on crops such as corn and cotton; also used for termite control	Under FIFRA: ■ No U.S. registrations; most uses canceled in 1969; all uses by 1987 ■ All tolerances on food crops revoked in 1986 No production, import, or export
Chlordane	Insecticide used on crops, including vegetables, small grains, potatoes, sugarcane, sugar beets, fruits, nuts, citrus, and cotton. Used on home lawn and garden pests. Also used extensively to control termites	Under FIFRA: ■ No US registrations; most uses canceled in 1978; all uses by 1988 ■ All tolerances on food crops revoked in 1986 No production (stopped in 1997), import, or export. Regulated as a hazardous air pollutant (CAA)
DDT	Insecticide used on agricultural crops, primarily cotton, and insects that carry diseases such as malaria and typhus	Under FIFRA: ■ No US registrations; most uses canceled in 1972; all uses by 1989 ■ Tolerances on food crops revoked in 1986 No US production, import, or export. DDE (a metabolite of DDT) regulated as a hazardous air pollutant (CAA). Priority toxic pollutant (CWA)
Endrin	Insecticide used on crops such as cotton and grains; also used to control rodents	Under FIFRA, no US registrations; most uses canceled in 1979; all uses by 1984. No production, import, or export. Priority toxic pollutant (CWA)
Mirex	Insecticide used to combat fire ants, termites, and mealybugs. Also used as a fire retardant in plastics, rubber, and electrical products	Under FIFRA, no US registrations; all uses canceled in 1977. No production, import, or export
Heptachlor	Insecticide used primarily against soil insects and termites. Also used against some crop pests and to combat malaria	Under FIFRA: ■ Most uses canceled by 1978; registrant voluntarily canceled use to control fire ants in underground cable boxes in early 2000 ■ All pesticide tolerances on food crops revoked in 1989. No production, import, or export
Hexachlorobenzene	Fungicide used for seed treatment. Also an industrial chemical used to make fireworks, ammunition, synthetic rubber, and other substances. Also unintentionally produced during combustion and the manufacture of certain chemicals. Also an impurity in certain pesticides	Under FIFRA, no US registrations; all uses canceled by 1985. No production, import, or export as a pesticide. Manufacture and use for chemical intermediate (as allowed under the Convention). Regulated as a hazardous air pollutant (CAA). Priority toxic pollutant (CWA)

Table 6.12 The Dirty Dozen persistent organic pollutant—cont'd

Pollutant	Global historical use/Source	Overview of US status
Polychlorinated bicphenyls (PCBs)	Used for a variety of industrial processes and purposes, including in electrical transformers and capacitors, as heat exchange fluids, as paint additives, in carbonless copy paper, and in plastics. Also unintentionally produced during combustion	Manufacture and new use prohibited in 1978 (TSCA). Regulated as a hazardous air pollutant (CAA). Priority toxic pollutant (CWA)
Toxaphene	Insecticide used to control pests on crops and livestock, and to kill unwanted fish in lakes	Under FIFRA: ▪ No US registrations; most uses canceled in 1982; all uses by 1990 ▪ All tolerances on food crops revoked in 1993. No production, import, or export. Regulated as a hazardous air pollutant (CAA)
Dioxins and furans	Unintentionally produced during most forms of combustion, including burning of municipal and medical wastes, backyard burning of trash, and industrial processes. Also can be found as trace contaminants in certain herbicides, wood preservatives, and in PCB mixtures	Regulated as hazardous air pollutants (CAA). Dioxin in the form of 2,3,7,8-TCDD is a priority toxic pollutant (CWA)

Source: US Environmental Protection Agency (2002). Persistent Organic Pollutants: A Global Issue, a Global Response; http://www.epa.gov/oia/toxics/pop.htm#thedirtydozen; accessed October 6, 2009.

Table 6.13 Pyrethroid pesticide mammalian toxicities

Pesticide	Rat oral LD_{50} (mg kg^{-1} body weight)	Rabbit dermal LD_{50} (mg kg^{-1} body weight)
Allethrin	860	11,332
Bifenthrin	375	>2000
Cyfluthrin	869–1271	>5000 (rat)
Cyhalothrin	79	632 (rat)
Cypermethrin	250	>2000
Deltamethrin	31–139 (female)	>2000
Esfenvalerate	451	2500
Fenpropathrin	70.6–164	>2000

(*Continued*)

Table 6.13 Pyrethroid pesticide mammalian toxicities—cont'd

Pesticide	Rat oral LD_{50} (mg kg^{-1} body weight)	Rabbit dermal LD_{50} (mg kg^{-1} body weight)
Fluvalinate	261–282	>20,000
Permethrin	430–4000	>2000
Resmethrin	1244 – >2500	>2500
Tefluthrin	969	>2000 (rat)
Tetramethrin	>5000	>2000
Tralomethrin	284	>2000

Source: F.M. Fischel (2005). Pesticide Toxicity Profile: Synthetic Pyrethroid Pesticides. Document PI-54, one of a series of the Agronomy Department, Florida Cooperative Extension Service, Institute of Food and Agricultural Sciences, University of Florida, Gainesville, FL.

Table 6.14 Pyrethroid pesticide wildlife toxicity ranges

Pesticide	Bird acute oral LD_{50} (mg/kg)[a]	Fish LC_{50} (ppm)[b]	Bee LD_{50}[c]
Allethrin	PNT	HT	HT
Bifenthrin	ST–PNT	HT	HT
Cyfluthrin	PNT	VHT	HT
Cyhalothrin	PNT	HT	HT
Cypermethrin	PNT	VHT	HT
Deltamethrin	PNT	HT	HT
Esfenvalerate	PNT	VHT	HT
Fenpropathrin	ST	VHT	HT
Fluvalinate	PNT	VHT	MT
Permethrin	PNT	VHT	HT
Resmethrin	PNT	VHT	HT
Tefluthrin	ST–PNT	VHT	HT
Tetramethrin	PNT	HT	—
Tralomethrin	—	VHT	HT

[a]*Bird LD_{50}: Practically nontoxic (PNT) = >2000; slightly toxic (ST) = 501–2000; moderately toxic (MT) = 51–500; highly toxic (HT) = 10–50; very highly toxic (VHT) = <10.*
[b]*Fish LC_{50}: PNT = >100; ST = 10–100; MT = 1–10; HT = 0.1–1; VHT = <0.1.*
[c]*Bee: HT = highly toxic (kills upon contact as well as residues); MT = moderately toxic (kills if applied over bees); PNT = relatively nontoxic (relatively few precautions necessary).*
Source: F.M. Fischel (2005). Pesticide Toxicity Profile: Synthetic Pyrethroid Pesticides. Document PI-54, one of a series of the Agronomy Department, Florida Cooperative Extension Service, Institute of Food and Agricultural Sciences, University of Florida, Gainesville, FL.

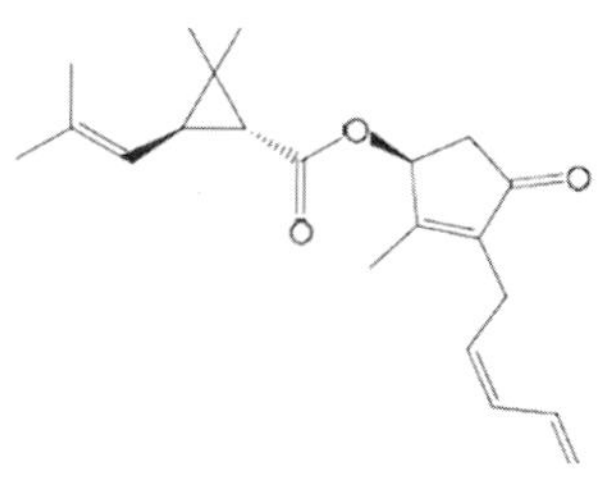

Pyrethrin I

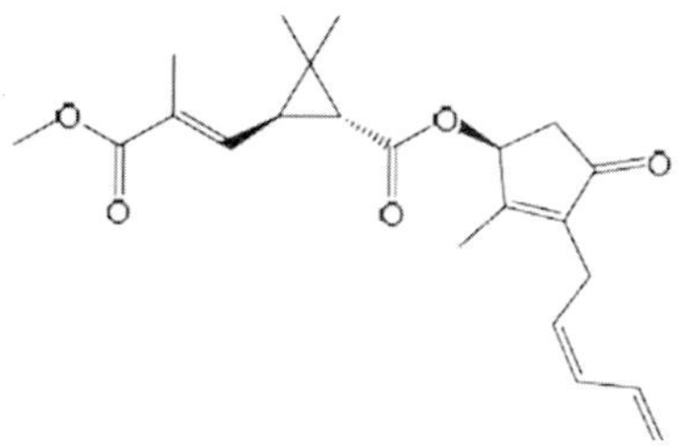

Pyrethrin II

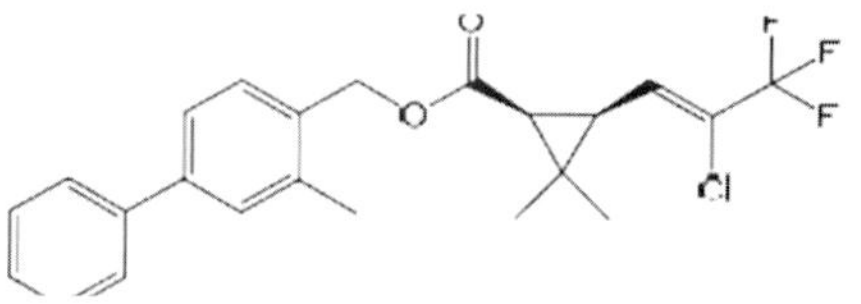

Bifenthrin

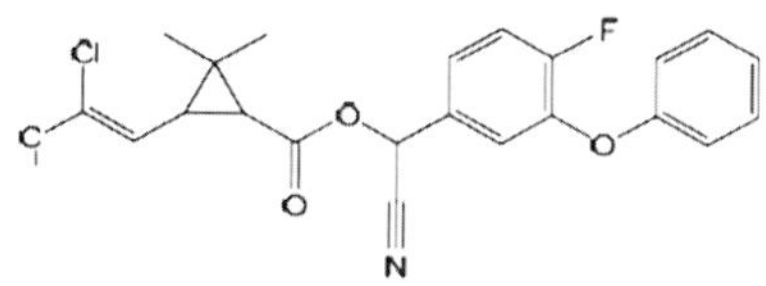

Cyfluthrin

FIGURE 6.14
Chemical structures of two pyrethrins (top) and two pyrethroids (bottom).
Source: US Department of Health and Human Services (2003). Agency for Toxic Substances and Disease Registry. Public Health Statement for Pyrethrins and Pyrethroids.

REVIEW QUESTIONS

Which differences in perspectives between the public and technical communities are most difficult to overcome? Why?

What are the best approaches for addressing uncertainty in biotechnologies?

What lessons can we draw from the Bhopal disaster to prevent problems with bioreactors and other biotechnological operations?

What are the key differences between a chemical plant disaster and a more subtle biotechnological problem, like release and flow of genetic material in the environment?

Give a biotechnological example of each of the five types of failure. How can each be prevented?

What are some of the advantages of using a Bayesian approach over conventional approaches (e.g. contingent probabilities) to characterize the potential for an environmental problem from a biotechnology? What are some disadvantages?

How might new analytical tools be used to improve risk assessment and predictions to support environmental decisions?

Are natural systems always less risky than synthetic systems? Give two examples to support your answer.

How do abiotic systems affect biotechnology? For example, list five physicochemical factors that can influence the risk of released genetically modified bacterial populations.

Consider the postulation that the working definition of a pollutant is that it is a resource that is simply out of place. Does this hold for natural microbial populations? How about genetically engineered microbial populations? Give reasons to support your answer, including a risk–benefit assessment.

NOTES AND COMMENTARY

1. United Kingdom Health and Safety Commission (1998). http://www.hse.gov.uk/nuclear/computers.pdf; accessed May 26, 2006. The Commission is responsible for health and safety regulation in Great Britain. The Health and Safety Executive and local government are the enforcing authorities who work in support of the Commission.
2. The concept of "reasonableness" is a constant theme of this book. Unfortunately, the means of determining reasonable designs are open to interpretation. The legal community uses the reasonable person standard. Professional ethics review boards may also indirectly apply such a standard. The reasonable person is akin to the Kantian categorical imperative. That is, if the act (e.g. the approach to the design) were universalized, would this be perceived to be a "good" or "bad" approach? This is the type of query where reviewers might ask whether the designer "knew or should have known" the adverse aspects of the design in advance.
3. United Kingdom Health and Safety Commission.

4. The term "intangible" is quite interesting in modern vernacular. First, it is one of those adjectives that have morphed into a noun. It is often used in sports analyses, such as when the University of North Carolina plays Duke in the Atlantic Coast Conference championship. UNC may have taller and more gifted players in a given year. They may also have home-court advantage, a deeper bench, and momentum. However, sportscasters may say that Duke has the "intangible advantage". This seems to consist of variables that defy quantitation or even objective analysis. If Duke goes on to win the game, the analysts may do their best retrospectively to deconstruct these intangibles (exam week, UNC players on the cover of *Sports Illustrated*, traffic on Highway 15-501, premature bonfires on Franklin Street, etc.). However, such weightings were not possible *before the game*.
This is similar to risk assessment and perception. The value of information may be that the perceptions were correct, but the actual process by which they led to the correct risk assessment are not well understood. The deconstruction of these processes can lead to better risk assessments next time (lessons learned) if the factors can be identified. Thus, a retrospective sensitivity analysis can inform future risk-based decision models.
Next time, we will know better *why* Duke won. In this sense, Coach K is a risk analyst who is applying sound science to decide how to handle the UNC Tar Heels' next invasion of the Cameron Stadium.
5. The following discussion is based on: US Environmental Protection Agency (2007). Appendix E: Risk Quotient Method and LOCs – Risks of Metolachlor Use to Federally Listed Endangered Barton Springs Salamander. Washington, DC.
6. US Environmental Protection Agency (2009). Technical Overview of Ecological Risk Assessment; http://www.epa.gov/oppefed1/ecorisk_ders/toera_risk.htm.
7. Ibid.
8. US Environmental Protection Agency (2003). Generic Ecological Assessment Endpoints (GEAEs) for Ecological Risk Assessment. Report No. EPA/630/P-02/004F. Washington, DC.
9. Ibid.
10. US Environmental Protection Agency (2009). ECOTOX Database; http://cfpub.epa.gov/ecotox/index.html; accessed October 7, 2009.
11. F. Basquero, J-L. Martínez and R. Cantón (2008). Antibiotics and antibiotic resistance in water environments. *Current Opinion in Biotechnology* 19: 260–265.
12. IUCN/SSC Invasive Species Specialist Group (ISSG): http://www.issg.org/database/species/search.asp?st=100ss&fr=1&;sts=; accessed April 20, 2005.
13. Environment Canada and US Environmental Protection Agency (2004). Lake Erie Lakewide Management Plan. Chicago, IL.
14. National Research Council (2004). *Biological Confinement of Genetically Engineered Organisms*. National Academies Press, Washington, DC.
15. Ibid.
16. A.R. Kapuscinski and T.J. Patronski (2005). Genetic methods for biological control of non-native fish in the Gila River basin: Development and testing of methods, potential environmental risks, regulatory requirements, multi-stakeholder deliberation, and cost estimates. Contract report to the US Fish and Wildlife Service (USFWS agreement number 201813N762). University of Minnesota, Institute for Social, Economic and Ecological Sustainability, St Paul, Minnesota. Minnesota Sea Grant Publication F 20.
17. US Food and Drug Administration, L. Bren (2001). Antibiotic resistance from down on the farm, *FDA Veterinarian* 16: (1): 2–4; and C. Richardson (2000). Ontario Ministry of Agriculture and Food: http://www.gov.on.ca/OMAFRA/english/livestock/sheep/facts/info_resist.htm.
18. F. Baquero, J.L. Martínez and R.Cantón (2008). Antibiotics and antibiotic resistance in water environments. *Biotechnology* 19: 260–265.
19. This is a type of "opportunity risk," such as when people worry about pesticides risks to the point where they eat less fresh fruit and vegetables, thereby increasing their risks to many diseases, including cancer. Another example of opportunity risk tradeoff is that of native populations, such as the Inuit who, because of long-range transport of persistent organic pollutants (so-called POPs) have elevated concentrations of polychlorinated biphenyls and other POPs in mother's milk. However, the ill effects of not breastfeeding (e.g. immunity, neonatal health, and colostrums intake) may far outweigh the long-term risks from exposures to POPs.
20. P.M. Schlosser (1997). Risk assessment: the two-edged sword: http://pw1.netcom.com/~drpauls/just.html; accessed August 25, 2009.
21. For a different, even contrary view, see http://www.brown.edu/Administration/George_Street_Journal/value.html. Richard Morin gives a thoughtful outline of the Allen Feldman's model and critique of the "willingness to pay" argument (very commonly used in valuation).
22. H. Bauer, A. Kasper-Giebl, F. Zibuschka, R. Hitzenberger, G.F. Kraus and Hans Puxbaum (2002). Determination of the carbon content of airborne fungal spores. *Analytical Chemistry* 74: 91–95.
23. US Environmental Protection Agency (1986). Guidelines for Carcinogen Risk Assessment, Report No. EPA/630/R-00/004. *Federal Register* 51(185): 33992–34003, Washington, DC.
24. A. Bradford Hill (1965). The environment and disease: association or causation?, *Proceedings of the Royal Society of Medicine, Occupational Medicine*, 58: 295; and A. Bradford Hill (1965). The environment and disease: association or causation? President's Address: *Proceedings of the Royal Society of Medicine*, 9, 295–300.
25. The derivation of the mythical Pandora's box is found in Chapter 2.
26. H.A.L. Fisher (1936). Preface, in *A History of Europe*. Edward Arnold, London, UK.
27. George Santayana (1905). *The Life of Reason*, Volume 1. New York, Scribner's Sons.
28. US Environmental Protection Agency (1986). The Proceedings of the United States Environmental Protection Agency Workshop on Biotechnology and Pollution Control. Report No. EPA-Z/566, Bethesda, MD.

29. Thomas Bayes, English preacher and mathematician, argued that knowledge of prior events is needed to predict future events. Thus Bayes, like Santayana for political thought, advocated for the role of memory in statistics. Bayes' theorem, which was published two years after his death in 1761 in *An Essay Towards Solving a Problem in the Doctrine of Chances*, introduced the mathematical approach to prediction, based on logic and history, the probability of an uncertain outcome. This is very valuable in science, i.e. it allows uncertainty to be quantified.
30. National Institutes of Health (2002). Guidelines for Research Involving Recombinant DNA Molecules; http://oba.od.nih.gov/oba/rac/guidelines_02/NIH_Guidelines_Apr_02.htm; accessed August 29, 2009.
31. K.M. Khleifat (2007). Biodegradation of phenol by *Actinobacillus sp.*: Mathematical interpretation and effect of some growth conditions. *Bioremediation Journal* 11 (3): 103–112.
32. A. Čejková, J. Masák, Vl. Jirků, M. Veselý, M. Pátek and J. Nešvera (2005). Potential of *Rhodococcus erythropolis* as a bioremediation organism. *World Journal of Microbiology and Biotechnology* 21 (3): 317–321.
33. D. Gronewold collaborated with the author in developing these scenarios.
34. J.D. Keasling and S.W. Bang (1998). Recombinant DNA techniques for bioremediation and environmentally-friendly synthesis. *Current Opinion in Biotechnology*, 9(2): 135–140.
35. Human Factors and Ergonomics Society (2006). HFES History: http://www.hfes.org/WEB/AboutHFES/history.html; accessed July 31, 2006.
36. The American Heritage Dictionary of the English Language, 4th Edition (2004). Houghton Mifflin Company, New York, NY; accessed at *Answers.com* on July 8, 2006.
37. J.K. Hammitt, E.S. Belsky, J.I. Levy and J.D. Graham (1999). Residential building codes, affordability, and health protection: a risk-tradeoff approach. *Risk Analysis* 19 (6): 1037–1058.
38. L.J. Morse, J.A. Bryan, J.P. Hurley, J.F. Murphy, T.F. O'Brien and T.F. Wacker (1972). The Holy Cross Football Team hepatitis outbreak. *Journal of the American Medical Association*, 219: 706–708.
39. The principal sources for this case are: M.W. Martin and R. Schinzinger (1996). Ethics in Engineering, 3rd Edition. McGraw-Hill, New York; and C.B. Fledderman (1999). *Engineering Ethics*. Prentice-Hall, Upper Saddle River, NJ.
40. Although the Bhopal incident is unprecedented and, thankfully, unique, spills and releases of toxic substances are all too common. Thus, the lessons of Bhopal can be applied to incidents and episodic events of smaller spatial extents and with, hopefully, more constrained impacts. For example, two freight trains collided in Graniteville, SC, just before 3:00 a.m. on January 6, 2005, resulting in the derailment of three tanker cars carrying chlorine (Cl_2) gas and one tanker car carrying sodium hydroxide (NaOH) liquids. The highly toxic Cl_2 gas was released to the atmosphere. The wreck and gas release resulted in hundreds of injuries and eight deaths.
41. Examples include improper cleaning due to mislabeled containers, mismatched transplants due to mislabeled blood types and injuries due to improper warning labels on power equipment.
42. Hydraulics and hydrology provide very interesting case studies in the failure domains and ranges, particularly how absolute and universal measures of success and failure are almost impossible. For example, a levee or dam breach, such as the recent catastrophic failures in New Orleans during and in the wake of Hurricane Katrina, experienced failure when flow rates reached cubic meters per second. Conversely, a hazardous waste landfill failure may be reached when flow across a barrier exceeds a few cubic centimeters per decade.
43. Aviation Safety Network (2002). http://aviation-safety.net/database/index.html.
44. Drivers of diesel-engine cars can still mistakenly pump gasoline into their cars, however.
45. S. Pfatteicher (2002). Learning from failure: terrorism and ethics in engineering education. *Technology and Society Magazine, IEEE* 21 (2): 8–12, 21.
46. D.E. Shea, Congressional Research Service (2006). Oversight of dual-use biological research: The National Science Advisory Board for Biosecurity, Updated July 10, 2006; Order Code RL33342.
47. L.M. Wein and Y. Liu (2005). Analyzing a bioterror attack on the food supply: the case of *Botulinum toxin* in milk. *Proceedings of the National Academy of Sciences of the United States of America* 102: 9984.
48. J. Kaiser (2005). ScienceScope. *Science* 309: 31; and A. McCook (2005). PNAS publishes bioterror paper, after all. *The Scientist* June 29, 2005.
49. K.C. Cole (1999). *The Universe and the Teacup: The Mathematics of Truth and Beauty.* Harcourt Brace, New York, NY.
50. F.P. Feynman and J. Robbins (1999). *The Pleasure of Finding Things Out.* Perseus Publishing Co., New York, NY.
51. D. Rumsfeld (2002). Press conference. US Department of Defense, February 12, 2002. This is in some ways similar to Socrates' advice that wisdom begins with knowing what one does not know. Henry David Thoreau also had a similar quote: "To know that we know what we know, and that we do not know what we do not know, that is true knowledge"; http://www.famousquotesandauthors.com/authors/henry_david_thoreau_quotes.html; accessed August 25, 2009.
52. US Environmental Protection Agency (2002). Interim Reregistration Eligibility Decision for Chlorpyrifos. Report No. EPA 738-R-01-007. Washington, DC.
53. Ibid.
54. US Department of Health and Human Services (2003). Agency for Toxic Substances and Disease Registry. Public Health Statement for Pyrethrins and Pyrethroids.
55. US Environmental Protection Agency (2002). Permethrin, Resmethrin, Sumithrin: Synthetic Pyrethroids for Mosquito Control.; http://www.epa.gov/opp00001/health/mosquitoes/pyrethroids4mosquitoes.htm#pyrethroids; accessed October 6, 2009.

CHAPTER 7

Applied Microbial Ecology: Bioremediation

Chances are, if one were to ask a group of environmental engineers to define "environmental biotechnology" there would be agreement among them that it has to do with putting biology to work to treat pollutants. Biotechnologists might agree, but may emphasize the "technology" over the "environment" in the application. Many scientists and engineers who apply life science principles would add that the field is more inclusive than simply solving problems, but also would use living organisms to find better ways to prevent pollution, such as innovative, green technologies like harnessing algae as energy sources. Certainly, these are correct.

As has been discussed thus far, even when biotechnologies work as designed there loom potential and actual downsides, and the environment includes numerous receptors of these downsides. This book focuses both on the environmental *applications* and the environmental *implications* of biotechnologies. Indeed, most bioengineering and scientific literature focuses on environmental applications. In fact, the likelihood is fairly good that if you pull a book from the library shelf with "environmental biotechnology" in its title it will largely if not exclusively deal in applications involving the degradation of pollutants.

Chapter 4 introduced a number of environmental applications, predominately using biology to provide indications of environmental conditions. This chapter focuses on the ways that biological principles, mainly microbiological principles, can be put to use to address pollution in a systematic way. This means that a pollutant and its spheres of influence must be considered in space and time, with attention to scale and complexity in all of their forms.

Indeed, natural and genetically modified organisms, including microbes (mainly bacteria, but also protozoa, fungi, and algae and even viruses), flora (i.e. large plants), and fauna (e.g. earthworms) degrade pollutants into simpler, less toxic forms. Organisms at the lower levels of biological organization transform substances in the direction of mineralization, i.e. toward inorganic compounds (those that lack carbon-to-carbon and carbon-to-hydrogen covalent bonds).

In comparison to the scale and complexity of medical and industrial applications of biotechnology, environmental biotechnologies usually address much larger volumes of chemicals, in both human populations and ecosystem, usually in the form of some type of waste or unwanted material.

The explosion in recent decades of the array of technologies and species and strains of organisms used in bioremediation and engineered biodegradation has been remarkable. Beginning with empirical evidence and expanding to general microbiological principles, bioengineers now deduce with confidence that treatment is possible to degrade a wide range of pollutants using biotechnologies. Figures 7.1 and 7.2 illustrate the predominant organic

Environmental Biotechnology

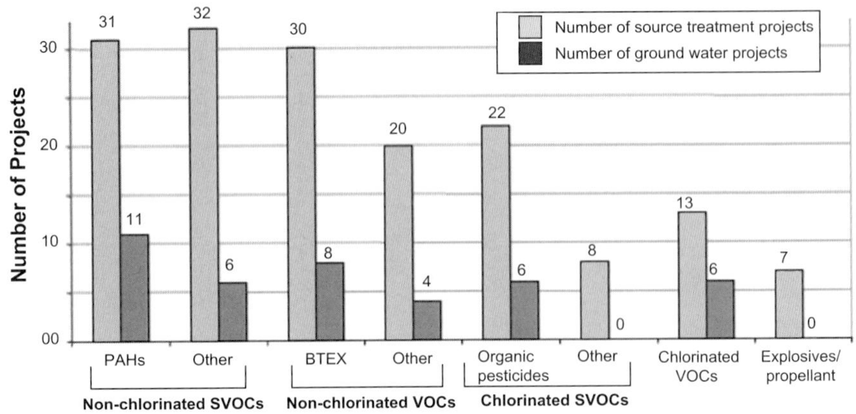

FIGURE 7.1
Some important chemical pollutants that may be treated using biotechnologies at abandoned waste site remediation projects (number shown at top of each bar) from 1982 through 1999. Some projects address more than one contaminant group. PAH = polycyclic aromatic hydrocarbons; VOCs = volatile organic compounds; SVOCs = semivolatile organic compounds; BTEX = gasoline remnants, benzene-toluene-ethylbenzene-xylenes.
Source: US Environmental Protection Agency (2001). Use of Bioremediation at Superfund Sites. Report No. EPA 542-R-01-019, Washington, DC.

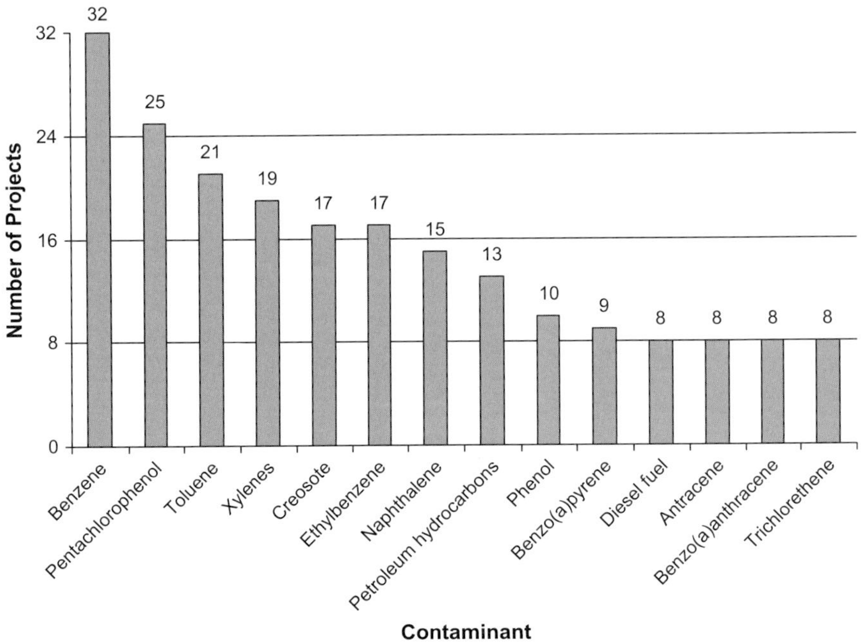

FIGURE 7.2
Some important chemical pollutants that may be treated using biotechnologies at abandoned waste site remediation projects (number shown at top of each bar) from 1982 through 1999. Some projects address more than one contaminant group.
Source: US Environmental Protection Agency (2001). Use of Bioremediation at Superfund Sites. Report No. EPA 542-R-01-019, Washington, DC.

pollutants that bioengineers must address, in these instances, at abandoned hazardous waste sites. However, similar success has been shown in all media, i.e. sediment, soil, surface and groundwater, air and biota.

Biological engineering applications were formerly covered under sanitary engineering and, later, environmental engineering courses with names like Biological Principles of Pollution Control, or Wastewater Treatment Engineering. Indeed, environmental biotechnology as it presently exists generally refers to enhancements of natural biological systems, particularly the use of microbes whose traits may or may not have been genetically modified to improve the efficiency of biodegradation of specific environmental contaminants.

Prior to World War II and the concomitant, dramatic increase in crude oil-based industry, most chemicals were relatively simple. They were either exactly the types produced in nature or only slightly different in chemical form. Thus, microbes were quite efficient at breaking down these chemicals as a source of food and metabolic processes. However, with the petrochemical revolution came more exotic chemical forms, especially organic compounds. When these found their way to soil and water, they presented major challenges to treatment, since the microbes had not previously come into contact with these compounds, so their ordinary metabolic processes of oxidation, reduction, hydrolysis, and other mechanisms lacked the ability to degrade these more complex chemical compounds.

Sanitary engineering pioneers, such as Ross McKinney at the University of Kansas, knew that microbes are highly adaptive, especially in conditions of food scarcity and stress. Microbes (often soil bacteria, e.g. *Pseudomonas spp.*) became acclimated to various chemicals by withholding their ordinary food sources, requiring the microbes to adapt their own metabolic processes to make use of the previously unknown chemicals.

Biotechnological advances, without doubt, have provided many benefits. Pollution is present throughout the globe. Chemicals that are useful in one place can find their way to the air, water, soil, and organisms where they present an array of problems. They can adversely affect the health of people exposed to them. They can upset balances and functions of ecosystems. As such, society demands that these problematic chemicals must be removed and that resources be decontaminated. Engineers have known for many decades that biological systems and processes provide key solutions to these problems. Biotechnological advances in recent decades have supercharged the natural processes used by engineers for pollution abatement.

One means of reducing the environmental and public health risks is to change the chemical structure of compounds so that they do not bind or block receptor sites on cells. Slight changes in structure, such as the addition or deletion of a single atom or the arrangement of the same set of atoms (i.e. an isomer), can substantially reduce the likelihood of toxic effects elicited by a compound. Chemical compounds that are relatively easily broken down by microbes may be completely treated by common techniques. More persistent compounds will require intensive treatment systems that enhance the degradation either chemically or microbiologically by bacteria, including species of *Pseudomonas, Rhodococcus,* and *Mycobacterium*. Concentrated wastes may require anaerobic treatment (where molecular oxygen is absent) or facultative treatment (where bacteria grow with or without molecular oxygen) to degrade the complex contaminant molecules into simpler compounds.

Pollutants can be transformed biologically by three specific mechanisms:

- use of the compound as an electron acceptor;
- use of the compound as an electron donor; or
- cometabolism.

These processes can occur together and simultaneously [1].

Use of a compound as an electron acceptor takes place under reducing conditions. Such biotransformation needs a source of carbon (i.e. the electron donor) for microbial growth and metabolism. For example, if the compound contains chlorine or other halogen, then reductive dehalogenation will occur. That is, the halogens that are bound to carbon and other elements are removed, usually reducing the toxicity of the compound. The electron donor carbon can be gained from various sources, both from natural and anthropogenically derived organic matter.

In an electron donor situation, the compound is used as the primary substrate (electron donor), so that the microbe gets its energy and organic carbon from the compound. This may occur under either aerobic and some anaerobic conditions. Lesser oxidized compounds (e.g., vinyl chloride, dichlorethylene, or 1,2-dichloroethane) are amenable to this mode of biotransformation.

Cometabolism is the interaction "between enzyme specificity and metabolic regulation, the metabolic interdependence of microorganisms, and co-substrate requirements in the catabolism of xenobiotic compounds" [2]. When a compound is broken down by cometabolism, the degradation is catalyzed by an enzyme that is fortuitously produced by the organisms for other purposes. The microbe does not directly benefit from the degradation of the compound.

The biotransformation of the compound could actually be harmful or can inhibit growth and metabolism a microbe. Cometabolism has been most often observed in aerobic environments, but has potential under anaerobic conditions, as well.

These three processes are evidence that biotechnology is essentially a means of manipulating systems. Biotechnology takes advantage of the physics, chemistry, and biology that occurs within various species. The fundamental building block in physics is the particle. For chemistry, it is the atom. For chemical reactions, the basic unit is the element. The fundamental unit in living systems is the cell [3]. Whether it is a self-contained organism, like that of bacteria or algae, or a part of a complex organism, like human beings, the cell is where the biochemical processes, for good or bad, take place. When operating effectively, cells are the factories that turn nutrients and energy into biomass through the processes of photosynthesis, metabolism, and ion exchanges in microbes and plants. In animals, the cell is the location of metabolism and respiration. These mechanisms unfortunately are often disrupted by environmental contaminants. They are also the processes that convert complex chemical contaminants into simpler, hopefully less toxic, forms through the process of *biotransformation*

SYSTEMATIC VIEW OF OXYGEN

A distinction in environmental engineering applications often begins with whether oxygen is the electron acceptor. That is, is a biodegradation pathway aerobic or anaerobic? Oxygen, particularly in its molecular form O_2, is an essential factor in environmental biotechnology, as it limits most life forms, all higher order organisms, and all environmental systems. In fact, one of the key indicators of an environmental system's conditions is the availability of ample O_2 concentrations. When O_2 concentrations are depleted, the system becomes *anoxic* and less capable of sustaining diverse populations. Like other types of pollution, oxygen depletion can be direct or indirect. Some agents elicit direct toxic responses; however, certain contaminants are particularly harmful because they use up resources in the environment at the expense of other organisms. So, it is important to consider rising and falling O_2 levels systematically. Microbes are directly and indirectly involved in changing oxygen budgets in all environmental systems. That is, the available molecular oxygen determines abundance and diversity of microbial populations, which in turn use oxygen as a function of the metabolic processes, which affect the overall diversity, productivity, and sustainability of an ecosystem or the health of the organism in which the microorganisms reside.

Interestingly, O_2 depletion affects ecosystems unevenly and often in nonlinear ways. For example, lowering dissolved O_2 in surface waters will first adversely affect more sensitive aquatic species, such as trout, salmon, and other game fish. The lower O_2 concentrations may actually benefit "rough" fish like carp and buffalo until the waters are almost completely devoid of dissolved O_2. Effects also vary according to life stages. Often, at lower oxygen levels, mature fish may live, but not be able to reproduce effectively. Figure 7.3 shows that the combination of vulnerabilities during these different life stages will determine the ability of an aquatic population to thrive, or at least survive (in this instance, in saltwater). For example, below 4 mg L^{-1} dissolved O_2 an adult trout may not suffer acute effects, but a trout larvae or young-of-the-year fish may not survive. Dissolved oxygen (DO) criteria apply to both continuous and cyclically depressed oxygen levels. If O_2 concentrations are continuously above the chronic criterion for growth (4.8 mg L^{-1}), the aquatic life at that location should not be harmed. When dissolved oxygen conditions at a site fall below the juvenile/adult survival criterion (2.3 mg L^{-1}), there is not a sufficient amount of oxygen to protect aquatic organisms. Thus, when conditions lead to persistently depressed oxygen levels, the DO concentrations will fall between the growth and survival levels. The duration and intensity of these depressed oxygen levels requires ongoing monitoring and, when necessary, actions may need to be taken to ensure that the aquatic ecosystem remains healthy [4]. Actions may include mechanical methods, e.g. aeration fountains, and decreasing the allowable amounts of oxygen depleting compounds in effluents reaching the water body.

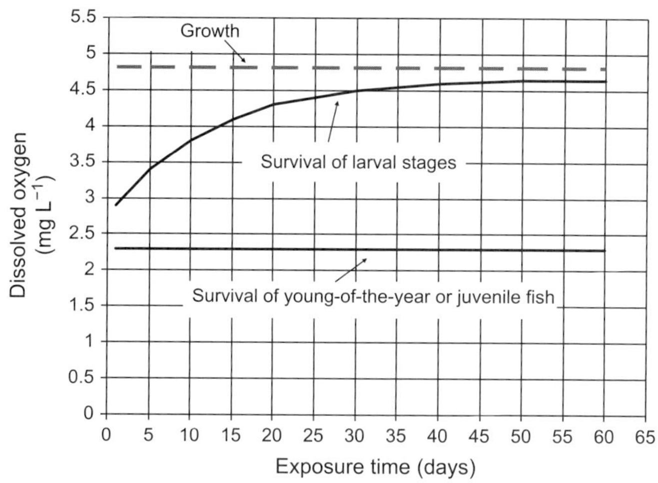

FIGURE 7.3
Summary of dissolved molecular oxygen (O_2) criteria for persistent exposure for a fish population. Shown are the lower bound limits on protective O_2 concentrations. The chronic growth limit may be violated for a specific number of days provided the chronic larval recruitment limit is not violated.
Source: US Environmental Protection Agency (2000). Ambient Aquatic Life Water Quality Criteria for Dissolved Oxygen (Saltwater): Cape Cod to Cape Hatteras. Report No. EPA-822-R-00-012. Washington, DC.

Sources of oxygen depletion are direct when they chemically react with oxygen, decreasing the dissolved oxygen (DO) content of the water. Sources of oxygen depletion are indirect if they allow for large growth of bacteria, fungi or algae that in turn uses up the oxygen. The total amount of oxygen used chemically (by agents other than living organisms) is known as chemical oxygen demand (COD). The amount used by microbes is known as biochemical oxygen demand (BOD). Naturally occurring organic matter, including organic wastes from sewage treatment plants, improperly operating septic systems, and runoff from agricultural and residential areas are the energy sources (i.e. "food") for water-borne bacteria. Bacteria decompose these organic materials using dissolved oxygen, thereby decreasing the DO present for aquatic life.

The BOD is the amount of oxygen that bacteria will consume in the process of decomposing organic matter under aerobic conditions. The BOD is measured by incubating a sealed sample of water for five days and measuring the loss of oxygen by comparing the O_2 concentration of the sample at time = 0 (just before the sample is sealed) to the concentration at time = 5 days (known specifically as BOD_5). Samples are commonly diluted before incubation to prevent the bacteria from depleting all of the oxygen in the sample before the test is complete [5].

BOD_5 is simply the measured DO at the beginning time, i.e. the initial DO (D), measured immediately after it is taken from the source minus the DO of the same water measured exactly five days after D_1, i.e. D_5:

$$\mathrm{BOD} = \frac{D_1 - D_5}{P} \tag{7.1}$$

where P = decimal volumetric fraction of water utilized and D units are in mg L^{-1}. If the dilution water is seeded, the calculation becomes:

$$\mathrm{BOD} = \frac{(D_1 - D_5) - (B_1 - B_5)f}{P} \tag{7.2}$$

where B_1= initial DO of seed control; B_5= final DO of seed control; and f= the ratio of seed in sample to seed in control ={% seed in $D1$/% seed in $B1$}. B units are in mg L^{-1}.

For example, to find the BOD_5 value for a 10 mL water sample added to 300 mL of dilution water with a measured DO of 7 mg L^{-1} and a measured DO of 4 mg L^{-1} five days later:

$$P = \frac{10}{300} = 0.03$$

$$BOD_5 = \frac{7-4}{0.03} = 100 \text{ mg L}^{-1}$$

Thus, the microbial population in this water is demanding 100 mg L^{-1} dissolved oxygen over the five-day period. So, if a conventional municipal wastewater treatment system is achieving 95% treatment efficiency, the effluent discharged from this plant would be 5 mg L^{-1}.

Chemical oxygen demand (COD) does not differentiate between biologically available and inert organic matter, and it is a measure of the total quantity of oxygen required to oxidize all organic material completely to carbon dioxide and water. COD values always exceed BOD values for the same sample. COD (mg L^{-1}) is measured by oxidation using potassium dichromate ($K_2Cr_2O_7$) in the presence of sulfuric acid (H_2SO_4) and silver. By convention, 1 g of carbohydrate or 1 g of protein accounts for about 1 g of COD. On average, the ratio BOD:COD is 0.5. If the ratio is <0.3, the water sample likely contains elevated concentrations of recalcitrant organic compounds, i.e. compounds that resist biodegradation [6]. That is, there are numerous carbon-based compounds in the sample, but the microbial populations are not efficiently using them for carbon and energy sources. This is the advantage of having both BOD and COD measurements. Sometimes, however, COD measurements are conducted simply because they require only a few hours compared to the five days for BOD.

Since available carbon is a limiting factor, the carbonaceous BOD reaches a plateau, i.e. *the ultimate carbonaceous BOD* (see Figure 7.4). However, carbonaceous compounds are not the only substances demanding oxygen. Microbial populations will continue to demand O_2 from the water to degrade other compounds, especially nitrogenous compounds, which account for the bump in the BOD curve. Thus, in addition to serving as an indication of the amount of molecular oxygen needed for biological treatment of the organic matter, BOD also provides a guide to sizing a treatment process, assigning its efficiency, and giving operators and regulators information about whether the facility is meeting its design criteria and is complying with pollution control permits.

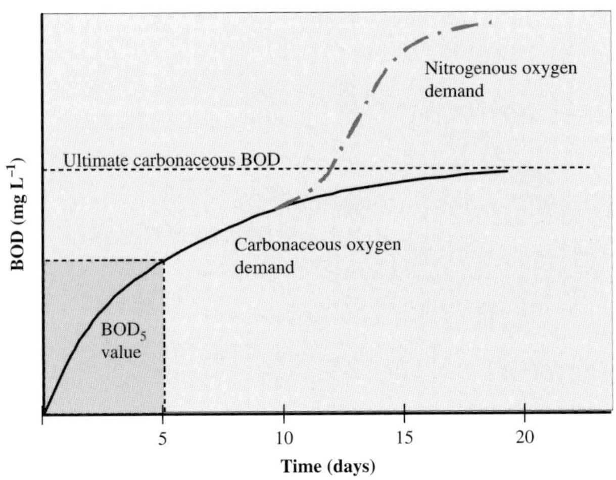

FIGURE 7.4
Biochemical oxygen demand (BOD) curve showing ultimate carbonaceous BOD and nitrogenous BOD. [*See color plate section*]
Source: Adapted from C.P. Gerba and I.L. Pepper (2009). Wastewater treatment and biosolids reuse. In: R.M. Maier, I.L. Pepper and C.P. Gerba (2009). *Environmental Microbiology*, 2nd Edition. Elsevier Academic Press, Burlington, MA.

DIGESTION

Biodegradation by aerobic microbes mimics natural systems. That is, those microbes that use oxygen as the electron acceptor are often the same as those found in natural systems and their survival and growth are similar to that in nature, albeit often at a more rapid rate.

Microbes degrade substrates and grow according to the Monod equation (Eq. 7.13 below). Usually, the Monod-constant K ranges between 1 and $10\,g\,m^{-3}$ for organic substrates, compared to $0.1\,g\,m^{-3}$ for oxygen. Whereas the liquids in bioreactors are frequently greater than the Monod-constant K, so they follow a zero-order growth curve (see Figure 7.11), if all other environmental conditions are equal (e.g. nutrient concentrations, pH, temperature, toxicities) [9].

Many biologically mediated processes that degrade contaminants are redox reactions that involve the transfer of electrons from an organic contaminant to an electron acceptor. In these cases, oxygen is the electron acceptor if the process is aerobic (i.e. sufficient molecular oxygen is available). In anaerobic microbial processes, Fe^{3+}, sulfate, and carbon dioxide can be the electron acceptors. The redox reactions for the commonly encountered contaminants benzene and substituted derivatives of benzene (i.e. xylene and ethylbenzene) are shown in Table 7.1.

The microbes will vary in dominance as the dissolved oxygen levels change. With falling dissolved oxygen concentrations, anaerobic bacteria will begin to dominate in degrading organic wastes. The transfer of electrons is the means by which the microbes receive energy from the food (i.e. the organic waste).

The process of adding nutrients to contaminated sites, such as nitrogen and phosphorus, oxygen, and other elements and compounds that serve as electron acceptors to stimulate the activity of microbial populations is known as *biostimulation* [10]. *Bioaugmentation* is the process of adding microbes to the subsurface environment. The microbes can come from "seed" microbes taken from the contaminated environment and mixing them with these elements and compounds in a reactor, and then reintroducing them to the contaminated soil or water, especially groundwater. Other times, special targeted and cultivated strains, including genetically modified microorganisms, with known abilities to degrade certain compounds, are injected into contaminated soil and groundwater. Many bioremediation experts favor biostimulation over bioaugmentation because almost every needed microbe is available in the subsurface, indigenous species have likely already developed enzyme production systems that will break down the contaminants, and the exogenous species that were augmented elsewhere may not survive the new, hostile environment after introduction.

A hybrid of biostimulation and vapor extraction is known as *bioventing*. This type of remediation stimulates the growth and metabolism of aerobic bacteria in contaminated soil by applying a vacuum at depth underground. Holes are drilled around the perimeter of contamination. Air is pulled from the surface to the lower pressure zone created by the vacuum pump, and flows through the contaminated soil. If the organic contaminants have a sufficiently high vapor pressure (i.e. volatile organic compounds), they will be pulled from the vadose (unsaturated) zone of soil or other unconsolidated matter, will flow upward and will enter a chamber where the vapors are treated. Obviously, the two limiting factors for bioventing are the vapor pressure of the contaminants and the permeability of the soil. Another variation is *biosparging*, wherein air is pumped into and through groundwater in the saturated zone to enhance volatilization and biodegradation.

The first-order biodegradation rate (R) can be calculated empirically from two representative points within a plume:

$$R = \ln\frac{C_d}{C_u}(Dv) \tag{7.3}$$

Table 7.1 Oxidation-reduction reactions for aromatic compounds

Type	Reaction	Electron acceptor
Benzene redox reactions:		
Oxidation	$C_6H_6 + 12H_2O \rightarrow 6CO_2 + 30H^+ + 30e^-$	
Reduction	$7.5O_2 + 30H^+ + 30e^- \rightarrow 15H_2O$	Oxygen
Reduction	$6NO_3^- + 36H^+ + 30e^- \rightarrow 3N_2 + 18H_2O$	Nitrate
Reduction	$15Mn^{4+} + 30e^- \rightarrow 15Mn^{2+}$	Manganese
Reduction	$30\ Fe^{3+} + 30e^- \rightarrow 30\ Fe^{2+}$	Iron
Reduction	$6SO_2^- + 37.5H^+ + 30e^- \rightarrow 3.75H_2S + 15H_2O$	Sulfate
Reduction	$3.75CO_2 + 30H^+ + 30e^- \rightarrow 3.75CH_4 + 7.5H_2O$	Methanogenic bacteria
Overall	$C_6H_6 + 7.5O_2 \rightarrow 6CO_2 + 3H_2O$	Oxygen
Overall	$C_6H_6 + 6H^+ + 6NO_3^- \rightarrow 6CO_2 + 3N_2 + 6H_2O$	Nitrate
Overall	$C_6H_6 + 15Mn^{4+} + 12H_2O\ ^- \rightarrow 6CO_2 + 30H^+ + 15Mn^{2+}$	Manganese
Overall	$C_6H_6 + 30Fe^{3+} + 12H_2O\ ^- \rightarrow 6CO_2 + 30H^+ + 0Fe^{2+}$	Iron
Overall	$C_6H_6 + 3.75SO_4^{2-} + 7.5H^+ \rightarrow 6CO_2 + 3.75H_2S + 3H_2O$	Sulfate
Overall	$C_6H_6 + 4.5H_2O \rightarrow 2.25CO_2 + 3.75CH_4$	Methanogenic bacteria
Xylene and ethylbenzene redox reactions:		
Oxidation	$C_8H_{10} + 16H_2O \rightarrow 8CO_2 + 42H^+ + 42e^-$	
Reduction	$10.5\ O_2 + 42H^+ + 42e^- \rightarrow 21H_2O$	Oxygen
Reduction	$8.4NO_3^- + 50.4H^+ + 42e^- \rightarrow 4.2N_2 + 25.2H_2O$	Nitrate
Reduction	$21Mn^{4+} + 42e^- \rightarrow 21Mn^{2+}$	Manganese
Reduction	$42Fe^{3+} + 42e^- \rightarrow 42Fe^{2+}$	Iron
Reduction	$5.25SO_2^- + 52.5H^+ + 42e^- \rightarrow 5.25H_2S + 21H_2O$	Sulfate
Reduction	$5.25CO_2 + 42H^+ + 42e^- \rightarrow 5.25CH_4 + 10.5H_2O$	Methanogenic bacteria
Overall	$C_8H_{10} + 10.5O_2 \rightarrow 8CO_2 + 5H_2O$	Oxygen
Overall	$C_8H_{10} + 8.4H^+ + 8.4NO_3^- \rightarrow 8CO_2 + 4.2N_2 + 9.2H_2O$	Nitrate
Overall	$C_8H_{10} + 21Mn^{4+} + 16H_2O^- \rightarrow 8CO_2 + 42H^+ + 21Mn^{2+}$	Manganese
Overall	$C_8H_{10} + 42Fe^{3+} + 16H_2O^- \rightarrow 8CO_2 + 42H^+ + 42Fe^{2+}$	Iron
Overall	$C_8H_{10} + 5.25SO_4^{2-} + 10.5H^+ \rightarrow 8CO_2 + 5.25H_2S + 5H_2O$	Sulfate
Overall	$C_8H_{10} + 5.5H_2O \rightarrow 2.75CO_2 + 5.25CH_4$	Methanogenic bacteria

Source: US Environmental Protection Agency (2003). Bioplume III Natural Attenuation Decision Support System: Users Manual, Version 1.0. Washington, DC.

where: C_d = Highest downgradient concentration of compound

C_u = Highest upgradient concentration of compound

D = Distance traveled

v = Plume velocity

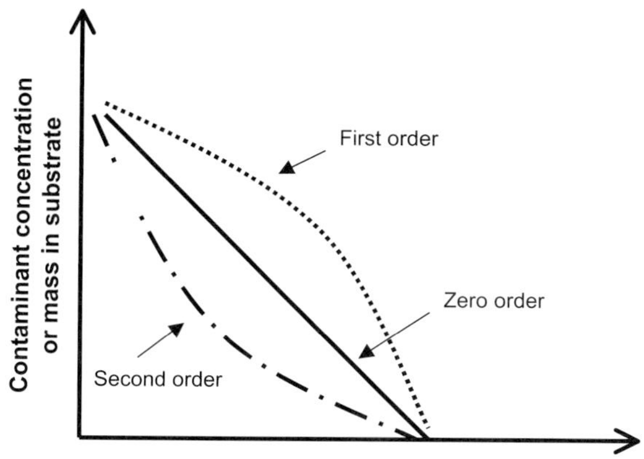

FIGURE 7.8
Prototypical decay curve for microbial degradation of organic contaminants.

Figure 7.8 illustrates the three prototypical microbial decay curves, where the beginning and ending amounts of contamination are the same but the rates or kinetics of the three systems are different. The mathematics of the three orders of decay kinetics have been derived from experimental data. While very useful in understanding the theory and first principles of degradation, especially biodegradation, they must be adapted to the heterogeneous conditions of the field, especially for *in situ* bioremediation, and the expected growth and metabolism of microbes that catalyze contaminants.

Biodegradation reactions follow the rate law:

$$aA + bB \rightarrow gG + hH, \text{ the rate } = k[A]^x[B]^y \tag{7.4}$$

where k represents the rate constant for molar concentrations of the reactants and x and y represent the reaction rate order for the reactants.

Summing all of the reactant orders gives the overall order of the reaction. A homogeneous (same physical state), abiotic example of decomposing the gas nitrogen pentoxide to the gases nitrogen dioxide and oxygen is:

$$2N_2O_5 \leftrightarrow 4NO_2 + O_2 \tag{7.5}$$

Laboratory studies have shown that rate law for this reaction is first order:

$$\text{Rate} = k[N_2O_5]^1 \text{or simply } k[N_2O_5]. \tag{7.6}$$

This means that for any first-order reaction for any chemical species, the rate is:

$$\text{Rate} = \frac{-d[A]}{dt} = \frac{\Delta[A]}{\Delta t} + k[A] \tag{7.7}$$

In a first-order reaction, doubling the concentration of chemical species A doubles the reaction rate, or a ten-fold increase in concentration of A means 10 times the reaction rate. Integrating the rate equation above yields:

$$\ln [A] = \ln[A]_0 - k\Delta t \tag{7.8}$$

Similarly, integrating the rate equation for a second-order reaction yields:

$$\frac{1}{[A]} = \frac{1}{[A]_0} + k\Delta t \tag{7.9}$$

However, living systems are more complicated than this. When the substrate is not limiting, i.e. a lot of contaminant (food) is available to the microbes, the contaminant is degraded as

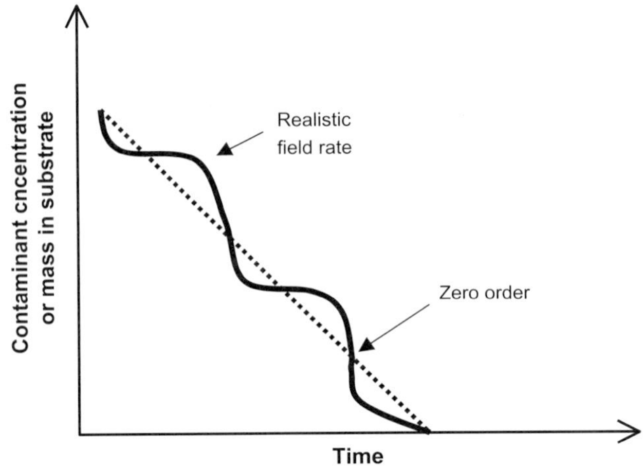

FIGURE 7.9
Rate changes within an overall zero-order decay rate. At various times during biodegradation, the orders will differ, depending on abiotic and biotic conditions.

a function of the logarithmic growth of the microbes, following zero-order kinetics (constant: log growth and log decay). When the rate of degradation of the chemical contaminant becomes directly proportional with the concentration of the contaminant, the decay follows first-order kinetics.

One of the more realistic biological kinetics scenarios is second order. That is, the first-order kinetics are related to microbial population density. Over time the realistic [11] biodegradation rate may cycle through various orders over the life of a remediation project, as shown in Figure 7.9. Modelers often consider microbial degradation rates to be nonlinear reactions, possibly because of the dearth of information available supporting stepwise degradation, especially in natural systems. Thus, they have developed biodegradation models to fit the results, expressed as:

$$-d[\mathrm{A}] = k[\mathrm{A}]^{\mathrm{n}}\frac{-d[\mathrm{A}]}{dt} = \mathrm{k}[\mathrm{A}]_{\mathrm{n}} \tag{7.10}$$

where n is the rate curve fitting parameter.

Numerous factors may account for why bioremediation does not follow the theoretical decay expected if degradation depended only on total concentration of the contaminant chemical species. Microbial populations not only must degrade soil concentrations that are (readily) available to the microbes, but also the water-borne fraction (e.g. in the water-filled pore spaces of the soil and unconsolidated media). For non-aqueous phase liquids (NAPLs), which comprise many of the organic contaminants, much of the contaminant is not in a soluble form. Many contaminants have strong affinity for soil particles due to their sorption coefficients, so they resist diffusion. They may also be physically encapsulated within the soil matrices. The bioengineer may need to consider conditioning the soil and pore water (e.g. with surfactants) to overcome some of these factors that limit contact between microbes and chemical contaminants.

Figure 7.10 illustrates an idealized biodegradation growth curve, which has well-defined stages. The extent and duration of each stage will of course vary according to the microbial species and environmental conditions.

Thus, engineered systems bring the chemical contaminant into contact with the microbes and enhance the degradation environment, such as by pumping nutrients and air (or pure oxygen) into the groundwater. Engineered systems may also add activated microbes (if indigenous microbes are not already breaking down the chemical) into the

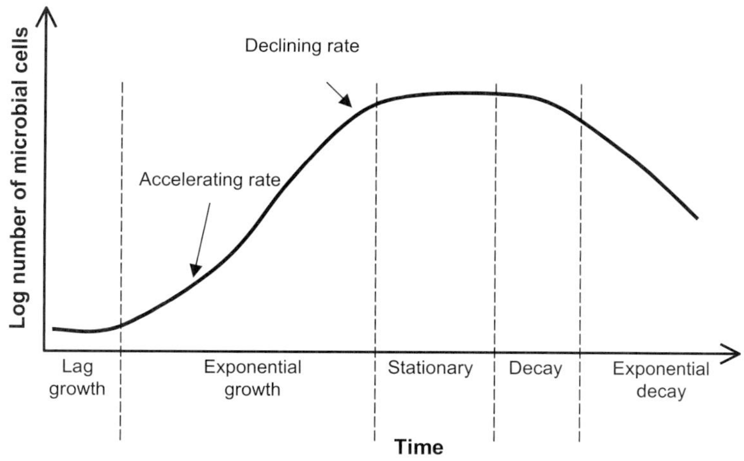

FIGURE 7.10
Prototypical growth and decay curve for bacteria.

vadose or saturated zones. As these conditions change, the microbes undergo a series of stages [12]:

1. *Lag phase:* Upon initial exposure of the microbes to the chemical contaminant, a period of time is needed for the organisms to become acclimated.
2. *Accelerated growth rate phase:* Following acclimation, the microbes propagate at an increasing rate. There are two major processes that allow for the degradation of chemical compounds by microbes:
   (A) The most effective is when the organisms use the contaminant as a food source for their growth, metabolism, and reproduction. This is accomplished by the microbe's ability to produce enzymes that catalyze the chemical contaminant as a carbon source (hence, that is why bioremediation is often a good choice for organic contaminants). In addition, the chemical contaminant is also a source of electrons that the microbe needs to extract for energy. That is, the chemical is the microbe's oxygen acceptor during respiration. The microbe produces enzymes that hasten the process of breaking chemical bonds and transferring electrons from the contaminant to an electron acceptor (oxygen for aerobic respiration, and metals (e.g. iron and manganese) and inorganic chemicals (e.g. nitrates and sulfates) for anaerobic respiration).
   (B) A second, less effective process is known as "secondary utilization." Microbes can transform chemical contaminants although the reaction provides no direct benefit to the microbial cell. Probably the most common, at least best-understood secondary utilization process is *cometabolism*, wherein the microbes break down chemicals coincidentally with enzymes that they normally synthesize for metabolism or detoxification. A case is the methane-oxidizing bacteria that happen to degrade chlorinated hydrocarbons, benzene, phenol, and toluene by producing enzymes needed to transfer electrons to methane (i.e., the bacterium's normal electron acceptor). The normal methane oxidation enzymes auspiciously degrade the chemical contaminants, even though the chemicals cannot serve as the primary food source for the bacteria.
3. *Exponential phase:* The cell mass and the number of cells are growing exponentially by binary fission.
4. *Declining growth phase:* Cell mass and numbers of cells continue to grow, but at a decreasing rate. This is usually due to depletion of the food and electron source (i.e. the contaminant, hopefully). Other limiting factors could also come into play, such as some of the newly generated chemicals ("degradates") inhibiting the growth of the microbes due to their toxicity.
5. *Stationary phase:* Cell decay is about equal to cell propagation during this time.

6. *Decay:* Cell decay now exceeds cell propagation.
7. *Exponential death phase:* Cells are dying exponentially as cells no longer grow or propagate. Under optimal conditions, this indicates successful bioremediation because the food source (toxic organics) is entirely used up.

The cycle will repeat if the microbes are again introduced to another slug of pollutants. This happens *ex situ*, for example a new batch of contaminated soil is introduced in the reactor, and *in situ*, with biostimulation (e.g. when the new mixes of nutrients are injected) or bioaugmentation (e.g. microbes are sent to another part of the aquifer).

The increase in microbial biomass in a bulk solution is inversely proportional to the decrease in food, if the food source is constant. It is also directly proportional to the biodegradation half-life of the contaminant. For example, if 400 kg of a chemical contaminant in a water solution must be treating using a reactor system, we can estimate the time it will take to degrade 375 kg of the contaminant, if we know the rate constant and the reaction order. In this instance, the contaminant's rate constant is 10^{-4} sec^{-1} and the reaction is first order.

By definition, thę half-life ($T_{1/2}$) is the time it takes for the concentration of a reactant to reach one-half of its initial value, so in this case it takes four half-lives (x – x/2 – x/4 – x/8 – x/16) to reach 25 kg.

The equation for a first-order half-life ($T_{1/2}$) is:

$$T_{1/2} = \ln(2)(k)^{-1} \tag{7.11}$$

where k = the rate constant.

Since the natural log of 2 is 0.693 and $k = 10^{-4}$ the half-life of the chemical is:

$$T_{1/2} = (0.693)\,(10^{-4})^{-1} = 6930 \text{ sec}$$

Thus, the time it takes to reach 25 kg of remaining chemical, i.e. four half-lives, can be calculated:

$$4(T_{1/2}) = 27720 \text{ sec or } \underline{7.7 \text{ hours}}$$

to destroy 375 kg of the initial 400 kg of the contaminant.

We still have a large amount of chemical (25 kg) to be treated. Assume that we need to reach 1 kg of the remaining chemical (i.e. degrade 24 kg). This will take an additional five $t_{1/2}$, so if the first-order reaction continues it will take 34,650 sec or 9.6 hours for remaining mass of the contaminant in the solution to fall below 1 kg. It is important to note that reaction rates can be affected by concentrations and other factors (see Figure 7.9). For example, when much reactant is available the reaction is not rate limited by concentration, but as the mass drops in the solution, the microbes may transition to non-exponential growth.

Given the assumptions, the total time needed to go from 400 kg to <1 kg of the waste is more than 17 hours. However, an engineer would likely want to achieve the higher removal rates found at higher mass (and concentrations). If this were not the only waste source for the chemical contaminant, it would be better to keep adding wastes to the reactor. In other words, treating the first half-life (400–200 kg) continuously takes 2 hours to destroy 200 kg of the contaminant. Conversely, treating at the ninth half-life (about 1.6–0.8 kg), it takes 2 hours to destroy 0.8 kg of the contaminant [13].

Degradation rates are often expressed empirically. For example, an enzyme-catalyzed reaction to degrade organic compounds follows the Michaelis–Menton equation:

$$\frac{dC}{dt} = -\frac{V_{\max}C}{K_m + C} \tag{7.12}$$

where C = concentration of the compound in soil; t = time; V_{max} = maximum reaction rate; and K_m = the Michaelis constant.

V_{max} is the reaction rate threshold above which the concentration of the compound is not a rate limiting factor. It includes both extracellular enzymatic activities in the soil and intracellular microbial activities in the soil. Thus, it changes with the concentrations of enzymes. K_m is the substrate concentration at which the reaction rate is one-half of the V_{max} [14].

With decreasing concentrations of an organic substrate, the rate of microbial biomass growth also decreases. This is expressed by the empirically-derived Monod equation:

$$\mu = \frac{\mu_{max} S}{K_s + S} \qquad (7.13)$$

where μ = the specific growth rate of the microbe, μ_{max} = maximum specific growth rate, and K_s = the Monod growth rate coefficient representing the substrate concentration at which the growth rate is half the maximum rate (see Figure 7.11). The μ_{max} is reached at the higher ranges of substrate concentrations. K_s is an expression of the affinity of the microbe for a nutrient, i.e. as K_s decreases the more affinity that microbe has for that particular nutrient (as expressed by the concomitantly increasing μ).

Thus, at low substrate concentrations, the reaction rates are often assumed to be first order with respect to both microbial population biomass and substrate concentration. Microbiologists and other biologists describe first-order processes using a half-life, whereas engineers commonly describe first-order processes using a first-order rate constant. This is because the rate of change relates directly to the constant. Also, in the presence of simultaneously operating first-order processes, such as biodegradation, the rate constant for the combined effect can be calculated as the sum of the individual rate constants. Transport and fate models can be evaluated using this property, since numerous processes can be combined into a single calibration parameter [15]. Figure 7.12 compares first-order rate constants to the respective half-life units.

There is some concern that the Monod equation is based solely on observational information and has not been fully explained theoretically, so there continues to be research to explain the mechanics underlying these empirical relationships between microbial populations and substrate degradation.

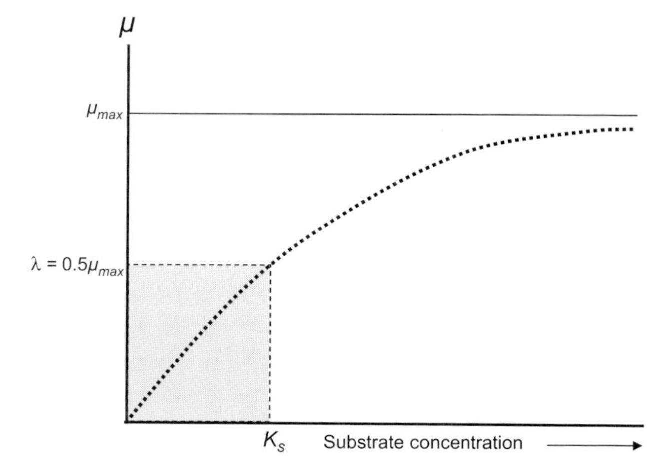

FIGURE 7.11
Empirical basis for deriving the Monod growth rate coefficient. This has been derived from numerous studies showing the hyperbolic relationship between the microbial growth rate and the carbon source (organic compound) supporting the microbial growth. Curve represents no inhibitions.
Source: Adapted from D.L. Russell (2006). *Practical Wastewater Treatment.* John Wiley & Sons, Inc. Hoboken, NJ.

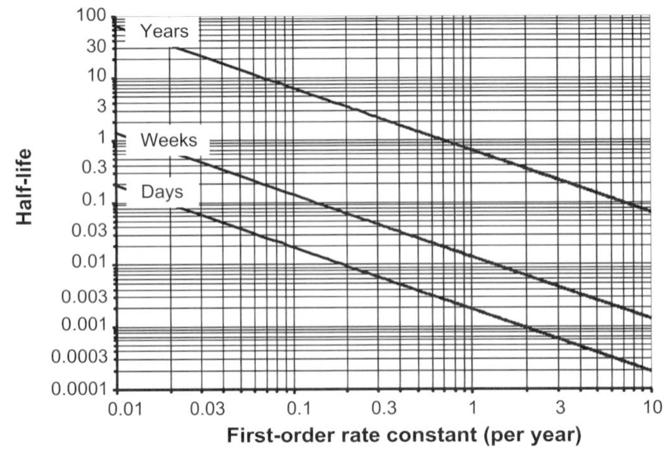

FIGURE 7.12
Relationship between a first-order rate constant in units of per year and half-lives in units of days, weeks, and years. *Source:* US Environmental Protection Agency (2008). Natural attenuation of the lead scavengers 1,2-Dibromoethane (EDB) and 1,2-Dichloroethane (1,2-DCA) at motor fuel release sites and implications for risk management. Report No. EPA 600/R-08/107. Ada, OK.

The biodegradable, organic feedstock coming into an aerobic digester is oxidized to CO_2 and water. The exiting material usually also includes untreated or converted biomass and nitrogenous compounds. To increase the rates of anaerobic digestion over those in nature, the surface area of biofilm must be greatly increased, so as to enhance O_2 exchange in fixed film processes (see Discussion Box: Biodynamic Films), especially trickling filters, and returning some of the activated sludge to the digester vessel.

Biodegradation by anaerobic digestion is initiated when microbes hydrolyze the organic matter entering the bioreactor. The rate of hydrolysis depends on the reactor conditions (retention time, mixing, temperature, pressure) and the complexity of the organic material. The feedstock can be mixtures containing simple carbohydrates to very lipophilic, organic compounds (e.g. polymers). *Acidogenic* microbes convert the feedstock sugars and amino acids into carbon dioxide, molecular hydrogen (H_2), ammonia (NH_3), and organic compounds with the carboxyl functional group (R-COOH), i.e. the organic acids. Next *acetogenic* microorganisms convert the organic acids into simpler acids, e.g. acetic acid, with the H_2 and NH_3 remaining in the mixture. The final step is *methanogenesis,* also known as *biomethanation.* This consists of methanogenic bacteria converting this mixture into methane (CH_4) and CO_2.

DISCUSSION BOX

Biochemodynamic Films

A key limitation of biodegradation and other biological processes is whether an organism comes into contact with the substance that needs to be transformed and degraded. This takes place in the biofilm, which is a thin layer of biota that lives on substrate, such as soil particles, unconsolidated materials like sand and gravel in aquifers, or media in bioreactors and other biological waste treatment systems. A biofilm consists of microorganisms, e.g. alga, fungi, bacteria, their byproducts and waste, water and air (see Figure 7.13).

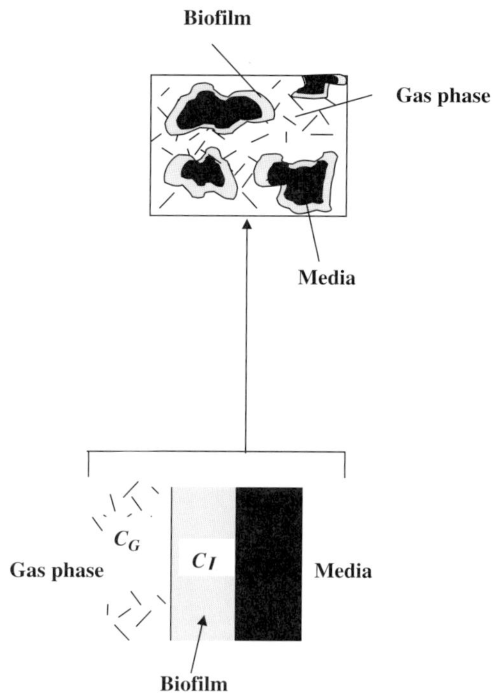

FIGURE 7.13
The structure of a biofilm. Substances in the gas phase (C_G) traverse porous media. The soluble fraction of the substances in the air stream partitions into the biofilm (C_L) according to Henry's law: $C_L = \frac{C_G}{H}$; where H is the Henry's law constant. *Source:* D.A. Vallero (2008). *Fundamentals of Air Pollution.* Elsevier Academic Press, Burlington, MA; adapted from S.J Ergas and K.A Kinney (2000). Biological control systems, in W.T. Davis (Ed.), *Air and Waste Management Association: Air Pollution Engineering Manual*, 2nd Edition. John Wiley & Sons, Inc., New York, NY.

During degradation, the mass or concentration of a chemical substance decreases with respect to time. This is the case for any type of degradation, whether abiotic chemical degradation or biodegradation. All other factors being equal in biodegradation and many abiotic chemical decay processes, the concentration dependence reflects a first-order reaction sequence. That is, the concentration of the product increases in proportion to the decrease in concentration of the single substance undergoing change:

$$\frac{dC_A}{dt} = kC_A \tag{7.14}$$

where C_A is the concentration of reactant A, t is time, and k is the first-order reaction constant, i.e., the fraction of A degrading per unit of time.

The reactions may occur in various combinations of physical phases and environmental compartments [16]. Reactions that take place in a single phase are known as *homogeneous reactions.* Film is the area between two different phases (see Figure 7.14).

Biofilms follow the same biochemodynamic principles of any film. The movement and change of substances (e.g. nutrients and contaminants) across phase boundaries and compartmental interfaces can be envisioned as the separation and joining of films. The *two-film model* depicted in Figure 7.15 demonstrates the relationship between the liquid and gas phases at a microscopic scale [17]. The model is designed under the assumption that the gas and liquid phases are in turbulent contact with each other, but segregated by an interface, where contaminants may cross in either direction. Each film is a mass transfer zone that comprises a small volume of the gas and liquid phases on either side of the interface.

The flow in the mass transfer zones is assumed to be laminar and the flow in the bulk liquid and gas regions is generally turbulent. Assuming complete mixing in the bulk phases and

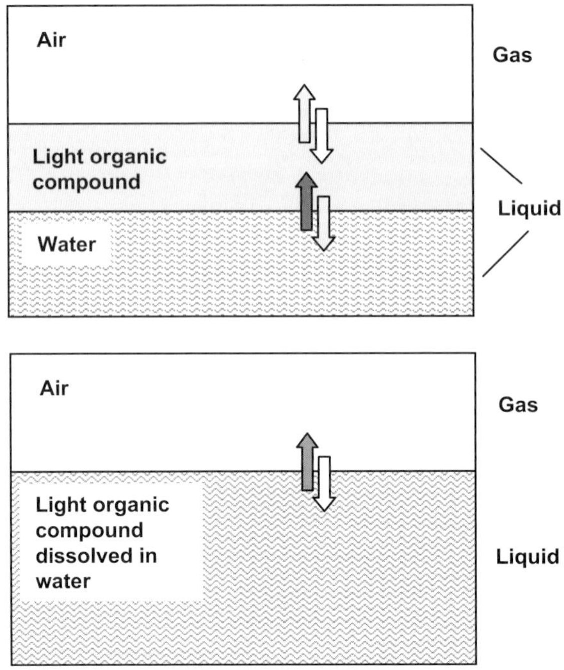

FIGURE 7.14

Top diagram: Transport and reactions occurring within and among two phases (liquid and gas) and three compartments (water, organic compound, and air) in an environmental system. *Bottom diagram:* Transport and reactions if the substances in the same phase are miscible (i.e., the organic compound is dissolved in the water).
Source: Adapted from W.J. Weber, Jr and F.A. DiGiano (1996). *Process Dynamics in Environmental Systems.* John Wiley & Sons, New York, NY.

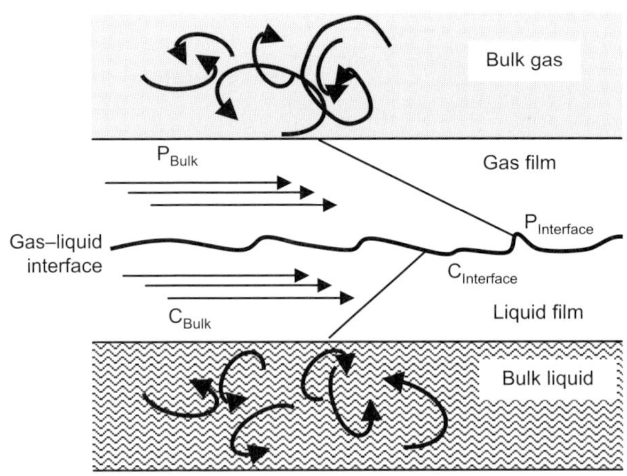

FIGURE 7.15

Two-film model. C_{Bulk} is the concentration of a substance in the bulk liquid and $C_{Interface}$ is the concentration of the substance at the gas–liquid interface. P_{Bulk} is the partial pressure of the substance in the bulk liquid and $P_{Interface}$ is the partial pressure of the substance at the gas–liquid interface. Flow lines indicate laminar flow in films and turbulent flow in bulk phases.
Source: Based on W.G. Whitman (1923). The two-film theory of gas absorption. *Chemical and Metallurgical Engineering* 29: 147.

equilibrium between the substance's molecule and the molecules of the interface, for a substance to move from the gas phase to the liquid phase, it must first migrate from the bulk gas phase into the gas film, then diffuse through the gas film (a function of the gas' partial pressure), before crossing the interface. After it crosses the interface, the substances must diffuse through the liquid film and mix into the bulk liquid (a function of the substance's concentration in the liquid). This means that all resistance to movement occurs as the substance's molecules diffuse through the films to reach the interface region. That is why the two-film theory is also referred to as the *double resistance theory.* In liquid-to-gas transport, the concentration in the liquid changes as it moves from the bulk liquid phase through the liquid film. The rate of mass transfer from one phase to the other then equals the product of the amount of the substance being transferred times the resistance the substance receives when moving through the films. Thus, the difference in a contaminant's partial pressure in the bulk gas and at the interface is known as the *gas-side impedance,* while the difference in a contaminant's concentration in the bulk liquid and at the interface is known as the *liquid-side impedance* [18]. Therefore, chemical reactions can occur in the bulk liquid, in the bulk solid, but also within the film. The reaction rates and characteristics depend on in which phase or mass transfer zone (film) the contaminant or nutrient is found.

The film models are also important for the transport of contaminants across various types of surfaces. For example, the biofilm behaves similarly to the two-film exchanges in tissue, such as between water and lipid diffusion between blood and gastro-intestinal tracts in animals (see Figure 7.16).

Biofilms are important considerations in any biological treatment system. For example, the traditional design of a trickling filter system illustrated in Figure 7.17 includes a bed of

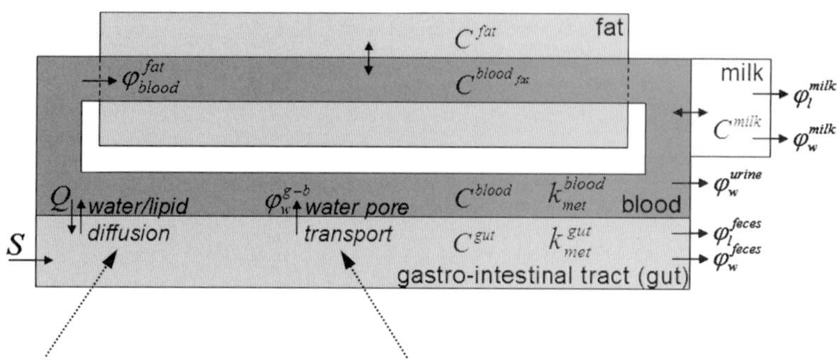

FIGURE 7.16
Mass balance model based on surface–film interfaces. Substances are transported across interfaces according to two-film and biofilm principles (see Figs. 7.12 and 7.14). [*See color plate section*]
Source: R. Rosenbaum, T.E. McKone and O. Jolliet (2005). Reducing uncertainties in biotransfer modeling in meat and milk. Society of Environmental Toxicology and Chemistry 26th Annual Meeting in North America. Baltimore, MD, November 13–18, 2005.

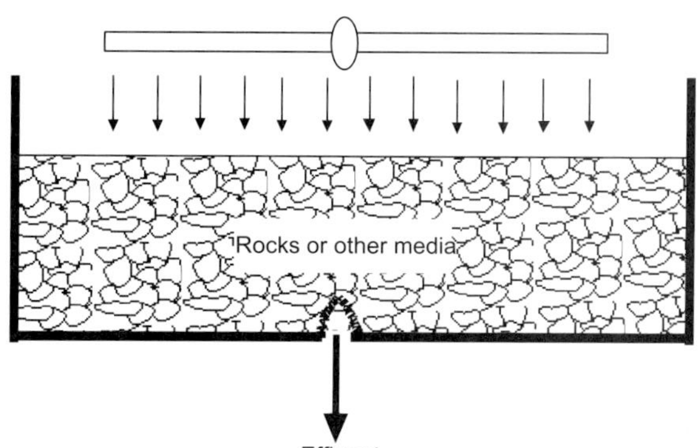

FIGURE 7.17
Trickling filter treatment system
Source: D. Vallero (2003). *Engineering the Risks of Hazardous Wastes.* Butterworth-Heinemann, Boston, MA.

fist-sized rocks, enclosed in a rectangular or cylindrical structure, through which is passed the waste of concern. Biofilms are selected from laboratory studies and encouraged to grow on the rocks; as the liquid waste moves downward with gravity through the bed the microorganisms comprising the biofilm are able to come into contact with the organic contaminant/food source and ideally metabolize the waste into relatively harmless $CO_2 + H_2O +$ microorganisms (+ energy ?). Oxygen is supplied by blowers from the bottom of the reactor and passes upward through the bed. The treated waste that moves downward through the bed subsequently enters a quiescent tank where the microorganisms and their biofilms are sloughed off of the media and then settled, collected, and ultimately disposed.

Trickling filters are actually considered to be mixed treatment systems because aerobic bacteria grow in the upper, higher oxygen layers of the media, while anaerobes grow in the lower, more reduced regions lower in the system. No matter the type of microbe, however, their only means of contact with the waste to be treated is the biofilm.

The same principles hold for microbial degradation in soil and in an aquifer (see Figure 7.18). As the contaminated water passes through the unsaturated zone, the microbes depend on the biofilm to come into contact with the food (the organic matter in the passing water). In fact, the zone immediately above the water table is the capillary fringe. Regardless of how densely soil particles are arranged, void spaces (i.e. pore spaces) will exist between the particles. By definition, the pore spaces below the water table are filled exclusively with water. However, above the water table, the spaces are filled with a mixture of air and water, so microbial contact depends on the biofilm around the individual particles. Figure 7.18 illustrates that the spaces between unconsolidated material (e.g. gravel, sand, or clay) are interconnected, and behave like small conduits or pipes in their ability to distribute water. Depending on the grain size and density of packing, the conduits will vary in diameter, ranging from large pores (i.e. macropores), to medium pore sizes (i.e. mesopores), to extremely small pores (i.e. micropores).

Biofilm transport is a limiting factor in biodegradation. Their properties can also interact with other substrate and media properties. These relationships can be put to use by the bioengineer. Lipophilic and hydrophobic compounds can be particularly difficult to treat. However, if the concentration gradient of these compounds is greater than that of the free water phase, the

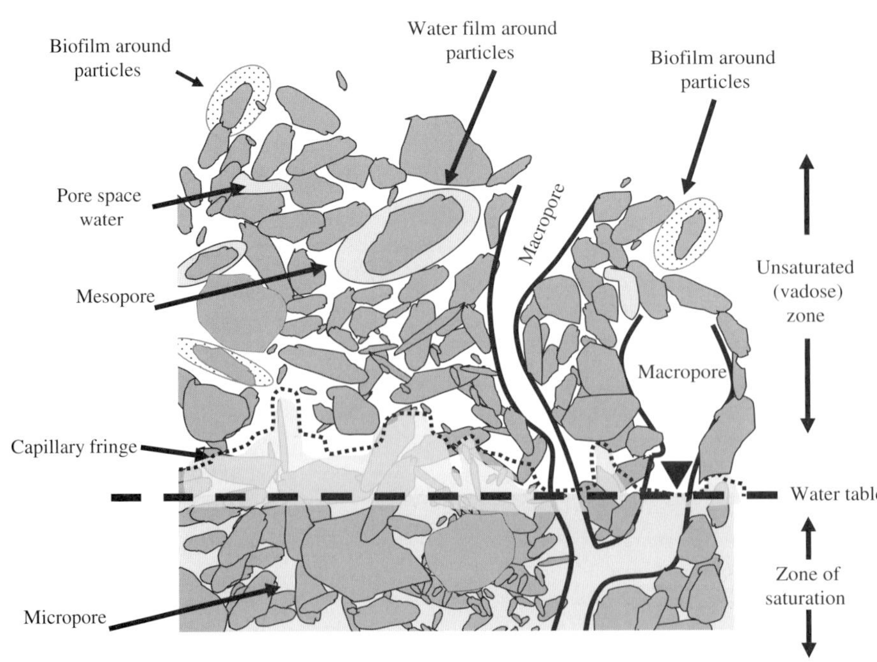

FIGURE 7.18
Capillary fringe above a water table. [*See color plate section*]

partitioning rate in the biofilm can be enhanced for those microbes that produce hair-like structures (fimbriae) that enhance the transport of these compounds toward the cells. The combination of these appendages and the biofilm are conducive to biodegradation. The question remains whether genetically enhancing microbial appendages and other properties to improve contact with contaminants carries its own risks.

AEROBIC BIODEGRADATION

Theoretically, if a substance is completely organic in structure, it should be able to be completely destroyed using principles based in microbiology with the engineering inputs and outputs summarized as:

$$\text{Hydrocarbons} + O_2 + \text{microorganisms (+energy)} \rightarrow CO_2 + H_2O + \text{microorganisms (+energy?)}$$

Organic wastes are mixed with oxygen and aerobic microorganisms, sometimes with an added energy source in the form of added nutrition for the microorganisms, and in seconds, hours, or possibly days the byproducts of gaseous carbon dioxide and water are produced which exit the top of the reaction vessel. Simultaneously, a solid mass of microorganisms is produced to exit the bottom of the reaction vessel [19].

On the other hand, if the waste of concern to the engineer contains other chemical constituents, in particular chlorine and/or heavy metals, and if in fact the microorganisms are able to withstand and flourish in such an environment and not shrivel and die, the simple input and output relationship is modified to:

$$\begin{aligned}&\text{Hydrocarbons} + O_2 + \text{microorganisms (+energy ?)} + \text{Cl or heavy metal(s)} + H_2O\\&+ \text{inorganic salts} + \text{nitrogen compounds} + \text{sulfur compounds} + \text{phosphorus compounds}\\&\rightarrow CO_2 + H_2O\text{ (+energy?)} + \text{chlorinated hydrocarbons or heavy metal(s) inorganic salts}\\&+ \text{nitrogen compounds} + \text{sulfur compounds} + \text{phosphorus compounds}\end{aligned}$$

A word of caution: if the microorganisms survive in this complex environment they can be transformed into potentially more toxic molecules that contain chlorinated hydrocarbons, higher heavy metal concentrations, as well as more mobile or more toxic chemical species of heavy metals.

All bioreactors have a number of common attributes. All rely on a population(s) of microorganisms to metabolize organic contaminant ideally into the harmless byproducts of $CO_2 + H_2O$ (+energy ?). In all of the systems the microorganisms must be either initially cultured in the laboratory to be able to metabolize the specific organic waste of concern, or target populations of microorganisms in the system must be given sufficient time, days, weeks, possible even years, to evolve to the point where the cumbersome food, that is the contaminant, is digestible by the microorganisms. However, biotechnology, including rDNA manipulation, has sped up these acclimation processes for certain microbial populations.

During all treatment processes the input waste must be monitored and possibly controlled to maintain environmental conditions that do not upset or destroy the microorganisms in the system. These monitoring and control requirements for each of the systems include but are not limited to:

temperature, possibly in the form of a heated building;
pH, possibly in the form of lime addition;
oxygen availability, possibly in the form of atmospheric diffusers that pump ambient atmosphere into the mixture of microorganisms and contaminant;

presence of additional food sources and/or nutrients, possibly in the form of a secondary carbon source for the microorganisms; and,
changes in the characteristics of the input hazardous waste, including hydrocarbon availability and chemicals that may be toxic to the microorganisms, possibly including holding tanks to homogenize the waste prior to exposure to the microorganisms.

The populations of microorganisms must be matched to the particular contaminant of concern. The engineer must plan for and undertake extensive and continual monitoring and fine-tuning of each microbiological processing system during its complete operation.

The advantages of the biotreatment systems include: (1) the potential for energy recovery; (2) volume reduction of the hazardous waste; (3) detoxification as selected molecules are reformulated; (4) the basic scientific principles, engineering designs, and technologies are well understood from a wide range of other applications including municipal wastewater treatment at facilities around the world; (5) application to most organic contaminants which as a group compose a large percentage of the total hazardous waste generated worldwide; (6) the possibility to scale the technologies to handle a single gallon/pound (liter/kilogram) of waste per day or millions of gallons/pounds (liters/kilograms) of waste per day; and (7) land areas that could be small relative to such other hazardous waste management facilities as landfills.

The disadvantages of these systems include: (1) the operation of the equipment requires very skilled operators and is more costly as input contaminant characteristics change over time and correctional controls become necessary; and (2) ultimate disposal of the waste microorganisms is necessary and particularly troublesome and costly if heavy metals and/or chlorinated compounds are found during the expensive monitoring activities.

Given these underlying principles of biotreatment systems four general guidelines are suggested whenever such systems are considered as a potential solution to any contaminant problem:

only liquid organic contaminants are true candidates;
chlorine-containing organic materials deserve special consideration if in fact they are to be biotreated at all, and special testing is required to match microbial communities to the chlorinated wastes, realizing that useful microbes may not be identifiable and even if they are the reactions may take years to complete;
hazardous waste containing heavy metals generally should not be bioprocessed; and
residual masses of microorganisms must be monitored for chemical constituents, and each residual must be addressed as appropriate so the entire bioprocessing system operates within the requirements of the local, state and federal environmental regulators.

The bottom line is that each application of biotechnology must be tailored to the specific characteristics of the contaminant under consideration, including the quantity of waste to be processed over the planning period as well as the physical, chemical, and microbiological characteristics of the waste also over the entire planning period of the project. Laboratory tests matching a given waste to a given bioprocessor must be conducted prior to the design and citing of the system.

Variations on three types of bioprocessors are available to the engineer: (1) trickling filter; (2) activated sludge; and, (3) aeration lagoons. As a group these three types of treatment systems represent a broad range of opportunities available to engineers searching for methods to control the risks associated with contaminants.

Trickling filter

The trickling filter is a time-tested, proven bioreactor system that has been widely used to treat municipal wastes and hazardous wastes. As mentioned, the classic design of a trickling filter system illustrated in Figure 7.17 includes packed media (often rocks, but there are numerous materials onto which microbes can adhere) through which liquids contacting the organic matter move by gravity after being sprayed onto the surface. The large amount of surface area provides ample contact between microbial populations and the liquid waste. That is, the microbes' biofilm comes into contact with an organic contaminant/food source. The concomitant microbial growth and metabolism result in degradation of the organic compounds, i.e. $CO_2 + H_2O$ + microorganisms (+energy ?).

Media bed depths can vary, but traditional trickling filter beds have been between about 1 to 3 meters. The media size ranges from about 25 to 60 mm in diameter (e.g. round gravel). However, various shapes and materials (e.g. polyvinyl chloride) have been used successfully. Waste is applied in a circular rotating manner to prevent pooling or uneven deposits. The interstices and openings between the media allow for air penetration throughout the system. In addition air can be supplied by blowers from the bottom of the reactor, migrating upward through the bed. This makes most of the trickling filter aerobic. The treated waste that moves downward through the bed subsequently enters a quiescent tank where the microorganisms that are sloughed off of the rocks are settled, collected, and ultimately disposed of. Even though much of the surface area is aerobic, trickling filters are actually considered to be mixed treatment systems because aerobic bacteria grow in the upper, higher oxygen layers of the media, while anaerobes grow in the lower, more reduced regions lower in the system, particularly if air is not pumped upward.

For particularly strong and recalcitrant compounds, and when the organic loading increases with time, a multiple filter system will be needed. For example, the first filter in a series may accept a high loading rate of organic matter, followed by subsequent filters in which their media come into contact with the partially treated waste liquids that broke through the first filter. Indeed, the process can be reversed so that the overloaded filter can recover from the loading (i.e. the second filter becomes the first filter, so that the original, first filter will receive less organic loading).

Activated sludge

The key to the activated sludge system (see Figure 7.19) is the enhancement of biodegradation due to greater numbers of microbes so that they can metabolize the contaminant/food source. This is accomplished by recycling the microbially rich biosolids (i.e. sludge), which enables this bioprocessing system to evolve over time as the microorganisms adapt to the changing characteristics of the influent. This evolution increases the potential for the microorganisms to be more efficient at metabolizing the waste stream of concern.

Liquid organic mixtures are injected with a mass of microorganisms into the bioreactor. Oxygen is supplied through the aeration basin as the microorganisms come in contact, sorb, and metabolize the waste ideally into $CO_2 + H_2O$ + microorganisms (+energy ?). The heavy, satisfied, microorganisms then flow into a quiescent tank where the microorganisms that are settled with gravity are collected, and ultimately disposed. Depending on the current operating conditions of the processor some or many of the settled and now active microorganisms are returned to the aeration basin where they are given another opportunity to metabolize the organic compounds in the tank. Liquid effluent from the activated sludge system may require additional microbiological and/or chemical processing prior to release into a receiving stream or city sewer system.

The activated sludge process in theory and in practice is a sequence of three distinct physical, chemical, and biological steps:

Sorption. The microorganisms come in contact with the food source, the organic material in the contaminant, and the food either is adsorbed to the cell walls or adsorbed through the cell walls of the microorganisms. In either case the food is now directly

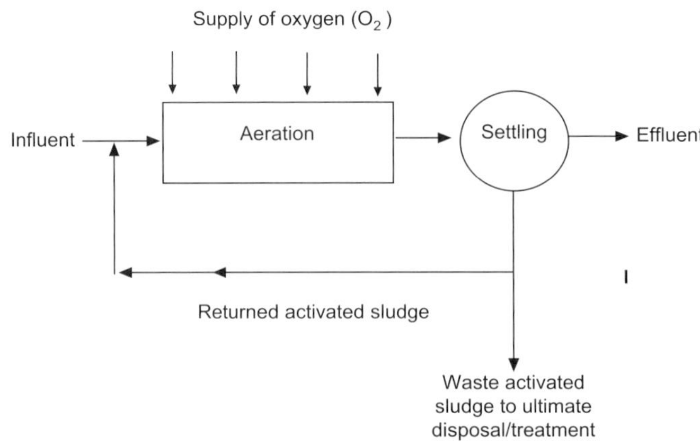

FIGURE 7.19

Activated sludge treatment system. An aerobic treatment approach breaks down toxic substances from household or manufacturing sources. The waste is combined with recycled biomass and aerated to maintain a target dissolved oxygen (DO) content. Organisms use the organic components, expressed as biochemical oxygen demand (BOD) of waste as food, decreasing the organic levels in the wastewater. Oxygen concentrations must be controlled to maintain optimal treatment efficiencies. One means of achieving optimal DO content is tapered aeration (shown in the diagram). The tapered system provides high concentrations of oxygen near the influent to accommodate the large oxygen demand from microbes as waste is introduced to the aeration tank (photo).

Source: Adapted from D. Vallero (2003). *Engineering the Risks of Hazardous Wastes.* Butterworth-Heinemann, Boston, MA.; photo courtesy of D.J. Vallero.

available to the individual microorganisms. In a correctly operated facility this sorption phase generally takes about 30 minutes.

Growth. The microorganisms metabolize the food and biochemically break down, or destroy, the hazardous organic molecules. This growth phase, during which individual organisms grow and multiply, may take up to hours or possibly days for complete metabolism of the hazardous constituents in the waste. Thus the design of the activated sludge system must include a basin with a detention time adequate for the correct amount of growth to take place.

Settling. Solid (the microorganisms) – liquid (the liquid remaining from the process) separation is achieved in a settling basin where the heavy and satisfied microorganisms sink to the bottom with gravity.

A critical design consideration of the activated sludge system is the *loading* to the aeration basin. *Loading* is defined as the food (F) to microorganism (M) ratio (F:M) at the start of the aeration process. Loading requires a balance between sufficient food available for the microbial population that is returned from the settling basin to the aeration tank. The operating engineer must fine-tune the F:M ratio by adjusting the number of returned microorganisms. This balancing act between the amount of food and the numbers of microorganisms is summarized in two extreme examples suggesting ranges of F:M ratios, aeration times, and treatment efficiencies:

	F to M Ratio +	Aeration Time →	Degree of Treatment
1.	lower	longer	higher
	(little food, large microbial population, long retention time in tank)		
2.	higher	shorter	lower
	(smaller tanks, shortened retention time, less efficient treatment)		

Sample loadings that are observed in practice range from 0.05 to greater than 2.0. The process of *extended aeration*, lasting up to greater than 30 hours, might have a loading of between 0.05 and 0.20 with an efficiency of hazardous waste removal in excess of 95%. The process of *conventional aeration*, closer to 6 hours for aeration, might have a loading between 0.20 and 0.50 with a treatment efficiency of possibly 90%. The process of *rapid aeration*, in the range of 1 to 3 hours for aeration, might have a loading between 1.0 and 2.0 with a removal efficiency closer to 85%. For each given problem, the engineer must design an individual activated sludge facility based on laboratory testing of a specific hazardous waste; the engineer must operate that facility and select different loadings through time based on ongoing laboratory tests of the facility's input, process variables, and outputs.

Variations of the classic activated sludge system summarized above exist to help process very specific and difficult to treat contaminants. These variations in the design and operation of such facilities include:

Tapered aeration (Figure 7.20): The oxygen that is supplied to the aeration basin is in greater amounts at the input end of the basin and in lesser amounts at the output end of the basin, with the goal of supplying more oxygen where it may be needed the most to address a specific hazardous waste problem.

Step aeration (Figure 7.21): The influent oxygen *and* food are supplied to the aeration basin in equal amounts throughout the basin with the goal of matching the oxygen demand to the location where it may be needed the most for a specific contaminant problem.

Contact stabilization or biosorption (Figure 7.22): The sorption and growth phases of the microbiological processing system are separated into different tanks with the goal of achieving growth at higher solids concentrations, saving tank space, and thus saving money.

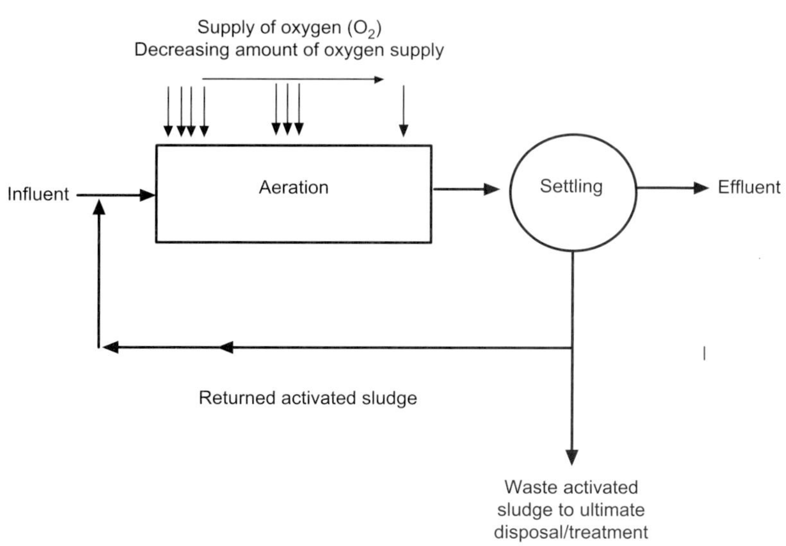

FIGURE 7.20
Tapered aeration activated sludge treatment system (greater amount of oxygen added closer to influent due to the large oxygen demand from microbes as waste is introduced to the aeration tank).
Source: D. Vallero (2003). *Engineering the Risks of Hazardous Wastes.* Butterworth-Heinemann, Boston, MA.

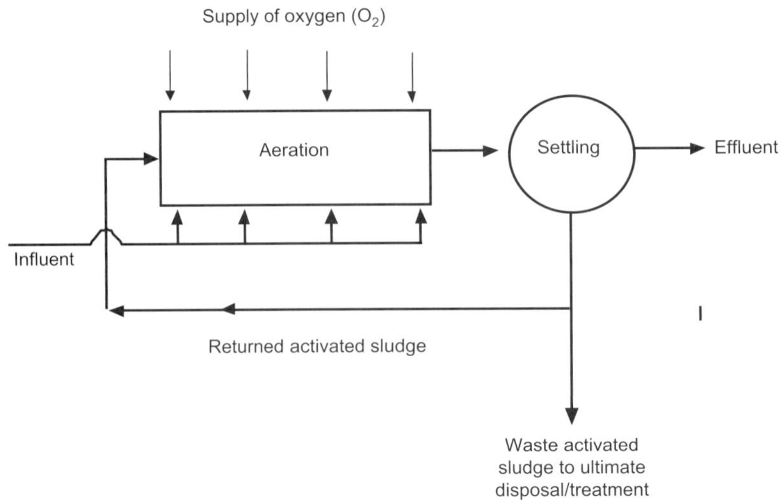

FIGURE 7.21
Step activated sludge treatment system.
Source: D. Vallero (2003). *Engineering the Risks of Hazardous Wastes.* Butterworth-Heinemann, Boston, MA.

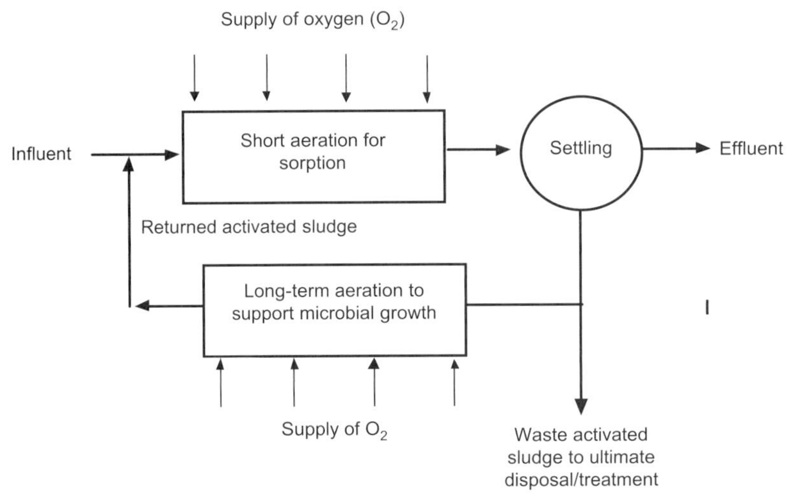

FIGURE 7.22
Contact stabilization activated sludge treatment system.
Source: D. Vallero (2003). *Engineering the Risks of Hazardous Wastes.* Butterworth-Heinemann, Boston, MA.

Aeration Ponds and Lagoons

Ponds like the one illustrated in Figure 7.23A treat liquid and dissolved contaminants over the long term, months to years. Persistent organic molecules, those not readily degraded in trickling filter or activated sludge systems, are potentially broken by certain microbes into $CO_2 + H_2O$ + microorganisms (+ energy ?) if given enough time. The ponds are open to the weather and ideally oxygen is supplied directly to the microorganisms from the atmosphere. Design decisions based on laboratory experiments and pilot studies include:

Design: Pond size: 0.5 to 20 acres
Design: Pond depth: 1 foot to 30 feet
Design: Detention time: days to months to possibly years
Operation: In series with other treatment systems, other ponds, or not
Operation: The flow to the pond is either continuous or intermittent
Operation: The supply of additional oxygen to the system through blowers and diffusers may be required (i.e. active systems).

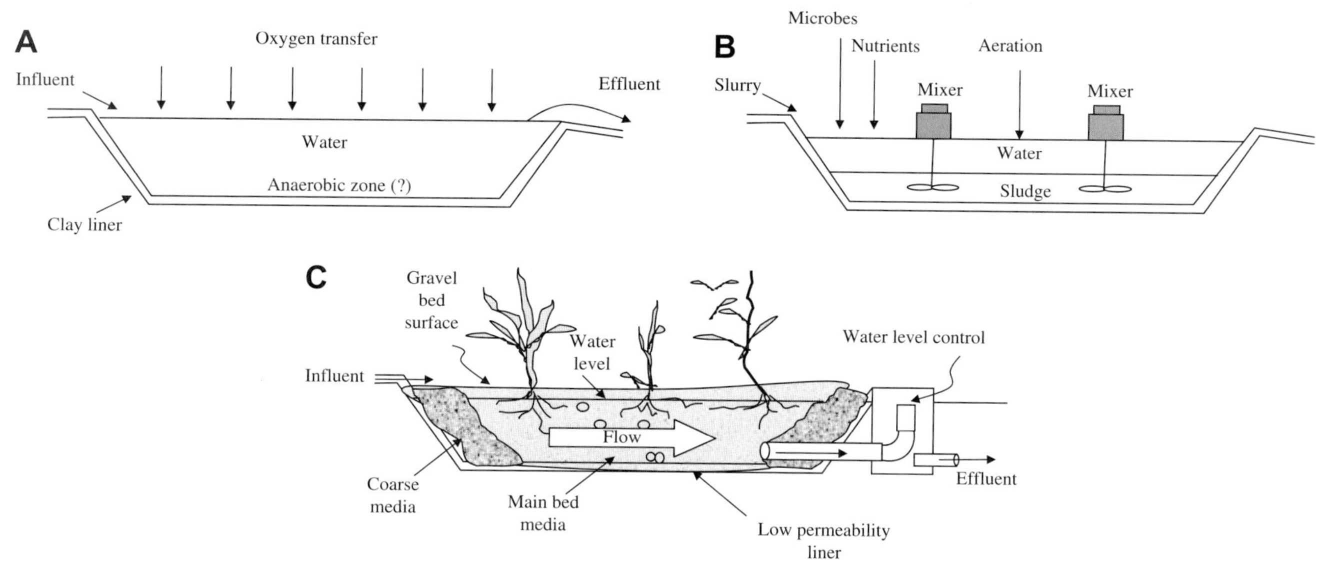

FIGURE 7.23
(A) Passive system: aeration pond. (B) Active system: *in situ* slurry-phase lagoon. (C) Combined active and passive system: engineered and constructed wetland. All three systems make use of microbes, nutrients and oxygen, but the active system increases the contact between the microbes and organic matter by mechanically mixing the sludge layer (which contains the organic contaminants to be degraded) with the water in the lagoon. The engineered wetland also incorporates plant processes in the degradation process.
Sources: US Environmental Protection Agency (1993). Pilot-scale demonstration of slurry-phase biological reactor for creosote-contaminated soil: Applications analysis report. Report No. EPA/540/A5-91/009. Cincinnati, OH; and S. Wallace (2004). Engineered wetlands lead the way. *Land and Water.* 48 (5); http://www.landandwater.com/features/vol48no5/vol48no5_1.html; accessed September 9, 2009.

A similar aerobic system is *slurry-phase lagoon activation*, where air and contaminated soil are made to come into contact with one another to promote biodegradation. Such a system can be used to treat an entire batch of sludge in single operation. These are usually less than 2 acres, with geometry and depth dependent upon the type of liner material, as well as the sludge characteristics and thickness. Larger systems require sectioning off into smaller lagoon compartments (se Figure 7.23B). Total solids content ranges from 5 to 20% [20].

Slurry systems are also used *ex situ* in tanks where higher degradation rates and greater engineering controls are needed (e.g. to treat soils and other media contaminated with very toxic and/or particularly recalcitrant organic contaminants). These stirred-tank bioreactors can receive sequenced batches or continuously fed sludge.

Another standing water biological treatment system (Figure 7.23C) is the engineered and constructed wetland, which operates as a type of biofilter. Wetlands naturally contain diverse and abundant microbial populations. They combine these with growth, nutrient extraction, photosynthesis, and ion exchange processes of plants (see Figure 7.24 and the Phytoremediation section later in this chapter). For example, gasoline-contaminated groundwater has been bioremediated using a radial-flow constructed wetland system in Casper, Wyoming. In this system, subsurface beds of crushed concrete reclaimed from a closed refinery were insulated with 6-inch layers of mulch. Above these layers bulrushes, switchgrass, and cordgrass were planted in separate sections of the wetland system. About 700,000 gallons of water are passively treated each day, with large reductions in benzene and other hydrocarbons. The rock/mulch insulated layers allow biological activity to continue throughout the year, even in the cold Wyoming winters. The passive system is also cost-effective (estimated construction costs were about 20% of a pump and treat system that can meet similar criteria, e.g. air stripping and catalytic oxidation) [21].

The crucial engineering concerns in the design and operation of ponds and other biotreatment facilities are the identification and maintenance of microbial populations that metabolize the specific contaminant of concern. Once selected, the conditions most favorable to the microbial populations can be determined and controlled.

The fluid dynamics and biological principles of the slurry-phase lagoon (Figure 7.23B) and the engineered wetland (Figure 7.23C and 7.24) can be put to use to treat air pollutants in a simple biofilter (Figure 7.25). However, polluted air rather than water is pumped into the system and mixed with water and pushed into the bottom of a 1-meter deep trench covered with unconsolidated material (soil or compost). The saturated air and water containing the organic contaminants move into the gravel and come into contact with microbial biofilm. The vapor pressure allows the more volatile compounds to move first through the media, but with time less volatile compound vapors will percolate through the unconsolidated material. During percolation, the air contacts the biofilm of microbes sorbed to the particles, which degrade the organic contaminants. A more intricate version is the packed bed biological control system (Figure 7.27) to treat volatile compounds.

Treatment optimization

In situ treatment literally means that the waste being treated is not moved, but treated where it is found. However, a number of treatment methods require that the waste, e.g. contaminated soil or water, be removed but not moved to a treatment facility. That is, they are on-site *ex situ* treatment approaches.

FIGURE 7.24
Engineered and constructed wetland system in San Diego, California, utilizing macrophytes, i.e. water hyacinths (*Eichhornia spp.*). The system takes advantage of both microbial and larger plant species (bioremediation and phytoremediation, respectively). [*See color plate section*]
Source: C.P Gerba and I.L. Pepper (2009).Wastewater treatment and biosolids reuse. In: R.M. Maier, I.L. Pepper and C.P. Gerba (2009). *Environmental Microbiology*, 2nd Edition. Elsevier Academic Press, Burlington, MA.

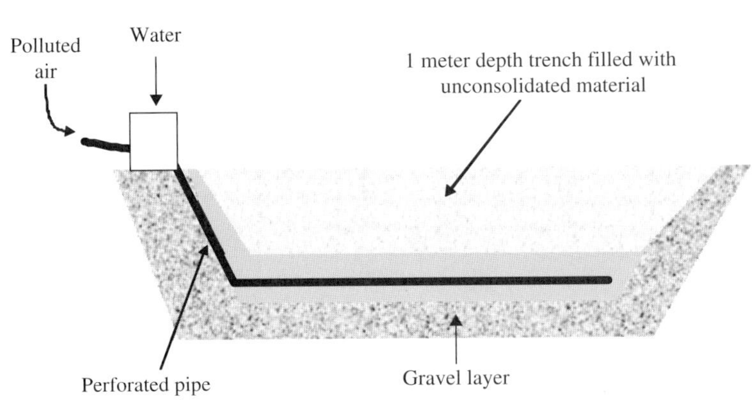

FIGURE 7.25
Biofilter system used to treat air contaminated with organic pollutants.
Source: Adapted from A. Scragg (2004). *Environmental Biotechnology*, 2nd Edition. Oxford University Press, Oxford, UK.

Anaerobic degradation consists of a series of steps whereby polysaccharides, proteins, fats, and other complex polymeric materials are hydrolyzed by the microbes. These microbial reactions, catalyzed by enzymes, generate products with greater aqueous solubility. The new compounds are secreted by the microorganisms into the biofilm where they can be transported, including diffusion across cellular membranes.

Ultimately, if the anaerobic biodegradation is successful, the production and ensuing escape of methane indicates that the organic material has been stabilized and degraded. The design of anaerobic systems needs to ensure that the retention time of the solids is sufficient for contact and reaction by the microbes with the substrate [24]. For example, if the input of relatively easily degradable organics is too rapid, this can lead to acidic and toxic conditions, with the buildup of organic acids, which can foul the reactor by inhibiting methanogenesis (i.e. instead of reaching the desired methane-water products, the system is stuck in the acid production steps).

Acclimatization of the microorganisms to a substrate can take several weeks. The biochemodynamics of the bioreactor affect anaerobic digestion rates, including temperature, pH, and concentration of toxic substances. The conventional anaerobic treatment process consists of a reactor containing waste and biosolids (sludge containing large microbial populations). As in the aerobic, activated sludge systems, these biosolids can be added continuously or in semi-batches, whereupon they are mixed in the bioreactor. In theory, the anaerobic digester is a once-through, completely mixed, reactor. If so, the hydraulic retention time (HRT) equals the solids retention time (SRT). This means that efficiency depends directly on the contact with the biosolids, i.e. the SRT. However, HRT and SRT can be decoupled for increased bioremediation efficiency. For example, an anaerobic upflow filter (AF) can substantially improve anaerobic degradation volumes and rates, since the filter can catch and sustain high concentrations of biosolids. By holding the solids, attenuated SRTs allow for much larger throughput which is needed to degrade low-strength organic wastes under feasible, environmental conditions (e.g. ambient temperatures and barometric pressures). A bed upflow reactor is another anaerobic system, but depends on the sorption of biomass on the surfaces of media. This is done by passing liquid solutions of the organic compounds to be treated upward through a bed of sand-sized particles at a velocity necessary to fluidize and partially expand the sand bed. More recently, the upflow anaerobic sludge blanket process (UASB) takes advantage of the inherent properties of flocculation and settling of anaerobic sludge, allowing for much higher HRT loadings and partitioning of gases (e.g. H_2, CH_4) from the sludge solids. The UASB is based on two biochemodynamic processes (see Figure 7.27): separation of solids and gases from the liquid; and degradation of biodegradable organic matter. If heavy mechanical agitation is

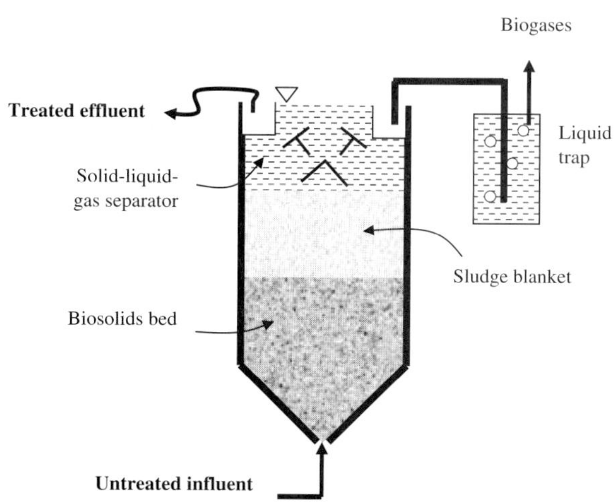

FIGURE 7.27
Schematic of upflow anaerobic sludge blanket system.
Source: Adapted from D.R. Christensen, J.A. Gerick and J.E. Eblen (1984). Design and operation of an upflow anaerobic sludge blanket reactor. *Journal of the Water Pollution Control Federation* 56 (9): 1059–1062.

avoided, unlike the other aerobic digesters, a separate settler with a biosolids return pump is not needed, so reactor volume can flow consistently through the systems. Unlike the fluidized bed reactor, high rate effluent recirculation and pumping are also eliminated. The biogas production enhances continuous contact between substrate and anaerobes. The UASB reactor, under optimal conditions, can be assumed to be a completely mixed reactor. The biosolids' contact with the incoming organic liquids is enhanced by the agitation caused by the release of gases from the biodegradation, along with an inlet that evenly distributes incoming materials in the lower level of the reactor.

MULTIMEDIA-MULTIPHASE BIOREMEDIATION

Waste streams containing volatile organic compounds (VOCs) may be treated with combinations of phases, i.e. solid media, gas and liquid flow in complete biological systems. These systems are classified as three basic types [25]:

Biofilters
Biotrickling filters
Bioscrubbers

Biofilms of microorganisms (bacteria and fungi) are grown on porous media in biofilters and biotrickling systems. The air or other gas containing the VOCs is passed through the biologically active media, where the microbes break down the compounds to simpler compounds, eventually to carbon dioxide (if aerobic), methane (if anaerobic), and water. The major difference between biofiltration and trickling systems is how the liquid interfaces with the microbes. The liquid phase is stationary in a biofilter (see Figure 7.28), but liquids move through the porous media of a biotrickling system (i.e. the liquid "trickles").

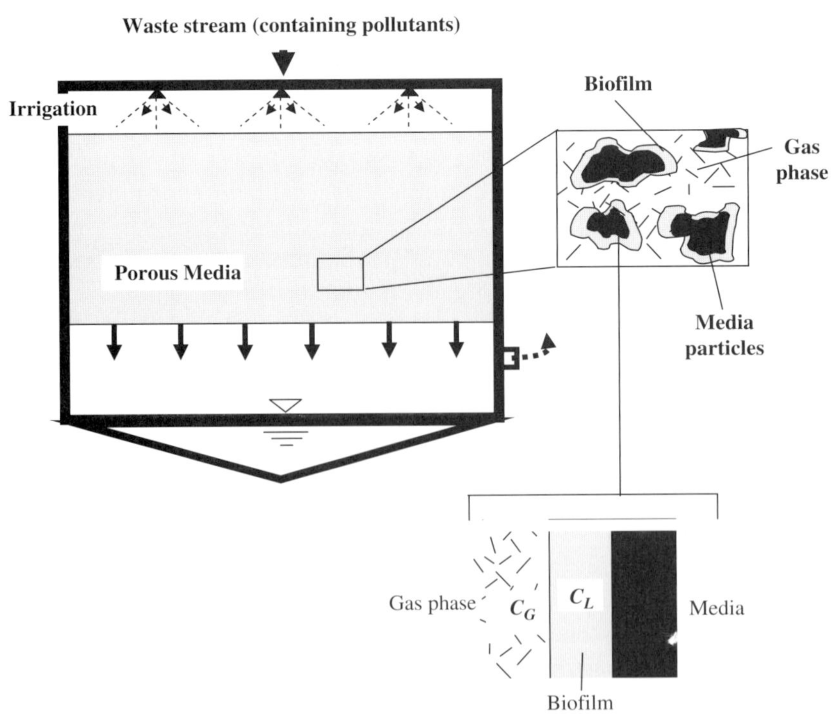

FIGURE 7.28
Schematic of packed bed biological control system to treat volatile compounds. Air containing gas phase pollutants (CG) traverse porous media. The soluble fraction of the volatilized compounds in the air stream partition into the biofilm (CL) according to Henry's law, $C_L = \{CG/H\}$; where H is the Henry's law constant.
Source: D.A. Vallero (2007). *Fundamentals of Air Pollution*, 4th Edition. Academic Press, Burlington, MA; adapted from S.J. Ergas and K.A. Kinney (2000). Biological control systems. In: W.T. Davis (Ed.), *Air and Waste Management Association, Air Pollution Control Manual*, 2nd Edition. John Wiley & Sons, Inc., New York, pp. 55–65.

A particularly novel biotechnological method in biofiltration (see Figure 7.29) uses compost as the porous media. Compost contains numerous species of beneficial microbes that are already acclimated to organic wastes. Industrial compost biofilters have achieved removal rates at the 99% level. Biofilters are also the most common method for removing VOCs and odorous compounds from air streams.

In addition to a wide assortment of volatile chain and aromatic organic compounds, biological systems have successfully removed vapor phase inorganics, such as ammonia, hydrogen sulfide, and other sulfides including carbon disulfide, as well as mercaptans. The operational key is the biofilm. The gas must interface with the film. In fact, this interface may also occur without a liquid phase (see Figure 7.12). According to Henry's law, the compounds partition from the gas phase (in the carrier gas or air stream) to the liquid phase (biofilm). Compost has been a particularly useful medium in providing this partitioning.

The bioscrubber is a two-unit setup. The first unit is an adsorption unit. This unit may be a spray tower, bubbling scrubber or packed column. After this unit, the air stream enters a bioreactor with a design quite similar to an activated sludge system in a wastewater treatment facility. Bioscrubbers are much less common in the United States than biofiltration systems [26].

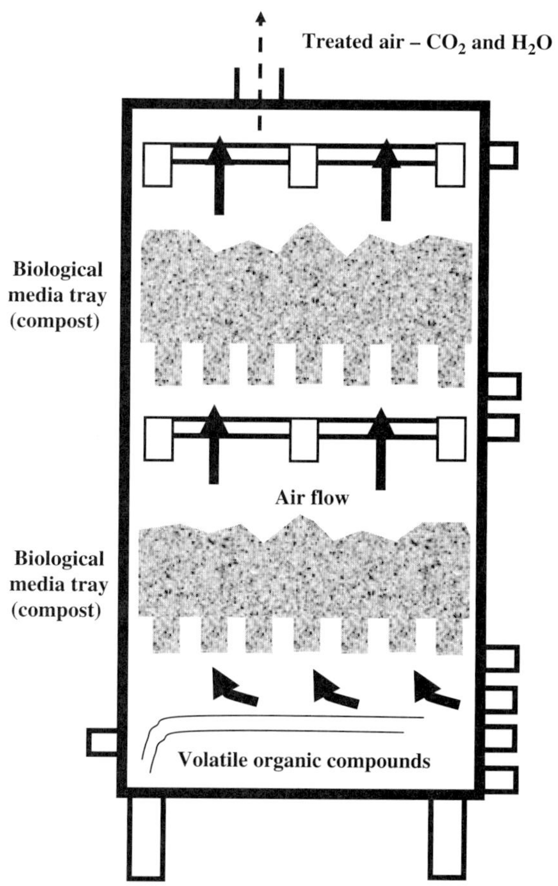

FIGURE 7.29
Biofiltration without a liquid phase used to treat vapor phase pollutants. Air carrying the volatilized contaminants upward through porous media (e.g. compost) containing microbes acclimated to break down the particular contaminants. The wastes at the bottom of the system can be heated to increase the partitioning to the gas phase. Microbes in the biofilm surrounding each individual compost particle metabolize the contaminants into simpler compounds, eventually converting them into carbon dioxide and water vapor.

All three types of biological systems have relatively low operating costs since they are operated near ambient temperature and pressure conditions. Power needs are generally for air movement and pressure drops are low ($<10\ cmH_2O\ m^{-1}$ packed bed). Other costs include amendments (e.g. nutrients) and humidification. Another advantage is the usual small amount of toxic byproducts, as well as low rates of emissions of greenhouse gases (oxides of nitrogen and carbon dioxide), compared to thermal systems.

Success is highly dependent on the degradability of the compounds present in the air stream, their fugacity and solubility needed to enter the biofilm (see Figure 7.12), and pollutant loading rates. Care must be taken in monitoring the porous media for incomplete biodegradation, the presence of substances that may be toxic to the microbes, excessive concentrations of organic acids and alcohols, and pH. The system should also be checked for shock and the presence of dust, grease or other substances that may clog the pore spaces of the media [27].

PHYTOREMEDIATION

As mentioned in Chapter 1, biotechnology in its broadest sense goes beyond microbial organisms in environmental applications. Plants are also used in remedying environmental problems. *Phytoremediation* is bioremediation by way of plant life. It is usually *in situ* and is almost always dependent on available air (i.e. it is almost exclusively an aerobic process).

Phytoremediation takes advantage of plants' absorption of CO_2 for photosynthesis, the process whereby plants convert solar energy into biomass and release O_2 as a byproduct. Thus, the essential oxygen is actually the waste product of photosynthesis and is derived from carbon-based compounds. Respiration generates carbon dioxide as a waste product of oxidation that takes place in organisms, so there is a balance between green plants' uptake of CO_2 and release of O_2 in photosynthesis and the uptake of O_2 and release of CO_2 in respiration by animals, microbes and other organisms.

Phytoremediation [28] can be used for a wide range of contaminants and soil types. It is frequently used to remediate metal-contaminated sites (e.g. nickel and its compounds). Phytoextraction (or *phytoaccumulation*) involves roots in the uptake and transfer of contaminants, i.e. translocation, to aboveground portions of the plants. Various species of plants, depending on the contaminant, soil, climate, and local conditions are planted and grown. Some are harvested like crops (e.g. grasses) and others maintained for removal and sequestration of the contaminants for longer time periods. The harvested materials can be treated and recycled by various methods, including composting and other bioremediation approaches and thermal processes, depending on the sequestered pollutants (e.g. content and form of halogenated and metallic compounds). This procedure may be repeated as necessary to decrease soil contaminant levels down to target cleanup concentrations. If thermally treated, e.g. incinerated, emissions and ash residues must meet regulatory requirements, including disposal as a possible hazardous waste.

Nickel, zinc, and copper are metals that have been particularly amenable to phytoextraction since hundreds of plant species have mechanisms for their uptake and absorption. This would seem to indicate numerous other metals would also be candidates since the micronutrient cycling and translocation mechanisms may be similar.

Rhizofiltration (*rhizo-* root) is the adsorption or precipitation of dissolved compounds onto plant roots or absorption into the roots. Rhizofiltration usually addresses contaminated groundwater, while phytoextraction targets contaminated soil. Thus, the plants are usually grown hypoponically in greenhouses and exposed to water from the contaminated site once the plants grow mature root systems. This acclimation step allows the plant's endogenous processes to become acclimated to the chemicals in the water. After acclimation, the plants are transplanted in soil above the contaminated aquifer or perched water table, and when their uptake is considered to have reached a threshold they are harvested and the biomass treated similarly to those from phytoextraction.

Phytostabilization immobilizes contaminants in the soil and groundwater by sorption and accumulation by roots, by adsorption onto roots, or by precipitation within the root zone of plants (*rhizosphere*). The biochemodynamics hinder transport, e.g. to the groundwater or air, as well as decreasing bioavailability of the contaminants.

In addition to these physical translocation and filtering processes, plants can degrade pollutants, using mechanisms similar to microbial degradation. *Phytodegradation* or *phytotransformation* is the process of degrading compounds after uptake by the plant's metabolism. It may also involve the breakdown of compounds externally by exogenous secretions, such as enzymes. Complex organic molecules follow similar degradation pathways to those of microbes, discussed in Chapter 3 and this chapter. That is, the complex molecules are broken down into simpler compounds to gain energy and to build plant biomass, similar to the mechanisms in Figure 3.1.

Rhizodegradation goes by a number of names, including *enhanced rhizosphere biodegradation, phytostimulation*, and *planted-assisted bioremediation/degradation*. However, it is actually the same microbial process as those described earlier in this chapter. That is, rhizodegradation is biodegradation by microbes using the contaminants as their energy and carbon sources, but in this case the degradation is conducted by a specific type of bacterium that has adapted to the root environment, where numerous chemical exudations occur (for example, see Figure 1.4). Such degradation is enhanced by both extracellular and intracellular enzymes.

Microbes degrade soil-borne contaminants microbially, but these mechanisms are enhanced by the biochemodynamics of the root zone (the *rhizosphere*). This process is slower usually than phytodegradation. This process can enhance cometabolism because plant roots release natural substances, e.g. sugars, alcohols, and acid, that are food for soil microorganisms. The fixation and release of nutrients are also a natural type of biostimulation for the microbes. In addition, the roots provide conduits and physical loosening of soil, improving microbial contact with oxygen and nutrients.

Phytovolatilization is both a transport and transformation process. A plant takes up a compound from the soil or aquifer and transpires it to the troposphere. This can be solely the parent compound, if it resists degradation (i.e. metabolism and growth products), or it may be the metabolic degradation products, or both. For example, a poplar may volatilize as much as 90% of the volatile organic compound taken up by the tree.

Phytoremediation has been used successfully on a variety of compounds in numerous locations (see Table 7.2). The local conditions and properties of the contaminants determine the degree of degradation.

BIOMARKERS

When a contaminant interacts with an organism, substances like enzymes are generated as a response. Thus, measuring such substances in fluids and tissues can provide an indication or "marker" of contaminant exposure and biological effects resulting from the exposure. The term *biomarker* includes any such measurement that indicates an interaction between an environmental hazard and a biological system [29]. In fact, biomarkers may indicate any type of hazard, chemical, physical, and biological. An exposure biomarker is often an actual measurement of the contaminant itself and/or any chemical substance resulting from the metabolism and detoxification processes that take place in an organism. For example, measuring total lead (Pb) in the blood may be an acceptable exposure biomarker for people's exposures to Pb. However, other contaminants are better reflected by measuring chemical byproducts [30].

Table 7.2 Examples of successful phytoremediation projects

Location	Application	Contaminants	Medium	Plant(s)
Edgewood, MD	Phytovolatilization rhizofiltration hydraulic control	Chlorinated solvents	Groundwater	Hybrid poplar
Fort Worth, TX	Phytodegradation phytovolatilization rhizodegradation hydraulic control	Chlorinated solvents	Groundwater	Eastern cottonwood
New Gretna, NJ	Phytodegradation hydraulic control	Chlorinated solvents	Groundwater	Hybrid poplar
Ogden, UT	Phytoextraction rhizodegradation	Petroleum hydrocarbons	Soil groundwater	Alfalfa, poplar juniper, fescue
Portsmouth, VA	Phytodegradation rhizodegradation	Petroleum	Soil	Grasses, clover
Portland, OR	Phytodegradation	PCP, PAHs	Soil	Ryegrass
Trenton, NJ	Phytoextraction	Heavy metals, radionuclides	Soil	Indian mustard
Anderson, SC	Phytostabilization	Heavy metals	Soil	Hybrid poplar, grasses
Chernobyl, Ukraine	Rhizofiltration	Radionuclides	Groundwater	Sunflowers
Ashtabula, OH	Rhizofiltration	Radionuclides	Groundwater	Sunflowers
Upton, NY	Phytoextraction	Radionuclides	Soil	Indian mustard, cabbage
Milan, TN	Phytodegradation	Explosives wastes	Groundwater	Duckweed parrotfeather
Beaverton, OR	Vegetative cover	Metals, nitrates, BOD	not applicable	Cottonwood
Texas City, TX	Vegetative cover rhizodegradation	PAHs	Soil	Mulbery
Amana, IA	Riparian corridor phytodegradation	Nitrates	Groundwater	Hybrid poplar

Source: US Environmental Protection Agency (1998). A Citizen's Guide to Phytoremediation. Report No. EPA 542-F-98-011. Washington, DC.

Exposure biomarkers are also useful as an indication of contamination of fish and wildlife in ecosystems. For example, measuring the activity of certain enzymes, e.g. ethoxyresorufin-O-deethylase (EROD) in fish *in vivo*, indicates that the organism has been exposed to planar halogenated hydrocarbons, PAHs, or other similar contaminants. The mechanism for EROD activity in the fish is the receptor-mediated induction of cytochrome P450-dependent monooxygenases when exposed to these contaminants [31].

BIOENGINEERING CONSIDERATIONS FOR GENETICALLY MODIFIED ORGANISMS

Bioremediation, whether enhanced by biotechnology or not, can occur in *in situ* treatments, i.e. where the water, soil or sediment is contaminated in the real world, or it can be *ex situ*, i.e. excavating and transporting the contaminated material to a treatment center. *Ex situ* bioremediation allows for more variety in techniques, from entirely "natural" approaches of land application of sludges and wastewater, to land farming of contaminated materials, to bioreactors making use of genetically modified microbes, cometabolism, and enzymatic processes. Up to a few decades ago, many of these processes were "black boxes," in which the result was rather well understood, but the specific mechanisms were not. As these mechanisms became better understood, engineers were able to replicate and to enhance the use of microbes in engineered systems, i.e. bioreactors. This imparts greater control over the degradation processes and, as a result, can expedite the rate of the biotechnological processes (e.g. specifically enhanced microbes) and physicochemical processes (adding nutrients and dissolved oxygen, changing redox) taking place in the bioreactors. Let us briefly consider some of the important environmental biotechnological applications.

The first step in applying biological principles to treat a waste requires that the microorganisms' can be cultured reliably. This is difficult in natural microbial populations, but since genetically engineered bacteria rarely function in the wild, tools are needed to characterize their actual and expected behavior in the environment. To address these challenges, bioengineers can apply the so-called "omics" tools, beginning with genomics, i.e. characterizing the complete genetic information of these organisms. Some currently used tools include methods to monitor electron acceptance and electron donation, enzyme probes to measure microbial functional activity in the environment, functional genomic and phylogenetic microarrays, metabonomics, proteomics, and quantitative approaches for genetic testing (e.g. quantitative real-time polymerase chain reaction (Q-PCR/qPCR) or kinetic polymerase chain reaction).

Figure 7.30 illustrates one conception of the use of omics tools to predict outcomes of environmental biotechnologies, especially in ecosystems. Genomics, including, functional genomics, proteomic and systems modeling characterize community population ecosystem structure, ecosystem function, and ecosystem dynamics. Usually, DNA extracted from an environmental sample is cloned from a library or affixed to beads, and then directly sequenced. Following assemblage of the sequence, the computational identification of marker

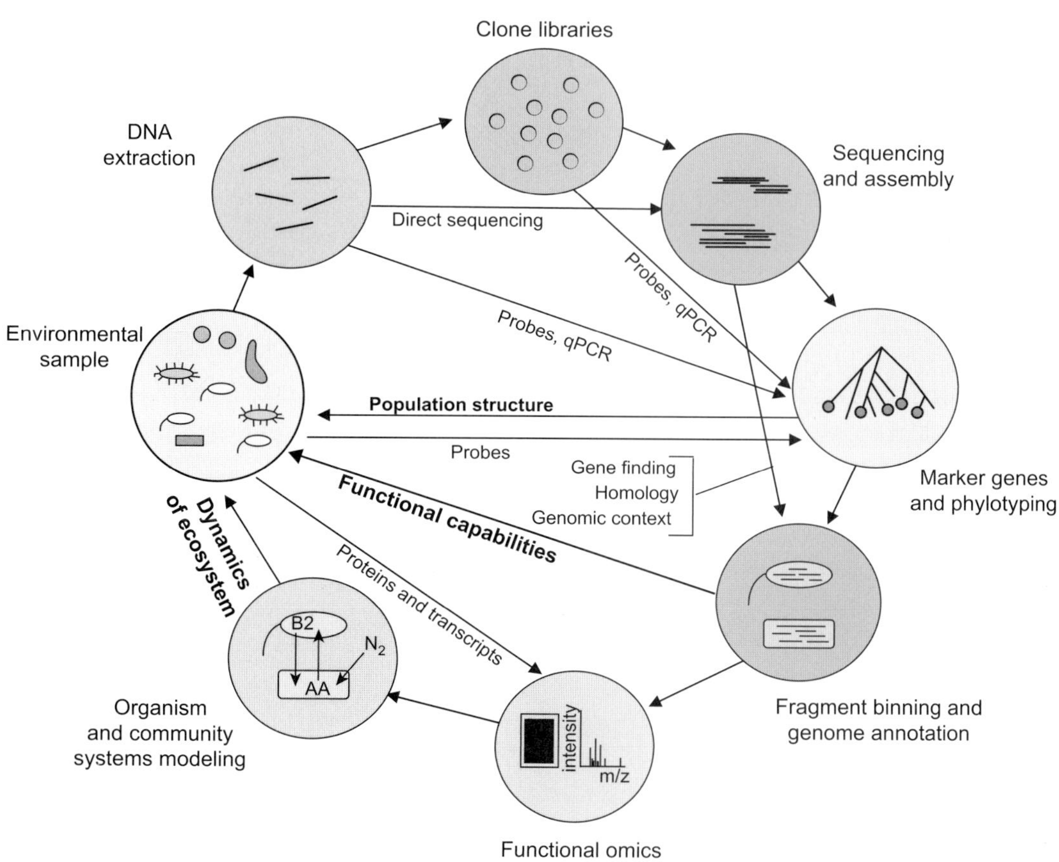

FIGURE 7.30
The application of "omics" technologies to characterize microbial behavior in ecosystems. [*See color plate section*]
Source: A.M. Deutschbauer, D. Chivian and A.P. Arkin (2006). Genomics in environmental microbiology. *Current Opinion in Biotechnology* 17 (3): 229–235.

genes allows for the members of the community to be identified and phylogenetically classified. Thereupon, probes can be designed for later population scale experiments (i.e. structure). Identification of the marker genes within fragments, along with other characterizations (e.g. G+C content bias and codon usage preferences) allows for assigning sequence fragments into groups that correspond to one type of organism (i.e."binning"). Functional capabilities of communities are described by computationally annotating genomes, i.e. predicting genes and assigning function using characterized homologs and genomic context. Information and knowledge pertaining to the presence of the genes also allows for the application of other "omics" approaches, including evaluation of the extracts of proteins and RNA transcripts from the environmental sample. Such investigations inform and enhance systems modeling efforts, which in turn can be applied to understanding and predicting ecosystem dynamics and lead the way to future investigations [32].

One of the challenges of environmental biotechnology is the ability to determine success or failure of systems, whether in a bioreactor (*ex situ*) or in the field (*in situ*). The "omics" techniques, such as those in Figure 7.31, certainly will help, as will improved models and field techniques (see Discussion Box: Measuring Biodegradation Success). Bioengineers must select the treatment system that meets the design requirements. Certainly, biotechnologies have emerged that provide a substantial part of these requirements.

Computational modeling can be both descriptive and predictive, within a spectrum ranging from highly specified and reductive approaches to abstracted and systematic methods (see Figure 7.32). The use of such models is an exercise in optimization. The US National Institute of General Medical Sciences aptly states:

> Most biological systems are too complex for even the most powerful computational models to capture all the system properties. A useful model, however, should be able to accurately conceptualize the system under study and provide reliable predictive values. To accomplish this, a certain level of abstraction may be required that focuses on the system behaviors of interest while neglecting some of the other details ... A useful model ... should be able to accurately conceptualize the system under study and provide reliable predictive values. To accomplish this, a certain level of abstraction may be required that focuses on the system behaviors of interest while neglecting some of the other details. [33]

Environmental systems are arguably even more complex than most medical systems, since they not only include human individuals, subpopulations and populations, but also those of other

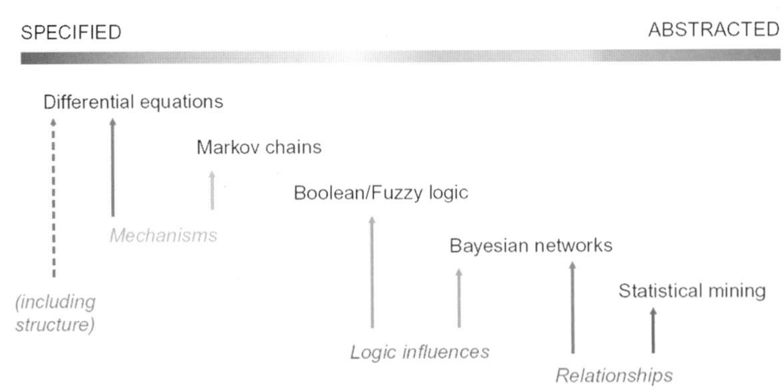

FIGURE 7.31
Computational modeling ranges from concrete output data for biological systems that have been well characterized and for which a substantial amount of reliable data exist to systems with little data and numerous "black boxes", such as the lack of information about interactions among factors and variability in time and space.
Source: B.A. Joughin, E. Cheung, R.K. Murthy Karuturi, J. Saez-Rodriguez, D.A. Lauffenburger and E.T. Liu (2009). Cellular regulatory networks. In: E.T. Liu and D.A. Lauffenburger (Eds), *Systems Biomedicine: Concepts and Perspectives.* Elsevier Academic Press, Amsterdam, The Netherlands.

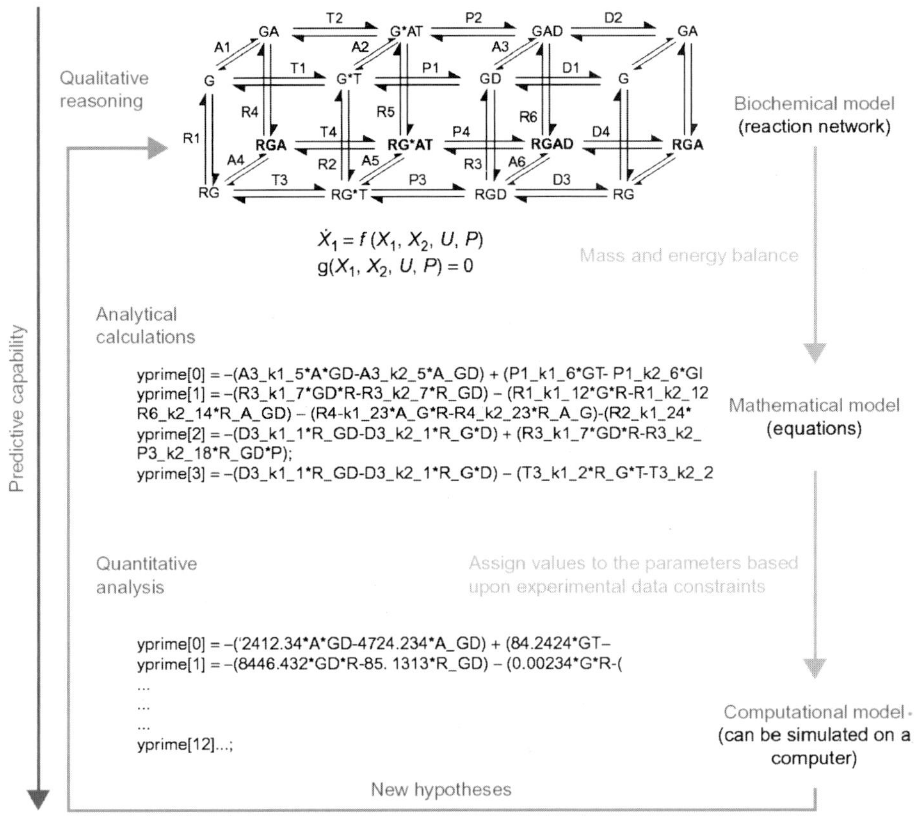

FIGURE 7.32
Flow of knowledge and increase in predictive reliability of systems in a three level modeling strategy, including feedback of knowledge to improve lower tier (e.g. biochemical) models. [*See color plate section*]
Source: M.R. Maurya and S. Subramaniam (2009). Computational challenges in systems biology. In: E.T. Liu and D.A. Lauffenburger (Eds), *Systems Biomedicine: Concepts and Perspectives.* Elsevier Academic Press, Amsterdam, The Netherlands.

species, as well as the abiotic environments that comprise ecosystems. Thus, abstraction becomes even more necessary, but the concomitant uncertainties can be increased as organismal and environmental system uncertainties are propagated. This spectrum in Figure 7.31 will be better characterized as better data are gathered. The advantage of computational approaches, however, is that they can be based on the first principles of physics, chemistry, and biology, whereas typical environmental models have often had to rely on probabilities and site-specific characteristics. That is to say, stochastic and deterministic models can benefit from the information that can be input from computational tools (e.g. metabonomics and proteomics). At present, however, the most commonly employed models rely on differential equations that characterize those biochemical and biophysical interactions among system components that are thought to represent a particular aspect of the system (e.g. dynamic properties of protein expression that would not be seen looking at individual proteins) [34].

Thus, appropriately scaled mathematical models that capture the underpinning principles at work in biochemical processes set the stage for computational models that continuously improve predictions of environmental consequences. This three-stage loop is shown in Figure 7.32. For example, the effect on ecosystem function and structure (Figure 7.30) from an introduced genetically modified microbe will be better characterized in the biochemical and mathematical models, but these systematic impacts ultimately depend on the computation approach. However, the improved systems biology and engineering better informs the biochemical model, which improves the mathematics and further enhances the computational model's predictive capacity.

In addition, improvements in the precision, accuracy, and reliability of the measurement tools in Figure 7.30 (e.g. marker genes, phytotyping, and environmental sampling) will improve all three modeling levels, since the real-world data both ground truths each model and provides actual data to allow for more and better extrapolations and interpolations about the biological systems. For example, combining chemical and biological exposure assessment tools like advanced molecular indicators of exposure and biosensors based on nanotechnologies will allow for measurements of biologically relevant exposures simultaneously with measurements of multiple real-world stressors. They may also link mechanistically traditional exposure metrics with *in vitro* assays [35].

DISCUSSION BOX

Measuring Biodegradation Success

Determining the amount and rate of degradation of toxic pollutants in soil and groundwater is difficult and often requires invasive techniques, such as deploying extensive monitoring well networks [36]. Even with these networks, degradation rates across entire systems cannot readily be extrapolated from the samples. When organic compounds are degraded by microbes, especially nitrifying bacteria, oxides of nitrogen (NO_x) are released to the atmosphere (see discussion of the biogeochemistry of nitrogen in Chapter 2, especially Figure 2.15). Thus, the flux of nitric oxide (NO) from the soil to the lower troposphere can be used to predict the rate at which organic compounds are degraded.

The measurement of NO emissions can be used as a screening tool for the success of *in situ* bioremediation, as well as an indirect measure of releases of toxic organic compounds to the atmosphere. This is done by tracking the production of NO results from both biotic and abiotic processes including nitrification, denitrification, chemo-denitrification, and other microbial processes [37]. In soils, however, most NO is produced biotically [38]. Several nitrifiers and denitrifiers have been reported to degrade aromatic hydrocarbons [39]. A recent study of a monoaromatic hydrocarbon contaminated aquifer found the concentration of nitrate to be inversely related to concentration of monoaromatic (i.e. single-ring) hydrocarbons, suggesting that nitrate reduction and degradation of organic ring compounds (i.e. aromatics) are related [40].

Toluene (C_7H_8) is designated to be a toxic air pollutant under Section 112 of the Clean Air Act Amendments. Its chemistry is similar to other aromatics in that it includes a substituted methyl group at one of the six positions of the benzene ring. It is also representative of a large group of contaminants that are frequently found near leaking underground storage tanks, i.e. the so-called "BTEX" group (i.e. benzene, toluene, ethylbenzene, and xylene). The BTEX chemicals are indicators of gasoline and other ubiquitous fuels that have been known to contaminate large areas of soil and groundwater. In addition, toluene is important from a biodegradation standpoint since it serves as a substrate for specific microbial populations [41]. Soil microorganisms exposed to toluene at varying concentrations respond differently to the toluene; hence the NO emissions from soil will be affected differently. Thus, toluene is a good candidate indicator of the degradation of a wide range of organic soil contaminants in soils, especially those that can be linked to NO releases.

Landfills and other storage, treatment, and disposal systems often employ microbial processes to degrade pollutants. The ultimate indication of success of aerobic microbial degradation is the production of carbon dioxide (CO_2) and water, and the indication of complete anaerobic microbial degradation is the production of methane (CH_4) and water. However, numerous abiotic and biotic reactions produce various products either as side reactions or as steps toward complete digestion. Managing *in situ* bioremediation projects requires reliable monitoring in the field so that adjustments can be made (e.g. injection of air or O_2, changing from exclusively natural to genetically enhanced microbial populations). Both passive biotechnological systems, such as natural attenuation, and active remediation systems, such as pump and treat, including those using genetically modified microbes, require extensive and elaborate monitoring for regulatory compliance, process control, and performance measurement. This can be measured directly by analysis of soil and groundwater samples in an analytical laboratory. These monitoring activities often include quarterly groundwater or soil sampling required for periods up to and exceeding 30 years. Unfortunately, these direct measures of bioremediation are often expensive and sample collection and analysis is often time-consuming [42]. In fact, measurement data are almost always extended using models, such as simple and direct interpolation between sampling points. Indirect measures of bioremediation are not meant to replace analytical testing of samples, but can be useful as surrogate indicators of bioremediation success. In this case, measuring NO emissions and modeling attendant toluene remaining in a contaminated site can decrease the number of samples requiring expensive analysis, and can allow for a real-time, integrated, and system-wide measure of remediation.

Nitric oxide as an indicator of degradation

Emissions of NO from soil are influenced by physical, chemical, and biological parameters of soil. Important variables in the production, transformation, and transport of NO from soils include pH, soil moisture content, temperature, physical and chemical soil properties, nutrient availability, atmospheric pressure, native microbial populations, and naturally occurring inhibitors of these microbial populations. NO emitted from soils represents a significant source of atmospheric NO. However, it is difficult to study the influence of soil variables on NO flux under field conditions because the net flux of NO between soil and atmosphere results from myriad processes that operate simultaneously and are regulated independently from each other, as well as other processes that are interdependent and affect one another (e.g. nutrient cycling, which affects the soil's ionic strength, which in turn affects pH and nitrogen speciation). Therefore, laboratory experiments are necessary in order to control these parameters and delineate the importance of various parameters on the NO flux from soils.

Soils consist of living and non-living components existing in complex and heterogeneous mixtures. Soil microorganisms play an important role in the breakdown and transformation of organic matter, with many species contributing to different aspects of soil fertility. Any long-term interference with these biochemical processes can interfere with nutrient cycling, which could alter the soil fertility. Transformation of carbon and nitrogen compounds occurs in all fertile soils. The pathways for transformation of carbon and nitrogen need to be characterized to monitor the bioremediation of organic contaminants when used as a carbon source for soil microorganisms, although the microbial communities responsible for these processes differ from soil to soil. Simulation models that describe these carbon and nitrogen pathways in the soil system in great detail can be important tools in this respect.

In 1998, over 100 million pounds (45 million kg) of toluene was released into the environment from 3801 facilities [43]. Elevated concentrations of toluene can be toxic to certain microorganisms, including eukaryotic cells in organisms and in prokaryotic bacteria [44], so the compound is also important to pretreatment and other environmental engineering processes. This is an instance of how certain genetically modified strains can help in bioremediation. For example, if the bacteria can gain resistance to toluene toxicity, the initial shock when the microbes come into contact with the wastes would be reduced. In spite of its toxicity, however, toluene has been demonstrated to be readily bioremediated in a manner similar to other substituted benzenes, which account for a large share of contaminants found in the environment [45]. This suggests that the structural activity relationships of toluene are indicative of those of other aromatic contaminants. So, perhaps, the indirect methods for estimating how much toluene remains in polluted sites can be used for quite a few more aromatic compounds.

To develop a model using indirect measurements first requires laboratory experiments of the NO flux from soil exposed to varying levels of toluene. Next, the lab findings must be analyzed and correlated to arrive at NO-toluene-soil microbial activity relationships to document the viability of using NO as a surrogate indicator of bioremediation in toluene-contaminated soil. Then, a mathematical model is adapted to characterize and estimate the NO-toluene-soil microbial relationships in a complex soil system. Finally, the laboratory results are compared to the results derived from the models, and the model appropriately adjusted.

In the laboratory, toluene and water were added to the soil to achieve approximately 30% water-filled pore space and five different toluene-soil amendment regimens: 0, 5, 30, 45, or 60 parts per million (ppm). After the toluene-contaminated soil was incubated at room temperature (22 °C) over varying time periods, the soil was then transferred to a dynamic test chamber (see Figure 7.33) for measurement of NO emissions, the concentration of toluene in soil and headspace vapor was measured, and the microbial activity was determined using fluorescent *in situ* hybridization (FISH).

(Continued)

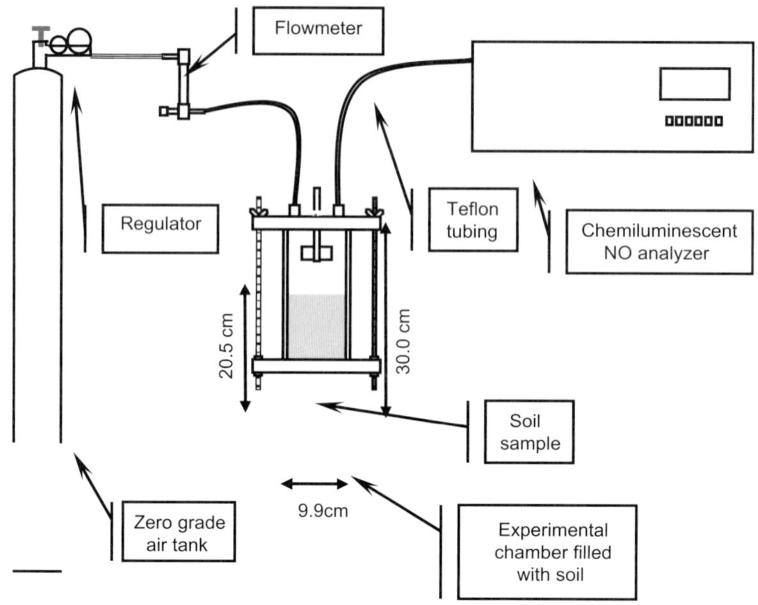

FIGURE 7.33
Dynamic soil to headspace flux test chamber set-up at Duke University.

Humility in biotechnological modeling

There are many methods for predicting the future. For example, you can read horoscopes, tea leaves, tarot cards, or crystal balls. Collectively, these methods are known as "nutty methods." Or you can put well-researched facts into sophisticated computer models, more commonly referred to as "a complete waste of time."

Scott Adams, US cartoonist [46]

This is a good time to remind ourselves that predicting environmental outcomes, using intricate models and scientific information requires a healthy dose of humility. Part of Adams' comic strip *Dilbert*'s appeal is that it often includes some underlying truth, however humorously exaggerated. This particular quote indeed contains an admonition to those of us who use models to try to get at the truth. Not to get too philosophical, but there is a scientific corollary to Plato's famous allegory of the cave, or St Paul's admonition that humans are quite limited in their view of how things work. Both warn us that what our limited senses tell us is at best incomplete, and sometimes misleading if we use the wrong set of metrics and assumptions. We may not understand much about a system, but it may seem to keep giving us reliable results. This is what scientists may call a "black box."

We often see correlations in complex systems, but cannot explain them. In fact, environmental systems are very complex and multivariable. We seldom have a lot of certainty about how to weight the variables and how to set up algorithms in models. In fact, environmental models are generally distinguished as to whether they are mainly *stochastic* or *deterministic.* A stochastic model is built under assumptions of randomness. Thus, it is statistically based. Conversely, deterministic models are built from an understanding of how things work. They are built on the assumption that there is indeed one solution; all we have to do is properly parameterize our model. That is, to find the solution, we must identify all the elements and factors that influence the outcome and assign values to them. So, if we have a strong understanding of microbial processes and the conditions that lead to degradation of a particular compound, we may be able to construct a deterministic model to estimate rates of biodegradation for that specific microbial population under prescribed environmental conditions.

Developing an indirect, chemical model of microbial activity

When discussing this model, it is important to remind ourselves about the uncertainties and the need for humility in its application. That said, we are ready to discuss how such a model can be built, starting with some background of how microbes get their energy.

All other things being equal, microbes with the most efficient metabolic mechanisms grow at the fastest rate, so these organisms will overwhelm the growth of microbes with less efficient redox systems. Thus, if O_2 is available in surplus, this will be the preferred reaction in the model. Once a system becomes anaerobic, nitrate is the most preferred redox reaction, followed by solid phase ferric iron, sulfate, and carbon dioxide (the least preferred redox reaction).

A thermodynamically dictated system would give preference, even exclusivity, to the reaction that provides the most energy, so a model that uses a sequential process does not allow the microbes to use any other less preferred electron acceptor until the more preferred acceptor is depleted. However, in reality, when monitoring wells are analyzed near plumes undergoing active biodegradation, the samples are seldom entirely depleted of one or more of these electron acceptors. There are seldom such "bright lines" in the field. For example, facultative aerobes, those that can shift from oxygen to anaerobic electron acceptors (especially nitrate), can change electron acceptors even when molecular oxygen is not completely depleted. This can be attributed to the fact that redox potentials for oxygen and nitrate are not substantially different (at pH 7, $O_2 = +820$ volts and $NO_3 = +740$ volts, compared to $CO_2 = -240$ volts). Also, the apparent divergence from pure thermodynamics in the field may simply be a sampling artifact, which can be attributed to the way monitoring is conducted. For example, monitoring wells do not collect water from a "point." Rather, the screens (the perforated regions of underground piping where water enters) are set at 1.5 to 3 m intervals, so waters will mix from different vertical horizons. Thus, if different reactions are occurring with depth, these are actually aggregated into a single water sample.

When a contaminant degrades sequentially, the slowest degradation step has the greatest influence on the time it takes the chemical to break down. If this most sensitive step can be sped up, the whole process can be sped up. Conversely, if an engineer or scientist devotes much time and effort to one of the faster steps in the degradation sequence, little or no enhancement to the degradation process may occur. Thus, bioengineers need to take care not to overgeneralize, assuming a contamination plume is limited by oxygen or even other redox conditions. Adding iron to an anaerobic system or pumping air into an aerobic stratum of an aquifer will help, but only so much. Figure 7.34 demonstrates a way to apply microbial kinetics limits to redox. The actual degradation can take a number of breakdown paths. For example, two possible anaerobic degradation pathways for benzene are shown in Figure 7.35.

Laboratory and the field measurements can differ regarding the presence of confounding chemical mixtures in real contamination scenarios. For example, leaking underground storage tanks (LUSTs) are a widespread problem. It is tempting to think that since these tanks contain refined fuels, that most spills will be similar. However, each compound has specific physicochemical properties that will affect its reactivity and movement in the environment. As evidence, the BTEX compounds of benzene, toluene, ethyl benzene, and xylenes usually comprise only a small amount (ranging from about 15 to 26%) of the mole fraction of gasoline or jet fuel [47]. However, largely because the BTEX compounds have high aqueous solubilities (152 to 1780 mg L^{-1}) compared to the other organic constituents (0.004 to 1230 mg L^{-1}) in these fuels, they often account for more than two-thirds of the amount of the contaminants that migrate away from the LUST. Also, soils are seldom homogeneous, so even if the contaminant is well characterized, how it will react and move is largely affected by the media's characteristics, such as their potential to sorb pollutants.

Ease of implementation and sensitivity are both important considerations when deciding how to address environmental problems. In some situations, steps that are readily available may be relatively insensitive to the intended outcome. In other situations, immediate and relatively inexpensive measures can be taken that are sensitive, such as pumping air and water to speed up biodegradation in an aquifer that has already shown natural attenuation.

(Continued)

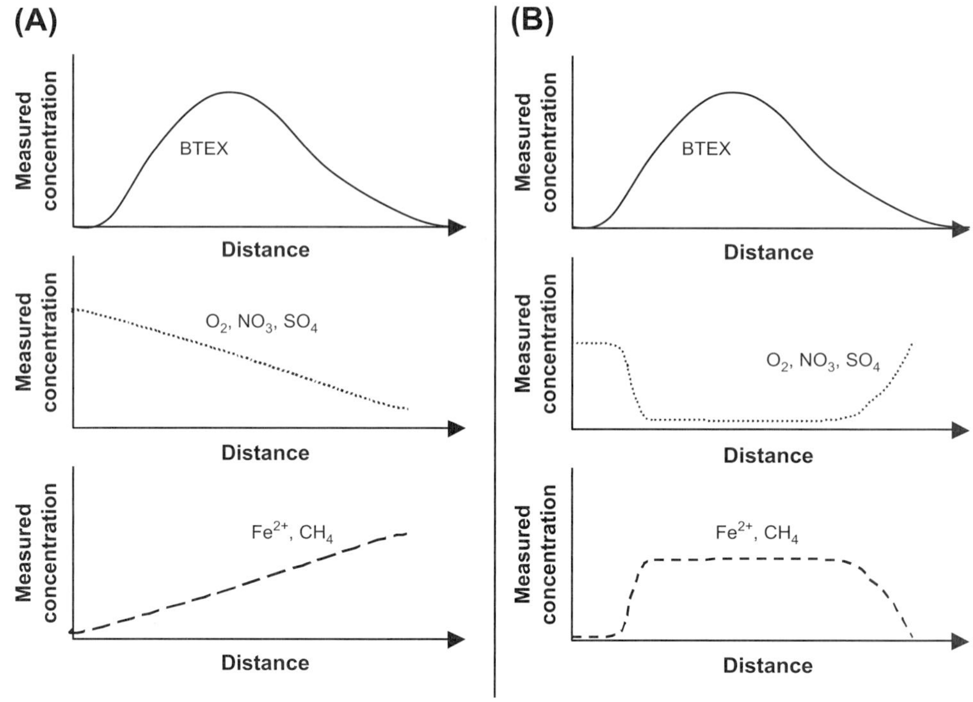

FIGURE 7.34

Two possible ways that microbes degrade benzene, toluene, ethyl benzene, and xylenes (BTEX): (A) Rate of biodegradation is limited by microbial kinetics: Concentrations of anaerobic electron acceptors (nitrate and sulfate) decrease at a constant rate downgradient from the pollutant source, with a concomitant increase in the concentrations of the byproducts of these anaerobic reactions (ferrous iron and methane). (B) Rate of biodegradation is relatively fast (days, not years, so compared to many groundwater replenishment rates, this can be characterized as instantaneous). Virtually all of the nitrate and sulfate anaerobic electron acceptors are depleted, while the iron and methane byproducts of these anaerobic reactions show the highest concentrations near the contaminant source. In both (A) and (B) the total concentrations of the byproducts are inversely related to the total concentrations of the principal electron acceptors in the anaerobic reactions overall. *Source:* Adapted from US Environmental Protection Agency (2003). Bioplume III Natural Attenuation Decision Support System: Users Manual; Version 1.0, Washington, DC.

The challenge for the bioengineer is to understand the entire system. Based on this understanding, solutions to environmental problems can be developed. For example, if a bioengineer locks into one microbe, either one existing in the soil or introduced (e.g. a genetically modified *Pseudomonas* bacterium) based on the assumption that the degradation will be limited by kinetics (A in Figure 7.34), but in fact it is limited by the presence of a non-BTEX compound, the microbial degradation will not be efficacious. Ideally, improvements can be made by focusing first on actions that bring about the most improvements, i.e. where the environmental responses are most sensitive to changes. Unfortunately, this works the other way as well. That is, some parts of the environment are highly sensitive to small changes. Small changes in surface water temperature, pH, or dissolved oxygen and other essential factors can greatly affect survival.

The highly interconnected carbon and nitrogen cycles can be modeled as a system (Figure 7.36). The soil organic matter (SOM) is divided into three pools that represent nitrogen, carbon, and toluene used as a different carbon source from outside of the system, respectively, while the inorganic nitrogen in the soil is divided into ammonium (NH_4^+) and nitrate (NO_3^-) species. To simplify the model, no distinction is made among the different bacterial colonies. All colonies are included in a single microorganism pool. Due to the high rate of nitrification, the presence of nitrite (NO_2^-) was below limits of detection, so is not reported in this model. The system with three SOM pools suggested agreement with the findings of Jenkinson (1990), Bolker *et al.* (1998), and Porporato *et al.* (2003) who studied models with different numbers of compartments

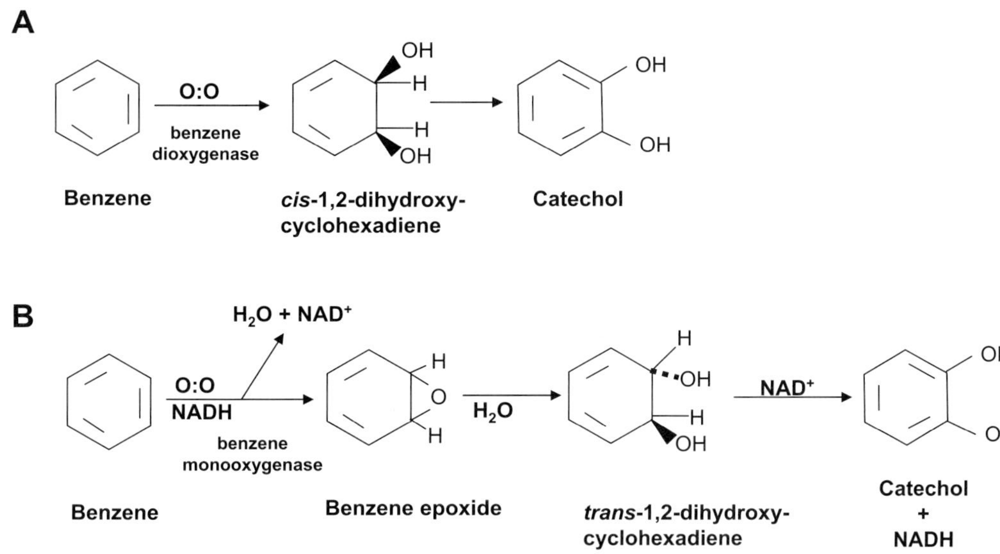

FIGURE 7.35

(A) Two proposed reaction sequences for aerobic biodegradation of benzene to form catechol via cis-1,2-dihydroxycyclohexadiene and a benzene epoxide intermediate. The reaction illustrates oxygen addition to benzene ring in the initial transformation steps. (B) Proposed pathway for anaerobic biodegradation of benzene to form the first intermediate phenol in the initial transformation step.

Source: S.A. Mancini, A.C. Ulrich, G. Lacrampe-Couloume, B. Sleep, E.A. Edwards and B.S. Lollar (2003). Carbon and hydrogen isotopic fractionation during anaerobic biodegradation of benzene. *Applied Environmental Microbiology* 69 (1): 191–198.

to describe SOM and suggested using more than one and less than five pools for SOM [48]. The use of a single pool for microorganisms is justified by the scarce quantitative information available on soil microbial biomass. Only the input of the added mineral fertilizer (MF) and the outputs of NO and N_2 through nitrification and denitrification are determinants of the external fluxes to the soil system for the SOM nitrogen pool. For the SOM carbon pool, there was one input source (toluene), and three external fluxes (vapor phase toluene, carbon dioxide [CO_2], and soil residual toluene). The C/N ratios of the pools that contain organic matter need to be characterized in terms of the dynamics of nitrogen and carbon cycling. Understanding the mechanisms and the interactions of the controlling factors provides a way to integrate and to expand the information from specific studies.

This integration has been achieved through the use of conceptual models. As additional conceptual models are developed they can be incorporated into numerical simulation models, as in this research. Table 7.3 summarizes the definition of the variables shown in Figure 7.36. Models using this same structure can be applied to other chemical compounds so long as there is a link to emitted gases other than the parent compound or even breakdown products from the parent compound. In this case, the actual metabolic and electron acceptance/donation need not necessarily be completely understood, since the model depends on a correlation between emitted gas mass and remaining mass of the parent compound to be biodegraded. In this sense, this model is less deterministic than a model that measures emitted metabolites or chemical degradation products, such as the degradations shown in Figures 7.35, 7.37 and 7.38.

To quantify the dynamics of carbon and nitrogen in this system, numerical models include multiple state variables corresponding to functionally and dynamically distinguishable ecosystem components. Construction of the model equation is based upon fundamental equations for production and transport through the soil.

(Continued)

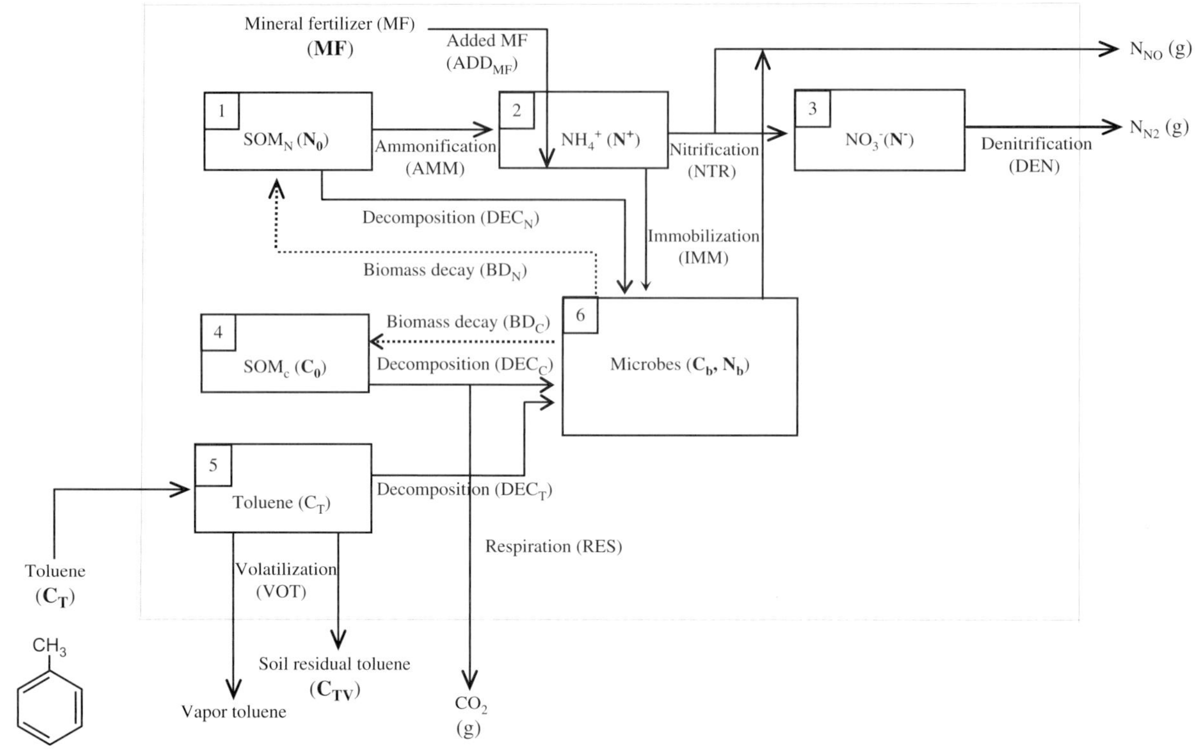

FIGURE 7.36

Potential model linking toxic compound biodegradation to NO emissions from contaminated soil. This model applies directly to one organic chemical species being degraded and another chemical species, in this instance inorganic NO, being emitted by microbes. This indirect method can be used to gauge the biodegradation of other compounds, so long as the emitted gas is correlated to the amount of parent compound being biodegraded. *Source:* D.A. Vallero, J. Peirce and K.D. Cho (2009). Modeling toxic compounds from nitric oxide emission measurements. *Atmospheric Environment* 43: 253–261.

Table 7.3 Definition of variables shown in model schematic diagram in Figure 7.36

Variable	Definition	Units
C_0	Carbon concentration in the indigenous carbon pool	mg-C kg^{-1} soil
C_T	Carbon concentration in the toluene pool	mg-C kg^{-1} soil
C_b	Carbon concentration in the soil microorganism pool	mg-C kg^{-1} soil
N_0	Organic nitrogen concentration in the nitrogen pool	mg-N kg^{-1} soil
N_b	Organic nitrogen concentration in the soil microorganism pool	mg-N kg^{-1} soil
N^+	Ammonium concentration in the soil	mg-N kg^{-1} soil
N^-	Nitrate concentration in the soil	mg-N kg^{-1} soil
N_{NO}	NO flux from soil	mg-N kg^{-1} soil

Indigenous carbon pool

According to the fluxes presented in Figure 7.36, the carbon balance equation for indigenous carbon can be described as:

$$\frac{dC_0}{dt} = -DEC_c + BD_C \tag{7.15}$$

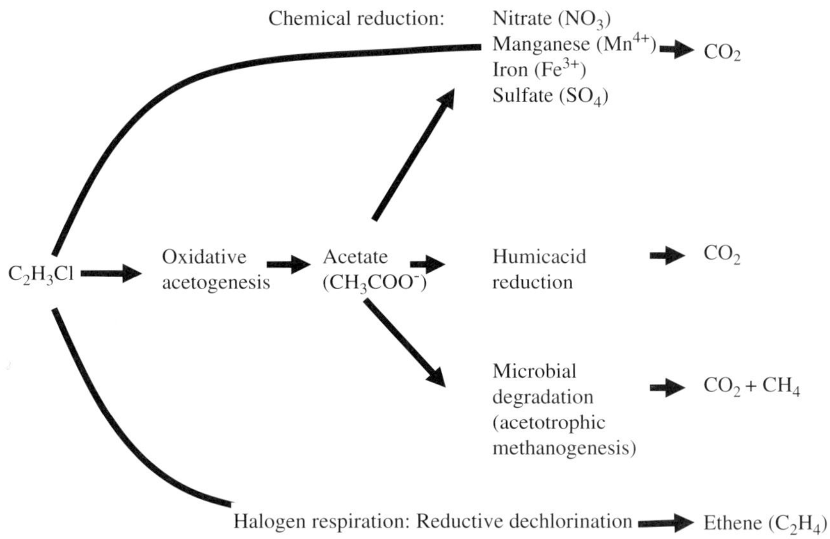

FIGURE 7.37
Degradation of vinyl chloride (C_2H_3Cl), which includes both abiotic degradation and biodegradation. In this instance, the model shown in Figure 7.36 could be applied to the oxidative acetogenesis–acetate–microbial degradation pathway, so long as reliable measures of methane (CH_4) emitted from the soil surface exist. Carbon dioxide is more difficult, since it is a relatively large component of the troposphere (>300 parts per million). In this case, the degradation of acetate (CH_3COO^-) would be compared to emitted CH_4. Samples of vinyl chloride and acetate taken at representative *in situ* locations could provide an overall C_2H_3Cl to CH_3COO^- ratio, from which the amount of remaining vinyl chloride can be estimated. Also, since CH_3COO^- is relatively volatile using the model in Figure 7.36 would provide an estimate of the rate of vinyl chloride degradation.

The term DEC_c represents the carbon output due to microbial decomposition. It is modeled using first order kinetics [49]:

$$DEC_c = k_c C_b C_0 \tag{7.16}$$

where k_c is the constant that defines the rate of decomposition for the carbon pool as a weighted average of the different organic carbon compounds. The term BD_c represents the rate at which carbon returns to the indigenous carbon pool due to the death of microorganisms. This can be simply expressed as a linear dependence on the amount of microorganisms:

$$BD_c = k_b C_b \tag{7.17}$$

The term *RES* is the respiration by soil microorganism that produce CO_2 into the outer system. The *RES* should be expressed as a function of DEC_c.:

$$RES = r_r DEC_c \tag{7.18}$$

The constant r_r defines the fraction of decomposed organic carbon that produces CO_2 by respiration. This value is usually estimated in the range of 0.6–0.8 [50].

Nitrogen pool

The nitrogen balance is similar to Eq. 7.15, but with each term divided by the C/N ratio of its respective pool:

$$\frac{dN_0}{dt} = -AMM - \frac{DEC_c}{(C/N)_c} + \frac{BD_c}{(C/N)_b} \tag{7.19}$$

(Continued)

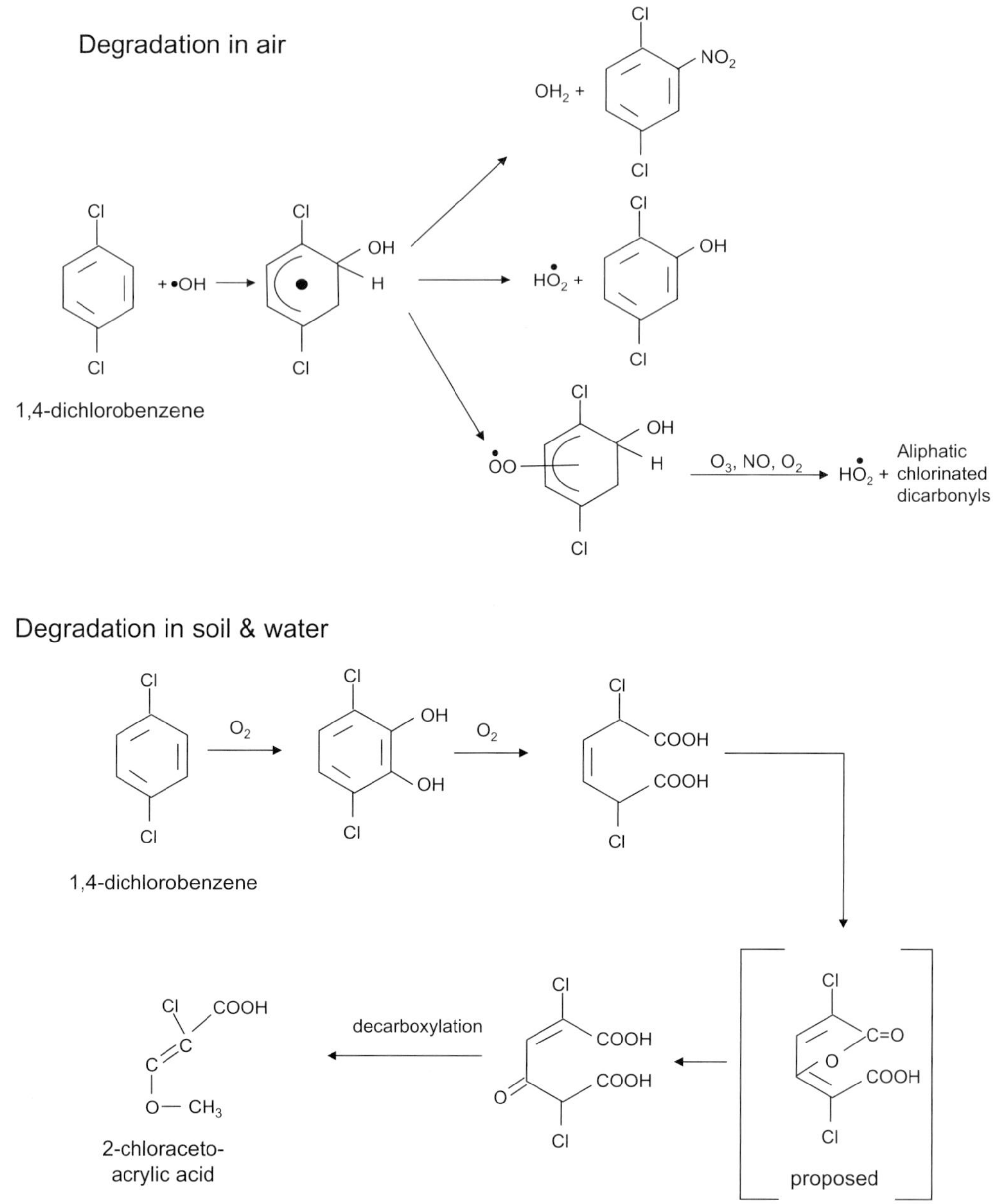

FIGURE 7.38
Degradation pathways of 1,4-dichlorobenzene reactions by a physicochemical process (photochemical) and biochemical processes (biodegradation). *Source:* Agency for Toxic Substances and Disease Registry (2006). Toxicological Profile for Dichlorobenzenes; http://www.atsdr.cdc.gov/toxprofiles/tp10-c6.pdf; accessed August 7, 2009.

The term AMM also can be simply described as a linear dependence on the nitrogen decomposition by microorganisms. Therefore, Eq. 7.19 can be rearranged:

$$\frac{dN_0}{dt} = -\alpha DEC_N - \frac{DEC_c}{(C/N)_c} + \frac{BD_c}{(C/N)_b} \tag{7.20}$$

$$\frac{dN_0}{dt} = \frac{-\alpha DEC_c}{(C/N)_c} - \frac{DEC_c}{(C/N)_c} + \frac{BD_c}{(C/N)_b} = \frac{-(\alpha+1)DEC_c}{(C/N)_c} + \frac{BD_c}{(C/N)_c} \tag{7.21}$$

The constant α defines the fraction of ammonification which converts the organic nitrogen from an organic form to an inorganic form. Approximately 1.5% to 3.5% of the soil organic nitrogen is converted to inorganic nitrogen via ammonification [51].

Toluene pool as an outer carbon source

The balance equation for toluene as an outer carbon source is given by

$$\frac{dC_T}{dt} = -r_T C_T - \left(\frac{DEC_T}{(H \times k_S) + C_T}\right) - (ADS - DES) \tag{7.22}$$

where the only input is represented by the fraction of r_T that undergoes volatilization, H is the dimensionless Henry's law partition coefficient, and K_s is the Monod half saturation constant for toluene. This study determined experimentally that coefficient r_T falls in the range of 0.14–0.32, depending on the incubation period. The second term uses the Monod equation to describe toluene consumption and the biomass growth [52]. The last term describes the adsorption and desorption of toluene in soil. Li and Gupta [53] measured the adsorption and desorption of toluene on clay minerals and reported 0.47 and 0.001 mg C hr^{-1} for adsorption and desorption coefficient, respectively. The output of decomposed toluene by microorganisms is modeled in the same way as carbon pool:

$$DEC_T = k_T C_b C_T \tag{7.23}$$

Soil microorganisms pool

The carbon balance in the soil microorganism pool can be described by:

$$\frac{dC_b}{dt} = (1 - r_r)DEC_c + \left(\frac{Y \times DEC_T}{(H \times k_S) + C_T}\right) - BD_c \tag{7.24}$$

where Y is the yield coefficient. The input is represented by the fraction of organic matter that is incorporated by the microorganisms from the carbon, nitrogen, and toluene decomposition. The only output is BD.

The balance of the nitrogen component in the soil microorganisms may be expressed as:

$$\begin{aligned}\frac{dN_b}{dt} &= \frac{(1 - r_r)DEC_c}{(C/N)_c} + \left(\frac{Y \times DEC_T}{(H \times k_S) + C_T}\right)\left(\frac{1}{(C/N)_T}\right) - \frac{BD_C}{(C/N)_b} \\ &+ DEC_c\left(\frac{1}{(C/N)_c} - \frac{1}{21}\right) + DEC_T\left(\frac{1}{(C/N)_T} - \frac{1}{21}\right)\end{aligned} \tag{7.25}$$

From the first to third terms, nitrogen balance for soil microorganisms simply divided by C/N ratio. The last two terms describe the input source by immobilization (IMM) from NH_4^+, which can be given by net nitrogen immobilization [54];

$$\varsigma_i = \varepsilon_i\left(\frac{1}{(C/N)_i} - \frac{1}{21}\right) \tag{7.26}$$

where ς_i is net nitrogen immobilization and ε_i is a decomposition rate of carbon residues. Immobilization is defined as the conversion of inorganic nitrogen ions into an organic form by soil microorganisms which incorporate mineral ions to synthesize cellular components. Since the application of mineral fertilizer provides a form of ammonium ion, it is assumed that immobilization is unrestricted, which means that the bacteria can meet their nitrogen requirement and decompose the organic matter at a maximum potential rate.

Inorganic nitrogen pool: ammonium (NH_4^+) and nitrate (NO_3^-)

The balance of ammonium and nitrate in the soil can be modeled respectively as:

$$\frac{dN^+}{dt} = ADD_{MF} + AMM - NTR - IMM \tag{7.27}$$

$$\frac{dN^-}{dt} = NTR_{N^+} - DEN \tag{7.28}$$

(Continued)

ADD_{MF} is the only input source from the framework, which is a constant value, and *AMM* and *IMM* have already been defined in Eqs. 7.7 and 7.14, respectively. *NTR* is a nitrification, which can be modeled as first order kinetics: [55]

$$NTR = k_n C_b f_{N^+} N^+ + k_n C_b (1 - f_{N^+}) \left(\frac{r_{\max} \times N^+}{K_{NO} + N^+} \right) = NTR_{N^+} + NTR_{NO} \quad (7.29)$$

where C_b expresses the dependence of nitrification on microbial activity, k_n is the rate of nitrification, K_{NO} is the half saturation constant for Monod-type production of NO from N^+ (that is, it corresponds to the concentration at which the specific growth rate coefficient is one-half of its maximum) m and f_{N^+} is the fraction of nitrification that produces nitrate (e.g. compartment #3 in Figure 7.36). Only about 0.1–0.4% of the NH_4^+ is being oxidized to NO_2^- and NO_3^- [56]. Therefore, the balance of ammonium can be finalized by:

$$\frac{dN^+}{dt} = ADD_{MF} + \frac{\alpha DEC_c}{(C/N)_c} - k_n C_b N^+ - \left[DEC_c \left(\frac{1}{(C/N)_c} - \frac{1}{21} \right) + DEC_T \left(\frac{1}{(C/N)_T} - \frac{1}{21} \right) \right] \quad (7.30)$$

The denitrification rate can also be modeled in the same way as the rate of nitrification by:

$$\frac{dN^-}{dt} = k_n C_b N^+ - k_d C_b N^- = C_b (k_n N^+ - k_d N^-) \quad (7.31)$$

where k_d is the rate of denitrification.

Finally, NO flux can be modeled based on the nitrification kinetics and Monod equation as follows:

$$\frac{dN_{NO}}{dt} = k_n \times C_b \times \left(\frac{r_{\max} \times N^+}{K_s + N^+} \right) \quad (7.32)$$

where, $r_{\max}$ is the maximum specific growth rate expressed as a NO flux and K_s is the Monod half saturation constant for the parent compound (in this instance, toluene). NO can be produced by both nitrification and denitrification from soil, however, previous research has suggested nitrification is a greater source of NO when compared to denitrification [57]. Therefore, the model approaches for NO flux studied in this research are based on nitrification as the major source of NO as shown in Eq. 7.32.

Model parameters

Conceptual development and construction of a toluene degradation model using carbon and nitrogen cycle require appropriate system parameters constrained by basic assumptions: (1) soil pH, temperature (22 °C), and moisture content (WFPS 30%) are constant throughout the time period; (2) all microbes in the population are active and their growth rate is a fixed proportion of carbon and nitrogen so that their C/N ratio remains constant (C/N ratio is 8); and (3) the presence of nitrite (NO_2^-) is very low and can be neglected because of the high rate of nitrification. These assumptions are reasonable and valid for the environmental conditions simulated by this model, but will likely not hold under extreme conditions of pH, temperature, and soil moisture content. The stochastic model of carbon and nitrogen cycle suggested above was validated using average rates derived from values reported in the literature within the constraints of general nitrogen and carbon dynamics, as well as from the experimental analysis and data.

The values of the constants k_T, k_c, and k_b for the first-order kinetics of indigenous carbon and toluene decomposition and of soil microorganisms are estimated using the steady-state solutions of Eqs. 7.15, 7.16, and 7.22, along with the average values of carbon in the indigenous, toluene, and soil microorganisms pool. When the temporal derivatives are set to zero and the values of the other variables are assigned the average conditions reported in Table 7.4, the solution of the linear system in these equations leads to the values of the k_T, k_c, and k_b.

Table 7.4 Assigned conditions to model parameters

Parameters	Units	Value
Added mineral fertilizer $(ADD)_{MF}$	mg-N kg^{-1} soil	8
C in carbon pool (C_c)	mg-C kg^{-1} soil	7800
C in soil microorganism pool (C_b)	mg-C kg^{-1} soil	20–250
C in toluene pool (C_T)	mg-C kg^{-1} soil	4.6–54.7
N in ammonium pool (N^+)	mg-N kg^{-1} soil	<28.6
N in nitrate pool (N^-)	mg-N kg^{-1} soil	<5.7
$(C/N)_c$	N/A	16.4
$(C/N)_T$	N/A	0.033
$(C/N)_b$	N/A	8
r_T	N/A	0.14–0.32
r_r	N/A	0.6
r_{SRT}	N/A	0.37
α	N/A	0.015–0.035
k_T	kg hr^{-1} mg-C	1.8×10^{-5}
k_c	kg hr^{-1} mg-C	1.5×10^{-4}
k_b	kg hr^{-1} mg-C	0.35
k_n	N/A	0.02
k_{oc}	mg C hr^{-1}	0.47
k_{ocd}	mg C hr^{-1}	0.001
H	N/A	0.261
Y	mg-biomass mg^{-1} C in toluene	0.35
K_s	N/A	6.04
r_{max}	mg-N kg^{-1} soil	6.0×10^{-6}

Model evaluation

Data collected and analyzed from laboratory experiments of toluene degradation, quantification of soil microorganisms, and NO emissions from soil were used to characterize the physical, chemical, and biological algorithms for carbon and nitrogen cycling in soil. In turn, this information provided the toluene degradation parameter from which soil microbial activity could be matched with NO emissions from soil. Thus, the predictive model was developed by comparing the laboratory studies to the model output using analysis of variance (ANOVA) for linear regressions. The laboratory results indicated that microbial populations grew with respect to toluene concentration significantly ($P \leq 5\%$ and most P values $\leq 1\%$), so this allowed for a mathematical relationship between microbial population size and toluene degradation [58]. Also, the mass of NO released from soil in chambers had a direct relationship to microbial population size. Thus, a numerical model could be developed to link toluene concentrations in soil with NO flux measurement. This led to the development of a numerical model using ordinary differential equations, which could

(Continued)

be evaluated from actual, direct measurements of the chemical undergoing degradation (toluene) and of the chemical being released by the microbes (NO).

The rate of NO production from soil and the toluene degradation in soil were measured directly. NO production rate and toluene degradation with different levels of toluene contamination are closely linked, which suggests that NO is a good surrogate measurement for toluene bioremediation progress in soil. NO is produced mainly by nitrifiers and denitrifiers (which are important hydrocarbon-degrading bacteria). Specific identification of the actual microbial species producing the NO, however, is not necessary to link the rates of decomposition of organic contaminants to NO production. Although there were good correlations between the NO production rate and toluene degradation, long-term NO production that accompanies toluene degradation should be measured in future research, since this may offer insights as to how nitrifying bacteria growth and metabolism can be associated with biodegradation. From the relationship between NO production rate and toluene degradation the data can be integrated to estimate contamination persistence, making it possible to deduce the contamination levels and duration by extrapolating from NO production in soil.

Model comparison to laboratory study for toluene degradation

Experimental data were correlated with the model values and found to be statistically significant ($P < 0.01$) for all levels of toluene concentration. However, the model values slightly overestimated experimental results from the 16 hour incubation period and the model did not predict the observed decrease in toluene degradation between 1 hour and 16 hours.

Typically, biodegradation followed a lag period, defined as the time period preceding significant chemical disappearance. The laboratory study and the model results were well matched because the model parameters such as fraction of volatilization of toluene (r_T) were determined in the laboratory, and therefore measured under similarly controlled laboratory conditions.

To compare the experimental data to model output, total cell counts (cells g^{-1} soil) derived from experimental data were converted to carbon content in soil microorganisms (mg-C kg^{-1} soil). Positive correlations were found between model outputs and experimental measurements across all initial toluene concentration ($P < 0.01$). The values of model output approximate those determined from the quantification of soil microorganisms using FISH, indicating that the numerical model studied here properly describes the experimental data.

The model developed for NO emission estimated NO flux well for both laboratory and field studies. The model reproduced the trends of the experimental measurements of NO flux from toluene contaminated soil, showing that the simple balance equations of the model are able to capture the effectual dynamics of nitrogen and carbon transformation during the experiments.

The overall nitrification process is controlled primarily by ammonium and oxygen concentrations, Total N gas emissions can be modeled as proportional to the rates of gross mineralization and fertilizer input (Eq. 7.27), and this suggestion has been supported by the large body of research showing increasing NO flux with increasing soil nitrogen and fertilization. The model output is largely affected by ammonium nitrogen conversion to NO production in soil.

As mentioned, nitrifiers are usually considered the major driver in most NO. Nitrifiers can survive and produce NO as a byproduct over a very wide range of pH, temperature, water content, soil texture, and nutrient availability. Diffusion and consumption of NO are also important factors affecting net release. Therefore, although denitrification may be a producer of NO, the factors of limited diffusion under denitrifying anaerobic conditions and consumption of NO by denitrifiers limit the contribution of denitrification to gross NO production. If nitrification is largely responsible for the net NO production as applied in this model, this would explain the favorable match between the studies and the model output.

The model output for the NO emissions at different levels of toluene amendment were also observed to follow the number of microorganisms trend observed at the same levels of toluene amendments. Therefore, the results support the hypothesis that NO emission is a good surrogate indicator of soil microbial activity in toluene-contaminated soil.

Carbon and nitrogen speciation and soil microbial activity can be used to predict the degradation of an organic toxic substance in soil. The model used an indirect, noninvasive metric (NO) to estimate rates of microbial population growth and the concomitant breakdown of toluene. The ratio of NO emission to toluene degradation was positively correlated across all levels of toluene concentrations and incubation times, meaning that NO emissions can be a good surrogate measure of toluene contamination in soil, if the duration of toluene incubation or initial soil toluene concentration is known. Therefore, NO flux from soil measured in the field may help to estimate the rate of toluene degradation in the soil and thus provide an ongoing metric of remediation success, an indication when soil remediation has met target concentrations, and a simple and integrated complement to post-closure monitoring (e.g. a spike in NO production and application of this model may indicate the presence of a new carbon source, which should be investigated). The simulated toluene concentration, number of soil microorganisms, and NO flux as a function of time showed statistical significance ($P < 0.01$) for all levels of toluene concentration with the experimental data. This finding supports the adequacy of the model for supporting the NO–toluene–soil microbial relationship. The mathematical model derived in this study suggests a novel approach for characterizing and predicting the fate of other toxic aromatic compounds in soil. Because the results of the investigation were based on experimental data and the analysis was conducted under specific environmental conditions, the emphasis throughout the research focused on specific rather than general features of the dynamics. However, the results can presumably be extended to other bioremediation systems with different environmental conditions, so long as these systems are evaluated using the protocol described here.

Measured toluene concentrations in soil, microbial population growth and NO fluxes in chamber studies based of carbon and nitrogen cycling provide useful information to those who manage toxic and criteria air pollutants. Nitric oxide may prove to be a valuable indicator of bioremediation of air toxic (i.e. toluene) concentrations. The model found that chemical concentration, soil microbial abundance, and NO production can be directly related to experimental results (significant at $P<0.01$) for all toluene concentrations tested. This indicates that the model may prove useful in monitoring and predicting the fate of toxic aromatic contaminants in a complex soil system.

This can be valuable in predicting the duration of such contaminants in soil sinks, but may also prove useful to environmental decision making. The model developed from this research may prove useful in predicting emissions of oxides of nitrogen from soils contaminated with organic compounds. Therefore, the model may help in predicting the interrelations of ozone precursors, such as changes in reservoirs of hydrocarbons and releases of various oxides of nitrogen. As such, the model may be a tool for decision makers in ozone nonattainment areas. For example, an ozone nonattainment area that is hydrocarbon-limited may benefit from the breakdown of organic compounds via nitrification or denitrification (i.e. lessening the sources of hydrocarbons stored and potentially available to be released to the atmosphere). Conversely, in an ozone nonattainment area that is NO_x limited, the emissions of NO may need to be limited to reduce ambient concentrations of ozone (i.e. perhaps increasing bioremediation efforts during winter months and decreasing them in summer, thereby lowering NO_x emissions when photochemical reaction rates are slower, thus limiting the production of ozone).

Future research should be conducted to characterize and to predict the fate of toxic aromatic compounds in soil and production of chemical and biological indicators of soil microbial activity to monitor the bioremediation processes: (1) fitting the model to compounds other than toluene, especially those with more complicated substitutions; (2) investigating other factors to estimate the degradation of toluene and other toxic aromatic compounds such as chemical dissolution and diffusion; and (3) performing additional studies of the NO_x emissions, other than NO, during the nitrification as additional potential indicators of important reactions occurring in soil and ground water, as well as their contributions as a tool to characterize potential sinks of hydrocarbons and to predict fluxes of ozone precursors and air toxics.

SEMINAR TOPIC

Contrasting the Risks from Experimental Microbiology and Bioremediation

The prototypical path to progress in engineering is from concept to design to construction to operation. Usually, a promising laboratory finding can lead to larger experiments and eventually to pilot facility. Each step involves an increasing number of variables, so that when enough is known about an engineering project it can be taken to full-scale operation. Genetic engineering of microbes to address environmental problems has, to some extent, taken this path. However, using recombinant organisms for environmental biotechnological applications varies from the laboratory to the field.

Rather than being propagated as a pure culture under highly controlled conditions and lacking excess nutrients, in a bioremediation project, genetically modified microbes (usually bacteria) are introduced into an environment of widely diverse biota where they must become established. The interactions within this field environment are unknown since the abiotic structure (e.g. soil, water, and air) differs in time and space. Extrapolations from previous settings will never be completely reliable. Add to this the microbial and macrobiotic population dynamics differ from place to place, due to different soil types, widely varied types and amounts of plant roots, presence of arthropods, population dynamics of other soil biota, etc.

These and other external factors can diminish the introduced microbes' establishment and survival rates and can adversely affect growth and metabolism, even if the introduced microbes survive. The European Union has provided a useful perspective on these challenges and how they differ from the containment issues associated with experimentation:

Thus, whereas contained applications are mainly based on a few well-characterized microorganisms such as *Escherichia coli*, *Bacillus subtilis*, *Saccharomyces cerevisiae*, and some cell lines that perform well in bioreactors, open applications are based on a more diverse range of organisms able to survive and perform in natural communities in the environment, such as *Pseudomonas*, *Alcaligenes*, etc. Efforts in the early 1980s focused on the development of new plasmid vectors based on broad host range replicons. However, these vectors suffered from the disadvantages generally common to plasmids. The specific characteristics of open biotechnological applications clearly necessitated the development of novel genetic tools and concepts to engineer new properties and meet the new challenges. Among others, these included stability without selection, minimal physiological burden, small size non-antibiotic selection markers, minimal lateral transfer of cloned genes to indigenous organisms, and traceability of specific genes and strains in complex ecosystems [59].

Success in containment is inversely related to success in biodegradation. That is, in bioremediation efforts, the bioengineer seeks the most rapid, widespread microbial growth since growth and metabolism are the processes by which the contaminants are used for energy and carbon sources. Conversely, attempting to keep a microbe from moving beyond a safe zone requires that transport, transformation, and fate be stifled beyond the containment area.

The risk assessment process is really not as stepwise as indicated in the typical flow chart, i.e. form hazard identification to dose-exposure response relationships to exposure assessment to effects assessment. Rather, many processes and feedbacks go on among these steps. For example, as new hazard information comes to the fore, it may well change the exposure pathways, e.g. the dermal pathway may have been thought to be low, but new information about cosolvation and dissolution may indicate higher dermal exposure potentials. This means the threshold values at which no effect is observed may also change, as do the uncertainty factors. These changes may lead to a new factor of safety (e.g. the reference dose may be decreased or increased).

As exposure and effects data increase in amount and quality due to research and after a substance is in the marketplace (e.g. monitoring the use of a pesticide after registration), the risk assessment will also change. An illustrative example is provided by the hormonally active pesticides, which include a wide variety of chemical structures, including OPs, organochlorines, and pyrethroids. For example, the US EPA has attempted to identify the chemicals that have the greatest potential to elicit endocrine disruption and to which people and ecosystems are likely to be exposed. The first group of chemicals are being screened in the Endocrine Disruptor Screening Program (EDSP) [60].

The Food Quality Protection Act of 1996 (FQPA) amended the Federal Food, Drug, and Cosmetic Act (FFDCA), requiring that the US EPA develop a chemical screening program based on validated test systems and other scientifically sound and relevant information to determine whether certain substances may have hormonal effects. Those chemicals that are suspected of being endocrine disruptive are further evaluated in what are called Tier I and Tier 2 testing protocols.

Tier 1 screening includes a battery of screening assays to identify substances with the potential to interact with the estrogen, androgen, or thyroid hormone systems, including [61]:

- *Amphibian (Frog) Metamorphosis* – involves the use of tadpoles to determine if chemicals affect the thyroid during metamorphosis and consequently result in developmental effects.
- *Receptor binding in vitro assays* – chemicals can affect the endocrine system by binding to hormone receptors to either mimic the action of the natural hormone or block access of the hormone to the site and thus block hormone controlled activity. The androgen receptor (AR) is involved in the development of male sexual characteristics and the estrogen receptor (ER) is involved in female maturation and reproductive function. Several receptor binding assays are being considered, including:

1. an AR binding assay that utilizes rat prostate cytosol to examine the ability of a test chemical to bind with androgen receptors;
2. an AR binding assay that utilizes a rat recombinant AR to examine the ability of a test chemical to bind with androgen receptors;

3. an ER binding assay that utilizes rat uterine cytosol to examine the ability of a test chemical to bind with estrogen receptors; and
4. an ER binding assay utilizes the alpha isoform of the human recombinant ER to examine the ability of a test chemical to bind with estrogen receptors.

- *Aromatase* – aromatase is an enzyme complex responsible for estrogen biosynthesis that converts androgens into estrogens, estradiol, and estrone. The aromatase *in vitro* assay focuses on this portion of the steroidogenic pathway to detect substances that inhibit aromatase activity.
- *Fish Screen* – screens for estrogenic and androgenic effects. The assay examines abnormalities associated with survival, reproductive behavior, secondary sex characteristics, histopathology, and fecundity (i.e., number of spawns, number of eggs/spawn, fertility, and development of offspring) of fish exposed to test chemicals.
- *Hershberger* – the Hershberger assay is designed to detect androgenic and anti-androgenic effects. In this *in vivo* assay, accessory sex gland weights, including several androgen-dependent tissues, are measured in castrated or immature male rats.
- *Pubertal Female* – involves the use of rats to screen for estrogenic and thyroid activity in females during sexual maturation. This assay examines abnormalities associated with sex organs and puberty markers, as well as thyroid tissue.
- *Pubertal Male* – involves the use of rats to screen for androgenic, anti-androgenic, and thyroid activity in males during sexual maturation. This assay examines abnormalities associated with sex organs and puberty markers, as well as thyroid tissue
- *Steroidogenesis* – detects interference with the body's production of male and female steroid sex hormones. A version of the assay using sliced testis as a source of steroidogenic enzymes was optimized by EPA to detect chemicals that inhibit synthesis of steroid hormones, but continued concerns about being able to distinguish between compounds that inhibit steroid hormone synthesis and chemicals that kill the cells responsible for testosterone synthesis led to a halt in further work on validating this assay. It is being replaced by a cell-based assay using the H295R human adrenocortical carcinoma cell line. The H295R cell line also holds promise in being able to detect inducers of enzymes responsible for steroid synthesis in addition to chemicals that inhibit it.
- *Uterotrophic* – The Uterotrophic assay involves the use of female rats to screen for estrogenic effects. In this *in vivo* assay, uterine weight changes are measured in ovariectomized or immature female rats.
- *15-day Adult Intact Male* – The Adult Male assay involves the use of rats to screen primarily for anti-androgenic and thyroid activity. The assay will screen for abnormalities associated with primary and secondary sex organs, systemic hormone concentrations, and thyroid.

Table 7.5 includes the draft list of Tier 1 compounds.

Table 7.5 **Tier 1 listing of pesticide active ingredients and high production volume (HPV) chemicals used as pesticide inert ingredients (also known as other ingredients)[a]**

Chemical name	CAS number	Pesticide active ingredient	HPV/Inert
2,4-D	94757	x	
4,7-Methano-1H-isoindole-1,3(2H)-dione,2-(2-ethylhexyl)-3a,4,7,7a-tetrahydro-	113484	x	
Abamectin	71751412	x	
Acephate	30560191	x	
Acetone	67641		x
Aldicarb	116063	x	
Allethrin	584792	x	
Atrazine	1912249	x	
Azinphos-methyl	86500	x	

(Continued)

Table 7.5 **Tier 1 listing of pesticide active ingredients and high production volume (HPV) chemicals used as pesticide inert ingredients (also known as other ingredients)[a]—cont'd**

Chemical name	CAS number	Pesticide active ingredient	HPV/Inert
Benfluralin	1861401	x	
Bifenthrin	82657043	x	
Butyl benzyl phthalate	85687		x
Captan	133062	x	
Carbamothioic acid, dipropyl-, S-ethyl ester	759944	x	
Carbaryl	63252	x	
Carbofuran	1563662	x	
Chlorothalonil	1897456	x	
Chlorpyrifos	2921882	x	
Cyfluthrin	68359375	x	
Cypermethrin	52315078	x	
DCPA (or chlorthal-dimethyl)	1861321	x	
Diazinon	333415	x	
Dibutyl phthalate	84742		x
Dichlobenil	1194656	x	
Dichlorvos	62737	x	
Dicofol	115322	x	
Diethyl phthalate	84662		x
Dimethoate	60515	x	
Dimethyl phthalate	131113		x
Di-sec-octyl phthalate	117817		x
Disulfoton	298044	x	
Endosulfan	115297	x	
Esfenvalerate	66230044	x	
Ethoprop	13194484	x	
Fenbutatin oxide	13356086	x	
Fenvalerate	51630581	x	
Flutolanil	66332965	x	
Folpet	133073	x	
Gardona (cis-isomer)	22248799	x	
Glyphosate	1071836	x	

Table 7.5 **Tier 1 listing of pesticide active ingredients and high production volume (HPV) chemicals used as pesticide inert ingredients (also known as other ingredients)[a]—cont'd**

Chemical name	CAS number	Pesticide active ingredient	HPV/Inert
Imidacloprid	138261413	x	
Iprodione	36734197	x	
Isophorone	78591		x
Linuron	330552	x	
Malathion	121755	x	
Metalaxyl	57837191	x	
Methamidophos	10265926	x	
Methidathion	950378	x	
Methiocarb	2032657	x	
Methomyl	16752775	x	
Methyl ethyl ketone	78933		x
Methyl parathion	298000	x	
Metolachlor	51218452	x	
Metribuzin	21087649	x	
Myclobutanil	88671890	x	
Norflurazon	27314132	x	
o-Phenylphenol	90437	x	
Oxamyl	23135220	x	
Permethrin	52645531	x	
Phosmet	732116	x	
Piperonyl butoxide	51036	x	
Propachlor	1918167	x	
Propargite	2312358	x	
Propiconazole	60207901	x	
Propyzamide	23950585	x	
Pyridine, 2-(1-methyl-2-(4-phenoxyphenoxy)ethoxy)-	95737681	x	
Quintozene	82688	x	
Resmethrin	10453868	x	
Simazine	122349	x	
Tebuconazole	107534963	x	
Toluene	108883		x

(Continued)

Table 7.5 Tier 1 listing of pesticide active ingredients and high production volume (HPV) chemicals used as pesticide inert ingredients (also known as other ingredients)[a]—cont'd

Chemical name	CAS number	Pesticide active ingredient	HPV/Inert
Triadimefon	43121433	x	
Trifluralin	1582098	x	

[a]*The US EPA notes that this is not a list of known or likely endocrine disruptors. Nothing in the approach for generating the initial list provides a basis to infer that any of the chemicals selected interferes with or is suspected to interfere with the endocrine systems of humans or other species.*
Source: US Environmental Protection Agency (2009). Overview of the April 2009 Final List of Chemicals for Initial Tier 1 Screening. http://www.epa.gov/scipoly/oscpendo/pubs/prioritysetting/final_listfacts.htm; accessed October 9, 2009.

Tier 2 assays have three goals:

determining whether a substance may cause endocrine-mediated effects through or involving estrogen, androgen, or thyroid hormone systems;
determining the consequences to the organism of the activities observed in Tier 1; and
establishing the relationship between doses of an endocrine-active substance administered in the test and the effects observed.

The Tier 2 tests are more protracted than Tier 1 tests and are designed to consider systematically the life stages and processes involved in a broad range of doses within relevant routes of exposure. That way, regulators hope to obtain a more comprehensive profile of the biological consequences of a chemical exposure and to identify the dose or exposure that caused the consequences, since the effects associated with endocrine disruption are often not expressed until later in the test subject's life. Endocrine disruption can also be *transgenerational*, i.e. the exposure may occur in one generation but the effects are not manifested until future generations, such as the women who were exposed to diethylstilbestrol (DES), but their daughters experienced increased incidences of cervical cancer (i.e. the "DES daughters").

Even within the same organism, the effects will likely be delayed and may not appear until the reproduction. Thus, Tier 2 tests usually encompass two generations and include effects on fertility and mating, embryonic development, sensitive neonatal growth and development, and transformation from the juvenile life stage to sexual maturity. The assays in current use for Tier 2 screening include [62]:

- *Amphibian Development, Reproduction* – involves the use of frogs to characterize dose-response characteristics and adverse reproductive and developmental effects.
- *Avian 2-Generation* – involves the use of Japanese quail to characterize dose-response characteristics and adverse reproductive and developmental effects.
- *Fish Lifecycle* – involves the use of fish to characterize dose-response characteristics and adverse reproductive and developmental effects.
- *Invertebrate Lifecycle* – involves the use of mysid shrimp to characterize dose-response characteristics and adverse reproductive and developmental effects.
- *Mammalian 2-Generation* – involves the use of rats to characterize dose-response characteristics and adverse reproductive and developmental effects.

Another extremely important tier is under consideration, i.e. *in utero* exposures through the lactation stage. This assay involves the use of pregnant rats to assess post-natal development of the neonate after *in utero* and lactational exposure. Obviously gestation and neonatal stages are highly vulnerable to the effects of endocrine disruptors, since tissue growth is very prolific and effects early in life can be extended and exacerbated with time.

The EDSP is actually a combination of all elements of the risk assessment process, as well as risk management factors. EPA will select 50 to 100 chemicals. The chemicals will be selected based on their relatively high potential for human exposure rather than using a combination of exposure- and effects-related factors. The scope of this first group of chemicals to be tested includes pesticide active ingredients and High Production Volume (HPV) chemicals used as pesticide inerts. This will allow EPA to focus its initial screening efforts on a smaller and more manageable universe of chemicals that emphasizes the early attention to the pesticide chemicals that Congress specifically mandated EPA to test for possible endocrine effects.

The US EPA considered a number of methods for screening chemicals that may elicit hormonal effects. These included quantitative approaches, such as the *quantitative structure activity relationship (QSAR)* method. This analysis uses computer simulations to estimate how a chemical behaves based on its structure. Regulators could use QSAR models to simulate and predict the likelihood of a chemical binding with estrogen and androgen receptors.

This prediction is based on a number of factors, especially the *agonistic* behavior of a substance, i.e. the ease with which a molecular structure of a certain chemical fits into the molecular structure of the estrogen and androgen receptors. Such a capacity to bind is known as the *binding affinity* of the chemical, which can be compared with the binding affinity of the natural hormone. For example, a chemical with a square-shaped structure trying to fit into a receptor with a smaller round structure would likely have a low binding affinity. This is the same process discussed in Chapter 3 regarding the specificity of an enzyme, i.e. an enzyme is limited in the kinds of substrate that it will catalyze. Thus, both a hormonally active molecule and an enzyme can bind a limited number of substrate molecules (see Figure 3.3). That is, the binding site is specific so that other compounds do not fit the specific three-dimensional shape and structure of the active site (analogous to a specific key fitting a specific lock).

Chemicals can affect the endocrine system in a number of ways other than through receptor binding; thus QSAR analyses provide limited information on the potential for a chemical to interfere with the endocrine system. Currently, QSAR analyses are not yet sufficiently developed for regulatory purposes.

The use of QSAR methods for screening endocrine disruptors is under review, but QSAR types of methods are indeed being applied to other aspects of risk assessment, including elements of transport, transformation, and fate, many which are listed in Table 2.9. Environmental biotechnology is particularly concerned about characterizing and predicting biodegradation, so QSAR applications to microbe-chemical degradation would be valuable.

Ideally, biodegradation can be estimated by distinguishing the initial alteration of the chemical structure of a chemical compound that results in the loss of a specific property of that compound (i.e. primary biodegradation) versus the complete conversion of the parent compound to water, carbon dioxide, and inorganic compounds (i.e. ultimate biodegradation).

In aerobic conditions, these are CO_2 and H_2O. In anaerobic conditions, ultimate biodegradation would be to CH_4 and H_2O. However, the few primary and ultimate biodegradation rates are not available for many of the compounds of concern.

The principal physicochemical processes [63] that affect persistence and biodegradation include the rates of atmospheric oxidation, aqueous hydrolysis, photolysis, sorption, and microbial growth and metabolism (electron acceptance and donation). Based on these processes, green engineering has provided a simple method for estimating biodegradation in the form of an index.

Biodegradation can be reflected by BOD, CO_2 production, loss of mass of parent compound, gain of mass of known product of degradation, etc. For example, one method is to designate qualitative characteristics (e.g. BR = biodegrades rapidly; BSA = biodegrades slowly even with acclimation). Other designations include microbial toxicity and temperature regimes [64]. These qualifiers go into a database, from which models can be developed and run to estimate persistence and biodegradation potential. For example, a database of 186 chemicals received summary evaluations of "biodegrades rapidly" and 109 chemicals designated "does not biodegrade rapidly". From this information, an indicator variable was established to categorize compounds, i.e. the rapid biodegradation category assigned a value of 1 and chemicals in the slow biodegradation category assigned a value of 0. The indicator variable was subsequently used as the dependent variable in multiple linear and nonlinear regressions against 37 independent variables. From these parameters, a regression model estimated the probability that a chemical would fall in the "biodegrades rapidly" category. Counts of structural fragments (i.e., the number of times a chemical substructure occurs in the molecule) formed the independent variables. The resulting index is [65]:

$$I = 3.199 + a_1 f_1 + a_2 f_2 + \ldots + a_n f_n + a_m \times M_w \qquad (7.33)$$

where I is an indicator of the aerobic biodegradation rate; a_n is the regression coefficient; f_n is the ultimate degradation biodegradation rate; M_w is the molecular weight; a_m is the regression coefficient for M_w.

Thus, this is an index, which is based on fitting coefficients to the regressions for models. In this case, coefficients for chemical functional groups and structural fragments can be fitted to regressions in models, as shown in Table 7.6. A value of 1 indicates that the compound is expected to degrade over hours; a value of 2 corresponds to a lifetime of days; 3, 4, and 5 correspond to weeks, months, and longer, respectively.

The parameter f_n is the number of groups of type n in the molecule, and a_n denotes the contributions of group n to degradation rate.

Thus, as the value of I increases, the chemical structure becomes less recalcitrant, and the greater the biodegradation rate (see Table 7.7).

For example, based on this approach, we can classify the biodegradation for the alcohol 1-propanol (molecular weight = 60), based on the molecule's aliphatic –OH group. Thus, from the ultimate coefficient column in Table 7.6, we find that an aliphatic –OH group = 0.160 and added to the ultimate coefficient × M_w product, we calculate 1-propanol's biodegradation index as:

$$I = 3.199 + 0.160 - 0.00221 \times 60 = 3.22$$

which implies a lifetime in the environment of weeks.

A heavier molecule with a more complex chemical structure would have a lower I value. For example, diphenyl ether (molecular weight = 170) contains an aromatic ether and two mono-aromatic rings. Its biodegradation index is:

$$I = 3.199 + 2 \times (0.022 - 0.058) - (0.00221 \times 170) = 2.81$$

This implies a lifetime of weeks; literature data indicate a lifetime of months.

Estimation Tools

This type of information has recently been incorporated into mathematical models. For example, BIOWIN™ estimates aerobic and anaerobic biodegradability of organic chemicals using seven different models [66]. Two of these are the original Biodegradation Probability Program (BPP™). The seventh and newest model estimates anaerobic

Table 7.6 Chemical substructures linked to coefficients in a biodegradation index

	BIODEG models			Survey models		
Fragment or parameter	Freq[a]	Linear coeff	Nonlinear coeff	Freq[a]	Primary coeff	Ultimate coeff
Equation constant		0.748	3.01		3.848	3.199
M_w	295	−0.000476	−0.0142	200	−0.00144	−0.00221
Unsubstituted aromatic (≤3 rings)	2	0.319	7.191	1	−0.343	−0.586
Phosphate ester	5	0.314	44.409	6	0.465	0.154
Cyanide/nitrile (C≡N)	5	0.307	4.644	11	−0.065	−0.082
Aldehyde (CHO)	4	0.285	7.180	5	0.197	0.022
Amide (C(=O)N or C(=S)N)	9	0.210	2.691	13	0.205	−0.054
Aromatic (C(=O)OH)	24	0.177	2.422	6	0.0078	0.088
Ester (C(=O)OC)	23	0.174	4.080	25	0.229	0.140
Aliphatic OH	34	0.159	1.118	18	0.129	0.160
Aliphatic NH_2 or NH	13	0.154	1.110	7	0.043	0.024
Aromatic ether	11	0.132	2.248	11	0.077	−0.058
Unsubstituted phenyl group (C_6H_5)	25	0.128	1.799	22	0.0049	0.022
Aromatic OH	46	0.116	0.909	21	0.040	0.056
Linear C4 terminal alkyl ($CH_2CH_2CH_2CH_3$)	44	0.108	1.844	26	0.269	0.298
Aliphatic sulfonic acid or salt	4	0.108	6.833	4	0.177	0.193
Carbamate	4	0.080	1.009	6	0.194	−0.047
Aliphatic (C(=O)OH)	33	0.073	0.643	10	0.386	0.365
Alkyl substituent on aromatic ring	36	0.055	0.577	36	−0.069	−0.075
Triazine ring	5	0.0095	−5.725	4	−0.058	−0.246
Ketone (CC(=O)C)	12	0.0068	−0.453	10	−0.022	−0.023
Aromatic F	1	−0.810	−10.532	1	0.135	−0.407
Aromatic I	2	−0.759	−10.003	2	−0.127	−0.045
Polycyclic aromatic hydrocarbon (≥4 rings)	6	−0.657	−10.164	2	−0.702	−0.799
N-nitroso (NN=O)	4	−0.525	−3.259	1	0.019	−0.385
Trifluoromethyl (CF_3)	1	−0.520	−5.670	2	−0.274	−0.513
Aliphatic ether	11	−0.347	−3.429	16	−0.0097	−0.0087
Aromatic NO_2	14	−0.305	−2.509	13	−0.108	−0.170
Azo group (N=N)	2	−0.242	−8.219	3	−0.053	−0.300
Aromatic NH_2 or NH	32	−0.234	−1.907	23	−0.108	−0.135
Aromatic sulfonic acid or salt	11	−0.224	−1.028	8	0.022	0.142

Table 7.6 Chemical substructures linked to coefficients in a biodegradation index—cont'd

	BIODEG models			Survey models		
Fragment or parameter	Freq[a]	Linear coeff	Nonlinear coeff	Freq[a]	Primary coeff	Ultimate coeff
Tertiary amine	10	−0.205	−2.223	10	−0.288	−0.255
Carbon with 4 single bonds and no H	9	−0.184	−1.723	32	−0.153	−0.212
Aromatic Cl	40	−0.182	−2.016	27	−0.165	−0.207
Pyridine ring	18	−0.155	−1.638	8	−0.019	−0.214
Aliphatic Cl	12	−0.111	−1.853	14	−0.101	−0.173
Aromatic Br	5	−0.110	−1.678	4	−0.154	−0.136
Aliphatic Br	5	−0.046	−4.443	2	0.035	0.029

[a]*Number of compounds in the training get containing the fragment.*
Source: R.S. Boethling, P.H. Howard, W. Meyian, W. Stiteler, J. Beauman and M. Tiradot (1994). Group contribution method for predicting probability and rate of aerobic biodegradation. *Environmental Science & Technology* 28: 459–465.

Table 7.7 Relative ranking of recalcitrance of chemical compounds

I Value	Biodegradability rank
5	Hours
4	Days
3	Weeks
2	Months
1	Years

Source: Based on modeling by R.S. Boethling, P.H. Howard, W. Meyian, W. Stiteler, J. Beauman and M. Tiradot (1994). Group contribution method for predicting probability and rate of aerobic biodegradation. *Environmental Science & Technology* 28: 459–465.

biodegradation potential. The Ecological Structure Activity Relationships (ECOSAR) Class Program is a computerized predictive system that estimates the aquatic toxicity of chemicals. The program estimates a chemical compound's toxicity to aquatic organisms such as fish, aquatic invertebrates, and aquatic plants by using *structure activity relationships (SARs)* [67].

The toxicity data used to build the SARs are collected from publicly available experimental studies and confidential submissions submitted to the US EPA New Chemicals Program. The SARs in ECOSAR express correlations between a compound's physicochemical properties and its aquatic toxicity within specific chemical classes. ECOSAR contains a library of class-based SARs for predicting aquatic toxicity, overlaid with an expert decision tree based on expert rules for selecting the appropriate chemical class for the compound. ECOSAR presently includes more than 120 chemical classes and allows access to over 440 SARs. The SARs estimate acute and chronic toxicity endpoints for fish, aquatic invertebrates, and green algae (species used in standard US EPA New Chemicals Program aquatic toxicity profiles) along with limited SARs for other salt water and terrestrial species, where data were available.

Besides ECOSAR, numerous exposure assessment tools use an estimation program interface (EPI) suite to run estimation tools [68]. These are:

- KOWWIN™: Estimates the log octanol-water partition coefficient, log K_{ow}, of chemicals using an atom/fragment contribution method.
- AOPWIN™: Estimates the gas-phase reaction rate for the reaction between the most prevalent atmospheric oxidant, hydroxyl radicals, and a chemical. Gas-phase ozone radical reaction rates are also estimated for olefins and acetylenes.

(Continued)

In addition, AOPWIN™ informs the user if nitrate radical reaction will be important. Atmospheric half-lives for each chemical are automatically calculated using assumed average hydroxyl radical and ozone concentrations.

- HENRYWIN™: Calculates the Henry's law constant (air/water partition coefficient) using both the group contribution and the bond contribution methods.
- MPBPWIN™: Melting point, boiling point, and vapor pressure of organic chemicals are estimated using a combination of techniques. Included is the subcooled liquid vapor pressure, which is the vapor pressure a solid would have if it were liquid at room temperature. It is important in fate modeling.
- BIOWIN™: Estimates aerobic and anaerobic biodegradability of organic chemicals using seven different models. Two of these are the original Biodegradation Probability Program (BPP™). The current model estimates anaerobic biodegradation potential.
- BioHCwin: Estimates biodegradation half-life for compounds containing only carbon and hydrogen (i.e. hydrocarbons).
- KOCWIN™: Formerly called PCKOCWIN™, this program estimates the organic carbon-normalized sorption coefficient for soil and sediment; i.e. K_{oc}. K_{oc} is estimated using two different models: the Sabljic molecular connectivity method with improved correction factors; and the traditional method based on log K_{ow}.
- WSKOWWIN™: Estimates an octanol-water partition coefficient using the KOWWIN™ program, then estimates a chemical's water solubility from this value and applicable correction factors if any.
- WATERNT™: Estimates water solubility directly using a "fragment constant" method similar to that used in the KOWWIN™ program.
- BCFBAF™: Formerly called BCFWIN™, this program estimates fish bioconcentration factor and its logarithm using two different methods. The first is the traditional regression based on log K_{ow} plus any applicable correction factors, and is analogous to the WSKOWWIN™ method. The second is the Arnot–Gobas method, which calculates BCF from mechanistic first principles. BCFBAF also incorporates prediction of apparent metabolism half-life in fish, and estimates BCF and BAF for three trophic levels.
- HYDROWIN™: Estimates aqueous hydrolysis rate constants and half-lives for the following chemical classes: esters, carbamates, epoxides, halomethanes, selected alkyl halides, and phosphorus esters. Estimates rate constants for acid- and base-catalyzed hydrolysis, but with the exception of phosphorus esters, not neutral hydrolysis. In addition, HYDROWIN™ identifies a variety of chemical structure classes for which hydrolysis may be significant (e.g. carbamates) and gives relevant experimental data.
- KOAWIN: Estimates KOA, the octanol/air partition coefficient, using the ratio of the octanol/water partition coefficient (K_{ow}) from KOWWIN™ and the dimensionless Henry's law constant (KAW) from HENRYWIN™. KOA has multiple uses in chemical assessment.
- AEROWIN™: Estimates the fraction of airborne substance sorbed to airborne particulates, i.e. the parameter phi (φ), using three different methods. AEROWIN™ results are also displayed with AOPWIN™ output as an aid in interpretation of the latter.
- WVOLWIN™: Estimates the rate of volatilization of a chemical from rivers and lakes; and calculates the half-life for these two processes from their rates. The model makes certain default assumptions with respect to water body depth, wind velocity, etc.
- STPWIN™: Using several outputs from EPI Suite™, this program predicts the removal of a chemical in a typical activated sludge-based sewage treatment plant. Values are given for total removal and three processes that may contribute to removal: biodegradation, sorption to sludge, and air stripping. The program assumes a standard system design and set of default operating conditions.
- LEV3EPI™: This program contains a level III multimedia fugacity model and predicts partitioning of chemicals among air, soil, sediment, and water under steady state conditions for a default model "environment". Some (but not all) system default values can be changed by the user.

EPI Suite™ provides users with screening-level estimates of physical/chemical and environmental fate properties. These properties are the building blocks of exposure assessment. Before using EPI Suite™, users should first determine whether any suitable data are available from the literature (e.g., Merck Index, Beilstein). This is facilitated by a database of >40,000 chemicals (called PHYSPROP©) that is included in the EPI Suite™ software. Dermwin™, a program that estimates the dermal permeability coefficient K_p, is included in EPI Suite™. ECOSAR™ is a program that predicts aquatic toxicity and is included in EPI Suite™.

So, let us compare the results for biodegradation using BIOWIN for our two original compounds, 1-propanol and diphenyl ether, using one of these models. Table 7.8 provides the output for 1-propanol and Table 7.9 provides this information for diphenyl ether. Note that BIOWIN predicts ultimate degradation for 1-propanol to be almost equal to our earlier prediction, i.e. the BIOWIN3 = 3.2263, which calls for weeks to reach ultimate degradation. Likewise, the BIOWIN3 value for diphenyl ether in Table 7.9 is very close to our original estimate based on the fragments, i.e. BIOWIN3 = 2.8089, with ultimate degradation expected to be reached in weeks.

We can compare these values to another compound that would be expected to be more recalcitrant. Perfluorooctane sulphonate (PFOS) is a synthetic surfactant that has been found around the world. PFOS and other perfluoride compounds have caused concern because they appear to be building up in the environment (e.g. bald eagle plasma concentrations >2 ppm, polar bear liver concentrations >3 ppm and

Table 7.8 **BIOWIN output for biodegradation estimates for 1-propanol**

```
SMILES (Simplified Molecular Input Line Entry System): OCCC
CHEM   : 1-Propanol
MOL FOR: C3 H8 O1
MOL WT : 60.10
--------------------------- BIOWIN v4.10 Results ---------------------------

   Biowin1 (Linear Model Prediction)     :  Biodegrades Fast
   Biowin2 (Non-Linear Model Prediction):  Biodegrades Fast
   Biowin3 (Ultimate Biodegradation Timeframe):  Weeks
   Biowin4 (Primary Biodegradation Timeframe):  Days
   Biowin5 (MITI Linear Model Prediction)     :  Readily Degradable
   Biowin6 (MITI Non-Linear Model Prediction):  Readily Degradable
   Biowin7 (Anaerobic Model Prediction):  Biodegrades Fast
   Ready Biodegradability Prediction:  YES

------+-----+--------------------------------------------+---------+---------
 TYPE | NUM |       Biowin1 FRAGMENT DESCRIPTION         |  COEFF  |  VALUE
------+-----+--------------------------------------------+---------+---------
 Frag |  1  |  Aliphatic alcohol  [-OH]                  | 0.1587  |  0.1587
 MolWt|  *  |  Molecular Weight Parameter                |         | -0.0286
 Const|  *  |  Equation Constant                         |         |  0.7475
============+============================================+=========+=========
   RESULT   |    Biowin1 (Linear Biodeg Probability)     |         |  0.8777
============+============================================+=========+=========

------+-----+--------------------------------------------+---------+---------
 TYPE | NUM |       Biowin2 FRAGMENT DESCRIPTION         |  COEFF  |  VALUE
------+-----+--------------------------------------------+---------+---------
 Frag |  1  |  Aliphatic alcohol  [-OH]                  | 1.1178  |  1.1178
 MolWt|  *  |  Molecular Weight Parameter                |         | -0.8534
============+============================================+=========+=========
   RESULT   |  Biowin2 (Non-Linear Biodeg Probability)   |         |  0.9635
============+============================================+=========+=========

 A probability greater than or equal to 0.5 indicates --> Biodegrades Fast
 A probability less than 0.5 indicates --> Does NOT Biodegrade Fast

------+-----+--------------------------------------------+---------+---------
 TYPE | NUM |       Biowin3 FRAGMENT DESCRIPTION         |  COEFF  |  VALUE
------+-----+--------------------------------------------+---------+---------
 Frag |  1  |  Aliphatic alcohol  [-OH]                  | 0.1600  |  0.1600
 MolWt|  *  |  Molecular Weight Parameter                |         | -0.1328
 Const|  *  |  Equation Constant                         |         |  3.1992
============+============================================+=========+=========
   RESULT   |  Biowin3 (Survey Model - Ultimate Biodeg)  |         |  3.2263
============+============================================+=========+=========

------+-----+--------------------------------------------+---------+---------
 TYPE | NUM |       Biowin4 FRAGMENT DESCRIPTION         |  COEFF  |  VALUE
------+-----+--------------------------------------------+---------+---------
 Frag |  1  |  Aliphatic alcohol  [-OH]                  | 0.1294  |  0.1294
 MolWt|  *  |  Molecular Weight Parameter                |         | -0.0867
 Const|  *  |  Equation Constant                         |         |  3.8477
============+============================================+=========+=========
   RESULT   |   Biowin4 (Survey Model - Primary Biodeg)  |         |  3.8905
============+============================================+=========+=========

   Result Classification:   5.00 -> hours     4.00 -> days    3.00 -> weeks
    (Primary & Ultimate)    2.00 -> months    1.00 -> longer
```

(Continued)

Table 7.8 BIOWIN output for biodegradation estimates for 1-propanol—cont'd

TYPE	NUM	Biowin5 FRAGMENT DESCRIPTION	COEFF	VALUE
Frag	1	Aliphatic alcohol [-OH]	0.1611	0.1611
Frag	1	Methyl [-CH3]	0.0004	0.0004
Frag	2	-CH2- [linear]	0.0494	0.0988
MolWt	*	Molecular Weight Parameter		-0.1788
Const	*	Equation Constant		0.7121
RESULT		Biowin5 (MITI Linear Biodeg Probability)		0.7937

TYPE	NUM	Biowin6 FRAGMENT DESCRIPTION	COEFF	VALUE
Frag	1	Aliphatic alcohol [-OH]	1.0041	1.0041
Frag	1	Methyl [-CH3]	0.0194	0.0194
Frag	2	-CH2- [linear]	0.4295	0.8590
MolWt	*	Molecular Weight Parameter		-1.7349
RESULT		Biowin6 (MITI Non-Linear Biodeg Probability)		0.9354

A probability greater than or equal to 0.5 indicates --> Readily Degradable
A probability less than 0.5 indicates --> NOT Readily Degradable

TYPE	NUM	Biowin7 FRAGMENT DESCRIPTION	COEFF	VALUE
Frag	1	Aliphatic alcohol [-OH]	0.1328	0.1328
Frag	1	Methyl [-CH3]	-0.0796	-0.0796
Frag	2	-CH2- [linear]	0.0260	0.0520
Const	*	Equation Constant		0.8361
RESULT		Biowin7 (Anaerobic Linear Biodeg Prob)		0.9413

A probability greater than or equal to 0.5 indicates --> Biodegrades Fast
A probability less than 0.5 indicates --> Does NOT Biodegrade Fast

Ready Biodegradability Prediction: (YES or NO)

Criteria for the YES or NO prediction: If the Biowin3 (ultimate survey model) result is "weeks" or faster (i.e. "days", "days to weeks", or "weeks" AND the Biowin5 (MITI linear model) probability is >= 0.5, then the prediction is YES (readily biodegradable). If this condition is not satisfied, the prediction is NO (not readily biodegradable). This method is based on application of Bayesian analysis to ready biodegradation data (see Help). Biowin5 and 6 also predict ready biodegradability, but for degradation in the OECD301C test only; using data from the Chemicals Evaluation and Research Institute Japan (CERIJ) database.

Source: US Environmental Protection Agency (2009). *On-Line BIOWIN™ User's Guide (v4.10)*; http://www.epa.gov/oppt/exposure/pubs/episuite.htm; accessed November 23, 2009.

Table 7.9 **BIOWIN output for biodegradation estimates for diphenyl ether (1,1'-oxybis-benzene)**

```
SMILES : O(c(cccc1)c1)c(cccc2)c2
CHEM   : Benzene, 1,1'-oxybis-
MOL FOR: C12 H10 O1
MOL WT : 170.21
--------------------------- BIOWIN v4.10 Results ----------------------------

   Biowin1 (Linear Model Prediction)     :  Biodegrades Fast
   Biowin2 (Non-Linear Model Prediction):  Biodegrades Fast
   Biowin3 (Ultimate Biodegradation Timeframe):  Weeks
   Biowin4 (Primary Biodegradation Timeframe):  Days-Weeks
   Biowin5 (MITI Linear Model Prediction)    :  Not Readily Degradable
   Biowin6 (MITI Non-Linear Model Prediction):  Readily Degradable
   Biowin7 (Anaerobic Model Prediction):  Does Not Biodegrade Fast
   Ready Biodegradability Prediction:  NO
```

TYPE	NUM	Biowin1 FRAGMENT DESCRIPTION	COEFF	VALUE
Frag	1	Aromatic ether [-O-aromatic carbon]	0.1319	0.1319
Frag	2	Unsubstituted phenyl group (C6H5-)	0.1281	0.2562
MolWt	*	Molecular Weight Parameter		-0.0810
Const	*	Equation Constant		0.7475
RESULT		Biowin1 (Linear Biodeg Probability)		1.0546

TYPE	NUM	Biowin2 FRAGMENT DESCRIPTION	COEFF	VALUE
Frag	1	Aromatic ether [-O-aromatic carbon]	2.2483	2.2483
Frag	2	Unsubstituted phenyl group (C6H5-)	1.7991	3.5982
MolWt	*	Molecular Weight Parameter		-2.4170
RESULT		Biowin2 (Non-Linear Biodeg Probability)		0.9984

A probability greater than or equal to 0.5 indicates --> Biodegrades Fast
A probability less than 0.5 indicates --> Does NOT Biodegrade Fast

TYPE	NUM	Biowin3 FRAGMENT DESCRIPTION	COEFF	VALUE
Frag	1	Aromatic ether [-O-aromatic carbon]	-0.0581	-0.0581
Frag	2	Unsubstituted phenyl group (C6H5-)	0.0220	0.0440
MolWt	*	Molecular Weight Parameter		-0.3761
Const	*	Equation Constant		3.1992
RESULT		**Biowin3 (Survey Model - Ultimate Biodeg)**		**2.8089**

(Continued)

Table 7.9 **BIOWIN output for biodegradation estimates for diphenyl ether (1,1'-oxybis-benzene)—cont'd**

```
------+-----+--------------------------------------------+---------+---------
 TYPE | NUM |        Biowin4 FRAGMENT DESCRIPTION          |  COEFF  |  VALUE
------+-----+--------------------------------------------+---------+---------
 Frag |  1  |  Aromatic ether  [-O-aromatic carbon]      | 0.0771  |  0.0771
 Frag |  2  |  Unsubstituted phenyl group (C6H5-)        | 0.0049  |  0.0098
 MolWt|  *  |  Molecular Weight Parameter                |         | -0.2456
 Const|  *  |  Equation Constant                         |         |  3.8477
============+============================================+=========+=========
   RESULT   |   Biowin4 (Survey Model - Primary Biodeg)  |         |  3.6891
============+============================================+=========+=========

 Result Classification:   5.00 -> hours     4.00 -> days    3.00 -> weeks
  (Primary & Ultimate)    2.00 -> months    1.00 -> longer

------+-----+--------------------------------------------+---------+---------
 TYPE | NUM |        Biowin5 FRAGMENT DESCRIPTION          |  COEFF  |  VALUE
------+-----+--------------------------------------------+---------+---------
 Frag |  1  |  Aromatic ether  [-O-aromatic carbon]      | 0.1952  |  0.1952
 Frag | 10  |  Aromatic-H                                | 0.0082  |  0.0822
 MolWt|  *  |  Molecular Weight Parameter                |         | -0.5064
 Const|  *  |  Equation Constant                         |         |  0.7121
============+============================================+=========+=========
   RESULT   |  Biowin5 (MITI Linear Biodeg Probability)  |         |  0.4832
============+============================================+=========+=========

------+-----+--------------------------------------------+---------+---------
 TYPE | NUM |        Biowin6 FRAGMENT DESCRIPTION          |  COEFF  |  VALUE
------+-----+--------------------------------------------+---------+---------
 Frag |  1  |  Aromatic ether  [-O-aromatic carbon]      | 1.3227  |  1.3227
 Frag | 10  |  Aromatic-H                                | 0.1201  |  1.2014
 MolWt|  *  |  Molecular Weight Parameter                |         | -4.9138
============+============================================+=========+=========
   RESULT   |Biowin6 (MITI Non-Linear Biodeg Probability)|         |  0.5339
============+============================================+=========+=========

 A probability greater than or equal to 0.5 indicates --> Readily Degradable
 A probability less than 0.5 indicates --> NOT Readily Degradable

------+-----+--------------------------------------------+---------+---------
 TYPE | NUM |        Biowin7 FRAGMENT DESCRIPTION          |  COEFF  |  VALUE
------+-----+--------------------------------------------+---------+---------
 Frag |  1  |  Aromatic ether  [-O-aromatic carbon]      | 0.1780  |  0.1780
 Frag |  2  |  Unsubstituted phenyl group (C6H5-)        | 0.2182  |  0.4364
 Frag | 10  |  Aromatic-H                                | -0.0954 | -0.9543
 Const|  *  |  Equation Constant                         |         |  0.8361
============+============================================+=========+=========
   RESULT   |   Biowin7 (Anaerobic Linear Biodeg Prob)   |         |  0.4962
============+============================================+=========+=========

 A probability greater than or equal to 0.5 indicates --> Biodegrades Fast
 A probability less than 0.5 indicates --> Does NOT Biodegrade Fast
```

Table 7.9 **BIOWIN output for biodegradation estimates for diphenyl ether (1,1'-oxybis-benzene)—cont'd**

```
Ready Biodegradability Prediction: (YES or NO)
----------------------------------------------

 Criteria for the YES or NO prediction: If the Biowin3 (ultimate survey
 model) result is "weeks" or faster (i.e. "days", "days to weeks", or
 "weeks" AND the Biowin5 (MITI linear model) probability is >= 0.5, then
 the prediction is YES (readily biodegradable). If this condition is not
 satisfied, the prediction is NO (not readily biodegradable). This method
 is based on application of Bayesian analysis to ready biodegradation data
 (see Help). Biowin5 and 6 also predict ready biodegradability, but for
 degradation in the OECD301C test only; using data from the Chemicals
 Evaluation and Research Institute Japan (CERIJ) database.
```

Source: US Environmental Protection Agency (2009). On-Line BIOWIN™ User's Guide (v4.10); http://www.epa.gov/oppt/exposure/pubs/episuite.htm; accessed on November 23, 2009.

mink liver concentrations >59 ppm) [69]. Table 7.10 provides the estimated biodegradation for PFOS, which indicates that the ultimate degradation will take many months, since the modeled value is 0.2887.

Seminar Questions

Estimate the biodegradation potential for the four pesticides in Figure 6.14. How do the natural pesticides compare to the synthesized versions? How do the I values compare to DDT and chlorpyrifos?

The methodologies presented in this chapter represent only a small fraction of possible environmental degradation pathways. Environmental biotechnology is more often than not "data poor." How can limited chemical and biological data regarding a released substance be compensated for when conducting an overall assessment of environmental persistence of these substances?

What are the likely weaknesses in using a model such as the one discussed here to predict biodegradation of endocrine disrupting compounds? How can a risk assessment be strengthened in light of these weaknesses?

How may the coefficients of degradation in Table 7.6 differ for natural microbial strains versus genetically modified strains in the same species? Give two examples to support your conjecture.

Table 7.10 **BIOWIN output for biodegradation estimates for perfluorooctanesulfonic acid**

```
SMILES : O=S(=O)(O)C(F)(F)C(F)(F)C(F)(F)C(F)(F)C(F)(F)C(F)(F)C(F)(F)C(F)(F)F
CHEM   : 1-Octanesulfonic acid, 1,1,2,2,3,3,4,4,5,5,6,6,7,7,8,8,8-heptadecafluoro-
MOL FOR: C8 H1 F17 O3 S1
MOL WT : 500.13
--------------------------- BIOWIN v4.10 Results ----------------------------

   Biowin1 (Linear Model Prediction)     :  Does Not Biodegrade Fast
   Biowin2 (Non-Linear Model Prediction):  Does Not Biodegrade Fast
   Biowin3 (Ultimate Biodegradation Timeframe):  Recalcitrant
   Biowin4 (Primary Biodegradation Timeframe):  Months
   Biowin5 (MITI Linear Model Prediction)     :  Not Readily Degradable
   Biowin6 (MITI Non-Linear Model Prediction):  Not Readily Degradable
   Biowin7 (Anaerobic Model Prediction):  Does Not Biodegrade Fast
   Ready Biodegradability Prediction:  NO
```

(Continued)

Table 7.10 BIOWIN output for biodegradation estimates for perfluorooctanesulfonic acid—cont'd

TYPE	NUM	Biowin1 FRAGMENT DESCRIPTION	COEFF	VALUE
Frag	7	Carbon with 4 single bonds & no hydrogens	-0.1839	-1.2875
Frag	1	Sulfonic acid / salt -> aliphatic attach	0.1084	0.1084
Frag	1	Trifluoromethyl group [-CF3]	-0.5204	-0.5204
MolWt	*	Molecular Weight Parameter		-0.2381
Const	*	Equation Constant		0.7475
RESULT		Biowin1 (Linear Biodeg Probability)		-1.1901

TYPE	NUM	Biowin2 FRAGMENT DESCRIPTION	COEFF	VALUE
Frag	7	Carbon with 4 single bonds & no hydrogens	-1.7232	-12.0624
Frag	1	Sulfonic acid / salt -> aliphatic attach	6.8331	6.8331
Frag	1	Trifluoromethyl group [-CF3]	-5.6696	-5.6696
MolWt	*	Molecular Weight Parameter		-7.1018
RESULT		Biowin2 (Non-Linear Biodeg Probability)		0.0000

A probability greater than or equal to 0.5 indicates --> Biodegrades Fast
A probability less than 0.5 indicates --> Does NOT Biodegrade Fast

TYPE	NUM	Biowin3 FRAGMENT DESCRIPTION	COEFF	VALUE
Frag	7	Carbon with 4 single bonds & no hydrogens	-0.2121	-1.4848
Frag	1	Sulfonic acid / salt -> aliphatic attach	0.1926	0.1926
Frag	1	Trifluoromethyl group [-CF3]	-0.5130	-0.5130
MolWt	*	Molecular Weight Parameter		-1.1052
Const	*	Equation Constant		3.1992
RESULT		**Biowin3 (Survey Model - Ultimate Biodeg)**		**0.2887**

TYPE	NUM	Biowin4 FRAGMENT DESCRIPTION	COEFF	VALUE
Frag	7	Carbon with 4 single bonds & no hydrogens	-0.1534	-1.0741
Frag	1	Sulfonic acid / salt -> aliphatic attach	0.1771	0.1771
Frag	1	Trifluoromethyl group [-CF3]	-0.2744	-0.2744
MolWt	*	Molecular Weight Parameter		-0.7216
Const	*	Equation Constant		3.8477
RESULT		Biowin4 (Survey Model - Primary Biodeg)		1.9548

Table 7.10 BIOWIN output for biodegradation estimates for perfluorooctanesulfonic acid—cont'd

```
 Result Classification:   5.00 -> hours     4.00 -> days    3.00 -> weeks
  (Primary & Ultimate)    2.00 -> months    1.00 -> longer

------+-----+--------------------------------------------+---------+---------
 TYPE | NUM |        Biowin5 FRAGMENT DESCRIPTION         |  COEFF  |  VALUE
------+-----+--------------------------------------------+---------+---------
 Frag |  7  |  Carbon with 4 single bonds & no hydrogens | 0.0676  |  0.4733
 Frag | 17  |  Fluorine   [-F]                           | 0.0174  |  0.2954
 MolWt|  *  |  Molecular Weight Parameter                |         | -1.4879
 Const|  *  |  Equation Constant                         |         |  0.7121
============+============================================+=========+=========
   RESULT   |  Biowin5 (MITI Linear Biodeg Probability)  |         | -0.0070
============+============================================+=========+=========

------+-----+--------------------------------------------+---------+---------
 TYPE | NUM |        Biowin6 FRAGMENT DESCRIPTION         |  COEFF  |  VALUE
------+-----+--------------------------------------------+---------+---------
 Frag |  7  |  Carbon with 4 single bonds & no hydrogens | 0.3990  |  2.7929
 Frag | 17  |  Fluorine   [-F]                           | -3.9878 |-67.7934
 MolWt|  *  |  Molecular Weight Parameter                |         |-14.4381
============+============================================+=========+=========
   RESULT   |Biowin6 (MITI Non-Linear Biodeg Probability)|         |  0.0000
============+============================================+=========+=========

 A probability greater than or equal to 0.5 indicates --> Readily Degradable
 A probability less than 0.5 indicates --> NOT Readily Degradable

------+-----+--------------------------------------------+---------+---------
 TYPE | NUM |        Biowin7 FRAGMENT DESCRIPTION         |  COEFF  |  VALUE
------+-----+--------------------------------------------+---------+---------
 Frag |  7  |  Carbon with 4 single bonds & no hydrogens | -0.3342 | -2.3396
 Frag |  1  |  Sulfonic acid / salt -> aliphatic attach  | 0.0000  |  0.0000
 Frag |  1  |  Trifluoromethyl group    [-CF3]           | 0.0000  |  0.0000
 Frag | 17  |  Fluorine   [-F]                           | 0.0000  |  0.0000
 Const|  *  |  Equation Constant                         |         |  0.8361
============+============================================+=========+=========
   RESULT   |   Biowin7 (Anaerobic Linear Biodeg Prob)   |         | -1.5035
============+============================================+=========+=========

 A probability greater than or equal to 0.5 indicates --> Biodegrades Fast
 A probability less than 0.5 indicates --> Does NOT Biodegrade Fast

Ready Biodegradability Prediction: (YES or NO)
----------------------------------------------
 Criteria for the YES or NO prediction: If the Biowin3 (ultimate survey
 model) result is "weeks" or faster (i.e. "days", "days to weeks", or
 "weeks" AND the Biowin5 (MITI linear model) probability is >= 0.5, then
 the prediction is YES (readily biodegradable). If this condition is not
 satisfied, the prediction is NO (not readily biodegradable). This method
 is based on application of Bayesian analysis to ready biodegradation data
 (see Help). Biowin5 and 6 also predict ready biodegradability, but for
 degradation in the OECD301C test only; using data from the Chemicals
 Evaluation and Research Institute Japan (CERIJ) database.
```

Source: US Environmental Protection Agency (2009). On-Line BIOWIN™ User's Guide (v4.10); http://www.epa.gov/oppt/exposure/pubs/episuite.htm; accessed November 23, 2009.

REVIEW QUESTIONS

Explain why a trickling filter facility is considered to be a mixed environmental bioreactor. How might this phenomenon be optimized when treating a halogenated compound? Could it work for an organometallic compound? Why or why not?

Consider the biochemodynamic information provided for the preliminary site investigation of an abandoned chemical plant (Table 7.11). Give reasons for expecting neither, either, or both sites to be amenable to biological treatment methods.

A database has been compiled during site characterizations for certain soil microbial classes to document intrinsic bioremediation (natural attenuation) of a chlorinated hydrocarbon. The data show that the biodegradation of this compound occurs in direct proportion to the compound's concentration. That is, it follows first-order kinetics.

Calculate the biodegradation rate constant for this compound if the highest measured concentration is 90 $\mu g\,L^{-1}$ upgradient (at point A) and 450 ng L^{-1} (corrected for dilution) downgradient (at point B 1330 m south of point A) in a groundwater plume moving at $10^{-5}\,m\,sec^{-1}$.

How does the life cycle of bacteria help to explain the changes in rate orders of *in situ* bioremediation?

Explain the shape of the curves in Figure 7.9. What does this mean to the bioengineer managing a bioremediation project?

If ethylbenzene, rather than toluene, degradation is being estimated using the model depicted in Figure 7.12, what changes would need to be made? Why?

If the model in Figure 7.12 showed that toluene was being degraded at a rate of 100 $\mu g\,L^{-1}\,d^{-1}$ based on NO emissions, would benzene degradation be faster or slower based on the same NO flux? Why?

The Monod equation is referred to as empirically derived. Why is it generally reliable at predicting degradation? What conditions might detract from its accuracy?

How does cometabolism work? Explain the role of electron acceptors in this process.

Why is the sludge blanket important in certain anaerobic reactors?

What does the oxygen sag curve tell us about designing an aerobic digester? ... an anaerobic digester?

You are working with a strain of bacteria to increase the biodegradation rate of the polycyclic aromatic hydrocarbon, benzo(a)pyrene. How would you use the information in Tables 2.9 and 7.6 to assist you in your environmental biotechnological research and the operations that will follow? How might your use of this information and your methodology and approach differ in designing an *in situ* bioremediation project, an on-site bioreactor system and off-site bioreactor system?

What is the difference between a Tier 1 and a Tier 2 assay for endocrine disrupting compounds? How would these assays differ for a neurotoxic compound? ... a reproduction/developmentally toxic compound?

Table 7.11 Comparison of site conditions from a hypothetical site investigation (hypothetical information)

Site ID	Contaminants detected to date	Soil type	Microbial enzyme type	Extent of contamination
1	Short chain, non-homogenate hydrocarbons and amines	Sandy	Primary	Highly concentrated within a 50 m^2 area
2	Multichlorinated C13–C20 hydrocarbons and PCBs	Clayey	Cometabolism	Widely dispersed and unevenly distributed pockets of contamination over a 0.5 km^2 area

Table 7.12 Selected examples of degradation efficiencies of dichloromethane removal in conventional (BF) and trickling (BTF) biofilters

Reactor configuration	Filter bed	Maximum EC ($g\,m^{-3}\,h^{-1}$)
BF	Compost–perlite–crushed oyster shell	10.3
BF	Peat	6.4
BTF	Ceramix Novalox saddles	157
BTF	Polypropylene packing	103.5
BTF	Polypropylene saddles	152
BTF	PVC	102
BTF	Lava rock	160

Source: C. Kennes, E.R. Rene and M.C. Veiga (2009). Bioprocesses for air pollution control. *Journal of Chemical Technology and Biotechnology* 84 (10): 1419–1436.

Table 7.12 gives the biodegradation rates observed for selected biotechnologies:

Treatment efficiency can be represented by measuring the amount of specific substance that is eliminated, in this case, dichloromethane. The elimination capacity (EC) in $g\,m^{-3}\,h^{-1}$ is calculated as:

$$EC = \frac{Q \times (S_{in} - S_{out})}{V} \tag{7.34}$$

where Q is the gas flow rate ($m^3\,h^{-1}$), V is the volume of the reactor (m^3) and S_{in} and S_{out} are, respectively, the inlet and outlet pollutant concentrations ($g\,m^{-3}$).

Give at least two possible reasons for the differences in elimination capacities. How might these rates be improved?

NOTES AND COMMENTARY

1. Battelle Pacific Northwest Laboratory (2009). Basics of *in situ* bioremediation (ISB). Battelle Chlorinated Solvent Bioremediation Design Service. http://bioprocess.pnl.gov/isb_defn.htm; accessed July 20, 2009.
2. L. P. Wackett (1996). Co-metabolism: is the emperor wearing any clothes? *Current Opinion in Biotechnology* 7(3): 321–325.
3. The principal source for the cell discussion is National Institutes of Health, National Institute of General Medical Science.
4. US Environmental Protection Agency (2000). *Fact Sheet: Dissolved Oxygen (Saltwater): Cape Cod to Cape Hatteras.* EPA-822-F-99-009, Washington, DC.
5. State of Georgia (2003). Watershed Protection Plan Development Guidebook.
6. C. P. Gerba and I. L. Pepper (2009). Wastewater treatment and biosolids reuse. In: R. M. Maier, I. L. Pepper and C. P. Gerba (Eds), *Environmental Microbiology* (2nd Edition). Elsevier Academic Press, Burlington, MA.
7. R. L. Raymond, V. W. Jamison and J. O. Hudson Jr. (1975). *Final Report on Beneficial Stimulation of Bacterial Activity in Groundwater Containing Petroleum Products.* American Petroleum Institute, Washington, DC.
8. N. P. Cheremisinoff (1996). *Biotechnology for Waste and Wastewater Treatment.* Noyes Publications, Westwood, NJ.
9. A. J. Dragt and J. van Ham. (Eds). (1991). *Biotechniques for Air Pollution Abatement and Odour Control Policies. Proceedings of an International Symposium.* Maastricht, The Netherlands, October 27–29, 1991.
10. S. S. Suthersan (1997). *Remediation Engineering: Design Concepts.* CRC Press, Inc., Boca Raton, FL, See pp.143–144.
11. See S. S. Suthersan (1997). *Remediation Engineering: Design Concepts.* CRC Press, Inc, Boca Raton, FL.
12. For example, see D. Grasso (1993). *Hazardous Waste Site Remediation: Source Control.* CRC Press, Inc., Boca Raton, FL
13. Bioengineers owe a debt to the economists for the concept of the Law of Diminishing Returns, which we see at work here.
14. M. Alexander (1994). *Biodegradation and Bioremediation* (2nd Edition). Academic Press, San Diego, CA.
15. US Environmental Protection Agency (2008). *Natural attenuation of the lead scavengers 1,2-Dibromoethane (EDB) and 1,2-Dichloroethane (1,2-DCA) at motor fuel release sites and implications for risk management.* Report No. EPA 600/R-08/107, Ada, OK.
16. The environmental literature is not completely consistent in its use of the term "phase." The term may be interpreted to mean exactly what physical chemists have defined, i.e., phase is the distinct state of matter in

a physical system. Such matter is identical in chemical composition and physical state and separated from other material by the phase boundary as represented in Figure 7.8. However, phase may be more akin in environmental contexts to "fractional solubility," meaning that a substance can be found in varying amounts in different solvents. One of the most common examples of fractional solubility is demarcation between the concentration of a contaminant in the "organic phase" and "aqueous phase" as represented by the octanol–water partition coefficient (K_{ow}), discussed in this chapter. Another less commonly applied connotation of environmental phases is its synonymous usage with environmental "media" or "compartments." For example, one may hear environmental engineers discussing the transport of a contaminant from the water phase to the air phase, or the soil phase to the water and air phases. Context, in fact, is important in many environmental discussions, and reading in context is often the only way to know what a publication means. Since environmental science, engineering, and technology draws from so many fields, numerous terms (e.g., particle) have various meanings which can only be understood within the context of the specific discussion. Furthermore, even within the environmental disciplines, uses of terms like phase will vary. For example, air experts may use the physical chemists' definition when discussing phase distributions within a stack gas, but may use the solubility fractionation definition when discussing the transformation and transport of a pollutant between a raindrop and the air (i.e., air-water partitioning). Likewise, soil and water scientists may apply the physicochemical definition of phases when in the laboratory (e.g., measuring the amount of a liquid or solid-phase analyte volatilizing to the gas phase from the gas chromatograph's column and carried as a gas to the detector). However, in the field they may refer to phase distribution as the movements among soil, water, and air compartments.

17. W. G. Whitman (1923). The two-film theory of gas absorption. *Chemical and Metallurgical Engineering* 29: 147.
18. For an excellent discussion of these theories, see Several other theories predict interphase transfer, including other film models, as well as penetration, and surface renewal models. Chapter 4 of W.J. Weber, Jr. and F.A. DiGiano (1996). *Process Dynamics in Environmental Systems.* John Wiley & Sons, New York, NY.
19. For decades books have been published that focus on the current understandings of the science, engineering, and technology of biological waste treatment. See, for example, Metcalf and Eddy as revised by G. Tchobanoglous and F. Burton (1991). *Wastewater Engineering.* McGraw-Hill, New York, NY; A. Gaudy and E. Gaudy (1988). *Elements of Bioenvironmental Engineering.* Engineering Press, Inc., San Jose, CA; and J. Peirce, R. Weiner and P. Vesilind (1998). *Environmental Pollution and Control.* Butterworth-Heinemann, Boston, MA. For a particular focus on the biotreatment of hazardous wastes see, for example, C. Haas and R. Ramos (1995). *Hazardous and Industrial Waste Treatment.* Prentice-Hall, Englewood Cliffs, NJ; and C. Wentz (1989). *Hazardous Waste Management.* McGraw-Hill, Inc., New York, NY.
20. US Environmental Protection Agency (1993). *Pilot-scale demonstration of slurry-phase biological reactor for creosote-contaminated soil: Applications analysis report.* Report No. EPA/540/A5-91/009, Cincinnati, OH.
21. S. Wallace (2004). Engineered wetlands lead the way. *Land and Water* 48 (5); http://www.landandwater.com/features/vol48no5/vol48no5_1.html; accessed September 9, 2009.
22. US Environmental Protection Agency (2008). *Technology Primer: Green Remediation: Incorporating Sustainable Environmental Practices into Remediation of Contaminated Site.* Report No. EPA 542-R-08-002, Washington, DC.
23. US Environmental Protection Agency (2009). Principles for greener cleanups; http://www.epa.gov/oswer/greencleanups/principles.html#attachment; accessed September 9, 2009.
24. A source for anaerobic system description is A. S. Bal and N. N. Dhagat (2001). Upflow anaerobic sludge blanket reactor – a review. *Indian Journal of Environmental Health* 43(2): 1–82.
25. This discussion is taken from D. Vallero and C. Brasier (2008). *Sustainable Design: The Science of Sustainability and Green Engineering.* John Wiley & Sons, Hoboken, NJ; and D.A. Vallero (2007). *Fundamentals of Air Pollution,* 4th Edition. Academic Press, Burlington, MA.
26. S. J. Ergas and K. A. Kinney (2000). Biological control systems. In: W. T. Davis (Ed), *Air and Waste Management Association. Air Pollution Control Manual* (2nd Edition). John Wiley & Sons, Inc., New York, NY.
27. Ibid.
28. The source of this discussion is US Environmental Protection Agency (1998). *A Citizen's Guide to Phytoremediation.* Report No. EPA 542-F-98-011, Washington, DC.
29. National Research Council (1989). *Biologic Markers in Reproductive Toxicology.* National Academies Press, Washington, DC.
30. Another connotation of biomarker is that some chemical denotes some ongoing or past biological activity, whether or not it represents a hazard. Geochemists may group compounds as indicators of past or present biota (e.g. crude oil or coal).
31. See T. Bucheli and K. Fent (1995). Induction of cytochrome P450 as a biomarker for environmental contamination in aquatic ecosystems. *Critical Reviews in Environmental Science & Technology* 25: 201–268; and J. Stegeman and M. Hahn (1994). Biochemistry and molecular biology of monooxygenases: Current perspectives on forms, functions, and regulation of cytochrome P450 in aquatic species. In: D. Malins and G. Ostrander (Eds), *Aquatic Toxicology: Molecular, Biochemical, and Cellular Perspectives.* CRC Press, Boca Raton, FL.
32. A. M. Deutschbauer, D. Chivian and A. P. Arkin (2006). Genomics in environmental microbiology. *Current Opinion in Biotechnology* 17(3): 229–235.
33. National Institutes of Health (2006). National Institute of General Medical Sciences. Systems Biology Center Request for Assistance; http://grants.nih.gov/grants/guide/RFA-files/RFA-GM-07-004.html; accessed October 1, 2009.
34 For example, see US Bhalla and R. Iyengar (1999). Emergent properties of networks of biological signaling pathways. *Science* 283: 381–387; and W.W. Chen, B. Schoeberl, P.J. Jasper *et al.* (2009). Input–output behavior of ErbB signaling pathways as revealed by a mass action model trained against dynamic data. *Molecular Systems*

Biology 5: 239. Chen's differential equation-based model, for instance, shows the variance between reductionist and systematic approaches, e.g. receptors that are highly sensitive to ligand concentration due to mitogen-activated protein kinase and phosphatidyl inositol 3-kinase cascades that enhance signals in a nonlinear way, but the individual cascades each behaved very differently in isolation.
35. E. A. Cohen Hubal, A. M. Richard, S. Imran, J. Gallagher, R. Kavlock, J. Blancato and S. Edwards (2008). Exposure science and the US EPA National Center for Computational Toxicology. *Journal of Exposure Science and Environmental Epidemiology*, doi: 10.1038/jes.2008.70.
36. This discussion box is primarily based on D. A. Vallero J. Peirce and K. D. Cho. (2009). Modeling toxic compounds from nitric oxide emission measurements. *Atmospheric Environment* 43: 253–261.
37. R. Conrad (1990). Flux of NOx between soil and atmosphere: importance of soil microbial metabolism. In: N. P. Reusbech and J. Sorenson (Eds), *Denitrification in Soil and Sediment* (pp. 105–128). Plenum, New York, pages.
38. G. L. Hutchinson, W. D. Guenzi, and G. P. Livingston (1993). Soil water controls on aerobic soil emissions of gaseous nitrogen oxides. Soil Biology and Biochemistry 25(1): 1–9; and S. Jousset, R.M. Tabachow and J.J. Peirce (2001). Soil nitric oxide emissions from nitrification and denitrification. *Journal of Environmental Engineering* 127(4): 322–328.
39. J. G. Leahy and R. R. Colwell (1990). Microbial degradation of hydrocarbons in the environment. *Microbiol. Rev.* 54(3): 305–315.
40. B. Zarda, G. Mattison, A. Hess, D. Hahn, P. Hohener and J. Zeyer (1998). Analysis of bacterial and protozoan communities in an aquifer contaminated with monoaromatic hydrocarbons. *FEMS Microbiology Ecology* 27(2): 141–152.
41. See, for example, A. S. Allard and A. H. Neilson (1997). Bioremediation of organic waste sites: a critical review of microbiological aspects. *International Biodeterioration and Biodegradation* 39: 253–285; P.J.J. Alvarez, P.J. Anid and T.M. Vogel (1991). Kinetics of aerobic biodegradation of benzene and toluene in sandy aquifer materials. *Biodegradation* 2(1): 43–51; L.R. Krumholz, E. Caldwell and J.M. Suflita (1996). Biodegradation of BTEX hydrocarbons under anaerobic conditions. In: R.L. Crawford and D.L. Crawford (Eds), *Bioremediation: Principles and Applications* (pp. 61–99). Cambridge University Press, Cambridge, UK.
42. D. X. Li and P. D. Lundegard (1996). Evaluation of subsurface oxygen sensors for remediation monitoring. *Ground Water Monitoring and Remediation* 16(1): 106–111.
43. National Safety Council (2000). *Toluene Chemical Background*. Washington, DC.
44. M. J. Huertas, E. Duque, L. Molina, R. Rossello-Mora, G. Mosqueda, P. Godoy, *et al.* (2000). Tolerance to sudden organic solvent shocks by soil bacteria and characterization of *Pseudomonas putida* strains isolated from toluene polluted sites. *Environmental Science & Technology* 34(16): 3395–3400.
45. US Environmental Protection Agency (2009). Technical Fact Sheet on: Toluene. EPA, Office of Water, Washington, DC; http://www.epa.gov/safewater/pdfs/factsheets/voc/tech/toluene.pdf; accessed July 31, 2009.
46. S. Adams (1998). *The Dilbert Future: Thriving on Business Stupidity in the 21st Century.* Harper Business, New York, NY.
47. See, for example P.C. Johnson, M. W. Kemblowski and J. D. Colthart (1990a) Quantitative analysis of cleanup of hydrocarbon-contaminated soils by in-situ soil venting. *Ground Water 28(3): 413–29;* P.C. Johnson, C.C. Stanley, M.W. Kemblowski, D.L. Byers, and J.D. Colthart (1990b). A practical approach to the design, operation, and monitoring of in site soil-venting systems. *Ground Water Monitoring and Remediation* Spring 159-178; and M.E. Stelljes and G.E. Watkin (1993). Comparison of environmental impacts posed by different hydrocarbon mixtures: a need for site specific composition analysis. In: P.T. Kostecki and E.J. Calabrese (Eds), *Hydrocarbon Contaminated Soils and Groundwater, vol. 3.* Lewis Publishers, Boca Raton, FL, 554.
48. D. Jenkinson (1990). The turnover of organic carbon and nitrogen in soil. Philosophical Transactions of the Royal Society B 329: 361–368; B.J. Bolker, S.W. Pacala and W.J. Parton (1998). Linear analysis of soil decomposition: Insights from the century model. *Ecological Applications* 8(2): 425–439; A. Porporato, P. D'Odorico, F. Laio and I. Rodriguez-Iturbe (2003). Hydrologic controls on soil carbon and nitrogen cycles: I. Modeling scheme. *Advances in Water Resources* 26: 45–58.
49. S.J. Birkinshaw and J. Ewen (2000). Nitrogen transformation component for SHETRAN catchment nitrate transport modeling. *Journal of Hydrology* 230: 1–17; A.J. Gusman and M.A. Marino (1999). Analytical modeling of nitrogen dynamics in soils and ground water. *Journal of Irrigation and Drainage Engineering* 125(6): 330–337; S. Hansen, H.E. Jensen and M.J. Shaffer (1995). Development in modeling nitrogen transformations in soils. In P.E. Bacon (ed.), *Nitrogen Fertilization in the Environment.* Marcel Dekker Inc., New York, NY.
50. N. C. Brady and R. R. Weil (2002). *The Nature and Properties of Soils* (13th Edition). Prentice-Hall, Upper Saddle River, NJ.
51. Ibid.
52. Y. H. El-Farhan, K. M. Scow, S. Fan and D. E. Rolston (2000). Kinetics of trichloroethylene cometabolism and toluene biodegradation: Model application to soil batch experiments. *Journal of Environmental Quality* 29: 778–786.
53. Y. C. Li and G. Gupta (1994). Adsorption of hydrocarbons by clay-minerals from gasoline. *Journal of Hazardous Materials* 38(1): 105–112.
54. S. Hansen, H. E. Jensen and M. J. Shaffer (1995). Development in modeling nitrogen transformations in soils. In: P. E. Bacon (Ed), *Nitrogen Fertilization in the Environment.* Marcel Dekker, Inc., New York, NY.
55. A. Porporato, P. D'Odorico, F. Laio and I. Rodriguez-Iturbe (2003). Hydrologic controls on soil carbon and nitrogen cycles: I. Modeling scheme. *Advances in Water Resources* 26: 45–58.
56. E. A. Davidson, P. A. Matson, P. M. Vitouset and J. M. Maass (1993). Processes regulating soil emissions of NO and N_2O in a seasonally dry tropical forest. *Ecology* 74: 130–139.

57. F. Slemr and W. Seiler (1984). Field measurements of NO and NO_2 emissions from fertilized and unfertilized soils. *Journal of Atmospheric Chemistry* 2: 1–24; I.C. Anderson and J.S. Levine (1987). Simultaneous field measurement of biogenic emissions of nitric oxide and nitrous oxide. *Journal of Geophysical Research* 92(D): 965–976; and G.L. Hutchinson, W.D. Guenzi and G.P. Livingston (1993). Soil water controls on aerobic soil emissions of gaseous nitrogen oxides. *Soil Biology and Biochemistry* 25(1): 1–9.
58. K. Cho and J. J. Peirce (2007). Nitric oxide emission and soil microbial activities in toluene contaminated soil. *Journal of Environmental Engineering* 133(2): 237–244.
59. V. de Lorenzo (2009). EC Sponsored Research on Genetically Modified Organisms – Bioremediation. Cleaning up polluted environments: how microbes can help: Introduction. Centro Nacional de Biotecnología, Madrid, Spain; http://ec.europa.eu/research/quality-of-life/gmo/05-bioremediation/05-intro.htm; accessed October 5, 2009.
60. US Environmental Protection Agency (2009). Overview of the April 2009 Final List of Chemicals for Initial Tier 1 Screening. http://www.epa.gov/scipoly/oscpendo/pubs/prioritysetting/final_listfacts.htm; accessed October 9, 2009.
61. US Environmental Protection Agency (2009). Assays under consideration. http://www.epa.gov/scipoly/oscpendo/pubs/assayvalidation/consider.htm.; accessed October 9, 2009.
62. US Environmental Protection Agency (2009). Assays under consideration. http://www.epa.gov/scipoly/oscpendo/pubs/assayvalidation/consider.htm; accessed October 9, 2009.
63. D. T. Allen and D. R. Shonnard (2002). *Green Engineering: Environmentally Conscious Design of Chemical Processes*. Prentice-Hall PTR, Upper Saddle River, NJ.
64. R. S. Boethling, P. H. Howard, W. Meyian, W. Stiteler, J. Beauman and M. Tiradot (1994). Group contribution method for predicting probability and rate of aerobic biodegradation. *Environmental Science & Technology* 28: 459–465.
65. Ibid.
66. US Environmental Protection Agency (2009). *On-Line BIOWIN™ User's Guide* (v4.10); http://www.epa.gov/oppt/exposure/pubs/episuite.htm; accessed November 23, 2009.
67. Ibid.
68. This discussion was taken from: US Environmental Protection Agency (2009). *On-Line BIOWIN™ User's Guide* (v4.10); http://www.epa.gov/oppt/exposure/pubs/episuite.htm; accessed November 23, 2009.
69. M. Houde, J. W. Martin, R. J. Letcher, K. R. Solomon, and D. C. Muir (2006). Biological monitoring of polyfluoroalkyl substances: A review. *Environmental Science & Technology* 40(11): 3463–3473.

CHAPTER 8

Biotechnological Implications: A Systems Approach

Scientists and engineers who engage in environmental processes must have a common understanding and application of living systems, i.e. biosystems. The previous chapter discussed the use of organisms to assist in the degradation and detoxification of substances that pose risks to human populations and the environment. This chapter extends that discussion in view of the fact that such biotechnologies present both challenges and opportunities for environmental science and engineering. That is, most environmental biotechnologies involve optimizing variables pertaining to the biota, the environment in which these organisms undergo metabolism, and the properties of the substances they are using as food sources.

To consider the environmental impacts of biotechnologies, much can be learned from two venues: successful biological treatment processes and biotechnologies in non-environmental applications. The former group has a storied history, with ongoing debates and interactions among experts in abiotic chemistry and the biological sciences. The latter group is quite eclectic, but most of the progress has been oriented toward the production of materials demanded by the marketplace, as will be discussed in greater detail in Chapter 9.

The agricultural, medical, industrial, and other biotechnologies provide lessons in both product-oriented and process-oriented systems. The life cycle of a product, for example, often incorporates and integrates the same biological, chemical, and physical processes and mechanisms as those needed for successful bioremediation, as introduced in Chapter 7. As evidence, the industrial bioreactor relies on sound science to characterize these processes and mechanisms systematically, i.e. information about the organism, the bioreactor's physical conditions, the chemicals being used as carbon sources and the iterative changes. This last feature, iterative changes, is very important. As conditions change (e.g. from aerobic to anaerobic as O_2 is consumed), what may have been a quite hospitable environment for a microbial population could become increasingly toxic. In fact, the system could become sterile (no microbial growth) or, as is often the case, hospitable for a completely different microbial population (e.g. latent spore forming anaerobes will be reproducing exponentially in a low pH, anoxic conditions, as organic acids are formed during fermentation in ethanol production plants, but also in a sanitary landfill cell).

Biotechnologies can be visualized as sets of biological reactions occurring at various scales in the environment [1]. The reactions can lead to desirable results, such as the chemical transformation and ultimate degradation of toxic substances into harmless compounds. Biological reactions may also lead to undesirable results, such as the introduction of genetically modified organisms to an ecosystem or the generation of toxic chemicals [2]. The challenge in predicting the beneficial and detrimental outcomes associated with using organisms in this way is complicated. In most cases, it is not whether biodegradation works, but whether the positive and desired outcome is accompanied by undesired and difficult-to-predict outcomes.

To some extent any biological outcome is unknowable before it occurs. One is very certain, but not completely certain, that a specific antimicrobial compound will be efficacious in treating a targeted organism, when the same compound has *almost always* worked in the past. Note that scientists are or at least should be circumspect about such matters. For example, it is possible that matching the antimicrobial compound to an organism does not work due to environmental conditions (e.g. the pH of the water changes the compound to render it less efficacious). It could also be that the microbial population has developed strains that are now resistant to the microbe. It could also be a combination of these factors. This makes prediction a humbling enterprise.

The scientific community relies heavily on models to extrapolate and to interpolate to build knowledge from what is known. For example, there is seldom enough reliable and relevant measurement data to determine hazards and risks. New tools are being developed to extend these extrapolations and interpolations. As evidence, environmental models are becoming increasingly computational; that is, they increasingly employ the first principles of science (e.g. computational fluid dynamics, computational chemistry and computational toxicology).

In environmental systems, thermodynamic processes occur over a broad domain; having scales ranging from just a few angstroms to global. For example, the processes that lead to a contaminant moving and changing within a bacterium may be very different from those processes that move and change microbial populations and communities. The processes differ further from the small niche in soils to larger systems, e.g. at the lake or river scale, which in turn are different from those processes that cause the contaminant's fate as it crosses the ocean.

In spite of such differences, all changes in human populations and ecosystems are simply manifestations of the first law of thermodynamics, i.e. energy or mass is neither created nor destroyed, only altered in form. This law also dictates that energy and mass within a system must be in balance; that which comes in must equal that which goes out. The mass and energy coming in and going out across two-dimensional surfaces in these systems are known as *fluxes*. These fluxes are measured and yield energy balances within a region in space through which a fluid travels. This region, i.e. the *control volume,* is where balances that occur can take many forms. With any control volume the calculated mass balance is:

$$\begin{bmatrix} \text{Quantity of} \\ \text{mass per unit volume} \\ \text{in a medium} \end{bmatrix} = [\text{Total flux of mass}] + \begin{bmatrix} \text{Rate of production or loss} \\ \text{of mass per unit volume} \\ \text{in a medium} \end{bmatrix} \quad (8.1)$$

Or, stated mathematically:

$$\frac{dM}{dt} = M_{in} - M_{out} \quad (8.2)$$

where M = mass and t = specified time interval.

If we are concerned about a specific chemical (e.g. environmental engineers worry about losing good ones, like oxygen, or forming bad ones, like the toxic dioxins), the equation needs a reaction term (R):

$$\frac{dM}{dt} = M_{in} - M_{out} \pm R \quad (8.3)$$

Biosystems include interrelationships of the *abiotic* (non-living) and *biotic* (living) components of the environment. Biotechnology has been a dramatic example of the manipulation and application of the concept of "trophic state" for much of human history. Organisms, including humans, live within an interconnected network or web of life. Ecologists attempt to understand the complex interrelationships between and within the compartments of the food webs and chains, and consider humans to be among the consumers [3]. Food chains illustrate the complexity of biosystems (see Figure 8.1). Species at a higher tropic level are predators of lower level species, so materials and energy flow downward. The transfer of mass and energy upwardly and downwardly between *levels of biological organization* can be measured and predicted, given certain initial and boundary conditions. However, the types and abundance of species and interaction rates vary in time and space. From a biotechnological standpoint, the introduction of modified species or changes in environmental conditions (e.g. introduction of nutrients and toxic byproducts) can change these trophic interrelationships.

All species living in biosystems consist of molecular arrangements of the elements carbon, oxygen, hydrogen, and most contain nitrogen. These four biophile elements have an affinity for each other so as to form complex organic compounds. In fact, up until less than two centuries ago, such organic compounds were thought only to be able to be produced within natural biological systems. The smallest biosystems, e.g. the viruses, bacteria, and other microbes, can be seen as biochemical factories. They are quite efficient in finding and using organic material as sources of energy and carbon, but for much of human history, the systems within microbes have been considered to be "black boxes". Indeed, to a large extent, they still are.

Biosystematic processes provide microbes with remarkable proficiencies to adapt to various hostile environments. Some produce spores; many have durable latency periods, all have the ability to reproduce in large numbers when environmental conditions become more favorable. The various systems that allow for this efficient survival have become better understood in recent decades, to the point that cellular and subcellular processes of uptake and absorption, nutrient distribution, metabolism, and product elimination have been characterized, at least empirically. That is, the black boxes are fewer and smaller, but are still present. More recently, the genetic materials of deoxyribonucleic acid (DNA) and the various forms of ribonucleic acids (RNA) have been mapped. As genes have become better understood, so has the likelihood of their being manipulated. Such manipulation is the stuff at the cutting edge of environmental biotechnology.

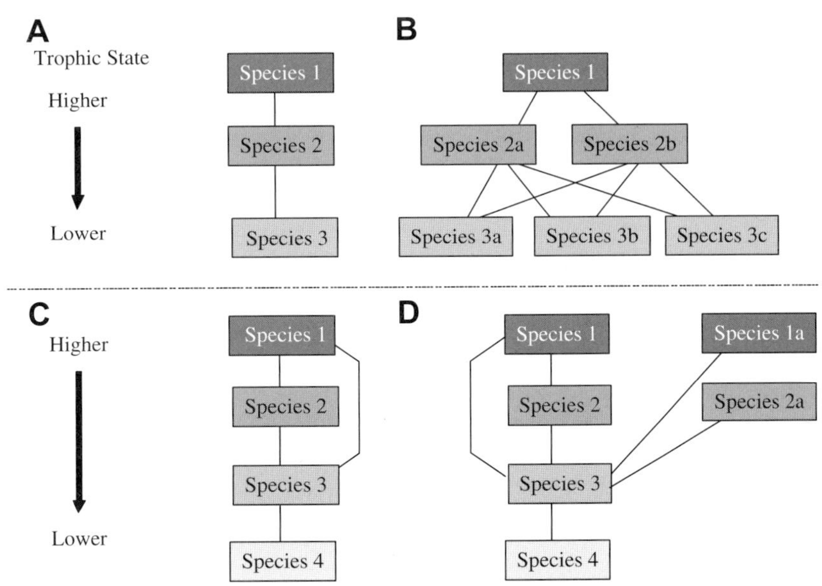

FIGURE 8.1
Flow of energy and matter in biosystems moves from higher trophic levels to lower trophic levels. Lines represent interrelationships among species. (A) Linear biosystem; (B) Multilevel trophic biosystem; (C) Omnivorous biosystem; and (D) Multilevel biosystem with predation and omnivorous behaviors. [*See color plate section*]
Source: Based on information from T.E. Graedel (1996). On the concept of industrial ecology. *Annual Review of Energy and the Environment* 21: 69-98.

SYSTEMATIC VIEW OF BIOTECHNOLOGICAL RISKS

Estimating and predicting the risks associated with the manufacture and use of biotechnologies is complicated by the diversity and complexity of the types of technologies available and being developed, as well as the seemingly limitless potential uses of these processes. As mentioned in Chapter 6, a risk assessment is the evaluation of scientific information regarding the hazardous properties of environmental agents, the dose-response relationship, and the extent of exposure of humans or environmental receptors to those agents. The beginning of risk assessment is a formulation of the problem and the product of the risk assessment is a statement regarding the probability that humans (populations or individuals) or other environmental receptors so exposed will be harmed and to what degree such harm will be manifested (i.e. the overall risk characterization). As more products and byproducts are developed using biotechnologies, the potential for environmental exposure has increased. Potential sources of biotechnological risk include direct and/or indirect releases to the environment from the manufacture and processing of biochemicals generated (e.g. proteins), as well the release of the modified organisms themselves and their biological products (e.g. spores and cysts).

The transfer of genetic material between separate populations, i.e. gene flow, is an example of the downstream risks from biotechnologies. Such transfer can be similar to that of chemical compounds in the environment, e.g. by dispersion of matter via advection. It can also take on very different transport mechanisms, such as biological transfer within and between levels of biological organization. An example is the organism-to-organism transport of genetic information via processes that resemble contagion behavior of disease transmission within a species.

Transfer of genetic material from modified strains is not always well understood. Transport models must continue to avail themselves to better science. Hopefully, increasingly improved transport mechanisms will be incorporated into risk assessments so that predictions of damage and risk become more reliable. For instance, the present commercialization of transgenic grasses is raising concerns about the ecological risks associated with future gene flow of modified species from high use agent contagion centers, such as golf courses adjacent to native and managed plant communities. One particular herbicide-tolerant creeping bentgrass (*Agrostis stolonifera*) has been the center of debate weighing the expected benefits of genetically modified turfgrasses, e.g. less pesticide use and lower labor costs, against potential ecological risks involved in the transfer of the herbicide-resistant trait to ecosystems (e.g. native grasses, introduced grasses, and conventional creeping bentgrass). The transgenic bentgrass populations may alter ecosystem function and structure in nearby ecosystems, e.g. wetlands. Reliable information is currently lacking, e.g. regarding the potential for gene flow to *A. perennans*, a common native perennial bentgrass. Complicating matters further, numerous native or introduced bentgrass species are able to hybridize with the herbicide-tolerant creeping bentgrass. Such genetically modified populations represent a potential environmental hazard in wetlands and other areas where invasive plants and weeds are managed by herbicides [4].

To date, most agricultural biotechnological products have come from microorganisms. The exact data requirements for each product have been developed on a case by case basis. All of the products have been proteins, either related to plant viruses or based on proteins from the common soil bacterium *Bacillus thuringiensis* (Bt). Generally, the types of data that need to be collected and provided to regulators have included product characterization, mammalian toxicity, allergenicity potential, effects on nontarget organisms, environmental fate and, for the Bt products in particular, the degree to which insects have become resistant to the Bt product after microbial spray applications and after plant incorporated protectants (PIPs) have been in use.

A transgene is an exogenous gene that has been introduced into the genome of another organism, and a transgenic species is one whose genome has been genetically altered. For instance, if a biotechnology is used as a PIP, the movement of transgenes from a host plant into weeds and other crops presents a concern that new types of exposures will occur. Bt corn and potato PIPs that have been registered to date have been expressed in agronomic plant species that, for the most part, do not have a reasonable possibility of passing their traits to wild native plants. Most of the wild species in the United States cannot be pollinated by these crops (corn and potato) due to differences in chromosome number, phenology and habitat. There is a possibility, however, of gene transfer from Bt cotton to wild or feral cotton relatives in Hawaii, Florida, Puerto Rico, and the US Virgin Islands. Where feral populations of cotton species similar to cultivated cotton exist, regulators have prohibited the sale or distribution of Bt cotton in these areas. These containment measures prevent the movement of the registered Bt endotoxin from Bt cotton to wild or feral cotton relatives [5].

Researchers have reviewed the potential for gene capture and expression of Bt plant-incorporated protectants (to date, only the colorado potato beetle control protein (Cry3A) has been introduced into potato) by wild or weedy relatives of cultivated potato in the United States, its possessions or territories [6]. Based on data submitted by the registrant and a review of the scientific literature, regulators have concluded that there is no foreseeable risk of unplanned pesticide production through gene capture and expression of the Cry3A in wild potato relatives in the US tuber-bearing *Solanum* species, including *S. tuberosum*. *S. tuberosum* cannot hybridize naturally with the non-tuber bearing *Solanum* species in the United States. Three species of tuber-bearing wild species of *Solanum* occur in the United States: *Solanum fendleri*, *Solanum jamesii*, and *Solanum pinnatisectum*. But, successful gene introgression into these tuber-bearing *Solanum* species is virtually excluded due to constraints of geographical isolation and other biological barriers to natural hybridization. These barriers include incompatible (unequal) endosperm balance numbers that lead to endosperm failure and embryo abortion, multiple ploidy levels, and incompatibility mechanisms that do not express reciprocal genes to allow fertilization to proceed. No natural hybrids have been observed between these species and cultivated potatoes in the United Species. The extent to which these findings will continue for the potato, or will be similar to those of other species, depends on the unique genomic characteristics of those species and the specific conditions of the ecosystems where they are introduced.

The constraints, drivers, and boundary conditions of the control volume wherein gene flow may occur must be understood to predict possible risks of genetically modifying these plant species. A systematic question is the extent to which such transfers present problems in microbial populations (e.g. genetically modified bacteria introduced in oil spills or hazardous waste sites to degrade toxic compounds; industrial applications of microbial reactors to produce pharmaceuticals and other chemicals that are released into the environment), and animal populations (i.e. the animals themselves, e.g. genetically modified fish released into the surface waters) or microbes that undergo transformation (e.g. similar to the recent H1N1 pandemic, but from a genetically modified microbe).

APPLIED THERMODYNAMICS

Biotechnology began as a passive approach. For example, sanitary engineers noted that natural systems, such as surface waters and soil, were able to break down organic materials. As they studied these processes, they realized that various genera of microbes had the ability to use detritus on forest floors, suspended organic material in water and organic material adsorbed onto soil particles as sources of energy need for growth, metabolism, and reproduction. The engineers correctly hypothesized that a more concentrated system could be fabricated to do the same thing with society's organic wastes. Thus, trickling filters, oxidation ponds, landfills, and other wastewater and solid waste treatment systems are merely supercharged versions of natural systems (see Discussion Box: Landfill Bioreactors).

DISCUSSION BOX

Landfill Bioreactors

No matter where one lives in most of the world, chances are good that a pile of solid waste is nearby. Greater than 230 million tons of municipal solid waste (MSW) is generated in the United States annually, of which 57 percent reaches a MSW landfill [7]. Most of the MSW landfills currently in service are regulated under Subtitle D of the Resource Conservation and Recovery Act, mainly to minimize risk to human health and the environment. Much of the potential risk from MSW landfill results from the migration of contaminated leachate and landfill gas (LFG). Thus, bioengineers are challenged to address contamination in two phases, liquid and solid, which are transported offsite by advection and other processes (see Table 2.9).

In North America, landfill regulations call for a system that minimizes liquid infiltration into the solid waste mass by controlling the amount of moisture allowed into these landfills. These so-called "dry tomb" landfill designs produce strata within the bioreactor system with low moisture content. While this decreases the amount of leachate, it also severely limits biodegradation, since moisture is a limiting factor for biofilm production and microbial metabolism. Thus, the likelihood of undegraded contaminants increases, as does their associated risk since the exposure integration time is protracted. These exposures include emissions from fugitive dust, combustion from the flare, microbial releases and the migration of contaminants through soil, ground, and surface waters. This is the rationale for long-term monitoring around these sites, e.g. current regulations require leachate and LFG emissions to be monitored for at least 30 years after closure of a landfill site or even longer if there is reason to believe the risks continue after that time.

A number of small-scale and large-scale projects have demonstrated that the rate of microbial degradation of solid wastes in these landfills can be enhanced, especially by increasing the moisture content of the waste. This can be by increasing the amount of water in the strata and by leachate re-circulation [8] (in most currently operated landfills the leachate is collected and treated – see Figure 7.6).

Waste arrives and is stored in a landfill on an ongoing basis so the age of the waste between the cells and within the cells (e.g. cells within layers) can vary considerably. Since degradation is a function of time, this means that the chemical conversion between and within cells will also vary, meaning that the chemical speciation within a landfill varies within two and three dimensions in space. That is, the age of the waste will affect the amount of oxidation-reduction, pH, and other important environmental conditions within each of the landfill's cells. As can be seen in Figure 7.6, the different landfill stabilization phases often overlap and can be viewed with regard to each phase's role within the system. The initial phase results in aerobic decomposition followed by four stages of anaerobic degradation. Thus, the majority of landfill decomposition by volume occurs under anaerobic conditions. Generally, biodegradation follows three basic stages [9]:

The organic material in solid phase (represented by chemical oxygen demand, i.e. COD_S) decays rapidly as larger organic molecules degrade into smaller molecules.

These smaller organic molecules in the solid phase undergo dissolution and move to the liquid phase (COD_L), with subsequent hydrolysis of these organic molecules.

The smaller molecules are transformed and volatilize as CO_2 and CH_4, with remaining biomass in solid and liquid phases.

During the first two phases, little material volume reaches the leachate. However, the biodegradable organic matter of the waste undergoes a rapid decrease in volume. Meanwhile, the leachate COD accumulates as a result of excesses of more recalcitrant compounds compared to the more reactive compounds in the leachate.

These three steps can be further grouped into five phases by which degradation occurs in a landfill bioreactor system, as shown in Figure 8.2. Successful conversion and stabilization of the waste depends on how well microbial populations function in *syntrophy* (i.e. an interaction of different populations that supply each other's nutritional needs).

Phase I: Initial adjustment

In environmental microbiology this phase is referred to as the lag phase. As the waste is placed in the landfill, the void spaces contain high volumes of molecular oxygen (O_2). As additional wastes are added and compacted, the O_2 content of the landfill bioreactor strata gradually decreases. With increasing moisture,

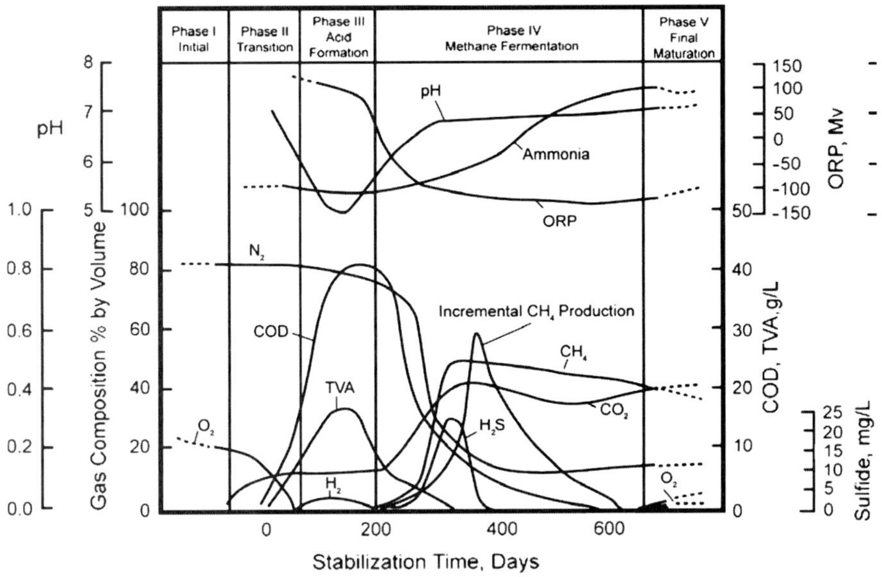

FIGURE 8.2
Phases of solid waste decomposition in a landfill, showing changes in released compounds and landfill conditions. Note: COD = chemical oxygen demand; TVA = total volatile acids; and ORP = oxidation-reduction potential.
Source: US Environmental Protection Agency (2007). National Risk Management Research Laboratory. Landfill Bioreactor Performance: Second Interim Report: Outer Loop Recycling & Disposal Facility – Louisville, Kentucky. Report No. EPA/600/R-07/060. Cincinnati, OH.

the microbial population density increases, initiating aerobic biodegradation, i.e. the primary electron acceptor is O_2.

Phase II: Transition
This phase is short-lived as the O_2 is rapidly degraded by the existing microbial populations. The decreasing O_2 results in a transition from aerobic to anaerobic conditions in the stratum. The primary electron acceptors during transition are nitrates and sulfates, since O_2 is rapidly displaced by CO_2 in the effluent gas.

Phase III: Acid formation
Hydrolysis of the biodegradable fraction of the solid waste begins in the acid formation phase, which leads to rapid accumulation of volatile fatty acids (VFAs) in the leachate. The increased organic acid content decreases the leachate pH from approximately 7.5 to 5.6 [10]. During this phase, the decomposition intermediate compounds like the VFAs contribute much COD. Long-chain volatile organic acids (VOAs) are converted to acetic acid ($C_2H_4O_2$), CO_2, and hydrogen gas (H_2). High concentrations of VFAs increase both the BOD and VOA concentrations, which initiates H_2 production by fermentative bacteria, which stimulates the growth of H_2-oxidizing bacteria. The H_2 generation phase is relatively short because it is complete by the end of the acid formation phase.

As seen in Figure 8.2, this phase also is accompanied by an increase in the biomass of acidogenic bacteria and rapid degradation of substrates and consumption of nutrients.

Since metals are generally more water soluble at lower pH, metallic compounds may become more mobile during this phase.

Phase IV: Methane fermentation
The acid formation phase intermediary products (e.g. acetic, propionic, and butyric acids) are converted to CH_4 and CO_2 by *methanogenic* microorganisms. As VFAs are metabolized by the methanogens, the landfill water pH returns to neutrality. The organic strength (i.e. oxygen demand) of the leachate decreases at a rapid rate as gas production increases in correspondence with increases in CH_4 and CO_2 gas production. This is the longest-lived waste decomposition phase.

Phase V: Final maturation and stabilization
The rate of microbiological activity slows during the last phase of waste decomposition as the supply of nutrients limits the chemical reactions, e.g. as bioavailable phosphorus becomes increasingly scarce.

(Continued)

CH_4 production almost completely disappears, with O_2 and oxidized species gradually reappearing in the gas wells as O_2 permeates downwardly from the troposphere. This transforms the oxidation–reduction potential (ORP) in the leachate toward oxidative processes. The residual organic materials may incrementally be converted to the gas phase, and as organic matter is composted; i.e. the organic matter is converted to humic-like compounds (although these processes have not yet been completely scientifically documented within a landfill system).

On-site degradation, such as that which occurs in a landfill, is carried out by different genera of microbes, making the kinetics difficult to predict for any facility. Such microbial populations occur naturally in the landfill (e.g. natural soil bacteria). However, varying temporal and spatial site-specific conditions to achieve optimal performance is often a heuristic process. The kinetics of microorganisms in landfill bioreactors have not been widely investigated, likely because these groups of microorganisms are much more difficult than aerobes to culture. Hopefully, the emergence of molecular-based non-culture techniques will increase this information. To date, landfill bioreactor performance has been evaluated and controlled by indirect evidence, e.g. waste stabilization is characterized by monitoring the outcome of the decomposition process. For example, re-circulation of leachate has been shown to shorten the initial lag phase, even though the microbial species responsible for this accelerated decomposition have not been precisely identified [11].

Bioengineering improvements to a landfill bioreactor requires attention to the sensitivity of variables. Finding methods to improve efficiency depends on optimizing microbiological metabolic processes. For example, the amount of leachate to be re-circulated affects the quantity of organic acids. If VFA concentrations are high, methanogenesis can be inhibited. This is not a direct inhibition of microbial metabolism, but a response of the microbial population to the lower pH induced by the VFAs. Thus, the volume of re-circulated leachate must be adjusted to minimize the accumulation of VFAs (see Figure 8.3).

Thus, effective degradation depends on matching the abiotic and biotic conditions in a bioreactor with the needs of the microbial population. Environmental biotechnologies are designed to enhance and control these relationships.

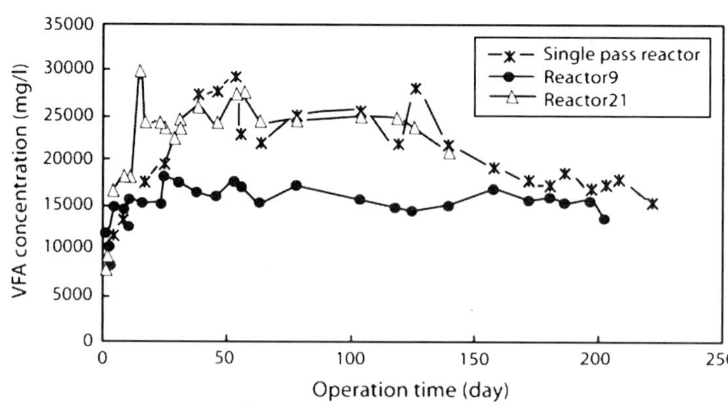

FIGURE 8.3
Effects of leachate re-circulation on volatile fatty acid (VFA) accumulation compared to VFA generation under conventional landfill management. The leachate in "Reactor 9" was re-circulated at 9 L day^{-1} (2.4 gallons per day – gpd) = 13% of the reactor volume. The leachate in "Reactor 21" was re-circulated at 21 L day^{-1} (5.5 gpd) = 30% of the reactor volume. The VFA buildup in the reactor with a higher leachate re-circulation rate of 21 L day^{-1} (5.5 gpd) was nearly as high as the VFAs generated in the single-pass (i.e., conventional) reactor. The 21 L day^{-1} (5.5 gpd) bioreactor experienced a spike = 30,000 mg $VFAL^{-1}$ within 30 days, which can be detrimental to methanogenic bacteria. *Sources:* US Environmental Protection Agency (2007). National Risk Management Research Laboratory. Landfill Bioreactor Performance: Second Interim Report: Outer Loop Recycling & Disposal Facility – Louisville, Kentucky. Report No. EPA/600/R-07/060. Cincinnati, OH; and D. Sponza and O. Agdag (2004). Impact of leachate re-circulation and re-circulation volume on stabilization of municipal solid wastes in simulated anaerobic bioreactors. *Process Biochemistry* 39(12): 2157–2165.

In such a passive system, the biotechnology makes use of the same microbes that have adapted metabolic processes to degrade an array of organic compounds in natural settings, but they have been allowed to acclimate to the organic material that needs to be broken down. The microbes' inherent or adaptive preference to have more easily and directly derived electron transfer (i.e. energy sources) must be overcome. Environmental biotechnologists have accomplished this by limiting the microbes to an exclusively available carbon and energy source. That is, the bioengineer only permits the microorganism population, under a control setting, to come into contact with the chemicals in the waste. Thus, the microbes adapt their biological processes to use these formerly unfamiliar compounds as their energy sources and, in the process, break them down into less toxic substances. Ultimately the microbes degrade complex organic waste to carbon dioxide and water in the presence of molecular oxygen. In the absence of molecular oxygen, the microbes degrade the organic waste into methane and water. These are known as aerobic and anaerobic digestion, respectively.

Numerous examples of passive systems have been put to use as society has become increasingly complex. Human systems have evolved to support the need for safe food supplies, clean water and air, better shelter, and urbanization. For example, passive biotechnologies were needed to allow for large-scale agriculture, including hybrid crops and nutrient cycling in agriculture. Likewise, biomedical advances, such as vaccines, have often led to other societal change, e.g. better living standards, and disease treatment and prevention.

Very recently, more active systems have been used increasingly to achieve such societal gains, but at an exponentially faster pace. In addition, scientists have developed biotechnologies that produce products that simply would not exist in passive systems.

The relationships between organisms and their environments reveal themselves in cycles of matter and energy into and out of the organism. This means that the organism itself is a thermodynamic control volume. In turn, the population of organisms is part of the larger control volumes (e.g. microbes in the intestine; the intestine in the animal; the herd as prey in a habitat; the habitat as part of an ecosystem's structure).

Smaller control volumes assimilate into larger ones. Within reactors are smaller-scale reactors (e.g. within the fish liver, on a soil particle, or in the pollutant plume or a forest). Thus, scale and complexity can vary by orders of magnitude in environmental systems. For example, the human body is a system, but so is the liver, and so are the collections of tissues through which mass and energy flow as the liver performs its function. Each hepatic cell in the liver is a system. At the other extreme, large biomes that make up large parts of the earth's continents and oceans are systems, both from the standpoint of biological organization and thermodynamics.

The interconnectedness of these systems is crucial to understanding biotechnological implications, since mass and energy relationships between and among systems determine the efficiencies of all living systems. For example, if a toxin adversely affects a cell's energy and mass transfer rates, it could have a cumulative effect on the tissue and organs of the organism. And, if the organisms that make up a population are less efficient in survival, then the balances needed in the larger systems, e.g. ecosystems and biomes, may be changed, causing problems at the global scale. Viewing this from the other direction, a larger system can be stressed, such as changes in ambient temperature levels or the increased concentrations of contaminants in water bodies and the atmosphere. This results in changes all the way down to the subcellular levels (e.g. higher temperatures or the presence of foreign chemicals at a cell's membrane will change the efficiencies of uptake, metabolism, replication, and survival). Thus, the changes at these submicroscopic scales determine the value of any biotechnology.

At very high levels of biological organization, i.e. the biomes, the very large are linked and interact with each other, such as the way that coral reef systems are inextricably connected to distant continental deserts, as will be addressed in this chapter's seminar discussion.

PREDICTING ENVIRONMENTAL IMPLICATIONS

A particular measure for emerging technologies' success is whether such technological advancements are accompanied by or induce human and ecological impacts. That is, the research must not only be efficient and effective within a short timeframe and for a specific purpose, but the technologies must be sustainable. This means that successful new technologies must not lead to unacceptable environmental risk and any risk will be minimized and justly distributed in time and space throughout society (see Discussion Box: Justice and Genetically Modified Organisms).

DISCUSSION BOX

Justice and Genetically Modified Organisms

> *Bioengineering is a rapidly evolving field that adapts to new knowledge, new technology, and the needs of society, it also draws on distinct roots that go back to the origins of civilization. Maintaining a linkage of the past with the future is fundamental to the rational and fact-based approaches that engineers use in identifying and confronting the most difficult issues.*
>
> *National Academy of Engineering (2004) [12]*

Biotechnologies present bioethical challenges. The National Academy's recommendation to link the past with the future is not always a positive experience. Adapting new knowledge and novel technologies, especially those that affect living things, has not always been exemplary. Arguably the nadir of 20th century biomedical research occurred between 1932 and 1972, with the US Public Health Service's (US PHS) investigation of the effects of syphilis (caused by the spirochete bacterium *Treponema pallidum*), i.e. the so-called Tuskegee Experiment. With a disturbing disregard for justice, beneficence, and respect for the human person, US PHS researchers deliberately denied full knowledge and treatment for the incapacitating sexually transmitted disease to 399 African-American men infected with *T. pallidum*. The researchers took advantage of a trusted African-American medical facility's reputation, ultimately leading to the deaths of 128 men, and allowing 59 spouses to contract the disease or their children to be born with syphilis.

Horrible as this case was, one may ask what lessons it has to offer biotechnology. And indeed, there are many. First, biotechnologies in every sector, industrial, medical, agricultural, and environmental, vary in the level of trust. The trust varies not just between researchers and the public, but within the scientific community. For example, scientists holding a more precautionary perspective on the advancement of science are more circumspect about potential risks from emerging technologies. Those scientists that see biotechnologies as merely extensions of a much larger knowledge base (a logical extension of centuries of genetics, hybridization, husbandry, botany, and microbiology) may prefer to address each specific nuance as a slight permutation, rather than a comprehensive threat. Hence, where some may see a genetically modified strain of grain as "Frankencorn", others see an almost routine variation of *Zea mays.*

Another important lesson from the Tuskegee Experiment is how to put science within a social context. Often, this presents a paradox to researchers. Most have not been trained in things sociological, let alone geopolitical. Their expertise is in an often esoteric area of research. The old adage says that to be a successful PhD-toting scientist requires us "to learn more and more about less and less, until we know everything about nothing." While extreme and *reductio ad absurdum*, it is somewhat accurate in pointing out that technical researchers often know much about the science, but too little about the social implications of that science (including whether to continue an investigation after evidence of ethical and justice issues – often even when such information becomes available, the scientist is not trained to recognize it).

Some parallels with biotechnologies have been drawn, as when Rebecca Bratspies of the City University of New York's Law School states:

> *Examining the degree to which environmental concerns have or have not been incorporated into the registration requirements of Bt crops, it becomes clear that the regulatory process suffers from many ills. In approving these genetically engineered crops for market, the United States Department of Agriculture (USDA) and Environmental Protection Agency (EPA) repeatedly disregarded significant but unresolved scientific questions about these GM crops. Seed*

companies agreed to environmentally protective measures in their crop registrations, but assumed no responsibility for implementing those measures ... No regulatory framework existed (or, for that matter, exists) to monitor and enforce these registration restrictions ... These serious regulatory deficiencies call into question the soundness of the entire biotechnology regulatory process, a question ultimately much broader than any particular GMO. [13]

Thus, biotechnological research scientists and biotechnologists work within an uncertain regulatory environment. Since the distribution of goods and services is so uneven, scientists may be tempted to "regress to the mean" and to satisfy some part of society by catering to a population within some level of variance around a measure of central tendency for that population (e.g. persons falling within ± 2 standard deviations of an arithmetic mean). That way, "most" people would be satisfied with the investigation. There are entire philosophical schools of thought surrounding the ethical treatment of populations. The most moral approach in science is to protect public health and the environment by ensuring that *all* persons are adequately protected.

The Reverend Martin Luther King put it this way: "Injustice anywhere is a threat to justice everywhere" [14]. By extension, if any group is disparately exposed to an unhealthy environment, then the whole nation is subjected to inequity and injustice. If a sensitive subpopulation is disproportionately harmed by a genetically modified organism, this is an example of biotechnological injustice. Put in a more positive way, researchers and biotechnologists can work to provide a safe and livable environment by including everyone, leaving no one behind. This is called *environmental justice.*

The concept of environmental justice has evolved over time. In the early 1980s the first name for the movement was *environmental racism*, followed by *environmental equity.* These transitional definitions reflect more than changes in jargon. When attention began to be paid to the particular incidents of racism, the focus was logically placed on eradicating the menace at hand, i.e. blatant acts of willful racism. This was a necessary, but not completely sufficient component in addressing the environmental problems of minority communities and economically disadvantaged neighborhoods, so the concept of equity was employed more assertively. Equity implies the need not only to eliminate the overt problems associated with racism, but to initiate positive change to achieve more evenly distributed environmental protection. 411

We now use the term environmental justice, which is usually applied to social issues, especially as they relate to neighborhoods and communities. The so-called environmental justice (EJ) communities possess two basic characteristics:

They have experienced historical (usually multi-generational) exposures to disproportionately [15] high doses of potentially harmful substances (the *environmental* part of the definition). These communities are home to numerous pollution sources, including heavy industry and pollution control facilities, which may be obvious by their stacks and outfall structures, or which may be more subtle, such as long buried wastes with little evidence on the surface of their existence. These sites increase the likelihood of exposure to dangerous substances. Exposure is preferred to *risk*, since risk is a function of the hazard and the exposure to that hazard. Even a substance with a very high toxicity (one type of hazard) that is confined to a laboratory of a manufacturing operation may not pose much of a risk due to the potentially low levels of exposure.

Environmental justice communities have certain, specified socioeconomic and demographic characteristics. EJ communities must have a majority representation of low socioeconomic status (SES), racial, ethnic, and historically disadvantaged people (the *justice* part of the definition).

Thus EJ is a system and calls for an integrated response to ensure justice. The first component is a sound scientific and engineering underpinning to decisions. The technical quality of designs and operations is vital to addressing the needs of any group. However, the engineering codes' call that engineers be "faithful agents" lends an added element of social responsibility to environmental practitioners [16]. No "blank slate" can be assumed for any biotechnological design. Historic disenfranchisement and even outright bias may well have put certain neighborhoods at a disadvantage. In fact, a metric to keep in mind when siting a biotechnological operation is why no one seems to be complaining. Such silence could be rooted in historical disenfranchisement and not necessarily a validation of the site selection criteria.

Bioscientific accountability and responsibility does not stop at sound science, but considers the social milieu, especially possible disproportionate impacts. The determination of disproportionate impacts,

(Continued)

e.g. changes to urban and neighborhood habitats and pollution-related impacts, is a fundamental step in ensuring environmental justice. Certainly, this step relies on the application of sound science. For example, a first step in assessing environmental insult may consist of epidemiological evidence, e.g. clusters of elevated exposures and effects in populations. For example, certain cancers, neurological, hormonal, and other chronic diseases have been found to be significantly higher in minority communities and in socioeconomically depressed areas. Acute diseases, as indicated by hospital admissions, may also be higher in certain segments of society, such as pesticide poisoning in migrant workers [17].

Disease incidence variation can be an example of *disparate effects*. To complicate such disparity, each person responds to an environmental insult uniquely and that person is affected differently at various life stages. For example, young children are at higher risk when exposed to certain proteins expressed in plants, leading to higher vulnerability to allergenicity. This is an example of *disparate susceptibility.* Indeed, subpopulations also can respond differently than the whole population, meaning that genetic differences seem to affect people's susceptibility to exposure to biological, chemical, and physical agents. Scientists are very interested in genetic variation, so that genomic techniques [18] (e.g. identifying certain polymorphisms) are a growing area of inquiry.

In a sense, historical characteristics constitute the "environmental" aspects of EJ communities, and socioeconomic characteristics entail the "justice" considerations. The two sets of criteria are mutually inclusive, so for a community to be defined as an EJ community, both of these sets of criteria must be present.

A recent report by the Institute of Medicine [19] found that numerous EJ communities experience a "certain type of double jeopardy." The communities must endure elevated levels of exposure to contaminants, while being ill equipped to deal with these exposures because so little is known about the exposure scenarios in EJ communities. The first problem (i.e. higher concentrations of contaminants) is an example of *disparate exposure.* The latter problem is exacerbated by the disenfranchisement from the political process that is endemic to EJ community members. This is a problem of *disparate opportunity* or even *disparate protection* [20]. The report also found large variability among communities as to the type and amount of exposure to toxic substances.

Toxicity is specific to the agent. For example, one of the most common exposures in EJ communities is to the metal lead (Pb) and its compounds. The major health problem associated with Pb is brain, as well as central and peripheral nervous system diseases, including learning and behavioral problems. Another common contaminant in EJ communities is benzene, as well as other organic solvents. These contaminants can also be neurotoxic, but also have very different toxicity profiles from neurotoxic metals like Pb. For example, benzene is a potent carcinogen, having been linked to leukemia and lymphatic tumors, as well as severe types of anemia. They also have very different exposure profiles. For example, Pb exposure is often in the home and yard, while benzene exposures often result from breathing air near a source (e.g. at work or near an industry, such as an oil refinery or pesticide manufacturer). The Institute's findings point to the need for improved approaches for characterizing human exposures to toxicants in EJ communities.

The double effect principle

A metric for environmental justice is the so-called *double effect*. In its most simple form, the principle holds that an act that leads to negative side effects is permitted, but deliberate harm (even for good causes) is wrong. Clearly the Tuskegee Experiment fails miserably in this regard.

Bioengineers and biotechnologists to some degree during their careers will face risk tradeoffs and double effects. For example, some time during the life cycle of a biotechnology some persons (e.g. immunocompromised) may be harmed by its use. If *all* of a certain group is harmed at the expense of another group's benefit, this could be conceived as intentional harm and an injustice. The way to mitigate this harm is to disclose fully the shortcomings of the technology and to work on improvements that will decrease the likelihood of harm. The technology must do inherent good (e.g. clean up a hazardous waste, provide food, generate insulin for diabetics or deliver drugs to ill patients). In other words, the designer must not actually intend to accomplish the bad effect (harming immunocompromised people). The ill effect is simply unavoidable in the effort to provide a needed good or service. If another approach would avoid the negative outcome and still provide the good or service, then such an option is the preferred and obligatory act.

Another provision of the double effect is that the good effect must be at least as directly an effect of the action as is the negative bad effect. In particular, the bad outcome must not *cause* the good effect. A biotechnological operation must not "use" any subpopulation as a commodity to achieve the intended result. Finally, the benefit of the bad outcome must not outweigh the benefit of the good (e.g. more people are harmed to provide a benefit to the few). Also, another ethically acceptable approach without the side effects must not be available. If it is, this is the approach that should be followed (e.g. a proven safer progenitor strain of a bacterium to degrade a toxic substance provides the same intended outcome as a less proven genetically modified strain).

A biotechnological example of the double effect is that of the vaccine. A government and manufacturer normally calculate an estimate of the population risk of administering a vaccine to the public. Most are expected to benefit, with a small number with adverse side effects. And, from this small group, a subset of vaccine recipients will die. The lives are saved as a result of the vaccine, not as a result of the deaths of those who die of side effects. The side effects do not advance any goals of the drug manufacturer. Thus, the side effects are not intended as a means to any other outcome. Finally, the proportion of lives saved compared to lives lost is very high, satisfying the requirement that benefits outweigh the negative outcomes.

It is unjust to produce and administer a vaccine with side effects if another means of preventing the disease is available. For example, a recent controversy arose over the use of a vaccine containing traces of mercury in a preservative. The preservative had been associated with increased incidences of autism, which is at least somewhat biologically plausible given that mercury is neurotoxic. Thus, a mercury-free preservative was sought and is being used in subsequent vaccines (e.g. the H1N1 virus vaccine).

Scales of justice

The justice aspects of environmental risks present issues at the local, national, and geopolitical scale. Environmental justice is a requirement for any socially responsible science and engineering endeavor, since the application of science without consideration of its social dimensions may lead to unfair outcomes. Environmental justice not only focuses on the need to avoid disproportionate risks (e.g. resulting from siting of hazardous waste sites) in certain disadvantaged subpopulations, e.g. minorities, but it is also concerned with inequitable distribution of environmental resources and services (e.g. flood control).

Bioethics is more than biomedical ethics. In fact, Van Rensselaer Potter II (1911–2001) originally coined the term *bioethics* to invoke the need to use integration and systematic thinking in decisions related to living things. Potter considered bioethics a bridge between the sciences and the humanities to serve the best interests of human health and to protect the environment. In his own words, Potter described this bridge:

> *From the outset it has been clear that bioethics must be built on an interdisciplinary or multidisciplinary base. I have proposed two major areas with interests that appear to be separate but which need each other: medical bioethics and ecological bioethics. Medical bioethics and ecological bioethics are non-overlapping in the sense that medical bioethics is chiefly concerned with short-term views: the options open to individuals and their physicians in their attempts to prolong life ... Ecological bioethics clearly has a long-term view that is concerned with what we must do to preserve the ecosystem in a form that is compatible with the continued existence of the human species. [21]*

The march of biological sciences has been justified as an overall benefit to humankind. However, this commitment and involvement calls for deliberate and serious considerations of actual and potential ethical issues. The President's Council on Bioethics [22] has summarized the dichotomy between the promise and ethical challenges:

> *... knowledge of how things work often leads to new technological powers to control or alter these workings, powers generally sought in order to treat human disease and relieve suffering. But, once available, powers sought for one purpose are frequently usable for others. The same technological capacity to influence and control bodily processes for medical ends may lead (wittingly or unwittingly) to non-therapeutic uses, including "enhancements" of normal life processes or even alterations in "human nature." Moreover, as a result of anticipated knowledge of genetics and developmental biology, these transforming powers may soon be able to transmit such alterations to future generations.*

(Continued)

Justice and genetically modified organisms

The United States' position on how to determine the safety and risks associated with genetically modified organisms differs from that of the European Union's. The US applies a *product-oriented approach*, whereas Europe uses a *process-oriented approach.*

A product-oriented approach is concerned with the benefits and risks of a particular product. That is, regulators are concerned with making sure that the products meet the criteria for risk and safety, not necessarily how the products are produced. The environmental objective is to produce products that are less harmful. From a green engineering perspective, a product-oriented approach is based on the assumption that the product will be produced, so the bioengineer needs to find ways to make it more sustainable. At this point, all phases of the product's life cycle are optimized, based on systematic thinking. Factors that go into such thinking include environmental and health risks associated with every material in the life cycle. It also considers the services involved in the process rather than apparatus to generate the product [23]. For example, a sustainable biotechnological operation to produce a solvent would not start with the design of a bioreactor or other equipment, but how best to produce a good solvent in a sustainable manner (e.g. it may not need a traditional bioreactor, but may be produced using modular bioreactors co-generationally near a number of chemical companies).

A process-oriented approach to environmental protection considers how products come to market, analyzing the input/output and material flow, ecological and economic factors, and risks to identify technical and organizational options to improve a process, including considerations on how to reduce the number of processes needed to bring a product to market. This includes a review of the internal cycles for auxiliary materials and how production wastes are introduced, hazardous substances are replaced and can be used more efficiently and safely, and how to introduce and apply innovative technologies [24]. Of course, biotechnologies comprise an important group of such innovations.

The differences in polices based on these two approaches can have profound impacts on international trade. In fact, the United States, Canada, and Argentina filed a complaint with the World Trade Organization (WTO) that mandates to apply the process-oriented, precautionary principles to genetically modified organisms constituted a threat to free trade. The precautionary approach relies on anticipatory outcomes of risks in the absence of sufficient evidence, when such risks are deemed to be serious and unreasonable. In a working paper, Paulette Kurzer of the University of Arizona succinctly summarizes this debate:

European officials claim that, in the absence of scientific proof of risk, nobody should assume the absence of risk. Therefore, officials should undertake proportionate measures to remove or reduce threats of serious harm. Not knowing what the long-term detriments are of GM seeds/food, the EU position is that we must assume that the product is not safe unless otherwise proven. US trade negotiators believe that this attitude prevents the unwarranted entrance of GMOs into the EU market. To them, this type of reasoning must conceal ulterior motivations. In the US, after all, hardly a debate has occurred on GMO. Yet Americans are very health conscious and are equally obsessed about "risks." [25]

In 2006, a panel of the World Trade Organization (WTO) ruled that the European Union had invoked a *de facto* moratorium on biotechnological products between June 1999 and August 2003 that led to "undue delay" that violated the WTO Agreement on the Application of Sanitary and Phytosanitary Measures. The panel also struck down some individual nations' bans that were not considered to be based on evidence-based risk assessments. The panel's ruling was quite narrow, so the biotechnology trade debate has continued. For example, the panel did not address the safety of genetically modified organisms, nor did it rule on the legality of the precautionary principle. [26]

Food is at the center of the controversy surrounding the environmental justice aspects of biotechnology. That hunger is directly related to poverty in the developing world is a function of poverty, not food scarcity. The world's food production has far outpaced population growth in recent decades. Rural areas account for 75% of the world's poor and undernourished people, in spite of the global urbanization trend [27]. Since biological diversity is a requirement for a sustainable and reliable global food supply, environmental justice in developing countries depends on a systematic viewpoint.

The risk of replacing indigenous crop varieties and diverse cultivation systems with monocultures is an environmental justice issue, since monocultures are more vulnerable to the ravages of pests and plant diseases, loss of soil fertility, and increased application of agrochemicals. As evidence, global food supplies now depend on merely 100 or so species of food crops, rather than the thousands of species and varieties that have been used locally for millennia. This consolidation is inextricably linked to international environmental justice. The individual farmer in developing countries cannot continue previously subsistent and sustainable farming practices, leading to more monocultures, and diminishing diversity.

Other environmental justice considerations may be similar to chemical risk assessment issues, such as the location of bioreactor landfills and other biotechnological operations in neighborhoods that are less likely to complain (e.g. the ones with high unemployment rates) or which have historically been disproportionately industrialized (e.g. neighborhoods that in the past had a high percentage of workers and their families, but that now mostly include people who do not reap the benefits of the operation, but moved there because of low property values). Siting an industrial biotechnology facility may be easier in these areas, but unjust if the exposures and risks are disproportionate to that of the larger population.

Thus, what may seem to be a scientific issue (e.g. growing crops that produce the most food or siting a bioreactor) in fact can also frequently be an issue of justice. In 1992 the US EPA created the Office of Environmental Justice to coordinate the agency's EJ efforts, and in 1994, President Clinton signed Executive Order 12898, ''Federal Actions to Address Environmental Justice in Minority and Low-Income Populations.'' This order directs that federal agencies attend to the environment and human health conditions of minority and low-income communities, and requires that the agencies incorporate EJ into their missions. In particular, EJ principles must be part of the federal agency's day-to-day operation by identifying and addressing ''disproportionately high and adverse human health and environmental effects of programs, policies and activities on minority populations and low-income populations'' [28].

This order has in many ways been an extension to the environmental impact statement process under the National Environmental Policy act–NEPA (see Chapter 1 and Appendix 1). The application to genetic engineering is sparse compared to that of chemical agents, so reconciling EJ with biotechnology remains a challenge.

Environmental "feedbacks" are crucial to environmental biotechnology, wherein bioengineers optimize the variables that lead to the intended products and the mechanisms needed to preserve these products (limit the effects) on the energy and mass balances [29]. Sometimes, the bioengineer must decide that there is no way to optimize both. In some instances, the bioengineer must recommend the "no go" option. That is, the potential downstream costs and risks are either unacceptable or the uncertainties of possible unintended, unacceptable outcomes are too high. Usually, though, scientists can model a number of permutations and optimize solutions from more than two variables (e.g. species diversity, productivity and sustainability, costs and feasibility, and bioengineered product efficiencies). The challenge is to know the extent to which the model represents the realities as they vary in time and space. However, every predictive model has uncertainties, so estimating outcomes in chaotic systems can lead to surprises, both pleasant (e.g. faster biodegradation) and unpleasant (slower kinetics or unexpected gene flows of genetic modified organisms).

Even beneficial biotechnologies can introduce hazards in time and space. The benefits of medical, industrial, agricultural, and environmental biotechnologies must be weighed against these possible hazards. As discussed in Chapter 4, scientists continue to develop models to characterize hazards. However, risk–benefit analyses are difficult since the science underpinning biotechnologies is emerging and is fraught with uncertainties. This is a principal biotechnological challenge for policy makers and regulators [30]. On the one hand, advances in industrial, agricultural, medical, and environmental biotechnologies must be supported, but on the other hand, any risks must be at least considered to determine whether they are acceptable in light of the possible benefits.

Stressor–receptor relationships have temporal and spatial dependencies. Localized stressors can result from episodes, such as a spill, immediate release or emergency situation. For example, carbon sources are highly variable, as are the receptors. The source may release organic forms, but these may undergo partitioning to form liquid and solid phases, as well as chemically react in the troposphere (see Figure 8.4), so that the receptor may be exposed to primary aerosols or products resulting from secondary chemical reactions in the atmosphere (e.g. alkanes and aromatic compounds).

Secondary organic aerosols (SOAs) are formed by oxidation reactions of gas-phase organic species. These species include alkanes, alkenes, aromatics, cyclic olefins, isoprene, and terpenes. These reactions can change the stressor-receptor relationship substantially. For example, the SOAs may have either lower the vapor pressure than the primary aerosols due to the addition of functional groups or reactions may increase vapor pressure when carbon–carbon bonds are cleaved. In addition, reactions of molecules sorbed to a particle may have their vapor pressures altered by oxidation or by the formation of higher-molecular-weight species. Such reactions can lead to the formation of *oligomers*, which have lower total vapor pressures than their component compounds or by the formation of more volatile products. Vapor pressure can also change as a result of ongoing reactions from the varied volatility of oxidation products [31].

The means by which the products from the atmospheric processes in Figure 8.4 come into contact with humans and ecosystem receptors is controlled by deposition from the atmosphere to surfaces in and on the receptor system. This can be envisioned as a flux (g sec^{-2}) across a two-dimensional surface which is calculated as the product of a concentration (g m^{-3}) times a deposition velocity (V_d), expressed in m sec^{-1}. V_d is inversely proportional to the resistance caused by atmospheric processes:

$$V_d = \frac{1}{R_a + R_b + R_c} \tag{8.4}$$

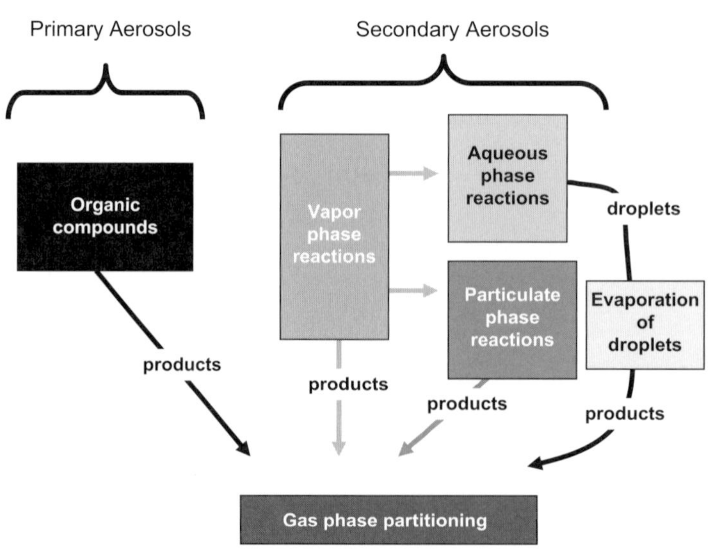

FIGURE 8.4
Source, transformation, and receptor relationships after release of an organic substance released to the atmosphere. [*See color plate section*]

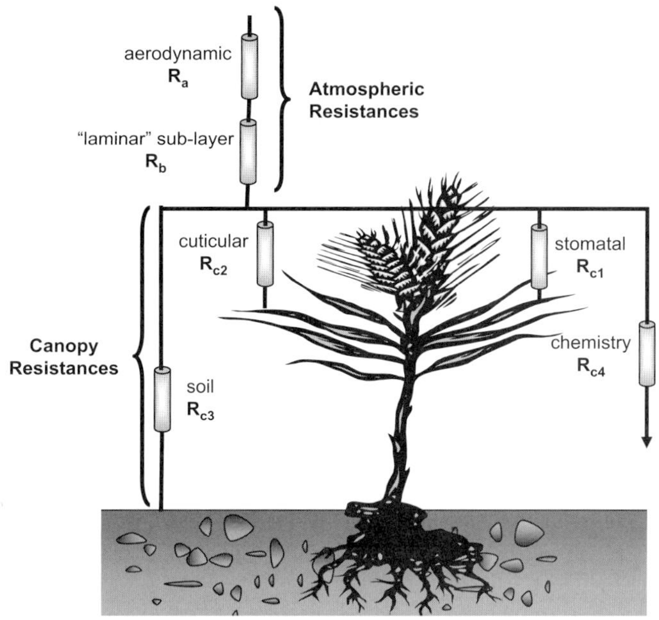

FIGURE 8.5
Resistance analogy for the deposition of particles from the atmosphere to the surface. This resistance-in-series analogy is a function of wind speed, solar radiation, plant characteristics, precipitation/moisture and soil/air temperature. Drawing by T. Peirce (2009). US Environmental Protection Agency, Research Triangle Park, North Carolina. [*See color plate section*]

where R_a represents the resistance from atmospheric turbulence, R_b the resistance due to transport in the fluid sublayer very near the elements of surface such as leaves or soil, and R_c the resistance to uptake of the surface itself (see Figure 8.5). This method is not appropriate for compounds with a substantial likelihood of being re-emitted after deposition.

Similarly, biological agents may also undergo changes after release, such as sorption to aerosols, spore formation and degradation.

Incremental or immediate stresses can spread in space and time, such as those due to the climatic changes that can result from incremental and continuous releases of greenhouse gases with expansive, global impacts on ecosystems and public health if global temperatures rise significantly and if in fact the concomitant changes that are expected as a result of mean temperature rises do materialize.

ENVIRONMENTAL IMPLICATIONS OF ENGINEERING ORGANISMS

Engineers never seem to be satisfied with things as they are at any given time. They restlessly seek improvements in efficiency and output. An engineer starts with a baseline and applies scientific principles to change things for the better, or at least for what the engineer and the scientific community perceive to be "better." They respect baselines and limits, but look for ways to improve them.

A mere half-century ago in 1953, shortly after Watson and Crick published their work on sequencing the DNA molecule, engineers saw an opportunity to address their dissatisfaction with baselines and thresholds; such as the amount of water and fertilizer needed for crops, the taste of food crops, the ability of plants to resist insults from pests, chemical processing rates for producing pharmaceuticals, and microbes' efficiencies at breaking down recalcitrant contaminants in the environment (see Discussion Box: Recalcitrance). They saw living things as means to various ends.

DISCUSSION BOX

Recalcitrance

Certain molecules resist biodegradation in the environment. They are chemically persistent, but there is something about their structure that makes them unattractive as electron acceptors and donors to microbes. Although they may be rich in carbon, microorganisms seek carbon from more familiar and usually chemically simpler sources unless drastic conditions induce degradation, e.g. these molecules are the only food that is available or the microbes' genetics are engineered to alter their carbon preferences.

The term *recalcitrance* is anthropogenic. Recalcitrant chemicals are stubbornly resisting human controls. The microbes are not being blamed for the lack of degradation, but the chemical compounds themselves are, especially when stereochemistry and structural activity *should* be amenable to biodegradation. In fact, however, recalcitrance is a function of both the chemical structure and the microorganisms' preferences for electron acceptance and carbon.

Actually, recalcitrance is not limited to microorganisms but is also applied to plants. Plants have the inherent capacity to degrade xenobiotic pollutants, but they generally lack the catabolic pathway to provide complete degradation, i.e. mineralization, relative to microbes. In fact, current research is being directed toward the transfer of genes involved in xenobiotic degradation from microbes to plants to enhance specific plant taxons' potential for remediating more recalcitrant compounds, e.g. trichloroethylene, pentachlorophenol, trinitrotoluene (TNT), glycerol trinitrate, atrazine, ethylene dibromide, metolachlor and hexahydro-1,3,5-trinitro-1,3,5-triazine [32].

In fact, phytodegradation and microbial biodegradation are interrelated and mutual. Plant evapotranspiration, plant-microbe rhizosphere degradation, and microbial metabolic processes are combined. For examble, *in situ* natural and engineered remediation projects take advantage of both plant life and microbial populations, with much of the degradation at the root-microbe interface (i.e. in the rhizosphere) in the soil (see Figure 8.6). Degradation is enhanced in plants through numerous processes. During phytoextraction, recalcitrant compounds are removed and stored. In fact, recalcitrant compounds may be stored without metabolism for protracted periods. Phytodegradation, on the other hand, occurs when plants can metabolize a compound. The metabolism during phytodegradation may resemble that of animal degradation of toxic substances [33].

The microbial-macrophytic rhizosphere stabilization and degradation are enhanced by the plant roots' release of cometabolites and by the roots' facilitating soil aeration (see Figure 8.6). Thus, the physical contact and the biochemistry needed for degradation is improved when roots and microbes are both available in the right environment.

The chemical-organism-environmental systematics inevitably leads to the question, then, as to what are the properties of molecules that render them recalcitrant? Why do some compounds that *should* be broken down in fact persist in the environment for so long? Indeed, not only large molecules are recalcitrant. Some relatively low molecular weight compounds can last for years in the environment.

In addition, certain environmental conditions add to the likelihood of molecular recalcitrance. Therefore, the time it takes to break down molecules is a function of the structure of the molecule, the carbon preference of the organism and the environmental conditions.

An environmental "rule of five?"

Pharmacologists use some standard screening approaches to characterize chemicals in respect of their likely biological activity based on chemical structures. For example, certain general molecular properties will drive the pharmacokinetics, i.e. the extent to which a chemical will be absorbed, distributed, metabolized, and eliminated after uptake into an organism. One model is the so-called "rule of five," proposed by C.A. Lipinski [34], which predicts that a compound is less likely to be absorbed and to permeate cellular membranes if:

The structure includes more than 5 H-bond donors (expressed as the sum of hydroxyl and amine groups, i.e. OHs + NHs);

The molecular weight is greater than 500;

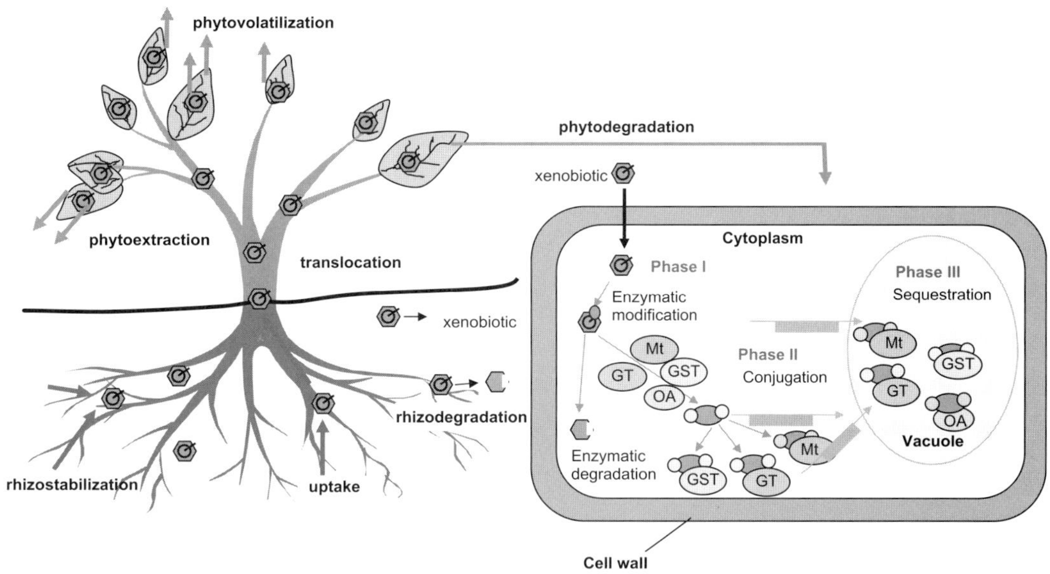

FIGURE 8.6
Attenuation and degradation mechanisms in macrophytic plants. A compound (xenobiotic) is stabilized or degraded in the rhizosphere, adsorbed or accumulated in to the roots and transported to the aerial parts, volatilized or degraded inside the plant tissue. Degradation generally involves enzymatic mediated metabolism (Phase I); conjugation (Phase II); and active sequestration (Phase III). Note: Active transporters are marked in green boxes (GST = glutathione *S*-transferases; GT = glucosyltransferases; Mt = Malonyltransferases; OA = organic acids. [*See color plate section*]
Source: P.C. Abhilash, S. Jamil and N. Singh (2009). Transgenic plants for enhanced biodegradation and phytoremediation of organic xenobiotics. *Biotechnology Advances* 27 (4): 474–488.

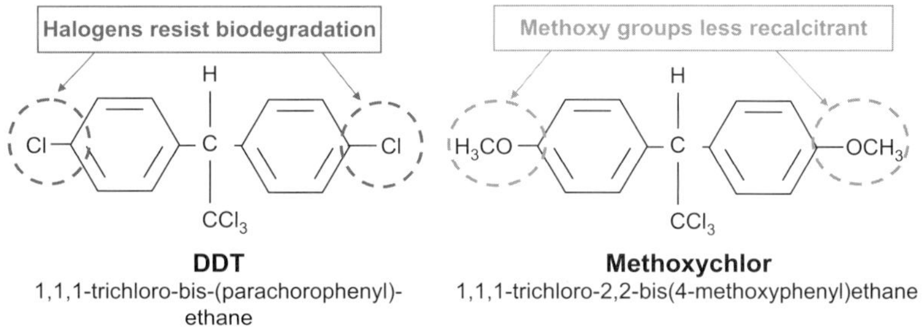

FIGURE 8.7
Structures of the insecticides and methoxychlor, showing the primary sites of attack. DDT is more recalcitrant than methoxychlor due to the chlorine substitutions, rather than the more easily degraded methoxy groups.

The log of the octanol-water coefficient (Log K_{ow}) is greater than 5;
There are more than 10 H-bond acceptors (expressed as the sum of Ns and Os); and,
Compound classes that are substrates for biological transporters are exceptions to the rule.

Thus, large, lipophilic molecules that promote electron acceptance will resist biological activity. This rule can be instructive for understanding recalcitrance. Both the chemical detoxification and chemical metabolism in cells of unicellular and multicellular organisms are biological processes. The degradation of a compound is a product of these biological processes. Therefore, the breakdown pathways may be predicted to some extent based on a substance's inherent chemical and physical properties. In fact, personal care products and pharmaceuticals usually have to be somewhat recalcitrant to be used in the marketplace; otherwise their shelf life would be very short. For example, drugs must resist the effects of oxidation, heat, photolysis, and pH.

(Continued)

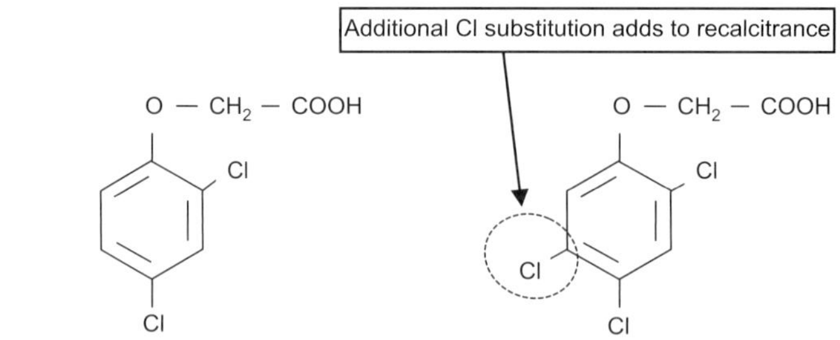

2-4-D
[2,4-dichlorophenoxyacetic acid]

2-4-5-T
[2,4,5-trichlorophenoxyacetic acid]

FIGURE 8.8
Greater chlorination of 2,4,5-T [2,4,5-trichlorophenoxyacetic acid] makes it more recalcitrant than the similarly structured 2-4-D [2,4-dichlorophenoxyacetic acid].

The insecticide DDT [1,1,1-trichloro-bis-(parachorophenyl)-ethane] is structurally quite similar to another insecticide, methoxychlor [1,1,1-trichloro-2,2-bis(4-methoxyphenyl)ethane], except for the substitution at the outside (para) positions of the molecules (see Figure 8.7). The DDT molecule is more recalcitrant because its para-substitutions are with chlorines (i.e. para-chloro substitution, whereas the methoxychlor molecule is vulnerable to dealkylation of the para-methoxy (H_3HO) groups. Most methoxychlor is likely from the air by wet and dry deposition processes in weeks. Although methoxychlor binds tightly to soil particles, it usually does not persist since it is readily biodegraded. However, it has been found to have a soil half-life as high as 120 days in some low oxygen soil conditions [35].

Methoxychlor's degradation products are generally detected in lower levels of soil, probably because these compounds are more mobile than the parent methoxychlor. Methoxychlor has an affinity for sediment, so coupled with its biodegradability potential is seldom protected above detection levels in ground and surface waters, except near sources of release, likely to a large extent due to its low aqueous solubility (about 1 mg L^{-1}). These factors mean that the less recalcitrant methoxychlor is much less likely to bioconcentrate in organisms compared to DDT, which has a range of bioaccumulation in fish, insects, and mammals [36].

Likewise, the herbicides 2-4-D [2,4-dichlorophenoxyacetic acid] and 2,4,5-T [2,4,5-trichlorophenoxyacetic acid] are quite similar, but 2,4,5-T is much more recalcitrant due to the additional halogen (Cl) substitution (see Figure 8.8). 2,4-D's half-life in soil is less than 7 days, predominantly due to microbial degradation [37]. In water, biodegradation rates increase with increased concentrations of nutrients, sediment load, and dissolved organic carbon. Under oxygenated conditions the sediment half-life is 1 week to several weeks. Despite its short half-life in soil and in aquatic environments, it has been detected in ground and surface water in the United States and groundwater in Canada [38]. The biodegradation rates for 2,4,5-T would be lower in most environments, due to the interference and added recalcitrance from the additional Cl atom.

The concept of recalcitrance does not exclusively apply to an entire molecule. In fact, larger molecules are likely to contain different components with varied recalcitrance. One part is susceptible to degradation while others are recalcitrant. For example, microbial enzymes can enhance the cleavage of one of a molecule's moieties (e.g. an aliphatic group) which may be readily mineralized. However, other moieties will be recalcitrant to the enzymatic attack. In fact, the remaining molecule could become even more recalcitrant and more toxic (i.e. a process known as *biological activation*).

Bioengineering researchers seek improved strains to address recalcitrance. The Ecological Society of American [39] points out that:

> *... the development of microbial GEOs for bioremediation remains an active field of research because cleanup of toxic sites is so difficult and costly, and because these sites are often isolated, restricted in access, and altered from their native condition. GE strategies have involved construction of biosensors to monitor pollutant concentrations; production of biosurfactants to increase pollutant uptake by other microbes; adding missing enzymes to complete biodegradation pathways;*

improving those enzymes, often by directed evolution; altering the regulation of biodegradation gene expression (e.g., to achieve constitutive, over-expression); and placing the biodegradation genes in a more suitable host. Primary targets are microbes that can be made to grow on PCBs, chlorinated ethylenes (PCE, TCE), explosives (TNT), and polynuclear aromatic hydrocarbons (PAHs), which are the most problematic environmental pollutants in the industrialized world.

In recent decades, engineering has gained some new disciplines, bioengineering among them. Bioengineers began asking whether genes could be altered in certain organisms to make them more drought tolerant, more nutritionally productive, better tasting, more pest resistant, more efficient biological factories, and more effective waste treatment systems. This was the birth of genetic engineering.

GENETIC ENGINEERING BASICS

Genetic engineering is no different than other types of engineering in the sense that scientific principles are being applied to improve a system's performance, whether it is to improve the cellular metabolic rates to enhance environmental cleanup or to provide better pharmaceuticals. The difference has resided in the scientific principles that were being applied. For a few decades now, this is being accomplished by novel ways of selecting certain traits (genetic expressions).

A genetically modified organism (GMO) is one whose genetic material has been changed in a way that does not occur under natural conditions through cross-breeding or natural recombination [40]. Thus, a genetically engineered (GE) organism's genetic material has been altered using technologies based on recombinant deoxyribonucleic acid (rDNA). It seems unsatisfying, even absurd, to say that the "only" difference between a GE organism and its natural counterpart is the rDNA, since this form of DNA does not exist in natural, unaltered organisms, but is engineered combining DNA sequences in novel ways.

In sequencing, the order of specific nucleotide bases (i.e. adenine, guanine, cytosine, and thymine) in the DNA molecule is manipulated to change the genetic expression. Sequencing has grown rapidly with improvements of analytical technology. In addition to the famous Human Genome Project, complete DNA sequences have become available for numerous organisms important to environmental biotechnology, especially those of microbial genomes.

Actually, the genetic engineering revolution is a continuum from conventional breeding approaches to modification with the introduction of foreign DNA to modification by the transfer of foreign DNA.

Conventional breeding approaches

Desirable traits to improve agricultural crops have been occurring for millennia. If the genetic information already resides in a subpopulation of the species, plants and animals with certain traits are bred by crossing a well-performing, survivable line with a line with the desired characteristics. When the desired trait does not reside within the species, but does exist in distantly related varieties, cell tissue culture can be used to obtain fertile generations from normally sterile crossings.

Such methods take advantage of innate DNA without insertion of foreign DNA. When a desirable trait is manifested, it is encouraged to reproduce and the strain is multiplied. This can take considerable time.

Modification of organisms without introducing foreign DNA

Occasionally, mutant lines with the desired characteristics are selected by chemical mutagenesis using various chemical substances such as alkylating agents (e.g. ethane-methyl-sulphonate and N-ethyl-N-nitrosourea) by irradiation, or by using naturally occurring transposable genetic elements, known as *transposons* [41]. Transposons are sometimes called "hopping genes" because of their unpredictable or random presence in species. Transposons that move about a genome via an RNA intermediate are known as retrotransposons. When stressed, transposons and retrotransposons can be "encouraged" to move. A copy of the transposon is made at the original site on the chromosome using reverse transcriptase. Such movement may stimulate genes that express the desired trait [42]. The various approaches exhibit different characteristics, e.g. ethane-methyl-sulphonate application produces primarily cytosine to thymine changes in the DNA, resulting in C/G to T/A transition mutations in the DNA.

These methods make use of the mobile capabilities of "otherwise inactive fragments" of an organism's own DNA, transposons, and retrotransposons. The genetic reshuffling of a chromosome's genome provides the changes sought by the genetic engineer.

Another approach is to perform *in situ* microsurgery to induce changes in transposons. In fact, the surgery is designed to mimic the random changes mentioned above, but by physically moving transposons. Transposons can be inserted midway through a coding sequence, so that a gene's function is destroyed or modified, leading to a new gene product. Another approach is to target the regulatory portion of the genome to increase or decrease gene expression. Thus, the organism will produce a protein not previously made by the organism (i.e. proteomics). Rarely, insertion of a transposon is made downstream from the regulatory promoter so that a dormant gene will be expressed [43].

Modification of organisms by introducing foreign DNA

The addition of relevant DNA into an existing organism's DNA changes the code so that certain traits are expressed. Novel genes are introduced to another organism's genome by various methods, but fall into two basic categories: transfection and infection using vectors.

TRANSFECTED DNA

Genetic material can be physically injected into a cell nucleus, i.e. transfection. Transfection can be enhanced using various methods. For example, the DNA can cross membranes through transient pores that open using electric pulses, i.e. electroporation. DNA can also be added to a cell by using polycations to neutralize the electrical charges on the DNA molecule and the cell surface that would ordinarily prohibit DNA uptake. Lipofection, i.e. enclosing the DNA in a lipid vesicle, can also be used to transmit the DNA through the lipophilic sites on the cell membrane, mimicking viral transfer. Sperm can also be used to carry foreign DNA, e.g. via intracytoplasmic sperm injection or by electroporation. Widely used approaches for transferring DNA for genetic modification include micro-injecting DNA into eggs or embryos, transfer via bacterial plasmids, and biolistic or so-called "shotgun" methods.

Biolistic impaction is another method of inserting foreign DNA into an organism. Particles, often of gold or titanium, are used as projectiles with the desired gene sorbed to the particle surface. Pressurized helium is used to project the particles into the organism's cells. Those that take up the inserted DNA are identified by a marker gene and cultured.

Microinjection of DNA into eggs or embryos has been used since 1980. The objective is to instill a certain trait from a known DNA sequence, often in animals, that induces the organism to produce a new protein. A promoter is a region of the gene where messenger RNA synthesis begins. The terminator is the region where RNA synthesis ends. So, when the RNA is formed, it travels to the cytoplasm wherein ribosomes produce the specific protein that gives the desired trait. It is not certain that the trait will be expressed, since the insertion into the chromosome is random. More commonly, insertion of the gene is facilitated by a DNA carrier, such as a plasmid or a virus. The embryo is implanted into the animal's uterus. This is known as a "knock-in" method.

VECTOR-BORNE DNA

Two types of vectors can deliver foreign DNA to a cell: viruses and plasmids. Retroviruses infect a cell by reverse transcription, wherein the viral RNA genome is copied into DNA, which is then integrated into the host cell DNA. At that point, the integrated (i.e. modified) DNA can be genetically expressed by the invaded cell's normal transcription mechanisms. This has been done to introduce genetic traits into animal somatic tissue and for germline modification (fish and birds, for example) [44].

Transposons provide another common method of genetic modification, taking advantage of bacteria's ability to penetrate cell walls and the cell's nuclear membrane, so that the gene with the desired trait will be incorporated into the organism's genome or chromosomes (see Figure 8.9). Transposons are not only used in microbes, but have increasingly been used to transmit DNA in plants and animals.

Plasmids are extra-chromosomal DNA molecules. They are often circular and double-stranded, but are separate from the DNA in the chromosome. Plasmids can independently replicate. These features have made plasmids the choice of vector for numerous genetic engineering applications.

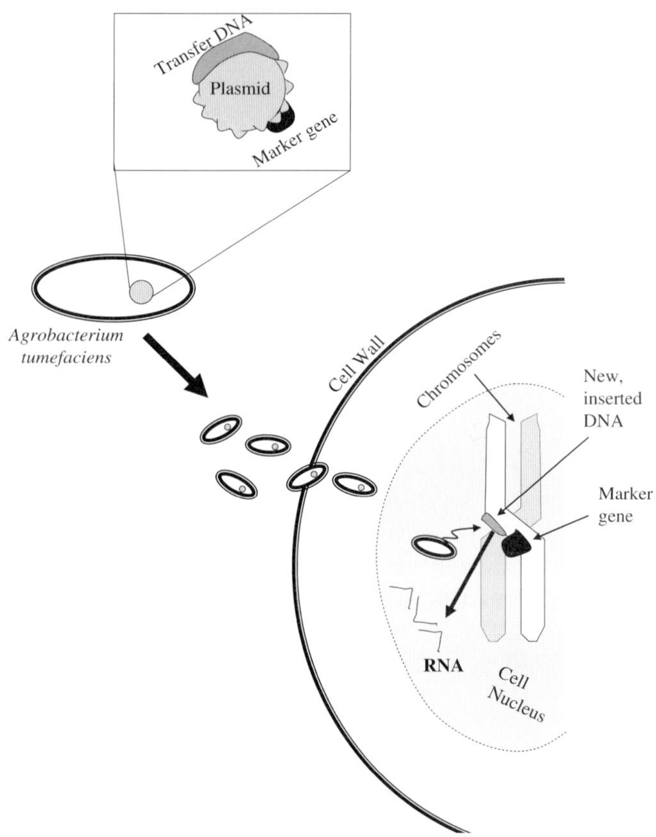

FIGURE 8.9
Soil bacterium *Agrobacterium tumefaciens* plasmid penetration of cell wall and nucleus to insert a genetic trait into an organism. With the help of proteins in the plasmid, transfer DNA is inserted randomly into the organism's chromosomes, accompanied by the marker gene.
Source: Adapted from T.F. Budinger and M.D. Budinger (2006). *Ethics of Emerging Technologies: Scientific Facts and Moral Challenges.* John Wiley & Sons, Inc. Hoboken, NJ.

As mentioned in Chapter 1, biotechnology is often a way of mimicking nature. This is certainly the case in vectors used for genetic modification. As evidence, biologists have observed the behavior of the soil bacterium *Agrobacterium tumefaciens*, leading them to some rather widely used approaches to seek biomimicry. Agrobacteria are natural plant parasites. Part of their survival depends on their ability to insert genes into plant hosts, eliciting a crown gall, i.e. proliferation of cells near the soil surface. The genetic information for infecting the plant's neoplasmic growth is encoded within the plasmid. The *Agrobacterium* infection entails the transfer of DNA (T-DNA or Ti) to a random site in the plant genome. This natural capability to transfer genes has been put to use by bioengineers in biotechnologies. That is, agrobacteria are used to send foreign genes into plants by cutting out the bacterial T-DNA of the plasmid and replacing it with the desired foreign gene. This has been done on various species of dicotyledonous plants.

Thus, plasmids are transferable genetic elements that can replicate autonomously within a host. In this sense they resemble the retroviruses, but differ in that the plasmids are actually "naked" forms of DNA. That is, plasmids do not encode genes necessary to encase the genetic material for transfer to a new host. The foreign DNA is introduced with this recombinant DNA technology and still must be transmitted through the germ line so that every cell has the same modified genetic material. Bioengineering's use of plasmid host-to-host for genetic modification relies on the transfer of genetic material directly by conjugation, i.e. modifying in-host gene expression so as to enhance the uptake of the genetic element by transformation [45]. Microbial transformation with plasmid DNA is not parasitic, nor is it symbiotic. It simply provides a mechanism for horizontal gene transfer within a population of microbes and typically provides a selective advantage under a given set of environmental conditions.

Engineering plasmids also can instill an ability of microbes to change nutrient cycling, such as bacteria fixing elemental nitrogen or to degrade persistent organic compounds by advantageous microbial metabolism and growth in hostile, nutrient-deprived conditions [46]. Genetic modification, whether by transfection or from transposons' ability to carry genes to cells, may well end up giving the modified organisms competitive advantages over natural species within an environmental niche. From a systematic, environmental perspective, this certainly can be good for environmental cleanup, since the advantage can increase biodegradation. However, it can be bad if the competitive advantage changes microbial diversity in a negative way. Also, some of the proteins produced may be toxic and disrupt the life cycles of other organisms.

ENVIRONMENTAL ASPECTS OF CISGENIC AND TRANSGENIC ORGANISMS

Genetically engineered organisms that have been modified exclusively by using the DNA from their own species are called *cisgenic species*. Cisgenesis in microbes may not be all that different from natural gene flow, since microorganisms reproduce so rapidly and microbial populations drift so readily that many of the DNA insertions probably have already occurred. For example, adaptation and acclimation for biodegradation of organic matter may result from the microbe's ability to "insert" DNA in a few organisms, which rapidly reproduce due to their selective competitive advantage.

Conversely, GE organisms that receive DNA from a different species than their own are known as *transgenic species*, so their competitive advantages may not resemble any in the past. In recent decades, genetic research has expanded the use of rDNA methodologies that permit the introduction of genes from distantly related species or even from different biologic kingdoms.

Foreign DNA in plants

Major scientific uncertainties regarding recombinant DNA remain. These are certainly more pronounced in animals than in plants and microorganisms. Many of the environmental concerns early on addressed plants.

Environmental concerns emerge when foreign DNA (such as a gene in plants that expresses pest-protection) is added in between T-DNA plasmid-encoded insertion sequences. This results in the foreign DNA sequences also being inserted into the plant's chromosome. That is, the DNA insertion is not always as precise as bioengineers may hope, resulting in the potential for uncertain effects from gene flow [47].

As evidenced by the National Institutes of Health guidelines [48] in 1978, concern arose quickly about possible release of genetically engineered organisms and the concomitant environmental and health risks. These concerns have continued, albeit under various scenarios, ever since. The NIH guidelines in 1978 prohibited the environmental release of genetically engineered organisms unless exempted by the NIH director. Eventually a number of court cases about a genetically modified species field trial required an environmental impact statement under NEPA (see Chapter 1). Simultaneously in the early 1980s, the US Congress began questioning the ability of federal agencies to address hazards to ecosystems in light of the uncertainties. In 1984, the Senate Committee on Environment and Public Works discussed the potential risks with representatives of the US EPA, NIH, and USDA, who held that existing statutes were sufficient to address the environmental effects of genetically engineered organisms. Also in 1984, a White House committee was formed under the auspices of the Office of Science and Technology Policy (OSTP) to propose a plan for regulating biotechnology [49].

The OSTP published the Coordinated Framework for the Regulation of Biotechnology in 1986 that is still in use. The overarching principle of the framework is that biotechnology is not inherently risky and, as such, should not be regulated as a *process*. Rather, the *products* of biotechnology should be regulated in the same way as any product. The coordinated framework outlined the roles and policies of the federal agencies and was built from the contention that existing laws were, on the whole, adequate for oversight of biotechnology products. The upshot of the rules was that, overall, the federal government's consideration of the products led it to believe that the behavior of genetically engineered organisms would not be expected to be fundamentally different from that of non-GMOs [50]. Of course, this might have had a different outcome had the full process been considered.

In 1987, the National Academy of Sciences prepared a white paper [51] with similar conclusions and recommended that the product, not the process, be regulated. It also stated genetically engineered organisms posed no new kinds of risks, that the risks were "the same in kind" as those presented by non-GE organisms. Shortly thereafter, the USDA reviewed and approved transgenic crop varieties for field trials under the Federal Plant Pest Act, which defined a plant pest as:

> ... any living stage of ... insects, mites, nematodes, slugs, snails, protozoa, or other invertebrate animals, bacteria, fungi, other parasitic plants or reproductive parts thereof, viruses, or any organisms similar to or allied with any of the foregoing, or any infectious substances, which can directly or indirectly injure or cause disease or damage in any plants or parts thereof, or any processed, manufactured, or other products of plants. [52]

Numerous international agencies have reached conclusions similar to those of the NAS white paper that the risks posed by transgenic organisms are expected to be the "same in kind" as those associated with the introduction of unmodified organisms and organisms modified by other methods. The types of genetic engineering included in most of the plant modification techniques are shown in Table 8.1.

This may well be true. However, as mentioned in previous chapters, ecosystems are quite complex. For example, the conclusion that plant biotechnological processes are the same and only the products differ needs to be challenged. The NAS report identified numerous examples of potential gene flow (see Table 8.2). Whereas hybridization takes place within the same

Table 8.1 National Academy of Sciences' summary of genetic basis of resistance traits that have been bred into cultivated plants using conventional and transgenic techniques!

Conventionally bred plants only	Polygenic traits – controlled by several interacting genes, usually selected without knowledge of which genes are involved
Both conventionally bred and transgenic plants	Single-gene traits from the same species or a related species Several single-gene traits that are not genetically linked and are therefore inherited independently Several single-gene traits that are physically linked and inherited as a unit; occasionally possible with conventional breeding, as when a chromosome segment bearing more than one resistance gene is transferred to the cultivar usually accompanied by extraneous DNA; transgenic methods allow several single-gene traits to be tightly linked without extraneous DNA Single-gene traits expressed only in particular tissues or at particular developmental stages because of specific promoters; occasionally possible with conventional breeding, but more flexible and precise with transgenic methods
Transgenic plants only	Single-gene traits found in the same species or a related species and modified by changes in the nucleotide sequence of the structural gene or the promoter to improve the plant's phenotypic characteristics Single-gene traits obtained from unrelated organisms (such as viruses, bacteria, insects, vertebrates, and other plants); sometimes modified by a change in the nucleotide sequence of the structural gene or the promoter to improve the plant's phenotypic characteristics Single-gene traits that can be induced by a chemical spray or by specific environmental conditions (such as threshold temperature), based on the action of specific promoters; (these traits may also occur naturally in non-transgenic plants, such as those with systemic acquired resistance, but have rarely been selected intentionally by conventional breeding)

Source: National Academy of Sciences (2000). National Research Council. *Genetically Modified Pest-Protected Plants: Science and Regulation.* The National Academies Press, Washington, DC.

genus, the ability of the genetic material of a genetically modified species to flow to other species should highlight the uncertainties about the conclusion that these biotechnological processes can be considered to be essentially identical, as evidence, the Academy notes the complexities of the gene flow processes:

> The frequency of a given crop gene in a wild population depends on many factors, including the rate at which it is introduced into the population; temporary fitness barriers, if any, in the first and early backcross generations; possible fitness costs associated with the gene itself; and possible benefits of the gene for the plant's survival and reproduction ... When several transgenes are inserted together as tightly linked traits (inherited as a unit), the combined ecological costs and benefits of these traits will determine whether and to what extent a wild relative's fitness is enhanced. [53]

Table 8.2 **Selected commercially important plant species able to hybridize with wild species in the continental United States**

Family and cultivated species	Wild relative
Apiaceae	
Apium graveolens (celery)	Same species
Daucus carota (carrot)	Same species (wild carrot)
Chenopodiaceae	
Beta vulgaris (beet)	*B. vulgaris* var. *maritima* (hybrid is a weed)
Chenopodium quinoa (quinua, a grain)	*C. berlandieri*
Compositae	
Chicorium intybus (chicory)	Same species
Helianthus annuus (sunflower)	Same species
Lactuca sativa (lettuce)	*L. serriola* (wild lettuce)
Cruciferae	
Brassica napus (oilseed rape; canola)b	Same species, *B. campestris*, *B. juncea*
Brassica rapa (turnip)	Same species (= *B. campestris*)
Raphanus sativus (radish)	Same species, *R. raphanistrum*
Cucurbitaceae	
Cucurbita pepo (squash)	Same species (= *C. texana*, Wild squash)
Ericaceae	
Vaccinium macrocarpon (cranberry)	Same species
Vaccinium angustifolium (blueberry)	Same species
Fabaceae	
Trifolium spp. (clover)	Same species
Medicago sativa (alfalfa)	Same species
Hamamelidaceae	
Liquidambar styraciflua (sweetgum)	Same species
Juglandaceae	
Juglans regia (walnut)	*J. hindsii*
Liliaceae	
Asparagus officinalis (asparagus)	Same species
Pinaceae	
Picea glauca (spruce)	Same species
Poaceae	
Avena sativa (oat)	*A. fatua* (wild oats)
Cynodon dactylon (bermuda grass)	Same species
Oryza sativa (rice)	Same species and others (red rice)
Saccharum officinarum (sugar cane)c	*S. spontaneum* (wild sugarcane)
Sorghum bicolor (sorghum)	*S. halepense* (johnsongrass)
Sorghum bicolor (sorghum)	Same species (shattercane)
Triticum aestivum (wheat)	*Aegilops cylindrica* (jointed goatgrass)e
Rosaceae	
Amelanchier laevis (serviceberry)	Same species
Fragaria sp. (strawberry)	*Fragaria virginiana*
Rubus spp. (raspberry, blackberry)	Same species
Salicaceae	
Populus alba x *P. grandidentata* (poplar)	*Populus spp.*

(*Continued*)

Table 8.2 Selected commercially important plant species able to hybridize with wild species in the continental United States—cont'd

Family and cultivated species	Wild relative
Solanaceae	
Nicotiana tabacum (tobacco)	Same species
Vitaceae	
Vitis vinifera (grape)	*Vitis spp*. (wild grape)

Source: National Academy of Sciences (2000). National Research Council. Genetically Modified Pest-Protected Plants: Science and Regulation. National Academies Press, Washington, DC.

Furthermore, the Academy concludes:

> Until better data are available, it will be necessary to rely on general ecological and agricultural knowledge to predict the consequences of commercial-scale, crop-to-wild gene flow from pest-protected plants. [54]

This apparent oversimplification of the differences between biotechnological processes and reliance on the similarities of the products illustrates a common problem of intermingling of risk assesment with risk management. The former is to be a scientifically based process, whereas the latter takes into account numerous perspectives, such as costs and feasibility. Some could argue the conclusion about the environmental acceptability of rDNA modification has jumped too quickly into the risk management milieu, without sufficient risk management. As evidence, the Academy states:

> Because of the uncertainties described above, it is premature to predict the ecological impacts of gene flow from transgenic pest-protected plants. Meanwhile, regulatory decisions must be made in a timely fashion. It seems unlikely that the transfer of one or two novel crop genes for pest-protection would transform a wild species into a problematic weed, although in some cases unwanted population increases of weedy species could result. Moreover, the cumulative effects of beneficial crop genes could potentially lead to expensive and ecologically damaging problems in weeds that are already difficult to control, such as Johnson grass (*Sorghum halepense*). In the future, additional phenotypic traits might include broad-spectrum resistance to insects or diseases and greater tolerance of cold, drought, salinity, nutrient scarcity, or acidic soils. Such traits could be more advantageous to wild relatives than those now in use. [55]

Indeed, few are arguing that biotechnology is needed, but many are not completely assured that the products *and* the process have undergone proper risk assessment underpinned by systematic scientific scrutiny.

Biochemodynamic flow of modified genetic material

Gene flow is the exchange of genes or genetic material between different populations within a species. An extreme form of hybridization involves gene flow between completely different species, which is one of the concerns with genetically modified organisms, wherein the rDNA flow can lead to downstream implications, including changes in diversity and uneven competition.

Risk assessments of gene flow are usually very limited in time and space. Large-scale studies of genetically modified crop plants, for example, are seldom studied epidemiologically. That is, they are not studied at the same temporal or spatial scales as they are actually grown (often greenhouse or test plot scales). This greatly limits their usefulness in application, since the processes at work may miss important synergistic, antagonistic, and chaotic outcomes, which can occur in agricultural and other

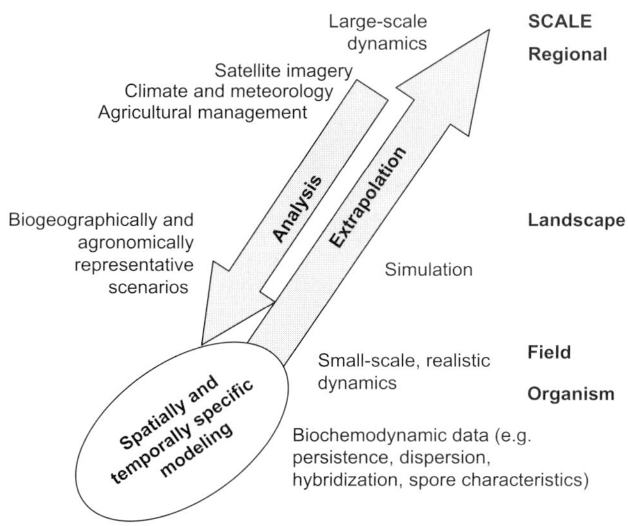

FIGURE 8.10
Upscaling and spatial extrapolation method. Landscape analysis based on spatial information results in the generation of representative scenarios for a region. Based on this information and biochemodynamic information about the genetically modified organism and the environmental conditions, realistic, site-specific models are developed. From these models, larger scale (e.g. landscape or regional) conditions are estimated based on indicators, and extrapolations are made to similar areas.
Source: Adapted from H. Rueter, G. Schmidt, W. Schröder, U. Middelhoff, H. Pehlke and B. Breckling (2009). Regional distribution of genetically modified organisms (GMOs) – Up-scaling the dispersal and persistence potential of herbicide resistant oilseed rape (*Brassisca napus*). *Ecological Indicators.* Article in press. doi:10.1016/j.ecolind.2009.03.007.

ecosystems. For example, experiments do not allow much certainty in how genetic material may integrate, persist and be dispersed. Recall from the discussions on risk assessment that public health and environmental risks involve very rare events. It is not all uncommon to attempt to predict an outcome (e.g. cancer case per population) of less than one event in a million (i.e. risk $<10^{-6}$).

Dispersion models may be useful in extrapolating from smaller scales (see Figure 8.10). One approach is to start with the smaller-scale studies, apply landscape characterization by overlaying satellite images, climate and agricultural data, and generate representative scenarios of GMO gene flows. Next, individual models are developed from biochemodynamic information (e.g. development, persistence and dispersal of spores and seeds, hybridization partners such as those in Table 8.2, and other site-specific information). With this information, various simulations can be generated from which indicators of a GE organism's gene flow can be extrapolated and indicators of the dynamics of the species can be estimated or predicted [56].

This approach can be used to analyze gene flow at the landscape and regional level and to estimate changes in neighboring areas' GMO content. For example, Figure 8.11 shows the landscape and neighborhood structural changes modeled for GM content in a German soil seedbank for oil seed rape (*Brassisca napus*), based on monitored fields on which no GM crops were grown during the simulation time. During the simulated 10-year period the number of times that a transgenic crop was grown on a directly adjacent field was entered into the model. In this case, the GM fraction on those fields declined from an average 1.25% when GM plants were grown on four neighboring fields to 0.46% with one neighbor. With no neighbor growing the GM *B. napus* crop, the mean GM content in the soil seedbank reached 0.07%, likely the result of long-distance pollen flow. Figure 8.11 illustrates the number of extrapolated GM oil rapeseeds per hectare for the whole region. The figure shows the regional increase in GM *B. napus* seed contents over the 10-year period. This indicates that gene flow can occur extensively and rapidly.

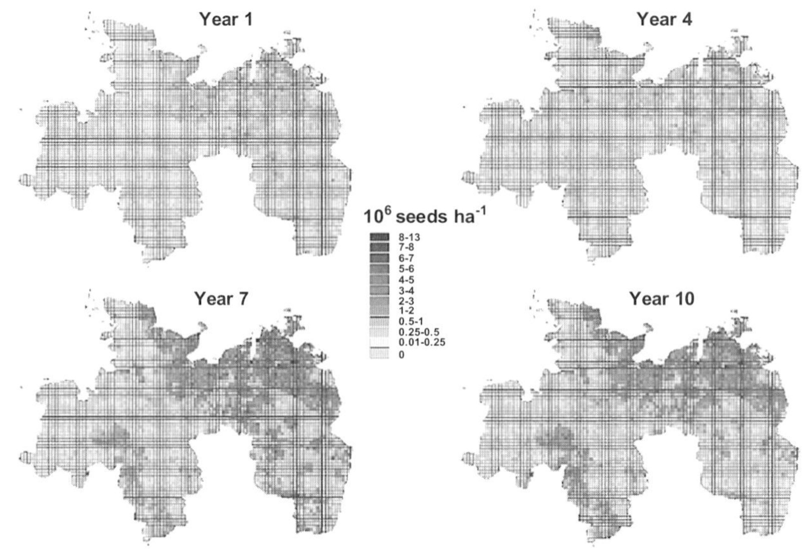

FIGURE 8.11
Indicator of gene flow. 10-year estimates for *Brassisca napus* seeds in a soil seedbank for a region in northern Germany, based on the methods described in Figure 8.10. [*See color plate section*]
Source: H. Rueter, G. Schmidt, W. Schröder, U. Middelhoff, H. Pehlke and B. Breckling (2009). Regional distribution of genetically modified organisms (GMOs) – up-scaling the dispersal and persistence potential of herbicide resistant oilseed rape (*Brassisca napus*). *Ecological Indicators*. Article in press.

Emerging analytical tools can be applied to gene flow. For example, the probability that a foreign pollen grain will result in a successful fertilization in an adjacent field can be calculated from gene-flow experiments falling into two design classes:

In fields situated next to each other (adjacent fields) wherein the probability of foreign pollination is measured as a function of the distance from the common edge.
In fields separated by some distance (non-adjacent fields) wherein the probability of foreign pollination is measured at a single or few locations within the field so that only a mean probability of foreign pollination can be determined for the field. [57]

With inherently complex environmental systems, such as agricultural ecosystems, reliable methods are needed to address ecosystem response, uncertainty, variability, and change. Bayesian statistics have been particularly effective in forecasting pollutant scenarios and should be useful for predicting potential environmental impacts from GMO gene flow. For example, Bayesian methods have been used to pull together multiple gene flow studies. These analyses have shown that increasing isolation distance appears to be more effective to reduce GM-pollen dispersal than the use of a buffer zone, especially for small recipient fields. Expanding the width of a recipient field, relative to the pollen donor field can greatly reduce the average level of fertilization by foreign pollen within the recipient field. The results indicate that a GM-pollination success decreases with isolation distance with the width of the non-GM-field.

The biochemodynamic processes described in Chapters 2 and 3 can form the basis for developing estimates of the temporal and spatial extent of gene flow and other movement of genetic materials within and among ecosystems. There is some value, at the screening level at least, to use simple Gaussian dispersion algorithms (see Figure 8.12) to estimate drift over simple terrain (e.g. low roughness index and low relief). This approach assumes that the organisms or their materials (e.g. spores) will be dispersed randomly according to wind vectors. That is, standard deviations of particles (microbes, spores, etc.) in the x, y and z axes are calculated to determine the location of the plume carrying these particles. However, new atmospheric dispersion methods may be applied to more complex ecosystems characterized by vertical venting in forest areas, channeling down canyons, and both horizontal and vertical

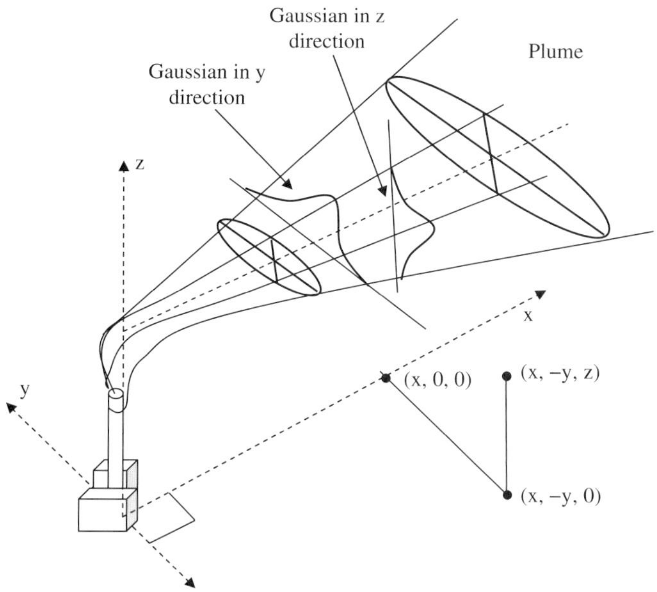

FIGURE 8.12
Atmospheric plume model based upon random (Gaussian) distributions in the horizontal and vertical directions.

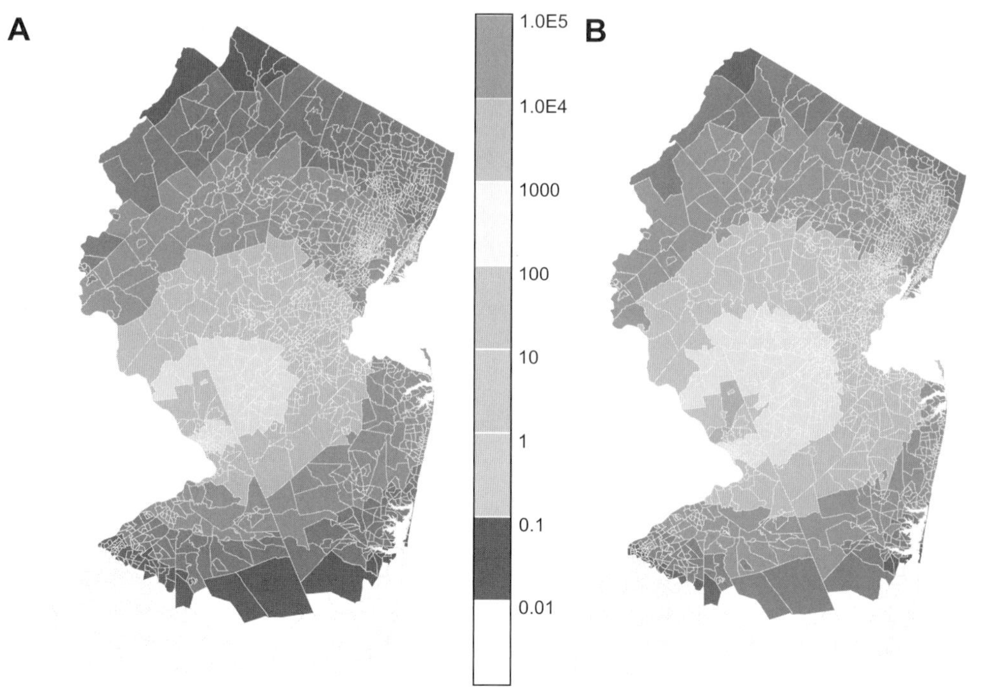

FIGURE 8.13
Estimated median (A) and 99th percentile (B) individual biological doses (spores/day) in each New Jersey census tract for a one-hour release scenario, based on a dispersion model (CALPUFF). [*See color plate section*] *Source:* D. Vallero, S. Isukapalli, P. Georgopoulos and P. Lioy (2009). Improved Assessment of Risks from Emergency Events: Application of Human Exposure Measurements. 4th Annual Interagency Workshop: Using Environmental Information to Prepare and Respond to Emergencies. July 14, 2009, New York, NY.

re-circulations that may occur at local sites. In fact, some of the recent computational models that have been developed for air pollution in complex airsheds may be put to use in ecological risk assessments of GE organisms.

Modeling biological agent transport: Examples

Dispersion models can also be applied to estimate exposures in human populations to releases of biotechnological toxic byproducts or to GE organisms or their products (e.g. spores). In fact, emergency response models are already being developed for biological agents, such as anthrax spores, which could be adapted to numerous GMO releases (see Figure 8.13).

The process for these estimates has consisted of a number of steps for a hypothetical scenario, e.g. the release of 100 g anthrax released over two time periods: one hour and 10 hours. This allows for a distinction of dispersion following an immediate release versus an attenuated release. In this case, the site was that of an anthrax letter mailing in New Jersey in 2001. The modeled release and dispersion consisted of eight steps:

- Step 1: Background spore level estimation from dispersion model: CALPUFF
- Step 2: Census tract (CT) level aggregation/averaging of CALPUFF results
- Step 3: Microenvironmental concentration estimates based on simple steady state mass balance
- Step 4: Characterization of populations – 500 individuals per CT (Census 2000)
- Step 5: Activity patterns – Comprehensive Human Activity Database (no change in activities due to release)
- Step 6: Inhalation rates calculated from published literature
- Step 7: Inhalation dosimetry – simple uptake modeling (deposition fractions based on ICRP data)
- Step 8: Dose-response – simple age-dependent dose-response model (multiple formulations)

The results for a one-hour release scenario are shown in Figure 8.12, which demonstrates that a biological agent can be dispersed extensively in a short time. Obviously, these steps address an immediate human health risk. However, it could be adapted to other scenarios, including those in ecosystems. A non-accidental release would be more continuous. In addition, the population characterization could be based on vulnerable species or habitat, the activity patterns could follow ecosystem functions, and inhalation rates and dosimetry could be replaced by uptake and cycling within an ecosystem. Dose-response could be replaced by another endpoint, such as biodiversity.

The transport of genetic materials can be complex. Some materials (e.g. spores) are transported as whole particles. Other genetic materials hitch a ride, so to speak, when they are sorbed to dust and soil particles.

The *endotoxin* lipopolysaccharide (LPS) is derived from the cell wall of gram-negative bacteria. The substance is very *antigenic*, i.e. it elicits an immune response. The likelihood of exposure to humans is high since LPS is widely distributed throughout the environment. Exposures [58] to levels of endotoxin $\geq$0.2 endotoxin unit (EU) per cubic meter dust have associated with acute respiratory diseases. Asthma, bronchitis, and other chronic respiratory diseases [59] have been associated with daily exposures $\geq$10 EU m^{-3}. The airborne LPS can be elevated near certain facilities, e.g. swine barns [60] (4385 EU m^{-3}) and composting plants [61] (0 to 400 EU m^{-3}). A major exposure pathway for endotoxins is by atmospheric transport, especially the inhalation of aerosols with sorbed or dissolved toxins. Bacterial concentrations in soil routinely exceed 10^8 per gram (with the majority being gram negative).

Whereas genetically engineered organisms share many characteristics of their non-modified counterparts, subtle changes can evoke major changes in any system, especially chaotic and complex environmental systems. The more reliable the tools being used, the better the estimates of gene flow and other indicators of potential biomaterial transport. With these improved estimates, the potential risk of GMOs and their byproducts can be better assessed.

Risk recommendations

The Ecological Society of America (ESA) has recently recommended ways to balance the environmental benefits and risks from genetically engineered organisms [62]. The society concluded that some genetically engineered organisms could play a positive role in sustainable agriculture, forestry, aquaculture, bioremediation, and environmental management. However, deliberate or inadvertent releases of genetically engineered organisms into the environment may induce negative ecological effects under certain circumstances. The potential risks include:

creating new or more vigorous pests and pathogens;
exacerbating the effects of existing pests through hybridization with related transgenic organisms;
harm to nontarget species, such as soil organisms, non-pest insects, birds, and other animals;
disruption of biotic communities, including agroecosystems;
irreparable loss or changes in species diversity or genetic diversity within species.

The ESA also concluded that many potential applications of genetic engineering extend beyond traditional breeding; encompassing viruses, bacteria, algae, fungi, grasses, trees, insects, fish, and shellfish. Genetically engineered organisms that present novel traits will need special scrutiny with regard to their environmental effects. As such, the ESA recommends the following:

Genetically engineered organisms should be designed to reduce environmental risks.
More extensive studies of the environmental benefits and risks associated with genetically engineered organisms are needed.
Environmental effects should be evaluated relative to appropriate baseline scenarios.
Environmental release of genetically engineered organisms should be prevented if scientific knowledge about possible risks is clearly inadequate.
In some cases, post-release monitoring will be needed to identify, manage, and mitigate environmental risks.
Science-based regulation should subject all transgenic organisms to a similar risk assessment framework and should incorporate a cautious approach, recognizing that many environmental effects are genetically engineered organism- and site-specific.
Ecologists, agricultural scientists, molecular biologists, and others need broader training and wider collaboration to address these recommendations.

Furthermore, the ESA recommends that risk evaluations of genetically engineered organisms focus on the phenotype or product rather than the process of genetic engineering, but that some genetically engineered organisms possess novel characteristics that require greater scrutiny than organisms produced by traditional techniques of plant and animal breeding. Also, the ESA points out that unlike commercialized crops or farm-raised fish, for a number of genetically engineered organisms there is little experiential information about breeding, release, and monitoring. Genetic engineering is different in both degree and kind compared to "traditional breeding, encompassing transgenic viruses, bacteria, algae, fungi, grasses, trees, insects, fish, shellfish, and many other nondomesticated species that occur in both managed and unmanaged habitats" [63].

The ESA states that "the environmental benefits and risks associated with genetically engineered organisms should be evaluated relative to appropriate baseline scenarios (e.g., transgenic vs. conventional crops), with due consideration of the ecology of the organism receiving the trait, the trait itself, and the environment(s) into which the organism will be introduced. Predicting impacts prior to commercialization is difficult, thus the ESA strongly recommends a "cautious approach to releasing such genetically engineered organisms into the environment" [64]. Scenarios in need of special concern include those where:

there is little prior experience with the trait and host combination;
the genetically engineered organism may proliferate and persist without human intervention;
genetic exchange is possible between a transformed organism and nondomesticated organisms; or
the trait confers an advantage to the genetically engineered organism over native species in a given environment. [65]

Thus, scientifically rigorous risk assessments should determine the likelihood and extent of biotechnological hazards, including:

creating new or more vigorous pests and pathogens;
exacerbating the effects of existing pests through hybridization with related transgenic organisms;
harm to nontarget species, such as soil organisms, non-pest insects, birds, and other animals;
disruptive effects on biotic communities;
irreparable loss or changes in species diversity or genetic diversity within species. [66]

According to the ESA, genetically engineered organisms should be evaluated in a transparent manner. The relevant regulatory policies must be evaluated and adjusted in time to accommodate emerging applications of genetic engineering and improved knowledge and scientific advances. As such, the ESA recommends the following research priorities [67]:

Early planning in genetically engineered organism development – Genetically engineered organisms should be designed to reduce unwanted environmental risks by incorporating specific genetic features, which might include sterility, reduced fitness, inducible rather than constitutive gene expression, and the absence of undesirable selectable markers.

Analyses of environmental benefits and risks – Rigorous, well-designed studies of the benefits and risks associated with genetically engineered organisms are needed. Ecologists, evolutionary biologists, and a wide range of other disciplinary specialists should become more actively involved in research aimed at quantifying benefits and risks posed by genetically engineered organisms in the environment. Because of the inherent complexity of ecological systems, this research should be carried out over a range of spatial and temporal scales. ESA further recommends that the government and commercial sectors expand their support for environmental risk assessment (including environmental benefits) and risk management research.

Preventing the release of unwanted genetically engineered organisms – Strict confinement of genetically engineered organisms is often impossible after large-scale field releases have occurred. Therefore, ESA recommends that large-scale or commercial release of genetically engineered organisms be prevented if scientific knowledge about possible risks is inadequate or if existing knowledge suggests the potential for serious unwanted environmental (or human health) effects.

Monitoring of commercial genetically engineered organisms – Well-designed monitoring will be crucial to identify, manage, and mitigate environmental risks when there are reasons to suspect possible problems. In some cases, post-release monitoring may detect environmental risks that were not evident in small-scale, pre-commercial risk evaluations. Because environmental monitoring is expensive, a clear system of adaptive management is needed so that monitoring data can be used effectively in environmental and regulatory decision-making.

Regulatory considerations – Science-based regulation should: (a) subject all transgenic organisms to a similar risk assessment framework, (b) recognize that many environmental risks are genetically engineered organism- and site-specific, and therefore that risk analysis should be tailored to particular applications, and (c) incorporate a cautious approach to environmental risk analysis.

Multidisciplinary training – Ecologists, agricultural scientists, molecular biologists, and others need broader training to address the above recommendations. The ESA strongly encourages greater multidisciplinary training and collaborative, multidisciplinary research on the environmental risks and benefits of genetically engineered organisms.

In summary, the ESA urges that sound science underpins the risk–benefit assessments of genetically engineered organisms that may be released into the environment, including attention to the potential environmental effects over large spatial scales and long timeframes. Further, the ESA considers genetically engineered organisms that are phenotypically similar to conventionally bred organisms to raise few new environmental concerns; however, numerous types of genetically engineered organisms are being considered for future development. These include baculoviruses that are engineered for more effective biological control, microorganisms that promote carbon storage, fast-growing fish, and fast-growing plants that tolerate cold, drought, or salinity. The ESA has stated its commitment to providing scientific expertise for evaluating and predicting ecological benefits and risks posed by field-released transgenic organisms. [68] Clearly, this rightfully calls for a systematic perspective in addressing environmental risks potentially posed by biotechnologies.

SEMINAR TOPIC

Biosystematic Perspective on Coral Reef Impairment

Coral reefs are some of the most productive ecosystems on earth (see Table 8.3). Threats to the function and structure of coral reef ecosystems have been associated with wastewater discharges to major rivers that ultimately drain into oceans, carrying nutrients and microbial populations. More recently, these discharges have been found to contain drugs, personal care products, and antibiotics that are not degraded or are incompletely degraded, leading to stresses on aquatic ecosystems. For example, the corals off the coast of Florida, the world's third largest barrier reef, have been highly stressed, with half of the live coral off the Florida coast lost in the past few years. An additional indication is that fish feeding on these corals are developing deformities and experience premature mortality [69].

The threats to coral reefs come in many chemical and biological forms. The US Coral Reef Task Force recently identified what it considers to be the most prominent threats that federal agencies and states must address to protect coral reefs in the United States [70]:

Pollution, including eutrophication and sedimentation from poor or overly intensive land use, chemical loading, oil and chemical spills, marine debris and invasive alien species.

Overfishing and exploitation of coral reef species for recreational and commercial purposes, and the collateral damage and degradation to habitats and ecosystems from fishing activities.

Habitat destruction and harmful fishing practices, including those fishing techniques that have negative impacts on coral reefs and associated habitats. This can include legal techniques such as traps and trawls used inappropriately, as well as illegal activities such as cyanide and dynamite fishing.

Dredging and shoreline modification in connection with coastal navigation or development.

Vessel groundings and anchoring that directly destroy corals and reef framework.

Disease outbreaks that are increasing in frequency and geographic range are affecting a greater diversity of coral reef species.

Global climate change and associated impacts including reduced rates of coral calcification, increased coral bleaching and mortality (associated with variety of stresses including increased sea surface temperatures), increased storm frequency, and sea level rise.

Whereas all seven threats relate to biotechnologies, the sixth in the list appears to be an implication of microbial infestation and infection. These microbes appear to be transported long distances in winds aloft. For example, some of the invasive bacteria that threaten coral reef habitats may be coming from Africa in the form of Saharan dust. Deserts commonly contain gravel and bedrock, along with some sand. The Sahara is the exception, with sand covering 20% of the spatial extent of the desert. This means that the Sahara often loses large amounts of dust by winds that advectively transport particles in plumes that can travel across the Atlantic Ocean, depositing dust along the way (see Figure 8.14). Saharan dust carries disease-causing bacteria and fungi that have been associated with the destruction of coral reefs in the Caribbean Sea.

Scientists from numerous disciplines are studying these phenomena and trying to evaluate the linkages, the threats, and possible interventions. The scientific goals associated with these threats, along with their respective ordinal (high, medium, and low) priorities for scientific research are shown in Figure 8.15. Note that invasive species can include numerous organisms, but recently algae and bacteria have been highlighted as a much larger concern than had previously been thought. The bacteria are likely arriving in the coral reef ecosystems by long-range, atmospheric transport. Note in Figure 8.16 that the actions recommended by the National Oceanic and Atmospheric Administration to address these threats vary by region.

Figures 8.15 and 8.16 indicate that coral reefs are threatened by an array of physical, chemical, and biological stressors. Ultraviolet radiation is an example of a physical stressor. UV radiation is a ubiquitous stressor that is very likely impacting human and ecological systems on a global scale. It has been implicated in observed shifts in polar plankton community composition, local and global declines in amphibian population abundance and diversity, coral bleaching syndrome, not to mention an increasing incidence of human skin cancer and other diseases. Disruption and loss of coral reef ecosystem communities due to coral bleaching/disease leads to coral mortality, changes in reef persistence and formation dynamics, as well as cascading reef community interactions [71].

An example of a chemical threat is tributyl tin, a potent endocrine disrupting compound that was used throughout much of the 20th century as an antifouling agent in coatings applied to watercraft. In other words, the tin compounds were added to paint for the stated purpose of preventing the growth of marine organisms. So, when the compounds leach into water, they continue this mode of action, including in coral reef habitats, decreasing the reproduction mechanisms in aquatic organisms. This was a major reason for the 2003 ban on this use of tributyl tin in the United States. This is also an example of an indirect biological threat, since tributyl tin first affects hormonal response at the cellular level, which in turn affects the overall structure and function of the coral community.

Biological agents can present an even greater and more ominous threat to coral reefs. The microbial ecology within a coral reef is complex. For example, some algae are symbiotic and others are parasitic to corals. Fungi are also widely distributed in calcium carbonate, i.e. they have the *endolithic* associations with corals [72]. Bacterial diseases are also appearing. Thus, various biological taxa of microorganisms present different, but possibly synergistic threats to coral reef systems.

Seminar Questions

How much of the coral reef damage and loss can be attributed to anthropogenic causes?

How much of the damage and loss can be attributed to long-range atmospheric transport of microbes?

Are the bacterial, fungal, and algal population changes in the reefs related to one another?

Are genetically modified organisms potentially a factor in the decline of coral reefs?

Table 8.3 Net primary productivity of ecosystems (grams dry organic matter per square meter per year)

Ecosystem type	Net primary productivity
Open ocean water	100
Coastal seawater	200
Desert	200
Tundra	400
Upwelling area	600
Rice paddy	340–1200
Freshwater pond	950–1500
Temperate deciduous forest	1200–1600
Cropland (cornfield)	1000–6000
Temperate grassland	≤1500
Cattail swamp	2500
Tropical rain forest	≤2800
Coral reef	4900
Sugarcane field	≤9400

Source: R.M. Maier, I.L. Pepper and C.P. Gerba (2009). *Environmental Microbiology*, 2nd Edition. Elsevier Academic Press, Burlington, MA.

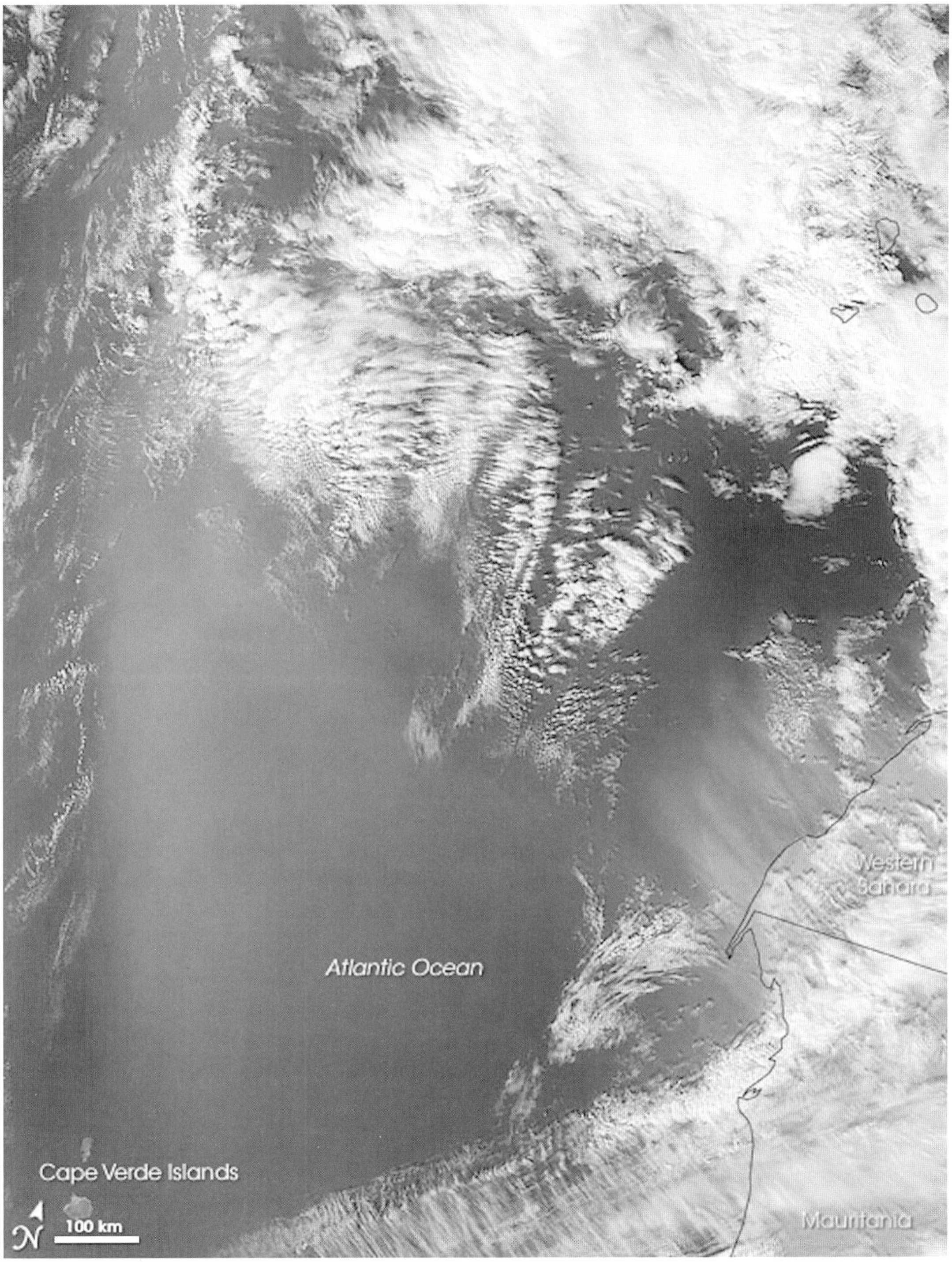

FIGURE 8.14
Plumes blowing off Africa's west coast on January 22, 2008. The plumes carry dust from the Sahara Desert across the Atlantic Ocean. The dust in the plumes contains bacteria, fungi, and their spores that have been associated with coral reef destruction. In this photo, the dust stayed airborne over the Atlantic Ocean for several days in mid-January 2008, before the National Air and Space Agency's Aqua satellite captured this image using Moderate Resolution Imaging Spectroradiometer (MODIS). [*See color plate section*]

KEY THREATS	GOALS												
	Understand coral reef ecosystems				Reduce adverse impacts of human activities on reefs								
	Map all U.S. coral reefs	Assess & monitor reef health	Conduct strategic research	Understand social & economic factors	Improve use of marine protected	Reduce impacts of fishing	Reduce impacts of coastal uses	Reduce pollution	Restore damaged reefs	Improve education & outreach	Reduce threats to international reefs	Reduce impacts from international trade	Improve coordination & accountability
Global warming/climate change	M	H	H	H	M	L	M	H	L	H	H	L	H
Diseases	M	H	H	L	L	M	L	M	L	L	L	L	M
Hurricanes/typhoons	L	L	M	L	L	L	L	L	M	L	L	L	L
Extreme biotic events	L	M	H	H	L	L	L	L	M	L	M	L	L
Overfishing	H	H	H	H	H	H	L	L	L	H	H	H	H
Destructive fishing practices	M	H	M	H	H	H	L	L	H	H	H	H	H
Habitat destruction	H	H	M	H	H	H	H	L	H	H	H	M	H
Invasive species	L	H	M	M	M	L	L	H	L	H	M	M	H
Coastal development	H	H	H	H	M	L	H	M	L	H	H	L	H
Coastal pollution	H	H	H	H	M	L	H	H	L	H	H	L	H
Sedimentation/runoff	M	H	H	H	M	L	H	H	L	H	H	L	H
Marine debris	L	M	L	L	L	H	L	H	L	H	M	L	M
Overuse from tourism or recreation	M	H	M	H	H	M	H	L	L	H	H	M	L
Vessel groundings	M	M	L	L	H	L	H	L	H	H	H	L	H
Vessel discharges	L	M	L	M	H	L	M	H	L	H	H	L	H

H = High priority action needed to address threat
M = Medium priority action to address threat
L = Low priority action to adderss threat

FIGURE 8.15
Long-term conservation measures included in the US Coral Reef Task Force's National Action Plan to increase understanding of coral reef ecosystems.
Source: National Oceanic and Atmospheric Administration (2002). National Coral Reef Action Strategy: Report to Congress on Implementation of the Coral Reef Conservation Act of 2000 and the National Action Plan to Conserve Coral Reefs.

GOALS	OBJECTIVES	REGIONS							
		Atlantic/Carribean			Polynesia			Micronesia	
		Florida	Puerto Rico	US Virgin Islands	Main Hawaiian Islands	NW Hawaiian Islands	American Samoa	Guam	N.Mariana Islands
Map U.S. coral reefs	Map all shallow reefs (<30m)	H	H	H	H	H	H	M	H
	Map selected deep reefs (>30m)+B8	M	L	H	L	L	M	L	L
Assess & Monitor Reef Health	Conduct rapid assessments & inventories	H	M	H	H	H	M	M	H
	Monitor coral, fish, and other living resources	H	H	H	H	H	H	H	H
	Assess water & substrate quality	H	H	H	M	L	H	H	H
	Assess global warming & bleaching	H	M	M	L	L	M	L	M
Conduct Strategic Research	Understand reef processes	H	M	H	M	L	L	M	M
	Understand reef diseases and bleaching	H	H	H	L	L	L	L	M
	Understand impacts of management actions	M	H	M	H	H	H	M	H
Understand Social and Economic Factors	Assess Human Uses of Reefs	H	H	H	H	M	M	H	H
	Assess Social/Economic impacts of reef management	L	H	H	H	M	L	H	M
	Assess value of reef resources	M	H	M	H	M	M	H	H
Improve Use of marine protected areas (MPAs)	Strengthen existing MPAs	H	H	H	H	H	H	H	H
	Identify gaps in MPA system	M	H	H	H	H	H	L	H
	Establish new MPAs	L	H	H	H	M	H	L	H
Reduce adverse impacts of fishing	Reduce Overfishing	H	H	H	H	M	H	L	H
	Reduce habitat destruction and other indirect impacts	M	H	M	H	H	M	M	H
Reduce impacts of coastal uses	Reduce dredging and other habitat impacts	H	H	M	L	L	M	L	H
	Reduce Impacts from Ocean Recreation	L	H	H	H	L	L	H	H
	Improve Vessel management	M	H	M	H	H	L	H	M
Reduce pollution	Reduce Sediment Pollution	M	H	H	H	L	H	H	H
	Reduce Nutrient Pollution	H	H	M	H	L	M	H	H
	Reduce Chemical Pollution	M	M	M	M	M	H	H	H
	Reduce Marine Debris	M	M	M	M	H	M	L	L
	Prevent and Control Invasive Species	M	L	L	H	H	L	L	L
Restore damaged reefs	Improve Response Capabilities	H	H	M	H	H	M	L	M
	Improve Restoration Techniques	M	H	M	M	L	L	M	M
	Restore Damaged Reefs	M	H	L	L	L	L	L	M
Improve Education & Outreach	Increase awareness	H	H	H	H	M	H	H	H
Reduce International Threats to Reefs	Increase Capability for Resource Management	L	H	H	H	H	H	L	H
	Support International Organizations and Institutions	L	L	M	L	L	M	H	M
	Support Project Development and Implementation	L	L	M	L	L	L	L	H
	Provide Technical Assistance	L	L	M	L	L	L	L	H
Reduce Impacts from International Trade	Reduce Destructive Fishing Practices	L	M	H	L	L	L	L	L
	Increase International Awareness	L	L	L	L	L	L	L	L
Improve Coordination and accountability	Improve coordination and accountability	H	M	M	H	H	H	L	H

H = High priority action needed to address threats
M = Medium priority action to address threats
L = Low priority action to address threats

FIGURE 8.16
Relative importance of the objectives under each goal outlined in the US Strategy and National Action Plan to address key threats to United States coral reef ecosystems by region.
Source: National Oceanic and Atmospheric Administration (2002). National Coral Reef Action Strategy: Report to Congress on Implementation of the Coral Reef Conservation Act of 2000 and the National Action Plan to Conserve Coral Reefs.

REVIEW QUESTIONS

How might the flow of matter and energy to and from trophic states (Figure 8.1) be used to predict the movement of genetic material through an ecosystem?

What changes need to be made to dispersion models that generally predict aerosol transport to be used for biological agent transport? What algorithms can remain the same? Why?

What factors may account for the differences in dispersion of genetic material and that of a biological warfare agent (e.g. would the dispersion follow a path similar to Figure 8.11 versus Figure 8.13)?

What are some of the inherent uncertainties in Figure 8.11 and Figure 8.13? How can these be addressed in a model?

List the most important challenges and obstacles to scaling up field-specific measurements. How can computational tools help to improve this process?

What makes a compound recalcitrant?

How might genetic engineering be used to address recalcitrant xenobiotics?

If a strain of a microbe is genetically modified to improve the *in situ* degradation rate of total polycyclic aromatic hydrocarbon content of river sediments by 25% over progenitor strains, how might you apply the ESA's recommendations for risk assessments to the decision to use this organism? What should be ignored or added to these recommendations? Explain your answer.

NOTES AND COMMENTARY

1. For an example of nanoscale issues, see US Environmental Protection Agency (2007). Nanotechnology White Paper, EPA 100/B-07/001.
2. US General Accountability Office (2004). Genetically Modified Foods: Experts View Regimen of Safety Tests as Adequate, but FDA's Evaluation Process Could Be Enhanced, GAO-02-566, 2002; and, Food and Agriculture Organization of the United Nations, The State of Food and Agriculture, 2003–2004. Agricultural Biotechnology – Meeting the Needs of the Poor?

3. T.E. Grandel (1996). On the concept of industrial ecology. *Annual Review of Energy and the Environment* 21: 69–98.
4. C.A. Auer (2008). Ecological risk assessment and regulation for genetically-modified ornamental plants. *Critical Reviews in Plant Sciences* 27: 255–271.
5. R.E. Evenson and V. Santaniello (Eds) (2004). *The Regulation of Agricultural Biotechnology*, CABI.
6. See, for example: US Environmental Protection Agency (2009). Regulating Biopesticides; http://www.epa.gov/pesticides/biopesticides/index.htm; accessed July 20, 2009.
7. US Environmental Protection Agency (2009). Summary of the EPA municipal solid waste program. http://www.epa.gov/reg3wcmd/solidwastesummary.htm; accessed October 10, 2009.
8. F.G. Pohland (1975). Sanitary Landfill Stabilization with Leachate Recycle and Residual Treatment. EPA-600/2-75-043. US Environmental Protection Agency. Cincinnati, OH.
9. Y. Long, Y-Y. Long, H-C. Liu and D-S. Shen (2009). Degradation of refuse in hybrid bioreactor landfill. *Biomedical and Environmental Sciences* 22: 303–310.
10. F. Pohland, W. Cross, J. Gloud and D. Reinhart (1993). Behavior and assimilation of organic and inorganic priority pollutants co-disposed with municipal refuse. Report No. EPA/600/R-93/137a. Risk Reduction Engineering Laboratory. Office of Research and Development. Cincinnati, OH.
11. R. Amman, W. Ludwig and K-H. Schleifer (1995). Phylogenetic identification and in situ detection of individual microbial cells without cultivation. *Microbiological Reviews* 59: 143–169; P. Hugenholtz, B. Goebel and N. Pace (1998). Impact of culture-independent studies on the emerging phylogenetic view of bacterial diversity. *Journal of Bacteriology* 180 (18): 4765–4774; and P. Jjemba (2004). Environmental Microbiology: Principles and Applications. Science Publishers, Enfield, NH.
12. National Academy of Engineering (2004). *The Engineer of 2020: Visions of Engineering in the New Century.* National Academies Press, Washington, DC, p. 49.
13. R. Bratspie (2002). The illusion of care: regulation, uncertainty and genetically modified food crops. *New York University Law Journal* 10: 297.
14. Letter from Birmingham Jail, in M.L. King (1963). *Why We Can't Wait.* HarperCollins, New York, NY.
15. Presidential Executive Order 12898 (1994). Federal Actions to Address Environmental Justice in Minority Populations and Low/Income Populations. February 11, 1994.
16. For example, this is the fourth canon of the American Society of Civil Engineers (1996). Code of Ethics, Adopted 1914 and most recently amended November 10, 1996, Washington, DC. This canon reads: "Engineers shall act in professional matters for each employer or client as faithful agents or trustees, and shall avoid conflicts of interest."

17. Even this is a challenge for environmental justice communities, since certain sectors of society are less likely to visit hospitals or otherwise receive early healthcare attention. This is not only a problem of assessment, but can lead to more serious, long-term problems compared to those of the general population.
18. W. Burke, D. Atkins, M. Gwinn, A. Guttmacher, J. Haddow, J. Lau, *et al.* (2002). Genetic test evaluation: information needs of clinicians, policy makers, and the public. *American Journal of Epidemiology* 156: 311–318.
19. Institute of Medicine (1999). *Toward Environmental Justice: Research, Education, and Health Policy Needs.* National Academies Press, Washington, DC.
20. This harkens back to the Constitution's requirement of equal protection.
21. V.R. Potter, II (1996). What does bioethics mean? *The Ag Bioethics Forum* 8 (1): 2–3.
22. The President's Council on Bioethics (2002). Working Paper 1, Session 4: Human Cloning 1: Human Procreation and Biotechnology, January 17.
23. Eionet – European Topic Centre on Sustainable Consumption and Production (2009). Waste prevention; http://scp.eionet.europa.eu/themes/waste/prevention/#product; accessed October 12, 2009.
24. Ibid.
25. P. Kurzer (2004). Working Paper: European Citizens Against Globalization: Public Health and Risk Perceptions. Lehigh University, Pennsylvania.
26. C.G. Gonzalez (2007). Genetically modified organisms and justice: the international environmental justice implications of biotechnology. *Georgetown International Environmental Law Review* 19: 584–642.
27. Ibid.
28. Presidential Executive Order 12898 (1994). Federal Actions to Address Environmental Justice in Minority Populations and Low-Income Populations. February 11, 1994.
29. R. Araujo (2007). US EPA, Research Triangle Park, NC; personal communication with author.
30. For example, see current challenges at: Organization for Economic Co-ordination and Development (2009). Directorate of Science, Technology and Industry, Biotechnology Policies; http://www.oecd.org/department/0,3355,en_2649_34537_1_1_1_1_1,00.html; accessed July 20, 2009.
31. J.H. Kroll and J.H. Seinfeld (2008). Chemistry of secondary organic aerosol: formation and evolution of low-volatility organics in the atmosphere. *Atmospheric Environment* 42: 3593–3624.
32. S. Eapen, S. Singh and S.F. D'Souza (2007). Advances in development of transgenic plants for remediation of xenobiotic pollutants. *Biotechnology Advances* 25 (5): 442–451.
33. P.C. Abhilash, S. Jamil and N. Singh (2009). Transgenic plants for enhanced biodegradation and phytoremediation of organic xenobiotics. *Biotechnology Advances* 27 (4): 474–488.
34. C.A. Lipinski, F. Lombardo, B.W. Dominy and P.J. Feeney (2001). Experimental and computational approaches to estimate solubility and permeability in drug discovery and development settings. *Advanced Drug Delivery Reviews* 46 (1-3): 2–26.

35. R.D. Wauchope, T.M. Buttler, A.G. Hornsby, P.W.M. Augustijn Beckers and J.P. Burt (1992). Pesticide properties database for environmental decision making. *Reviews in Environmental Contaminant Toxicology* 123: 1–157.
36. US Department of Health and Human Services (2002). Toxicological Profile of Methoxyclor. Public Health Service. Agency for Toxic Substances and Disease Registry. Atlanta, GA.
37. Wauchope *et al.*, Pesticide properties database; and P.H. Howard (Ed.) (1991). Pesticides. In: *Handbook of Environmental Fate and Exposure Data for Organic Chemicals.* Lewis Publishers, Chelsea, MI, pp. 7–21.
38. US Environmental Protection Agency (1992). Pesticides in Ground Water Database: A Compilation of Monitoring Studies, 1971-1991: National Summary. Washington, DC.
39. A.A. Snow, D.A. Andow, P. Gepts, E.M. Hallerman, A. Power, J.M. Tiedje and L.L. Wolfenbarger (2005). Genetically engineered organisms and the environment: current status and recommendations. *Ecological Applications* 15 (2): 377–404.
40. European Union (2001). Directive 2001/18/EC of the European Parliament and of the Council of 12 March 2001 on the deliberate release into the environment of genetically modified organisms and repealing Council Directive 90/220/EEC – Commission Declaration. Article 2.
41. E.J. Kok, J. Keijer, G.A. Kleter, and H.A. Kuiper (2008). Comparative safety assessment of plant-derived foods. *Regulatory Toxicology and Pharmacology* 50: 98–113.
42. T.F. Budinger and M.D. Budinger (2006). *Ethics of Emerging Technologies: Scientific Facts and Moral Challenges.* John Wiley & Sons, Inc., Hoboken, NJ.
43. Ibid.
44. National Academy of Sciences (2002). National Research Council. *Animal Biotechnology: Science Based Concerns.* The National Academies Press, Washington, DC.
45. G. Lipps (Ed.) (2008). *Plasmids: Current Research and Future Trends.* Caister Academic Press, Norwich, UK.
46. National Academy of Sciences (2000). National Research Council. *Genetically Modified Pest-Protected Plants: Science and Regulation.* The National Academies Press, Washington, DC.
47. Lipps, *Plasmids.*
48. National Institutes of Health (1978). NIH Guidelines for Research Involving Recombinant DNA Research. 43 Federal Register. 60108. Bethesda, MD; and National Institutes of Health (1976). Recombinant DNA research: Guidelines. 41 Federal Register. 27901.
49. National Academy of Sciences (2000). *National Research Council. Genetically Modified Pest-Protected Plants: Science and Regulation.* The National Academies Press, Washington, DC.
50. Ibid.
51. National Academy of Sciences (1987). *Introduction of Recombinant DNA-Engineered Organisms into the Environment: Key Issues.* National Academies Press, Washington, DC.

52. US Congress (1957). Federal Plant Pest Act, 7 USC. § 150aa-jj, as amended.
53. National Academy of Sciences, *Genetically Modified Pest-Protected Plants.*
54. Ibid.
55. Ibid.
56. H. Rueter, G. Schmidt, W. Schröder, U. Middelhoff, H. Pehlke and B. Breckling (2009). Regional distribution of genetically modified organisms (GMOs) – Up-scaling the dispersal and persistence potential of herbicide resistant oilseed rape (*Brassisca napus*). *Ecological Indicators.* Article in press. doi:10.1016/j.ecolind.2009.03.007.
57. C. Damgaard and G. Kjellsson (2005). Gene flow of oilseed rape (*Brassica napus*) according to isolation distance and buffer zone. *Agriculture, Ecosystems & Environment* 108 (4): 291–301.
58. O. Michel (2003). Role of lipopolysaccharide (LPS) in asthma and other pulmonary conditions. *Journal of Endotoxin Research* 9: 293–300.
59. S.A. Olenchock (2001). Airborne endotoxin. In: C.J. Hurst, R.L. Crawford, G.R. Knudsen, M.J. McInerney and L.D. Stetzenbach (Eds), 2nd Edition. ASM Press, Washington, DC, pp. 814–826.
60. C. Duchaine, P.S. Thorne, A. Merizux, Y. Grimard, P. Whitten and Y. Cormier (2001). Comparison of endotoxin exposure assessment by bioaerosol impinger and filter-sampling methods. *Applied Environmental Microbiology* 67: 2775–2780.
61. C.S. Clark, R. Rylander and L. Larsson (1983). Levels of gram-negative bacterial, *Aspergillus fumigatus*, dust and endotoxin at compost plants. *Applied Environmental Microbiology* 45: 1501–1505.
62. A.A. Snow, D.A. Andow, P. Gepts, E.M. Hallerman, A. Power, J.M. Tiedje and L.L. Wolfenbarger (2005). Genetically engineered organisms and the environment: Current status and recommendations. *Ecological Applications* 15 (2): 377–404.
63. Ibid.
64. Ibid.
65. Ibid.
66. Ibid.
67. Ibid.
68. Ibid.
69. *New Scientist* (2002). Future of corals is going down. August 10, 2002.
70. National Oceanic and Atmospheric Administration (2002). National Coral Reef Action Strategy: Report to Congress on Implementation of the Coral Reef Conservation Act of 2000 and the National Action Plan to Conserve Coral Reefs.
71. W.R Munns, Jr., R. Kroes, G. Veith, G.W. Suter II, *et al.* (2003). Approaches for integrated risk assessment. *Human and Ecological Risk Assessment* 9 (1): 267–273.
72. S. Golubic, G. Radtke and T. Le Campion-Alsumard (2005). Endolithic fungi in marine ecosystems. *Trends in Microbiology* 13: 229–235.

CHAPTER 9

Environmental Risks of Biotechnologies: Economic Sector Perspectives

Biotechnology is a prominent part of numerous aspects of contemporary society, so this book addresses biotechnology from a number of environmental perspectives. Indeed, biotechnology is a spectrum and, as such, the use of living things to provide for society's need has been categorized using colors (see Figure 9.1). During a meeting between the European Union and the United States to discuss international perspectives on biotechnology, Dr Rita Colwell, at that time the Director of the US National Science Foundation, made a very apropos statement:

> If we could weave a Flag of Biotechnology, some say, it would feature three colors: red for medical applications, green for agricultural and white for industrial. In fact this flag may accrue even more colors over time as environmental and marine biotechnology and other applications add their stripes.

In this chapter, the first three priorities will be addressed with respect to their actual and potential environmental implications, i.e. medical, agricultural, and industrial biotechnologies. Others have been and will be addressed throughout the text.

Both the intended and unintended products of biotechnology affect myriad sectors of society, especially applications in industry, medicine, and agriculture. The public is mixed in its support and fear of biotechnologies, depending on the application. For example, in a 2005 poll, Europeans were less optimistic about biotechnology than information technology, renewable energy sources and mobile phones, but more optimistic about biotechnology than space exploration, nanotechnology, and nuclear energy (see Figure 9.2). This is complicated when factors interrelate. For example, Europeans are more sanguine about biotechnologies if they improve the availability of sustainable products (see Figure 9.3).

Some of the most well known applications in the early stages of biotechnology consisted of advances in medical applications, especially drugs. However, a decade after these advances began, industrial applications began to increase. Actually, industrial biotechnology is not completely distinct from the other biotechnology fields. Medicine accounts for about 17% of the US economy and agriculture touches quite a few other economic sectors, such as petrochemical (pesticides, fertilizers and other agricultural chemicals) and pharmaceuticals (veterinarian prescriptions). Although most medical biotechnology has involved microbes, other organisms are also of interest. For example, in recent decades pharmaceutical interests

Table 9.1 Examples of economic sectors that apply biotechnologies

Industrial sector	Description	Example application
Fine chemical production	Biocatalysis using selectivity of enzymes for one of the enantiomers of a chiral molecule, i.e. one enantiomer of a racemate is unaffected and the other enantiomer is converted into the desired, pure chemical	Hydrolases are most prominent enzyme class used in production of fine chemicals by biocatalytic resolution [1]
Bulk chemical production	Microbial production of 1,3-propanediol	*Klebsiella pneumoniae* fermentation of glycerol in the sodium cellulose sulfate/poly-dimethyl-diallyl-ammonium chloride (NaCS/PDMDAAC) microcapsule to produce 1,3-propanediol [2]
Ethanol production	Microbial production of ethanol and other alcohols from sugar	Respiration deficient strain *Saccharomyces cerevisiae* ATCC 24553 fermentation of pineapple cannery waste [3]
Chiral compound synthesis	Biochemicals from genetically enhanced microbes can efficiently resolve racemic amines using ethylmethoxyacetate as acylating agent in a lipase-catalyzed reaction	Chiral compounds yields have been enhanced by reaction of 1-phenyethylamine with ethylmethoxyacetate in the presence of a lipase from *Burkholderia plantarii* [4]
Pharmaceutical manufacturing/ Processing	Modification of bacteria to produce biochemicals, including those previously only produced endogenously	Insulin was one of the first biotechnologically produced substances. Artificial genes for each of insulin's two protein chains. The artificial genes were then inserted … into plasmids … among a group of genes that are activated by lactose. Thus, the insulin producing genes were also activated by lactose. The recombinant plasmids were inserted into *Escherichia coli* [5]
Synthesis of vanillin and other food flavor agents	Recombinant strains harboring hybrid plasmid can grow on alcohols as carbon sources	Insertion of vanillyl alcohol oxidase gene from *Penicillium spp* [6]
Biopolymers/Plastics	Enzymes or whole cell systems use sugars as feedstock for product manufacturing	Microbial emulation of fossil fuel processes
Ethanol production	Feed stock is cellulosic biomass (e.g. corn ears and stalks, wheat straw, or switchgrass)	Recent advances in cellulase enzymes have improved efficiencies [7]
Nutritional oil production	Genetically enhanced biomass (e.g. soybeans) to yield oil with improved properties, especially functional and nutritional quality	Increasing concentration of beta-conglycinin, a seed storage protein; decreases linolenic acid content to improve oil stability and require less hydrogenation, reducing trans fatty acids [8]
Oil and gas biodesulfurization	Uses bacteria as the catalyst to remove sulfur from the feedstock (e.g. coal or crude oil)	Organosulfur compounds, e.g. dibenzothiophene and its alkylated homologues, are oxidized with genetically engineered microbes; removing sulfur as aqueous soluble sulfate salts [9]
Leather degreasing	Developing proteases for use in soaking, de-hairing and bating processes	Proteases from *Aspergillus tamarii* and *Alcaligenes faecalis* can loosen hair without chemical assistance. Alkaline protease produced from *Rhizopus oryzae* through solid-state fermentation de-hairs the skins completely; use of enzymes for de-hairing; bacterial cultures have keratinolytic activity [10]

Table 9.1 Examples of economic sectors that apply biotechnologies—cont'd

Industrial sector	Description	Example application
Biofilms	Biofilms used in production of ethanol, butanol, lactic acid, acetic acid/ vinegar, succinic acid, fumaric acid and other industrial chemicals. Biofilm forms when microbial cells attach to support particles without using exogenous chemicals forming a dense layer around the particle	Treatment of various types of wastewater, using numerous species of microorganisms
Biohydrogen production	H_2 reactions catalyzed by either nitrogenase or hydrogenase enzymes	*E. coli*, *Enterobacter aerogenes*, and *Clostridium butyricum* use multienzyme systems. Can continuously produce H_2 photochemically and non-photochemically. Nitrogenase enzymes from *Rhodopseudomonas palustris* and *Rhodobacter sphaeroides* generate H_2 under N-limited conditions. Electrons derived from the breakdown of organic compounds. Cyanobacteria can do same as by-product of nitrogen fixation [11]
Chemical/biological warfare agent decontamination	Enzymatic processes can speed the decomposition of organophosphate nerve agents and other warfare agents	Bacterial enzymes catalyze hydrolysis from bacteria genetically modified to express protein variants, e.g. phosphotriesterase and organophosphorus anhydrolase [12]
Pulp and paper bleaching	Xylanase is applied before bleaching, replacing Cl-containing compounds in the first stage of the five-stage bleaching sequence. White rot fungus (*Phanerochaete chyrsosporium*) degrades lignin in bioreactor: wood chips injected with fungus and a growth medium, incubate for 2 weeks, and followed by traditional chemical or mechanical processes	Enzyme replaces traditional Cl addition. Biotechnology process reduces the amount of Cl-containing compounds by more than 10% Bioreactor method reduces *bleaching-related energy requirements by 40%*, with concomitant pollution reduction [13]
Textiles	Transgenic *Bacillus thuringiensis* cotton species have improved pest resistance. Colored silks are developed using genetically engineered silkworms (genome has been mapped)	New advances may result with the synthesis of tractile fiber protein by *Bacillus coli* [14]
Antibiotic biosynthesis	Soil and other bacteria can be modified to produce production-grade antibiotics	Plasmids and phage vectors for cloning Streptomyces to produce antibiotics. DNA introduced by polyethylene glycol-assisted transformation or transfection of protoplasts. Expression of genes coding for antibiotic biosynthesis also in *E. coli*. *Saccharopolyspora erythraea* produces polyketide erythromycin A. Biology can be based on genomes of microbes with similar biosynthesis capabilities [15]
Electroplating/metal cleaning	Enzymes make degreasing/metal cleaning. Fungi can be used to treat metal-laden wastes	Proteases may be similar to those listed for leather degreasing (see above). Mycelial biomass *Aspergillus japonicus* used to sorb metal ions, e.g. Fe(II), Ni(II), Cr(VI) and Hg(II) [16]

Source: Left column adapted from BIO, Biotechnology Industrial, The Third Wave in Biotechnology: A Primer on Industrial Biotechnology; http://www.bio.org/ind/background/thirdwave.asp; accessed July 23, 2009.

Red	Medical
Yellow	Food Biotechnology
Green	Agriculture
Blue	Aquatic
White	Gene-based industry
Grey	Fermentation
Brown	Arid
Gold	Nanotechnology/Bioinformatics
Purple	Intellectual
Dark	Bioterrorism/Warfare

FIGURE 9.1
Classification of biotechnologies. [*See color plate section*]
Source: Adapted from E.J. DiSilva (2004). The colours of biotechnology: science, development and humankind. *Electronic Journal of Biotechnology* 7 (3): doi: 10.4067/S0717-34582004000300001.

have been *bioprospecting* for plants with medicinal traits, such as the hallucinogen *Ayahuasca* in the Amazon rainforests. Although used by native peoples for centuries for religious ceremonies, pharmacologists are interested in potential psychotropic value. So, the push toward biotechnologies has been a mutual aspiration among economic sectors. Industrial research and development have happily responded, so that today, numerous biotechnologies are being applied in a wide swath of manufacturing and operational settings (see Table 9.1).

INDUSTRIAL BIOTECHNOLOGY

The growth of biotechnologies has occurred in just a few decades and signals the private and public sectors' embrace of bioscientifc applications. This trend also demonstrates the array of possible opportunities for failure and potential hazards, many of which are quite subtle and obscure. Prospective risk assessments are difficult in that as a technology evolves unforeseen

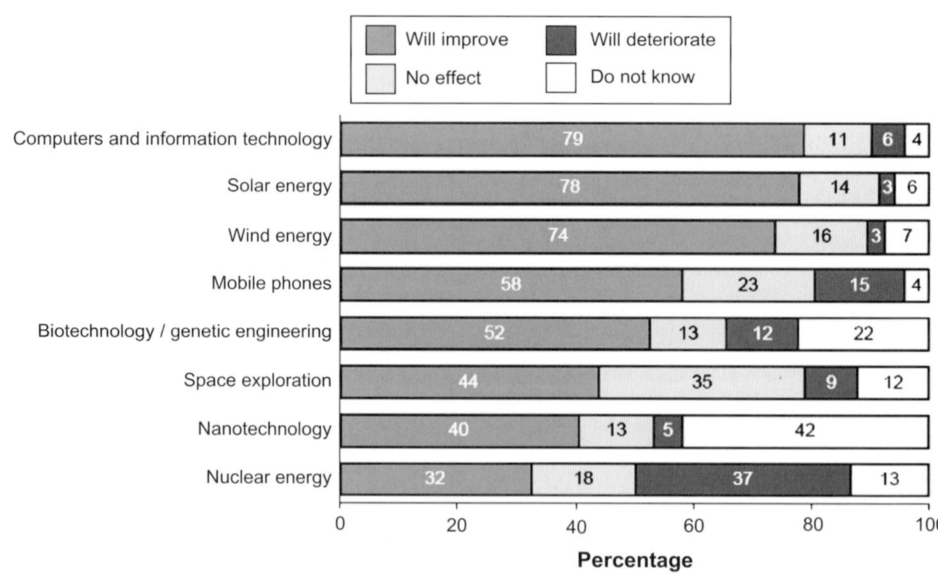

FIGURE 9.2
Levels of optimism in the year 2005 reported by Europeans regarding various technologies. [*See color plate section*]
Source: G. Gaskell, A. Alansdottir, N. Allum, C. Corchero, C. Fischler, J. Hampel, *et al.* (2006). Europeans and biotechnology in 2005: Patterns and trends. Eurobarometer 64.3. Report to the European Commission's Directorate-General for Research.

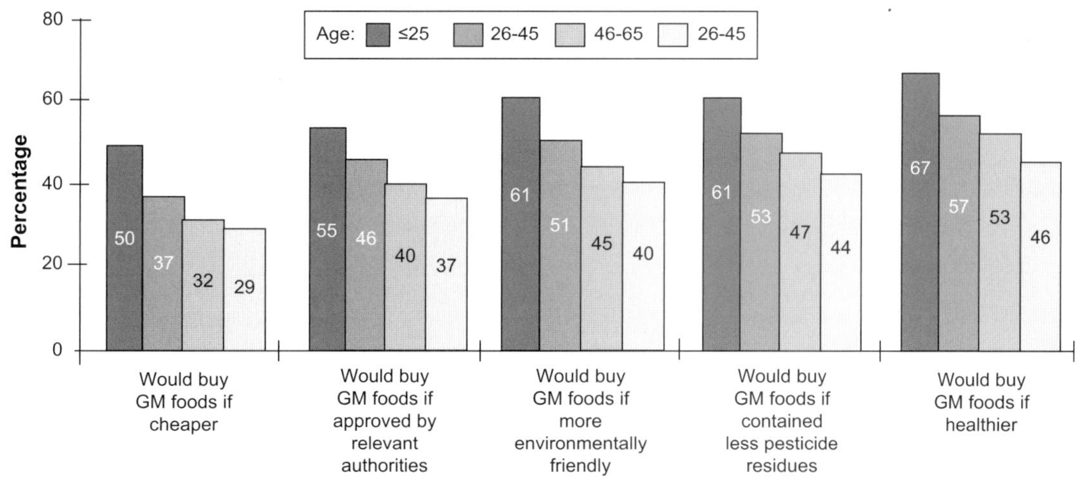

FIGURE 9.3
Intention by age groups to purchase genetically modified food in the year 2005 reported by Europeans regarding various technologies. [*See color plate section*] *Source:* G. Gaskell, A. Alansdottir, N. Allum, C. Corchero, C. Fischler, J. Hampel, *et al.* (2006). Europeans and biotechnology in 2005: Patterns and trends. Eurobarometer 64.3. Report to the European Commission's Directorate-General for Research.

events lead to unforeseen subsequent events and, ultimately, to unforeseen outcomes. In fact, the uncertainties can propagate in time and space (see Chapter 6).

Numerous laws and regulations cover the possible implications of biotechnology. For example, the Toxic Substances Control Act (TSCA) is the primary law in the United States addressing new chemicals (see Table 9.2). As such, it regulates:

> ... persons conducting commercial research and development activities or persons manufacturing, importing, or processing for commercial purposes intergeneric microorganisms used for a TSCA purpose. [17]

PRODUCTION OF ENZYMES [18]

Enzymes are crucial components of metabolism and other biological processes that are applied in biotechnology (see Table 9.1). Certainly the white and grey sectors in Figure 9.1 are not the exclusive domain of enzymatic optimizations. In fact, the microbial and plant processes involved in biodegradation discussed in the two previous chapters include crucial enzymatic steps. Thus, the enzymatic activity discussed within the context of environmental applications, particularly bioremediation, are instructive for many biotechnological processes in all sectors in Figure 9.1. However, in light of their importance in the industrial sectors, it is worth discussing these reactions within the more controllable conditions of industrial bioreactors [19].

Microbial industrial production of enzymes is usually a stepped, aerobic reaction process within a submerged culture in a stirred tank reactor. Figure 9.4 illustrates the flow in a typical enzyme production process. The organism, the media, and feedstock (raw material) are all limiting factors in the efficiency of enzyme fermentation.

The organism

The enzyme biochemistry is driven by transcription, translation, and posttranslational processing. Much variability exists in metabolic processes of diverse classes of organisms. The enzyme molecules themselves have large ranges in molecular mass, number of polypeptide chains, isoelectric point, and degree of glycosylation, i.e. a saccharide's reaction with a hydroxyl or amino group to form a glycoside.

The selection of microbes as candidates for fermentation depends on process characteristics (e.g. viscosity or recoverability), legal approval of the use, and the state of knowledge about the selected organism. Cellular enzyme expression is strongly influenced by the microbe's

Table 9.2 Biotechnological activities regulated under the toxic substances control act

Regulated activity category	Examples of regulated entities
Biotechnology research and development activities involving commercial funds	Persons conducting commercial research using intergeneric microorganisms for biofertilizers; biosensors; biotechnology reagents; commodity or specialty chemical production; energy applications; waste treatment or pollutant degradation; and other TSCA subject uses
Commercial biotechnology products	Persons manufacturing, importing or processing products for commercial purposes; intergeneric microorganisms for biofertilizers; biosensors; biotechnology reagents; commodity or specialty chemical production; energy applications; waste treatment or pollutant degradation; and other TSCA subject uses

regulatory mechanism. The enzyme synthesis system can be managed either by changing the structural characteristics (e.g. strain improvement), or by optimizing environmental conditions. Sugars comprise the principal feedstock for microbial processes (carbon and energy sources). Feedstocks include molasses, unrefined sugar, sulfite liquor from cellulose production plants, hydrolysates of wood and starch, or fruit juices, such as the grape juice used in wine making. Often, starches from cereals or tubers may be preferred, due to costs of refined products. Thus, these raw sources therefore contain other compounds besides sugars. This can be beneficial, because vegetative materials almost invariably contain nitrogen, phosphorus, and potassium, important nutrients to maintain microbial growth. The additional biomass can also be detrimental if the additional compounds are toxic or interfere with microbial growth and metabolism. For the purer feedstocks, the nutrients are added to the reactor as inorganic compounds such as ammonium compounds, phosphates, and potassium chloride. Organic supplements include meal, fish meal, cotton seed, low-quality protein materials such as casein or its hydrolysates, millet, stillage, and corn steep liquor. In addition, these

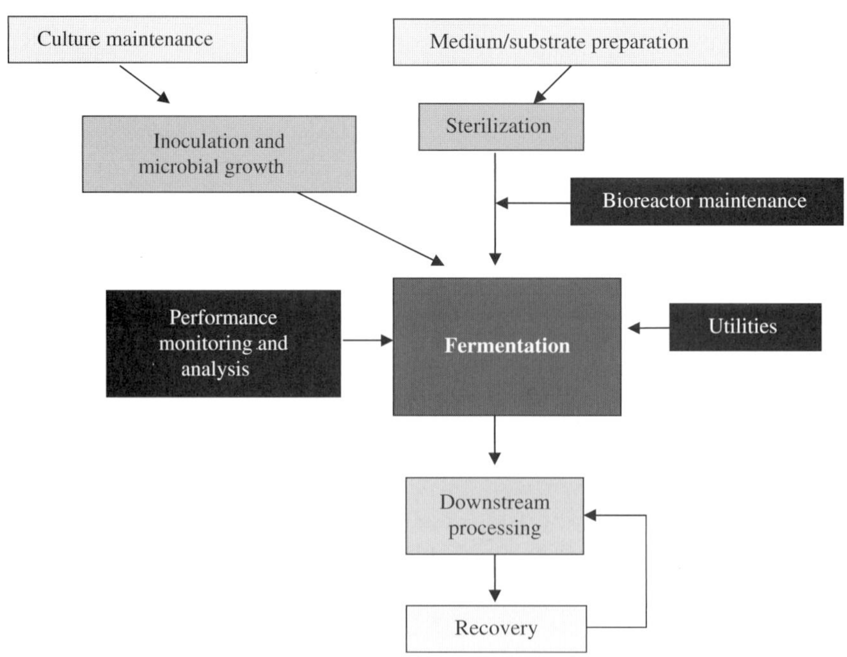

FIGURE 9.4

Steps in industrial fermentation process. [*See color plate section*]

Source: Adapted from European Commission and Federal Environment Agency Austria (2002). Collection of Information on Enzymes. Final Report. Contract No B4-3040/2000/278245/MAR/E2; http://www.agronavigator.cz/attachments/enzymerepcomplete.pdf; accessed August 10, 2009.

chemically complicated mixtures must include micronutrients, i.e. trace elements and growth promoters, which are limiting factors.

In general, the raw materials are dissolved or suspended in water; then the medium is heated, filtered, and sterilized. For downstream processing (harvest, concentration, and purification) or for analytical assays during the process, additional pre-treatment of the raw material can reduce unwanted side reactions.

Contamination in the bioreactor interferes with efficiency, so sterilization is essential. Solid substrates must be segregated at elevated temperatures ($>100\,^{\circ}C$) for specified time periods, whereas liquids are sterilized *in situ* in the vessel or in separate chambers. Steam treatment of a concentrate is another method of assuring aseptic conditions. The media pH is also a limiting factor in sterilization since it affects the viability of microbes and their spores. Injected air for aerobic processing is often filtered to improve controls.

The quantity of active cell culture added in the inoculation step depends on the size of the batch and characteristics of the microbes. For example, fungal inoculates require premoistening of the spores by adding small amounts of surfactants to the broth. Bacterial spores are activated thermally prior to inoculation. Bioprocess cells can be harvested during the exponential growth phase for subsequent inoculations.

The typical industrial bioreactor is large, with an operational volume range of 20 to 200 m^3. Aerobic fermentation has higher efficiencies than anaerobic processes. There are numerous "black boxes" since the biochemodynamics of the system, even these highly controlled reactors, are rarely known completely. The relationship between synthesis and growth rates is dependent upon the presence of molecules to start genetic expression in the microbes (inducers), or the absence proteins that regulate the microbes' expression by decreasing the rate of transcription (repressors). This is accomplished by fine-tuning the reactor's physical and chemical conditions. This is one of the reasons genetically modified strains that optimize these rates are preferred. The total enzyme synthesis rate depends on both the microbial growth rate *and* the concentration of biomass.

Enzymes are eclectic and present in all living organisms. Replicating or mimicking these natural processes is the first step in purification, followed by downstream processing via enzymes at the industrial scale. Figure 9.5 shows the sequence of steps involved in the recovery of enzymes.

From the perspective of an enzyme-producing efficiency in the industrial sectors, genetic modifications are often preferred for the following reasons [20]:

Exploitation of new types of enzymes and new source organisms, even enzymes from organisms that are difficult to handle or previously non-culturable (e.g. extremophiles).
Abbreviated development times from screening to marketing.
Large cost reductions in the development and production process of enzymes.
Genetic engineering a prerequisite for the optimization of enzyme molecular properties by protein engineering.
Potentially improved product safety and fewer production risks, due to the production of enzymes from various source organisms in a well defined population of well-characterized microbes.

Table 9.3 provides some of the rationale for industrial preferences for genetically modified organisms in enzyme production genetic engineering techniques.

The expectation is that enzyme production can be more efficient and better controlled by either multiplying the enzyme gene or by constructing an *artificial expression system*. Both approaches aim at increasing the transcription/translation of the enzyme gene into proteins on the cellular level. Multiplying enzyme genes can be achieved by amplifying the copies of the

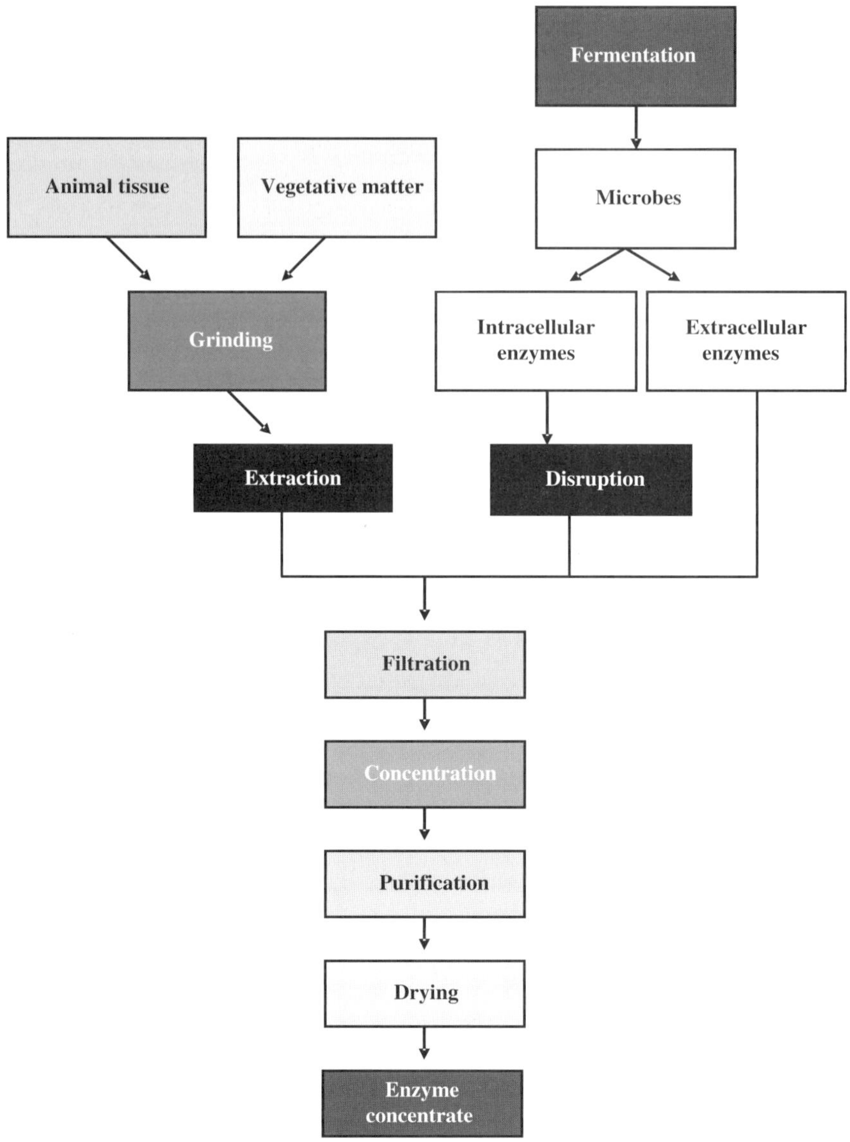

FIGURE 9.5
Steps in industrial fermentation process. [*See color plate section*]
Source: Adapted from European Commission and Federal Environment Agency Austria (2002). Collection of Information on Enzymes. Final Report. Contract No B4-3040/2000/278245/MAR/E2; http://www.agronavigator.cz/attachments/enzymerepcomplete.pdf; accessed August 10, 2009.

enzyme gene in source organisms [21]. Alternatively, the gene could be isolated from the source organism, delivering the rDNA onto a plasmid which is then introduced in a production strain. Normally a microbe has only a single set of genomic genes (two sets in the case of diploid organisms). Nearly 1000 copies of plasmids and consequently of plasmid-related enzyme genes can exist in a cell.

Genetic modification has led to a rapid increase in the availability of enzymes. This has grown with the continuous improvements in polymerase chain reaction (PCR) techniques in the past few decades. Between 1993 and 1997 more than 130 fungal enzyme genes were cloned by one company, Novozymes, including the lipase Lipolase. This enzyme was originally produced by *Humicola lanuginose*, but was found to be more economically isolated and expressed in *Aspergillus oryzae*. It is now a bulk enzyme used in detergents.

Another illustrative example of industrial enzymes is that of metabolic engineering, which involves genetically engineering a microbe to contain all the enzyme steps for a series of

Table 9.3 Rationale for selecting genetically modified organisms over naturally occurring strains to produce enzymes at the industrial scale

Goals of Innovation	Technical approaches due to genetic engineering
Reduction of manufacturing costs	Increase of enzyme yield by *increasing enzyme expression* in the production organism No need for a *de novo* design of a production process for a new production organisms. Instead, the enzyme gene of interest is cloned into a well-known production strain
New enzymes	Increase of accessibility of new enzymes especially from extremophiles Isolation of the respective genes and expression in known production strains
Improved enzyme properties	Rational protein engineering/directed molecular evolution
Improved product safety	Use of well-characterized production strains instead of new less characterized strains and sometimes strains that might be less safe

Source: European Commission and Federal Environment Agency Austria (2002). Collection of Information on Enzymes. Final Report. Contract No B4-3040/2000/278245/MAR/E2; http://www.agronavigator.cz/attachments/enzymerepcomplete.pdf; accessed August 10, 2009.

reactions leading to a particular product and then uses the cell metabolism to drive the reaction. The cell effectively becomes a highly efficient micro-reactor that synthesizes the product. Hoffman La-Roche, for example, now uses a metabolically engineered microbe to produce Vitamin B_2. Metabolic engineering has changed the previous six-step chemical process to a one-step biological process. Also, this process is greener, since the use of non-renewable raw materials has decreased by 75%, volatile organic compound releases to air and water have decreased by 50% and operating costs improved by 50% [22].

Health and safety regulations

From a regulatory perspective industrial enzymes are chemical substances, so they need to adhere to premanufacture notification rules. Many nations require information upfront before approving substances that may affect people's health or ecosystem condition. For example, under the Toxic Substances Control Act (TSCA) in the United States, the manufacture or import of any new chemical substance for a non-exempt commercial purpose requires that the company provide the US EPA with a premanufacture notice (PMN) at least 90 days prior to the manufacture or import of the chemical. The EPA evaluates chemistry, hazard, and exposure data to perform a risk assessment following a process similar to that described in Chapter 5. If, based on the assessment of the potential exposures and releases associated with the new chemical, the government determines that the new substance may pose an unreasonable risk to human health or the environment, testing and restrictions may be required:

chemistry review
hazard (toxicity) evaluation
exposure evaluation
risk assessment/risk management

Note that this process follows directly the risk assessment process for chemicals discussed in Chapters 6 and 7. Accordingly, the US EPA categorizes PMN chemicals in relation to chemical and toxicological properties (45 categories), including those described in Chapter 3, e.g. molecular structure, the log of the octanol/water partition coefficient (log K_{OW}), aqueous solubility, as well as standard hazard and fate tests. When a new substance is designated to be a member of a category, the chemical is evaluated in the context of the potential health or environmental concerns.

Most enzymes were listed in the TSCA Inventory in 1979. The decision on equivalence of enzymes is predominately based on substrate specificity but any other distinguishing characteristic of the enzyme can also drive the decision. TSCA's specific characteristics of the enzymes are the same whether the enzyme is produced by genetically modified or naturally occurring microorganisms. The government uses numerous technical approaches to address the commonly lacking or uncertain data available in the risk assessment, including structure–activity relationships (SARs), quantitative structure–activity relationships (QSARs), and various types of predictive models. Recently, the regulators have increasingly considered the function of the enzyme rather than simply the chemical structure in decisions regarding the health and safety of a proposed enzyme.

Environmental implications

Enzymes are not the only chemicals of concern related to industrial biotechnology, but they illustrate the uncertainties and the complications in assessing their risks. Biotechnology is a prominent part of contemporary industrial development. Most of the processes appear to be boons to efficiency and smooth production processes. However, these benefits must be compared to the potential risks, now and possibly in the future, as more industries make use of genetically modified organisms in their daily operations.

The question remains whether risk assessment of complex industrial biotechnologies can be sufficiently reliable to inform environmental decisions. The problem of physical containment of genetically modified organisms varies by industrial sector and the application within the sector. For example, the life cycle for a well-defined and circumscribed bioreactor where every step from raw materials to wastes and recycling is characterized and controlled (e.g. a small-scale fermentation vessel with known mass and energy balances) is likely easier to contain than a multi-vessel, multi-feedstock, multi-product system that has numerous opportunities for chemicals and organisms to be transformed and released.

Containment can be viewed as a biochemodynamic problem. For example, all control volumes described in Chapter 3 are open systems. As shown in Figure 3.10, genetic material has an opportunity to flow through environmental media, to be taken up with a compartment, be transported to other compartments, with the possibility of transformation before reaching its ultimate fate (e.g. a microbial population downstream). Thus, the release initiates a complex process that involves many systems as various scales. Unfortunately, models are not usually designed to predict the fate and transport of genetic material. In particular, the assumption that a microbe will behave like a single chemical or that the material will not undergo relationships with other living and abiotic systems after release is a gross and often inaccurate oversimplification that will likely lead to poorly defined environmental decisions.

The United Kingdom's Advisory Committee for Releases to the Environment (ACRE) illustrates the problem of assessing the risks posed by industrial biotechnologies. ACRE was formed under the Environmental Protection Act in 1990 to have the formal responsibility to assess actual and potential risks to human health and the environment in the event that a genetically modified organism is released [23]. From January to December 2008, ACRE reviewed merely two research and development applications for releases, one for a trial of potatoes resistant to potato cyst nematodes and the other for a vaccine against two respiratory diseases. Besides that, ACRE issued three notifications to place genetically modified organisms on the market, processed 11 requests to market food and feed products; and issued six licenses to release non-native species [24]. The overwhelming venue for release appears to be in agricultural operations, with the one exception of the organism for the vaccine. No other industrial sector seems to be at risk of release.

This limited venue may reflect a major flaw in methodology and information gathering. This has been articulated recently by those recommending that mistakes made in biotechnologies not be repeated in other emerging technologies, especially nanotechnology:

> As the only established mechanism for the regulatory support of GM releases this advisory body became the de facto political authority on GM releases, backed by the UK commitment, backed by the UK Government's commitment to "sound science."

> However, ACRE was concerned solely with the risks of individual GM crops. In seeking to address specific risks on a case-by-case basis, the risk assessment template came to be structurally built on past knowledge, rather than taking account of the potential for new types of hazards that might arise in unknown forms. [25]

ACRE's former chair pointed to the flaws of a non-systematic, reductionist approach to GMO release risks, saying:

> ... it was really very easy to give approval, say, for GM maize as is being done at the moment. You could not see any human risks, you couldn't really see any serious environmental ones, and as was proven in the farm trials, it's actually slightly better than traditional treatment in terms of wildlife ... [W]e can do this for one crop, one manipulation. But when all crops are being manipulated, every effect becomes additive ... Where is the mechanism to put this together?
>
> ... You'd need to know something about the interrelationships of those genes if they come together ... [26]

It is important to note that these observations came from the head of the advisory group responsible for the release of GMOs from *all economic sectors* in the UK. Some may liken this to continuing to use a hammer because it worked well driving a nail. Indeed the hammer is an excellent tool for that purpose, but not for other uses like clamping, driving screws or leveling. Single perspective risk assessments have worked reasonably well for single chemical releases into well characterized media, but not for complex biological manipulations and subsequent releases into unknown environmental compartments with highly uncertain and variable interactions.

In addition, risk assessment of biotechnology may not follow the traditional approach often used for single chemical exposures. In part, this is because, as shown in Figures 9.4 and 9.5, numerous chemicals and mixtures may be part of the industrial biotechnological set-up, operation, and maintenance. Furthermore, organisms are involved in most of these processes, so the agent of concern may well be biological, rather than or in addition to chemical. Also, the risks may well not be presented in a step-wise manner, but become problematic during different stages of the product and process life cycle.

The life cycle of a product, for example, often incorporates and integrates the same biological, chemical, and physical processes and mechanisms as those needed for successful bioremediation, as introduced in Chapter 7. As evidence, the industrial bioreactor relies on sound science to characterize these processes and mechanisms systematically, i.e. information about the organism, the bioreactor's physical conditions, the chemicals being used as carbon sources and the iterative changes. This last feature, iterative changes, is very important. As conditions change (e.g. from aerobic to anaerobic as O_2 is consumed), what may have been a quite hospitable environment for a microbial population could become increasingly toxic. In fact, the system could become sterile (no microbial growth) or, as is often the case, hospitable for a completely different microbial population (e.g. latent spore-forming anaerobes will be reproducing exponentially in a low pH, anoxic conditions, as organic acids are formed during fermentation in ethanol production plants, but also in a sanitary landfill cell).

Biotechnologies can be visualized as sets of biological reactions occurring at various scales in the environment [27]. The reactions can lead to desirable results, such as the chemical transformation and ultimate degradation of toxic substances into harmless compounds. Biological reactions may also lead to undesirable results, such as the introduction of genetically modified organisms to an ecosystem or the generation of toxic chemicals [28]. The challenge in predicting the beneficial and detrimental outcomes associated with using organisms in this way is complicated. In most cases, it is not whether biodegradation works, but whether the positive and desired outcome is accompanied by undesired and difficult to predict outcomes.

To some extent any biological outcome, even under highly controlled industrial process conditions, cannot completely be understood and thus all important variables cannot be completely controlled in time and space. A specific microbial population may have proven time and again to be efficacious in fermentation, but there is always a possibility that matching a set of reactor conditions to an organism will not work properly due to slight variations in conditions (e.g. the alkalinity of upstream substrates may have be altered rendering a more hostile environment for the desired microbial population, rendering the biotechnological operation less efficacious).

As mentioned in Chapter 8, the United States generally has determined the safety and risks associated with genetically modified organisms using a product oriented approach. Conversely, Europe has generally applied a process-oriented approach. The US regulators are often concerned with ensuring that manufactured products meet the criteria for risk and safety, and are less concerned about the products that are produced. Chemicals must be assessed a priori for their likely exposure and potential risks to human populations and ecosystems. However, the General Accountability of Office has found that the one of the major chemical databases is deficient [29]:

> (The) Integrated Risk Information System (IRIS) – a database that contains EPA's scientific position on the potential human health effects of exposure to more than 540 chemicals – is at serious risk of becoming obsolete because the agency has not been able to complete timely, credible assessments or decrease its backlog of 70 ongoing assessments. Overall, EPA has finalized a total of only 9 assessments in the past 3 fiscal years. As of December 2007, 69 percent of ongoing assessments had been in progress for more than five years, and 17 percent had been in progress for more than 9 years. In addition, EPA data as of 2003 indicated that more than half of the 540 existing assessments may be outdated. Five years later, the percentage is likely to be much higher.
>
> ... Since EPA estimates that the assessment process for complex chemicals such as dioxin could take 6 to 8 years to complete, the public in the meantime will likely remain at risk. Other toxic chemicals with widespread human exposure whose assessments have been in progress for 10 or more years include formaldehyde, trichloroethylene, and tetrachloroethylene.

In response to these deficiencies, the US EPA has recently articulated six principles for managing chemical risks, most of which support a life cycle perspective (see Discussion Box: Managing Chemical Risks in the United States).

DISCUSSION BOX

Managing Chemical Risks in the United States

The US EPA has recently articulated the need to update the way that chemicals are evaluated prior to manufacture and use, especially how regulations under the Toxic Substances Control Act (TSCA) can be revised to reduce risks of chemicals used in commerce. In particular, the regulators are looking for improving methods to ensure that chemicals do not endanger the public health and welfare of consumers, workers, and especially sensitive subpopulations such as children, or the environment.

The following goals are designed to give the government certain mechanisms and authorities to expeditiously target chemicals of concern and promptly assess and regulate new and existing chemicals [30].

- Principle No. 1: Chemicals should be reviewed against safety standards that are based on sound science and reflect risk-based criteria protective of human health and the environment.

EPA should have clear authority to establish safety standards that are based on scientific risk assessments. Sound science should be the basis for the assessment of chemical risks, while recognizing the need to assess and manage risk in the face of uncertainty.

- Principle No. 2: Manufacturers should provide EPA with the necessary information to conclude that new and existing chemicals are safe and do not endanger public health or the environment.

Manufacturers should be required to provide sufficient hazard, exposure, and use data for a chemical to support a determination by the Agency that the chemical meets the safety standard. Exposure and hazard assessments from manufacturers should be required to include a thorough review of the chemical's risks to sensitive subpopulations.

Where manufacturers do not submit sufficient information, EPA should have the necessary authority and tools, such as data call in, to quickly and efficiently require testing or obtain other information from manufacturers that is relevant to determining the safety of chemicals. EPA should also be provided the necessary authority to efficiently follow up on chemicals that have been previously assessed (e.g., requiring additional data or testing, or taking action to reduce risk) if there is a change that may affect safety, such as increased production volume, new uses or new information on potential hazards or exposures. EPA's authority to require submission of use and exposure information should extend to downstream processors and users of chemicals.

- Principle No. 3: Risk management decisions should take into account sensitive subpopulations, cost, availability of substitutes and other relevant considerations.

EPA should have clear authority to take risk management actions when chemicals do not meet the safety standard, with flexibility to take into account a range of considerations, including children's health, economic costs, social benefits, and equity concerns.

- Principle No. 4: Manufacturers and EPA should assess and act on priority chemicals, both existing and new, in a timely manner.

EPA should have authority to set priorities for conducting safety reviews on existing chemicals based on relevant risk and exposure considerations. Clear, enforceable, and practicable deadlines applicable to the Agency and industry should be set for completion of chemical reviews, in particular those that might impact sensitive subpopulations.

- Principle No. 5: Green chemistry should be encouraged and provisions assuring transparency and public access to information should be strengthened.

The design of safer and more sustainable chemicals, processes, and products should be encouraged and supported through research, education, recognition, and other means. The goal of these efforts should be to increase the design, manufacture, and use of lower risk, more energy efficient and sustainable chemical products and processes.

TSCA reform should include stricter requirements for a manufacturer's claim of Confidential Business Information (CBI). Manufacturers should be required to substantiate their claims of confidentiality. Data relevant to health and safety should not be claimed or otherwise treated as CBI. EPA should be able to negotiate with other governments (local, state, and foreign) on appropriate sharing of CBI with the necessary protections, when necessary to protect public health and safety.

- Principle No. 6: EPA should be given a sustained source of funding for implementation.

Implementation of the law should be adequately and consistently funded, in order to meet the goal of assuring the safety of chemicals, and to maintain public confidence that EPA is meeting that goal. To that end, manufacturers of chemicals should support the costs of Agency implementation, including the review of information provided by manufacturers.

Discussion

The second principle appears to increase attention to processes and the life cycle of a chemical, including manufacture, use and post-use, compared to the traditional chemical risk assessment process based on evidence of possible and actual hazards of the chemical substance (i.e. hazard identification and dose-response, followed by exposure and effects assessments).

It is interesting to compare the US EPA principles with those of trade associations. For example, the Consumer Specialty Products Association (CSPA) has its own essential principles for a regulatory framework aimed at managing chemicals [31]:

CSPA supports a chemicals management program based on sound scientific risk assessment.
CSPA supports company-performed safety-based assessments of consumer products – prior to the marketing of a product – that take into consideration all of the phases of a product's life cycle.

(Continued)

CSPA supports initiatives that encourage manufacturers of consumer products to continuously evaluate and improve their internal product safety assessment management systems.

CSPA supports appropriate use-restrictions and/or substitution for chemical ingredients when scientific based assessments indicate that they cannot be used safely in a consumer product or use application.

CSPA supports initiatives that minimize unnecessary and duplicative chemical screening processes, data development and unnecessary animal testing.

CSPA supports initiatives among companies, government, and interested parties to promote consumer awareness of the importance of reading and following label instructions for safe product use, storage, and disposal.

CSPA supports initiatives that leverage information submitted through the Inventory Update Rule (IUR), Health Canada, and Environment Canada.

CSPA supports initiatives to encourage manufacturers to voluntarily develop and make health and safety information public for chemicals in commerce (e.g. under the ICCA's Global Product Strategy, EHPV, or other government or industry initiatives).

CSPA supports the multinational collaboration of regulatory harmonization for existing chemical substances demonstrated through the Chemical Assessment and Management Program (ChAMP).

CSPA supports initiatives to encourage collaboration between EPA and state and international agencies.

There is substantial overlap in the principles, especially with respect to green chemistry and life cycle perspectives. However, in such a far-reaching policy debate, it is difficult to predict the importance of subtle differences in terminology. For example, what is meant by encouraging voluntary actions?

Among the likely and important challenges will be how to deal with uncertainty. Some of the tools needed to prioritize chemicals have been enhanced, but still need to advance to be reliable screening and priority-setting methodologies. For example, the quantitative structural activity relationships (QSARs) have only been performed on a handful of chemicals and there is no consensus within the scientific community as to QSARs' usefulness. Other approaches, including computational and ''omics'' tools, are rapidly advancing, but just how they will support chemical risk prioritization remains to be seen.

Chemical transport, transformation, and fate models will also be needed. For example, a computational tool, known as MetaPath, is being designed to predict the metabolic pathways of chemicals [32]. MetaPath includes a database of metabolic pathways and associated metadata, constructed primarily from rat metabolic *in vivo* studies of pesticides to support critical analysis and interpretation of data by risk assessment and to advance research to form hypotheses critical to the understanding of metabolic activation after chemical uptake. MetaPath is a software system with chemical structure/substructure search queries to identify commonalities and differences in metabolites among chemicals, species, dosing regimes, and other biochemodynamic information. The database supports an expert system to predict metabolite formation. Metabolic activation (i.e. biological activation) of chemicals results in potentially hazardous transformation products from parent chemicals. MetaPath is designed to characterize these processes.

An initial version of a metabolic simulator, which is under development, uses a library of more than 340 functional-group transformations that target *in vitro* and *in vivo* mammalian liver metabolism. Linking the simulator to exposure and toxic effects models is expected to support a scientific approach for prioritizing large chemical lists, especially categorizing chemicals that need additional exposure and toxicity evaluations, since many have a paucity of reliable data. It also support risk assessors with systematic tools to characterize hazards posed by parent chemicals and their potentially bioactive metabolites.

The simulator is expected to be an enhancement to other chemical characterization and screening tools, such as those that estimate the likelihood that a chemical will be persistent, will bioaccumulate, and will present toxic endpoints if released (see Chapter 3). In other words, the chemical risk screening has been based in part on whether a compound is a persistent, bioaccumulating toxic substance, or a ''PBT.'' For example, the US EPA's ''PCB Profiler'' calculates an atmospheric half-life by determining the importance of a chemical's reaction with two of the most prevalent atmospheric oxidants, hydroxyl radicals and ozone.

Expected persistence is often expressed as a half-life, which is calculated from rate constants in the environmental compartments. These rate constants are obtained from a database of measured values or, if no experimental values are available, the environmental half-lives ($t_{1/2}$) for each process is calculated based on the "Category for Persistent, Bioaccumulative, and Toxic New Chemical Substances" under TSCA [33]. Such multimedia fate models require compartmental half-lives for air, water, soil and sediment, which cannot necessarily be interpreted as half-lives for any specific process such as biodegradation. Data on air half-lives for input to models would be either measured or derived. Half-lives in bulk soil may be assumed for screening purposes to be about the same as for surface water, and that sediment half-lives may be assumed to be 3–4 times longer. The US EPA's suggested approach to finding water half-life has been to use the Ultimate Survey Model (USM) in the EPI BIOWIN program. Estimation of bulk compartment half-lives from the USM derived data requires several assumptions, including that (1) biodegradation is the only significant fate process in water, soil, sediment; (2) water and soil half-lives are the same; and (3) sediment is dominated by anaerobic conditions and therefore sediment half-life is four times longer than water half-life.

New chemicals identified as potential PBTs are assessed on a case-by-case basis. Regulatory agencies in many countries, to varying extents, control commercial activities involving a new chemical substance for which available information is often insufficient to permit a reasoned evaluation of potential health and environmental effects. In the United States, TSCA allows the US EPA to control new chemicals if the government can prove either (1) that the manufacture, importation, processing, distribution in commerce, use, or disposal of the substance may present an unreasonable risk of injury to health or the environment (known as a "risk-based" finding) or (2) that the substance is or will be produced in substantial quantities, and that the substance either enters or may reasonably be anticipated to enter the environment in substantial quantities or there is or may be significant or substantial human exposure to the substance (known as an "exposure-based" finding). Regulators consider P and B and T attributes individually and collectively, along with exposure in making risk-based judgments.

Risk, specific to substance as well as its risk relative to substitutes currently on the market, is predicted as a function of the potential hazard of the substance and the expected exposure. Otherwise, the US EPA may determine that a new substance will be produced in substantial quantities and may reasonably be anticipated to enter the environment in substantial quantities or there is or may be significant or substantial human exposure to the substance, and that the available information is insufficient to determine the effects of the substance. For such exposure-based determinations on suspected PBT new chemicals, the Regulators can apply a case-by-case approach beyond the quantitative classifications of persistence and bioaccumulation potential, beyond tightly defined toxicity or physicochemical properties. Regulators may also consider persistence and bioaccumulation potential as factors to require actions at a lower production volume or for expected release and exposure levels than would ordinarily be prescribed by general guidelines. To date, companies have not been explicitly prevented from developing and using new substances that are judged to be potential PBT chemicals, but this may soon change.

To be identified as a PBT new chemical based on a risk-based finding, all three criteria must be satisfied. Regulators have adopted a 1 to 3 rating system for each of P, B, and T. If a chemical has a low K_{ow} [i.e., B1, with an estimated bioconcentration factor (BCF) <1000], the B1 rating does not support the new chemical's identification as a potential "PBT chemical" under TSCA. For example, certain surfactants could be rated as P3-B1-T3; that is, they are highly persistent in the environment and chronically toxic to organisms, but with low bioaccumulation potential. However, regulations may still be taken under TSCA on chemicals not meeting all of the PBT criteria, so long as these chemicals otherwise meet the risk or exposure-based elements of TSCA section 5(e). Similarly, calcium would also not be considered a PBT chemical, as it would be ranked P3-B3-T1; i.e. it is persistent in the environment, it bioaccumulates, but it is not considered toxic. Although the Regulators have not promoted the environmental release of more persistent materials, the environmental "desirability" of a given chemical often depends on a systematic approach that balances various factors, including toxicity and ability of the chemical to bioaccumulate. As in the surfactant example, regulators may choose to take

(Continued)

action on a P3-B3-T1 chemical (not necessarily calcium), but most likely under its exposure-based authority.

The toxicity rating for a PBT chemical applies to repeated exposures that result in human or environmental toxicity, including systemic toxicity, mutagenic damage, reproductive toxicity, or developmental toxicity. Organotins, for example, present chronic toxicity in aquatic organisms (endocrine disruption, e.g. impo-sexual response in gastropods) when exposed to contaminated marine environments (see Figure 9.6). This exposure scenario ultimately led to a highly restricted use of tributyltin in marine anti-fouling paints. The nature of recalcitrance and bioconcentration would drive repeated exposures from a PBT chemical after it has been released into the environment, in contaminated water, sediments, or food (atmospheric exposures may also be long-term from a continuous or resuspended source). The prototypical PBT problems (i.e. PCBs and DDT and its metabolites) have been often an expression of food chain contamination (see Figure 9.7).

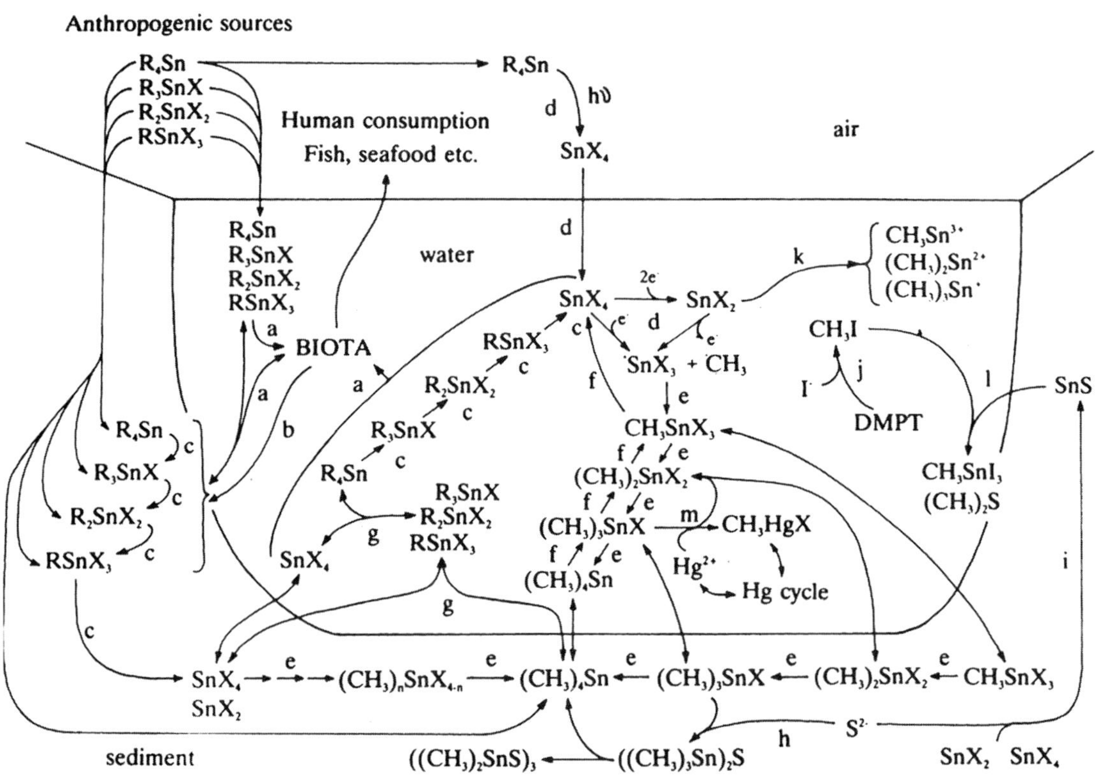

FIGURE 9.6

Food chain and biogeochemical cycling of tin (Sn) compounds. Tributyltin compounds are included in the nomenclature R_3Sn *X*. The influence of the anionic radical (*X*) on breakdown is not well understood. The important processes in the pathways are: a. bioaccumulation; b. deposition or release from biota on death or other processes; c. biotic and abiotic degradation; d. photolytic degradation and resultant free radical production; e. biomethylation; f. demethylation; g. disproportionation reactions; h. sulphide-mediated disproportionation reactions; i. SnS formation; j. formation of methyl iodide by reaction of dimethyl b-propiothetin (DMPT) with aqueous iodide; k. CH_3I methylation of Sn *X2* ; and m. transmethylation reactions between organotins and mercury.

Source: G.M. Gadd (2000). Microbial interactions with tributyltin compounds: detoxification, accumulation, and environmental fate. *The Science of the Total Environment* 258: 119–127.

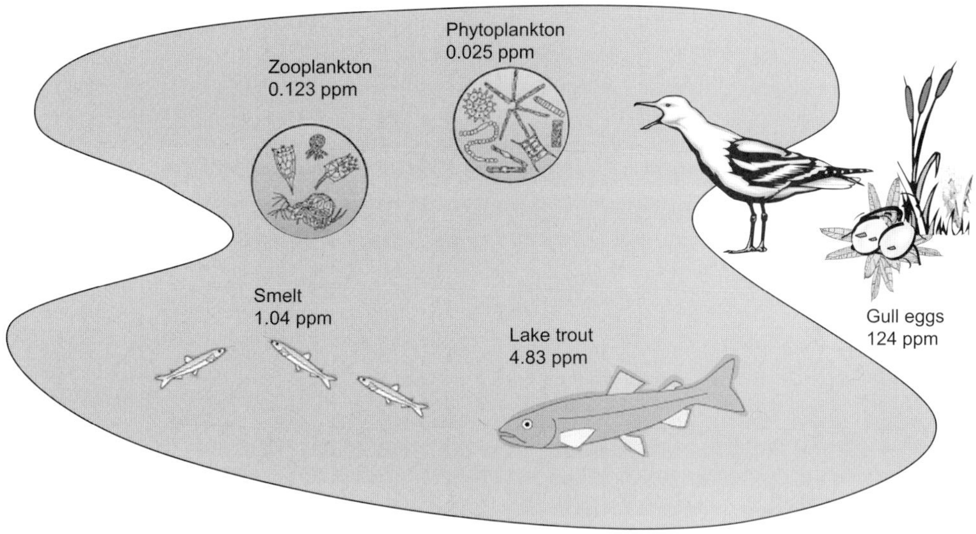

FIGURE 9.7
Persistence, bioaccumulation, and toxic substances in a food chain. In this instance, polychlorinated biphenyls (PCBs) concentrate in each level of the Great Lakes aquatic food chain, i.e. PCB concentrations (ppm) are shown for various levels of biological organization. The highest levels are reached in the eggs of piscivorous (fish-eating) birds such as herring gulls. [*See color plate section*]
Source: US Environmental Protection Agency (2009). The Great Lakes Today: Concerns. http://www.epa.gov/glnpo/atlas/glat-ch4.html; accessed October 16, 2009.

The environmental objective of a product-oriented risk analysis is to ensure the products themselves are not hazardous under various use and exposure scenarios. In fact, a green design approach would also look for ways to reduce their risks; including new formulations of products that provide the same benefit (for example, see the "rule of five" under the Recalcitrance Discussion Box in Chapter 8). From a green engineering perspective, a product-oriented approach is based on the assumption that the product will be produced, so the bioengineer needs to find ways to make it more sustainable. At this point, all phases of the product's life cycle are optimized, based on systematic thinking. Factors that go into such thinking include environmental and health risks associated with every material in the life cycle. It also considers the services involved in the process rather than apparatus to generate the product [34]. For example, a sustainable biotechnological operation to produce a solvent would not start with the design of a bioreactor or other equipment, but how best to produce a good solvent in a sustainable manner (e.g. it may not need a traditional bioreactor, but may be produced using modular bioreactors co-generationally near a number of chemical companies).

A process-oriented approach to environmental protection considers how products come to market, analyzing the input/output and material flow, ecological and economic factors, and risks to identify technical and organizational options to improve a process, including considerations on how to reduce the number of processes needed to bring a product to market. This includes a review of the internal cycles for auxiliary materials and how production wastes are introduced, hazardous substances are replaced and can be used more efficiently and safely, and how to introduce and apply innovative technologies [35]. Of course, biotechnologies comprise an important group of such innovations.

MEDICAL BIOTECHNOLOGY

> Several scientists working in the field of gene therapy are appalled that genetic technology is being applied to food, which exposes our entire population and ecosystem to unnecessary risks. Gene therapy or GM medicine, on the other hand, may limit risk to just those individuals who agree in advance. I invite you to evaluate the other genetic technologies on a case-by-case basis.
>
> **J.M. Smith (2003) [36]**

This quote from an advocate opposed to genetically engineered food reflects one of the current policy positions distinguishing medical biotechnologies from other types of biotechnology. It may also reflect Nelson Mandela's adage that "where you stand depends on where you sit." Medical applications perhaps represent the paragon of biotechnological success. Most insulin is now produced from genetically modified bacteria, as are numerous other pharmaceuticals. However, it is also one of the most controversial sectors, when one considers the bioethical challenges of cloning, use of embryonic stem cells, animal welfare, and the value of biological information (see Discussion Box: Patenting Life). These indeed have relationships to environmental risks, but since the major issues and dilemmas center around respect for human life and ethics, it is best to consult texts that address these crucial issues directly (including the author's own: D.A. Vallero (2007). *Biomedical Ethics for Engineers: Ethics and Decision Making in Biomedical and Biosystem Engineering.* Elsevier Academic Press, Burlington, MA).

DISCUSSION BOX

Patenting Life

Bioprospecting, the search for natural substances of medicinal value, is a very divisive topic in bioethical debates. In November of 1999, the US Patent and Trademark Office rescinded a patent on the plant species *Banisteriopsis caapi* held by a Californian since 1986. The plant is sacred to tribal communities living in the Amazon basin and is the source of the hallucinogen *ayahuasca*, used in their religious rituals. In addition to being a harbinger of the complications of religious and cultural respect, it presages the looming, bitter debates about the extent to which biological materials can be "owned."

In fact, in one form or another, humankind has been in the bioprospecting business for millennia. Like many bioethical issues, emerging technologies and research have changed the landscape (literally and figuratively) dramatically. And, powerful interests, such as pharmaceutical companies, see natural materials (including certain genes) as lucrative ventures that need to be harnessed for profit. For example, the biotechnology firm Diversa Inc. entered into an agreement with the National Park Service to find efficacious and beneficial microbes in the geysers and springs in Yellowstone National Park. However, the agreement was suspended by a federal court ruling.

As controversial as the subject of patents on plant genetic material is, it pales in comparison to the bioethical debates surrounding that of animals. This is in part because patenting animals' genetic materials is linked to cloning. The larger bioethical issue is captured well by the Church of Scotland's Society, Religion and Technology Project:

> *Many people would also say that knowledge of a genetic sequence itself is part of the global commons and should be for all to benefit from. To patent parts of the human genome as such, even in the form of "copy genes", would be ethically unacceptable to many in Europe. In response it is argued that patenting is the legal assessment of patent claims, and should not be confused with ethics. But patenting is already an ethical activity, firstly in that it expresses a certain set of ethical values of our society; it is a response to a question of justice, to prevent unfair exploitation of inventions. Secondly a clause excluding inventions "contrary to public order and decency" is part of most European patent legislation – an extreme case of something like a letter bomb would be excluded as immoral. But now we have brought cancerous mice and human genetic material in the potential frame of intellectual property, ethics has moved to a much more central position, where it sits uncomfortably with the patenting profession. They do not like the role of ethical adjudicator to be thrust upon them by society. [37]*

Keith Douglas Warner of Environmental Studies Institute at Santa Clara University states:

> *The privatization of germplasm formerly considered the common heritage of humankind is incompatible with notions of the common good and economic justice. The scrutiny that life industries have been receiving is well deserved, although most of this attention has been focused on the potential threats to human and ecosystem health. The economic implications of the biotechnology patent regime are less obvious because they do not impact individuals, but rather social groups. The public appears less interested in this dimension of the biotechnology revolution. Nevertheless, addressing this patent regime through the lens of the common good is a better strategy for critics of agricultural biotechnology, who will likely be more successful in slowing down the expansion of corporate control over germplasm by addressing economic issues. [38]*

The biotechnical revolution has improved crop yields and greatly increased the world's food supply in recent centuries. Along with these improvements, there have been "human and ecosystem health" tradeoffs associated with these benefits, many of which cannot be quantified. One must ask, then, whether it is morally preferable to engage in "slowing down the expansion of corporate control over germplasm" and other genetic materials?

The biomedical sector shares numerous aspects with the other sectors, e.g. those biotechnical operations described in Table 9.1. In addition, medicine has some unique challenges. For example, a growing concern is the presence of personal care and biomedical products released into the environment. Pharmaceutical and medicine manufacturing has led to diagnostic, preventive, and therapeutic drugs that save the lives of millions of people from various diseases and will likely continue to improve humans' ability to recover from diseases. There are also substantial veterinary drugs produced from biotechnologies. Like most other biotechnologies, drugs are produced by first inserting a nucleotide sequence (either natural or synthetic) into a vector, which is then introduced into a host organism which expresses the desired gene effect [39].

In fact, biotechnological products can differ substantially from abiotically-generated organic compounds. Synthetic drugs, like aspirin and oligopeptides, are readily synthesized into small molecules, whereas hemi-synthetic drugs (e.g. anticancer drugs and steroids) often need active stereo-isomerization. Even more complex are the extraction biologicals that mimic endogenous production in animals (e.g. heparins and insulin) and humans (e.g. albumin, coagulation agents, and human growth hormones). Biotechnologies allow for improved yields and safety compared to other techniques. Also, some drugs are entirely impossible to produce without biotechnologies (i.e. extraction is not possible), such as interferon and interleukin (see Table 9.4).

Again, various microbes and compounds can be released during numerous stages in medical biotechnologies, from research and testing to prototypes to manufacturing and marketing, and ultimately to use and post-use (see Figure 9.8). Note that the fermentation processes can be the same as those for other sectors (Figures 9.4 and 9.5), but the release–event–outcome causal chain can be more complex following the manufacture and operation steps.

Typically the manufacture of drugs is the final stage in a protracted process beginning with exploratory research [40]. Thus, many potential drugs do not pan out. Those that survive the process of models, combinatorial chemistry, and high-throughput screening (HTS) must meet premanufacture notification and scrutiny requirement (e.g. the US Food and Drug Administration's regulatory review and approval process can take years). However, these processes are usually most concerned with side effects and other medical concerns, with little interest in the environmental fate of genetically modified organisms or the chemicals they produce (this is more the mandate of the US Department of Agriculture, and as mentioned in the discussion regarding industrial enzymes, regulated by the US EPA).

The purification stage is a particularly hazardous step in biotechnological processes. During this stage, each of the numerous steps has its own hazards, including those during precipitation, filtration, and chemical separation. The genetically modified microbes themselves must not be released into the environment, since their proliferation can upset delicate ecological population balances, or they may exhibit characteristics harmful to human and animal populations, or to ecosystems. Adventitious agents are possibly present in original cells or somewhere in the master cell bank. Also, adventitious viruses can be introduced during this step. Hazardous chemicals are also used during this stage, such as cyanogen bromide, various heavy metals, organic solvents, and antibiotics. Infectious disease agents may also be present, such as yeasts, viruses, and mycoplasmas [41].

Table 9.4 Comparison of processes needed to produce biomedical compounds conventionally, compared to biotechnologically

Conventional production	Biotechnological production
Abiotic processes	Biological processes (e.g. by prokaryotic or eukaryotic microbes that have been genetically modified)
Made by formulation	Made by purification
Well defined starting material and reagents	Variable starting material, reagents, catalysts (enzymes)
In-process testing infrequent	Numerous in-process tests needed
Simple products	Complex products
Process specific to type of products, so generalizations are usually direct	Process is usually specific to one product, so generalizations are seldom possible (must have new process for even slightly different compounds, since there are many endogenous ''black boxes'')
Relatively large batch sizes (several kilograms of end product generated each time)	Small batch sizes (e.g. less than one gram of product produced in certain processes each time)

Source: Department of Life Sciences, Fu Jen Catholic University, Sinjhuang, Taiwan: www.bio.fju.edu.tw/handout/bio/7.ppt; accessed July 27, 2009.

Mycoplasmas are bacteria that lack a cell wall. Since many of the antibiotics bind to cell walls, coupled with mycoplasma's rapid mutation rates, mycoplasmas present a particular risk when released to the environment, since they will be resistant to many antibiotics. Mycoplasma's other daunting characteristics include their ability to pass through 0.2 μm "sterilizing grade" filters, their ability not to thrive in the typical conditions of bioreactors, their knack for being able to assume the characteristics of their hosts and to be isolated from plants, animals, and humans, their capacity to cause occult infections, and their effects on host cell metabolism and expression [42].

Most firms devote a substantial portion of their R&D budgets to applied research, using scientific knowledge to develop a drug targeted to a specific use. For example, an R&D unit may focus on developing a compound that will effectively slow the advance of breast cancer. If the discovery phase yields promising compounds, technical teams then attempt to develop a safe and effective product based on the discoveries.

The approval process can be likened to a series of sieves. The first sieve has the largest holes, so it only removes the biggest rocks. However, the biggest rocks are also often the heaviest, so the first screen usually removes the largest mass of rocks. The successive screens continue to remove large pebbles, then sand, the silt, until only the finest clays are left. Likewise, the "screening" of potential drugs removes many compounds early in the process. For example, the first screen of an antibiotic may consist of culturing a sample in a petri dish (i.e., an *in vitro* analysis). If the antibiotic passes this test, it may be tested in studies on animals (i.e., an *in vivo* analysis) that have the disease [43]. Laboratory animals also are used to study the safety and efficacy of the new drug. A new drug is selected for testing on humans only if it promises to have therapeutic advantages over drugs already in use, or is safer. Drug screening is an incredibly risky, laborious, and costly process – only 1 in every 5000 to 10,000 compounds screened eventually becomes an approved drug.

After laboratory screening, firms conduct clinical investigations, or "trials," of the drug on human patients. Human clinical trials normally take place in three phases. First, medical scientists administer the drug to a small group of healthy volunteers to determine and adjust dosage levels, and monitor for side effects. If a drug appears useful and safe, additional tests are conducted in two more phases, each phase using a successively larger group of volunteers or carefully selected patients, sometimes upwards of 10,000 individuals.

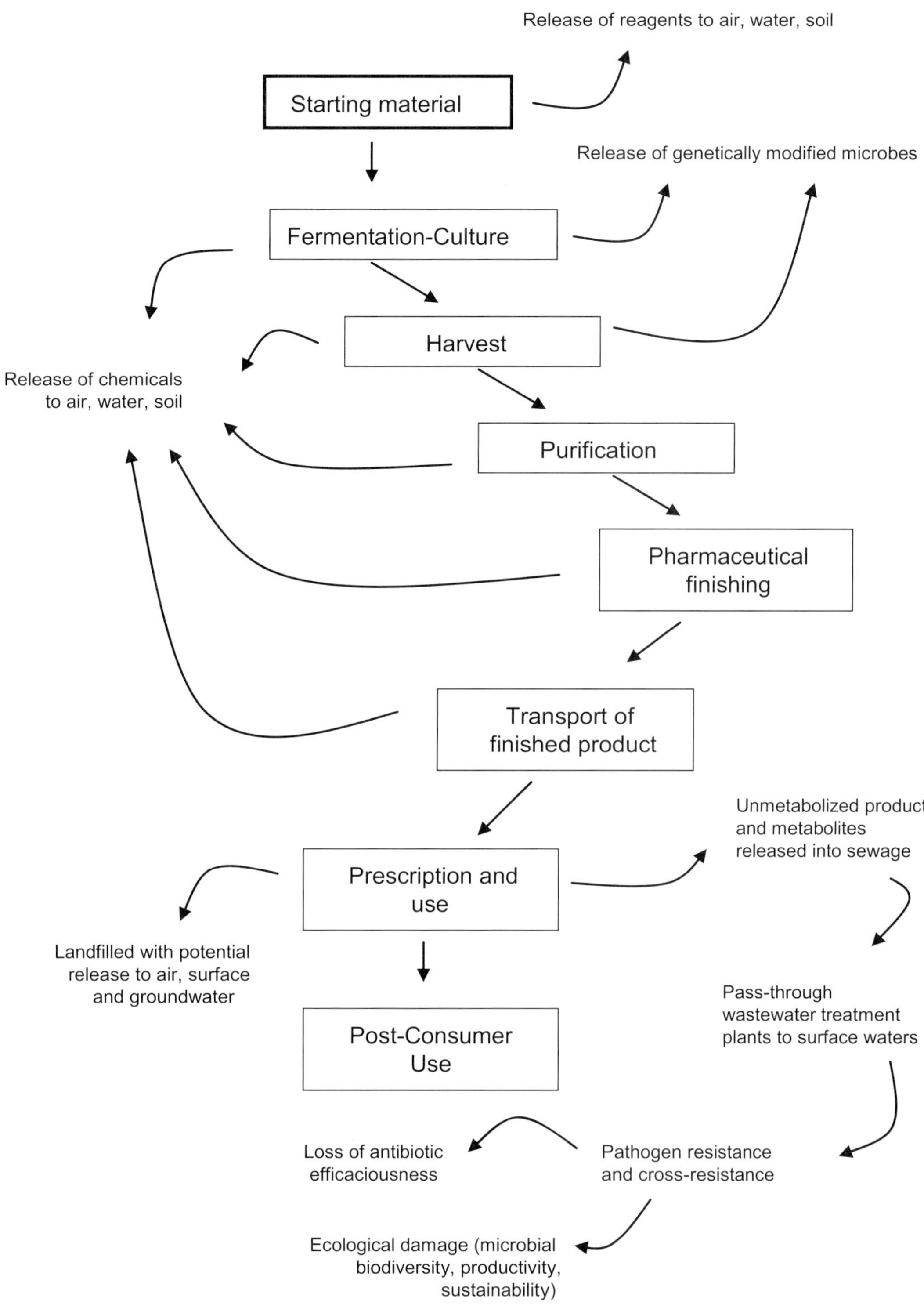

FIGURE 9.8
Flow chart of possible environmental implication from medical biotechnologies.

After a drug successfully passes animal and clinical tests, the US Food and Drug Administration's (FDA) Center for Drug Evaluation and Research (CDER) must review the drug's performance on human patients before approving the substance for commercial use. The entire process, from the first discovery of a promising new compound to FDA approval, can take over a decade and cost hundreds of millions of dollars.

After FDA approval, problems of production methods and costs must be worked out before manufacturing begins. If the original laboratory process of preparing and compounding the ingredients is complex and too expensive, pharmacists, chemists, chemical engineers, packaging engineers, and production specialists are assigned to develop a manufacturing process economically adaptable to mass production. After the drug is marketed, new production methods may be developed to incorporate new technology or to transfer the manufacturing operation to a new production site.

Advances in biotechnology are transforming drug discovery and development. Bioinformatics uses information technologies to evaluate myriad forms of biological data. Advances in technology and the knowledge of how cells work will allow pharmaceutical discovery processes to improve. These same tools can be used to predict possible interactions within containment (e.g. fermentation), during application, and post-use (including releases to the environment).

Biosystem engineering is vital to both biomedical and environmental engineering whose lexicons are full of terms with the prefix "bio." Often, this is meant to distinguish a process initiated, mediated, and sited in living systems, especially at molecular and cellular levels. This is similar to the definitions of biotechnology discussed in Chapter 1.

The interface between medicine and ecosystems deploys myriad physical and chemical processes that differ in varying degrees between those that take place in *abiotic* systems (e.g. sand and air) from those that occur in *biotic* systems, such as populations (human and forest), organism (humans and trees), tissue (liver and leaf), cellular (liver cell and leaf cell), or a receptor molecule (e.g. on the leaf cell or liver cell membrane). Thus, additional discussion beyond the introduction in Chapter 3 is needed for certain "bio" terms: bio-effective dose (bio-exposure); bio-uptake; bioactivation; bioaccumulation; biosequestration; bioconcentration; biotransformation; biodegradation; biomagnification; and bio-depuration (elimination).

Bio-uptake and bioaccumulation

Upon entering an organism a chemical compound moves and changes as a result of several processes, especially accumulation, metabolism, and excretion. All organisms share the pharmacokinetic processes of absorption, distribution, and excretion. Bioaccumulation is a function of these three processes. However, the type of chemicals able to be processed, the time that each mechanism takes, and the ultimate change to the compound after uptake vary significantly among species, or even strains of the same species. Thus bioaccumulation is a "species-dependent" factor.

The mass of the chemical substance that ultimately is accumulated by an organism is known as the organism's *body burden*. Bioaccumulation is another equilibrium condition. As shown in Figure 9.9, the organism goes through a stage, perhaps even before birth, where it begins uptake of a chemical substance. The rate of uptake is greater than the rate of elimination during the toxicokinetics phase. Eventually, the accumulation reaches equilibrium with its surrounding environment, so that the body burden remains constant. Through treatment or with the elimination of the source and release of the chemical substance from its fatty tissues or other storage sites (e.g. the liver), the process of bio-depuration may result in a reduced body burden.

One a substance has been taken up by an organism, for an adverse effect, or any effect for that matter, to occur the substance must interact with its cells. The interaction sites may be on the cell's surface, such as when an endocrine disruptor mimics a hormone by linking with a hormone receptor site on the cell's surface. The interaction may also occur within a cell, such as when carcinogenic contaminant enters a cell's nucleus and interferes with normal DNA sequencing. The contaminant may also interact within an organism's extracellular spaces. So, for plants, a chemical substance may interact with root cells, stomata, vascular tissue, and cuticle tissues. Animal interactive sites include skin and stomach tissue. The lung for land animals and the gills for fish are also sites that interact with contaminants that have been taken up.

If the dose-response and biological gradient relationships hold, and they usually do, the intensity of an adverse effect from exposure to an environmental contaminant must depend on

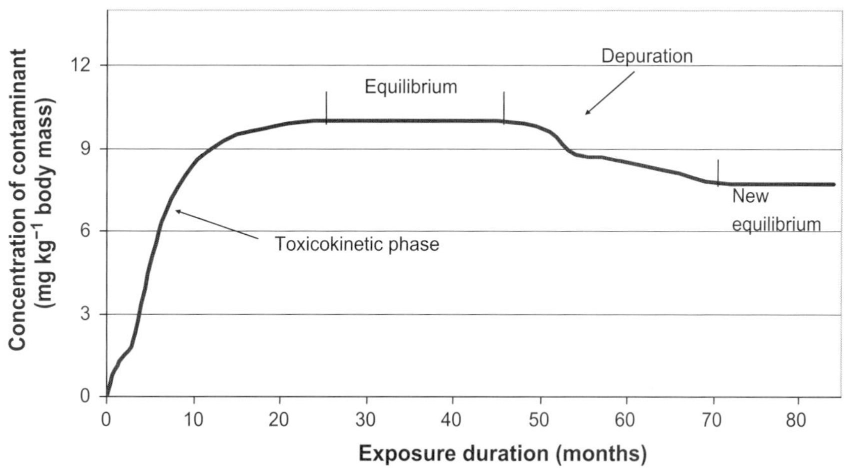

FIGURE 9.9
Bioaccumulation in an organism. During the toxicokinetic stage, uptake of the chemical substance is greater than elimination. At equilibrium, the uptake and elimination processes are equal. During detoxification or depuration, elimination is greater than uptake, so the body burden of the organism is reduced until a new equilibrium is established.

the concentration of the contaminant. If a contaminant persists in an organism, it is more likely to elicit toxicity, particularly if the contaminant stays at the ultimate site of action. For example, if a neurotoxic contaminant is stored in fat reserves, but does not find its way to the central nervous system or any other nerve site, the organism will not exhibit neural dysfunction. However, once the neurotoxin is released and distributed to a nerve site, the neurotoxicity will be manifested. And, if the contaminant finds its way to a nerve cell, the longer it remains at this site of action, the more likely the cell will be damaged. Since the endogenous target molecule in the cell is the site of action, a contaminant's chemical reactions with that molecule represents the initiation of toxicity in the organism. For example, a dioxin molecule may react with a receptor molecule on the cell's surface. This reaction may signal the feminine or masculine responses (e.g. hair growth, testis or ova development) much like a hormone would do (see Figure 9.10). In other words, the dioxin and the natural hormone both bind to the cell's receptor (see Discussion Box: Hormonally Active Agents). They are both *ligands*, i.e. molecules that travel through the bloodstream as chemical messengers that will bind to a target cell's receptor. Or, the new polypeptide that is formed from this receptor-contaminant interaction may react with DNA in the nucleus. The former reaction is an example of an endocrine response, while the latter may lead to mutagenicity or cancer.

The same chemical substance can elicit different responses. In our example above, dioxins have been shown to be both endocrine disruptors by binding with or interfering with cellular receptors, and they are carcinogenic and mutagenic because of the reactions that they or their metabolites have with the DNA molecule. Contaminants may also react with a wide range of molecules besides receptors and DNA, including lipids and microfilamental proteins. Contaminants may also enter into catalytic reactions, where enzymes are involved. Enzymes are important in the metabolism of cells, whether in a unicellular bacterium or a multicellular human being.

Absorption is the process whereby a substance moves from the site of exposure (e.g. the skin, lung tissue, or stomach) to the circulatory system. The principal mechanism for transferring a substance that has entered an organism is diffusion, i.e. movement of the substance from high to low concentrations. Most substances travel across epithelial barriers to find their way to blood capillaries via diffusion. So, if a chemical mass is high enough (i.e. sufficient rate of exposure) and the chemical can be readily dissolved into the bloodstream, then absorption will occur. Absorption also depends on area of exposure, the type of epithelial layers, microcirculation intensity in the subepithelial regions, and the properties of the substance [44].

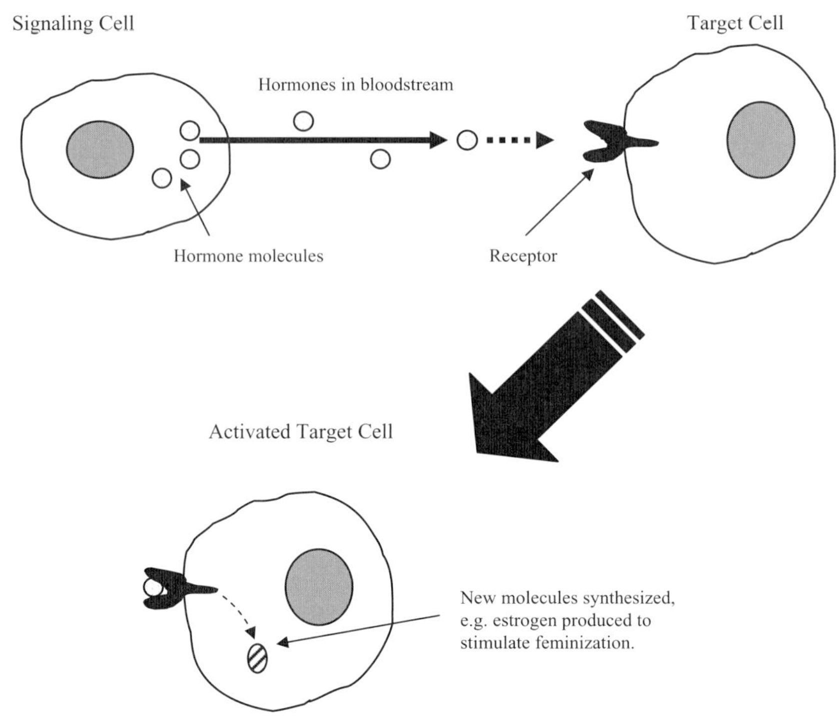

FIGURE 9.10
Schematic of the process for endocrine signals between cells. The signaling cell releases hormones into the bloodstream that reach the receptor of the target cell. When the receptor binds to the hormone, new molecules are synthesized in the activated target cell.

It is possible for some substances to be eliminated before even being absorbed. This process is known as "presystemic elimination" and can take place while the substance is being transferred from the exposure site (e.g. the outer layer of the skin or the gastrointestinal (GI) tract). As a substance moves through the GI mucosal cells, lungs, or liver, much of the substance may be eliminated. The heavy metal manganese (Mn) can be eliminated during uptake by the liver, even before it is absorbed into the bloodstream. Presystemic elimination, however, does not necessarily mean that an organism experiences no adverse effect. In fact, in the example above, Mn exposure can damage the liver without ever being absorbed into the bloodstream. This is also one of the complications of biomarkers (to be discussed later), since the body is protected against Mn toxicity by low rates of absorption or by the liver's presystemic Mn elimination [45].

Distribution [46] is the step where substances move from the point of entry and/or absorption to other locations in an organism. The principal mechanism for distribution is circulation of fluids. The absorbed substance first moves through cell linings of the absorbing organ, e.g. the skin or GI tract. After this, the substance enters that organ's interstitial fluid, i.e. the fluid that surrounds cells. About 15% of the human body mass is interstitial fluids. The substance may continue to be distributed into intracellular fluids, which account for about 40% of body mass. The substance can move to more remote locations in blood plasma (about 8% of body mass). Interstitial and intracellular fluids are stationary, i.e. they remain in place, so while the substance resides in these fluids they are not mechanically transported. Only after entering the bloodstream does distribution become rapid. A substance can leave the interstitial fluids by entering cells of local tissue, by flowing into blood capillaries and the blood circulatory system, and by moving into the lymphatic system.

A substance's distribution is largely influenced by its affinity for binding to proteins, e.g. albumin, in the blood plasma. When a substance binds to these proteins it is no longer available for potential cell interactions. In the bloodstream, only the bound fraction of the substance is in equilibrium with the free substance. Only the free (unbound) fraction may pass through the capillary membranes. The portion of the substance that is bound to proteins, therefore, determines the substance's biological half-life and toxicity. Passive diffusion of the toxicant to and from fluids is the result of the chemical substance's concentration gradient. The diffusive processes follow the same Fickian principles as those discussed in previous chapters. The apparent volume of distribution (V_D) is the total volume of fluids (units = liters) in the body to which the chemical substance has been distributed:

$$V_D = \frac{m}{[C]_{plasma}} \tag{9.1}$$

where m is the mass or pharmacological dose (mg) of the chemical substance and $[C]_{plasma}$ is the concentration of the chemical substance in the plasma (mg L^{-1}).

Chemical substances distributed exclusively in the blood will have higher values of V_D, while those distributed in several fluid types (blood and the interstitial and intracellular fluids) will be more diluted and would have lower V_D values. These values can be influenced by a chemical substance's rates of sequestration, biotransformation, and elimination.

The value is a good indication of just how widely a chemical substance is distributed within an organism. It is also a key factor in calculating the chemical substance *body burden* (mg):

$$\text{Body burden} = [C]_{plasma} \cdot V_D \tag{9.2}$$

So, if a person is exposed to 30 mg Contaminant A and has a blood plasma concentration of 3 mg L^{-1}, the volume of distribution is the quotient of the dose and the concentration in the plasma. That person's $V_D = 30/3 = 10$ L of Contaminant A. If another person is exposed to 9 mg of Contaminant B, but has a plasma concentration of 3 mg L^{-1}, then that person's $V_D = 9/3 = 3$ L of Contaminant B.

The body burden is the product of the plasma concentration and the volume of distribution, so the first person's body burden = $3 \times 10 = 30$ mg of Contaminant A, and the second person's body burden = $3 \times 3 = 9$ mg of Contaminant B.

Therefore, in this example, Contaminant B is distributed less than A (only 30%). Also, this has caused the first person to have a greater body burden of A than the second person does of B. It is important to keep in mind, however, that numerous factors can affect distribution and body burden. For example, the sex and age of a person can influence how rapidly a chemical substance is distributed. In fact, if men on average distribute these chemical substances 3.3 times more rapidly than women, then A and B could be the same chemical substance (all other factors, such as age, being equal).

The route of exposure is an important factor that can affect the concentration of the parent chemical substance, or its metabolites, within the blood or lymph regions. This can be important since the degree of biotransformation, storage, elimination, and ultimately, toxicity can be influenced by the time and path taken by the chemical substance within the body. For example, if the contaminant goes directly to the liver before it travels to other parts of the body, most of the contaminant mass can be biotransformed rapidly. This means that "downstream" blood concentrations will be muted or entirely eliminated, which obviates any toxic effects. This occurs when chemical substances become absorbed through the gastrointestinal (GI) tract. The absorbed chemical substance mass that enters the vascular system of the GI tract is carried by the blood directly to the liver via the portal system. Blood from the liver subsequently travels to the heart and then on to the lung, before being distributed to other organs. Thus, contaminants that enter from the GI tract are immediately available to be biotransformed or excreted by the liver and eliminated by the lungs. This is known as the "first-pass effect." For example, if the first-pass biotransformation of a contaminant is 75% via the oral exposure route, the contaminant-blood concentration is only about 25% of that of a comparable dose administered intravenously.

The routes of exposure follow the same principles discussed in earlier discussions. For example, respiratory exposures to contaminant gases are a function of gas diffusion. Recall that Fick's law expresses gas flux as:

$$J_{Diffusion} = -D\frac{dC}{dx} \tag{9.3}$$

This may be reordered, and values added [47] for the contaminant and the lung:

$$J_{Diffusion} = -D \times \frac{S}{MW^{1/2}} \times \frac{A}{d} \times (p_a \times p_b) \tag{9.4}$$

where $J_{Diffusion}$= diffusion rate (mass per length^2 per time)

D = diffusion coefficient for the chemical substance (area per time)

S = solubility of the chemical substance gas in the blood (mass per volume)

MW = molecular weight of the chemical substance (dimensionless)

A = surface area of membrane in contact with the chemical substance (length^2)

d = membrane thickness (length)

p_a = partial pressure of chemical substance gas in inhaled air (pressure units)

p_b = partial pressure of chemical substance gas in blood (pressure units)

The Fickian relationship shows that so long as p_a is larger than p_b, the diffusion rate is positive and the chemical substance is taken up (i.e. is more likely to reach the target organ). As the partial pressure in the blood increases and becomes greater than that in the air, the gradient reverses and the chemical substance moves out of the lung. Also, note that for a highly soluble compound, the rate of diffusion is rapid. Obviously, the slowest processes (smallest variable in the numerators, largest variable in the denominators) will be rate limiting. Aerosols (particles) will effectively diffuse if the chemical substance is lipophilic. Particle size is a major limiting factor, and is inversely proportional to dose. Currently, particles with diameters $\leq$2.5 m are considered to be most effective in passing by the nasopharyngeal region and penetrating to the tracheobronchial region and being deposited in alveoli. Larger particles are filtered physically and are considered to be less problematic.

Fundamental chemical principles apply to the oral route. For example, the pH varies among the fluids found in different organs, lowest in the stomach (pH near 1.0) and highest in some urines (pH about 7.8). Blood is also basic, with a pH of 7.4, while the small intestines are slightly acidic (pH about 6.5). This means that the acid-base relationships described in our discussions of chemical reactions are very important to the oral exposure route. For example, lipophilic organic acids and bases will be absorbed by passive diffusion only when they are not in an ionized form, so the Henderson–Hasselbach equation is a determinant in the amount of organic acids absorbed:

$$pK_a = pH + \log\frac{[HA]}{[A^-]} \tag{9.5}$$

Chemical substances absorbed through the inhalation or dermal routes will enter the blood and go directly to the heart and systemic circulation. Therefore, the chemical substance is distributed to other organs of the body before it finds its way to the liver, and is not subject to this first-pass effect. Also, a chemical substance entering the lymph of the intestinal tract will not first travel to the liver. Rather, the chemical substance will slowly enter the circulatory system. The proportion of a chemical substance that moves via lymph is much smaller than that amount carried in the blood.

The chemical substance blood concentration also depends on the rate of biotransformation and excretion. Some chemical substances are rapidly biotransformed and excreted while others are slowly biotransformed and excreted.

Disposition is the mechanism that integrates the processes of distribution, biotransformation, and elimination. Disposition (kinetic) models describe how a chemical substance moves within the body with time. The disposition models are named for the number of compartments of the body where a chemical substance may be transported. Important compartments include blood, fat (adipose) tissue, bone, liver, kidneys, and brain.

Kinetic models may be a one-compartment open model, a two-compartment open model, or a multiple-compartment model. The one-compartment open model describes the disposition of a substance that is introduced and distributed instantaneously and evenly in the body, and eliminated at a rate and amount that is proportional to the amount left in the body (see Figure 9.11). This is known as a "first-order" rate, and represented as the logarithm of concentration in blood as a linear function of time.

The half-life of the chemical that follows a one-compartment model is simply the time required for half the chemical to no longer be found in the plasma. Only a few contaminants adhere to simple, first-order conditions of the one-compartment model.

For most chemicals, it is necessary to describe the kinetics in terms of at least a two-compartment model (see Figure 9.12). This model assumes that the chemical substance enters and distributes in the first compartment, usually the blood. From there, the chemical substance is distributed to another compartment from which it can be eliminated or it may return to the first compartment. Concentration in the first compartment declines continuously over time. Concentration in the second compartment increases, peaks, and subsequently declines as the chemical substance is eliminated from the body.

A half-life for a chemical whose kinetic behavior fits a two-compartment model is often referred to as the "biological half-life." This is the most commonly used measure of the kinetic behavior of a trace chemical substance.

Frequently the kinetics of a chemical within the body cannot be adequately described by either of these models since there may be several peripheral body compartments that the chemical may go to, including long-term storage. In addition, biotransformation and elimination of a chemical may not be simple processes but subject to different rates as the blood levels change.

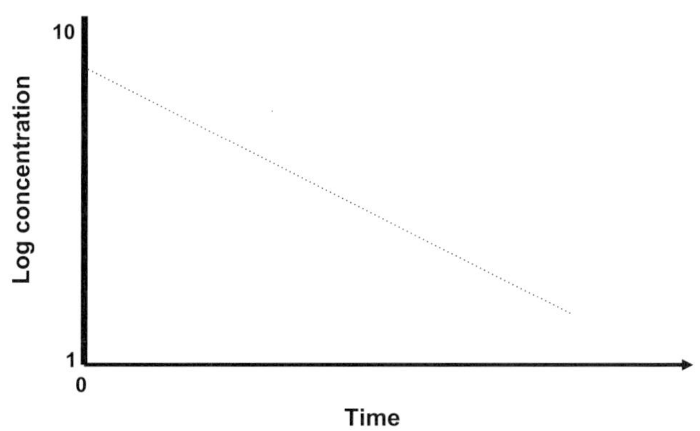

FIGURE 9.11
One-compartment toxicokinetic model. The decline of the chemical substance concentration (−dC/dt) is determined by a single first-order process, e.g. elimination. Thus, log C declines as a straight line on semi-log paper.

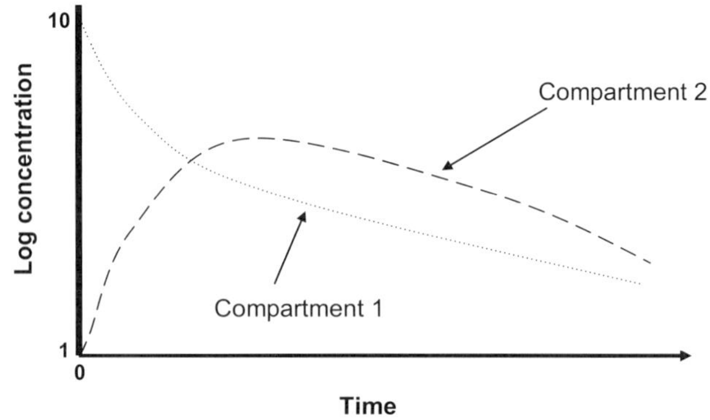

FIGURE 9.12
Two-compartment toxicokinetic model. A multi-compartmental model involves more than one endogenous process, e.g. elimination plus distribution. These multiple processes change the concentrations by moving the chemical substance away from the vascular space. Thus, $-dC/dt$ depends upon more than one interaction of processes, so the sum of more than one straight line is curvilinear on semi-log paper.

DISCUSSION BOX

Hormonally Active Agents [48]

In animals and plants, the endocrine system is actually a chemical messaging network. Hormones interact with ligands to initiate biochemical processes. Endocrine disrupting chemicals (EDCs) provide a unique challenge for environmental biotechnology, since EDCs can mimic hormones, antagonize normal hormones, alter the pattern of synthesis and metabolism of natural hormones, or modify hormone receptor levels [49]. Anthropogenic EDCs that are of concern in water and wastewater include pesticide residues (e.g. DDT, endosulfan, methoxychlor), PCBs, dioxin, alkyphenols (e.g. nonylphenol), plastic additives (e.g. bisphenol A, diethyl phthalate), PAHs, and pharmaceutical hormones (e.g. 17β-estradiol, ethinylestradiol) [50]. When microbes and plants and higher animals take part in biotechnological endeavors, implications to these chemical messages need to be considered.

Recent research has shown that many EDCs are present in the environment at levels capable of negatively effecting wildlife. One of the first EDCs heavily researched was DDT [51]. Throughout the 1980s, exposure to this pesticide was associated with abnormal sexual differentiation in seagulls, as well as thinning and cracking of bald eagle eggs [52]. Sharp decreases in the numbers of male alligators were observed in Lake Apopka, Florida, following a large spill of a DDT-laden pesticide. The alligator population also experienced feminization and the loss of fertility in the remaining males [53]. Since then, other pesticides and chemicals have been associated with endocrine-related abnormalities in fish and wildlife, including the inducement of feminine traits in males. One of the most dramatic observations has been the male secretion of the egg-laying hormone, vitellogenin, downstream of treatment facilities. This phenomenon has been observed in numerous aquatic species [54]. Birds and terrestrial animals are also affected by EDCs [55]. Recently, these problems have found their way to humans, exposed to halogenated compounds and pesticides [56]. A recent nationwide survey of pharmaceuticals in US surface water found EDCs at ng L^{-1} levels in 139 stream sites throughout the United States. Several of these EDCs were found at even $\mu g\ L^{-1}$ levels, including nonylphenol (40 $\mu g\ L^{-1}$), bisphenol A (12 $\mu g\ L^{-1}$), and ethinylestradiol (0.831 $\mu g\ L^{-1}$) [57]. Many of these compounds are extremely persistent in the environment, so their removal before entering environmental media is paramount to reducing exposures.

The search for the specific chemical structure moiety responsible for inducing the estrogenic response is a key area of endocrine disruption researchers. Phenolic rings appear to be a major chemical structure involved in the estrogenicity of EDCs [58]. Figure 9.13 shows how several known EDCs compare structurally with estrogen, the hormone they are thought to mimic.

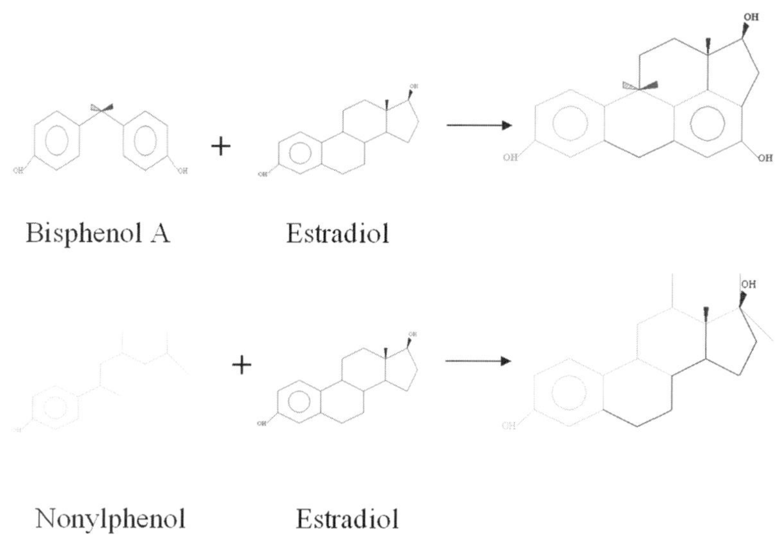

FIGURE 9.13
Comparison of the structure of bisphenol A and nonylphenol with estradiol, showing their overlap in the combined structures. [*See color plate section*]

Determining estrogenicity

Several bioassays are currently being developed and tested for their ability to predict the estrogenicity of various compounds. These assays work in various ways, but all have the common goal of identifying compounds that will cause responses similar to estrogen in various organisms. Some of these bioassays include the Yeast Estrogen Screen (YES), human cell reporter gene construct (ER-CALUX), MCF-7 cell proliferation (E-Screen), vitellogenin induction in fish, and developmental studies of fish with specific endpoints.

As an example, the YES is an assay based on yeast cells modified to harbor the human estrogen receptor. When activated, this receptor binds to the estrogen response element of some plasmid DNA that is engineered to produce S-galactosidase. When estrogens are present, S-galactosidase is excreted by the cells into the culture medium where it reacts and liberates a red dye. The resulting color change is measured with a spectrophotometer, and the responses have been calibrated based on the response of actual estrogen. This method has been widely used to determine the ''estrogenicity'' (in terms of ability to bind with the estrogen receptor and produce a response) of many compounds as well as mixtures of compounds, of known and unknown composition. Table 9.5 displays the relative binding affinity for several suspected EDCs as compared with estrogen (17β-estradiol).

Environmental fate of endocrine disrupting compounds

By examining the physical and chemical properties of EDCs, it is possible to examine where in the environment a threat from the chemicals will occur. Table 9.6 displays physical data and major uses of three EDCs of particular concern to human health, bisphenol A (BPA), 17β-estradiol (E2), and 17β-ethinylestradiol (EE2). These three compounds are xeno-estrogens, natural or synthetic compounds that act to mimic the effect of estrogens.

Their low vapor pressures mean that these chemical substances are not regularly found in the atmosphere unless associated with particles. Similarly, due to their hydrophobic nature, they will more readily associate with organic solvents or particles within a liquid water phase. However, portions of the compounds do exist in the aqueous phase, and this proportion can be greater at higher pH values, especially for BPA. Also, due to the hormonal nature of these compounds, their effects can be felt at extremely low concentrations (on the order of ng L^{-1}). Thus, treatment technologies to remove these chemical substances in drinking water to levels below their active concentrations must be found and utilized in order to protect human health. Example 1 examines the possible impacts of an industrial spill of an EDC, even when a viable treatment scheme exists to protect against such an accident.

(Continued)

Table 9.5 Relative binding affinity compared to estrogen (YES assay)

Test compound	Relative estrogenic potency
17β-estradiol (E2)	1.0
17β-ethinylestradiol (EE2)	0.7
Diethylstilbestrol (DES)	1.1
Nonylphenol (NP)	7.2×10^{-7}
Bisphenol A (BPA)	6.2×10^{-5}

Source: Data from E. Silva, N. Rajapakse and A. Kortenkamp (2002). Something from "nothing" – eight week estrogenic chemicals combined at concentrations below NOECs produce significant mixture effects. *Environmental Science & Technology* 36 (8): 1751–1756; and L. Folmar, *et al.* (2002). A comparison of the estrogenic potencies of estradiol, ethinylestradiol, diethylstilbestrol, nonylphenol and methoxychlor in vivo and in vitro. *Aquatic Toxicology* 60: 101–110.

Table 9.6 Physical data for and major products containing BPA, EE2, and E2

Compound	Melting point(°C)	Vapor pressure (mm Hg)	Solubility (mg/L)	Log K_{ow}	Uses
BPA	153	4×10^{-8}	129 (25°C)	3.32	Plasticizer (adhesives, paints, CDs, baby bottles)
EE2	183	NA	11.3 (27°C)	3.67	Synthetic estrogen (birth control pills)
E2	178.5	NA	3.6 (27°C)	4.01	Natural estrogen

Source: Physical data from Chemfinder.com.

Let us consider a hypothetical bioengineering case. A chemical plant that produces polycarbonate for baby bottles recently spilled one ton of BPA into a wastewater stream with a flow of 1 MGD that discharges its effluent into the Ohio River. The plant has the capability to feed 10 mg/L Powdered Activated Carbon (PAC) into the wastewater stream and enough holding capacity to achieve 4 hours of contact time. To remove this PAC and other solids, the plant also has the capability to filter solid particles down to 1 μm from their wastewater in an emergency. If it is assumed that the spill is evenly dispersed throughout one day, and equilibrium conditions are achieved in the water stream, what is the final concentration of the water being discharged from the plant into the Ohio River? Also, according to the YES assay, how "estrogenic" is the wastewater stream due to the BPA?

- Step 1: Find the concentration of BPA in the waste stream before any treatment.

If 1 ton of solid BPA is spilled into 1 million gallons (assume even dispersion through the waste stream for 1 day), upon unit conversion, a concentration of 237 mg/L would be achieved if all BPA dissolved in water. However, Table 9.6 shows that the solubility of BPA in water is only 129 mg L^{-1}, implying this is the maximum concentration of BPA in water at 25 °C. The rest of the BPA remains as solid particles in the water (assumed greater than 1 μm in diameter.

- Step 2: Determine the concentration after PAC addition and filtration.

For the conditions given (10 mg/L PAC with a contact time of 4 hours removing BPA), approximately 4% of the original concentration of BPA remains in solution. Also, the filtration step will remove all PAC, plus any undissolved BPA, implying the final concentration in the wastewater stream will be approximately 5.16 mg/L.

- Step 3: How "estrogenic" is this stream?

According to the YES data given in Table 9.5, BPA displays a relative potency of 6.2e-5 as compared to estrogen. This means the concentration of 5.16 mg/L displays the "estrogenic" response of 0.32 μg/L of

17β-estradiol. This is equivalent to an estrogen concentration capable of inducing estrogenic responses in all of the bioassays.

Note that the wastewater will be substantially diluted when it enters the Ohio River. However, it is naive to assume that a wastewater stream dumping into a larger water body will disperse widely, a conclusion supported by a United States Geological Survey study examining wastewater discharge from Las Vegas into Lake Meade, which used the vitellogenin bioassay to show elevated levels of EDCs greatly affecting male carp, fish that prefer sheltering near large underwater objects, including wastewater effluent pipes [59].

Treatment of EDCs in drinking water – UV applications

Because of their proven ability to interfere with normal endocrine function of so many aquatic species at low concentrations, and their presence in waters used as drinking water sources, inclusion of a treatment technology capable of removing or destroying EDCs in a drinking water treatment train may be imperative to the goal of protecting human health. Current treatment technologies that have been tested for their efficacy regarding removal or degradation of EDCs include conventional biological treatment, chlorination, activated carbon (GAC, PAC), membranes, and several oxidative techniques, with mixed success. Of great concern are several recent reports that have indicated that chlorination, a treatment process utilized by nearly every water utility in the United States, may react with certain EDCs to produce products that exhibit greater estrogenic activity than their parent compounds. These studies were performed regarding the chlorination of bisphenol A and nonylphenol [60], two persistent EDCs (see Figure 9.10).

A novel approach to removing synthetic estrogens involves using emerging ultraviolet light (UV)-based water treatment technology, currently used to disinfect microbial contaminants in drinking water. Ultraviolet radiation water treatment has proved very effective in removing threats presented by pathogenic organisms, and is being installed in many treatment facilities throughout the world. The use of UV radiation for destruction of contaminants, an area of increasing interest, may also present a viable alternative for effective treatment of EDCs in water supplies throughout the world.

Direct UV photolysis is governed by two main parameters, the molar absorption coefficient, and the quantum yield. Both parameters are chemical-specific and describe the interaction of the chemical with UV radiation. The molar absorption coefficient describes the amount of radiation at a specific wavelength that a compound within solution will absorb. With units of $M^{-1}\,cm^{-1}$, the UV absorbance due to a solution of the compound at a specific concentration is described by using the molar absorption coefficient. Figure 9.11 shows the relative absorption at each wavelength from 200 to 300 nm of BPA, E2, and EE2. All three compounds exhibit a multi-modal absorption spectrum over this range, and each exhibits an absorption minimum at approximately 250 nm. This is significant because LP lamps emit UV radiation only at approximately 254 nm, which corresponds closely to the minimum absorption of each contaminant, while MP lamps emit radiation throughout the UV range.

The first law of photochemistry states that only radiation that is absorbed can produce a photochemical effect. Thus, direct UV treatment of contaminants is effective only if the UV radiation emitted by a UV lamp is absorbed by the contaminant. The emission spectrum for the MP lamps overlaps much of the major absorbance features of the contaminants under study. Therefore, it is expected that a MP lamp will destroy the EDC more rapidly simply because more radiation is absorbed by the compounds.

Another important factor in understanding direct UV treatment is the quantum yield (Φ). The quantum yield is a measure of the photon efficiency of a photochemical reaction. It is defined as the number of moles of reactant removed per Einstein (mole of photons) absorbed by the chemical. There are no simple rules to predict reaction quantum yields from chemical structure, so Φ values need to be determined experimentally for each compound. Additionally, the wavelength dependence of Φ must be considered when using polychromatic radiation sources. Quantum yield values can be approximated as wavelength-independent, at least over the wavelength range of a given absorption band, corresponding to one mode of excitation. If, as in the case of our EDCs, multiple light absorption bands are displayed, quantum yields

(Continued)

may have to be determined for various wavelengths to accurately predict the transformation rate of a given compound.

When peroxide is added to the solution before irradiation with UV, the direct photolysis process for the target chemical is augmented by an indirect process through the production of the hydroxyl radical. Addition of UV energy in the presence of H_2O_2 is known as an advanced oxidation process (AOP). AOPs can be generated via a number of scenarios including vacuum UV in water, ozone, ozone/peroxide, UV/ozone/peroxide, UV/TiO_2, UV/NO_3, fenton processes, and photo-fenton processes. AOPs are characterized by the formation of a highly reactive, oxidative-intermediate species, such as the hydroxyl radical.

When UV radiation hits a H_2O_2 molecule, the molecule splits apart into two OH radicals.

$$H_2O_2 + h\nu \rightarrow 2\ OH \tag{9.6}$$

Although the stoichiometry of this reaction implies two radicals per parent H_2O_2 molecule, due to recombining and inefficiencies in the process, only one OH radical is formed per photon of light absorbed. Therefore, the quantum yield of the process is unity. This means that in the bulk solution, for every mole of photon of light absorbed by H_2O_2, one mole of hydroxyl radical is formed. Once the hydroxyl radical is formed, it will rapidly undergo an oxidation reaction with almost any species present, including the contaminant of interest. OH radicals will also react quickly with carbonate species (HCO_3^-, CO_3^{-2}), natural organic matter (NOM), other organic compounds present, chloride ion, and even H_2O_2. Given this non-selective nature of OH radicals, water quality must be accounted for when determining the effectiveness of the process towards degrading a specific contaminant of concern. Table 9.7 displays the second-order rate constants of OH radical with several organic contaminants of concern, as well as with carbonate species and NOM.

Modeling the UV/H_2O_2 process

Because of the unselective nature of the OH radical, the concentration of the species can often be considered constant and relatively low (10^{-14} to 10^{-12} M) when compared to the levels of other species in the water. Using these assumptions, the steady state model for destruction involving the OH radical has been developed. This model assumes the OH radical concentration at a constant level throughout the process, thus reducing the second-order rate equation:

$$\frac{d[M]}{dt} = k[OH][M] \tag{9.7}$$

Table 9.7 Second-order rate constants of OH radical with several organic contaminants and inorganic species

Compound	Second-order rate constants	Source [61]
Atrazine ($M^{-1} s^{-1}$)	3×10^9	Acero (2000)
MTBE ($M^{-1} s^{-1}$)	1.6×10^9	Huber (2003)
Ethinyl estradiol ($M^{-1} s^{-1}$)	9.8×10^9	Huber (2003)
HCO_3^- ($M^{-1} s^{-1}$)	8.5×10^6	Buxton (1988)
CO_3^{2-} ($M^{-1} s^{-1}$)	3.9×10^8	Buxton (1988)
DOM (L (mg C)$^{-1}$ s^{-1})	2.5×10^4	Larson and Zepp (1988)
H_2O_2 ($M^{-1} s^{-1}$)	2.7×10^7	Buxton (1988)

Table 9.8 Water quality parameters for a sample of natural water

H_2O_2 (ppm)MW = 34 g/mol	$[EE2]_i$ ($\mu g\ L^{-1}$)	DOM (excluding EE2) mg/L	pH	Alkalinity (mg L^{-1} as $CaCO_3$)
15	50	4.92	7.35	24.8

to a pseudo-first-order rate equation

$$\frac{d[M]}{dt} = k'[M] \tag{9.8}$$

where k' is the product of the second-order rate constant and the steady state OH radical concentration.

The steady state OH radical concentration is influenced by many parameters in the UV/H_2O_2 process, including intensity of the UV radiation, concentration of H_2O_2, and water quality. Eq. 9.9 is used to calculate $[OH]_{ss}$ for a low pressure lamp with a known hydrogen peroxide concentration.

$$[OH]_{ss} = \frac{I_{ave}\Phi\varepsilon[H_2O_2]}{\sum k_s[S]} \tag{9.9}$$

where I_{ave} is the average UV irradiance (Es/s)

Φ is the quantum yield of OH radical formation from H_2O_2 (1 mol/Es)

ε is the molar absorption coefficient of H_2O_2 (17.9 $M^{-1}\ cm^{-1}$ at 254 nm)

$[H_2O_2]$ is the initial concentration of hydrogen peroxide (M)

$\Sigma k_s[S]$ is the sum of the second-order rate constants times the concentration of all scavenger species present

Let us consider another hypothetical case to illustrate how this mode can show photodegradation of an EDC in natural waters using the UV/H_2O_2 process. Given the second-order rate constant for the reaction between EE2 and OH radical in Table 9.8, find the time required to degrade EE2 by 2 logs (99%) using LP UV/H_2O_2 process (average irradiance = 0.015 mEs/s, $[H_2O_2]_i$ = 15 mg/L) in water described by the water quality parameters given in Table 9.8.

- Step 1: Find the OH radical steady state concentration.

First, the molar concentration of all scavenger species must be known. The scavengers in this case are H_2O_2, DOM, HCO_3^-, and CO_3^{2-} (the initial concentration of EE2 can be neglected because it is significantly less than the concentration of other organics in the water). HCO_3^- and CO_3^{2-} are calculated using the pH and alkalinity. A simplified version of the alkalinity equation is:

$$Alk = [OH^-] - [H^+] + [HCO_3^-] + [CO_3^{-2}] \tag{9.10}$$

And the carbonate species are related through acid/base chemistry:

$$K_a = \frac{[CO_3^{2-}][H^+]}{HCO_3^-} \tag{9.11}$$

where K_a is the second acid dissociation constant for the carbonate system ($K_a = 10^{-10.3}$)

By manipulating these equations and solving for the molar concentrations of the other species, the concentrations of the scavenging species are as follows.

$$[H_2O_2] = 4.4e\text{-}4\ M,\ DOM = 4.92\ mg/L,\ [HCO_3^-] = 5.0e\text{-}4\ M,\ [CO_3^{-2}] = 5.6e\text{-}07$$

The second-order OH radical rate constants for all of these species can be found in Table 9.7 and so Eq. 9.8 can be solved to find the steady state OH radical concentration of 8.5×10^{-13} M.

(Continued)

- Step 2: Integrate the pseudo first-order rate equation.

To find the time necessary for a reaction to occur, an integrated rate expression must be found. In this case, separation of variable and integration of both sides of Eq. 9.8 yields the following integrated rate equation.

$$\frac{C}{C_o} = e^{-k't} \tag{9.12}$$

- Step 3: Solve the integrated rate equation to find the time needed for 2 log removal.

2 logs of removal implies 99% removal, so if 0.01 is input for the left-hand side of Eq. 9.12, and k_{EE2} from Table 9.7 (9.8×10^9) is multiplied by $[OH]_{ss}$ to find k', a time of 553 seconds, or 9 minutes and 13 seconds is needed to achieve the desired removal.

As a final note, complete mineralization of organic contaminants can be achieved by utilizing the UV/H_2O_2 advanced oxidation process. However, complete transformation to mineral acids, H_2O, and CO_2 takes long exposure times and high concentrations of H_2O_2. As is the case in most chemical treatment situations, incomplete destruction of the contaminants will occur with the UV/H_2O_2 processes. As such, a variety of destroyed but not mineralized byproducts will remain in the treated water. These products are likely to be more polar and smaller than the original pollutant. Both the identities and the toxicity of these compounds must be determined to evaluate the true effectiveness of any degradation process. The ultimate guiding question that needs to be answered when determining the effectiveness of the treatment process for destruction of EDCs is, "Does this treatment process solve the problem, exacerbate it, or cause new problems?" By utilizing various bioassays, including the YES assay, the E-Screen assay, or a developmental fish assay, future research will attempt to examine the relative toxicity of the byproducts of the destruction processes, in an attempt to determine the effectiveness of the UV treatment process not only in regards to destruction of EDCs, but ultimately in regards to protecting the water supply from these contaminants and the possibility of estrogenic behavior of their degradation products.

Environmental implications

Medical processes and products have the same potential challenges as many other industrial applications, i.e. potential release of genetically modified organisms from confined operations into the environment and public facilities (e.g. drinking water supplies). In addition, human and veterinary medicines can introduce biomedical and environmental problems of cross-resistance as drugs and personal care products pass through treatment facilities and leave animal feeding operations (see Chapter 6).

The complex interactions between genetically modified and non-modified organisms may be even more complicated for medical biotechnologies since the biological agents and the chemicals they release are, by design, expected to be biologically active. Thus, they trigger cellular responses, many of which are enzymatic, that may affect microbial populations in humans, wildlife, and ecosystems.

ANIMAL BIOTECHNOLOGY

The medical and agricultural sectors share the application of animal biotechnologies. While recognizing that there "might be many benefits," the National Academy of Sciences highlighted the need to consider the environmental concerns that can accompany the exponential growth in modifying animals' genetic material:

> Potential impacts on the environment from the escape or release of genetically engineered organisms was the committee's greatest science-based concern associated with animal biotechnology, in large part due to the uncertainty inherent in identifying environmental problems early on and the difficulty of remediation once a problem

> has been identified ... The committee based its assessment on principles of risk analysis that are general in their application and not limited to currently developed biotechnology ...
>
> Any analysis of GE organisms and their potential impact on the environment needs to distinguish between organisms engineered for deliberate release and those that are engineered with the intention of confinement but escape or are inadvertently released. The discussion in this report focuses primarily on the latter category, but the committee recognized the possibility of intentional release of GE organisms into the environment and expressed a high level of concern about it ...
>
> Consideration of environmental concerns posed by GE animals must be based on an understanding of key concepts underlying the science and practice of ecologic risk assessment. [62]

According to the Academy, genetically engineered animals for biomedical purposes have included reducing allergenicity, xenotransplantation, and control of the spread of pests and diseases. Insects that have been modified have predominantly been vectors (especially mosquitoes) to eliminate parasite transmissions. Genetically modified animals may overcome viability disadvantages if other fitness components are enhanced, such as mating success, fecundity, or age at sexual maturity. Introgression of genes decreases fitness, which may introduce "a near-term demographic risk to small receiving populations (i.e., small populations might not remain viable until the transgene is selected out, which poses a risk if a threatened or endangered or otherwise valued population is at issue)." The potential phenotypic change by transgenesis "could exceed that of conventional breeding or natural mutations." Physiological changes can be much faster and more substantial in transgenic organisms than in naturally occurring mutations (e.g. dwarfism or gigantism in mammals and poultry). Transgenic fish, for example, have shown as much as 11 times normal size [63].

Ecosystem disruptions by the dispersal of domesticated species have had dramatic impacts when released, such as those shown in Table 9.9. The Academy states that:

> ... if wild or feral populations exist locally, the escaped transgenic organisms could breed with those and spread the transgene into populations that otherwise are well adapted to the local environment. If the GE animal is released into an area where a native wild or feral population of the same species exists, mates might be readily available, and the transgene could spread via mating. Even in areas where the GE species does not exist, it might breed with members of a closely related species with which it is reproductively compatible (e.g., transgenic rainbow trout, Oncorhynchus mykiss, with native cutthroat trout) ... [64]

Thus, the likelihood for a transgenic animal to disrupt ecological resources depends on its ability to escape and its fitness in the environment that it enters. In one scenario, the genetically modified organism may displace its near relative. In another, both the transgenic and non-transgenic genotypes persist in the environment. If a more fit (e.g. larger) transgenic genotype survives, it could upset the predator–prey and other ecological mechanisms, making the ecosystem less stable [65].

Therefore, in addition to safety concerns, numerous environmental concerns accompany medical biotechnologies.

AGRICULTURAL BIOTECHNOLOGY

> Think of a farmer: how patiently he waits for the precious fruit of the ground until it has had the autumn rains and the spring rains!
>
> **Letter of James 5.7**

Table 9.9 Factors contributing to level of concern for species transformed

	Factor contributing to concern					
Animal	Number of citations[1]	Ability to become feral[2]	Likelihood of escape captivity[3]	Mobility[4]	Community disruptions reported[5]	Level of concern[6]
Insects[8]	1804	High	High	High	Many	High
Fish[7]	186	High	High	High	Many	↓
Mice/Rats	53	High	High	High	Many	↓
Cat	160	High	High	Moderate	Many	↓
Pig	155	High	Moderate	Low	Many	↓
Goat	88	High	Moderate	Moderate	Some	↓
Horse	93	High	Moderate	High	Few	↓
Rabbit	8	High	Moderate	Moderate	Few	↓
Mink	16	High	High	Moderate	None	↓
Dog	11	Moderate	Moderate	Moderate	Few	↓
Chicken	11	Low	Moderate	Moderate	None	↓
Sheep	27	Low	Low	Low	Few	↓
Cattle	16	Low	Low	Low	None	Low

[1]*Number of scientific papers dealing with feral animals of this species.*
[2]*Based on number of feral populations reported.*
[3]*Based on ability of organism to evade confinement measures by flying, digging, swimming, or jumping ability for any of the life stages.*
[4]*Relative dispersal distance by walking, running, flying, swimming, or hitchhiking in trucks, trains, boats, etc.*
[5]*Based on worldwide citations reporting community damage and extent of damage.*
[6]*A ranking based on the four contributing factors.*
[7]*Did not include shellfish, some of which (such as zebra mussel and Asiatic clam) have proven highly invasive.*
[8]*Limited to gypsy moth and Africanized honeybee.*
Source: National Academy of Sciences (2002). *Animal Biotechnology: Science-Based Concerns.* National Academies Press, Washington, DC.

Agriculture requires patience. In the millennia that preceded modern agriculture, farmers from throughout the globe painstakingly and incrementally selected seeds and livestock generation after generation. Biotechnologies that emerged in the latter part of the 20th century took these efforts to a new level.

Much of the discussion in this and previous chapters has already addressed agricultural biotechnologies, especially the risks associated with genetic material leaving containment, such as confined and targeted fields, within specific organisms and other systems. The biochemodynamic and geopolitical complexities of agricultural biotechnology are illustrated by the push and pull of crops. Society demands that sufficient food is available. Society also demands that the food be safe.

In the medical biotechnology discussion, we considered the similarities of possible environmental implications from animal biotechnologies. Fish populations appear to be particularly vulnerable to these disruptions (see Chapter 6, especially Case Study Box: Genetic Biocontrols of Invaders). In fact, genetically modified salmon provide a dramatic illustration of the complexity of predicting environmental risks from animal biotechnologies. The modified salmon appear to have 40 times the growth hormone of non-modified salmon.

Fish farms have been associated with increased concentrations of certain contaminants, including mercury (Hg) and polychlorinated biphenyls (PCBs), whether or not the fish are genetically engineered. Both of these substances are neurotoxic. Both born and unborn children are vulnerable to neurotoxicity, as well as to hormonally active agents. Given that the immune, neural, and endocrine system are strongly interconnected, the question for risk assessors is what is the effect of a large increase in both the endogenous chemicals (growth hormones) and xenobiotic compounds?

Most of the attention in agricultural biotechnologies, however, resides in plant life. Agricultural chemicals, especially fertilizers and pesticides, are part of modern agriculture's arsenal for a stable food supply, but at the same time, they are part of the threat to the safety of that food supply. Thus, biotechnologies have been enlisted as an important means to optimize crops to provide adequate and safe food supplies (see Case Discussion: King Corn or Frankencorn).

CASE DISCUSSION

King Corn or Frankencorn?

Bt corn has been evaluated thoroughly by EPA, and we are confident that it does not pose risks to human health or to the environment.
Stephen L. Johnson, EPA Administrator (October 16, 2001) [66]
Once again, the EPA has taken the interests of a few corporations over public health and the environment.
Matt Rand, Washington-based National Environmental Trust (October 16, 2001) [67]

No discussion of the perceived risks of agricultural biotechnology would be complete without discussing corn. The grain has been the focus of two major biotechnological issues: biofuels and GMOs.

Biofuels

Corn is a particularly important and vulnerable crop. This became painfully obvious to the biotechnological community when it was recently touted as a source for biofuels. In his 2007 State of the Union Address, US President George W. Bush set a two-part goal:

- Setting a mandatory standard requiring 35 billion gallons of renewable and alternative fuels in the year 2017, which is approximately five times the 2012 target called for in current law. Thus, in 2017, alternative fuels will displace 15% of projected annual gasoline use.
- Reforming the corporate average fuel economy (CAFE) standards for cars and extending the present light truck. Thus, in 2017, projected annual gasoline use would be reduced by up to 8.5 billion gallons, a further 5% reduction that, in combination with increasing the supply of renewable and alternative fuels, will bring the total reduction in projected annual gasoline use to 20%.

Ethanol production has been politically popular in corn-growing states. In fact, since the presidential proclamation, dedicated corn crops and ethanol refining facilities in these states have emerged. Conversely, geopolitical impacts, i.e. the food versus fuel dilemma, are being raised. Scientific challenges to any improved efficiencies and actual decreases in the demand for fossil fuels have also been voiced. Some have accused advocates of ethanol fuels of using "junk science" to support the "sustainability" of an ethanol fuel system.

From a thermodynamics standpoint (see Table 9.10), the nation's increased ethanol use could actually increase demands for fossil fuels, such as the need for crude oil-based infrastructures, including farm chemicals derived from oil, farm vehicle, and equipment energy use (planting, cultivation, harvesting, and transport to markets) dependent on gasoline and diesel fuels, and even embedded energy needs in the ethanol processing facility (crude oil-derived chemicals needed for catalysis, purification, fuel mixing, and refining).

(Continued)

Table 9.10 Energy use and net energy value per gallon of ethanol produced with coproducts energy credits

	Milling process		
	Dry	Wet	Weighted average
Production phase		*Btu per gal*	
Corn production	21,803	21,430	21,598
Corn transport	2,284	2,246	2,263
Ethanol conversion	48,772	54,239	51,779
Ethanol distribution	1,588	1,588	1,588
Total energy used	74,447	79,503	77,228
Net energy value	9,513	4,457	6,732
Energy ratio	1.11	1.04	1.08

Source: H. Shapouri, J.A. Duffield and M.I. Wang (2002). The Energy Balance of Corn Ethanol: An Update. Agricultural Economic Report No. 813. US Department of Agriculture, Office of the Chief Economist, Office of Energy Policy and New Uses.

No matter how politically attractive or favorable to society, a biofuel must comport with the conservation of mass and energy. Further, each step in the life cycle (e.g. extraction of raw materials, value-added manufacturing, use, and disposal) must be considered in any benefit–cost or risk–benefit analysis.

The challenge of the bioengineer and policy maker is to sift through the myriad data and information to ascertain whether ethanol truly presents a viable alternative fuel. There is always the risk of mischaracterizing the social good or costs, a common problem with the use of benefit–cost relationships.

Biomass-based fuel efficiencies are evaluated in terms of net energy production that is based on thermodynamics (first and second laws):

$$\text{Efficiency} = \frac{E_{in} - E_{out}}{E_{in}} \times 100 \quad (9.13)$$

where E_{in} = Energy entering a control volume, and

E_{out} = Energy exiting a control volume

The numerator includes all energy losses. However, these are dictated by the specific control volume. This volume can be of any size from molecular to planetary. To analyze energy losses related to alternative fuels, every control volume of each step of the life cycle must be quantified.

The first two laws of thermodynamics drive this step (see Chapter 3). For example, if a life cycle for ethanol fuels begins with the corn arriving at the ethanol processing facility, none of the fossil fuel needs on the farm or in transportation will appear. Entropy is ever present. Losses must always occur in conversions from one type of energy (e.g. mechanical energy of farm equipment ultimately to chemical energy of the fuel). Thus, Eq. 9.13 is actually a series of efficiency equations for the entire process, with losses at every step. This study by the US Department of Agriculture shows a net positive efficiency (i.e. more energy produced from ethanol than lost to fossil fuels needed to produce the ethanol [68]. Other studies have found both negative and positive efficiencies. The bottom line, though, is that it takes a substantial amount of fossil fuels currently to produce ethanol from corn.

On the surface, the choice of whether to pursue a substantial increase in ethanol production is a simple matter of benefits versus costs. Is it more or less costly to generate ethanol than other fuels, especially those derived from crude oil? Engineers make much use of the benefit–cost ratio (BCR), owing to a strong affinity for objective measures of successes. Thus, *usefulness* is an engineering measure of success. Such utility is indeed part of any successful engineering enterprise. After all, engineers are expected to provide reasonable and useful products. Two useful engineering definitions of utilitarianism (Latin *utilis*, useful) are imbedded in BCR and life cycle analysis (LCA):

The belief that the value of a thing or an action is determined by its utility.
The ethical theory … that all action should be directed toward achieving the greatest happiness for the greatest number of people.

The BCR is an attractive metric due to its simplicity and seeming transparency. To determine whether a project is worthwhile, one need only add up all of the benefits and put them in the numerator and all of the costs (or risks) and put them in the denominator. If the ratio is greater than 1, its benefits exceed its costs.

One obvious problem is that some costs and benefits are much easier to quantify than others. Some, like those associated with quality of life, are nearly impossible to quantify and monetize accurately. Further, the comparison of action versus no-action alternatives cannot always be captured within a BCR. Opportunity costs and risks are associated with taking no action (e.g. loss of an opportunity to apply an emerging technology may mean delay or nonexistent treatment of diseases). Simply comparing the *status quo* to costs and risks associated with a new technology may be biased toward no action. Costs in time and money are not the only reasons for avoiding action. The greater availability of ethanol may introduce unforeseen risks that, if not managed properly, could interfere with quality of life of distant and future populations, and could add costs to the public (e.g. air pollutants and topsoil loss) with little net benefit. So, it is not simply a matter of benefits versus cost, it is often one risk being traded for another.

Often, addressing contravening risk is a matter of optimization, which is a proven analytical tool in engineering. However, the greater the number of contravening risks that are possible, the more complicated such optimization routines become. The product flows, critical paths, and life cycle inventory (LCI) can become quite complicated for complex issues.

Risk tradeoff is likely to occur in biofuel decisions. For example, if the government mandates more ethanol usage, it will also have to enforce new air pollution laws associated with the fuel. These added regulations can be associated with indirect, countervailing risks. For example, the costs of construction of new facilities and the price of feedstock (especially corn) may increase safety risks via "income" and "stock" effects. The income effect results from pulling money away from other fuel ventures to pay the capital costs associated with ethanol, making it more difficult for a company or backers to invest in other services that would have provided improved fuel efficiency. The stock effect results when the capital costs increase to a point where companies have to wait to purchase new facilities, so they are left with substandard manufacturing. Thus, the engineer is frequently asked to optimize for two or more conflicting variables in many situations.

The success of ethanol in displacing fossil fuels depends on the efficiency with which it can be produced and used. Complicating matters, the use of fossil fuels in their production and/or operation is part of ethanol production, as it is for all biofuels. Societal benefits and costs are tied to ethanol's energy balances.

Genetic modification

"Frankencorn" is the moniker of genetically modified corn. The term's notoriety grew when genetically engineered corn started showing up in US food supplies. Some of the fear of biosciences and engineering that was captured by Mary Shelley in *Frankenstein (The Modern Prometheus)* remains ensconced in most contemporary societies. The analogy to biotechnology is that we cannot know the downstream effects and artifacts when creatures are manipulated.

In 2000, the lay press reported frequently on possible ecosystem effects from genetically modified corn perspective (see Chapter 1, Discussion Box: Little Things Matter in a Chaotic World). A specific genetically modified corn, known as Starlink, which was not approved for human consumption, had also found its way into the food supply. This led to food distribution disruptions and law suits, some stemming from the concern that Starlink and other genetically altered varieties may have greater allergenicity than the progenitor varieties. In fact, gene flow of the modified corn species indeed seems to be occurring. For example, in 2001, a specific gene from a genetically modified corn produced in the US was isolated in native corn in Mexico. This was a controversial study because the authors attributed the flow to *introgression*

(Continued)

(i.e. incorporation of a gene from one organism complex into another organism complex by means of hybridization) [69]. This is not the likely mechanism. More important, however, is the concern that the genetic material had found its way to the traditional crops.

In addition, the whole area of allergenicity is a modern conundrum. Why do food allergies seem to be on the increase? The desire was to introduce the Bt property of releasing the toxin Cry9c to protect corn from the European corn borer, *Ostrinina nubilalis*, which is a major pest of corn and other crops. Starlink corn thus contains in its genome a modified sequence of the *cry9Ca1* bacterial gene (derived from strain BTS02618A of *Bacillus thuringiensis* serovar *tolworthi*) [70].

The knowledge gaps about Starlink were pronounced. It was obvious that the regulatory agencies fell woefully short when it came to predicting and modeling allergenicity posed by the newly expressed proteins. This required not only knowing the chemistry of the proteins, but also the sequence homology with known allergens (see Figure 9.14). Segregating the genetically engineered strains was also obviously not happening.

One of the metaphors for biotechnology is that it can be a two-edged sword. For example, the author was having lunch with his wife recently, when she mentioned severe food allergies. She asked, ''Could all this biotechnology be the reason for the increase in extent and severity of food allergies?'' The author's first reaction is usually to ask something like how good are the data and is this simply because the means of diagnosing and treating these allergies have improved over recent decades. Also, could another variable, such as less breast feeding or high exposures to allergens and other agents in high density environments (e.g. day care centers), be exposure and risk factors? However, in this case, even before researching it, these factors were not very satisfying. One could add the seeming increase in gluten and lactose intolerance. Could there have been gene mixing or flow into nearby fields of some test crops? Could the desired expression be accompanied by an unintended allergenic expression of certain enhanced plants?

A data search indicated sources that seemed to look to genetic engineering solutions to the allergies, but few that posited the potential that they were a part of the cause. It appears there is more fear of the type II error than the type I error. That is, bioscientists' greater interest in seeing biotechnologies as solutions to allergies may stem from their fear of rejecting a true hypothesis (i.e. saying biotechnology is not the solution

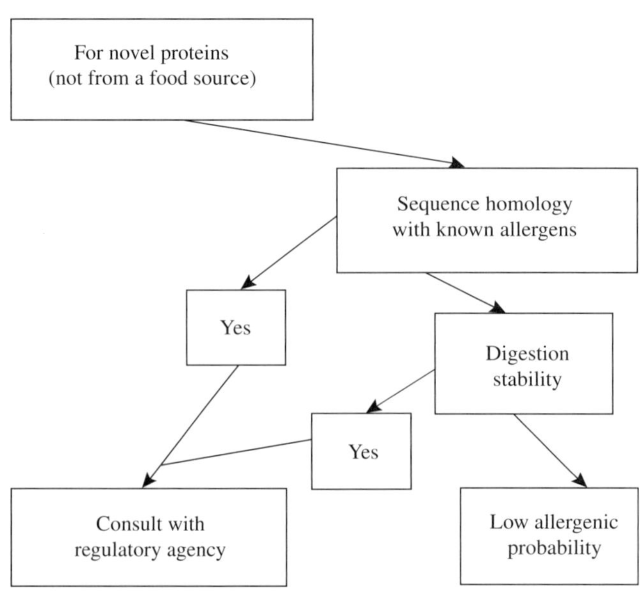

FIGURE 9.14
Decision tree for assessing potential for allergenicity associated with novel proteins form sources other than a food source. *Source:* L. Bucchini and L.R. Goldman (2002). Starlink corn: a risk analysis. *Environmental Health Perspectives* 110 (1): 5–13.

when biotechnology really is the solution) than in accepting an incorrect one (i.e. saying that biotechnology is not the problem when biotechnology really is the problem). Researchers seem to care more about doing something that hinders the advancement of science (e.g. finding the rDNA that makes peanuts less allergenic) than precaution (determining whether biotechnology has had some role in the increase in allergenic foods).

The advocacy group Institute for Responsible Technology argues that the major emphasis should be placed on reducing type II error:

> *The huge jump in childhood food allergies in the US is in the news often, but most reports fail to consider a link to a recent radical change in America's diet. Beginning in 1996, bacteria, virus and other genes have been artificially inserted to the DNA of soy, corn, cottonseed and canola plants. These unlabeled genetically modified (GM) foods carry a risk of triggering life-threatening allergic reactions, and evidence collected over the past decade now suggests that they are contributing to higher allergy rates. [71]*

The three principal types of biomaterials used in agriculture are biochemical pesticides, microbial pesticides, and plant-incorporated protectants (PIPs). In the United States, the US EPA regulates natural and engineered microbial pesticides, PIPs, and biochemical pesticides. Microbial pesticides can be naturally occurring or genetically engineered. Plant-incorporated protectants are biocidal substances that are produced in a living plant along with the genetic material necessary to produce the substance, where the substance is intended for use in the living plant. EPA includes the genetic material necessary to produce the substance in the definition of a plant-incorporated protectant because the genetic material introduced into the plant will ultimately result in a toxic effect. The genetic material may be responsible for the spread of the biocidal trait in the environment to related plant relatives [72].

Regulations for notification prior to small-scale field testing of engineered microbial pesticides were made final in 1994. In 1994, EPA proposed a set of regulations aimed at establishing the Agency's scope for PIPs and proposing to exempt several classes of PIPs that had very low risk. As of 2002, the majority of these proposed regulations were finalized.

Genetically engineered microorganisms are regulated using essentially the same data requirements used for naturally occurring microbial pesticides (40 CFR part 158.740). In addition, data may be required about the genetic engineering process used and the results from that process. EPA requires notification prior to small-scale field testing of genetically engineered microorganisms to allow EPA to determine whether an Experimental Use Permit is needed (40 CFR part 172 subpart C). When testing 10 acres or more, EPA requires an Experimental Use Permit before field testing naturally occurring or genetically engineered microorganisms. Under FIFRA, microbial biotech products, as with all other pesticides, must be evaluated for their risks and benefits. The approval process considers any potential adverse effects to nontarget organisms, environmental fate of the microorganism, and the potential pathogenicity and infectivity of the microorganism to humans.

All plants have survival mechanisms, including the production of substances that repel or kill fauna (e.g. insects) and pathogens such as bacteria and fungi. Plants that have more of these natural protections are less susceptible to pests than plants that have fewer of them.

Genes that express these traits have been inserted into other species as plant-incorporated protectants. Such manipulations are regulated in the United States under the Federal Insecticide, Fungicide and Rodenticide Act (FIFRA). In 1994, the EPA began registering plant-pesticides under FIFRA and regulating residues of plant-pesticides under FFDCA. The name plant-incorporated protectants or PIPs was selected after EPA took comments on a substitute name for plant-pesticides as requested in the public comments in response to the 1994 proposed rules.

In 2001, EPA published final rules exempting from FIFRA requirements (except for an adverse effects reporting requirement) pesticidal substances produced through conventional breeding of sexually compatible plants.

(Continued)

The residues of these pesticidal substances and of all nucleic acids that are part of a PIP are exempt from the Federal Food, Drug, and Cosmetic Act (FFDCA) pesticide residue requirements, based on a long history of human dietary exposure to these naturally occurring plant compounds, and epidemiological studies showing the health benefits of consuming foods that can contain low levels of these substances. Residues of nucleic acids that are part of PIPs were exempted as nucleic acids are common to all life forms, have always been present in human and domestic animal food, and are not known to cause adverse health effects when consumed as part of food. PIPs that have moved into plants from other organisms, including from plants not sexually compatible with the recipient plant, continue to be regulated, because the new exposure to the PIP is now expressed in the new host plant.

As mentioned in Chapter 8, the insecticidal protein from the common soil bacteria *Bacillus thuringiensis* (Bt) provides an illustrative example of the complexities associated with this type of PIP. The protein was introduced into a plant to induce protection against lepidopteran (caterpillar) pests. In 1994, EPA proposed exempting this use, as well as that of three other categories of PIPs. However, in 2001 EPA decided to solicit further guidance from the public on the three proposed exemptions because of the wide range of varied comments received on the proposals. In addition, EPA placed these rules in the public record, as well as asking for public comments on the information, analyses, and conclusions of the 2000 National Academy of Sciences study entitled "Genetically Modified Pest-Protected Plants: Science and Regulation" that recommended EPA reconsider its proposed categorical exemption of viral coat proteins, PIPs that act by primarily affecting the plant, and PIPs produced by using biotechnology to move genes between sexually compatible plants.

Generally, PIP data are based on those for microbial pesticides since to date most PIP products are produced by microbes. All of the products have been proteins, either related to plant viruses or based on proteins from Bt.

The quality of any risk assessment depends on physicochemical characterization, mammalian toxicity (including allergenicity potential), effects on nontarget organisms, extent of transport, and environmental fate. In addition, like chemical pesticides, there is always the concern of resistance. For example, the microbial population may adapt to the PIP-express pesticidal properties, similar to the problems encountered by farmers using traditional chemical pesticides. The risk assessment must, therefore, consider all biochemodynamic aspects on a systematic basis. This includes physical, chemical, and biological complexities and their spatial and temporal influences on the environment.

Recent studies by the EPA tend to show that using PIPs does appear to reduce the use of other pesticides.

Gene flow

In addition to the previously discussed problems of horizontal gene transfer (e.g. biodiversity), the movement of PIP transgenes from the host plant into weeds and other crops presents the possibility of novel exposures to the biocidal agent after release.

The Federal Insecticide, Fungicide, and Rodenticide Act directs the US EPA to examine all potentially adverse environmental impacts, including those that may arise from gene flow of PIPs to wild or feral populations of sexually compatible plants. In addition to this mandate, the Federal Food, Drug, and Cosmetic Act requires the issuance of a food tolerance or exemption from the requirement of a tolerance for all biocidal substances entering the food supply whether through seed mixing or cross pollination.

To date, only a handful of PIPs in crop species have been registered and all have received exemptions from the requirement of a tolerance. Bt corn, cotton, and potato were reviewed for their potential to hybridize with wild and feral relatives of sexually compatible plants. The government concluded that, with the conditions of registration in place, no significant risk of gene capture and expression is expected for any Bt endotoxin by wild or weedy relatives [73].

The Bt corn and potato PIPs that have been registered to date have been expressed in agronomic plant species that, for the most part, do not have a reasonable possibility of passing their traits to wild native plants. Most of the wild species in the United States cannot be pollinated by these crops (corn and potato) due to differences in chromosome number, phenology, and habitat. Gene transfer from Bt cotton to wild or feral cotton relatives may be possible in Hawaii, Florida, Puerto Rico, and the US Virgin Islands, where feral populations of cotton species are similar to cultivated cotton exist. Thus, the government does not allow the sale or distribution of Bt cotton in these locations, to prevent the movement of the registered Bt endotoxin from Bt cotton to wild or feral cotton relatives.

Movement of transgenes from crop plants to related or unrelated species of plants or other organisms can be a concern, as is the possibility of horizontal gene transfer. Horizontal transfer of several traits might be of concern for the current Bt crops, e.g. Cry genes and antibiotic resistance genes. Evidence to date does not support that horizontal gene transfer occurs from plants to microbes or bacteria to bacteria. Bt genes that naturally occur in many soils have never demonstrated horizontal gene transfer.

Industrial, medical, and agricultural biotechnologies must be considered in light of possible and actual environmental risks. In many instances, these risks are difficult to assess, given the lack of reliable and complete biochemodynamic and genetic data. However, as analytical and assessment tools improve, the genetically modified strains will be better characterized, distinguished from non-genetically modified strains, and considered from a multidisciplinary, multipathway and multicompartmental, i.e. systematic, perspective.

SEMINAR TOPIC

Vaccines from Genetically Modified Organisms

Joachim Frey of the Institute of Veterinary Bacteriology in Switzerland has stated:

> *... novel molecular methods enable the development of genetically modified organisms (GMOs) targeted to specific genes that are particularly suited to induce attenuation or to reduce undesirable effects in the tissue in which the vaccine strains can multiply and survive. Since live vaccine strains (attenuated by natural selection or genetic engineering) are potentially released into the environment by the vaccines, safety issues concerning the medical as well as environmental aspects must be considered. These involve (i) changes in cell, tissue and host tropism, (ii) virulence of the carrier through the incorporation of foreign genes, (iii) reversion to virulence by acquisition of complementation genes, (iv) exchange of genetic information with other vaccine or wild-type strains of the carrier organism and (v) spread of undesired genes such as antibiotic resistance genes. Before live vaccines are applied, the safety issues must be thoroughly evaluated case-by-case. Safety assessment includes knowledge of the precise function and genetic location of the genes to be mutated, their genetic stability, potential reversion mechanisms, possible recombination events with dormant genes, gene transfer to other organisms as well as gene acquisition from other organisms by phage transduction, transposition or plasmid transfer and cis- or trans-complementation. For this, GMOs that are constructed with modern techniques of genetic engineering display a significant advantage over random mutagenesis derived live organisms. The selection of suitable GMO candidate strains can be made under* in vitro *conditions using basic knowledge on molecular mechanisms of pathogenicity of the corresponding bacterial species rather than by* in vivo *testing of large numbers of random mutants. This leads to a more targeted safety testing on volunteers and to a reduction in the use of animal experimentation. [74]*

P.P. Pastoret of the University of Liège in Belgium adds:

> *New biotechnology has had a major impact on the design and development of new medicinal products, especially vaccines. In some cases, use of biotechnology has lead to results which would have been impossible to obtain through classical or conventional routes, for example, defining the complete sequence of human hepatitis C virus, which still cannot be grown in cell culture. Biotechnology has also helped to develop products which are safer than conventional ones; one notable example is the use in humans of growth hormone produced in genetically engineered bacteria instead of natural growth hormone, the use of the latter being responsible for the transmission of Creutzfeldt–Jakob Disease. Another example is the use of a recombinant vaccinia-rabies virus for vaccinating foxes against rabies, which is more efficacious and safer than the conventional attenuated SAD B19 strain. This strain, which is still used, is pathogenic for some non-target mammals and since its*

(Continued)

safety for man is unknown, people in contact with it must undergo post-exposure anti-rabies treatment.

Medicinal products are among the most stringently regulated products in the market place. They are evaluated for quality and efficacy but, above all, for safety, and if they consist of a genetically modified organism they must comply with specific regulations. If a medicinal product is developed using a biotechnological procedure it must be evaluated at a European level by the relevant scientific committee using the so-called Centralised Procedure: the Committee for Proprietary Medicinal Products (CPMP) deals with human products and the Committee for Veterinary Medicinal Products (CVMP) with products intended for use in animals; both are part of the European Agency for the Evaluation of Medicinal products (EMEA), based in London, UK.

Recombinant pharmaceutical products including genetically modified organisms are often developed primarily for veterinary use, taking advantage of the fact that efficacy and safety can be studied experimentally directly in the target species, and that the environmental impact for non-target species can also be studied experimentally.

DNA vaccination is the new frontier in vaccinology and, thanks to the existence of already available biosafety research results, the Immunological Working Party (IWP) of the CVMP was able to produce guidance notes on DNA vaccines for use by pharmaceutical companies and the relevant competent authorities.[75]

Octavio Guerrero-Andrade of the Center for Research and Advanced Studies in Guanajuato, Mexico, inserted a gene from the Newcastle Disease Virus, a serious threat to poultry, into corn DNA. Chickens that consumed the genetically modified (GM) corn produced antibodies against the virus. In addition, the corn provided a level of protection against infection comparable to that of commercially available vaccines [76]. This is evidence that genetically modified crops could be a tool for delivering the vaccine to millions of poor poultry farmers throughout the world.

So, then, what constitutes a threshold of safety for vaccines? Given recent public concerns over actual and perceived risks of influenza vaccinations (both seasonal and H1N1 strains), the issues surrounding risk and safety have become ones of popular and medical concern. How might this be exacerbated if vaccine concerns are combined with GMO concerns? Do GMO derived vaccines differ from traditional techniques with regard to the type and focus of risk assessments? According to a recent report in *Meat & Poultry*:

Researchers at Iowa State University seem to be involved in a medical/agricultural GMO "risk hybrid". They are inserting flu vaccines into the genetic makeup of corn, with hopes that eventually this will allow both livestock and humans to be vaccinated when consuming corn or corn products. ISU professor Hank Harris has stated that his team is "trying to figure out which genes from the swine influenza virus to incorporate into corn so those genes, when expressed, would produce protein," and "when the pig consumes that corn, it would serve as a vaccine" [77].

Is this an example of poor timing, given the controversies surrounding flu vaccines, or is this simply an example of the advancement of science in the face of ideological and popular sentiment?

There have been a number of research investigations recently into the safety and risk associated with GMO-mediated vaccines. For example, vaccine strains may persist in recipients and, if the target species is a food-producing animal, the strains and genetic material may find their way into the food chain. The second one is that administration of a vaccine may trigger long-term adverse effects in normal or immunodeficient recipients. Effects can be discarded by using a recombinant alphavirus, i.e. the Semliki Forest Virus (SFV), in mice, chicken, and sheep. The vaccine virus does not persist more than seven days after vaccination. Potential immunological problems associated with this new vaccination technology include unexpected immunopathological reactions or tolerance. Studies have shown that the purity of injected DNA vaccines can affect the inflammatory response and that the type of vaccination can influence the nature of this effect. This kind of work is of primary importance in establishing a scientific basis for proper regulation of DNA vaccines, but also to limit ill-informed opinions impeding clinical trials, which is detrimental both to the vaccine industry and also for human and animal health [78].

Hank Harris, the ISU researcher, goes on to say that "The big question is whether or not these genes will work when given orally through corn." He added, "That is the thing we've still got to determine" [79]. He considers that corn vaccine's stability and safety are advantages, claiming that once the corn with the vaccine is grown, it can be stored for long-term without a substantial loss in potency. This would allow corn to be shipped to potential pandemic sources early during a breakout, and since corn grain is a food and feed stock, extensive vaccine purification, an expensive process, can be avoided.

Seminar Questions

Are the ISU researchers correct that the biggest issue associated with a plant based vaccine is whether or not the genetic insertion will work and that the desired traits will be expressed to produce a vaccine?

What are the possible health and environmental concerns with using genetic engineering to produce vaccines in general and the combination of rDNA insertion in a food crop specifically?

How may justice issues (e.g. providing stability to poor farmers) be balanced with potential risks (e.g. release of genetic material from vaccination recipients into the food chain)?

REVIEW QUESTIONS

What do industrial, medical, agricultural, and environmental biotechnologies have in common?

How do industrial, medical, agricultural, and environmental biotechnologies differ?

If biotechnological operations were held to the same release restrictions as research under the National Institutes of Health's guidance, would reactor designs have to be changed? How?

Do you agree with the criticisms of ACRE in protecting the United Kingdom from releases of GMOs and genetic material? Support your answer and suggest ways of improving the risk assessment and management processes.

Explain the pros and cons of PIPs. What factors of safety are needed before they are applied on a larger scale? How should uncertainties be addressed?

How can fate and transport simulation models be improved to address biotechnologies, especially with regard to environmental persistence, bioaccumulation, and toxicity?

Compare the advantages and disadvantages of generating vaccines from genetically modified organisms.

NOTES AND COMMENTARY

1. B. Schulze and M.G. Wubbolts (1999). Biocatalysis for industrial production of fine chemicals. *Current Opinion in Biotechnology* 10 (6): 609–615.
2. Y-N. Zhao, G. Chen and S-J. Yao (2006). Microbial production of 1,3-propanediol from glycerol by encapsulated *Klebsiella pneumoniae. Biochemical Engineering Journal* 32 (2): 93–99.
3. J.N. Nigam (1999). Continuous ethanol production from pineapple cannery waste. *Journal of Biotechnology* 72: 197–202.
4. F. Balkenhohl, K. Ditrich, B. Hauer and W. Ladner (1997). Optically active amines via lipase-catalyzed methoxyacetylation. *Advanced Synthesis and Catalysis* 339 (4): 381–384.
5. D.V. Goeddel, D.G. Kleid, F. Bolivar, H.L. Heyneker, D.G. Yansura, R. Crea, *et al.* (1979). Direct expression in *Escherichia coli* of a DNA sequence coding for human growth hormone. *Nature* 281: 544–548.
6. M.W. Fraaije and W.J.H. van Berkel (2006). Flavin-containing oxidative biocatalysts. In: R.N. Patel (Ed.), *Biocatalysis in the Pharmaceutical and Biotechnology Industries.* Marcel & Dekker, New York, NY.
7. M. Knauf and M. Moniruzzama (2004). Lignocellulosic biomass processing: a perspective. *International Sugar Journal* 106 (1263): 147–150.
8. Susan Harlander (2009). Biotechnology's Possibilities for Soyfoods and Soybean Oil. United Soybean Board. http://www.soyconnection.com/newsletters/soy-connection/health-nutrition/article.php/Biotechnology's+Possibilities+for+Soyfoods+and+Soybean+Oil?id=64; accessed August 27, 2009.
9. US Department of Energy, Energy Administration Administration (2001). The Transition to Ultra-Low-Sulfur Diesel Fuel: Effects on Prices and Supply. Report No. SR-OIAF/2001-01; http://www.eia.doe.gov/oiaf/servicerpt/ulsd/uls.html; accessed August 27, 2009.
10. P. Thanikaivelan, J.R. Rao, U. Balachandran, U. Nair and T. Ramasami (2004). *Trends in Biotechnology* 22 (4): 181–188.
11. US Department of Energy, Office of Science (2009). Systems Biology for Energy and Environment: Biohydrogen Production; http://genomicsgtl.energy.gov/benefits/biohydrogen.shtml; accessed August 27, 2009.
12. A. Richardt and M-M. Blum (Eds) (1997). *Decontamination of Warfare Agents: Enzymatic Methods for the Removal of B/C Weapons.* Wiley-VCH, Weinheim, Germany.
13. M.B. Roncero, A.L. Torres, J.F. Colom and T. Vidal (2005). The effect of xylanase on lignocellulosic components during the bleaching of wood pulps. *Bioresource Technology* 96 (1): 21–30.
14. J. Chen, Q. Wang, Z. Hua and G. Du (2007). Research and application of biotechnology in textile industries in China. *Enzyme and Microbial Technology* 40 (7): 1651–1655.
15. J.F. Martín and José A. Gil (1984). Cloning and expression of antibiotic production gene. *Nature Biotechnology* 2: 63–72.
16. S.S. Ahluwalia and D. Goyal (2007). Microbial and plant derived biomass for removal of heavy metals from wastewater. *Bioresource Technology* 98 (12): 2243–2257.
17. US Environmental Protection Agency (1997). Microbial Products of Biotechnology: Final Regulation under the Toxic Substances Control Act. *Federal Register* 62 (70): 17910–17958.
18. The source for this section is: European Commission and Federal Environment Agency Austria (2002). Collection of Information on Enzymes. Final Report. Contract No B4-3040/2000/278245/MAR/E2; http://www.agronavigator.cz/attachments/enzymerepcomplete.pdf; accessed August 10, 2009.
19. It should be noted that environmental biotechnologies can be just as controlled as those of industrial biotechnologies. For example, a bioreactor that breaks down recalcitrant compounds (see Chapter 8) may use identical microbial populations and similar conditions as an industrial fermentation chamber. The product of the former is a pathway of compounds that are less hazardous; the product of the latter is an intermediate or final product for the marketplace.

20. K. Menrad, D. Agrafiotis, C.M. Enzing, L. Lemkow and F. Terragni (1999). Scientific and technological development. In: K. Menrad, *et al.* (Eds), *Future Impacts of Biotechnology on Agriculture, Food Production and Food Processing.* Physica-Verlag, Heidelberg, Germany, pp. 286–329.
21. Self-cloning consisting in the removal of nucleic acid sequences from a cell of an organism which may or may not be followed by reinsertion of all or part of that nucleic acid (or a synthetic equivalent) with or without prior enzymatic or mechanical steps, into cells of the same species or into cells of phylogenetically closely related species which can exchange genetic material by natural physiological processes where the resulting microorganism is unlikely to cause disease to humans, animals or plants. (Council Directive 98/81/EC of 26 October 1998 amending Directive 90/219/EEC on the contained use of genetically modified micro-organisms (1998). *Official Journal of the European Communities.* No L 330 P. 0013 - 0031, Annex II, Part A: Techniques or methods of genetic modification yielding micro-organisms to be excluded from the Directive.)
22. Organization for Economic Cooperation and Development (2001). *The Application of Biotechnology to Industrial Sustainability – A Primer.* OECD, Paris, France.
23. Advisory Committee on Releases to the Environment (2009). Reports, Advice and Other Publications; http://www.defra.gov.uk/environment/acre/about/index.htm; accessed August 30, 2009.
24. Advisory Committee on Releases to the Environment (2009). Annual Report: Number 15; http://www.defra.gov.uk/environment/acre/pdf/acre-annrpt15.pdf; accessed August 30, 2009.
25. P. Macnaghtan (2008). From bio to nano: learning the lessons, interrogating the comparisons. In: K. David and P.B. Thompson (Eds), *What Can Nanotechnology Learn from Biotechnology?* Elsevier Academic Press, Amsterdam, The Netherlands, p. 110.
26. J. Berringer (2008). Quoted in P. Macnaghtan (2008). From bio to nano: learning the lessons, interrogating the comparisons. In: K. David and P.B. Thompson (Eds), *What Can Nanotechnology Learn from Biotechnology?* Elsevier Academic Press, Amsterdam, The Netherlands, p. 110.
27. For an example of nanoscale issues, see US Environmental Protection Agency (2007). Nanotechnology White Paper, EPA 100/B-07/001.
28. US General Accountability Office (2004). Genetically Modified Foods: Experts View Regimen of Safety Tests as Adequate, but FDA's Evaluation Process Could Be Enhanced, GAO-02-566, 2002; and, Food and Agriculture Organization of the United Nations (2004), The State of Food and Agriculture, 2003–2004, "Agricultural Biotechnology – Meeting the needs of the poor?"
29. US General Accountability Office (2009). High Risk Series: An Update, GAO-09-271, 22–24.
30. US Environmental Protection Agency (2009). Essential Principles for Reform of Chemicals Management Legislation. http://www.epa.gov/oppt/existingchemicals/pubs/principles.html; accessed October 16, 2009.
31. Consumer Specialty Product Association (2009). http://www.cspa.org/infocenter/our-issues/principles-for-chemicals-management-policy/; accessed October 16, 2009.
32. W.J. Jones, P. Schmieder, R. Kolanczyk and O. Mekenyan (2009). Development of a searchable metabolite database and simulator of xenobiotic metabolism. Project Description. US Environmental Protection Agency. Office of Research and Development, National Exposure Research Laboratory, Athens, GA.
33. *Federal Register,* 64 (213), November 4, 1999.
34. Eionet – European Topic Centre on Sustainable Consumption and Production (2009). Waste prevention; http://scp.eionet.europa.eu/themes/waste/prevention/#product; accessed October 12, 2009.
35. Ibid.
36. J.M. Smith (2003). *Seeds of Deception: Exposing Industry and Government Lies about the Safety of the Genetically Engineered Foods You're Eating.* Yes! Books, Fairfield, IO.
37. Church of Scotland (2006). Society, Religion and Technology Project, Patenting Life: An Introduction to the Issues; http://www.srtp.org.uk/scsunpat.shtml; accessed September 17, 2006.
38. K.D. Warner (2002). Are life patents ethical? Conflict between Catholic social teaching and agricultural biotechnology's patent regime. *Journal of Agricultural and Environmental Ethics* 14 (3): 301–319.
39. Department of Life Sciences, Fu Jen Catholic University, Sinjhuang, Taiwan: www.bio.fju.edu.tw/handout/bio/7.ppt; accessed July 27, 2009.
40. The principal source for this section is: Bureau of Labor Statistics, US Department of Labor, Career Guide to Industries, 2008–09 Edition, Pharmaceutical and Medicine Manufacturing; http://www.bls.gov/oco/cg/cgs009.htm; accessed July 23, 2009.
41. Ibid.
42. R.J. Hay, M.L. Macy and T.R. Chen (1989). Mycoplasma infection of cultured cells. *Nature* 229: 487–488.
43. Incidentally, these animals are increasingly genetically selected, such as Harvard University's famous (infamous?), genetically modified "oncomouse." The biotechnology allows the mouse to carry an activated "oncogene", which is a specific gene that increases the mouse's susceptibility to cancer.
44. The source of this and the following discussions on contaminant toxicity mechanisms is Z. Gregus and C. Klaasen (1996). Mechanisms of toxicity. In: C. Klaasen (Ed.), *Casarett and Doull's Toxicology: The Basic Sciences of Poisons,* 5th Edition. McGraw-Hill, New York, NY. The whole edition is an excellent source of information on most aspects of toxicology.
45. For example, see J. Greger (1998). Dietary standards for manganese: overlap between nutritional and toxicological studies. *Journal of Nutrition* 128 (2): 368S–371S.
46. The general source for the distribution and toxicokinetic modeling discussion is the National Library of Medicine's Toxicokinetics Tutor program.

47. The discussions on pharmacological subjects are based upon discussions in S. Zakrewski (1991). *Principles of Environmental Toxicology.* American Chemical Society, Washington, DC. This is an excellent introduction to toxicology as it applies to public health and environmental assessments.
48. This discussion, including the cases and examples, was prepared by E. Rosenfeldt and K. Linden, both formerly of Duke University's Department of Civil and Environmental Engineering. The Duke University UV research lab studies pathogen disinfection and UV photochemical treatment of drinking water, through direct UV and advanced oxidation processes. For more information, visit the website for the International Ultraviolet Association at www.iuva.org.
49. C. Sonnenschein and A.M. Soto (1998). An updated review of environmental estrogen and androgen mimics and antagonists. *Journal of Steroid Biochemistry and Molecular Biology* 65 (1-6): 143.
50. United States Environmental Protection Agency (2001). Removal of Endocrine Disruptor Chemicals Using Drinking Water Treatment Processes. EPA/625/R-00/015, Washington, DC.
51. D. Fry and C. Toone (1981). DDT-induced feminization of gull embryos. *Science* 213 (4510): 922.
52. S. Weimeyer, *et al.* (1984). Organochlorine, pesticide, polychlorobiphenyl, and mercury residues in bald eagle eggs, 1969–79 – and their relationships to shell thinning and reproduction. *Archives of Environmental Contamination and Toxicology* 13 (5): 529.
53. L. Guillette, *et al.* (1994). Developmental abnormalities of the gonad and abnormal sex-hormone concentrations in juvenile alligators from contaminated and control lakes in Florida. *Environmental Health Perspectives* 102 (8): 680.
54. See, for example: C. Purdom, *et al.* (1994). Estrogenic effects from sewage treatment works. *Chem. Ecol.* 8: 275; and S. Joblin, *et al.* (1996). Inhibition of testicular growth in rainbow trout (*Oncorhynchus mikiss*) exposed to estrogenic alkyphenolic chemicals. *Environmental Toxicology and Chemistry* 15 (2): 194.
55. G. Fox (2001). Effects of endocrine disrupting chemicals on wildlife in Canada: past, present and future. *Water Quality Research Journal of Canada* 36 (2): 233.
56. See E.K. Sheiner, *et al.* (2003). Effect of occupational exposures on male fertility: literature review. *Industrial Health* 41 (2): 55; and P. Guzelian (1982). Comparative toxicology of chlordecone (kepone) in humans and experimental – animals. *Annual Reviews of Pharmacology and Toxicology* 22: 89; and T. Hayes, *et al.* (2002). Hermaphroditic, demasculinized frogs after exposure to the herbicide atrazine at low ecologically relevant doses. *Proceedings of the National Academy of Sciences of the United States of America* 99 (8): 5476.
57. D. Koplin, *et al.* (2002). Pharmaceuticals, hormones, and other organic wastewater contaminants in US streams, 1999–2000: A national reconnaissance. *Environmental Science and Technology* 36 (11): 1202.
58. See T. Schultz, *et al.* (2000). Estrogenicity of benzophenones evaluated with a recombinant yeast assay: comparison of experimental and rules-based predicted activity. *Environmental Toxicology and Chemistry* 19 (2): 301; T. Schultz (2002). Structure-activity relationships for gene activation oestrogenecity: evaluation of a diverse set of aromatic chemicals. *Environmental Toxicology* 17(1): 14; Y. Tabira (1999). Structural requirements of par-alkylphenols to bind to estrogen receptor. *European Journal of Biochemistry* 262 (1): 240; and C. Waller, *et al.* (1996). Ligand-based identification of environmental estrogens. *Chemical Research in Toxicology* 9 (8): 1240.
59. H. Bevans, *et al.* (1996). USGS, Water-Resources Investigations Report 96–4266.
60. J. Hu, *et al.* (2002). Products of aqueous chlorination of 4-nonylphenol and their estrogenic activity. *Environmental Toxicology and Chemistry* 21 (10): 2034.
61. References to this table are: J. Acero, K. Stemmler, and U. Von Gunten (2000). Degradation kinetics of atrazine and its degradation products with ozone and OH radicals: a predictive tool for drinking water treatment. *Environmental Science and Technology* 34 (4): 591–597; M.M. Huber, *et al.* (2003). Oxidation of pharmaceuticals during ozonation and advanced oxidation processes. *Environmental Science and Technology* 37(5): 1016; G. Buxton, *et al.* (1988). Critical review of data constants for reactions of hydrated electrons, hydrogen atoms and hydroxyl radicals in aqueous solutions. *Journal of Physical and Chemical Reference Data* 17: 513–886; and, R. Larson and R. Zepp (1988). Reactivity of the carbonate radical with aniline derivatives. *Environmental Toxicology and Chemistry* 7: 265–274.
62. National Academy of Sciences (2002). *Animal Biotechnology: Science-Based Concerns.* National Academies Press, Washington, DC.
63. Ibid.
64. Ibid.
65. Ibid.
66. D. Knight (2001). EPA reregisters Bt Frankencorn despite widespread criticism. *Inter Press Service* October 16, 2001.
67. Ibid.
68. H. Shapouri, J.A. Duffield and M.I. Wang (2002). The Energy Balance of Corn Ethanol: An Update. Agricultural Economic Report No. 813. U.S. Department of Agriculture, Office of the Chief Economist, Office of Energy Policy and New Uses.
69. D. Quist and I.H. Chapela (2001). Transgenic DNA introgression into traditional maize landraces in Oazaca, Mexico. *Nature* 414 (6863): 541–543.
70. B. Lambert, L. Buysse, C. Decock, S. Jansens, C. Piens, B. Saey, *et al.* (1996). A *Bacillus thuringiensis* insecticidal crystal protein with a high activity against members of the family Noctuidae. *Applied Environmental Microbiology* 62: 80–86.
71. J.M. Smith (2007). Institute for Responsible Technology. Genetically engineered food may cause rising food allergies. Organic Consumers Association: http://www.organicconsumers.org/articles/article_5296.cfm; accessed August 31, 2009.

72. The source for discussions on PIPs and the US EPA regulatory process is US Environmental Protection Agency (2009). Regulating Pesticides; http://www.epa.gov/pesticides/biopesticides/; accessed August 31, 2009.
73. USDA/APHIS made this same determination under its statutory authority under the Plant Pest Act.
74. J. Frey (2006). Biological safety concepts of genetically modified live bacterial vaccines. *Vaccine* 25 (30): 5598–5605.
75. P.P. Pastoret (2009). Vaccines: GM technology for better vaccines. European Union. Research Areas. http://ec.europa.eu/research/quality-of-life/gmo/08-vaccines/08-intro.htm; accessed October 16, 2009.
76. A.C. Baumer (2006). Genetically modified maize vaccine can aid farmers in developing nations. *GMO Food for Thought*. http://www.gmofoodforthought.com/2006/08/genetically_modified_maize_vac.html; accessed October 16, 2009.
77. Quote found in B. Savage (2009). Scientists experiment with vaccinations in GMO corn. *Meat & Poultry* May 11, 2009.
78. Ibid.
79. Ibid.

CHAPTER 10

Addressing Biotechnological Pollutants

Progress in environmental technologies has been remarkable. The environment has benefited from numerous scientific and engineering advances in chemistry and biology. However, as in most engineering endeavors, biotechnological operations will produce pollutants during their life cycles. Traditionally pollutants of any type and in any environmental compartment have been addressed on a contaminant-by-contaminant control basis. During the 1980s, this command and control paradigm added pollution prevention, along with waste minimization features. All of these, along with newer life cycle approaches, have been aimed at reducing risks to acceptable levels.

Application of physical and natural sciences began by addressing so-called conventional pollutants, like suspended solids and biochemical oxygen demand, but toward the end of the 20th century moved to the treatment of toxic substances. Both of these applications have continued, but the present approaches' increased attention to pollution prevention and green engineering have been looking to biotechnologies to address environmental problems. Greener approaches that incorporate natural processes, and mimic and borrow from those processes, have received attention from researchers and practitioners alike.

This chapter applies the principles introduced in Chapter 3 to provide a scientifically sound approach to present and future environmental problems caused or exacerbated by biotechnologies. The discussion begins with proven engineering interventions. These interventions formerly were considered to be sanitary engineering practice, but presently fall under the category of environmental engineering. The discussion highlights unique challenges presented by biotechnologies that are not ordinarily found in environmental engineering practice. This is followed by considerations of the proper design and implementation of environmental sampling and monitoring programs. Again, these are based on the sound application of physical, chemical, and biological processes, with an eye toward achieving feasible and representative data to support risk assessment and other environmental decision making. Ironically, or at least interestingly, the environmental biotechnologies mentioned in Chapter 7 can be used to address pollution, especially that posed by organic contaminants (that is, resulting from biotechnological operations in every economic sector described in Figure 9.1).

Environmental Biotechnology

CLEANING UP BIOTECHNOLOGICAL OPERATIONS

As an extension of the discussion in Chapter 7 on how the use biotechnology to treat wastes, let us discuss the wastes potentially generated from biotechnological operations and how these may be treated. The life cycle viewpoint is used in waste audits, environmental management systems, and green engineering to envisage the sources and strength of pollutants and the points at which such pollutants are likely to be released to the environment. This perspective can also be effectively applied to reducing toxicity, controlling exposures and treating pollutants after they have entered the environment. Five steps in sequence define an event that results in environmental contamination of the air, water, or soil pollution. These steps individually and collectively offer opportunities for engineers to apply biotechnologies to intervene and to control the risks associated with hazards and thus protect public health and the environment. The steps address the presence of waste at five points in the life cycle (see Figure 10.1).

As a first step, the contaminant source must be identified and characterized. Every genetically modified organism or hazardous substance has a source from which is will be or has been released. However, in most biotechnological operations, numerous steps along the way are potential sources of contaminants. In fact, there are potential workplace scenarios like that in Figure 10.1 within the operation, such as worker exposure. In occupational scenarios, each step would have unique source, release, transport, and receptor profiles.

Next, the organism or substance moves through the water, air, sediment or soil environment, before reaching human, animal, or plant receptors in a measurable dose, and the receptor must have a quantifiable detrimental response (i.e. stress on some biological function, such as adsorption or metabolism). Intervention can occur at any one of these steps to control the risks to public health and to the environment. Of course, any intervention scheme and subsequent control by the bioengineer must be justified in terms of scientific evidence, sound engineering design, technological practicality, economic realities, ethical considerations, and the laws of local, state and national governments.

INTERVENTION AT THE SOURCE OF CONTAMINATION

A source of the biological or chemical agent must be known and properly characterized, either in the form of an industrial facility that generates waste byproducts, a bioreactor and its various stages, a surface or subsurface land storage/disposal facility, or an accidental spill into a water, air or soil. The intervention must minimize or eliminate the risks to public health and the environment by utilizing technologies at this source that are economically acceptable and

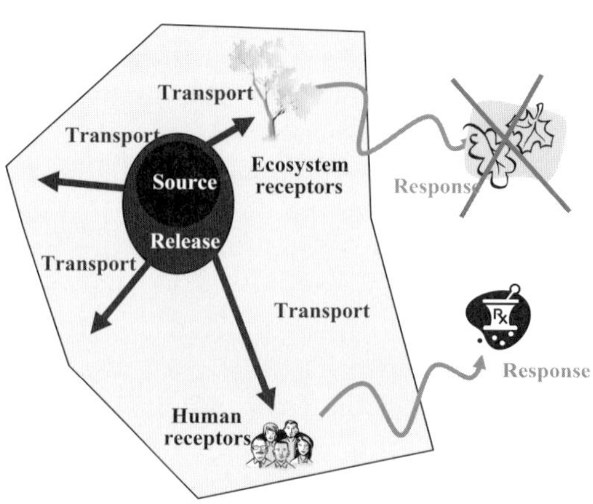

FIGURE 10.1
Points of intervention for a pollutant. The bioengineer has numerous points during a process or product life cycle to address pollution, however, the opportunities tend to lose effectiveness when moving from pre-release to release to transport to receptor to response. [*See color plate section*]

based on applicable scientific principles and sound engineering designs. Numerous source data sets are available, but often do not provide sufficient detail for an individual audit of a community or facility. The Toxic Release Inventory (TRI), for example, is a publicly available EPA database that contains information on toxic chemical releases and waste management activities reported annually by a number of industries and federal facilities. TRI is authorized under the Emergency Planning and Community Right-To-Know Act (EPCRA), for the purposes of supporting state and local planning for chemical emergencies, providing for notification of emergency releases of chemicals, and addressing communities' right-to-know about toxic and hazardous chemicals. An example of TRI retrieval reports for Orange County, North Carolina, and Madison County, Illinois, are shown in Table 10.1.

The TRI data are limited to chemicals, so biological agents are not reported. However, certain chemicals may be indicators of biological processes. Acetaldehyde, for example, is released during yeast fermentation processes. This requires expert elicitation however, since acetaldehyde is a byproduct of numerous industrial processes, not just biotechnologies.

Source identification and characterization can be difficult for biotechnologies, since some of the organism's characteristics may have been significantly altered, making it difficult to predict the feasibility of intervention (e.g. how rapidly they reproduce, how persistent and resistant they are, and what their sources of food and energy may be).

In the case of an industrial facility producing biological or chemical agents as necessary byproducts of a profitable item, bioengineers can take advantage of the growing body of knowledge that has become known as life cycle analysis (LCA). Software such as SimaPro has been used to predict implications of enzyme production, in which the life cycle consists of processes prior to manufacture (e.g. extraction): fermentation – pure culture microbial growth in a liquid medium; recovery; extracellular enzyme separation from biomass; formulation; preservation and standardization of enzyme products and addition of formulation chemicals and biomass treatment; and inactivation of microbes and drying the biomass for use as soil additive in agriculture (see Figure 10.2) [1]. Ultimately, the enzymes and their degradation products will either be completely degraded in the land application process, or find their way to food supplies and ecosystems.

Under the LCA method of intervention the bioengineer considers the environmental impacts that could be incurred during the entire life cycle of (1) all of the resources that go into the product; (2) all the materials that are in the product during its use; and (3) all the materials that are available to exit from the product once it or its storage containers are no longer economically useful to society. Few simple examples exist that describe how *life cycle analysis* is conducted but consider for now any one of a number of household cleaning products. Consider that a particular cleaning product, a solvent of some sort, must be fabricated from one of several basic natural resources. Assume for now that this cleaning product currently is petroleum-based. The engineer could intervene at this initial step in the life cycle of this product, as the natural resource is being selected, and consequently the engineer could preclude the formation of a source of hazardous waste by suggesting instead the production of a water-based solvent. For biotechnologies, solvents are used in various stages, such as the purification stages of bioreactions. If a safer (e.g. more biodegradable and less persistent) solvent is available, the overall environmental quality of the bioreaction process is enhanced.

A cursory review of Figure 10.2 would indicate that this life cycle is at a gross scale. Any of the arrows or boxes would need to be expanded to assess adequately a biotechnological operation. For example, Table 10.2 summarizes the ingredients in one enzyme production process. Each of these ingredients has its own life cycle. Figure 10.3 illustrates an abbreviated life cycle for some of the agricultural inputs from one source (i.e. potatoes) to supply ingredients to enzyme production (starch, protein and sugar), which are co-produced with other agricultural products [2].

Similarly, intervention at the production phase of this product's life cycle can preclude the formation of a source of certain contaminants from the outset. The life cycle approach can also

Table 10.1 Toxic Release Inventory 2007 retrieval report for on-site and off-site reported disposed of or otherwise released (pounds) for facilities in Orange County, NC, Carolina and in Madison County, IL

Row #	Chemical	Total on-site disposal or other releases	Total off-site disposal or other releases	Total on- and off-site disposal or other releases
Orange County, NC				
1	Antimony compounds	271	1770	2041
2	Certain glycol ethers	5800	0	5800
3	Copper	37	4	41
4	Lead	0	.	0
5	Styrene	32	85	117
6	Zinc compounds	2856	9786	12,642
	Total	**8997**	**11,645**	
Madison County, Illinois				
1	1,2,4-trimethylbenzene	20,046	35	20,081
2	1,3-butadiene	677	0	677
3	Aluminum (fume or dust)	0	0	0
4	Aluminum oxide (fibrous forms)	0	0	0
5	Ammonia	169,683	8	169,691
6	Anthracene	2	60	62
7	Antimony	457	504	961
8	Antimony compounds	435	5080	5515
9	Arsenic	197	161	358
10	Barium compounds	84,568	2087	86,655
11	Benzene	78,260	4907	83,167
12	Benzo(g,h,i)perylene	121	0	121
13	Biphenyl	927	11	938
14	Calcium cyanamide	38	.	38
15	Carbon disulfide	14,322	54	14,376
16	Carbonyl sulfide	7903	0	7903
17	Certain glycol ethers	14,752	0	14,752
18	Chlorine	5	0	5
19	Chromium	15	6	21
20	Chromium compounds (except chromite ore mined in the transvaal region)	27,549	21,652	49,201
21	Cobalt	4	215	219
22	Cobalt compounds	116	595	711

Table 10.1 **Toxic Release Inventory 2007 retrieval report for on-site and off-site reported disposed of or otherwise released (pounds) for facilities in Orange County, NC, Carolina and in Madison County, IL—cont'd**

Row #	Chemical	Total on-site disposal or other releases	Total off-site disposal or other releases	Total on- and off-site disposal or other releases
23	Copper	30,628	531,139	561,767
24	Copper compounds	4163	1672	5835
25	Cresol (mixed isomers)	587	7	594
26	Cumene	1206	0	1206
27	Cyanide compounds	1453	0	1453
28	Cyclohexane	47,585	0	47,585
29	Dibenzofuran	76	36	112
30	Dibutyl phthalate	0	0	0
31	Diethanolamine	2910	0	2910
32	Diisocyanates	49	0	49
33	Dimethyl phthalate	1386	0	1386
34	Dioxin and dioxin-like compounds	**	0	**
35	Ethylbenzene	12,342	27	12,369
36	Ethylene	26,668	0	26,668
37	Ethylene glycol	6587	0	6587
38	Fluorine	.	.	0
39	Hydrochloric acid (1995 and after "acid aerosols" only)	70,841	0	70,841
40	Hydrogen cyanide	2683	0	2683
41	Hydrogen fluoride	80,585	0	80,585
42	Lead	1478	308,430	309,908
43	Lead compounds	125,346	177,202	302,548
44	Lithium carbonate	0	0	0
45	Manganese	1162	5383	6545
46	Manganese compounds	939,301	380,453	1,319,754
47	Mercury	0	0	0
48	Mercury compounds	145	16	161
49	Methanol	21,454	0	21,454
50	Molybdenum trioxide	992	0	992
51	N-butyl alcohol	4168	0	4168
52	N-hexane	134,895	0	134,895
53	Naphthalene	6694	527	7221

(*Continued*)

Table 10.1 Toxic Release Inventory 2007 retrieval report for on-site and off-site reported disposed of or otherwise released (pounds) for facilities in Orange County, NC, Carolina and in Madison County, IL—cont'd

Row #	Chemical	Total on-site disposal or other releases	Total off-site disposal or other releases	Total on- and off-site disposal or other releases
54	Nickel	1189	26,798	27,987
55	Nickel compounds	5226	28,725	33,951
56	Nitrate compounds	491,289	24,909	516,198
57	Nitric acid	0	0	0
58	Nitroglycerin	0	0	0
59	Peracetic acid	0	.	0
60	Phenanthrene	320	144	464
61	Phenol	3986	4	3990
62	Polychlorinated alkanes	0	5	5
63	Polycyclic aromatic compounds	1342	165	1507
64	Propylene	12,910	0	12,910
65	Pyridine	3	8	11
66	Quinoline	0	9	9
67	Sodium nitrite	0	0	0
68	Styrene	1118	95	1213
69	Sulfuric acid (1994 and after "acid aerosols" only)	427,763	0	427,763
70	Tetrachloroethylene	608	0	608
71	Toluene	45,422	867	46,289
72	Vanadium compounds	32,697	29,133	61,830
73	Xylene (mixed isomers)	45,042	243	45,285
74	Zinc (fume or dust)	67,999	97,630	165,629
75	Zinc compounds	5,115,117	1,966,345	7,081,463
	Total	**8,197,492**	**3,615,347**	**11,812,839**

Note: In the Madison County table above, asterisks are shown to indicate that data for dioxin and dioxin-like compounds in grams (as required by EPA) was reported by the facility. EPA has converted these data into pounds and included them in the table totals (in pounds). Refer to the dioxin and dioxin-like compounds table below for reported amounts of dioxin and dioxin-like compounds in grams. Grams can be converted to pounds by multiplying by 0.002205.

TRI on-site and off-site reported disposed of or otherwise released (in grams), for facilities in all Industries, dioxin and dioxin-like compounds, Madison County, IL, 2007

Row #	Chemical	Total on-site disposal or other releases	Total off-site disposal or other releases	Total on- and off-site disposal or other releases
1	Dioxin and dioxin-like compounds	0.8223	0.0000	0.8223
	Total	0.8223	0.0000	0.8223

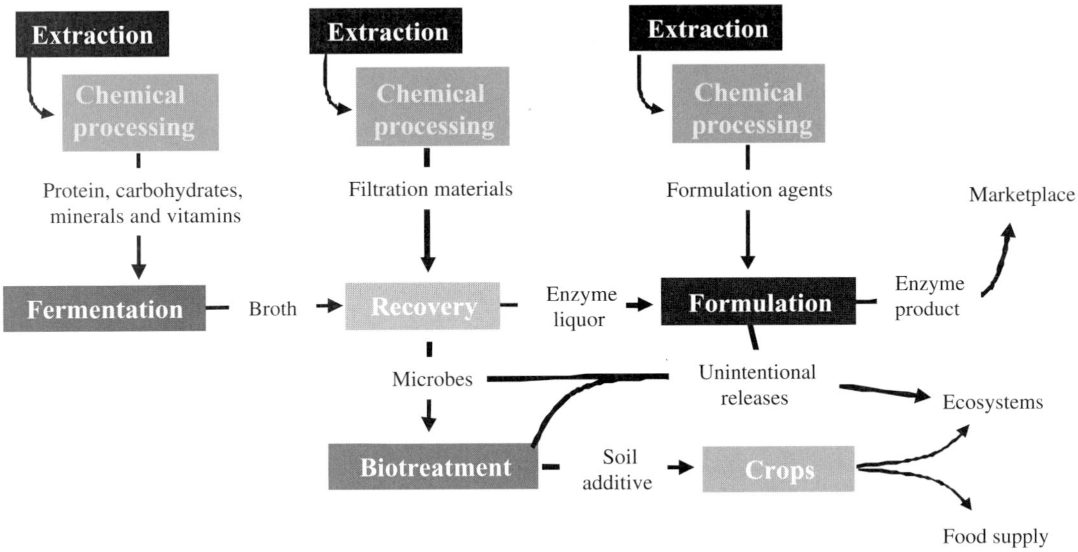

FIGURE 10.2
Life cycle of industrial enzyme production. There is also commensurate energy life cycle (e.g. heat, electricity) and water life cycle. Neither is shown here. [*See color plate section*] *Source:* Portions of the figure are adapted from P.H. Nielsen and W.H. Oxenbøll (2007). Cradle-to-gate environmental assessment of enzyme products produced industrially in Denmark by Novozymes A/S. *International Journal of Life Cycle Analysis* 12 (6): 432–438.

suggest "greener" manufacturing techniques that can replace less sustainable, conventional processes. For example, "co-generation" allows for simultaneous and serial sharing of products and byproducts by two manufacturing facilities when they are located near one another. What would be a "waste" for one facility becomes a "resource" for the other. This can be applied to physical, chemical, and biological wastes. An example would be to locate a bioreactor near a chemical plant with materials needed for the various reactions, so that the needed materials generated by the chemical plant can be piped to the nearby bioreactor, allowing for greater controls and quality assurance than trucking in these same materials. Conversely, an alcohol waste from a bioreactor fermentation process could serve as a source for chemical processes at another biotechnology or chemical facility.

A biological example may be if excess enzyme broth is produced from one biotechnology it may be used in a nearby facility. In fact, the microbial population, used to produce enzyme products for industry, could be a source of microbes to degrade similar type substances in environmental technologies. For example, if a strain of microorganisms has successfully broken down starches in a fermentation bioreactor, this microbial population may be more readily acclimated to a spill of starches or similar hydrocarbons than the microbes living in the soil or water where the spill has occurred.

The design process must account for possible waste streams long before any switches are flipped and valves turned. For example, a particular cleaning product may result in unintended human exposure to barrels of solvent mixtures that fumigate the air in a bioreactor facility or may pollute the town's sewers as the barrels of liquid are flushed down a drain. In this way, LCA is a type of systems engineering where a critical path is drawn, and each decision point considered.

Using a sustainable design approach (e.g. Design for the Environment) requires that the disposal of this solvent's containers must be incorporated as a design constraint from a long-term risk perspective. The challenge is that every potential and actual environmental impact of a product's fabrication, use, and ultimate disposal must be considered. This is seldom, if ever, a "straight line projection."

INTERVENTION AT THE POINT OF RELEASE

Once the source has been identified, the possible point of remedy is at the release to the environment. This could be at the top of a stack or vent from the reactor to the atmosphere, or it could be a more indirect release, such as from the bottom-most layer of a clay liner in a waste

Table 10.2 Summary of ingredients used in the life cycle described in Figure 10.2

Process	Ingredient
Fermentation	Corn starch
	Sucrose
	Glucose/maltose
	Corn steep powder
	Soy bean meal
	Potato protein
	Phosphoric acid
	Glucose syrup
	Ammonia
Recovery	Kiselgur supplier (mined)
	Perlite supplier (mined)
	Sodium chloride
Formulation	Sodium sulphate supplier (mined)
	Cellulose powder
	Palm oil
	Wheat starch
	Kaolin
	Calcium carbonate
	Titanium dioxide
	Sodium chloride
	Sucrose
	Calcium chloride
	Acetic acid

Source: P.H. Nielsen and W.H. Oxenbøll (2007). Cradle-to-gate environmental assessment of enzyme products produced industrially in Denmark by Novozymes A/S. *International Journal of Life Cycle Analysis* 12 (6): 432–438.

landfill connected to surrounding soil material. Similarly this point of release could be a series of points as a contaminant is released along a shoreline from a plot of land into a river or through a plane of soil underlying a storage facility (i.e. a so-called "non-point source").

INTERVENTION DURING TRANSPORT

Wise site selection of biotechnology facilities is the first step in preventing or reducing the likelihood that released materials will move. For example, the distance from a source to a receptor is a crucial factor in controlling the quantity and characteristics of waste as it is transported.

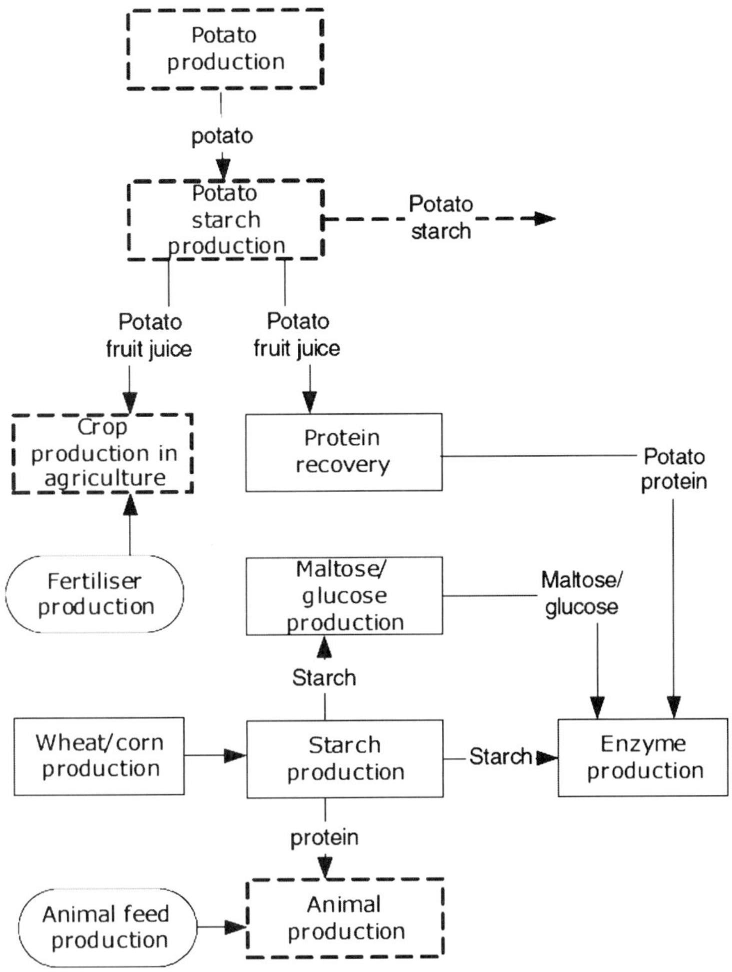

FIGURE 10.3
Life cycle analysis system expansions related to potato protein, starch, and maltose/glucose production. Note: Boxes refer to production processes and arrows refer to material streams. Processes indicated with dotted boxes may not be included in a life cycle assessment if they are independent of demand for enzyme products. Rounded boxes refer to displaced processes.
Source: P.H. Nielsen and W.H. Oxenbøll (2007). Cradle-to-gate environmental assessment of enzyme products produced industrially in Denmark by Novozymes A/S. *International Journal of Life Cycle Analysis* 12 (6): 432–438.

Meteorology is a primary determinant of the opportunities to control the atmospheric transport of contaminants, just as soil porosity and texture and aquifer permeability are determinants of transport in ground and surface water. For example, biotechnology and ancillary processing and storage facilities must avoid areas where specific local weather patterns are unpredictable or where problems are persistent. These avoidance areas include locations that experience ground-based inversions, elevated inversions, valley winds, shore breezes, and city heat island effects. In each of these venues, the pollutants become locked into air masses with little of no chance of moving out of the respective areas. Ground water recharge areas and flood plains can also allow for unacceptable transport of bioengineered substances and their byproducts. In the soil environment the engineer has the opportunity to site facilities in areas of great depth-to-groundwater, as well as in soils (e.g. clays) with very slow rates of transport. Thus the concentrations of the toxins, microbes, and other bioengineering substances can quickly and greatly pose risks to public health and the environment. In this way, engineers and scientists must work closely with city and regional planners early in the site selection phases [3].

INTERVENTION TO CONTROL THE EXPOSURE

The receptor of contamination can be a human, other fauna in the general scheme of living organisms, flora, or materials or constructed facilities. In the case of humans, as we discussed previously, the contaminant can be ingested, inhaled, or dermally contacted. Such exposure can be direct with human contact to, for example, microbes or aerasols laden with microbes and spores or volatile byproducts like alcohols that are present in inhaled indoor air. Such exposure also can be indirect as in the case of human ingestion of the cadmium and other heavy metals found in the livers of beef cattle that were raised on grasses receiving nutrition from cadmium-laced municipal wastewater treatment biosolids (commonly known as "sludge").

Biological agents, heavy metals or chlorinated hydrocarbons similarly can be delivered to domestic animals and animals in the wild. Isolating potential receptors from exposure to hazardous substances gives the bioengineer an opportunity to control the risks to those receptors.

The opportunities to control exposures to contaminants are directly associated with the ability to control the amount of hazardous substances transported to the receptor through source control and siting of biotechnology and attendant facilities. The processes underlying this transport, especially advection, dispersion and diffusion, are described in Chapter 3.

In the 1970s a popular saying in the environmental protection business was: "Dilution is not the solution to pollution." This is sometimes accurate, but not always. However, it runs contrary to current policies for most, non-carcinogenic substances. In fact, most regulated substances are considered hazardous only in certain quantities, i.e. doses. Below some dose, i.e. the so-called "no observable adverse effect level," no harm would be expected. The challenge for biomaterials that have not been used before or that are newly released to the environment or human populations is that this safe level is unknown and must be extrapolated from computer models, animal studies, and observations of the characteristics of similar substances (e.g. by *quantitative structure-activity relationships*). Thus, one solution to environmental contamination could be to increase their dilution in the water, air or soil environments. However, given the rapid doubling of microbes, such a solution may be less compelling for biological agents than for traditional chemical pollutants.

INTERVENTION AT THE POINT OF RESPONSE

Opportunities for intervention are grounded in basic scientific principles, bioengineering designs and processes, and applications of proven and developing biotechnological and other technologies to control the risks associated with contaminants. The discussion begins with the application of thermal processes to control the release and reduce the risks of contaminants generated by biotechnologies.

Thermal treatment of biotechnological wastes

Residue materials from biotechnological operations usually consist predominantly of organic compounds. All organic compounds have heat value and are subject to complete destruction by applying thermodynamic principles. The destruction follows a conversion of matter and energy:

$$\text{Hydrocarbons} + O_2(+\text{energy}) \rightarrow CO_2 + H_2O(+\text{energy}) \quad (10.1)$$

The biomaterials are mixed with oxygen, sometimes in the presence of an external energy source, and in fractions of a second or, at most, in several seconds, the byproducts of gaseous carbon dioxide and water exit the top of the reaction vessel while the solid ash that is produced exits the bottom of the reaction vessel [4]. Energy may also be produced during the reaction and the heat may be recovered. Conversely, if the biomaterials contain other chemical constituents, in particular chlorine and/or heavy metals, the original simple input and output relationship is modified to a very complex situation:

$$\text{Hydrocarbons} + O_2(+\text{energy?}) + \text{Cl or heavy metal(s)} + H_2O + \text{inorganic salts} + \text{nitrogen compounds} + \text{sulfur compounds} + \text{phosphorus compounds} \rightarrow CO_2 + H_2O\,(+\text{energy?}) + \text{chlorinated hydrocarbons or heavy metal(s) inorganic salts} + \text{nitrogen compounds} + \text{sulfur compounds} + \text{phosphorus compounds} \quad (10.2)$$

With these contaminants the potential exists for destruction of the initial contaminant, but actually the problem is exacerbated by generating more hazardous off-gases containing chlorinated hydrocarbons and/or ashes containing heavy metals are produced (e.g. the improper incineration of certain chlorinated hydrocarbons can lead to the formation of the highly toxic chlorinated dioxins, furans, and hexachlorobenzene). All of the thermal systems discussed below have common attributes. All require the balancing of the three "*T*'s" of the science, engineering, and technology of incineration of any substance:

*T*ime of incineration
*T*emperature of incineration
*T*urbulence in the combustion chamber

The advantages of thermal systems include: (1) the potential for energy recovery; (2) volume reduction of the contaminant; (3) detoxification as selected molecules are reformulated; (4) the basic scientific principles, engineering designs, and technologies are well understood from a wide range of other applications including electric generation and municipal solid waste incineration; (5) application to most organic contaminants that compose a large percentage of the total contaminants generated worldwide; (6) the possibility to scale the technologies to handle a single gallon per pound (liter per kilogram) of waste or millions of gallons per pound (liter per kilogram) of waste: and (7) land areas that are small compared to many other facilities (e.g. landfills). An additional advantage for biotechnological operations is that the time, temperature and turbulence needed to thermally destroy chemical compounds is usually sufficient to kill most microorganisms.

Each system design must be customized to address the specific biotechnological operation, including the quantity of waste to be processed over the planning period as well as the physical, chemical, and microbiological characteristics of the waste over the planning period, operational life, and closure/post-closure of the operations. The space required for the incinerator itself ranges from several square yards to possibly the back of a flat bed truck to several acres to sustain a regional incinerator system. Laboratory testing and pilot studies matching a given waste to a given incinerator must be conducted prior to the design, citing, and construction of each incinerator. Generally, the same reaction applies to most thermal processes, i.e. gasification, pyrolysis, hydrolysis, and combustion [5]:

$$C_{20}H_{32}O_{10} + x_1O_2 + x_2H_2O \underset{\Delta}{\rightarrow}$$

$$y_1C + y_2CO_2 + y_3CO + y_4H_2 + y_5CH_4 + y_6H_2O + y_7C_nH_m \quad (10.3)$$

The coefficients x and y balance the compounds on either side of the equation. The delta under the arrow indicates heating. In many thermal reactions, C_nH_m includes the alkanes, C_2H_2, C_2H_4, C_2H_6, C_3H_8, C_4H_{10}, C_5H_{12}, and benzene, C_6H_6. Of all of the thermal processes, incineration is the most common process for destroying organic contaminants in industrial wastes. Incineration simply is heating wastes in the presence of oxygen to oxidize organic compounds (both toxic and non-toxic). The principal incineration steps are shown in Figure 10.4.

It is important to note that incineration alone does not "destroy" some wastes. Some elements, like the metals, are actually concentrated in the ash and other incineration residues. Thermal treatment changes the valence of the metals. In fact, incineration can increase the leachability of metals via oxidation, although processes like slagging (operating at sufficiently high

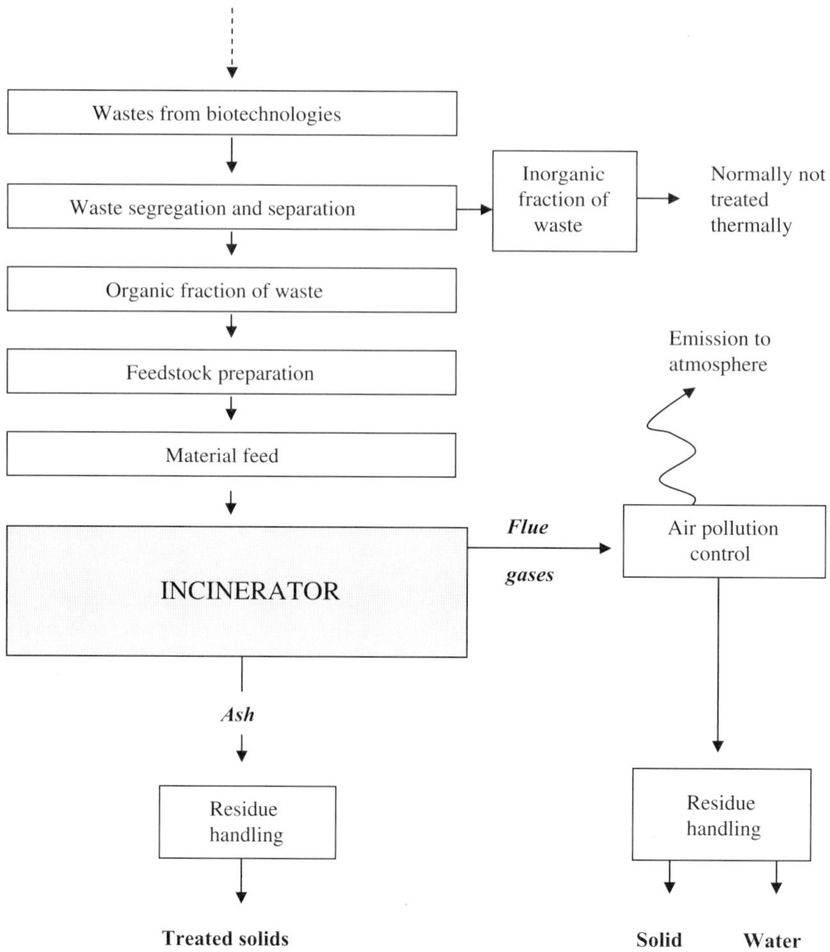

FIGURE 10.4
Steps in the incineration of contaminants from biotechnological operations.
Source: Adapted from US Environmental Protection Agency (2003). Remediation Guidance Document, EPA-905-B94-003 Chapter 7.

temperatures to melt and remove incombustible materials) or vitrification (producing nonleachable, basalt-like residue) can substantially reduce the mobility of many metals.

Leachability is a measure of the ease with which compounds in the waste can move into the accessible environment. The increased leachability of metals would be problematic if the ash and other residues are to be buried in landfills or stored in piles. The leachability of metals is generally measured using the toxicity characteristic leaching procedure (TCLP) test, discussed earlier. Incinerator ash that fails the TCLP must be disposed of in a waste facility approved for hazardous wastes. Enhanced leachability would be advantageous only if the residues are engineered to undergo an additional treatment step of metals. Again, the engineer must see incineration as but one component within a systematic approach for any contaminant treatment process.

There are a number of places in the incineration flow of the contaminant through the incineration process where new compounds may need to be addressed. As mentioned, ash and other residues may contain high levels of metals, at least higher than the original feed. The flue gases are likely to include both organic and inorganic compounds that have been released as a result of temperature-induced volatilization and/or newly transformed products of incomplete combustion with higher vapor pressures than the original contaminants.

The disadvantages of incinerators include: (1) the equipment is capital-intensive, particularly the refractory material lining the inside walls of each combustion chamber that must be replaced as cracks form whenever a combustion system is cooled and/or heated; (2) the operation of the equipment requires skilled operators and fuel must be added to the system; (3) ultimate disposal of the ash is necessary and particularly troublesome and costly if heavy metals and/or chlorinated compounds are found during the expensive monitoring activities; and (4) air emissions and control equipment discharges (e.g. scrubber sludge) may be hazardous and thus must be monitored for chemical constituents and controlled.

Given these underlying principles of incineration, seven general guidelines emerge:

Liquid phase, nearly pure organic contaminants are best candidates for combustion.
Chlorine-containing organic materials deserve special consideration if in fact they are to be incinerated at all; special materials used in the construction of the incinerator, long (many seconds) of combustion time, high temperatures (>1600 °C), with continuous mixing if the contaminant is in the solid or sludge form.
Feedstock containing heavy metals generally should not be incinerated.
Sulfur-containing organic material will emit sulfur oxides which must be controlled.
The formation of nitrogen oxides can be minimized if the combustion chamber is maintained above 1100 °C.
Destruction depends on the interaction of a combustion chamber's temperature, dwell time and turbulence.
Off-gases and ash must be monitored for chemical constituents, each residual must be treated as appropriate so the entire combustion system operates within the requirements of the local, state, and federal environmental regulators, and hazardous components of the off-gases, off-gas treatment processes, and the ash must reach ultimate disposal in a permitted facility.

CALCULATING DESTRUCTION REMOVAL

Federal hazardous waste incineration standards require that hazardous organic compounds meet certain destruction efficiencies. These standards require that any hazardous waste undergo 99.99% destruction of all hazardous wastes and 99.9999% destruction of extremely hazardous wastes like dioxins. The destruction removal efficiency (DRE) is calculated as:

$$\mathrm{DRE} = \frac{W_{in} - W_{out}}{W_{in}} \times 100 \qquad (10.4)$$

where W_{in} = Rate of mass of waste flowing into the incinerator

W_{out} = Rate of mass of waste flowing out of the incinerator.

For example, let us calculate the DRE if during a stack test the mass of pentachlorodioxin is loaded into an incinerator at the rate of 10 mg min^{-1} and the mass flow rate of the compound measured downstream in the stack is 200 pg min^{-1}. Is the incinerator up to code for the thermal destruction of this dioxin?

$$\mathrm{DRE} = \frac{W_{in} - W_{out}}{W_{in}} \times 100 = \frac{10\ \mathrm{mg\ min^{-1}} - 200\ \mathrm{pg\ min^{-1}}}{10\ \mathrm{mg\ min^{-1}}} \times 100$$

Since 1 pg = 10^{-12} g and 1 mg = 10^{-3}, then 1 pg = 10^{-9} mg. So:

$$\frac{10\ \mathrm{mg\ min^{-1}} - 200 \times 10^{-9}\ \mathrm{mg\ min^{-1}}}{10\ \mathrm{mg\ min^{-1}}} \times 100 = 999999.98\%\ \text{removal}$$

Even if pentachlorodioxin is considered to be "extremely hazardous," this is better than the "rule of six nines" so the incinerator is operating up to code.

If we were to calculate the DRE during the same stack test for the mass of tetrachloromethane (CCl_4) loaded into the incinerator at the rate of 100 L min^{-1} and the mass flow rate of the compound measured downstream is 1 ml min^{-1}, is the incinerator up to code for CCl_4? This is a lower removal rate since 100 L are in and 0.001 are leaving, so the DRE = 99.999%. This is acceptable, i.e. better removal efficiency than 99.99% by two orders of magnitude, so long as CCl_4 is not considered an extremely hazardous compound. If it were, then it would have to meet the rule of six nines (it only has five).

By the way, both of these compounds are chlorinated. As mentioned, special precautions must be taken when dealing with such halogenated compounds, since even more toxic compounds than those being treated can end up being generated. Incomplete reactions are very important sources of environmental contaminants. For example, these reactions generate products of incomplete combustion (PICs), such as dioxins, furans, carbon monoxide (CO), polycyclic aromatic hydrocarbons (PAHs), and hexachlorobenzene.

OTHER THERMAL STRATEGIES

High-temperature incineration may not be needed to treat many biotechnology wastes, including most volatile organic compounds (VOCs). Also, in soils with heavy metals, high-temperature incineration will likely increase the volatilization of some of these metals into the combustion flue gas. The presence of high concentrations of volatile trace metal compounds in the flue gas poses complicated pollution control strategies.

When successful in decontaminating soils to the necessary treatment levels, thermally desorbing contaminants from substrates has benefits over incineration, including lower fuel consumption, no formation of slag, less volatilization of metal compounds and less complicated air pollution control demands. Beyond monetary costs and ease of operation, a less energy- (heat-)intensive system can be more advantageous in terms of actual pollutant removal efficiency.

Pyrolysis is the process of chemical decomposition induced in organic materials by heat in the absence of oxygen. It is practicably impossible to achieve a completely oxygen-free atmosphere; so pyrolytic systems run with less than stoichiometric quantities of oxygen. Because some oxygen will be present in any pyrolytic system, there will always be a small amount of oxidation. Also, desorption will occur when volatile or semivolatile compounds are present in the feed.

During pyrolysis [6] organic compounds are converted to gaseous components, along with some liquids and coke, i.e. the solid residue of fixed carbon and ash. Gas phase compounds that are commonly produced and emitted include CO, H_2, and CH_4 and other hydrocarbons are produced. If these gas phase hydrocarbons cool and condense, liquids will form and leave oily tar residues and water with high concentrations of total organic carbon (TOC). Pyrolysis generally takes place well above atmospheric pressure at temperatures exceeding 430 °C. The secondary gases need their own treatment, such as by a secondary combustion chamber, by flaring, and partial condensation. Particulates must be removed by additional air pollution controls, e.g. fabric filters or wet scrubbers.

Conventional thermal treatment methods, such as rotary kiln, rotary hearth furnace, or fluidized bed furnace, are used for waste pyrolysis. Kilns or furnaces used for pyrolysis may be of the same design as those used for combustion (i.e. incineration) discussed earlier, but operate at lower temperatures and with less air than in combustion.

Pyrolysis allows for separating organic contaminants from various wastes and may be used to treat a variety of organic contaminants that chemically decompose when heated (i.e. "cracking"). Pyrolysis is not effective in either destroying or physically separating inorganic compounds that coexist with the organics in the contaminated medium.

Other promising thermal processes include high-pressure oxidation and vitrification [7]. High-pressure oxidation combines two related technologies, i.e., wet air oxidation and supercritical water oxidation, which combine high temperature and pressure to destroy organics. Wet air oxidation can operate at pressures of about 10% of those used during supercritical water oxidation, an emerging technology that has shown some promise in the treatment of PCBs and other stable compounds that resist chemical reaction. Wet air oxidation has generally been limited to conditioning of municipal wastewater sludges, but can degrade hydrocarbons (including PAHs), certain pesticides, phenolic compounds, cyanides, and other organic compounds. Oxidation may benefit from catalysts.

Vitrification uses electricity to heat and destroy organic compounds and immobilize inert contaminants. A vitrification unit has a reaction chamber divided into two sections: the upper section to introduce the feed material containing gases and pyrolysis products, and the lower section consisting of a two-layer molten zone for the metal and siliceous components of the waste. Electrodes are inserted into the waste solids, and graphite is applied to the surface to enhance its electrical conductivity. A large current is applied, resulting in rapid heating of the solids and causing the siliceous components of the material to melt as temperatures reach about 1600 °C. The end product is a solid, glass-like material that is very resistant to leaching.

As mentioned, the environmental biotechnologies discussed in Chapter 7 can certainly be applied to waste generated by medical, agricultural, and industrial biotechnologies.

Nitrogen and sulfur problems

Bioreactors and other biotechnologies may be sources of nitrogen and sulfur since they are essential nutrients in the microbes and other biota that are part of most biotechnologies. The oxidized chemical species [e.g. sulfur dioxide (SO_2) and nitrogen dioxide (NO_2)] form acids when they react with water. The lowered pH is responsible for numerous environmental problems (i.e. acid deposition). Many compounds contain both nitrogen and sulfur along with the typical organic elements (carbon, hydrogen, and oxygen). The reaction for the combustion of such compounds, in general form, is:

$$C_aH_bO_cN_dS_e + (4a + b - 2c) \rightarrow aCO_2 + \left(\frac{b}{2}\right)H_2O + \left(\frac{d}{2}\right)N_2 + eS \quad (10.5)$$

Reactions 10.3 and 10.5 demonstrate the incremental complexity as additional elements enter the reaction. In the real world, pure reactions are rare. The environment is filled with mixtures. Reactions can occur in sequence, parallel or both. For example, a feedstock to a municipal incinerator contains myriad types of wastes, from garbage to household chemicals to commercial wastes, and even small (and sometimes) large industrial wastes that may be illegally dumped. For example, the nitrogen-content of typical cow manure is about 5 kg per metric ton (about 0.5%). If the fuel used to burn the waste also contains sulfur along with the organic matter, then the five elements will react according to the stoichiometry of reactions 10.3 and 10.5.

Certainly, combustion specifically and oxidation generally are very important processes that lead to nitrogen and sulfur pollutants. But they are certainly not the only ones. At this point, a more detailed explanation of oxidation is in order. In the environment, oxidation *and* reduction occur. The formation of sulfur dioxide and nitric oxide by acidifying molecular sulfur is a redox reaction:

$$S(s) + NO_3^-(aq) \rightarrow SO_2(g) + NO(g) \quad (10.6)$$

The designations in parentheses give the physical phase of each reactant and product: "s" for solid, "aq" for aqueous, and "g" for gas.

The oxidation half-reactions for this reaction are:

$$S \rightarrow SO_2 \quad (10.7)$$

$$S + 2H_2O \rightarrow SO_2 + 4H^+ + 4e^- \quad (10.8)$$

The reduction half-reactions for this reaction are:

$$NO_3^- \rightarrow NO \quad (10.9)$$

$$NO_3^- + 4H^+ + 3e^- \rightarrow NO + 2H_2O \quad (10.10)$$

Therefore, the balanced oxidation-reduction reactions are:

$$4NO_3^- + 3S + 16H^+ + 6H_2O \rightarrow 3SO_2 + 16H^+ + 4NO + 8H_2O \quad (10.11)$$

$$4NO_3^- + 3S + 4H^+ \rightarrow 3SO_2 + 4NO + 2H_2 \quad (10.12)$$

A reduced form of sulfur that is highly toxic and an important pollutant is hydrogen sulfide (H_2S). Certain microbes, especially bacteria, reduce nitrogen and sulfur, using the N or S as energy sources through the acceptance of electrons. For example, sulfur-reducing bacteria can produce hydrogen sulfide (H_2S), by chemically changing oxidized forms of sulfur, especially sulfates (SO_4). To do so, the bacteria must have access to the sulfur, i.e. it must be in the water, which can be in surface or groundwater, or the water in soil and sediment. These sulfur-reducers are often anaerobes, i.e. bacteria that live in water where concentrations of molecular oxygen (O_2) are deficient. The bacteria remove the O_2 molecule from the sulfate leaving only the S, which in turn combines with hydrogen (H) to form gaseous H_2S. In groundwater, sediment, and soil water, H_2S is formed from the anaerobic or nearly anaerobic decomposition of deposits of organic matter, e.g. plant residues. Thus, redox principles can be used to treat H_2S contamination. Strong oxidizers, like molecular oxygen and hydrogen peroxide, most effectively oxidize the reduced forms of S, N or any reduced compound.

Often, the biotechnologically efficient approach is to use both aerobic and anaerobic processes. This is the case for tetrachloroethylene, which is often biodegraded in two steps. The first step is anaerobic, wherein the oxygen is removed, and the anaerobes then remove two of the chlorines, leaving dichloroethylene. Next, the soil or sediment media are again aerated and the aerobes allow for mineralization to carbon dioxide and water. Microbes that can break down hydrocarbons (e.g. after an oil spill) are ubiquitous in the environment. In fact, more than 30 distinct genera of oil-degrading bacteria and fungi have been identified for both *in situ* and *ex situ* bioremediation purposes [8]. This has made them attractive to biotechnological enhancement, since even with this wealth of available microbes ready to break down hydrocarbons, scientists still need, or at least want, to increase their kinetics and want to break down some very recalcitrant compounds.

SAMPLING AND ANALYSIS

The first step in controlling contaminants is to know as precisely and accurately as possible where they are and at what concentrations they exist. So, we will begin with an overview of environmental contaminant sampling and analysis.

Environmental monitoring

The terms "monitoring" and "sampling" are frequently used interchangeably by environmental scientists, however, monitoring is a more inclusive term. Environmental monitoring is dependent upon the quality of sample collection, preparation, and analysis. Sampling is a statistical term, and usually a geostatistical term.

An environmental sample is a small portion of air, water, soil, biota, or other environmental media (e.g. paint chips, food, etc. for indoor monitoring) that represents a much larger entity. For example, a sample of air may consist of a canister or bag that holds a defined quantity of air that will be subsequently analyzed. The sample is representative of a portion of an air mass. Ideally, a sufficient number of samples are collected and their results aggregated to ascertain with defined certainty the quality of an air mass. So, more samples will be needed for the New York City air shed than for that of a small town. However, intensive sampling is often needed for highly toxic contaminants and for sites that may be particularly critical, e.g. a national park, a hazardous waste site, or an "at risk" neighborhood (such as one near a chemical manufacturing facility). Like other statistical measures, the sample is used to infer the condition of the larger population or larger area (in the case of geostatistics).

A simple example of the representativeness of a sample is illustrated by attempting to characterize the amount of trichloroethane (TCA) that may have been released from a biotechnological operation into a lake. The bioengineer gathers a single 500 mL sample in the middle of the lake that contains 1 million liters of water. Thus, the sample represents only 5×10^{-7} of the lake's water. In addition, the collected sample is limited in location vertically and horizontally, so there is much uncertainty. However, if 10 samples are taken at 10 spatially distributed sites, the inferences are improved. Furthermore, if the samples were taken in each season, then there would be some improvement to understanding of intra-annual variability. And, if the sampling is continued for several years, the inter-annual variability is better characterized. Such sampling can be conducted on any environmental media.

Before the samples are collected and arrive at the laboratory, the general monitoring plan, including quality assurance provisions, must be in place. The plan describes the procedures to be employed to examine a particular site. These procedures must be strictly followed to investigate the extent of contamination of an environmental resource. The plan describes in detail the kinds of samples to be taken (e.g. real-time probes, sample bags, bottles, and soil cores), the number of samples needed, methods for collection, sample handling, and transportation. The quality and quantity of samples are determined by data quality objectives (DQOs), which are defined by the objectives of the overall contaminant assessment plan. DQOs are qualitative and quantitative statements that translate non-technical project goals into scientific and engineering outputs needed to answer technical questions [9].

Quantitative DQOs specify a required level of scientific and data certainty, while qualitative DQOs express decisions goals without specifying those goals in a quantitative manner. Even when expressed in technical terms, DQOs must specify the decision that the data will ultimately support, but not the manner in which the data will be collected. DQOs guide the determination of the data quality that is needed in both the sampling and analytical efforts. The US Environmental Protection Agency has listed three examples of the range of detail of quantitative and qualitative DQOs [10]:

Example of a less detailed, quantitative DQO: Determine with greater than 95% confidence that contaminated surface soil will not pose a human exposure hazard.

Example of a more detailed, quantitative DQO: Determine to a 90% degree of statistical certainty whether or not the concentration of mercury in each bin of soil is less than 96 ppm.

Example of a detailed, qualitative DQO: Determine the proper disposition of each bin of soil in real-time using a dynamic work plan and a field method able to turn around lead (Pb) results on the soil samples within two hours of sample collection.

In other words, if all one needs to know is the seasonal change in pH near a fish hatchery, only a few samples using simple pH probes would be defined as the DQO. However, if the environmental assessment calls for the characterization of year-round water quality for trout in the stream, the sampling plan's DQO may dictate that numerous samples at various points be continuously sampled for inorganic and organic contaminants, turbidity, nutrients, and ionic strength. This is even more complicated for biotechnological operations than most environmental assessments, since biotechnologies always involve organisms. This may mean that microbiological monitoring (e.g. most probable numbers, etc.) will need to be conducted and interpreted to meet the specific DQOs.

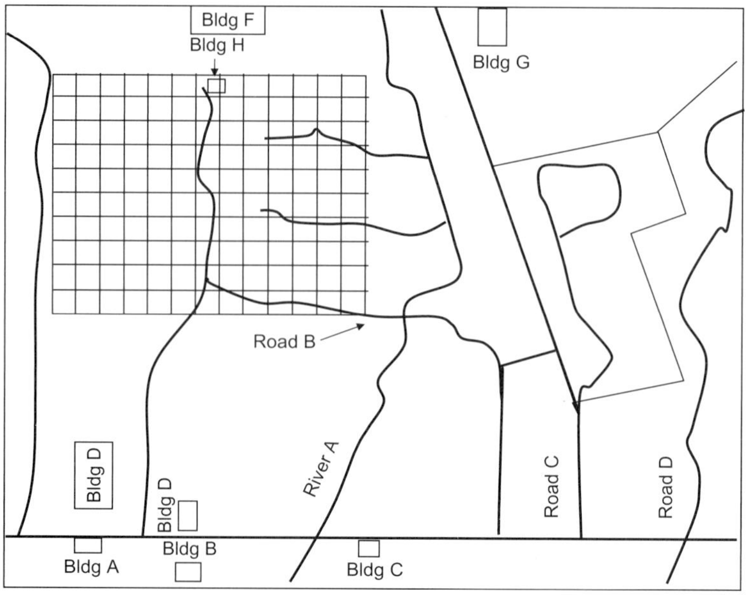

FIGURE 10.5
Environmental assessment area delineated by map boundaries.
Source: US Environmental Protection Agency (2002). Guidance for the Data Quality Objectives Process, EPA QA/G-4, EPA/600/R-96/055, Washington, DC.

The sampling plan must include all environmental media, e.g. soil, air, water, and biota, that are needed to characterize the exposure and risk of any biotechnological operation. The sampling and analysis plan should explicitly point out which methods will be used. For example, if toxic chemicals are being monitored, the US EPA specifies specific sampling and analysis methods [11].

The geographic area where data are to be collected is defined by distinctive physical features such as volume or area, e.g. metropolitan city limits, the soil within the property boundaries down to a depth of 6 cm, a specific water body, length along a shoreline, or the natural habitat range of a particular animal species. Care should be taken to define boundaries. For example, Figure 10.5 indicates a study area by a grid, wherein each cell is sampled [12].

The target population may be divided into relatively homogeneous subpopulations within each area or subunit. This can reduce the number of samples needed to meet the tolerable limits on decision errors, and to allow more efficient use of resources.

Time is another essential parameter that determines the type and extent of monitoring needed. Conditions vary over the course of a study due to changes in weather conditions, seasons, operation of equipment, and human activities. These include seasonal changes in groundwater levels, seasonal differences in farming practices, daily or hourly changes in airborne contaminant levels, and intermittent pollutant discharges from industrial sources. Such variations must be considered during data collection and in the interpretation of results.

The US EPA's Guidance for the Data Quality Objectives Process identifies the following examples of temporally sensitive parameters:

- measurement of lead in dust on windowsills may show higher concentrations during the summer when windows are raised and paint/dust accumulates on the windowsill;
- terrestrial background radiation levels may change due to shielding effects related to soil dampness;
- measurement of pesticides on surfaces may show greater variations in the summer because of higher temperatures and volatilization;

- instruments may not give accurate measurements when temperatures are colder; or
- measurements of airborne particulate matter may not be accurate if the sampling is conducted in the wetter winter months rather than the drier summer months.

Thus, the population and optimal timeframe for collecting data are crucial considerations in the monitoring plan. After ensuring that the scientific (i.e. biochemodynamic) criteria are met, feasibility should also be considered. This includes gaining legal and physical access to the properties, equipment acquisition and operation, and environmental conditions, times and conditions when sampling is prohibited (e.g. freezing temperatures, high humidity, and noise).

Siting an environmental monitoring study: an example

The steps needed to plan an environmental monitoring study can be illustrated by a recent study design to determine the amount of air toxics being released near a major thoroughfare in Las Vegas, NV. Although this is not a biotechnological study, a similar process may be used to design monitoring studies for bioreactor and other biotechnology facilities. The objective of the study is to determine mobile source air toxics (MSAT) concentrations and variations in concentrations as a function of distance from the highway and to establish relationships between MSAT concentrations as related to highway traffic flows, including traffic count, vehicle types and speeds, and meteorological conditions such as wind speed and wind direction. As such, the MSAT study is expected to provide data detailing concentrations and distributions of motor vehicle emitted pollutants, including regulated gases, air toxics, and particulate matter (see Table 10.3). Specifically, the data will be used to address the following goals [13]:

Identify the existence and extent of elevated air pollutants near roads.
Determine how vehicle operations and local meteorology influence near-road air quality for criteria and toxic air pollutants.
Collect data that will be useful in evaluating and refining, if necessary, models used to determine the emissions and dispersion of motor vehicle related pollutants near roadways.

The broad science needs of the study are to improve the understanding of: (1) ambient air concentrations in a near-road environment; (2) exposures and uncertainties of exposures

Table 10.3 Site selection process steps

1	Determine site selection criteria	Monitoring protocol	Developed by FHWA
2	Develop list of candidate sites	Geographic information system (GIS) data; on-site visit(s)	Additional sites added as information is developed
3	Apply coarse site selection filter	Team discussions, management input	Eliminate sites below acceptable minimums
4	Site visit	Field trip	Application of fine site selection filter
5	Select candidate site(s)	Team discussions, management input	
6	Obtain site access permissions	Contact property owners	If property owners do not grant permission, then the site is dropped from further consideration
7	Site logistics (i.e., physical access, utilities – electrical and communications)	Site visit(s), contact utility companies	

Source: S. Kimbrough, D. Vallero, R. Shores, A. Vette, K. Black and V. Martinez (2008). Multi-criteria decision analysis for the selection of a near road ambient air monitoring site for the measurement of mobile source air toxics. *Transportation Research, Part D: Transport and Environment* 13 (8): 505–515.

to a population living in a near-road environment; and (3) health risks, impacts, and uncertainties of a population living in a near road environment. Specifically this study will provide input to item 1 above, improving the understanding of ambient air concentrations in a near-road environment. An example of a sampling and analysis management structure is shown in Figure 10.6. A complex monitoring effort requires management and technical staff with a diversity of skills that can be brought to bear on the implementation of this project. This diverse skill set includes: program management, contracts administration, field monitoring experience, laboratory expertise, and quality assurance oversight.

The purpose of any site selection process is to gather and analyze sufficient data that would lead one to draw informed conclusions regarding the selection of the most appropriate site for the monitoring that will be performed in Las Vegas, NV. Moreover, the site selection process needs to include programmatic issues to ensure an informed decision is reached.

Initially, the targeted site to conduct the monitoring in Las Vegas, NV was at a particular elementary school since this would represent vulnerable subpopulations, i.e. a combination of high exposure and sensitivity (children are more sensitive to the effects of air toxics than the general population). This was one of the three schools named in a settlement agreement issued requiring that better environmental sampling be conducted to provide better data to be used to determine risks from MSATs. However, after further investigation and analysis it became apparent that a more formalized site selection process to either confirm or deny the suitability of this site for our project would be required. During this process, the school became increasingly less acceptable as the optimal site owing to the presence of large sound walls (>15 feet in height), very poor quality wind rose data (prevailing winds channeled down roadway as opposed to across), lack of access for site installation (10 meter site – no access at roadside due to sound walls), and the roadway being below grade. In essence, the school site is on top of an urban canyon. Thus, it became necessary to expand the search for a more optimal

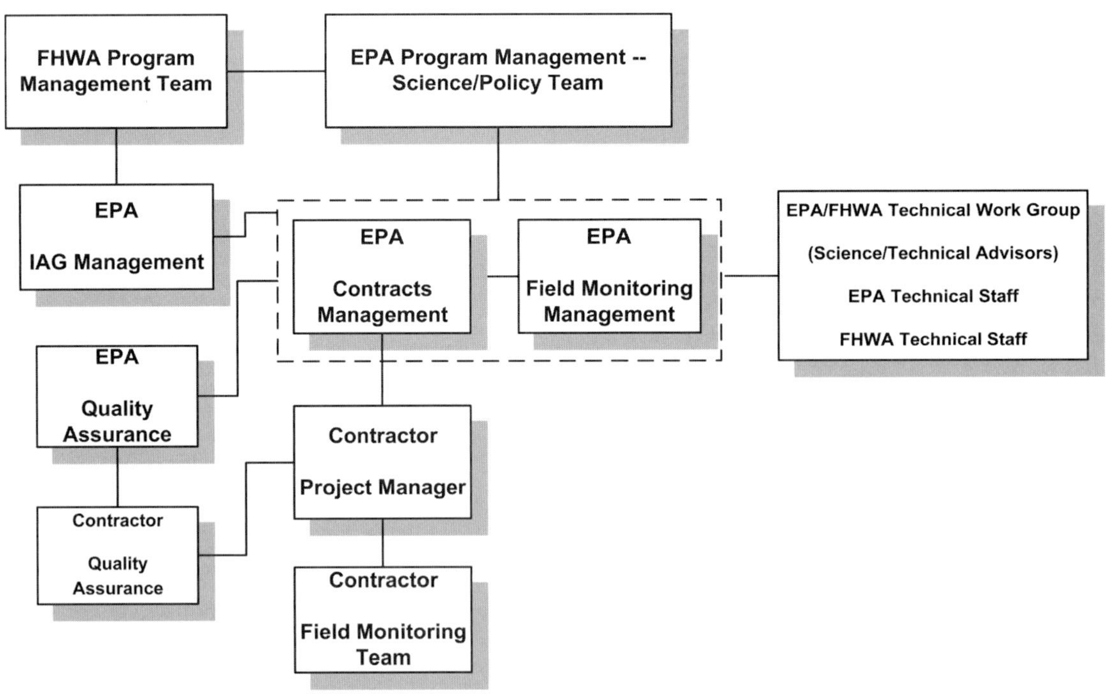

FIGURE 10.6
Sampling and analysis project roles and responsibilities in a multi-agency investigation. Note: FHWA = Federal Highway Administration; EPA = Environmental Protection Agency; IAG = Interagency Agreement (a formal assistance agreement that transfers funds between two different federal agencies).
Source: S. Kimbrough, D. Vallero, R. Shores, A. Vette, K. Black and V. Martinez (2008). Multi-criteria decision analysis for the selection of a near road ambient air monitoring site for the measurement of mobile source air toxics. *Transportation Research, Part D: Transport and Environment* 13 (8): 505–515.

site through a more formal process. For purposes of this project, the site selected must meet the requirements of the monitoring protocol. As is often the case, adjustments to selection criteria need to be made as a study unfolds.

The site selection process utilized for this project consisted of a series of steps as shown in Table 10.3 and Figure 10.7. Each of these seven steps has varying degrees of complexity due to "real-world" issues. The first step was to determine site selection criteria (see Table 10.4). The follow-on steps include: (2) develop list of candidate sites and supporting information; (3) apply site selection filter ("coarse" and "fine"); (4) site visit; (5) select candidate site(s) via team discussion; (6) obtain site access permission(s); and (7) implement site logistics.

A list of candidate sites was developed using the monitoring protocol's site selection criteria. *Geographic information system* (GIS) data, tools, and techniques and on-site visits were used by project team members as a means of developing supporting information regarding each potential site. The Nevada Department of Transportation (NDOT) provided annual average daily traffic (AADT) counts and their associated spatial coordinate locations. Other types of spatial data (e.g., street network) were downloaded from the Clark County GIS website as well as other relevant websites. Non-spatial data (meteorological data) were downloaded from the National Climatic Data Center for Las Vegas, NV. ArcGIS 9.2 was used to create the maps for the site selection process. WRPLOT View by Lakes Environmental was used to create wind rose plots for the meteorological data. In addition, on-site visits provided information not readily available elsewhere or provided information not easily gained from site maps.

As mentioned, it is quite common to need to adjust even a well-designed environmental monitoring plan. For example, in this study, the investigators learned that sound barriers would be constructed along most of US 95, especially the area of concern *vis-à-vis* the

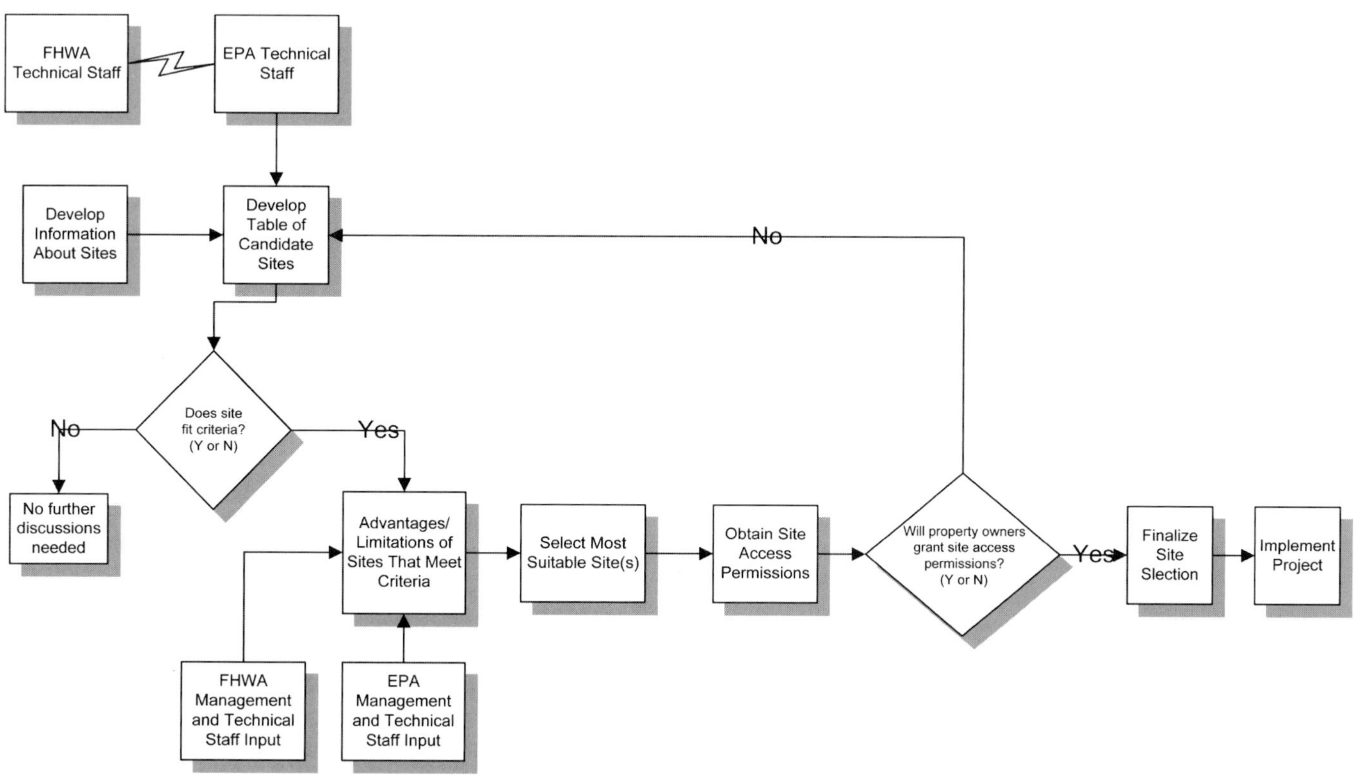

FIGURE 10.7
Monitoring location selection decision flow chart.
Source: S. Kimbrough, D. Vallero, R. Shores, A. Vette, K. Black and V. Martinez (2008). Multi-criteria decision analysis for the selection of a near road ambient air monitoring site for the measurement of mobile source air toxics. *Transportation Research, Part D: Transport and Environment* 13 (8): 505–515.

Table 10.4 Example selection considerations and criteria

Selection considerations	Monitoring protocol criteria
AADT (>150,000)	Only sites with more than 150,000 annual average daily traffic (AADT) are considered as candidates
Geometric design	The geometric design of the facility, including the layout of ramps, interchanges, and similar facilities, will be taken into account. Where geometric design impedes effective data collection on MSATs and $PM_{2.5}$, those sites will be excluded from further consideration. All sites have a "clean" geometric design
Topology (i.e., sound barriers, road elevation)	Sites located in terrain making measurement of MSAT concentrations difficult or that raise questions of interpretation of any results will not be considered. For example, sharply sloping terrain away from a roadway could result in under representation of MSAT and $PM_{2.5}$ concentration levels on monitors in close proximity to the roadway simply because the plume misses the monitor as it disperses
Geographic location	Criteria applicable to representing geographic diversity within the U.S. as opposed to within any given city
Availability of data (traffic volume data)	Any location where data, including automated traffic monitoring data, meteorological or MSAT concentration data, are not readily available, or instrumentation cannot be brought in to collect such data, will not be considered for inclusion in the study
Meteorology	Sites will be selected based on their local climates to assess the impact of climate on dispersion of emissions and atmospheric processes that affect chemical reactions and phase changes in the ambient air
While not explicitly included in the Monitoring Protocol, the following selection criteria were deemed important to the selection process and were included:	
Downwind sampling	Any location where proper siting of downwind sampling sites is restricted due to topology, existing structures, meteorology, etc., may exclude otherwise suitable sites for consideration and inclusion in this study
Potentially confounding air pollutant sources	The presence of confounding emission sources may exclude otherwise suitable sites for consideration and inclusion in this study
Site access (admin/physical)	Any location where site access is restricted or prohibited either due to administrative or physical issues, will not be considered for inclusion in the study

Source: S. Kimbrough, D. Vallero, R. Shores, A. Vette, K. Black and V. Martinez (2008). Multi-criteria decision analysis for the selection of a near road ambient air monitoring site for the measurement of mobile source air toxics. *Transportation Research, Part D: Transport and Environment* 13 (8): 505–515.

settlement agreement. In addition, these areas also turned out to be the areas with the highest daily traffic counts. This was unfortunate since, when it comes to environmental measurements, worst-case scenarios can provide important insights into how sources of pollution are connected to diseases and other adverse effects in populations. In this case, however, such siting would be highly problematic as it would violate a number of the site selection criteria of the monitoring protocol, particularly the need to avoid complex topology, and poor meteorology.

From this process, a list of 19 of the 22 sites was developed (see Table 10.5). Three additional sites were added during an on-site visit to Las Vegas during the spring of 2007. This list contains sites that are located along interstate or US highways, state highways or major streets that would be of interest to a project of this nature. At this point, sites were included that might fall below certain minimum requirements (e.g., AADT <100,000).

After applying our site selection criteria as a set of "filters" most of the candidate sites were eliminated. For example, the most obvious first filter eliminated sites with low AADT

Table 10.5 **All sites considered for intensive monitoring in a near-road environmental study**

		Selection considerations						
			Topology					
Candidate locations		AADT (2006)	Sound barriers	Road elevation	Meteorology	Traffic volume data	Downwind sampling	Nearby sources
Interstate/US highway								
1	O.K. Adcock School	Y	Y	BG	NW	Y	S	N
2	Fyfe/Western Schools	Y	Y	AtG/BG	NW	Y	S	N
3	Sunset/Lake Meade Interchange – US95	N	N	AtG/BG	SW	N	S	CM/S
4	I-215 (between Warm Springs/Robindale)	Y	Y	AG	SW	N	S	N
5	I-215 (vicinity of E. Pebble Rd)	N	P	AtG/AG	SW	N	S	N
6	I-215 (east of I-15)	Y	N	BG	SW	N	S	M
14	Flamingo & I-15	Y	N	AtG	SW	N	C	UT
15	I-215 (Eastern & Pebble)	N	N	AG	SW	N	S	N
16	US95 (Kelso Dunes/Auto Mall)	N	N	AtG	SW	N	S	CM/S
17	US95 (Gibson/Sunset Area)	N	N	AtG	SW	N	S	S
18	US95 (Sunset & I-515)	N	N	AtG	SW	N	S	S
19	I-15 (Martinez School)	N	N	AtG	NW		C	RS
20	I-15 (vicinity of Ensworth)	Y	N	BG / Modest Cut	SW	Y	S	M
21	US95/Lake Meade Blvd	N	N	AG	W	N	CM/S	UT
22	I-215 (vicinity of Jones Rd)	N	N	AtG	W/SW	N	S	N
Major street								
7	E Flamingo Rd	N	N	AtG	SW	N	CM	UT
8	W Flamingo Rd/S Decatur Blvd	N	N	AtG	SW	N	CM	UT
State highway								
9	W Summerlin Pkwy (1)	N	Y	AtG	NW	N	R	N/C
10	W Summerlin Pkwy (2)	N	Y	AtG	NW	N	R	N/C
11	US95 east of Rancho Dr.	N	P	BG	NW	N	R/S	N/C
12	W Summerlin Pkwy (3)	N	P	AtG	NW	N	R/S	N/C
13	W Summerlin Pkwy (4)	N	P	AtG	NW	N	R/S	N/C

AADT: >150,000 = Y (Yes); otherwise N (No)
Topology: Sound barriers – Y = Yes; N = No; P = Partial; Road elevation – AG = above grade; BG = below grade; AtG = at grade
Meteorology: SW = McCarran – SW winds; NW = North LV – NW winds; W = westerly winds
Traffic volume data: Operational = Y (Yes); otherwise N (No)
Downwind sampling: R = residential; C = complex (mixed commercial); S = semi-open fields
Nearby sources: N = none; UT = urban traffic; M = McCarran Airport; S = sand/gravel; RS = railroad/scrap yards; C = construction/possible construction; CM = commercial; R = residential
Source: S. Kimbrough, D. Vallero, R. Shores, A. Vette, K. Black and V. Martinez (2008). Multi-criteria decision analysis for the selection of a near road ambient air monitoring site for the measurement of mobile source air toxics. *Transportation Research, Part D: Transport and Environment* 13 (8): 505–515.

(i.e., <100,000). The next filter, the presence of extensive sound barriers, eliminated additional sites. Other filters, complex geometric design or lack of available traffic volume data, eliminated additional sites. Thus, we eliminated most of these 22 sites. Other criteria of interest, while not explicitly stated in the monitoring protocol, include restricted downwind sampling, presence of confounding air pollutant sources and site access (administrative and physical).

On-site visits are crucial. During the first site visit to Las Vegas, it was found that what were expected to be the highest priority sites were considered to be unsuitable based on previously unforeseen factors that had not been readily obvious from our earlier analysis. These unsuitable factors included the roadway being above/below grade and presence of sound barriers. Site 19, which was also visited, was deemed to be unsuitable for several reasons: confounding winds, adjacent railroad (main line), and the presence of confounding air pollutant sources (nearby vehicle scrapping plants). Moreover, the NDOT staff provided information indicating that this location, which happened to be in the vicinity of I-15, is the focus of an upcoming "design-build" highway project. The NDOT staff indicated that contracts were about to be implemented and that subsequent construction activities would effectively place our monitors within the construction zone for the next 18 months to 2 years, thus making this location unsuitable. During this site visit the remaining, available candidate sites were also eliminated. It was also during this visit that three additional sites were added to "List of Candidate Sites" (see Table 10.5). These additional sites were added based on "real-world" observations at these locations and how well these locations met the site selection criteria. It is important to note that it is not always possible to determine the most suitable sites without an actual site visit. Thus, one may add sites as well as delete sites based on factors not previously known. In this case, these additional candidate sites were Sites 20, 21, and 22. Site 20 was deemed the most promising for several reasons: high AADT (>190,000), lack of sound barriers, road being at or near grade, acceptable downwind sampling, and acceptable meteorology.

An important component of "ground truthing" or site visit is to obtain information from local sources. For purposes of this project, both NDOT and Clark County Department of Air Quality and Environmental Management (DAQEM) staff provided information about local road and meteorological conditions that would not have been known otherwise. (Too often, local resources are overlooked during a decision process such as this, which can in turn lead to poor decision making.)

The use of spatial tools in decision processes is increasing and will continue to increase in the future [14]. Historically, the use of spatial tools (i.e., GIS) in decision processes has been somewhat problematic in part due to the magnitude of the data required by a GIS, perception, and reality of operating GIS software, and the level-of-knowledge required by end-users to manipulate data in a GIS. In the last 15 years, GIS data have become more readily available in both quantity and quality, while the availability of a GIS in a Windows operating system environment, availability of low-cost computer hardware, and the development and implementation of easy to use GIS tools (in a Windows environment) has made implementation of GIS-based decision support tools more practical.

A typical example of the use of a GIS as a decision support tool is its use for siting a landfill. Typical data layers that are required are the location of suitable soils, wells, surface water sources, residential areas, schools, airports, roads, etc. From these data layers queries are formulated to provide the most suitable site(s). For example, a landfill should not be in the vicinity of an airport due to safety issues (i.e., aircraft striking birds) but would require suitable soil (i.e., soils with low permeability). A landfill should not be in the vicinity of wells or other water sources due to the issue of landfill leaching. Typically, quantitative weighting criteria are associated with the siting criteria as well as elements of the data layers (e.g., certain types of soils would be more suitable than others and thus would have applicable quantitative values) [15]. The data layers relevant to this project are shown in Table 10.6. Based on an evaluation of these weightings, Site 20 was considered the optimal site. The

Table 10.6 Example of spatial and non-spatial data inputs

Data input	Source	Comments
Spatial data		
AADT	Nevada DOT	http://www.nevadadot.com/reports_pubs/traffic_report/2005/pdfs/Clark.pdf
		Excel spreadsheet with X, Y coordinates of AADT station locations and AADT counts.
Topology	Clark County GIS website	http://gisgate.co.clark.nv.us/gismo/gismo.htm
	EPA/FHWA personnel	Site visits by EPA/FHWA personnel
Potentially confounding air pollutant sources	EPA/FHWA personnel	Site visits by EPA/FHWA personnel
	Clark County GIS website	http://gisgate.co.clark.nv.us/gismo/gismo.htm
Street data	Clark County GIS website	http://gisgate.co.clark.nv.us/gismo/gismo.htm
Points of interest		
Administrative boundaries		
Schools		
Aerial imagery	GlobeXplorer	ImageConnect Service (ArcGIS)
	Google Earth	http://earth.google.com/
Non-spatial Data		
Selection criteria	Settlement agreement	http://www.fhwa.dot.gov/environment/airtoxicmsat/setagree.pdf
	Monitoring protocol	http://www.fhwa.dot.gov/environment/airtoxicmsat/FinalDMPJune.pdf
Geometric design, geographic location	Aerial photos – Digital Globe – October 2005	Aerial images downloaded using ArcGIS tools
Availability of traffic volume data	Nevada DOT	Conference calls, site visit by EPA/FHWA personnel
Meteorology	National Climatic Data Center	http://cdo.ncdc.noaa.gov/CDO/cdo
	Clark County Air Quality	http://www.ccairquality.org/archives/index.html
Downwind sampling	EPA/FHWA personnel	Site visits by EPA/FHWA personnel

Source: S. Kimbrough, D. Vallero, R. Shores, A. Vette, K. Black and V. Martinez (2008). Multi-criteria decision analysis for the selection of a near road ambient air monitoring site for the measurement of mobile source air toxics. *Transportation Research, Part D: Transport and Environment* 13 (8): 505–515.

significant difference between the near-road site selection process and the landfill example is that the former did not explicitly assign quantitative values to our selection criteria. However, during our site visit and subsequent discussions quantitative values were assigned to the selection criteria. For example, sites with high AADT ($>$150,000) were more highly "valued" than sites with a lower AADT, sites without sound barriers were "valued" more highly than sites with sound barriers, etc. This is an example of combining qualitative and quantitative DQOs in a sampling and analysis plan.

Appropriately siting downwind sampling locations is an important criterion for an environmental monitoring study. Any location where proper siting of downwind sampling sites is restricted due to topology, existing structures, and meteorology may exclude otherwise suitable sites. Site selection must account for proper meteorological conditions (e.g., wind direction). For example, meteorological conditions in Las Vegas can be problematic as the city is located in a valley surrounded by mountains that channel the wind. This channeling of wind can present technical challenges in site selection and achieving proper wind flow from the source to the detector (i.e., gas analyzers). The topographic and meteorological conditions are shown in Figure 10.8. As shown in Figure 10.9, the wind direction for Site 20 does have acceptable meteorological conditions. Site 20, on the other hand, is 1 km west of a nearby airport (a potential source of air toxics) and is slightly below grade (modest cut) as shown in Figure 10.10. Another feature of note at this location is a spur line of the Union Pacific Railroad (UPRR). This is a commodity line that runs approximately 12 miles from Las Vegas, NV to Henderson, NV and passes this site twice a day (once in the morning and returns in the afternoon).

Activities can also impact site selection. For example, a construction project in the vicinity of Site 20 involves the conversion of the inner shoulders and median to express lanes (lanes in either direction of travel). Based on information from the NDOT design engineer, there would be minimal impact to the monitoring project for the following reasons: (1) construction will involve the addition of center lanes – the sampling project is on the shoulder behind a guardrail to the east of any construction activity; (2) the segment of roadway carries >200,000 vehicles per day – diesel emissions from construction equipment would be overwhelmed by the sheer volume of the on-highway vehicles; and (3) construction vehicles will not be operating in "front of monitoring station" 24-hours per day – construction activity will be occurring along a 5–6 mile stretch of the freeway. Therefore, while not perfect, Site 20 is still a viable site.

As mentioned, a key consideration is site access. For example, Site 20 would require that the investigators obtain right-of-way access from both Nevada DOT and the UPRR. Other property owners in the vicinity of our location were reluctant to permit access to their property. Property owners may be reluctant to grant ROW access for a variety of reasons: liability, financial issues, suspicion of government activities, etc.

Whereas Site 21 has minimally acceptable meteorology and minimally acceptable downwind sampling, the road is above grade (US 95 overpass) and lacks sound barriers. Moreover, Site 21 is somewhat problematic with regards to nearby sources. Site 21 is in close

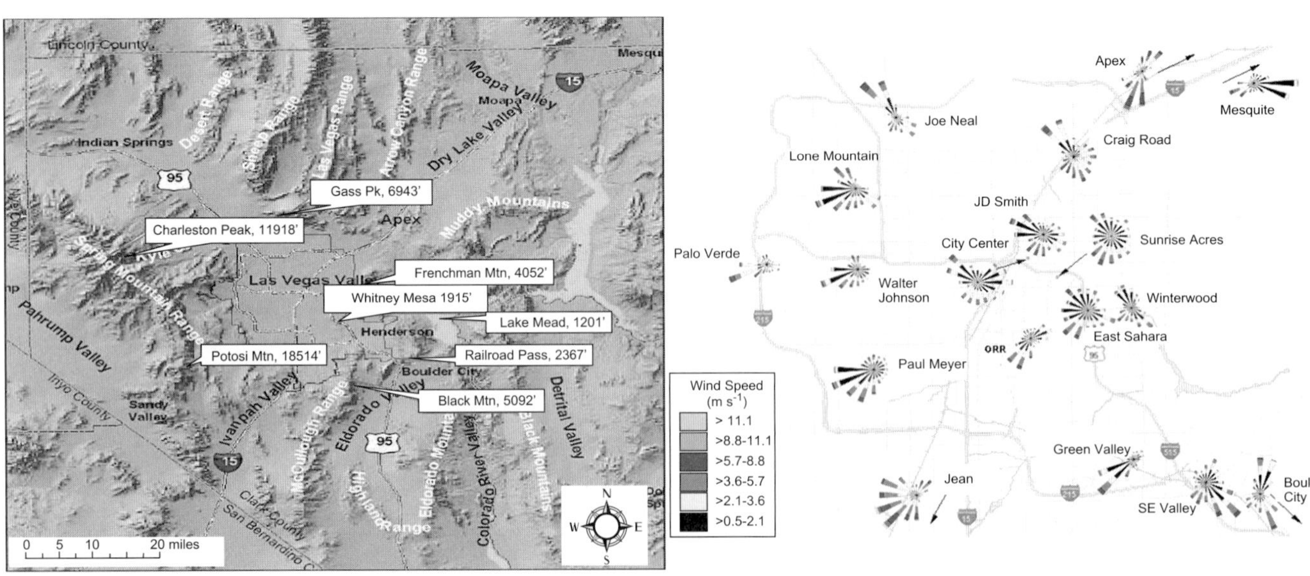

FIGURE 10.8

Topographic and meteorological conditions in Las Vegas, NV. [*See color plate section*]

Source: Clark County DAQEM 2007 with permission from Clark County Department of Air Quality and Environmental Management Officials.

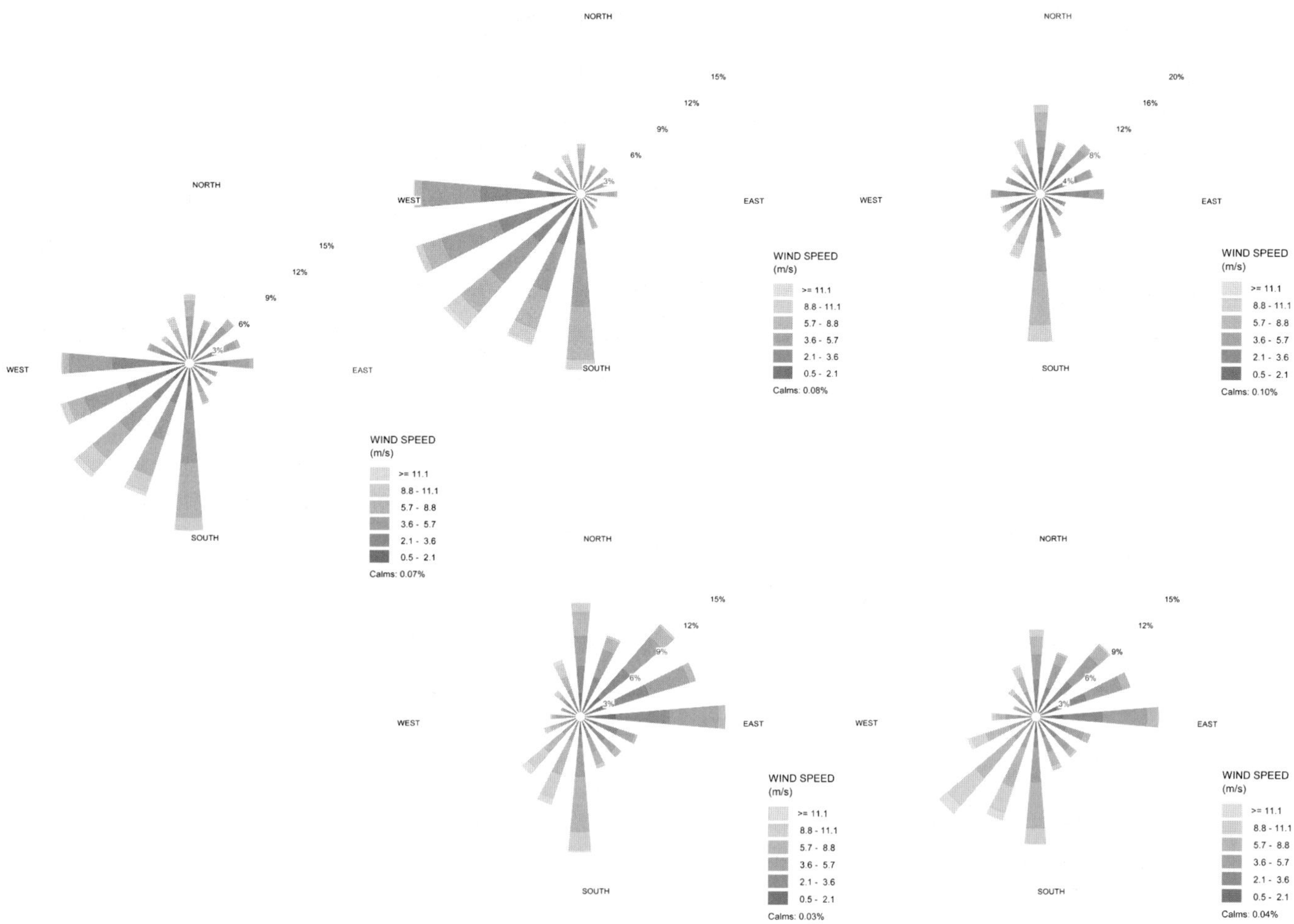

FIGURE 10.9
Wind roses for air toxic monitoring site in Las Vegas.
Source: S. Kimbrough, D. Vallero, R. Shores, A. Vette, K. Black and V. Martinez (2008). Multi-criteria decision analysis for the selection of a near road ambient air monitoring site for the measurement of mobile source air toxics. *Transportation Research, Part D: Transport and Environment* 13 (8): 505–515.

proximity to Rainbow Boulevard (<200 meters), a major north–south street with approximately 60,000–75,000 AADT. Highway vehicle emissions from this nearby source could and would confound sampling data – it would not be possible to differentiate the contribution of vehicle emissions from Rainbow Boulevard and the contribution of vehicle emissions from Site 21 (i.e., US 95). Site 21 is described as having minimally acceptable meteorology and minimally acceptable downwind sampling. Downwind sampling at this location was deemed to be problematic as well. The area surrounding Site 21 is mixed use residential and commercial. Thus, it would be probable that our downwind sampling site (100 meters downwind of the road) would be in a parking lot or in the yard of a private residence. Typically, commercial property owners do not want parking space consumed by non-revenue producing activities (i.e., non-customers) and private residence owners do not want yard space consumed by an 8′ × 16′ trailer that would require regular on-site visits to operate and maintain the ambient air monitoring equipment that would be housed in the trailer. Traffic monitoring equipment is planned for this location but is not currently installed and operating in the vicinity of this location.

Site 22 is an at grade roadway, lacks sound barriers, acceptable downwind sampling, minimally acceptable meteorology and has no known nearby sources. However, further investigation demonstrated that Site 22 lacked sufficient AADT (<100,000) to be considered

FIGURE 10.10
Photos of site ultimately selected for monitoring of air toxics in Las Vegas, NV. [*See color plate section*]
Photos by J.L. Vallero, used with permission.

minimally acceptable. Moreover, traffic monitoring equipment is not currently installed or planned for this location.

There is seldom a "perfect" monitoring site; compromises often have to be made. It is a question of balancing benefits with risks and costs. The selection is further complicated by external constraints and drivers. The principal constraint is the legal mandates of the settlement agreement, especially the data that must be derived pursuant to the monitoring protocol. Few, if any, design decisions can be made exclusively from a single perspective. These decisions can be visualized as attractions within a force field, where the center of the diagram represents the initial condition with a magnet placed in each sector at points equidistant from the center of the diagram as shown in Figure 10.11(a). If the factors are evenly distributed and weighted, the diagram might appear as that in Figure 10.11(b). But, as the differential in magnetic force increases, that factor will progressively drive the decision. In the present case study, the decision is most directly influenced by legal requirements, but which also needs to be scientifically credible and economically feasible (Figure 10.11c).

There is also the question of the best use of resources for this project. For example, a site could be chosen that would call for additional monitoring (and concomitantly additional resources) to overcome certain physical constraints (i.e., above/below grade, sound walls). Or, a site could be chosen that has some other issue such as low AADT or where traffic monitoring equipment would have to be installed. Some sites that would otherwise be favorable are near open areas, such as desert land that is prone to fugitive dust (common in unpaved or un-irrigated areas in Las Vegas).

Ultimately, Site 20 (I-15) was deemed the most suitable site. This site has the most advantages and fewest disadvantages. The site has high AADT (206,000 AADT for 2006), no sound barriers, meteorological and traffic data availability, manageable site logistics including right-of-way access, "clean geometric design," and favorable wind direction. "Clean geometric design" is defined as a facility that does not impede the effective data

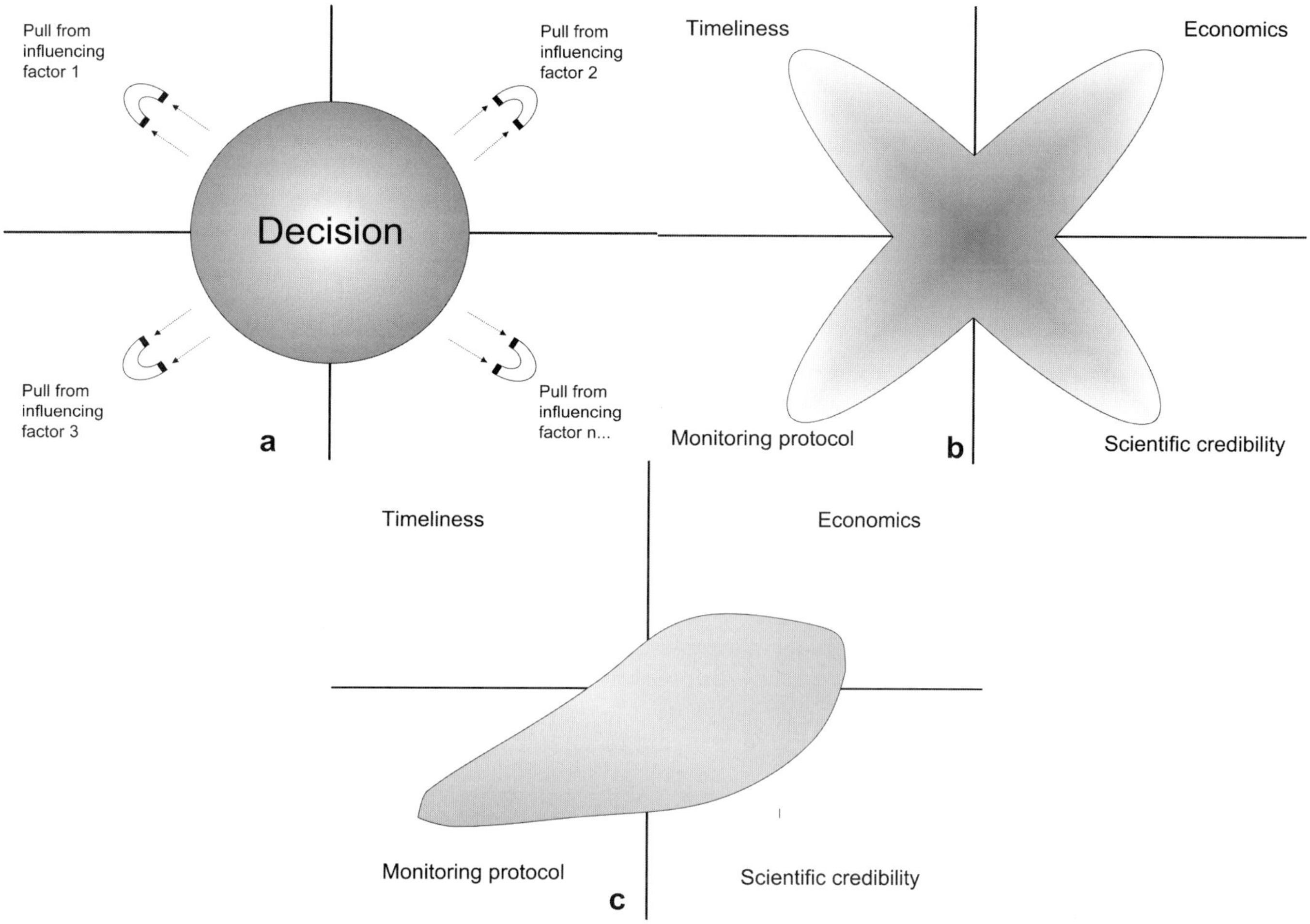

FIGURE 10.11

(a) Decision force field. The initial conditions will be driven toward influences. The stronger the influence of a factor (e.g. high AADT) the greater the decision will be drawn to that perspective. (b) Decision force field where a number of factors have nearly equal weighting in a design decision. For example, if the monitoring protocol is somewhat ambiguous, a number of alternatives are available, costs are flexible, and scientific credibility is minimally impacted, the design has a relatively large degree of latitude and elasticity. (c) Force field for a decision that is most strongly influenced by legal constraints and drivers. Note that all factors drive the decision, but that the monitoring protocol and other legal instruments have the greatest "magnetic pull." In fact, the scientific, economic, and timeliness factors may indeed be embedded in the legal instruments, as is the case in this paper.
Source: D. Vallero and C. Brasier (2008). *Sustainable Design: The Science of Sustainability and Green Engineering.* John Wiley & Sons, Inc. Hoboken, NJ.

collection of MSATs and $PM_{2.5}$. For example, a "clean geometric design" would be a site that does not include multiple on/off ramps, interchanges, or other complicating facility characteristics. Of the disadvantages, Site 20 has a modest road cut at that location and is only for a short distance. In other words, the roadway passes under a railroad bridge (see Figure 10.10) and returns to at/near grade conditions. Thus, this site is much more suitable than the urban canyon situation that exists with the elementary school site. The airport, a source of vehicle and aircraft emissions, is approximately 1 km due east of Site 20; however the predominant winds at this location generally maintain the monitoring site upwind of the airport. The wind blows predominately from the southwest quadrant until approximately 3:00 pm. Beginning at about 3:00 pm in the afternoon, the winds shift and become more variable. In the late afternoon and early evening (6–9:00 pm), the winds are predominately from the east and northeast. Then, after 9:00 pm the winds shift again into a more variable pattern. After midnight the winds return to blow predominately from the southwest.

An additional issue about Site 20 is related to construction that will be occurring along I-15 in the vicinity of the monitoring location. This construction involves center lane resurfacing and median replacement to accommodate the new express lanes between Russell Rd and Sahara Ave., with most of the work taking place north of the monitoring site. The construction work in the immediate area of the monitoring site will consist of mainly restriping and tapering. These construction activities (as planned) will overlap with the monitoring times of this project.

Site logistics includes, but is not limited to, gaining access to electrical power, communications connectivity, arranging for security fencing, etc. Site logistics, while not explicitly included in the monitoring protocol, is mission critical. Any location where site logistics, is restricted or prohibited either due to administrative or physical issues, is highly problematic and would eliminate a site from further development. For this specific project, obtaining the proper electrical feed, communications connectivity, and being able to establish security fencing is mission critical.

Sampling approaches

The environmental sampling plan defines the kinds of samples to be collected. A grab sample is simply a measurement at a site at a single point in time. Composite sampling physically combines and mixes multiple grab samples (from different locations or times) to allow for physical, instead of mathematical, averaging. The acceptable composite provides a single value of contaminant concentration measurement that can be used in statistical calculations. Multiple composite samples can provide improved sampling precision and reduce the total number of analyses required compared to non-composite sampling [16] (e.g. "grab" or integrated soil sample of x mass or y volume), the number of samples needed (e.g. for statistical significance), the minimum acceptable quality as defined by the quality assurance (QA) plan and sampling standard operating procedures (SOPs), and sample handling after collection. A weakness of composite sampling is the false negative effect. For example,

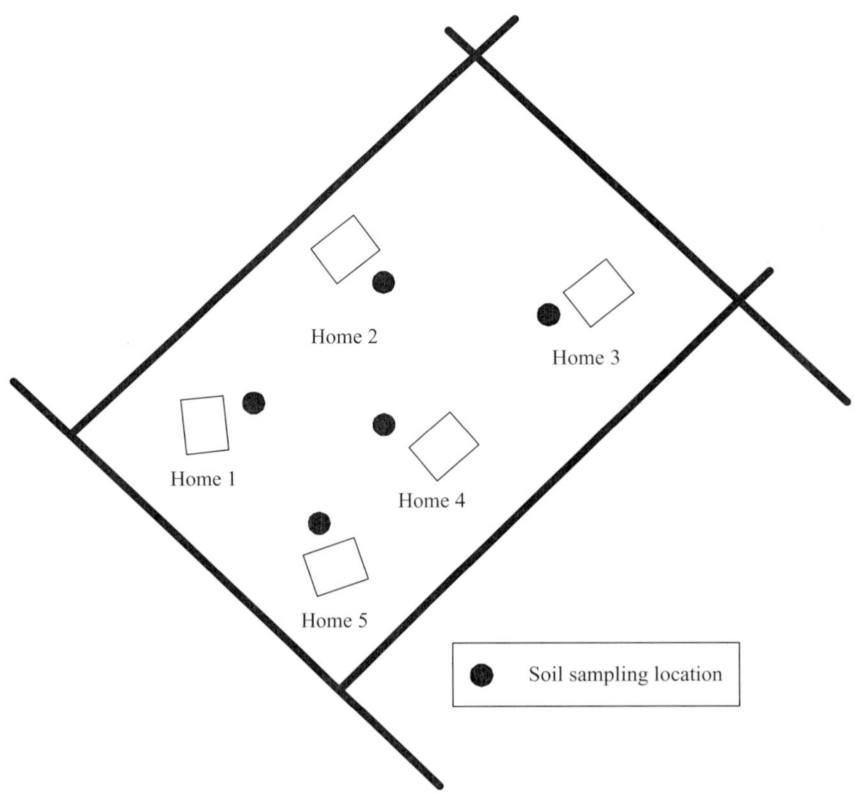

FIGURE 10.12 Composite sampling grid for a neighborhood.

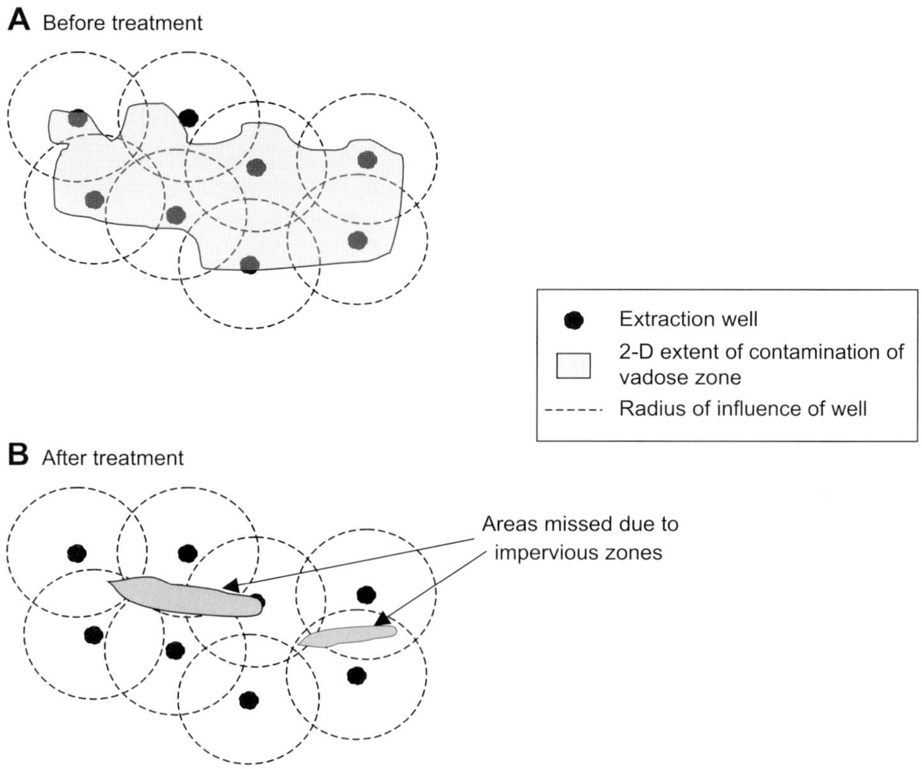

FIGURE 10.13
Extraction well locations on a geometric grid, showing hypothetical cleanup after 6 months.

samples are collected from an evenly distributed grid of homes to represent a neighbor exposure to a contaminant, as shown in Figure 10.12, where action is needed above the threshold of 5 mg L^{-1}. The assessment found the values are 3, 1, 2, 12, and 2 mg L^{-1}, the mean contamination concentration is only 4 mg L^{-1}, so it would be reported below the threshold level. However, the fourth home is well above the safety level. This could also have a false positive effect. For example, if the mean concentration were 6 mg L^{-1} in the example, the whole neighborhood may not need cleanup if the source is isolated to a confined area in the yard of home 5. Another example of where geographic composites may not be representative is in cleaning up and monitoring the success of cleanup actions. For example, if a grid is laid out over a contaminated groundwater plume (Figure 10.13), it may not take into account horizontal and vertical impervious layers, unknown sources (e.g. tanks), and flow differences among strata, so that some of the plume is eliminated but pockets are left (as shown in Figure 10.13B).

Thus, it is often good practice to assume that a contaminated site will have a heterogeneous distribution of contamination. Sampling methods and considerations on their use are as follows [17]:

- *Random sampling:* While it has the value of statistical representativeness, with a sufficient number of samples for the defined confidence levels (e.g. x samples needed for 95% confidence), random sampling may lead to large areas of the site being missed for sampling due to chance distribution of results. It also neglects prior knowledge of the site. For example, if maps show an old tank that may have stored contaminants, a purely random sample will not give any preference to samples near the tank.
- *Stratified random sampling:* By dividing the site into areas and randomly sampling within each area, avoiding the omission problems of random sampling alone.

- *Stratified sampling:* Contaminants or other parameters are targeted. The site is subdivided and sampling patterns and densities varied in different areas. Stratified sampling can be used for complex and large sites, such as mining.
- *Grid or systematic sampling:* The whole site is covered. Sampling locations are readily identifiable, which is valuable for follow-on sampling, if necessary. The grid does not have to be rectilinear. In fact, rectangles are not the best polygon to use if the value is to be representative of a cell. Circles provide equidistant representation, but overlap. Hexagons are sometimes used as a close approximation to the circle. The US Environmental Monitoring and Assessment Program (EMAP) has used a hexagonal grid pattern, for example.
- *Judgmental sampling:* Samples are collected based upon knowledge of the site. This overcomes the problem of ignoring sources or sensitive areas, but is vulnerable to bias of both inclusion and exclusion. Obviously, this would not be used for spatial representation, but for pollutant transport, plume characterization, or monitoring near a sensitive site (e.g. a day care center).

At every stage of monitoring from sample collection through analysis and archiving, only qualified and authorized persons should be in possession of the samples. This is usually assured by requiring chain-of-custody manifests. Sample handling includes specifications on the temperature range needed to preserve the sample, the maximum amount of time the sample can be held before analysis, special storage provisions (e.g. some samples need to be stored in certain solvents), and chain-of-custody provisions (only certain, authorized persons should be in possession of samples after collection).

Each person in possession of the samples must require that recipient to sign and date the chain-of-custody form before transferring the samples. This is because samples have evidentiary and forensic content, so any compromising of the sample integrity must be avoided.

Laboratory analysis

Analytical chemistry is an essential part of an environmental assessment of a biotechnological operation. When a sample arrives at the laboratory, the next step may be "extraction." Extraction is needed for two reasons. First, the environmental sample may be in sediment or soil, where the chemicals of concern are sorbed to particles and must be freed for analysis to take place. Second, the actual collection may have been by trapping the chemicals onto sorbants. So, to analyze the sample, the chemicals must first be freed from the sorbant matrix. Again, dioxins provide an example. Under environmental conditions, dioxins are fat soluble and have low vapor pressures, so they may be found on particles, in the gas phase, or in the water column suspended to colloids (and very small amounts dissolved in the water itself, especially as a result of co-solvation – see chapter 3). Therefore, to collect the gas phase dioxins, the standard method calls for trapping it on polyurethane foam (PUF). Thus, to analyze dioxins in the air, the PUF and particle matter must first be extracted, and to analyze dioxins in soil and sediment, those particles must also be extracted.

Extraction makes use of physics and chemistry. For example, many compounds can be simply extracted with solvents, usually at elevated temperatures. A common solvent extraction is the Soxhlet extractor, named after the German food chemist Franz Soxhlet (1848–1913). The Soxhlet extractor (the US EPA Method 3540) removes sorbed chemicals by passing a boiling solvent through the media. Cooling water condenses the heated solvent and the extract is collected over an extended period, usually several hours. Other automated techniques apply some of the same principles as solvent extraction, but allow for more precise and consistent extraction, especially when large volumes of samples are involved. For example, supercritical fluid extraction (SFE) brings a solvent, usually carbon dioxide to the pressure and temperature near its critical point of the solvent, where the solvent's properties are rapidly altered with very slight variations of pressure [18]. Solid phase extraction (SPE), which uses a solid and a liquid phase to isolate a chemical from a solution, is often used to clean up a sample before analysis. Combinations of various extraction methods can enhance the extraction efficiencies, depending upon the chemical and the media in which it is found. Ultrasonic and microwave extractions may be used alone or in combination with solvent extraction. For example, the US EPA Method 3546 provides a procedure for extracting hydrophobic (that is, not soluble in water) or slightly water-soluble organic compounds from particles such as soils, sediments, sludges, and solid wastes. In this method, microwave energy elevates the temperature and pressure conditions (i.e., 100–115 °C and 50–175 psi) in a closed extraction vessel containing the sample and solvent(s). This combination can improve recoveries of chemical analytes and can reduce the time needed compared to the Soxhlet procedure alone.

Not every sample needs to be extracted. For example, air monitoring using canisters and bags allows the air to flow directly into the analyzer. Water samples may also be directly injected. Surface methods, such as fluorescence, sputtering, and atomic absorption, require only that the sample be mounted on specific media (e.g. filters). Also, continuous monitors like the chemiluminescent system mentioned in the next section provide ongoing measurements.

Chromatography consists of separation and detection. Separation makes use of the chemicals' different affinities for certain surfaces under various temperature and pressure conditions. The first step, injection, introduces the extract to a "column." The term column is derived from the time when columns were packed with sorbents of varying characteristics, sometimes meters in length, and the extract was poured down the packed column to separate the various analytes. Today, columns are of two major types, gas and liquid. Gas chromatography (GC) makes use of hollow tubes ("columns") coated inside with compounds that hold organic chemicals. The columns are in an oven, so that after the extract is injected into the column, the temperature is increased, as well as the pressure, and the various organic compounds in the extract are released from the column surface differentially, whereupon they are collected by a carrier gas (e.g. helium) and transported to the detector. Generally, the more volatile compounds are released first (they have the shortest retention times), followed by the semi-volatile organic compounds. So, boiling point is often a very useful indicator as to when a compound will come off a column. This is not always the case, since other characteristics such as polarity can greatly influence a compound's resistance to be freed from the column surface. For this reason, numerous GC columns are available to the chromatographer (different coatings, interior diameters, and lengths). Rather than coated columns, liquid chromatography (LC) makes use of columns packed with different sorbing materials with differing affinities for compounds. Also, instead of a carrier gas, LC uses a solvent or blend of solvents to carry the compounds to the detector. In the high performance LC (HPLC), pressures are also varied.

Detection is the final step for quantifying the chemicals in a sample. The type of detector needed depends upon the kinds of pollutants of interest. Detection gives the "peaks" that are used to identify compounds (see Figure 10.14). For example, if hydrocarbons are of concern, GC with flame ionization detection (FID) may be used. GC-FID gives a count of the number of carbons, so for example, long chains can be distinguished from short chains. The short chains come off the column first and have peaks that appear before the long-chain peaks. However, if pesticides or other halogenated compounds are of concern, electron capture detection (ECD) is a better choice.

A number of detection approaches are also available for LC. Probably the most common is absorption. Chemical compounds absorb energy at various levels, depending upon their size, shape, bonds, and other structural characteristics. Chemicals also vary in whether they will absorb light or how much light they can absorb depending upon wavelength. Some absorb very well in the ultraviolet (UV) range, while others do not. Diode arrays help to identify compounds by giving a number of absorption ranges in the same scan. Some molecules can be

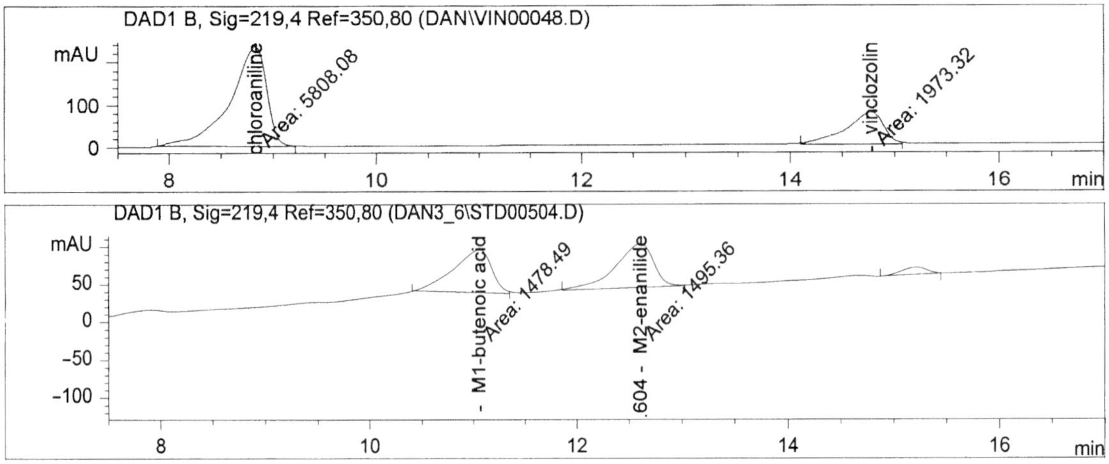

FIGURE 10.14
High performance liquid chromatograph/ultraviolet detection peaks for standard acetonitrile solutions: 9 mg L^{-1} 3,5-dichloroaniline and 8 mg L^{-1} the fungicide vinclozolin (*top*); and 7 mg L^{-1} M1 and 9 mg L^{-1} M2 (*bottom*). Note: mAU = milli-absorprtion units; 1mAU = 10^{-3}AU.
Source: D. Vallero (2003). *Engineering the Risks of Hazardous Wastes.* Butterworth-Heinemann, Boston, MA.

excited and will fluoresce. The *Beer–Lambert law* tells us that energy absorption is proportional to chemical concentration:

$$A = eb[C] \tag{10.13}$$

where, *A* is the absorbency of the molecule, *e* is the molar absorptivity (proportionality constant for the molecule), *b* is the light's path length, and *[C]* is the chemical concentration of the molecule. Thus, the concentration of the chemical can be ascertained by measuring the light absorbed, which is expressed in absorption units (AU) and more commonly in milli-absorption units (mAU = 10^{-3}AU).

One of the most popular detection methods is mass spectrometry (MS), which can be used with either GC or LC separation. The MS detection is highly sensitive for organic compounds and works by using a stream of electrons to consistently break apart compounds into fragments. The positive ions resulting from the fragmentation are separated according to their masses. This is referred to as the "mass to charge ratio" or m/z. No matter which detection device is used, software is used to decipher the peaks and to perform the quantitation of the amount of each contaminant in the sample.

For inorganic substances and metals, the additional extraction step may not be necessary. The actual measured media (e.g. collected airborne particles) may be measured by surface techniques like atomic absorption (AA), X-ray fluorescence (XRF), inductively coupled plasma (ICP), or sputtering. As for organic compounds, the detection approaches can vary. For example ICP may be used with absorption or MS. If all one needs to know is elemental information, for example to determine total lead or nickel in a sample, AA or XRF may be sufficient. However, if speciation is required (i.e. knowing the various compounds of a metal), then significant sample preparation is needed, including a process known as "derivatization." Derivatizing a sample is performed by adding a chemical agent that transforms the compound in question into one that can be recognized by the detector. This is done for both organic and inorganic compounds, for example, when the compound in question is too polar to be recognized by MS.

The physical and chemical characteristics of the compounds being analyzed must be considered before visiting the field and throughout all the steps in the laboratory. Although it is beyond the scope of this book to go into detail, it is worth mentioning that the quality of results generated about contamination depends upon the sensitivity and selectivity of the analytical equipment. Table 10.7 defines some of the most important analytical chemistry threshold values.

Table 10.7 Expressions of chemical analytical limits

Type of limit	Description
Limit of detection (LOD)	Lowest concentration or mass that can be differentiated from a blank with statistical confidence. This is a function of sample handling and preparation, sample extraction efficiencies, chemical separation efficiencies, and capacity and specifications of all analytical equipment being used (see IDL below)
Instrument detection limit (IDL)	The minimum signal greater than noise detectable by an instrument. The IDL is an expression of the piece of equipment, not the chemical of concern. It is expressed as a signal to noise (S:N) ratio. This is mainly important to the analytical chemists, but the engineer should be aware of the different IDLs for various instruments measuring the same compounds, so as to provide professional judgment in contracting or selecting laboratories and deciding on procuring for appropriate instrumentation for all phases of remediation
Limit of quantitation (LOQ)	The concentration or mass above which the amount can be quantified with statistical confidence. This is an important limit because it goes beyond the ''presence-absence'' of the LOD and allows for calculating chemical concentration or mass gradients in the environmental media (air, water, soil, sediment, and biota)
Practical quantitation limit (PQL)	The combination of LOQ and the precision and accuracy limits of a specific laboratory, as expressed in the laboratory's quality assurance/quality control (QA/QC) plans and standard operating procedures (SOPs) for routine runs. The PQL is the concentration or mass that the engineer can consistently expect to have reported reliably

Source: D. Vallero (2003). *Engineering the Risks of Hazardous Wastes.* Butterworth-Heinemann, Boston, MA.

EXAMPLE

Calculating a Chemical Gradient

You have received a report that shows that the level of detection (LOD) for naphthalene is 5 $\mu g\ L^{-1}$. The LOQ for the same chemical is 10 :g Ll^{-1}. The report shows that the soil concentrations of naphthalene every 10 m from a shed to a roadway 50 m away:

6 $\mu g\ L^{-1}$ closest to the shed
7 $\mu g\ L^{-1}$
8 $\mu g\ L^{-1}$
8 $\mu g\ L^{-1}$
9 $\mu g\ L^{-1}$
9 $\mu g\ L^{-1}$ at the roadway

Can a gradient be calculated from these data? Why or why not? If yes, what would the average concentration gradient be? If not, how might these data be useful (Hint: Consider what an LOD tells you about the data and the presence of naphthalene).

What if the report showed the following concentrations?

10 $\mu g\ L^{-1}$ closest to the shed
11 $\mu g\ L^{-1}$
11 $\mu g\ L^{-1}$
12 $\mu g\ L^{-1}$
13 $\mu g\ L^{-1}$
13 $\mu g\ L^{-1}$ at the roadway

(Continued)

Can a gradient be calculated from these data? Why or why not? If yes, what is the average concentration gradient?

If these data are truly representative and accurate, is the shed the likely source of naphthalene contamination? Explain.

Solution and discussion

Since all of the values in the first data set are above the LOD, we can say that naphthalene is present in the soil. However, since the values are below the LOQ, we cannot confidently tell whether they vary amongst themselves, so we cannot calculate a gradient.

Since all values in the second data set are at or above the LOQ, we can calculate a gradient. For example, in a span of 50 m the naphthalene concentrations increased by 3 $\mu g\ L^{-1}$ from the shed to the roadway. So, the gradient is 3 $\mu g\ L^{-1}$ 50 m^{-1} = 0.6 $\mu g\ L^{-1}\ m^{-1}$.

Since the concentrations of naphthalene are decreasing from the road to the shed, the roadway is more likely to be the location of the source rather than the shed.

Source identification

If soil naphthalene concentrations similar to those found above are also found along the roadway for a distance of 50 m where there are no buildings and the land in undeveloped, what is a likely source of the naphthalene? Why were only naphthalene results reported?

Discussion

Was there a spill, or is there a continuous source of the naphthalene at a distant site that is depositing it on the soil? An investigation is in order.

An environmental assessment of one compound may be the result of an investigation requested by a company, town or regulatory agency due to a complaint, for example, anonymously via a ''hot line'' call or a formally registered complaint or inquiry from a citizen or group. Or, the analysis could also be part of a research study by a laboratory or university. Another possibility is that a whole suite of compounds was studied. In this instance, a number of polycyclic aromatic compounds were targeted, but only naphthalene was detected.

It behooves the scientist and engineer to find out which, if any, of these is the reason for only one identified contaminant.

Fluorescent in-situ hybridization (FISH): an environmental monitoring biotechnology

Many sophisticated instruments and procedures are available to monitor for the presence and concentration of contaminants in water, air, and soil. These instruments with their attendant procedures offer engineers many opportunities to better understand the magnitude of the associated risks. For example these instruments and techniques are used to determine the magnitude of a pollution problem, answering such questions as: how much toluene is in the soil surrounding a hazardous waste landfill? The same instruments and techniques are used to track the successes and/or failures of the engineers' attempts to control the risks associated with hazardous waste, answering such questions as: how much has this pump and treat technology, that was used for the past 6 months at a cost of $1,200,000, reduced the levels of toluene in the soil surrounding this hazardous waste landfill?

Two examples of measurement and monitoring instruments and procedures illustrate the recent advances in efforts to determine levels of hazardous contaminants in water, air and soil. Chemiluminescence and fluorescent in-situ hybridization (FISH) are state-of-the-art examples of the science, engineering, and technology of measurement instrumentation and techniques available to address pollution problems.

Soils at sites nationwide are contaminated with dense non-aqueous phase liquids (DNAPLs) that sink in the groundwater, non-aqueous phase liquids (NAPLs) that float within the groundwater, as well as heavy metals. These sites have become a nationwide public health and economic concern. Such active soil remediation systems such as pump and treat and such passive remediation systems as *natural attenuation* require elaborate, expensive, decades-long monitoring for process control, performance measurement, and regulatory compliance. Under current monitoring practices many liquid and/or soil samples must be collected from contaminated sites, packaged, transported and analyzed in a certified laboratory, all at great expense and with great time delay to the owner of the site and to regulatory agencies at the state and federal levels. In addition, waste that is generated by sample collection and by sample analyses properly must be disposed, again at great cost. Nationwide monitoring of contaminated soils will continue into the future in order to protect public health and the environment.

Consider the extremely complex and only partially understood biogeochemical nitrogen cycle. The processes in soil that traditionally are suggested to contribute to the levels of NO emissions are, in order of general importance, autotrophic nitrification, respiratory denitrification, chemo-denitrification and heterotrophic nitrification. Except for chemo-denitrification, all of these mechanisms are microbially mediated transformations performed by such bacteria as *Nitrosolobus* and *Nitrobacter* genera in autotrophic nitrification and *Pseudomonas* and *Alcaligenes* genera in respiratory denitrification. Autotrophic nitrification and then respiratory denitrification are suggested to be the principal sources of NO in the cycle, while heterotrophic nitrification and chemo-denitrification can be important NO sources under extraordinary soil pH and other conditions [19].

As levels of contamination change in a soil, chemiluminescence monitoring of NO emissions from contaminated soil can indicate to the engineer the absence or presence, including the level, of contamination of a pollutant in the soil. This monitoring of NO emissions from soil may be used as a surrogate indicator of the level of contamination in soils during remediation and post-remediation activities at contaminated soil sites and thus could assist in determining when expensive soil pollution remediation activities may cease.

Historically NO concentrations in ambient air have been determined using chemiluminescence analyzers that are inexpensive, durable, accurate, and precise. For example these analyzers are used widely by the state and federal environmental agencies to measure NO concentrations as precursors to ozone formation in cities and towns nationwide contributing to decades of successful ambient air quality monitoring programs.

Chemiluminescence analyzers convert NO to electronically excited NO_2 (indicated as NO_2^*) when O_3 is supplied internally by the analyzer as summarized:

$$O_3 + NO \rightarrow NO_2^* + O_2 \qquad (10.14)$$

These excited NO_2^* molecules emit light when they move to lower energy states as:

$$NO_2^* \rightarrow NO_2 + h\nu \; (590 < \lambda < 3000 \text{ nm}) \qquad (10.15)$$

The intensity of the emitted light is proportional to the NO concentration and is detected and converted to a digital signal by a photomultiplier tube that is recorded.

Dynamic test chambers and systems are available to measure the NO flux from soil. The mass balance for NO in the chamber is summarized by:

$$\frac{dC}{dt} = \left(\frac{Q[C]_0}{V} + \frac{JA_1}{V}\right) - \left(\frac{LA_2}{V} + \frac{Q[C]_f}{V}\right) + R \qquad (10.16)$$

where A = surface area of the soil

V = volume of the chamber

Q = air flow rate through the chamber

J = emission from the soil (flux)

C = NO concentration in the chamber

$[C]_o$ = NO concentration at the inlet of the chamber

$[C]_f$ = NO concentration at the outlet of the chamber

L = loss of NO on the chamber wall assumed first order in [C]

R = chemical production/destruction rate for NO in the chamber.

The NO emissions from soil to the headspace are calculated as J.

The so-called "FISH" method identifies microorganisms by using fluorescently labeled oligonucleotide probes homologous to target strains or groups of microorganisms and viewing by epifluorescent microscope in samples of soil studied in the laboratory and in the field. This technique was first applied to activated sludge cultures in 1994 but is continually undergoing modifications building on the understandings of procedures and oligonucleotide probes designed and applied to identify nitrifying bacteria in wastewater treatment systems. Methods of FISH application to soil samples are and will continue to evolve, as does every method of contaminant monitoring and measurement [20].

Historically, the FISH techniques applied to the study of microbial communities in soil have not been as well developed as have the FISH techniques applied to the study of microbial communities in water or slurried sediment samples. The classification of active soil bacteria using FISH is a challenging research topic that appears to be developing almost exclusively outside the United States. The FISH techniques for identifying bacteria extracted from soils generally are particularly difficult to perform because of: (1) the high background fluorescence signals from soil particles; (2) the exclusion of bacteria associated with soil particles; (3) the nonspecific attachment of the fluorescent probes to soil debris; (4) probing microorganisms that are entrapped in soil solids; and (4) determining the optimal stringency of hybridization.

Other obstacles to the application of the general FISH method to soils include difficulties in sequence retrieval, finding rRNA sequences of less common organisms, nonspecific staining, low signal intensity, and target organism accessibility. In addition, cells that are in the stationary phase often do not contain a sufficient cellular rRNA content to produce a detectable fluorescent image with FISH. These challenges can be overcome with the development of a variety of directed modifications to the general FISH methodologies, including altering experimental procedures for extraction and filtration of soil microbes, different selection and sequencing of oligonucleotide probes, and improving detection instrumentation, particularly the software to analyze the images obtained on a microscope with epifluorescent capability. Example FISH probes are presented in Table 10.8.

For different soils from different contaminated sites having different levels of different contaminants, microbial activity and consequently NO production will be affected during remediation and post-remediation activities in the field. Consider, for example, a site where the soil is contaminated with non-aqueous phase liquids (NAPLs). At sampling locations at

Table 10.8 Examples of oligonucleotide probes

Probe	Target bacteria	Applicability in NO studies
EUB338	Eubacteria	All bacteria
ALF1B	α proteobacteria	*Pseudomonas* and *Nitrobacter*
BET42A	β proteobacteria	*Nitrosomas*
GAM42A	γ proteobacteria	*Pseudomonas* (for example *P. putida*)

this site observed NO emissions measurements that are lower than representative background levels of NO emissions from the soil could indicate depressed levels of microbial activity due to high levels of contamination that are toxic to the microorganisms in the soil. On the other hand, when NO emissions equilibrate to background soil NO concentrations, this may indeed indicate a successful cleanup. That is, a more normal level of microbial activity is indicated by lower NO emissions, meaning that the remediated site has acceptably low concentrations of contamination, e.g. the formerly contaminated soil now has concentrations of organic compounds that would have been observed prior to the contamination.

Depending on the type of contaminant, the level of contamination, and the physical, chemical, and microbiological characteristics of the soil itself, chemiluminescence NO emissions monitoring at different locations at the site could indicate the presence, absence, or level of contamination in the soil.

Emissions of NO from soil are seen as a direct indicator of microbiological activity in the soil, which in turn can suggest the presence, absence and/or concentration of different contaminants in soil. For example, laboratory measurements of NO emissions from uncontaminated soil and soil that has been contaminated with toluene have shown that the toluene-contaminated soil can produce ten times more NO than the uncontaminated soil. The additional production of NO is suggested to be the result of increased microbial activity in the contaminated soil. (see Discussion Box: Measuring Biodegradation Success, in chapter 7)

SOURCES OF UNCERTAINTY

Contaminant assessments have numerous sources of uncertainty. There are two basic types of uncertainty: Type A and Type B. Type A uncertainties result from the inherent unpredictability of complex processes that occur in nature. These uncertainties cannot be eliminated by increasing data collection or enhancing analysis. The scientist and engineer must simply recognize that Type A uncertainty exists, but must not confuse it with Type B uncertainties, which can be reduced by collecting and analyzing additional scientific data.

The first step in an uncertainty analysis is to identify and describe the uncertainties that may be encountered. Sources of Type B uncertainty take many forms [21]. There can be substantial uncertainty concerning the numerical values of the attributes being studied (e.g. contaminant concentrations, wind speed, discharge rates, groundwater flow, and other variables). These should be represented as standard deviations or confidence intervals. Modeling generates its own uncertainties, including errors in selecting the variables to be included in the model, such as surrogate contaminants that represent whole classes of compounds (e.g. how well does benzene represent the behavior or toxicity of other aromatic compounds?).

In other words, the data themselves have uncertainties, but the use of data introduces its own uncertainties. When data are entered into models, this may lead to unacceptable uncertainty, even if the data themselves have individual uncertainties that are tolerable uncertainty. Thus, models and other data manipulation techniques can propagate uncertainties, especially when ambiguity arises regarding the data's meaning. For example, a decision rule is a statement about which alternative will be selected, e.g. for cleanup, based on the characteristics of the decision situation. A "decision-rule uncertainty" occurs when there are disagreements or poor specification of objectives (i.e. is our study really addressing the client's needs?).

Variability and uncertainty must not be confused. Variability consists of measurable factors that differ across populations such as soil type, vegetative cover, or body mass of individuals in a population. Uncertainty consists of unknown or not fully known factors that are difficult to measure, such as the inability to access an ideal site that would be representative because it is on private property. Thus, uncertainty represents a lack of knowledge about the factors of the subject matter under investigation, whereas variability is the heterogenity of the values pertaining to those factors.

Modeling uncertainties, for example, may consist of extrapolations from a single value to represent a whole population, .i.e. a point estimate (e.g. 70 kg as the weight of an adult male). Such estimates can be typical values for a population or an estimate of an upper end of the population's value, e.g. 70 years as the duration of exposure used as a "worst-case" scenario. Another approach is known as the Monte Carlo technique (Figure 10.15). The Monte Carlo-type exposure assessments use probability distribution functions, which are statistical distributions of the possible values of each population characteristic according to the probability of the occurrence of each value. These are derived using iterations of values for each population characteristic. While the Monte Carlo technique may help to deal with the point estimate limitation, it can suffer from confusing variability with uncertainty.

Other data interpretation uncertainties can result from the oversimplification of complex entities. For example, assessments consist of an aggregation of measurement data, modeling, and combinations of sampling and modeling results. However, these complicated models are providing only a snapshot of highly dynamic human and environmental systems. The use of more complex models does not necessarily increase precision, and extreme values can be improperly characterized. For example, a 50th percentile value can always be estimated with more certainty than a 99th percentile value.

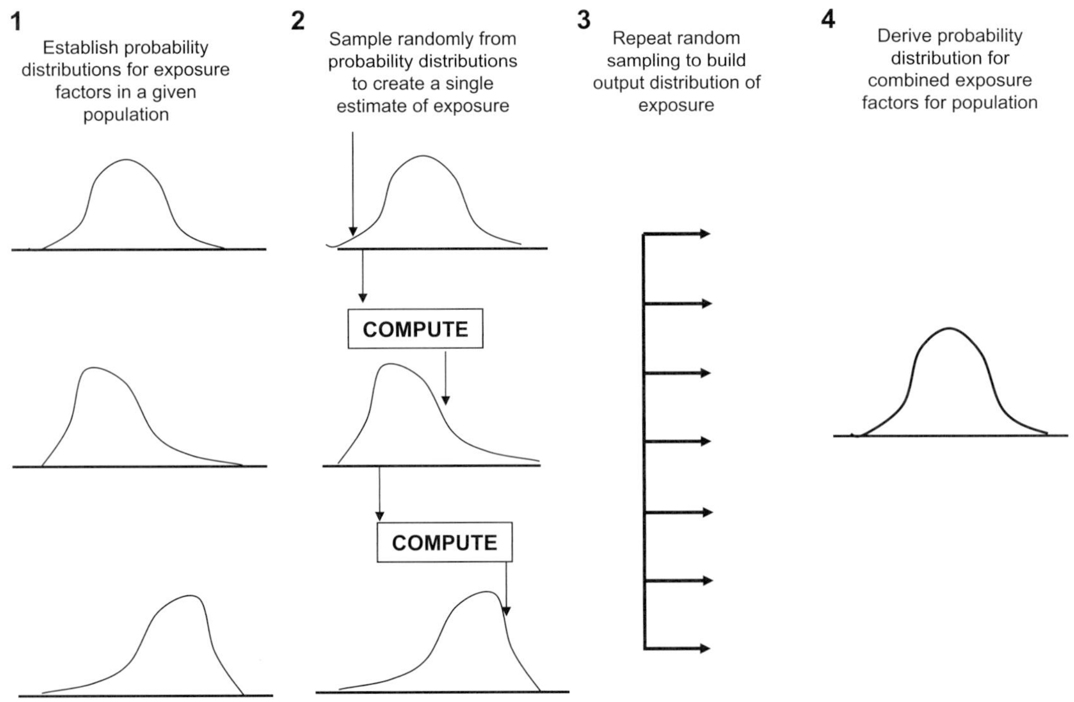

FIGURE 10.15
Principles of the Monte Carlo method for aggregating data.

The bottom line is that uncertainty is always present in sampling, analysis, and data interpretation, so the monitoring and data reduction plan should be systematic and rigorous. The uncertainty analysis must be addressed for each step of the contaminant assessment process, including any propagation and enlargement of cumulative error (e.g. an incorrect pH value that goes into an index where pH is weighted heavily, and then used in another algorithm for sustainability). The characterization of the uncertainty of the assessment includes selecting and rejecting data and information ultimately used to make environmental decisions, and includes both qualitative and quantitative methods (see Table 10.9).

In discussing reference doses and concentrations (RfDs and RfCs, respectively), recall that uncertainty factors (UF) were applied to address both the inherent and study uncertainties upon which to establish safe levels of exposure to contaminants. These include 10-fold factors, used to derive the RfD and RfC from experimental data. The UFs consider the uncertainties resulting from the variation in sensitivity among the members of the populations, including inter-human and intra-species variability, the extrapolation of animal data to humans

Table 10.9 **Example of an uncertainty table for exposure assessment**

	Effect on exposure[a]		
Assumption	Potential magnitude for over-estimation of exposure	Potential magnitude for under-estimation of exposure	Potential magnitude for over- or under-estimation of exposure
Environmental sampling and analysis			
Sufficient samples may not have been taken to characterize the media being evaluated, especially with respect to currently available soil data			Moderate
Systematic or random errors in the chemical analyses may yield erroneous data			Low–High
Exposure parameter estimation			
The standard assumptions regarding body weight, period exposed, life expectancy, population characteristics, and lifestyle may not be representative of any actual exposure situation			Moderate
The amount of media intake is assumed to be constant and representative of the exposed population	Moderate		
Assumption of daily lifetime exposure for residents	Moderate to high		

[a]*As a general guideline, assumptions marked as 'low', may affect estimates of exposure by less than one order of magnitude; assumptions marked 'moderate' may affect estimates of exposure by between one and two orders of magnitude; and assumptions marked 'high' may affect estimates of exposure by more than two orders of magnitude.*

Source: Australian Department of Health and Ageing (2002). Environmental Health Risk Assessment: Guidelines for Assessing Human Health Risks from Environmental Hazards.

(i.e. inter-species variability), the extrapolation from data gathered in a study with less-than-lifetime exposure to lifetime exposure, i.e., extrapolating from acute or subchronic to chronic exposure; the extrapolation from different thresholds, such as the LOAEL rather than from a NOAEL; and the extrapolation from an incomplete database. Note that most of these sources of uncertainty have a component associated with measurement and analysis.

The numerical value uncertainties are directly related to the quality and representativeness of the sampling design and the analytical expressions described in Table 10.9. When these values are input into models, they are known as "parameter uncertainties." So imprecision and inaccuracy associated with the measurement and analytical equipment can combine with insufficient sample size (random error) and systemic weaknesses in data gathering (bias). The net effect is diminished reliability and usefulness of environmental investigations.

SEMINAR TOPIC

Comparing Study Designs to Assess Releases of Biotechnology Agents with Those of Chemical Agents in the Environment

This chapter includes a number of recommendations of how to design an environmental study. However, these have been extrapolated from investigations of chemical agents. Medical geographic studies and epidemiological diseases studies have been designed to address the migration of pathogenic microbes in time and space. Such studies have included substantial environmental parameters, e.g. those aimed at documenting sources and migration of *Cryptosporidium spp.* during and after outbreaks. These have usually been related to a particular legislative or regulatory mandate, such as those to protect drinking water from pathogens.

Recent site-specific studies have gone a step further. For example, one study was designed to investigate the contribution of a landfill to bioaerosol exposures [22]. The investigators sampled airborne bacteria and fungi using soy and malt extract agar, collected with an impactor to determine seasonal and diurnal variations in these biological agents. They found that concentrations of culturable bacteria and fungi were higher in winter than in other seasons. The study also found substantial differences in bioaerosol concentrations and fungal percentages in the air around active versus closed landfills. These study designs seem to indicate that biological agents may be monitored in a similar manner as chemical agents if the biological agents and their materials are sorbed to aerosols.

Environmental studies often include multiple compartments, e.g. investigations in and around hazardous waste sites include measurements of numerous compounds in air, soil, sediment, ground and surface water, as well as biota (e.g. uptake of contaminants in plants and animals). The designs are complicated, and are usually aggregations of individual methods within the media. For example, the air monitoring may follow separate state protocols and standards for measuring air pollutants, water pollutants, and soil pollutants; each with their own quality assurance plan, sampling procedures, and analytical protocols. The results are later presented, either compartment-by-compartment or as an assimilated report. The actual integration of the results may not occur until the risk management phases of a project (e.g. preparation of executive summaries and public information documents).

Biological agents may be measured in a similar way, but it is best to keep in mind how the information will be presented *before* designing the monitoring study. For example, will the data be input into a geographic information system (GIS)? Will the sampling locations be selected using geostatistical techniques? These questions need to be answered as early as possible in the study design stage so that the measured data will be most useful, e.g. as input to exposure predictive models (see Figure 10.16).

Measuring biological materials may involve combining methods, for example, a recent risk assessment [23] of a genetically modified strain of oil rapeseed (*Brassica napus*) combined volumetric spore traps and passive traps to investigate movement and concentrations of airborne pollen (see Figure 10.17). The objectives of the assessment were to study gene flow, as indicated by pollen movement by:

Characterizing *B. napus* cv. Marinka using molecular markers.
Elucidating the distance traveled by the pollen by biotic dispersal.
Elucidating the distance traveled by the pollen by abiotic dispersal.
Elucidating pollination and seed set at various distances from a source crop using male sterile bait plants.
Developing risk assessment/containment strategies.

The collected pollen was identified using molecular markers to trace insect and wind dispersal of genetic material from *B. napus*. In most of the sampling, the volumetric spore traps proved to be more reliable given the fixed air volume of air in each trap. This allowed for a direct calculation of pollen and spore concentrations [24]. Passive traps do not need pumps and usually work under the principles of Fick's law of diffusion, so they have the advantages of being very portable, low maintenance, simple and inexpensive to construct and use. Such tradeoffs are common in monitoring studies, so the investigator must prioritize and optimize with respect to the required data quality

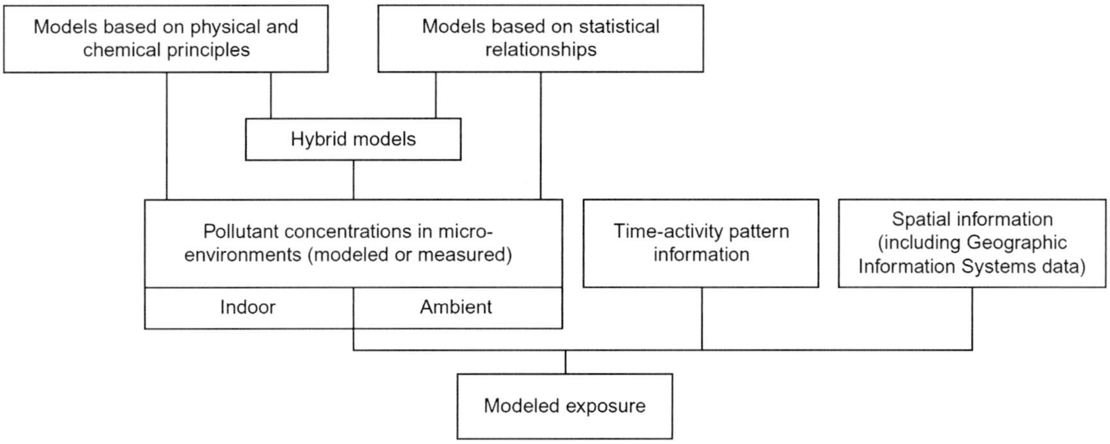

FIGURE 10.16
Measurement and modeling steps in environmental exposure assessments.
Source: Includes information from US National Research Council (1990). Chapter 6: Models. In: Committee on Advances in Assessing Human Exposure to Airborne Pollutants and Committee on Geosciences, Environment, and Resources. *Human Exposure Assessment for Airborne Pollutants: Advances and Applications*. National Academies Press, Washington, DC.

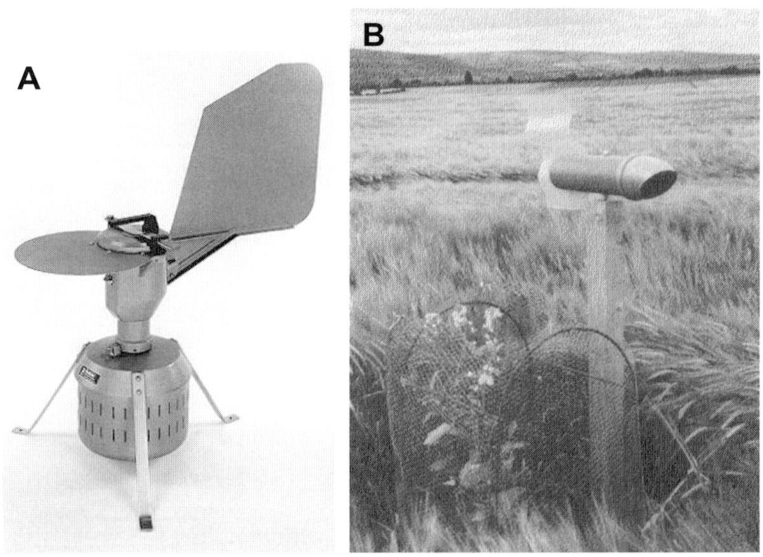

FIGURE 10.17
Traps used to collect airborne pollen: (A) volumetric spore trap; (B) passive spore trap. [*See color plate section*]
Source: M.L. Flannery, F.J.G. Mitchell, T.R. Hodkinson, T. Kavanagh, P. Dowding, S. Coyne and J.I. Burke (2004). *Environmental Risk Assessment of Genetically Modified Crops: The Use of Molecular Markers to Trace Insect and Wind Dispersal of Brassica napus Pollen.* Irish Agriculture and Food Development Authority, Dublin, Ireland.

needed. The results of the two-year study are provided in Figures 10.18 and 10.19.

Monitoring microbial gene flow

Environmental biotechnology approaches microbial survival rates dichotomously. On the one hand, the microbial populations are encouraged not only to survive, but to rapidly grow and thrive in order to break down pollutants. On the other hand, the population dynamics must be limited to the site of contamination, since the effects of gene flow are not known, but may be detrimental, e.g. changing diversity and predator–prey relationships in invaded habitats. Malcolm S. Shields of the University of West Florida and Stephen C. Francesconi of the US EPA [25] aptly captured this dilemma:

> *In virtually all scenarios, the anticipated utility of GEMs (genetically engineered microorganisms) in deliberate release hinges on their continued viability in order to provide a source of enzyme to transform the pollutant compound. Because released cells are alive, they can grow, proliferate, and even spread their genetic material to native organisms. By the very*

(Continued)

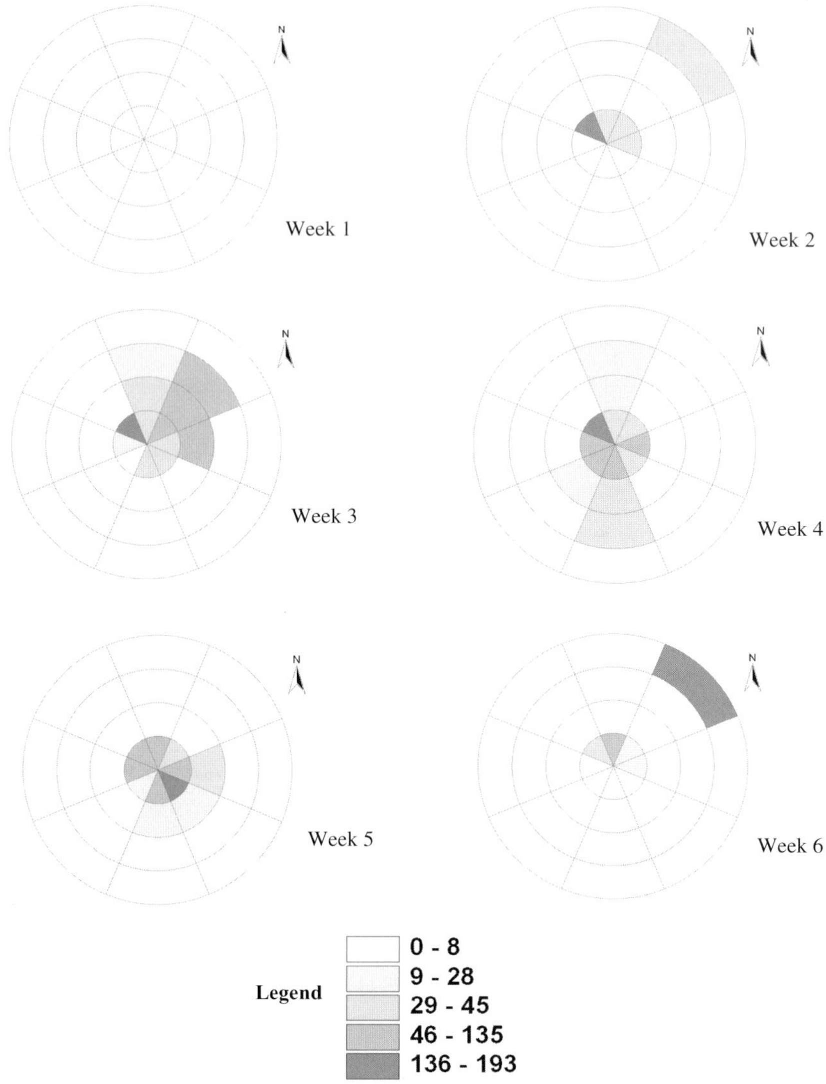

FIGURE 10.18

Mean concentration of Brassicaceae pollen during 2001 (values in pollen grains m^{-3}). Rings indicate locations of traps: 0 m, 50 m, 100 m, and 200 m. A geographic information system, i.e. ArcView GIS 3.2 (Environmental Systems Research Institute Inc.) was used to generate charts. [*See color plate section*]

Source: M.L. Flannery, F.J.G. Mitchell, T.R. Hodkinson, T. Kavanagh, P. Dowding, S. Coyne and J.I. Burke (2004). *Environmental Risk Assessment of Genetically Modified Crops: The Use of Molecular Markers to Trace Insect and Wind Dispersal of Brassica napus Pollen.* Irish Agriculture and Food Development Authority, Dublin, Ireland.

nature of the genetic manipulations, the GEMs are phenotypically distinct, creating the possibility of altering native organisms as well. Many worry that these recombinants may change the microbial ecology where they are applied, and even spread outside the treatment area. While such strains may appear harmless in the laboratory, it is virtually impossible to be certain that they would remain so in the field.

This statement lends credence to the need for reliable measurements of the migration of genetically modified microbes on and adjacent to treatment sites. Such measurements should document changes from baseline conditions and be input into models to predict microbial migrations. Engineered microbes have been modified to include a so-called *hok* (''host killing'') trait. However, this has not proven to be a fail-safe control measure. For example, when hok cells have been grown under selective pressure and inducement the populations experience an initial decline. However, in time the microbial populations recover and return to log growth [26]. Soil bacteria, which are a common source of genetically modified organisms used in environmental treatments, have shown *hok* resistance. Inexplicably, the

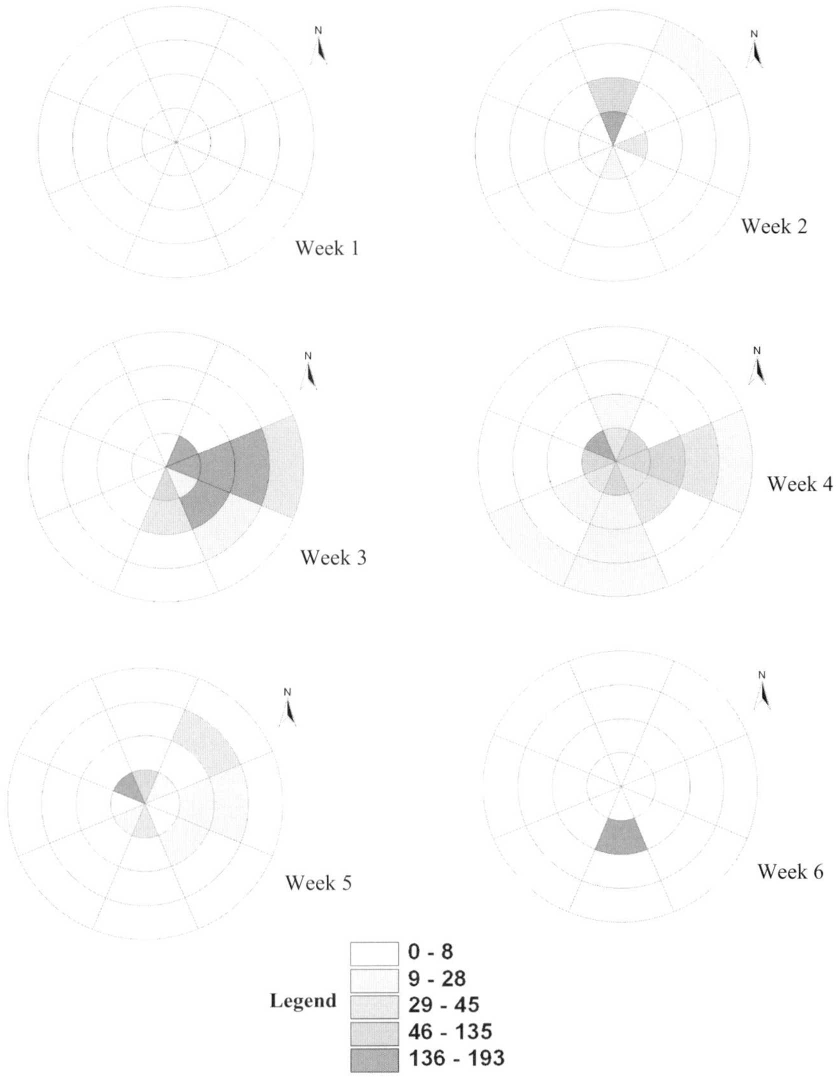

FIGURE 10.19
Mean concentration of Brassicaceae pollen during 2002 (values in pollen grains m^{-3}). Rings indicate locations of traps: 0 m, 50 m, 100 m, and 200 m. A geographic information system, i.e. ArcView GIS 3.2 (Environmental Systems Research Institute Inc.) was used to generate charts. [*See color plate section*]
Source: M.L. Flannery, F.J.G. Mitchell, T.R. Hodkinson, T. Kavanagh, P. Dowding, S. Coyne and J.I. Burke (2004). *Environmental Risk Assessment of Genetically Modified Crops: The Use of Molecular Markers to Trace Insect and Wind Dispersal of Brassica napus Pollen.* Irish Agriculture and Food Development Authority, Dublin, Ireland.

research related to *hok* peptides appears to have ended abruptly in the mid-1990s [27].

The bottom line is that treatment sites using modified microbial populations need not only to measure the changes in concentrations of parent and degradation chemical compounds, but to measure the actual and predict the potential off-site migration of the modified organisms and their genetic material and markers. Ideally, measurements at various locations along transects could be taken within and away from the treatment site. Such monitoring should distinguish between the genetically modified organisms and the related progenitor and native strains, presented in a format similar to that shown in Figure 10.20.

Seminar Questions

How important are the differences between the characteristics of a biological agent versus a chemical agent in terms of monitoring and modeling?

What other materials need to be measured besides the organisms themselves and the biological materials (e.g. spores and pollen) in a risk assessment? ... for routine environmental compliance monitoring?

How may these differ from laboratory containment studies?

What lessons do plant gene flow studies provide for microbial gene flow from bioremediation sites?

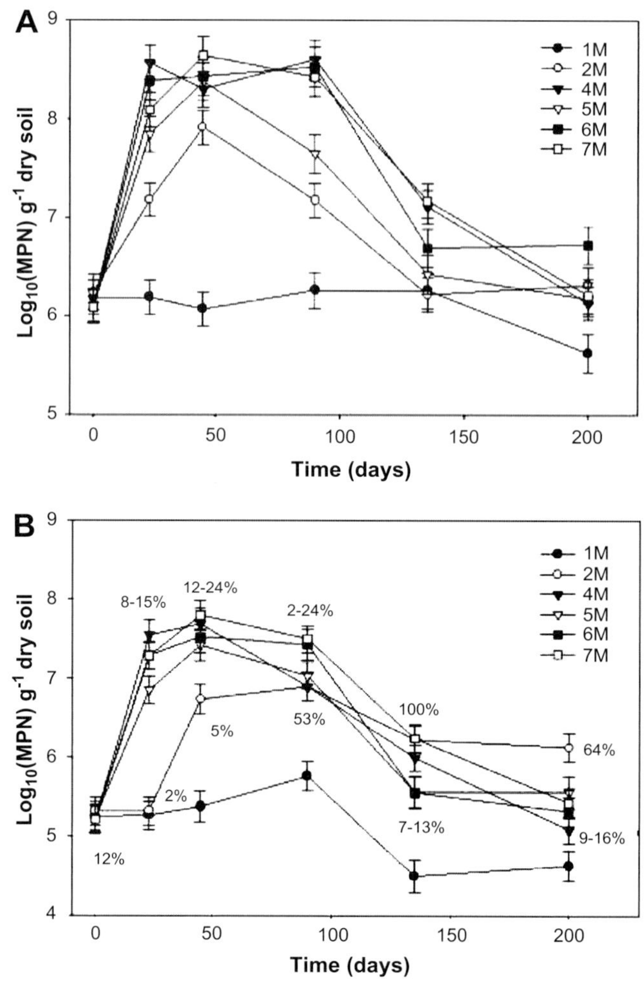

FIGURE 10.20

Example of measurement demonstrating a change in microbial population diversity with time: (A) heterotrophic bacterial populations and (B) polycyclic aromatic hydrocarbon (PAH)-degrading populations in soil treatments over the course of 200 days of incubation. Autoclaved soil presented a microbial population only after 135 days of incubation (10 to 100 most probable number (MPN) g [dry weight] of soil^{-1}). Boldface percentages apply to the PAH-degrading population for the basic treatment, while the other values apply to the PAH-degrading populations for the nutrient treatments and untreated soil.

Source: M. Viñas, J. Sabaté, M.J. Espuny and A.M. Solanas (2008). Bacterial community dynamics and polycyclic aromatic hydrocarbon degradation during bioremediation of heavily creosote-contaminated soil. *Applied Environmental Microbiology* 71(11): 7008–7018.

REVIEW QUESTIONS

How are biotechnology wastes similar to organic chemical wastes in terms of cleanup and pollution prevention? How do they differ, if at all?

At which point in the life cycle are genetic material interventions most effective? What might be a problem in treating a genetically modified organism-related waste using bioremediation?

Why are halogens and metals problematic in thermal treatment processes? What can be done to ameliorate these problems?

How is intervention at the point of use similar to the way agricultural biotechnologies are regulated? Decide whether this is the best way to manage the risks or if and where other interventions may be better.

Consider the biochemodynamics cycles of nitrogen and sulfur. How can these cycles be put to use in addressing potential wastes from bioreactors?

How might nutrient cycling be used to prevent problems with PIPs and GMOs?

How are the "3 Ts" of incineration similar to a bioreactor process? How do they differ?
What are the processes involved in the release of nitric oxide from soil? How can these be used to predict and to characterize microbial activity in the soil?
Find a data set of diseases that may have an environmental etiology. How do these data compare to the criteria in Table 10.5?
Find an environmental data set (e.g. Toxic Release Inventory) and match the releases to the diseases in Question 9. What are the key uncertainties involved in this exercise? Can a Monte Carlo approach assist in this endeavor?

NOTES AND COMMENTARY

1. P.H. Nielsen and W.H. Oxenbøll (2007). Cradle-to-gate environmental assessment of enzyme products produced industrially in Denmark by Novozymes A/S. *International Journal of Life Cycle Analysis* 12 (6): 432–438.
2. Ibid.
3. Obviously, the engineer should be certain that the planned facility adheres to the zoning ordinances, land use plans, and maps of the state and local agencies. However, it behooves all of the professionals to collaborate, hopefully before any land is purchased and contractors are retained. Councils of Government (COGs) and other "A-95" organizations can be rich resources when considering options on siting. They can help avoid the need for problems long before implementation, to say nothing of contentious zoning appeal and planning commission meetings and perception problems at public hearings!
4. Numerous textbooks address the topic of incineration in general and hazardous waste incineration in particular. For example, see C.N. Haas and R.J. Ramos (1995). *Hazardous and Industrial Waste Treatment.* Prentice-Hall, Englewood Cliffs, NJ; C.A. Wentz (1989). *Hazardous Waste Management.* McGraw-Hill, Inc., New York, NY; and J.J. Peirce, R.F. Weiner and P.A. Vesilind (1998). *Environmental Pollution and Control.* Butterworth-Heinemann, Boston, MA.
5. Biffward Programme on Sustainable Resource Use (2003). *Thermal Methods of Municipal Waste Treatment*; http://www.biffa.co.uk/pdfs/massbalance/Thermowaste.pdf.
6. Federal Remediation Technologies Roundtable (2002). *Remediation Technologies Screening Matrix and Reference Guide*, 4th Edition.
7. A principal source for all of the thermal discussions is: US Environmental Protection Agency (2003). Remediation Guidance Document, EPA-905-B94-003 Chapter 7.
8. V. Lehmann (1998). Bioremediation: A solution for polluted soils in the South? *Biotechnology and Development Monitor*, No. 34, pp. 12–17.
9. A useful resource is the Data Quality Objectives Guidance (US EPA): http://www.epa.gov/quality/qs-docs/g4-final.pdf. It provides guidance for project managers, technical staff, regulators, stakeholders, and others who apply the data quality objective (DQO) process to plan data collection efforts and to develop an appropriate sampling and analytical design to support decision making.
10. US Environmental Protection Agency, D. Crumbling (2001). Clarifying DQO Terminology Usage to Support Modernization of Site Cleanup Practice, EPA 542-R-01-014.
11. For sampling and analyzing dioxins and furans in soil and water, a good place to start is US Environmental Protection Agency (1994). "Method 1613," Tetra-through octa-chlorinated dioxins and furans by isotope dilution HRGC/HRMS(Rev. B). Office of Water, Engineering and Analysis Division, Washington, DC; as well as US Environmental Protection Agency (1994). "RCRA SW846 Method 8290," Polychlorinated dibenzodioxins (PCDDs) and polychlorinated dibenzofurans (PCDFs) by high resolution gas chromatograph/high resolution mass spectrometry (HRGC/HRMS). Office of Solid Waste, Washington, DC. For air, the best method is the PS-1 high-volume sampler system described in US Environmental Protection Agency (1999). "Method TO-9A" in Compendium of Methods for the Determination of Toxic Organic Compounds in Ambient Air, Second Edition, EPA/625/R-96/010b.
12. US Environmental Protection Agency (2002). Guidance for the Data Quality Objectives Process, EPA QA/G-4, EPA/600/R-96/055, Washington, DC.
13. The source for this section is S. Kimbrough, D. Vallero, R. Shores, A. Vette, K. Black and V. Martinez (2008). Multi-criteria decision analysis for the selection of a near road ambient air monitoring site for the measurement of mobile source air toxics. *Transportation Research, Part D: Transport and Environment* 13 (8): 505–515.
14. J. Malczewski (1999). *GIS and Multicriteria Decision Analysis.* J. Wiley & Sons, New York, NY; and V.R. Sumathi, U. Natesan and C. Sarkar (2009). GIS-based approach for optimized siting of municipal solid waste landfill. *Waste Management* in Press.
15. ESRI (1995). *Understanding GIS: The Arc/Info Method*, 3rd Edition. Redlands, CA.
16. US Environmental Protection Agency (2003). Test Methods: Frequently Asked Questions: http://www.epa.gov/cgi-bin/epaprintonly.cgi.
17. Australian Department of Health and Ageing (2002). Environmental Health Risk Assessment: Guidelines for Assessing Human Health Risks from Environmental Hazards.
18. See M. Ekhtera, G. Mansoori, M. Mensinger, A. Rehmat and B. Deville (1997). Supercritical fluid extraction for remediation of contaminated soil. In: M. Abraham and A. Sunol (Eds), *Supercritical Fluids: Extraction and Pollution Prevention.* ACSSS, Vol. # 670, pp. 280–298, American Chemical Society, Washington, DC.

19. Developing research in the area of FISH applications to the microbial populations in water and soil includes G.A. Kowalchuk, J.R. Stephen, W. De Boer, J.I. Prosser, T.M. Embley and J.W. Woldendorp (1997). Analysis of *B*-proteobacteria ammonia-oxidising bacteria in coastal sand dunes using denaturing gradient gel electrophoresis and sequencing of PCR amplified 16S rDNA fragments. *Applied Environmental Microbiology* 63: 1489–1497; W.R. Manz, R. Amann, M. Wagner and K.-H. Schleifer (1992). Phylogenetic oligonucleotide probes for the major subclasses of Proteobacteria: problems and solutions. *Systematic and Applied Microbiology* 15: 593–600; B. Nogales, E.R.B. Moore, E. Llobet-Brossa, R. Rossello-Mora, R. Amann and K.N. Timmis (2001). Combined use of 16S ribosomal DNA and 16S RNA to study the bacterial community of polychlorinated biphenyl-polluted soil. *Applied and Environmental Microbiology* 67 (4): 1874–1884; M. Wagner, G. Rath, H.-P. Koops, J. Flood and R. Amann (1996). In situ analysis of nitrifying bacteria in sewage treatment plants. *Water Science and Technology* 34 (1–2): 237–244.
20. See A. Finkel (1990). *Confronting Uncertainty in Risk Management: A Guide for Decision-Makers.* Center for Risk Management, Resources for the Future, Washington, DC.
21. Developing research in the area of nitric oxide emissions from soil includes F. Chase, C. Corke and J. Robinson (1968). Nitrifying bacteria in the soil. In: T.R.G. Gray and D. Parkinson (Eds), *Ecology of Soil Bacteria.* University of Liverpool Press, Liverpool; H. Christensen, M. Hansen and J. Sorensen (1999). Counting and size classification of active soil bacteria by fluorescence in situ hybridization with an rRNA oligonucleotide probe. *Applied Environmental Microbiology* 65 (4): 1753–1761; I. Galbally (1989). Factors controlling NO emissions from soils. In: M. O. Andreae and D. S. Schimel (Eds), *Exchange of Trace Gases Between Terrestrial Ecosystems and the Atmosphere: The Dahlem Conference.* John Wiley, New York, NY; S. Jousset, R. Tabachow and J. Peirce (2001). Nitrification and denitrification contributions to soil nitric oxide emissions. *Journal of Environmental Engineering* 127 (4): 322–328; J. Peirce and V. Aneja (2000). Laboratory study of nitric oxide emissions from sludge amended soil. *Journal of Environmental Engineering* 126 (3): 225–232; and D. Rammon and J. Peirce (1999). Biogenic nitric oxide from wastewater land application. *Atmospheric Environment* 33: 2115–2121.
22. C-Y. Huang, C.C. Lee, F-C. Li, Y-P. Ma and H-J. Su (2002). The seasonal distribution of bioaerosols in municipal landfill sites: a 3-yr study. *Atmospheric Environment* 36 (27): 4385–4395.
23. M.L. Flannery, F.J.G. Mitchell, T.R. Hodkinson, T. Kavanagh, P. Dowding, S. Coyne and J.I. Burke (2004). *Environmental Risk Assessment of Genetically Modified Crops: The Use of Molecular Markers to Trace Insect and Wind Dispersal of Brassica napus Pollen.* Irish Agriculture and Food Development Authority. Dublin, Ireland.
24. Ibid.
25. M.S. Shields and S.C. Francesconi (1996). Molecular techniques in bioremediation. In: L. Crawford and Don L. Crawford (Eds), *Bioremediation: Principles and Applications.* Cambridge University Press, Cambridge, UK.
26. Ibid.
27. National Research Council (2004). *Biological Confinement of Genetically Engineered Organisms.* National Academies Press, Washington, DC.

CHAPTER 11

Analyzing the Environmental Implications of Biotechnologies

PREDICTING AND MANAGING OUTCOMES

The second law of motion tells us that for every action, there is an equal and opposite reaction. The laws of chemistry remind us that what we put into a reaction always leads to a specific product. Thus, science tells us that so long as we understand the variables and parameters of any system, the outcome is predictable. There is the rub; in living systems, we seldom fully understand the variables and parables, so outcomes are never completely predictable. That said, however, there are numerous ways to identify, characterize, and analyze existing and planned applications of biotechnology for their potential environmental impacts. The key to any analytical, evaluative or assessment approach is that it accurately describes the events and situation at hand, considers the important associated and causative factors, fully articulates the decisions that are to be made, and that it be completely objective, so that it results in comprehensive, fair and consistent decision-making that leads to appropriate actions, outcomes and consequences.

Predicting and managing outcomes based on extrapolations and interpolations from the known to the less known goes by many names, such as risk management, science-based decision making, and forward trajectories. Conversely, looking at present conditions to determine the importance of all contributing factors to outcomes also goes by various names, such as root cause analysis, decision trees, deconstruction/reconstruction, and failure analysis.

Cumulative environmental impacts

Simply because a bioengineer is doing a "good" thing does not mean that there are not potential "bad" outcomes to go along with the desired outcome. Even if every bioengineer on a project is doing a good job, there can still be an unacceptable cumulative outcome. In other words, the reductionism is needed, but not sufficient to ensure systematic success.

This lesson was driven home to the author three decades ago, when he worked for a regulatory agency. The agency's mission was to protect the environment. Two of the ways that it did so was by issuing permits to keep releases below a scientifically based pollution level and by paying municipalities to build facilities to treat wastes so that their releases met scientifically based

Environmental Biotechnology

pollution standards. This meant that the effort to clean up surface waters was strongly mandated by the public and supported by science-based regulations, in addition to being well funded. The public entrusted the regulatory agency to protect and to enhance the environment. Aggregately and additively, each facility met an expectable measure of success, but antagonisms within the system led to an overall diminished outcome (i.e. each component was positive but when combined led to some negative impacts).

The author was responsible for writing environmental impact statements (see Chapter 1 and Appendix 1); and such statements are designed to take a systematic view. One might assume that this agency's attitude toward environmental assessment would be better than that of other agencies that had missions other than protecting the environment. That may have been so, but when it came to assessing this environmental agency's own projects, the agency was often just as project-oriented and seemingly myopic, especially when it needed to analyze possible downsides of its own actions in an objective manner. Some at the time may have characterized the situation that "perception was reality." That is, by definition, since the agency had "environmental" in its name, it *had to be doing good things for the environment* and could never do bad things when it came to environmental protection. Experience shows this was not the case. Indeed, the overall objectives of writing the permits and building treatment plants were noble and, and on the whole, necessary. Nevertheless, these actions were not without their own environmental implications.

The difference between the reductionist and systems view is well illustrated by site-specific needs and cumulative impacts brought on by revisions to the Clean Water Act (the Federal Water Pollution Control Act Amendments of 1972) in the United States. For example, heat exchange and balances were changing conditions of receiving water bodies. In fact, in the Midwestern and Western regions of the United States, the value of a fishable stream was directly related to water temperature. This is because water temperature is directly proportional to dissolved oxygen content, which is a limiting factor of the type of fish communities that can be supported by a water body (see Tables 11.1 and 11.2). For example, a trout (*Salma, Oncorhynchus and Salvelinus spp.*) stream is a highly valued resource that is adversely impacted if mean temperatures increase. Rougher, less valued fish (e.g. carp, Order Siluriformes) can live at much lower ambient water body temperatures than can salmon, trout, and other cold water fish populations.

Cumulative heat exchange provides a useful example of the need for a systematic approach to evaluate potential outcomes. The reasons behind whether an environment is supportive or hostile to organisms may be direct or indirect. The aforementioned relationship between increased temperatures in surface waters and game fish populations provides an illustrative example. The added heat may directly stress the game fish population. That is, the fish simply cannot tolerate higher temperatures.

The stress may also be indirect, as the stress is derived as a result of the increased temperature, such as the resulting drop in dissolved oxygen (DO) concentrations in the water (see Figures 11.1 and 11.2), which deems the water body hostile to the fish. Even this derived stress is uneven. For example, the adult fish may do well at the reduced DO levels, but their reproductive capacities decreases. Or, the reproduction is not adversely affected, but the survival of juvenile fish can be reduced.

The stress may be further derived as a type of "second-order" change. For example, the increased temperature leads to greater concentrations of dissolved metals (see Figure 3.23), since aqueous solubility is directly proportional to temperature. Thus, a number of these newly formed metallic compounds are not only more available since they are in solution, but they can also be highly toxic. This is a first-order change. However, the reduced DO in combination with the increased concentrations of dissolved metals makes for redox conditions whereby the metals may be reduced. These reduced metal species can be particularly toxic to fish and other aquatic fauna. The organometallic species that are formed may also have an increased potential to bioaccumulate and biomagnify in the food chain, meaning they will find their way to human

Table 11.1 **Relationship between water temperature and maximum dissolved oxygen (DO) concentration in water (at 1 atm)**

Temperature (°C)	DO (mg L^{-1})	Temperature (°C)	DO (mg L^{-1})
0	14.60	23	8.56
1	14.19	24	8.40
2	13.81	25	8.24
3	13.44	26	8.09
4	13.09	27	7.95
5	12.75	28	7.81
6	12.43	29	7.67
7	12.12	30	7.54
8	11.83	31	7.41
9	11.55	32	7.28
10	11.27	33	7.16
11	11.01	34	7.16
12	10.76	35	6.93
13	10.52	36	6.82
14	10.29	37	6.71
15	10.07	38	6.61
16	9.85	39	6.51
17	9.65	40	6.41
18	9.45	41	6.41
19	9.26	42	6.22
20	9.07	43	6.13
21	8.90	44	6.04
22	8.72	45	5.95

Source: US Environmental Protection Agency (1997). *Volunteer Stream Monitoring Methods Manual.* Report No. EPA 841-B-97-003. Chapter 5. Monitoring and assessing water quality: 5.2. Dissolved oxygen and biochemical oxygen demand. *[See color plate section]*

food supplies. Thus, increased temperature drives numerous aspects in a system. The change in temperature drives the systematic effects. That is, with the resulting decrease in DO, and increasing metal concentrations, both ecosystem function and human health are threatened. The synergistic impact of combining the hypoxic conditions with presence of reduced metal compounds leads to larger problems than if either condition existed alone (see Figure 11.3).

Additional "orders" of stress can continue to occur, which may be predictable and modeled. If the underlying scientific principles are understood, this model may be deterministic. If probabilities of events and outcomes are known (e.g. from laboratory studies or observations of previous scenarios), they can be applied to each stress. From these event probabilities, decision trees and belief networks, such as those described in Chapter 6, can be used to predict the likelihoods of various outcomes. These networks can be perturbed to predict how the outcomes change under various scenarios (e.g. disallowing heated effluents entirely, controlling the effluent to meet certain levels, controlling metal inputs and pH, etc.).

The first law of thermodynamics requires, for example, that allowing heated water to be released in any amount, even the permitted level, would increase the overall temperature of the

Table 11.2 **Normal temperature tolerances of aquatic organisms**

Organism	Taxonomy	Range in temperature tolerance (°C)	Minimum dissolved oxygen (mg L^{-1})
Trout	*Salma, Oncorhynchus* and *Salvelinus spp.*	5–20	6.5
Smallmouth bass	*Micopterus dolomieu*	5–28	6.5
Caddisfly larvae	*Brachycentrus* spp.	10–25	4.0
Mayfly larvae	*Ephemerella invaria*	10–25	4.0
Stonefly larvae	*Pteronarcys spp.*	10–25	4.0
Catfish	Order Siluriformes	20–25	2.5
Carp	Cyprinus spp.	10–25	2.0
Water boatmen	*Notonecta spp.*	10–25	2.0
Mosquito larvae	Family Culicidae	10–25	1.0

Source: Data from Vernier Corporation (2009). Computer 19: Dissolved Oxygen in Water. http://www2.vernier.com/sample_labs/BWV-19-COMP-dissolved_oxygen.pdf; accessed October 19, 2009. *[See color plate section]*

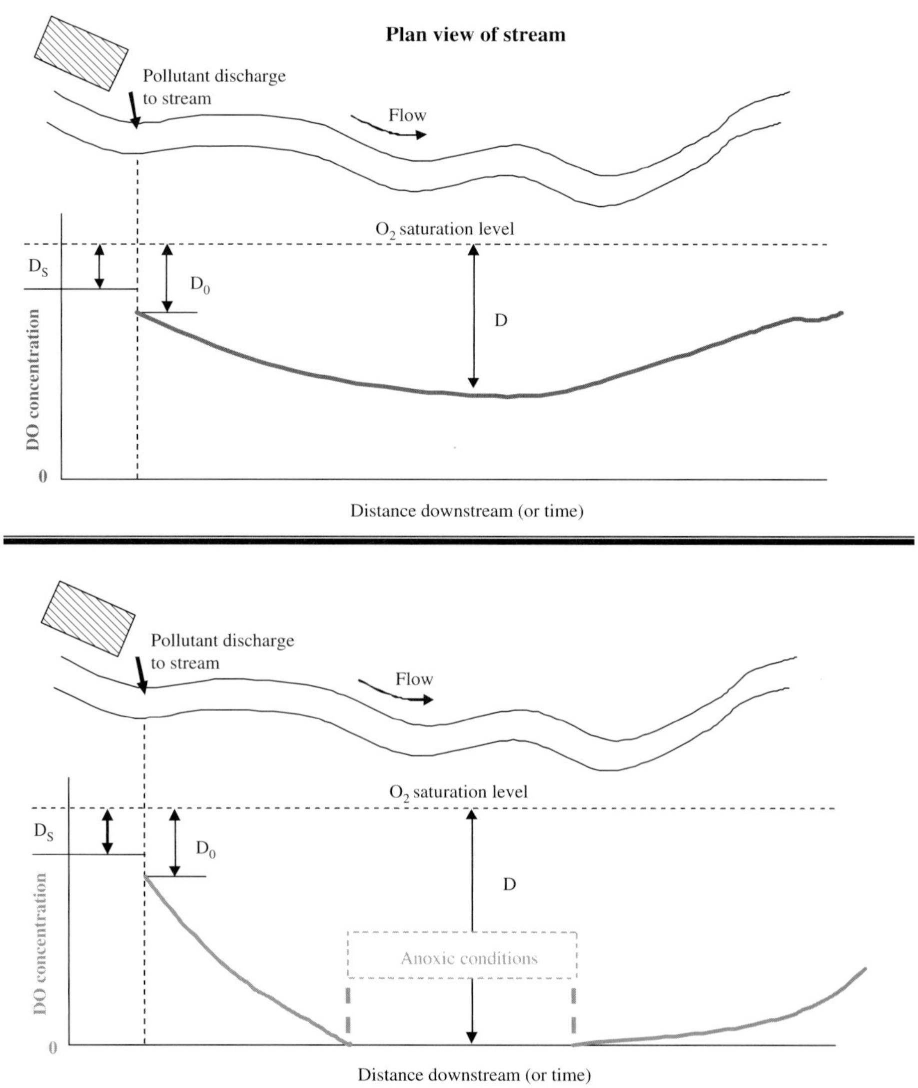

FIGURE 11.1
Drop in dissolved oxygen (DO) downstream from a heated effluent. The increased temperature can result in an increase in microbial kinetics (see Figures 7.5 and 7.9), as well as more rapid abiotic chemical reactions, both consuming DO. The concentration of dissolved oxygen in the top curve remains above 0, so although the DO decreases, the overall system DO recovers. The bottom sags where dissolved oxygen falls to 0, and anaerobic conditions result and continue until the DO concentrations begin to increase. D_S is the background oxygen deficit before the pollutants enter the stream. D_0 is the oxygen deficit after the pollutant is mixed. D is the deficit for contaminant A which may be measured at any point downstream. The deficit is overcome more slowly in the lower curve (less slope) because the re-oxygenation is dampened by the higher temperatures and changes to the microbial system, which means the system has become more vulnerable to another insult, e.g. another downstream source could cause the system to return to anoxic conditions (see Figure 11.2).

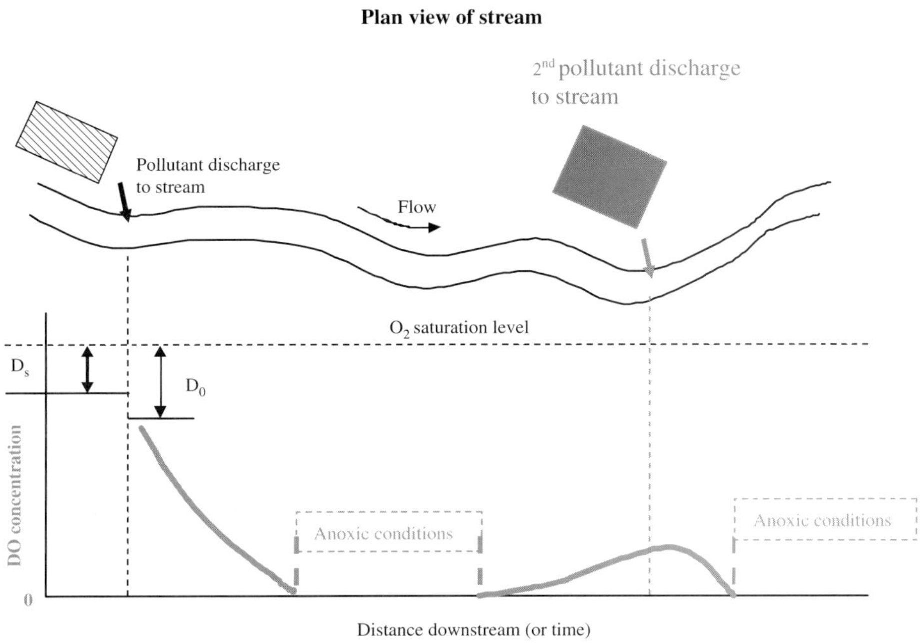

FIGURE 11.2
Cumulative effect of a second heat source, causing an overall system to become more vulnerable. The rate of re-oxygenation is suppressed, with a return to anoxic conditions.

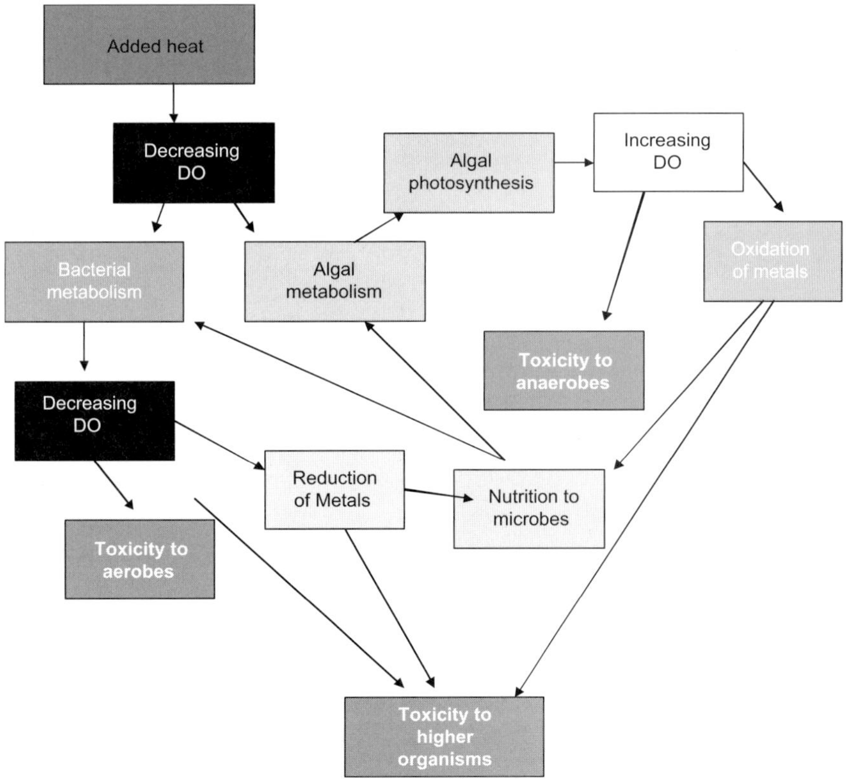

FIGURE 11.3
Adverse effects in the real world usually result from a combination of conditions. In this example, the added heat results in an abiotic response (i.e. decreased DO concentrations in the water). This first-order abiotic effect then results in an increased microbial population. However, the growth and metabolism of the bacteria results in decreasing the DO levels, but the growth of the algae both consume DO for metabolism and produce DO by photosynthesis. Meanwhile a combined abiotic and biotic response occurs with the metals. The increase in temperature increases their aqueous solubility and the decrease in DO is accompanied by redox changes, e.g. formation of reduced metal species, such as metal sulfides. This is also being mediated by the bacteria, some of which will begin reducing the metals as the oxygen levels drop (reduced conditions in the water and sediment). However, the opposite is true in the more oxidized regions, i.e. the metals are forming oxides. The increase in the concentration of bioavailable metal compounds combined with the reduced DO, combined with the increased temperatures can act synergistically to make the conditions toxic for higher animals, e.g. a fish kill.

receiving stream. Up to the 1970s, every power plant along the major rivers of the United States was releasing heated water to a stream (see Figure 11.4). This meant that the incremental effect of all the permitted releases would lead to a cumulative increase in the temperature, with a possible impact on aquatic life. In the late 1970s, once-though cooling, i.e. letting water pass through turbines and then discharged to adjacent streams, was no longer allowed in US waters (Figure 11.4A). Other cooling systems, e.g. cooling towers and cooling lakes, had to be installed and operated, which meant power plant water systems became more closed, both from a fluid dynamic and thermodynamic perspective (Figure 11.4B).

The National Pollution Discharge Elimination System was the Clean Water Act system under which the state and federal regulars were to control pollution released from point sources. These were issued for both private and public dischargers. Publicly funded wastewater treatment facilities had their own challenges under this program. The good news was that the plants were taking out large amounts of materials that would have found their way to the surface waters. The

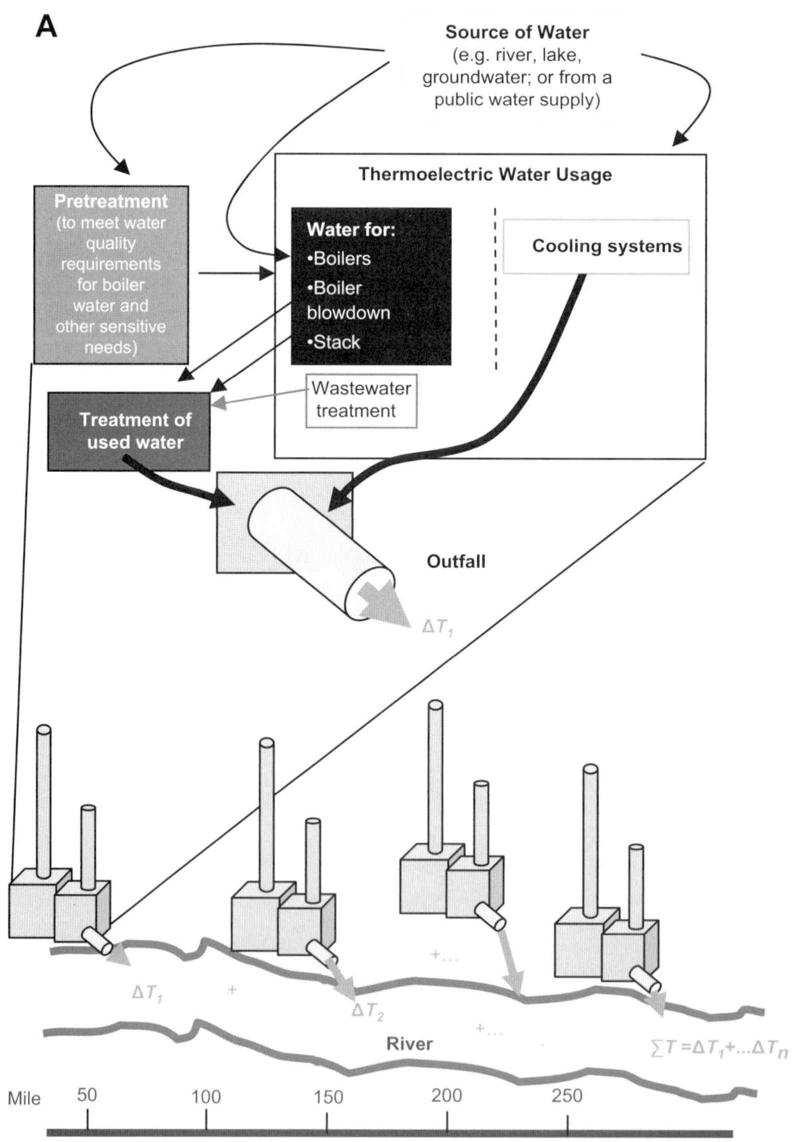

FIGURE 11.4
Difference in cumulative heat contribution to a river using a once-through cooling system (A) versus using a cooling water return system (B). The cumulatively added heat is greatly reduced with the closed water return systems compared to the once-through cooling systems. [*See color plate section*]

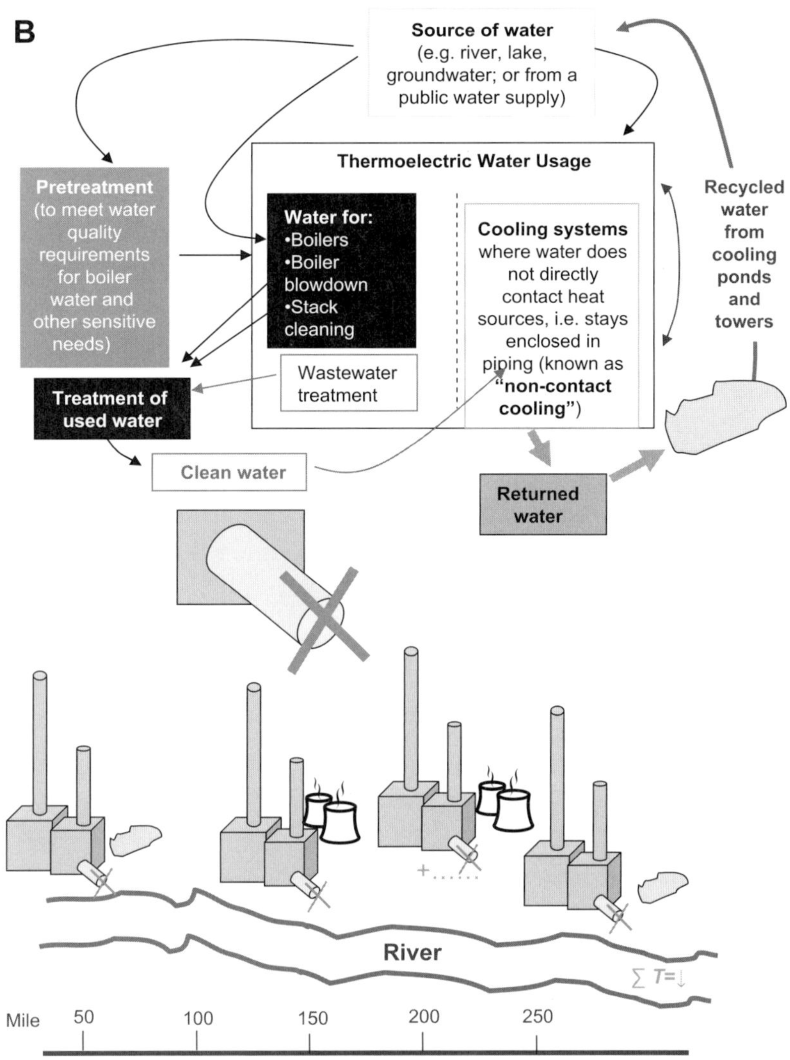

FIGURE 11.4—cont'd

bad news was these materials ended up in sludge (now more euphemistically called "biosolids"), with tons of concentrated chemical and microbial wastes. This does not by any means suggest that the emphases on permits and treatment facilities of the early 1970s were not needed. Indeed, the biosolids/sludge problem remains today, and is a reminder that a problem does not necessarily end with one phase of treatment; in this case, treating the wastewater moves some of the problem to another medium, i.e. sludge. This is known as "cross media transfer."

The lesson is that, no matter how essential a new technology, all of its possible impacts must be considered. As discussed in Chapter 10, the entire life cycle must be considered before a technology can be properly deemed as environmentally acceptable.

As in any scientific analysis, the first step is descriptive. This includes an analysis of all relevant data and any other descriptive materials that will help identify possible events that may lead to other events. As such, this is a practice in chaos. So, first what we need is a complete and accurate telling of the proposed or actual biotechnology and its uses.

Biological processes are often not value-free. That is, the same process under one condition can be a benefit, whereas the same process under a different condition is harmful. Two of the major concerns about biotechnology are the release of genetically engineered organisms and attendant materials into the environment, and the release of chemicals from biotechnological operations. Figure 11.5 demonstrates the similarities between the events that lead to an

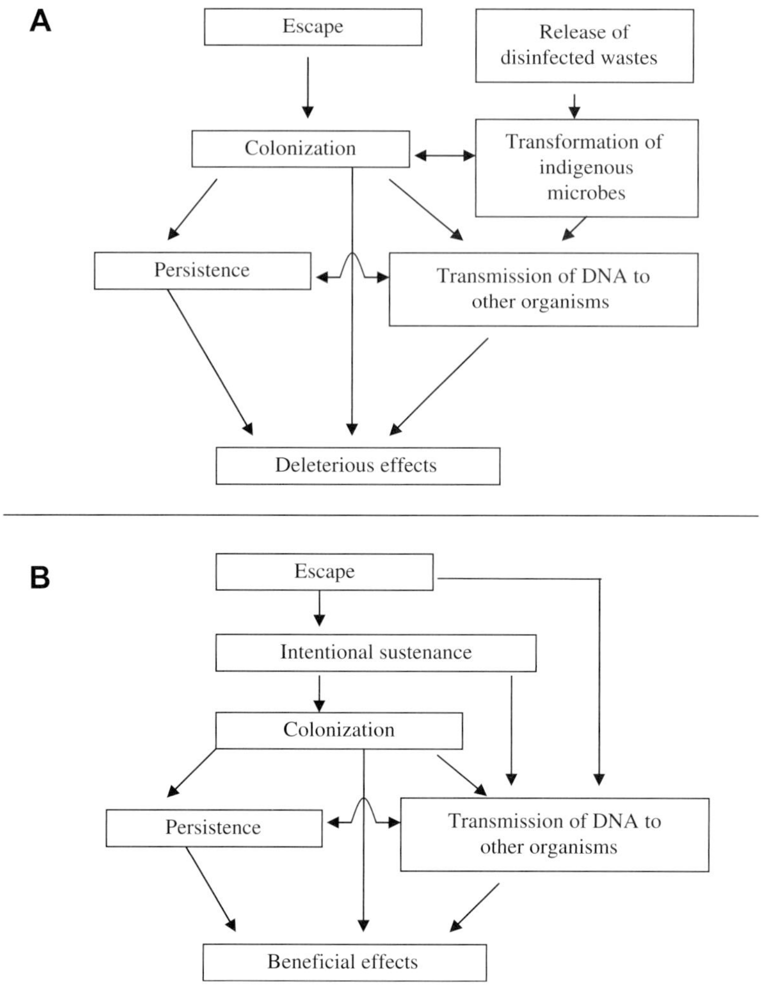

FIGURE 11.5
Critical paths of a microbial disaster scenario (A) versus a bioremediation success scenario (B). In both scenarios, a microbial population (either genetically modified or non-genetically modified) is released into the environment. The principal difference is that the effects in A are unwanted and deleterious, whereas the effects in B are desired and beneficial.

environmental disaster and those that lead to a successful environmental treatment (e.g. bioremediation). The event trees demonstrate that it is not the act of release (escape, if it is unintended) that renders the chain of events as deleterious or beneficial, but the system in its entirety. In fact, the critical path seldom ends at the final steps shown in Figure 11.5, but will have continuing impacts. In the case of a deleterious effect in Figure 11.5(A), additional impacts can ensue, such as if the released microbes change the biodiversity of an ecosystem, and if these microbial populations' genetic material drifts into standing adjacent wild strains of the same microbial species, which in turn change conditions that adversely affect agricultural crops and ultimately lead to threats to the food supply. Or, if the deleterious effect of the microbial release is a virulent form of a bacterium, which not only causes a health effect from those who consume contaminated drinking water, but which may lead to cross-resistant pathogens, so that existing antibiotics become less effective in treating numerous other diseases.

Despite a successful bioremediation project, such as that shown in Figure 11.5(B), other unexpected events can occur after the specific success. For example, genetically modified bacteria have been widely and successfully used to degrade oil spills. If the bacteria, which have a propensity to break down hydrocarbons, do not stop at the spill, but begin to degrade asphalt roads (i.e. the "bugs" do not distinguish between the preferred electron acceptors and donors in an oil spill or asphalt), this is a downstream, negative impact of a successful bioremediation

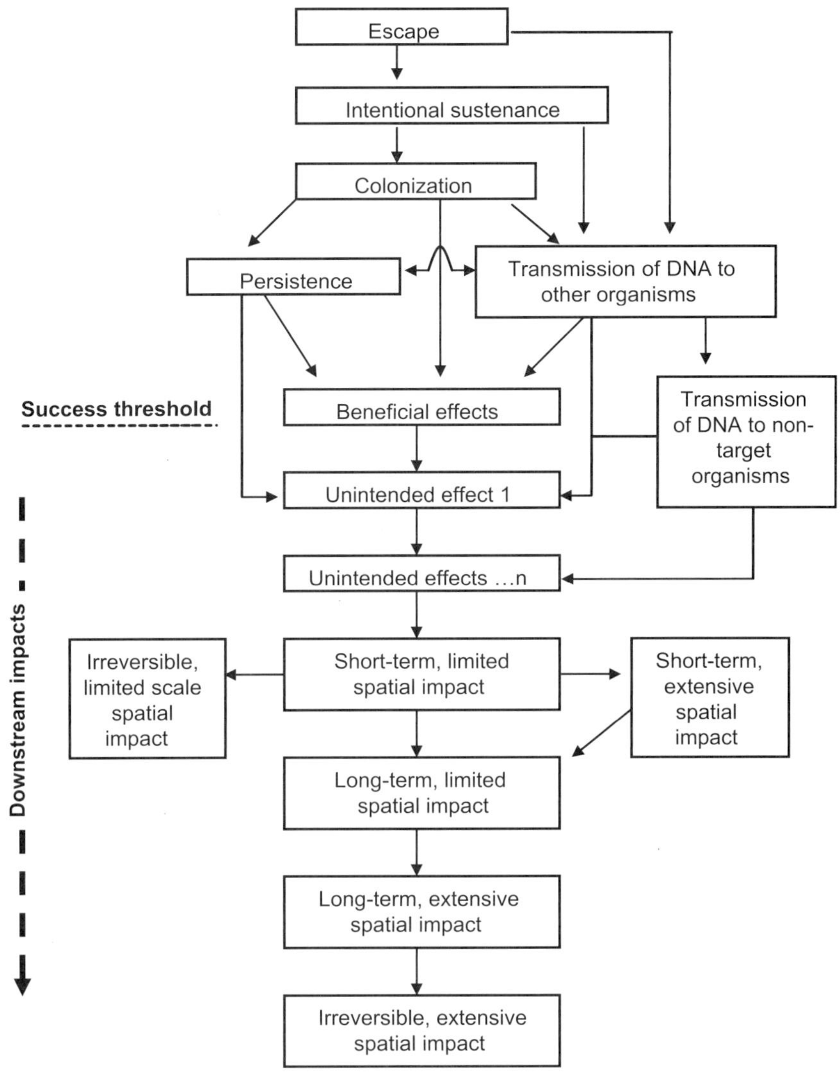

FIGURE 11.6
Same scenario as Figure 11.5(A), but with subsequent or coincidental, unexpected and unremediated events, leading to downstream, environmental or public health impacts.

effort (see Figure 11.6). Further, if the microbes do not follow the usual script, where the next generation are completely sterile, but are able to reproduce and become part of the formerly exclusively natural microbial species, the traits of the population may be altered in unknown ways.

Figure 11.7 provides a similar scenario for chemical (abiotic) releases (treatment, if intentional). This is actually an example of addressing a risk, but at the same time, introducing another, so-called contravening risk. For example, if thousands of people are intentionally exposed to a toxic substance, is this an immoral act? Sometimes, of course, it is. For example, exposing people intentionally to Sarin gas is an act of terrorism; but exposing people to a very toxic pesticide is an act of protecting public health. In the latter scenario, the risks of not applying the pesticide had to outweigh the risks of applying it.

Figures 11.5 through 11.7 illustrate that biological and chemical agents flow systematically under both desired and undesired scenarios. This is crucial to bioengineering and biotechnology since an advance in genetic engineering is not an isolated endeavor. In fact, each genome is a system where the molecules (DNA, proteins, and other biomolecules) interact within the cellular environment. Thus, these interactions and feedback mechanisms within the genome, within the cell, and between the cell and the biotic and abiotic conditions of the organism's environment call for an "ecosystem" perspective. Biological systems are complex.

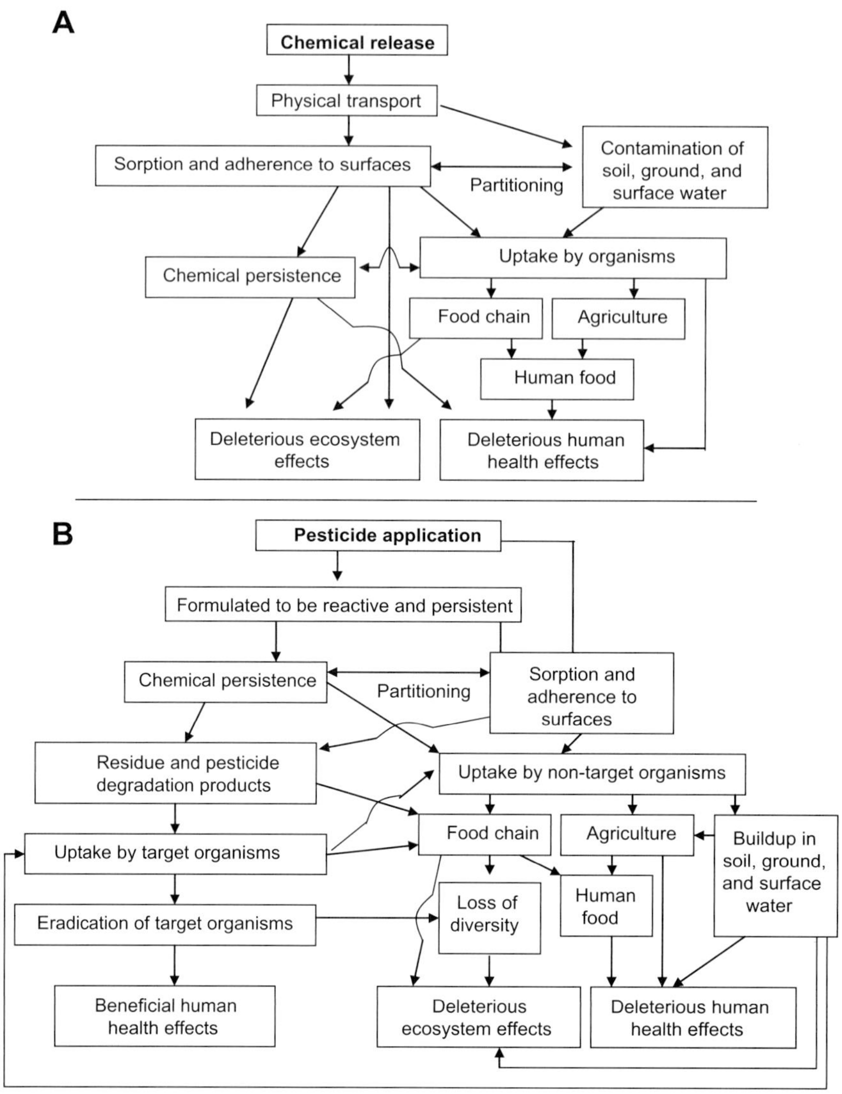

FIGURE 11.7
Critical paths of a chemical disaster scenario (A) versus a public health pesticide application scenario (B). In both scenarios, a chemical compound is released into the environment. The principal difference is that the effects in (A) are unwanted and deleterious, whereas the left side of (B) shows effects that are desired and beneficial, which can be eradication of a vector (e.g. mosquito) that carries disease (e.g. malaria).The same critical path can be applied to herbicides for weed control, rodenticides for rodents that carry disease, and fungicides for prevention of crop damage. The right side of (B) is quite similar to the chemical disaster scenario in (A); that is, the pesticide that would be useful in one scenario is simply a chemical contaminant in another. Examples include pesticide spills (e.g. in dicofol in Lake Apopka, Florida), or more subtle scenarios, such as the buildup of persistent pesticides in sediment for years or decades.

The interactions, modes of action, and mechanistic behaviors are incompletely and poorly understood. Much of what we know about biology at the subcellular level is more empirical and descriptive than foundational and predictive. Thus, there are so many uncertainties about these processes that even a seemingly small change ("tweaking") of any part of the genomic system can induce unexpected consequences [1].

Assessment uncertainties and complexities

The law of unintended consequences should be respected when it comes to predicting possible environmental outcomes from genetic manipulations. Biological agents elicit myriad effects; for example, molds have intricate mechanisms to ward off predators that may produce substances that are toxic (see Discussion Box: Biological Agent: *Stachybotrys*). A slight modification of the transcription could lead to unwelcome results. Making matters worse, bioengineered genes "are typically inserted into random positions in the receiving organism's genome" [2]. This is not exactly the precision that engineers desire. The less certainty about the location, the less control an engineer has over outcomes. The location in the genome is not precisely known, but the traits are being selected empirically, as are other traits of which the biotechnologist may or may not be aware. Some have likened the need for a life cycle view and the need for humility to that of ecosystem failures:

> Inserting genes is similar to ecological practices that we thought we understood well, but which held unexpected consequences, such as introducing industrial chemicals to the environment (consider DDT, PCBs), or such as introducing alien species (consider Purple Loosestrife, Kudzu, Starlings) ... Regardless of our fundamental ignorance of the genetic mechanisms mentioned above, no one has properly studied the ecological and health ramifications of releasing so many GMOs into farms and grocery stores. [3]

DISCUSSION BOX

Biological Agent: *Stachybotrys*

Fungi comprise the kingdom of organisms that includes about 250,000 species, only about 200 of which have been identified as pathogenic [4]. Molds are fungi that live on numerous surfaces, including indoor walls and fixtures, as well as outdoors in soil, on plants, and on detritus. Over 1000 species of molds have been found in indoor environments. Mold growth is usually increased with increasing temperature and humidity under environmental conditions, but this does not mean molds cannot grow in colder conditions. Like other fungi, molds reproduce by producing spores that are emitted into the atmosphere. Living spores are disseminated to colonize growth wherever conditions allow. Most ambient air contains large amounts of so-called "bioaerosols," i.e. particles that are part of living or once-living organisms. In this instance the bioaerosols are live mold spores, meaning that inhalation is a major route of exposure.

Some molds produce toxic substances called mycotoxins. There is much uncertainty related to possible health effects associated with inhaling mycotoxins over a long time period. Extensive mold growth may cause nuisance odors and health problems for some people. It can damage building materials, finishes, and furnishings and, in some cases, cause structural damage to wood. Sensitive persons may experience allergic reactions, similar to common pollen or animal allergies, flu-like symptoms, and skin rash. Molds may also aggravate asthma. Rarely, fungal infections from building-associated molds may occur in people with serious immune disease. Most symptoms are temporary and eliminated by correcting the mold problem, although much variability exists on how people are affected by mold exposure. Infants, children, the elderly, pregnant women and persons with compromised immune systems and allergies are particularly sensitive subpopulations.

Health hazards from exposure to environmental molds and their metabolites relate to four broad categories of chemical and biological characteristics: (1) irritants, (2) allergens, (3) toxins, and rarely (4) pathogens. Risks from exposure to a particular mold species vary depending on a number of factors. Uncertainty is increased with the lack of information on specific human responses to well-defined mold contaminant exposures. In combination, these knowledge gaps make it impossible to set simple exposure standards for molds and related contaminants.

A useful method for interpreting microbiological results is to compare the kinds and levels of organisms detected in different environments. Usual comparisons are indoors to outdoors or complaint areas to areas where no complaints have been made. Specifically, in buildings without mold problems, the qualitative diversity of airborne fungi indoors and outdoors are expected to be similar. On the other hand, dominance of one or a few species of fungi indoors and their absence outdoors may indicate a moisture problem and degraded air quality. Also, the consistent presence of certain fungi species, including *Stachybotrys chartarum, Aspergillus versicolor*, or various *Penicillium* species in counts above background concentrations may indicate the conditions conducive to their growth (i.e. moisture and ventilation problems). Total bacterial levels indoors versus outdoors may not be as useful as with fungi, because bacteria reservoirs exist in both. However, specific strains of bacteria that are present may help in apportioning potential building-related sources [5].

RISK TRADEOFFS

The uncertainties and complexities also indicate that controlling and managing these agents often entails risk tradeoffs, since every one of the scenarios, even the decision not to take any action, has contravening risks. For example, the persistent organic pollutants (POPs) all have helped to meet society's needs, such as disease control (1,1,1-trichloro-2,2-bis-(4-chlorophenyl)-ethane, DDT), food supply (aldrin, dieldrin, hexachlorobenzene), and distribution of electricity (polychlorinated biphenyls, PCBs). But these uses were always accompanied by contravening risks. Disputes between the pros and cons of DDT, for example, not only centered around environmental and public health risks versus the commercial rewards, however, the arguments were between public health risks and public health rewards, e.g. the tradeoff between possible chronic diseases – cancer, endocrine disruption, etc. – and acute diseases – malaria. In addition, people are justifiably concerned that even though the use of a number of pesticides, including DDT, has been banned in Canada and the United States, they may still be exposed by importing food that has been grown where these pesticides are not banned. In fact, Western nations may continue to allow the pesticides to be formulated at home, but do not allow their application and use. But in the long-run (short-term and long-term, extensive spatial impact boxes in Figure 11.5), the pesticide comes back in the important products treated with the domestically banned pesticide. This is known as the "circle of poisons."

Does this analogy hold for genetically modified organisms? Do they return in places that one would not expect or want? Do the chemicals generated by these organisms differ in subtle physical, chemical, and biological characteristics from the abiotically generated chemicals, so that safety and health criteria (material safety data sheets, premanufacture notification data, etc.) may no longer be representative of hazard or risk?

Risks *versus* risks can also come into play. In other words, it is not simply a matter of taking an action, e.g. banning worldwide use of DDT, which leads to many benefits, e.g. less eggshell thinning of endangered birds and less cases of cancer. No, it sometimes comes down to trading off one risk for another. Since there are yet to be reliable substitutes for DDT in treating disease-bearing insects, policy makers must decide between ecological and wildlife risks and human disease risk. As mentioned, since DDT has been linked to some chronic effects like cancer and endocrine disruption, how can these be balanced against expected increases in deaths from malaria and other diseases where DDT is part of the strategy for reducing outbreaks? Is it appropriate for economically developed nations to push for restrictions and bans on products that can cause major problems in the health of people living in developing countries? Some have even accused Western nations of "eco-imperialism" when they attempt to foist temperate climate solutions onto tropical, developing countries. That is, we are exporting fixes based upon one set of values (anti-cancer, ecological) that are incongruent with another set of values of other cultures (primacy of acute diseases over chronic effects, e.g. thousands of cases of malaria are more important to some than a few cases of cancer, and certainly more important than threats to the bald eagle from a global reservoir of persistent pesticides).

Finding substitutes for chemicals that work well on target pests can be very difficult. This is the case for DDT. In fact, the chemicals that have been formulated to replace it have been found either to be more dangerous, e.g. aldrin and dieldrin (which have also been subsequently banned), or much less effective, especially in the developing world (e.g. pyrethroids). Most developing nations are in the tropics and subtropics, where mosquitoes and other disease vectors can be very difficult to treat. This makes finding substitutes for DDT difficult. In temperate climates DDT use has been almost completely replaced by other pesticides, such as the pyrethroids. However, buildings and other structures can be quite different in tropical regions than those in temperate climates. Huts that have been effectively sprayed with DDT often are not effectively treated with pyrethroids and other pesticide formulations. This is likely largely due to the greater chemical reactivity of the pyrethroids compared to DDT. Once the DDT is sprayed onto the mud in these hut structures it remains for long periods of time.

Conversely, sprayed pyrethroids break down much more rapidly. Persistence is a desirable trait in this instance for pesticide efficacy, but would be an undesirable trait in terms of environmental recalcitrance.

The POPs provide abundant lessons about risk tradeoffs. First, the engineer must ensure that recommendations are based upon sound science. While seemingly obvious, this lesson is seldom easy to put into practice. Sound science can be trumped by perceived risk, such as when a chemical with an ominous sounding name is uncovered in a community, leading the neighbors to call for its removal. However, the toxicity may belie the name. The chemical may have very low acute toxicity, has never been associated with cancer in any animal or human studies, and is not regulated by any agency. This hardly allays the neighbors' fears. The engineer's job is not done by declaring that removal of the chemical is not necessary, even though the declaration is absolutely right. The community deserves clear and understandable information before we can expect any capitulation.

Second, removal and remediation efforts are entirely never risk-free. To some extent, they always represent risk shifting in time and space. A spike in exposures is particularly likely during the early stages of removal and treatment. During these early stages, chemical compounds may have been shielded from contact with oxygen, water, microbial populations, and other conditions that had inhibited their release. Also, the substance may have been in a chemical form that is less likely to be released, e.g. low aqueous solubility, high vapor pressure. However, with the changing conditions brought on by removal and treatment, the compound may undergo chemical changes that make it more likely to be released. In other words, had no action been taken, the compound's release would have been lower than what occurred as a result of the cleanup actions. Due in part to this initial exposure, the concept of "natural attenuation" has recently gained greater acceptance within the environmental community. However, the engineer should expect some resistance from the local community when the public is informed that the proposed best solution is to do little or nothing but to allow nature (i.e. indigenous microbes) to take its course (doing nothing could be interpreted as intellectual laziness!).

Third, the mathematics of benefits and costs is inexact. Finding the best engineering solution is seldom captured with a benefit/cost ratio. Opportunity costs and risks are associated with taking no action (e.g. the recent Hurricane Katrina disaster presents an opportunity to save valuable wetlands and to enhance a shoreline by *not developing and not rebuilding* major portions of the Gulf region). The costs in time and money are not the only reasons for avoiding an environmental action. Constructing the new wetland or adding sand to the shoreline could inadvertently attract tourists and other users who could end up presenting new and greater threats to the community's environment.

Arguably, biotechnological solutions, such as the use of genetically modified organisms to produce pesticide substitutes, are even more complicated than dealing with abiotic systems. As such, the bioengineer will need tools to optimize the risk management.

LIFE CYCLE AS AN ANALYTICAL METHODOLOGY

Discussions in the previous chapters have given examples of the benefits of using biology to solve societal problems, with an eye toward possible hazards from these solutions. The life cycle perspective is very valuable in identifying possible problems, even for some very beneficial environmental applications.

Environmental systems consist of intricately interconnected components that are in balance at many levels, i.e. thermodynamic, fluid dynamic, trophic, and physical-chemical-biological cycling. Introducing change, sometimes seemingly small, can have profound impacts on these systems.

This life cycle view also is the first step toward preventing problems. For example, if we consider all of the possible contaminants of concern, we can compare which of these must be avoided completely, which are acceptable with appropriate safeguards and controls, and which are likely to present hazards beyond our span of control. We may also ascertain certain processes that generate none of these hazards. Obviously, this is a preferable way to prevent problems. This is a case where we would be applauded for thinking first "inside the box." We can then progress toward thinking outside the box, or better yet in some cases, get rid of the box completely by focusing on function rather than processes.

REVISITING FAILURE AND BLAME

Sometimes, environmental impacts result from weaknesses in the assessment and management of what should have been manageable risks. Quite commonly, to paraphrase *Cool Hand Luke*, "what we have here is failure to communicate." In deconstructing environmental failures, it is quite common that events were worsened and protracted because of poor risk communications, whether intentional or unintentional. The good news is that bioengineers are becoming more skillful in communications, bolstered by courses in engineering curricula and continuing education.

The short but dramatic history of biotechnology certainly includes examples of human failure, coupled with or complicated by physical realities. Water flows down hill. Air moves in the direction of high to low pressure. A rock in a landslide falls at an increasing rate of speed. Do we blame the water, the air, or the rock? Generally, no; we blame the engineer, the ship captain, the government inspector, or whomever we consider to have been responsible. And we hold them accountable. If we ignore physical, chemical, and biological principles, we have failed in a fundamental component of risk assessment. Risk assessment must be based in sound science. If we fail to apply these principles within the societal context and political, legal, economic, and policy milieu, we have failed in a fundamental component of risk management. And, if we fail to share information and include all affected parties in every aspect of environmental decisions, we have failed in a fundamental component of risk communication. Thus, environmental decisions can be likened to a three-legged stool. If any of the legs is missing or weak, our decision is questionable, no matter how strong the other two legs are.

Failure analysis is an important role of every biomedical and engineering discipline, including environmental science, planning, and engineering. When there is a problem, especially if it is considered to be a disaster, considerable attention is given to the reasons that damages occurred. This is primarily an exercise in what historians refer to as "deconstruction" of the steps leading to the negative outcomes, or what engineers call a "critical path." We turn back time to see which physical, chemical, and biological principles dictated the outcomes. Science and engineering are ready-made for such retrospective analyses. Factor A (e.g. gravity) can be quantified as to its effect on Factor B (e.g. stress on a particular material with a specified strength), which leads to Outcome C (e.g. a hole in the stern of a ship).

The severity of the outcome of an environmental event also affects the actual and perceived failure (see Table 11.3), i.e. the greater the severity of the consequences the more intense the blame for those expected to be responsible. The people thought to have caused it will assume more blame if they are professionals, e.g. engineers and physicians.

Solutions to technical problems should lead to different results than the consequences from the outcome expected from the *status quo*. In fact, throughout this text, there are ample examples of why a biotechnology was needed at all, such as speeding up a reaction to degrade a recalcitrant pollutant or a reaction to provide a needed drug. The biotechnological value to society is better agriculture, more efficacious medicine, greener industries, and unique products that would not be possible without genetic manipulations.

If a biotechnology site is contaminated with a carcinogen, the engineer can select from numerous interventions, all with different outcomes. A prototypical curve for an engineered facility that contains or caps the pollutant may reduce exposure to contaminants and, therefore, reduce health risks in a manner similar to the curve in Figure 11.8, with relatively high risk reduction early or with the initial expenditure of resources, and diminishing returns

Table 11.3 Risk matrix comparing frequency to consequences of a failure event

Frequency	Consequence				
	Disastrous	Severe	Serious	Considerable	Insignificant
Very likely	Unacceptable	Unacceptable	Unacceptable	Unwanted	Unwanted
Likely	Unacceptable	Unacceptable	Unwanted	Unwanted	Acceptable
Occasional	Unacceptable	Unwanted	Unwanted	Acceptable	Acceptable
Unlikely	Unwanted	Unwanted	Acceptable	Acceptable	Negligible
Very unlikely	Unwanted	Acceptable*	Acceptable	Negligible	Negligible

*Note: Depending on the wording of the risk objectives, it may be argued that risk reduction should be considered for all risks with a consequence assessed to be ''severe,'' and thus be classified as ''unwanted'' risks, even for a very low assessed frequency.
Source: S.D. Eskesen, P. Tengborg, J. Kampmann and T.H. Veicherts (2004). Guidelines for Tunnelling Risk Management: International Tunnelling Association, Working Group No. 2. *Tunnelling and Underground Space Technology* 19: 217–237.

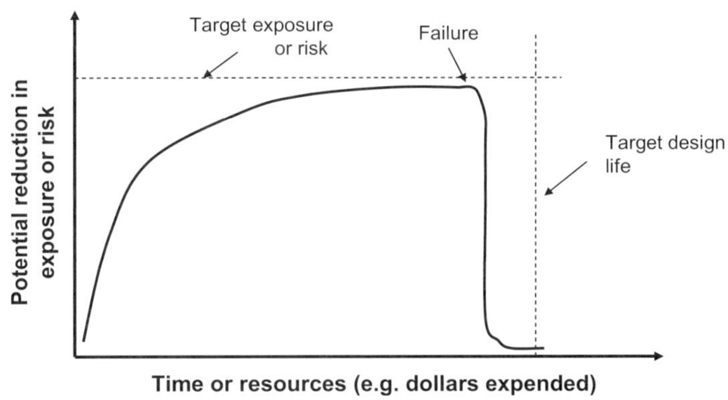

FIGURE 11.8
Hypothetical change in risk in relation to time and resources expended for exposure reduction (e.g. containing a release of a microbe or chemical within a defined perimeter), without any reduction in the amount (mass) of the contaminant. Since the pollutant is still present, there is the potential for a catastrophic failure, followed by an increase in contaminant exposure and elevated risk.
Source: Adapted from National Research Council (2003). *Environmental Cleanup at Navy Facilities: Adaptive Site Management*. Committee on Environmental Remediation at Naval Facilities. National Academies Press, Washington, DC.

thereafter. The exposure or risk reduction is a measure of engineering effectiveness. The figure also depicts a catastrophic failure. This failure does not necessarily have to occur all at once, but could be an incremental series of failures that lead to a disaster, such as the containment and capping of hazardous wastes at the Love Canal, New York, site or the slow hybridization and gene flow from a genetically modified plant into a native population.

So far, biotechnologies have not had to deal with a site-specific problem on the scale of Love Canal, but this infamous case demonstrates the difficulty in labeling actions and decisions as distinctly good or bad. Actions, including some environmentally irresponsible ones, were taken. Among them was the burial of wastes in the first place and then capping the landfill. The outcome of these mistakes was not immediately apparent. Eventually, however, the failures of these engineered systems became obvious in terms of health endpoints, e.g. birth defects, cancer, and other diseases, as well as measurements of contamination in the air, water, and

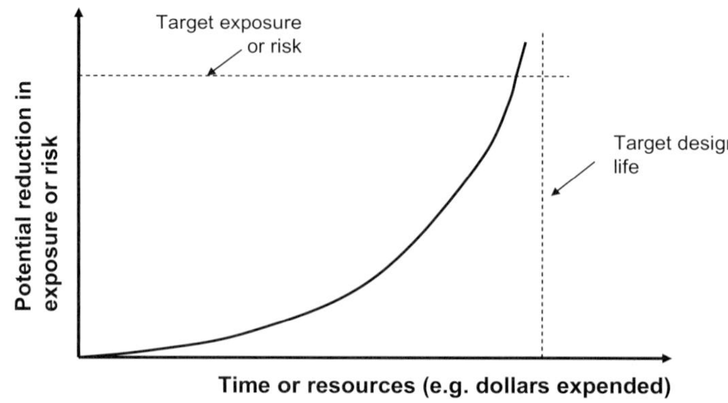

FIGURE 11.9

Hypothetical change in risk in relation to time and resources expended for exposure reduction from an aggressive cleanup action. In this depiction, the cleanup achieves the targeted risk reduction (e.g. less than one additional cancer per 10,000 population, i.e. cancer risk = 10^{-4}), within the specified project life (i.e. target cleanup date).
Source: Adapted from National Research Council (2003). *Environmental Cleanup at Navy Facilities: Adaptive Site Management.* Committee on Environmental Remediation at Naval Facilities. National Academies Press, Washington, DC.

soil. Whether or not the facilities reach catastrophic failure, the curve becomes asymptotic, i.e. virtually no additional risk reduction, even with drastically increased costs. The target design life for persistent chemical and nuclear wastes can be many decades, centuries, even millennia. Any failure before this target is a design failure.

The design life for containment of microbes is until they can no longer move into unwanted areas. This is when all of the genetically modified strains have died and their spores and genetic material are no longer viable when favorable conditions return.

Another possible situation is where aggressive measures are taken, such as treating any biotechnology's released contaminants where they are found (i.e. *in situ*), such as pump and treat for volatile organic compounds (VOCs) or chemical oxidation of dense non-aqueous phase liquids (DNAPLs) like phthalates (see Figure 11.9). The actual relationship of risk reduction with time and expended resources varies according to a number of factors, such as recalcitrance of the contaminant, ability to access the pollutant (e.g. in sediment or groundwater), matching the treatment technology to the pollution, microbial and other biological factors, and natural variability, such as variability in meteorological and hydrological conditions (see Curves A and B in Figure 11.10).

Problems can result if the life of a project is shorter than what is required by the environmental situation. For example, "high maintenance" engineering solutions may provide short-term benefits, i.e. rapid exposure reduction, but when the project moves to the operation and maintenance (O&M) stage, new risks are introduced (see Curve C in Figure 11.10). This is particularly problematic when designing environmental solutions in developing countries or even in local jurisdictions with little technical capacity. For example, if the local entities must retain expensive human resources or "high-tech" programs to achieve environmental and public health protection, there is a strong likelihood that these systems will not achieve the planned results and may even be abandoned once the initial incentives are gone. Certain engineering and environmental *pro bono* enterprises have recognized this and encourage "low-tech" systems that can be easily adopted by local people.

APPLYING KNOWLEDGE AND GAINING WISDOM

Technical analyses require not only knowing how to solve problems, but also having the wisdom to decide when conditions warrant one solution over another and where one solution is workable and another is not. For example, the engineer is called upon to foresee which, if any, of the curves in Figure 11.10 applies to the situation at hand. Intuition has always been an asset for environmental engineers, and its value is increasing. The term "intuition" is widely

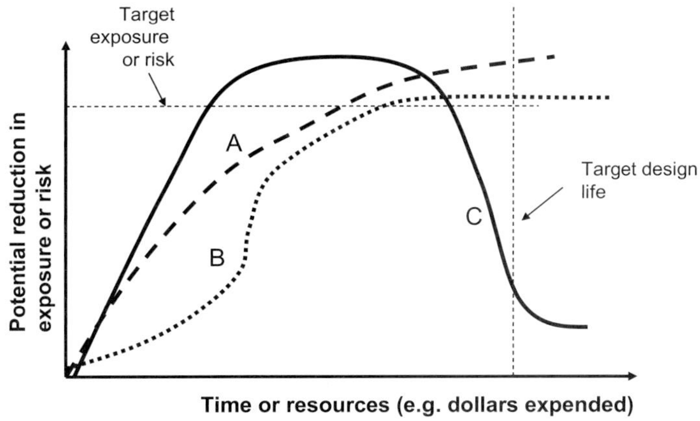

FIGURE 11.10
Hypothetical change in risk in relation to time and resources expended for exposure reduction from various actions, including natural attenuation (i.e. allowing the microbial populations to acclimate themselves to the pollutant and, with time, degrading the contaminations). For example, Curve A could represent an *in situ* treatment process. Curve B may represent natural attenuation, which lags the engineered approach, but the rate of biodegradation increases as the microbial populations become acclimated. Curve C is a situation where controls are effective up to a point in time, but the effectiveness of the risk reduction approach plateaus and then becomes increasingly ineffective. The risk increases either due to physical limits of the treatment system, e.g. in bioremediation operations that pull in water from other aquifers that may be polluted with substances toxic to the microbes, or when treatment technologies are high maintenance.
Source: Adapted from National Research Council (2003). *Environmental Cleanup at Navy Facilities: Adaptive Site Management.* Committee on Environmental Remediation at Naval Facilities. National Academies Press, Washington, DC.

used in a number of ways, so it needs to be defined here so that we are clear about what we mean by intuition, and more importantly, what engineering intuition is *not*. One of the things that set apart engineers from most other scientists is the way that engineers process information. There are two ways of looking at data to derive information and, one hopes, to gain knowledge. These are deductive and inductive reasoning. When we "deduce," we use a general principle or fact to give us information about a more specific situation. This is the nature of scientific inquiry. We use general theories, laws, and experiential information to provide accurate information about the problem or the situation we are addressing. A classic example in environmental engineering is deducing from a cause to the effect. Low dissolved oxygen levels in a stream will not support certain fish species, so we reason that the fish kill is the result of low O_2. This demonstrates a product of deductive reasoning, i.e. "synthesis." A biotechnological example of deduction may be the extrapolation of possible ecosystem damage (e.g. loss of microbial diversity in detritus) due to gene flow from a genetically modified bacterium used to treat a waste in another sector of the ecosystem.

Biologists, engineers, and other technical professionals also engage in inductive reasoning or "analysis." When we induce, we move from the specific to the general and from the effect to the cause. We attribute the fish kill to the low dissolved oxygen levels in a stream that results from the presence of certain substances that feed microbes that, in turn, use up the O_2. We conduct experiments in microcosms that allow us to understand certain, well-defined and well-controlled aspects of a system. We induce larger meaning from these observations. Thus, inductive reasoning helps to form larger principles beyond the data derived from the collective scientists' specific studies. The new data either reinforce existing knowledge or call for new interpretations.

The peril of induction is that any conclusion must be limited [6]. For example, a biotechnological safety experiment may show a direct relationship between an independent and dependent variable, but one does not know just how far to extend the relationship beyond the controlled environment of the laboratory. Could the product or process be deemed "safe" prematurely? We may show that increasing X results in growth of Y, but what happens in the presence of A, B, C, and Z? This is why biotechnologists must overcome their advocacy of a particular process, even though it has great potential for good, and must be arbiters between the stated benefit versus possible implications in time and space in real-world settings.

So, like other scientists, biotechnologists build up a body of information and knowledge from deductive and inductive reasoning. They must rigorously apply scientific theory (deduction) and extend specific laboratory and field results (induction). Over time, the engineer's comfort level increases. To observe the decision making of a seasoned engineer might well lead to the conclusion that the engineer is using a lot of "intuition." Bioengineers, for example, learn about how their designs and plans will work in two ways:

Their formal and continuing education, i.e. what others tell them; and
What they have experienced personally.

The bioengineer learns both subject matter, i.e. "content," and processes, i.e. "rules." The scientific and practical content is what each engineer has learned about the world. Facts and information about matter and energy and the relationships between them are the content of engineering. Rules are the sets of instructions that each engineer has written (literally and figuratively) over time of how to do things [7].

The accumulation of content and rules over one's academic experience and professional practice leads to intuition. Thus, intuition can be explained as the lack of awareness of why or how professional judgments have come to be. Kenneth Hammond [8], a psychologist who has investigated intuitive processes, says that intuition is, in fact, "a cognitive process that somehow produces an answer, solution, or idea without the use of a conscious, logically defensible step-by-step process." So, intuition is an example of something that we know occurs, and probably quite frequently, but it is not deliberative, nor can it be explained explicitly after it occurs. This book's author has argued that it is really a collective memory of the many deductive and inductive lessons learned (content), using a system to pull these together, sort out differences, synthesize, analyze, and come to conclusions (rules). The more one practices, the more content that is gathered and the more refined and tested the rules become.

Thus, the right solution in one instance may be downright dangerous in another. Or, as the National Academy of Engineering puts it, "engineering is a profoundly creative process" [9]. However, like all engineers, bioengineers must always design solutions to problems within constraints and tolerances called for by the problem at hand. In environmental risk decision making, this is a balance between natural and artificial systems. This balance depends on data from many sources. Good data make for reliable information. Reliable information adds to scientific and societal knowledge. Knowledge, with time and experience, leads to wisdom. Environmental assessment and protection need to include every step in the "wisdom cascade" (see Figure 11.11).

Building a structure such as a bioreactor may be part of the solution to a particular problem. At all times, the solution calls for a process that may or may not require the design and construction of a structure. Certainly, the engineer must design a structure that is the best of the options considered. However, the structure alone is not the solution. It only becomes the solution with proper operation and maintenance (O&M) and when life cycle analysis (LCA) dictates the type of structure *vis-à-vis* the systems of processes, of which the structure is a key component. Indeed, a single process or series of processes may represent the entire solution to the environmental problem, such as instituting recycling or pollution prevention based entirely on "virtual" systems like waste clearinghouses. Such solutions do not require a standalone structure. They do not even very closely follow the traditional concept-to-construction paradigm. The form still follows the desired function, but the "form" may be more intellectual than structural.

This thinking has gained currency in that it is a vital part of sustainable design, which applies to all engineering disciplines, not just environmental engineering. For biotechnologies, it may have

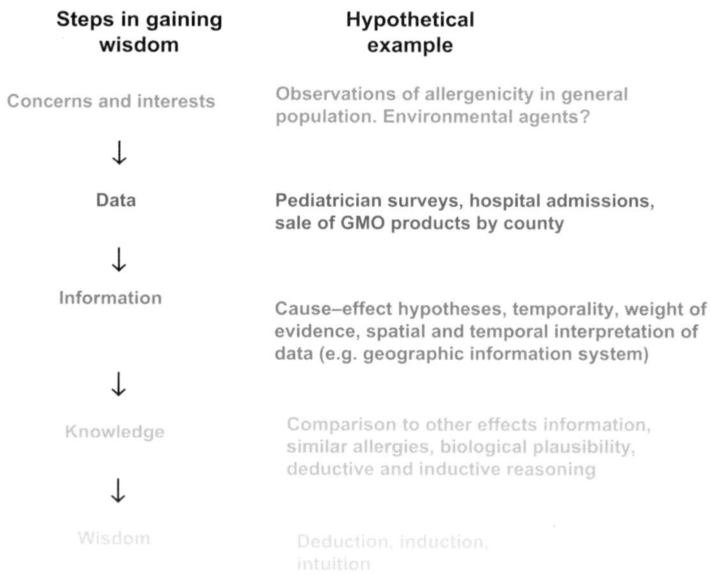

FIGURE 11.11
Value-added chain from data to knowledge and, with experience, professional wisdom. [*See color plate section*]
Source: Adapted from D.A. Vallero (2004). *Environmental Contaminants: Assessment and Control.* Elsevier Academic Press, Burlington, MA.

an added benefit that such materials, including genetically modified organisms, have been previously tested under similar conditions as those being proposed. Standard practice in many engineering disciplines now embodies sustainable design; for example, we now expect engineers to design for the environment (DFE), design for recycling (DFE), and design for disassembly (DFD), as well as to consider ways to reduce the need for toxic chemicals and substances and to minimize the generation of wastes when they conceive of new products and processes [10]. Environmental engineering seldom, if ever, can rely exclusively on a single scientific solution, but is always a choice among many possible solutions dictated by the particular environmental conditions. The design of environmentally sound and socially acceptable biotechnologies calls for the application of all of the biological and physical sciences.

ENVIRONMENTAL ENGINEERING

Throughout the first half of the 20th century, when the field was predominantly considered "sanitary engineering," structural considerations were paramount. Yet, even then, operational conditions had to include chemistry and biology, as well as fluid mechanics and other physical considerations. This amalgam of science grew more complex as those who designed pollution control systems earned the designation of "environmental engineers."

All engineers apply scientific principles. Every engineering discipline to some extent applies chemical principles; certainly the "life science engineers," biomedical and environmental engineers are steeped in chemistry. Most importantly, biomedical and environmental engineers must also account for biology. In the case of environmental engineering, the concern for biology ranges across all kingdoms, phyla, species, subspecies, and stains. Engineers use biological principles and concepts to solve problems, for example bacteria and fungi adapted to treat wastes, algae and macrophytes to generate biofuels, macrophytic flora to extract and degrade contaminants (i.e. phytoremediation) and to restore wetlands, and benthic organisms to help to clean contaminated sediments, including benthic and soil fauna to degrade pollutants, and organisms from all kingdoms as indicators of environmental conditions.

In fact, environmental engineers and scientists use a wide array of organisms as indicators of levels of contamination (e.g. algal blooms, species diversity, and abundance of top predators and other so-called "sentry species") to act as our "canaries in the coal mine" to give early warning about stresses to ecosystems and public health problems. And, arguably most important, environmental engineers study organisms as endpoints in themselves.

Environmental engineers, like all engineers, care principally about human safety, health, and well-being. The particular area of biology that addresses health outcomes and is so important to environmental engineers is known as "toxicology," which deals with the harmful effects of substances on living organisms. Usually, when toxicology is not further specified, it deals with the harmful effects of substances on human beings. However, there are toxicological subdisciplines, such as ecotoxicology, which addresses harm to components of ecosystems, and these subdisciplines are further categorized into even more specific fields, such as aquatic toxicology, which is concerned with harm to those organisms living in water.

Interestingly, what has traditionally accounted for the lion's share of biological applications in environmental systems is what is now often called "natural" (e.g. natural attenuation is an example of using extant microbial populations *in situ* to biodegrade hazardous wastes in time). This is presently contrasted with "engineered" systems. In these, natural microbes may not degrade substances rapidly enough, if at all, so their or other microbes' genetic material is modified so that they can be introduced to the system to improve efficiency. As we have discussed in previous chapters, such technologies are welcomed so long as all of the physical and biological factors have been appropriately considered.

SCIENCE AS A SOCIAL ENTERPRISE

So, then, how is it possible to appropriately consider *all* of the factors that may lead to the desired outcome, without introducing the possibilities of negative outcomes? First we must remember that the purpose of scientific research is to understand natural systems and add to the knowledge about nature. This means that a scientist first believes that something is going to add to the knowledge base about some area of scientific endeavor. Engineering scientists would probably add that the scientific knowledge would address some societal need.

The decision to study a particular aspect of a biological system is itself evidence that a least one person, i.e. that particular scientist, believes it is important. In this way, science is a social enterprise. The reason we know more about many aspects of the environment today is that the scientific community has decided or been forced to decide to give attention to these matters [11].

Engineers have devoted entire lifetimes to ascertaining how a specific scientific or mathematical principle should be applied to a given event (e.g. why compound X evaporates more quickly, although compound Z under the same conditions remains on the surface). Such research is more than academic. For example, once we know why something does or does not occur, we can use it to prevent problems (e.g. choosing the right materials and designing a reactor correctly) as well as to respond to problems after they occur. A case in point may be that compound X may not be as problematic in a spill as compound Z if during the event that compound Z does not evaporate in a reasonable amount of time (i.e. Z's vapor pressure $<<$X's vapor pressure). However, compound X may be very dangerous if it is toxic and if people nearby are breathing air contaminated with compound X. In other words, in one scenario the higher vapor pressure is advantageous and in the other it is detrimental. Such factors affect what the US Coast Guard, fire departments, and other first responders should do when they encounter these compounds. The release of volatile compound X may call for an immediate evacuation of human beings, whereas a spill of compound Z may be a bigger problem for fish and wildlife (it stays in the ocean or lake and makes contact with plants and animals).

Thus, when deconvoluting the events that lead to an adverse outcome to determine responsibility and to hold the right people accountable, one must look at events in time and space and how these events react with one another. Biological factors have all of these challenges, and more. For example, genetically modified organisms introduce wholly different and important

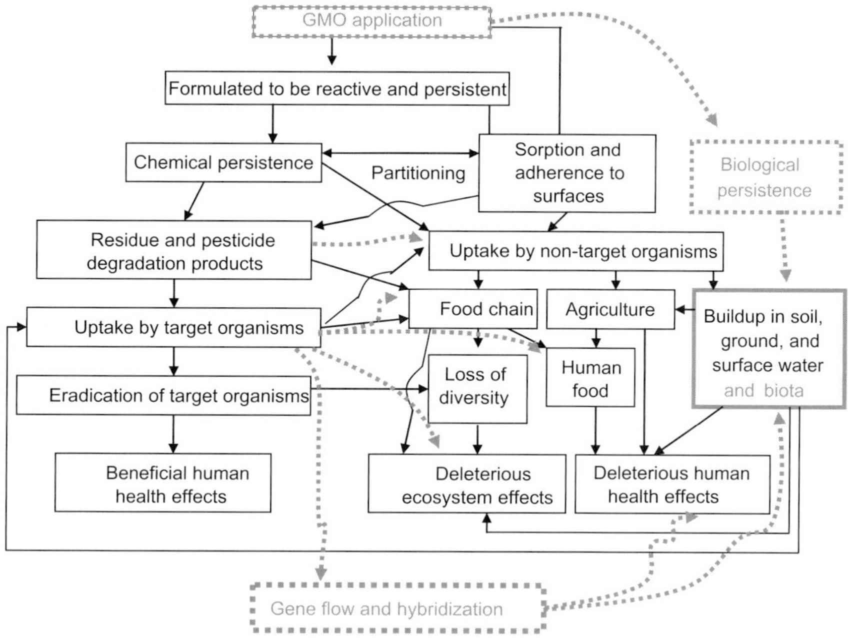

FIGURE 11.12
Added complexities for biological agents (bold dashed arrows) compared to chemical pesticides described in Figure 11.7B.

factors to Figure 11.7B (e.g. varied virulence, reproduction, potential for disease transmission or gene flow). Some of the added complexities for biological agents are depicted in Figure 11.12.

Arguably, the compartment that the majority of engineers and scientists are most comfortable with is the "physical" compartment. This is the one all engineers know the most about, or at least engineers can trust the existence of physical laws in natural systems. From these principles, engineers know how to measure things. Engineers can even use models to extrapolate from what they actually have measured. They can also fill in the blanks between points of measurement by interpolation. Thus, engineers can assign values to important scientific features and can extend the meaning of what they find in space and time. For example, if a bioengineer applies sound methods and uses statistics correctly, measurements of the amount of spilled crude oil that has found its way to the feathers of a few ducks can reveal much about the extent of an oil spill's impact on waterfowl in general. More oil on more feathers reveals even more. And, good models can even begin to divulge how the environment will change with time (e.g. is the oil likely to be broken down by microbes and, if so, how fast?).

This is not to imply that the physical applications in biotechnologies are easily mastered. Such applications are often very complex and fraught with uncertainty. Every event is somewhat unique. There is always inherent and extrinsic variability. But physics is the engineer's domain.

Missions of government agencies, such as the US Department of Homeland Security, the US Environmental Protection Agency, the Agency for Toxic Substances and Disease Registry, the National Institutes of Health, the US Food and Drug Administration, the US Public Health Service, the World Health Organization, and the United Nations Environmental Programme, devote considerable effort in just getting the science right. Universities and research institutes are collectively adding to the knowledge base to improve the science and engineering that underpins the physical principles that themselves underpin public health and environmental consequences from pollutants, whether these be intentional or by happenstance, whether biological or chemical agents, and whether they cause acute or chronic effects.

Systematic bioengineering solutions must consider "anthropogenic" aspects of environmental problems, that is, the portion of the problem that can be attributed to human activities (*anthropo* denotes human and *genic* denotes origin). This consideration includes the *gestalt* of humanity, taking into account all of the factors that society imposes down to the things that drive an individual or group. For example, the anthropogenic contribution to a problem would include the factors that led to a ship captain's failure to stay awake. However, it must also include why the fail-safe mechanisms did not kick in. That is why, for example, standard operating procedures for genetically modified organism containment need to consider human error.

Anthropogenic failures do have physical factors that drive them: for example, a release valve may have rusted shut or the alarm's quartz mechanism failed because of a power outage, but there is also frequently a more important human failure, if past disasters are indications. For example, one common theme in many disasters is that the safety procedures are often adequate in and of themselves, but the implementation of these procedures was insufficient. Often, failures have shown that the safety manuals and data sheets were properly written and available and contingency plans were adequate, but the workforce was not properly trained and inspectors failed in at least some crucial aspects of their jobs, leading to horrible consequences.

The bottom line is that an understanding of the physical factors is necessary to understand and to begin to point to solutions to environmental problems, but most certainly this understanding is not sufficient to solve the problem. In this age of specialization in technical professions, one negative side effect is the increased likelihood that no single person can understand all of the physical and human factors needed to prevent a problem. Thus, each individual scientist and engineer may be following best practices and sound science, but if their work is not integrated with the others' there is a strong possibility that the systematic solution will not be realized.

Preventative systems provide an example of the problems of not adequately considering environmental scales and complexities. Such systems are often needed as early warnings and contingencies, so these systems must be tested and inspected continuously. Every step in the critical path that leads to failure is important. In fact, the more seemingly "mundane" the task, the less likely people are to think a lot about it. So, these small details may be the largest areas of vulnerability. One can liken this to the so-called "butterfly effect" of chaos theory, where the flapping of a butterfly's wings under the right conditions in a certain part of the world can lead to a hurricane. One of the adages of the environmental movement is that "everything is connected." A loss of a small habitat can lead to endangering a species and altering the entire diversity of an ecosystem. A seemingly safe reformulation of a pesticide can alter the molecule to make it toxic or even cancer-causing. Preventing an environmental disaster may rest on how well these details are handled.

Many of the cases in engineering failure owe their origin or enlarged effect in part to a failure of fundamental checks and balances. Often, these requirements have been well documented, yet ignored. A lesson going forward is the need to stay vigilant. One of the major challenges for safety and health enterprises is that human beings tend to be alert to immediacy. If something has piqued their interest they are more than happy to devote attention to it. However, their interest drops precipitously as they become separated from an event in space and time. Psychologists refer to this phenomenon as a memory extinction curve. For example, we may learn something, but if we have no application of what we have learned we will forget it in a relatively short time. Even worse, if we have never experienced something (e.g. a real spill, a fire, or leak), we must further adapt our knowledge of a simulation to the actual event. We never know how well we will perform under actual emergency conditions.

ENVIRONMENTAL ACCOUNTABILITY

One's area of responsibility and accountability is inclusive. The buck stops with the professional. The credo of the professional is *credat emptor*, i.e. let the client trust. Environmental and public health professionals are charged with responsibilities to protect the public and ecosystems. When failures occur, the professionals are accountable. When a manufacturing, transportation, or other process works well, the professional can take pride in its success. The professional is responsible for the successful project. That is why we went to school and are highly trained in our fields. We accept the fact that we are accountable for a well-running system. Conversely, when things go wrong, we are also responsible and must account for every step, from the largest and seemingly most significant to those we perceive to be the most minuscule, in the system that was in place. Professional responsibility cannot be divorced from accountability. The Greeks called this *ethike areitai*, or "skill of character." It is not enough to be excellent in technical competence. Such competence must be coupled with trust gained from ethical practice.

One of the difficult tasks in writing and thinking about failures is the temptation to assign "status" to key figures involved in the episodes. Most accounts in the media and even in the scientific literature readily assign roles of villains and victims. Sometimes, such assignments are straightforward and enjoy a consensus. However, often such classifications are premature and oversimplified.

Granted, case studies have their limits, but one can strongly argue that failure has common elements, whether a reactor is nuclear, such as the meltdown in Chernobyl, or chemical, such as arguably the worst reactor disaster on record, the Bhopal toxic cloud in India. Whereas neither was a bioreactor or biotechnological system disaster, there are profound lessons to be learned.

The Bhopal release killed thousands of people and left many more thousands injured; yet there are still unresolved disagreements about which events leading up to the disaster were most critical. In addition, the incident was fraught with conflicts of interest that must be factored into any thoughtful analysis. In fact, there is no general consensus on exactly how many deaths can be attributed to the disaster, especially when trying to ascertain mortality from acute exposures versus from long-term, chronic exposures. Certainly, virtually all of the deaths that occurred within hours of the methylisocynate (MIC) release in nearby villages can be attributed to the Bhopal plant. However, with time, the linkages between deaths and debilitations to the release become increasingly indirect and more obscure. Also, lawyers, politicians, and business people have reasons beyond good science for including and excluding deaths. Frequently, the best one can do is say that more deaths than those caused by the initial, short-term MIC exposure can be attributed to the toxic cloud. But, just how many more is a matter of debate and speculation. One important lesson to environmental biotechnology, however, is that consequences of inadequate design failure can be very protracted and persistent.

LIFE CYCLE APPLICATIONS [12]

The complexity of LCA ranges from cursory attention paid to inputs and outputs of materials and energy (Figure 11.13), to multifaceted decision fields extending deeply into time and space. The latter is preferable for decisions involving large scales, such as the cumulative build-up of greenhouse gases, or those with substantially long-term implications, such as the release of genetically altered microbes into the environment. Complex LCAs are also favored over cursory models when the effects are extensive, such as externalities and artifacts resulting in geopolitical impacts. Biotechnologies may fall into any or all of these three categories.

Recall from the Case Discussion – King Corn or Frankencorn – in Chapter 9, that the decision to increase the use of ethanol as a fuel additive and a reformulated fuel is such a decision. Thus, the recent proclamation by the United States government to increase ethanol's share of refined fuel to 10% by the year 2012 provides a case study of the application of LCA, from both design and pedagogical perspectives. Ethanol had been increasingly touted as an alternative to crude oil-based fuels. This interest has been diverse, with coverage in the national media and in professional and research journals. In his 2007 State of the Union Address US President George W. Bush set a two-part goal:

- Setting a mandatory standard requiring 35 billion gallons of renewable and alternative fuels in the year 2017, which is approximately five times the 2012 target called for in current law. Thus, in 2017, alternative fuels will displace 15% of projected annual gasoline use.

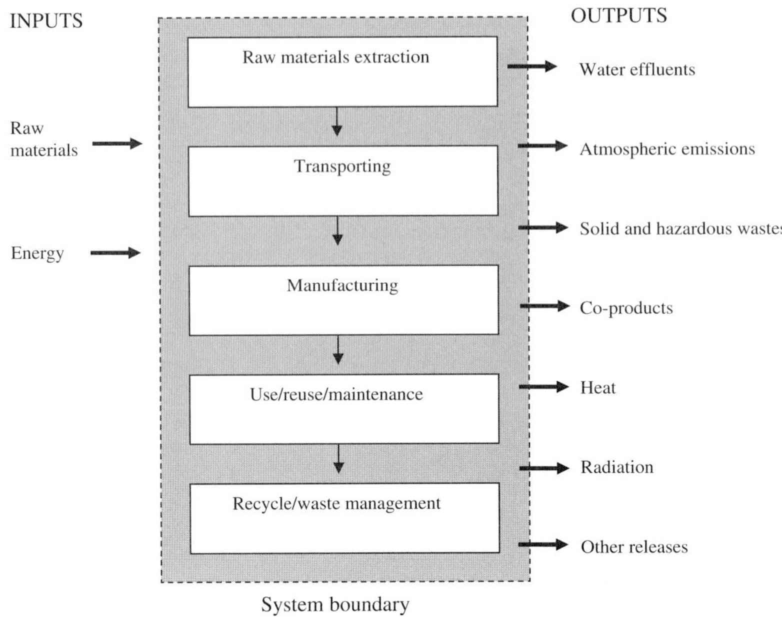

FIGURE 11.13
Life cycle stages of a process must follow the conservation law, with material and energy balances. For biotechnologies, the manufacturing box consists of numerous components from which outputs can result (see Figure 3.10).
Source: US Environmental Protection Agency (1993). *Life Cycle Assessment: Inventory Guidelines and Principles.* EPA/600/R-92/245. Office of Research and Development, Cincinnati, OH.

- Reforming the corporate average fuel economy (CAFE) standards for cars and extending the present light truck. Thus, in 2017, projected annual gasoline use would be reduced by up to 8.5 billion gallons, a further 5% reduction that, in combination with increasing the supply of renewable and alternative fuels, will bring the total reduction in projected annual gasoline use to 20%.

These and other alternative fuel standards have met with skepticism and even dissent. In particular, the viability of ethanol is being challenged from scientific and policy standpoints. Corn-based ethanol is indeed a biotechnology. In fact, since the presidential proclamation, dedicated corn crops and bioreactors in these states have emerged. On the other hand, geopolitical impacts, such as food versus fuel dilemmas, are being raised. Scientific challenges to any improved efficiencies and actual decreases in the demand for fossil fuels have also been voiced. Some have accused advocates of ethanol fuels of using "junk science" to support the "sustainability" of an ethanol fuel system. Notably, some critics contend that ethanol is not even renewable, since its product life cycle includes a large number of steps that depend on fossil fuels. The metrics of success are often deceptively quantitative. For example, the two goals for increasing ethanol use include firm dates and percentages. However, the means of accountability can be quite subjective. For example, the 2017 target *could* be met, but if overall fossil fuel use were to increase dramatically the percentage of total alternative use could be quite small, i.e. not near the 15%. Thus, *both* absolute and fractional metrics are needed.

Another accountability challenge is how accurately energy and matter losses are included in calculations. From a thermodynamics standpoint, the nation's increased ethanol use could actually increase demands for fossil fuels, such as the need for crude oil-based infrastructures, including farm chemicals derived from oil, farm vehicle, and equipment energy use (planting, cultivation, harvesting, and transport to markets) dependent on gasoline and diesel fuels, and even embedded energy needs in the ethanol processing facility (crude oil-derived chemicals needed for catalysis, purification, fuel mixing, and refining). A comprehensive LCA is a vital tool for ascertaining the actual efficiencies.

The questions surrounding ethanol can be addressed using a three-step methodology. First, the efficiency calculations must conform to the physical laws, especially those of thermodynamics and motion. Second, the "greenness," as a metric of sustainability and effectiveness can be characterized by life cycle analyses. Third, the policy and geopolitical options and outcomes can be evaluated by decision force field analyses. In fact, these three approaches are sequential. The first must be satisfied before moving to the second. Likewise, the third depends on the first two methods. No matter how politically attractive or favored by society, an alternative fuel must comport with the conservation of mass and energy. Further, each step in the life cycle (e.g. extraction of raw materials, value-added manufacturing, use, and disposal) must be considered in any benefit–cost or risk–benefit analysis. Finally, the societal benefits and risks must be viable for an alternative fuel to be accepted. Thus, even a very efficient and effective fuel may be rejected for societal reasons (e.g. religious, cultural, historical, or ethical).

The challenge of the scientist, engineer, and policy maker is to sift through the myriad data and information to ascertain whether ethanol truly presents a viable alternative fuel. Of the misrepresentations being made, some clearly violate the physical laws. Many ignore or do not provide correct weights to certain factors in the life cycle. There is always the risk of mischaracterizing the social good or costs, a common problem with the use of benefit–cost relationships.

Biomass-based fuel efficiencies are evaluated in terms of net energy production that is based on thermodynamics (first and second laws). Recalling from our discussions in Chapter 1, energy balances can be calculated from the first law of thermodynamics:

$$\text{Accumulation} = \text{creation rate} - \text{destruction rate} + \text{flow in} - \text{flow out} \quad (11.1)$$

Stated quantitatively as efficiency, Eq. 11.1 is:

$$\mathit{Efficiency} = \frac{E_{in} - E_{out}}{E_{in}} \times 100 \quad (11.2)$$

where E_{in} = Energy entering a control volume, and

E_{out} = Energy exiting a control volume.

The numerator includes all energy losses. However, these are dictated by the specific control volume. This volume can be of any size, from molecular to planetary. To analyze energy losses related to alternative fuels, every control volume of each step of the life cycle must be quantified.

The first two laws of thermodynamics drive this step. First, the conservation of mass and energy requires that every input and output be included. Energy or mass can neither be created nor destroyed, only altered in form. For any system, energy or mass transfer is associated with mass and energy crossing the control boundary within the control volume. If mass does not cross the boundary, but work and/or heat do, the system is a "closed" system. If mass, work, and heat do not cross the boundary, the system is an isolated system. Too often, open systems are treated as closed, or closed systems include too small control volume (Figure 11.14). A common error is to assume that the life cycle begins at an arbitrary point conveniently selected to support a benefit–cost ratio. For example, if a life cycle for ethanol fuels begins with the corn arriving at the ethanol processing facility, none of the fossil fuel needs on the farm or in transportation will appear.

The second law is less direct and obvious than the first. In all energy exchanges, if no energy enters or leaves the system, the potential energy of the state will always be less than that of the initial state. The tendency toward disorder, i.e. entropy, requires that external energy is needed to maintain any energy balance in a control volume, such as a heat engine, waterfall, or an ethanol processing facility. Entropy is ever present. Losses must always occur in conversions from one type of energy (e.g. mechanical energy of farm equipment ultimately to chemical energy of the fuel). Thus, Eq. 11.2 is actually a series of efficiency equations for the entire process, with losses at every step.

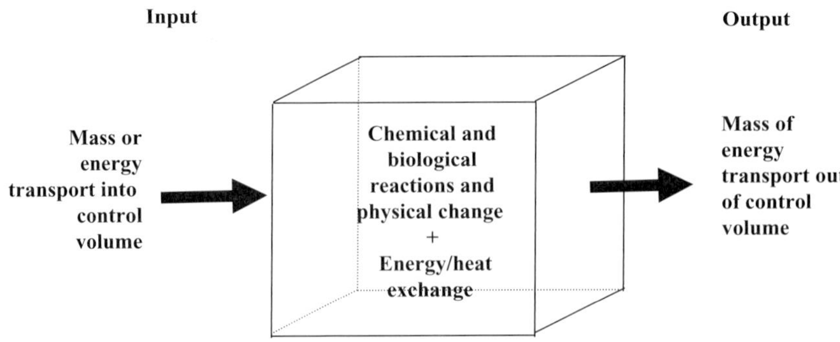

FIGURE 11.14
Control volume showing input, change, and output. The process applies to both mass and energy balances.

These equations illustrate the importance of reliable inventories. In the LCA process, the information is provided by the life cycle inventory (LCI). In fact, the LCA process uses the LCI data to assess the environmental implications associated with a product, process, or service, by compiling an inventory of relevant energy and material inputs and environmental releases. From this LCI, the potential environmental impacts are evaluated. The results aid in decision making. Thus, the LCA process is a systematic, four-component process:

Goal Definition and Scoping – Define and describe the product, process or activity. Establish the context in which the assessment is to be made and identify the boundaries and environmental effects to be reviewed for the assessment.

Inventory Analysis – Identify and quantify energy, water, and materials usage and environmental releases (e.g., air emissions, solid waste disposal, wastewater discharges).

Impact Assessment – Assess the potential human and ecological effects of energy, water, and material usage and the environmental releases identified in the inventory analysis.

Interpretation – Evaluate the results of the inventory analysis and impact assessment to select the preferred product, process or service with a clear understanding of the uncertainty and the assumptions used to generate the results. [13]

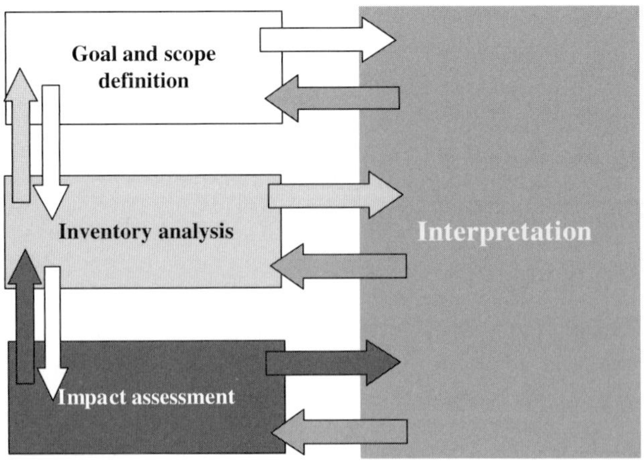

FIGURE 11.15
Life cycle assessment framework consists of: (1) a specifically stated purpose of boundaries of the study (Goal and scope definition); (2) an estimate of the energy use and raw material inputs and environmental releases associated with each stage of the life cycle (life cycle inventory); (3) an interpretation of the results of the inventory to assess the impacts on human health and the environment (impact assessment); and (4) an evaluation of ways to reduce energy, material inputs, or environmental impacts along the life cycle (interpretation).
Source: US Environmental Protection Agency (2009). Systems Analysis Research: Program Brief – Life Cycle Analysis; http://www.epa.gov/nrmrl/std/sab/lca/lca_brief.htm; accessed July 29, 2009.

Note that these steps, as illustrated in Figure 11.15, track closely with the life cycle stages dictated by the laws of physics (see Figure 11.14).

This stepwise process can be used to evaluate biotechnologies. First, a life cycle inventory (LCI) is constructed to define the boundaries of the possible effects of a technology (e.g. microbial populations, genetically modified organisms, and toxic chemical releases). If the technology is hypothetical, this can be done by analogy with a similar conventional process. Next, experts can participate in an expert panel to find the driving forces involved (this is known as "expert elicitation"). Then, scenarios can be constructed from these driving forces to identify which factors are most important in leading to various outcomes. This last step is known as a sensitivity analysis. The greater the weight of the factor the greater will be the change in the outcome. For example, if one genetic alteration of a microbe leads to very little gene flow, but another variant leads to 100 times the gene flow in the same type of ecosystem, then gene flow is 100 times more sensitive to the latter genetically modified microbe than the former in the prescribed ecosystem.

UTILITY AND THE BENEFIT–COST ANALYSIS

As discussed in Chapter 9, selecting a biotechnology is complicated, beginning with the subtle and difficult-to-measure benefits and costs, as expressed in the benefit–cost ratio (BCR). The BCR is attractive to technologists since it is, or at least appears to be, a quantitative measure of engineering success. Thus, it can also be an expression of a biotechnology's success. This utilitarian perspective is also attractive since it can compare one project to another, e.g. a BCR of 2 in Project 1 means that it is a more worthwhile endeavor than Project 2, which has a BCR of 1.5. But is that true? This is why it is advantageous to use more than one metric of success and why a combined BCR and LCA approach can be quite useful in environmental biotechnology. Certainly, the utility of a project is crucial to the decision to go forward, according to at least two criteria:

- The project has value based on its utility.
- In pursuing this project it must provide the most benefit for the greatest number (e.g. people, ecosystems, etc.).

A BCR value is more of a screening tool than an absolute measure of potential or value. If the benefits are far outweighed by the costs (i.e. $BCR << 1$), this is likely not a worthwhile endeavor, so long as the factors are representative and properly weighted. Conversely, a very large BCR may indicate that there is great potential for that project.

As mentioned in the discussion in Chapter 9, however, certain costs and benefits are much easier to quantify than others. Some, like those associated with the social sciences and humanities are nearly impossible to quantify and monetize accurately when compared to those that are aligned with the physical sciences. Recall also that the comparison of action versus no-action alternatives cannot always be captured within a BCR. This means that the so-called opportunity costs and risks that are associated with taking no action (e.g. what is lost by not implementing the biotechnology project) are not included in the calculations. Thus, comparisons of the *status quo* to costs and risks associated with a new technology may be biased toward the *status quo*. Thus, the decision to embrace or avoid a project is complex, involving factors that are readily quantifiable and monetizable and those that are almost impossible to assign a concrete value. This is particularly challenging for values that are long-term (e.g. overall ecological sustainability for future populations) and for those decisions where two societal values are clashing (e.g. the DDT/cancer/malaria conundrum mentioned previously).

The means of finding the best biotechnological approach is a matter of optimization, which is complicated for projects that have numerous contravening risks and many possible solutions. Life cycle analysis (LCA) can build from the BCR screening effort as it considers product flows, critical paths, and life cycle inventories (LCIs). If properly conducted, the LCA can be used to compare various biotechnological options, including the *status quo*, passive approaches and more aggressive endeavors, not only in terms of the principal project objective (e.g. contamination cleanup), but also from the standpoint of various ecosystem, public health and other societal values. In other words, the view from the past (e.g. material extraction) to the future (e.g. post-project impacts) is a critical path for each option that can be evaluated objectively.

The success of a biotechnology displacing conventional manufacturing depends on the efficiency and safety with which a product can be produced and used. Complicating matters, certain materials and costs may be embedded in the "alternative" technology. For example, fossil fuels are used in almost every step in ethanol production and/or operation, as is the case for all biofuels.

The underlying assumption in any environmental assessment of biotechnology is that we can identify "cause and effect" with regard to the credible science needed to connect exposure to a biotechnological hazard and a negative outcome. Scientists frequently "punt" on this issue. We have learned from introductory statistic courses that association and causation are not synonymous. We are taught, for example, to look for the "third variable." Something other than what we are studying may be the reason for the relationship. In statistics classes, we are given simple examples of such occurrences:

> Studies show that people who wear shorts in Illinois eat more ice cream.
> Therefore, wearing shorts induces people to eat more ice cream.

The first statement is simply a measurement. It is stated correctly as an association. However, the second statement contains a causal link that is clearly wrong for most occurrences [14]. Something else is actually causing both variables, i.e. the wearing of shorts and the eating of ice cream. For example, if one were to plot ambient average temperature and compare it to either the wearing of shorts or the eating of ice cream, one would see a direct relationship between the variables. That is, as temperatures increase so does short wearing and so does the rate of ice cream eating.

As mentioned, scientists usually "punt" regarding causality. Punting is not inherently a bad thing (Ask the football coach who decides to go for the first down on fourth and inches and whose team comes up a half-inch short. He would have likely wished he had asked for a punt!). It is only troublesome when we use the association argument invariably. (The football coach who always punts on fourth and short might be considered to lack courage.) People want to know what our findings mean. Again the medical science community may help us deal with the causality challenge. The best that science usually can do in this regard is to provide enough weight-of-evidence to support or reject a suspicion that a substance causes a disease. The medical research and epidemiological communities use a number of criteria to determine the strength of an argument for causality, but the first well-articulated criteria were Hill's Causal Criteria [15] (see Table 11.4). Some of Hill's criteria are more important than others. Interestingly, the first criterion is, in fact, association.

The LCA is useful, but it is not a panacea for biotechnological impact assessment. For example, an analysis is only as good as the availability and quality of the life cycle inventory data. Uncertainties in the inventory and in the impact assessment methodology can lead to omitted options and unforeseen impacts. The lack of agreement among elements of the impact assessment methodology, i.e. internal inconsistencies, can lead to additional uncertainties and errors. Finally, many of the factors are qualitative and subjective so that differences in LCA problem formulation may in fact be due to differences in weights placed on factors. Such weighting can be biased according to the values of those conducting the LCA.

Table 11.4 Criteria for causality

Factors to be considered in determining whether a biotechnology elicits an effect:

- **Criterion 1: Strength of Association.** For a chemical exposure to cause an effect, the exposure must be associated with that affect. Strong associations provide more certain evidence of causality than is provided by weak associations. Common epidemiological metrics used in association include risk ratio, odds ratio, and standardized mortality ratio
- **Criterion 2: Consistency.** If the chemical exposure is associated with an effect consistently under different studies using diverse methods of study of assorted populations under varying circumstances by different investigators, the link to causality is stronger. For example, the carcinogenic effects of Chemical X is found in mutagenicity studies, mouse and Rhesus monkey experiments, and human epidemiological studies, there is greater consistency between Chemical X and cancer than if only one of these studies showed the effect
- **Criterion 3: Specificity.** The specificity criterion holds that the cause should lead to only one disease and that the disease should result from only this single cause. This criterion appears to be based in the germ theory of microbiology, where a specific strain of bacteria and viruses elicits a specific disease. This is rarely the case in studying most chronic diseases, since a chemical can be associated with cancers in numerous organs, and the same chemical may elicit cancer, hormonal, immunological and neural dysfunctions
- **Criterion 4: Temporality.** Timing of exposure is critical to causality. This criterion requires that exposure to the chemical must precede the effect. For example, in a retrospective study, the researcher must be certain that the manifestation of a disease was not already present before the exposure to the chemical. If the disease were present prior to the exposure, it may not mean that the chemical in question is not a cause, but it does mean that it is not the sole cause of the disease (see ''Specificity'' above)
- **Criterion 5: Biologic Gradient.** This is another essential criterion for chemical risks. In fact, this is known as the ''dose-response'' step in risk assessment. If the level, intensity, duration, or total level of chemical exposure is increased a concomitant, progressive increase should occur in the toxic effect
- **Criterion 6: Plausibility.** Generally, an association needs to follow a well-defined explanation based on known biological system. However, ''paradigm shifts'' in the understanding of key scientific concepts do change. A noteworthy example is the change in the latter part of the 20th century of the understanding of how the endocrine, immune, neural systems function, from the view that these are exclusive systems to today's perspective that in many ways they constitute an integrated chemical and electrical set of signals in an organism [16]
- **Criterion 7: Coherence.** The criterion of coherence suggests that all available evidence concerning the natural history and biology of the disease should ''stick together'' (cohere) to form a cohesive whole. By that, the proposed causal relationship should not conflict or contradict information from experimental, laboratory, epidemiologic, theory, or other knowledge sources
- **Criterion 8: Experimentation.** Experimental evidence in support of a causal hypothesis may come in the form of community and clinical trials, *in vitro* laboratory experiments, animal models, and natural experiments
- **Criterion 9: Analogy.** The term analogy implies a similarity in some respects among things that are otherwise different. It is thus considered one of the weaker forms of evidence

Source: Adapted from A. Bradford Hill (1965). The environment and disease: association or causation? *Proceedings of the Royal Society of Medicine, Occupational Medicine* 58: 295.

PREDICTING ENVIRONMENTAL DAMAGE

The life cycle approach depends on reliable tools for predicting possible outcomes. The following discussion is an introduction of some of these tools, but is by no means an exhaustive listing.

Analysis of biotechnological implications

Tom and Miriam Budinger have provided an elegant approach, the "Four As", for approaching a scientific dilemma, which can be used to address possible environmental impacts that can result from a biotechnology [17]. First, one should *acquire* the facts, including the uncertainties associated with the technology. Second, the *alternative* solutions are listed and compared in parallel. Third, each solution is *assessed* with respect to principles (the authors were mainly concerned with moral theories, but other scientific theories can also be used as benchmarks). This includes a thorough risk analysis, where appropriate. Finally, a decision is made on which *action* to take. This includes a comprehensive action plan, keeping the other alternatives available if needed, and with continuous adaptation and improvement, and eye toward new options (since the nature of emergent technologies is that things change very rapidly).

Problem analysis is a step-wise process, such as that which follows.

STEP 1 – SCENARIO DESCRIPTION

Key characters and events pertinent to the case are identified. This description includes narrative, tables, figures, maps, organization charts, critical path diagrams, and photographs that are needed to place the case in ethical context. The storyboard must be both accurate and complete. This can be challenging since almost any biotechnological endeavor includes numerous perspectives, so each perspective must be described adequately. Since this is the *descriptive stage*, no judgments about what is right need be made; that is done in the next steps. Completeness means that the science and societal concepts are fully understood. For example, if the project is designed to degrade a recalcitrant chemical compound, it must be described completely in all matters that could affect the environmental decision (e.g. electron donation/acceptance of microbes in previous studies, similarities in chemical structure to compounds that have been successfully bioremediated, and problems encountered in similar projects).

A key consideration of Step 1 is assigning responsibility and accountability.

STEP 2 – DEDUCTIVE ARGUMENTS

Based upon the findings in Step 1, the validity of the decisions or lack thereof is analyzed. The syllogism includes a factual premise, a connecting fact-value premise, and an evaluative premise to reach an evaluative conclusion. Many moral (and scientific) arguments fail because of weaknesses in any of these components of the logical argument, i.e. the syllogism. Depending on the case, numerous arguments must be evaluated.

STEP 3 – PROBLEM-SOLVING ANALYSIS

Once the facts are identified, articulated and sufficiently explained, any issues must be categorized as to whether they are factual, conceptual, or related to human factors [18].

Human factors would include constraints or drivers that are not based on physical science, but more related to perceptions and expectations of the people potentially affected by the project (e.g. historical, cultural, financial).

From descriptions in Step 1, the depth of each type of issue can be assessed. Factual issues are those that are known. This can sometimes be apparent just by reading the events, but in certain cases the facts may not be so clear (e.g. two scientists may agree on the "fact" that carbon dioxide is a radiant gas, but may disagree on whether the build-up of CO_2 in the troposphere will lead to increased global warming). Agreement on first principles of science and even the data being used may still be followed by large disagreements about the relative weightings in indices and models. This leads to a need to ascribe causality, a very difficult problem indeed.

Step 3a – Application of Hill's criteria

To begin to evaluate whether a biotechnology is appropriate to the problem at hand, oftentimes the best that science usually can do in this regard is to provide enough weight-of-evidence between a cause and an effect. The medical research and epidemiological communities use a number of criteria to determine the strength of an argument for causality, but the first well-articulated criteria were Hill's Causal Criteria [19] (see Table 11.4). Depending on the biotechnology, some of Hill's criteria are more relevant and important than others.

Conceptual issues involve different ways that the meaning may be understood. For example, what one considers to be "pollution" or "good lab practices" may vary (although the scientific community strives to bring consensus to such definitions).

Many engineers and scientists believe it is the job of technical societies and other collectives to try to eliminate factual and conceptual disagreements. Most agree on first principles (e.g. fundamental physical concepts like the definitions of matter and energy), but unanimity fades as the concepts drift from first principles. For example, John Ahearnes, the former President of Sigma Xi, the Scientific Research Society, recently told an audience of engineers that we should not be disagreeing about the facts [20]. The progress of research and knowledge helps to resolve factual issues (eventually), and the consensus of experts aids in resolving conceptual issues. But since complete agreements are not generally possible even for the factual and conceptual aspects of a case, the moral or ethical issues are further complicated.

Step 3b – Force fields

A simple "polar diagram" that illustrates the "forces" that pull or push the key individuals or groups toward decisions can be quite useful, at least in identifying the various stakeholders (Figure 11.16). The shape and size of the resulting diagram give an idea of what are the principal driving factors that lead to decisions. Envision a source in the outer middle of each sector pulling against the shape. A force field diagram can be drawn as a subjective assessment of each decision and for each decision maker. For example, lawyers may proceed in one direction, while engineers another, and the land owners another, all because of different forces.

Step 3c – Net goodness analysis

This is a subjective analysis of whether a decision will be moral or less than moral. It puts the case into perspective, by looking at each factor driving a decision from three perspectives: (1) how good or bad would the consequence be; (2) how important is decision; and (3) how likely is it that the consequence would occur. These factors are then summed to give the overall net goodness of the decision:

$$NG = \sum(\text{goodness of each consequence}) \times (\text{importance}) \times (\text{likelihood}) \qquad (11.3)$$

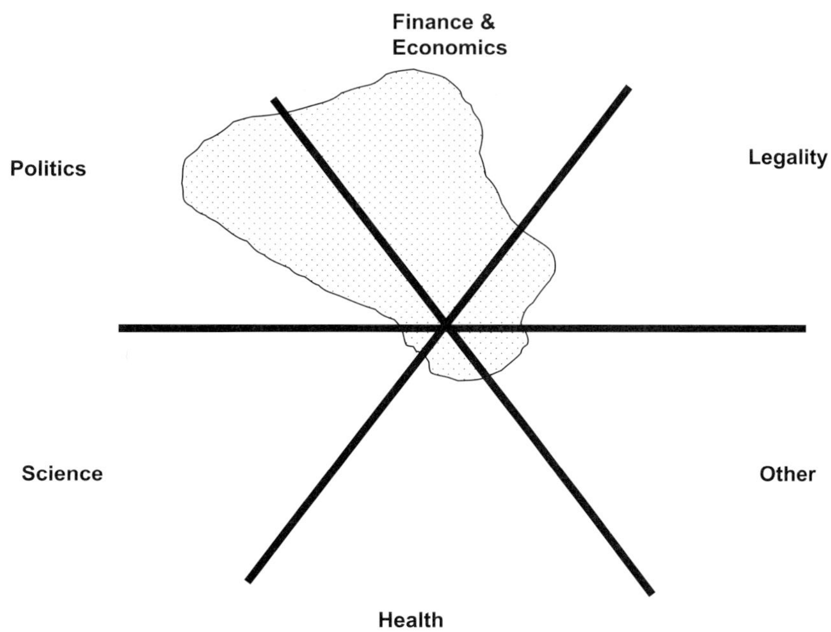

FIGURE 11.16
Example of a force field diagram for a biotechnological "Go or No Go" decision. In this instance, the decision is clearly driven by political and financial factors. This should give the scientist pause on whether the decision is properly weighted.

Thus, this can be valuable in decisions that have not yet been made, as well as in evaluating what decisions "should" have been made in a case. For example, these analyses sometimes use ordinal scales, such as 0 through 3, where 0 is nonexistence (e.g. zero likelihood or zero importance) and 1, 2, and 3 are low, medium, and high, respectively. Thus, there may be many small consequences that are near zero in importance and, since NG is a product, the overall net goodness of the decision is driven almost entirely by one or a few important and likely consequences.

There are two cautions in using this approach. First, although it appears to be quantitative, the approach is very subjective. Second, as we have seen many times in cases involving health and safety, even a very unlikely but negative consequence is unacceptable.

Step 3d – Line drawing

Graphical techniques like line drawing, flow charting, and event trees are very valuable in assessing a case. Line drawing is most useful when there is little disagreement on what the moral principles are, but when there is no consensus about how to apply them. The approach calls for a need to compare several well-understood cases for which there is general agreement about right and wrong and show the relative location of the case being analyzed. Two of the cases are extreme cases of right and wrong, respectively; that is, the positive paradigm is very close to being unambiguously moral and the negative paradigm unambiguously immoral:

Next, our case (T) is put on a scale showing the positive paradigm (PP) and the negative paradigm (NP), as well as other cases that are generally agreed to be less positive than PP but more positive than NP. This shows the relative position of our case T:

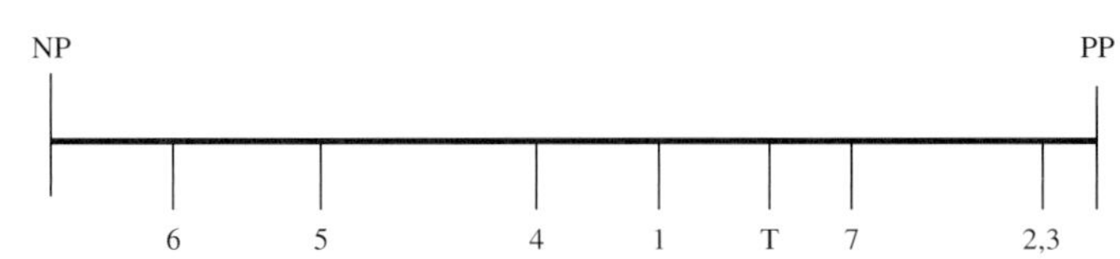

This gives us a sense that our case is more positive than negative, but still short of being unambiguously positive. In fact, two other actual, comparable cases (2 and 3) are much more morally acceptable. This may indicate that we consider taking an approach similar to these if the decision has not yet been made. If the decision has been made, we will want to determine why the case being reviewed was so different from these.

Although being right of center means that our case is closer to the most moral than to the most immoral approach, other factors must be considered, such as feasibility and public acceptance. Like risk assessment, ethical analysis must account for tradeoffs (e.g. security versus liberty).

Step 3e – Flow charting

Critical paths, PERT charts, and other flow charts are commonly used in design and engineering, especially computing and circuit design. They are also useful in ethical analysis if sequences and contingencies are involved in reaching a decision, or if a series of events and ethical and factual decisions lead to the consequence of interest. Thus, each consequence and the decisions that were made along the way can be seen and analyzed individually and collectively. Fleddermann [21] shows a flow chart for the Bhopal incident. This flow chart (Figure 11.17) deals with only one of the decisions involved in the incident, i.e. where to site the plant. Other charts need to be developed for safety training, the need for fail-safe measures, and proper operation and maintenance. Thus, a "master flow chart" can be developed for all of the decisions and subconsequences that ultimately led to the disaster.

Step 3f – Event trees

Event trees or fault trees allow us to look at possible consequences from each decision. A straightforward example is provided in Figure 11.18.

The event tree can build from all of the other analytical tools, starting with the timeline of key events and list of key actors. What are their interests and why were the decisions made? The event tree allows us to visualize a number of different paths that could have been taken that could have led to better or worse decisions. We would do this for every option and suboption that should have been considered in our case, comparing each consequence. It may be, for example, that even in a disaster, there may have been worse consequences than what actually occurred. Conversely, even though something did not necessarily turn out all that badly, the event tree could point out that the outcomes were simply fortunate! In fact, the fault tree approach applies a probability to each option and suboption.

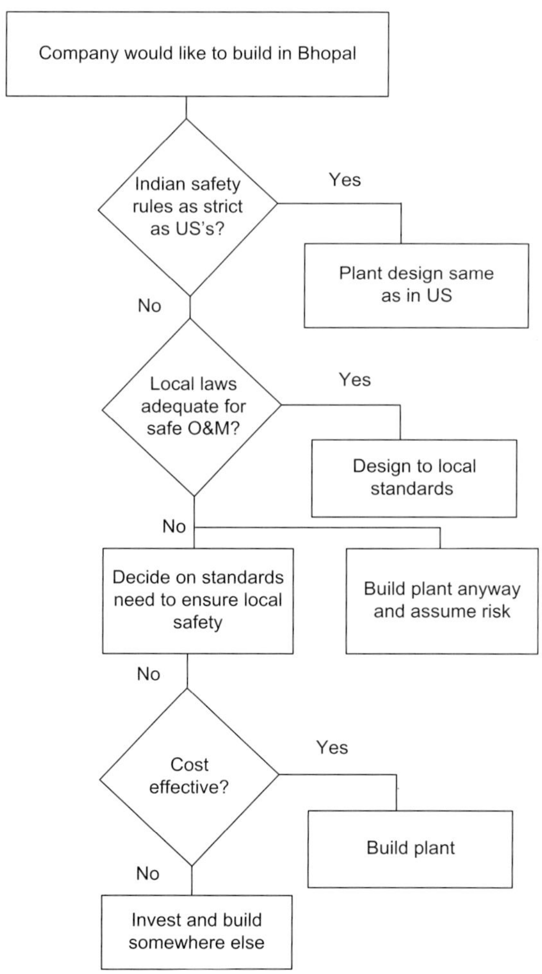

FIGURE 11.17
Flow chart on decision to locate pesticide plant in Bhopal, India.
Source: Adapted from Fleddermann [21].

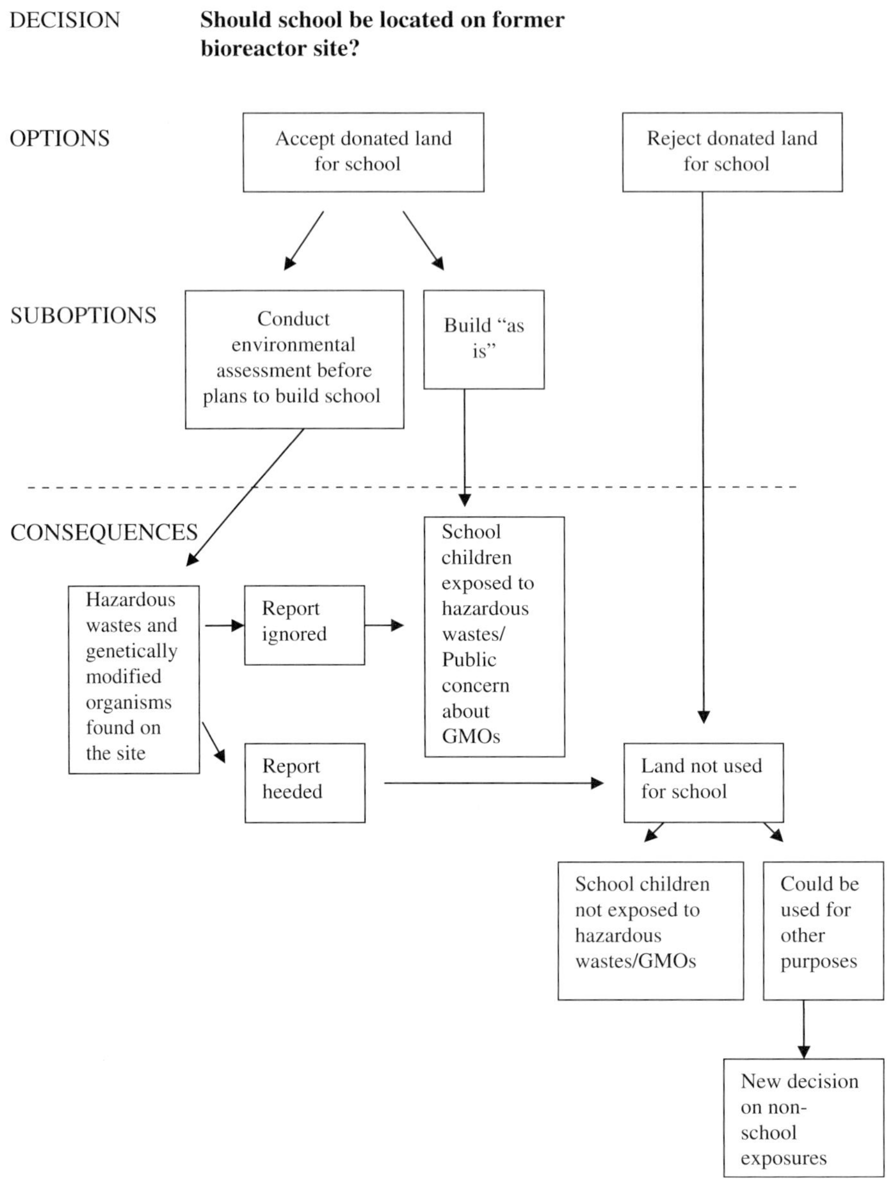

FIGURE 11.18
Event tree for hypothetical siting decision

STEP 4 – SYNTHESIS

Using the information from the steps above, we can begin to decide about how "right" the decision is. This is tantamount to a moral decision. That is, is the implementation of the biotechnological enterprise a moral or immoral decision compared to the other approaches that could have been taken, including the so-called "no action" alternative? If the decision has not yet been made, then the alternatives can be compared before choosing the best one.

Most environmental decisions are moral decisions. That is, they must be based on sound scientific and engineering principles, but there is usually some value being placed on one alternative compared to that of another. Thus, few environmental decisions are *amoral* (i.e. devoid of ethical content). For example, deciding whether one particular method of nutrient addition to improve bioremediation is better than another is predominantly an amoral decision, so long as it is completely based on undisputed facts. However, if the science is not completely driving the decision (e.g. the best nutrient addition is "too" expensive), the decision is taken on moral relevance. Deciding how to address the moral aspects of a scientific decision has been addressed by professional societies and scientific groups. One approach has been proposed by the National Academy of Engineering [22], based on work by Swazey and Bird, Weil, and Velasquez [23].

Checklist for Ethical Decision Making

- Recognize and define the ethical issues (i.e., identify what is [are] the problem[s] and who is involved or affected).
- Identify the key facts of the situation, as well as ambiguities or uncertainties, and what additional information is needed and why.
- Identify the affected parties or "stakeholders" (i.e., individuals or groups who affect, or are affected by, the problem or its resolution). For example, in a case involving intentional deception in reporting research results, those affected include those who perpetrated the deception, other members of the research group, the department and university, the funder, the journal where the results were published, other researchers developing or conducting research on the findings, etc.
- Formulate viable alternative courses of action that could be taken, and continue to check the facts.
- Assess each alternative (i.e., its implications; whether it is in accord with the ethical standards being used, and if not, whether it can be justified on other grounds; consequences for affected parties; issues that will be left unresolved; whether it can be publicly defended on ethical grounds; the precedent that will be set; practical constraints, e.g., uncertainty regarding consequences, lack of ability, authority or resources, institutional, structural, or procedural barriers).
- Construct desired options and persuade or negotiate with others to implement them.
- Decide what actions should be taken and in so doing, recheck and weigh the reasoning in steps 1–6.

It should be noted that these track quite well with the Budingers' Four A's and steps noted in this chapter. However, it is not enough to be right; the biotechnologist must be able to communicate with and convince others that the selected approach is best and needed (and be sufficiently open-minded to possible improvements and needed adjustments to the proposed approach, whether from other professionals or the lay public). Thus, based on the analysis, the findings and arguments (including all the necessary facts and figures) must be placed in the context of the stakeholders in an understandable way. The audience will vary, so one size definitely does not fit all when it comes to presenting information on possible environmental risks.

SEMINAR TOPIC

Public Participation in Environmental Biotechnology

A recent focus group [24] investigated the range of perceptions regarding plant biotechnologies. The five focus groups who participated were diverse:

Participants were students studying environmental policy in the Department of Resource Development at a land grant college, Michigan State University (MSU). Only one was not a graduate student.

Participants were a mix of graduate and undergraduate students from MSU. Their areas of study included: zoology, agricultural economics, and environmental policy.

Participants were all active members (i.e. not simply dues paying) in non-profit environmental groups in the local area.

Participants were either farmers or worked in a farm-related profession. All had used transgenic crops or were familiar with them.

Participants were all researchers in the field of plant biotechnology or plant breeding at MSU.

The focus group members shared some interesting insights:

> *Many advocates of biotechnology have argued that any public unease with GMOs could be overcome through improved communication (e.g. standardizing terminology) and education campaigns that explain the technology and its societal benefits ... The results of this study suggest, however, that, though a better-informed public is needed, merely promoting the benefits of plant biotechnology is in itself unlikely to persuade those who have concerns about the technology.*

The findings continued with:

> *... concerns were expressed about the potential impacts of releasing GMOs into the environment. The Losey, Rayor, and Carter [25] study of the possible negative side effects to monarch butterflies from Bt [*Bacillus thuringiensis*] corn pollen*

was mentioned in every focus group, although most were unclear regarding the details. There were concerns expressed about enhancing an organism's ability to become an invasive species, the potential for new diseases to emerge and gene flow from herbicide-resistant crops to wild species.

The regulatory and industry decision-making processes associated with plant biotechnology drew fire. The biotechnology companies were criticized for not being more open or receptive to questions. Their size and wealth raised concerns about their influence on the regulatory system, their control of world food systems, including intellectual property, and their impact on the credibility of industry-funded research. The perceived secrecy arising from the commercial nature of plant biotechnology raised warning flags for many participants.

A related issue was that of trust. A perceived lack of openness on the part of the biotechnology companies appeared to have diminished their trustworthiness in the eyes of many participants. Discussions about testing, the regulatory process and scientific findings revolved about the issue of who had sufficient credibility to be trusted by the public. There were frequent calls for more openness in the regulatory process.

The investigators concluded that:

... despite public opinion polls showing significant American consumer support for food biotechnology, the two focus group studies suggest that support may be fragile. The ordinary citizen has little awareness or understanding of food biotechnology. When confronted with new information, their existing opinions may not prove to be robust. If the general public is to have a meaningful role to play in influencing policy, a concerted effort is needed to transform today's uninformed and superficial public opinions into soundly reasoned public judgments.

This would not be an easy task. Scientific illiteracy and lack of public awareness of the presence of GMOs in the American food system are impediments to an engaged society. An intensive effort would be needed to overcome the knowledge gaps and make the topic of plant biotechnology and its place in society accessible to the general public. It would require an approach that not only addressed scientific issues but all of the associated socio-economic, value and ethical issues raised by the study's participants. On the basis of this study, it is likely that any such public debate would focus as much, and likely more so, on issues other than the science of food biotechnology; with issues of choice, control and trust likely to dominate.

There appears to be a problem of trust, especially when it comes to potential widespread and irreversible systematic outcomes. Food seems to be a triggering mechanism for concerns about biotechnological advances. This has not been observed, at least to nearly as great an extent, in environmental biotechnologies, i.e. enhanced biodegradation.

Seminar Questions

It would appear that environmental biotechnologies do not elicit the advocacy and concern to agricultural and food biotechnologies? Why not?

How can complex issues like gene flow for microbes versus those for higher organisms be explained in a generally understandable way? In other words, if people ate microbes as food sources, would environmental biotechnologies be under the same level of scrutiny as plant biotechnologies?

But people do indeed fear pathogenic microbes. Why has that fear not permeated risk perceptions of microbial bioremediation?

The focus group findings seem to indicate that better risk communication, while needed, will not eliminate many of these concerns. This would seem to indicate that the concerns may be more substantial than merely problems of perception. Where do the most important plant biotechnological problems fall in the risk matrix in Table 11.3?

REVIEW QUESTIONS

Explain the similarities and difference between an accidental release and a bioremediation project.

Conduct an LCA for a process that will extract DNA from the fictional flower Abner's Daisy, and insert it into a cabbage plant for mythical *Cavola mangia* moth control. The flower is listed as threatened on the endangered species list and is one of the few food sources for the larvae of a *Cavola* eating wasp. The DNA extract requires the use of a retrovirus vector. What other critical information is needed? State all important assumptions and advise the regulators whether this GMO should be approved for agricultural use.

In the scenario above, if the company proposing the genetic modification just completed a thorough and scientifically credible assessment of all of the chemicals used in the process and showed the risks to be negligible, would this likely be approved in the United States? In the UK? Use a decision force field to show the differences, if any, between the two countries' review and approval processes.

What if instead of the modification being used for agriculture the same DNA insertion were used for a new medicine to treat Alzheimer's patients? Construct an event tree for both the agricultural and medical biotechnology. How do they differ?

Construct a line drawing, conduct a net goodness analysis, and draw a flow chart diagram for the two biotechnologies above. Do these indicate factors not evident in the decision tree? Why might they show up using an ethical decision tool and not with a risk assessment (e.g. chemical hazard-exposure-effect) method?

Are most environmental biotechnology decisions chaotic? Support your answer.

Give an example of an amoral versus moral decision regarding a biotechnology? Which one is easier to resolve? Give the reasons, including any uncertainties, for this conclusion.

NOTES AND REFERENCES

1. C. Picone, D. Andresen, G. Thomas, D. Griffith, *et al.* (1999). Say No to GMOs! (Genetically Modified Organisms). *Agenda.* University of Michigan Chapter of the New World Agriculture and Ecology Group, May/June, pp. 6–8.
2. Ibid. See also: Anonymous (1999). Unpalatable truths. *New Scientist* April 17; and P.C. Ronald (1997). Making rice disease-resistant. *Scientific American* 277: 100–105.
3. Picone, *et al.* Say No to GMOs!.
4. S. Reid (2002). State of the Science on Molds and Human Health. Centers for Disease Control and Prevention, Statement for the Record before the Subcommittees on Oversight and Investigations and Housing and Community Opportunity, US House of Representatives, Washington, DC.
5. More information is available at:
   - Field Guide for the Determination of Biological Contamination (stock #227-RC-96), American Industrial Hygiene Association (AIHA), www.aiha.org.
   - Report of Microbial Growth Task Force (stock #456.EQ-01), AIHA, www.aiha.org.
   - Listing of AIHA Laboratory Quality Assurance Program Environmental Microbiology Laboratory Accreditation Program (LQAP EMLAP) accredited laboratories, AIHA, www.aiha.org.
   - Bioaerosols: Assessment and Control, American Conference of Governmental Industrial Hygienists (ACGIH), www.acgih.org.
   - IICRC S500, Standard and Reference Guide for Professional Water Damage Restoration, Institute of Inspection, Cleaning, and Restoration Certification, www.iicrc.org.
   - Mold Remediation in Schools and Commercial Buildings (EPA 402-K-01-001), Environmental Protection Agency (EPA), www.epa.gov/iaq/molds/index.html.
   - Draft Guideline for Environmental Infection Control in Healthcare Facilities (especially sections I.C.3, I.C.4, I.F, II.C.1, and Appendix B), Centers for Disease Control (CDC), www.cdc.gov/ncidod/hip/enviro/env_guide_draft.pdf.
   - EPA and FEMA (Federal Emergency Management Agency) Flood Clean-Up Guidelines, www.epa.gov/iaq/pubs/flood.html and www.fema.gov/hazards/floods/.
   - Centers for Disease Control and Prevention (CDC), www.cdc.gov/nceh/airpollution/mold/default.htm.
   - California Indoor Air Quality Program, www.cal-iaq.org/iaqsheet.htm.
   - New York City Department of Health, Guidelines on Assessment and Remediation of Fungi in Indoor Environments, www.nyc.gov/html/doh/html/epi/moldrpt1.html.
   - American College of Occupational and Environmental Medicine guideline: Adverse Human Health Effects Associated with Molds in the Indoor Environment, www.acoem.org/guidelines/pdf/mold-10-27-02.pdf.
6. Inductive reasoning is also called "abstraction," because it starts with something concrete and forms a more abstract ideal. Philosophers have argued for centuries regarding the value of inductive reasoning. Induction is the process that takes specific facts, findings or cases and then generally applies them to construct new concepts and ideas. Abstraction leaves out specific details, unifying them into a whole based on a defined principle. For example, a brown-feathered chicken, a white-feathered chicken, and a polka-dot-feathered chicken can all be integrated because each is a chicken, albeit with differences. The feather color, then, can be eliminated under the principle or criterion of being a chicken (i.e. "chickenness"), i.e. color is not "relevant." A brown chicken, brown bear, and brown paper bag can be integrated under the criterion of having brown color. The other aspects besides "brownness" of each item's characteristics are not relevant in this case, so they are omitted.
   In the 18th century, the Scottish philosopher David Hume postulated the so-called "problem of induction." To paraphrase, Hume was asking "Why should things that we may be observing on a regular basis continue to hold in the future?" In other words, there is no justification in using induction, because there is no reason that the conclusion of any inductive argument is valid. Like the scientific revolutionaries a couple of centuries earlier, Hume rejected *a priori* reason, since humans are incapable of fully and directly comprehending the laws of nature.

This can only be accomplished *a posteriori*, through experience.
Hume would have a problem with this inductive syllogism:

Every time I add nickel to my activated sludge, the bacteria grow more rapidly.

Therefore, the next time I add nickel to the sludge, my bacteria's growth rate will increase.

Although engineers can think of many reasons why the Ni addition may not lead to increased growth, e.g. different strains may not have adapted an enzymatic need for Ni, temperature changes may induce changed behaviors that render the Ni ineffective, and incomplete mixing does not allow the microbes access to the Ni, we also know that under the regular (expected?) conditions in the plant that the fact it has worked every time is a strong indicator that it will work again. Mathematicians may have a harder time with this expectation, but is it really any different than pressing your brake pedal and expecting the car to stop? Yes, there is always a probability (hopefully very low) that a leak in the master cylinder or brake line could cause the hydraulics to fail and the car would not stop when the brake pedal is depressed, but such probabilities do not render, in my opinion, inductive reasoning useless.

7. The discussion on intuition draws upon R.M. Hogarth (2001). *Educating Intuition*. University of Chicago Press, Chicago, IL.
8. Ibid. and K. Hammond (1996). *Human Judgment and Social Policy: Irreducible Uncertainty, Inevitable Error, Unavoidable Injustice.* Oxford University Press, New York, NY.
9. National Academy of Engineering (2004). *The Engineer of 2020: Visions of Engineering in the New Century.* The National Academies Press, Washington, DC.
10. For example, see S.B. Billatos and N.A. Basaly (1997). *Green Technology and Design for the Environment.* Taylor & Francis Group, London, UK.
11. For example, see D.E. Stokes (1997). *Pasteur's Quadrant.* Brookings Institute Press, Washington, DC; and H. Brooks (1979). Basic and applied research. In: *Categories of Scientific Research.* National Academies Press, Washington, DC, pp. 14–18.
12. The principal source for this discussion is: D. Vallero and C. Brasier (2008). Teaching green engineering: the case of ethanol lifecycle analysis. *Bulletin of Science, Technology & Society* 28 (3): 236–243.
13. US Environmental Protection Agency (2006). Life Cycle Assessment: Principles and Practice. Report No. EPA/600/R-06/060.
14. This is a typical way that scientists report information. In fact, there may be people who, if they put on shorts, will want to eat ice cream even if the temperature is $-30°$. These are known as "outliers". The term outlier is derived from the prototypical graph that plots the independent and dependent variables (i.e. the variable that we have control over and the one that is the outcome of the experiment, respectively). Outliers are those points that are furthest from the line of best fit that approximates this relationship. There is no standard for what constitutes an outlier, which is often defined by the scientists who conduct the research, although statistics and decision sciences give guidance in such assignments.
15. A. Bradford Hill (1965). The environment and disease: Association or causation? *Proceedings of the Royal Society of Medicine, Occupational Medicine* 58: 295.
16. For example, Candace Pert, a pioneer in endorphin research, has espoused the concept of mind/body, with all the systems interconnected, rather than separate and independent systems.
17. T.F. Budinger and M.D. Budinger (2006). *Ethics of Emerging Technologies: Scientific Facts and Moral Challenges.* John Wiley & Sons, Inc. Hoboken, NJ.
18. See C.E. Harris, Jr., M.S. Pritchard and M.J. Rabins (2000). *Engineering Ethics, Concept and Cases.* Wadsworth Publishing Co., Belmont, CA.
19. Bradford Hill, The environment and disease.
20. Ahearne's comments were made at the National Academy of Engineers' workshop on emerging technologies and ethics held in Washington, DC in November 2003.
21. C.B. Fleddermann (2004). *Engineering Ethics*, 2nd Edition. Pearson Education, Inc. Upper Saddle River, NJ.
22. National Academy of Engineering (2004). Emerging Technologies and Ethical Issues in Engineering: Papers from a Workshop, October 14–15, 2003. National Academies Press, Washington, DC.
23. From J.P. Swazey and S.J. Bird (1995). Teaching and learning research ethics. *Professional Ethics* 4: 155–178; M. Velasquez (1992). *Business Ethics*, 3rd Edition. Englewood Cliffs, NJ, Prentice-Hall; and V. Weil (1993). Teaching ethics in science. In: *Ethics, Values, and the Promise of Science.* Research Triangle Park, NC: Sigma Xi, pp. 243–248.
24. J.A. Beckwith, T. Hadlock and H. Suffron (2003). Public perceptions of plant biotechnology – a focus group study. *New Genetics and Society* 22 (2): 125–141.
25. J.E. Losey, L.S. Rayor and M.E. Carter (1999). Transgenic pollen harms monarch larvae. *Nature* 399: 214.

CHAPTER 12

Responsible Management of Biotechnologies

> Science is a wonderful thing if one does not have to earn one's living at it.
>
> **Albert Einstein [1]**

Responsible bioengineering and sound bioscience are defined in two dimensions. First, the bioengineering profession and the bioscience community have the privilege and responsibility to serve society by ensuring that organism-based designs and biotechnologies are in the public's best interest. Second, the individual engineer has agreed, either formally or tacitly, to a specific set of moral obligations to the public and the client. The formal agreements are codified in codes of professional practice. The tacit agreements are defined not only by one's specific professional community, but also by larger societal norms.

Likewise, bioscientists and biotechnologists have specific responsibilities to the public and their clients. The scientist's and technologist's "clients" are diverse, but all should be treated, to some extent, with what the Sigma Xi, the Scientific Research Society, considers to be "honor in science." The Society [2] so states:

> Semantic arguments apart, there remain two fundamental reasons why scientists should be concerned with the ethics of their research. The first reason is that without the basic principles of truthfulness – the assumption that we can rely on other people's words – the whole scientific research enterprise is liable to grind to a halt. Truthfulness may or may not be the cement that holds together society as a whole, but certainly it is essential to science. Secondly, whereas truthfulness in a wider context can be maintained and enforced by the institutions of the society we live in, scientific research is a specialized activity, each scientist working largely on individual experiments and analysis on the fringes of knowledge. Truthfulness – honesty – therefore has to depend primarily on individual scientists themselves.

In Chapter 1, biotechnology as a unique discipline was discussed; not merely the manner of conducting research, but also how that research is or will be applied. Certainly, the research that underpins environmental biotechnology is "on the fringes of knowledge," but so are the practical applications. It can be argued that every bioremediation project is an experiment that is pushing the envelopes of science.

There are seldom "bright lines" between what has worked in the past or in a highly controlled laboratory setting and the conditions at hand in an actual biotechnological operation. After all, these are living things that are being manipulated. Biological systems with myriad "black boxes" are always messier than abiotic systems. As such, all engineers and scientists must ensure that their work is of the utmost integrity and that it exceeds the norms

Environmental Biotechnology

of acceptability, since biotechnology's impacts can be widespread and irreversible. Any profession's role is the assurance that each of its members adheres to a defined set of ethical expectations. That is certainly the case for bioengineering and its allied professions and research disciplines.

BIOENGINEERING PERSPECTIVES

> Ethical responsibility ... involves more than leading a decent, honest, truthful life, as important as such lives certainly remain. And it involves something much more than making wise choices when such choices suddenly, unexpectedly present themselves. Our moral obligations must ... include a willingness to engage others in the difficult work of defining what the crucial choices are that confront technological society and how intelligently to confront them.
>
> **Langdon Winner (1990)[3]**

The engagement that Langdon speaks of necessitates both the bottom-up and the top-down approaches to the conception, design, realization and operation of biotechnological enterprises. This engagement must account for sound technical principles and the ethical and social dimensions of biotechnologies. This engagement holds doubly true for environmental aspects of bioengineering. The organisms involved in solving problems can, in other venues, be hazardous. In addition, environmental protection is one of the modern expectations of responsible engineering and science. It is one of the indications of public trust in a profession. Any enterprise is evaluated on how much insult it inflicts on the environment. What would be a worthy enterprise may be too distasteful for the public if it adds too much pollution in its achieving these otherwise commendable ends.

The responsibilities of the bioengineer are prospective and systematic. They are prospective in that they must adapt to a changing set of rules, both scientific and societal. They are systematic in that everything in biosystems affects every other thing. The "thing" can be a material substance (e.g. a xenobiotic toxin) or an organism (both the agent and the receptor, e.g. a microbe and a human host for a contaminant, or a microbe and ecosystem in a remediation project). The "thing" may also be a process or mechanism, such as those processes discussed in Chapter 3 of this book. K.W. Miller, Editor of the *IEEE Technology and Society Magazine*, captured this prospective, systematic perspective, adding society's expectation of scientific vigilance:

> From stone spearheads to nanotubes, our artifacts can change how we live and, ultimately, who we are. The social significance of technological change requires us to take responsibility for the design, implementation, and deployment of the things we make. ... [I]t is clear that technology can change society. But sometimes we lose sight of the idea that society can change technology. [4]
>
> As technological changes come at us thick and fast, we can be overwhelmed. Either consciously or unconsciously, we may start to accept some degree of technological determinism or Chandler's inevitability thesis. The idea that technology is going to happen no matter what we do is both tempting and highly dangerous, as many have pointed out. We have to keep reminding ourselves that we not only *can* steer technology, but that we *should.* Engineers especially must remember that part of our professional responsibility is to shape technology for the benefit of the public at large.

As discussed in the three previous chapters, identifying, characterizing, controlling, and reducing risks is an apparent cacophony or symphony (depending on one's point of view) of multi-faceted efforts to solve ill-posed problems. There is seldom a singular critical path to success. And, being human, even a well-designed, properly focused path to success is seldom followed as conceived, especially when living things are involved. Even the smallest of organisms does not completely behave as expected, particularly when a project is *scaled up* from the laboratory to the meso-scale pilot to the prototype and finally, in the real world.

Error, variability, and unexpected contingent events are missed or mischaracterized in a setting with additional and changing variables, even after applying the lessons from each scale-up step.

The challenge is to achieve something of importance, e.g. medical, agricultural, manufacturing and environmental breakthroughs, without undue harm in space and time. Spatial harm is a well-documented phenomenon. A contaminant is released into the environment and causes immediate harm within a defined distance of the release. As discussed in Chapters 3 and 4, it may be transported to places where it causes additional harm and it may be transformed by abiotic and biotic processes to form compounds that cause harm within their own spheres of influence.

Temporal harm can take the form of short-term impacts. A chemical or biological agent is released into the environment, with an immediate, acute impact. These releases can range from the imperceptible (e.g. a release of a highly reactive substance in sufficiently low quantities and distance between the release and the receptor so that it breaks down long before causing any harm), to disastrous (an immediate release of sufficiently large quantities of a substance that reaches the receptor and elicits effects to a large population of receptors). The insult may be either isolated or episodic (e.g. a one-time event like a hurricane or violation in containment of a genetically modified bacterium), or it may be continuous (e.g. a leak that has released a chemical or biological agent over decades into the groundwater). Temporal harm may also take the form of longer-term impacts, such as the continuous release of a harmful agent until some point in time when a threshold is reached whereupon an adverse effect occurs in a population or ecosystem. Another example of long-term temporal harm is that of the latency period, where the exposure to an agent may not cause immediate harm, but after prolonged *latency period*, the adverse effect is manifested (eg. cancers commonly do not occur until years after the exposure to the carcinogen).

The measure of temporal harm is *persistence*. The measure of environmentally acceptable designs and operations is *sustainability*. For example, a recalcitrant chemical compound will have a high measure of persistence. A well-designed biotechnology must be sustainable, i.e. it has sufficient *design integrity* to ensure that unacceptable, negative impacts will not arise down the road. If fact, to be truly sustainable, an enterprise must improve with time.

CODES OF CONDUCT

All engineering professions and many scientific disciplines have embraced sustainability as what they consider to be acceptable research and practice. The American Society of Civil Engineers (ASCE) was the first of the engineering disciplines to codify this into the norms of professional practice. The first canon of the ASCE code of ethics now reads:

> Engineers shall hold paramount the safety, health and welfare of the public *and shall strive to comply with the principles of sustainable development* in the performance of their professional duties [emphasis added]. [5]

The ASCE code's most recent amendment, on November 10, 1996, incorporated the principle of sustainable development. As a subdiscipline of civil engineering, much of the environmental engineering mandate is encompassed under engineering professional codes in general and more specifically in the ASCE. The Code mandates four principles that engineers abide by to uphold and to advance the "integrity, honor, and dignity of the engineering profession":

Using their knowledge and skill for the enhancement of human welfare and the environment.
Being honest and impartial and serving with fidelity the public, their employers, and clients.
Striving to increase the competence and prestige of the engineering profession.
Supporting the professional and technical societies of their disciplines.

The code further articulates seven fundamental canons, beginning with the just mentioned first canon:

> Engineers shall hold paramount the safety, health, and welfare of the public and shall strive to comply with the principles of sustainable development in the performance of their professional duties.
> Engineers shall perform services only in areas of their competence.
> Engineers shall issue public statements only in an objective and truthful manner.
> Engineers shall act in professional matters for each employer or client as faithful agents or trustees, and shall avoid conflicts of interest.
> Engineers shall build their professional reputation on the merit of their services and shall not compete unfairly with others.
> Engineers shall act in such a manner as to uphold and enhance the honor, integrity, and dignity of the engineering profession.
> Engineers shall continue their professional development throughout their careers, and shall provide opportunities for the professional development of those engineers under their supervision.

The first canon is a direct mandate for the incorporation of sustainability principles in any design. It is a call to evaluate the design and operation under consideration for potential applications of *green engineering*. The remaining canons prescribe and proscribe activities to ensure trust. It is important to note that the Code applies to all civil engineers, not just environmental engineers. Thus, even a structural engineer must "hold paramount" the public health and environmental aspects of any project, and must seek ways to ensure that the structure is part of an environmentally sustainable approach.

The engineering professions' increasing concern that designs be sustainable is matched by other disciplines engaged in biotechnologies. As evidence, the American Society for Microbiology (ASM) revised its code of conduct in 2005 to reflect biotechnological vulnerabilities of its membership, when they added a sixth guiding principle to address the problem of outright bioterrorism, but also potential dual use dilemmas. The principles now read [6]:

> ASM members aim to uphold and advance the integrity and dignity of the profession and practice of microbiology.
> ASM members aspire to use their knowledge and skills for the advancement of human welfare.
> ASM members are honest and impartial in their interactions with their trainees, colleagues, employees, employers, clients, patients, and the public.
> ASM members strive to increase the competence and prestige of the profession and practice of microbiology by responsible action and by sharing the results of their research through academic and commercial endeavors, or public service.
> ASM members seek to maintain and expand their professional knowledge and skills.
> ASM members are obligated to discourage any use of microbiology contrary to the welfare of humankind, including the use of microbes as biological weapons. Bioterrorism violates the fundamental principles upon which the Society was founded and is abhorrent to the ASM and its members. ASM members will call to the attention of the public or the appropriate authorities misuses of microbiology or of information derived from microbiology.

The ASM's first two principles, it can be argued, would cover most of the nefarious activities addressed in Principle 6. Indeed, the repercussions of biotechnologies are problematic and potentially irreversible; it seems the ASM needs to remind its members that human welfare is vulnerable to such misuse. Just as the first mandate of all engineers, public trust, calls for sustainability, the first mandate of microbiologists requires vigilance against abuses and misuses of the knowledge they unveil.

A recurring theme throughout this book is that the environment is both a beneficiary and victim of biotechnology. This trust, then, is most certainly not deferred exclusively to the so-called "environmental" professions but is truly an overarching mandate for all professions (including medical, legal, and business-related professionals). Gaining and keeping trust in science is more complicated than merely avoiding improprieties. It invokes a social contract between the purveyors and users of science. In *Pandora's Picnic Basket*, University of California professor Alan McHughen [7] states:

> The final decision on accepting or rejecting GM (genetic modification) technology demands consideration of social, ethical, political, and other aspects. But because so much of the debate is founded on science and scientific interpretation, this book concentrates on the science issues. With the scientific and factual basis, you may then incorporate the other aspects to build your position, confident in a stable and solid foundation.

All environmental decisions, particularly those related to emerging technologies with little track record, must incorporate a wide array of perspectives, while simultaneously being based in sound science. The first step in this inclusive decision making process, then, is to ensure that every stakeholder sufficiently understands the data and information gathered when assessing environmental contamination.

Examples of a possible disconnect between scientists and the consuming public is the perception of three prominent risks associated with genetically modified food: (1) allergenicity; (2) toxicity; and (3) anti-nutrition [8]. The US General Accountability Office (GAO) has succinctly summarized these threats (see Table 12.1).

The GAO [9] succinctly encapsulates the challenge:

Beyond these technical challenges, however, lies a more fundamental problem. Because these new technologies are more sensitive, they may identify a flood of differences between conventional and GM food products that existing tests could not detect. Not all of these differences will stem from genetic modification. Some of the differences will stem from the tremendous natural variations in all plants caused by factors such as the maturity of the plants and a wide range of environmental conditions, such as temperature, moisture, amount of daylight, and unique soil conditions that vary by region of the country. For example, there can be a tenfold difference in the level of key compositional elements, such as nutrients, depending on the region in which soybeans are grown. Thus, according to a biotechnology company expert, it will be difficult to differentiate naturally occurring changes from the effects of deliberate genetic modifications.

One of the scientific problems in distinguishing between actual and perceived hazards and risks is the need for benchmarks of what is good or bad and a baseline from which to make such comparisons. The question that the consuming public wants answered is: What constitutes the difference in effects between unmodified and genetically modified organisms? GAO [10] adds:

Industry and university scientists have expressed strong concerns about the problem of interpreting the potential significance of these differences. They believe that the new technologies will be of limited value unless baseline data on the natural variations of nutrients and other compositional values for each of the major food crops can be developed. However, experts disagree on the difficulty of developing this baseline. Some experts, including those at FDA, assert that developing the baseline will be difficult because of the extreme sensitivity of plants to environmental variations. Other experts, especially those pioneering the new techniques, state that a baseline can definitely be established in the next few years.

Numerous scientific tests, assays, and evaluations are being applied to the products of biotechnology (see Figure 12.1). These are generally designed to predict adverse outcomes (see Table 12.2). For example, the regimen usually includes an analysis of the source of

Table 12.1 Risks associated with genetically modified (GM) food

Risks of genetically modified food	Description
Allergenicity	An allergic reaction is an abnormal response of the body's immune system to an otherwise safe food. Some reactions are life threatening, such as anaphylactic shock (a severe allergic reaction that can lead to death). To avoid introducing or enhancing an allergen in an otherwise safe food, the biotechnology food industry evaluates genetically modified (GM) foods to determine whether they are "as safe as" their natural counterparts. For example, in 1996 FDA reviewed the safety assessment for a GM soybean plant that can produce healthier soybean oil. As part of a standard safety assessment, the GM soybean was evaluated to see if it was as safe as a conventional soybean. Although soybeans are a common food allergen and the GM soybean remained allergenic, the results showed no significant difference between its allergenicity and that of conventional soybeans. Specifically, serums (blood) from individuals allergic to the GM soybean showed the same reactions to conventional soybeans.
Toxic reaction	A toxic reaction in humans is a response to a poisonous substance. Unlike allergic reactions, all humans are subject to toxic reactions. Scientists involved in developing a GM food aim to ensure that the level of toxicity in the food does not exceed the level in the food's conventional counterpart. If a GM food has toxic components outside the natural range of its conventional counterpart, the GM food is not acceptable. To date, GM foods have proven to be no different from their conventional counterparts with respect to toxicity. In fact, in some cases there is more confidence in the safety of GM foods because naturally occurring toxins that are disregarded in conventional foods are measured in the pre-market safety assessments of GM foods. For example, a naturally occurring toxin in tomatoes, known as "tomatine," was largely ignored until a company in the early 1990s developed a GM tomato. FDA and the company considered it important to measure potential changes in tomatine. Through an analysis of conventional tomatoes, they showed that the levels of tomatine, as well as other similar toxins in the GM tomato, were within the range of its conventional counterpart.
Anti-nutritional effects	Anti-nutrients are naturally occurring compounds that interfere with absorption of important nutrients in digestion. If a GM food contains anti-nutrients, scientists measure the levels and compare them to the range of levels in the food's conventional counterpart. If the levels are similar, scientists usually conclude that the GM food is as safe as its conventional counterpart. For example, in 1995 a company submitted to FDA a safety assessment for GM canola. The genetic modification altered the fatty acid composition of canola oil. To minimize the possibility that an unintended anti-nutrient effect had rendered the oil unsafe, the company compared the anti-nutrient composition of its product to that of conventional canola. The company found that the level of anti-nutrients in its canola did not exceed the levels in conventional canola. To ensure that GM foods do not have decreased nutritional value, scientists also measure the nutrient composition, or "nutrition profile," of these foods. The nutrient profile depends on the food, but it often includes amino acids, oils, fatty acids, and vitamins. In the example previously discussed, the company also presented data on the nutrient profile of the GM canola and concluded that the significant nutrients were within the range of those in conventional canola.

Source: US General Accountability Office (2002). Genetically Modified Foods: Experts View Regimen of Safety Tests as Adequate, but FDA's Evaluation Process Could Be Enhanced. Report No. GAO-02- 566.

transferred genetic material, specifically whether the source of the transferred gene has a history of causing allergic or toxic reactions or containing anti-nutrients, along with the degree of similarity between the amino acid sequences in the newly introduced proteins of the GM food and the amino acid sequences in known allergens, toxins, and anti-nutrients. The tests should also include information about *in vitro* digestibility (i.e., the ease with which proteins break down in simulated digestive fluids), along with a comparison of the severity of individual allergic reactions to the GM product and its conventional counterpart as measured through blood (serum) screening – when the conventional counterpart is known to elicit allergic reactions or allergenicity concerns remain. Predicted hazards also benefit from reliable information regarding endogenous responses, e.g. any changes in

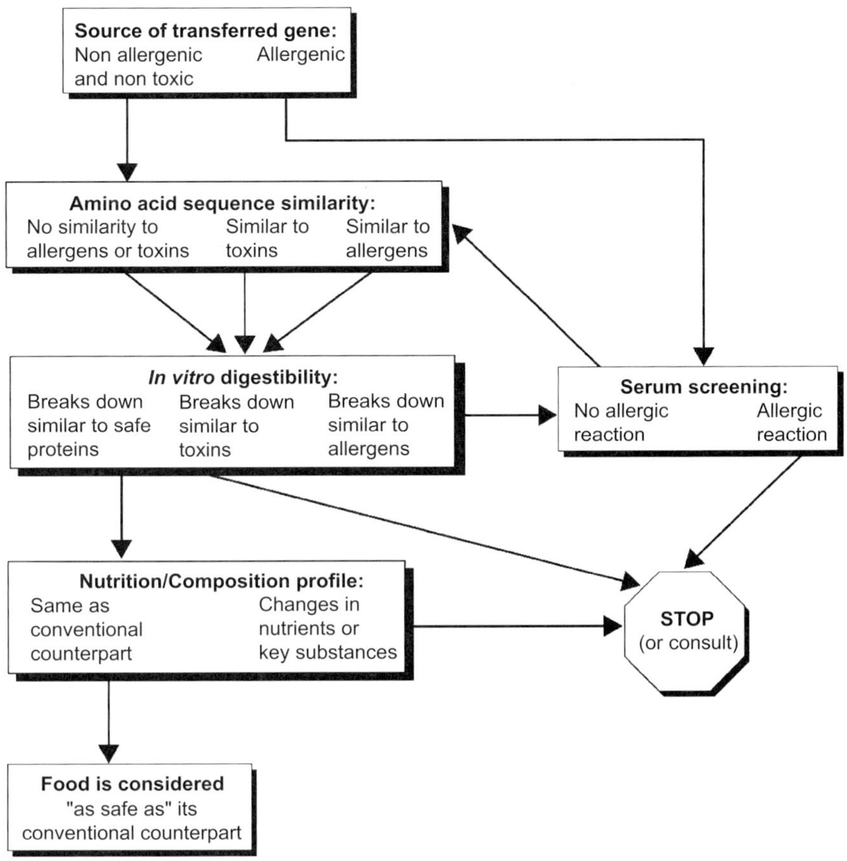

FIGURE 12.1
Example of the regimen of tests and critical path to predict safety of genetically modified foods. Anti-nutrients are tested as a subset of toxicity. In addition, they are often measured with a simple nutrition/composition profile. If a company transfers genetic material from an allergenic source and undertakes serum screening tests, it does not have to go through serum screening again if *in vitro* digestibility tests uncover a similarity to an allergen. At such a point, it would be assessed by amino acid sequence similarity and *in vitro* digestibility tests for potential toxicity.
Source: US General Accountability Office (2002). Genetically Modified Foods: Experts View Regimen of Safety Tests as Adequate, but FDA's Evaluation Process Could Be Enhanced. Report No. GAO-02- 566.

nutrient substances, such as vitamins, proteins, fats, fiber, starches, sugars, or minerals due to genetic modification. Occasionally, the regimen of tests also includes animal studies for toxicity [11].

The testing is designed to build trust with the consuming public and provides lessons to practitioners of environmental biotechnologies. Trust depends on more than assuring the public that risks will be minimal and, hopefully, acceptable. For example, with the emergence of a newer, greener era, the public is increasingly expecting, indeed prompting, companies and agencies to look beyond ways to treat pollution to find better processes to prevent the pollution in the first place. In fact, the adjective "green" has been showing up in front of many disciplines, e.g. green chemistry, green engineering, and green architecture, as has the adjective "sustainable." Increasingly, companies have come to recognize that improved efficiencies save time, money, and other resources in the long run. Hence, companies are thinking systematically about the entire product stream in numerous ways:

- Applying sustainable development concepts, including the framework and foundations of "green" design and engineering models.
- Applying the design process within the context of a sustainable framework: including considerations of commercial and institutional influences.
- Considering practical problems and solutions from a comprehensive standpoint to achieve sustainable products and processes.

Table 12.2 Tests with potential to meet the US Food and Drug Administration's policy regarding the assessment of the safety of genetically modified foods (see Figure 12.1)

Test or assay	Description
Source of the transferred genetic material	Examining the source of the transferred genetic material is the starting point in the regimen of tests for safety assessments. Two principles of allergenicity assessment underlying the regimen of tests that contribute to adequate safety are that assessments scientists (1) avoid transferring known allergenic proteins and (2) assume all genes transferred from allergenic sources create new food allergies until proven otherwise. If the source contains a common allergen or toxin, industry scientists must prove that the allergenic or toxic components have not been transferred. However, as a practical matter, biotechnology companies repeatedly state that if the conventional food is considered a major food allergen, they will not transfer genes from that source. Accordingly, experts from FDA and the biotechnology industry agree that the probability of introducing a new allergen, enhancing a toxin, or enhancing an anti-nutrient is very small.
Amino acid sequence similarity	The next step involves a comparison between the amino acid sequences of the transferred proteins of the GM food plant and those of known allergens, toxins, or anti-nutrients. If scientists detect an amino acid sequence in a GM food identical or similar to one in an allergen, toxin, or anti-nutrient, then there is a likelihood that the GM food poses a health risk. Overall, sequence similarity tests are very useful in eliminating areas of concern and revealing areas for further evaluation.
Digestibility tests	*In vitro* digestibility tests are a primary component of all GM food safety assessments. These tests analyze the breakdown of a GM protein in simulated human digestive or gastric fluids. The quick breakdown of a GM protein in these fluids indicates a very high likelihood that the protein is not allergenic or toxic. Safe dietary proteins are almost always rapidly digested, while allergens and toxins are not.
Serum screening	If a gene raises allergenicity concerns, a company can include serum screening tests in its safety assessment of a GM food. Serum screening is used only for allergenicity assessment. Serum screening involves evaluating the reactivity of antibodies in the blood of individuals with known allergies to the plant that was the source of the transferred gene. Antibody reactions suggest the presence of an allergenic protein. Serum screening tests are valuable because they can expose allergens whose presence was only suggested in amino acid sequence similarity tests. Since there are neither abundant, appropriate stored serums nor many suitable human test subjects, these tests cannot always be used.
Nutritional and compositional profile	Scientists also create a nutritional and compositional profile of the GM food to assess whether any unexpected changes in nutrients, vitamins, proteins, fibers, starches, sugars, minerals, or fats have occurred as a result of the genetic modification. While changes in these substances do not pose a risk of allergenicity, toxicity, or anti-nutrient effects to human health, creating a nutritional and compositional profile further ensures that the GM food is comparable to its conventional counterpart.
Animal (*in vivo*) studies	Biotechnology companies occasionally use animal studies to confirm the results of prior toxicity tests. For the most part, these studies have involved feeding extraordinarily high doses of the modified protein from a GM food to mice. The doses of the modified protein are often hundreds to thousands of times higher than the likely dose from human diets. Scientists perform these studies to determine if there are any toxic concerns from the GM food. Animal studies also have the potential to predict allergenicity in humans, although scientists have not yet identified an animal that suffers from allergic reactions the same way that humans do. The brown Norway rat has provided the closest approximation to human allergic reactions to several major food allergens. However, animal models – as predictors of allergenic responses in humans – are not scientifically accepted at this time.

Source: US General Accountability Office (2002). Genetically Modified Foods: Experts View Regimen of Safety Tests as Adequate, but FDA's Evaluation Process Could Be Enhanced. Report No. GAO-02- 566.

- Characterizing waste streams resulting from designs.
- Understanding how first principles of science, including thermodynamics, must be integral to sustainable designs in terms of mass and energy relationships, including reactors, heat exchangers, and separation processes.
- Applying creativity and originality in group product and building design projects.

Ironically, and perhaps predictably, many biotechnologies are justified for their "greenness." Biofuel sources and refining, microbial manufacturing, and numerous other biotechnologies are compared to conventional processes, often favorably, in terms of sustainable solutions to big social problems. Biofuels are seen as better than fossil fuels; microbial manufacturing is preferable to conventional chemical processes. Thus, the benefits in the benefit-to-cost ratio can be quite compelling. Thus, biotechnologies can be justified, so long as they do not introduce comparatively high and different types of risks with unacceptable uncertainties. Such decisions are not exclusively scientific, but are also ethical.

ETHICS AND DECISIONS IN ENVIRONMENTAL BIOTECHNOLOGY

Philosopher Immanuel Kant's *categorical imperative* may provide insight into how to balance the benefits and risks of biotechnological research, development and operations [12]. This imperative, the most famous of Kant's maxims, can be briefly summarized to say that to determine whether something is ethical, one should consider what would happen if that act were a law adopted by everyone. If the biotechnology is designed to meet some environmental ideal, while showing complete respect for humankind, the act is moral; if, on the other hand, it hurts others, the act is immoral. Kant embraced the categorical imperative as the theoretical underpinning for duty ethics (a philosophical and ethical school of thought, known as *deontology*).

This can be a sticking point for scientists and engineers; that is, most would not engage in any enterprise that does not have a noble endpoint, albeit far removed from one's particular project. However, the categorical imperative has two parts, the work itself *and* the debt to duty. A biotechnology can fail to meet either or both of these requirements.

Any biotechnology that is poorly conceived, designed, and operated fails the test of professional duty, even if the stated purpose is noble. An example would be to miss some key detrimental traits of a strain of microbes that effectively degrades a recalcitrant pollutant. The endpoint, destruction of a pollutant, meets half of the categorical imperative (noble objective), but fails with regard to universalization. That is, if all bioengineers behaved this way, the world would be a much riskier place to live.

Conversely, even a well-conceived, designed, and operated biotechnology can fail to meet the categorical imperative if the means for carrying it out are suspect. Staying with the bioremediation example, if the bioengineer did not follow appropriate safety protocols (e.g. physical containment) or if he or she violated research norms (cooking, trimming or forging data) to genetically engineer a strain of microorganisms that effectively degrades the recalcitrant pollutant, this would be an unethical act. In research and practice, noble ends do not justify immoral means.

The third scenario is a well-conceived, designed, and operated biotechnology, but one that is detrimental to the public's safety, health, and welfare. The classic examples in history are instruments of torture. The contemporary biotechnological examples include genetically engineered microbes for bioterrorism. These blatant immoralities are obviously blotches on past and present scientific research. However, there are elements that are more subtle that need to be avoided by researchers and practitioners. Due diligence requires that one considers all possible good and bad outcomes of a project. The categorical imperative requires that even a good biotechnology will need to have commensurate safety and security measures to ensure that it is not misused. Even though one's research is not related to bioterrorism, the imperative requires that those who know the most consider the points of vulnerability, so that these points will not be misused and adapted for harmful purposes (i.e. the so-called *dual use* quandary in an emergent technological research).

It must be noted that Kant's imperative is not universally accepted by philosophers and scientists. John Stuart Mill's utilitarian axiom of "greatest good for the greatest number of people" is moderated by his *harm principle* which holds that even if an act can be good for the majority, it may still be unethical if it causes undue harm to individuals. In this way, the sensitive subpopulations (e.g. highly allergic children) are better targets for risk reduction than are the general population. For ecosystems, the corresponding sensitive subpopulations would be threatened and endangered species and vulnerable habitats. This also is consistent with contractarian ethics (i.e. those spawned by Thomas Hobbes' social contract theory). For example, John Rawls has moderated the social contract with the *veil of ignorance* as a way to consider the perspective of the most vulnerable members of a system [13].

Although the end-versus-duty balance varies among schools of thought, most call for some modicum of a systematic view and a balance between a needed outcome and an ethical pathway to achieve it. Indeed, this is the substance of Aristotle's "golden mean." That is, the best solution to a problem optimizes variables to achieve a balance that achieves the societal objective and simultaneously allows the individual engineer to act responsibly.

The golden mean can be likened to an environmental index (see Chapter 4), which includes numerous physical, chemical and biological variables. Some of these variables are essential to a healthy system. Some are desirable. Those that are detrimental are to be avoided. Those that are essential are to be selected and encouraged. Some variables are minimum operators. That is, if a value falls below a threshold, the whole system crashes. Some are maximum operators, that if their value is above a certain level, the whole system crashes.

UNINTENDED CONSEQUENCES

All biotechnological decisions *can* lead to unanticipated consequences. One could argue that every decision in fact *does* lead to unanticipated outcomes. For example, moving from the circumscribed and controlled conditions of the laboratory is always different than moving to the prototype stage, which is always different from moving to actual conditions in the environment. We are never completely certain about the extent of how things will change once they reach the environment. This is chaos. Initial conditions may be relatively well defined, but almost immediately after, new factors appear.

Some argue that we may need to proceed carefully by following a "precautionary principle" which states that if the consequences of an action, such as the application of a new technology, are unknown but the possible scenario is sufficiently devastating, then it is prudent to *avoid the action*. However, the precautionary approach must also be balanced against opportunity risks. In other words, by our extreme caution are we missing opportunities that would better serve the public and future generations? The key to balancing these connections can be a complete and accurate characterization of risks and benefits. Unfortunately, biotechnological design decisions are often not fully understood until after the fact (and viewed through the prism of lawsuits and media coverage).

SYSTEMATIC BIOTECHNOLOGY AND THE STATUS QUO

The interdependence of science, engineering, and technology are important in environmental biotechnology, as in almost any technical field. Interestingly, thermodynamic terms are frequently used, and sometimes abused by the lay public and by the scientific community. The system and its boundary, surroundings, constraints, and drivers must be known in order to engage in biotechnological design. We must be aware of other connotations of the word "system," especially that used by the lay public for a mental construct that influences the way that one thinks. Designing to improve a process or to solve a problem requires that we keep in mind how best to measure success "systematically".

The two metrics that are most commonly applied to a system are efficiency and effectiveness. The first is a thermodynamics term. The second is a design term. Efficiency and effectiveness refer to whether the design is conducive to the purpose for which it was created, and whether the design function performs some task, respectively. As mentioned in Chapter 11, unlike mathematics, engineering, and architecture are not exclusively deductive endeavors. Bioengineers also base knowledge on experience and observation. Design professions first generate rules based on observations (i.e., the laws of nature: chemistry, physics, biology, etc.) of the world around them; the way things work. Once they have this understanding, they may apply it by using the rule to create something, some technology, designed to reach some end, from manufacturing insulin to making a crop more resistant to pests to degrading a toxic waste. According to the National Academy of Engineering:

> Technology is the outcome of engineering; it is rare that science translates directly to technology, just as it is not true that engineering is just applied science. [14]

It is worth pondering why most bioscientists are never satisfied with the *status quo*, why they risk failure to advance science in such seemingly small doses. Part of the answer is innovation, and the need to find out why something is not able to work in certain venues. The playing field is not even. The *status quo*, by definition, works against innovation. At the other end are the extremely risk averse who prefer the *status quo* rather than introduce any risk. Some say that is the extreme position of the precautionary principle. However, some of the resistance to adoption is the sheer difference between what the public perceives to be "normal" and "abnormal." Granted, in the 18th century it was not normal to treat a waste before releasing it into a river. At that time, it may not have even been all that harmful, since the natural microbial population and the rest of the biological system could degrade the relatively dilute wastes efficiently. Also, most of the chemical compounds released to the river were not synthetic, so the microbes had already adapted capacities to biodegrade them. Once the petrochemical revolution and industrial expansion took hold, the natural processes were no longer able to withstand the onslaught of chemical loading. Thus, the "normal" became the "no longer acceptable."

Bioengineering and biotechnology represented a paradigm shift, the term coined in the late 20th century by Thomas S. Kuhn. He changed the meaning of the word "paradigm," extending the term to an accepted specific set of scientific practices. The scientific paradigm is made up of what is to be observed and analyzed, the questions that arise pertaining to this scientific subject matter, to whom such questions are to be addressed, and how the results of the investigations into this subject matter will be interpreted. The paradigm can be harmful if incorrect theories and information become accepted by the scientific and engineering communities. Excessive comfort with the status quo, i.e. xenophobia, can be sometimes attributed to a community's well-organized protections against differences, i.e. groupthink [15].

Innovations in design occur when a need or opportunity arises. "Necessity is the mother of invention." Environmental bioscience and technology have followed a progression similar to other research advances since the mid-20th century. In 1944, Vannevar Bush, Franklin D. Roosevelt's director of the wartime Office of Scientific Research and Development, was asked to consider the role of science in peacetime. As reported by science policy expert Donald Stokes, Bush did this through his work *Science, the Endless Frontier*, through two aphorisms. The first aphorism was that "basic research is performed without thought of practical ends." According to Bush, basic research is to contribute to "general knowledge and an understanding of nature and its laws." Seeing an inevitable conflict between research to increase understanding and research geared towards use, he held that "applied research invariably drives out pure" [16].

Today, Bush's "rugged individual approach" has been largely supplanted by a paradigm of teamwork. Certainly, any emerging technology thrives on the synergies of teams. Here the emphasis in design has evolved toward a cooperative approach. The paradigm recognizes that, to be effective, we need not only groups of people who are technically competent but also who are good at collaborating with one another in order to realize a common objective [17]. If we are to succeed by the new paradigm, we have to act synergistically.

Basic research is defined by the fact that it seeks to widen the understanding of the phenomena of a scientific field – it is guided by the quest to further knowledge. Numerous influential works of research are in fact driven by both these goals. A prime biotechnological example is the work of Pasteur, who both sought to understand the microbiological processes he discovered, but also sought to apply this understanding to the prevention of the spoilage of vinegar, beer, wine, and milk [18].

The disparity between basic and applied research is captured in the "linear model" of the dynamic form of the postwar paradigm. It is important to keep in mind though that in the dynamic flow model, each of the successive stages depends upon the stage before it (Figure 12.2).

There is an interesting parallel between the advancement of science and the application of science. For example, does science always lead to engineering that subsequently drives the need for technology? Of course, this may be the default, but is not the exclusive transition. Engineering has driven basic science (e.g. bioscience's "black boxes" that progressively, but never completely, become understood by bioscience researchers). Technology has driven both science and engineering. A new device allows scientists to embark on whole new areas of research (e.g. the PCR in rDNA research) and engineers to conceive new designs (e.g. the DNA markers allow for enhanced bioremediation projects).

This simple model of scientific advances has come to be called technology *transfer*, as it describes the movement from basic science to technology. The first step in this process is basic research, which charts the course for practical application, eliminates dead ends, and enables the applied scientist and engineer to reach their goal quickly, and economically. Then, applied research involves the elaboration, and the application of the known. Here, scientists convert the possible into the actual. The final stage in the technological sequence, development, is the stage where scientists systematically adapt research findings into useful materials, devices, systems, methods, processes, etc. [19].

The characterization of evolution from basic to applied science has been criticized for being too simple an account of the flow from science to technology. In particular, the one-way flow from scientific discovery to technological innovation does not seem to fit with 21st century science. The supposition that science exists entirely *outside* of technology is rather absurd in today's way of thinking. In fact, throughout history there is seen a reverse flow, a flow from technology to the advancement of science. The innovation of the calculus and the inventions of the microscope and telescope, and later examples of fractal dimensions and rDNA illustrate that science has progressively become more technology-derived [20]. Biotechnology is a prime example of the technology advancing the science and vice versa.

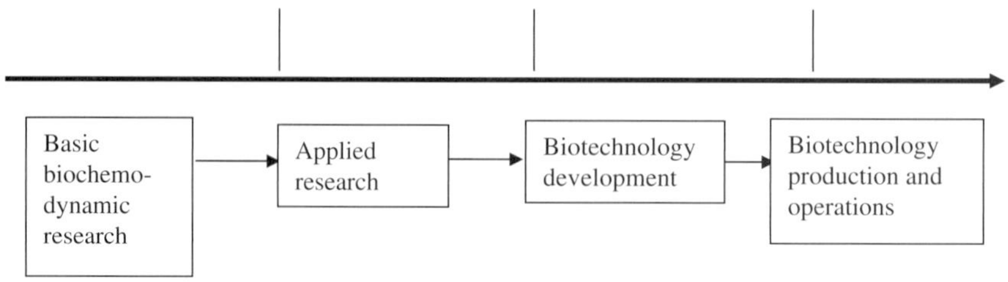

FIGURE 12.2
Progression from basic research to product/system realization.

The relationship between basic and applied sciences is not universally held within the bioscience and bioengineering communities. Some agree that:

> The terms basic and applied are, in another sense, not opposites. Work directed toward applied goals can be highly fundamental in character in that it has an important impact on the conceptual structure or outlook of a field. Moreover, the fact that research of such a nature can be applied does not mean that it is not also basic. [21]

Biosystematic design based in sound science is actually a synthesis of the goals of understanding and use. Good design, then, is the marriage of theory and practice. One could argue that Louis Pasteur was among the first bioengineers, optimizing biological theory and utility. The one-dimensional model of Figure 12.2 consists of a line with "basic research" on one end and "applied research" on the other (as though the two were polar opposites). Pasteur's worldview could be force fit into this model, by placing his design paradigms at the center of the flow in Figure 12.2. However, Pasteur's equal and strong commitments to *understanding* the theory (microbiological processes) and to practice (controlling the effects of these processes), would cover the entire line segment. Arguably, two points within a spectrum better represent Pasteur: one at the "basic research" end of the spectrum and another at the "applied research" end of the spectrum. This placement led Stokes to suggest a different model, i.e. a matrix rather than a flow chart, that reconciles the shortcomings of this one-dimensional model (see Figure 12.3).

This model can also be applied to the entities conducting biotechnology research and development. For example, research that takes place within a university can be compartmentalized in a manner similar to that in Figure 12.4. The science departments are concerned with knowledge-building, the engineering departments with applied knowledge to understand how to solve society's problems, and the university designer is interested in finding innovative ways to use this knowledge. For example, the medical doctor at the university's medical center may know what research has led to a particular medical procedure and the devices used in that procedure, but may want to "figure out" better designs in terms of ease of application, improved recovery time, and better drug delivery. The physician is behaving much like Thomas Edison, who was most interested in utility, and less interested in knowledge for knowledge sake. In addition, the physician must work closely with the health administrators of the university, who purchase the devices and maintain the systems. This is not to say that innovations do not come from the southwest box in Figure 12.4, because they clearly do. It simply means that their measures of success at the university stress operation and maintenance. In fact, the quadrants must all have feedback loops to one another.

This view can also apply to symbiotic relationships among institutions. Duke University is located at one of the points of Research Triangle in North Carolina. The other two points are

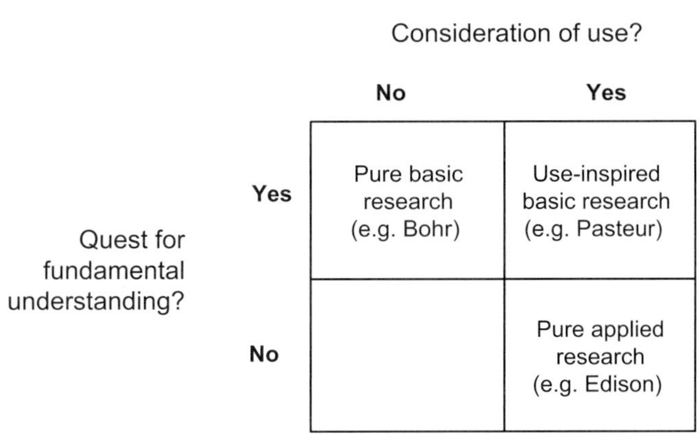

FIGURE 12.3
Research categorized according to knowledge and utility drivers.
Source: Adapted from D.E. Stokes (1997). *Pasteur's Quadrant.* The Brookings Institution, Washington, DC.

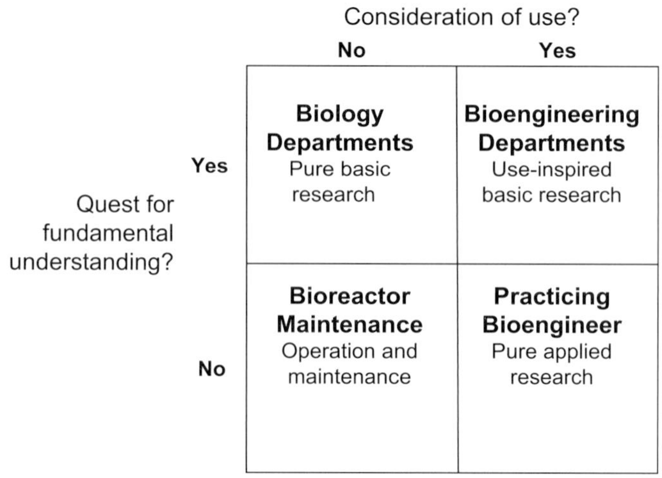

FIGURE 12.4
University research categorized according to knowledge and utility drivers.

the University of North Carolina – Chapel Hill and North Carolina State University. All three schools have engineering programs, but their emphasis differs somewhat. Duke is recognized as a world leader in basic research, but its engineering school tends to place a greater emphasis on application of these sciences. For example, there is much collaboration between Duke's schools of medicine and environment with the biomedical and environmental engineering programs in Duke's Pratt School of Engineering. The University of North Carolina also has a world-renowned medical school, but its environmental engineering program is housed in the School of Public Health. So, this engineering research tends to advance health by addressing environmental problems. North Carolina State is the first place that the State of North Carolina looks for designers, so the engineers graduating from NC State are ready to design as soon as they receive their diplomas. However, NC State also has an excellent engineering research program that applies the basic sciences to solve societal problems. All of this occurs within the scientific community of the Research Triangle, with a variety of research taking place in Research Triangle Park (RTP). The RTP research is supported by both private and public entities with particular, e.g. biotechnological research is driven by needs expressed by the environmental, agricultural, industrial and medical communities. In this way, the RTP researchers are looking for new products and better processes. The RTP can be visualized as the "Edison" of the Triangle, although research in the other two quadrants is ongoing in the RTP labs. This can be visualized in an admittedly oversimplified way, as in Figure 12.5.

The degree to which a given body of research seeks to expand understanding is represented on the vertical access, and the degree to which the research is driven by considerations of use is represented on the horizontal axis. A body of research that is equally committed to potential utility *and* to advancing fundamental understanding is represented as "use-inspired" research [22].

In an area as complex as environmental biotechnology, arriving at the right amount of shift is a challenge. For example, not changing is succumbing to groupthink, but changing the paradigm beyond what needs to be changed can be unprincipled, lacking in scientific rigor. Groupthink is a complicated concept. As evidence, recently an undergraduate team in a Duke Professional Ethics course considered the role of groupthink in technical decisions and concluded that it is a worthwhile mechanism. Quite possibly, some of the students either did not read or disagreed with the assigned text's discussion of the matter, but nonetheless made some good points about the value of the members of a group thinking in similar ways (see Figure 12.6). Consensus views are often very valuable, but lack of dissent can lead to overreliance on existing approaches, even in the face of contrasting evidence.

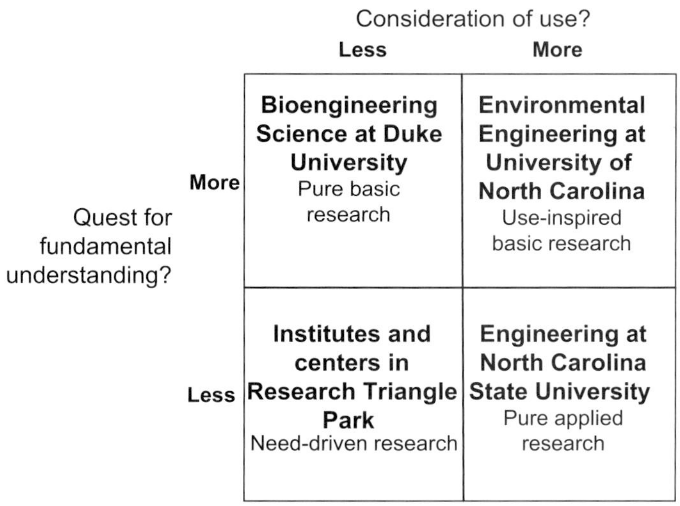

FIGURE 12.5
Simple differentiation of the knowledge and utility drivers in the design-related research ongoing at institutions of the Research Triangle, North Carolina.

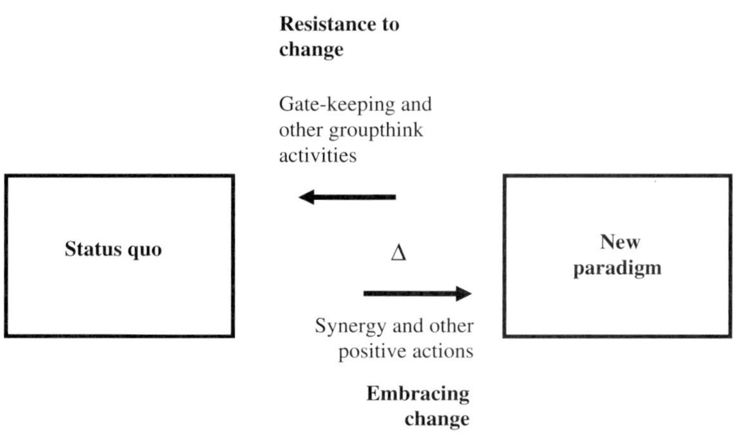

FIGURE 12.6
Resistance and openness to change: the difference between groupthink and synergy.

During the last quarter of the 20th century, advances and new environmental applications of bioscience, bioengineering, and their associated biotechnologies began to coalesce into a whole new way to see the world, at least new to most of Western Civilization. Ancient cultures on all continents, including the Judeo-Christian belief systems, had warned that humans could destroy the resources bestowed upon us unless the view as stewards and caretakers of the environment were taken seriously. Scientifically based progress was one of the major factors behind the exponential growth of threats to the environment. Environmental controls grew out of the same science, which is now part of a widely accepted environmental ethos.

It is difficult to place biotechnologies in any one ethos. Most practitioners and researchers immersed in biosciences see their work as beneficial and self-justified. In other words, their specific work is justified by the general societal need for better medicine, food, environmental quality, better products, and more efficient manufacturing. As discussed, such rationalizations can be dangerous, since even noble goals can be achieved by ignoble means. In addition, such single-mindedness does not allow for considerations of possible side effects, contravening risks and unintended consequences. Biotechnology has its share of ends justifying the means, as recent cases in cooking, trimming and forging data can attest. Some who have engaged in these unethical acts saw, or at least claimed to have seen, their deceptions as being necessary to advance science for an overall good. This is in opposition to sound science and public trust. C.P. Snow [23] put it this way:

> If we do not penalize false statements made in error, we open up the way, don't you see, for false statements by intention. And of course a false statement of fact, made deliberately, is the most serious crime a scientist can commit.

The societal demands can be very compelling. That is why Snow likely saw the need to warn most scientists about a problem about which most scientists would reflexively agree, i.e. truth-telling is essential to the scientific method. After all, it all seems quite straightforward. Science is the explanation of the physical world, while engineering encompasses applications of science to achieve results. Thus, what we have learned about the environment by trial and error has incrementally grown into what is now standard practice of environmental science and engineering.

This heuristically attained knowledge base has come at a great cost in terms of the loss of lives and diseases associated with mistakes, poorly informed decisions, and the lack of appreciation of environmental effects. That is why scientists must be careful to guard the public trust. Science and engineering are not popularity contests. That is one worrisome aspect of many public debates.

Scientists can be recruited, sometimes unknowingly, as advocates for one cause or another. Many of these causes are worthwhile. For example, among the most prominently studied environmental problems are those associated with climate change. Many studies are quite good and specifically directed toward a tightly defined research objective. The reality is that research funding follows both scientific and policy-directed agendas. The temptation is to apply for funds and to write research proposals that fit what policy makers want; even if that means that the better and more relevant research would be in another area. It is seldom that the scientific community has 100% consensus on anything except the basic principles (and even these are suspect in quantum mechanics and mathematics). It is scary when one hears that the "issue is settled," as we have heard recently regarding anthropogenic global warming. Indeed there is arguably consensus, but nothing near unanimity, in any scientific discipline at this time. Indeed, the thoughtful scientist must ask what exactly is supposed to be "settled". Is it the overall agreement that CO_2 and other greenhouse gases are a factor in how the earth is warmed by incoming solar radiation? Few scientists would disagree, since this is basic thermodynamics. Beyond these basics, however, the so-called settlement becomes friable.

This creeping advocacy is present in biotechnology, including environmental applications. The general consensus within the environmental engineering community is that genetic engineering appears to work. That is, a genetically enhanced microbe's biodegradation rates can be objectively compared to the rates of an unenhanced strain of the same microbe. However, there is disagreement about the difference. Sometimes, in environmental assessments, all one can say is that something is better or worse. It is an ordinal comparison, perhaps with adjective descriptions like fast, medium, and slow reactions. Stronger quantitation is more difficult and takes many studies and field experiences before empirical equations can be derived.

The general consensus begins to erode when risks are discussed. The genetically enhanced microbe may indeed have a faster rate than a progenitor strain, but is this rate justification enough to ignore certain risks? For example, if the genetically modified organism degrades tricholoethylene 10% faster in soil than natural strains of bacteria acclimated to the TCE, is this sufficient justification to use the new strain, given an unlikely horizontal gene transfer off-site? What if it is 50% faster, meaning that a site will be contaminated half as long with the GM degradation than with natural stains? What if the GM strain can degrade another cancer-causing compound that is almost 100% recalcitrant under environmental conditions? The risk of the microbes would likely be much less than the attendant risk of the chemical in the environment, where it can lead to exposures (e.g. drinking water) that will cause the population's cancer risks to increase. Even in this last scenario, however, the scientist and engineer must disclose completely even unlikely and uncertain risks. It is understandable that technical experts fear even bringing up very unlikely outcomes, since the public is likely to cease on these and not hear the experts' comparative information.

Good risk communication must do better than simply saying the risk of gene flow is 10^{-9} (one in a billion chance) and the risk of cancer from the carcinogen in drinking water is 10^{-6}. First, engineering notation is not generally understood. More importantly, however, is that these risks are not completely comparable (different receptors, different values, ecological versus human). This would be a problem for a technical audience, let alone a diverse one.

At least at first, risk communications about biotechnologies will require more listening than speaking by the experts. This calls for presenting a systematic view of benefits and risks. The good news is that popular culture has come to appreciate the systematic relationship between the biological sciences, engineering, and technologies. As evidence is the concept of "spaceship earth," i.e. our planet consists of a finite life support system and that our air, water, food, soil, and ecosystems were not infinitely elastic in their ability to absorb humanity's willful disregard. The other good news is that our preconceived ideas about how great our project is may cause us to revisit some previously neglected details that, in the long run, may reduce risks and enhance the project from a straightforward, off-the-shelf biotechnology to a site-specific, targeted solution tailored to the explicit environmental problem.

A FEW WORDS ABOUT ENVIRONMENTAL ETHICS

The book of nature is one and indivisible: it takes in not only the environment but also life, sexuality, marriage, the family, social relations: in a word, integral human development.

> Our duties towards the environment are linked to our duties towards the human person, considered in himself and in relation to others. It would be wrong to uphold one set of duties while trampling on the other. Herein lies a grave contradiction in our mentality and practice today: one which demeans the person, disrupts the environment and damages society.
>
> **Pope Benedict XVI [24]**

Since biotechnology is at the center of so much ethical dialogue and debate, it is important to consider concepts that underpin the ways that society places values on environmental resources. This book has addressed the technical aspects of the applications and the possible environmental implications of biotechnologies. Most of the discussion has admittedly been incomplete. The issues surrounding the pursuit of biotechnologies to advance societal problems are complicated and often do not lend themselves completely to a rational, scientific decision framework. They are almost always laden with ethical content. That is, few biotechnological decisions are entirely *amoral*. More often, they involve decisions on whether a decision is morally permissible or impermissible.

Environmental ethics is the set of morals, those actions held to be right and wrong, about how people interact with the environment. A few major ethical viewpoints dominate environmental literature: anthropocentrism; biocentrism; ecocentrism; and sentientism (see Figure 12.7).

Anthropocentrism is the philosophy or decision framework entailing that all and only humans have moral value. Nonhuman species and ecological resources have value only in respect to that associated with human values (known as instrumental value). In this view, biotechnology's value is completely ascertained from the utility that it provides to humans. The arguments within anthropocentrism are about the focus of the biotechnology: for example, justice issues about who will benefit and who will be at risk (certain subpopulations may be susceptible to biotechnologically-derived medicines and vaccines).

Conversely, biocentrism is a systematic and comprehensive account of moral relationships between humans and other living things. The biocentric view requires an acceptance that all living things have inherent moral value, so that respect for nature is the ultimate moral

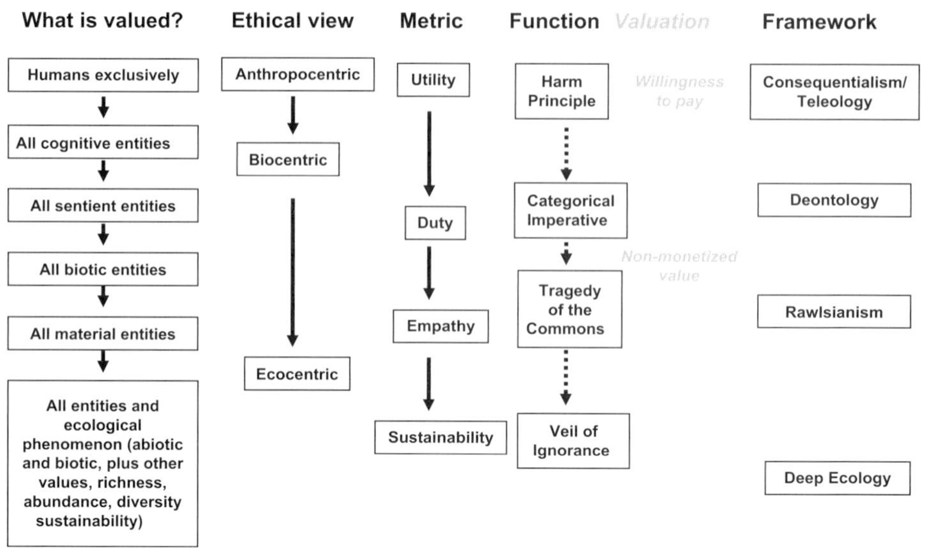

FIGURE 12.7
Continuum of ethical viewpoints.
Source: Adapted from D.A. Vallero (2007). *Biomedical Ethics for Engineers: Ethics and Decision Making in Biomedical and Biosystem Engineering.* Elsevier Academic Press, Burlington, MA.; some information from R.B. Meyers (2003). Environmental Values, Ethics and Support for Environmental Policy: A Heuristic, and Psychometric Instruments to Measure their Prevalence and Relationships. International Conference on Civic Education Research. November 16–18, 2003. New Orleans, Louisiana.

attitude. It is encapsulated by Albert Schweitzer's "reverence for life" [25]. Here, arguments about the value of biotechnologies may have to consider the costs to certain species. For example, is it acceptable to manipulate a fish's genetic material so that it changes color in the presence of certain chemical compounds. Clearly, this has great potential value for human health, i.e. it has anthropocentric utility (e.g. an early warning system of possible drinking water contamination at a municipal water plant), but from the biocentric perspective, this utility must be weighed against the effects this action has on other species.

By extension of the biocentric view, ecocentrism is based on the notion that the whole ecosystem, rather than just single species, has moral value. In the ecocentric view, the biotechnologies are only acceptable if they do not harm entire systems. Thus, even if genetically modifying an organism provides some benefit to humans (e.g. produces a needed medication), and it does not seem to adversely affect the organism that produces the benefit (e.g. a microbe's mitochondrial changes do not harm the individual bacterium), it may still be unacceptable from a systematic perspective (e.g. the change in population ecology by introducing the altered species somehow affects the function or structure of the ecosystem).

Sentient-centered ethics falls between *anthropocentric* (human-centered) and *biocentric* (i.e. broad concern for all living things) ethical frameworks. This approach suggests that all creatures with a nervous system are entitled to moral regard. That said, this view would cause many scientists to do whatever possible, or at least practicable, to prevent or reduce suffering in these other species that share pain. This view demonstrates that from a purely neurological and physiological perspective, the difference between humans and animals is a continuum, as indicated by the development of the nervous system and other physiological metrics.

It is important to note that it can be problematic to apply a single ethical perspective in every circumstance. For example, extreme anthropocentrism can lead to animal cruelty and destruction of habitat since it ignores the critical interdependence among myriad organisms, including humans, and the environment. In fact, anthropocentrism is always utilitarian, since humans are the only organisms making decisions about the usefulness of a new technology or a new application of a technology. In other words, the only utility is the extent to which the technology is good for humans. Thus, the only value placed on the technology is *instrumental,* i.e. how does it serve a certain human need.

Ecologists and others have argued that even an anthropocentric view requires a systematic perspective. That is, even if one cares only about humans, the human support systems of clean air, water, soil, sediment, biota, and even social issues like animal welfare all must be

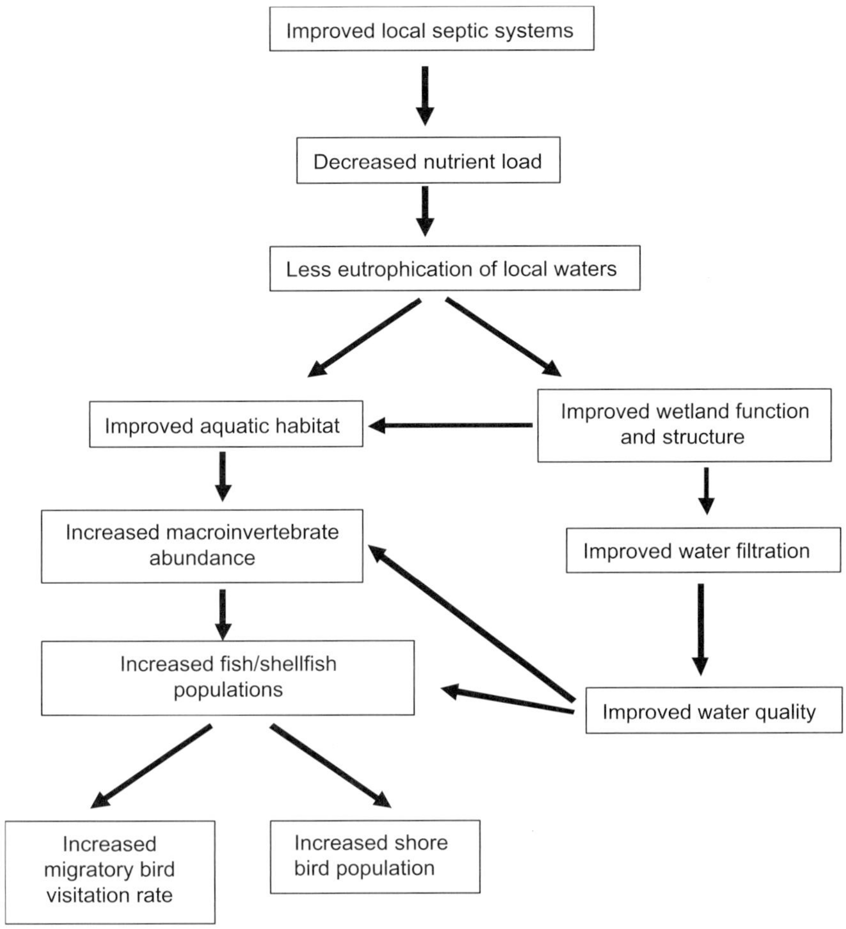

FIGURE 12.8
Ecosystem service flow chart showing need to control nutrient loads to local surface and ground water, by protecting adjoining wetlands. The protection results in ecosystem improvements, in this case indicated by improved aquatic life and bird diversity and abundance. However, the protection is justified, not from the need to protect the wetlands' inherent, ecological value, but their anthropocentric utility (instrumental values).
Source: US Environmental Protection Agency (2002). *A Framework for Economic Assessment of Ecological Benefits.* Washington, DC.

protected. Without these systems, the human population would suffer. Thus, in our role as stewards of the environment, we cannot neglect the intricacies, interconnections, and interrelationships that exist between humans and their environment.

Recently, the systematic, anthropocentric view has manifested itself in the form of *ecosystem services.* In this context, the value of an ecosystem is in the form of processes by which the environment provides resources, often not fully appreciated, such as keeping water clean, providing materials, e.g. timber, preserving habitat for fisheries, and keeping agriculture productive by protecting pollinating insects, as well as preserving esthetics of parks and national monuments. In other words, simply ecosystems are not just inherently valuable, but instrumentally valuable to humans. Interestingly, our kindergartners often have less difficulty in seeing these values than some extremely anthropocentric bioengineers or their bosses! A specific example of protecting an aquifer or a surface water body from septic tank infiltration is shown in Figure 12.8.

Problems also emerge at the other extreme. Exclusive biocentrism could completely eliminate genetic engineering, notwithstanding its usefulness (e.g. genetically modified microbes to treat hazardous wastes). If "harm" is done to a single microbial species, e.g. having some of its population's genetic material manipulated to enhance the degradation of a toxic compound, this would be unacceptable at the extreme. Again, the lack of acceptance is not that it could harm humans down the road; indeed, that would be an anthropocentric view. The genetic modification is unacceptable to the biocentrist simply because of its impact on the organisms, no matter the positive human value.

Ecocentrism is attractive since it is comprehensive and systematic at its core, but extreme versions can deprive humans of value and dignity (e.g. deep ecology sometimes implies that humans can be "parasites" or "vermin" and that there are "excessive" numbers of people).

Thus, placing value on a particular biotechnology crosses a number of ethical frameworks. But, where do these values come from? Psychologists argue that moral development takes a predictable and stepwise progression as the result of social interactions over time. For example, according to Lawrence Kohlberg and his followers [26], people first behave according to authority, then in accordance with social norms, before finally maturing to the point where they are genuinely interested in the welfare of others and in upholding philosophical notions like justice. This model can be directly applied to bioengineers. The most basic (bottom tier) actions are pre-conventional. That is, decisions are made primarily to stay out of trouble. While proscriptions against unethical behavior at this level are effective, the training, mentorship and other opportunities for professional growth try to push the engineer to higher ethical expectations. With experience as guided by observing and emulating ethical role models, the engineer can move to higher professional stages.

In the second level, the bioengineer dutifully acts within a range of expectations prescribed by the profession, i.e. the engineering convention. Thus, the engineering practice *is* the convention as articulated in codes of ethics. At this level, the engineer is charged with being a loyal and faithful agent to the clients. Researchers are beholden to their respective universities and institutions. Engineers working in companies and agencies are required to follow mandates to employees (although never in conflict with their obligations to the engineering profession). Thus, engineers must stay within budget, use appropriate materials, and follow best practices as they concern their respective designs. For example, if an engineer is engaged in work that would benefit from collaborating with another company working with similar genetic material, the engineer must take precautionary steps to avoid breeches in confidentiality, such as trade secrets and intellectual property.

The highest level of bioengineering development has a number of aspects. Many of the research and development projects address areas that could greatly benefit society, but may lead to unforeseen costs. The bioengineer is called to consider possible contingencies, such as those in Figures 1.2 and 1.3. For example, if an engineer is selecting microbial traits, is there a possibility that self-replication mechanisms in the cell could be modified to lead to potential adverse effects, such as generating mutant pathological cells, toxic byproducts, or changes in genetic structure not previously expected? Thus, this highest level of professional development is often where the previously mentioned *risk tradeoffs* must be considered. In the case of our example, the risk of adverse genetic outcomes must be weighed against the loss of advancing the state of environmental science (e.g. improved bioremediation rates). The engineer must design a solution that optimizes outcomes; that is, not too much risk while advancing the state-of-the-science. This is done using a number of tools, such as benefit–cost ratios, best practice guidelines, and transparency in terms of possible downstream impacts, costs, side effects and interactions.

Going beyond the conventional stages, the truly effective bioengineer makes decisions based on the greater good of society, sometimes even at personal costs. In our zeal to advance science, we must not ignore some of the larger, albeit low-probability societal repercussions of our research and operations.

Research introduces a number of challenges that must be approached at all three ethical levels. At the most basic level, laws, rules, regulations and policies dictate certain behaviors.

In other words, the expectations are not yet at the conventional level, so only the fear of punishment (e.g. sanctions, job loss) and desire for rewards (e.g. monetary, recognition) prescribe behavior. For example, genetic research, especially that which receives federal funding, is controlled by rules overseen by federal and state agencies. Such rules are often proscriptive, that is, they tell you what *not to do*, but are less clear on what actually *to do*.

The engineering profession and engineering education standards require attention to both the macro and micro dimensions of ethics. Macroethics are those articulated for the profession as a whole, whereas microethics are those ascribed to the individual engineer. The Accreditation Board for Engineering and Technology, Inc. (ABET) includes a basic microethical requirement for engineering education programs, identified as "(f) an understanding of professional and ethical responsibility," along with macroethical requirements that graduates of these programs should have "(h) the broad education necessary to understand the impact of engineering solutions in a global and societal context" and "(j) a knowledge of contemporary issues" [27]. These are tall orders for the next generation of bioengineers and those who teach them.

BIOTECHNOLOGY DECISION TOOLS

A common human failing is that we tend to keep doing things the way we have always done things. Maybe this is not so much a "failing" as an adaptive skill of the survival of the species. If we suffered no ill effects after eating that berry, we will eat more of that particular species. But, even a similar looking species may be toxic, so we are careful to eat only the species that did not kill us. So, we have a paradox as professionals. We do not want to expose our clients or ourselves to unreasonable risks, but we must to some extent "push the envelope" to find better ways of doing this. So, we tend to suppress new ways of looking at problems. Sometimes, the facts and theories may be so overwhelmingly convincing that we must change our worldview. Recall that Thomas S. Kuhn refers to this as a "paradigm shift." According to Kuhn, there are essentially two types of paradigm shifts: those that result from a discovery caused by encounters with anomaly, and those that result from the invention of new theories brought about by failures of existing theories to solve problems the theory defines. Both apply to biotechnology. In the case of a paradigm shift brought about by discovery, the first step in shifting said paradigm is the discovery of the anomaly itself. Once the paradigm has been adjusted so that the anomalous becomes the norm or at least the expected, it is said that the paradigm change is complete. In the case of a paradigm shift that results from the invention of new theories caused by the failure of existing theory, the first step is the failure itself (when the system in place fails, the creation of a new system is necessary). Several things can bring about this failure: observation of discrepancies between the theory and the fact, changes in the surrounding social or cultural climate, and academic criticism of the existing theory. The scientific community is highly resistant to adopting new paradigms. In the early stages of a paradigm, theoretical alternatives can be proposed that are merely adaptations of the existing paradigm, but once the new paradigm becomes well-established, these theoretical alternatives are strongly resisted [28].

For engineers, society evaluates our success and failure in accordance with the performance of a design. Does the resulting product or system "work" (effectiveness) as designed? Is it the best way to reach the end for which we strive (efficiency)? Next, we consider whether or not it will likely continue to "work" (reliability) and our due diligence causes us to reflect on what adverse implications we might face (risk). Thus the "risk" associated with a design is used to refer to the possibility and likelihood of undesirable and possibly harmful effects. Errors in engineering can range from those that are merely annoying, such as when a concrete building develops cracks that mar it as it settles, to those that are seemingly unforgivable, such as the collapse of a bridge which causes human death [29]. This can be illustrated by the case of the Ford Pinto, a subcompact car produced by Ford Motor Company between 1971 and 1980. The car's fuel tank was poorly placed in such a way that increased the probably of a fire from fuel spillage by a rear collision. The ensuing adverse implications manifested in the series of injuries that resulted from the defect. Engineers in such instances have been criticized for missing key facts, or worse yet, knowing the facts but allowing a dangerous design to go to production. In the bioethical arena, similar criticisms have been lodged against designs of devices and systems.

In his book, *To Engineer is Human*, Henry Petroski discusses success and failure in engineering as part of an ongoing process of trial and error that leads to innovation. For example, he discusses the effects of the growth of the railroads on engineering structures. As the network of railroads in the United States was quickly expanding, there were more incentives for heavier trains to travel progressively faster on more rugged terrain. But soon enough, it was obvious that the strength that was required for earlier railroads to facilitate the earlier trains was no match for the new trains. As collapses occurred, "each defective bridge resulted in demands for excess strength in the next similar bridge built, and thus the railroad bridge evolved through the compensatory process of trial and error" [30]. As the processes of engineering has become progressively more scientific in nature over the decades, engineers have had to deal with "success" and "failure" more explicitly. Thus the real *trial and error* that has been passed down to today's engineers by their predecessors is that of "mind over matter." According to Petroski, while no one *wants* to learn by mistakes, they are an integral part of the process. As engineers perpetually strove to employ new concepts to create lighter, more cost-effective structures, each new structure can be seen as a trial of sorts [31].

CHARACTERIZING SUCCESS AND FAILURE

Academic and professional disciplines characterize success and failure in unique ways. For example, economists view them from two perspectives. First, how well is the distribution of goods and services and management of economic systems working? What is the best way to allocate scarce resources? When we have found the best solution to any problem of allocation or distribution, we are at *Pareto* optimality, the point where no one else can be made any better off without making someone worse off. Therefore "success" in the case of this task involves attaining Pareto optimality.

The second economic perspective of success and failure is based on how an individual or organization makes economic decisions. This can follow a version of ethical egoism, hedonism, and utilitarianism. How do we optimize *our* profits, or lower *our* individual costs? Here, our goal becomes to make ourselves or our individual firms as well off as we can, given the constraints of the marketplace, including legal constraints. Since such a view is not concerned with the degree of improvement or degradation of others, Pareto optimality is not the goal. The duty-based aspects of the engineering profession are in conflict with such a system. For example, in the consumer model, the goal is to maximize the level of utility a consumer achieves, while subject to the constraints of the budget (essentially the total amount of money we have to spend), but most engineering codes of ethics require that the more expansive "public" be considered. In fact, engineering norms dictate that the public is *more important* than the individual and the organization.

The free market, economic model is based purely on maximizing a predefined utility. In the producer model, firms optimize by maximizing profits, given the limitations they face due to the wage (the price of labor) and the rental rate (the price of capital).

This is not to say that engineers can operate without regard to economics. To the contrary, we are to be faithful agents to our clients. Thus, optimization for engineers involves an economic element, as reflected in the use of benefit–cost ratios (BCRs), discussed in Chapter 11. Essentially, engineering involves design under constraint [32]. Engineers may strive to optimize the level of elegance of a given structure while subject to the constraints of fixed physical and mental resources, as well as the demands of the marketplace. At the same time, engineers additionally face the greatest constraint of all: safety. This constraint becomes more important than the esthetic and economic aspects because "the loss of a single life due to structural collapse can turn the most economically promising structure into the most costly and can make the most beautiful one ugly" [33].

Engineering success is somewhat different. Perhaps, the best way to address what it means to have a successful design and project is to explore the ways we fail.

Accountability

The level and type of accountability for success and failure of a design is affected by the setting in which the engineer works. However, in every engineering office or department there is a designated "engineer in responsible charge" whose job it is to make sure that every project is completed successfully and within budget. Biotechnological and other life science institutions are no different. This responsibility is often indicated by the fact that the engineer in responsible charge places his or her professional engineering seal on the design drawings or the final reports. By this action the engineer is telling the world that the drawings or plans or programs or whatever are correct, accurate, and that they will work. (In some countries, not too many years ago, the engineer in responsible charge of building a bridge was actually required to stand under the bridge while it was tested for bearing capacity!) In sealing drawings or otherwise accepting responsibility the engineer in charge places his or her professional integrity and professional honor on the line. There is nowhere the engineer in charge can hide if something goes wrong. If something *does* go wrong, saying "One of my younger engineers screwed up" is not a defense because it is the engineer in charge who is supposed to have overseen the calculations or the design.

For very large projects where the responsible engineer may not even know all of the engineers working on the project, much less be able to oversee their calculations, this is clearly impossible. In a typical engineering office the responsible engineer depends on a team of senior engineers who oversee other engineers, who oversee others and so on down the line. How can the responsible engineer at the top of the pyramid be confident that the product of collective engineering skills meets the client's requirements?

The rules governing this activity are fairly simple. Central is the concept of truthfulness in engineering communication. Such technical communication up and down the organization requires an uncompromising commitment to tell the truth no matter what the next level of engineering wants to hear.

What we value influences the attention to detail. For example, an engineer in the lower ranks may develop spurious data, lie about test results, or generally manipulate the basic design components. Such information might not be readily detected by supervisory engineers if the bogus information is beneficial to the completion of the project. If the information is not beneficial, on the other hand, everyone along the chain of engineering responsibility will give it a hard critical look. Therefore the inaccurate information, if it is the desired information, can steadily move up the engineering ladder because at every level the tendency is not to question good news. The superiors at the next level also want good news, and want to know that everything is going well with the project. They do not want to know that things may have gone wrong somewhere at the basic level.

The only correcting mechanism in engineering exists at the very end of the project if failure occurs: the contamination persists; the bioreactor fouls; the genes drift. And then the search begins for what went wrong. Eventually the truth emerges, and often the problems can be traced to the initial level of engineering design, the development of data and the interpretation of test results.

It is one thing to make a mistake (we all do), but it is totally another thing to use misinformation in the design. Fabricated or spurious test results can lead to catastrophic failures because there is an absence of a failure detection mechanism in engineering until the project is completed. Without trust and truthfulness in engineering, the system will fail. In the words of scientist and philosopher Jacob Bronowski [34]:

> All engineering projects are communal; there would be no computers, there would be no airplanes, there would not even be civilization, if engineering were a solitary activity.

What follows? It follows that we must be able to rely on other engineers; we must be able to trust their work. That is, it follows that there is a principle that binds engineering together, because without it the individual engineer would be somewhat rudderless. It also follows that technical collaborations can only work effectively if they be fact-based and honest. These are elements of trust. The trustworthiness of the practicing engineer within the engineering community must first be met before the engineering profession can expect to be trusted by our clients and the public.

Value

The various definitions of success and failure require a means for determining success. We need ways to measure the level of success of an engineering design in terms of effectiveness. For instance, assuming that pump Y has been designed to deliver chemical X to a fermentation unit. We can measure the efficiency of chemical delivery, by a typical mass balance, such as the mass of X administered compared to the mass reaching the active fermentation site. Thus, efficiency is relatively straightforward. Effectiveness, however, may not be. For example, even if the total efficiency is high (e.g. 99.99% of the mass of X reaches the target site), it may be ineffective (non-efficacious), because microbial metabolism changes X to an oxidized metabolite X′ before it reaches the site that is not effective in catalyzing the reaction. Thus, effectiveness assumes some definition of *value.*

In engineering, one might consider "value" through the idea of value engineering. This concept was created at General Electric Co. during World War II. As the war caused shortages of labor and materials, the company was forced to look for more accessible substitutes. Through this process, they saw that the substitutes often reduced costs, or improved a product [35]. Consequently, they turned the process into a systematic procedure called "value analysis."

Value engineering consists of assessing the value of goods in terms of *function*. Value, as a ratio of function to cost, can be improved in various ways. Oftentimes, value engineering is done systematically through the Job Plan [36]. This method involves four basic steps: information gathering, alternative generation, evaluation, and presentation. In the information gathering step, engineers consider what the requirements for the object are. This step also includes function analysis, which attempts to determine what functions or performance characteristic are important. In the next step, alternative generation, value engineers consider the possible alternative ways of meeting requirements. Next, in evaluation, the engineers assess the alternatives in terms of functionality and cost-effectiveness. Finally, in the presentation stage, the best alternative is chosen and presented to the client for the final decision [37].

In considering the realm of economics, "value" is considered to be the worth of one commodity in terms of other commodities (or currency). There are three main value theories in economics. The first, an intrinsic theory of value is a theory which holds that the value of an object, good or service is contained in the item itself. These theories tend to consider the costs associated with the process of producing an item when assigning the item value. For example, the labor theory of value, a model developed by David Ricardo, holds that the value of a good is derived from the effort of its production, reduced to the two inputs in the production frontier, labor and capital [38]. In this model, if a lamp is produced in 5 hours by 3 men, then the lamp is worth $3 \times 5 = 15$ man-hours. On the other hand, the subjective theory of value holds that goods have no *intrinsic* value, outside the desire of individuals to have the items. Here, value of an item becomes a function of how much an individual is willing to give up in order to have that item [39].

Similarly, the marginal theory of value takes into account both the scarcity and desirability of a good, holding that the utility rendered by the last unit consumed determines the total value of a good. The main difference between David Ricardo's labor theory of value (and the concept of intrinsic value) and marginal theory of value is that the former takes into account a form of value derived from utility – from satisfying human desire [40].

Furthermore, consider the meaning of the term "value" to the common member of society. When a person says that they *value* something, this generally means that the item is of importance to them for one reason or another. At the same time, to say something is "valuable" would mean that it costs a lot of money, meaning that a high demand for the item in society has driven the prices up. Therefore, person A might *value* her beaded necklace because she derives pleasure from it, but her diamond necklace would be considered *valuable,* because of the large number of other members of society who also feel they would derive pleasure from it (which causes a high demand, which raises the price).

This leads to the discussion of the diamond–water paradox, which is a noteworthy example of the role of scarcity in economic value theory. Diamonds, which have relatively little use, only esthetic value, have an extremely high price when compared to water, which is essential to life itself. The example illustrates the importance of scarcity in economic value. Here, as diamonds are far scarcer than is water, they have the higher price. However, this is situational. If one were to find oneself in the middle of the desert, where water is extremely scarce, one would almost certainly be willing to pay more money for water than for diamonds. Thus, the person would *value* water more at this point.

Informing decisions

Recall from Chapter 11 that there are numerous ways to identify, characterize, and analyze success and failure of decisions after they are made. It is more difficult to predict these for decisions yet to be made. The key is to have an accurate and objective set of facts from which valid arguments are made:

Step 1 – Factual Description: During the descriptive stage, we are not ready yet to discuss whether a decision is right or wrong. We are gathering information that will be needed in subsequent steps.

Step 2 – Arguments: Drivers and limiting factors must include all biochemodynamic variables, as well as legal, social and financial considerations.

Step 3 – Problem-Solving Analysis: Once the facts and reasons for the biotechnological operation are sufficiently identified and explained, the issues must be classified as to whether they are factual, conceptual or ethical in nature [41].

- Step 3a – Identify Possible Causal Relationships: Frequently, weight-of-evidence between a cause and an effect is the determining criterion [42] (see Chapter 11). Depending on the case, some of Hill's criteria are more important than others.

Conceptual issues involve different ways that the meaning may be understood. For example, two scientists may not completely agree on the meaning of "pollution" or "good lab practices" (although the scientific community strives to bring consensus to such definitions). Most scientists agree on first principles (e.g. fundamental physical concepts like the definitions of matter and energy), but unanimity fades as the concepts drift from first principles.

As mentioned in Chapter 11, decisions are seldom made exclusively from physical scientific principles and the influence of these other factors can be likened to the force fields created by the pull of a magnet (see Figure 12.9). Thus, the decision being considered will be pulled in the direction of the strongest magnetic force. The stronger the magnet the more the decision that will actually be made will be pulled in that direction. For example, lawyers may proceed in one direction, while engineers another, and the business clients another, all applying different forces that "deform" the decision.

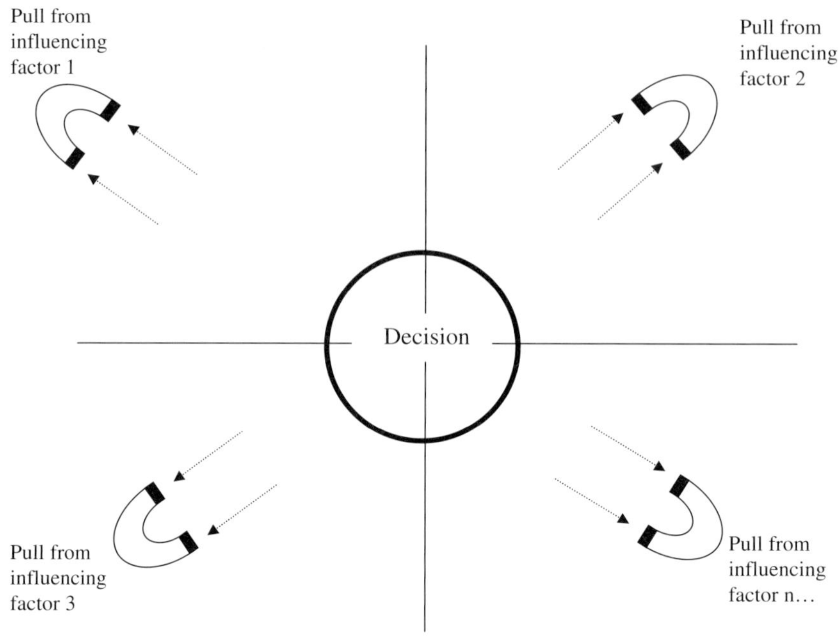

FIGURE 12.9
Decision force field.

A decision that is almost entirely influenced by strong science will appear something like the force field in Figure 12.10. However, if other factors are present, such as a plant closing or the possibility of new jobs being attracted at the expense of environmental, geological, or other scientific influences, the decision would migrate toward these stronger influences, as shown in Figure 12.11.

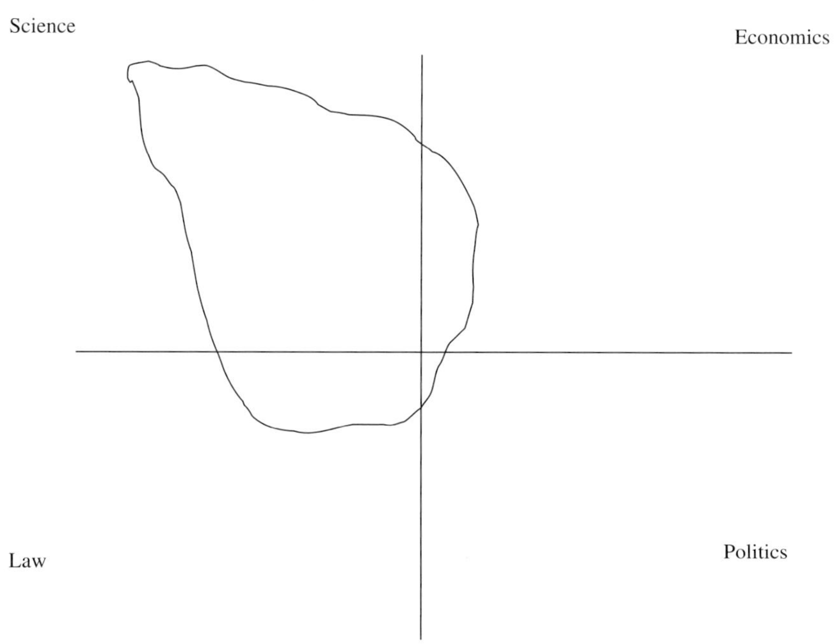

FIGURE 12.10
Decision force field for a scientifically based decision, although other factors (legal, financial, and political) are influencing the final outcome. For example, the data are clear that there is no way to contain genetically modified fish if they are introduced to a water body. Thus, the likelihood of severe ecosystem damage is high, which overshadows local economics, diminishes the likelihood of lawsuits and may even drive the regulations for more protection against invading species.

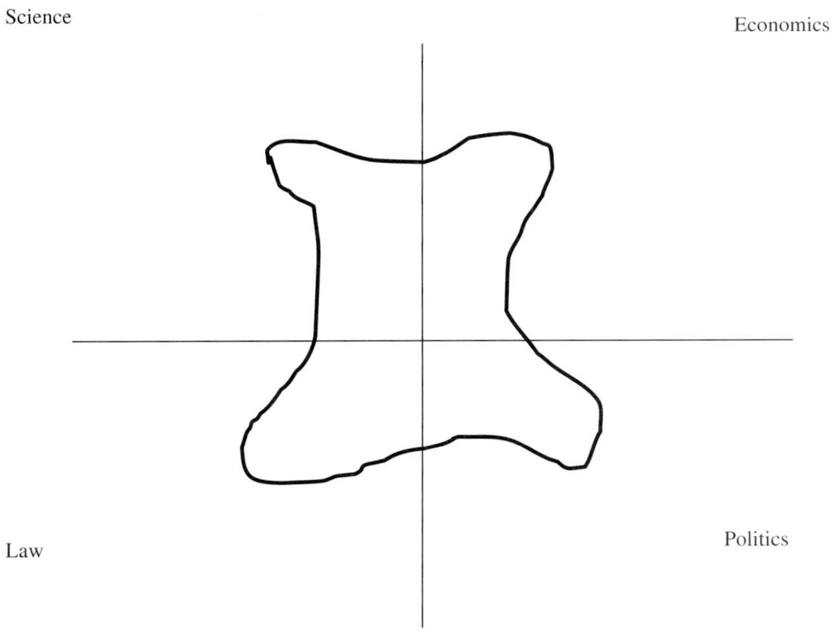

FIGURE 12.11
Biotechnology decision force field for complex decision with multiple influences with about equal influence on the outcome, e.g. whether to use in agriculture a GMO that would enhance food supply, but may introduce gene flow problems to an ecosystem. Thus, lawsuits and regulation may drive the decision from being purely scientific to being multi-perspective.

Thus, bioengineers make decisions under risk and uncertainty (need for factors of safety). The bioengineering risk management process is informed by the quantitative results of the risk assessment process, but not exclusively. The shape and size of the resulting decision force field diagram give an idea of what are the principal driving factors that lead to decisions. While helpful, the force field step is merely subjective.

Biotechnological net goodness analysis

As mentioned in Chapter 11, this is also a subjective analysis of each factor driving and limiting a decision from three perspectives: (1) what the severity of the consequence might be; (2) the importance of the decision; and (3) the probability that the consequence will happen. Recall that these factors are then summed to give the overall net goodness of the decision:

$$\text{NG} = \Sigma(\text{goodness of each consequence}) \times (\text{importance}) \times (\text{likelihood}) \quad (12.1)$$

This can be modified from a purely ethical decision making tool to a risk management tool. For example, Figure 12.12 depicts a decision on whether to use Bt in a particular application. The decision is based on the likelihood of various beneficial and adverse outcomes, with ranked importance to three receptors: the environment, public health, and food production. As the net goodness analysis, this is qualitative, but can be an effective means of identifying the important factors, as well as potential downstream impacts and artifacts of an immediate decision. The difficulty will be to arrive at probabilities to fill the "likelihood" column. Sometimes these are published, but often will have to be derived from focus groups and expert elicitation. Often, likelihood is presented as an ordinal scale (e.g. high, medium, or low – or 1, 2, 3). As such, this is not a fault tree, but an event tree.

Thus, this can be valuable in decisions that have not yet been made, as well as what decisions "should" have been made in a case. For example, these analyses sometimes use ordinal scales, such as 0 through 3, where 0 is nonexistence (e.g. zero likelihood or zero importance) and 1, 2, and 3 are low, medium, and high, respectively. Thus, there may be many small consequences that are near zero in importance and, since NG is a product, the overall net goodness of the decision is driven almost entirely by one or a few important and likely consequences.

Note that even a very unlikely but negative consequence is unacceptable. In fact, the most unacceptable environmental biotechnological outcomes are rare events. For that matter, many

FIGURE 12.12
Decision tree and net goodness analysis of a decision to insert *Bacillus thuringiensis* (Bt) expressed crops near an ecosystem. Data are hypothetical. Outcomes can be expanded, both within the first-order outcome groups and by adding other first-order outcomes.

environmental pollution problems that are of the highest priority, such as chronic diseases (e.g. cancer) have low probabilities in a population. For example, an elevated population risk of cancer is one in a million. But in larger populations, these do occur.

Flow charts, event trees, and fault trees display the paths to possible consequences from each decision. Figure 12.13 provides a simple, environmental biotechnological example.

The event tree draws on the other analytical tools, starting with the timeline of key events, analyzing a number of different paths that can be taken. This is done for every option and suboption that should have been considered, comparing each consequence. It can be retrospective; for example, even in a disaster, there may have been worse consequences than what actually occurred. Conversely, even though something did not necessarily turn out all that badly, the event tree could point out that with one or more changed factors, the outcome could have been disastrous. In fact, the fault tree approach applies a probability to each option and suboption.

The event tree above depicts an unusually dichotomous decision, i.e. whether or not to use a toxic substance. Often, however, decisions are fraught with advantages and disadvantages no matter what is decided. For example, not using the Bt crop could result in food shortages that may lead to other environmental impacts (e.g. greater use of commercially available chemical pesticides), as well as geopolitical complications, such as destabilizing economies. There are also opportunity risks when choosing the "no action" alternatives.

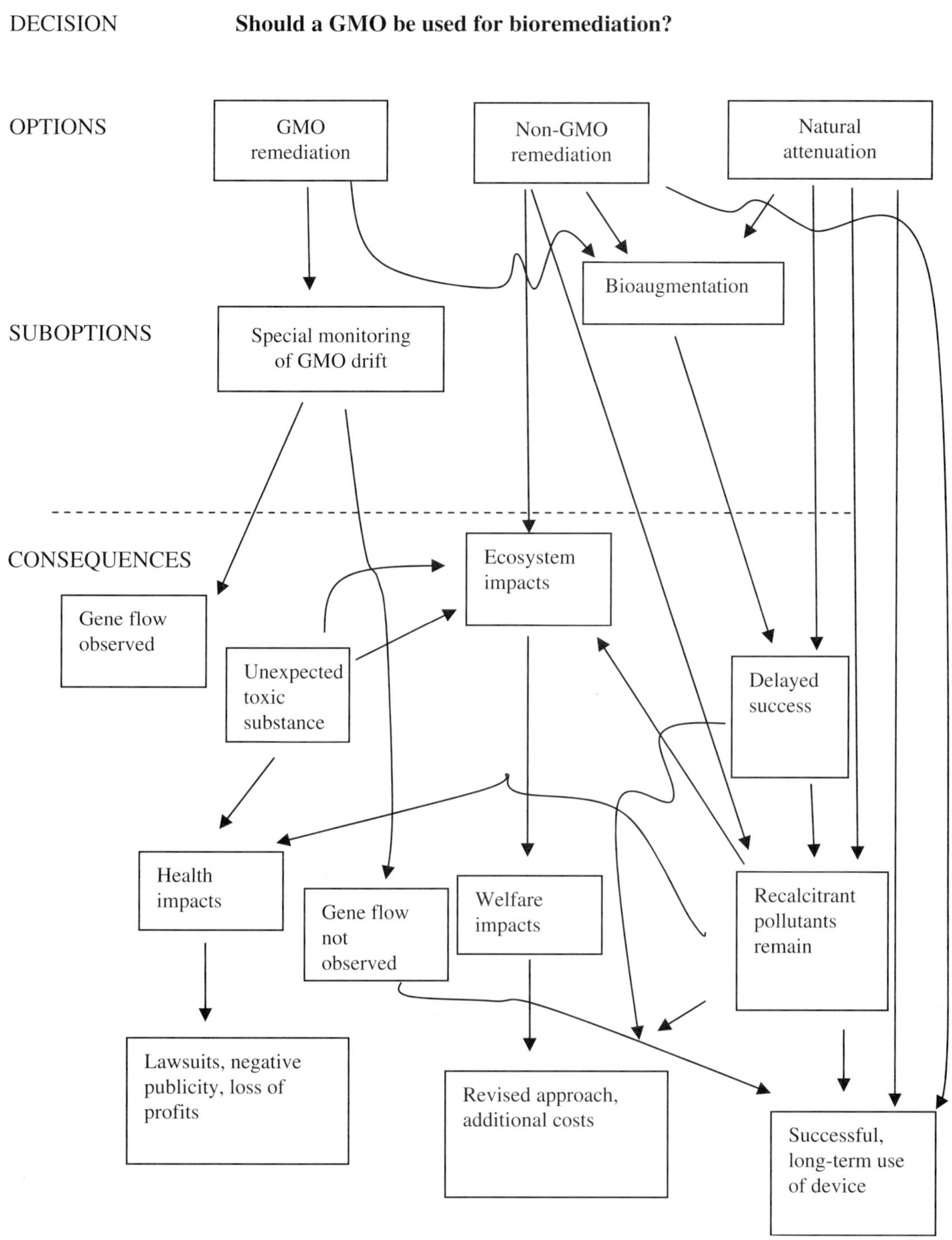

FIGURE 12.13
Event tree on whether to use a genetically modified bacterial strain in remediating an abandoned hazardous waste site.

The current challenge in assessing impacts is the lack of comprehensive data and reliable models. For example, most studies are at very small scales (laboratory and single fields). As such, risk calculations may appear to be more precise than they actually are.

GREEN ENGINEERING AND BIOTECHNOLOGY

One of the most recent and best ways to gauge the success of any environmental endeavor, including biotechnology, is its sustainability. That is, how does this new technology or enhanced application of an existing technology support green engineering? Engineers and technologists are successful when their designs are implemented so that desired results are achieved. In recent decades, engineers have increasingly been asked to design buildings, devices, and systems that are sustainable. Many see the need for greener manufacturing, operations, and products being met to some extent by biotechnologies. Sustainable approaches are outward looking; that is, they are designed to provide benefits not only to the present users, but to do so in a way that future people will not be harmed by present benefits. According to the National Academy of Engineering [43]:

> It is our aspiration that engineers will continue to be leaders in the movement toward the use of wise, informed, and economical sustainable development. This should begin in our educational institutions and be founded in the basic tenets of the engineering profession and its actions.

Environmental conscientiousness evolved in the 20th century from a peculiar interest of a few design professionals to an integral part of every engineering disciple. In fact, one of the most important macroethical challenges for engineers is to provide sustainable designs. The US Environmental Protection Agency defines "green engineering" as:

> ... the design, commercialization and use of processes and products that are feasible and economical while: Reducing the generation of pollution at the source. Minimizing the risk to human health and the environment. [44]

Green engineering asks the bioengineer to incorporate "environmentally conscious attitudes, values, and principles, combined with science, technology, and engineering practice, all directed toward improving local and global environmental quality" [45]. However, the design must also be feasible and must adhere to the first canon of engineering practice; holding paramount the safety, health, and welfare of the public. One of the principles of "green engineering" is the recognition of the importance of sustainability. Green engineering and sustainable design are key aspects of environmental biotechnology that have been addressed often in this book.

Environmental biotechnology has always been a form of biomimicry. This was true even before it was called biotechnology. Nature presents a viable model for innovation worthy of imitation. The biomimicry model starts, like any scientific investigation, by observation. Natural systems are the subject matter and a pedagogy rather than simply a natural resource commodity to be extracted from the earth:

> ... [N]ature would provide the models: solar cells copied from leaves, steely fibers woven spider-style, shatterproof ceramics drawn from mother-of-pearl, cancer cures compliments of chimpanzees, perennial grains inspired by tallgrass, computers that signal like cells, and a closed-loop economy that takes its lessons from redwoods, coral reefs, and oak-hickory forests. [46]

Nature elegantly demonstrates how scientific principles such as optimization and the laws of thermodynamics, hydrodynamics, and aerodynamics are evident and interwoven in the natural environment, which is a network of diverse and cooperative systems. This is evidenced in the principles of biomimicry (Table 12.3).

Innovations in material science have accelerated over the past few years with new materials that are built on science and engineering discoveries. These are outright examples of biomimicry, with so many innovations in material science drawing inspiration from natural systems. For example, the study of the lotus petal's ability to repel rain water is now finding application in "biometic paint" and surface treatment for concrete that absorbs pollution from the air. What can the orb-weaver spider teach today's architects, engineers and material scientists? Sanitary engineers, before they were called environmental engineers, respected the bacterium as a marvelously efficient chemical factory. The study of this organism and a closer look at the chemistry underlying the transformation of flies and crickets into materials that are five times stronger per ounce than steel and done at room temperature could lead to a new way of conceiving and manufacturing materials and assembling them to create more sustainable environments.

Table 12.3 Principles of biomimicry

Nature runs on sunlight
Nature uses only the energy it needs
Nature fits form to function
Nature recycles everything
Nature rewards cooperation
Nature banks on diversity
Nature demands local expertise
Nature curbs excesses from within
Nature taps the power of limits

Source: J.M. Benyus (1997). *Biomimicry*. William Morrow and Company, New York, NY.

BIOENGINEERING SAFETY

Safety is a key expectation of all biotechnological endeavors. If not stated as one of the outright goals of a new biotechnology, it is one of the underpinning criteria of whether the biotechnology will be deemed a success or failure. Biotechnological safety is not only important in the design and manufacture of the biotechnology or its consequent products, its success is also dependent upon its use. Medical, engineering, and other professional practices will have their own unique professional obligations associated with the use and application of the new technologies. Safety and risk are also critical components of whether a professional or research approach is considered sound. When patients, subjects or the public are exposed to *unacceptable* risk, the practitioner or researcher is deemed to have performed in an unethical manner.

The fifth principle of medical ethics of the American Medical Association (AMA) states:

> A physician shall continue to study, apply, and advance scientific knowledge, maintain a commitment to medical education, make relevant information available to patients, colleagues, and the public, obtain consultation, and use the talents of other health professionals when indicated. [47]

The seventh AMA principle states:

> A physician shall recognize a responsibility to participate in activities contributing to the improvement of the community and the betterment of public health. [48]

Likewise, the first fundamental canon of the engineering profession, as articulated by the National Society of Professional Engineers, requires that "engineers, in the fulfillment of their professional duties, shall hold paramount the safety, health and welfare of the public" [49]. To emphasize this professional responsibility, the engineering code includes this same statement as the engineer's first rule of practice.

These principles and canons indicate that safety and risk are not only technical concepts, but are also ethical concepts. Professionals cannot behave ethically and simultaneously ignore the direct and indirect risks stemming from their practice. Thus, competence *and* character demand an appreciation of risks, both real and perceived. For example, perceived risks may be much greater than actual risks, or they may be much less. So then, how is it possible to square technical facts with public fears? Like so many engineering concepts, timing and scenarios are crucial. What may be the right manner of saying or writing something in one situation may be very inappropriate in another. Communication approaches will differ according to whether we need to motivate people to take action, alleviate undue fears, or simply share our findings clearly, no matter whether they are good news or bad. For example, some have accused certain businesses of using public relations and advertising tools to lower the perceived risks of their products. The companies may argue that they are simply presenting a counterbalance against unrealistic perceptions.

Engineering success or failure is in large measure determined by what we do compared to what our profession "expects" us to do. Safety is a fundamental facet of our professional duties. Thus, we need a set of criteria that tells us when our designs and projects are sufficiently safe. Four safety criteria are applied to test engineering safety [50]:

- The design must comply with applicable laws.
- The design must adhere to "acceptable engineering practice."
- Alternative designs must be sought to see if there are safer practices.
- Possible misuse of the product or process must be foreseen.

The first two criteria are easier to follow than the third and fourth. The well-trained designer can look up the physical, chemical, and biological factors to calculate tolerances and factors of safety for specific designs. Laws have authorized the thousands of pages of regulations and guidance that demark when acceptable risk and safety thresholds are crossed; meaning that the design has failed to provide adequate protection. Engineering standards of practice go a step further. Failure here is difficult to recognize. Only other engineers with specific expertise can judge whether the ample margin of safety as dictated by sound engineering principles and practice has been provided in the design. Identifying alternatives and predicting misuse requires quite a bit of creativity and imagination.

But, can risks really be quantified? Risk assessors and actuary experts would answer with a resounding "yes." Medical practitioners routinely share risks with patients in preoperative preparations, usually in the form of a probability. However, the general public's perception is often that one person's risk is different from another's and that risk is in the "eye of the beholder." Some of the rationale appears to be rooted in the controversial risks of tobacco use and daily decisions, such as choice of modes of transportation. What most people perceive as risks and how they prioritize those risks is only partly driven by the actual objective assessment of risk, i.e. the severity of the hazard combined with the magnitude, duration, and frequency of the exposure to the hazard. For example, young student smokers may be aware that cigarette smoke contains some nasty compounds, but is not directly aware of what these are (e.g. polycyclic aromatic hydrocarbons and carcinogenic metal compounds). They have probably read the conspicuous warning labels many times as they held the pack in their hands, but these really have not "rung true" to them. They may have never met anyone with emphysema or lung cancer, or they may not be concerned (yet) with the effects on the unborn (i.e. *in utero* exposure) [51]. Psychologists also tell us that many in this age group have a feeling of invulnerability. Those who think about it may also think that they will have plenty of time to end the habit before it does any long-term damage. Thus, we should be aware that what we are saying to people, no matter how technically sound and convincing to us as engineers and scientists, may be simply a din to our targeted audience.

The converse is also true. We may be completely persuaded based upon data, facts, and models, that something clearly does not cause significant damage, but those we are trying to convince of this finding may not buy it. They may think we have some vested interest, or that they find us guilty by association with a group they do not trust, or that we are simply "guns for hire" for those who are sponsoring our research or financially backing the product development. The target group may not understand us because we are using jargon and are not clear in how we communicate the risks. So, do not be surprised if the perception of risk does not match the risk you have quantified.

Engineers add value when we decrease risk, a crucial concern of bioethics. By extension, reliability tells us and everyone else just how well designs are performing by reducing overall risk. What we design must continue to serve its purpose throughout its useful life.

As it is generally understood, risk is the chance that something will go wrong or that some undesirable event will occur. Every time we go skating, for example, we are taking a risk that we might be in an accident and damage property (e.g. go through a store window pane), get hurt, injure others, or even die (unlikely, but possible, as in the window pane example). The understanding of the factors that lead to a risk is called *risk analysis* and the reduction of this risk (for example, by wearing equipment and skating only in controlled environments, such as rinks) is *risk management.* Risk management is often differentiated from *risk assessment,* which is comprised of the scientific considerations of a risk [52]. Risk management includes the policies, laws, and other societal aspects of risk.

An overarching theme of this book is that bioscientists and bioengineers constantly engage in risk analysis, assessment and management, which translates into the need to provide safe and reliable products and process. They must consider the interrelationships among factors that put people at risk, suggesting that we are risk analysts. Scientists and engineers provide decision makers with thoughtful studies based upon the sound application of the physical sciences and, therefore, are risk assessors by nature. Scientists and engineers control and characterize variables and, as such, are risk managers. Engineers, specifically, are held responsible for designing safe products and processes, and the public holds them accountable for its health, safety, and welfare. The public expects engineers to "give results, not excuses" [53]. Engineers design systems to reduce risk and look for ways to enhance the reliability of these systems. Every engineer deals directly or indirectly with risk and reliability. Thus, risk and reliability are accountability measures of biotechnological success [54].

RELIABILITY OF BIOTECHNOLOGIES

> Probable impossibilities are to be preferred to improbable possibilities.
>
> **Aristotle**

Aristotle was not only a moral philosopher and natural philosopher (the forerunner to "scientist"); he was also a risk assessor. Biotechnology presents both "probable impossibilities" and "improbable possibilities." Risk is an expression of probability. At some level, Aristotle recognized this, but it was incrementally refined with the advance of science (see Discussion Box: Probability and Biotechnology). People, at least intuitively, assess risks and determine the reliability of their decisions every day. We want to live in a "safe" world. But, *safety* is a relative term. The "safe" label requires a value judgment and is always accompanied by uncertainties, but engineers frequently characterize the safety of a product or process in objective and quantitative terms. Factors of safety are a part of every design. Biological safety is usually expressed by its opposite term, risk.

DISCUSSION BOX

Probability and Biotechnology

Probability is the likelihood of an outcome. The outcome can be bad or good, desired or undesired. The history of probability theory, like much modern mathematics and science, is rooted in the Renaissance. Italian mathematicians considered some of the contemporary aspects of probability as early as the 15th century, but did not see the need to or were unable to devise a generalized theory. Blaise Pascal and Pierre de Fermat, the famous French mathematicians, developed the theory after a series of letters in 1654 considering some questions posed by the nobleman Antoine Gombaud, Chevalier de Méré, regarding betting and gaming. Other significant Renaissance and post-Renaissance mathematicians and scientists soon weighed in, with Christian Huygens publishing the first treatise on probability, *De Ratiociniis in Ludo Aleae*, which was specifically devoted to gambling odds. Jakob Bernoulli (1654–1705) and Abraham de Moivre (1667–1754) also added to the theory. However, it was not until 1812 with Pierre LaPlace's publication of *Théorie Analytique des Probabilités*, that probability theory was extended beyond gaming to scientific applications [55].

(Continued)

Probability is now accepted as the mathematical expression that relates a particular outcome of an event to the total number of possible outcomes. This is demonstrated when we flip a coin. Since the coin has only two sides, we would expect a 50–50 chance of either a heads or a tails. However, scientists must also consider rare outcomes, so there is a very rare chance (i.e. highly unlikely, but still possible) that the coin could land on its edge, i.e. the outcome is neither a heads nor a tails, A "perfect storm" of a confluence of unlikely events is something that engineers must always consider, such as the combination of factors that led to major disasters like Hurricane Katrina and Bhopal, India (see Chapter 6), or the introduction of a seemingly innocuous opportunistic species that devastates an entire ecosystem (e.g. Iron Gates Dam in Europe, see Chapter 1), or the interaction of one particular congener of a compound in the right cell in the right person leads to cancer. As engineers, we also know that the act of flipping or the characteristics of the coin may tend to change the odds. For example, if for some reason the heads is heavier than the tails side or the aerodynamics is different, then the probability could change.

The total probability of all outcomes must be unity, i.e. the sum of the probabilities must be 1. In the case of the coin standing on end rather than being a heads or tails, we can apply a quantifiable probability to that rare event. Let us say that laboratory research has shown that one in a million times ({1/1,000,000 }= $0.000001 = 10^{-6}$), the coin lands on edge. By difference, since the total probabilities must equal 1, the other two possible outcomes (heads and tails) must be $1 - 0.000001 = 0.999999$. Again, we are assuming that the aerodynamics and other physical attributes of the coin give it an equal chance of being either heads or tails, then the probability of heads = 0.4999995 and the probability of tails = 0.4999995.

Stated mathematically, an event (e) is one of the possible outcomes of a trial (drawn from a population). All events – in our coin toss case, heads, tails, and edge – together form a finite "sample space," designated as $E = [e_1, e_2, \ldots, e_n]$. The lay public is not generally equipped to deal with such rare events, so by convention, they usually ignore them. For example, at the beginning of overtime in a football game a tossed coin determines who will receive the ball, and thus have the first opportunity to score and win. When the referee tosses the coin there is little concern about anything other than "heads or tails." However, the National Football League undoubtedly has a protocol for the rare event of the coin not being a discernible heads or tails. In environmental studies, *e* could represent a case of cancer. Thus, if a population of 1 million people is exposed to a biologically derived pesticide over a specific time period, and one additional cancer is diagnosed that can be attributed to that pesticide exposure; we would say that probability of *e*, i.e. $p\{e\} = 10^{-6}$. Note that this was the same probability that we assigned to the coin landing on its edge.

Returning to our football example, the probability of the third outcome (a coin on edge) is higher than "usual" since the coin lands in grass or artificial turf, compared to landing on a hard flat surface. Thus, the physical conditions increase the relative probability of the third event. This is analogous to a person who may have the same exposure to a carcinogen as the general population, but who may be genetically predisposed to develop cancer. The exposure is the same, but the probability of the outcome is higher for this "susceptible" individual. Thus, risk varies by both environmental and individual circumstances.

Events can be characterized a number of ways. Events may be discrete or continuous. If the event is forced to be one of a finite set of values (e.g. six sides of a die) the event is discrete. However, if the event can be any value, e.g. size of tumor (within reasonable limits), the event is continuous. Events can also be independent or dependent. An event is independent if the results are not influenced by previous outcomes. Conversely, if an event is affected by any previous outcome then the event is a dependent event.

Joint probabilities must be considered and calculated since, in most environmental scenarios, events occur in combinations. So, if we have *n* mutually exclusive events as possible outcomes from *E* that have probabilities equal to $p\{e_i\}$, then the probability of these events in a trial equals the sum of the individual probabilities:

$$p\{e_i \text{ or } e_2 \cdots \text{or } e_k\} = p\{e_1\} + p\{e_2\} + \cdots + p\{e_k\} \quad (12.2)$$

Further this helps us to find the probabilities of events e_i and g_1 for two independent sets of events, *E* and *G*, respectively:

$$p\{e_i \text{ or } g_i\} = p\{e_i\}p\{gi\} \quad (12.3)$$

For example, a company record book indicates that a waste site has 10 unlabeled buried chemical drums: five drums that contain mercury (Hg), two drums that contain chromium (Cr), and three drums that contain tetrachloromethane (CCl_4). We can determine the probability of pulling up one of the drums that contains a metal waste (i.e. Hg or Cr). The two possible events (Hg drum or Cr drum), then, are mutually exclusive and come from the same sample space. So we can use Equation 12.1:

$$p\{\text{Hg or Cr}\} = p\{\text{Hg}\} + p\{\text{Cr}\} = \frac{5}{10} + \frac{2}{10} = \frac{7}{10}$$

Thus we have a 70% probability of pulling up a metal-containing drum.

If we have another waste site that also has 10 unlabeled, buried drums, three drums that contain dichloromethane (CH_2Cl_2), and seven drums that contain trichloromethane ($CHCl_3$), we calculate the probability of pulling up a chromium drum from our first site and a $CHCl_3$ drum from the second site. Since the two trials are independent, we can use Equation 12.2:

$$p\{\text{Cr and CH}_2\ \text{Cl}_2\} = p\{\text{Hg}\} + p\{\text{Cr}\} = \frac{2}{10} \times \frac{3}{10} = \frac{6}{100}$$

Thus we have 6% probability of extracting a chromium and dichloromethane drum on our first excavation.

Another important concept for environmental data is that of conditional probability. If we have two dependent sets of events, *E* and *G*, the probability that event e_k will occur if the dependent event *g* has previously occurred is shown as $p\{e_k|g\}$, which is found using Bayes' theorem:

$$p\{ek|g\} = \frac{p\{e_k \text{ and } g\}}{p\{g\}} = \frac{p\{g|e_k\}p\{e_k\}}{\sum_{i=1}^{n} p\{g|e_i\}p\{e_i\}} \qquad (12.4)$$

A review of this equation shows that conditional probabilities are affected by a cascade of previous events. Thus, the probability of what happens next can be highly dependent upon what has previously occurred. For example, the cumulative risk of cancer depends on the serial (dependent) outcomes. Similarly, reliability can also be affected by dependencies and prior events. Thus, characterizing any risk or determining how reliable our systems are expressions, at least in part, of probability.

Put another way, one can characterize risk and reliability using a "probability density function" (PDF) for data. The PDF is created from a probability density, that is, when the data are plotted in the form of a histogram, as the amount of data increases, the graph increases its smoothness, i.e. the data appear to be continuous. The smooth curve can be expressed mathematically as a function, $f(x)$. This is the PDF. The probability distribution can take many shapes, so the $f(x)$ for each will differ accordingly. For example, in environmental matters, distributions commonly seen are normal, log-normal, and Poisson. The normal (Gaussian) distribution is symmetrical and is best known as the "bell curve". The log-normal distribution is also symmetrical, but its *x*-axis is plotted as a logarithm of the values.

The Poisson distribution is a representation of events that happen with relative infrequency, but regularly [56]. Stated mathematically, the Poisson distribution function expresses the probability of observing various numbers of a particular event in a sample when the mean probability of that event on any one trial is very small. So, the Poisson probability distribution characterizes discrete events that occur independently of one another during a specific period of time. This is useful for risk assessments, since exposure-related measurements can be expressed as a rate of discrete events, i.e., the number of times an event happens during a defined time interval (e.g. the frequency (times per week) that a person eats shellfish that contain polychlorinated biphenyls (PCBs) in concentration fish containing a methyl mercury concentration greater than 12.0 mg L^{-1}). The Poisson distribution describes events that take place during a fixed period of time (i.e., a rate), so long as the individual events are independent of each other.

As the expected number of events or counts increases (i.e., the event rate increases), so does variability. Obviously, if we expect a count to equal 1, then we should have little trouble picturing an observation of 2 or 0. If we expect a count equal to 50,000, counts of 49,700 and 50,300 are within reason. The range and

(Continued)

variance of the latter, however, is much larger. The Poisson equation needed to compute the probability of a specific number of counts being observed over a defined time interval is:

$$P_\lambda = \frac{e^{-\pi}\lambda^n}{n!} \tag{12.5}$$

where λ = average or expected counts or events per unit time

n = number of encounters

Thus, the Poisson distribution is useful in a risk assessment to estimate exposures. It may be used to characterize the frequency with which a person (or animal or ecosystem) comes into contact with a substance (e.g. the number of times per day a person living near a wood treatment facility is exposed to pentachlorophenol). Assuming that, based on existing data, the expected number of encounters is two per day, applying Equation 12.4 with $\lambda = 2$, there is a 9% chance that an individual will have four (i.e., $n = 4$) encounters with pentachlorophenol on a given day.

Risk itself is a probability (i.e. the chance of an adverse outcome). So, any calculation of environmental insult will likely be based on some use of probabilities. The challenge of biotechnology can reside in the very low probabilities (small likelihood) of important outcomes. After all of the contingent, joint probabilities are calculated and after all of the posterior distributions are considered, is the biotechnology indeed worth the risk?

Engineering success or failure is in large measure determined by comparing what we do to what our profession "expects" of us. Safety is a fundamental facet of our engineering duties. Thus, we need a set of criteria that tells us when our designs and projects are sufficiently safe. Recall from Chapter 1 that the four major safety criteria that need to be applied to biotechnology are [57]:

The design must comply with applicable laws.
The design must adhere to "acceptable engineering practice."
Alternative designs must be sought to see if there are safer practices.
Possible misuse of the product or process must be foreseen.

Complying with health and safety laws, regulations and guidelines is often a technical endeavor, e.g. are physical containment facilities sufficiently leak proof and do the organisms meet specific hazard or exposure criteria? If not, the bioengineer has failed. Acceptable engineering practice is a bit more subjective, but is also a matter of meeting technical guidance in a manner which would be with a consensus of fellow bioengineers. The margin of safety and acceptable risk are dictated by sound bioengineering principles and practice. The key is in finding alternatives and predicting misuse and mistakes calls for prospective viewpoints, such as complete critical paths and event trees to indicate even unlikely, but unacceptable outcomes.

If one were to query a focus group as to whether risk can be quantified, the group is usually divided. At first thought, most respondents consider risk not to be quantifiable. Risk is in the "eye of the beholder." Some of the rationale appears to be rooted in the controversial risks of tobacco use and daily decisions, such as choice of modes of transportation. Like most decisions, risk decisions consist of five components:

An inventory of relevant choices
An identification of potential consequences of each choice
An assessment of the likelihood of each consequence actually occurring
A determination of the importance of these consequences
The synthesis of this information to decide which choice is the best. [58]

These perceptions change with age, as a result of experiences and physiological changes in the brain. However, like the risks associated with the lack of experience in driving an automobile, the young person may do permanent damage while traversing these developmental phases. In fact, this mix of physiological, social, and environmental factors in decision making is an important variable in characterizing hazards. In addition, the hazard itself influences the risk perception. For example, whether the hazard is intense or diffuse, or whether it is natural or human-induced (see Figure 12.14) is a determination of public acceptance of the risks associated with the hazard. People tend to be more accepting of hazards that are natural in origin, voluntary, and concentrated in time and space [59].

Other possible explanations are risk mitigation and sorting of competing values. The biker may well know that smoking is a risky endeavor and is attempting to mitigate that risk by other positive actions, such as exercise and clean water. Or, she may simply be making a choice that the freedom to smoke outweighs other values like a healthy lifestyle (students have reported that biking may well simply be a means of transportation and not a question of values at all).

It is likely that all of these factors affect different people in myriad ways, illustrating the complexities involved in risk management decisions.

Demographics is a determinant of risk perception, with certain groups more prone to "risk taking" and averse to authority. Teenagers, for example, are often shifting dependencies, e.g. from parents to peers. Later, the dependencies may be transitioning to greater independence, such as that found on college campuses. Eventually, these can lead to interdependent, healthy relationships. Engineers have to deal with these dynamics as a snapshot. Although the individual is changing, the population is often more static. There are exceptions, for example, if the mean age of a neighborhood is undergoing significant change (e.g. getting younger), then there may a concomitant change in risk acceptance and acceptance of controls (e.g. changes to zoning and land use).

People seem to have their own "mathematics" when it comes to risk. If you visit a local hospital, you are likely to see a number of patients gathered near the hospital entrance in designated smoking areas. Here they are hooked up to IVs, pumps, and other miracles of medical technology and simultaneously engaging in one of the most potent health hazards,

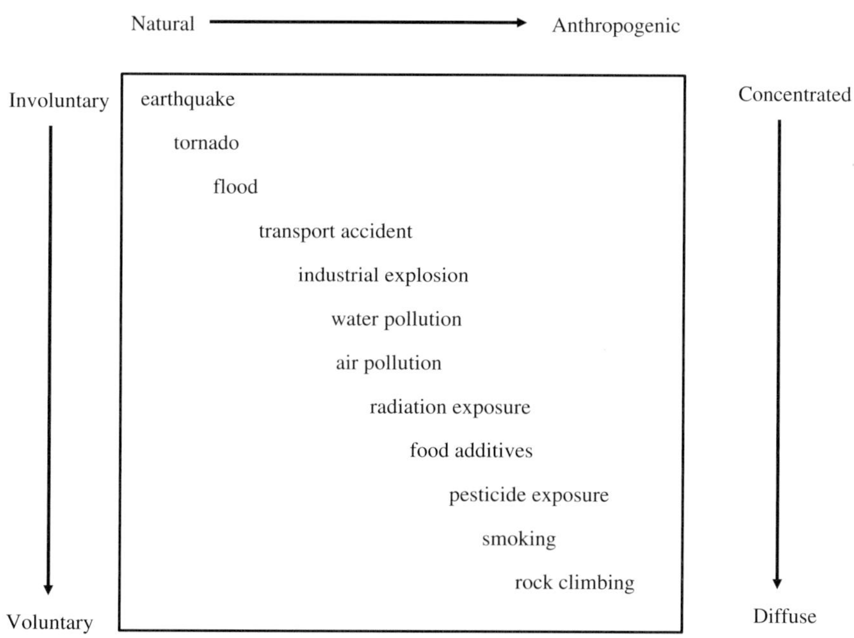

FIGURE 12.14
Spectrum of hazards.
Source: D.A. Vallero (2005). *Paradigms Lost: Learning from Environmental Mistakes, Mishaps, and Misdeeds.* Butterworth-Heinemann, Burlington, MA; adapted from K. Smith (1992). *Environmental Hazards: Assessing Risk and Reducing Disaster.* Routledge, London.

smoking. Of course, this is ironic. On the one hand we are assigning the most talented (and expensive) professionals to treat what is ailing them, yet they have made a personal decision to engage in very unhealthful habits. It seems analogous to training and equipping a person at great social costs to drive a very expensive vehicle, all the time knowing that the person has a nasty habit of tailgating and weaving in traffic. Even if the person is well trained and has the best car, this habit will increase the risk. However, there is another way to look at the smoking situation: that is, the person has decided (mathematically) that the "sunken costs" are dictating the decision. Intuitively, the smoker has differentiated long-term risks from short-term risks. The reason the person is in the hospital (e.g. heart disease, cancer, emphysema, etc.) is the result of risk decisions the person made years, maybe decades ago. The exposure is long in duration and the effect is chronic. So the person may reason that the effects of today's smoking will only be manifested 20 years hence and has little incentive to stop engaging in the hazardous activity. Others see the same risk and use a different type of math. They reason that there is X odds that they will live a somewhat normal life after treatment, so they need to eliminate bad habits that will put them in the same situation 20 years from now. They both have the same data, but reach very different risk decisions.

Another interesting aspect of risk perception is how it varies in scale and scope. The scope of a problem colors what we consider to be of value. For example, how do we value the life of an animal? Indeed, biotechnology is all about engaging the instrumental value of organisms. Biotechnological advances are presently highly dependent on information gathered from comparative biology. We use models to extrapolate potential hazards of new drugs, foods, pesticides, cosmetics and other household products. We study the effects of devices as prototypes in animals to determine their risks and efficacy prior to introducing them to humans. When we do this, we are placing a value on the utility of the results of the animal research, that is, we are applying a utilitarian ethical model. If the "greater good" is served, we conclude that the animal studies are worthwhile. Critics of this approach are likely to complain that the animals' welfare is not given the appropriate weight in such a model. For example, the suffering and pain experienced by animals with nervous systems like ours and clear capacities for higher level social behavior are unjustifiable means to an end, albeit one that serves science well.

We do not have go to the laboratory, however, to consider valuation as it pertains to non-human animals. Many communities are overrun with deer populations because suburban developments have infringed on the habitats of deer and their predators, changing the ecological dynamics. Certainly, the deer population increases present major safety problems, especially during the rutting season when deer will enter roadways and the likelihood of collisions increase. The deer also present nuisances, such as their invasions of gardens. They are even part of the life cycle of disease vectors when they are hosts to ticks that transmit Rocky Mountain spotted fever and Lyme's disease. In this sense, we may see deer as a "problem" that must be eradicated. However, when we come face to face with the deer, such as when we see a doe and her fawns, we can appreciate the majesty and value of those individual deer.

> Recently, I have been trying desperately to prevent the unwelcome visits of deer to my Chapel Hill garden. I have had little success (they particularly like eating the tomatoes before they ripen and enjoy nibbling young zucchini plants). In the process, I have uttered some rather unpleasant things about these creatures. On a recent occasion, however, I was traveling on a country highway and noticed an emergency vehicle assisting a driver who had obviously crashed into something. The driver and responder were in the process of leaving the scene. Driving 20 meters further, I noticed a large doe trying to come to her feet to run into the woods adjacent to the highway, but she could not lift her back legs. I realized that the people leaving the scene must have concluded that the "emergency" was over, without regard to the deer that had been struck by the car. Seeing the wounded creature was a reminder that the individual, suffering animal had an intrinsic value. Such a value is lost

when we see only the large-scale problem, without consideration of the individual. Incidentally, by the time I had turned around another person had already called the animal control authorities. Coming into personal contact with that deer makes it *our* deer; just as knowing that there are too many dogs and cats in the nation does not diminish our devotion to *our* own dog or cat [60]. When we do this, we move from a utilitarian model to one of empathy. This certainly applies to most matters of bioethics. Our ability to empathize about the risks and benefits of a given drug, device, or system is a tool for predicting the necessary moral action.

Another lesson here is our need to be aware of what we are saying to people. No matter how technically sound and convincing to us as engineers, our communications of efficacy and risk may be simply a din to our targeted audience, or even an affront to the values they cherish. Their value systems and the ways that they perceive risk are the result of their own unique blend of experiences.

The converse is also true. We may be completely persuaded based upon data, facts, and models that something clearly does not cause significant harm, but those we are trying to convince of this finding may not buy it. They may think we have some vested interest, or that they find us guilty by association with a group they do not trust, or that we are simply hired guns. They may not understand us because we are using jargon and are not clear in how we communicate the risks.

Thus, it requires considerable effort to ensure that the perception of risk of one's clients and the public matches the actual risks calculated and quantified by the bioscientist and bioengineer.

RELIABILITY OF BIOTECHNOLOGICAL SYSTEMS

Reliability, like risk, is an expression that incorporates probability, but instead of conveying something bad, it expresses the likelihood of a good, or at least *desired*, outcome. Reliability is the extent to which something can be trusted. A system, process or item is reliable so long as it performs the designed function under the specified conditions during a certain time period. In most environmental biotechnological applications, reliability means that the designed systems will not fail during their proposed operating periods. Or, stated more positively, reliability is the mathematical expression of success; that is, reliability is the probability that something that is in operation at time *0* (t_0) will still be operating until the designed life (time $t = (t_t)$). As such, it is also a measure of the engineer's social accountability. Neighborhoods, municipalities and other clients want to be assured that the bioremediation project will work as designed and will not fail to treat the toxic waste.

The probability of a failure per unit time is the "hazard" rate, but many engineers may recognize it as a "failure density," or $f(t)$. This is a function of the likelihood that an adverse outcome will occur, but note that it is not a function of the severity of the outcome. The $f(t)$ is not affected by whether the outcome is very severe (such as pancreatic cancer and loss of an entire species) or relatively benign (muscle soreness or minor leaf damage). The likelihood that something will fail at a given time interval can be found by integrating the hazard rate over a defined time interval:

$$P\{t_1 \leq T_f \leq t_2\} = \int_{t_1}^{t_2} f(t)dt \tag{12.6}$$

where T_f = time of failure.

Thus, the reliability function $R(t)$, of a system at time t is the cumulative probability that the system has not failed in the time interval from t_0 to t_t:

$$R(t) = P\{T_f \geq t\} = 1 - \int_0^t f(x)dx \tag{12.7}$$

Engineers must be humble, since everything we design *will* fail. We can improve reliability by extending the time (increasing t_t), thereby making the system more resistant to failure. For example, proper engineering design of a stent allows for improved blood flow. The stent may perform well for a certain period of time, but loses efficiency when it becomes blocked. A design flaw occurs if the wrong material is used, such as one that allows premature sorption of cholesterol or other substances tô the interior lining of the stent. However, even when the proper materials are used, the failure is not completely eliminated. In our case, the sorption still occurs, but at a slower rate, so 100% blockage (i.e. $R(t) = 0$) would still be reached eventually. Selecting the right materials simply protracts the time before the failure occurs (increases T_f) [61]. Ideally, if possible the reliability of a device or system reflects an adequate margin of safety. So, if the stent fails after 100 years, few reasonable people would complain. If it fails after 10 years, and was advertised to prevent blockage for five years, there may still be complaints, but the failure rate and margin of safety were fully disclosed. If it fails after 10 years, but its specifications require 20 years, this may very well constitute an unacceptable risk. Obviously, other factors must be considered, such as sensitive subpopulations. For example, it may be found that the stent performs well in the general population, but in certain subgroups (e.g. blood types) the reliability is much less. The engineer must fully disclose such limitations. These lessons can be applied to biotechnologies, such as deficiencies in bioreactor vessel geometry, improper environment for targeted microbes, lack of monitoring and adjustments improper levels of nutrients and build-up of toxic substances and using the wrong statistics (e.g. mean nutrient needs rather than 90th percentile) to predict optimal conditions for microbial growth.

Equation 12.7 illustrates built-in vulnerabilities, such as the inclusion of inappropriate design criteria, like cultural bias, the time of failure is shortened. If we do not recognize these inefficiencies upfront, we will pay by premature failures (e.g. law suits, unhappy clients, and a public that has not been well served in terms of our holding paramount their health, safety, and welfare). So, if we are to have reliable engineering we need to make sure that whatever we design, build, and operate is done so with fairness and openness. Otherwise, these systems are, by definition, unreliable.

Reliability engineering, a discipline within engineering, considers the expected or actual reliability of a process, system or piece of equipment to identify the actions needed to reduce failures, and once a failure occurs, how to manage the expected effects from that failure. Thus, reliability is the mirror image of failure. Since risk is really the probability of failure (i.e. the probability that our system, process, or equipment will fail), risk and reliability are two sides of the same coin. The most common graphical representation of engineering reliability is the so-called "bathtub" curve (Figure 12.15). The U-shape indicates that failure will more likely occur at the beginning (infant mortality) and near the end of the life of a system, process, or equipment. Actually, the curve indicates engineer's common proclivity to compartmentalize. We are tempted to believe that the process only begins after we are called on to design a solution. Indeed, failure can occur even before infancy. In fact, many problems in environmental justice occur during the planning and idea stage. A great idea may be shot down before it is born.

Biotechnological decisions must be inclusive. For example, some environmental and public health decisions can have a greater negative impact on certain subpopulations (sensitivity, allergenicity, etc.). These are known as disparities in exposure and in outcomes. Injustices can gestate even before the engineer becomes involved in the project. This "miscarriage of justice" follows the physiological metaphor closely. Certain groups of people have been historically excluded from preliminary discussions, so that if and when they do become involved they are well beyond the "power curve" and have to play catch up. The momentum of a project, often being pushed by the project engineers, makes participation very difficult from some groups. So, we can modify the bathtub distribution accordingly (see Figure 12.16). The figure shows that the rate of failure is highest during gestation. This may or may not be the case, since identifying the number of premature failures is extremely difficult to document with any modicum of certainty.

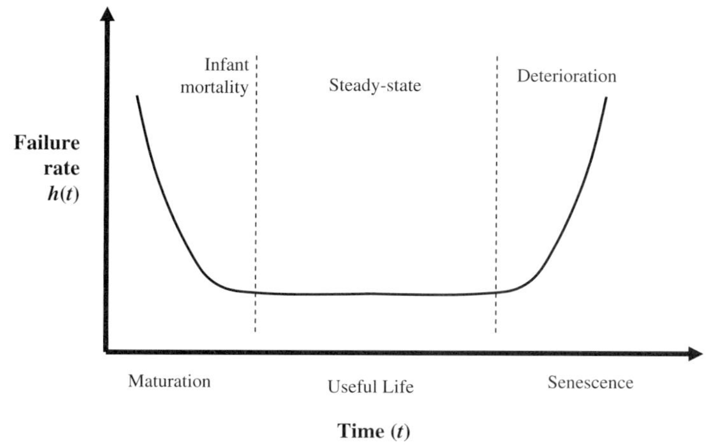

FIGURE 12.15
Prototypical reliability curve, i.e. the bathtub distribution. The highest rates of failure, $h(t)$, occur during the early stages of adoption (infant mortality) and when the systems, processes or equipment become obsolete or begin to deteriorate. For well-designed systems, the steady-state period can be protracted, e.g. decades.
Source: D.A. Vallero (2005). *Paradigms Lost: Learning from Environmental Mistakes, Mishaps, and Misdeeds.* Butterworth-Heinemann, Burlington, MA.

Another way to visualize reliability as it pertains to biotechnology is to link potential causes to effects. Cause and effect diagrams (also known as *Ishikawa diagrams*) identify and characterize the totality of causes or events that contribute to a specified outcome event. The "fishbone" diagram (see Figure 12.17) arranges the categories of all causative factors according to their importance (i.e. their share of the cause). The construction of this diagram begins with the failure event to the far right (i.e. the "head" of the fish), followed by the spine (flow of events leading to the failure). The "bones" are each of the contributing categories. This can be a very effective tool in explaining failures to clients and the public. Even better, the engineer may construct the diagrams in "real time" at a design meeting. This will help to open up the design for peer review and will help get insights, good and bad, early in the design process.

The premise behind cause and effect diagrams like the fishbones and fault trees is that all the causes have to connect through a logic gate. This is not always the case, so another more qualitative tool may need to be used, such as the *Bayesian belief network* (BBN). Like the fishbone, the BBN starts with the failure (see Figure 12.18). Next, the most immediate contributing causes are linked to the failure event. The next group of factors that led to the immediate causes are then identified, followed by the remaining contributing groups.

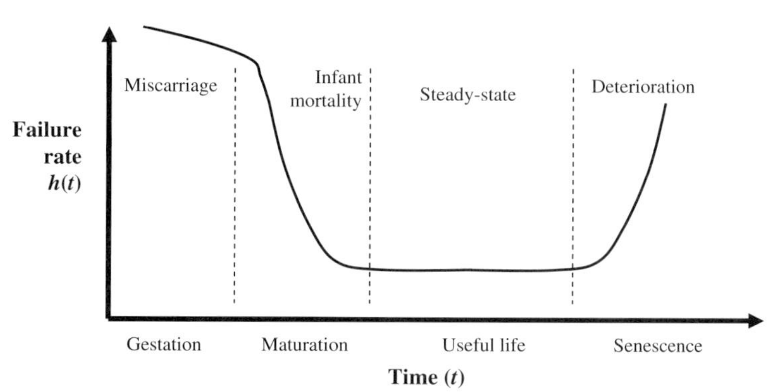

FIGURE 12.16
Prototypical reliability curve with a gestation (e.g. idea) stage. The highest rate of failure, $h(t)$, can occur even before the system, process, or equipment has been made a reality. Exclusion of people from decision making or failure to get input about key scientific or social variables can create a high hazard.
Source: D.A. Vallero (2005). *Paradigms Lost: Learning from Environmental Mistakes, Mishaps, and Misdeeds.* Butterworth-Heinemann, Burlington, MA.

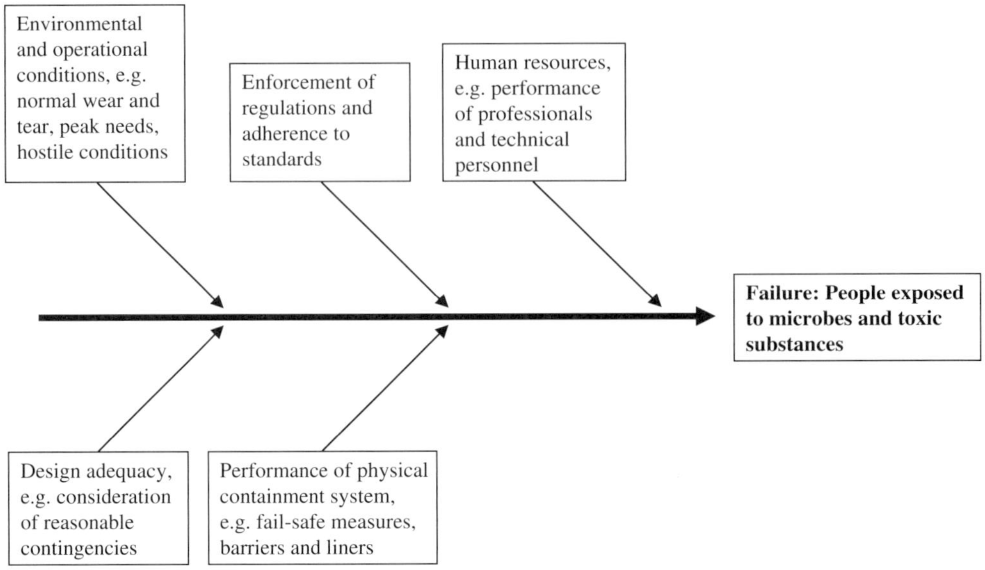

FIGURE 12.17
Fishbone reliability diagram showing contributing causes to an adverse outcome (exposure to products of biotechnological operations).

This diagram helps to catalog the contributing factors and also compares how one group of factors impacts the others. Note the similarities between the BBN and Figures 6.11 and 6.12.

The engineering and scientific communities often use the same terms for different concepts. This is the case for reliability. Environmental engineering and other empirical sciences also use the term "reliability," to indicate quality, especially for data derived from measurements, including health data. In this use, reliability is defined as the degree to which measured results are dependable and consistent with respect to the study objectives, e.g. stream water quality. This specific connotation is sometimes called "test reliability" in that it indicates how consistent measured values are over time, how these values compare to other measured values, and how they differ when other tests are applied. Test reliability, like engineering reliability, is a matter of trust. As such, it is often paired with test validity, that is, just how near to the true value (as indicated by some type of known standard) the measured value is. The less reliable and valid the results, the less confidence scientists and engineers have in interpreting and using them. This is very important in engineering communications generally, and risk communications specifically.

The engineer must know just how reliable and valid the data are. And, the engineer must properly communicate this to clients and the public. This means, however discomfiting, the

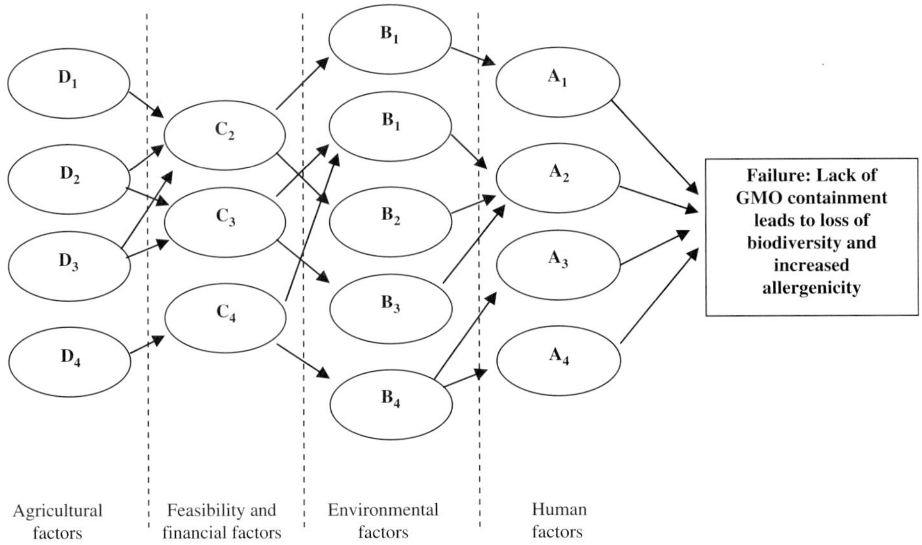

FIGURE 12.18
Bayesian belief network (BBN), with three groups of contributing causes leading to a failure.

engineer must "come clean" about all uncertainties. Uncertainties are ubiquitous in risk assessment. The engineer should take care not to be overly optimistic nor overly pessimistic about what is known and what needs to be done. Full disclosure is simply an honest rendering of what is known and what is lacking for those listening to make informed decisions. Part of the uncertainty involves conveying the meaning; we must clearly communicate the potential risks. A word or phrase can be taken many ways. Engineers should liken themselves to physicians writing prescriptions. Be completely clear, otherwise confusion may result and lead to unintended, negative consequences.

The concept of safety is laden with value judgments. Thus, ethical actions and decisions must rely on both sound science and quantifiable risk assessment, balanced with an eye toward social fairness.

RISK HOMEOSTASIS AND THE THEORY OF OFFSETTING BEHAVIOR

Any engineer's paramount concern is design failure. Product liability is one of the public's responses to past failures and a means to induce researchers and designers to build in safeguards to prevent future failures. One example of product liability oversight is that of the US Food and Drug Administration (FDA). Figure 12.19 provides a simplified model of FDA's licensing process.

The reasons that products fail are two-fold. Either they have flaws or they are misused. Good engineering practice prevents the first, but what about the second? At what point are engineers to blame for misuse of their products? This is even more complicated for bioengineering than many other endeavors, since biotechnologies involve living organisms, which are always more complicated than abiotic systems.

The response to biotechnological failure can be dramatic and severe. Baruch Fishhoff, a psychologist, and Jon Merz, an engineer [62] have noted that people process failures both cognitively and motivationally:

> Cognitively, injured parties see themselves as having been doing something that seemed sensible at the time, and not looking for trouble. As a result, any accident comes as a surprise. If it was to be avoided, then someone else needed to provide the missing expertise and protection. Motivationally, no one wants to feel responsible for an accident. That just adds insult to injury, as well as forfeiting the chance for emotional and financial redress.
>
> Of course, the natural targets for such blame are those who created and distributed the product or equipment involved in an accident. They could have improved the design to prevent accidents. They should have done more to ensure that the product would

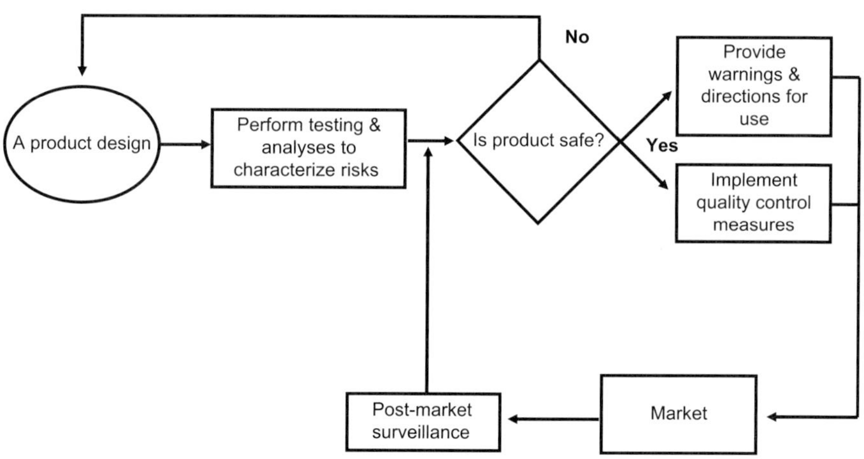

FIGURE 12.19
Licensing process of the US Food and Drug Administration.
Source: B. Fishhoff and J.F. Merz (1994). The inconvenient public: behavioral research approaches to reducing product liability risks. In: National Academy of Engineering, *Product Liability and Innovation: Managing Risk in an Uncertain Environment*. National Academies Press, Washington, DC.

> not fail in expected use. They could have provided better warnings and instructions in how to use the product. They could have sacrificed profits or forgone sales, rather than let users bear (what now seem to have been) unacceptable risks.

Another explanation of misuse may fall under the first definition as a design flaw. It is known as *risk homeostasis* [63]. Basically, users defeat built-in factors of safety by asserting new ways to use the products. If such aggressiveness occurs, it may simply be because consumers want more from their products. Some argue that the instructions and warnings accompanying products should be sufficient consideration, and if a user willfully ignores this information this constitutes the user's autonomy in a rational decision made with "informed consent." Others disagree, holding that liability extends beyond labeling and reasonable use [64]. This is particularly problematic for complicated biosystems like those employed by biotechnologists.

Economists have found that policies designed to protect the public may inadvertently lead to "attenuation and even reversal of the direct policy effect on expected harm ... because of *offsetting behavior* (OB) by potential victims as they reduce care in response to the policy" [65]. Much of the research has been related to transportation, especially road safety. For example, drivers with anti-lock brake systems tend to tailgate more closely and the use of helmets has not been commensurate with expected injury prevention in cycling. It is logical that such behavior would also extend to biotechnologies, since they involve human factors and are sensitive to even microscopic changes in bioreactor conditions.

Events such as the presence of genetically modified foodstuffs found in human diets indicate that society's response is colored by offsetting behavior. For example, many did not question the use of genetically modified corn as food for animals. Is it possible that those feeding animals relied on safeguards too blindly and did not recognize when genetically modified corn found its way into human diets? Relying on conventional safeguards rather than instituting biotechnological safeguards can result in offsetting behaviors.

CHAOS AND ARTIFACTS

Chaos is defined as a dynamical system that is extremely sensitive to its initial conditions. Small differences early in a process can profoundly and extensively change the final state of the system even over small timescales. Such unintended change can be considered an artifact of the system. Bioengineers are vulnerable to artifacts. That is, even well-conceived designs may lead to impacts down the road. Biotechnological risk assessment should not be limited to bioengineers. For example, biotechnological enterprises can be harmed by failure in any engineering discipline, as well as by other professions.

In the case of genetically modified corn in the human diet, offsetting behavior was not that of the consumer, but of the gatekeepers expected to keep GMOs separate from regular grain. These gatekeepers may have become overly reliant on the backup and redundant systems, to the extent that normal operations were not adequately monitored.

The engineering canon that an engineer must be competent within an area of expertise is not an excuse not to consider possible flaws that result down the road. In fact, just the opposite is true. That is, part of the competence requires consideration of offsetting behavior. So, then, what actions can lead to artifacts? The FDA and others offer some insights about how to avoid offsetting behaviors [66]:

- Beware of overconfidence in computer systems – Engineers must overcome the temptation to assume that problems are usually about hardware.
- Do not confuse reliability with safety – Software can work thousands of times without error, but even these probabilities are unacceptable when consequences are potentially widespread and long-term.

Include a defensive design – Redundancies as well as self-checks, trouble-shooting, and error detection and correction systems are vital. A worst-case design scenario should be identified.

Address root causes – Causes may be misidentified and the corrections may solve problems other than the ones leading to the real failures.

Avoid complacency – In many engineering failures the contingency that led to the artifact was part of a routine process. Continued success does not mean due diligence can be short-cut.

Conduct realistic risk assessments – The failure analysis should not assume independence and miss key factors. Seemingly mundane or benign technologies can have adverse effects.

Thus, a biotechnology itself may be ethically neutral, but the risk homeostasis and offsetting behaviors that it spawns are not. This is another type of the consequence that must be considered in a bioengineer's design. In fact, it is akin to and supportive of *root cause analysis,* which is a method to address a problem, retrospectively retracing events, i.e. a "reverse event tree." The goal is to learn lessons and not repeat the same mistakes, but in a very orderly step-wise manner. An excellent tool to display this retracing is the fishbone diagram (Figure 12.17). Another means at getting at the root cause is the Bayesian inference (see Chapter 6).

Engineering, by its nature, attempts to be as exact as possible in addressing inexact problems. That is part of the reason that this book is more about guidance than about "plug and chug" tools. Any endeavor that involves living things, any endeavor that applies biological principles, and any endeavor using terminology that begins with "bio" is going to be imprecise and uncertain. Environmental biotechnology, in both its applications and implications, is simultaneously exciting and frightening. The famous engineer Norm Augustine has articulated the bioengineering challenge:

> The bottom line is that the things engineers do have consequences, both positive and negative, sometimes unintended, often widespread, and occasionally irreversible ... Engineers who make bad decisions often don't know they are confronting ethical issues. [67]

Biotechnology comes with opportunities and peril, Augustine's *caveat* reminds us that the bioscientist, bioengineer and biotechnologist have been given the responsibility to manage these promising and daunting prospects with care and trust.

SEMINAR TOPIC

Risk Tradeoffs in Environmental Biotechnology

In environmental biotechnology, problems are seldom presented as "good versus bad." More often, they are a choice between "good versus good." It is good to clean up a hazardous waste site. It is also good to contain microbial populations so as to protect neighboring habitats.

The overall objective is to render hazardous compounds nontoxic. Empirical evidence has shown that microbial growth and metabolism, as well as that of larger organisms, can be very efficient at degradation. This is an essential part of natural systems, and environmental biotechnologists are quite good at observing natural systems and transferring these processes to engineered systems. Traditionally, indigenous soil bacteria have been encouraged to use some rather exotic chemical compounds as food, especially when it is the sole source of carbon and electrons (i.e. acclimation).

We also know that some compounds are recalcitrant, refusing to be the carbon and energy sources for most microbes. Sometimes, the solution lies in finding a different type of organism. The keen observations of environmental engineers and biologists led to the logical use of fungi, for example, since these organisms are quite proficient at degrading the otherwise recalcitrant natural polymers, especially lignin in wood. Bacteria have a much harder time degrading these substances, but the bioscientists reckoned that different orders of organisms could be put to use. One observation was that thousands of strains and species of white rot fungi (numerous species, including *Phanerochaete chryosporium*) have been shown to degrade lignin, cellulose, and hemicellulose. Taking this into account, *P. chryosporium* has been put to work in increasingly larger reactors (from lab benches to full scale *ex situ* operations). In addition to aromatic hydrocarbons and other organic compounds, fungi have also been used to extract and detoxify metallic contaminants.

(Continued)

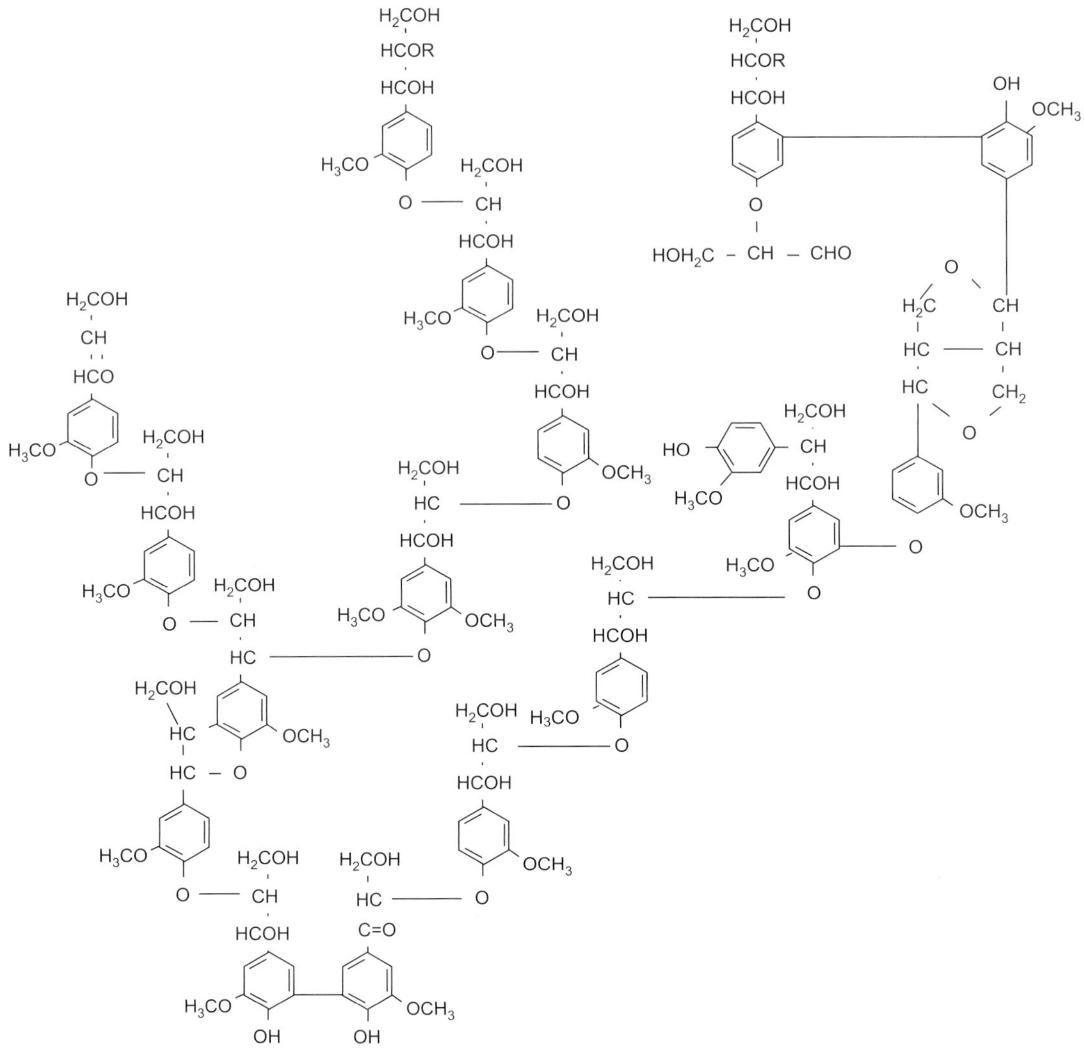

FIGURE 12.20
An example of possible configurations of a lignin polymer.
Source: Institute of Biotechnology and Drug Research, *Environmental Biotechnology and Enzymes.* Kaiserslautern, Germany; adapted from E. Adler (1977). Lignin chemistry – past, present and future. *Wood Science Technology* 11: 169–218.

Both cellulose and lignin are polymers, which are large organic molecules comprised of repeated subunits (i.e. monomers). However, lignin is naturally recalcitrant, owing to its covalent bonds to hemicellulose and cross-linking to polysaccharides, which holds woody tissue together and provides the rigidity of wood compared to non-lignin-containing plants (see Figure 12.20). Lignin fills the spaces in a woody plant's cell wall between cellulose and two other compounds, hemicellulose and pectin.The monomers that comprise lignin polymers can vary, depending on the sugars from which they are derived. In fact, lignin polymers contain so many random couplings that the exact chemical structure is seldom known [68].

Lignin is not a source of energy for white rot fungi, so their biodegradation of toxic pollutants follows the prototypical *cometabolic pathway*, i.e. they need other substrates (e.g. cellulose). This need, however, makes fungi particularly attractive to green engineering solutions (e.g. composting). That is to say, silage and other vegetative waste material from agricultural operations can be degraded (e.g. composted), and in the process, enhance degradation rates of xenobiotic contaminants in polluted material, either *in situ* or *ex situ* at polluted sites. All plants contain cellulose, but woody plants also contain lignin.

The systematic perspective takes into account that white rot and other fungi branch and form filaments, which indicates that their growth is especially conducive to soil treatment. The principal mechanism for white rot fungus degradation is enzymatic, especially the fortuitous lignin degradation system of enzymes. Extracellular *lignin modifying enzymes* (LMEs) are not very substrate-specific, which enables them to mineralize numerous, otherwise recalcitrant contaminants, particularly those with chemical structures similar to lignin [69]. The predominant LMEs are lignin peroxidase [70], manganese (Mn)-dependent peroxidase [71], and laccase [72]. The extracellular enzymatic mechanism allows for contact with the numerous organic contaminants with low aqueous solubility (e.g. chlorinated and multiple ring aromatic compounds) not allowed when restricted to intracellular processes. For example, catalysis by cytochrome P450 would first require that the contaminant compound be transferred across the microbe's membrane, followed by contact with the cytochrome enzymes (see Figure 12.21). These are predominantly localized to the endoplasmic reticulum (ER), and can also be found in other subcellular compartments, such as mitochondria, plasma membrane, and lysosomes [73].

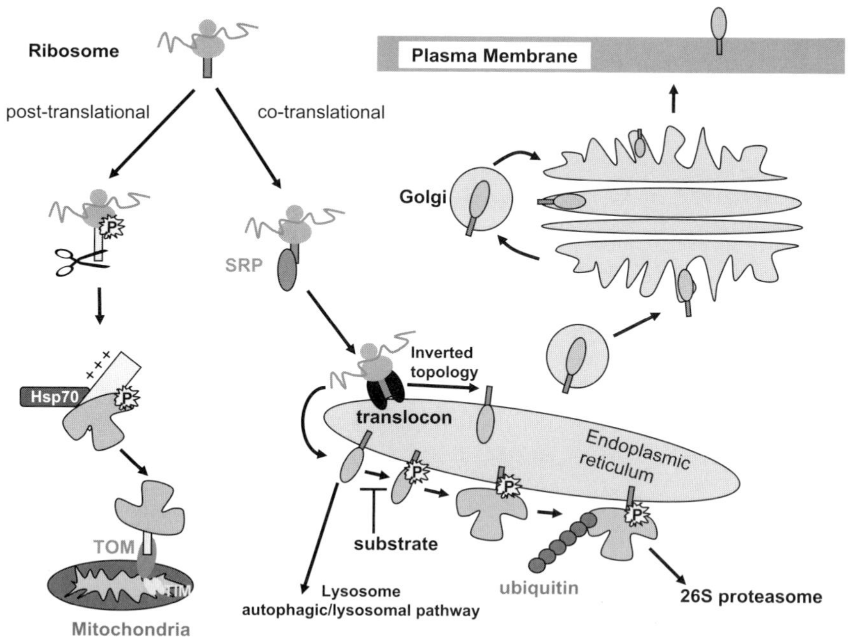

FIGURE 12.21
Intracellular targeting, transport, and degradation of microsomal cytochrome P450. Normally, cytochromes P450 are targeted to the endoplasmic reticulum (ER) via a signal recognition particle (SRP) process. A very small quantity of the enzymes may be inserted inversely and transported via the Golgi apparatus to the plasma membrane. Cytochromes P450 are degraded via the lysosomal pathway or after phosphorylation (*P*) or covalent modification by the proteasomal pathway. The presence of a substrate can prevent phophosporylation and subsequent proteosomal degradation. Modification of the signal sequence by phosphorylation or proteolytic processing (shown as scissors) prevents efficient binding of the SRP that leads to translation of the entire protein in the cytoplasm and after interaction with chaperones (Hsp70) to mitochondrial targeting and import. TOM = translocase of the outer membrane; TIM = translocase of the inner membrane. Cytochromes P450 can be degraded via the lysosomal pathway or after phosphorylation (*P*) or covalent modification by the proteasomal pathway. The presence of a substrate may prevent phophosporylation and subsequent proteosomal degradation. [*See color plate section*]
Source: E.P.A. Neve and M. Ingelman-Sundberg (2008). Intracellular transport and localization of microsomal cytochrome P450. *Analytical & Bioanalytical Chemistry* 392 (6): 1075–1084.

Cometabolism is an excellent illustration of a biotechnological systems approach. On its own, each species in a successful bioremediation project may not solve the problem. However, the synergies among species and strains of microbes can lead to successes not possible from reductionism. As mentioned in Chapter 7, the metabolism of a microbe that does not need our targeted contaminant as a carbon source will nonetheless lead indirectly to detoxification due to degradation that is catalyzed by an enzyme that is fortuitously produced by the organisms for other purposes. The microbe does not directly benefit from the degradation of the compound [74]. In fact, the biotransformation of the compound could actually be harmful or can inhibit growth and metabolism of a microbe. This process would make little sense from a reductionist viewpoint, but is perfectly reasonable from the systematic perspective.

Thus, bioengineers have been able to add a second or additional carbon sources (e.g. unsubstituted alkanes) that stimulate biodegradation of the targeted pollutant. For example, halogenated compounds that are inherently more recalcitrant than their non-halogenated counterparts can be degraded by stimulating soil bacteria with the addition of O_2 and CH_4 to the subsoil strata.

The systematic view is further reinforced by the relationships among abiotic and biotic features in the environment. For example, chemical compounds have varying affinities to certain types of soils, depending on surface area, sorptive properties, functional groups on the compounds, redox conditions, and moisture. In other words, from a reductionist perspective, it would be impossible to predict the rates of degradation of a compound solely from its inherent physicochemical properties.

Consider the explosive trinitrotoluene (TNT). Obviously, the compound degrades extremely fast when ignited, i.e. the explosion from the addition of oxygen in the presence of a heat source. This extremely rapid oxidation produces CO_2, H_2O, and oxides of nitrogen. Scientists also observed that in the environment TNT can be reduced slowly. In laboratory studies, the majority of the TNT degrades to monoaminodinitrotoluene and diaminonitrotoluene isomers within a few days (see Figure 12.22).

(Continued)

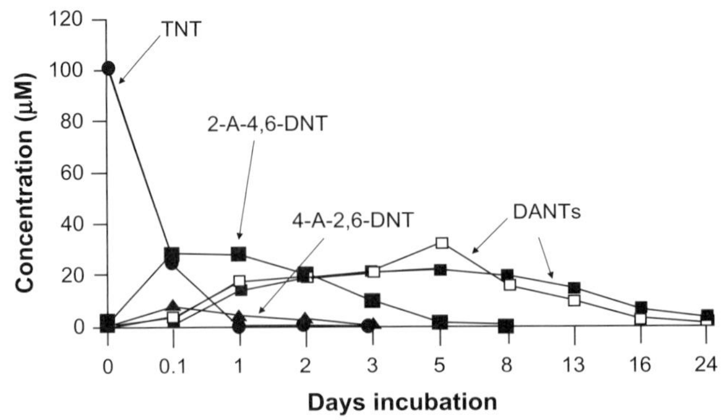

FIGURE 12.22
Degradation of trinitrotoluene (TNT) biodegradation in serum bottles incubated under reduced (methanogenic) conditions. Note: 2-A-4,6-DNT = 2-amino-4,6-dinitrotoluene; 4-A-2,6-DNT = 4-amino-2,6-dinitrotoluene; and DANTs = diaminonitrotoluene isomers.
Source: P. Hwang, T. Chow and N.R. Adrian (1998). Transformation of TNT to triaminotoluene by mixed cultures incubated under methanogenic conditions. US Army Corps of Engineers. USACERL Technical Report No. 98/116.

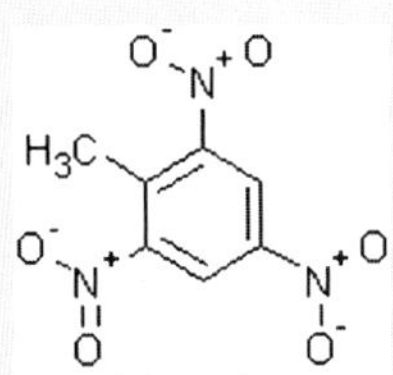

Trinitrotoluene

The persistence of TNT in the environmental systems is quite different. When released into the atmosphere it is degraded by predominantly direct photolysis, with an estimated atmospheric half-life ranging from 18.4 days to 184 days [75]. These $T_{1/2}$ estimates are based on the expected reactions with hydroxyl radicals in the atmosphere. The transformation of TNT in surface waters by microbial metabolism is much slower than photolysis. Under anaerobic and aerobic environments, the predicted biodegradation $T_{1/2}$ of TNT in surface water ranges between 1 and 6 months [76].

These rates do not represent environmental biodegradation, however, since environmental fate depends on more than mere abiotic chemistry. In addition to redox reactions, the TNT is sorbed to clay and organic molecules (e.g. humic acid) in the soil, with the intermediate degradates formed along the way. That is, TNT initially undergoes oxidation to form a variety of reduction products, culminating in the formation of triaminotoluene, which can be irreversibly adsorbed to soil's clay and organic matter content (see Figure 12.23). These chemical and physical mechanisms, i.e. redox and soil binding, call for a series of aerobic, then anaerobic degradation processes (see Figure 12.24) to completely remove and degrade TNT when it is in soil. The reduction of one nitro (-NO_2) group of TNT is quite fast by aerobes. Conversely, the reduction of 2-amino-4,6-dinitrotoluene requires a lower redox potential, and reduction of 2,4-diamino-6-nitrotoluene requires a very low redox potential (i.e. < -200 mV), due to the electron-donating properties of the amino (–NH_2) group's decreasing the electron deficiency of the molecule [77].

TNT is biodegraded aerobically by a number of organisms, including fungi (*Phanerochaete chryosporium* and *Irpex lacteus*), yeasts (*Candida* and *Geotrichum spp.)* and bacteria (e.g. *Actinomycetes spp.*, *Pseudomonas spp.* and *Alcaligenes* 1-15), and anaerobically by bacteria (e.g. *Methanococcus spp.* B strain [78], *Desufovibrio spp.*, *Clostridium pasteurianum* and *Moorella thermoacetica* [79]).

Incidentally, cometabolism appears to enhance TNT's degradation rates (see Table 12.4). The cosubstrates' presence may enhance the TNT degradation rates by serving as H_2 donors. Complex substrates can be fermented by microorganisms in a bacterial consortium, thus generating H_2 as one of the products, which in turn becomes available to the TNT-degrading bacteria [80].

In all of these biodegradation pathways, existing processes have been enhanced to achieve a societal need, i.e. the degradation of a hazardous and otherwise recalcitrant pollutant. There is little debate in the enhancement of indigenous bacteria, yeast or fungi, to break down these unwanted contaminants. Adding oxygen, nutrients, and water to improve rates of biodegradation is good engineering practice. But, does this hold when we change the genetic material within the biota? The answer is most likely that "it depends."

Indeed, when indigenous bacteria are acclimated to use otherwise unattractive food sources, the bioengineer is changing the genetic makeup of the microbial population. So, then, what is the difference if genetic engineering, e.g. insertion of recombinant DNA, does what is seemingly the same thing? Bioengineers are often unprepared for such questions in the environmental fields, but these confrontations are not at all uncommon in other areas of biotechnology especially, as discussed in Chapter 9, when it comes to food supply.

One pivotal concern is that of containment. There are numerous dimensions to containment, but they either have to do with the biological agent or the environment, and usually both. If the agent does not elicit real or perceived hazards, containment would not be an issue. And, even if an agent is considered hazardous, if there is no

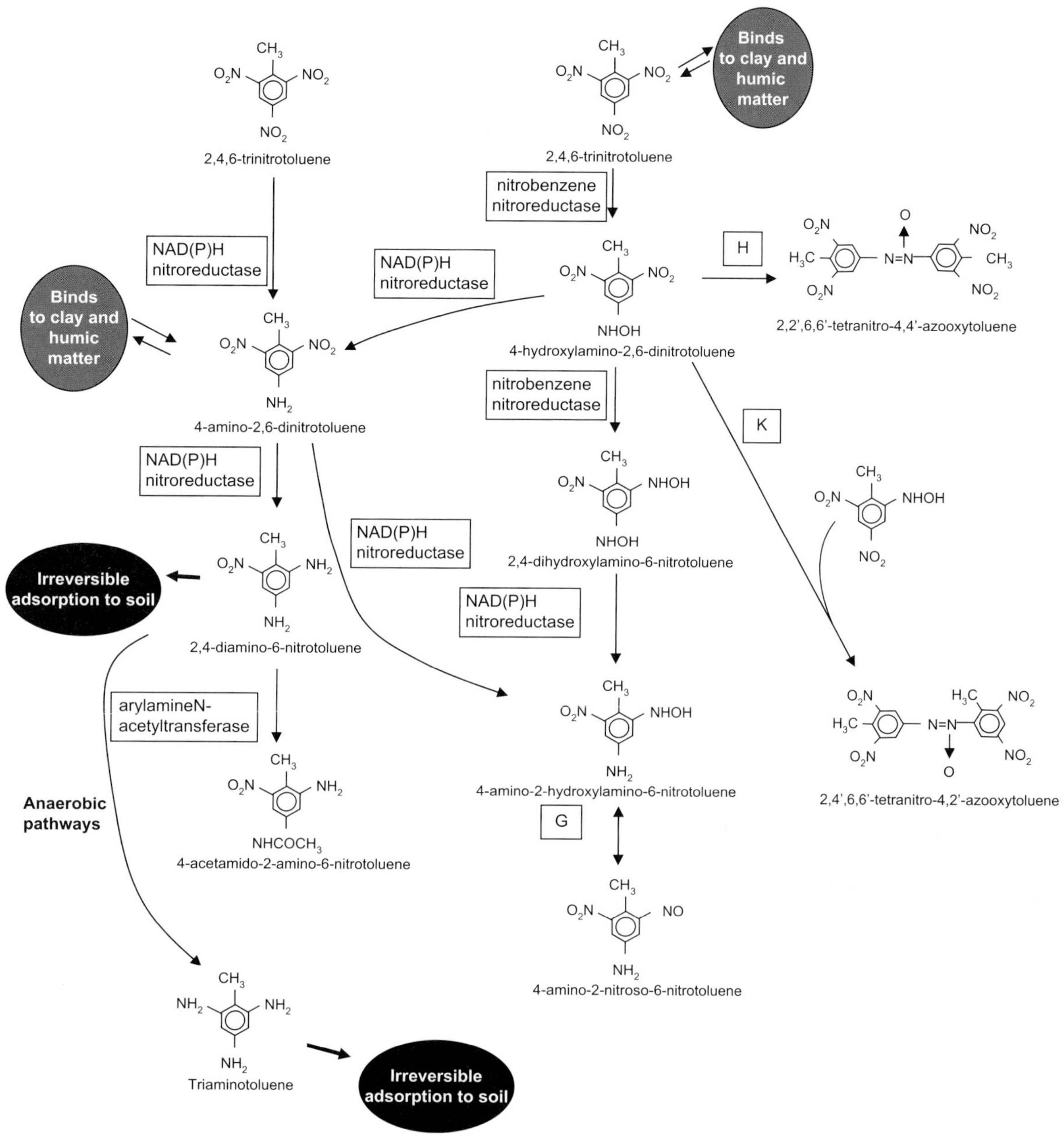

FIGURE 12.23
One set of aerobic (e.g. *Pseudomonas savastanoi*) biodegradation pathways of trinitrotoluene, including influence of soil binding. Anaerobic pathway details are shown in Figure 12.22.
Sources: A. Scragg (2004). *Environmental Biotechnology*, 2nd Edition. Oxford University Press, Oxford; and S. McFarlan (2009). 2,4,6-Trinitrotoluene Pathway Map. University of Minnesota Biocatalysis/Biodegradation Database.

means by which it will move away from a "safely" localized site, it is unlikely to be of much concern.

Thus, when it comes to the compounds that receive the most headlines in technical and lay literature, e.g. carcinogens, neurotoxins, endocrine disruptors, teratogens, and other toxic substances, the technical and lay communities are likely to consider the risk tradeoffs acceptable, so long as the best engineering practices are followed. For example, if PCBs are found in sediments that are likely to leach into surface waters with which human beings and their food sources may come into contact, a well-designed bioremediation project is likely to be considered a necessary cleanup action. Let us consider three scenarios.

(Continued)

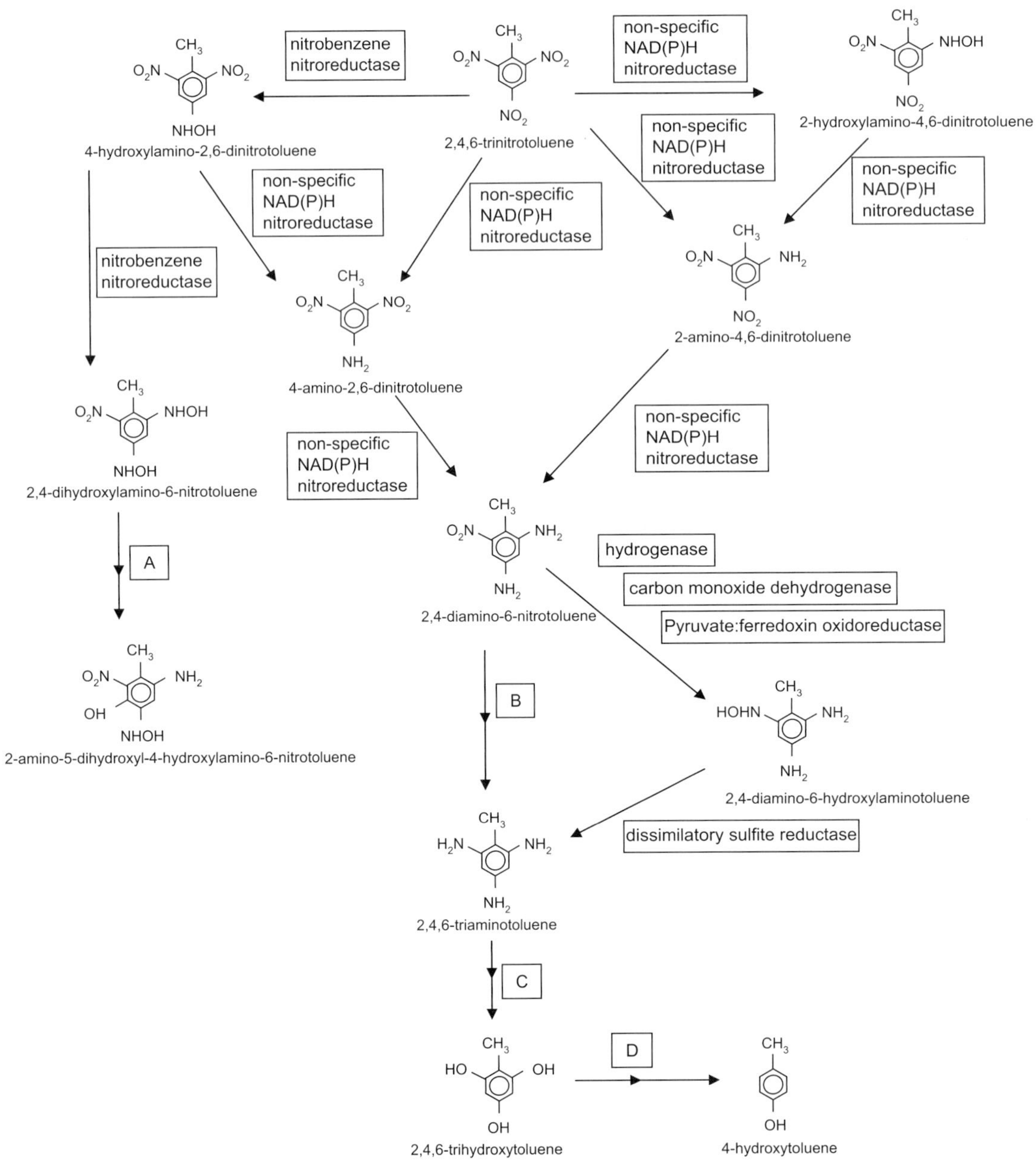

FIGURE 12.24

Anaerobic biodegradation pathway for trinitrotoluene. Most reactions are catalyzed by non-specific NAD(P)H-dependent nitroreductases. The last reduction steps to produce triaminotoluene occur only under anaerobic conditions catalyzed by enzymes in *Desufovibrio spp.*, *Clostridium pasteurianum*, and *Moorella thermoacetica*. These nitro group reductions are also catalyzed by purified xenobiotic reductase enzyme.
Source: S. McFarlan and G. Yao (2009). Anaerobic Trinitrotoluene Pathway Map. University of Minnesota Biocatalysis/Biodegradation Database.

Scenario 1: Different Outcome between Indigenous and Genetically Modified Strains

Further, if it can be shown that the cleanup of certain PCBs would not occur *unless* a genetically modified microorganism is added as a biostimulation step, the genetically modified organism is likely to be accepted. This is because the risk of being exposed to these PCBs far outweighs the risks associated with introducing a heretofore nonexistent bacterial strain.

Scenario 2: Same Outcome between Indigenous and Genetically Modified Strains but Protracted Remediation Time

But, what if the PCBs would be degraded, but at a much slower rate by indigenous bacteria acclimated to them? We now have a scenario where the outcome is the same, but the path to the outcome is longer for the natural microbes than the genetically modified strains. Likely, in a public meeting, if the difference is in

Table 12.4 **Enhancement of biodegradation of trinitrotoluene (TNT) by addition of cosubstrates**

Electron donor	Trinitrotoluene degradation rate (μM day^{-1})
None	2.2
Acetate	2.7
Ethanol	4.2
Glucose	6.3

Source: P. Hwang, T. Chow and N.R. Adrian (1998). Transformation of TNT to triaminotoluene by mixed cultures incubated under methanogenic conditions. US Army Corps of Engineers. USACERL Technical Report No. 98/116.

years, there would probably be little difference between this and Scenario 1.

However, there may be differences within the various technical communities and advocacy groups. For example, some ecologists may see the introduction of any unknown species as potentially irreversible, and from a precautionary perspective, should be avoided if another organism or method will do the job. In this instance, abiotic degradation (e.g. *ex situ* incineration) may be more attractive to some. To others, the additional chemical risks of shipping and treating the hazardous sediments may render this option unacceptable (e.g. PCBs that are improperly incinerated can generate compounds that are more toxic than even the PCB, like dioxins and furans).

Clearly, though, the less the difference in time and costs between native and genetically modified strains the greater the advantage of native over the modified biota. If there is no difference, there is no need to use genetically modified organisms, but at what point do conditions favor one or the other? This "tipping point" will very likely be different between the lay public and the technical community. Indeed, it will likely be very different for the various disciplines within the scientific community.

Scenario 3: Same Outcome between Indigenous and Genetically Modified Strains, but by Different Pathways

Sometimes, there are not only different rates but qualitatively different pathways of degradation between native versus modified strains. If modifying an organism's genetic material results in new enzymatic changes and production of proteins not in the native pathways, these products have to be considered in the overall risk assessment. That is, the addition (and to some extent the elimination) of an intermediate compound introduces the need for a new risk–benefit calculation.

If all that is estimated and modeled is the rate of degradation of the parent compound, this new risk would not even enter into the risk assessment. The worst case would be that the genetically modified organism's biodegradation rate is significantly faster than that of the native organism, but the modified strain produces toxic intermediates not present in the native strain's degradation pathway.

Risk Tradeoffs and Optimization

The ideal scenario is one in which the genetically modified organism's pathway eliminates toxic intermediates *and* provides more rapid ultimate degradation (i.e. to CO_2 and water), *and* is completely contained within the project area during and after remediation (spatial and temporal). The rates and products can vary between the best and worst cases, so the selection of the best process requires that all the variables be optimized.

Engineering solutions always involve a combination of actions. For example, in addition to biostimulation with inoculants of microbial strains, the physical conditions of the targeted cleanup site will be changed. Surfactants may be added to facilitate the desorption of otherwise, low aqueous-soluble contaminants (e.g. chlorinated and large molecules) from the solid matrices in soil and sediment.

As mentioned in the TNT discussion, soil provides absorbents for the retention of organic compounds, with most sorption surfaces in clay and organic matter. Cation exchange capacity, charge, and the nature of cations on the exchange complexes will determine the amount of sorption of organic compounds. Double layer phenomena often control sorption at the soil particle solid–liquid interface. These thin films around soil particles have thicknesses of 1–10 nm and are composed of ions of opposite charge to the particle. The hydrolysis can be acid or base-catalyzed, depending upon the pH of the soil water and other soil conditions. For example, by adding titanium dioxide particles or dissolved natural organic matter particles to a solution, surface catalyzed, neutral hydrolysis can dominate across a large pH range, even when acid or base catalysis had been observed. Such a relationship is a determinant in controlling remediation.

When additives like surfactants are present in soil *micellar* solutions they can also affect hydrolysis and modify the kinetics of the compound–soil–water system. The *micelles* are aggregated surfactant molecules that become colloidally suspended in the water (see Figure 12.25). The micellar head is relatively polar and the tail is relatively nonpolar, meaning that the normally hydrophobic, lipophilic compounds will be dissolved in the head, resulting in a much larger amount of these organic

(Continued)

compounds in the aqueous phase. Thus, the surfactants allow the desorbed molecules to be dispersed, improving contact with the microbes with a concomitant improvement in degradation.

The value-added of surfactants is the extent to which they increase the ease with which otherwise nonpolar substances can reach the aqueous phase and come into contact with the biofilms of microbes. This is determined by the structure of the chemical compounds and the concentration as a multiple of *critical micellar capacity* (CMC). Another systematic consideration is that even though synthetic surfactants can be quite efficient in desorbing recalcitrant compounds, they or the structures they form can be toxic to microorganisms at the concentrations needed for desorption [81].

Thinking systematically, however, this could present a considerable problem in terms of transport (see Figure 12.26A). In other words, since the contaminants are now more dispersed throughout the groundwater and soil (Figure 12.26B), have we not made them more mobile and weakened containment? Again, the solution must be systematic. With more of the contaminant dispersed, additional engineering measures must be deployed, such as slurry walls and pumps (see Figure 12.26C). The resulting treatment will be much better, so long as *all of the systems are operating as designed.*

This discussion illustrates the complexities that are involved in natural systems and the possibilities that engineering actions taken to solve one problem can exacerbate another or even introduce a problem that would not have existed had the engineering action not be taken. More positively, these linkages can also be leveraged to provide better results than if each action were taken separately. The key is to apply reliable information to fit the constraints, needs, and opportunities of each system.

Seminar Questions

Are there ever instances where a reductionist approach is favored over a systematic one in environmental biotechnologies? Why or why not?

How can abiotic and biotic uncertainties in the TNT degradation pathway be properly addressed?

Are there ways to ensure that a bioengineer does not make matters worse, e.g. adding surfactants only when the circumstances and controls are appropriate?

How can aerobic and anaerobic processes be used for halogenated compounds?

Is it possible to create an engineered and constructed system that uses numerous types of organisms in series and succession, e.g. fungi, aerobic and anaerobic bacteria, and larger plants? Where can you go to find prototypes of such complex systems?

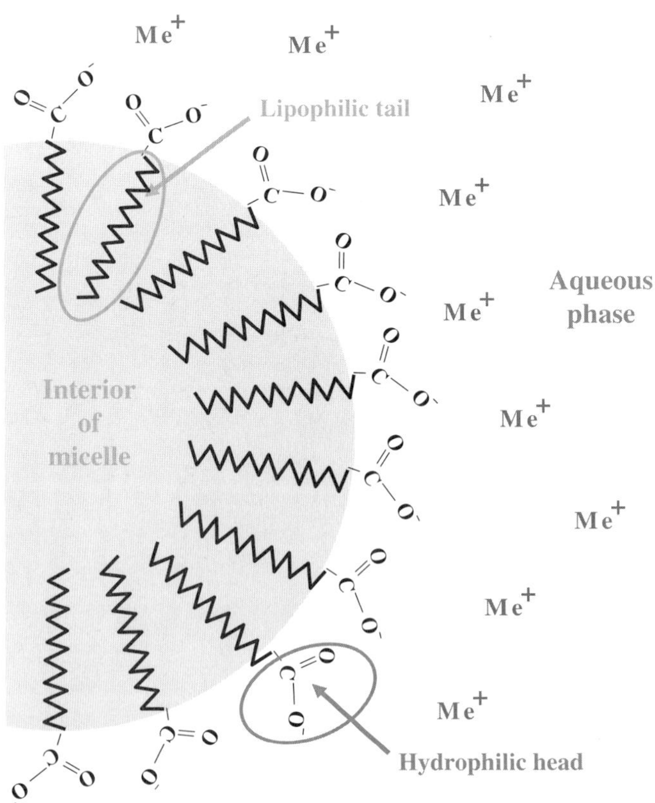

FIGURE 12.25

Structure of micelle in an aqueous solution forming an aggregate with the hydrophilic heads in contact with surrounding water and its dissociated substances (e.g. cations, Me^{+}), and sequestering the lipophilic tails in the micelle's interior. For example, the nonpolar chains of the fatty acids become oriented toward the interior of the micelle, and the polar carboxylate groups keep the micelles in colloidal suspension in water. This allows otherwise nonpolar substances, e.g. halogenated hydrocarbons, to be suspended and dispersed within the aqueous phase. [*See color plate section*]

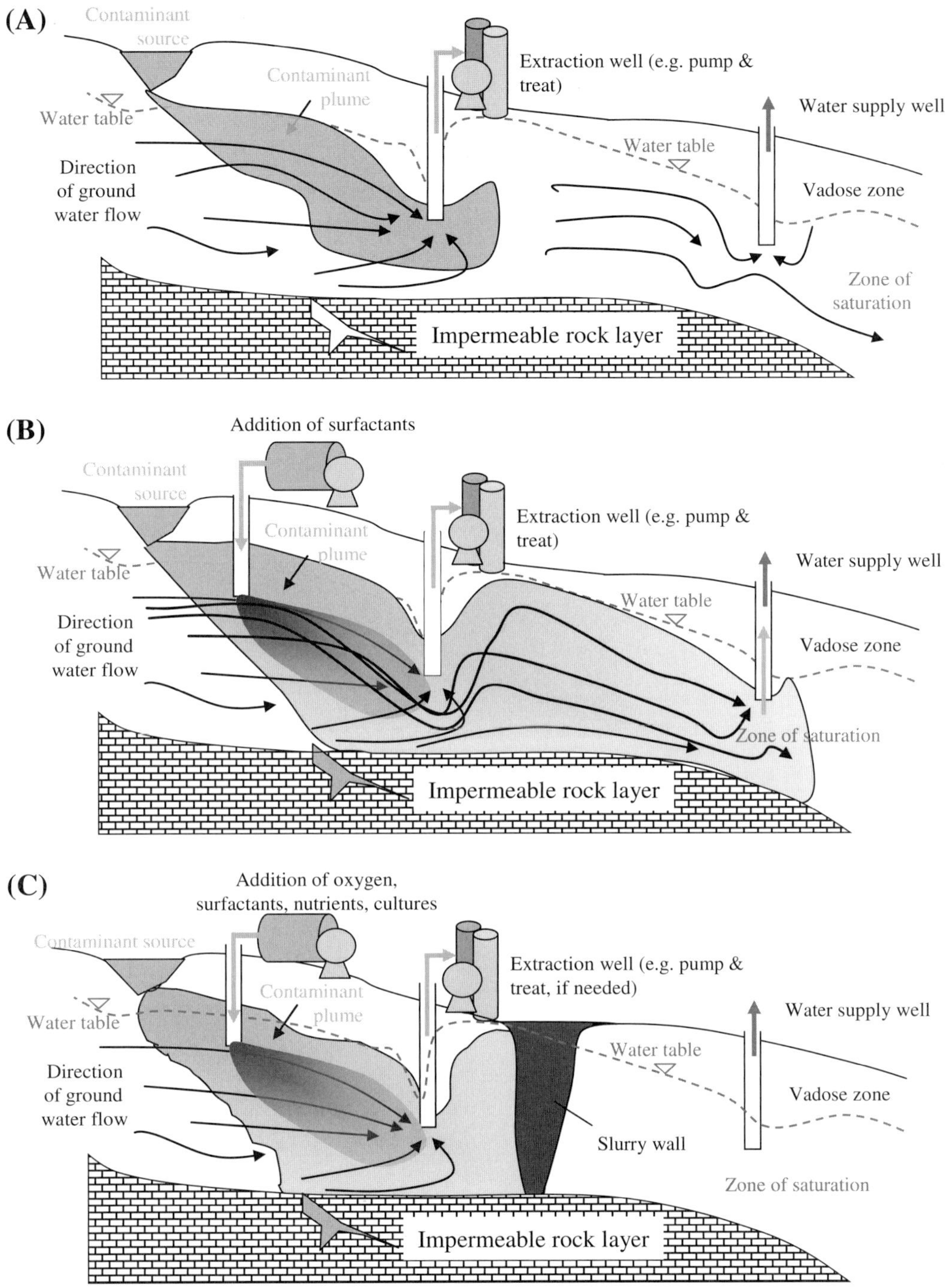

FIGURE 12.26
Changes in contaminant plume by engineering actions. (A) The plume moves advectively with flow lines toward the extraction well, but is not yet influenced by the downgradient drinking water well. (B) The plume becomes more dispersed as the contaminant becomes desorbed from the clay and organic matter in soil by the addition of surfactants, allowing the pollutant to be in the cone of depression of the drinking water well (i.e. contaminating the water supply). (C) The systematic control strategy takes advantage of the desorption and entry of the contaminant into the aqueous phase with a slurry wall or other system used to capture and degrade the pollutant *in situ* or in combination with pump and treat and excavation to *ex situ* treatment. In addition, the barrier protects the drinking water supply. [*See color plate section*]

REVIEW QUESTIONS

What are some of the advantages and disadvantages of taking a solely utilitarian view of agricultural biotechnologies? How does this compare and contrast to environmental, commercial and medical utility?

How are decisions about the environmental impacts of biotechnologies made in universities? ... in government? ... in industry? Do the decision makers fall mainly within the use or knowledge quadrants in Figures 12.3–12.5? Give an example of decisions made in each quadrant.

Read Garrett Hardin's "Tragedy of the commons". Science 1968; 162: 1243–1248. How might this apply to the current approach to biotechnology? Give a specific example of a biotechnology that is "in the commons" (or perhaps, Buckminster Fuller's concept of "Spaceship Earth") and how it has affected and/or will affect the environment.

Construct a fishbone reliability diagram and a Bayesian belief network diagram showing contributing factors that led to regulatory decisions about Bt corn (see Chapter 9). Identify points where interventions could have ameliorated certain problems.

Identify points in the FDA approval process (Figure 12.19) that may not track well with biotechnologies. Discuss examples of how a product may be good for medicine but detrimental for the environment.

Discuss the similarities and differences between the engineering and medical discipline codes of ethics as they apply to environmental decisions. What is the primary focus of each?

Explain the relationship between risk and reliability as they drive environmental biotechnological decisions. Give an example of a biotechnology that seems to be unbalanced between risks and reliability. How does this comport with evidence-based and precaution-based decision making?

A recommendation to improve health care in the United States is to make more health records electronic and to use data mining and other informatics to enhance the physician's information about the patient. "Interoperability" is a term that embodies the ability of different data bases and systems to be joined to serve a specified purpose, e.g. wellness. Luis Kun [82] has said that an "individual's life may be saved a few times by having the right information in the right place at the right time. The quality of life may improve, and the number of potential medication or allergy errors may be eliminated. The knowledge stored could also provide information related to environmental factors (e.g., quality of the water and air), as well as short- and long-term effects of diet, exercise, and vaccines." Comment on whether this statement is correct and how environmental information about genetically modified organisms and gene flow may or may not be part of the electronic health record.

Do you agree with Albert Einstein's quote at the beginning of the chapter? Why or why not?

NOTES AND COMMENTARY

1. Albert Einstein in a letter to an admirer. Quoted in H. Dukas and B. Hoffman (Eds) (1981). *Albert Einstein, The Human Side.* Princeton University Press, Princeton, NJ, p. 43.
2. Sigma Xi (1986). *Honor in Science.* Sigma Xi, the Scientific Research Society, Inc. Research Triangle Park, North Carolina, p. 39.
3. L. Winner (1990). Engineering ethics and political imagination. In: P.T. Durbin (Ed.), *Broad and Narrow Interpretations of Philosophy of Technology.* Kluwer Academic Publishers, Dordrecht, The Netherlands, pp. 53–64. Reprinted in D.G. Johnson (Ed.) (1991). *Ethical Issues in Engineering.* Prentice-Hall, Englewood Cliffs, NJ.
4. K.W. Miller (2009). While we weren't paying attention. *IEEE Technology and Society Magazine* 28 (1): 4.
5. American Society of Civil Engineers (1996). Code of Ethics, Adopted 1914 and most recently amended November 10, 1996, Washington, DC.
6. American Society for Microbiology (2005). Code of Ethic (Revised and Approved by Council 2005); http://www.asm.org/ccLibraryFiles/FILENAME/000000001596/ASMCodeofEthics05.pdf; accessed October 23, 2009.
7. A. McHughen (2000). *Pandora's Picnic Basket: The Potential and Hazards of Genetically Modified Foods.* Oxford University Press, Oxford, UK, p. 10.
8. US General Accountability Office (2002). Genetically Modified Foods: Experts View Regimen of Safety Tests as Adequate, but FDA's Evaluation Process Could Be Enhanced. Report No. GAO-02- 566. 10–11.
9. Ibid. pp. 28–29.
10. Ibid. pp. 29.

11. Ibid. pp. 11–12.
12. I. Kant (1785). *Groundwork of the Metaphysics of Morals* (trans. H.J. Paton, Routledge Reprint, 1991). Harper-Collins, San Francisco, CA.
13. J. Rawls (1971). *A Theory of Justice.* Belknap Press Reprint (1999), Cambridge, MA.
14. National Academy of Engineering (2004). *The Engineer of 2020: Visions of Engineering in the New Century.* National Academies Press, Washington, DC.
15. I. Janus (1982). *Groupthink: Psychological Studies of Policy Decisions and Fiascoes,* 2nd Edition. Houghton Mifflin Co., Boston, MA.
16. D.E. Stokes (1997). *Pasteur's Quadrant.* The Brookings Institution, Washington, DC.
17. J. Fernandez (2002). Understanding group dynamics. *Business Line* December 2, 2002.
18. Stokes, *Pasteur's Quadrant,* p. 12.
19. Ibid., pp. 10–11.
20. Ibid., pp. 18–21.
21. H. Brooks (1967). Applied science and technological progress. *Science* 156 (3783): 1706–1712.
22. Stokes, *Pasteur's Quadrant,* pp. 70–73.
23. C.P. Snow (1959). *The Search.* Charles Scribner's Sons. New York, NY.
24. Pope Benedict XVI (2009). Encyclical Letter: Caritas in Veritate (Truth in Love). The Vatican.
25. A. Schweitzer (1933). *Out of My Life and Thought* (trans. A.B. Lemke, 1990 edition). Henry Holt & Co., New York, NY.
26. A. Colby and L. Kohlberg (1987). *The Measurement of Moral Judgment, Volume 2: Standard Issue Scoring Manual.* Cambridge University Press, Cambridge, UK.
27. ABET (2003). Criterion 3: Program Outcomes and Assessment.
28. T.S. Kuhn (1962). *The Structure of Scientific Revolutions.* University of Chicago Press, Chicago, IL.
29. H. Petroski (1985). *To Engineer Is Human: The Role of Failure in Successful Design.* St Martin's Press, New York, NY.
30. Ibid., p. 58.
31. Ibid., pp. 62–63.
32. National Academy of Engineering (2004). *The Engineer of 2020: Visions of Engineering in the New Century.* National Academies Press, Washington, DC.
33. Petroski, *To Engineer Is Human,* p. 41.
34. Quoted by I. Jackson (1984). *Honor in Science,* p. 7. Sigma Xi, Research Triangle Park, NC; from J. Bronowski (1956). *Science and Human Values.* Messner, New York, NY.
35. Value engineering. OD Value Engineering Program, February 24, 2006; http://ve.ida.org/ve/ve.html; accessed March 16, 2006.
36. What is the value method? Systemic Analytic Methods and Innovations; http://www.value-engineering.com; accessed November 25, 2009.
37. The Value Engineering (VE) process. US Department of Transportation Federal Highway Administration, March 11, 2005; http://www.fhwa.dot.gov/ve/veproc.htm; accessed March 16, 2006.
38. Value in economics, *The Columbia Encyclopedia,* 6th Edition, 2001-05; http://www.bartleby.com/65/va/value2.html; accessed May 2, 2006.
39. D. Ricardo (1821). *On the Principles of Political Economy and Taxation.* John Murray, London, UK; http://www.econlib.org/library/Ricardo/ricP.html; accessed March 15, 2006.
40. Value in economics, *Columbia Encyclopedia.*
41. See C.E. Harris, Jr., Pritchard, M.S. and Rabins, M.J. (2000). *Engineering Ethics, Concept and Cases.* Wadsworth Publishing Co., Belmont, CA.
42. A. Bradford Hill (1965). The environment and disease: Association or causation? *Proceedings of the Royal Society of Medicine, Occupational Medicine* 58: 295.
43. National Academy of Engineering, *The Engineer of 2020,* pp. 50–51.
44. US EPA (2009). What is Green Engineering? http://www.epa.gov/oppt/greenengineering/pubs/whats_ge.html; accessed November 25, 2009.
45. Virginia Polytechnic Institute and State University (2008). Green Engineering Definition; http://www.eng.vt.edu/green/program.php; accessed May 30, 2008.
46. J.M. Benyus (1997). *Biomimicry.* William Morrow and Co., New York, NY, p. 3.
47. American Medical Association (2004). Code of Medical Ethics: Current Opinions with Annotations, 2004–2005. American Medical Association, Chicago, IL.
48. Ibid.
49. National Society of Professional Engineers (2003). NSPE Code of Ethics for Engineers; http://www.nspe.org/ethics/eh1-code.asp, accessed August 21, 2005.
50. C.B. Fleddermann (1999). Safety and risk. *Engineering Ethics,* Chapter 5. Prentice-Hall, Upper Saddle River, NJ.
51. An episode of the popular television series *The Sopranos* included a scene where the teenage girlfriend of the main character's son is eating fruit and smoking a cigarette at a wedding reception. She is offered an *hors d'oeuvre* which contains seafood. She rejects the offer after puffing on the cigarette, saying that she does not eat fish because it "contains toxins."
52. A segregation of risk assessment (science) from risk management (feasability) began in the 1980s, in response to some dramatic cases, notably chemical contamination in Times Beach, Missouri and Love Canal, New York, where the intermingling of assessment and management led to distrust in the reliability of the science.
53. C. Mitcham and R.S. Duval (2000). Responsibility in engineering. *Engineering Ethics,* Chapter 8. Engineer's Toolkit Series, Prentice-Hall, Upper Saddle River, NJ.

54. Although presented within the context of how risk is a key aspect of environmental justice, the information in this chapter is based on two principal sources: D.A. Vallero (2004). *Environmental Contaminants: Assessment and Control.* Elsevier Academic Press, Burlington, MA; and D.A. Vallero (2005). *Paradigms Lost: Learning from Environmental Mistakes, Mishaps, and Misdeeds.* Butterworth-Heinemann, Burlington, MA.
55. T.M. Apostol (1969). *Calculus, Volume II*, 2nd edition. John Wiley & Sons, New York, NY.
56. This discussion and examples are based on: United Nations Environmental Program (2000). International Labor Organisation, International Programme on Chemical Safety, Environmental Health Criteria 214: Human Exposure Assessment. Geneva, Switzerland.
57. Fleddermann, Safety and risk.
58. R. Beyth-Marom, B. Fischhoff, M. Jacobs-Quadrel and L. Furby (1991). Teaching decision making in adolescents: a critical review. In: J. Baron and R.V. Brown (Eds), *Teaching Decision Making to Adolescents.* Lawrence Erlbaum Associates, Hillsdale, NJ, pp. 19–60.
59. K. Smith (1992). Hazards in the environment. *Environmental Hazards: Assessing Risk and Reducing Disaster,* Chapter 1. Routledge, London, UK.
60. This calls to mind Jesus' parable of the lost sheep (Gospel of Matthew, Chapter 18). In the story, the shepherd leaves, abandons really, 99 sheep to find a single lost sheep. Some might say that if we as professionals behaved like that shepherd, we would be acting irresponsibly. However, it is actually how most of us act. We must give our full attention to the patient or client one at a time. There are a number of ways to interpret the parable, but one is that there is an individual value to each member of society and the value of society's members is not mathematically divisible. In other words, a person in a population of one million is not one-millionth of the population's value. The individual value is not predicated on the group's value. This can be a difficult concept for those of us who are analytical by nature, but it is important to keep in mind when estimating risk.
61. Hydraulics and hydrology provide very interesting case studies in the failure domains and ranges, particularly how absolute and universal measures of success and failure are almost impossible. For example, a levee or dam breach, such as the catastrophic failures in New Orleans during and in the wake of Hurricane Katrina, experienced failure when flow rates reached cubic meters per second. Conversely, a hazardous waste landfill failure may be reached when flow across a barrier exceeds a few cubic centimeters per decade.
62. B. Fishhoff and J.F. Merz (1994). The inconvenient public: behavioral research approaches to reducing product liability risks. In: National Academy of Engineering. *Product Liability and Innovation: Managing Risk in an Uncertain Environment.* National Academies Press, Washington, DC.
63. G.J.S. Wilde (1982). The theory of risk homeostasis: implications for safety and health. *Risk Analysis* 2: 209–225.
64. See, for example, P. Slovic and B. Fischhoff (1983). Targeting risks: comments on Wilde's "Theory of Risk Homeostasis." *Risk Analysis* 2: 227–234.
65. J. C. Hause (2006). Offsetting behavior and the benefits of safety regulations. *Economic Inquiry* 44 (4): 689–698.
66. N. Leveson (1995). Medical Devices: The Therac-25. In: *Safeware: System Safety and Computers.* Addison-Wesley Professional Publications, New York, NY.
67. N. Augustine (2002). Ethics and the second law of thermodynamics. *The Bridge* 32 (3): 4–7.
68. E. Adler (1997). Lignin chemistry – past, present and future. *Wood Science Technology* 11: 169–218.
69. T. Cajthaml, M. Möder, P. Kacer, V. Šašek and P. Popp (2002). Study of fungal degradation products of polycyclic aromatic hydrocarbons using gas chromatography with ion trap mass spectrometry detection. *Journal of Chromatography A* 974: 213–222; M. Mansur, E. Arias, J.L. Copa-Patino, M. Flärdh and A. E. González (2003). The white-rot fungus *Pleurotus ostreatus* secretes laccase isozymes with different substrate specificities. *Mycologia* 95 (6): 1013–1020; S.B. Pointing (2001). Feasibility of bioremediation by white-rot fungi. *Applied Microbiology Biotechnology* 57: 20–33; and, E. Veignie, C. Rafin, P. Woisel and F. Cazier (2004). Preliminary evidence of the role of hydrogen peroxide in the degradation of benzo[a]pyrene by a non-white rot fungus *Fusarium solani. Environmental Pollution* 129: 1–4.
70. See, for example, R. Manimekalai and T. Swaminathan (1999). Optimization of lignin peroxidase production from *Phanerochaete chrysosporium* using response surface methodology. *Bioprocess Engineering* 21: 465–468.
71. See, for example, O.V. Koroleva, E.V. Stepanova, V.P. Gavrilova, N.S. Yakovleva, E.O. Landesman, I.S. Yavmetdinov *et al.* (2002). Laccase and Mn-peroxidase production by *Coriolus hirsutus* strain 075 in ajar fermenter. *Journal of Biosciences and Bioengineering* 93(5): 449–455.
72. See, for example, J. Hess, C. Leitner, C. Galhaup, K.D. Kulbe, B. Hinterstoisser, M. Steinwender *et al.* (2002). Enhanced formation of extracellular laccase activity by the white-rot fungus *Trametes multicolor. Applied Biochemistry and Biotechnology* 98-100: 229–241; N. Hatvani and I. Mecs (2001). Production of laccase and manganese peroxidase by *Lentinus edodes* on malt-containing by-product or the brewing process. *Process Biochemistry* 37: 491–496; and I. Herpoel, S. Moukha, L. Lesage-Meessen, J-C. Sigoillot and M. Asther (2000). Selection of *Pycnoporus cinnabarinus* strains for laccase production. *FEMS Microbiological Letters* 183: 301–306.
73. Intracellular transport and localization of microsomal cytochrome P450. *Analytical & Bioanalytical Chemistry* 392 (6): 1075–1084.
74. L.P. Wackett (1996). Co-metabolism: is the emperor wearing any clothes? *Current Opinion in Biotechnology* 7(3): 321–325.
75. Agency for Toxic Substances and Disease Registry (1996). US Department of Health and Human Services. *Toxicological Profile for 2,4,6 Trinitrotoluene.* Atlanta, GA.
76. Ibid.
77. J.C. Spain (1995). Biodegradation of nitroaromatic compounds. *Annual Review of Microbiology* 49: 523–555.
78. Agency for Toxic Substances and Disease Registry (1996). H-Y. Kim and H.G. Song (2003). Transformation and mineralization of 2,4,6-trinitrotoluene by the white rot fungus *Irpex lacteus. Applied Microbiology and Biotechnology*

61 (2): 151–156; and, A.M. Ziganshin, A.V. Naumov, E.S. Suvorova, E.A. Naumenko and R.P. Naumova (2007). Hydride-mediated reduction of 2,4,6-trinitrotoluene by yeasts as the way to its deep degradation. *Microbiology* 76 (76): 676–682.

79. S. McFarlan and G. Yao (2009). Anaerobic trinitrotoluene pathway map. University of Minnesota Biocatalysis/ Biodegradation Database.
80. P. Hwang, T. Chow and N.R. Adrian (1998). Transformation of TNT to triaminotoluene by mixed cultures incubated under methanogenic conditions. US Army Corps of Engineers. USACERL Technical Report No. 98/116.
81. I.V. Robles-González, F. Fava and H.M. Poggi-Varaldo (2008). Review on slurry bioreactors for bioremediation of soils and sediments. *Microbial Cell Factories* 7(5); http://www.microbialcellfactories.com/content/7/1/5; accessed November 26, 2009.
82. L. Kun (2007). Interoperability: the cure for what ails us. *IEEE Engineering in Medicine and Biology* 26 (1): 87–90.

APPENDIX 1

Background Information on Environmental Impact Statements

SUMMARY OF THE COUNCIL ON ENVIRONMENTAL QUALITY GUIDANCE FOR COMPLIANCE WITH THE NATIONAL ENVIRONMENTAL POLICY ACT OF 1969

Title of guidance	Summary of guidance	Citation	Relevant regulation/ documentation
Forty Most Often Asked Questions Concerning CEQ's National Environmental Policy Act Regulations	Provides answers to 40 questions most frequently asked concerning implementation of NEPA	46 FR 18026, dated March 23, 1981	40 CFR Parts 1500–1508
Implementing and Explanatory Documents for Executive Order 12114, Environmental Effects Abroad of Major Federal Actions	Provides implementing and explanatory information for EO 12114. Establishes categories of Federal activities or programs as those that significantly harm the natural and physical environment. Defines which actions are excluded from the order and those that are not	44 FR 18672, dated March 29, 1979	EO 12114, Environmental Effects Abroad of Major Federal Actions
Publishing of Three Memoranda for Heads of Agencies on: – Analysis of Impacts on Prime or Unique Agricultural Lands (Memoranda 1 and 2)	1/2 Discusses the irreversible conversion of unique agricultural lands by Federal Agency action (e.g., construction activities, developmental grants, and federal land management). Requires identification of and cooperation in retention of important agricultural lands in areas of impact of a proposed agency action. The agency must identify and summarize existing or proposed agency policies, to preserve or mitigate the effects of agency action on agricultural lands	45 FR 59189, dated September 8, 1980	1/2 Farmland Protection Policy Act (7 U.S.C. §4201 et seq.) 3 The Wild and Scenic Rivers Act of 1965 (16 U.S.C. §1271 et seq.)
– Interagency Consultation to Avoid or Mitigate Adverse Effects on Rivers in the Nationwide Inventory (Memorandum 3)	3 "Each Federal agency shall, as part of its normal planning and environmental review process, take care to avoid or mitigate adverse effects on rivers identified in the Nationwide Inventory prepared by the Heritage Conservation and Recreation Service in the Department of the Interior." Implementing regulations includes determining whether the proposed action: affects an Inventory River; adversely affects the natural, cultural and recreation values of the Inventory river segment; forecloses options to classify any portion of the Inventory segment as a wild, scenic or recreational river area, and incorporates avoidance/mitigation measures into the proposed action to maximum extent feasible within the agency's authority		
Memorandum for Heads of Agencies for Guidance on Applying Section 404(r) of the Clean Water Act at Federal Projects Which Involve the Discharge of Dredged or Fill Materials into Waters of the US Including Wetlands	Requires timely agency consultation with US Army Corps of Engineers (COE) and the US Environmental Protection Agency (EPA) before a Federal project involves the discharge of dredged or fill material into US waters, including wetlands. Proposing agency must ensure, when required, that the EIS includes written conclusions of EPA and COE (generally found in Appendix)	Council on Environmental Quality, dated November 17, 1980	Clean Water Act (33 U.S.C. §1251 et seq.) EO 12088, Federal Compliance with Pollution Control Standards

Scoping Guidance	Provides a series of recommendations distilled from agency research regarding the scoping process. Requires public notice; identification of significant and insignificant issues; allocation of EIS preparation assignments; identification of related analysis requirements in order to avoid duplication of work; and the planning of a schedule for EIS preparation that meshes with the agency's decisionmaking schedule	46 FR 25461, dated May 7, 1981	40 CFR Parts 1500–1508
Guidance Regarding NEPA Regulations	Provides written guidance on scoping, CatEx's, adoption regulations, contracting provisions, selecting alternatives in licensing and permitting situations, and tiering	48 FR 34263, dated July 28, 1983	40 CFR Parts 1501, 1502, and 1508
National Environmental Policy Act (NEPA) Implementation Regulations, Appendices I, II, and III	Provides guidance on improving public participation, facilitating agency compliance with NEPA and CEQ implementing regulations. Appendix I updates required NEPA contacts, Appendix II compiles a list of Federal and Federal-State Agency Offices with jurisdiction by law or special expertise in environmental quality issues; and Appendix III lists the Federal and Federal-State Offices for receiving and commenting on other agencies' environmental documents	49 FR 49750, dated December 21, 1984	40 CFR Part 1500
Incorporating Biodiversity Considerations into Environmental Impact Analysis under the National Environmental Policy Act	Provides for "acknowledging the conservation of biodiversity as national policy and incorporates its consideration in the NEPA process"; encourages seeking out opportunities to participate in efforts to develop regional ecosystem plans; actively seeks relevant information from sources both within and outside government agencies; encourages participating in efforts to improve communication, cooperation, and collaboration between and among governmental and nongovernmental entities; improves the availability of information on the status and distribution of biodiversity, and on techniques for managing and restoring it; and expands the information base on which biodiversity analyses and management decisions are based	Council on Environmental Quality, Washington, DC, dated January 1993	Not applicable

(*Continued*)

Title of guidance	Summary of guidance	Citation	Relevant regulation/ documentation
Pollution Prevention and the National Environmental Policy Act	Pollution-prevention techniques seek to reduce the amount and/or toxicity of pollutants being generated, promote increased efficiency of raw materials and conservation of natural resources and can be cost-effective. Directs Federal agencies that to the extent practicable, pollution prevention considerations should be included in the proposed action and in the reasonable alternatives to the proposal, and to address these considerations in the environmental consequences section of an EIS and EA (when appropriate)	58 FR 6478, dated January 29, 1993	EO 12088, Federal Compliance with Pollution Control Standards
Considering Cumulative Effects under the National Environmental Policy Act	Provides a "framework for advancing environmental cumulative impacts analysis by addressing cumulative effects in either an environmental assessment (EA) or an environmental impact statement". Also provides practical methods for addressing coincident effects (adverse or beneficial) on specific resources, ecosystems, and human communities of all related activities, not just the proposed project or alternatives that initiate the assessment process	January 1997	40 CFR §1508.7
Environmental Justice Guidance Under the National Environmental Policy Act	Provides guidance and general direction on Executive Order 12898 which requires each agency to identify and address, as appropriate, "disproportionately high and adverse human health or environmental effects of its programs, policies, and activities on minority populations and low-income populations"	Council on Environmental Quality, Washington, DC, dated December 10, 1997	EO 12898, Federal Actions to Address Environmental Justice in Minority Populations and Low-Income Populations

FORMAT OF AN ENVIRONMENTAL IMPACT STATEMENT

Cover Sheet (see next table for information to be included)

EXECUTIVE SUMMARY

TABLE OF CONTENTS

LIST OF ABBREVIATIONS AND ACRONYMS

MEASUREMENT CONVERSION TABLES

CHAPTERS:

1. PURPOSE AND NEED FOR THE ACTION
2. DESCRIPTION AND COMPARISON OF ALTERNATIVES
   - Description of proposed action and each reasonable alternative, including No-Action
   - Brief description of alternatives not considered in detail; explain why
   - Summary of environmental impacts of proposed action and reasonable alternatives, including No-Action
3. DESCRIPTION OF THE AFFECTED ENVIRONMENT
   - Appropriate-level descriptions of the physical, natural, and socioeconomic aspects of the environment that will be impacted, including, but not limited to, air quality, historical/cultural resources, threatened or endangered species and habitats, wetlands, floodplains, and other sensitive/protected resources
4. ENVIRONMENTAL CONSEQUENCES
   - Impact analyses for the proposed action and reasonable alternatives, including No-Action
   - Mandatory subsections
   - Relationship Between Short-term Use of the Human Environment and the Maintenance and Enhancement of Long-term Productivity
   - Irreversible and Irretrievable Commitments of Resources
5. MITIGATION AND MONITORING (optional; can be incorporated into Chapter 4 if appropriate)
6. REFERENCES
7. LIST OF PREPARERS
8. AGENCIES, ORGANIZATIONS, AND INDIVIDUALS CONSULTED
   - Consulting Agencies
   - Distribution List
9. INDEX

Appendices (Final EIS must have a ''Response to Comments'' chapter; as either an appendix or in a separate volume.)

Source: National Aeronautics and Space Administration (2001). Implementing the National Environmental Policy Act and Executive Order 12114, Chapter 6.

REQUIRED COVER SHEET FOR AN ENVIRONMENTAL IMPACT STATEMENT

Note: Boxes are to be completed by a Federal representative (NASA EIS preparer).

POPULAR NAME of PROPOSAL

(includes type, e.g., draft or final)

Lead Agency:	NASA, state name of Sponsoring Entity; name(s) of cooperating agency(ies) if appropriate
Point of Contact for Information:	Name, title, address, and phone number of NASA point of contact
Date:	Date of issuance (recommend using month and year)
Abstract:	Succinct statement of proposed action; brief abstract of the EIS, stating proposed action, alternatives examined, and summary of key findings (the abstract may be printed on a separate page, if necessary)

Source: National Aeronautics and Space Administration (2001). Implementing the National Environmental Policy Act and Executive Order 12114, Chapter 6.

APPENDIX 2

Cancer Slope Factors

Slope factors are expressed in inverse exposure units since the slope of the dose-response curve is an indication of risk per exposure. Thus, the units are the inverse of mass per mass per time, usually $(\text{mg kg}^{-1}\,\text{day}^{-1})^{-1}. = \text{kg day}^{-1}\,\text{mg}^{-1}$. This means that the product of the cancer slope factor and exposure, i.e. risk, is unitless. The SF is the toxicity value used to calculate cancer risks. SF values are contaminant-specific and route-specific (e.g. via inhalation, through the skin, or via ingestion). Inhalation and oral cancer slope factors are shown in Table A2.1. Note that the more potent the carcinogen, the larger the slope factor (i.e. the steeper the slope of the dose-response curve). For example, arsenic and benzo(a)pyrene are quite carcinogenic, with slope factors of 1.51 and 3.10, respectively. Their cancer potency is three orders of magnitude greater than aniline, bromoform, and chloromethane, for example. Also note that the SF is based on the linear portion of the curve.

The route of exposure can greatly influence the cancer slope. Note, for example, that the carcinogenicity of 1,2-dibromo-3-chloropropane is three orders of magnitude steeper via the oral route than from breathing vapors. Conversely, the cancer slope factor for chloroform is more than an order of magnitude greater from inhalation than from oral ingestion. Such information is important in deciding how to protect populations from exposure to contaminants. For example, if an industrial facility is releasing vinyl chloride, both inhalation and oral ingestion must be considered as possible routes of exposures for people living nearby. Both the inhalation and oral slope factors are high, i.e. 3.00×10^{-1} and $1.90\,\text{kg day}^{-1}\,\text{mg}^{-1}$, respectively. In addition, if the vinyl chloride finds its way to the water supply, not only the amount in food and drinking water must be considered, but also indirect inhalation routes, e.g. showering, since vinyl chloride is volatile and can be released and inhaled. The physical and chemical characteristics, e.g. vapor pressure and Henry's law constants, of vinyl chloride coupled with its marked toxicity via multiple routes of exposure, make it a particularly onerous contaminant.

The table also indicates that the structure of a compound greatly affects its biological activity. For example, comparing halogen substitutions indicates that the greater the number of chlorine atoms on a molecule, the steeper the slope of the dose-response curve. Unsubstituted ethane is not carcinogenic (no slope factor). A single chlorine substitution in chloromethane renders the molecule carcinogenic, with a slope factor of 2.90×10^{-3}. Adding another chlorine atom to form 1,2-dichloroethane increases the slope to 9.10×10^{-2}. Completely halogenated ethane, i.e. hexachloroethane, has seen its cancer slope factor increase to 1.40×10^{-2}. Also, where the chlorine or bromine substitutions occur on the molecule will affect the cancer potential. For example, the isomers of tetracloroethane have different slope factors; 1,1,1,2-tetrachloroethane's slope factor is 1.40×10^{-2}, but 1,1,2,2-tetrachloroethane's slope factor is 2.03×10^{-1}. This seemingly small difference in molecular structure leads to an order of magnitude greater cancer potency.

Table A2.1 Cancer slope factors for selected environmental contaminants*

Contaminant	Inhalation slope factor (kg day^{-1} mg^{-1})	Oral slope factor (kg day^{-1} mg^{-1})
Acephate	1.74×10^{-2}	8.70×10^{-3}
Acrylamide	4.55	4.50
Acrylonitrile	2.38×10^{-1}	5.40×10^{-1}
Aldrin	1.71×10^{1}	1.70×10^{1}
Aniline	5.70×10^{-3}	5.70×10^{-3}
Arsenic	1.51×10^{1}	1.50
Atrazine	4.44×10^{-1}	2.22×10^{-1}
Azobenzene	1.09×10^{-1}	1.10×10^{-1}
Benzene	2.90×10^{-2}	2.90×10^{-2}
Benz(a)anthracene	3.10×10^{-1}	7.30×10^{-1}
Benzo(a)pyrene	3.10	7.30
Benzo(b)fluoranthene	3.10×10^{-1}	7.30×10^{-1}
Benzo(k)fluoranthene	3.10×10^{-2}	7.30×10^{-2}
Benzotrichloride	1.63×10^{1}	1.30×10^{1}
Benzyl chloride	2.13×10^{-1}	1.70×10^{-1}
Beryllium	8.40	Not given
Bis(2-chloroethyl)ether	1.16	1.16
Bis(2-chloroisopropyl)ether	3.50×10^{-2}	1.10×10^{-2}
Bis(2-ethyl-hexyl)phthalate	1.40×10^{-2}	7.00×10^{-2}
Bromodichloromethane	6.20×10^{-2}	6.20×10^{-2}
Bromoform	3.85×10^{-3}	7.90×10^{-3}
Cadmium	Not given	6.30
Chlordane	3.50×10^{-1}	3.50×10^{-1}
Captan	7.00×10^{-3}	3.50×10^{-3}
Chlorodibromomethane	8.40×10^{-2}	8.40×10^{-2}
Chloroethane (Ethylchloride)	2.90×10^{-3}	2.90×10^{-3}
Chloroform	8.05×10^{-2}	6.10×10^{-3}
Chloromethane	3.50×10^{-3}	1.30×10^{-2}
Chromium(VI)	3.50×10^{-3}	Not given
Chrysene	3.10×10^{-3}	7.30×10^{-3}
DDD	2.40×10^{-1}	2.40×10^{-1}
DDE	3.40×10^{-1}	3.40×10^{-1}
DDT	3.40×10^{-1}	3.40×10^{-1}

Table A2.1 **Cancer slope factors for selected environmental contaminants*—cont'd**

Contaminant	Inhalation slope factor (kg day^{-1} mg^{-1})	Oral slope factor (kg day^{-1} mg^{-1})
Dibenz(a,h)anthracene	3.10	7.30
Dibromo-3-chloropropane,1,2-	2.42×10^{-3}	1.40
Dichlorobenzene,1,4-	2.20×10^{-2}	2.40×10^{-2}
Dichlorobenzidine,3,3-	4.50×10^{-1}	4.50×10^{-1}
Dichloroethane,1,2-	9.10×10^{-2}	9.10×10^{-2}
Dichloroethene (mixture),1,1-	1.75×10^{-1}	6.00×10^{-1}
Dichloromethane	7.50×10^{-3}	1.64×10^{-3}
Dichloropropane,1,2-	6.80×10^{-2}	6.80×10^{-2}
Dichloropropene,1,3-	1.30×10^{-1}	1.75×10^{-1}
Dieldrin	1.61×10^{1}	1.61×10^{1}
Dinitrotoluene,2,4-	6.80×10^{-1}	6.80×10^{-1}
Dioxane,1,4-	2.20×10^{-2}	1.11×10^{-2}
Diphenylhydrazine,1,2-	7.70×10^{-1}	8.00×10^{-1}
Epichlorohydrin	4.20×10^{-3}	9.90×10^{-3}
Ethyl acrylate	6.00×10^{-2}	4.80×10^{-2}
Ethylene oxide	3.50×10^{-1}	1.02
Formaldehyde	4.55×10^{-2}	Not given
Heptachlor	4.55	4.50
Heptachlor epoxide	9.10	9.10
Hexachloro-1,3-butadiene	7.70×10^{-2}	7.80×10^{-2}
Hexachlorobenzene	1.61	1.60
Hexachlorocyclohexane, alpha	6.30	6.30
Hexachlorocyclohexane, beta	1.80	1.80
Hexachlorocyclohexane, gamma (lindane)	1.30	1.30
Hexachloroethane	1.40×10^{-2}	1.40×10^{-2}
Hexahydro-1,3,5-trinitro-1,3,5-traizine (RDX)	2.22×10^{-1}	1.11×10^{-1}
Indeno(1,2,3-cd)pyrene	3.10×10^{-1}	7.30×10^{-1}
Isophorone	9.50×10^{-4}	9.50×10^{-4}
Nitrosodi-*n*-propylamine, *n*-	7.00	7.00
Nitrosodiphenylamine, *n*-	4.90×10^{-3}	4.90×10^{-3}
Pentachloronitrobenzene	5.20×10^{-1}	2.60×10^{-1}
Pentachlorophenol	1.20×10^{-1}	1.20×10^{-1}

(*Continued*)

Table A2.1 Cancer slope factors for selected environmental contaminants*—cont'd

Contaminant	Inhalation slope factor (kg day^{-1} mg^{-1})	Oral slope factor (kg day^{-1} mg^{-1})
Phenylphenol,2-	3.88×10^{-3}	1.94×10^{-3}
Polychlorinated biphenyls (Arochlor mixture)	3.50×10^{-1}	2.00
Tetrachlorodibenzo-*p*-dioxin,2,3,7,8	1.16×10^{5}	1.50×10^{5}
Tetrachloroethane,1,1,1,2-	2.59×10^{-2}	2.60×10^{-2}
Tetrachloroethane,1,1,2,2-	2.03×10^{-1}	2.03×10^{-1}
Tetrachloroethene (PCE)	2.00×10^{-3}	
Tetrachloroethylene	2.03×10^{-3}	5.20×10^{-2}
Tetrachloromethane	5.25×10^{-2}	1.30×10^{-1}
Toxaphene	1.12	1.10
Trichloroethane,1,1,2-	5.60×10^{-2}	5.70×10^{-2}
Trichloroethene (TCE)	6.00×10^{-3}	1.10×10^{-2}
Trichlorophenol,2,4,6-	1.10×10^{-2}	1.10×10^{-2}
Trichloropropane,1,2,3-	8.75	7.00
Trifluralin	3.85×10^{-3}	7.70×10^{-3}
Trimethylphosphate	7.40×10^{-2}	3.70×10^{-2}
Trinitrotoluene,2,4.6- (TNT)	6.00×10^{-2}	3.00×10^{-2}
Vinyl chloride	3.00×10^{-1}	1.90

These values are updated. If a carcinogen is not listed in the table, visit: http://risk.lsd.ornl.gov/tox/rap_toxp.shtml.

Sources: US Environmental Protection Agency (2002). Integrated Risk Information System; US EPA (1994). Health Effects Summary Tables, 1994.

Dermal exposures are generally extrapolated from the other two major routes. For example, the dermal slope factor for Aroclor 1254, the polychlorinated biphenyl (PCB) mixture (21% $C_{12}H_6Cl_4$, 48% $C_{12}H_5Cl_5$, 23% $C_{12}H_4Cl_6$, and 6% $C_{12}H_3Cl_7$) is dermal exposure to soil or food is 2.22 kg day^{-1} mg^{-1}. Keep in mind that this is the dose-response slope associated with the handling or other skin contact with the contaminant, not the actual ingestion. The Aroclor 1254 dermal slope factor for exposure to water is 4.44 kg day^{-1} mg^{-1}. Both of these dermal slopes have been extrapolated from a gastrointestinal absorption factor of 0.9000. (This information was obtained from the Risk Assessment Information System of the Oak Ridge National Laboratory, 2003.) The dermal slope factors shown in Table A2.2 have been extrapolated from other routes. The GI tract absorption rate is also given, since these are often used to extrapolate slope factors for dermal and other routes of exposure. Note that the larger the GI absorption decimal, the more completely the contaminant is absorbed. For complete absorption, the value equals 1.

The absorption factor is not only important for extrapolating slope factors, but it is a variable in calculating certain exposures. The air (both particle and gas) and water exposure equations include an absorption factor. The dermal exposure equation does not include an absorption factor, but since dermal cancer slope factors are extrapolated from the inhalation or ingestion slopes, by extension, the absorption factor is part of the dermal risk calculations. So, all other factors being equal, a contaminant with a larger absorption factor will have a larger risk. This is evident when considering the pathway taken by a chemical after it

Table A2.2 Gastrointestinal absorption rates and dermal cancer slope factors for selected environmental contaminants*

Contaminant	GI absorption	Inhalation slope factor (kg day^{-1} mg^{-1})
Acephate	0.5	1.74×10^{-2}
Acrylamide	0.5	9.00
Acrylonitrile	0.8	6.75×10^{-1}
Aldrin	1	1.72×10^{1}
Aniline	0.5	1.14×10^{-3}
Arsenic	0.95	1.58×10^{1}
Atrazine	0.5	4.44×10^{-1}
Azobenzene	0.5	2.20×10^{-1}
Benzene	0.9	3.22×10^{-2}
Benz(a)anthracene	0.5	1.46
Benzo(a)pyrene	0.5	1.46×10^{1}
Benzo(b)fluoranthene	0.5	1.46
Benzo(k)fluoranthene	0.5	1.46×10^{-1}
Benzotrichloride	0.8	1.63×10^{1}
Benzyl chloride	0.8	2.13×10^{-1}
Beryllium	0.006	Not given
Bis(2-chloroethyl)ether	0.98	1.13
Bis(2-chloroisopropyl)ether (DEHP)	0.8	8.75×10^{-2}
Bis(2-ethyl-hexyl)phthalate	0.5	2.80×10^{-2}
Bromodichloromethane	0.98	6.37×10^{-2}
Bromoform	0.75	1.05×10^{-2}
Cadmium	0.044	Not given
Captan	0.5	7.00×10^{-3}
Chlordane	0.8	4.38×10^{-1}
Chloroethane (Ethylchloride)	0.8	1.28
Chloroform	1	6.10×10^{-3}
Chloromethane	0.8	1.63×10^{-2}
Chromium(VI)	0.013	Not given
Chrysene	0.5	1.46×10^{-2}
DDD, 4,4-	0.8	3.00×10^{-1}
DDE, 4,4-	0.8	4.25×10^{-1}
DDT, 4,4-	0.8	4.25×10^{-1}
Dibenz(a,h)anthracene	0.5	1.46×10^{1}

(Continued)

Table A2.2 Gastrointestinal absorption rates and dermal cancer slope factors for selected environmental contaminants*—cont'd

Contaminant	GI absorption	Inhalation slope factor (kg day^{-1} mg^{-1})
Dibromo-3-chloropropane,1,2-	0.5	1.12×10^{-1}
Dichlorobenzene,1,4-	1	2.40×10^{-2}
Dichlorobenzidine,3,3-	0.5	9.00×10^{-1}
Dichloroethane,1,2- (EDC)	1	9.10×10^{-2}
Dichloroethene,1,1-	1	6.00×10^{-1}
Dichloropropane,1,2-	1	6.80×10^{-2}
Dichloropropene,1,3-	0.98	1.84×10^{-1}
Dieldrin	1	1.60×10^{1}
Dinitrotoluene,2,4-	1	6.80×10^{-1}
Dioxane,1,4-	0.5	2.20×10^{-2}
Diphenylhydrazine,1,2-	0.5	1.60
Epichlorohydrin	0.8	1.24×10^{-2}
Ethyl acrylate	0.8	6.00×10^{-2}
Ethylene oxide	0.8	1.28
Formaldehyde	0.5	Not given
Heptachlor	0.8	5.63
Heptachlor epoxide	0.4	2.28×10^{1}
Hexachloro-1,3-butadiene	1	7.80×10^{-2}
Hexachlorobenzene	0.8	2.00
Hexachlorocyclohexane, alpha	0.974	6.47
Hexachlorocyclohexane, beta	0.907	1.99
Hexachlorocyclohexane, gamma (lindane)	0.994	1.31
Hexachloroethane	0.8	1.75×10^{-2}
Hexahydro-1,3,5-trinitro-1,3,5-traizine (RDX)	0.5	2.22×10^{-1}
Indeno(1,2,3-cd)pyrene	0.5	1.46
Isophorone	0.5	1.90×10^{-3}
Nitrosodi-*n*-propylamine, *n*-	0.475	1.47×10^{1}
Nitrosodiphenylamine, *n*-	0.5	9.80×10^{-3}
Pentachloronitrobenzene	0.5	5.20×10^{-1}
Pentachlorophenol	0.5	2.40×10^{-1}
Phenylphenol,2-	0.5	3.88×10^{-3}
Polychlorinated biphenyls (Arochlor mixture)	0.85	2.35

Table A2.2 **Gastrointestinal absorption rates and dermal cancer slope factors for selected environmental contaminants*—cont'd**

Contaminant	GI absorption	Inhalation slope factor (kg day^{-1} mg^{-1})
Tetrachlorodibenzo-*p*-dioxin,2,3,7,8	0.9	1.68×10^{5}
Tetrachloroethane,1,1,1,2-	0.8	3.25×10^{-2}
Tetrachloroethane,1,1,2,2-	0.7	2.86×10^{-1}
Tetrachloroethene (PCE)	1	5.20×10^{-2}
Tetrachloromethane	0.85	1.53×10^{-1}
Toxaphene	0.63	1.75
Trichloroethane,1,1,2-	0.81	7.04×10^{-2}
Trichloroethene (TCE)	0.945	1.16×10^{-2}
Trichlorophenol,2,4,6-	0.8	2.20×10^{-2}
Trichloropropane,1,2,3-	0.8	8.75
Trifluralin	0.2	3.85×10^{-3}
Trimethylphosphate	0.5	7.40×10^{-2}
Trinitrotoluene,2,4.6- (TNT)	0.5	6.00×10^{-2}
Vinyl chloride	0.875	2.17

These values are updated. If a carcinogen is not listed in the table, visit: http://risk.lsd.ornl.gov/tox/rap_toxp.shtml.
Sources: US Environmental Protection Agency (2002). Integrated Risk Information System; US EPA (1994). Health Effects Summary Tables, 1994.

enters an organism. As shown in Figure A2.1, the potential dose in a dermal exposure is what is available before coming into contact with the skin, but after this contact (i.e. the applied dose), it crosses the skin barrier and is absorbed. The absorption leads to the biological effectiveness of the contaminant when the chemical reaches the target organ, where it may elicit the effect (e.g. cancer). The absorption factor is the first determinant of the amount of the contaminant that reaches the target organ. For example, although the dermal slope factors for 1,4-dioxane and 1,4-dichlorobenzene are nearly the same (2.20×10^{-2} and 2.40×10^{-2} respectively), all of the dichlorobenzene is expected to be absorbed (i.e. absorption = 1), while only half of the dioxane will be absorbed (i.e. absorption = 0.5). This means that if all other factors are equal, the risk from dichlorobenzene is twice that of dioxane.

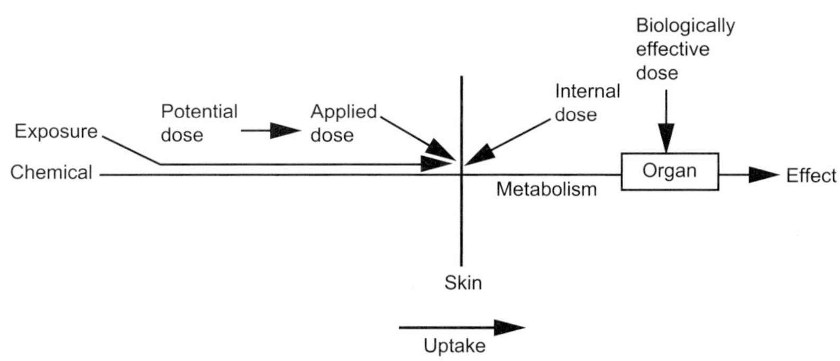

FIGURE A2.1
Pathway of a contaminant from ambient exposure through health effect.
Source: US Environmental Protection Agency, and D. Vallero (2003). *Engineering the Risks of Hazardous Wastes.* Butterworth-Heinemann, Boston, MA.

Exposure route can influence the steepness of the slope. Note for example, that the very lipophilic PCBs have a dermal slope that is an order of magnitude steeper than their inhalation slope. The absorption rate and, hence, the dermal slope is also affected by the contaminant's chemical structure. For example, trichloroethene (TCE) with its double bond between the carbon atoms has an absorption rate of 0.945, while 1,1,2-trichloroethane with only single bonds has a much lower absorption rate of 0.81, even though both have three chlorine substitutions.

Verification Method for Rapid Polymerase Chain Reaction Systems to Detect Biological Agents

The US Environmental Protection Agency's Advanced Monitoring Systems (AMS) Center applied the following approach [1] to evaluate the performance, in terms of its accuracy, specificity, number of false positive/negative responses, precision, interferences, ease of use, and sample throughput, of rapid polymerase chain reaction (PCR) systems to detect biological agents and pathogens in water for the Invitrogen Corporation's PathAlert™ Detection Kits for the detection of *Francisella tularensis* (*F. tularensis*), *Yersinia pestis* (*Y. pestis*), and *Bacillus anthracis* (*B. anthracis*).

Performance test (PT) samples, drinking water (DW) samples, and quality control (QC) samples were used in the verification test for each bacterium. PT samples included individual bacteria spiked into American Society of Testing and Materials (ASTM) Type II deionized (DI) water at 2, 5, 10, and 50 times the vendor-stated method limit of detection (LOD), as well as the infective/lethal dose for each contaminant. PT samples also included potential interferent samples containing a single concentration (10 times the method LOD) of the contaminant of interest in the presence of fulvic and humic acids [at 0.5 milligram (mg)/liter (L) each and 2.5 mg/L each] spiked into ASTM Type II DI water. Interferent samples also were analyzed without the addition of any bacteria. DW samples consisted of chlorinated filtered surface water, chloraminated filtered surface water, chlorinated filtered groundwater, and chlorinated unfiltered surface water collected from four geographically distributed municipal sources. DW samples were analyzed without adding contaminant and after fortification with each individual bacterium at a single concentration level (10 times the vendor-stated method LOD). QC samples included method blank samples and positive (both internal and external) and negative controls, as supplied with each PathAlert™ Detection Kit. For all contaminants, plate enumerations were performed in triplicate to confirm the concentrations of the stock solutions of each bacterium prior to testing.

For the purposes of this test, 1×10^4 colony forming units per milliliter (cfu/mL) were used to calculate the concentration levels of *F. tularensis* and *B. anthracis* spiked into the PT and DW samples; 100 cfu/mL were used to calculate levels of *Y. pestis* spiked in the PT and DW samples. These vendor-provided concentration levels were anticipated to be the levels for the entire experimental process at which quantifiably reproducible positive results could be obtained from a raw water sample. These concentration levels are referred to as the "method LOD" for a particular assay. The method LOD incorporates the sensitivities and uncertainties of not only the PathAlert™ Detection Kit, but also the deoxyribonucleic acid (DNA) purification step; and, as such, it is an experimental detection limit rather than an instrument or reagent-specific detection limit. As mentioned previously, the method LOD provided by the vendor was used specifically as a guideline in calculating sample concentration ranges for use with the PathAlert™ Detection Kit and all other components used in this verification test to analyze a sample, and it should be noted that Invitrogen Corporation does not claim this to be the true LOD of the PathAlert™ Detection Kit alone. The vendor claims the absolute LOD (the least amount of target DNA that would generate a positive result) for the PathAlert™ Detection Kit alone is as low as 1 to 10 copies of DNA, depending on the assay. This information was not verified in this test.

Samples were spiked with *F. tularensis* and *B. anthracis* at 2×10^4 colony-forming units (cfu)/milliliter (mL), 5×10^4 cfu/mL, 1×10^5 cfu/mL, and 5×10^5 cfu/mL for PT samples and 1×10^5 cfu/mL for interferent and DW samples. Samples were spiked with *Y. pestis* at 2×10^2 cfu/mL, 5×10^2 cfu/mL, 1×10^3 cfu/mL, and 5×10^3 cfu/mL for PT samples and 110^3 cfu/mL for interferent and DW samples. The infective/lethal dose of each contaminant was determined by calculating the concentration at which ingestion of 250 mL of water is likely to cause the death of a 70-kilogram person based on human LD_{50} or ID_{50} data. The infective/lethal doses for *F. tularensis, Y. pestis,* and *B. anthracis* were 4×10^5 cfu/mL, 0.28 cfu/mL, and 200 cfu/mL, respectively.

Samples were prepared in 1 mL quantities and tested blindly by trained Battelle operators who had prior PCR experience. To test a 1 mL liquid sample for the presence or absence of *F. tularensis, Y. pestis,* or *B. anthracis,* DNA was extracted and purified from the sample using the Roche High Pure PCR Template Preparation Kit, assays were prepared using the PathAlert™ Detection Kit reagents, PCR was performed using a MJ Research DNA Engine® (PTC-200™) Peltier Thermal Cycler, and the amplified products were analyzed using the Agilent 2100 Bioanalyzer instrument along with the 2100 Bioanalyzer DNA 500 chips and reagent kit and associated 2100 Bioanalyzer software.

The kit was exclusively tested for one bacterium at a time. All samples were analyzed in quadruplicate from the same batch of purified DNA. The PathAlert™ Detection Kit was evaluated for qualitative results only by monitoring the internal positive control (IPC) along with the bacteria-specific peaks in the 2100 Bioanalyzer electropherogram output. Only positive, negative, and inconclusive results were recorded. Inconclusive results occurred when not all of the bacteria-specific peaks were present in the electropherogram. Quality assurance oversight of verification testing was provided by Battelle and EPA. Battelle QA staff conducted a technical systems audit and a data quality audit of 10% of the test data [2].

NOTES

1. US Environmental Protection Agency and Battelle National Laboratory (2004). ETV Joint Verification Statement. Rapid Polymerase Chain Reaction: Detecting Biological Agents and Pathogens in Water; http://www.epa.gov/ordnhsrc/pubs/vsInvitrogen121404.pdf; accessed September 24, 2009.
2. For more information, visit: http://www.epa.gov/nrmrl/std/etv/pubs/600etv09003.pdf; accessed September 24, 2009.

APPENDIX 4

Summary of Persistent and Toxic Organic Compounds in North America, Identified by the United Nations as Highest Priorities for Regional Actions

Compound	Biochemodynamic Properties	Persistence/Fate	Toxicity*
Aldrin 1,2,3,4,10,10-Hexachloro-1,4,4a,5,8,8a-hexahydro-1,4-endo,exo-5,8-dimethanonaphthalene ($C_{12}H_8Cl_6$)	Solubility in water: 27 μg L^{-1} at 25 °C; vapor pressure: 2.31×10^{-5} mmHg at 20 °C; log K_{ow}: 5.17–7.4	Readily metabolized to dieldrin by both plants and animals. Biodegradation is expected to be slow and it binds strongly to soil particles, and is resistant to leaching into groundwater. Classified as moderately persistent with $T_{1/2}$ in soil ranging from 20 to 100 days	Toxic to humans. Lethal dose for an adult estimated to be about 80 mg kg^{-1} body weight. Acute oral LD_{50} in laboratory animals is in the range of 33 mg kg^{-1} body weight for guinea pigs to 320 mg kg^{-1} body weight for hamsters. The toxicity of aldrin to aquatic organisms is quite variable, with aquatic insects being the most sensitive group of invertebrates. The 96-h LC_{50} values range from 1 to 200 μg L^{-1} for insects, and from 2.2–53 μg L^{-1} for fish. The maximum residue limits in food recommended by the World Health Organization varies from 0.006 mg kg^{-1} milk fat to 0.2 mg kg^{-1} meat fat. Water quality criteria between 0.1 to 180 μg L^{-1} have been published
Dieldrin 1,2,3,4,10,10-Hexachloro-6,7-epoxy-1,4,4a,5,6,7,8,8a-octahydroexo-1,4-endo-5,8-dimethanonaphthalene ($C_{12}H_8Cl_6O$)	Solubility in water: 140 μg L^{-1} at 20 °C; vapor pressure: 1.78×10^{-7} mmHg at 20 °C; log K_{ow}: 3.69–6.2	Highly persistent in soils, with a $T_{1/2}$ of 3–4 years in temperate climates, and bioconcentrates in organisms	Acute toxicity for fish is high (LC_{50} between 1.1 and 41 mg/L) and moderate for mammals (LD_{50} in mouse and rat ranging from 40 to 70 mg kg^{-1} body weight). Aldrin and dieldrin mainly affect the central nervous system but there is no direct evidence that they cause cancer in humans. The maximum residue limits in food recommended by WHO varies from 0.006 mg kg^{-1} milk fat and 0.2 mg kg^{-1} poultry fat. Water quality criteria between 0.1 to 18 μg L^{-1} have been published
Endrin 3,4,5,6,9,9-Hexachloro-1a,2,2a,3,6,6a,7,7a-octahydro-2,7:3,6-dimethanonaphth[2,3-b]oxirene ($C_{12}H_8Cl_6O$)	Solubility in water: 220–260μg L^{-1} at 25 °C; vapor pressure: 7×10^{-7} mmHg at 25 °C; log K_{ow}: 3.21–5.34	Highly persistent in soils ($T_{1/2}$ of up to 12 years have been reported in some cases). Bioconcentration factors of 14 to 18,000 have been recorded in fish, after continuous exposure	Very toxic to fish, aquatic invertebrates, and phytoplankton; the LC_{50} values are mostly less than 1 μg L^{-1}. The acute toxicity is high in laboratory animals, with LD_{50} values of 3–43 mg kg^{-1}, and a dermal LD_{50} of 6–20 mg kg^{-1} in rats. Long-term toxicity in the rat has been studied over two years and a NOEL of 0.05 mg kg^{-1} bw/day was found
Chlordane 1,2,4,5,6,7,8,8-Octachloro-2,3,3a,4,7,7a-hexahydro-4,7-methanoindene ($C_{10}H_6Cl_8$)	Solubility in water: 180 μg L^{-1} at 25 °C; vapor pressure: 0.3×10^{5} mmHg at 20 °C; log K_{ow}: 4.4–5.5	Metabolized in soils, plants, and animals to heptachlor epoxide, which is more stable in biological systems and is carcinogenic. The $T_{1/2}$ of	Acute toxicity to mammals is moderate (LD50 values between 40 and 119 mg kg^{-1} have been published). The toxicity to aquatic organisms is higher and LC_{50} values down to 0.11 μg L^{-1} have been found for pink shrimp. Limited information is available on the effects in humans and

		heptachlor in soil in temperate regions is 0.75–2 years. Its high partition coefficient provides the necessary conditions for bioconcentrating in organisms	studies are inconclusive regarding heptachlor and cancer. The maximum residue levels recommended by FAO/WHO are between 0.006 mg kg^{-1} milk fat and 0.2 mg kg^{-1} meat or poultry fat
Dichlorodiphenyltrichloroethane (DDT) 1,1,1-Trichloro-2,2-bis-(4-chlorophenyl)-ethane ($C_{14}H_9Cl_5$)	Solubility in water: 1.2–5.5 μg L^{-1} at 25 °C; vapor pressure: 0.02×10^{-5} mmHg at 20 °C; log K_{ow}: 6.19 for pp-DDT, 5.5 for pp-DDD and 5.7 for pp-DDE	Highly persistent in soils with a $T_{1/2}$ of about 1.1 to 3.4 years. It also exhibits high bioconcentration factors (in the order of 50,000 for fish and 500,000 for bivalves). In the environment, the parent DDT is metabolized mainly to DDD and DDE	Lowest dietary concentration of DDT reported to cause egg shell thinning was 0.6 mg kg^{-1} for the black duck. LC_{50} of 1.5 mg L^{-1} for largemouth bass and 56 mg L^{-1} for guppy have been reported. The acute toxicity of DDT for mammals is moderate with an LD_{50} in rat of 113–118 mg kg^{-1} body weight. DDT has been shown to have an estrogen-like activity and possible carcinogenic activity in humans. The maximum residue level in food recommended by WHO/FAO, ranges from 0.02 mg kg^{-1} milk fat to 5 mg kg^{-1} meat fat. Maximum permissible DDT residue levels in drinking water (WHO) is 1.0 μg L^{-1}
Toxaphene polychlorinated bornanes and camphenes ($C_{10}H_{10}Cl_8$)	Solubility in water: 550 μg L^{-1} at 20 °C; vapor pressure: 0.2–0.4 mmHg at 25 °C; log KOW: 3.23–5.50	Half-life in soil from 100 days up to 12 years. It has been shown to bioconcentrate in aquatic organisms (BCF of 4247 in mosquito fish and 76,000 in brook trout)	Highly toxic in fish, with 96-hour LC_{50} values in the range of 1.8 μg L^{-1} in rainbow trout to 22 μg L^{-1} in bluegill. Long-term exposure to 0.5 μg L^{-1} reduced egg viability to zero. The acute oral toxicity is in the range of 49 mg kg^{-1} body weight in dogs to 365 mg kg^{-1} in guinea pigs. In long-term studies NOEL in rats was 0.35 mg kg^{-1} bw/day, LD_{50} ranging from 60 to 293 mg kg^{-1} bw. For toxaphene, there exists strong evidence of the potential for endocrine disruption. Toxaphene is carcinogenic in mice and rats and is of carcinogenic risk to humans, with a cancer potency factor of 1.1 mg kg^{-1}/day for oral exposure

(*Continued*)

Compound	Biochemodynamic Properties	Persistence/Fate	Toxicity*
Mirex 1,1a,2,2,3,3a,4,5,5a,5b,6-Dodecachloroacta-hydro-1,3,4-metheno-1H-cyclobuta[cd] pentalene ($C_{10}Cl_{12}$)	Solubility in water: 0.07 $\mu g\ L^{-1}$ at 25 °C; vapor pressure: 3×10^{-7} mmHg at 25 °C; log K_{ow}: 5.28	Among the most stable and persistent pesticides, with a $T_{1/2}$ in soils of up to 10 years. Bioconcentration factors of 2600 and 51,400 have been observed in pink shrimp and fathead minnows, respectively. Capable of undergoing long-range transport due to its relative volatility (VPL = 4.76 Pa; H = 52 Pa m^{-3}/mol).	Acute toxicity for mammals is moderate with an LD_{50} in rat of 235 mg kg^{-1} and dermal toxicity in rabbits of 80 mg kg^{-1}. Mirex is also toxic to fish and can affect their behavior (LC_{50} (96 h) from 0.2 to 30 mg L^{-1} for rainbow trout and bluegill, respectively). Delayed mortality of crustaceans occurred at 1 $\mu g\ L^{-1}$ exposure levels. There is evidence of its potential for endocrine disruption and possibly carcinogenic risk to humans
Hexachlorobenzene (HCB) (C_6H_6)	Solubility in water: 50 $\mu g\ L^{-1}$ at 20 °C; vapor pressure: 1.09×10^{-5} mmHg at 20 °C; log K_{ow}: 3.93–6.42	Estimated 'field half-life' of 2.7–5.7 years. HCB has a relatively high bioaccumulation potential and long $T_{1/2}$ in biota	LC_{50} for fish varies between 50 and 200 $\mu g\ L^{-1}$. The acute toxicity of HCB is low with LD_{50} values of 3.5 mg/g for rats. Mild effects of the [rat] liver have been observed at a daily dose of 0.25 mg HCB/kg bw. HCB is known to cause liver disease in humans (porphyria cutanea tarda) and has been classified as a possible carcinogen to humans by IARC
Polychlorinated biphenyls (PCBs) ($C_{12}H_{(10-n)}Cl_n$, where n is within the range of 1–10)	Water solubility decreases with increasing chlorination: 0.01 to 0.0001 $\mu g\ L^{-1}$ at 25 °C; vapor pressure: 1.6–0.003 $\times 10^{-6}$ mmHg at 20 °C; log K_{ow}: 4.3–8.26	Most PCB congeners, particularly those lacking adjacent unsubstituted positions on the biphenyl rings (e.g., 2,4,6-, 2,3,6- or 2,3,6-substituted on both rings) are extremely persistent in the environment. They are estimated to have a $T_{1/2}$ ranging from 3 weeks to 2 years in air and, with the exception of mono- and dichlorodiphenyl, more than 6 years in aerobic soils and sediments. PCBs also have extremely long $T_{1/2}$ in adult fish, for example, an 8-year study of eels found that the $T_{1/2}$ of CB153 was more than 10 years	LC_{50} for the larval stages of rainbow trout is 0.32 $\mu g\ L^{-1}$ with a NOEL of 0.01 $\mu g\ L^{-1}$. The acute toxicity of PCB in mammals is generally low and LD_{50} values in rat of 1 g/kg bw. IARC has concluded that PCBs are carcinogenic to laboratory animals and probably also for humans. They have also been classified as substances for which there is evidence of endocrine disruption in an intact organism

Polychlorinated dibenzo-p-dioxins (PCDDs) and Polychlorinated dibenzofurans (PCDFs) ($C_{12}H_{(8-n)}Cl_nO_2$) and PCDFs ($C_{12}H_{(8-n)}Cl_nO$) may contain between 1 and 8 chlorine atoms. Dioxins and furans have 75 and 135 possible positional isomers, respectively	Solubility in water: in the range 550–0.07 ng L^{-1} at 25 °C; vapor pressure: 2–0.007 × 10^{-6} mmHg at 20 °C; log K_{ow}: in the range 6.60–8.20 for tetra- to octa-substituted congeners	PCDD/Fs are characterized by their lipophilicity, semi-volatility and resistance to degradation ($T_{1/2}$ of TCDD in soil of 10–12 years) and to long-range transport. They are also known for their ability to bio-concentrate and biomagnify under typical environmental conditions	Toxicological effects reported refers to the 2,3,7,8-substituted compounds (17 congeners) that are agonist for the AhR. All the 2,3,7,8-substituted PCDDs and PCDFs plus dioxin-like PCBs (DLPCBs) (with no chlorine substitution at the ortho positions) show the same type of biological and toxic response. Possible effects include dermal toxicity, immunotoxicity, reproductive effects and teratogenicity, endocrine disruption and carcinogenicity. At the present time, the only persistent effect associated with dioxin exposure in humans is chloracne. The most sensitive groups are fetus and neonatal infants. Effects on the immune systems in the mouse have been found at doses of 10 ng kg^{-1} bw/day, while reproductive effects were seen in rhesus monkeys at 1–2 ng kg^{-1} bw/day. Biochemical effects have been seen in rats down to 0.1 ng kg^{-1} bw/day. In a re-evaluation of the TDI for dioxins, furans (and planar PCB), the WHO decided to recommend a range of 1–4 TEQ pg/kg bw, although more recently the acceptable intake value has been set monthly at 1–70 TEQ pg/kg bw
Atrazine 2-Chloro-4-(ethlamino)-6-(isopropylamino)-s-triazine ($C_{10}H_6Cl_8$)	Solubility in water: 28 mg L^{-1} at 20 °C; vapor pressure: 3.0 × 10^{-7} mmHg at 20 °C; log K_{ow}: 2.34	Does not adsorb strongly to soil particles and has a lengthy $T_{1/2}$ (60 to >100 days). Atrazine has a high potential for groundwater contamination despite its moderate solubility in water	Oral LD_{50} is 3090 mg kg^{-1} in rats, 1750 mg kg^{-1} in mice, 750 mg kg^{-1} in rabbits, and 1000 mg kg^{-1} in hamsters. The dermal LD_{50} in rabbits is 7500 mg kg^{-1} and greater than 3000 mg kg^{-1} in rats. Atrazine is practically nontoxic to birds. The LD_{50} is greater than 2000 mg kg^{-1} in mallard ducks. Atrazine is slightly toxic to fish and other aquatic life. Atrazine has a low level of bioaccumulation in fish. Available data regarding atrazine's carcinogenic potential are inconclusive

(*Continued*)

Compound	Biochemodynamic Properties	Persistence/Fate	Toxicity*
Hexachlorocyclohexane (HCH) 1,2,3,4,5,6-hexachlorocyclohexane (mixed isomers) ($C_6H_6Cl_6$)	γ-HCH (lindane): solubility in water: 7 mg L^{-1} at 20° C; vapor pressure: 3.3×10^{-5} mmHg at 20 °C; log K_{ow}: 3.8	Lindane and other HCH isomers are relatively persistent in soils and water, with half-lives generally greater than 1 and 2 years, respectively. HCHs are much less bioaccumulative than other organochlorines of concern because of their relatively low lipophilicity. On the contrary, their relatively high vapor pressures, particularly of the α-HCH isomer, determine their long-range transport in the atmosphere	Lindane is moderately toxic for invertebrates and fish, with LC_{50} values of 20–90 μg L^{-1}. The acute toxicity for mice and rats is moderate with LD_{50} values in the range of 60–250 mg kg^{-1}. Lindane resulted to have no mutagenic potential in a number of studies but an endocrine disrupting activity
Chlorinated paraffins (CPs) Polychlorinated alkanes ($C_xH_{(2x-y+2)}Cl_y$) Manufactured by chlorination of liquid *n*-alkanes or paraffin wax and contain from 30 to 70% chlorine. The products are often divided in three groups depending on chain length: short chain (C_{10}–C_{13}), medium (C_{14}–C_{17}) and long (C_{18}–C_{30}) chain lengths	Properties largely dependent upon the chlorine content. Solubility in water: 1.7 to 236 μg L^{-1} at 25 °C; vapor pressure (pure chemical in solid phase) ranges from 2.8×10^{-9} mmHg to 4.0×10^{-2} mmHg at 25 °C; however VP varies considerably by amount of chlorine substitution and molecular weight log K_{ow}: in the range from 5.06 to 8.12	May be released into the environment from improperly disposed metal-working fluids or polymers containing chlorinated paraffins. Loss of chlorinated paraffins by leaching from paints and coatings may also contribute to environmental contamination. Short chain CPs with less than 50% chlorine content seem to be degraded under aerobic conditions. The medium and long chain products are degraded more slowly. CPs are bioaccumulated and both uptake and elimination are faster for the substances with low chlorine content	Acute toxicity of CPs in mammals is low with reported oral LD_{50} values ranging from 4 to 50 g/kg bw, although in repeated dose experiments, effects on the liver have been seen at doses of 10–100 mg kg^{-1} bw/day. Short-chain and mid-chain grades have been shown, in laboratory tests, to show toxic effects on fish and other forms of aquatic life after long-term exposure. The NOEL appears to be in the range of 2–5 μg L^{-1} for the most sensitive aquatic species tested

Chlordecone or Kepone Chemical name: 1,2,3,4,5,5,6,7,9,10,10-dodecachlorooctahydro-1,3,4-metheno-2H-cyclobuta(cd)pentalen-2-one ($C_{10}Cl_{10}O$)	Solubility in water: 7.6 mg L^{-1} at 25 °C; vapor pressure: less than 3×10^{-5} mmHg at 25 °C; log K_{ow}: 4.50	Estimated $T_{1/2}$ in soils is between 1–2 years, whereas in air is much higher, up to 50 years. Not expected to hydrolyze, biodegrade in the environment. Also direct photodegradation and vaporization from what not significant. General population exposure to chlordecone mainly through the consumption of contaminated fish and seafood	Workers exposed to high levels of chlordecone over a long period (more than one year) have displayed harmful effects on the nervous system, skin, liver, and male reproductive system (likely through dermal exposure to chlordecone, although they may have inhaled or ingested some as well). Animal studies with chlordecone have shown effects similar to those seen in people, as well as harmful kidney effects, developmental effects, and effects on the ability of females to reproduce. There are no studies available on whether chlordecone is carcinogenic in people. However, studies in mice and rats have shown that ingesting chlordecone can cause liver, adrenal gland, and kidney tumors. Very highly toxic for some species, such as Atlantic menhaden, sheepshead minnow, or Donaldson trout, with LC_{50} between 21.4 and 56.9 mg L^{-1}
Endosulfan 6,7,8,9,10,10-hexachloro-1,5,5a,6,9,9a-hexahydro-6,9-methano-2,4,3-benzodioxathiepin-3-oxide ($C_9H_6Cl_6O_3S$)	Solubility in water: 320 μg L^{-1} at 25 °C; vapor pressure: 0.17×10^{-4} mmHg at 25 °C; log K_{ow}: 2.23–3.62	Moderately persistent in soil, with a reported average field $T_{1/2}$ of 50 days. The two isomers have different degradation times in soil ($T_{1/2}$ of 35 and 150 days for α- and β-isomers, respectively, in neutral conditions). It has a moderate capacity to adsorb to soils and it is not likely to leach to groundwater. In plants, endosulfan is rapidly broken down to the corresponding sulfate, on most fruits and vegetables, 50% of the parent residue is lost within 3 to 7 days	Highly to moderately toxic to bird species (mallards: oral LD_{50} 31–243 mg kg^{-1}) and it is very toxic to aquatic organisms (96-hour LC_{50} rainbow trout 1.5 μgL^{-1}). It has also shown high toxicity in rats (oral LD_{50}: 18–160 mg kg^{-1}, and dermal: 78–359 mg kg^{-1}). Female rats appear to be 4–5 times more sensitive to the lethal effects of technical-grade endosulfan than male rats. The α-isomer is considered to be more toxic than the β-isomer. There is a strong evidence of its potential for endocrine disruption

(*Continued*)

Compound	Biochemodynamic Properties	Persistence/Fate	Toxicity*
Pentachlorophenol (PCP) (C_6Cl_5OH)	Solubility in water: 14 mg L^{-1} at 20 °C; vapor pressure: 16 × 10^{-5} mmHg at 20 °C; log K_{ow}: 3.32–5.86	Photodecomposition rate increases with pH ($T_{1/2}$ 100 h at pH 3.3 and 3.5 h at pH 7.3). Complete decomposition in soil suspensions takes >72 days, other authors report $T_{1/2}$ in soils of about 45 days. Although enriched through the food chain, it is rapidly eliminated after discontinuing the exposure ($T_{1/2}$ 10–24 h for fish)	Acutely toxic to aquatic organisms. Certain effects on human health. 24-h LC_{50} values for trout were reported as 0.2 mg L^{-1}, and chronic toxicity effects were observed at concentrations down to 3.2 μg L^{-1}. Mammalian acute toxicity of PCP is moderate-high. LD_{50} oral in rat ranging from 50 to 210 mg kg^{-1} bw have been reported. LC_{50} ranged from 0.093 mg L^{-1} in rainbow trout (48 h) to 0.77–0.97 mg L^{-1} for guppy (96 h) and 0.47 mg L^{-1} for fathead minnow (48 h)
Hexabromobiphenyl (HxBB) ($C_{12}H_4Br_6$) A congener of the class polybrominated biphenyls (PBBs)	Solubility in water: 11 μg L^{-1} at 25 °C; vapor pressure 5.2 × 10^{-8} at 25 °C; log K_{ow}: 6.39	Strongly adsorbed to soil and sediments and usually persist in the environment. Resists chemical and biological degradation. Found in sediment samples from the estuaries of large rivers and has been identified in edible fish	Few toxicity data are available from short-term tests on aquatic organisms. The LD_{50} values of commercial mixtures show a relatively low order of acute toxicity (LD_{50} range from >1 to 21.5 g/kg body weight in laboratory rodents). Oral exposure of laboratory animals to PBBs produced body weight loss, skin disorders, and nervous system effects, and birth defects. Humans exposed through contaminated food developed skin disorders, such as acne and hair loss. PBBs exhibit endocrine disrupting activity and possible carcinogenicity to humans
Polybrominated diphenyl ethers (PBDEs) ($C_{12}H_{(10-n)}Br_nO$, where n = 1–10). As in the case of PCBs the total number of congeners is 209, with a predominance in commercial mixtures of the tetra-, penta-, and octa-substituted isomers	Solubility in water: mg L^{-1} at 25 °C; vapor pressure: 3.85 up to 13.3 × 10^{-3}mmHg at 20–25 °C; log K_{ow}: 4.28–9.9	Biodegradation does not seem to be an important degradation pathway, but that photodegradation may play a significant role. Have been found in high concentrations in marine birds and mammals from remote areas. The half-lives of PBDE components in rat adipose tissue vary between 19 and 119 days, the higher values being for the more highly brominated congeners	Lower (tetra- to hexa-) PBDE congeners likely to be carcinogens, endocrine disruptors, and/or neurodevelopmental toxicants. Studies in rats with commercial penta BDE indicate a low acute toxicity via oral and dermal routes of exposure, with LD_{50} values >2000 mg kg^{-1} bw. In a 30-day study with rats, effects on the liver could be seen at a dose of 2 mg kg^{-1} bw/day, with a NOEL at 1 mg kg^{-1} bw/day. The toxicity to *Daphnia magna* has also been investigated and LC_{50} was found to be 14 μg L^{-1} with a NOEC of 4.9 μg L^{-1}. Although data on toxicology is limited, they have potential endocrine disrupting properties, and there are concerns over the health effects of exposure

Polycyclic Aromatic Hydrocarbons (PAHs) A group of compounds consisting of two or more fused aromatic rings	Solubility in water: 0.00014–2.1 mg L^{-1} at 25 °C; vapor pressure varies considerably depending on molecular weight, e.g. relatively high for naphthalene (0.087 mmHg at 25 °C) and quite low for benzo(a)pyrene (5.6×10^{-9} mmHg at 25 °C) and dibenzo[a,h]anthracene (1.0×10^{-10} mmHg at 25 °C); log K_{ow}: 4.79–8.20	Persistence of the PAHs varies with their molecular weight. The low molecular weight PAHs are most easily degraded. The reported $T_{1/2}$ of naphthalene, anthracene and benzo(e)pyrene in sediment are 9, 43, and 83 hours, respectively, whereas for higher molecular weight PAHs $T_{1/2}$ are up to several years in soils and sediments. The BCFs in aquatic organisms frequently range between 100 and 2000 and it increases with increasing molecular size. Due to their wide distribution, the environmental pollution by PAHs has aroused global concern	Acute toxicity of low PAHs is moderate with an LD_{50} of naphthalene and anthracene in rat of 490 and 18,000 mg kg^{-1} body weight respectively, whereas the higher PAHs exhibit higher toxicity and LD_{50} of benzo(a)anthracene in mice is 10 mg kg^{-1} body weight. In *Daphnia pulex*, LC_{50} for naphthalene is 1.0 mg L^{-1}, for phenanthrene 0.1 mg L^{-1} and for benzo(a)pyrene is 0.005 mg L^{-1}. The critical effect of many PAHs in mammals is their carcinogenic potential. The metabolic actions of these substances produce intermediates that bind covalently with cellular DNA. IARC has classified benz[a]anthracene, benzo[a]pyrene, and dibenzo[a,h]anthracene as probable carcinogenic to humans. Benzo[b]fluoranthene and indeno[1,2,3-c,d]pyrene were classified as possible carcinogens to humans
Phthalates Includes a wide family of compounds. Among the most common contaminants are: dimethylphthalate (DMP), diethylphthalate (DEP), dibutylphthalate (DBP), benzylbutylphthalate (BBP), di(2-ethylhexyl)phthalate (DEHP)($C_{24}H_{38}O_4$) and dioctylphthalate (DOP)	Properties of phthalic acid esters vary greatly depending on the alcohol moieties. log K_{ow} 1.5 to 7.1	Ubiquitous pollutants, in marine, estuarine, and freshwater sediments, sewage sludges, soils and food. Degradation ($T_{1/2}$) values generally range from 1 to 30 days in freshwaters	Acute toxicity of phthalates is usually low: the oral LD_{50} for DEHP is about 26–34 g/kg, depending on the species; for DBP reported LD_{50} values following oral administration to rats range from 8 to 20 g/kg body weight; in mice, values are approximately 5–16 g/kg body weight. In general, DEHP is not toxic for aquatic communities at the low levels usually present. In animals, high levels of DEHP damaged the liver and kidney and affected the ability to reproduce. There is no evidence that DEHP causes cancer in humans but they have been reported as endocrine disrupting chemicals. The EPA proposed a Maximum Admissible Concentration (MAC) of 6 μg L^{-1} of DEHP in drinking water

(*Continued*)

Compound	Biochemodynamic Properties	Persistence/Fate	Toxicity*
Nonyl- and Octyl-phenols NP: $C_{15}H_{24}O$; OP: $C_{14}H_{22}O$	log K_{ow}: 4.5 (NP) and 5.92 (OP)	NP and OP are the end degradation products of APEs under both aerobic and anaerobic conditions. Therefore, the major part is released to water and concentrated in sewage sludges. NPs and t-OP are persistent in the environment with $T_{1/2}$ of 30–60 years in marine sediments, 1–3 weeks in estuarine waters and 10–48 hours in the atmosphere. Due to their persistence they can bioaccumulate to a significant extent in aquatic species. However, excretion and metabolism is rapid	Acute toxicity values for fish, invertebrates and algae ranging from 17 to 3000 $\mu g\ L^{-1}$. In chronic toxicity tests the lowest NOEC are 6$\mu g\ L^{-1}$ in fish and 3.7 $\mu g\ L^{-1}$ in invertebrates. The threshold for vitellogenin induction in fish is 10 $\mu g\ L^{-1}$ for NP and 3 $\mu g\ L^{-1}$ for OP (similar to the lowest NOEC). Alkylphenols are endocrine disrupting chemicals also in mammals
Perfluorooctane sulfonate ($C_8F_{17}SO_3$)	Solubility in water: 550 mg L^{-1} in pure water at 24–25 °C; the potassium salt of PFOS has a low vapor pressure, 2.5 × 10^{-6} mmHg at 20 °C. Due to the surface-active properties of PFOS, the Log K_{ow} cannot be measured	Does not hydrolyze, photolyze or biodegrade under environmental conditions. It is persistent in the environment and has been shown to bioconcentrate in fish. It has been detected in a number of species of wildlife, including marine mammals. Animal studies show that PFOS is well absorbed orally	Moderate acute toxicity to aquatic organisms, the lowest LC_{50} for fish is a 96-h LC_{50} of 4.7 mg/l to the fathead minnow (*Pimephales promelas*) for the lithium salt. For aquatic invertebrates, the lowest EC_{50} for freshwater species is a 48-h EC_{50} of 27 mg L^{-1} for *Daphnia magna* and for saltwater species, a 96-h LC_{50} value of 3.6 mg L^{-1} for the Mysid shrimp (*Mysidopsis bahia*). Both tests were conducted on the potassium salt. The toxicity profile of PFOS is similar among rats and monkeys. Repeated exposure results in hepatotoxicity and mortality; the dose-response curve is very steep for mortality. PFOS has

	and distributes mainly in the serum and the liver. The half-life in serum is 7.5 days in adult rats and 200 days in Cynomolgus monkeys. The half-life in humans is, on average, 8.67 years (range 2.29–21.3 years, SD = 6.12)	shown moderate acute toxicity by the oral route with a rat LD_{50} of 251 mg kg^{-1}. Developmental effects were also reported in prenatal developmental toxicity studies in the rat and rabbit, although at slightly higher dose levels. Signs of developmental toxicity in the offspring were evident at doses of 5 mg kg^{-1}/day and above in rats administered PFOS during gestation. Significant decreases in fetal body weight and significant increases in external and visceral anomalies, delayed ossification, and skeletal variations were observed. A NOAEL of 1 mg kg^{-1} day and a LOAEL of 5 mg kg^{-1} day for developmental toxicity were indicated. Studies on employees conducted at PFOS manufacturing plants in the US and Belgium showed an increase in mortality resulting from bladder cancer and an increased risk of neoplasms of the male reproductive system, the overall category of cancers and benign growths, and neoplasms of the gastrointestinal tract

Lethal dose to 50% of tested organism = LD_{50}
Lethal concentration to 50% of tested organism = LC_{50}
Bioconcentration factor = BCF
No observable effect level = NOEL
No observable effect concentration = NOEC
Chemical half-life = $T_{1/2}$
Source: United Nations Environmental Programme (2002). Chemicals: North American Regional Report, Regionally Based Assessment of Persistent Toxic Substances, Global Environment Facility.

APPENDIX 5

Sample Retrieval from ECOTOX Database for Rainbow Trout (*Oncorhynchus mykiss*) Exposed to DDT and its Metabolites in Freshwater

Chemical name	Response site description	Exposure duration (days)	Conc. (μg/L)	Test location	Author	Title	Source
1,1′-(2,2,2-Trichloroethylidene)bis-[4-chlorobenzene]	Whole organism	30	128.8	LAB	D.C.G. Muir and A.L. Yarechewski	Dietary accumulation of four chlorinated dioxin congeners by Rainbow Trout and fathead minnows	*Environ Toxicol Chem* 1988; 7(3): 227–236
1,1′-(2,2,2-Trichloroethylidene)bis-[4-chlorobenzene]	Not reported	30	20	LAB	F.L Mayer, Jr.	Pesticides as pollutants	In: B.G. Liptak (Ed.), *Environmental Engineer's Handbook*, Chilton Book Co., Radnor, PA; 1974: 405–418 (Publ in part as 6797)
1,1′-(2,2,2-Trichloroethylidene)bis-[4-chlorobenzene]		14	10	LAB	J. Miyamoto, Y. Takimoto and K. Mihara	Metabolism of organophosphorus insecticides in aquatic organisms, with special emphasis on fenitrothion	In: M.A.Q. Khan, J.J. Lech and J.J. Menn (Eds), *Pesticide and Xenobiotic Metabolism in Aquatic Organisms*, ACS (Am. Chem. Soc.) Symp. Ser. 1979; 99: 3–20
1,1′-(2,2,2-Trichloroethylidene)bis-[4-chlorobenzene]	Whole organism	4	0.0205	LAB	D.C.G. Muir, B.R. Hobden and M.R. Servos	Bioconcentration of pyrethroid insecticides and DDT by Rainbow Trout: uptake, depuration, and effect of dissolved organic carbon	*Aquat. Toxicol.* 1994; 29 (3/4): 223–240
1,1′-(2,2,2-Trichloroethylidene)bis-[4-chlorobenzene]	Whole organism	4	0.0193	LAB	D.C.G. Muir, B.R. Hobden and M.R. Servos	Bioconcentration of pyrethroid insecticides and DDT by Rainbow Trout: uptake, depuration, and effect of dissolved organic carbon	*Aquat. Toxicol.* 1994; 29 (3/4): 223–240
1,1′-(2,2,2-Trichloroethylidene)bis-[4-chlorobenzene]	Whole organism	4	0.0205	LAB	D.C.G. Muir, B.R. Hobden and M.R. Servos	Bioconcentration of pyrethroid insecticides and DDT by Rainbow Trout: uptake, depuration, and effect of dissolved organic carbon	*Aquat. Toxicol.* 1994; 29 (3/4): 223–240
1,1′-(2,2,2-Trichloroethylidene)bis-[4-chlorobenzene]	Whole organism	4	0.0193	LAB	D.C.G. Muir, B.R. Hobden and M.R. Servos	Bioconcentration of pyrethroid insecticides and DDT by Rainbow Trout: uptake, depuration, and effect of dissolved organic carbon	*Aquat. Toxicol.* 1994; 29 (3/4): 223–240
1,1′-(2,2,2-Trichloroethylidene)bis-[4-chlorobenzene]	Blood	20	NR	LAB	P.V. Hodson, B.R. Blunt, U. Borgmann, C.K. Minns and S. Mcgaw	Effect of fluctuating lead exposures on lead accumulation by Rainbow Trout (*Salmo gairdneri*)	*Environ. Toxicol. Chem.* 1983; 2(2): 225–238

1,1′-(2,2,2-Trichloroethylidene)bis-[4-chlorobenzene]	Whole organism	4	0.0631	LAB	D.C.G. Muir, B.R. Hobden and M.R. Servos	Bioconcentration of pyrethroid insecticides and DDT by Rainbow Trout: uptake, depuration, and effect of dissolved organic carbon	*Aquat. Toxicol.* 1994; 29(3/4): 223–240
1,1′-(2,2,2-Trichloroethylidene)bis-[4-chlorobenzene]	Whole organism	4	0.064	LAB	D.C.G. Muir, B.R. Hobden and M.R. Servos	Bioconcentration of pyrethroid insecticides and DDT by Rainbow Trout: uptake, depuration, and effect of dissolved organic carbon	*Aquat. Toxicol.* 1994; 29(3/4): 223–240
1,1′-(2,2,2-Trichloroethylidene)bis-[4-chlorobenzene]	Whole organism	4	0.0631	LAB	D.C.G. Muir, B.R. Hobden and M.R. Servos	Bioconcentration of pyrethroid insecticides and DDT by Rainbow Trout: uptake, depuration, and effect of dissolved organic carbon	*Aquat. Toxicol.* 1994; 29(3/4): 223–240
1,1′-(2,2,2-Trichloroethylidene)bis-[4-chlorobenzene]	Whole organism	4	0.0497	LAB	D.C.G. Muir, B.R. Hobden and M.R. Servos	Bioconcentration of pyrethroid insecticides and DDT by Rainbow Trout: uptake, depuration, and effect of dissolved organic carbon	*Aquat. Toxicol.* 1994; 29(3/4): 223–240
1,1′-(2,2,2-Trichloroethylidene)bis-[4-chlorobenzene]	Whole organism	4	0.0638	LAB	D.C.G. Muir, B.R. Hobden and M.R. Servos	Bioconcentration of pyrethroid insecticides and DDT by Rainbow Trout: uptake, depuration, and effect of dissolved organic carbon	*Aquat. Toxicol.* 1994; 29 (3/4): 223–240
1,1′-(2,2,2-Trichloroethylidene)bis-[4-chlorobenzene]	Whole organism	5	0.106	LAB	D.C.G. Muir, A.L. Yarechewski and G.R.B. Webster	Bioconcentration of four chlorinated dioxins by Rainbow Trout and fathead minnows	In: R.C.Bahner and D.J.Hansen (Eds), *Aquatic Toxicology and Hazard Assessment*, 8th Symposium, ASTM STP 891, Philadelphia, PA; 1985: 440–454
1,1′-(2,2,2-Trichloroethylidene)bis-[4-chlorobenzene]	Whole organism	4	0.0638	LAB	D.C.G. Muir, B.R. Hobden and M.R. Servos	Bioconcentration of pyrethroid insecticides and DDT by Rainbow Trout: uptake, depuration, and effect of dissolved organic carbon	*Aquat. Toxicol.* 1994; 29(3/4): 223–240
1,1′-(2,2,2-Trichloroethylidene)bis-[4-chlorobenzene]	Multiple Tissue/ Organ	168	NR	LAB	K.J. Macek, C.R. Rodgers, D.L. Stalling and S. Korn	The uptake, distribution and elimination of dietary 14C-DDT and 14C-dieldrin in Rainbow Trout	*Trans. Am. Fish. Soc.* 1970; 99 (4): 689–695
1,1′-(2,2,2-Trichloroethylidene)bis-[4-chlorobenzene]	Not reported	NR	NR	FIELD N	E.B. Welch and J.C. Spindler	DDT persistence and its effect on aquatic insects and fish after an aerial application	*J. Water Pollut. Control Fed.* 1964; 36 (10): 1285–1292
1,1′-(2,2,2-Trichloroethylidene)bis-[4-chlorobenzene]	Not reported	NR	NR	FIELD N	W.R. Bridges, B.J. Kallman and A.K. Andrews	Persistence of DDT and its metabolites in a farm pond	*Trans. Am. Fish. Soc.* 1963; 92: 421–427

(Continued)

Chemical name	Response site description	Exposure duration (days)	Conc. (μg/L)	Test location	Author	Title	Source
1,1′-(2,2,2-Trichloroethylidene)bis-[4-chlorobenzene]	Not reported	1	NR	LAB	C.F. Peters and D.D. Weber	DDT: effect on the lateral line nerve of Steelhead Trout	In: F.J. Vernberg, A. Calabrese, F.P. Thurberg and W.B. Vernberg (Eds), *Physiological Responses of Marine Biota to Pollutants*. Academic Press Inc., NY; 1977: 75–91
1,1′-(2,2,2-Trichloroethylidene)bis-[4-chlorobenzene]	Not reported	1	NR	LAB	D. Ludemann and H. Neumann	Acute toxicity of present contact insecticides for freshwater animals (Versuche uber die Akute Toxische Wirkung Neuzeitlicher Kontaktinsektizide auf Susswassertiere)	*Z. Angew. Zool.* 1961; 48: 87–96 (GER) (ENG ABS)
1,1′-(2,2,2-Trichloroethylidene)bis-[4-chlorobenzene]	Not reported	12	500	FIELD N	O.B. Cope	Effects of DDT spraying for Spruce Budworm on fish in the Yellowstone River System	*Trans. Am. Fish. Soc.* 1961; 90 (3): 239–251
1,1′-(2,2,2-Trichloroethylidene)bis-[4-chlorobenzene]	Not reported	344	NR	FIELD N	O.B. Cope	Effects of DDT spraying for Spruce Budworm on fish in the Yellowstone River System	*Trans. Am. Fish. Soc.* 1961; 90 (3): 239–251
1,1′-(2,2,2-Trichloroethylidene)bis-[4-chlorobenzene]	Whole organism	140	1000	LAB	B.F. Grant and P.M. Mehrle	Pesticide effects on fish endocrine function	In: Resour. Publ. No. 88, Prog. Sport Fish. Res. 1969, Div. Fish. Res., Bur. Sport Fish. Wildl., USDI, Washington, DC; 1970: 13–15
1,1′-(2,2,2-Trichloroethylidene)bis-[4-chlorobenzene]	Whole organism	30	20	FIELD U	O.B. Cope	Contamination of the freshwater ecosystem by pesticides	*J. Appl. Ecol.* 1966; 3: 33–44 (Publ in part as 6797)
1,1′-(2,2,2-Trichloroethylidene)bis-[4-chlorobenzene]	Serum	140	1000	LAB	B.F. Grant and P.M. Mehrle	Pesticide effects on fish endocrine function	In: Resour. Publ. No. 88, Prog. Sport Fish. Res. 1969, Div. Fish. Res., Bur. Sport Fish. Wildl., USDI, Washington, DC; 1970: 13–15
1,1′-(2,2,2-Trichloroethylidene)bis-[4-chlorobenzene]	Residual, remnant, carcass	140	200	LAB	K.J. Macek, C.R. Rodgers, D.L. Stalling and S. Korn	The uptake, distribution and elimination of dietary 14C-DDT and 14C-dieldrin in Rainbow Trout	*Trans. Am. Fish. Soc.* 1970; 99(4): 689–695

1,1′-(2,2,2-Trichloroethylidene)bis-[4-chlorobenzene]	Residual, remnant, carcass	140	1000	LAB	K.J. Macek, C.R. Rodgers, D.L. Stalling and S. Korn	The uptake, distribution and elimination of dietary 14C-DDT and 14C-dieldrin in Rainbow Trout	*Trans. Am. Fish. Soc.* 1970; 99 (4): 689–695
1,1′-(2,2,2-Trichloroethylidene)bis-[4-chlorobenzene]	Serum	140	1000	LAB	B.F. Grant and P.M. Mehrle	Pesticide effects on fish endocrine function	In: Resour. Publ. No. 88, Prog. Sport Fish. Res. 1969, Div. Fish. Res., Bur. Sport Fish. Wildl., USDI, Washington, DC; 1970: 13–15
1,1′-(2,2,2-Trichloroethylidene)bis-[4-chlorobenzene]	Liver	140	1000	LAB	B.F Grant and P.M. Mehrle	Pesticide effects on fish endocrine function	In: Resour. Publ. No. 88, Prog. Sport Fish. Res. 1969, Div. Fish. Res., Bur. Sport Fish. Wildl., USDI, Washington, DC; 1970: 13–15
1,1′-(2,2,2-Trichloroethylidene)bis-[4-chlorobenzene]	Whole organism	140	1000	LAB	B.F. Grant and P.M. Mehrle	Pesticide effects on fish endocrine function	In: Resour. Publ. No. 88, Prog. Sport Fish. Res., 1969, Div. Fish. Res., Bur. Sport Fish. Wildl., USDI, Washington, DC; 1970: 13–15
1,1′-(2,2,2-Trichloroethylidene)bis-[4-chlorobenzene]	Liver	7	10000	LAB	J.L. Newsted and J.P. Giesy	Effect of 2,3,7,8-tetrachlorodibenzo-p-dioxin (TCDD) on the epidermal growth factor receptor in hepatic plasma membranes of Rainbow Trout (*Oncorhynchus mykiss*)	*Toxicol. Appl. Pharmacol.* 1993; 119: 41–51
1,1′-(2,2,2-Trichloroethylidene)bis-[4-chlorobenzene]		140	1	LAB	P.M. Mehrle, D.L. Stalling and R.A. Bloomfield	Serum amino acids in Rainbow Trout (*Salmo gairdneri*) as affected by DDT and dieldrin	*Comp. Biochem. Physiol. B* 1971; 388: 373–377
1,1′-(2,2,2-Trichloroethylidene)bis-[4-chlorobenzene]	Not reported	2.958333333	NR	LAB	P.G. McNicholl and W.C. Mackay	Effect of DDT on discriminating ability of Rainbow Trout (*Salmo gairdneri*)	*J. Fish. Res. Board Can.* 1975; 32(6): 785–788
1,1′-(2,2,2-Trichloroethylidene)bis-[4-chlorobenzene]	Not reported	1	5000	LAB	C. MacPhee, and R. Ruelle	Lethal effects of 1888 chemicals upon four species of fish from Western North America	Bull. No. 3, Forest, Wildl. and Range Exp. Stn., Univ. of Idaho, Moscow, ID; 1969: 112 pp.

(Continued)

Chemical name	Response site description	Exposure duration (days)	Conc. (μg/L)	Test location	Author	Title	Source
1,1′-(2,2,2-Trichloroethylidene)bis-[4-chlorobenzene]	Multiple tissue/organ	16	NR	LAB	R.D. Campbell, T.P. Leadem and D.W. Johnson	The *in vivo* effect of p,p′DDT on Na+ – K+ – activated ATPase activity in Rainbow Trout (*Salmo gairdneri*)	*Bull. Environ. Contam. Toxicol.* 1974; 11 (5): 425–428
1,1′-(2,2,2-Trichloroethylidene)bis-[4-chlorobenzene]	Not reported	30	128.8	LAB	D.C.G. Muir and A.L. Yarechewski	Dietary accumulation of four chlorinated dioxin congeners by Rainbow Trout and Fathead Minnows	*Environ. Toxicol. Chem.* 1988; 7(3): 227–236
1,1′-(2,2,2-Trichloroethylidene)bis-[4-chlorobenzene]		56	150000	LAB	P.O. Fromm and K.R. Olson	Industrial and municipal wastes: action of some water soluble pollutants on fish	Office of Water Res. and Technol., Michigan State University, East Lansing, MI; 1973: 24 pp. (US NTIS PB-237428) (Author communication used)
1,1′-(2,2,2-Trichloroethylidene)bis-[4-chlorobenzene]		56	150000	LAB	P.O. Fromm and K.R. Olson	Industrial and municipal wastes: action of some water soluble pollutants on fish	Office of Water Res. and Technol., Michigan State University, East Lansing, MI; 1973: 24 pp. (US NTIS PB-237428) (Author communication used)
1,1′-(2,2,2-Trichloroethylidene)bis-[4-chlorobenzene]	Liver	213.08	25000	LAB	J.D. Hendricks	Neoplasia in fish: tumor and mechanism studies in trout	In: Technical Report 9306, Compendium of the FY1988 & FY1989 Research Reviews for the Research Methods Branch, US Army Biomedical Research & Development Lab., Ft Detrick, Frederick, MD; 1993: 61–74, 103–113 (US NTIS AD-A272667)
1,1′-(2,2,2-Trichloroethylidene)bis-[4-chlorobenzene]	Not reported	0.010416667	NR	FIELDU	O.B. Cope, C.M. Gjullin and A. Storm	Effects of some insecticides on trout and salmon in Alaska, with reference to blackfly control	*Trans. Am. Fish. Soc.* 1949: 77: 160–177
1,1′-(2,2,2-Trichloroethylidene)bis-[4-chlorobenzene]	Not reported	1	NR	LAB	D. Ludemann and H. Neumann	Uber die Wirkung der Neuzeitlichen Kontaktinsektizide auf die Tiere des Subwassers	*Anz. Schaedlingskd. Pflanzenschutz*; 1962: 35: 5–9 (GER)

1,1′-(2,2,2-Trichloroethylidene)bis-[4-chlorobenzene]	Not reported	1	NR	LAB	D. Ludemann and H. Neumann	Acute toxicity of present contact insecticides for freshwater animals (Versuche uber die Akute Toxische Wirkung Neuzeitlicher Kontaktinsektizide auf Susswassertiere)	*Z. Angew. Zool.* 1961; 48: 87–96 (GER) (ENG ABS)
1,1′-(2,2,2-Trichloroethylidene)bis-[4-chlorobenzene]	Not reported	1	NR	FIELD N	O.B. Cope, C.M. Gjullin and A. Storm	Effects of some insecticides on trout and salmon in Alaska, with reference to blackfly control	*Trans. Am. Fish. Soc.* 1949; 77: 160–177
1,1′-(2,2,2-Trichloroethylidene)bis-[4-chlorobenzene]	Not reported	7	NR	FIELD N	W.R. Bridges, B.J. Kallman and A.K. Andrews	Persistence of DDT and its metabolites in a farm pond	*Trans. Am. Fish. Soc.* 1963; 92: 421–427
1,1′-(2,2,2-Trichloroethylidene)bis-[4-chlorobenzene]	Not reported	0.041666667	40000	LAB	K. Matsuo and T. Tamura	Laboratory experiments on the effect of insecticides against blackfly larvae (*Diptera*: Simuliidae) and fishes	*Sci. Pest Control/Botyu-Kagaku* 1970; 35(4): 125–130
1,1′-(2,2,2-Trichloroethylidene)bis-[4-chlorobenzene]	Not reported	0.5	1000	LAB	J. Mayhew	Toxicity of seven different insecticides to Rainbow Trout *Salmo gairdnerii* (Richardson)	*Proc. Iowa J. Acad. Sci.* 1955; 62: 599–606
1,1′-(2,2,2-Trichloroethylidene)bis-[4-chlorobenzene]	Not reported	0.5	500	LAB	J. Mayhew	Toxicity of seven different insecticides to Rainbow Trout *Salmo gairdnerii* (Richardson)	*Proc. Iowa J. Acad. Sci.* 1955; 62: 599–606
1,1′-(2,2,2-Trichloroethylidene)bis-[4-chlorobenzene]	Not reported	0.666666667	100	LAB	J. Mayhew	Toxicity of seven different insecticides to Rainbow Trout *Salmo gairdnerii* (Richardson)	*Proc. Iowa J. Acad. Sci.* 1955; 62: 599–606
1,1′-(2,2,2-Trichloroethylidene)bis-[4-chlorobenzene]	Not reported	0.666666667	1000	LAB	J. Mayhew	Toxicity of seven different insecticides to Rainbow Trout *Salmo gairdnerii* (Richardson)	*Proc. Iowa J. Acad. Sci.* 1955; 62: 599–606
1,1′-(2,2,2-Trichloroethylidene)bis-[4-chlorobenzene]	Not reported	0.666666667	250	LAB	J. Mayhew	Toxicity of seven different insecticides to Rainbow Trout *Salmo gairdnerii* (Richardson)	*Proc. Iowa J. Acad. Sci.* 1955; 62: 599–606
1,1′-(2,2,2-Trichloroethylidene)bis-[4-chlorobenzene]	Not reported	0.666666667	500	LAB	J. Mayhew	Toxicity of Seven Different Insecticides to Rainbow Trout *Salmo gairdnerii* (Richardson)	*Proc. Iowa J. Acad. Sci.* 1955; 62: 599–606
1,1′-(2,2,2-Trichloroethylidene)bis-[4-chlorobenzene]	Not reported	1	100	LAB	J. Mayhew	Toxicity of seven different insecticides to Rainbow Trout *Salmo gairdnerii* (Richardson)	*Proc. Iowa J. Acad. Sci.* 1955; 62: 599–606

(*Continued*)

Chemical name	Response site description	Exposure duration (days)	Conc. (µg/L)	Test location	Author	Title	Source
1,1′-(2,2,2-Trichloroethylidene)bis-[4-chlorobenzene]	Not reported	1	1000	LAB	J. Mayhew	Toxicity of seven different insecticides to Rainbow Trout *Salmo gairdnerii* (Richardson)	*Proc. Iowa J. Acad. Sci.* 1955; 62: 599–606
1,1′-(2,2,2-Trichloroethylidene)bis-[4-chlorobenzene]	Not reported	1	250	LAB	J. Mayhew	Toxicity of seven different insecticides to Rainbow Trout *Salmo gairdnerii* (Richardson)	*Proc. Iowa J. Acad. Sci.* 1955; 62: 599–606
1,1′-(2,2,2-Trichloroethylidene)bis-[4-chlorobenzene]	Not reported	1	500	LAB	J. Mayhew	Toxicity of seven different insecticides to Rainbow Trout *Salmo gairdnerii* (Richardson)	*Proc. Iowa J. Acad. Sci.* 1955; 62: 599–606
1,1′-(2,2,2-Trichloroethylidene)bis-[4-chlorobenzene]	Not reported	1	10000	LAB	C. MacPhee and R. Ruelle	Lethal effects of 1888 chemicals upon four species of fish from Western North America	Bull. No. 3, Forest, Wildl. and Range Exp. Stn., Univ. of Idaho, Moscow, ID; 1969: 112 pp.
1,1′-(2,2,2-Trichloroethylidene)bis-[4-chlorobenzene]	Not reported	1	5000	LAB	C. MacPhee and R. Ruelle	Lethal effects of 1888 chemicals upon four species of fish from Western North America	Bull. No. 3, Forest, Wildl. and Range Exp. Stn., Univ. of Idaho, Moscow, ID; 1969: 112 pp.
1,1′-(2,2,2-Trichloroethylidene)bis-[4-chlorobenzene]	Not reported	3	150000	LAB	State of Washington Dept of Fisheries	Toxic effects of organic and inorganic pollutants on young salmon and trout	*Res. Bull.* 1960; No. 5: 1–161
1,1′-(2,2,2-Trichloroethylidene)bis-[4-chlorobenzene]	Not reported	43	180000	LAB	P.M. Mehrle, F.L. Mayer and W.W. Johnson	Diet quality in fish toxicology: effects on acute and chronic toxicity	In: F.L. Mayer and J.L. Hamelink (Eds), *Aquatic Toxicology and Hazard Evaluation*, 1st Symposium, ASTM STP 634, Philadelphia, PA; 1977: 269–280 (Publ in part as 6797)
1,1′-(2,2,2-Trichloroethylidene)bis-[4-chlorobenzene]	Not reported	8	5	LAB	F.L.J. Mayer, J.C. Street and J.M. Neuhold	DDT intoxication in Rainbow Trout as affected by dieldrin	*Toxicol. Appl. Pharmacol.* 1972; 22(3): 347–354
1,1′-(2,2,2-Trichloroethylidene)bis-[4-chlorobenzene]	Not reported	30	128.8	LAB	D.C.G. Muir and A.L. Yarechewski	Dietary accumulation of four chlorinated dioxin congeners by Rainbow Trout and Fathead Minnows	*Environ. Toxicol. Chem.* 1988; 7(3): 227–236

1,1′-(2,2,2-Trichloroethylidene)bis-[4-chlorobenzene]	Nervous tissue	1	NR	LAB	C.F. Peters and D.D. Weber	DDT: Effect on the lateral line nerve of Steelhead Trout	In: F.J. Vernberg, A. Calabrese, F.P. Thurberg and W.B. Vernberg (Eds), *Physiological Responses of Marine Biota to Pollutants.* Academic Press Inc., NY; 1977: 75–91
1,1′-(2,2,2-Trichloroethylidene)bis-[4-chlorobenzene]		0.083333333	NR	LAB	T.G. Bahr and R.C. Ball	Action of DDT on the evoked and spontaneous activity from the Rainbow Trout lateral line nerve	*Comp. Biochem. Physiol. A* 1971; 38(2): 279–284
1,1′-(2,2,2-Trichloroethylidene)bis-[4-chlorobenzene]		0.208333333	140	LAB	C.R. Lunn, D.P. Toews and D.J. Pree	Effects of three pesticides on respiration, coughing, and heart rates of Rainbow Trout (*Salmo gairdneri* Richardson)	*Can. J. Zool.* 1976; 54(2): 214–219
1,1′-(2,2,2-Trichloroethylidene)bis-[4-chlorobenzene]		0.208333333	350	LAB	C.R. Lunn, D.P. Toews and D.J. Pree	Effects of three pesticides on respiration, coughing, and heart rates of Rainbow Trout (*Salmo gairdneri* Richardson)	*Can. J. Zool.* 1976; 54(2): 214–219
1,1′-(2,2,2-Trichloroethylidene)bis-[4-chlorobenzene]		0.25	15000	LAB	T.G. Bahr and R.C. Ball	Action of DDT on the evoked and spontaneous activity from the Rainbow Trout lateral line nerve	*Comp. Biochem. Physiol. A* 1971; 38(2): 279–284
1,1′-(2,2,2-Trichloroethylidene)bis-[4-chlorobenzene]		0.208333333	140	LAB	C.R. Lunn, D.P. Toews and D.J. Pree	Effects of three pesticides on respiration, coughing, and heart rates of Rainbow Trout (*Salmo gairdneri* Richardson)	*Can. J. Zool.* 1976; 54(2): 214–219
1,1′-(2,2,2-Trichloroethylidene)bis-[4-chlorobenzene]		0.208333333	52.5	LAB	C.R. Lunn, D.P. Toews and D.J. Pree	Effects of three pesticides on respiration, coughing, and heart rates of Rainbow Trout (*Salmo gairdneri* Richardson)	*Can. J. Zool.* 1976; 54(2): 214–219
2,4′-DDD	Liver	NR	NR	LAB	S. Benguira, V.S. Leblond, J.P. Weber and A. Hontela	Loss of capacity to elevate plasma cortisol in Rainbow Trout (*Oncorhynchus mykiss*) treated with a single injection of o,p′-dichlorodiphenyldichloroethane	*Environ. Toxicol. Chem.* 2002; 21(8): 1753–1756
2,4′-DDD	Blood	NR	NR	LAB	S. Benguira, V.S. Leblond, J.P. Weber and A. Hontela	Loss of capacity to elevate plasma cortisol in Rainbow Trout (*Oncorhynchus mykiss*) treated with a single injection of o,p′-dichlorodiphenyldichloroethane	*Environ. Toxicol. Chem.* 2002; 21(8): 1753–1756

(Continued)

Chemical name	Response site description	Exposure duration (days)	Conc. (μg/L)	Test location	Author	Title	Source
2,4′-DDD	Whole organism	NR	NR	LAB	S. Benguira, V.S. Leblond, J.P. Weber and A. Hontela	Loss of capacity to elevate plasma cortisol in Rainbow Trout (*Oncorhynchus mykiss*) treated with a single injection of o,p′-dichlorodiphenyldichloroethane	*Environ. Toxicol. Chem.* 2002; 21(8): 1753–1756
2,4′-DDD	Plasma	NR	NR	LAB	S. Benguira, V.S. Leblond, J.P. Weber and A. Hontela	Loss of capacity to elevate plasma cortisol in Rainbow Trout (*Oncorhynchus mykiss*) treated with a single injection of o,p′-dichlorodiphenyldichloroethane	*Environ. Toxicol. Chem.* 2002; 21(8): 1753–1756
2,4′-DDD	Gonad(s)	NR	NR	LAB	S. Benguira, V.S. Leblond, J.P. Weber and A. Hontela	Loss of capacity to elevate plasma cortisol in Rainbow Trout (*Oncorhynchus mykiss*) treated with a single injection of o,p′-dichlorodiphenyldichloroethane	*Environ. Toxicol. Chem.* 2002; 21(8): 1753–1756
2,4′-DDD	Liver	NR	NR	LAB	S. Benguira, V.S. Leblond, J.P. Weber and A. Hontela	Loss of capacity to elevate plasma cortisol in Rainbow Trout (*Oncorhynchus mykiss*) treated with a single injection of o,p′-dichlorodiphenyldichloroethane	*Environ. Toxicol. Chem.* 2002; 21(8): 1753–1756
1,1′-(Dichloro-ethenylidene)bis-(4-chlorobenzene)	Whole organism	7	0.013	LAB	B.G. Oliver and A.J. Niimi	Bioconcentration factors of some halogenated organics for Rainbow Trout: limitations in their use for prediction of environmental residues	*Environ. Sci. Technol.* 1985; 19(9): 842–849
1,1′-(Dichloro-ethenylidene)bis-(4-chlorobenzene)	Whole organism	7	0.0013	LAB	B.G. Oliver and A.J. Niimi	Bioconcentration factors of some halogenated organics for Rainbow Trout: limitations in their use for prediction of environmental residues	*Environ. Sci. Technol.* 1985; 19(9): 842–849
1,1′-(Dichloro-ethenylidene)bis-(4-chlorobenzene)	Whole organism	21	0.013	LAB	B.G. Oliver and A.J. Niimi	Bioconcentration factors of some halogenated organics for Rainbow Trout: limitations in their use for prediction of environmental residues	*Environ. Sci. Technol.* 1985; 19(9): 842–849
1,1′-(Dichloro-ethenylidene)bis-(4-chlorobenzene)	Whole organism	21	0.0013	LAB	B.G. Oliver and A.J. Niimi	Bioconcentration factors of some halogenated organics for Rainbow Trout: limitations in their use for prediction of environmental residues	*Environ. Sci. Technol.* 1985; 19(9): 842–849

1,1′-(Dichloro-ethenylidene)bis-(4-chlorobenzene)	Whole organism	35	0.013	LAB	B.G. Oliver and A.J. Niimi	Bioconcentration factors of some halogenated organics for Rainbow Trout: limitations in their use for prediction of environmental residues	*Environ. Sci. Technol.* 1985; 19(9): 842–849
1,1′-(Dichloro-ethenylidene)bis-(4-chlorobenzene)	Whole organism	50	0.0013	LAB	B.G. Oliver and A.J. Niimi	Bioconcentration factors of some halogenated organics for Rainbow Trout: limitations in their use for prediction of environmental residues	*Environ. Sci. Technol.* 1985; 19(9): 842–849
1,1′-(Dichloro-ethenylidene)bis-(4-chlorobenzene)	Whole organism	75	0.0013	LAB	B.G. Oliver and A.J. Niimi	Bioconcentration factors of some halogenated organics for Rainbow Trout: limitations in their use for prediction of environmental residues	*Environ. Sci. Technol.* 1985; 19(9): 842–849
1,1′-(Dichloro-ethenylidene)bis-(4-chlorobenzene)	Whole organism	75	0.013	LAB	B.G Oliver and A.J. Niimi	Bioconcentration factors of some halogenated organics for Rainbow Trout: limitations in their use for prediction of environmental residues	*Environ. Sci. Technol.* 1985; 19(9): 842–849
1,1′-(Dichloro-ethenylidene)bis-(4-chlorobenzene)	Whole organism	50	0.013	LAB	B.G. Oliver and A.J. Niimi	Bioconcentration factors of some halogenated organics for Rainbow Trout: limitations in their use for prediction of environmental residues	*Environ. Sci. Technol.* 1985; 19(9): 842–849
1,1′-(Dichloro-ethenylidene)bis-(4-chlorobenzene)	Whole organism	35	0.0013	LAB	B.G. Oliver and A.J. Niimi	Bioconcentration factors of some halogenated organics for Rainbow Trout: limitations in their use for prediction of environmental residues	*Environ. Sci. Technol.* 1985; 19(9): 842–849
1,1′-(Dichloro-ethenylidene)bis-(4-chlorobenzene)	Whole organism	96	0.0013	LAB	B.G. Oliver and A.J. Niimi	Bioconcentration factors of some halogenated organics for Rainbow Trout: limitations in their use for prediction of environmental residues	*Environ. Sci. Technol.* 1985; 19(9): 842–849
1,1′-(Dichloro-ethenylidene)bis-(4-chlorobenzene)	Whole organism	96	0.013	LAB	B.G. Oliver and A.J. Niimi	Bioconcentration factors of some halogenated organics for Rainbow Trout: limitations in their use for prediction of environmental residues	*Environ. Sci. Technol.* 1985; 19(9): 842–849
1,1′-(Dichloro-ethenylidene)bis-(4-chlorobenzene)	Whole organism	96	NR	LAB	B.G. Oliver and A.J. Niimi	Bioconcentration factors of some halogenated organics for Rainbow Trout: limitations in their use for prediction of environmental residues	*Environ. Sci. Technol.* 1985; 19(9): 842–849
1,1′-(Dichloro-ethenylidene)bis-(4-chlorobenzene)	Not reported	NR	NR	FIELD N	J.L. Hamelink and R.C. Waybrant	DDE and lindane in a large-scale model lentic ecosystem	*Trans. Am. Fish. Soc.* 1976; 1: 125–134

(Continued)

Chemical name	Response site description	Exposure duration (days)	Conc. (μg/L)	Test location	Author	Title	Source
1,1′-(Dichloro-ethenylidene)bis-(4-chlorobenzene)	Not reported	108	NR	FIELD N	J.L. Hamelink and R.C. Waybrant	DDE and lindane in a large-scale model lentic ecosystem	*Trans. Am. Fish. Soc.* 1976; 1: 125–134
1,1′-(Dichloro-ethenylidene)bis-(4-chlorobenzene)	Liver	NR	30000	LAB	R.M. Donohoe and L.R. Curtis	Estrogenic activity of chlordecone, o,p′-DDT and o,p′-DDE in juvenile Rainbow Trout: induction of vitellogenesis and interaction with hepatic estrogen binding sites	*Aquat. Toxicol.* 1996; 36 (1/2): 31–52
1,1′-(Dichloro-ethenylidene)bis-(4-chlorobenzene)	Blood	42	90000	LAB	R.M. Donohoe and L.R. Curtis	Estrogenic ativity of clordecone, o,p′-DDT and o,p′-DDE in juvenile Rainbow Trout: induction of vitellogenesis and interaction with hepatic estrogen binding sites	*Aquat. Toxicol.* 1996; 36 (1/2): 31–52
1,1′-(Dichloro-ethenylidene)bis-(4-chlorobenzene)	Blood	42	45000	LAB	R.M. Donohoe and L.R. Curtis	Estrogenic activity of chlordecone, o,p′-DDT and o,p′-DDE in juvenile Rainbow Trout: induction of vitellogenesis and interaction with hepatic estrogen binding sites	*Aquat. Toxicol.* 1996; 36 (1/2): 31–52
1,1′-(Dichloro-ethenylidene)bis-(4-chlorobenzene)	Blood	NR	NR	LAB	R.M. Donohoe and L.R. Curtis	Estrogenic activity of chlordecone, o,p′-DDT and o,p′-DDE in juvenile Rainbow Trout: induction of vitellogenesis and interaction with hepatic estrogen binding sites	*Aquat. Toxicol.* 1996; 36 (1/2): 31–52
1,1′-(Dichloro-ethenylidene)bis-(4-chlorobenzene)	Serum	9	100000	LAB	H.R. Andersen, A.M. Andersson, S.F. Arnold, H. Autrup, M. Barfoed, N.A. Beresford, P. Bjerregaard, and L.B. Christiansen	Comparison of short-term estrogenicity tests for identification of hormone-disrupting chemicals	*Environ. Health Perspect.* 1999; 107 (Suppl. 1): 89–108
1,1′-(Dichloro-ethenylidene)bis-(4-chlorobenzene)	Liver	2	50000	LAB	M. Machala, P. Drabek, J. Neca, J. Kolarova and Z. Svobodova	Biochemical markers for differentiation of exposures to nonplanar polychlorinated biphenyls, organochlorine pesticides, or 2,3,7,8-tetrachlorodibenzo-p-dioxin in trout liver	*Ecotoxicol. Environ. Saf.* 1998; 41: 107–111

1,1′-(Dichloro-ethenylidene)bis-(4-chlorobenzene)	Liver	2	NR	LAB	M. Petrivalsky, M. Machala, K. Nezveda, V. Piacka, Z. Svobodova and P. Drabek	Glutathione-dependent detoxifying enzymes in Rainbow Trout liver: search for specific biochemical markers of chemical stress	*Environ. Toxicol. Chem.* 1997; 16 (7): 1417–1421
1,1′-(Dichloro-ethenylidene)bis-(4-chlorobenzene)	Not reported	1	5000	LAB	V.C. Applegate, J.H. Howell, A.E. Hall Jr., and M.A. Smith	Toxicity of 4,346 chemicals to larval lampreys and fishes	*Spec. Sci. Rep. Fish.* No. 207, Fish Wildl. Serv., USDI, Washington, DC; 1957: 157 pp.
1,1′-(Dichloro-ethenylidene)bis-(4-chlorobenzene)	Liver	42	45000	LAB	R.M. Donohoe and L.R. Curtis	Estrogenic activity of chlordecone, o,p′-DDT and o,p′-DDE in juvenile Rainbow Trout: induction of vitellogenesis and interaction with hepatic estrogen binding sites	*Aquat. Toxicol.* 1996: 36 (1/2): 31–52
1,1′-(Dichloro-ethenylidene)bis-(4-chlorobenzene)	Liver	42	90000	LAB	R.M. Donohoe and L.R. Curtis	Estrogenic activity of chlordecone, o,p′-DDT and o,p′-DDE in juvenile Rainbow Trout: induction of vitellogenesis and interaction with hepatic estrogen binding sites	*Aquat. Toxicol.* 1996; 36 (1/2): 31–52
1,1′-(Dichloro-ethenylidene)bis-(4-chlorobenzene)	Liver	NR	NR	LAB	R.M. Donohoe and L.R. Curtis	Estrogenic activity of chlordecone, o,p′-DDT and o,p′-DDE in juvenile Rainbow Trout: induction of vitellogenesis and interaction with hepatic estrogen binding sites	*Aquat. Toxicol.* 1996; 36 (1/2): 31–52
1,1-Bis(ethylphenyl)-2,2-dichloroethane	Not reported	0.583333333	5000	LAB	V.C. Applegate, J.H. Howell, A.E. Hall Jr. and M.A. Smith	Toxicity of 4,346 chemicals to larval lampreys and fishes	*Spec. Sci. Rep. Fish.* No. 207, Fish Wildl. Serv., USDI, Washington, DC; 1957: 157 pp.
1-Chloro-2-(2,2,2-trichloro-1-(4-chlorophenyl)ethyl) benzene	Blood	42	45000	LAB	R.M. Donohoe and L.R. Curtis	Estrogenic activity of chlordecone, o,p′-DDT and o,p′-DDE in juvenile Rainbow Trout: induction of vitellogenesis and interaction with hepatic estrogen binding sites	*Aquat. Toxicol.* 1996; 36 (1/2): 31–52
1-Chloro-2-(2,2,2-trichloro-1-(4-chlorophenyl)ethyl)-benzene	Blood	42	NR	LAB	R.M. Donohoe and L.R. Curtis	Estrogenic activity of chlordecone, o,p′-DDT and o,p′-DDE in juvenile Rainbow Trout: induction of vitellogenesis and interaction with hepatic estrogen binding sites	*Aquat. Toxicol.* 1996; 36 (1/2): 31–52

(Continued)

Chemical name	Response site description	Exposure duration (days)	Conc. (μg/L)	Test location	Author	Title	Source
1-Chloro-2-(2,2,2-trichloro-1-(4-chlorophenyl)ethyl)-benzene	Plasma	NR	NR	LAB	L.B. Christiansen, K.L. Pedersen, S.N. Pedersen, B. Korsgaard and P. Bjerregaard	*In vivo* comparison of xenoestrogens using Rainbow Trout vitellogenin induction as a screening system	*Environ. Toxicol. Chem.* 2000; 19 (7): 1867–1874
1-Chloro-2-(2,2,2-trichloro-1-(4-chlorophenyl)ethyl)-benzene	Serum	9	50000	LAB	H.R. Andersen, A.M. Andersson, S.F. Arnold, H. Autrup, M. Barfoed, N.A. Beresford, P. Bjerregaard, and L.B. Christiansen	Comparison of short-term estrogenicity tests for identification of hormone-disrupting chemicals	*Environ. Health Perspect.* 1999; 107 (Suppl. 1): 89–108
1-Chloro-2-(2,2,2-trichloro-1-(4-chlorophenyl)ethyl)-benzene	Blood	NR	NR	LAB	R.M. Donohoe and L.R. Curtis	Estrogenic activity of chlordecone, o,p′-DDT and o,p′-DDE in juvenile Rainbow Trout: induction of vitellogenesis and interaction with hepatic estrogen binding sites	*Aquat. Toxicol.* 1996; 36 (1/2): 31–52
1-Chloro-2-(2,2,2-trichloro-1-(4-chlorophenyl)ethyl)-benzene	Liver	42	45000	LAB	R.M. Donohoe and L.R. Curtis	Estrogenic activity of chlordecone, o,p′-DDT and o,p′-DDE in juvenile Rainbow Trout: induction of vitellogenesis and interaction with hepatic estrogen binding sites	*Aquat. Toxicol.* 1996; 36 (1/2): 31–52
1-Chloro-2-(2,2,2-trichloro-1-(4-chlorophenyl)ethyl)-benzene	Liver	42	90000	LAB	R.M. Donohoe and L.R. Curtis	Estrogenic activity of chlordecone, o,p′-DDT and o,p′-DDE in juvenile Rainbow Trout: induction of vitellogenesis and interaction with hepatic estrogen binding sites	*Aquat. Toxicol.* 1996; 36 (1/2): 31–52
1-Chloro-2-(2,2,2-trichloro-1-(4-chlorophenyl)ethyl)-benzene	Liver	NR	NR	LAB	R.M. Donohoe and L.R. Curtis	Estrogenic activity of chlordecone, o,p′-DDT and o,p′-DDE in juvenile Rainbow Trout: induction of vitellogenesis and interaction with hepatic estrogen binding sites	*Aquat. Toxicol.* 1996; 36 (1/2): 31–52
o,p′-DDE	Blood	42	45000	LAB	R.M. Donohoe and L.R. Curtis	Estrogenic activity of chlordecone, o,p′-DDT and o,p′-DDE in juvenile Rainbow Trout: induction of vitellogenesis and interaction with hepatic estrogen binding sites	*Aquat. Toxicol.* 1996; 36 (1/2): 31–52

o,p′-DDE	Blood	42	NR	LAB	R.M. Donohoe and L.R. Curtis	Estrogenic activity of chlordecone, o,p′-DDT and o,p′-DDE in juvenile Rainbow Trout: induction of vitellogenesis and interaction with hepatic estrogen binding sites	*Aquat. Toxicol.* 1996; 36 (1/2): 31–52
o,p′-DDE	Blood	NR	NR	LAB	R.M. Donohoe and L.R. Curtis	Estrogenic activity of chlordecone, o,p′-DDT and o,p′-DDE in juvenile Rainbow Trout: induction of vitellogenesis and interaction with hepatic estrogen binding sites	*Aquat. Toxicol.* 1996; 36 (1/2): 31–52
o,p′-DDE	Liver	42	90000	LAB	R.M. Donohoe and L.R. Curtis	Estrogenic activity of chlordecone, o,p′-DDT and o,p′-DDE in juvenile Rainbow Trout: induction of vitellogenesis and interaction with hepatic estrogen binding sites	*Aquat. Toxicol.* 1996; 36 (1/2): 31–52
o,p′-DDE	Liver	42	45000	LAB	R.M. Donohoe and L.R. Curtis	Estrogenic activity of chlordecone, o,p′-DDT and o,p′-DDE in juvenile Rainbow Trout: induction of vitellogenesis and interaction with hepatic estrogen binding sites	*Aquat. Toxicol.* 1996; 36 (1/2): 31–52
o,p′-DDE	Liver	42	45000	LAB	R.M. Donohoe and L.R. Curtis	Estrogenic activity of chlordecone, o,p′-DDT and o,p′-DDE in juvenile Rainbow Trout: induction of vitellogenesis and interaction with hepatic estrogen binding sites	*Aquat. Toxicol.* 1996; 36 (1/2): 31–52
o,p′-DDE	Liver	NR	NR	LAB	R.M. Donohoe and L.R. Curtis	Estrogenic activity of chlordecone, o,p′-DDT and o,p′-DDE in juvenile Rainbow Trout: induction of vitellogenesis and interaction with hepatic estrogen binding sites	*Aquat. Toxicol.* 1996; 36 (1/2): 31–52

Note: Columns from the actual retrieval have been removed that did not add value to this sample (e.g. non-reported denotations, repeated information such as taxonomy, common name and freshwater versus salt water).
Source: US Environmental Protection Agency (2009). ECOTOX Database; http://cfpub.epa.gov/ecotox/index.html; accessed October 7, 2009.

GLOSSARY

These terms are the author's operational definitions. The sources are numerous and many terms have been modified from their original definitions. However, the resources used to augment this glossary are listed at the end. In addition, the sources cited in the Notes and Commentary following each chapter of this book contain useful definitions of terms used in these specific instances.

AB model: Two-stage microbiological model wherein the B portion of a toxin is responsible for toxin binding to a cell but does not directly harm it. Thereafter, the A portion enters the cell and disrupts its function.

AB toxin: Structure and activity of many *exotoxins* based on the *AB model.*

Abiotic: Description of chemical and physical processes occurring without the involvement of living organisms. In some cases, such processes do not involve microorganisms or plants at all, whereas in other cases, biological and abiotic processes occur simultaneously and/or serve to enhance each other.

Absorbed dose: 1. Amount of a substance that enters the body of an organism, e.g. through the eyes, skin, stomach, intestines, or lungs. 2. Amount of active ingredient crossing exchange boundaries of a test organism or human (same as "internal dose").

Absorption: 1. Process wherein a substance permeates another substance; a fluid is sorbed into a particle. 2. After uptake, the process by which a substance moves to tissues in an organism, e.g. absorption of a substance into the bloodstream. 3. Process by which incident radiated energy is retained in a medium; e.g. shortwave radiation from the sun is absorbed by soil and re-radiated as longer wave radiation; e.g. infrared radiation (see *greenhouse effect*).

Acceptable daily intake (ADI): The amount of a chemical a person can be exposed to on a daily basis over an extended period of time (usually a lifetime) without suffering deleterious effects.

Acceptable engineering practice: 1. Amount of engineering-related work needed, in years, to meet one of the minimal criteria to sit for the Professional Engineer (PE) examination. 2. Reasonably expected professional performance by an engineer demonstrating competence, especially adhering to codes for design, construction, and operation.

Acclimation: Adapting a microbe to food sources for carbon and energy in an attempt to enhance biodegradation (e.g. in wastewater treatment).

Accuracy: Degree of agreement between a measured value and the true value; usually expressed as $+/-$ percent of full scale. Compare to *precision*.

Acetogenic: Describing a *prokaryotic microbe* that uses carbonate as a terminal *electron acceptor,* producing acetic acid as waste.

Acetyl coenzyme A (acetyl Co-A): Energy-rich combination of acetic acid and coenzyme A, produced by numerous catabolic pathways and the substrate for the tricarboxylic acid cycle, fatty acid biosynthesis, and other pathways.

Acid fast: Description of bacteria, e.g. *mycobacteria*, which cannot be easily decolorized with acid alcohol after being stained with dyes like basic fuchsin.

Acid rain: Precipitation with depressed pH due to increases in concentrations of acid-forming compounds, such as oxides of sulfur and oxides of nitrogen, in the atmosphere. Term is usually limited to atmospheric deposition with pH < 5.6, which is about the mean value for natural water (mainly due to carbonic acid content).

Acidogenic: Describing microbes that convert sugars and amino acids into carbon dioxide, molecular hydrogen (H_2), ammonia (NH_3), and organic compounds with the carboxyl functional group (R-COOH), i.e. the organic acids.

Acidophile: Microbe that survives best in pH ≤ 5.5. Preferred scientific term is acid deposition.

Actinobacterium: Member of group of gram-positive bacteria, including *actinomycetes* and their high *G 1 C* relatives.

Actinomycete: Aerobic, gram-positive bacterium that forms branching filaments (*hyphae*) and asexual spores. Includes many but not all members of the order *Actinomyceales.*

Actinorhizae: Associations between actinomycetes and plant roots.

Action level: 1. Concentration threshold above which actions must be taken, e.g. remediation, removal, treatment, or use restrictions and closures (e.g. beaches). 2. Regulatory level recommended by the US Environmental Protection Agency for enforcement by Food and Drug Administration and United States Department of Agriculture when pesticide residues occur in food or feed commodities for reasons other than the direct application of the pesticide. Set for inadvertent residues resulting from previous legal use or accidental contamination. Compare to *tolerance.*

Activated carbon: Carbon with a very high ratio of surface area to mass, rendering it a strong *adsorbent.* In granulated form, known as granulated activated carbon (GAC).

Activated sludge: Product of mixing primary *effluent* with microbe-laden sludge and then agitated and aerated to promote biological treatment, speeding the breakdown of organic matter in raw sewage undergoing secondary waste treatment. The microbes readily use dissolved organic substrates and transform them into additional microbial cells and carbon dioxide.

Activation: See *bioactivation.*

Activation energy: Energy needed to bring all molecules in one mole of a substance to their reactive state at a given temperature.

Active ingredient: Compound in a pesticide formulation that provides the biocidal mechanism of action. All other components are known as *inerts.*

Active site: Portion of an enzyme that binds the substrate to form an enzyme-substrate complex and catalyze the reaction. Also referred to as the catalytic site.

Acute: Describing exposures, diseases or responses with short durations.

Acute exposure: A single exposure to a substance that results in severe biological harm or death. Acute exposures are usually characterized as lasting no longer than a day, as compared to longer, continuing *exposure* over a period of time.

Acute toxicity: Any adverse effect that occurs within a short period of time following *exposure,* usually up to 24–96 hours, resulting in biological harm and often death.

Adaptive management (AM): Also known as "adaptive resource management" (ARM), a structured, iterative process of optimal decision making in the face of uncertainty, with an aim to reducing uncertainty over time via system monitoring. In this way, decision making simultaneously maximizes one or more resource objectives and, either passively or actively, accrues information needed to improve future management. AM is often characterized as "learning by doing."

Additive effect: Response to exposure to multiple substances that equals the sum of responses of all the individual substances added together (compare with *antagonism* and *synergism*).

Adenine: Purine derivative, 6-aminopurine, found in nucleosides, nucleotides, coenzymes, and nucleic acids.

Adenosine diphosphate (ADP): The nucleoside diphosphate usually formed upon the breakdown of ATP when it provides energy for work. See *photosynthesis*.

Adenosine 5′-triphosphate (ATP): The triphosphate of the nucleoside adenosine, which is a high energy molecule or has high phosphate group transfer potential and serves as the cell's major form of energy currency. See *photosynthesis*.

Adsorption: Process wherein a substance permeates another substance; a fluid is sorbed onto the surface of a particle. Compare to *absorption* (*1*).

Advanced waste treatment: Physical, chemical or biological treatment beyond that achieved by secondary treatment (e.g. additional removal of nutrients and solids). Synonymous with and preferred over *tertiary treatment*.

Advection: Transport of a substance along with the flow of a fluid, e.g. transport of a solute by the bulk motion of flowing groundwater or particles in the flow streams of an air mass.

Adverse effect: A biochemical change, functional impairment, or pathologic lesion that affects the performance of the whole organism, or reduces an organism's ability to respond to an additional environmental challenge.

Aerobe: 1. Microorganism that can survive in the presence of molecular oxygen (O_2). 2. Microorganism that requires sufficient concentrations of O_2 to survive.

Aerobic: Conditions for growth or metabolism in which the organism is sufficiently supplied with molecular oxygen.

Aerobic respiration: Metabolic process whereby microorganisms use oxygen as the final *electron acceptor* to generate energy.

Aerobic treatment: Process by which microbes decompose complex organic compounds in the presence of oxygen and use the liberated energy for reproduction and growth.

Aerodynamic diameter: The diameter of a sphere with unit density that has aerodynamic behavior identical to that of the particle in question; an expression of aerodynamic behavior of an irregularly shaped particle in terms of the diameter of an idealized particle. Particles having the same aerodynamic diameter may have different dimensions and shapes. See also *Stokes diameter*.

Aerosol: A suspension of liquid or solid particles in air.

Affinity: Attraction, e.g. between antigen and an antibody, or for a polar compound to the aqueous compartments of the environment.

Agar: Polysaccharide complex gel derived from marine algae used to grow microbiological cultures. Gelling temperature usually ranges between 40 and 50 °C.

Agrobacterium: Genus of bacteria that includes several plant pathogenic species, causing tumor-like symptoms.

Air stripping: Removal of *volatile* compounds from solution by passing a higher concentration solution into an air stream of lower concentration. This process uses *Henry's law* as a means of removing pollutants from soil and groundwater.

Akinete: Resting cell with a thick wall, found in *algae* and *cyanobacteria*.

ALARA: As low as reasonably achievable.

Albedo: Reflectivity of light; inverse of light absorption.

Algae: Phototrophic *eukaryotic* microbes that can be either unicellular or multicellular. Plural of alga.

Algal bloom: Masses of *algae,* plants and other organisms that form scum at the top of surface waters; usually attributed to large inputs of nutrients to the waters.

Aliphatic compounds: Acyclic or cyclic, saturated or unsaturated carbon compounds, excluding aromatic compounds.

Alkalophile: Organism that prefers elevated pH levels, i.e. up to 10.5.

Alkane: Single bonded carbon chains or branched structures.

Alkene: Carbon chains or branched structures that contain at least one double bond.

Alkyne: Carbon chains or branched structures that contain at least one triple bond.

Allele: One of several alternative forms of a gene that occupies a given locus on a chromosome.

Allergen: Substance that causes an allergic reaction.

Allergenicity: Capacity or potential that a substance will elicit an allergic reaction.

Allosteric: Describing a site on an enzyme other than the active site to which a nonsubstrate binds, which could result in blocking the normal substrate binding.

Alpha-proteobacteria: Purple bacteria and relatives. One of five subgroups of *proteobacteria*, each with distinctive 16s rRNA sequences. Compare to *beta-proteobacteria.*

Alum: 1. Flocculant, $K_2SO_4Al_2(SO_4)_3 \cdot 2H_2O$. 2. Aluminum sulfate alum, used to precipitate hydroxides for coagulation.

Ambient: 1. Outdoor (e.g. ambient air). 2. Describing general environmental conditions, as contrasted with effluent or emission, e.g. ambient measurements versus effluent measurements. 3. Surrounding conditions.

Amendment: Substrate introduced to stimulate the in situ microbial processes (vegetable oils, sugars, alcohols, etc.).

Amensalism: One organism's production of a substance that inhibits another organism, e.g. fungi's exudations that inhibit the growth of bacteria.

Amino acid: Any of 20 basic building blocks of proteins with a free amino (-NH_2) and a free carboxyl (-COOH) group, and having the basic formula $NH_2 - C_R - COOH$. According to the side group R, they are subdivided intopolar or hydrophilic (serine, threonine, tyrosine, asparagine and glutamine); nonpolar or hydrophobic (glycine, alanine, valine, leucine, isoleucine, proline, phenylalanine, tryptophan and cysteine); acidic (aspartic acid and glutamic acid) and basics (lysine, arginine, hystidine). The sequence of amino acids determines the shape, properties, and the biological role of a protein.

Amino group: $-NH_2$ attached to carbon structures (e.g. in amines and amino acids).

Amoral: Lacking any moral characteristics. An amoral act is neither morally good nor morally bad; it simply exists. Contrast with moral or immoral.

Amphiphilic: Describing a chemical compound that has both hydrophilic and lipophilic properties.

Amphoteric: Able to react as either a weak acid or a weak base.

Amplification: 1. Treatment (e.g., use of chloramphenicol) designed to increase the proportion of plasmid DNA relative to that of bacterial (host) DNA. 2. Replicating a gene library in bulk. 3. Duplication of genes within a chromosomal segment. 4. Creation of numerous copies of a segment of DNA by the *polymerase chain reaction.*

Anabolism: Synthesis of complex molecules from simpler molecules with the input of energy. Compare to *catabolism.*

Anaerobe: 1. Microorganism that cannot survive in the presence of molecular oxygen (O_2). 2. Microorganism that requires electron acceptors other than O_2 to survive.

Anaerobic respiration: Process whereby microorganisms use a chemical other than oxygen as an electron acceptor. Common substitutes for oxygen are nitrate, sulfate, iron, carbon dioxide, and other organic compounds (fermentation).

Analogy: Comparison of similarities between two things to a conclusion about an additional attribute common to both things. This is a type of inductive reasoning (see *inductive reasoning*). One of *Hill's criteria.*

Analyte: 1. Substance measured in a scientific study. 2. Chemical for which a sample (e.g. water, air, or blood) is tested in a laboratory, e.g. to determine the amount of cadmium in soil, the specified amount of soil to be collected and analyzed in the laboratory.

Analytic epidemiology: Evaluation of associations between exposure to physical, chemical, and biological agents and disease by testing scientific hypotheses.

Anion: Negatively charged ion, e.g. ammonium [NH_4^+].

Anion exchange capacity (AEC): Total of exchangeable anions that can be sorbed by a soil (units = centimoles of negative charge per kg soil).

Anisotropy: Conditions under which one or more hydraulic properties of an aquifer vary with direction.

Anoxic: An environment where there is no free oxygen and where microbial and chemical reactions use other chemicals in the environment to accept electrons.

Antagonism: 1. Effect from a combination of two agents is less than the sum of the individual effects from each agent ($1+1<2$). Contrast with *synergism* and *additive effect.* 2. *Amensalism.*

Anthropocentrism: Philosophy or decision framework based on human beings. View that all and only humans have moral value. Nonhuman species and abiotic resources have value only in respect to that associated with human values. Contrast with *biocentrism* and *ecocentrism*.

Anthropogenic: 1. Made, caused, or influenced by human activities. Contrast with *biogenic*. 2. Derived from human activities, as opposed to those occurring in natural environments without human influences.

Antibody: Protein, e.g. immunoglobulin, manufactured by lymphocytes (a type of white blood cell) to neutralize an *antigen* or foreign protein. Microbes, pollens, dust mites, molds, foods, and other substances contain antigens which will trigger antibodies.

Antigen: Foreign substance (e.g. protein, nucleoprotein, polysaccharide) to which lymphocytes respond. When immune system responds, known as an immunogen.

Anti-nutritional: Factor, when present, stifles metabolism and growth of an organism (especially humans in the context of food biotechnologies). One of three major risks associated with food biotechnology, along with toxicity and *allergenicity.*

Apoptosis: Programmed cell death.

Applied mathematics: Mathematical techniques typically used in the application of mathematical knowledge to domains beyond mathematics itself.

Aqueous solubility: Maximum concentration of a substance that will dissolve in pure water at a reference temperature. See *hydrophilicity.*

Aquifer: A porous underground bed or layer of earth, sand, gravel, or porous stone that contains water. Geologic formation, group of formations or part of a formation containing saturated permeable material that yields sufficient, economical quantities of groundwater.

Artificial expression system: Cell system into which an expression vector has been artificially introduced and that contains all the enzyme systems needed for translation of *messenger RNA.*

Association: Relationship, not necessarily causal, between two variables. The antecedent variable comes before and is associated with an outcome; however, it may or may not be the cause of the outcome. For example, mean birth weight of minority babies is less than that of babies of the general population. Ethnicity is an antecedent of low birth weight, but not the cause. Other factors, e.g., nutrition, smoking status, and alcohol consumption, may be the causal agents.

Attached growth process: Treatment process wherein microbes are attached to media in a reactor, with the wastes to be treated flowing over the media. Examples include *trickling filter*, bio-tower, and rotating biological contactor (RBC).

Attenuation rate: The rate at which a contaminant is removed. This is not a rate constant but a rate, with typical units of $\mu g\, L^{-1}\, year^{-1}$.

Attenuation: The process by which a chemical compound's concentration decreases with time, through sorption, degradation, *dilution*, and/or *transformation*. The term applies to both destructive and non-destructive contaminant removal.

Attributable risk: The rate of a disease in exposed individuals that can be attributed to the *exposure*. This measure is derived by difference between the rate (usually *incidence* or *mortality*) of the disease among persons not exposed to the suspected agent and the corresponding rate among exposed individuals.

Autosome: Any chromosome other than the sex chromosomes or the mitochondrial chromosome.

Autotroph: Organisms feeding on inorganic minerals, producing complex organic compounds from simple inorganic molecules using energy by *photosynthesis* or by inorganic chemical reactions.

Average daily dose (ADD): Dose rate averaged over a pathway-specific period of exposure expressed as a daily dose on a per-unit-body-weight basis. The ADD is usually expressed as $mg\, kg^{-1}\, day^{-1}$ or other mass-time units.

Ayahuasca: Hallucinogen used by Amazonian indigenous people for religious rituals.

B-oxidation pathway: Major pathway of fatty acid oxidation to produce NADH, FADH2, and *acetyl coenzyme A*.

Bacillus: Rod-shaped bacterium.

***Bacillus thuringiensis* (Bt):** Bacterium that repels or kills insects; a major component of the microbial pesticide industry.

Bacteria: Unicellular microorganisms that exist either as free-living organisms or as parasites, ranging from beneficial to harmful to humans.

Bacteriophage: Virus (phage) that infects a bacterium.

Bacteriophage lambda (λ): Virus that infects *Escherichia coli*; often used as a genetic *vector* or *cloning* vehicle. Recombinant phages can be made in which certain non-essential λDNA is removed and replaced with the DNA of interest. The phage can accommodate a DNA insert of about 15-20 *kilobases*. Replication of that virus will thus replicate the investigated DNA.

Base composition: Proportion of the total bases of DNA or other molecule that consist of guanine plus either cytosine or thymine, plus base pairs. Usually expressed as G+C value, e.g. 59% G+C.

Batch culture: Microbial growth in which the logarithmically increasing phase is proportional to mass of the microbes.

Bayesian: Statistical approach, named after Thomas Bayes (*An Essay towards Solving a Problem in the Doctrine of Chances*, 1763), to decision making and inferential statistics that deals with probability inference (i.e., using knowledge of prior events to predict future events). In a Bayesian network, priors are updated by additional data that yield posterior probabilities that are often more robust than classical probabilities.

Bayesian belief network (BBN): Cause and effect tool represented by a probabilistic graphic model of a set of random variables and their conditional independencies.

Benefit–cost analysis (or cost–benefit analysis): Method designed to determine the feasibility or utility of a proposed or existing project. Yields a benefit–cost ratio. See *benefit–cost ratio (BCR)*.

Benefit–cost ratio (BCR): Weighted benefits divided by weighted costs; used to compare and differentiate among project alternatives. Gross BCR < 1 is undesirable. The greater the BCR, the more acceptable the alternative.

Benthic: Pertaining to the bottom of a body of water. Often used to distinction bottom organisms from those that swim and float.

Best available control technology: A limitation on an *emission* (including a visible emission standard), based on the maximum degree of reduction for each pollutant subject to regulation under the Clean Air Act, which would be emitted from any proposed major stationary source or major modification which the Administrator, on a case-by-case basis, taking into account energy, environmental, and economic impacts and other costs, determines is achievable for such source or modification through application of production processes or available methods, systems, and techniques, including fuel cleaning or treatment or innovative fuel combustion techniques for control of such pollutant.

Best management practice (BMP): Methods that have been determined to be the most effective, practical means of preventing or reducing pollution from nonpoint sources.

Best practice: 1. Optimal service to the client. 2. Treatment is appropriate, accepted, and widely used according to expert consensus; embodies an integrated, comprehensive, and continuously improving approach to care (medicine). It is morally obligatory that health care practitioners provide patients with the best practice (also known as standard therapy or standard of care).

Beta-proteobacteria: One of five subgroups of *proteobacteria*, each with distinctive 16S *rRNA* sequences. Similar to the *alpha-proteobacteria* metabolically, but tend to use substances that diffuse from organic matter decomposition in anaerobic zones.

Bias: 1. Systematic error in one direction; such as the positive bias of a scale that reads 1 mg too high (instrument error) or the negative bias in interpretations of lesions reported by a physician performing the procedure (operator bias) that consistently miss some lesions. Bias makes the reported values less accurate. 2. Any difference between the true value and that measured due to all causes other than sampling variability.

Binding site: 1. Location on cellular DNA to which other molecules and ions can bind. Typically, binding sites may be in the vicinity of genes and involved in activating *transcription* of that gene (i.e. promoter elements), in enhancing transcription of that gene (enhancer elements), or in reducing transcription of that gene (silencers). 2. Region on a protein where *ligands* can bind.

Bio-: Prefix indicating "life" (Greek).

Bioaccumulation: The process whereby certain substances build up in living tissues.

Bioaccumulation factor (BAF): Ratio of a tissue concentration of substance in an organism to its concentration in the environment (usually water) where the organism lives. BAF indicates a compound's potential to accumulate in tissue through exposure to both food and water. Compare to *biomagnification* and *bioconcentration factor.*

Bioactivation: Process by which an organism's metabolic or other endogenous processes increases the toxicity of a substance after uptake.

Bioassay: Test to assess the effects of certain substances on animals. Often used to estimate *acute toxicity.*

Bioaugmentation: Addition of beneficial microorganisms into groundwater to increase the rate and extent of *biodegradation.* Part of a *bioremediation* strategy.

Bioavailability: The degree to which a substance becomes available to the target tissue after administration or exposure.

Biocatalysis: Mediation of chemical reactions by biological systems, e.g. microbial communities, whole organisms or cells, cell-free extracts, or purified enzymes.

Biocentrism: View that all life has moral value. Contrast with *anthropocentrism*.

Biochemical oxygen demand (BOD): A standard test to assess wastewater pollution due to organic substances. BOD_5 is based on measurement of the oxygen used under controlled conditions of temperature (20 °C) and time (5 days). Compare to *chemical oxygen demand*.

Biochemodynamics: The physical, chemical, and biological processes that transport and transform substances.

Bioconcentration factor (BCF): Ratio of the concentration of substance in an organism's tissue versus its concentration in the environment (usually water) in situations where the organism is exposed exclusively to water. BCF measures the compound's potential to accumulate in tissue through direct uptake from water (excludes uptake from food). Compare to *bioaccumulation factor.*

Biocontrol: Managing pests by using other organisms or other natural agents, e.g. predators, parasites, and competition, rather than chemical controls.

Biodiversity: Number and variety of organisms in given system (e.g. wetland or forest). Usually, lower biodiversity indicates that the system is stressed and in poor condition.

Biodegradation: Breakdown of a contaminant, usually catalyzed by enzymes produced by organisms. Term usually applies only to microorganisms. Compare to *phytodegradation*.

Bioenergetics: Energy flow and transformation through living systems. Can be within an organism, e.g. cellular energetics, or within and between levels of *biological organization*, e.g. *trophic states* (energy transfer from producers to first-order consumers, to second-order consumers, to and from decomposers, etc.).

Bioengineering: See *biological engineering*.

Bioethics: 1. Inquiry into ethical implications of biological research and applications. 2. Ethical inquiry into matters of life, especially biomedical and environmental ethics.

Biofilm: Organized microbial system consisting of layers of microbial cells associated with surfaces, often with complex structural and functional characteristics. Influence microbial metabolic processes. Site where chemical degradation occurs both via extracellular enzymatic activity and by intracellular microbial processes. *Pseudomonas* and *Nitrosomonas* strains are notable for their ability to form a strong biofilm.

Biogenic: Made, caused, or influenced by natural processes. Contrast with *anthropogenic*.

Biogeochemical reductive dechlorination: Process that involves both biological and chemical reactions to effect the reduction of chlorinated solvents, such as trichloroethene and tetrachloroethene. Indigenous sulfate-reducing bacteria are stimulated through the addition of a labile organic and sulfate, if not already present at high concentrations. The stimulated bacteria produce reductants that react in conjunction with minerals in the aquifer matrix. Moreover, the reducing conditions necessary to produce such reactions most often are created as a result of microbial activity.

Biogeochemistry: 1. Study of the fluxes, cycles and other chemical and biological processes at various scales on earth. 2. Study of microbially mediated chemical transformations, especially with regard to nutrient cycling (N, P, S, and K, for example).

Bioinformatics: Management and analysis of data using advanced computing techniques applied to biological research and inquiry.

Biolistic gun (particle gun): Method used to modify genes by directly shooting genetic information into a cell. DNA is bound to tiny particles of gold or tungsten and subsequently inserted into tissue or single cells under high pressure. The accelerated particles penetrate both the cell wall and membranes, slowing down upon impact. The DNA separates from the metal and can be integrated into the genetic material inside the nucleus.

Biolitic: Formed by living organisms or their remains (e.g. sedimentary rocks).

Biological containment: Use of organisms that have reduced ability to survive or reproduce in the environment. Compare to *physical containment*.

Biological control: Method of addressing problematic organisms by using a biochemical product or bioengineered or naturally-occurring organism; e.g. introducing the European beetle (*Nanophyes marmoratus*) that feeds exclusively on the highly invasive purple loofestrife (*Lythrum salicaria*).

Biological criteria: Measures of the condition of an environment, e.g. incidence of cancer in benthic fish species.

Biological engineering: Combination of biomedical and biosystem engineering (see *biomedical engineering* and *biosystem engineering*) to develop useful biology-based technologies that can be applied across a wide spectrum of societal needs, including diagnosis, treatment, and prevention of disease, design and fabrication of materials, devices, and processes, and enhancement and sustainability of environmental quality.

Biological half-life: The time required for a biological system (such as a human or animal) to eliminate, by natural processes, half the amount of a substance (such as a radioactive material) that has been absorbed into that system.

Biological magnification: See *biomagnification*.

Biological organization: Levels of living things, from *biomolecules* to planetary. The levels generally representing biological systems aremolecule, *cell*, *tissue*, *organ*, organ system, *organism*, *population*, *community*, *ecosystem*, and *biosphere*.

Biological response: Manner and type of effect in an organism (e.g., disease, change in metabolism, and homeostasis).

Biologically based dose response (BBDR) model: Predictive model that describes biological processes at the cellular and molecular level linking the target organ dose to the adverse effect.

Biomagnification (biological magnification): Process whereby certain substances such as substances move up the *food chain*, work their way into rivers or lakes, and are eaten by aquatic organisms such as fish, which in turn are eaten by large birds, animals or humans. The substances become increasingly concentrated in tissues or internal organs as they move up the chain.

Biomarker: 1. Chemical, physical, or biological measurement that indicates biological condition. The biomarker may be a chemical to which an organism is exposed (e.g., lead in blood), a metabolite of the chemical (e.g., cotinine in blood as an indication of exposure to nicotine), or a biological response (e.g., an increase in body temperature as a result of exposure to a pathogen). 2. In geochemistry, organic compounds that are remnants of former living creatures (e.g. the suite of compounds that indicate the processes by which coal or petroleum has formed).

Biomass: Material produced by the growth of microorganisms.

Biomedical engineering: Application of engineering principles to medicine, including drug delivery systems, therapeutic systems, and medical devices.

Biomedical testing: Investigations to determine whether a change in a body function might have occurred because of exposure to a hazardous substance.

Biomethanation: See *methanogenesis*.

Biomolecule: Building block compounds of life that perform essential functions in living organisms, e.g. amino acids, carbohydrates, lipids, polysaccharides, proteins and nucleic acids.

Biomonitoring: 1. Measuring and investigating organisms as indications of environmental quality. 2. Collection and analysis of samples of tissue, fluid, and other components of humans to establish *biomarkers* from which exposure, effects, and risks in populations and groups of individuals can be estimated and inferred; e.g. measuring concentrations of lead from a sample of children to estimate lead exposure to children in a populations, or measuring a metabolite of an organic solvent to estimate a population's exposure to that solvent.

Biophile: An element, arranged in myriad ways, that provides the structure for all living systems; e.g. oxygen, carbon, hydrogen and nitrogen.

Bioprospecting: Search for novel products from organisms in the natural habitats, usually plants and microbes. In its negative connotation, known as biopiracy.

Bioreactor: Vessel, container or other system that uses microorganisms in attached or suspended biological systems to degrade contaminants. In suspended biological systems, such as activated sludge, fluidized beds, or sequencing batch reactors, contaminated water is circulated in an aeration basin where microbes aerobically degrade organic matter and produce carbon dioxide, water, and biomass. In attached systems, such as rotating biological contactors and trickling filters, a microbial population is established on an inert support matrix. The cells form a sludge, which is settled out in a clarifier and is recycled to the aeration basin and disposed of.

Bioremediation: Treatment processes that use microorganisms such as bacteria, yeast, or fungi to break down hazardous substances into less toxic or nontoxic substances. Bioremediation can be used to clean up contaminated soil and water. *In situ* bioremediation treats contaminated soil or groundwater in the location in which it is found. For *ex situ* bioremediation processes, contaminated soil is excavated or groundwater is pumped to the surface before they can be treated.

Bioscientist: One who studies and engages in the structure and behavior of living organisms.

Biosecurity: Measures to control the transmission of microorganisms into or out of a specified area or population, including *biological* and *physical containment*.

Biosensor: a portable device that uses living organisms, such as microbes, or parts and products of living organisms, such as enzymes, tissues, and antibodies, to produce reactions to specific chemical contaminants.

Bioslurping: The adaptation of vacuum-enhanced dewatering technologies to remediate hydrocarbon-contaminated sites. Bioslurping combines elements of both bioventing and free-product recovery to simultaneously recover free product and bioremediate soils in the vadose zone.

Biosolids: 1. See *sludge*. 2. Organic product of wastewater treatment that can be beneficially used.

Biosparging: *In situ remediation* that combines soil vapor extraction and *bioremediation*.

Biosphere: Earth's zone that includes *biota*, extending from ocean sediment to mountaintops.

Biostatistics: Application of statistical tools to interpret biological and medical data.

Biostimulation: Adding chemical amendments, such as nutrients or electron donors, to soil or groundwater to support bioremediation.

Biosystem: Living organism or a system of living organisms that are able to interact with other organisms directly or indirectly.

Biosystem or biosystems engineering (bioengineering): 1. Application of biological sciences to achieve practical ends. 2. Integration of physical, chemical, or mathematical sciences and engineering principles for the study of biology, medicine, behavior, or health to advance fundamental concepts, to create knowledge for the molecular to the organ systems levels, and to develop innovative biologics, materials, processes, implants, devices, and informatics approaches for the prevention, diagnosis, and treatment of disease, for patient rehabilitation, and for improving health.

Biota: 1. Any living creature, plant (flora), animal (fauna), or microbial. 2. Total of the living organisms of any designated area.

Biotechnologist: One who applies biological systems, especially living organisms, to address societal needs.

Biotechnology: Use of living creatures to produce things of value to humans (e.g., hazardous waste cleanup, production of drugs and improving agriculture and food supplies).

Bioterrorism: Use of living agents to cause intentional harm to people (e.g., anthrax spores or pathogenic viruses) and society (e.g., agricultural pests).

Biotic: Related to living systems.

Bio-tower: Attached culture system consisting of a tower filled with a medium similar to plastic rings in which air and water are forced up a counterflow movement in the tower.

Biotransformation: Biologically catalyzed transformation of a chemical to some other product.

Bio-uptake: 1. Process by which a compound enters an organism. 2. Amount of a substance that enters an organism. Often referred to simply as *uptake*.

Bioventing: An *in situ* remediation technology that stimulates the natural biodegradation of aerobically degradable compounds in soil by the injection of oxygen into the subsurface. Bioventing has been used to remediate releases of petroleum products, such as gasoline, jet fuels, kerosene, and diesel fuel. Bioventing stimulates the aerobic bioremediation of hydrocarbon-contaminated soils and vacuum-enhanced free-product recovery extracts light non-aqueous phase liquids (LNAPL) from the capillary fringe and the water table.

Biphasic reactor: Two-phase partitioning bioreactor.

Black body radiator: Idealized object that absorbs all electromagnetic radiation that reaches it. The earth behaves like a black body radiator when it absorbs incoming solar radiation (e.g. shortwave and visible light) and re-emits it at longer wavelengths (e.g. infrared heat). See *greenhouse effect.*

Blastocyst: Early stage embryo that consists of cells enclosing a fluid-filled cavity.

Blotting: Technique for detecting one RNA within a mixture of RNAs (a Northern blot) or one type of DNA within a mixture of DNAs (a Southern blot).

Body burden: Total amount of a specific substance in an organism, including the amount stored, the amount that is mobile, and the amount *absorbed.*

Bottom-up: View where fundamental components are first considered, working upward to larger perspectives. Contrast with *top-down*.

Broth: Liquid microbial growth medium.

Brownfield: Abandoned, idled, or under-used industrial and commercial site where expansion or redevelopment is complicated by real or perceived environmental contamination. Generally applied to such sites that have been or are expected to be re-used.

BTEX: Term used for benzene, toluene, ethylbenzene, and xylene, which are volatile aromatic compounds typically found in petroleum products such as gasoline and diesel fuel.

Bulk density: The mass of a soil per unit bulk volume of soil; mass is measured after all water has been extracted and the volume includes the volume of the soil particles and pores.

Butterfly effect: Sensitive dependence on initial conditions. Metaphor for the extreme sensitivity of chaotic systems (see *chaos theory*), in which small changes or perturbations lead to drastically different outcomes. The phrase is derived from a butterfly flapping its wings in California, and thereby initiating a change in weather patterns that results in the formation of a thunderstorm in Nebraska (from Edward Lorenz in his 1963 article "Deterministic Nonperiodic Flow," Journal of Atmospheric Sciences 20130–41, although in his presentation to the New York Academy, it was not a butterfly but a seagull's flapping of the wing that was posited as the initial condition; later, in 1972, Lorenz used the butterfly in the example).

Calvin cycle: Dominant pathway for the fixation (or reduction and incorporation) of CO_2 into organic material by *photoautotrophs* during *photosynthesis*. Also is found in *chemolithoautotrophs*.

Cancer: Disease of heritable, somatic mutations affecting cell growth and differentiation, characterized by an abnormal, uncontrolled growth of cells.

Capillary force: Interfacial force between immiscible fluid phases, resulting in pressure differences between the two phases. Force due to capillary action that "pulls" water and/or waterborne contaminants toward a substance that attracts them, leading to the production of thin trails of contamination and the incorporation of contamination into the inner windings of a soil particle.

Carbonaceous biochemical oxygen demand (CBOD): Measure of the dissolved oxygen used for biological oxidation of C-containing compounds in a sample. See *biochemical oxygen demand.*

Carcinogen: Physical, chemical or biological agent that induces cancer.

Carcinogenesis: Origin or production of a benign or malignant tumor. The carcinogenic event modifies the genome and/or other molecular control mechanisms of the target cells, giving rise to a population of altered cells.

CAS registration number: An organization that indexes information published in Chemical Abstracts by the American Chemical Society and provides index guides by which information about particular substances may be located in the Abstracts when needed. CAS numbers identify specific chemicals, so they can be important in queries about chemical toxicity and environmental characteristics.

Case study: Evaluation of an actual occurrence of events to describe specific environmental and health conditions and past exposures.

Case-control study: An epidemiologic study contrasting those with the disease of interest (cases) to those without the disease (controls). The groups are then compared with respect to exposure history, to ascertain whether they differ in the proportion exposed to the chemical(s) under investigation.

Catabolism: Metabolism in which larger, more complex molecules are broken down into smaller, simpler molecules with the release of energy. Compare to *anabolism*.

Catalyst: Substance that is not affected by a reaction by helps to initiate or accelerate it.

Categorical imperative: Central theme of Immanuel Kant's deontological ethics (see *deontology* or *deontological ethics*) that sets one principle from which all specific moral imperatives are derived:"Act only according to that maxim by which you can at the same time will that it should become a universal law" (*Groundwork of the Metaphysic of Morals* [*Grundlegung zur Metaphysik der Sitten*], 1785).

Cation: Positively charged ion.

Cation exchange capacity (CEC): Ability of soil, sediment or other solid matrix to exchange *cations* with a fluid. Very important measure of soil productivity and root behavior.

Causation (causality): Relationship between causes and effects. Contrast with association.

Cell: Basic unit of life; autonomous, self-replicating unit that either constitutes a unicellular organism or is a subunit of a multicellular organism; the lowest denomination of life.

Cell envelope: Cell *membrane* and cell wall configuration of a microbe that dictates its behavior in the environment. Two most common envelope architectures are *gram negative* and *gram positive.*

Central nervous system: Portion of the nervous system that consists of the brain and the spinal cord.

Chaos theory: Exposition of the apparent lack of order in a system that nonetheless obeys specific rules. Condition discovered by the physicist Henri Poincaré around the year 1900 that refers to an inherent lack of predictability in some physical systems (i.e., Poincaré's concept of dynamical instability).

Chemical oxygen demand (COD): Measure of the amount of oxygen required for the chemical oxidation of carbonaceous (organic) material in a waste, using inorganic dichromate or permanganate salts as oxidants in a 2-hour test. Compare to *biochemical oxygen demand.*

Chemisorption: Type of *adsorption* process wherein an adsorbate is held on the surface of an adsorbent by chemical bonds.

Chemolithoautotroph: Microbe that oxidizes reduced inorganic compounds to derive energy and electrons; major carbon source is CO_2.

Chemotroph: Organism that derives energy from inorganic reactions.

Chimera: Organism, usually animal, that is a mixture of cells from two different embryonic sources.

Chlorinated ethene: Chemical substances, such as trichloroethene and tetrachloroethene that have been used in industry as solvents.

Chlorinated solvent: Organic compounds with chlorine substituents that commonly are used for industrial degreasing and cleaning, dry cleaning, and other processes.

Chloromethanes: Chemical substances, such as carbon tetrachloride and chloroform that have been used in industry as solvents.

Chloroplast: Organelle containing chlorophyll, which carries out *photosynthesis* in plants and green algae.

Chromosome: Structure within a cell's nucleus consisting of strands of deoxyribonucleic acid (DNA) coated with specialized cell proteins, and duplicated at each mitotic cell division. Chromosomes transmit the genes of the organism from one generation to the next.

Chronic: Having a persistent, recurring or long-term nature. Contrast with *acute.*

Chronic effect: An adverse effect on a human or animal in which symptoms recur frequently or develop slowly over a long period of time.

Chronic exposure: Multiple exposures occurring over an extended period of time, or a significant fraction of the animal's or the individual's life-time.

Chronic toxicity: Capacity of a substance to cause long-term adverse effects (usually applied to humans and human populations).

Ciliate: Class of protozoans distinguished by short hairs on all or part of their bodies.

Cisgenesis: Process by which genes are artificially transferred between organisms that could be conventionally bred (i.e. closely related). Compare to *transgenesis.*

Clarification: Removal of suspended solids. Preferred terms are *sedimentation* or settling.

Clay: Soil particle <0.002 mm in diameter. Compare to *silt* and *sand.*

Cleanup: Actions taken to address a release or threat of release of a pollutant. Often used synonymously with *remediation*, but also can consist of pollutant removals and other corrective actions that do not necessarily require *degradation* and *detoxification*.

Clone: Line of cells genetically identical to the originating stem cell; group of genetically identical cells or organisms derived by asexual reproduction from a single parent. Act of generating these organisms is known as cloning.

Closure: Procedure following a *remediation* project or the useful life of a landfill, e.g. installing a permanent cap.

Coccus: Bacterial cell that is roughly spherical.

Code of ethics: Established set of moral expectations of a group, especially of professional societies.

Coefficient of determination (r2): Proportion of the variance of one variable predictable from another variable. The ratio of the explained variation to the total variation, which represents the percentage of the data nearest to the line of best fit. For example, if $r = 0.90$, then $r^2 = 0.81$, meaning that 81% of the total variation in one variable (y) can be explained by the linear relationship between the two variables (x and y) as described by the regression equation. Thus, the remaining 19% of the total variation is unexplained.

Coenzyme: Loosely bound non-protein component of an enzyme required for catalytic activity that often dissociates from the enzyme active site after product has been formed.

Coherence: Criterion for causality (i.e. *Hill's criteria*) based on the amount and degree of agreement among studies linking cause to effect; especially among various types of studies (e.g., animal testing, human epidemiological investigations, and *in vitro* studies).

Cohort study: Epidemiologic study comparing those with an exposure of interest to those without the exposure. These two cohorts are then followed over time to determine the differences in the rates of disease between the exposure subjects. Also called a prospective study.

Coliform: Gram-negative, non-sporing, facultative rod that ferments lactose with gas formation within 48 hours at 35°C.

Colloid: Fine solid (<0.002 mm and $>0.000\,001$ mm) that does not readily settle; intermediate between true solutions and suspensions.

Colony forming units (CFUs): Number of microorganisms that can form colonies when cultured using spread plates or pour plates; an indication of the number of viable microorganisms in a sample.

Combinational biology: Introduction of genes from one microorganism into another microorganism to synthesize a new product or a modified product, especially in relation to antibiotic synthesis.

Cometabolism: A reaction in which microorganisms transform a contaminant even though the contaminant cannot serve as an energy source for growth, requiring the presence of other compounds (primary substrates) to support growth.

Commensalism: Symbiosis where an organism lives on or within another organism with neither a positive nor negative effect on the other organism.

Comminutor: Shredding device to reduce the size of materials entering a waste treatment system.

Community: Assemblage of two or more biotic populations of different species that reside in the same spatial area.

Comparative risk: An expression of the risks associated with two (or more) actions leading to the same goal; may be expressed quantitatively (e.g. ratio of 1.5) or qualitatively (one risk greater than another risk). Any comparison among the risks of two or more hazards with respect to a common scale.

Compartmental model: 1. Model that predicts or characterizes the transport and fate of a compound within an organism (e.g. moving from blood to tissues, transformed by metabolism and detoxified or *bioactivated* during the path to elimination). 2. Similar model accounts for transport and fate, usually at a larger scale in the environment, e.g. the quantity and change of a chemical as it moves from the air to the water to the sediment to aquatic biota and within the *food chain*.

Compartmentalization: Viewing a system by its individual components. This can be problematic when an engineer does not consider the system as a whole (e.g., when the structural engineer and soil engineer do not collaborate on selecting the best and safest combination of materials and structures suited to a soil type, or when a biomedical engineer does not work closely with various specialized health care professionals in a clinical setting to adapt a realistic device to the comprehensive needs of the patient). Compartmentalization can be good when it allows the engineer to focus adequate attention on the components (see *bottom-up*), so long as the design is properly built into a system.

Competence: 1. Skill in practice. For professionals, competence is requisite to ethical practice. 2. Sufficient velocity for a fluid to carry a load (especially a stream's ability to carry solids).

Complementary DNA (cDNA): DNA copy of an RNA molecule.

Complexity: Relative measure of uncertainty in achieving functional requirements or objectives. Designers are frequently expected to reduce the complexity of engineered systems.

Compliance monitoring: Collection of data needed to evaluate the condition of the contaminated media against standards such as soil and or water quality regulatory standards, risk-based standards of remedial action objectives.

Composite sample: 1. Series of samples taken over a given period of time and weighted by flow rate or by other means to represent a concentration integrated with respect to time. 2. Soil sample that consists of soil taken from various depths or various locations.

Compost: Organic material produced from microbial degradation that is useful as soil conditioners and fertilizers. Process to produce such matter is known as composting.

Concentration: 1. Quantity of substance per unit volume (fluid) or per unit weight (solid matrix, e.g. soil, sediment, tissue). 2. Method of increasing the dissolved solids per unit volume of solution, e.g. via evaporation of the liquid. 3. Increasing suspended solids per unit volume of *sludge* via sedimentation or dewatering.

Conceptual site model (CSM): A hypothesis about how releases occurred, the current state of the source zone, and current plume characteristics (plume stability).

Cone of depression: Lowering of an aquifer's water table (or potentiometric surface) shaped like an inverted cone that develops around a vertical discharge well.

Confidence, confidence level: 1. Client's trust in a professional. 2. Amount of certainty that a statistical prediction is accurate. Physical sciences may differ from social sciences in what is considered acceptable confidence, e.g., the former may require 99% while social scientific research may consider 95% to be acceptable. Depending on the application, engineering research ranges in acceptable confidence level (e.g., structural fatigue research may require higher confidence levels than environmental research).

Confounder: Factor that distorts or masks the true effect of risk factors in an epidemiologic study. A condition or variable that is a risk factor for disease and is associated with an exposure of interest. This association between the exposure of interest and the confounder (a true risk factor for disease) may make it falsely appear that the exposure of interest is associated with disease. For example, a study of low birth weight children in low-income families must first address the confounding effects of tobacco smoking before ascribing the actual risk associated with income.

Confounding factor: Variable that may introduce differences between cases and controls, which do not reflect differences in the variables of primary interest. Factors that must be considered in *epidemiological* studies to ensure that the experimental variables are indeed the cause of an outcome (e.g. smoking can be a confounder in most cancer studies).

Conjugation: In *genetic engineering*, transferring genetic material between bacteria through direct cell-to-cell contact, or through a bridge between the two cells.

Consequentialism: Ethical theory with the perspective that the value of an action derives solely from the value of its consequences. Consequentialists hold that the consequences of a particular action form the basis for any valid moral judgment about that action, so that a morally right action is an action that produces good consequences. One of three major theories of normative ethics (see *normative ethics*), along with virtue ethics and deontological ethics (see *deontology* or *deontological ethics*).

Constitutive: Quality of enzyme meaning that is always synthesized and ready.

Contact stabilization: Enhanced *activated sludge* process by adding a period of contact between wastewater and *sludge* for rapid removal of soluble *biochemical oxygen demand* by adsorption, followed by a longer period of aeration in a separate tank so that the sludge is oxidized and new biosolids are synthesized.

Contamination: Contact with an admixture of an unnatural agent, with the implication that the amount is measurable. Increase in harmful or otherwise unwanted material in the environment. Often used synonymously with pollution.

Contingency plan: Document setting out an organized, planned, and coordinated course of action to be followed in case of an emergency or episodic event that threatens public health or the environment (e.g. an oil spill, toxic release or natural disaster).

Contingent probability: Probability that an event will occur as a result of one or more previous events. Also known as conditional probability.

Continuous culture: Microbial growth that is limited, and the effect of limiting the substrate or nutrients can be described by *Monod equation*.

Continuous sample: A flow of water from a particular place in a plant to the location where samples are collected for testing; may be used to obtain grab or composite samples.

Control group: Group used as the baseline for comparison in epidemiologic studies or laboratory studies. This group is selected because it either lacks the disease of interest (case control group) or lacks the exposure of concern (cohort study). Also known as a reference group.

Control volume: Arbitrary volume in which the mass of the fluid remains constant at steady state, so that as a fluid moves through, the mass entering the control volume is equal to the mass leaving the control volume.

Controlled liquid waste: Waste that meets the definition of a liquid waste and is in a container or piping system; a waste stream that can be shut off without a release to the environment.

Correlation coefficient (r): Statistical measurement of the strength and the direction of a linear relationship (association) between two variables.

Co-solvation: Process by which a substance is first dissolved in one solvent and then the new solution is mixed with another solvent.

Cost–benefit analysis: A formal quantitative procedure comparing costs and benefits of a proposed project or act under a set of preestablished rules. To determine a rank ordering of projects to maximize rate of return when available funds are unlimited, the quotient of benefits divided by costs is the appropriate form; to maximize absolute return given limited resources, benefits–costs is the appropriate form.

Criteria: Descriptive factors in setting standards for various pollutants. These factors are used to determine limits on allowable concentration levels, and to limit the number of violations per year (plural of criterion).

Critical micellar capacity (CMC): Likelihood of a substance to form *micelles* as a function of chemical structure, concentration, and other factors.

Critical path: Systems engineering of activities, decisions, and actions that must be completed on schedule and at a sufficient level of quality for the entire project to be successful.

Cross-resistance: Mutation of a microbe that has mutated in such a way that it loses its susceptibility to more than one antibiotic simultaneously, not just the one to which it has been directly exposed.

Cross-sectional study: Epidemiological study of observations representing a particular point in time. Contrast with *longitudinal study.*

Culture: Intentional organic growth.

Cyanobacteria: Large group of bacteria that carry out oxygenic *photosynthesis* using a system similar to that of photosynthetic *eukaryotes*.

Cyst: Specialized microbial cell enclosed in a wall; formed by protozoa and a few bacteria. They may be dormant, resistant structures formed in response to adverse conditions or reproductive cysts that are a normal stage in the life cycle.

Cytochrome: Heme protein that carry electrons, usually as members of electron transport chains.

Cytochrome P450 (CYP): Enzymes that use iron to oxidize substances, often as part of the body's strategy to dispose of potentially harmful substances by making them more water-soluble. Varying versions of CYP are used to identify different enzymatic activities, so are important in *toxicodynamics* and *toxicokinetics modeling.*

Cytokine: Nonantibody protein released by a cell in response to inducing stimuli, which are mediators that influence other cells. Produced by lymphocytes, monocytes, macrophages, and other cells.

Dalton's law: Total pressure exerted by a mixture of gases is equal to the sum of the pressures that would be exerted if each of the individual gases were to occupy the same volume by itself.

Darcy's law: An empirically derived equation for the flow of fluids through porous media; based on assumptions that flow is laminar and inertia can be neglected. States that the specific discharge, q, is directly proportional to the hydraulic conductivity and the hydraulic gradient.

Dark field: Microscope's optical system that makes small, clear and colorless particles (e.g. many microbes) visible, by illuminating the object at an angle such that no light enters the microscope system except that which is diffracted by particles.

Data: Plural of datum. Gathered facts from which conclusions can be drawn.

Decision tree: Diagram indicating various steps to different outcomes. Supports the optimal course of action in situations where several possible alternatives have uncertain outcomes.

Declining growth phase: Period of time in microbial population dynamics between the *log growth phase* and the *endogenous phase,* where the amount of food is in short supply, leading to incrementally slowing growth rates.

Decomposer: Organism that degrades complex materials into simpler ones.

Deductive reasoning: A conclusion is necessitated by previously known facts. If the premises are true, the conclusion must be true. Starting from general knowledge and moving to specifics (e.g., from cause to effects). Contrast with inductive reasoning.

Deep ecology: Environmental movement initiated in 1972 by Norwegian philosopher Arnie Naess, that advocates radical measures to protect the natural environment irrespective of their effect on the welfare of humans (opposite of *anthropocentrism*).

Degradation: Process of breaking down larger molecules into smaller molecules. See *biodegradation.*

Dehydrohalogenation: A process by which a halogenated alkane loses a halogen from one carbon atom and a hydrogen from the adjacent carbon atom, producing the alkene and an acid (e.g., 1,1,2,2-tetrachloroethane dehydrohalogenates to produce trichloroethene and HCl).

Demand: Quantity of a good or service that society chooses to buy at a given price.

Denitrification: Reduction of nitrate to gas products, primarily nitrogen gas (N_2), during anaerobic respiration.

Dense, non-aqueous-phase liquid (DNAPL): An immiscible organic liquid that is denser than water (e.g., tetrachloroethene).

Density: Mass per unit volume.

Deontology or deontological ethics: Ethical theory basing right and wrong on duty (Greek: *deon* meaning obligation).

Deoxygenation constant: Expression of the rate of the biochemical oxidation of organic matter under aerobic conditions. Value depends on the time unit involved (often 1 day) and varies with temperature and other environmental conditions.

Deoxyribonucleic acid (DNA): Double-stranded nucleic acid containing genetic information; polynucleotide composed of deoxyribonucleotides connected by phosphodiester bonds.

Depuration: Cleansing of a previously dosed organism by ending the dosing completely.

Dermal exposure: Contact between a chemical and the skin.

Dermal toxicity: The ability of a pesticide or toxic chemical to poison people or animals by contact with the skin.

Descriptive epidemiology: Study of the amount and distribution of a disease in a specified population by person, place, and time. Compare to *analytic epidemiology.*

Desiccation: 1. Removal of moisture; e.g. drying of a soil sample before analysis. 2. Loss of water, dehydration.

Design integrity: Quality of a design's function, performance, material quality and other metrics are accurately documented by its requirements, design, and support specifications.

Desorption: The converse of sorption, i.e., when a compound slowly releases from a surface(s) that it has previously accumulated upon or within.

Destruction and removal efficiency (DRE): Percentage of compound removed or destroyed by a process, usually thermal.

Detection limit: See *limit of detection.*

Detoxification: Process of making a substance less toxic. For example, removing chlorine atoms can render a molecule less toxic (e.g. less carcinogenic). Contrast with *bioactivation.*

Detritus: Dead biota matter, usually at varying degrees of *decomposition* by microbes.

Developmental toxicity: Adverse effects on developing organism that may result from exposure prior to conception (either parent), during prenatal development, or postnatally until the time of sexual maturation. The major manifestations of developmental toxicity include death of the developing organism, structural abnormality, altered growth, and functional deficiency.

Device (medical): Diagnostic or therapeutic instrument, apparatus, implement, machine, contrivance, implant, *in vitro* reagent, or other similar or related article, including any component, part, or accessory, that is intended for use in the diagnosis of disease or other conditions, or in the cure, mitigation, treatment, or prevention of disease, in humans or animals, or intended to affect the structure of any function of the body and that does not achieve its primary intended purpose through chemical action and that is not depended upon being metabolized for the achievement of its primary intended purposes. The last sentence of this definition helps to distinguish a device from a drug. Both are regulated, but differently, by the US Food and Drug Administration.

Dewater: 1. Remove or separate a portion of the water in a sludge or slurry to dry the sludge so it can be handled and disposed. 2. Remove or drain the water from a tank or trench.

Diatom: Algal protist with siliceous cell wall (frustule); constitute a substantial group of phytoplankton.

Dichlorodiphenyltrichloroethane (DDT): Organochlorine pesticide, banned in many parts of the world due to associations with eggshell thinning, endocrine effects and human effects. Still used to control mosquitoes and other disease *vectors*.

Diffused aeration: Injection of air through submerged porous plates, perforated pipes, or other devices to form small air bubbles from which oxygen is transferred to the liquid as the bubbles rise to the water surface.

Diffusion: 1. Process of net transport of solute molecules from a region of high concentration to region of low concentration caused by their molecular motion and not by turbulent mixing. Graham's law of diffusion states that a gas diffuses at a rate inversely proportional to its density. Liquids also diffuse as a result of net spontaneous and random movement of molecules or particles from a region in which they are at a high concentration to a region of lower concentration. Diffusion will continue for a fluid until a uniform concentration is achieved throughout the region of the system. Diffusion is not a major mechanism of mass transport in rapidly flowing systems, such as air and surface waters, but is quite important in more quiescent systems, such as across cellular *membranes*, and in slow moving regions of the environment, e.g. covered *sediment* and groundwater. 2. Synonym for *dispersion*.

Digester: 1. System where *biosolids* are decomposed by microbes. 2. Tank where such decomposition occurs.

Digestion: 1. In environmental biotechnology, the process of decomposing organic matter by microbial growth and metabolism. As such, organic matter is transformed and transferred to *sludge*, resulting in partial liquefaction, mineralization, and volume reduction. 2. Actions that occur in a *digester*.

Dilution: A reduction in solute concentration caused by mixing with water at a lower solute concentration.

Dimer: Molecule that consists of two identical simpler molecules e.g. NO_2 can form the molecule NO_2–O_2N or simply N_2O_4 that consists of two identical simpler NO_2 molecules.

Dinoflagellate: Algal protist characterized by two flagella used in swimming in a spinning pattern; many are bioluminescent and an important group of marine phytoplankton. A few species are important marine pathogens.

Dioxin: Highly *toxic*, *recalcitrant*, and *bioaccumulating* product of incomplete combustion and chlorination processes with a structure of two phenyl rings bonded by two oxygen atoms, with chlorine substitution. Most toxic form is 2,3,7,8-*tetrachlorodibenzo-para-dioxin*.

Diploid: Cell with normal amount of DNA per cell; i.e. two sets of chromosomes or twice the *haploid* number.

Direct filtration: A method of treating water that consists of the addition of coagulant chemicals, flash mixing, coagulation, minimal flocculation, and filtration. Sedimentation is not used.

Direct runoff: Water that flows over the ground surface or through the ground directly into streams, rivers, and lakes.

Disaster: A relative term meaning a catastrophic event that wreaks great destruction. However, the term is not exclusive to large-scale events, such as hurricanes or earthquakes, but can also include small-scale events with highly negative consequences, such as an engineering or medical failure where one or a few people are impacted but that has other implications (malpractice, bad publicity, blame, etc.).

Discharge: 1. Flow (Q) in a stream or canal or the outflow of a fluid from a source. Used in calculating liquid effluent from a facility or particulate or gaseous emissions into the air through designated venting mechanisms. 2. Any flow in an open or closed conveyance.

Disease: Abnormal and adverse condition in an organism.

Disparate effect: Health outcome, usually negative, that is disproportionately high in certain members of a population, such as an increased incidence of certain cancers in minority groups.

Disparate exposure: *Exposure* to a physical, chemical, or biological agent that is disproportionately high in certain members in a population, such as the higher than average exposure of minority children to lead.

Disparate susceptibility: Elevated risk of certain members of a population (e.g., genetically predisposed) to the effects of a physical, chemical, or biological agent; can lead to disparate effects. See *disparate effect*.

Dispersion: The spreading of a solute from the expected groundwater flow path as a result of mixing of groundwater.

Dispersion model: Prediction tool of how a substance will behave after release.

Dissolution: Act of going into solution; dissolving.

Dissolved oxygen (DO): Concentration of molecular O_2 in water.

DNA ligase: Enzyme that joins two DNA fragments together through the formation of a new phosphodiester bond.

DNA marker: Cloned chromosomal locus with allelic variation that can be followed directly by a DNA-based assay such as Southern blotting or *PCR*.

Dose: Amount of a substance available for interactions with metabolic processes or biologically significant receptors after crossing the outer boundary of an organism. Potential dose is the amount ingested, inhaled, or applied to the skin. Applied dose is the amount presented to an absorption barrier and available for absorption (although not necessarily having yet crossed the outer boundary of the organism). Absorbed dose is the amount crossing a specific absorption barrier (e.g., the exchange boundaries of the skin, lung, and digestive tract) through uptake processes. Internal dose is a more general term denoting the amount absorbed without respect to specific absorption barriers or exchange boundaries. The amount of the chemical available for interaction by any particular organ or cell is termed the delivered or biologically effective dose for that organ or cell.

Dose-effect: The relationship between dose (usually an estimate of dose) and the gradation of the effect in a population, that is a biological change measured on a graded scale of severity, although at other times one may only be able to describe a qualitative effect that occurs within some range of exposure levels.

Dose-response: 1. Relationship between a quantified exposure (dose) and the proportion of subjects demonstrating specific biologically significant changes in incidence and/or in degree of change (response). 2. Correlation between a quantified exposure (*dose*) and the proportion of a population that demonstrates a specific *effect* (response).

Dose-response assessment: Process of characterizing the relationship between dose of an agent and the effect elicited by that dose.

Double effect: Doctrine or ethical principle stating that an otherwise immoral act is acceptable provided a proportional good effect will accrue, so long as the immorality is not intended and that humans are not used as objects. For example, a vaccine may be morally acceptable, even if 500 people in a population die, so long as the benefits of the vaccine cannot be gained in a way in which fewer people would die and the 500 people (or any number, for that matter) are not used to achieve the good result (e.g. humans morally cannot be "harvested" for the greater good; rather the 500 people were not expected to die when they received the vaccine).

Double resistance theory: See *two-film model.*

Doubling time: Time required for a population, e.g. bacterial, or cells to double in number or biomass.

Downgradient: The direction that groundwater flows; analogous to "downstream" for surface waters.

Drawdown: Lowering of water table of an unconfined aquifer or the potentiometric surface of a confined aquifer caused by pumping of groundwater from wells. Vertical distance between the original water level and the new water level. See *cone of depression.*

Drug: Substance intended for use in the diagnosis, cure, mitigation, treatment, or prevention of disease, which is regulated by the US Food and Drug Administration. Contrast with device and nutritional supplement.

Dry lab: 1. *In silico* research (contrast with wet lab). 2. Walkthrough prior to actual laboratory work (step preceding wet lab). 3. Unethical practice of forging (making up) data.

Dual use: 1. Science, engineering, and technology designed to provide both military and civilian benefits. 2. Research and technology that simultaneously benefit and place society at risk (e.g., biotechnological advances that improve vaccines but also increase the risks of bioterrorism).

Dynamical instability: See *chaos theory.*

Ecocentrism: Perspective based on the whole ecosystem rather than a single species. Contrast with *anthropocentrism.*

Ecological community: See *community.*

Ecological impact: Total effect of an environmental change, natural or of human origin, on the community of living things.

Ecological indicator: A characteristic of the environment that, when measured, quantifies magnitude of stress, habitat characteristics, degree of exposure to a stressor, or ecological response to exposure. The term is a collective term for response, exposure, habitat, and stressor indicators.

Ecological risk assessment: The application of a formal framework, analytical process, or model to estimate effects of human action(s) on a natural resource and to interpret the significance of those effects in light of the uncertainties identified in each component of the assessment process.

Ecology: Science dealing with the relationship of all living things with each other and with their environment.

Ecosystem: The interacting system of a biological community and its nonliving surroundings.

Ecosystem function: Processes and interactions that operate within an ecosystem, including energy flow, nutrient cycling, filtering and buffering of contaminants, and regulation of populations.

Ecosystem service: Benefit to humans derived from ecosystems. *Anthropocentric* and *instrumental* value provided by natural ecosystems.

Ecosystem structure: Attributes related to instantaneous physical state of an ecosystem; examples include species population density, species richness or evenness, and standing crop biomass.

Ecotone: A habitat created by the juxtaposition of distinctly different habitats; an edge habitat; or an ecological zone or boundary where two or more ecosystems meet.

Effect: A biological change caused by an exposure.

Effectiveness: Measure of the extent and degree to which a design achieves a goal (compare to *efficacy* and *efficiency*).

Efficacy: A measure of the probability and intensity of beneficial effects.

Efficiency: Ratio of total energy or mass output to total energy or mass input, expressed as a percentage. Treatment or removal efficiency is the product of the contaminant mass prior to treatment (I) times the contaminant mass after treatment (E) divided by I. To express efficiency as a percentage, these values are multiplied 100 times:

$$\frac{I \times E}{I} \times 100$$

Effluent: Waste material discharged into the environment, treated or untreated. Generally refers to surface water pollution (analogous to *emission* in air pollution).

Effusion: Escape of fluid into a body space or tissue. Effusion of a gas is inversely proportional to the square root of either the density or molecular weight of the gas. Compare to *diffusion*.

Electron: A negatively charged subatomic particle that may be transferred between chemical species in chemical reactions.

Electron acceptor: 1. Oxidant. 2. Chemical substance, such as oxygen, nitrate, sulfate, and iron, which receives the electrons during microbial and chemical reactions. Microorganisms need these compounds to obtain energy. For MNA and EA, these electron acceptors often compete with chlorinated solvents and reduce the attenuation rates.

Electron donor: 1. Reductant. 2. Chemical substance, such as molecular hydrogen or organic substrate, which yields an electron as it is oxidized, producing energy to sustain life and for the subsequent degradation of other chemicals, in this case, chlorinated solvents.

Electron-transport chain: Final steps of reactions that occur in biological oxidation; composed of series of oxidizing agent (i.e. *electron acceptors*) arranged in sequence by increasing strength and terminating with oxygen (the strongest oxidizer).

Electrophoresis: See *gel electrophoresis.*

Embden–Meyerhof–Parnas (EMP) pathway: Biochemical pathway that degrades glucose to pyruvate; the six-carbon stage converts glucose to fructose-1,6-bisphosphate, and the three-carbon stage produces ATP while changing glyceraldehyde-3-phosphate to pyruvate. Also called Embden–Meyerhof pathway and glycolytic pathway.

Emission: Release of a pollutant or other substance to the atmosphere (analogous to *effluent* in water pollution).

Endergonic: Describing a reaction that does not spontaneously go to completion as written; the standard free energy change is positive, and the equilibrium constant is less than one.

Endocrine disruptor: Exogenous chemical compound that mimics or disrupts hormones in the endocrine system. Synonymous with *hormonally active agent.*

Endocrine system: Chemical messaging system in organisms used for regulation by secretion of hormones by glands that are sent through the circulatory system to cells where the hormones bind to receptors.

Endogenous phase: Microbial population growth period dominated by endogenous respiration.

Endoplasmic reticulum: Organelle consisting of a network of membranes within the cytoplasm of cells, where proteins and lipids are synthesized.

Endospore: Seed-like structure; formed by a microbe to survive during hostile conditions.

Endothermic: Requires energy (e.g. an endothermic reaction).

Endotoxin: Heat-stable lipopolysaccharide in the outer membrane of the cell wall of gram-negative bacteria that is released when the bacterium lyses, or during growth, and is toxic to the host.

Endpoint: Observable or measurable biological event or chemical concentration (e.g., metabolite concentration in a target tissue) used as an index of an effect of a chemical.

Engineering: 1. Application of scientific and mathematical principles to practical ends, especially design, manufacture, and operation of structures, machines, processes, and systems. 2. The profession that implements these applications.

Enhanced attenuation: Any type of intervention that might be implemented in a source-plume system to increase the magnitude of attenuation by natural processes beyond that which occurs without intervention. Enhanced attenuation is the result of applying an enhancement that manipulates a *natural attenuation* process, leading to an increased reduction in mass of contaminants.

Enhanced bioremediation: An engineered approach to increasing biodegradation rates in the subsurface.

Enteric bacteria: 1. Members of the family Enterobacteriaceae, (i.e. gram-negative, peritrichous or nonmotile, facultatively anaerobic, straight rods with simple nutritional requirements). 2. Bacteria that live in the intestinal tract.

Environmental assessment (EA): Investigation whether a proposed action will adversely affect the environment. If so, in the United States, such action is usually followed by a formal *environmental impact statement.* If not, the agency will issue a "finding of no significant impacts" document.

Environmental engineering: Subdiscipline of engineering (usually civil engineering) concerned with applications of scientific principles and mathematics to improve the condition of the environment.

Environmental impact statement (EIS): Document prepared by a government agency detailing the potential effects resulting from a major action being considered by that agency. In the United States, the EIS is required under the National Environmental Policy Act, and is usually preceded by an *environmental assessment.*

Environmental justice: Concern for the fair treatment of all people in environmental decisions.

Environmental science: Systematic study of the environment and its components and processes (e.g., nutrient cycling, pollutant transport, and adverse effects).

Environmentalism: 1. Advocacy in the protection of the environment. 2. Philosophy underpinning this advocacy. Such advocacy may or may not be scientifically based (i.e., differs from environmental science and environmental engineering).

Enzyme: Protein catalyst with specificity for both the reaction catalyzed and its substrates.

Enzyme-linked immunosorbent assay (ELISA): A technique used for detecting and quantifying specific antibodies and antigens.

Epidemic: Disease outbreak that occurs simultaneously or nearly simultaneously in a large area or in large percentage of a population.

Epidemiology: 1. Study of the causes, distribution, and control of disease in populations. 2. The study of the distribution and determinants of health-related states or events in specified populations.

Epigenetics: Concern with mechanisms that regulate gene activity.

Episome: Plasmid that can exist either independently of the host cell's chromosome or be integrated into it.

Epistasis: Modification of properties of one gene by one or more genes at other loci. Also known as a genetic interaction, but epistasis refers to the statistical properties of the phenomenon.

Equilibrium: Condition in which a reaction is occurring at equal rates in its forward and reverse directions, so that the concentrations of the reacting substances do not change with time.

Equilibrium constant: Value representing relationship between a compound in a system that has reached equilibrium. Also known as *partition coefficient*, e.g. *octanol–water coefficient, bioconcentration factor* and *Henry's law constant.*

Equilibrium vapor pressure: Pressure of a vapor in thermodynamic equilibrium with its condensed phases in a closed container.

Error: 1. Mistake. 2. In statistics, the difference between a reported value and the actual value. See *bias*.

Estimated exposure dose (EED): The measured or calculated dose to which humans are likely to be exposed considering all sources and routes of exposure.

Ethics: 1. Set of moral principles. 2. Study of morality and moral decision making.

Ethylenediaminetetraacetic acid (EDTA): Chelating agent that binds to and makes unavailable metal ions in a solution; because certain cations are essential for many enzymes to function, EDTA is applied to halt enzymatic and cellular activity (as such, is a common preservative).

Eukaryote: Organism whose cell contains a distinct, membrane-bound nucleus.

Eutrophication: Process by which water bodies receive excess nutrients, primarily nitrogen and phosphorus, which stimulate excessive algal and plant growth.

Event: Set of outcomes that are preceded and linked to an earlier set of outcomes (probability theory).

Event tree: Diagram of the flow of events following an initial event, showing subsequent possible events toward different outcomes. Each event has its own possible outcomes, so that the critical path chosen will result in numerous potential outcomes.

***Ex situ*:** Moved off-site (e.g., contaminated soil transported to an incinerator for treatment).

***Ex vivo*:** Outside the body, frequently the equivalent of *in vitro* (see *in vitro*).

Exergonic: Describing a reaction that spontaneously goes to completion as written; the standard free energy change is negative, and the equilibrium constant is greater than one.

Exothermic: Liberates energy (e.g. an exothermic reaction).

Exotoxin: Heat-labile, toxic protein produced by a bacterium as a result of its normal metabolism or because of the acquisition of a plasmid or prophage that redirects its metabolism; usually released into the bacterium's surroundings.

Experiment: Investigation to support or reject a hypothesis or to increase knowledge about a phenomenon.

Exponential growth phase: Microbial growth during a period of cell doubling, e.g., at a constant percentage per unit time; this occurs during the *logarithmic growth phase.*

Exposure assessment: Identification and evaluation of the human population exposed to a toxic agent, describing its composition and size, as well as the type, magnitude, frequency, route and duration of exposure.

Exposure: Contact made between a chemical, physical, or biological agent and the outer boundary of an organism. Exposure is quantified as the amount of an agent available at the exchange boundaries of the organism (e.g., skin, lungs, gut).

Exposure scenario: Set of facts assumptions, and inferences regarding how exposure occurs to support risk assessors in evaluating, estimating, or quantifying exposures.

Expression (genetic expression): Effect on cell resulting from the gene's instructions in transcription.

Extended aeration: Enhancement of activated-sludge process using long aeration periods to promote aerobic digestion of the biological mass by endogenous respiration; includes stabilization of organic matter under aerobic conditions and disposal of the gaseous end products into the air. Effluent contains both dissolved and fine, suspended matter.

Extrapolation: 1. Estimate of the extent of conditions from measured data. 2. Estimate of the response at a point below the range of the experimental data, generally through the use of a mathematical model. Compare to *interpolation*.

Extremophile: Microbe that grows under harsh environmental conditions, e.g. very high temperatures or extreme pH values.

Fact: That which can be shown to be true, to exist, or to have occurred.

Facultative anaerobe: Microorganism, usually a bacterium, which grows equally well under *aerobic* and *anaerobic* conditions.

Facultative pond: Most common type of treatment pond in current use. Upper portion (*supernatant*) is *aerobic* and the bottom layer is *anaerobic*. *Algae* supply most of the oxygen to the supernatant.

Failure: Lack of success as indicated by design specifications and measures of success.

Failure rate [$f(t)$]: See *hazard rate.*

False negative: Finding of the absence of a condition (e.g., disease) in a test when, in fact, the disease is present (e.g., a lung cancer screen shows that the patient has no cancer, but at a level of detection below the screen, the cancer has cancer cells in the lung). See *type II error.*

False positive: Positive finding of a test when, in fact, the true result was negative (e.g., a drug screen shows that a person has used opiates, even though the person has not). See *type I error.*

Fatty acid: Class of aliphatic monocarboxylic acids forming part of a lipid molecule that can be derived from fat by hydrolysis; simple molecules formed around a series of carbon atoms linked together in 12 to 22 carbon atom chains.

Fault tree analysis (FTA): Failure analysis in which an undesired state of a system is analyzed by combining a series of lower-level events. Mainly used in the field of safety engineering to find probability of a safety hazard.

Fauna: All animal life in a specific geographic region. See *biota*.

Fecal coliform: Coliform with the intestinal tract as its normal habitat and that can grow at 44.5 °C.

Feedstock: Material entering a *reactor.*

Fermentation: Energy-yielding process in *microbes* oxidize an energy substrate without an exogenous electron acceptor. Usually organic molecules serve as both electron donors and acceptors.

Fimbria: Hair-like protein appendage on certain gram-negative bacteria for adhesion to surfaces.

Finding of No Significant Impact (FONSI): Statement by an agency after completing an *environmental assessment* that a proposed action will not lead to significant impacts, so an *environmental impact statement* would not be required for this action.

Fishbone diagram: See *Ishikawa diagram.*

Fixed phase reactor: System in which solid-phase particles are fixed in position as the fluid phase passes through.

Flagellum: Threadlike appendage on numerous prokaryotic and eukaryotic cells responsible for their motility.

Flare: Device that combusts gaseous materials exiting a system, e.g. a landfill flare burns methane and a chemical processing facility includes a flare backup system in the event of unplanned releases of otherwise toxic substances.

Floc: Particles, including cells, which adhere to one another loosely to form clusters.

Flocculation: Process by which *flocs* are formed; often enhanced in water treatment processes by addition of flocculant, e.g. alum – aluminum sulfate [$Al_2(SO_4)_3$].

Flora: All plant life in a specific geographic region. See *biota.*

Flow cytometer: Instrument with a laser detector and a very small orifice through which particles (including microbes) flow through one at a time. As they pass through a laser beam, biochemicals may be determined on a per cell basis.

Fluidized bed reactor: System that suspends small solid particles in an upwardly flowing stream of fluid. Fluid velocity must be sufficient to suspend the particles, but not so high as to transport them out of the vessel. Mixing occurs as the solid particles swirl around the bed rapidly. The material fluidized is nearly always solid and the fluidizing medium can be either a liquid or a gas.

Flux: Rate of flow of fluid, particles, or energy through a given two-dimensional surface.

Food chain: A biological system in which individuals in a trophic level feed on organisms in the trophic level below theirs. As such, energy is transferred from level to level.

Food web: Complex of interrelated *food chains* in an ecological community.

Forcing: Effectiveness of a gas to warm or cool the atmosphere.

Free energy (ΔG): Intrinsic energy in a substance available to do work, especially to drive chemical transformations. Also known as Gibbs free energy.

Freundlich sorption isotherm: See *sorption isotherm.*

Fugacity: Tendency of a substance to prefer one phase over another, and tendency to flee or escape one compartment (e.g. water) to join another (e.g. atmosphere).

Fugacity capacity constant (Z): Term in the *fugacity* equation which relates fugacity (f) to concentration (C); i.e. $C = Zf$.

Fungus: Achlorophyllous, heterotrophic, spore-bearing eukaryotes with absorptive nutrition; often with a walled thallus.

Fuzzy logic: System dealing with the partial truths, assigning with values ranging from completely true to completely false.

Game theory: Decision making under conditions of uncertainty and interdependence; taking into account the characteristics of players, strategies, actions, payoffs, outcomes at equilibrium.

Gamete: Reproductive cell (i.e., an egg or a sperm).

Gamma-proteobacteria: One of the five subgroups of proteobacteria, each with distinctive 16S *rRNA* sequences. This is the largest subgroup and is very diverse physiologically; many important genera are facultatively anaerobic chemoorganotrophs.

GANTT chart: Graphical display of a project or program showing each task as a horizontal bar with its length being proportional to time needed for completion.

Gas-side impedance: In two-film theory, difference in a substance's partial pressure in the bulk gas and at the interface with liquid phase.

Gel: Inert polymer, often agarose or polyacrylamide, that separates macromolecules, e.g. nucleic acids or proteins, during electrophoresis.

Gel electrophoresis: Method of analyzing size of DNA and RNA fragments, based on the speed of movement, i.e. when exposed to electric field, larger fragments of DNA and RNA more slowly through a gel than smaller fragments.

Gel shift: Method for detecting the interaction of a nucleic acid (DNA or RNA) with a protein; if the protein binds to the nucleic acid, the complex migrates more slowly in the gel.

Gene: Ordered sequence of nucleotides located in a particular position on a particular chromosome, representing the fundamental unit of heredity. DNA segment or sequence that codes for a polypeptide, *rRNA*, or *tRNA*.

Gene flow: Movement of genes from one population to another.

Gene silencing: Inactivation of genes without changing the corresponding DNA sequence.

Gene targeting: Method to allow site-specific genomic modifications by precisely switching defined genes on or off.

Generation time: See *doubling time.*

Genetic engineering (GE): Modification of the structure of genetic material in a living organism; involving the production and use of *recombinant DNA*.

Genetic material: *Deoxyribonucleic acid* and *ribonucleic acid.*

Genetically modified microbe (GMM): Subdivision of *genetically modified organisms* that includes bacteria, fungi, algae and other microbes whose genetic material has been altered.

Genetically modified microbial pesticide: Bacteria, fungi, viruses, protozoa, or algae, whose DNA has been modified to express pesticidal properties. The modified microorganism generally performs as a pesticide's active ingredient. For example, certain fungi can control the growth of specific types of weeds, while other types of fungi can kill certain insects.

Genetically modified organism (GMO): Organism whose genetic material has been changed in a way that does not occur under natural conditions through cross-breeding or natural recombination.

Genetics: Scientific investigation of heredity.

Genome: Entire genetic complement, i.e. all of the hereditary material possessed by an organism.

Genomics: 1. Study of genes, including their functions. 2. Study of the molecular organization of genomes, their information content, and the gene products they encode.

Genotype: Combination of alleles, situated on corresponding chromosomes, that determines a specific trait of an individual.

Geographic information system (GIS): Mapping system that uses computers to collect, store, manipulate, analyze, and display data. For example, GIS can show the concentration of a contaminant in an ecosystem with respect to land cover, water depth, and potential sources of the contaminant.

Germ theory: Paradigm that diseases are caused by singular, proximate, pathogenic microbes. Displaced miasma theory (see *miasma*) in late nineteenth century.

Gestalt theory (Germanform): View that perception and other psychological phenomena must be understood for their overall patterns and forms, as opposed to the individual components.

Gestation: Time from fertilization of the ovum to birth. Also known as uterogestation.

Gibbs free energy: See *free energy.*

Glycocalyx: Extracellular layer in bacteria composed primarily of polysaccharide but which can contain proteins and even nucleic acids. Also known as slime layer (if diffuse and irregular) or capsule (if more defined and distinct). Resulting sticky layer protects against *desiccation*, predation, phagocytosis, and chemical toxicity (e.g. from antimicrobials), and acts as a means of attachment to surfaces. Thus, glycocalyx producing bacteria, e.g. *Pseudomonas* spp., are often found associated with microbial mats and *biofilms* in *trickling filters* and other attached media used to treat wastes.

Glycogen: Highly branched polysaccharide containing glucose, which is used to store carbon and energy.

Glycolysis: Anaerobic conversion of glucose to lactic acid via the *Embden–Meyerhof–Parnas pathway.*

Gycolytic pathway: See *Embden–Meyerhof–Parnas pathway.*

Glyoxylate cycle: Modification of Kreb's cycle in certain bacteria wherein *acetyl coenzyme A* is generated directly by oxidization of light lipids (e.g. *fatty acids*).

Grab sample: Single sample of environmental material collected without regard to flow or time.

Gram negative: Describing a bacterial cell that retains crystal violet during staining processes, and is then colored by a counterstain, e.g. *Thiobacillus* and *Pseudomonas.*

Gram positive: Describing a bacterial cell that loses crystal violet during staining processes, e.g. *Bacillus.*

Gram stain: Differential staining procedure that divides bacteria into gram-positive and gram-negative groups based on their ability to retain crystal violet when decolorized with an organic solvent such as ethanol.

Green engineering: Design, commercialization, and use of processes and products, which are feasible and economical while minimizing the generation of pollution at the source and the risk to human health and the environment.

Greenhouse effect: Physical process by which incoming solar radiation is re-radiated as infrared wavelengths from the earth's surfaces. In turn, the heat is retained by radiant gases (i.e. *greenhouse gases*), so that the earth stays warm. Without these gases virtually all of the heat would be returned to space, so that the diurnal heat variations would range from extremely hot during day and extremely cold at night. Thus, the greenhouse effect is absolutely essential to life on earth. However, with the increase in greenhouse gas concentrations in the troposphere is causing concern within the scientific community, with fears of global warming and other changes in global climate.

Greenhouse gas (GHG): Gas released to the atmosphere that in turn retains heat that has been radiated from the earth.

Grey goo scenario: Doomsday scenario related to *nanotechnology* in which an "extinction technology" is created from the cell's unchecked ability to exponentially replicate itself if part of their design is to be completely "omnivorous," using all organic matter as food. No other "life" on earth would exist if this "doomsday" scenario were to occur.

Guanine: Purine derivative, 2-amino-6-oxypurine, found in nucleosides, nucleotides, and nucleic acids.

Half-life (T1/2): Time needed for half the quantity of a substance taken up by a living organism to be metabolized and eliminated by normal biological processes. Also called biological half-life.

Halophile: Microbe that requires high levels of sodium chloride for growth ($\leq$ 25% NaCl).

Halorespiration: The use of halogenated compounds (e.g., trichloroethene) as electron acceptors. This is the essential processes of biological reductive dechlorination.

Haploid: Cell with only one complete set of chromosomes.

Harm: Damage to another person or creature.

Harm principle: John Stuart Mill's recommendation that utilitarianism's premise of greatest good (see *utilitarianism*) is restricted (e.g., by law) if an act that is good for the majority causes undue harm to individuals.

Hazard: Potential source of harm. Term is often formally defined and distinguished from non-hazards by regulatory agencies, e.g. hazardous and non-hazardous wastes.

Hazard assessment: The process of determining whether exposure to an agent can cause an increase in the incidence of a particular adverse health effect (e.g., cancer, birth defect) and whether the adverse health effect is likely to occur in humans.

Hazard characterization: A description of the potential adverse health effects attributable to a specific environmental agent, the mechanisms by which agents exert their toxic effects, and the associated dose, route, duration, and timing of exposure.

Hazard rate: Probability of a failure per unit time. Synonymous with failure density.

Henry's law: Relationship between the partial pressure of a compound and its equilibrium concentration in a dilute aqueous solution through a constant of proportionality, i.e. Henry's law constant. Expression of *fugacity*.

Herbicide-tolerant crop: Crop that contains new genes that allow the plant to tolerate herbicides. The most common herbicide-tolerant crops (cotton, corn, soybeans, and canola) are those that are resistant to glyphosphate, an effective herbicide used on many species of grasses, broadleaf weeds, and sedges.

Heterocyst: Specialized cell produced by cyanobacteria; sites of nitrogen fixation.

Heterogeneous reaction: Reaction in which the reagents and products involved include more than one physical state of matter.

Heterologous encapsidation (transcapsidation): Generation of "new" viruses by surrounding one virus with the envelope protein of another virus; a natural process that can occur when plants are co-infected by different strains of viruses.

Heterotroph: Organism need carbon-energy compounds.

Heterozygous: Having different alleles at a genetic locus.

Hill's criteria: Minimal conditions necessary to establish causal relationship between two items; presented by British medical statistician Sir Austin Bradford Hill (1897–1991) as a means of finding causal links between a specific factor (e.g., exposure to air pollution) and specific adverse effects (e.g., asthma). These criteria, originally recommended for occupation setting but now applied to numerous health and environmental problems, are meant to be guidelines rather than inviolable rules of *epidemiology*.

Histone: Small basic protein with large amounts of lysine and arginine; associated with eukaryotic DNA in chromatin.

Homeostasis: Ability of an organism to self-regulate functions; inherent trend toward stability.

Homogeneous reaction: Reaction in which the reagents and products involved include only one physical state of matter (e.g. aqueous or gaseous).

Homozygous: Having two identical alleles of a gene.

Horizontal gene transfer: Process by which an organism incorporates genetic material from another organism without being the offspring of that organism.

Hormonally active agent: Substances that possess hormone-like activity, regardless of mechanism. Synonymous with *endocrine disruptor.*

Hormone: Chemical released by glands of the endocrine system (see *endocrine system*).

Host: Organism that harbors another organism; microenvironment that shelters and supports the growth and multiplication of a parasitic organism.

Hours of retention (HRT): Common unit of time that a substance is held in a *bioreactor.*

Humus: Dark-colored organic material in soil and sediment that is a product of plant material decomposition.

Hybrid: Offspring of genetically dissimilar parents or stock.

Hybridization: 1. Act of mixing different species or varieties of organisms to produce hybrids. 2. Reaction by which pairing of complementary strands of nucleic acid occurs. Usually double-stranded, when the strands of DNA are separated they will re-hybridize under the appropriate conditions. Hybrids can form between DNA-DNA, DNA-RNA or RNA-RNA.

Hydraulic conductivity: A measure of the capability of a medium to transmit water.

Hydraulic gradient: The change in hydraulic head (per unit distance in a given direction) typically in the principal flow direction.

Hydrolysis: Decomposition of a chemical compound by reaction with water, such as the dissociation of a dissolved salt or the catalytic conversion of starch to glucose.

Hydrophilic: Describing a polar substance with a strong affinity for water (i.e. high aqueous solubility).

Hydrophilicity: Tendency of a substance to attract water or to be capable of completely dissolving in water. *Hydrophilic* substances are readily soluble in polar solvents, especially water. See *aqueous solubility* and *polarity.*

Hydrophobicity: Tendency of a substance to repel water or to be incapable of completely dissolving in water. Hydrophobic substances are readily soluble in many nonpolar solvents, such as octanol, but only sparingly soluble in water, a polar solvent. That is, most hydrophobic compounds are also *lipophilic.* The hydrophobicity of an organic contaminant influences the fate of the contaminant in the environment. In general the more hydrophobic a contaminant is, the greater the likelihood it will be associated with nonpolar organic matter such as humic substances and lipids (fats). Can be predicted fairly well by the *octanol-water coefficient.*

Hypothetico-deductive method: Method of logical deduction, attributed to Karl Popper (*The Logic of Scientific Discovery,* 1934), limiting scientific discovery to that which is testable; requiring an approach that formulates hypotheses, a priori, with the intent of rejecting these hypotheses. The method assumes that a hypothesis can never truly be proved, but at best can be corroborated.

Hysteresis: 1. Changes that occur depending on the direction taken in a pathway; e.g. a material may behave differently in the same temperature range when cooled than when heated in that same range. In mechanics, the changes of a body as it returns to its original shape after being stressed. 2. Failure to return to previous condition, such as due to an energy loss that always occurs under cyclical loading and unloading of a spring, proportional to the area between the loading and unloading load-deflection curves within the elastic range of a spring (engineering), or the failure of a variable to return to its initial equilibrium after a temporary shock (economics).

Ideal gas law: The product of pressure and volume of a gas is equal to the product of amount of gas and temperature:

$$\frac{n}{V} = \frac{p^{\circ}}{RT};$$

where V = volume of the container; n = number of moles of chemical; R = molar gas constant;

$$\frac{n}{V}$$

is the gas phase concentration (moles L^{-1}) of the chemical.

Imhoff tank: Two-story vessel for both settling and *digestion* occur in a waste treatment system, with one compartment below the other. This allows for a stepped and separate *aerobic* and *anaerobic* treatment.

***In silico*:** Based on information, usually using computational methods, rather than using actual materials being studied. To some extent, *in silico* research is an alternative to *in vivo* and *in vitro* research (see *in vivo* and *in vitro*), which is desirable in the case of limiting animal research and in reducing risks in humans who undergo *in vivo* procedures.

***In situ*:** Taking place where it is found (e.g., bioremediation of a hazardous waste site where it exists, rather than moving the materials off-site).

***In utero*:** In the womb (e.g., fetal alcohol syndrome results from the unborn child's exposure to alcohol and its metabolites during gestation). See *gestation*.

***In vitro*:** Outside of the organism (literally: "in glass," i.e., in a test tube).

***In vivo*:** Inside the organism (e.g., experiments within a rat to observe biochemical responses to a chemical dose).

Incidence or incidence rate: Number of new cases in a defined population within a period of time (compare to *prevalence*).

Index of biological integrity (IBI): Method of indicating the quality of aquatic systems. Usually, the total number of organisms and the number of different species present are inventoried, followed by the application of an index, or scale, that lists organisms according to their sensitivity to pollution.

Indicator organism: Organism whose abundance indicates the condition of a substance or environment. For example, the potential presence of pathogens in fecal pollution is indicated by *coliforms*.

Inducible: Feature of an enzyme indicating that it is not synthesized or activated until needed.

Inductive reasoning: Starting from a specific experience and drawing inferences from the specific set of facts or instances to a larger number of instances (generalization). Conclusions are drawn from the perspective that all individuals of a kind have a certain character on the basis that some individuals of the kind have that character. Contrast with *deductive reasoning*.

Industrial ecology: Study of industrial system that focuses on material cycling, energy flow, and the ecological impacts of such systems.

Inert ingredient (inert): 1. Non-reactive ingredient. 2. Any ingredient in a pesticide formulation other than the ones that provide the mechanisms of biocidal action. An inert ingredient may or may not be reactive or may or may not be toxic. See *active ingredient*.

Inference: Reasoning that one statement (the conclusion) is derived from one or more other statements (the premises). See *syllogism*.

Informatics: Application of computational and other technologies to access and to enhance information; one means of turning data into information (see *data* and *information*).

Information: Processed and organized *data*. Value-added data as a step toward knowledge.

Initiation: The first stage of carcinogenesis.

Inoculum: Microbial culture used to initiate growth and introduced to a container.

Inorganic compound: A compound that is not based on covalent carbon bonds, including most minerals, nitrate, phosphate, sulfate, and carbon dioxide.

Instrumental value: Worth based on usefulness. In biomedical ethics, the perspective of whether a human life has value that depends on usefulness (Will the baby be loved? Will the elderly person continue to enjoy life?) is an instrumental viewpoint. In environmental ethics, use of the term environmental resource implies that ecological value is based on the utility of the ecosystem (e.g., wetlands as breeding area for game fish, as retention areas to prevent floods, and as sinks for carbon to prevent global warming). Contrast with *intrinsic value*.

Integrated mass flux (IMF): The total quantity of a migrating substance that moves through a planar transect within the system of interest and oriented perpendicular to the direction of movement. If the transect is at the entry point to the system, the integrated mass flux is the loading. If the transect is at the exit point from the system, the integrated mass flux is the discharge. Note that these terms have units of mass per time (kg $year^{-1}$, g day^{-1}, etc.) and represent an extension of the traditional engineering definition of flux (e.g. kg $year^{-1}$ m^{-2}) in which the transect area is accounted for to allow mass balance calculation of plume or system-scale behavior.

Integrated pest management (IPM): Combination of various strategies to reduce pests, rather than simply relying on application of pesticides; e.g. including use of natural predators, physical removal of breeding areas (e.g. standing water for mosquitoes) and introduction of organisms that lead to sterile offspring, i.e. interruption of a pest or vector's life cycle.

Interferent: Substance which interferes with analytical procedure, leading to analytical error.

Intergenic: Between two genes; e.g. intergenic DNA is DNA found between two genes.

Interspecies dose conversion: The process of extrapolating from animal doses to human equivalent doses.

Intervention: Direct involvement of corrective action to change existing condition for the better.

Intrinsic value: Worth based on existence, not usefulness. All humans have intrinsic value in contemporary morality. In biomedical ethics, however, there is no unanimity of thought about the intrinsic value of an embryo or a fetus, or a person nearing end of life. Those subscribing to sanctity of life viewpoints see intrinsic value of any human being (beginning with the human zygote and ending in natural death). In environmental ethics, there is no unanimity of thought about nonhuman species. For example, the loss of a species is morally wrong based on the value of the existence of the species, not its actual or potential value (e.g., as a cure for cancer or as a food source for a food species). Contrast with *instrumental value*.

Introgression: Incorporation of a gene from one organism complex into another organism complex by means of hybridization.

Intuition: Direct perception of meaning without conscious reasoning. Compare to *deductive* and *inductive reasoning*.

Ion exchange: Transfer of ions between two electrolytes or between an electrolyte solution and a complex. Physicochemical process in which ions held electrostatically on the surface of a solid phase are exchanged with ions of similar charge in a solution (e.g. drinking water).

Ionic strength: Solution's total concentration and valences of ions.

Irreversible sorption: A hysteresis effect in which a chemical species becomes more strongly bound over time. The term sometimes appears to be used to describe a situation where, once sorbed, the contaminant is removed from the plume and remains associated with the soil.

Ishikawa diagram: Graphical technique for identifying cause-and-effect relationships among factors in a given situation or problem. (Also known as a fishbone diagram.)

Isolation: Separation of specific microorganisms from cultures.

Junk science: Term applied to questionable data or information used to support advocacy positions; or factually correct data and information misapplied to support advocacy positions.

Kilobase (kb): Unit of 1000 nucleotide bases, either RNA or DNA.

Kinase enzyme: Enzyme that adds phosphate (PO_4^{3-}) groups to molecules.

Kinetics: Rates of reactions and processes.

Knock-in: Targeted mutation in which an alteration in gene function other than a loss-of-function allele is produced.

Knock-out: Targeted mutation in which a loss-of-function (often a null allele) is produced.

Knowledge: Familiarity, awareness, or understanding gained through experience or study. A necessary step toward wisdom.

Kreb's cycle: Oxidative pathway in respiration by which pyruvate, via *acetyl coenzyme A*, is decarboxylated to form CO_2.

Lag phase: In microbial growth, period after inoculation and before exponential (log) growth.

Lagoon: Shallow surface water system used for wastewater treatment; often aerated mechanically to increase *aerobic* decomposition of waste material.

Laminar: Describing flow of a viscous fluid in which particles of the fluid move in parallel layers, each of which has a constant velocity but is in motion relative to its neighboring layers. Also called streamline or viscous flow.

Land farming: Addition of waste material, e.g. organic compound-laden waste, to the soil surface for biodegradation. The soil may be moistened or mixed to stimulate the desired degradation process.

Latency period: The time between first exposure to an agent and manifestation or detection of a health effect of interest. Period of time between disease occurrence and detection, sometimes used interchangeably with induction.

Law of diminishing returns: Economic principle espoused by Thomas Malthus (*Essay on the Principle of Population*, 1798) stating that when a fixed input is combined in production with a variable input, using a given technology, increases in the quantity of the variable input will eventually depress the productivity of the variable input. Malthus proposed this as a law from his pessimistic idea that population growth would force incomes down to the subsistence level.

Law of supply and demand: Economic principle stating that, in equilibrium, prices are determined so that demand equals supply; thus changes in prices reflect shifts in the demand or supply curves.

Leachate: 1. Percolated liquid through solid waste or other permeable material. 2. Extracted materials from this liquid.

Lethal concentration (LC_x): Concentration of a substance in air in which X% of test animals die, e.g. the median lethal concentration is LC_{50}: a common measure of *acute toxicity*.

Lethal dose (LD_x): Amount of a substance delivered to a test animal in a single dose that kills X%, e.g. the median lethal dose is LC_{50}: a common measure of *acute toxicity*.

Lichen: Organism composed of a fungus and either green algae or cyanobacteria in a symbiotic association.

Life: Period from onset (i.e. conception) to end (i.e. death) of a unique organism. Antonym can be either death or nonliving.

Life cycle analysis; life cycle assessment (LCA): Consideration and quantification of total environmental and energy impact of a product or process, beginning at or before raw material extraction through use, disposal, recycling, and post-use.

Lifetime average daily dose (LADD): Estimated dose to an individual averaged over a lifetime of 70 years; used in assessments of *carcinogenic* risk.

Ligand: Molecule that travels through the bloodstream as chemical messenger that will bind to a target cell's *receptor.*

Ligase: Enzyme, e.g. T4 DNA ligase, able to link pieces of DNA together.

Ligation: Process of *splicing* two pieces of DNA.

Lignin: Organic polymer stored in plant cell walls of woody plants; an aromatic hydrocarbon compound forming a three-dimensional structural matrix.

Lignin modifying enzymes (LMEs): Extracellular enzymes released by organisms, especially fungi, that enhance and/or accelerate mineralization of lignin, but which also increase *biodegradation* of other substances with chemical structures similar to lignin.

Limit of Detection (LOD): Lowest concentration of a chemical that can reliably be distinguished from a zero concentration (also known as detection limit).

Linearity: Following the mathematical equation for a line ($y = mx + b$; where m is the slope and b is the y intercept). Also used to describe the degree to which data points approximate the line of best fit (linear regression).

Liner: Clay or manufactured material that serves as a barrier against the movement of *leachate.* Liners have very low *hydraulic conductivity.*

Lipophilicity: Tendency of a chemical compound to dissolve in fats, oils, lipids, and non-polar solvents. See *hydrophilicity.*

Liquid-side impedance: In two-film theory, difference in a substance's concentration in the bulk liquid and at the interface with the gas phase.

Load: Quantity of a substance discharged into a system (e.g. water body).

Loading: In wastewater treatment, food (F) to microorganism (M) ratio (F:M) at the entry point of the aeration basin. See also: *mass loading, source loading, hysteresis* (2), and *total maximum daily load.*

LOAEL: See *lowest observed adverse effect level.*

Log growth phase: Also known as trophophase.

Logic: Branch of philosophy addressing inference (e.g., using a *syllogism* to determine the validity of an ethical argument).

Longitudinal study: Epidemiological study using data gathered at more than one point in time, e.g., after an exposure or a medical intervention. Contrast with *cross-sectional study.*

Lowest observed adverse effect level (LOAEL): Lowest exposure level at which there are biologically significant increases in frequency or severity of adverse effects between the exposed population and its appropriate control group. Compare to *NOAEL.*

Lysozyme: Enzyme that degrades peptidoglycan by hydrolyzing the glycosidic bond between N-acetylmuramic acid with C4 of N-acetylglucosamine.

Macroethics: Expectations of an entire profession, e.g., the engineering profession's positions regarding emerging technologies, social justice, or sustainability.

Macroscopic property: see *property*.

Malaria: Serious infectious disease caused by the parasitic protozoan *Plasmodium*. Characterized by bouts of high chills and fever that occur at regular intervals.

Marker gene: Genes used to identify the number of cells that have successfully acquired the new gene; transferred alongside the gene of interest (the target or beneficial gene). Most commonly used marker genes are antibiotic or herbicide resistance genes. All work by making the modified cells detoxify substances that would otherwise be fatal to them.

Mass balance: Assessment includes a quantitative estimation of the mass loading to the dissolved plume from various sources, as well as the mass attenuation capacity for the dissolved plume.

Mass loading: Contaminant released to the environment from the source material.

Mass transfer: The irreversible transport of solute mass from the non-aqueous phase (i.e., DNAPL) into the aqueous phase, the rate of which is proportional to the difference in concentration.

Mechanism of action: Specific biochemical interaction through which a substance produces its intended effect in the case of a pesticide or drug, or toxic reaction in the case of a toxic substance. The mechanism is usually characterized by specific molecular targets to which the substance binds, such as to an enzyme or to a receptor. For example, numerous pesticides' mechanism of action is by inhibition of neurotransmitters, e.g. acetylcholine. Organochlorine pesticides often alter movement of ions across the nerve cell membranes, changing the ability of the nerve to fire. Organophosphate and carbamate pesticides act primarily at the synapses, altering the regulation of the transmission of signal between neurons.

Medium: 1. Substance in which organisms are grown. 2. Material in a bioreactor, e.g. *trickling filter*, on which microbes grow and produce *biofilm*. 3. Environmental compartment (e.g. air, water, soil, sediment or biota).

Meiosis: Process by which diploid germ cell precursors segregate their chromosomes into haploid nuclei within eggs and sperm.

Membrane: 1. Film with pores. 2. Cellular amphiphilic layer that encloses the cell or separates parts within a cell.

Membrane bioreactor (MBR): System that combines suspended growth with solids separation using ultrafine porous membranes; often follows aeration step in *activated sludge* treatment.

Mendelism: Heredity theory underlying classical genetics, proposed by Roman Catholic monk and scientist Gregor Mendel in 1866.

Messenger RNA (mRNA): RNA containing sequences coding for a protein.

Mesophile: Microbe with optimal growth between 20 °C to 45 °C, a minimum of 15 °C to 20 °C, and a maximum $\leq$45 °C.

Metabolism: Act of a living organism converting and degrading a substance from one form to another (known as a metabolite). Chemical reactions in living cells that convert food sources to energy and new cell mass. See *catabolism* and *anabolism*.

Metabolonomics: Same as *metabonomics*, although recently metabolonomics is more interested in comprehensive metabolic profiling, whereas metabonomics is more interested in metabolic changes resulting from a perturbation.

Metabonomics: 1. Using changes in metabolite levels in cells, tissues, and fluids to estimate or characterize an exposure to a substance. 2. Determining the genome responsible for the changes observed in 1. See *metabolonomics*.

Methanogen: Strictly anaerobic Archaeabacteria able to use only a very limited substrate spectrum (e.g., molecular hydrogen, formate, methanol, carbon monoxide, or acetate) as substrates for the reduction of carbon dioxide to methane.

Methanogenesis: Breakdown of organic compounds by *anaerobic* microbes to form methane (CH_4).

Miasma: Theory believed from the Middle Ages to the late 19th century that smells emanating from decomposing material were the causes of a disease.

Micellar: Referring to *micelles.*

Micelle: Aggregated *surfactant* molecule that has become colloidally suspended.

Michaelis constant: Kinetic constant for an enzyme reaction equal to the substrate concentration required for the enzyme to operate at half maximal velocity.

Microalgae: *Algae* too small to be seen individually with the naked eye.

Microarray: A multifaceted tray or array of DNA material. Microarrays are expected to revolutionize medicine by helping pinpoint a very specific disease or the susceptibility to it. Sometimes called "biochips," microarrays are commonly known as "gene chips."

Microbe: Microorganism.

Microbial ecology: Study of microorganisms in their natural environments, with a major emphasis on physical conditions, processes, and interactions that occur on the scale of individual microbial cells.

Microbiology: Study of *microorganisms* (those too small to be seen with the naked eye). Special techniques are required to isolate and grow such organisms.

Microcosm: A batch reactor used in a bench-scale experiment designed to resemble the conditions present in the groundwater environment.

Microethics: Expectations of the individual professional practitioner or researcher. Compare to *macroethics.*

Microinjection: Process by which, DNA or other materials are injected into fertilized eggs or blastocysts.

Microorganism: An organism of microscopic or submicroscopic size, including bacteria.

Mineralization: The complete degradation of an organic compound to carbon dioxide and other inorganic compounds, such as water and chloride ions.

Minimal risk level (MRL): Estimate of daily human exposure to a hazardous substance at or below which that substance is unlikely to pose a measurable risk of harmful, noncancerous effects. Calculated for a route of exposure (inhalation or oral) over a specified time period (*acute,* intermediate, or *chronic*). Not recommended as predictors of adverse health effects.

Minimax theorem: Key convention of game theory holding that lowest maximum expected loss in a two-person zero-sum game equals the highest minimum expected gain. It is a useful technique to address uncertainties in decision making.

Miscibility: Chemical property where two or more liquids or phases readily dissolve in one another, such as ethanol and water.

Mitochondrion: Eukaryotic organelle that is the site of electron transport, oxidative phosphorylation, and pathways such as the Krebs cycle; it provides most of a nonphotosynthetic cell's energy under aerobic conditions. Constructed of an outer membrane and an inner membrane, which contains the electron transport chain.

Mitosis: Process in the nucleus of a eukaryotic cell that results in the formation of two new nuclei, each with the same number of chromosomes as the parent.

Mixed acid fermentation: Process carried out by members of the family Enterobacteriaceae in which ethanol and a complex mixture of organic acids are produced.

Mixed liquor: Activated sludge and water containing organic matter undergoing *activated sludge* treatment in aeration tank.

Mixed liquor suspended solids (MLSS): Volume of *suspended solids* in *mixed liquor.*

Mixed liquor volatile suspended solids (MLVSS): Volume of organic solids from *mixed liquor* that will evaporate at relatively low temperatures (e.g. 55 °C); MLVSS is an indicator that microbial populations are active.

Mixotrophic: Characteristic of microorganisms that combine autotrophic and heterotrophic metabolic processes (i.e. use both inorganic electron sources and organic carbon sources).

Mode of action: Overall manner in which a substance acts, e.g. the way a pesticide kills an insect at the tissue or cellular level, or the way that a drug works at the cellular level.

Model: A mathematical function with parameters that can be adjusted so the function closely describes a set of empirical data. A mechanistic model usually reflects observed or hypothesized biological or physical mechanisms, and has model parameters with real-world interpretation. In contrast, statistical or empirical models selected for particular numerical properties are fitted to data; model parameters may or may not have real-world interpretation. When data quality is otherwise equivalent, extrapolation from mechanistic models (e.g., biologically based dose-response models) often carries higher confidence than extrapolation using empirical models (e.g., logistic model).

Modifying factor (MF): A factor used in the derivation of a reference dose or reference concentration. The magnitude of the MF reflects the scientific uncertainties of the study and database not explicitly treated with standard uncertainty factors (e.g., the completeness of the overall database). A MF is greater than zero and less than or equal to 10, and the default value for the MF is *1*. Use of a modifying factor was generally discontinued in 2004. Compare to *uncertainty factor (UF)*.

Molecular pharming: Use of *genetically modified organisms* to produce pharmaceuticals. Application of genetic engineering that introduces genes, primarily of human and animal origin, into plants or farm animals to produce medicinal substances. Premise is using plants as efficient chemical factories for producing antibodies, vaccines, blood proteins, and other therapeutically valuable proteins.

Monod equation: Empirically-derived expression of the rate of microbial biomass:

$$\mu = \frac{\mu_{\max} S}{K_s + S}$$

where μ = the specific growth rate of the microbe; $\mu_{\max}$ = maximum specific growth rate; and K_s = the Monod growth rate coefficient representing the substrate concentration at which the growth rate is half the maximum rate. The $\mu_{\max}$ is reached at the higher ranges of substrate concentrations. K_s is an expression of the affinity of the microbe for a nutrient, i.e. as K_s decreases the more affinity that microbe has for that particular nutrient (as expressed by the concomitantly increasing μ). Named in honor of French researcher, Jacques Monod.

Monte Carlo technique: Repeated random sampling from the distribution of values for each of the parameters in a calculation (e.g., lifetime average daily exposure), to derive a distribution of estimates (e.g. of exposures) in the population.

Moral: 1. Pertaining to the judgment of goodness or evil of human action and character. 2. Often, an adjective for goodness or ethically acceptable actions (opposite of immoral).

Morality: Distinction between what is right and wrong.

Morbidity: State of disease.

Mortality rate: Proportion of a population that dies during a specified time period. Also called death rate.

Mosaic: Individual consisting of cells of two or more *genotypes*.

Most probable number (MPN): Estimation of the probable population in a liquid by diluting and determining endpoints for microbial growth and conducting statistical tests.

Motile: Capable of movement, e.g. used to characterize microorganisms.

Mucociliary escalator: Mechanism by which mucous fluid is moved upwardly by microscopic projections (i.e. fiber cilia) on the surfaces of cells lining the respiratory tract.

Mutagen: Substance that can induce an alteration in the structure of DNA.

Mutagenesis: Generation of mutations; breeding whereby random mutations are induced in cell's DNA using chemicals or ionizing radiation.

Mutualism: Symbiosis in which both partners gain from the association and are unable to survive without it. The mutualist and the host are metabolically dependent on each other.

Mycelium: Branching hyphae found in fungi and some bacteria.

Mycoplasma: Bacteria that are members of the class Mollicutes and order Mycoplasmatales, lacking cell walls and unable to synthesize peptidoglycan precursors; most require sterols for growth. Smallest organisms capable of independent reproduction.

Myxobacteria: Gram-negative, aerobic soil bacteria characterized by gliding motility, a complex life cycle with the production of fruiting bodies, and the formation of myxospores.

Nano-scale: Having at least one dimension <100 nm.

Nanotechnology: Science and engineering technologies addressing the design and production of extremely small (<100 nanometers diameter in at least one dimension) devices and systems fabricated from individual atoms and molecules.

Natural attenuation: Naturally occurring processes in soil and groundwater environments that act without human intervention to reduce the mass, toxicity, mobility, volume, or concentration of contaminants in those media. When analyzing data from a natural attenuation site, a key question often is whether the mechanisms that destroy or immobilize contaminants are sustainable for as long as the source area releases them to the groundwater; more specifically, whether the rates of the protecting mechanisms will continue to equal the rate at which the contaminants enter the groundwater may be a concern. Sustainability is affected by the rate at which the contaminants are transferred from the source area and whether or not the protecting mechanisms are renewable.

Navier–Stokes equations: Equations that explain motion of fluids, usually nonlinear partial differential equations.

Negative paradigm: Most unacceptable or unethical action or case possible. In line drawing, the negative paradigm is the polar opposite of the positive paradigm (compare to *positive paradigm*).

Nephelometry: Measurement of turbidity using light scattered at an angle to the incident beam; particularly sensitive at low turbidity.

Nerve: Enclosed, cable-like bundle of nerve fibers or axons.

Net primary productivity: Organisms' generation of organic compounds from carbon dioxide. Rate at which an ecosystem accumulates biomass and energy, but excluding the energy used for respiration.

Niche: Function of an organism in a complex system, including place of the organism, the resources used in a given location, and the time of use.

Nicotinamide adenine dinucleotide (NAD): Coenzyme for dehydrogenases; reduced form is *nicotinamide adenine dinucleotide phosphate*. Formerly called DPN (diphosphopyridine nucleotide) and Coenzyme I.

Nicotinamide adenine dinucleotide phosphate (NADP): Coenzyme for dehydrogenases; reduced form is NADPH. Formerly called TPN (triphosphopyridine nucleotide) and Coenzyme II.

Nitrification: Oxidation of reduced forms of nitrogen, e.g. ammonia to nitrate.

Nitrifying bacteria: *Chemolithotrophic*, gram-negative members of the family Nitrobacteriaceae that convert ammonia to nitrate and nitrite to nitrate.

Nitrogenase: Enzyme that catalyzes biological *nitrogen fixation*.

Nitrogen fixation: Metabolic process by which atmospheric molecular nitrogen is reduced to ammonia; carried out by cyanobacteria, Rhizobium, and other nitrogen-fixing bacteria.

Nitrogen oxygen demand (NOD): Demand for oxygen in sewage treatment, caused by nitrifying microorganisms.

NOAEL: See *no observed adverse effect level*.

Nongovernmental organization (NGO): Entity that advocates or represents positions, including scientific, legal, and medical perspectives, without a governmental mandate. Examples include Doctors without Borders, Resources for the Future, and Engineers without Borders.

Nonpoint source: Pollution discharged over an expansive area, not from one specific location. Compare to *point source*.

No observed adverse effect level (NOAEL): Highest exposure level where there are no biologically significant increases in the frequency or severity of adverse effect between the exposed population and its appropriate control. Compare to *LOAEL*.

Nosocomial: Describing infection acquired in a medical facility.

Nuclear transfer cloning: Transfer of a nucleus into an enucleated egg cell.

Nuclease: An enzyme which degrades nucleic acids.

Nucleotide: Combination of ribose or deoxyribose with phosphate and a purine or pyrimidine base; a nucleoside plus one or more phosphates.

Nucleus: Eukaryotic organelle enclosed by a double-membrane envelope that contains the cell's chromosomes.

Null mutation: The complete elimination of the function of a gene.

Nutritional supplement: A dietary supplement intended to be ingested in pill, capsule, tablet, or liquid form, not represented for use as a conventional food or as the sole item of a meal or diet, and labeled accordingly. It is regulated by the US Food and Drug Administration as a food and not as a drug.

Obligate: Absolutely required, e.g. an obligate *aerobe* can live only in the presence of molecular oxygen and an obligate *anaerobe* can live only in the absence of molecular oxygen.

Ockham's Razor: Principle espoused by medieval nominalist William of Ockham that entities are not to be multiplied beyond necessity. The principle encourages asking whether any proposed kind of entity is necessary.

Octanol-water coefficient (K_{ow}): Ratio of the concentration of a chemical in octanol and in water at equilibrium and at a specified temperature. Octanol is used as a surrogate for natural organic matter. K_{ow} is used to help determine the fate of chemicals in the environment, e.g. to predict the extent a contaminant will *bioaccumulate* in aquatic biota.

Odds ratio: Ratio of the odds of disease among the exposed compared with the odds of disease among the unexposed. For rare diseases, such as cancer, the odds ratio can provide an estimate of relative risk.

Offsetting behavior: Inadvertent attenuation or reversal of a risk management action due to a reduction of care by those targeted for risk reduction. For example, if actions are taken to reduce bacterial growth in food supplies, and consumers consider the foods to be safer to the extent that they are less careful in food preparation, this behavior offsets at least some of the safety margin of the risk reduction efforts.

Oligomer: Category between monomer and polymer, defining a compound that consists of between 5 and 100 monomers.

Oligonucleotide: Short DNA sequence of up to 1000 nucleotides.

Omics: Shorthand term for computational, biological subfields for describing very large-scale data collection and analysis, all with the suffix "omics" (e.g., genomics, proteomics, and metabonomics).

Onus: Burden of responsibility.

Opportunity risk: Likelihood that a better opportunity will present itself after an irreversible decision has been made (e.g., prohibiting research in an emerging technology may prevent exposure to a toxic substance to a few, but in the process the cure for a disease may be lost).

Optimal range: Range of success, below and above which are unacceptable (e.g., trivalent chromium must be taken within the optimal range because intake at too low a dosage leads to a nutritional deficiency and too high a dosage leads to toxicity).

Optimization: Selecting the best design for the conditions. The "best" is determined by the designer based on one or more variables (e.g., a heart valve may have three key variables: flow rate, reliability, and durability; the engineer would design the valve by optimizing these three variables to achieve the best performance).

Oral slope factor: Upper bound, approximating a 95% confidence limit, on the increased cancer risk from a lifetime oral exposure to an agent. This estimate, usually expressed in units of proportion of a population affected per mg kg^{-1} day^{-1}, is generally reserved for use in the low-dose region of the dose-response relationship, that is, for exposures corresponding to risks less than 1 in 100.

Organ: Completely differentiated unit of an organism that provides a certain, specialized function.

Organelle: Structure within or on a cell that performs specific functions.

Organism: Living entity; consisting of one or more cells (unicellular and multicellular organisms, respectively).

Organophosphate (OP): Phosphorous-containing synthetic pesticide active ingredient that acts on the nervous system by inhibiting acetylcholinesterase. Irreversible inhibition is characteristic of many OPs. Azamethiphos, chlorpyrifos, diazinon, dichlorvos, and malathion are OPs.

Organotroph: Organism that uses reduced organic compounds as its electron source.

Osmosis: Movement of fluid through a partially (selectively) permeable membrane separating solutions of different concentrations. Movement of a fluid across a selectively permeable membrane from a dilute solution to a more concentrated solution. Cellular *membranes* and root systems, for example, take advantage of the separation by osmosis for nutrient transport.

Outfall structure: Outlet where a sewer, drain, or stream discharges to a water body; or structure through which reclaimed water or treated effluent ultimately reaches a receiving water body.

Outlier: Value that is markedly smaller or larger than other values in a data set. Can be problematic for researchers since it decreases the coefficient of determination (i.e., r^2).

Oxidation: Loss of electrons from a compound.

Oxidation-reduction potential (ORP): Degree of completion of chemical reaction expressed as the ratio of reduced ions to oxidized ions.

Oxidation-reduction reaction: Reaction involving electron transfer; the reductant donates electrons to an oxidant. Also called redox reaction.

Ozonation: Addition of ozone (O_3) to water and other *media* for disinfection and other oxidative processes.

PAH: See *polycyclic aromatic hydrocarbon.*

Pandora's box: Metaphor for a prolific source of problems. (Greek: All gifted; from mythology of a box given to Pandora by Zeus who ordered that she not open it. Pandora succumbed to her curiosity and opened it; all the miseries and evils flew out to afflict humankind.)

Paradox: Argument appearing to justify a self-contradictory conclusion by using valid deductions from acceptable premises.

Parametrics: Descriptors of an entire population, without the need of inference. Compare to *statistics.*

Pareto efficiency: Resource allocation wherein there is no rearrangement that can make anyone better off without making someone else worse off.

Partial pressure: Pressure exerted by a single gas in a mixture of gases.

Particulate matter: Solid or liquid phase particles suspended in a gas or liquid (compare to *aerosol*).

Partition coefficient: Quotient of the concentration of a substance in one phase divided by the concentration of the substance in a different phase in a heterogeneous system wherein the two phases have reached equilibrium.

Partitioning: Chemical equilibrium condition where a chemical's concentration is apportioned between two different phases according to the partition coefficient, which is the ratio of a chemical's concentration in one phase to its concentration in the other phase, e.g. *octanol-water coefficient.*

Passive flux meter: Sampling device that uses the absorption and desorption properties of the sampling media to collect and measure the movement of contaminants through the device over a set period of time. These results are then used to estimate the rate at which the contaminants will move through the associated groundwater system for an extended period of time.

Pathogen: Microbe capable of producing disease.

PBT: Chemical that is persistent, bioaccumulates, and is toxic.

PCB: See *polychlorinated biphenyl.*

PCR: See *polymerase chain reaction.*

Pedagogy: Instruction techniques used to promote learning.

Peptide: 1. Chain formed by two or more *amino acids* linked through *peptide bonds*: dipeptide = two amino acids, oligopeptide = small number of amino acids. 2. Molecule formed by peptide bonds covalently linking two or more amino acids. Larger peptides (i.e. polypeptides) are usually *expressed* from recombinant DNA.

Peptide bond: Covalent bond between two *amino acids,* in which the carboxyl group of one amino acid (X_1-COOH) and the amino group of an adjacent amino acid (NH_2-X_2) react to form X_1-CO-NH-X_2 plus H_2O.

Perception: Information and knowledge gained through the senses.

Performance monitoring: The collection of information which, when analyzed, evaluates the performance of the system on the environmental contamination.

Permeability: Ease at which a fluid moves through a substance.

Permeable reactive barriers: Subsurface walls composed of reactive materials that will either degrade or alter the state of a contaminant when that contaminant in a groundwater plume passes through the wall.

Permissible exposure level (PEL): Occupational limit for a contaminant.

Persistence: Resistance to degradation in the environment.

Persistent organic pollutant (POP): *Recalcitrant* organic compounds, especially those listed under international treaties and agreement, e.g. Stockholm Convention on Persistent Organic Pollutants, in an effort to eliminate or restrict their production and use.

Pesticide: Substance used to control pesticide by using its toxic properties.

pH: Negative logarithmic measure of the hydrogen ion concentration in water, ranging from 0 to 14 (acidic to basic).

Pharmacodynamics: Manner in which a substance exerts its effects on living organisms (compare to *toxicodynamics*).

Pharmacokinetics: Behavior of substances within an organism, especially by absorption, distribution, biotransformation, storage, and excretion (compare to *toxicokinetics*).

Phenotype: Physical manifestation of the *genotype*.

Phosphorylation: Addition of phosphate monoester to macromolecule, catalyzed by a specific *kinase enzyme*.

Photoautotroph: An organism, especially plants and *algae*, that uses light as its primary energy source.

Photolithotrophic autotroph: Organism that uses light energy, an inorganic electron source (e.g., H_2O, H_2, H_2S), and CO_2 as a carbon source.

Photoorganotrophic heterotroph: Microbe that uses light energy and organic electron donors, and simple organic molecules rather than CO_2 as their carbon source.

Photoreactivation: Increase in survival rate or reduction in the frequency of mutation of a microbial population previously irradiated with ultraviolet light by exposure to light of 300–450 nm wavelength.

Photosensitization: Increased sensitivity of microbes to oxygen and light by applying certain stains to cells (e.g., acridine orange, methylene blue).

Phototaxis: Microbial movement (especially photosynthetic bacteria) toward light.

Phototroph: Organism that can use light as an energy source. Green algae (and higher plants) produce oxygen. Photosynthetic bacteria produce sulfur (for instance). Blue-green algae are sometimes considered algae (Cyanophyta) and sometimes bacteria (Cyanobacteria).

Phycology: Study of algae.

Physical containment: Use of good work practices, equipment and installation design to prevent the spread of organisms away from the location of their intended use. Compare to *biological containment*.

Physiologically based pharmacokinetic (PBPK) model: Model used to characterize pharmacokinetic behavior of a chemical. Available data on blood flow rates and on metabolic and other processes that the chemical undergoes within each compartment are used to construct a mass-balance framework for the PBPK model.

Phytodegradation: Process by which plants metabolically degrade a contaminant to a nontoxic form in roots, stems, or leaves.

Phytoextraction: Removal of a substance from soils and groundwater surrounding the roots of a plant through that plant's vascular system.

Phytoplankton: *Flora* in the plankton *community*.

Phytoremediation: Use of plants to clean up contamination.

Phytostabilization: Using plants to immobilize contaminants in soil and ground water, especially via sorption in and on roots and precipitation within root zone.

Phytovolatilization: Plants translocate contaminants into the atmosphere via normal transpiration.

Pilot plant: Small-scale production process (following laboratory scale) used to develop a subsequent full-scale process.

Plankton: Small, mainly microscopic, members of animal and plant communities in aquatic system. See *phytoplankton* and *zooplankton*.

Plant-incorporated protectants (PIPs): Proteins and other chemicals introduced to plants either through the conventional breeding of sexually compatible plants or through techniques of modern biotechnology, e.g. transferring specific genetic material from a bacterium to a plant to induce the plant to produce pesticidal proteins or other chemicals that the plant could not previously produce.

Plasmid: Extra-chromosomal DNA molecule separate from the chromosomal DNA capable of replicating independently of the chromosomal DNA; found in bacteria and protozoa. Since it is capable of replicating autonomously, it is often used for insertion of genetic material.

Plasmolysis: Water loss with concomitant shrinkage of the cell contents and cytoplasmic membrane resulting from high osmotic pressure in a medium.

Plating: Culturing microorganisms on Petri plates (*in vitro*).

Pleiotropy: Phenomenon in which one gene can influence two or more independent characteristics.

Ploidy: Number of complete sets of chromosomes in a cell.

Plume: A zone of dissolved contaminants, originating from a source and extending in the direction of flow. Volume of a substance moving from its source away from the source.

PMN: See *premanufacture notice*.

Point of departure: The dose-response (see *dose-response*) point that marks the beginning of a low-dose extrapolation. This point is most often the upper bound on an observed incidence or on an estimated incidence from a dose-response model.

Point source: Pollution released from a single source, e.g. a pipe or *outfall structure*. Compare to *nonpoint source*.

Polarity: Extent to which a molecule can bond together due to dipole–dipole intermolecular forces between a whole or part of a molecule with asymmetrical charge distribution to a molecule which also has asymmetrical charge distribution; e.g. water is a highly polar molecule.

Polychlorinated biphenyl (PCB): Highly toxic molecule of two benzene rings bonded to each other with chlorine substituents (209 structural variations, known as congeners); presently banned but manufactured by Monsanto for much of the 20th century.

Polycyclic aromatic hydrocarbon (PAH): Class of products of incomplete combustion consisting of fused aromatic rings. A number of them are suspected carcinogens (e.g., benzo(a)pyrene).

Polymerase: Enzyme that links individual nucleotides together into a long strand, using another strand as a template.

Polymerase chain reaction (PCR): Technique that enables the *in vitro* amplification of target DNA sequences.

Pool: An accumulation of a liquid phase chemical substance in porous media above a capillary barrier. Commonly, this is used for substances that are denser than water, such that they are expected to be found in lower layers of an *aquifer*.

POP: See *persistent organic pollutant*.

Population: 1. Statistical term for the entire aggregation of items from which samples are taken from which inferences can be made. 2. Number of individuals in a given area.

Porosity: Percentage of void space in a solid matrix (e.g. soil, sediment, or gravel).

Positive paradigm: Most acceptable or ethical action or case possible. In line drawing, the positive paradigm is the polar opposite of the negative paradigm (see *negative paradigm*).

Precautionary principle: Risk management approach taken when scientific knowledge is incomplete and the possible consequences could be substantial and irreversible (e.g., global climate change). The principle holds that scientific uncertainty must not be accepted as an excuse to postpone cost-effective measures to prevent a significant problem.

Precision: Exactness and reproducibility. Usually represented by the number of significant figures.

Premanufacture notice; premanufacture notification (PMN): Notice prior to manufacture or importation of substances to allow regulators to evaluate whether the substance poses a threat to human health or ecosystems. For example, Section 5 of the Toxic Substances Control Act requires that industries submit a PMN or any proposed new chemical.

Prevalence: Proportion of cases that exist within a population at a specific point in time, relative to the number of individuals within that population at the same point in time.

Primary treatment: Clarification of wastewater influent (i.e. removal of suspended solids) by *sedimentation*.

Probability: Measurement of the likelihood that an event will occur; ranging from 0 (no likelihood whatsoever) to 1 (absolute certainty).

Process: Thermodynamics term for the description of the change of a system from one state (e.g. equilibrium) to another.

Process monitoring: The collection of information documenting the operation of a system's engineered components.

Profession: Group (e.g., physicians or engineers) with a common mission, requiring substantial education and training, self determination of professional requirements to enter, organized into an identifiable professional body, and requiring the adherence to standards of conduct.

Progenitor strain: Original strain prior to *hybridization* or genetic modification. Unmodified strain.

Program evaluation review technique (PERT) chart: Diagram depicting project tasks and their interrelationships.

Prokaryote: Organism lacking a true nucleus and other membrane-bound cellular compartments, and containing a single loop of stable chromosomal DNA in the nucleoid region and cytoplasmic structures.

Promoter: An agent that is not carcinogenic itself, but when administered after an initiator of carcinogenesis, stimulates the clonal expansion of the initiated cell to produce a neoplasm.

Property: Thermodynamic term for a quantity that is either an attribute of an entire system or is a function of position which is continuous and does not vary rapidly over microscopic distances, except possibly due to immediate changes at boundaries between phases of the system; e.g. temperature, pressure, volume, concentration, surface tension, and viscosity.

Proteobacteria: Bacteria, primarily gram-negative, that 16S *rRNA* sequence comparisons show to be phylogenetically related; proteobacteria contain the purple photosynthetic bacteria and their relatives and are composed of *a, b, g, d,* and *e* subgroups.

Protist: Eukaryote with unicellular organization, either in the form of solitary cells or colonies of cells lacking true tissues.

Proteome: Complete collection of proteins that an organism produces.

Proteomics: Study of proteins in the body, especially the protein complement of the genome (see *omics*).

Protoplast: Cell bounded by cytoplasmic membrane, yet lacking rigid layer.

Protozoa: *Motile* microbes that consume bacteria as carbon and energy sources.

Psychrophilic: Describing a microbe with optimal temperatures between 0° and 20 °C.

Public, the: Whole collection of people comprising a society.

Pure culture: Cell population with all members identical because they arise from a single cell.

Purine: Basic, heterocyclic, nitrogen-containing molecule with two joined rings that occurs in nucleic acids and other cell constituents; most purines are oxy- or amino-derivatives of the purine skeleton. The most important purines are adenine and guanine.

Pyrethroid: Natural (from the chrysanthemum family) or synthetic pesticide, of varying chemical structure, which acts on the nervous system by interfering with nerve conduction. Permethrin and d-phenothrin are synthetic pyrethroids.

Quantitative polymerase chain reaction (qPCR): Determination of a polynucleotide by including a known amount of readily distinguished template as an internal standard to compensate for variation in efficiency of amplification.

Quantitative structure-activity relationship (QSAR): Understanding and application of chemical structure to estimate the behavior of a compound, e.g. *toxicity* or *persistence.*

Quantum yield: Number of photons required per molecule of carbon dioxide converted to sugar (or per oxygen molecule produced).

Raoult's law: A dissolved substance will lower the partial pressure of the solvent proportionally to the mole fraction of the dissolved substance.

Rate constant: Term that quantifies the speed of a chemical reaction.

Reactor: System where physical, chemical, and biological reactions occur. Traditionally, these have been thought of vessels or other engineered systems, but reactors can be quite large and occur in natural systems (e.g. a wetland is a reactor in which *aerobic, anaerobic,* and mixed reactions occur). At the other end of the scale, any living cell is a reactor.

Reasonable person standard: Position (legal, engineering, etc.) expected to be held by a hypothetical person in society who exercises average care, skill, and judgment in conduct.

Reasoning: Derivation of a conclusion from premises.

Rebound: After contaminant concentrations in groundwater have been reduced through *in situ* treatment and the treatment is terminated or reduced, concentrations return to elevated levels from the continued release of mass from a source zone beyond the natural attenuation capacity of the groundwater system.

Recalcitrance: 1. Resistance of a compound to degradation. 2. Inverse of degradability.

Receptor: 1. Molecule in a cell or on its surface that binds to a specific substance and causes a specific physiologic effect in the cell. 2. The potentially or actually affected entity exposed to a physical, chemical or biological agent.

Recombinant: Describing material produced by *genetic engineering.*

Recombinant DNA (rDNA): Genetically engineered DNA prepared by transplanting or splicing genes from one individual to another, including from the cells of one species into the cells of a host organism of a different species. Thereafter, the rDNA becomes part of the host's genetic makeup and is replicated.

Redictable: Ridiculously predictable (coined by D.J. Vallero and A.C.V. Randall). For example, if a microbe has been genetically modified to degrade crude oil spilled from a tanker, the likelihood that the same microbe will degrade asphalt in adjacent roads is redictable.

Reductionism: 1. Understanding complex phenomena by reducing them to the interactions of their parts, or to simpler or more fundamental processes and components. 2. Perspective that a complex system is merely the sum of its parts, and that an account of it can be reduced to accounts of individual constituents. Compare reductionist to *systematic.*

Reductive dechlorination: The removal of chlorine from an organic compound and its replacement with hydrogen. Often part of a two-step degradation process for *recalcitrant* halogenated compounds, i.e. the anaerobic step to remove halogens and the aerobic step to break aromatic rings and otherwise reach *ultimate degradation.*

Reference concentration (RfC): An estimate (with uncertainty spanning perhaps an order of magnitude) of a continuous inhalation exposure to the human population (including sensitive subgroups) that is likely to be without an appreciable risk of deleterious effects during a lifetime. It can be derived from a *NOAEL, LOAEL,* or benchmark concentration, with uncertainty factors generally applied to reflect limitations of the data used. Generally used in noncancer health assessments. Durations include *acute, short-term, subchronic,* and *chronic.*

Reference dose (RfD): An estimate (with uncertainty spanning perhaps an order of magnitude) of a daily oral exposure to the human population (including sensitive subgroups) that is likely to be without an appreciable risk of deleterious effects during a lifetime. It can be derived from a *NOAEL, LOAEL,* or benchmark dose, with uncertainty factors generally applied to reflect limitations of the data used. Generally used in noncancer health assessments. Durations include *acute, short-term, subchronic,* and *chronic.*

Relative risk: Ratio of the risk of disease or death among the exposed segment of the population to the risk among the unexposed. The relative measure of the difference in risk between the exposed and unexposed populations in a cohort study. The relative risk is calculated as the rate of an outcome in an exposed divided by the rate of the disease among the unexposed; e.g. a relative risk of 2 means that the exposed group has twice the disease risk as the unexposed group.

Reliability: Probability that a device or system will perform its specified function, without failure under stated environmental conditions, over a required lifetime.

Remediation: Measure to correct environmental damage, especially to prevent exposures to hazardous substances. See *bioremediation.*

Replicase: DNA-duplication catalyzing enzyme.

Replicon: Genomic unit that contains an origin for the initiation of replication and in which DNA is replicated.

Reporter gene: Gene coupled to another gene of interest to make its activity detectable; used to create gene products that are easily detectable and are non-toxic for the organism.

Resilience: Attribute of ecosystem stability, expressing the systems ability to recover after disturbance.

Respiration: Energy-yielding process in which the energy substrate is oxidized by means of an exogenous or externally derived electron acceptor.

Response boundary (control plane): A location within the source area, or immediately downgradient of the source area, where changes in the plume configuration are anticipated due to the implementation of the *in situ* bioremediation DNAPL source zone treatment.

Restriction enzyme: Enzyme that cleaves DNA molecules at a precisely defined site; provides ubiquitous defense proteins in bacteria that cut DNA strands at specifically defined sequences.

Return activated sludge (RAS): Settled material collected in secondary clarifier and returned to the aeration basin to be mixed with incoming wastewater to be treated.

Reverse osmosis: High-pressure filtration to separate extremely fine particle and ions.

Reverse transcriptase: Enzyme that synthesizes a strand of DNA complementary to the base sequence of an RNA template.

Reynolds number: Dimensionless number associated with fluid flow that determines the transition point from laminar to turbulent flow. It represents the ratio of the momentum forces to the viscous forces in the fluid flow.

Rhizodegradation: Plants promote a soil environment suitable for microbes that can degrade or sequester contaminants.

Rhizofiltration: Sorption, filtering and other mechanisms by roots to remove contaminants.

Rhizosphere: Narrow region of soil directly influenced by soil microbes and plant root secretions.

Ribonucleic acid (RNA): Nucleic acid molecule similar to deoxyribonucleic acid, but containing ribose rather than deoxyribose.

Ribosomal RNA (rRNA): Any of several RNAs that become part of the ribosome, and thus are involved in translating *mRNA* and synthesizing proteins.

Risk: 1. Likelihood of an adverse outcome. 2. Probability of adverse effects resulting from exposure to an environmental agent or mixture of agents.

Risk assessment: The evaluation of scientific information on the hazardous properties of environmental agents (hazard characterization), the dose-response relationship (dose-response assessment), and the extent of human exposure to those agents (exposure assessment). The product of the risk assessment is a statement regarding the probability that populations or individuals so exposed will be harmed and to what degree (risk characterization).

Risk–benefit analysis: Comparison of risks of various options to their benefits (e.g., health and wildlife risks of applying a pesticide compared to the benefits of crop protection).

Risk characterization: The integration of information on hazard, exposure, and dose-response to provide an estimate of the likelihood that any of the identified adverse effects will occur in exposed people.

Risk homeostasis: Defeat of built-in factors of safety by asserting new way to use the products.

Risk management: Decision making process that accounts for political, social, economic, and engineering implications together with risk-related information in order to develop, analyze, and compare management options and select the appropriate managerial response to a hazard.

Risk quotient (RQ) method: Using the ratio of predicted exposure concentration to predicted no effect concentration or some other quality criterion to express the risk posed by a substance.

Risk ratio (RR): See *relative risk*.

Risk shifting: Taking an action that changes and reduces the risk of one population, but increases the risk in a different population (e.g., banning DDT reduces the risk of cancer in developed nations but increases the risk of malaria in tropical and subtropical developing countries).

Risk tradeoff: Eliminating or reducing one risk, but introducing or increasing a countervailing risk (e.g., reducing the pain of a headache by taking aspirin, but increasing the risk of Reye's syndrome; or removing mold in buildings may increase worker exposure to asbestos).

Root cause analysis: Analytical method for deciding causes to an existing problem based on a comprehensive, systematic reconstruction of events that contributed to the outcome; retrospectively retracing events, i.e. a "reverse event tree."

Runoff: 1. Fraction of precipitation or other water that appears in uncontrolled surface streams, rivers, drains or sewers. 2. Total discharge described in (1) during a specified period of time. 3. Depth to which a drainage area would be covered if all runoff for a given time were uniformly distributed over that area. Water running overland, often carrying solids and other pollutants to surface waters; important vehicle in nonpoint pollution.

Sanctity of life: View that human life is precious from conception to natural death, since it is created in the image of the Creator.

Sand: Soil particle between 0.05 and 2.0 mm in diameter.

Sanitary engineer: One who applies the physical, chemical, and biological sciences to protect and improve public health, often involving the design of structures, e.g. wastewater treatment facilities, sewer systems, and sanitary landfills.

Saprophyte: Organism that derives its nutrients from decomposing organic matter.

Saturated zone: See *zone of saturation*.

Saturation: Ratio of the volume of a single fluid in the pores to pore volume expressed as a percentage or a fraction.

Scale: Spatial extent from very small (molecular) to very large (planetary).

Science: Systematic investigation, through experiment, observation, and deduction, in an attempt to produce reliable explanations of the physical world and its processes.

Scientific method: Progression of inquiry: 1. to identify a problem you would like to solve; 2. to formulate a hypothesis; 3. to test the hypothesis; 4. to collect and to analyze the data; and 5. to draw valid conclusions.

Secondary treatment: Clarification followed by a biological process with separate sludge collection and handling.

Sedimentation: Process in surface waters or in engineered systems wherein particles fall in direct proportion to a particle's mass and indirect proportion to flow. Most particle mass in a wastewater treatment facility is collected during *primary treatment*.

Senescence: Organism's aging process.

Sensitivity: 1. Ability of a test to detect a condition when it is truly present. 2. Smallest change in a physical quantity or parameter that can be detected by a measuring system. Determined by signal to noise ratio, system amplification or quantizing limit. Compare to *specificity*.

Septic tank (soil absorption system): Localized and residential scale treatment system that includes: waste component separation via settling and scum formation of grease and *fatty acids*; microbial decomposition and liquefaction; baffling to increase retention time; flow of liquids to soil adsorption ("leach") field of parallel trenches of gravel; soil filtering and plant uptake of high-nutrient product that remains.

Sewage: Untreated, predominantly liquid, waste in need of treatment; the fluid being transported to a wastewater treatment plant. Preferred term is wastewater.

Short-term: Describing an *exposure* to an agent that occurs multiple times or continuously for one week (or other designated short time period).

Silage: Fermented plant material with increased palatability and nutritional value for animals; often can be stored for extended periods.

Silt: Soil particle with a diameter between 0.002 and 0.05 mm. Compare to *clay* and *sand*.

Sink: Site where matter or energy is lost in a system; e.g. a wetland can be a sink for CO_2, but it may not be a net sink for carbon, for example, if it releases CH_4 from anaerobic decomposition, i.e. it would be a net carbon sink only if more is lost and stored than is released. If it releases more carbon than it takes up and stores it is a net source.

Slime mold: Organism that produces spores but moves with amoeba-like gliding motility; phenotypically similar to fungi and protozoa. Phylogenetically, slime molds are more closely related to amoeboid protozoa than to fungi. Cellular slime molds are composed of single amoeboid cells during their vegetative stage. Vegetative acellular slime molds are made up of plasmodia, amorphic masses of protoplasm.

Slope factor, cancer: Dose-response curve for a substance indicating cancer potency (units = mass per body mass per time; e.g., mg kg^{-1} day^{-1}). An upper bound, approximating a 95% confidence limit, on the increased cancer risk from a lifetime exposure to an agent. This estimate, usually expressed in units of proportion of a population affected per mg kg^{-1} day^{-1}, is generally reserved for use in the low-dose region of the dose-response relationship, that is, for exposures corresponding to risks less than 1 in 100.

Sludge: Muddy aggregate at the bottom of tanks generated as particles settle. Also known as biosolids when a sludge contains large amounts of organic materials, e.g. microbes and their remnants.

Slurry: Liquid mixture of aqueously insoluble matter, e.g. a lime (CaO, $CaCO_3$) slurry.

Slurry wall: Barrier used for containment of a plume in soil and groundwater consisting of a trench filled with slurry.

Social ecology: View that environmental problems are firmly rooted in human social interactions. Social ecologists believe that an ecologically sustainable society can still be socially exploitative.

Society: Collection of human beings that is distinguished from other groups by shared institutions and a common culture.

Solids retention time (SRT): Average time of retention of suspended solids in a treatment system, equal to the total weight of suspended solids exiting the system, per unit time.

Somatic cell nuclear transfer (SCNT): Method of cloning by transferring the nucleus from a donor somatic cell into an enucleated egg to produce a cloned embryo.

Sorption: The uptake of a solute by a solid.

Sorption isotherm: Curves based on properties of both the chemical and the soil (or other matrix) that determine how and at what rates the molecules partition into the solid and liquid phases. Examples are the Freundlich Sorption Isotherms.

Source: Site from which matter or energy originates. Contrast with *sink*.

Source loading: The flux of a substance leaving the original disposal location and entering the water migrating through the soil and aquifer.

Source zone: The subsurface zone containing a contaminant reservoir sustaining a plume in groundwater. The subsurface zone is or was in contact with DNAPL. Source zone mass can include sorbed and aqueous-phase contaminant mass as well as DNAPL.

Sparger: Air diffuser designed to generate large bubbles, used singly or in combination with mechanical aeration devices. Increases vapor pressure allowing volatile compounds to be collected and treated.

Spatial scale: Geographic extent of a resource or problem. Global scale examples include pandemics, changes in climate, or nuclear threats. Continental scale examples are shifting biomes and border control between nations. Regional scale examples include the contamination of rivers or polluting the air. Local scale examples include crime, job loss, hazardous waste sites, and landfills.

Specialization: Degree to which an individual professional concentrates his or her practice into a narrow range of expertise and activities.

Species: Subdivision of a genus having members differing from other members of the same genus in minor details.

Specific gravity: 1. Property of a substance defined by its density relative to that of water. 2. Ratio of the mass of a body to the mass of an equal volume of water at a specific temperature, typically 20 °C.

Specific heat: Property of a substance defined by the heat required to raise its temperature of one gram of a substance one degree centigrade.

Specific volume: Property of a substance defined by its volume of unit mass; reciprocal of density.

Specificity: Ability of a test to exclude the presence of a condition when it is truly not present. Compare to *sensitivity*.

Spirochete: Spiral-shaped bacterium with periplasmic flagella.

Splicing (gene splicing): Excising specific areas of messenger RNA in cells of higher organisms.

Spore: Differentiated, specialized form that can be used for dissemination, for survival of hostile conditions because of its heat and desiccation resistance or for reproduction. Highly variable, and usually unicellular. May develop into vegetative organisms or gametes and can be produced asexually or sexually.

Stakeholder: A person other than regulators, owners, or technical personnel involved in the environmental activity of concern, who has a vested interest in decisions related to those particular activities.

Standard deviation: Measure of the spread in a data set; wider spread means larger standard deviation.

Standard mortality ratio (SMR): Ratio of the number of deaths observed in the study group to the number of deaths expected based on rates in a comparison population, multiplied by 100.

Stationary phase: Microbial growth period after rapid growth, in which cell multiplication is balanced by cell death.

Statistical significance: The probability that a result is not likely to be due to chance alone. By convention, a difference between two groups is usually considered statistically significant if chance could explain it only 5% of the time or less. Study design considerations may influence the a priori choice of a different level of statistical significance.

Statistics: Mathematics concerned with collecting and interpreting quantitative data and applying probability theory to estimate conditions of a universe or population from a sample.

Stoichiometric: Related to chemical proportions exactly needed in a reaction.

Stokes diameter: Diameter of a spherical particle with the same density and settling velocity as the particle of interest.

Stokes law: At low velocities, frictional force on a spherical body moving through a fluid at constant velocity equals 6π times the product of the velocity, the fluid viscosity, and the radius of the sphere.

Strain: 1. Geometrical expression of deformation caused by the action of stress (see *stress*) on a physical body. 2. In biological taxonomy, group of organisms bred within a closed colony to maintain certain defining characteristics.

Stratified random sample: Separation of a sample into several groups and randomly assigning subjects to those groups.

Stress: 1. Applied force or system of forces that are apt to strain (see *strain*) or deform a physical body. 2. The internal resistance of a physical body to such applied force or system of forces.

Structure-activity relationships (SAR): Applying the chemical structure of a molecule to its expected behavior and effects in the environment or in a biological system.

Subchronic exposure: Repeated exposure by the oral, dermal, or inhalation route for more than 30 days, up to approximately 10% of the life span in humans (more than 30 days up to approximately 90 days in typically used laboratory animal species). (See also *chronic exposure.*)

Substrate: Molecule that can transfer an electron to another molecule. Substance upon which an enzyme or microorganism acts. For example, organic compounds, such as lactate, ethanol, or glucose, are commonly used as substrates for bioremediation of chlorinated ethenes.

Sulfate reducer: A microorganism that exists in anaerobic environments and reduces sulfate to sulfide.

Supernatant: 1. Liquid stratum in a bioreactor, e.g. a sludge digester. 2. Layer of liquid above a precipitate (e.g. sediment) after settling.

Superoxide dismutase: Enzyme that catalyzes the reaction of superoxide (O_2^-)$2\,O_2^- + 2\,H+ \rightarrow H_2O_2 + O_2$.

Supply: Quantity of a good or service that a seller would like to sell at a particular price.

Supply curve: Relationship between a good's quantity supplied and the good's price.

Surfactant: Surface-active agent that concentrates at interfaces, forms micelles, increases solution, lowers surface tension, increases adsorption and disperses otherwise sorbed substances into the aqueous phase. All have common chemical structure, i.e. a *hydrophobic* and a *hydrophilic* moiety. The hydrophobic part usually consists of an alkyl chain, which is then linked to a hydrophilic group. Each part of the surfactant molecule interacts differently with water. The hydrophilic group is surrounded by water molecules, leading to enhanced aqueous solubility. Simultaneously, the hydrophobic moieties are repulsed by strong interactions between the water molecules. This combination allows for otherwise insoluble substances to be desorbed from soil and sediment.

Susceptibility: 1. Extent to which an individual is prone to the effects of an agent. 2. Increased likelihood of an adverse effect, often discussed in terms of relationship to a factor that can be used to describe a human subpopulation (e.g., life stage, demographic feature, or genetic characteristic).

Susceptible subgroups: May refer to life stages, for example, children or the elderly, or to other segments of the population, for example, asthmatics or the immune-compromised, but are likely to be somewhat chemical-specific and may not be consistently defined in all cases.

Suspended solids (SS): 1. Floating solids with low aqueous solubility. Common measure of water pollution and treatment efficiency, e.g. 20 mg SS L-2 prior to treatment and 1 mg SS L-1 after treatment.

Sustainability: 1. Processes and activities that are currently useful and that do not diminish these same functions for future generations. 2. The ability of a system to maintain the important attenuation mechanisms through time. In the case of reductive dechlorination, sustainability might be limited by the amount of electron donor, which might be used up before remedial goals are achieved.

Sustainable design: Application of principles of sustainability (see *sustainability*) to structures, products, and system; an aspect of green engineering.

Sustainable enhancement: An intervention action that continues until such time that the enhancement is no longer required to reduce contaminant concentrations or fluxes.

Syllogism: Argument according to Aristotle's logical theory that includes a major premise, a minor premise, and a conclusion. Ethical syllogisms have factual premise, a connecting premise, an evaluative premise, and a moral conclusion.

Symbiosis: Two dissimilar organisms living together or in close association.

Synergism: Effect from a combination of two agents is greater than the sum of the additive effects from both agents ($1 + 1 > 2$). Contrast with *antagonism*.

System: 1. Combination of organized elements comprising a unified whole. 2. In thermodynamics, a defined physical entity containing boundaries in space, which can be open (i.e., energy and matter can be exchanged with the environment) or closed (no energy or matter exchange).

Systematic: Perspective that includes relationships of numerous factors simultaneously. Compare to *reductionist*.

Systemic effect: Toxic effect as a result of absorption and distribution of a toxicant to a site distant from its entry point. Also known as systemic toxicity.

Systems biology: 1. Discipline that seeks to study the relationships and interactions between various parts of a biological system (e.g. metabolic pathways, organelles, cells, and organisms) and to integrate this information to understand how biological systems function. 2. Treatment of biological entities as systems composed of defined elements interacting in distinct ways to enable the observed function and behavior of that system. Properties of such systems are embedded in a quantitative model that guides further tests of systems behavior.

Tapered aeration: Supply of air in increments into a treatment system, increasing from the inlet and decreasing with distance from the inlet. Tapering is usually adjusted to meet the *biological oxygen demand* of the *mixed liquor*.

Target organ: The biological organ(s) most adversely affected by exposure to a chemical, physical, or biological agent.

Taxonomy: Classification system.

Technology: 1. Application of scientific knowledge. 2. The apparatus that results from such applications.

Temporal scale: Range of complexity associated with time. Extremely short temporal scale may be measured in nanoseconds, e.g., nuclear reactions, whereas long temporal scales may be measured in millions of years, e.g., fossilization of plants to coal.

Temporality: Criterion for causality requiring that the cause (e.g., exposure to an infectious agent) precede the effect (e.g., disease). One of *Hill's criteria*.

Teratogen: Substance that causes defects in development between conception and birth. A teratogen is a substance that causes a structural or functional birth defect.

Teratogenic: Structural developmental defects due to exposure to a chemical agent during formation of individual organs.

Terrorism: Unlawful use of, or threatened use of, force or violence against individuals or property to coerce or intimidate governments or societies, often to achieve political, religious, or ideological objectives.

Tertiary treatment: Removal of nutrients, especially phosphorus and nitrogen, along with most suspended solids; generally synonymous with *advanced waste treatment*, which is becoming the preferred term.

Tetrachlorodibenzo-para-dioxin (TCDD): Most toxic *dioxin* form, especially when 2,3,7,8-TCDD.

Texture: Particle size classification of soil in terms of the U.S. Department of Agriculture system, which uses the term loam for a soil having equal properties of *sand*, *silt*, and *clay*. The basic textural classes, in order of their increasing proportions of fine particles are sand, loamy sand, sandy loam, loam, silt loam, silt, sandy clay loam, clay loam, silty clay loam, sand clay, silty clay, and clay. The sand, loamy sand, and sandy loam classes may be further divided by decreasing size, i.e. coarse, fine, or very fine.

Thallus: Body devoid of root, stem, or leaf; characteristic of certain algae, many fungi, and lichens.

Thermodynamics: Principles addressing the physical relationships between energy and matter, especially those concerned with the conversion of different forms of energy.

Thermophilic: Describing microbes that thrive in temperatures >45 °C, e.g. *Bacillus licheniformis.*

Threshold: The dose or exposure below which no deleterious effect is expected to occur. An example is the *no-observed-adverse-effect level (NOAEL)*.

Threshold limit value (TLV): Occupational standard of the concentration of airborne substances to which a healthy person may be exposed during a 40-hour workweek with adverse effects.

Tissue: Collection of interconnected cells that carry out a similar function in an organism.

Tolerance: Established concentration of a substance (e.g. in pesticide residues) occurring as a direct result of proper usage. Compare to *action level*.

Top-down: Starting at the upper levels of organization and working downward to the details. Contrast with *bottom-up*.

Total dissolved solids (TDS): Common measure of the aqueously soluble content in water or wastewater, estimated from electrical conductivity, since pure water is a poor conductor and the solids are often electrolytes and other conducting substances.

Total maximum daily load (TMDL): Maximum amount of a pollutant that a water body can receive and still meet water quality standards, and an allocation of that amount to the pollutant's sources; varies according to specific watersheds. Sum of the allowable loads of a single pollutant from all contributing *point* and *nonpoint sources*. Must include a margin of safety to ensure that the water body can be used for the purposes the State has designated. Calculation must also account for seasonal variation in water quality.

Total suspended solids (TSS): Fraction of *suspended solids* collected by a filter in water.

Toxic Release Inventory (TRI): Database of annual releases from specified manufacturers in the United States, which includes almost 400 chemicals, as part of the Community Right to Know regulations. Data are self-reported by the manufacturers annually.

Toxic substance: A chemical, physical, or biological agent that may cause an adverse effect or effects to biological systems.

Toxicity: 1. Deleterious or adverse biological effects elicited by a chemical, physical, or biological agent. 2. Extent and degree of biological harm of a chemical, physical, or biological agent, ranging from *acute* to *chronic*.

Toxicodynamics: The determination and quantification of the sequence of events at the cellular and molecular levels leading to a toxic response to an environmental agent. Often used synonymously with the broader term *pharmacodynamics*, but toxicodynamics is exclusively applied to adverse agents rather than the efficacy of substance.

Toxicogenomics: Investigation of genetic influences on how the organisms respond to toxic substances.

Toxicokinetics: The determination and quantification of the time course of absorption, distribution, biotransformation, and excretion of chemicals. Often used synonymously with the broader term *pharmacokinetics*, but toxicokinetics is exclusively applied to adverse agents rather than the efficacy of substance.

Toxicology: Study of harmful interactions between chemical, physical, or biological agents and biological systems.

Tragedy of the Commons: Term coined by Garrett Hardin [*Science* (1968), volume 162] characterizing the degradation of commonly held resources as a result to selfish self-interest in maximizing utility of each individual using the resource.

Transcapsidation: See *heterologous encapsidation*.

Transcription: Transfer of information in DNA sequences to produce complementary *messenger RNA (mRNA)* sequences. It is the beginning of the process by which the genetic information is translated to functional peptides and proteins.

Transduction: Gene transfer between bacteria by bacteriophages.

Transfection: Method by which experimental DNA is inserted into a cultured mammalian cell.

Transfer RNA (tRNA): Small RNA that binds an amino acid and delivers it to the ribosome for incorporation into a polypeptide chain during protein synthesis, using an mRNA as a guide.

Transect: 1. Cross-section through which groundwater flows. 2. Straight line placed on a surface along which measurements are taken.

Transformation: 1. Chemical change to a substance. 2. Process by which bacterium acquires a plasmid and gains antibiotic resistance. Commonly refers to bench procedure that introduces experimental plasmids into bacteria. 3. Change in cell morphology and behavior, especially related to carcinogenesis. Transformed cell sometimes referred to as transformed phenotype.

Transgene: Integrated sequences of exogenous DNA.

Transgenic: Referring to cells or organisms containing integrated sequences of cloned DNA transferred using techniques of *genetic engineering*.

Translation: Decoding of *messenger RNA (mRNA)* occurs after transcription to produce a specific polypeptide according to the rules specified by the genetic code.

Translocation: 1. Transfer by *capillary force*, usually in plants, of compounds from the *rhizosphere* to roots, and ultimately to other tissue, especially leaves. 2. Chromosome rearrangement in which two nonhomologous chromosomes are each broken and then repaired in such a way that the resulting chromosomes each contain material from the other chromosome (a reciprocal translocation).

Transmissivity: Rate at which a fluid of the prevailing kinematic *viscosity* is transmitted through a unit width of the aquifer under a unit hydraulic gradient.

Transpiration: Process by which water is released to the atmosphere by plants, usually through leaves.

Transposon (transposable element): "Jumping genes" or mobile genetic elements that can spontaneously move from one position in the *genome* to another, usually randomly; can be responsible for noticeable mutations by jumping into coding sequences of genes. Although rare, transposons are present in all organisms.

Tricarboxylic acid cycle (TAC): Cycle of the oxidation of *acetyl coenzyme A* to CO_2, and the generation of NADH and FADH2 for oxidation in the electron transport chain; the cycle also supplies carbon skeletons for biosynthesis.

Trickling filter: Beds of rock and other media covered *biofilm* that aerobically degrades organic waste during secondary sewage treatment.

Trophic state: Level of *biological organization*.

Truth: Conformity to fact.

Turbidity: Scattering and adsorption of light in a fluid, usually caused by suspended matter.

Turbulence (turbulent flow): 1. Fluid property characterized by irregular variation in the speed and direction of movement of individual particles or elements of the flow. 2. State of flow of which the fluid is agitated by cross currents and eddies, as contrasted with *laminar* flow.

Two-film model: Resistance model in which dissolved compound moves from the bulk fluid through a liquid film to the evaporating surface, and then diffuses through a stagnant air film to well-mixed atmosphere above; assumes that chemical is well mixed in the bulk solution below the liquid film and that mass transfer across each film is proportional to the concentration difference.

Type I error: Error of rejecting a true null hypothesis. Contrast with *type II error.*

Type II error: Error of accepting (not rejecting) a false null hypothesis. Contrast with *type I error.*

Ultimate biochemical oxygen demand (BODu): 1. Total quantity of oxygen needed for complete degradation of organic material in the first-stage of biochemical oxygen demand. 2. Quantity of oxygen required to completely degrade all biodegradable material in a wastewater.

Ultimate degradation: The final breakdown products of reactions of contaminants; usually carbon dioxide and water, but also methane and water for anaerobic systems.

Uncertainty: Difference between what is known and what is actually the truth. Scientific uncertainty includes error and unknowns, such as those resulting for selecting variables, undocumented variability, and limitations in measurements and models. In science, there is almost always uncertainty. The goal is to decrease uncertainty and to document known uncertainties.

Uncertainty factor (UF): One of several, generally 10-fold, default factors used in operationally deriving the *RfD* and *RfC* from experimental data. The factors are intended to account for: 1. variation in *susceptibility* among the members of the human population (i.e., inter-individual or intraspecies *variability*); 2. uncertainty in extrapolating animal data to humans (i.e., interspecies uncertainty); 3. uncertainty in extrapolating from data obtained in a study with less-than-lifetime exposure (i.e., extrapolating from subchronic to chronic exposure); 4. uncertainty in extrapolating from a *lowest-observed-adverse-effect level* rather than from a *no-observed-adverse-effect level,* and 5. uncertainty associated with extrapolation when the database is incomplete.

Unicellular: Describing an organism that consists of a single cell.

Unit risk: The upper-bound excess lifetime cancer risk estimated to result from continuous exposure to an agent at a concentration of 1 μg/L in water, or 1 $\mu g/m^3$ in air. The interpretation of unit risk would be as follows if unit risk $= 2 \times 10^{-6}$ per μg/L, 2 excess cancer cases (upper bound estimate) are expected to develop per 1,000,000 people if exposed daily for a lifetime to 1 μg of the chemical per liter of drinking water.

Upper bound: A plausible upper limit to the true value of a quantity. This is usually not a true statistical confidence limit.

Uptake: See *bio-uptake.*

Utilitarianism: Theory proposed by Jeremy Bentham (*Principles of Morals and Legislation,* 1789) and James Mill (*Utilitarianism,* 1863) that action should be directed toward achieving the greatest happiness for the greatest number of people (if applied to nonhuman species, this is referred to merely as "greatest number").

Utility: 1. Useful outcome. 2. Level of enjoyment an individual attains from choosing a certain combination of goods.

Vadose zone: Unsaturated zone of soil or other unconsolidated material above the water table (i.e. above the *zone of saturation*). Includes root zone, intermediate zone, and capillary fringe. Pore spaces contain water, as well as air and other gases at less than atmospheric pressure. Saturated bodies, such as perched groundwater, may exist in the unsaturated zone, and water pressure within these may be greater than atmospheric. Also known as unsaturated zone.

Valuation: Quantifying or otherwise placing value on goods and services. Monetized valuation uses monetary currency (e.g., gross domestic product), whereas many environmental and quality of life resources are not readily conducive to monetized valuation, e.g., old growth forests have non-monetized value (e.g., habitat, ecological diversity) but little monetized value since they are not used for timber.

Value: 1. Principle, standard, or quality that is good for a person to hold. 2. Worth.

Value engineering (VE): Systematic application of recognized techniques by a multi-disciplinary team to identify the function of a product or service, establish a worth for that function, generate alternatives through the use of creative thinking, and provide the needed functions to accomplish the original purpose of the project, reliably, and at the lowest life-cycle cost without sacrificing safety, necessary quality, and environmental attributes of the project.

Valve: Device in a pipe that controls the magnitude and direction of flow.

Vapor: 1. Gas. 2. Gas phase of compound that under standard conditions would not be a gas.

Vapor pressure: Pressure exerted by a vapor in a confined space.

Vaporization: Change of a liquid or solid to the vapor phase.

Variability: 1. True heterogeneity or diversity. For example, among a population that is exposed to airborne pollution from the same source and with the same contaminant concentration, the risks to each person as a result of breathing the polluted air will vary; or among a population that drinks water from the same source and with the same contaminant concentration, the risks from consuming the water may vary; i.e. different people drinking different amounts of water and having different body weights, different exposure frequencies, and different exposure durations. Variability exists in every aspect of environmental data, e.g. in transport, fate, exposures, and effects. Overall variability may be due to differences in exposure (i.e., different people drinking different amounts of water and having different body weights, different exposure frequencies, and different exposure durations) as well as differences in response (e.g., genetic differences in resistance to a chemical dose). Differences among individuals in a population are referred to as inter-individual variability, differences for one individual over time is referred to as intra-individual variability. 2. In modeling, the differences in responses and other factors among species, i.e. interspecies variability, and within the same species, i.e. intraspecies variability, are part of the *uncertainty factors*, important to calculating *reference dose and concentrations.*

Variety: Distinct population within a species with distinguishing, heritable traits.

Vector: 1. Vehicle by which genetic material is inserted into a cell (usually a *plasmid* or *virus*). 2. Agent that carries *pathogens* among hosts (e.g. mosquito). 3. In physics, straight line segment with length representing magnitude and orientation in space representing direction.

Vegetative: In a growth mode; e.g. a vegetative bacterial cell is one that is growing and feeding, in contrast with its *spore.*

Veil of ignorance: John Rawls's postulation that the morality of an act can be based on the perspective of the most vulnerable members of a system. It is a thought experiment whereby the decision maker imagines that societal roles have been completely re-fashioned and redistributed, and that from behind the "veil" one does not know the role to which the person making the decision will be reassigned.

Vertical gene transfer: Crossing of organisms sexually and passing their genes on to following generations (usually called "crossing"). For example, gene transfer via pollen between plants of the same or related species takes place in the wild, so the transfer of disease and pest resistance from cultivated plants to related wild species and vice versa takes place irrespectively to the resistance genes initially acquired.

Vesicle: Abruscular mycorrhizal fungi's intracellular structures, usually spherical in shape.

***Vibrio*:** 1. *Anaerobes* of the genus *Vibrio.* 2. Rod-shaped, curved bacterial cell.

Virology: Branch of microbiology that is concerned with viruses and viral diseases.

Virulence: Degree of pathogenicity, i.e. the greater the virulence the more pathogenic the microbe is.

Virus: Sub-microscopic organism typically containing a protein coat surrounding a nucleic acid core, only able to grow in a living cell.

Viscosity: Molecular attractions with a fluid that evokes a tendency to deform under applied force. Internal friction within a fluid that causes it to resist flow. Absolute or dynamic viscosity is a measure of a fluid's resistance to tangential or shear stress (typical units are centipoises). Kinematic viscosity is the ratio of dynamic viscosity to mass density, obtained by dividing dynamic viscosity by the fluid density (typical units are centistokes).

Volatile: Capacity to change to the vapor phase, often expressed as *vapor pressure.*

Volatile acid: Fatty acid containing ≤ 6 carbon atoms, with relatively high *aqueous solubility.* Often produced by *anaerobes.*

Volatile organic compound (VOC): Organic compound that readily evaporates under environmental conditions, e.g. benzene, methylene chloride.

Volatile solids (VS): Materials, generally organic, that can be driven off from a sample by heating, usually to 550 °C (1022 °F); nonvolatile inorganic solids (ash) remain.

Volatilization: The transfer of a chemical from its liquid phase to the gas phase.

Vulnerability: 1. Condition of an individual or population determined by physical, social, economic, and environmental variables, wherein susceptibility to a hazard increases (e.g., asthmatics are more vulnerable to the effects of some pollutants than is the average person). 2. Recently, the term has been a combination of *susceptibility* and high-end exposure; i.e. vulnerable subpopulations are those that are both susceptible and receive sufficient levels of exposure for an adverse outcome.

Waste activated sludge: Solids taken from the activated sludge systems to prevent accumulation and an in sludge return to add microbes to increase waste degradation.

Watershed: Topographic area drained by surface water, wherein all outflows are discharged through a single outlet.

Weight-of-evidence: Strength of data and information supporting a conclusion. When little reliable data are available the weight of evidence is lacking, whereas when numerous studies provide reliable information to support a particular position (e.g., exposure to a chemical associated with a health effect), the weight of evidence is strong. Used by regulatory Agencies to characterize the extent to which the available data support a hypothesis, e.g. whether a specific agent causes cancer in humans. Under the US EPA's 1986 risk assessment guidelines, the WOE was described by categories "A through E," i.e. Group A for known human carcinogens through Group E for agents with evidence of noncarcinogenicity. The approach outlined in EPA's guidelines for carcinogen risk assessment (2005) considers all scientific information in determining whether and under what conditions an agent may cause cancer in humans, and provides a narrative approach to characterize carcinogenicity rather than categories. Five standard weight-of-evidence descriptors are used as part of the narrative.

Wetland: Lands inundated by water at least part of the year, indicated by plants that require or at least tolerate saturated soil conditions for substantial time periods.

White rot fungus: Numerous species of fungi that attack lignin, cellulose, and hemicellulose; frequently used in *aerobic biodegradation.*

Wild type: Appearance of an organism that exists most frequently in nature. Contrasted with transgenic organisms; term commonly used in genetics to denote "normal" version of a mutated organism.

Wilderness: A large, natural (or nearly natural) region, not controlled by humans.

Willingness to pay: Economic concept meaning the most money that people will give for a good or service; depicted by the total area under a demand curve.

Wisdom: Insight, erudition, and enlightenment resulting from the accumulation of knowledge and the ability to discern what is meaningful from what is not.

Xenobiotic: 1. Compound that is not normally found in natural systems. 2. A general term for an *anthropogenic* substance that is not easily degraded (see *recalcitrant*) by native microbial populations.

Xenotransplantation: Transplantation of cells, tissue, and organs between non-related species.

Xerophile: Organism adapted to dry conditions.

Yeast: Unicellular fungus with a single nucleus that reproduces either asexually by budding or fission, or sexually through spore formation.

Z value: See *fugacity capacity constant.*

Zone of saturation: Underground layers below the water table where void space is filled with water. See *aquifer.* Compare to *vadose zone.*

Zooplankton: Fauna in the *plankton* community.

Zygote: *Diploid* cell that results from the fertilization of an egg cell by a sperm cell.

RESOURCES

Enhanced Attenuation: Chlorinated Organics Team. The Risk Assessment Information System Glossary: http://rais.ornl.gov/homepage/glossary.shtml#Committed%20effective%20dose%20equivalent.

Fox Chase Cancer Center, Glossary of Ethics Terms; http://www.fccc.edu/ethics/Glossary_of_Ethics_Terms.html.

L.M. Hinman, Ethics Update – Glossary; http://ethics.sandiego.edu/Glossary.html.

The Interstate Technology & Regulatory Council (2008). *Enhanced Attenuation: Chlorinated Organics.* EACO-1. Washington, DC: Interstate Technology & Regulatory Council.

R.N. Johnson, A Glossary of Standard Meanings of Common Terms in Ethical Theory: http://web.missouri.edu/~philrnj/eterms.html.

US Department of Health and Human Services, Agency for Toxic Substances and Disease Registry; http://www.atsdr.cdc.gov/glossary.html.

US Department of Health and Human Services, Office of Research Integrity; http://ori.dhhs.gov/education/products/rcradmin/glossary.shtml.

US Environmental Protection Agency, Argonne National Laboratory and US Army Corps of Engineers (2009). The Brownfields and Land Revitalization Support Center. Glossary: http://www.brownfieldstsc.org/glossary.cfm; accessed on September 9, 2009.

US Environmental Protection Agency, Glossary of IRIS Terms; http://www.epa.gov/iris/gloss8.htm.

US Environmental Protection Agency. Integrated Risk Information System Glossary: http://www.epa.gov/NCEA/iris/help_gloss.htm.

US Environmental Protection Agency, Terms of Environment: Glossary, Abbreviations and Acronyms; http://www.epa.gov/OCEPAterms/.

INDEX

AADT, *see* Annual average daily traffic
Absorption, 118
ACRE, *see* Advisory Committee for Releases to the Environment
Activated sludge, 349–351
Activation energy, 104
Adsorption, 118
Advection, 151–154
Advisory Committee for Releases to the Environment (ACRE), 452
Aerobic biodegradation, *see* Bioremediation
Aerosol, *see also* Particulate matter
  definition, 36
  detection limit, 50–53
AEROWIN, bioremediation estimation, 388
Agricultural biotechnology
  corn, 479–484
  gene flow, 484–485
  overview, 477–479
Air–water partitioning, 122–123
ALARP, *see* As low as reasonably practical
Algal index value, 210
AMA, *see* American Medical Association
American Medical Association (AMA), ethics, 607
American Society for Microbiology (ASM), code of conduct, 580
American Society of Civil Engineers (ASCE), code of ethics, 579
Ames test, 244
Ammonification, 144
Amphibian metamorphosis, screening assays, 380, 384
Amphibilic, 108
Anabolism, 102,103
Anaerobic biodegradation, *see* Bioremediation
Anaerobic upflow filter, 357
Androgen receptor (AR), binding assays, 380
Animal biotechnology, overview, 476–477
Annual average daily traffic (AADT), 511–512, 514–518
Anthrax, *see Bacillus anthracis*
Anthropocentrism, 593–594
Antibiotic resistance
  cross-resistance, 288–289
  transfer between bacteria, 35–36
AOPWIN, bioremediation estimation, 387–388
Applied research, 587–591
Aquatic toxicity
  fish, 214
  green algae, 214
AR, *see* Androgen receptor
Aromatase, activity assay, 381
Aromatic compounds, 74–75
Artificial expression system, 449
As low as reasonably practical (ALARP), 275
ASCE, *see* American Society of Civil Engineers
ASM, *see* American Society for Microbiology
Atmospheric oxidation, 213
Atomic absorption, biotechnological waste analysis, 524

Bacillus anthracis
  contamination and decontamination, 221–225
  detection, 226
  limit of detection, 53–55
Bacillus thuringiensis (Bt)
  fish toxicity, 169–170
  hazard/risk classification, 236
  mechanism of toxicity, 171
BACT, *see* Best achievable control technologies
Bacteria, *see also specific bacteria*
  classification, 61, 64
  distinction between progenitor and genetically modified bacteria, 299–305
  gene flow monitoring, 533–535
  genera for persistent organic contaminant degradation, 104
  Risk Group, 2, 300–301
Basic research, 587–591
Bayes' theorem, 301
BBDR model, *see* Biologically-based dose–response model
BCF, *see* Bioconcentration factor
BCFBAF, bioremediation estimation, 388
BCR, *see* Benefit–cost ratio
Beer–Lambert law, 524
Benefit–cost ratio (BCR)
  biofuel, 479–481
  overview, 565–566
Best achievable control technologies (BACT), 258
Bhopal disaster, 309–312
Bioaccumulation, 464–469
Bioaugmentation, 262
Bioavailability, 128–131
Biocentrism, 593–594
Biochemical oxygen demand (BOD), 175
Biochemodynamics
  bioremediation, 332–334
  environmental biochemodynamics, 65–72
  genetic material, 428–431
  nitrogen, 87–91
  overview, 2–3
  pharmaceuticals, 287–290
  phytoremediation, 13
  sulfur, 87–91
  transport
    advection, 151–154
    diffusion, 157–158
    dispersion, 154–157
    genetic materials in environment, 161–163
    load, 142–145
    models, 159–161
    overall effect of fluxes, sinks and sources, 159
    total maximum daily load, 145–151
Bioconcentration, 213, 243
Bioconcentration factor (BCF), 129–130, 140
Biodegradation, 213, 331–332, *see also* Bioremediation
Biodegradation Probability Program (BPP), bioremediation estimation, 385
Biodegradation rate, 335–336
Biodegradation rate constant, 126–127
Bioengineering
  codes of conduct, 579–585
  ethics, 579
Biofilm, bioremediation, 342–347
Biofuel, 479–481
Biohazardous agent, classification, 300
BioHCwin, bioremediation estimation, 388
Biological indicator, 202–205
Biological oxygen demand (BOD), 329
Biological warfare, exposure factors, 60
Biologically-based dose–response (BBDR) model, 3
Biomagnification, 129
Biomarkers
  bioremediation, 361–362
  databases and exposure reconstruction, 3–5

Biomimicry
    environmental, 56–58
    principles, 506
Biomonitoring, 204
Biopharming, 241
Biophile
    cycling, 72–75
    definition, 58
    elements, 58–59
Bioprospecting, 460
Bioremediation
    aerobic biodegradation
        activated sludge, 349–351
        optimization, 354–356
        overview, 347–348
        ponds and lagoons, 352–354
        trickling filter, 348–349
    anaerobic biodegradation, 342, 356–358
    applications, 325–326
    bacteria growth curve, 338–340
    biochemodynamic film, 342–347
    biochemodynamics, 332–334
    biomarkers, 361–362
    chemical modeling of microbial activity, 369–379
    digestion, 335–347
    genetically modified organism bioengineering, 362–366
    goal, 331
    mechanisms, 327–328
    multimedia-multiphase bioremediation, 358–360
    off-site treatment, 334
    oxygen systematic view, 328–331
    phytoremediation, 360–361
    rates, 335–338, 340–341
    redox reactions, 336
    risk analysis, 380–393
    success measurement, 366–367
Biosensor, 210–211
Biosorption, 351
Biotechnological waste
    clean-up overview, 492
    contamination source interventions, 492–497
    exposure control interventions, 500
    fluorescent in-situ hybridization for environmental monitoring, 526–529
    nitrogen, 505–506
    point of release interventions, 497–498
    sampling and analysis, 506–529
    sulfur, 505–506
    thermal treatment, 500–505
    transport interventions, 498–499
    uncertainty in assessment, 529–532
Biotechnology
    definition, 21
    historical perspective, 10, 15
    risks and reliability, 17–20
    technology ranking for developing country health improvement, 16, 18
    uncertainties, 16–17
Bioterrorism, 314–316
Biowall, 355
BIOWIN, bioremediation estimation, 385, 388–395, 457
Bisphenol A (BPA), environmental fate, 471–473
BOD, *see* Biochemical oxygen demand
BOD, *see* Biological oxygen demand
Body burden, 464
    definition, 115
    reconstruction, 2
BPA, *see* Bisphenol A
BPP, *see* Biodegradation Probability Program
Bt, *see Bacillus thuringiensis*
Bulk attenuation rate constant, 127–128
Bush, Vennevar, 587
Butterfly Effect, 13–14

CAA, *see* Clean Air Act
CAFO, *see* Combined animal feeding operation
Cancer risk, calculation, 255–256
Cancer slope factors
    inhalation slope factors, 642–647
    oral slope factors, 642–647
    overview, 641, 647–648
Carbohydrates, chemical indicators of biological agents, 293–295
Carbon
    biogeochemistry, 75, 78–87
    chemical modeling of microbial activity, 372–373
    cycle, 73
    elemental versus organic, 294
    nanostructures, 56
Carbon dioxide, 84
Catabolism, 102–103
Catalyst, 104–106
Causality, *see* Hill's criteria for causality
Cause–effect–event chain, 303–304
CDER, *see* Center for Drug Evaluation and Research
Center for Drug Evaluation and Research (CDER), 463
CERCLA, *see* Comprehensive Environmental Response, Compensation, and Liability Act
Chaos, 620
Chemical gradient, calculation, 525–526
Chemical oxygen demand (COD), 329–330
Chemisorption, 118
Chlorophyll a, environmental indicator, 209
Chlorophyll, environmental indicator, 207–209
Clean Air Act (CAA), 10, 51, 258
Clean Water Act (CWA), 145
Climate change, genetically modified algae sensitivity, 92–95
COD, *see* Chemical oxygen demand
Combined animal feeding operation (CAFO), 287
Cometabolism, bioremediation, 327
Comprehensive Environmental Response, Compensation, and Liability Act (CERCLA), 34
Consumer Specialty Products Association (CSPA), 455–456
Contaminated site, environmental clean-up, 33–34
Control volume, 402
Coral reef, biosystematic perspective on impairment, 435–439
Corn
    biofuel, 479–481
    genetic modification, 481–484
Cross-resistance, 288–289
CSPA, *see* Consumer Specialty Products Association
Cuyahoga basin, total maximum daily load, 145–149
CWA, *see* Clean Water Act

Data quality objective (DQO), 507–509
DDT
    persistence, 134
    rainbow trout effects of DDT and metabolites, 664–677
    risks versus benefits, 550
Deontology, 585
Deposition, advective transport, 153
Deposition velocity, 416
DES, *see* Diethylstilbesterol
Design for Disassembly (DFD), 214, 557
Design for Recycling (DFR), 214, 557
Design for the Environment (DFE), 214, 557
Destruction removal efficiency (DRE), calculation, 503–504
Detection limit
    aerosols, 51–53
    definitions, 46, 525
    determination, 47–50
    mercury and methylmercury, 47
    microbes, 53–55
Deterministic approach, risk assessment, 29, 31
DFD, *see* Design for Disassembly
DFE, *see* Design for the Environment
DFR, *see* Design for Recycling
Diethylstilbesterol (DES), 287
Diffusion, 157–158
Diffusion rate, 468
Dispersion, 154–157
Dissolved oxygen (DO), 175, 328–331, 540–542
DNA, structure, 100
DO, *see* Dissolved oxygen
Dose–response, assessment, 242, 244–248
DQO, *see* Data quality objective
DRE, *see* Destruction removal efficiency
Dual-use, 315
D value, 141

EA, *see* Environmental assessment
Ecocentrism, 593–596
Ecological indicator, 205, 207–209
Ecological Society of America (ESA), 432–435
Ecological Structure–Activity Relationships (ECOSTAR), bioremediation estimation, 387

ECOSTAR, *see* Ecological Structure–Activity Relationships
Ecosystem services, 595
ECOTOX database, rainbow trout effects of DDT and metabolites, 664–677
EDC, *see* Endocrine disrupting chemical
EJ, *see* Environmental justice
Endergonic reaction, 101
Endocrine disrupters, 137–138
Endocrine disrupting chemical (EDC), 470–476
Endothermic reaction, 106
Environmental assessment (EA), 6–8
Environmental biotechnology
- ethics and decisions, 585
- green engineering relationship, 212–221, 605–606
- historical perspective, 24–25
- public participation, 573–574
- scope, 15–16

Environmental impact statement (EIS)
- cover sheet, 640
- format, 639
- guidance sources, 636–638
- overview, 3, 6–7
- steps, 7

Environmental impact, biotechnology
- applying knowledge and gaining wisdom, 554–557
- benefit–cost ratio, 565–566
- cumulative impact, 539–548
- damage prediction steps, 567–573
- environmental accountability, 560–561
- environmental engineering, 557–558
- failure analysis, 552–554
- life cycle analysis, 551–552
- life cycle applications, 561–565
- science as social enterprise, 558–560
- uncertainty and complexity assessment, 548–551

Environmental justice (EJ), 410–415
Environmental medium, 151
Environmental persistence, 123, 125–126, 317–319
Environmental Protection Agency (EPA), 10, 21, 34, 384, 451, 454–457
Enzyme
- life cycle of industrial production, 497–498
- overview, 104–105
- production
  - environmental implications, 452–454, 459
  - organisms, 445, 448–451
  - overview, 445
  - regulatory control, 451–452

EPA, *see* Environmental Protection Agency
Equilibrium vapor pressure, 113
Equilibrium, thermodynamics, 100–101, 136, 139
ER, *see* Estrogen receptor
Error, types, 291
ESA, *see* Ecological Society of America
Estimated environmental concentration (EEC), 278
Estradiol, environmental fate, 471–473
Estrogen receptor (ER), binding assays, 380
Ethics
- anthropocentrism, 593–595
- biocentrism, 593–595
- ecocentrism, 593–596
- environmental ethics, 593–597
- sentientism, 593

Ethinylestradiol, environmental fate, 470–472
Eukaryotic cell, 99
Event tree
- bioremediation decision making, 605
- environmental damage analysis, 571–572

Exothermic reaction, 106
Exposure
- estimation, 248–255
- human exposure factors, 254
- pathway, 248

Exposure, risk, 241
Extraction, biotechnological waste for analysis, 522–523

Failure density, 313, 615
False negative, 248
False positive, 248, 290
FDA, *see* Food and Drug Administration
Federal Food, Drug, and Cosmetic Act (FFDCA), 380, 484
Federal Insecticide, Fungicide, and Rodenticide Act (FIFRA), 483
FFDCA, *see* Federal Food, Drug, and Cosmetic Act
Fick's law, 157, 468
FID, *see* Flame ionization detection
FIFRA, *see* Federal Insecticide, Fungicide, and Rodenticide Act
FISH, *see* Fluorescent in-situ hybridization
Fish
- *Bacillus thuringiensis* toxicity, 169–170
- invasive species control, 282–286
- life cycle, 286
- rainbow trout effects of DDT and metabolites, 664–677
- screening assays, 380

Flame ionization detection (FID), biotechnological waste analysis, 523
Flammability, 213
Fluorescent in-situ hybridization (FISH), environmental monitoring, 526–529
Flux density, 158
Food and Drug Administration (FDA), 463
Food Quality Protection Act (FQPA), 247, 380
Food web, compartmental model, 191–198
FQPA, *see* Food Quality Protection Act
Francisella tularensis, limit of detection, 53–55
Free energy, *see* Gibbs free energy
Freundlich sorption isotherm, 120
Fuel efficiency, biofuel, 479–481
Fugacity, 115–116, 185–189
Fugacity-based mass balance modeling, 189–191
Fugacity capacity constant, 139–142, 185

GAC, *see* Granulated activated carbon
Gas chromatography (GC), biotechnological waste analysis, 523–524
GC, *see* Gas chromatography
Gene flow
- agricultural biotechnology, 484–485
- genetically modified organisms, 428–430
- monitoring, 533–535

Genetic engineering, *see also* Genetically modified organism
- cisgenic versus transgenic organisms, 424
- conventional breeding, 421
- environmental implications, 417, 421
- modification without foreign DNA, 422
- transfected DNA, 422–423
- vector-borne DNA, 423–424

Genetically modified food
- risks, 581–582
- testing and safety, 583–584

Genetically modified organism (GMO)
- algae and climate sensitivity, 92–95
- bioengineering for bioremediation, 362–366
- definition, 421
- environmental justice, 410–415
- European Federation of Biotechnology risk classification, 11
- gene flow, 428–432
- plants, 424–428
- risk recommendations, 432–435
- risks, 27
- vaccine production, 485–486

Genomics, 168
Geographic information system (GIS), environmental monitoring, 511–515
Gibbs free energy
- microbial metabolism importance, 102–106
- overview, 101–102

GIS, *see* Geographic information system
GMO, *see* Genetically modified organism
Granulated activated carbon (GAC), 333–334
Green engineering, environmental biotechnology relationship, 212–221
Greenhouse gas
- overview, 83–85
- sequestration
  - soil, 85–87
  - active sequestration, 87

Half-life, 340
- biochemodynamic persistence and half-life, 134–136
- overview, 123, 126

Halide-resistant crop (HRC), 268
Hazard identification, 241–242
Hazard index (HI), 257
Hazard quotient (HQ), 257
Hazard rate, 313, 615
Hazard
- risk, 241
- spectrum, 613

Henry's law, 122, 140, 213
HENRYWIN, bioremediation estimation, 388
Hershberger assay, 381
HFE, *see* Human factors engineering
HHM, *see* Hierarchical holographic modeling
HI, *see* Hazard index
Hierarchical holographic modeling (HHM), 161
Hill's criteria for causality, 264, 296, 568–569
HQ, *see* Hazard quotient
HRC, *see* Halide-resistant crop
HRT, *see* Hydraulic retention time
Human factors engineering (HFE)
 failure types
  critical path, 309–313
  extraordinary natural circumstances, 308–309
  lack of imagination, 314
  mistakes and miscalculations, 308
  negligence, 313–314
 overview, 306–307
 utility as measure of success, 307–308
Hydraulic retention time (HRT), 357
Hydrogen peroxide, endocrine disrupting chemical treatment in drinking water, 473–476
Hydrophilic, 107
Hydrophobic, 108
HYDROWIN, bioremediation estimation, 388
Hysteresis, 192

IBI, *see* Index of biological integrity
Ideal gas law, 101, 111, 140
IDL, *see* Instrument detection limit
Incineration, *see* Thermal treatment
Index of biological integrity (IBI)
 alternative metrics, 176–178
 calculation and interpretation, 179
 original metrics, 175
Industrial sectors, biotechnology applications, 446–447
Instrument detection limit (IDL), 525
Intake, equation, 252
Integrated Risk Information System (IRIS), 454
Invasive Species Specialist Group (ISSG), 282
Ion exchange, 118
IRIS, *see* Integrated Risk Information System
Iron, nanotechnology for groundwater treatment, 57–58
ISSG, *see* Invasive Species Specialist Group

Kinetics, 136, 139
KOAWIN, bioremediation estimation, 388
KOCWIN, bioremediation estimation, 388
KOWWIN, bioremediation estimation, 387

LADD, *see* Lifetime average daily dose
Lagoon, aerobic biodegradation, 352–354
Landfill bioreactor
 overview, 406
 phases
  acid formation, 407
  final maturation and stabilization, 407–408
  initial adjustment, 406–407
  methane fermentation, 407
  transition, 407
LC, *see* Lethal concentration
LC, *see* Liquid chromatography
LCA, *see* Life cycle analysis
LD, *see* Lethal dose
Lethal concentration (LC), 243
Lethal dose (LD), 243
LEV3EPI, bioremediation estimation, 388
Level of concern (LOC), 278–281
Life cycle analysis (LCA), 11, 493, 499, 551–552, 556, 561–565
Lifetime average daily dose (LADD), 253–256, 259–261
Limit of detection, *see* Detection limit
Limit of quantitation (LOQ), 525
Lipophilic, 108
Lipophobic, 108
Liposome, 55
Liquid chromatography (LC), biotechnological waste analysis, 523–524
Load, 142–145
LOC, *see* Level of concern
LOEL, *see* Lowest observed effect level
LOQ, *see* Limit of quantitation
Lowest observed effect level (LOEL), 246

MACT, *see* Maximally achievable control technologies
Manganese, elimination, 466
Mass balance
 concentration-based mass balance modeling, 181–183
 control volume calculation, 402
 fugacity-based mass balance modeling, 189–191
 overview, 113
Mass spectrometry (MS), biotechnological waste analysis, 524
Maximally achievable control technologies (MACT), 258
Maximum daily dose (MDD), 253, 257
Maximum exposed individual (MEI), 256
MDD, *see* Maximum daily dose
MDL, *see* Method detection limit
Medical biotechnology
 environmental implications, 476
 hormonally-active agents, 470–476
 overview, 459–464
 uptake and bioaccumulation, 464–470
MEI, *see* Maximum exposed individual
Mercury, detection limit, 47
Metabolonomics, 168
Methane, 84
Methanogenesis, 342
Method detection limit (MDL)
 definition, 46
 determination, 47–50
Methyl isocyanate (MIC)
 Bhopal disaster, 309–312
 properties, 311
 synthesis, 310
Methylmercury, detection limit, 47
MIC, *see* Methyl isocyanate
Minimum limit (ML), 46
Mitigation measures, hypothetical impact on outcomes, 9–10
ML, *see* Minimum limit
Mobile source, 143
Mobile source air toxics (MSAT), 509
Monitoring
 biotechnological waste, 506–520
 definition, 506
Monod equation, 341
MPBWIN, bioremediation estimation, 388
MS, *see* Mass spectrometry
MSAT, *see* Mobile source air toxics

Nanotechnology
 carbon structures, 56
 iron nanotechnology for groundwater treatment, 57–58
 life cycle perspective, 57
National Environmental Policy Act (NEPA), 3, 6, 8, 10
Negative paradigm (NP), 570
NEPA, *see* National Environmental Policy Act
Net goodness (NG)
 analysis, 603–605
 calculation, 569–570
Net primary productivity, 195
NG, *see* Net goodness
Nitric oxide (NO), 88–89
 biodegradation marker, 367
 monitoring, 527–529
Nitrification, 144
Nitrogen
 biochemodynamics, 87–91
 biotechnological waste, 505–506
 chemical modeling of microbial activity, 373–376
Nitrogen dioxide, 88–89
Nitrogenous biological oxygen demand, 330
Nitrous oxide, 85
NOEL, *see* No observed effect level
Noncancer risk, calculation, 256
Nonpoint source, 142
No observed effect level (NOEL), 245–248, 279
NP, *see* Negative paradigm

Octanol–water partition coefficient, 108, 123, 126, 179
Offsetting behavior, 619–620
Organic carbon partition coefficient, 119–121
Organic compound
 classification, 74–75
 definition, 73
 physicochemical properties and hazards, 78
 structures, 75–78

Organophosphate pesticides
mechanism of action, 316–317
persistence, 318–319
Oxygen, systematic view, 328–331

PAD, *see* Population adjusted dose
Partial pressure, 111
Particulate matter (PM), *see also* Aerosol
diameter approximation, 37
environmental impact, 37–38
respiratory system interactions, 39–40
sampling, 52–53
size classification, 37
tropospheric particles, 36
Partition coefficient, 119–120
Patent, organisms, 460–461
Pathogen-derived resistance (PDR), 269
PBT, *see* Persistent bioaccumulating toxic substance
PBTK modelPhysiologically-based toxicokinetic model
PCR, *see* Polymerase chain reaction
PDF, *see* Probability density function
PDR, *see* Pathogen-derived resistance
Permeable reactive barrier (PRB), 355
Permethrin, 317
Persistent bioaccumulating toxic substance (PBT), 130–131, 457–458
Persistent organic pollutant (POP)
overview, 131–133
table of compounds by priority, 652–661
Pharmacodynamics, 115
Phase partitioning, 107–110
Physiologically-based toxicokinetic (PBTK) model, 3
Phytoaccumulation, 360
Phytoremediation, 13, 360–361
Phytostabilization, 361
PIP, *see* Plant-incorporated protectant
Plant-incorporated protectant (PIP), 483–484
Plant pathogens, European Federation of Biotechnology classification, 12
PM, *see* Particulate matter
PMN, *see* Premanufacture notice
Point decay rate constant, 127–128
Point source, 142–143
Poisson distribution, 611
Polymerase chain reaction (PCR)
overview, 450
verification in biological agent detection, 649–650
Pond, aerobic biodegradation, 356
POP, *see* Persistent organic pollutant
Population adjusted dose (PAD), 247
PPPositive paradigm
PQL, *see* Practical quantitation limit
Practical quantitation limit (PQL), 525
PRB, *see* Permeable reactive barrier
Premanufacture notice (PMN), 233, 451
Probability
biotechnology, 609–611
risk, 230
Probability density function (PDF), 611
Prokaryotic cell, 99
Proteomics, 168
Pyrethoids
pesticides, 317
structures, 321
toxicity, 320
Pyrolysis, *see* Thermal treatment

QSAR, *see* Quantitative structure–activity relationship
Quantitative structure–activity relationship (QSAR), 179, 230, 384, 456

Rainbow trout, effects of DDT and metabolites, 663–677
Rate constant, 126–128
Reaction rate, 136, 159
Recalcitrance, chemicals, 418
Record of Decision (ROD), environmental impact statement, 7
Red tide, 237
Reference dose (RfD), 245–247
Regulatory costs, 292
Resilience, systems, 192, 195–196
Resmethrin, 317
Respiratory system
anatomy, 38–39
particulate behavior, 39–40
RfD, *see* Reference dose
Rhizodegradation, 361
Rhizofiltration, 360–362
Risk
causes, 295–300
classes, 231
communication techniques, 264–266
genetically modified organism recommendations, 433–436
green transgenesis, 267–270
homeostasis and theory of offsetting behavior, 619–620
social definitions, 237
systematic view, 404–405
technical definitions, 237
tradeoffs, 550–551
tradeoffs in environmental biotechnology, 621–628
Risk analysis, definition, 19, 229
Risk assessment
definition, 229, 237
ecological risk framework, 232–233
paradigm, 230
principles, 262
risk perception comparison, 278
Risk-based cleanup standards, 258–269
Risk management
acceptable risk, 28–29, 31–35
definition, 21, 237, 307
exposure and effect process risk management, 30
paradigm, 230
Risk quotient (RQ), 278–281, 317
Risk reduction
biosystemic intervention, 281–293
bioterrorism, 314–316
chemical indicators of biological agents, 293–295
human factors engineering
failure types
critical path, 309–313
extraordinary natural circumstances, 308–309
lack of imagination, 314
mistakes and miscalculations, 308
negligence, 313–314
overview, 306–307
utility as measure of success, 307–308
level of concern, 279–281
overview, 275–278
risk causes, 295–300
risk quotient, 278–281
Rock Creek watershed, total maximum daily load, 149–151
ROD, *see* Record of Decision
Ross River virus, , ecosystem and human disease, 23–24
RQ, *see* Risk quotient
Rule of five, chemical adsorption, 418–421
Runoff, 142

Safety, bioengineering, 607–609
Sampling
overview, 520–521
types, 523–529
Scale, biosystems, 196–201
Sedimentation
concentration-based mass balance modeling, 183
fugacity-based mass balance modeling, 189
Sentientism, 593–594
SEPA, *see* State Environmental Policy Act
Severity, risk, 230
Sink, 114–115, 142
Sinorhizobium meliloti RMBPC-2
alfalfa yield impact, 234
antibiotic resistance, 234
commercialization risks, 232–233, 235–236
environmental fate, 234–235
exposure assessment, 234
nodulation, 234
risk characterization, 235
Slope factors, cancer, 641–648
Soil
carbon sequestration, 85–87
oxygen and carbon dioxide, 86
texture classification, 85
Solids retention time (SRT), 357
Solubility
overview, 106–107
phase partitioning, 107–110
polarity effects, 107
Sorption, 118–121
SRT, *see* Solids retention time
Stachybotrys, health hazards, 549

State Environmental Policy Act (SEPA), North Carolina, 8
Stationary source, 143
Step aeration, 351–352
Stokes' law, 36
STPWIN, bioremediation estimation, 388
Success, engineering
accountability, 599–600
criteria, 612
informing decisions, 601–603
net goodness analysis, 603–604
overview, 598
value, 599–601
Sulfide ion, oxidation, 90
Sulfur
biochemodynamics, 87–91
biotechnological waste, 504–505
Sumithrin, 317
Systems
biological indicators, 202–210
biosensor, 210–211
biotechnological systems
overview, 167–171, 586–593
reliability, 615–617
concentration-based mass balance modeling, 181–183
food web compartmental model, 191–195
fugacity, 185–188
fugacity-based mass balance modeling, 189–191
indices, 174–181
overview, 167
pollution control and prevention perspective, 171–174
resilience, 192, 195–196
scale in biosystems, 197–202
synergy and biotechnical analysis, 201

Tapered aeration, 350
TCE, *see* Trichloroethene
TCLP, *see* Toxicity characteristic leaching procedure
Terrorism, *see* Bioterrorism
Tetrachloroethene (TPE), biotransformation, 64–65
Thermal treatment, biotechnological waste
destruction removal efficiency calculation, 503–504
guidelines, 503
incineration steps, 501–502
overview, 500–501
pyrolysis, 504
vitrification, 504–505
Tier 1 screening, 380–385
Tier 2 screening, 384
TMDL, *see* Total maximum daily load
TNT, *see* Trinitrotoluene
Toluene, chemical modeling of microbial activity, 375, 378–379
Total maximum daily load (TMDL), 145–151
Toxicity characteristic leaching procedure (TCLP), 502
Toxic organic compounds, table of compounds by priority, 652–661
Toxic Release Inventory (TRI), 493–496
Toxic Substances Control Act (TSCA), 233–236, 447–448, 451, 457
Toxicokinetic modeling, dose and environmental exposure, 240
Toxin, biological toxin characteristics, 62–63
TPE, *see* Tetrachloroethene
Transcriptinomics, 168
Transformation, chemical, 113
TRI, *see* Toxic Release Inventory
Trichloroethene (TCE), biotransformation, 64–65
Trickling filter, 348–349
Trinitrotoluene (TNT), bioremediation, 623–628
TSCA, *see* Toxic Substances Control Act
Type I error, 291
Type II error, 291, 483

UF, *see* Uncertainty factor
Ultimate carbonaceous biological oxygen demand, 330
Ultraviolet light
carcinogenesis, 248
disinfection, 258
endocrine disrupting chemical treatment in drinking water, 473–476
Uncertainty
contamination assessment, 529–532
environmental impact of biotechnology, 548–551
types, 529
Uncertainty factor (UF), dose–response assessment, 247
Urea, synthesis, 72
Uterotrophic assay, 381
Utilitarianism, 275, 307

Vaccine, production from genetically modified organisms, 485–486
Vapor pressure, 111–112, 212
Vaporization
concentration-based mass balance modeling, 183
fugacity-based mass balance modeling, 189
Vitrification, *see* Thermal treatment
Volatilization, 122–128

Waste, *see* Biotechnological waste
Water quality criteria, 203
Water vapor, greenhouse gas, 83
WATERNT, bioremediation estimation, 388
WSKOWWIN, bioremediation estimation, 388
WVOLWIN, bioremediation estimation, 388

X-ray fluorescence, biotechnological waste analysis, 524

Yeast estrogen screen (YES), 471
Yersinia pestis, limit of detection, 53–55
YES, *see* Yeast estrogen screen

Z value, *see* Fugacity capacity constant

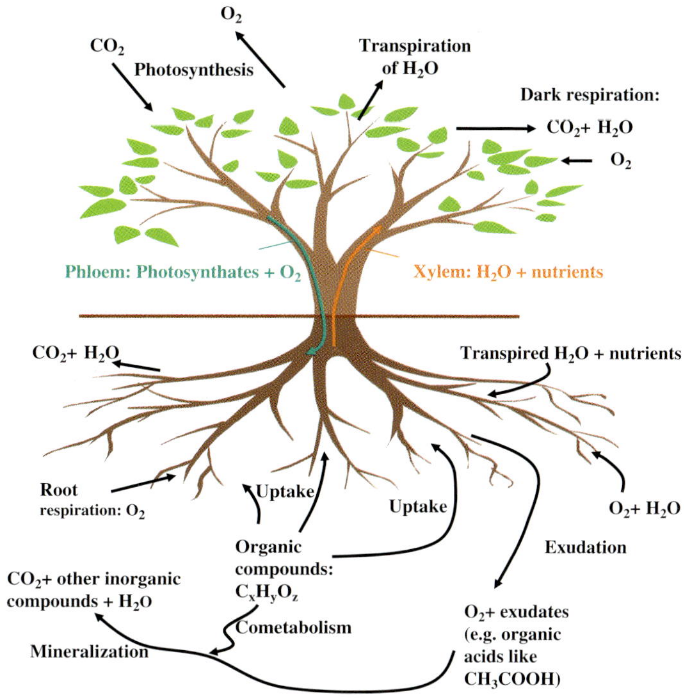

FIGURE 1.4

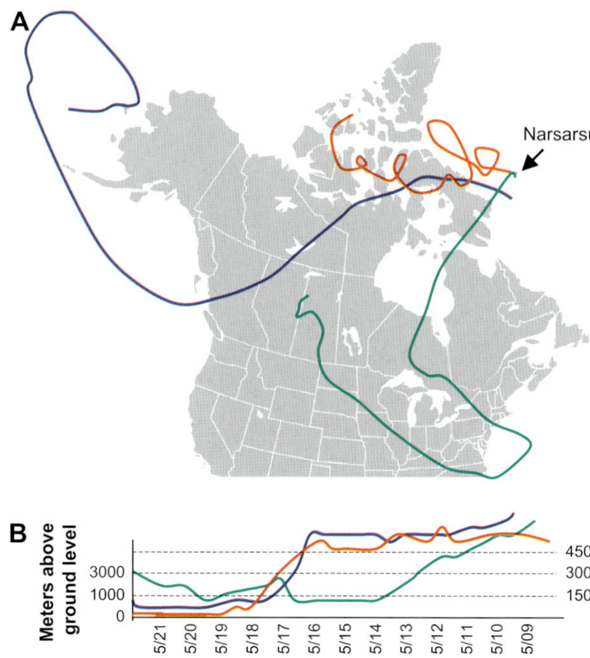

FIGURE 2.17

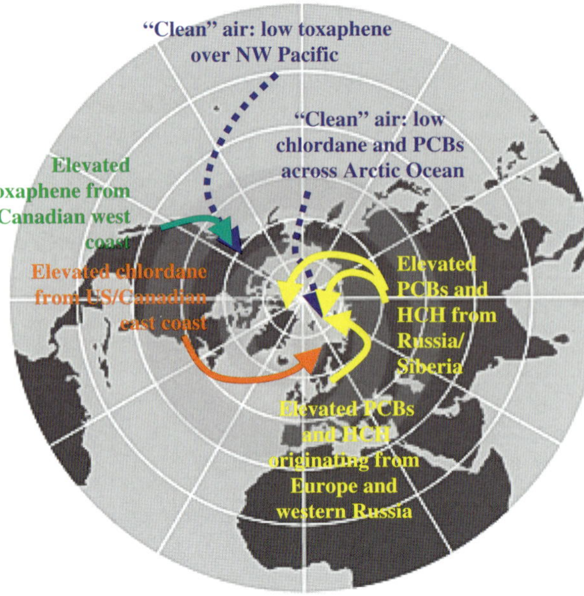

FIGURE 3.21

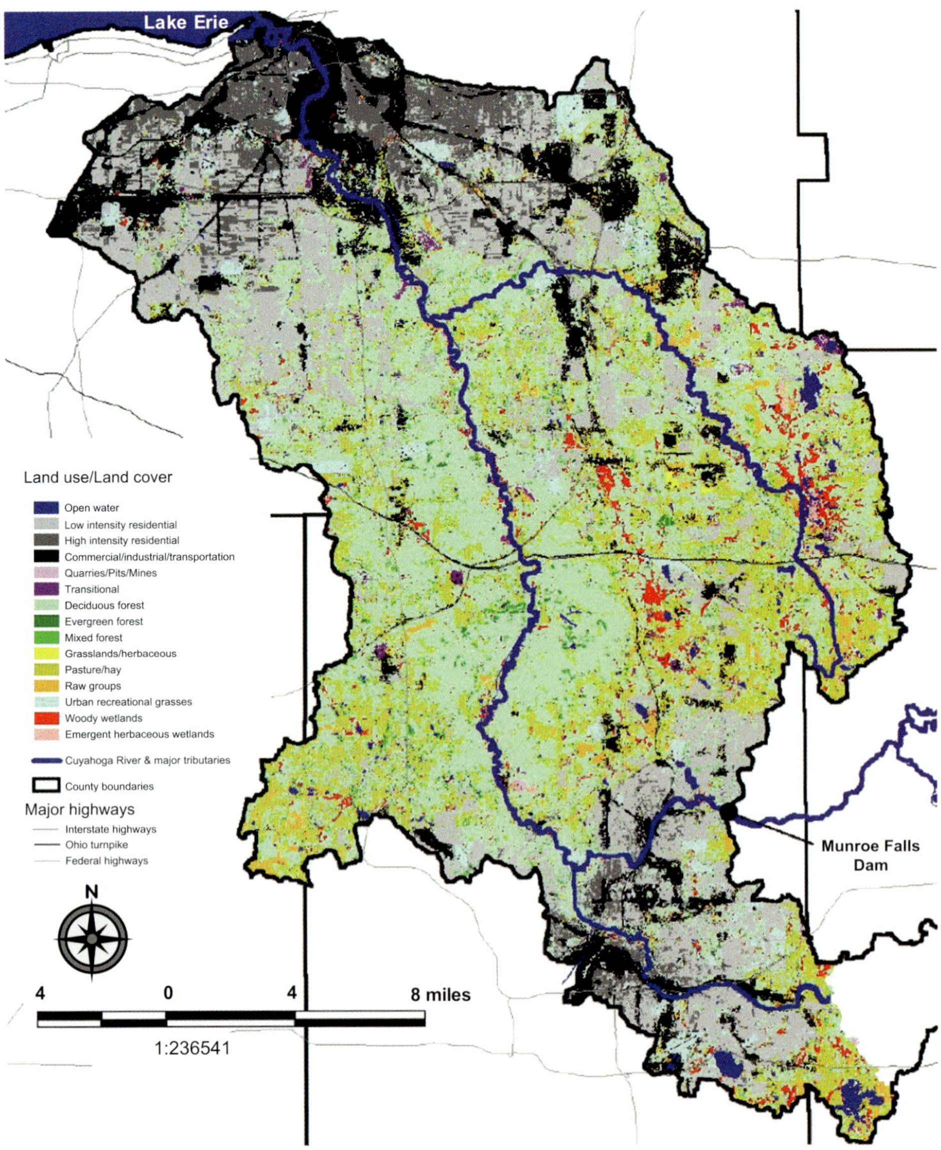

FIGURE 3.28

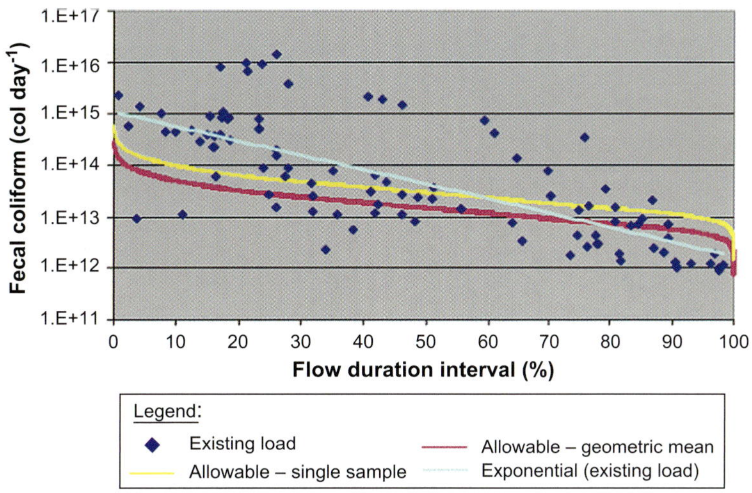

FIGURE 3.30

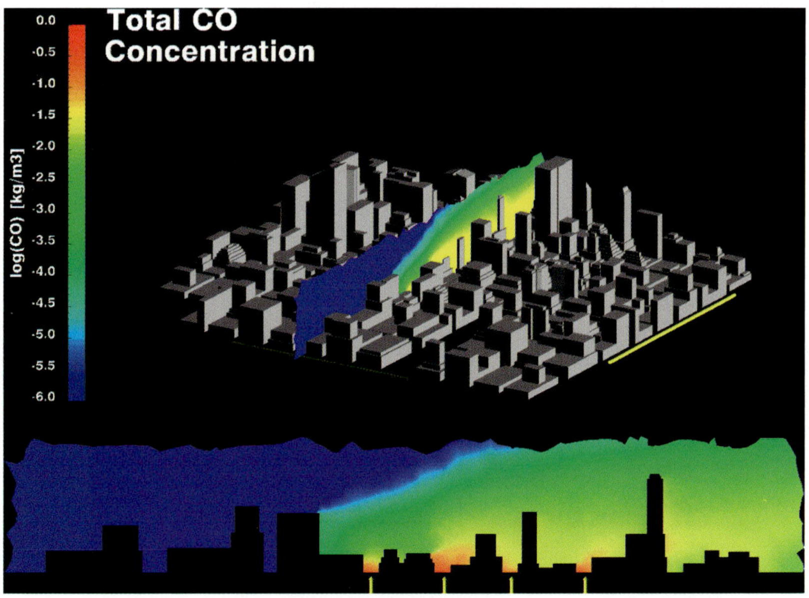

FIGURE 3.34

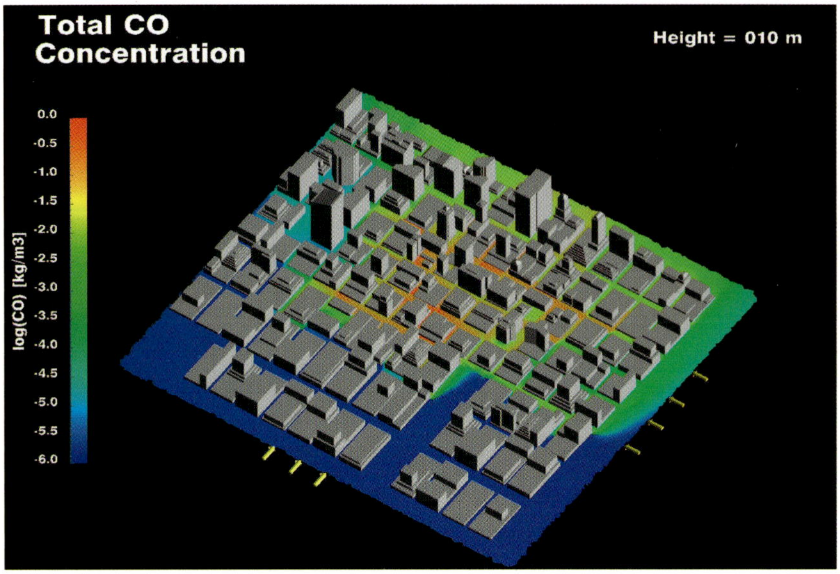

FIGURE 3.35

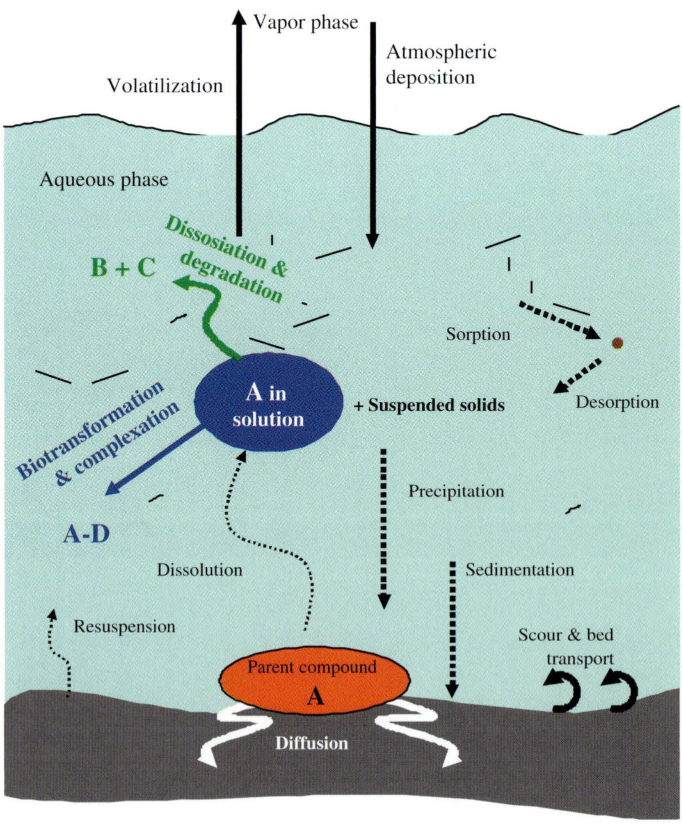

FIGURE 4.6

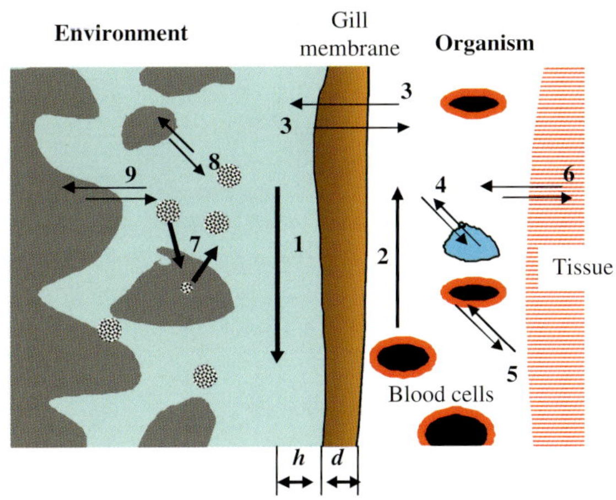

FIGURE 4.7

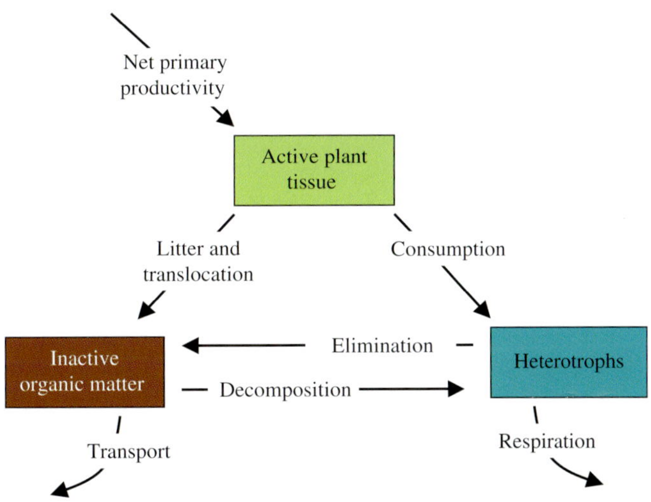

FIGURE 4.8

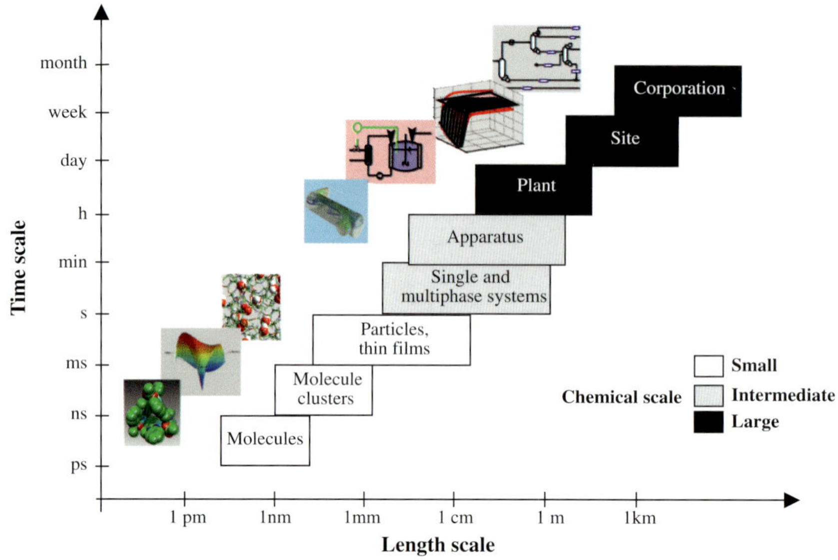

FIGURE 4.11

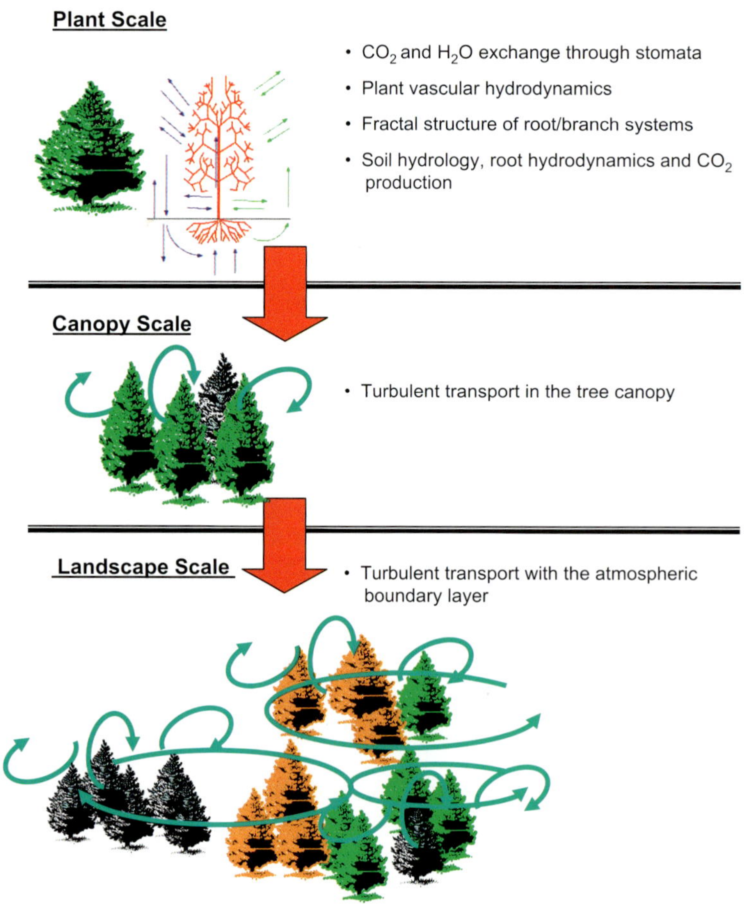

FIGURE 4.12

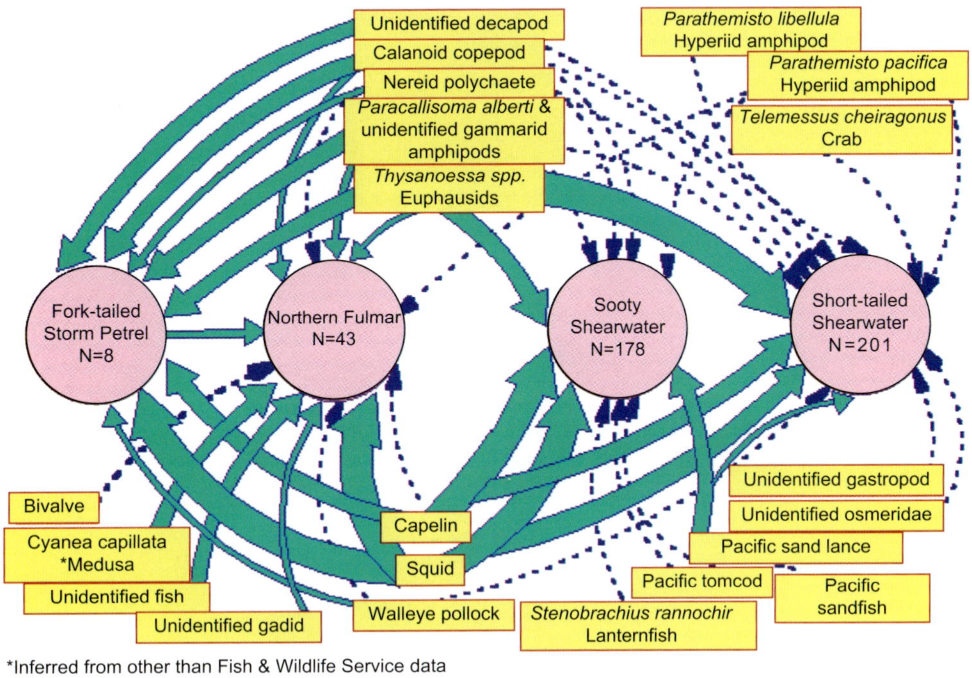

FIGURE 4.13

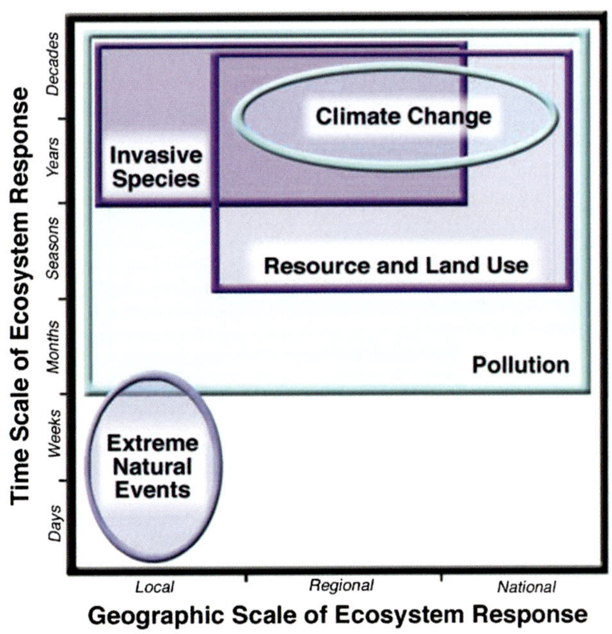

FIGURE 4.14

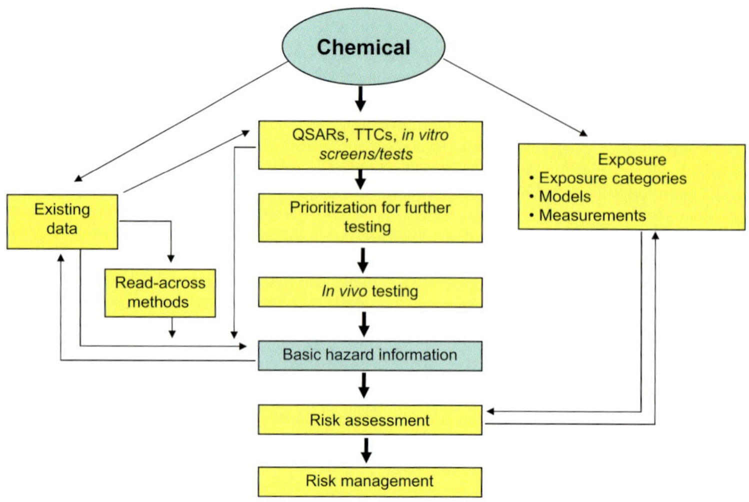

FIGURE 4.16

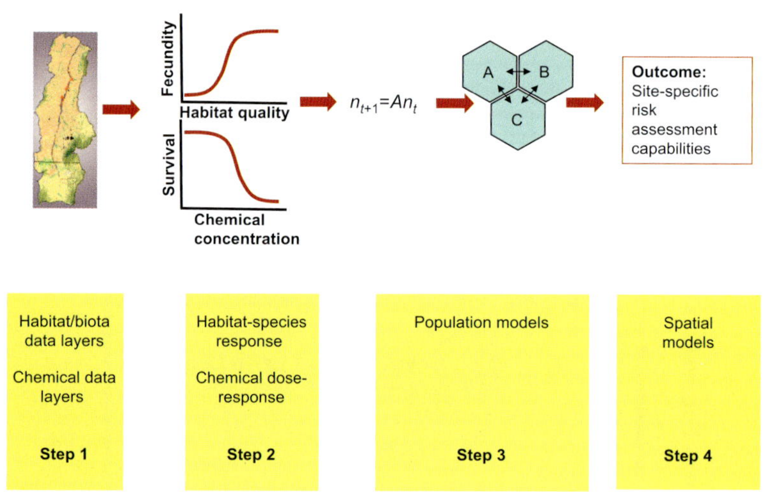

FIGURE 4.17

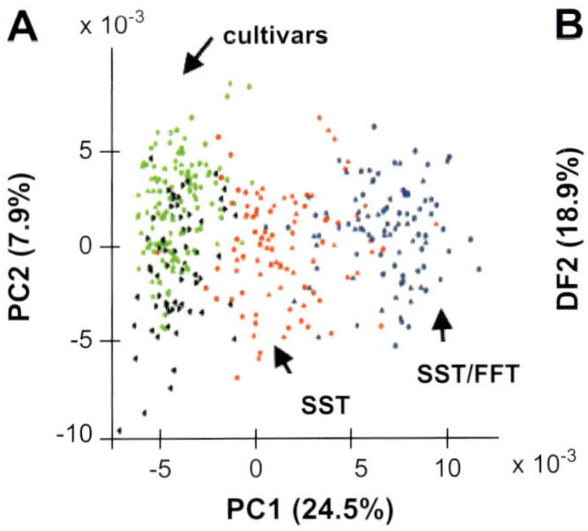

FIGURE 4.19

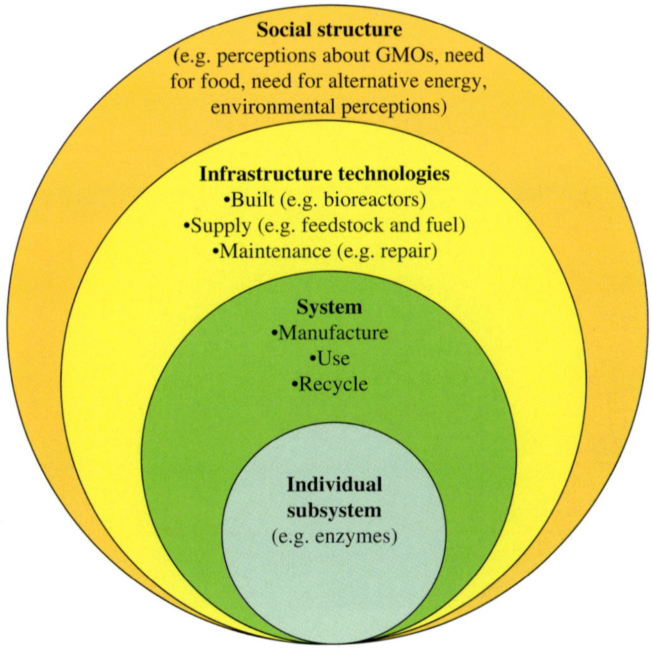

FIGURE 4.26

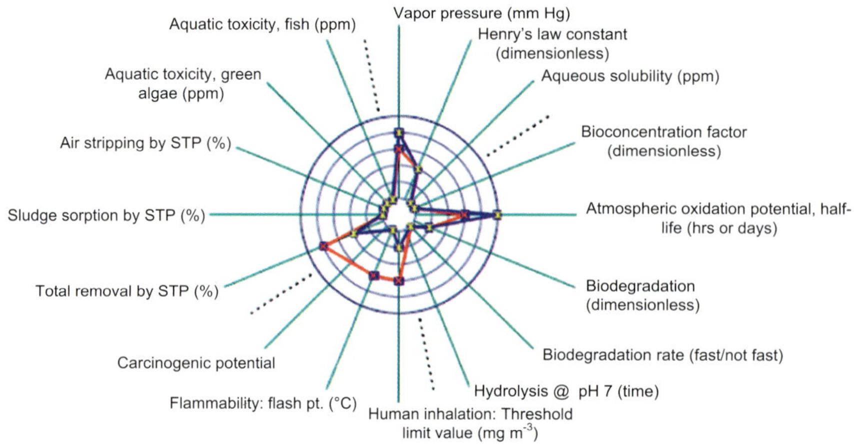

FIGURE 4.27

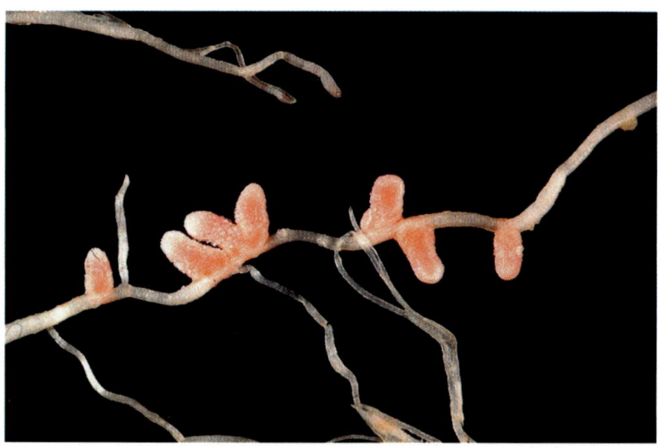

FIGURE 5.3

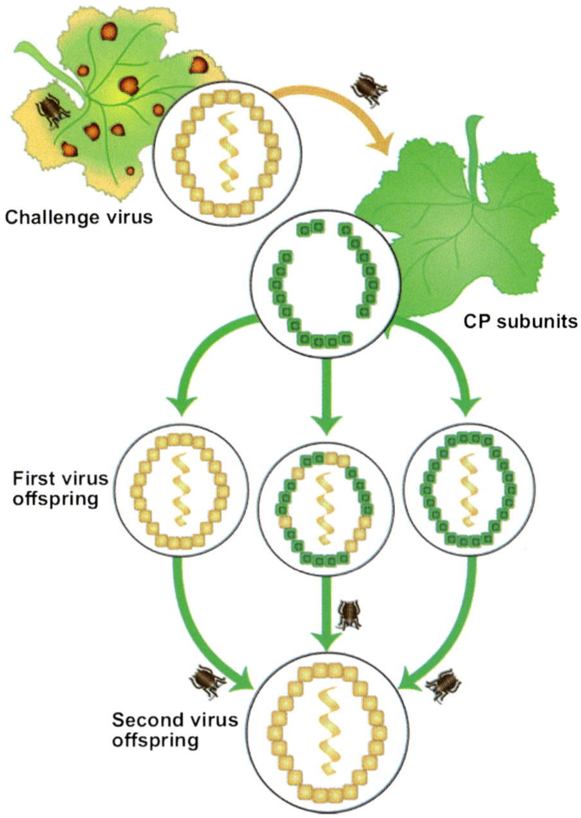

FIGURE 5.12

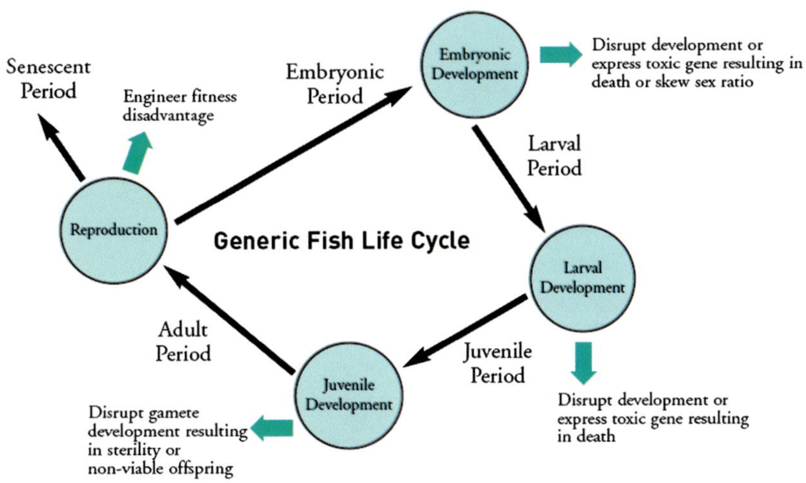

FIGURE 6.4

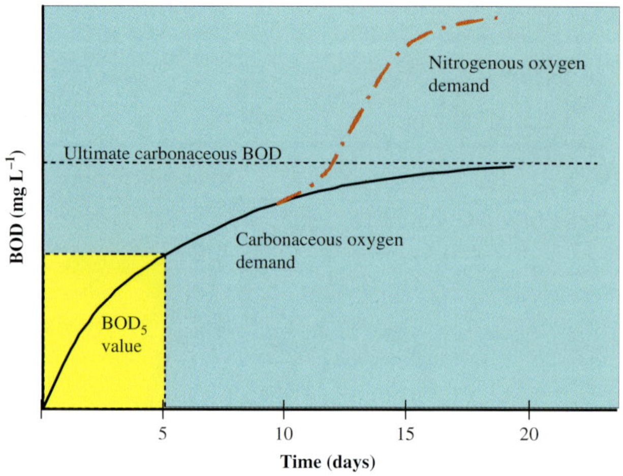

FIGURE 7.4

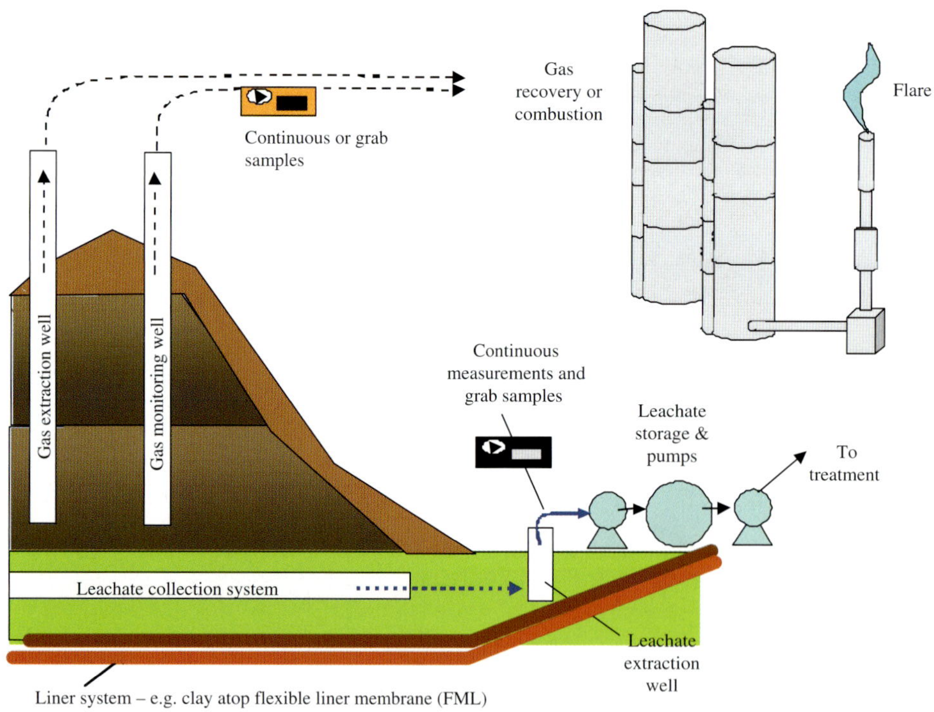

FIGURE 7.6

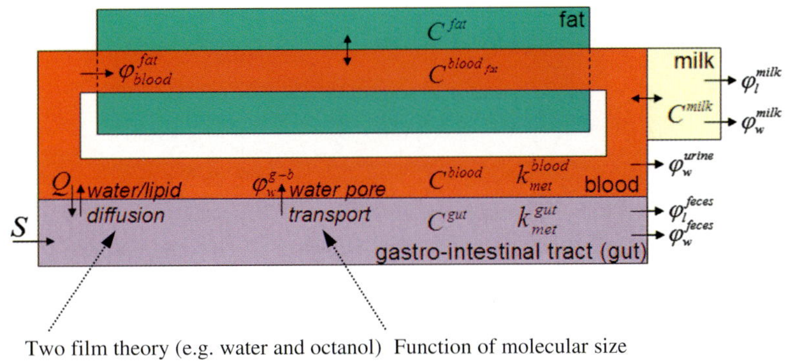

FIGURE 7.16

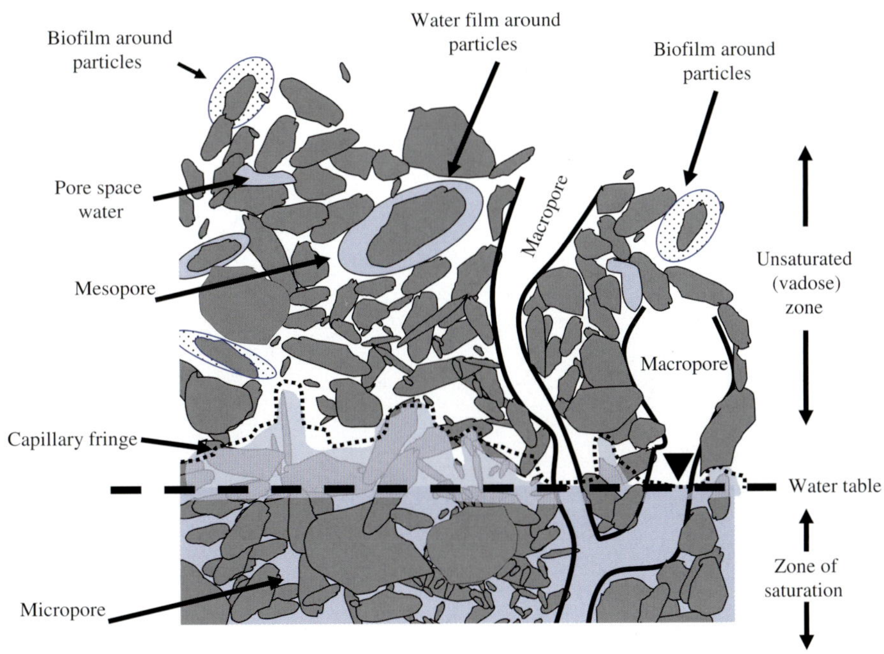

FIGURE 7.18

FIGURE 7.24

FIGURE 7.26

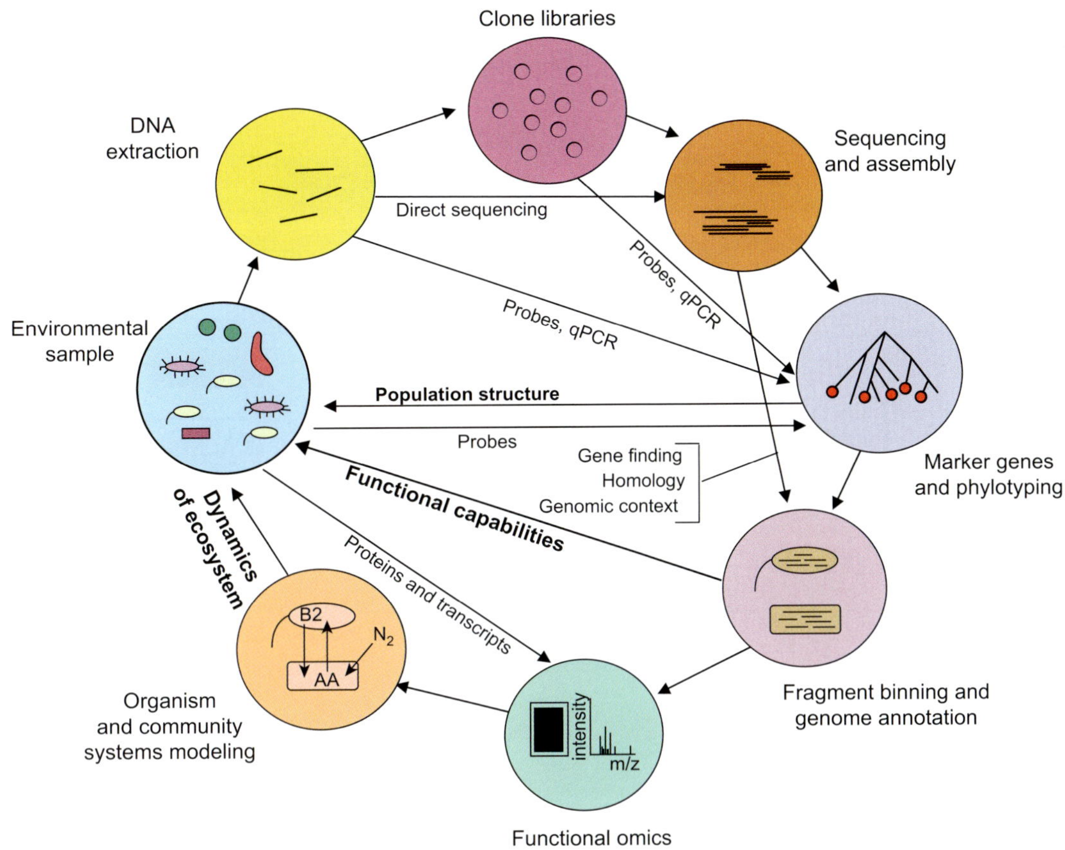

FIGURE 7.30

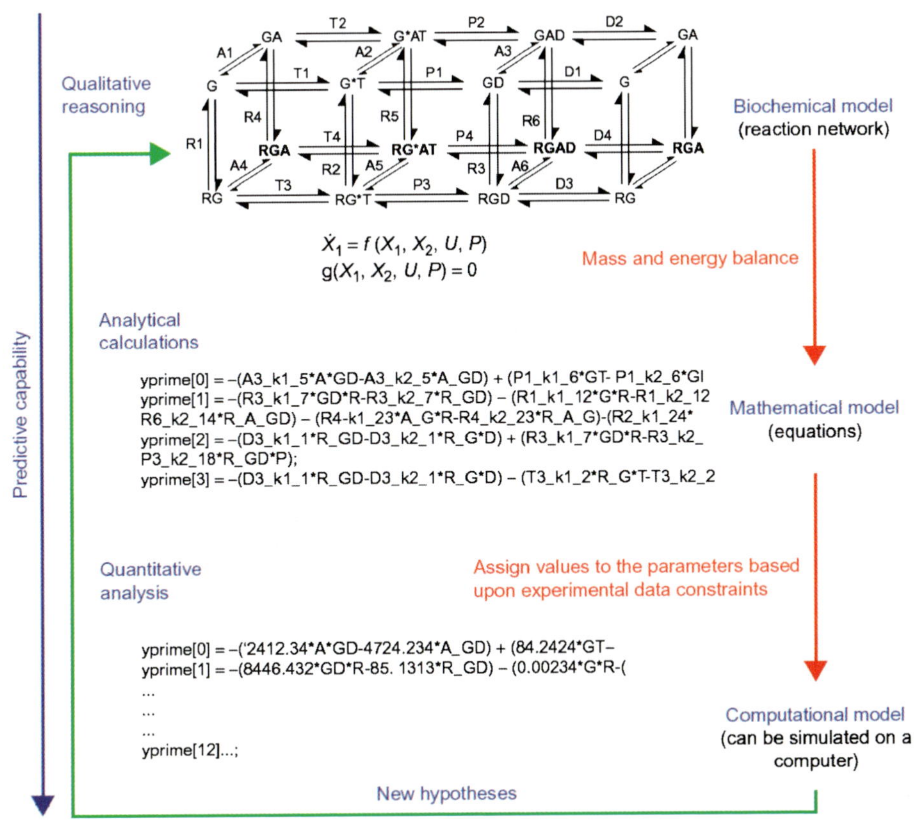

FIGURE 7.32

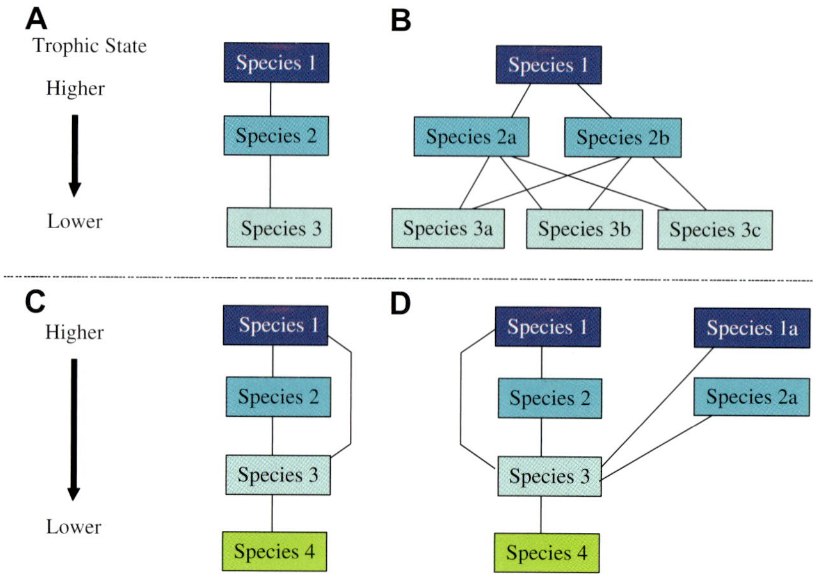

FIGURE 8.1

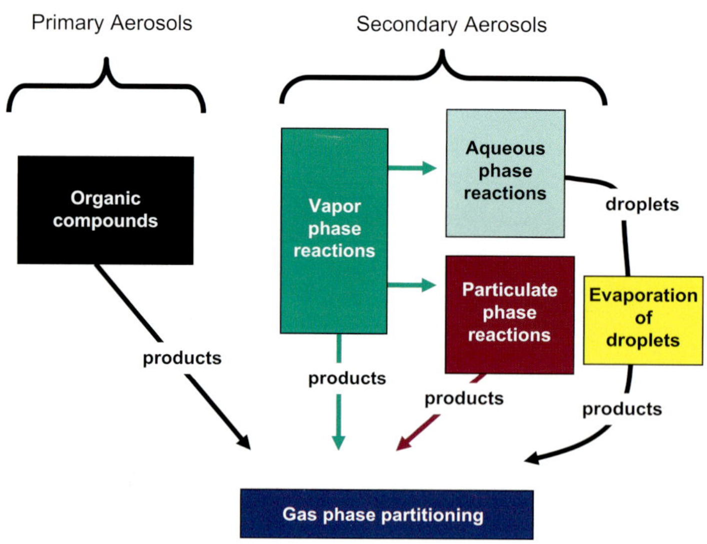

FIGURE 8.4

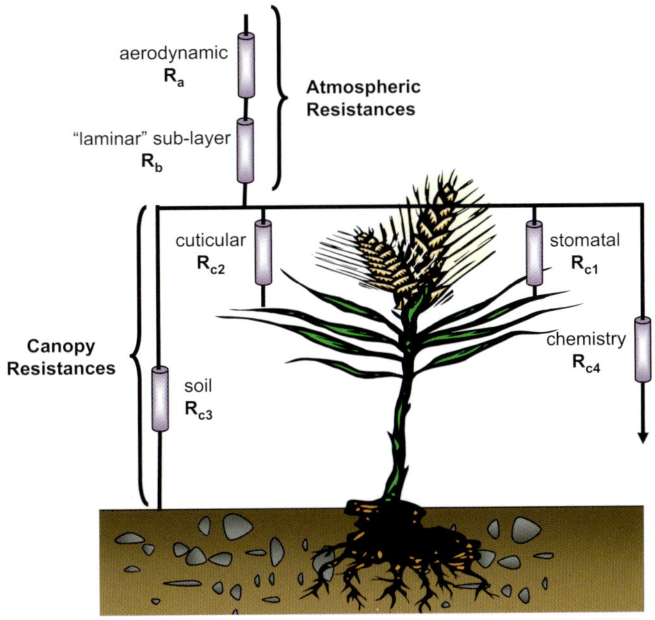

FIGURE 8.5

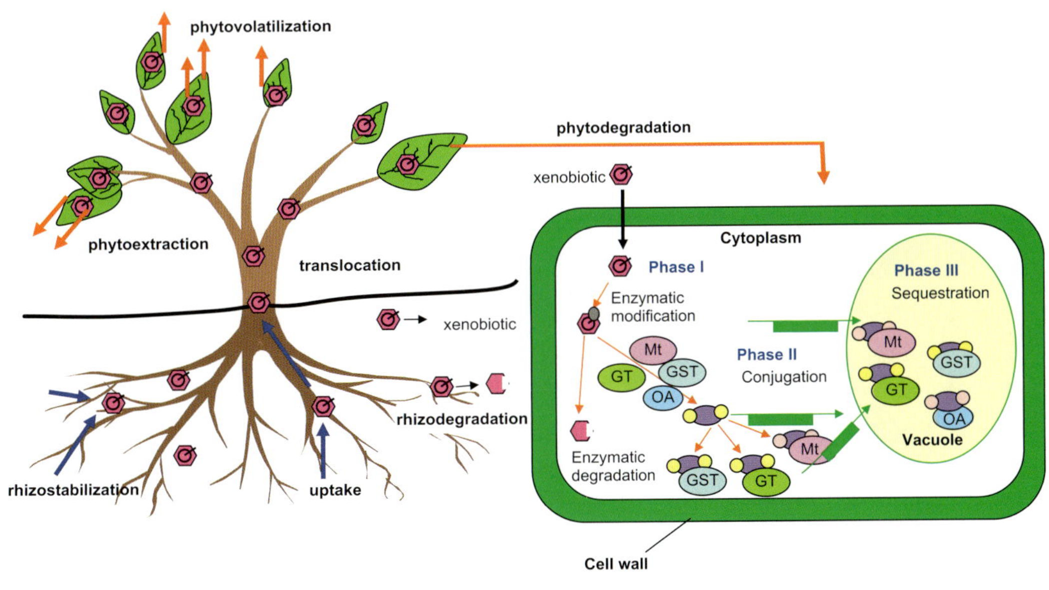

FIGURE 8.6

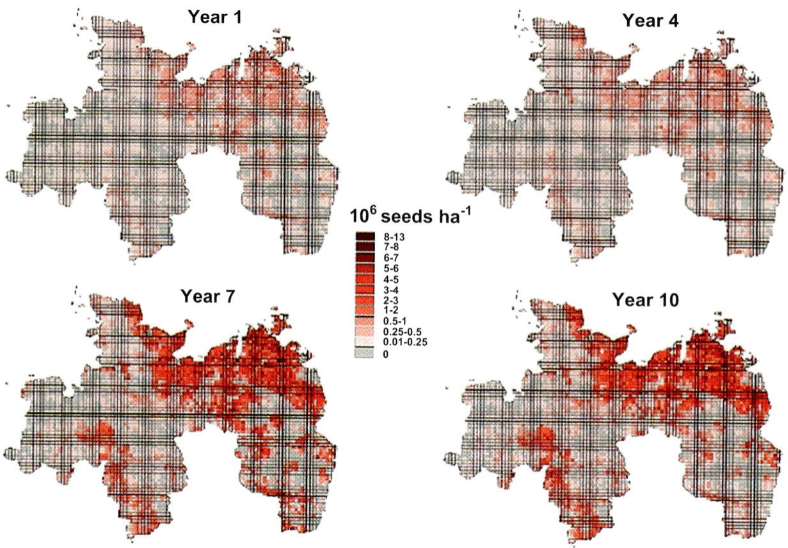

FIGURE 8.11

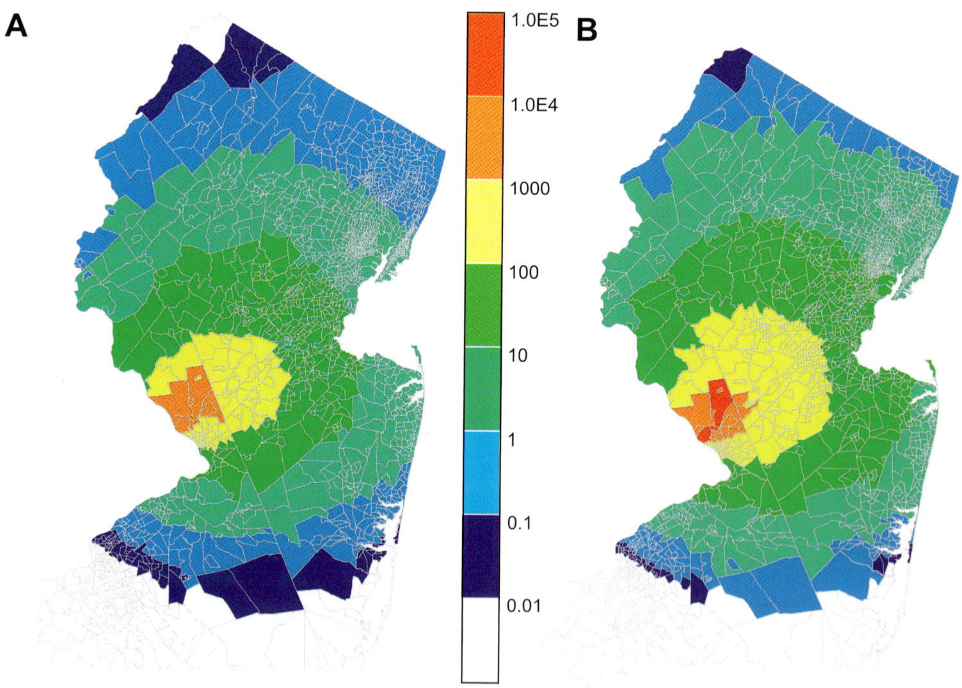

FIGURE 8.13

Western
Sahara
Atlantic Ocean
Cape Verde Islands
100 km
N
Mauritania

FIGURE 8.14

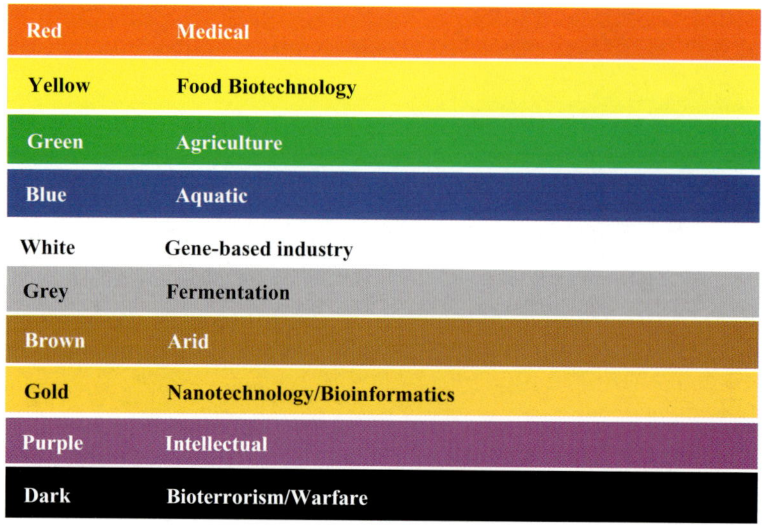

FIGURE 9.1

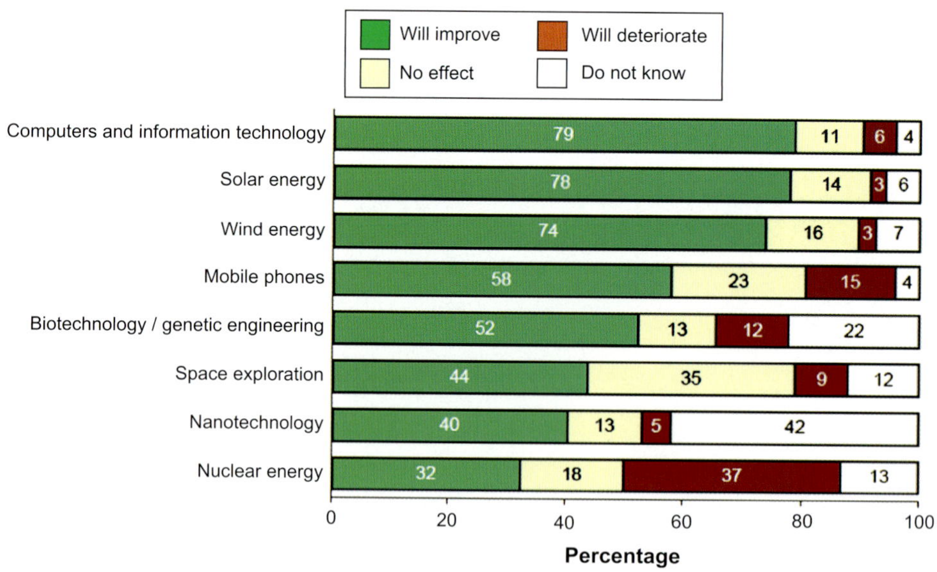

FIGURE 9.2

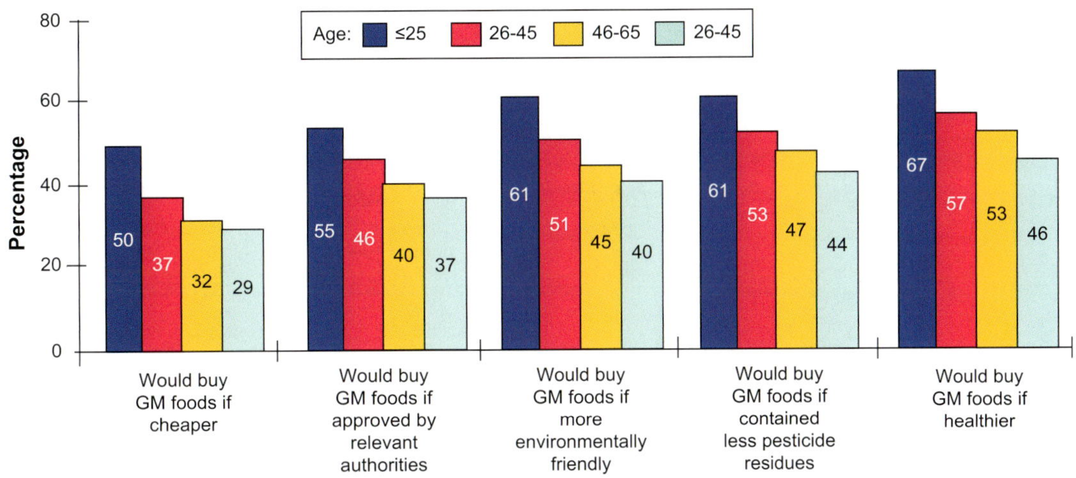

FIGURE 9.3

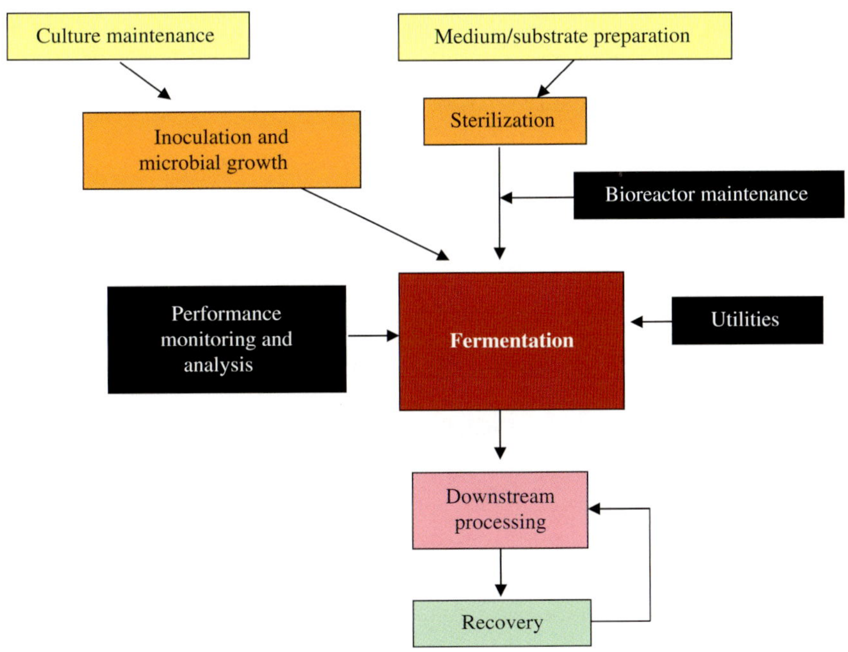

FIGURE 9.4

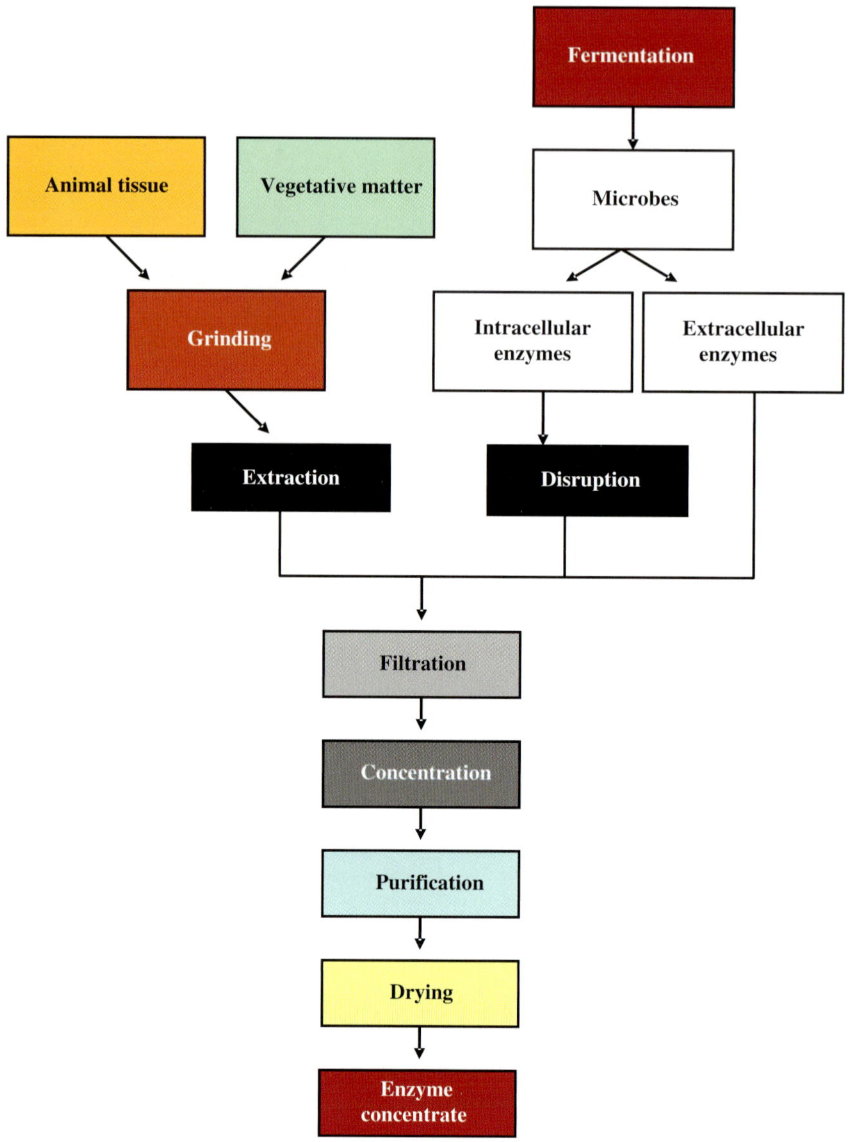
Fermentation
Animal tissue
Vegetative matter
Microbes
Grinding
Intracellular enzymes
Extracellular enzymes
Extraction
Disruption
Filtration
Concentration
Purification
Drying
Enzyme concentrate

FIGURE 9.5

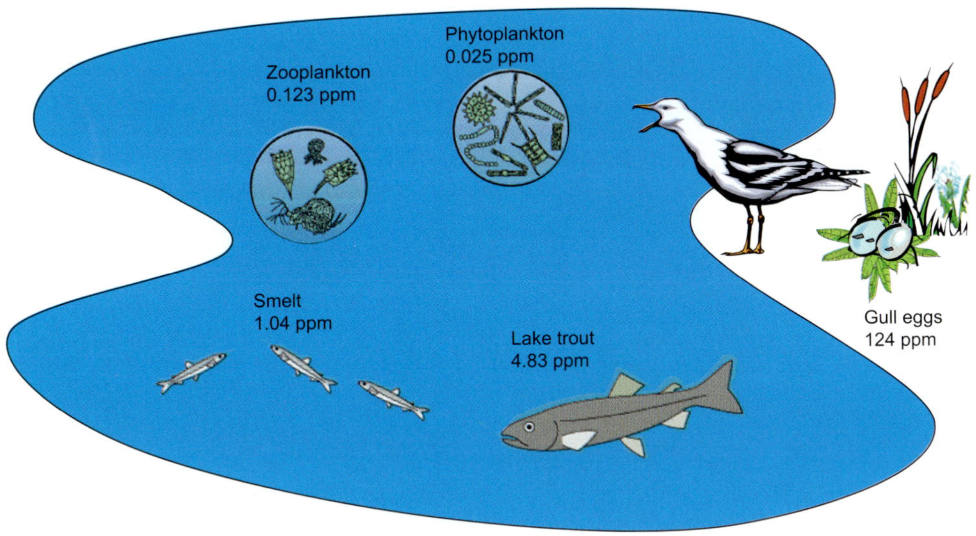

FIGURE 9.7

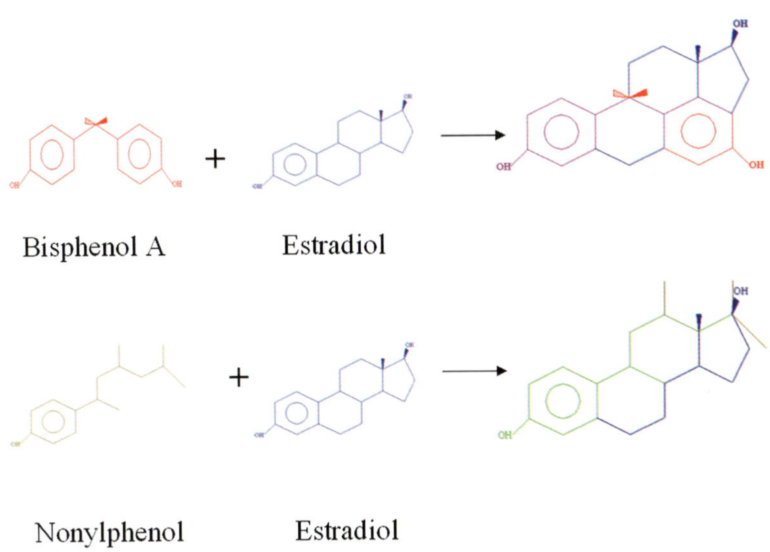

FIGURE 9.13

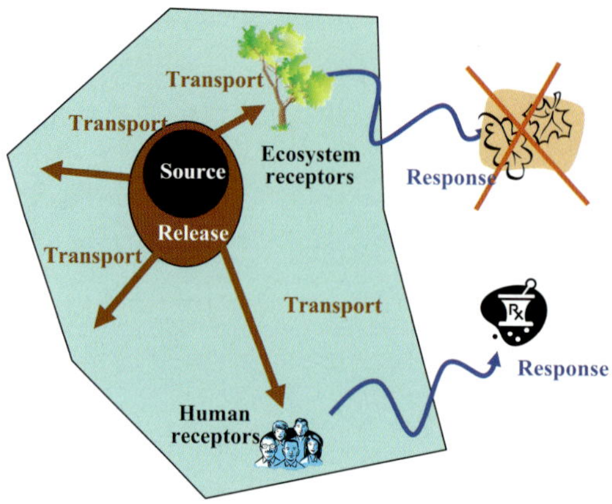

FIGURE 10.1

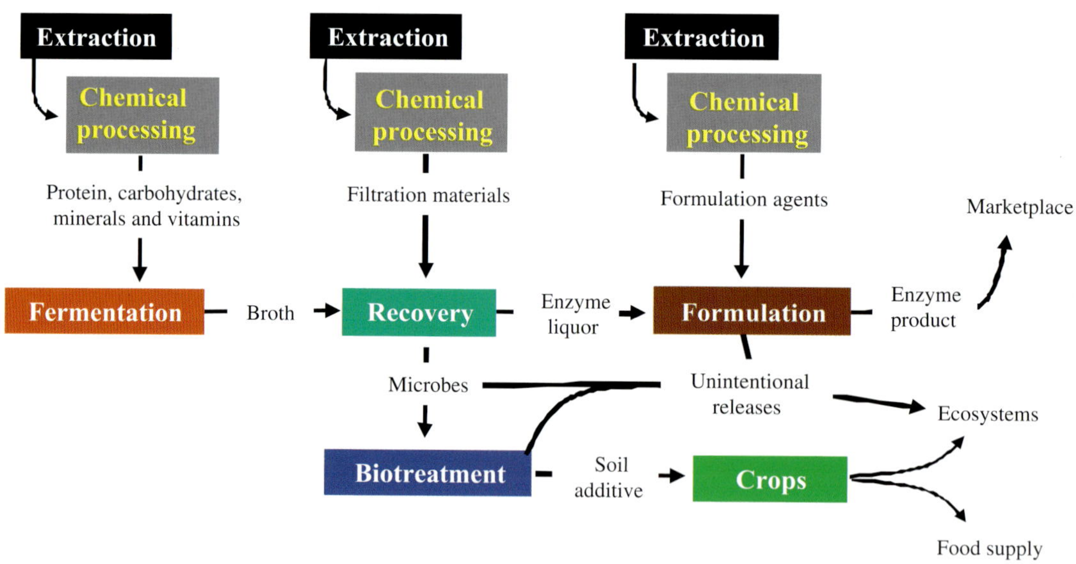

FIGURE 10.2

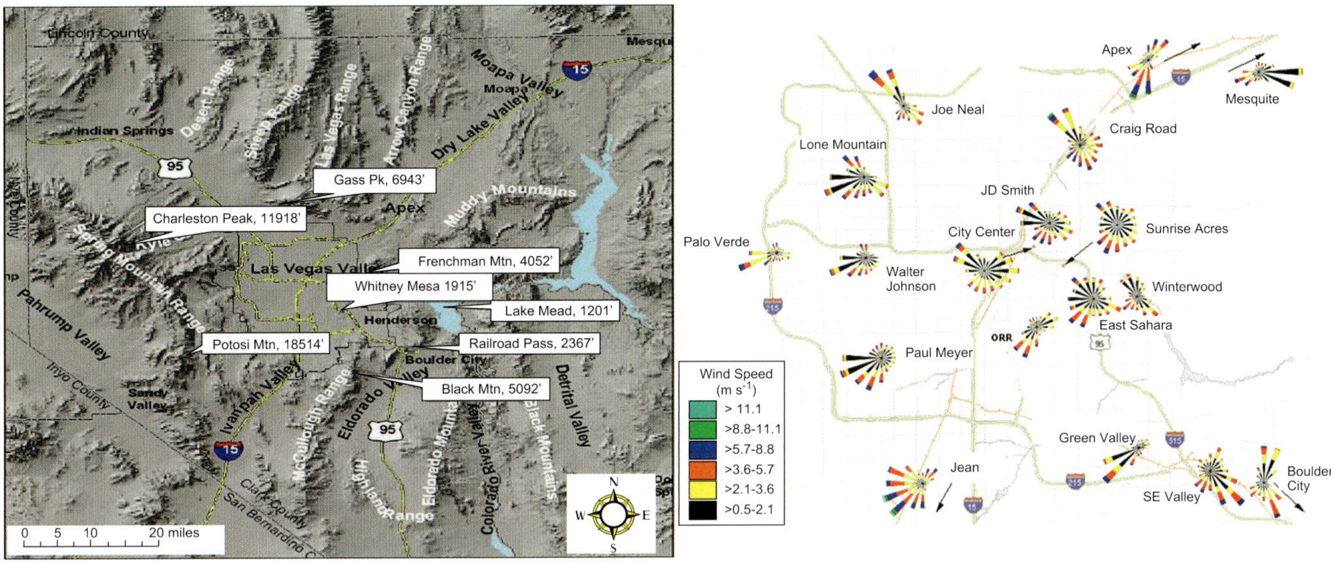

FIGURE 10.8

FIGURE 10.10

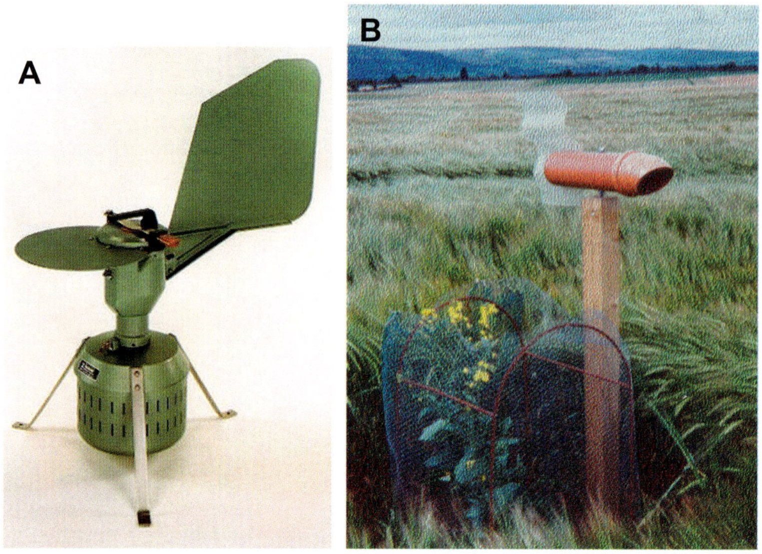

FIGURE 10.17

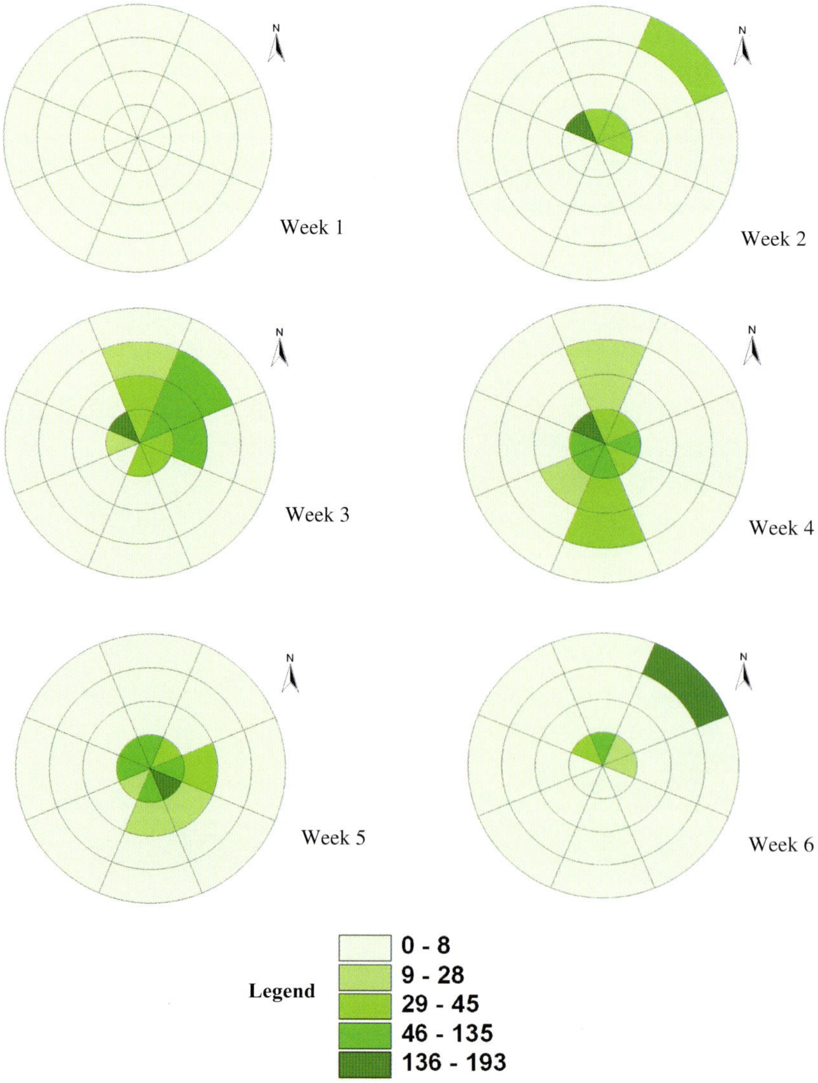

FIGURE 10.18

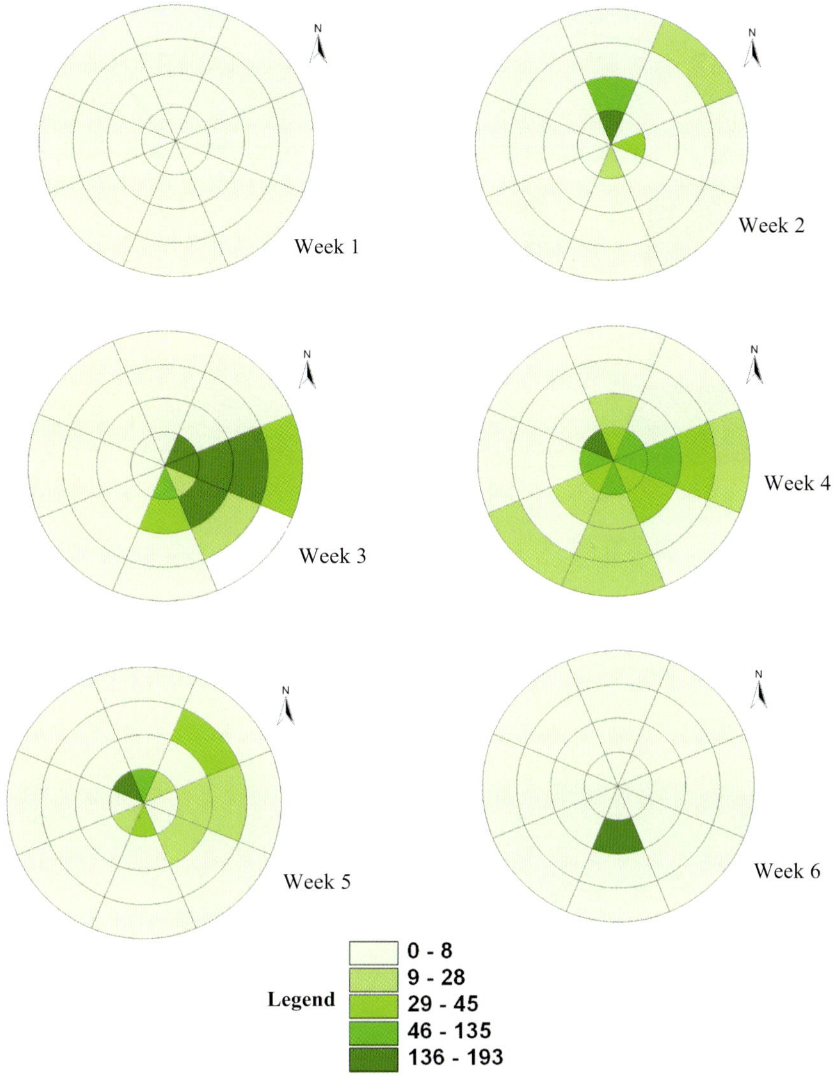

N
Week 1
N
Week 2
N
Week 3
N
Week 4
N
Week 5
N
Week 6
Legend
0 - 8
9 - 28
29 - 45
46 - 135
136 - 193

FIGURE 10.19

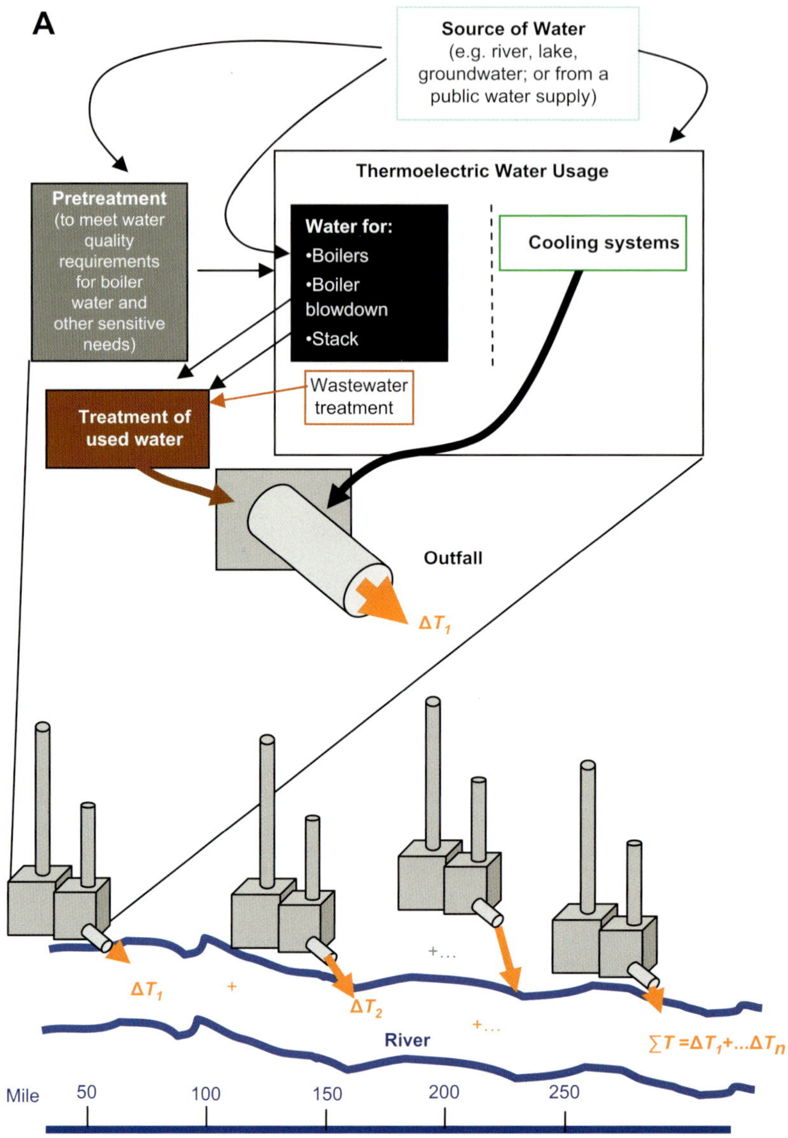
A
Source of Water
(e.g. river, lake,
groundwater; or from a
public water supply)
Pretreatment
(to meet water
quality
requirements
for boiler
water and
other sensitive
needs)
Thermoelectric Water Usage
Water for:
•Boilers
•Boiler
blowdown
•Stack
Cooling systems
Wastewater
treatment
Treatment of
used water
Outfall
ΔT1
ΔT1
+
ΔT2
+...
+...
River
∑T =ΔT1+...ΔTn
Mile
50
100
150
200
250

FIGURE 11.4

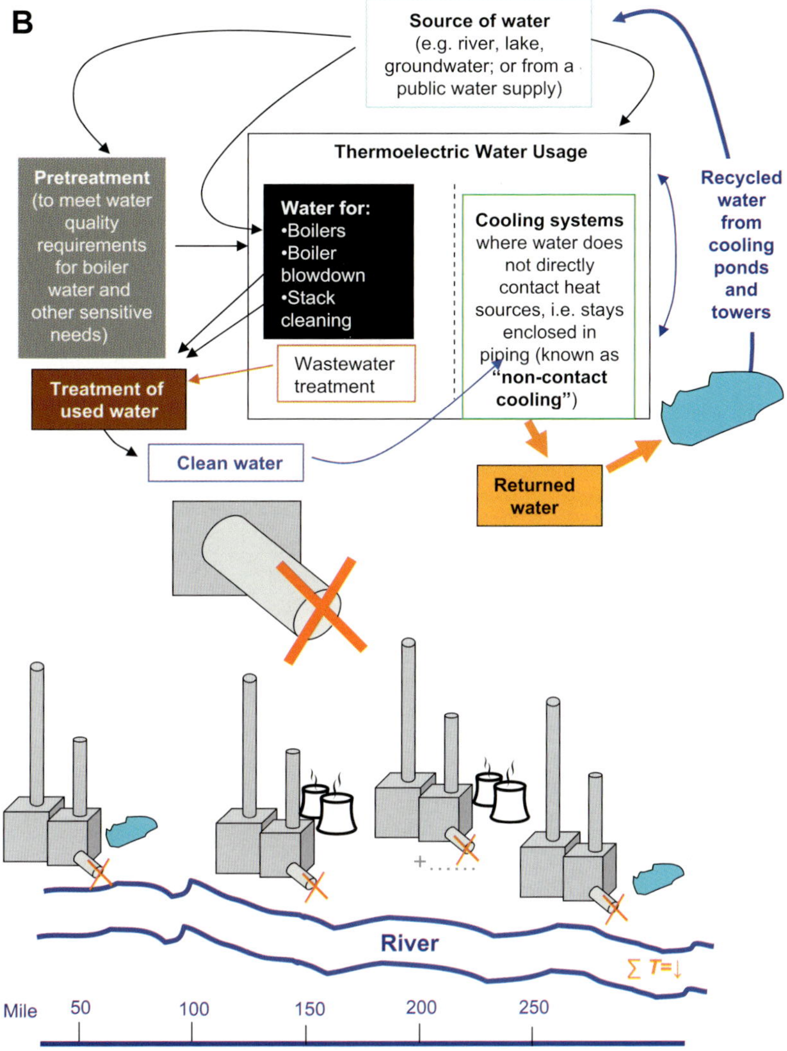

FIGURE 11.4—cont'd

Steps in gaining wisdom	Hypothetical example
Concerns and interests	Observations of allergenicity in general population. Environmental agents?
↓	
Data	Pediatrician surveys, hospital admissions, sale of GMO products by county
↓	
Information	Cause–effect hypotheses, temporality, weight of evidence, spatial and temporal interpretation of data (e.g. geographic information system)
↓	
Knowledge	Comparison to other effects information, similar allergies, biological plausibility, deductive and inductive reasoning
↓	
Wisdom	Deduction, induction, intuition

FIGURE 11.11

Temperature (°C)	DO (mg L^{-1})	Temperature (°C)	DO (mg L^{-1})
0	14.60	23	8.56
1	14.19	24	8.40
2	13.81	25	8.24
3	13.44	26	8.09
4	13.09	27	7.95
5	12.75	28	7.81
6	12.43	29	7.67
7	12.12	30	7.54
8	11.83	31	7.41
9	11.55	32	7.28
10	11.27	33	7.16
11	11.01	34	7.16
12	10.76	35	6.93
13	10.52	36	6.82
14	10.29	37	6.71
15	10.07	38	6.61
16	9.85	39	6.51
17	9.65	40	6.41
18	9.45	41	6.41
19	9.26	42	6.22
20	9.07	43	6.13
21	8.90	44	6.04
22	8.72	45	5.95

TABLE 11.1

Organism	Taxonomy	Range in temperature tolerance (°C)	Minimum dissolved oxygen (mg L^{-1})
Trout	*Salma, Oncorhynchus* and *Salvelinus spp.*	5–20	6.5
Smallmouth bass	*Micopterus dolomieu*	5–28	6.5
Caddisfly larvae	*Brachycentrus spp.*	10–25	4.0
Mayfly larvae	*Ephemerella invaria*	10–25	4.0
Stonefly larvae	*Pteronarcys spp.*	10–25	4.0
Catfish	Order Siluriformes	20–25	2.5
Carp	Cyprinus spp.	10–25	2.0
Water boatmen	*Notonecta spp.*	10–25	2.0
Mosquito larvae	Family Culicidae	10–25	1.0

TABLE 11.2

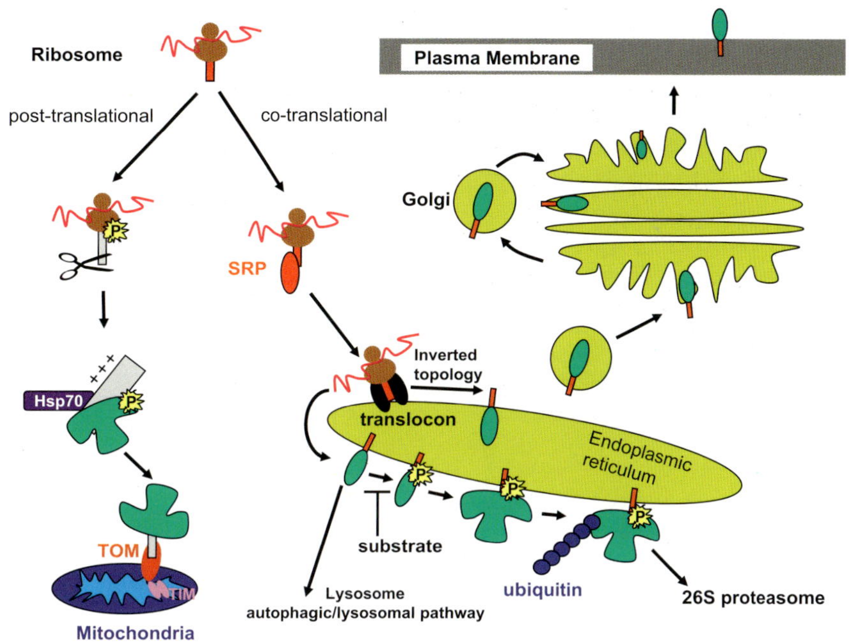

FIGURE 12.21

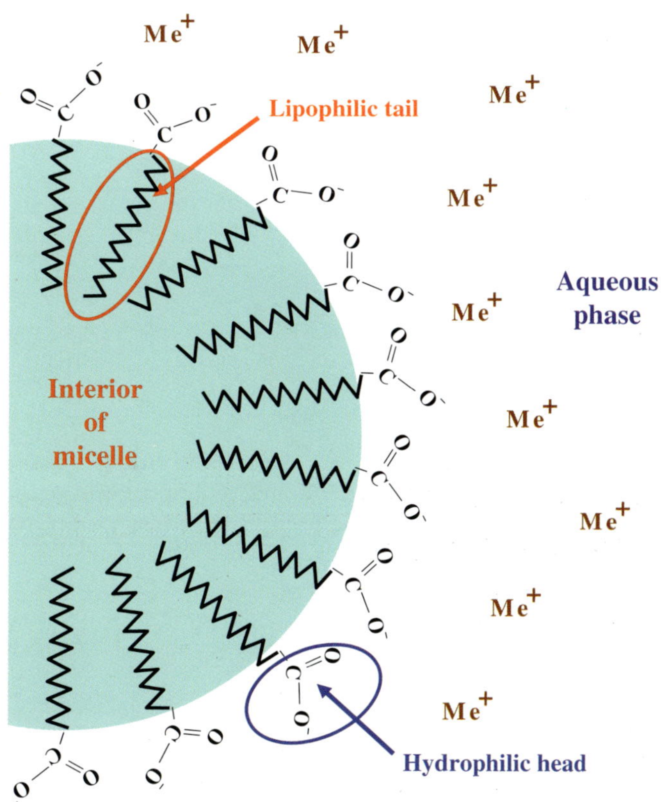

FIGURE 12.25

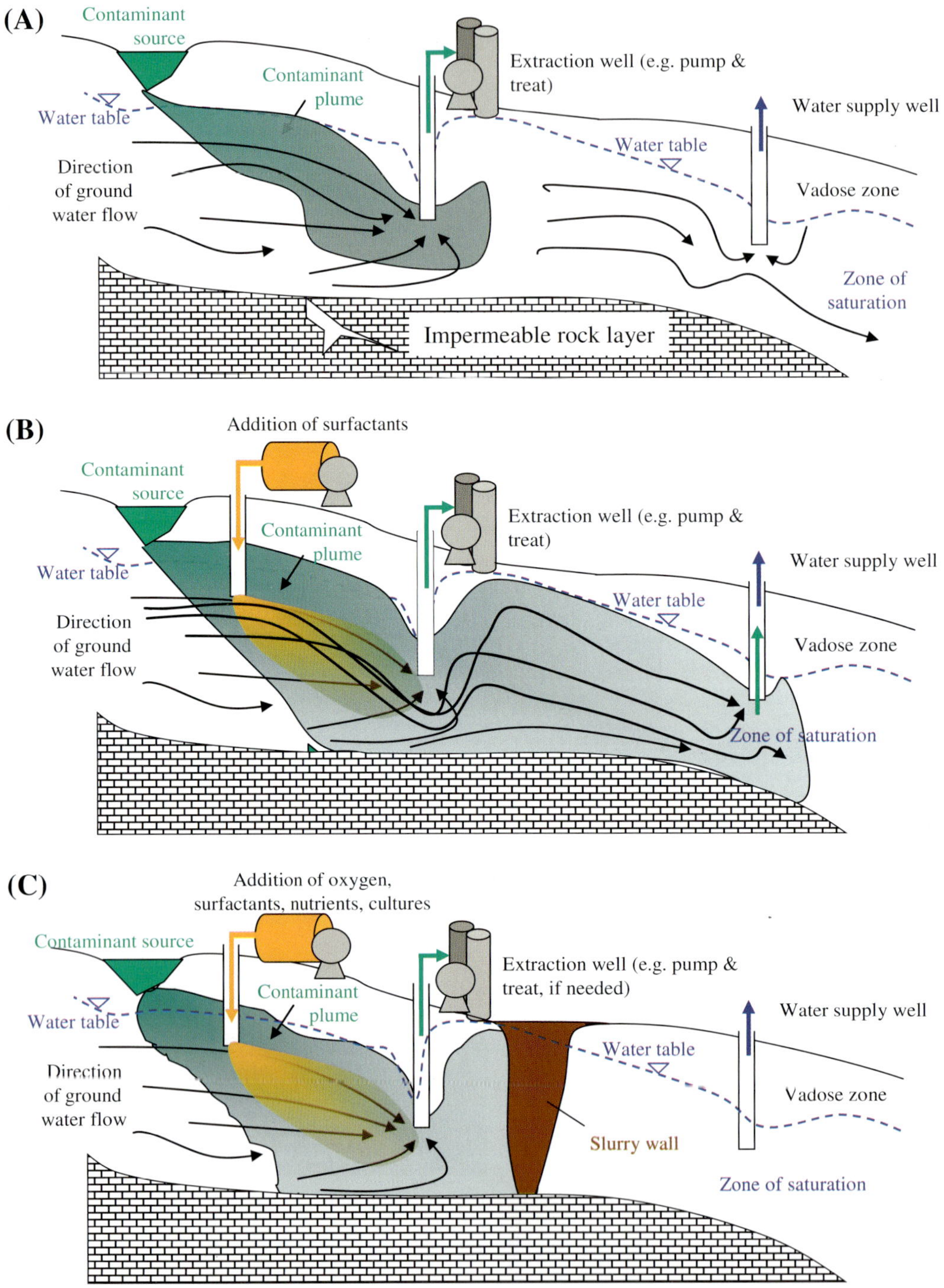
(A)
Contaminant source
Contaminant plume
Extraction well (e.g. pump & treat)
Water supply well
Water table
Water table
Direction of ground water flow
Vadose zone
Zone of saturation
Impermeable rock layer
(B)
Addition of surfactants
Contaminant source
Contaminant plume
Extraction well (e.g. pump & treat)
Water supply well
Water table
Water table
Direction of ground water flow
Vadose zone
Zone of saturation
(C)
Addition of oxygen, surfactants, nutrients, cultures
Contaminant source
Contaminant plume
Extraction well (e.g. pump & treat, if needed)
Water supply well
Water table
Water table
Direction of ground water flow
Vadose zone
Slurry wall
Zone of saturation

FIGURE 12.26